Geometry

Examination Guide

A complete course for students of all abilities provides everything you need—

- **Traditional proofs and constructions** help you teach synthetic and coordinate geometry.

  Chapters 2-8 include both two-column (pp. 48, 150, 192) and paragraph proofs (p. 236). Synthetic and coordinate representations show how to examine geometry from an algebraic perspective (Chapter 10).

- **Exercises** progress in complexity to give you a choice of assignments.

  Exercises are plentiful with Classroom Exercises (p. 155) for oral discussion and Written Exercises (p. 155) that progress from easy to more challenging. Examples correspond to each type of written exercise for levels A and B (p. 154).

- **Reviews, quizzes, and tests** make it easy to integrate assessment with instruction.

  Each lesson includes a Prerequisite Quiz (p. 77) and a post-lesson Checkpoint (p. 79) in the side column of the *Annotated Teacher's Edition* and as blackline masters in the Quick Quizzes of the testing package. Assessments in the textbook include Mixed Reviews (pp. 47, 52, 56) and an Algebra Review (p. 47) as well as Chapter Reviews and Tests (pp. 86, 88) and College Prep Tests (p. 89) at the end of each chapter and Cumulative Reviews in alternate chapters (p. 90).

- **Extensive features and worksheets** help you meet the NCTM *Standards* for the 90s!

  - * **Technology—calculators and computers**
    pp. 120, 214, 233, 347
    Technology Worksheets—
    Teacher's ResourceBank™

  - * **Math connections to algebra and life**
    In Teacher's side column for each lesson
    Applications features (pp. 204, 235, 349)
    Application Worksheets—
    Teacher's ResourceBank™

  - * **Communication about math**
    Focus on Reading (pp. 18, 217, 325)
    Oral explanations (p. 228)
    Written Exercises—essays (p. 239, #15)

  - * **Models, manipulatives, projects**
    Investigations that begin each chapter
    Project Worksheets and Manipulatives
    Worksheets—*Teacher's ResourceBank*™

  - * **Spatial and spherical geometry**
    pp. 2, 31, 93, 138, 352, 554

  - * **Reflections, translations, rotations and transformations**
    Chapter 15 (p. 598)

  - * **Problem solving strategies**
    pp. 58, 84, 133, 184, 256
    Problem Solving Worksheets—
    Teacher's ResourceBank™

  - * **Other valuable timesavers for you**
    Reteaching and Practice Worksheets and
    Tests booklet with Tests, Form A and
    Form B—*Teacher's ResourceBank*™
    Instructional Tr_____ _____ _
    Solution Key

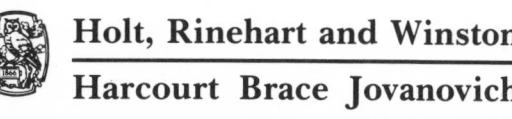 **Holt, Rinehart and Winston**

 Harcourt Brace Jovanovich

GEOMETRY

Annotated Teacher's Edition

Eugene D. Nichols
Mervine L. Edwards
E. Henry Garland
Sylvia A. Hoffman
Albert Mamary
William F. Palmer

 **Holt, Rinehart and Winston, Inc.**

Harcourt Brace Jovanovich, Inc.

Austin · Orlando · San Diego · Chicago · Dallas · Toronto

Acknowledgments

PHOTO CREDITS

Abbreviations used: (l)left; (t)top; (b)bottom; (tl)top left; (bl)bottom left; (ml)middle left.

Page xii-A, HRW photo by Dennis Fagan; xii-J, HRW photo by Dennis Fagan

Chapter 1: page 1, © California Institute of Technology, 1961; 5, HRW Photo by Bryan Tumlinson; 15, Lewis Portnoy/Spectra-Action Inc.; 31, NASA.

Chapter 2: page 48, Dave Stock; 83, HRW Photos by Bryan Tumlinson.

Chapter 3: page 106, Russ Kinne/Comstock; 119(t) Mark Antman/The Image Works; 119(b), Bill Varie/The Image Bank.

Chapter 4: page 147, Audrey Gibson; 152, Tony Freeman/PhotoEdit; 153, Phil Degginger/TSW-Click/Chicago.

Chapter 5: page 177, Cameramann International Ltd.; 182, Robert Brown/Allsport USA; 215, HRW Photos by Eric Beggs.

Chapter 6: page 229, Mark Antman/The Image Works; 230, IBM/PhotoEdit; 235, Tony Freeman/PhotoEdit; 239, HRW Photo by Eric Beggs; 247 HRW Photos by Eric Beggs.

Chapter 7: page 261, Frank Cezus/TSW-Click/Chicago; 283(ml)(b), HRW Photos by Eric Beggs; 283(t), Patrick Fischer Photography; 285, Patrick Fischer Photography.

Printed in the United States of America

ISBN: 0-03-005408-7

123456   036   98765432

About the Authors

Eugene D. Nichols
Distinguished Professor
of Mathematics Education
Florida State University
Tallahassee, Florida

Mervine L. Edwards
Chairman, Department of Mathematics
Shore Regional High School
West Long Branch, New Jersey

E. Henry Garland
Head of Mathematics Department
Developmental Research School
DRS Professor
Florida State University
Tallahassee, Florida

Sylvia A. Hoffman
Resource Consultant in Mathematics
Illinois State Board of Education
State of Illinois

Albert Mamary
Superintendent of Schools for Instruction
Johnson City Central School District
Johnson City, New York

William F. Palmer
Professor of Education and Director
Center for Mathematics and Science Education
Catawba College
Salisbury, North Carolina

Contents

Using *Geometry* .... xii-A
Implementing the NCTM *Standards* with *Geometry* .... xii-K
Manipulatives and Models .... xii-K
Cooperative Learning Groups .... xii-L
Problem-Solving Strategies .... xii-M
Critical Thinking Questions .... xii-N
Reading, Writing, and Discussing Math .... xii-N
Technology .... xii-O
Summary .... xii-P
Symbol List xiii

1 GEOMETRIC FIGURES — xiv

Overview .... xiiiA
Planning Guide .... xiiiB
1.1 Introduction to Geometric Figures 1
1.2 Distance, Segments, and Rays 5
1.3 Congruent Segments and Constructions 10
1.4 Angle Measurement and Constructions 15
1.5 Adjacent Angles and Angle Bisectors 21
1.6 Supplementary and Complementary Angles 26
1.7 Logic: Conjunction and Disjunction 32
1.8 Graphing Conjunctions and Disjunctions 36

Special Features

Focus on Reading 3, 18
Computer Investigation: Supplementary and
  Complementary Angles 20
Application: Surveying 25
Application: Astronomical Units 31
Application: Electric Circuits 35

Review and Testing

Algebra Review 4
Midchapter Review 19
Chapter 1 Review 38
Chapter 1 Test 40
College Prep Test 41

2 PROOF — 42

Overview 41A
Planning Guide 41B
2.1 Drawing Conclusions 43
2.2 Introduction to Proof 48
2.3 Writing Proofs in Geometry 53
2.4 Proofs and More Complex Figures 57
2.5 Conditional Statements 63
2.6 Deductive and Inductive Reasoning 67
2.7 Proving Theorems about Angles 71
2.8 Vertical Angles 77
2.9 Postulates and Theorems: Points, Lines, Planes 82

Special Features

Application: Tilt of the Earth 56
Problem Solving Strategies: Drawing a Diagram 62
Focus on Reading 65, 73
Computer Investigation: Vertical Angles 76

Review and Testing

Algebra Review 47
Midchapter Review 61
Chapter 2 Review 86
Chapter 2 Test 88
College Prep Test 89
Cumulative Review 90

3 PARALLELISM 92

Overview 91A
Planning Guide... 91B
3.1 Parallel and Skew Lines.. 93
3.2 Transversals and Special Angle Relationships 97
3.3 Proving Lines Parallel.. 100
3.4 Introduction to Indirect Proof................................ 106
3.5 Converses and the Parallel Postulate 109
3.6 The Angles of a Triangle..................................... 115
3.7 Exterior and Remote Interior Angles of a Triangle 121
3.8 Negation, Contrapositive, Inverse, Biconditional 126
3.9 Parallel Lines and Planes 130

Special Features

Computer Investigation: Angles of a Triangle 120
Focus on Reading 123
Application: Light Rays 125
Problem Solving Strategies: Using an
    Alternate Approach 133

Review and Testing

Midchapter Review 114
Algebra Review 114
Chapter 3 Review 134
Chapter 3 Test 136
College Prep Test 137

4 CONGRUENT TRIANGLES 138

Overview 137A
Planning Guide.. 137B
4.1 Triangle Classifications 139
4.2 Congruence of Triangles 143
4.3 The SAS Postulate 147
4.4 SSS and ASA Congruence Proofs 153
4.5 Congruence in Complex Figures............................... 158
4.6 The Third Angle and AAS Theorems 161
4.7 Applying Congruence: Corresponding Parts.................... 167

Special Features

Focus on Reading 141, 164
Computer Investigation: Constructing Triangles 146
Application: Angles in Sand Piles 152

Review and Testing		
Midchapter Review		157
Algebra Review		166
Chapter 4 Review		170
Chapter 4 Test		172
College Prep Test		173
Cumulative Review		174

5 CONGRUENCE 176

Overview		175A
Planning Guide		175B
5.1	Isosceles Triangles	177
5.2	Introduction to Overlapping Triangles	182
5.3	Congruence and Overlapping Triangles	184
5.4	Congruence of Right Triangles	188
5.5	Altitudes and Medians of Triangles	194
5.6	Perpendicular Bisectors	199
5.7	Inequalities in a Triangle	205
5.8	The Triangle Inequality Theorem	210
5.9	Inequalities for Two Triangles	215

Special Features		
Computer Investigation: Altitudes and Medians of a Triangle		193
Focus on Reading		202, 217
Application: Congruent Relationships		204

Review and Testing		
Midchapter Review		192
Algebra Review		209, 219
Chapter 5 Review		220
Chapter 5 Test		222
College Prep Test		223

6 POLYGONS 224

Overview		223A
Planning Guide		223B
6.1	Introducing Polygons	225
6.2	Interior Angles of Polygons	230
6.3	Exterior Angles of Polygons	236
6.4	Quadrilaterals and Parallelograms	241
6.5	Quadrilaterals That Are Parallelograms	247
6.6	The Midsegment Theorem	252
6.7	Lines Parallel to Many Lines	257

Special Features		
Application: Tiling and Soccer Ball		235
Application: Mirrors		239
Computer Investigation: Properties of a Parallelogram		240
Problem Solving Strategies: Making a Table		256

Review and Testing		
Algebra Review		235
Midchapter Review		246
Chapter 6 Review		262
Chapter 6 Test		264
College Prep Test		265
Cumulative Review		266

7 SPECIAL QUADRILATERALS 268

Overview 267A
Planning Guide 267B
7.1 Proving Quadrilaterals Congruent 269
7.2 Necessary Conditions: Rectangles, Rhombuses,
      and Squares 273
7.3 Sufficient Conditions: Rectangles, Rhombuses,
      and Squares 278
7.4 Trapezoids 285
7.5 Isosceles Trapezoids 290
7.6 Constructing Quadrilaterals 295

Special Features

Focus on Reading 275
Application: Conditions for Quadrilaterals 283
Computer Investigation: Necessary Conditions 284

Review and Testing

Algebra Review 272
Midchapter Review 282
Chapter 7 Review 298
Chapter 7 Test 300
College Prep Test 301

8 SIMILARITY 302

Overview 301A
Planning Guide 301B
8.1 Ratio and Proportion 303
8.2 Similar Polygons 308
8.3 Similar Triangles 313
8.4 The Triangle Proportionality Theorem 318
8.5 SAS and SSS Similarity Theorems 323
8.6 Segments in Similar Triangles 328

Special Features

Application: Pitch of a Roof 307
Problem Solving Strategies: Using a Formula to Find the
      Golden Ratio 312
Application: Finding the Height of a Tree 317
Computer Investigation: Similar Triangles 322
Focus on Reading 325

Review and Testing

Midchapter Review 316
Algebra Review 327
Chapter 8 Review 332
Chapter 8 Test 334
College Prep Test 335
Cumulative Review 336

9 RIGHT TRIANGLES 338

Overview 337A
Planning Guide. 337B
9.1 Right Triangle Similarity Properties 339
9.2 The Pythagorean Theorem 344
9.3 Converse of the Pythagorean Theorem 350
9.4 Two Special Types of Right Triangles 354
9.5 The Sine Ratio 360
9.6 Other Trigonometric Ratios 366
9.7 Applying Trigonometric Ratios 370

Special Features
Application: Baseball "Diamond" 349
Focus on Reading 357, 367
Computer Investigation: A Special Ratio in Right Triangles 359
Using The Calculator 365

Review and Testing
Algebra Review 353
Midchapter Review 358
Chapter 9 Review 374
Chapter 9 Test 376
College Prep Test 377

10 COORDINATE GEOMETRY 378

Overview 377A
Planning Guide. 377B
10.1 Coordinate Systems and Distance 379
10.2 The Midpoint Formula 384
10.3 Slope of a Line 388
10.4 Equation of a Line 393
10.5 Parallel or Perpendicular Lines 397
10.6 Proofs with Coordinates 403

Special Features
Focus on Reading 381
Computer Investigation: Equation of a Line 392
Application: Vector Methods 402

Review and Testing
Algebra Review 387, 407
Midchapter Review 391
Chapter 10 Review 408
Chapter 10 Test 410
College Prep Test 411
Cumulative Review 412

11 CIRCLES 414

Overview **413A**
Planning Guide **413B**
11.1 Circles and Spheres: Basic Definitions **415**
11.2 Properties of Chords **420**
11.3 Special Properties of Tangents to Circles **425**
11.4 Arcs and Central Angles **432**
11.5 Arcs, Chords, and Central Angles **438**
11.6 Inscribed Angles **442**
11.7 Angles Formed by Secants and Chords **447**
11.8 Angles Formed by Tangents and Secants **452**
11.9 Lengths of Segments Formed by Secants, Chords, and
         Tangents **457**
11.10 Circles and Constructions **462**
11.11 Equations of Circles **466**

Special Features

Focus on Reading **417, 435, 463**
Computer Investigation: Arcs and Related Angles **431**
Application: Geometric Designs **465**

Review and Testing

Algebra Review **437**
Midchapter Review **446**
Chapter 11 Review **470**
Chapter 11 Test **472**
College Prep Test **473**

12 AREA 474

Overview **473A**
Planning Guide **473B**
12.1 Standard Units **475**
12.2 Areas of Rectangles and Squares **478**
12.3 Areas of Parallelograms **483**
12.4 Areas of Triangles **488**
12.5 Areas of Trapezoids **493**
12.6 Measuring the Regular Polygons **498**
12.7 Areas and Perimeters of Similar Polygons **504**
12.8 Circumferences and Areas of Circles **510**
12.9 Measuring Arcs and Sectors of Circles **515**

Special Features

Focus on Reading **476**
Problem Solving Strategies: More than One Way **487**
Application: Areas of Irregularly-Shaped Regions **497**
Application: Polygons in Structures **503**
Application: Enlarging Drawings **508**
Computer Investigation: Estimating Pi **509**
Application: Estimating Area **520**

Review and Testing

Midchapter Review **497**
Algebra Review **503**
Chapter 12 Review **522**
Chapter 12 Test **524**
College Prep Test **525**
Cumulative Review **526**

13 LOCI 528

Overview 527A
Planning Guide. 527B
13.1 Finding Locations 529
13.2 Multiple Conditions for Loci 533
13.3 Loci in Space 538
13.4 Concurrent Bisectors in Triangles 541
13.5 Concurrent Altitudes and Medians 547

Special Features
Application: Site for a House. 540
Focus on Reading 544
Computer Investigation: Concurrence in Triangles 546

Review and Testing
Midchapter Review 540
Chapter 13 Review 551
Chapter 13 Test 552
College Prep Test 553

14 FIGURES IN SPACE 554

Overview 553A
Planning Guide. 553B
14.1 Polyhedra and Rigidity 555
14.2 Prisms and Cylinders 560
14.3 Areas of Prisms and Cylinders 564
14.4 Volumes of Prisms and Cylinders 569
14.5 Areas of Pyramids and Cones 574
14.6 Volumes of Pyramids and Cones 579
14.7 Areas and Volumes of Spheres 584
14.8 Coordinates in Space 589

Special Features
Focus on Reading 558
Application: Manufacturing Paper Cups. 583

Review and Testing
Midchapter Review 573
Algebra Review 588
Chapter 14 Review 592
Chapter 14 Test 594
College Prep Test 595
Cumulative Review 596

Overview **597A**
Planning Guide **597B**
15.1 Reflections **599**
15.2 Translations **606**
15.3 Rotations **612**
15.4 Transformations and Coordinates **618**
15.5 Dilations and Other Transformations **622**

**Special
Features**

Computer Investigation: Reflections **605**
Application: Designs **611**
Application: Stadium Lights **621**
Focus on Reading **623**

**Review
and Testing**

Midchapter Review **617**
Chapter 15 Review **626**
Chapter 15 Test **628**
College Prep Test **629**

Postulates, Theorems, and Corollaries **630**
Table of Roots and Powers **642**
Table of Trigonometric Ratios **643**
Glossary **644**
Additional Answers **649**
Index **668**

GEOMETRY

Helping You Meet the Standards of Tomorrow— Starting Today

It has been predicted that today's students will live their lives in a world far more technologically advanced than ours. With continual progress in the field of mathematics, this generation will be required to know more mathematics than ever before and will need to apply this knowledge to their daily lives and future careers.

To help students become adept problem solvers and to prepare them for the demands of tomorrow's world, the National Council of Teachers of Mathematics has developed the *Curriculum and Evaluation Standards for School Mathematics*. In the introduction of the *Standards*, NCTM recognizes that educational reforms are needed to keep pace with today's world.

> *Calls for reform in school mathematics suggest that new goals are needed. All industrialized countries have experienced a shift from an industrial to an information society, a shift that has transformed both the aspects of mathematics that need to be transmitted to students and the concepts and procedures they must master if they are to be self-fulfilled, productive citizens in the next century.*
>
> (NCTM Standards, p. 3)

The following lesson has been chosen to show you how Holt, Rinehart and Winston's *Geometry* helps you move towards the NCTM curriculum standards, while also providing you with a practical text that meets *your* standards.

A variety of supplementary materials has been developed that incorporates the newest technology and current trends in mathematics education into a traditional mathematics program. These complete components allow you to provide your students with the highest quality education available without spending hours of your time in researching and planning additional classroom activities.

The role that today's teachers play in the education of students will affect our society and our world for generations to come. At Holt, Rinehart and Winston, we recognize the importance of education and are dedicated to providing today's teachers and students with the finest textbooks available.

Flexible teaching strategies allow you to customize instruction.

The teaching strategies, which are interleaved in the *Annotated Teacher's Edition* before each chapter, help you focus lessons on problem-solving strategies, computer and calculator technology, and special features that enrich and extend students' learning.

These alternative methods of instruction will require the teacher's role to shift from dispensing information to facilitating learning, from that of director to that of catalyst and coach.

(*NCTM Standards*, p. 128)

In both the **Overview** and the **Objectives**, learning goals are defined in terms of what students will *do* — emphasizing the value of active, not passive, learning.

5 CONGRUENCE

OVERVIEW

In this chapter, students prove and apply theorems about isosceles triangles. They write proofs for overlapping triangles and identify parts of right triangles. They use the Hypotenuse-Leg Theorem, identify altitudes and medians of triangles, and prove theorems about them. They prove and apply theorems about perpendicular bisectors and construct perpendicular segments. Students prove and apply The Triangle Inequality Theorem, and, given the lengths of 3 sides decide whether a triangle can be constructed. They are also asked to write paragraph proofs.

OBJECTIVES

- To solve problems involving isosceles triangles
- To draw conclusions from diagrams with labeled data
- To construct a line perpendicular to a given line and passing through a given point on the line
- To determine whether a triangle can be constructed, given lengths of three sides
- To relate the measures of sides and angles of triangles with the symbol =, >, or <
- To prove relationships between segments and angles in complex figures
- To identify the hypotenuse, legs, altitudes, and median in a right triangle

PROBLEM SOLVING

Throughout Chapter 5, the problem solving strategy of Making a Model can be used to solve problems. Concrete models can help answer questions involving overlapping triangles (Lesson 5.2) by showing each triangle separately. Mathematical models can be used to discover theorems such as the Triangle Inequality Theorem (Lesson 5.8). Lessons 5.1 and 5.8 provide exercises that give students an opportunity to use algebraic equations to solve problems.

TECHNOLOGY

Computer: A Computer Investigation on page 193 enables students to discover properties of altitudes and medians of triangles. This activity can also be extended to introduce Lesson 5.6 on perpendicular bisectors.

Calculator: You may want to encourage students to use a calculator in Lesson 5.8, Exercises 23 and 24, to facilitate computations.

SPECIAL FEATURES

Mixed Review pp. 181, 183, 187, 198, 204, 209, 214, 219
Brainteaser pp. 187, 214
Midchapter Review p. 192
Focus on Reading pp. 202, 217
Application: Congruent Relationships p. 204
Computer Investigation: Altitudes and Medians of a Triangle p. 193
Algebra Review pp. 209, 219
Key Terms p. 220
Key Ideas and Review Exercises pp. 220, 221
Chapter 5 Test p. 222
College Prep Test p. 223

175A

Problem Solving identifies the strategies that occur in the chapter.

Lessons where the computer and calculator can be used to explore mathematical concepts and facilitate routine computations are included in **Technology**.

A **Special Features** list allows you to decide in advance which applications and additional activities to incorporate into the lessons.

Time-saving Planning Guides help you organize instruction.

A quality curriculum requires versatile and complete lesson plans, but lesson planning should not keep you from doing what you do best — teaching. Therefore, lesson plans are organized in a format that is both functional and convenient.

PLANNING GUIDE

Lesson	Basic	Average	Above Average	Resources
5.1 pp. 179–181	CE all WE 1–17 odd	CE all WE 1–24 odd	CE all WE 1–27 odd	Reteaching p. 67 Practice p. 68
5.2 pp. 183	CE all WE 1–6	CE all WE 1–21 odd	CE all WE 1–21	Reteaching p. 69 Practice p. 70
5.3 pp. 186–87	CE all WE 1–9 Brainteaser	CE all WE 1–17 odd Brainteaser	CE all WE 1–19 odd Brainteaser	Reteaching p. 71 Practice p. 72
5.4 pp. 190–92	CE all WE 1–8 Midchapter Review CI all	CE all WE 1–18 odd Midchapter Review CI all	CE all WE 1–21 odd Midchapter Review CI all	Reteaching p. 73 Practice p. 74
5.5 pp. 197–98	CE all WE 1–13 odd	CE all WE 1–18 odd	CE all WE 1–21 odd	Reteaching p. 75 Practice p. 76
5.6 pp. 202–04	FR all CE all WE 1–5 Application all	FR all CE all WE 1–8 Application all	FR all CE all WE 1–10 Application all	Reteaching p. 77 Practice p. 78
5.7 pp. 208–09	CE all WE 1–9 AR all	CE all WE 1–12 AR all	CE all WE 1–17 AR all	Reteaching p. 79 Practice p. 80
5.8 pp. 212–14	CE all WE 1–13 Brainteaser	CE all WE 1–19 odd Brainteaser	CE all WE 1–23 odd Brainteaser	Reteaching p. 81 Practice p. 82
5.9 pp. 217–19	FR all CE all WE 1–9 AR all	FR all CE all WE 1–13 AR all	FR all CE all WE 1–15 AR all	Reteaching p. 83 Practice p. 84
Chapter Review pp. 220–21	all	all	all	
Chapter Test p. 222	1–11, 13, 14	1–11	all	

CE = Classroom Exercises   WE = Written Exercises   FR = Focus on Reading   AR = Algebra Review   CI = Computer Investigation
NOTE: For each level, all students should be assigned all Mixed Review exercises.

175B

The **Planning Guide** recommends appropriate classroom and written exercises for students of all ability levels, suggests when to incorporate special features into the lessons, and reminds you of the availability of other components.

Attractive Chapter Openers connect math to the world outside the classroom.

Each **Chapter Opener** contains a colorful illustration of a famous person who has made a difference in the world by contributing to our knowledge of the theory and applications of mathematics.

The **Investigation** is a hands-on activity that helps students give concrete meaning to abstract mathematical concepts. Students work cooperatively, in small groups, acquiring interactive problem-solving skills they will use for a lifetime. Follow-up questions help students form conjectures and draw conclusions about the activity.

■■■INVESTIGATION

Manipulatives: Each pair of students makes one proportional divider using two strips of posterboard of equal length (8 inches by $\frac{1}{2}$ inch is a good size). Punch a hole in each strip two inches from the end and join the two strips with a metal fastener or brad.

Working with partners, open the strips to form an angle. Place the divider on a sheet of paper and trace the angle formed by the longer legs onto the paper. With a straightedge, complete the drawing of the triangle on the paper. Repeat this by changing the opening of the dividers and drawing several different triangles. Measure all three angles of each triangle. This activity can help the students to discover Theorem 5.1

Cut out three of the triangles. On one triangle, fold the triangles to find the midpoints of the sides, and then draw the medians.

| medians | perpendicular bisectors | angle bisectors |

On another triangle, fold to find the midpoints of the sides and then use a protractor to draw the perpendicular bisectors of all three sides. On the third triangle, fold to find the angle bisectors. These investigations lead to the relationships of Lesson 5.6.

5 CONGRUENCE

$$E=mc^2$$

Albert Einstein (1879–1955)—Although he failed algebra, Einstein developed the Theory of Relativity that forms the foundation for modern physics and space exploration. He played the violin to relax.

More about the Mathematician

While in school, Einstein acquired the questioning attitude towards assumptions that helped him develop his general theory of relativity. This theory says that the speed of light is absolute while space and time intervals are relative. Einstein used the example of a moving train, relative to a passenger and an observer outside the train, to explain his theory. He was a self-taught mathematical genius whose statement that the energy content (E) of a body can be measured by its mass (m) times the square of the speed of light (c) is the most famous equation of the 20th century. He also developed the idea of the existence of a curved universe and of time as the fourth dimension.

176

More About the Mathematician provides in-depth background information about the mathematician featured in the **Chapter Opener**. The accomplishments and innovations of this person provide an opportunity to discuss the value and usefulness of mathematics in our world.

Easy-to-follow lessons communicate learning goals clearly and intelligibly.

Easy-to-follow lessons make mathematics accessible to your students by featuring succinct objectives, brief introductions, and clear-cut examples.

5.1 Isosceles Triangles

Objectives
To prove and apply theorems about isosceles triangles
To solve problems involving isosceles triangles

Recall that an isosceles triangle has at least two congruent sides. Each of these sides is called a leg. The angle formed by the legs is called the **vertex angle**. The side opposite the vertex angle is called the **base**. The angles opposite the legs are called the **base angles**.

It can be proved that base angles are congruent. The drawing of an auxiliary segment within an isosceles triangle allows the proof to be easily completed.

Theorem 5.1
If two sides of a triangle are congruent, then the angles opposite these sides are congruent. (The base angles of an isosceles triangle are congruent.)

Given: $\triangle ABC$, $\overline{AC} \cong \overline{BC}$
Prove: $\angle A \cong \angle B$

Plan
Construct $\overline{CD}$, the angle bisector of $\angle ACB$.
Then prove that $\triangle ADC \cong \triangle BDC$.

Proof

Statement	Reason
1. $\overline{AC} \cong \overline{BC}$	(S) 1. Given
2. Draw $\overline{CD}$, the bisector of $\angle ACB$, intersecting $\overline{AB}$ at D.	2. Every $\angle$, except a st $\angle$, has exactly one bis.
3. $\angle 1 \cong \angle 2$	(A) 3. Def of $\angle$ bis
4. $\overline{CD} \cong \overline{CD}$	(S) 4. Reflex Prop
5. $\triangle ADC \cong \triangle BDC$	5. SAS
6. $\therefore \angle A \cong \angle B$	6. CPCTC

Corollary
If a triangle is equilateral, then it is also equiangular, and the measure of each angle is 60.

5.1 Isosceles Triangles **177**

Teaching Resources

Manipulative Worksheet 9
Problem Solving Worksheet 5
Project Worksheet 5
Quick Quizzes 34
Reteaching and Practice Worksheets, pp. 67, 68
Transparencies 6A, 6B, 6C

▰▰ GETTING STARTED

Prerequisite Quiz
Solve each equation.

1. $2x + 6 = 20$ 7
2. $x + (2x - 10) + (2x + 10) = 180$ 36
3. The measures of two angles of a triangle are 50 and 50. Find the measure of the third angle.   80
4. The three angles of a triangle have equal measures. Find the measure of each angle.   60
5. The measure of one angle of a triangle is 40. Find the measure of each of the other two angles if they are congruent.   70

Motivator

Have your students construct triangle *ABC* with $AB = AC = 6$ cm and $BC = 4$ cm.

Ask them what appears to be true about the measures of $\angle$B and $\angle$C. Then have them verify by measuring these angles with a protractor.

Have them complete the statement of the following theorem suggested by this construction: If two sides of a triangle are congruent, then _____. the angles opposite these sides are congruent.

Teaching Resources provides a list of all supplementary components available for use with the lesson. These materials contain practice, review, reteaching, and enrichment activities.

The **Prerequisite Quiz** helps you assess what students have learned in previous lessons and determine the overall readiness of the class for the lesson.

The **Motivator** helps students form conjectures about the lesson, investigate their conjectures with manipulatives and models, and then form conclusions.

Examples lead students through the problem-solving process by showing them how to formulate a **Plan** and then guides them step-by-step through the proof.

177

In mathematics, as in any field, knowledge consists of information plus know-how. Know-how in mathematics that leads to mathematical power requires the ability to use information to reason and think creatively and to formulate, solve, and reflect critically on problems.

(NCTM Standards, p. 205)

Teaching Suggestions integrate a variety of topics into the lessons.

The **Teaching Suggestions** relate each lesson to current issues in mathematics education. Located in the side-column of the text, these features provide you with a convenient way to enrich the lessons and make them meaningful to students.

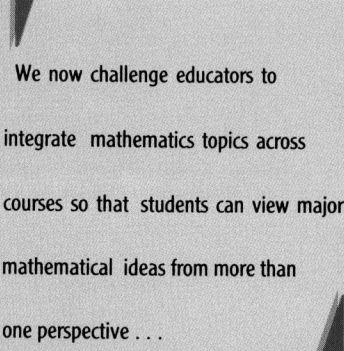

We now challenge educators to integrate mathematics topics across courses so that students can view major mathematical ideas from more than one perspective . . .

(NCTM Standards, p. 252)

The **Lesson Note** helps prepare students for the exercises by identifying concepts and skills from earlier lessons that may need to be reviewed.

Math Connections relates geometry to previously taught material, other areas of mathematics, real-life situations, and the history of mathematics.

Critical Thinking Questions help students develop higher-order thinking skills such as analysis, synthesis, and evaluation.

Common Error Analysis diagnoses potential problem areas and prescribes corrective strategies that focus students' attention on correct procedures.

▰▰ TEACHING SUGGESTIONS

Lesson Note

Review the concept of "converse," which is used in the theorems of this lesson. Use the Motivator to lead students to discover Theorem 5.1. It may be necessary to review the solution of linear equations.

Math Connections

Algebra: Lesson 5.1 introduces the properties of isosceles triangles and connects them with algebra by using algebraic equations to solve for the length of a missing side.

Critical Thinking Questions

Logic: What can you conclude about a given triangle $\triangle ABC$, if $\angle A \cong \angle B$? $\triangle ABC$ is isosceles and $\overline{AC} \cong \overline{BC}$.

Common Error Analysis

Error: Students soon discover that if the measure of the vertex angle of an isosceles triangle is known, they can find the measure of each base angle without using algebra. They subtract the given measure from 180 and then take half of the result. However, they sometimes forget the last step.

Suggest that they add the three angles of the triangle to verify that the sum is 180, before continuing to the next exercise.

EXAMPLE 1 The measure of a base angle of an isosceles triangle is 10 less than twice the measure of the vertex angle. Find the measure of a base angle.

Plan Draw a diagram. Let x = measure of vertex angle. Then $2x - 10$ = measure of a base angle.

Solution
$$m \angle A + m \angle B + m \angle C = 180$$
$$x + (2x - 10) + (2x - 10) = 180$$
$$5x - 20 = 180$$
$$5x = 200$$
$$x = 40$$
$$2x - 10 = (2 \cdot 40) - 10 = 70$$

Thus, the measure of a base angle is 70.

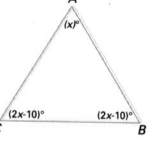

The converse of Theorem 5.1 is given below as Theorem 5.2.

Theorem 5.2 If two angles of a triangle are congruent, then the sides opposite these angles are congruent. (A triangle with two congruent sides is isosceles.)

The following corollary can be proved by using Theorem 5.2.

Corollary If a triangle is equiangular, then it is also equilateral.

The **perimeter** of a triangle is the sum of the lengths of its three sides. Example 2 applies the properties of isosceles triangles to a perimeter problem.

EXAMPLE 2 In $\triangle ABC$, $AC = BC$, $AB = 6$, and the perimeter is 20. Find AC.

Plan Draw $\triangle ABC$. Let x = length of $\overline{AC}$ and of $\overline{BC}$. Use the given perimeter to write an equation.

Solution
$$AC + BC + AB = 20$$
$$x + x + 6 = 20$$
$$2x + 6 = 20$$
$$x = 7$$

Thus, $AC = 7$.

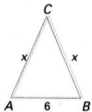

178 Chapter 5 Congruence

Additional Example 1

The measure of the vertex angle of an isosceles triangle is 20. Find the measure of each base angle. 80

Additional Example 2

In $\triangle PQR$, $\angle P \cong \angle R$, $PR = 16$ and the perimeter is 36. Find QP. 10

Classroom Exercises help students learn through success.

The **Classroom Exercises** reinforce the material in the lesson by providing students with the opportunity to practice newly learned skills.

EXAMPLE 3 Given: $\angle 3 \cong \angle 4$, D is the midpoint of $\overline{EC}$; $\angle 1 \cong \angle 2$
Prove: $\overline{EA} \cong \overline{CB}$

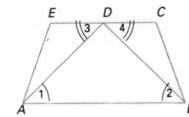

Plan Show $\overline{EA} \cong \overline{CB}$ by proving $\triangle EDA \cong \triangle CDB$. $\angle 1$ and $\angle 2$ are *not* angles of those triangles; however, since $\angle 1 \cong \angle 2$, $\overline{AD} \cong \overline{BD}$.

Proof

Statement	Reason
1. $\angle 3 \cong \angle 4$	(A) 1. Given
2. D is midpt of $\overline{EC}$.	2. Given
3. $\overline{ED} \cong \overline{CD}$	(S) 3. Def of midpt
4. $\angle 1 \cong \angle 2$	4. Given
5. $\overline{AD} \cong \overline{BD}$	(S) 5. Sides opp $\cong$ $\angle$s of a $\triangle$ are $\cong$.
6. $\triangle EDA \cong \triangle CDB$	6. SAS
7. $\therefore \overline{EA} \cong \overline{CB}$	7. CPCTC

Classroom Exercises

1. Name the vertex angle, base, legs, and base angles of the isosceles triangle. V: $\angle R$; B: $\overline{QP}$; L: $\overline{RQ}$, $\overline{RP}$; B $\angle$s: $\angle Q$, $\angle P$

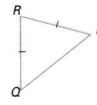

Complete.

2. If $\overline{FG} \cong \overline{CG}$, then $\underline{\angle 1} \cong \underline{\angle 4}$.

3. If $\overline{EG} \cong \overline{ED}$, then $\underline{\angle 7} \cong \underline{\angle 3}$.

4. If $\angle 6 \cong \angle 5$, then $\underline{\overline{AG}} \cong \underline{\overline{BG}}$.

5. If $\angle 2 \cong \angle 7$, then $\underline{\overline{DG}} \cong \underline{\overline{DE}}$.

6. If $\overline{EG} \cong \overline{DG}$, then $\underline{\angle 2} \cong \underline{\angle 3}$.

7. If $\overline{AG} \cong \overline{BG}$, then $\underline{\angle 6} \cong \underline{\angle 5}$.

8. If $\angle 1 \cong \angle 4$, then $\underline{\overline{FG}} \cong \underline{\overline{CG}}$.

5.1 Isosceles Triangles **179**

Checkpoint

Refer to the figure below.

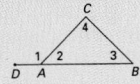

1. Given: $\overline{AC} \cong \overline{BC}$, m $\angle 4$ = 100. Find m $\angle 2$. 40
2. Given: $\overline{AC} \cong \overline{BC}$, AB = 4, and the perimeter of $\triangle ABC$ is 24. Find BC. 10
3. Given: $\overline{AC} \cong \overline{BC}$, m $\angle 1$ = 150. Find m $\angle 4$. 120
4. Given: $\overline{AC} \cong \overline{BC}$, m $\angle 4$ is three times m $\angle 3$. Find m $\angle 3$. 36
5. Given: $\angle 3$ is a supplement of $\angle 1$. Prove: $\overline{AC} \cong \overline{BC}$

Statement	Reason
1. $\angle 3$ supp $\angle 1$	1. Given
2. $\angle 2$ supp $\angle 1$	2. If outer rays of 2 adj $\angle$s form a straight $\angle$, the $\angle$s are supp.
3. $\angle 3 \cong \angle 2$	3. Supp of the same $\angle$ are $\cong$.
4. $\overline{AC} \cong \overline{BC}$	4. Sides opp $\cong$ $\angle$s of a $\triangle$ are $\cong$.

Closure

Ask your students to explain which angles of an isosceles triangle are the base angles and which sides of an isosceles triangle are the legs. Draw a triangle *PYT* given *PY* = *YT* and ask them what conclusion they can give based upon the theorem in this lesson. Have them explain how they would find m $\angle P$ if m $\angle P$ = $2x - 10$ and m $\angle T$ = $x + 30$.

■■ FOLLOW UP

Guided Practice
Classroom Exercises 1–7

Independent Practice

Ⓐ Ex. 1–17, Ⓑ Ex. 18–24, Ⓒ Ex. 25–27
Basic: WE 1–17 odd
Average: WE 1–24 odd
Above Average: WE 1–13 odd, 19–27 odd

Checkpoint provides a quick way to determine what students have learned from the lesson and what concepts and skills need to be reinforced.

Closure provides questions that help students summarize, communicate, and form conclusions about the major concepts of the lesson.

Follow-up suggests *Guided Practice* exercises that students can do in class to ensure a successful homework experience.

Independent Practice groups exercises according to A, B, and C difficulty levels for basic, average, and above-average students.

Additional Example 3

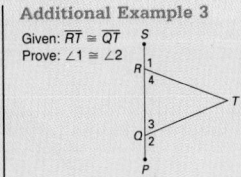

Given: $\overline{RT} \cong \overline{QT}$
Prove: $\angle 1 \cong \angle 2$

Proof: $\overline{RT} \cong \overline{QT}$ (Given); $\angle 3 \cong \angle 4$ ($\angle$s opp $\cong$ sides of a $\triangle$ are $\cong$); $\angle 1$ supp $\angle 4$, $\angle 3$ supp $\angle 2$ (if outer rays of 2 adj $\angle$s form a st $\angle$, the $\angle$s are supp); $\angle 1 \cong \angle 2$ (supp of $\cong$ $\angle$s are $\cong$)

179

Completely worked-out **Additional Examples** may be used for reteaching or as a quiz to evaluate students' understanding of the lesson.

Written exercises clarify and reinforce newly-learned concepts and skills.

An extensive variety of exercises covers all topics from the lesson and challenges students of all ability levels.

Additional Answers provides added convenience by supplying solutions which could not be reproduced on the pupil's page in the side column of the *Annotated Teacher's Edition*.

Additional Answers

Written Exercises

14. Statement	Reason
1. $\overline{AB} \cong \overline{BC}$	1. Given
2. $\angle 3 \cong \angle 1$	2. $\angle$s opp $\cong$ sides are $\cong$.
3. $\angle 2 \cong \angle 1$	3. Vert $\angle$s are $\cong$.
4. $\angle 3 \cong \angle 2$	4. Trans Prop of Congr

15. Statement	Reason
1. $\angle 4 \cong \angle 2$	1. Given
2. $\angle 1 \cong \angle 2$	2. Vert $\angle$s are $\cong$.
3. $\angle 4 \cong \angle 1$	3. Trans Prop of Congr
4. $\overline{AB} \cong \overline{AC}$	4. Sides opp $\cong$ $\angle$s are $\cong$.

18. Statement	Reason
1. $\overline{BA} \parallel \overline{CD}$, $\overline{CB} \cong \overline{DB}$	1. Given
2. $\angle C \cong \angle D$	2. $\angle$s opp $\cong$ sides are $\cong$.
3. $\angle 1 \cong \angle C$	3. If lines are $\parallel$, then corr $\angle$s are $\cong$.
4. $\angle 2 \cong \angle D$	4. If lines are $\parallel$, then alt int $\angle$s are $\cong$.
5. $\angle 1 \cong \angle 2$	5. Trans Prop of Congr

19. Statement	Reason
1. $\overline{OV} \cong \overline{RW}$, $\overline{PQ} \cong \overline{PR}$, $\overline{PS} \cong \overline{PT}$	1. Given
2. $\angle Q \cong \angle R$	2. $\angle$s opp $\cong$ sides are $\cong$.
3. $PQ - PS = PR - PT$	3. Equations may be subtracted.
4. $PQ - PS = SQ$, $PR - PT = TR$	4. Seg Add Post
5. $SQ = TR$	5. Sub
6. $\triangle SQV \cong \triangle TRW$	6. SAS
7. $\angle 1 \cong \angle 2$	7. CPCTC

20. Statement	Reason
1. $\overline{DA} \cong \overline{DB}$, $\angle 1 \cong \angle 2$	1. Given
2. m $\angle DAB =$ m $\angle DBA$	2. $\angle$s opp $\cong$ sides are $\cong$.
3. m $\angle 1 +$ m $\angle DAB =$ m $\angle 2 +$ m $\angle DBA$	3. Equations may be added.
4. m $\angle 1 +$ m $\angle DAB =$ m $\angle CAB$, m $\angle 2 +$ m $\angle DBA =$ m $\angle CBA$	4. Angle Add Post
5. m $\angle CAB =$ m $\angle CBA$	5. Sub
6. $\overline{AC} \cong \overline{BC}$	6. Sides opp $\cong$ $\angle$s are $\cong$.

Written Exercises

Find the measure of each angle of the triangle.

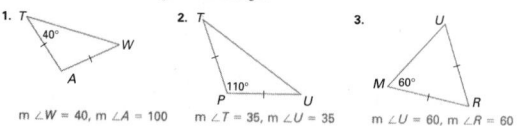

1. m $\angle W = 40$, m $\angle A = 100$
2. m $\angle T = 35$, m $\angle U = 35$
3. m $\angle U = 60$, m $\angle R = 60$

In $\triangle ABC$, $AC = BC$. Find the measure of each angle of the triangle.
(Exercises 4–5)

4. m $\angle C = x$, m $\angle A = 2x - 20$ 44, 68, 68
5. m $\angle C = 2x + 20$, m $\angle A = 6x + 10$ 40, 70, 70
6. The measure of the vertex angle of an isosceles triangle is twice the measure of a base angle. Find the measure of a base angle. 45
7. The measure of the vertex angle of an isosceles triangle is 20 less than twice the measure of a base angle. Find the measure of the vertex angle. 80

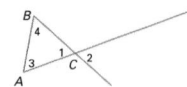

8. Given: $\overline{BF} \cong \overline{CF}$, m $\angle 5 = 130$. Find m $\angle 4$. 25
9. Given: $\overline{BF} \perp \overline{FC}$, $\overline{BF} \cong \overline{FC}$. Find m $\angle 2$. 45
10. Given: $\overline{BC} \cong \overline{FC}$, m $\angle 4 = 30$. Find m $\angle 1$. 105
11. Given: $\overline{FB} \cong \overline{FC}$, $FB = 10$, $BC = 15$. Find the perimeter of $\triangle FBC$. 35
12. Given: $\angle 2 \cong \angle 3$, $BF = 7\frac{1}{2}$, $BC = 11$. Find the perimeter of $\triangle FBC$. 26
13. Given: $\overline{BF} \cong \overline{CF}$, $BC = 20$, the perimeter of $\triangle BFC$ is 48. Find FC. 14

14. Given: $\overline{AB} \cong \overline{BC}$
    Prove: $\angle 3 \cong \angle 2$
15. Given: $\angle 4 \cong \angle 2$
    Prove: $\overline{AB} \cong \overline{AC}$

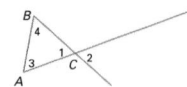

16. The perimeter of an equilangular triangle is 45 cm. Find the measure of each side. 15 cm
17. $\triangle XYZ$ is equilangular and XY is 27 cm. Find the perimeter of $\triangle XYZ$. 81 cm

180   Chapter 5 Congruence

Enrichment

Draw this figure on the chalkboard.

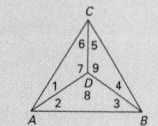

Have the students tell whether each set of given information is sufficient to prove that $\triangle ABC$ is equilateral. (If the information is not sufficient, the students should show counterexamples.)
1. $\overline{CD} \cong \overline{DA} \cong \overline{DB}$ ns
2. $\angle 1 \cong \angle 6$, $\angle 2 \cong \angle 3$, $\angle 4 \cong \angle 5$ ns
3. $\angle 7 \cong \angle 8 \cong \angle 9$ ns
4. $\angle 1 \cong \angle 2 \cong \angle 3 \cong \angle 4 \cong \angle 5 \cong \angle 6$ s
5. $\overline{AD} \cong \overline{DB}$, $\angle 1 \cong \angle 2$, $\angle 4 \cong \angle 3$ ns

180

The **Enrichment** problem is a challenging exercise for students who have mastered the lesson content. The solution is included.

Frequent review opportunities help you evaluate students' understanding.

The **Mixed Review**, which occurs at the end of most lessons, and the **Midchapter Review** reinforce newly learned concepts, as well as maintain skills from previous lessons. For the convenience of the student, exercises are referenced to the chapter and lesson where the related skill or concept was taught.

C-level exercises, numbered in red, require students to use critical-thinking skills to solve these challenging problems.

The **Mixed Review** reinforces problem-solving strategies, concepts, and skills from the material from earlier lessons.

The main purpose of evaluation, as

described in these standards, is to help

teachers better understand what students

know and make meaningful

instructional decisions.

(NCTM Standards, p. 189)

---

18. Given: $\overline{BA} \parallel \overline{CD}$,
$\overline{CB} \cong \overline{DB}$
Prove: $\angle 1 \cong \angle 2$

19. Given: $\overline{PQ} \cong \overline{PR}$,
$\overline{PS} \cong \overline{PT}$,
$\overline{QV} \cong \overline{RW}$
Prove: $\angle 1 \cong \angle 2$

20. Given: $\overline{DA} \cong \overline{DB}$,
$\angle 1 \cong \angle 2$
Prove: $\overline{AC} \cong \overline{BC}$

21. Prove the corollary to Theorem 5.1.
22. Prove Theorem 5.2. (HINT: Construct an auxiliary segment.)
23. $\triangle GHK$ is isosceles, with $\overline{GH}$ as the base. Point L is on $\overline{GH}$ such that $\overline{KL}$ bisects $\angle K$. m $\angle G = 40$. Find m $\angle GKL$ and m $\angle GLK$.   m $\angle GKL = 50$, m $\angle GLK = 90$
24. Prove the corollary to Theorem 5.2.
25. Prove: In an isosceles triangle, the bisector of an exterior angle of the vertex angle is parallel to the base.
26. Given: $\overline{BA}$ is perpendicular to every line in plane $\mathcal{M}$ passing through A; $\angle ACB \cong \angle ADB$
Prove: $\triangle BCD$ is isosceles.
27. Given: $\overline{AE} \perp \overline{CD}$, E is the midpoint of $\overline{CD}$; $\overline{BA}$ is perpendicular to every line in plane $\mathcal{M}$ passing through A.
Prove: $BC = BD$

Mixed Review

1. The measure of an angle is twice the measure of its complement. Find the measure of the angle.   1.6   60
2. $\angle A$ and $\angle B$ are vertical angles. m $\angle A = 4x - 40$, m $\angle B = 2x + 60$. Find m $\angle A$.   2.7   160
3. The measure of an exterior angle of a triangle is 150. The measure of one of the two remote interior angles is $\frac{2}{3}$ the measure of the other. Find the measure of each remote interior angle.   3.7   60, 90

5.1 Isosceles Triangles   **181**

21.

Statement	Reason
1. $\triangle ABC$ is equilateral.	1. Given
2. $\overline{AB} \cong \overline{AC}$, $\overline{CA} \cong \overline{CB}$	2. Def of equil $\triangle$
3. $\angle C \cong \angle B$, $\angle B \cong \angle A$	3. $\angle$s opp $\cong$ sides are $\cong$.
4. $\angle C \cong \angle A$	4. Trans Prop of Congr
5. m $\angle C$ + m $\angle B$ + m $\angle A = 180$	5. Sum of meas of $\angle$s of $\triangle = 180$.
6. m $\angle B$ + m $\angle B$ + m $\angle B = 180$, or $3 \times$ m $\angle B = 180$	6. Sub
7. m $\angle B = 60$	7. Div Prop of Eq
8. m $\angle A = 60$, m $\angle C = 60$	8. Sub

22.

Statement	Reason
1. m $\angle A$ = m $\angle B$	1. Given
2. m $\angle ACD$ = m $\angle BCD$	2. Def $\angle$ bis
3. $CD = CD$	3. Reflex Prop of Eq
4. $\triangle ACD \cong \triangle BCD$	4. AAS
5. $\overline{AC} \cong \overline{BC}$	5. CPCTC

24.

Statement	Reason
1. $\triangle ABC$ is equiangular.	1. Given
2. m $\angle A$ = m $\angle B$, m $\angle A$ = m $\angle C$	2. Def equiangular $\triangle$
3. $CA = CB$, $CB = AB$	3. Sides opp $\cong$ $\angle$s are $\cong$.
4. $CA = AB$	4. Sub
5. $\triangle ABC$ is equil	5. Def of equil $\triangle$

See page 651 for the answers to Ex. 25–27.

181

Expanded side columns make it possible to print proofs in a two-column format, making them easier to read when teaching a lesson at the chalkboard or grading papers.

A complete selection of ancillary materials enriches the lessons.

The supplementary materials described below provide additional activities that incorporate important topics in mathematics education. Calculator and computer activities allow students to feel comfortable using these tools and to realize the importance of technology in today's world.

Nor do we believe that textbooks should drive instruction. Rather, other materials, . . . such as manipulatives and courseware, must be developed, in addition to new textbooks.

(*NCTM Standards*, p. 252)

Teacher's ResourceBank™

The *Teacher's ResourceBank™* contains an array of materials that provide additional practice, review exercises, and an extensive testing program.

*Activity Worksheets
(Blackline Masters with Answer Key)

The **Activity Worksheets** help you integrate the following topics into your lessons.

- Problem Solving Worksheets
- Manipulative Worksheets
- Project Worksheets
- Application Worksheets
- Technology Worksheets
- Technology Software

*Also available for sale separately.

*Reteaching and Practice Worksheets
(Blackline Masters with Answer Key)

For each lesson, there is one *Reteaching* and one *Practice Worksheet*. These can be used together to reinforce the lesson and provide additional practice opportunities, or used separately to individualize instruction.

*Tests
(Blackline Masters with Answer Key)

The *Tests* provide opportunities for assessment at each step of the learning process.

- Prerequisite Quiz
- Checkpoint Quiz
- Chapter Tests (Forms A and B)
- Cumulative Tests (Forms A and B)

Other Ancillary Materials

Instructional Transparencies and Teacher's Notes

The **Instructional Transparencies** provide convenience for teachers who use an overhead projector in their classroom. Over forty colorful transparencies (at least one transparency per chapter) provide visual representations of abstract mathematical concepts such as proofs, coordinate graphs, and overlapping triangles.

Solution Key

The Solution Key contains worked-out answers for all A, B, and C level exercises in the text and end-of-chapter activities.

Implementing The NCTM *Standards* with *Geometry*

As a geometry teacher of the 1990s, you are probably involved in implementing the *Curriculum and Evaluation Standards for School Mathematics*, published by the National Council of Teachers of Mathematics. The professional articles in this section are designed to help you use *Geometry* to integrate the NCTM *Standards* into your classroom.

NCTM's vision of the future "sees students studying much of the same mathematics currently taught but with quite a different emphasis" and builds on the premise that "*what* a student learns depends to a great degree on *how* he or she has learned it." (p. 5, NCTM *Standards*)

How will your students learn geometry? The six articles that follow are designed to help you apply new approaches in your classroom.
- Manipulatives and Models
- Cooperative Learning Groups
- Problem-Solving Strategies
- Critical Thinking Questions
- Reading, Writing, and Discussing Math
- Technology

These articles can help you achieve these five general goals for all students as listed in the NCTM *Standards*.
1. Learning to value mathematics
2. Becoming confident in one's own ability
3. Becoming a mathematical problem solver
4. Learning to communicate mathematically
5. Learning to reason mathematically

Rest assured, however, that the NCTM has not discarded the traditional content of your geometry course. *Standard 7: Geometry from a Synthetic Perspective* states (in part) that the curriculum should include topics that teach students to
- classify figures in terms of congruence and similarity and apply these relationships;
- deduce properties of, and relationships between, figures from given assumptions;
- develop an understanding of an axiomatic system through investigating and comparing various geometries.

You will also find ample material that relates directly to these topics as you survey *Geometry*, as well as topics that fulfill the requirements stated in *Standard 8: Geometry from an Algebraic Perspective.* ■

Manipulatives and Models

What are manipulatives and models?

Manipulatives and models are physical representations of such geometric figures as lines, angles, planes, and triangles. Manipulatives and models give students hands-on experiences that help them visualize the abstractions they write about in problems and proofs.

In most situations, new mathematical ideas should continue [in grades 9-12] to be introduced at the concrete level. (p. 131, NCTM *Standards*)

Hands-on experiences with lines and angles (by tearing off the corners of triangles), with area and volume (by building a cube), and with using tools such as the ruler, the compass, and the straightedge, provide a concrete introduction to the abstractions of geometry.

At least thirty percent of the students in geometry classes are at the concrete operational level of cognitive development. (p. 239, Learning and Teaching Geometry, K-12, 1987 *NCTM Yearbook*) Manipulatives and models play a role in enabling these students to advance to the formal operational stage, where they can hypothesize and reason deductively—some of the key cognitive abilities needed for success in geometry.

How do I use manipulatives and models with my students?

Take every opportunity to use physical objects to illustrate the mathematics of the lesson. Use marbles or pieces of gravel to represent points; use pencils, pens, straws, or toothpicks for lines, rays, and segments; use the desktop, walls, notebook paper, or cardboard for planes; use cans for cylinders, boxes for rectangular solids, paper drinking cones for right circular cones. Encourage the students to bring models to class and to make their own models.

For more formal experiences, use the Manipulative Worksheets in the *Teacher's ResourceBank*™. These worksheets provide a list of needed materials and detailed step-by-step instructions. After you distribute the materials and the worksheets, walk among the students and help them keep on task by asking questions and encouraging them to keep trying. In most cases there is a visible outcome (such as a model, diagram, or data sheet) which you can use to verify that the students successfully completed the activity. Either on the worksheet or in the *Annotated Teacher's Edition* you will find closure for the activity in the form of questions that encourage students to formulate conjectures and generalizations.

How will models help me implement the *Standards*?

The *Standards* "call for a shift in emphasis from a curriculum dominated by memorization of isolated facts and procedures and by proficiency with paper-and-pencil skills to one that emphasizes conceptual understandings, multiple representations and connections, mathematical modeling, and mathematical problem solving." (p. 125, NCTM *Standards*)

Manipulatives and models help you and your students take the step from doing everything with paper and pencil to using dimensional aids to help create conceptual understandings. These devices help stimulate that all-important student comment, so rewarding to the teacher, "Oh, now I see!" ■

Cooperative Learning Groups

What are cooperative learning groups?

The *Standards* point out that "instructional settings that encourage investigation, cooperation, and communication foster problem posing as well as problem solving." (p. 138, NCTM *Standards*) Cooperative learning groups are such an instructional setting.

Cooperative learning means more than simply working in teams or groups. Some of the distinguishing characteristics of true cooperative learning are these.

- Groups are heterogeneous to increase tolerance and understanding.
- Students are responsible for their own participation and for making sure that others participate.
- Help is sought within the group and the teacher is asked only when the whole group agrees to ask a question.
- Consensus on the answer is required from the whole group.
- The group evaluates their own strategies and ideas rather than relying on the teacher for this evaluation.
- The teacher provides help by means of questions rather than giving hints.
- The students who learn at a faster pace do NOT do the task alone and then help other students.
- One student does NOT do all the work and then the others sign off on it.
- Participants encourage each other to explain answers and how they arrived at them.
- Ideas, not people, are criticized.
- The logic of an idea, not peer pressure or majority rule, determines its value.

Rather than working in competition with other students or as an individual working for a personal best, students in cooperative learning groups are working together towards shared goals; the success of each student in the group depends on the success of the group as a whole, which in turn depends on the success of each member. Each individual is accountable for mastering the material, and also for helping everyone in the group to master the material. Thus, cooperative learning groups foster interdependence.

The benefits of this learning structure incorporate those of peer tutoring (in which the tutor also learns by helping), increased understanding of diverse viewpoints, and simulation of the way work is done on the job.

How do I use cooperative learning groups with my students?

Students may not be used to interdependent groups in a school setting, though they will usually have experienced the team approach in games. The first time you organize cooperative learning groups, discuss a set of procedural rules for the groups based on the characteristics in the foregoing list. The first time, let students form their own groups of four or five. Later, after they have begun to learn how to work together, you can use random methods to form groups that are heterogeneous.

For the first try, assign a familiar task such as producing study notes for a test. Tell the students that the goal of the activity includes learning the process (cooperation) as well as producing the product (the notes). Encourage each student to take responsibility for

- participating and contributing,
- ensuring that all other members of the group participate,
- staying on task, and
- producing a high quality final product.

Once the procedures are understood (though the groups will probably need much practice before they can adhere effectively to them), use the appropriate *Activity Worksheets* (Projects, Applications, Manipulatives, Problem Solving, and Technology), and the Investigations that begin each chapter, to provide geometry-related tasks for the groups.

You can also use cooperative learning groups with the Quick Quizzes before the lesson to verify prerequisite skills (Prerequisite Quiz, also in the *Annotated Teacher's Edition*) and after the lesson to verify understanding (Checkpoint, also in the *Annotated Teacher's Edition*). The Classroom Exercises that follow each lesson can also be done in cooperative groups to prepare students for the Written Exercises.

Conclude each session by a discussion about how the procedure worked, and how they can improve the process.

How will groups help me implement the *Standards*?

Research indicates that "cooperative learning experiences tend to promote higher achievement than do competitive and individualistic learning experiences." (p. 15, *Circles of Learning: Cooperation in the Classroom*, by Johnson, Johnson, Holubec, and Roy) This results in part because the discussion process promotes the development of critical thinking skills and the group setting increases student motivation. The peer support and the oral repetition of information that occurs in the group also contribute to efficient learning.

The use of cooperative learning groups clearly belongs in an effective variety of instructional methods as called for by the *Standards*.

A variety of instructional methods should be used in classrooms in order to cultivate students' abilities to investigate, to make sense of, and to construct meanings from new situations; to make and provide arguments for conjectures; and to use a flexible set of strategies to solve problems from both within and outside mathematics.
(p. 125, 128, NCTM *Standards*) ■

Problem-Solving Strategies

What are problem-solving strategies?

The NCTM *Standards* recommend increased attention to "problem solving as a means as well as a goal of instruction." (p. 129 NCTM *Standards*) As a means of instruction, problem-solving strategies are taught overtly so that students can choose from many possible ways to begin solving a given problem.

Some students, especially those who have been unsuccessful in mathematics, have developed, on their own, only one problem-solving strategy—guess and give up. Teaching problem-solving strategies overtly shows students that there are many procedures used by successful problem solvers to solve a problem. When you use problem solving as a means of instruction, students realize that getting the right answer is not a matter of luck, innate talent, or magic. They begin to see that they, too, can experience success at mathematics.

In order to create an atmosphere of successful learning, change the emphasis in your classroom from correct answers to effective process and procedures. Pay attention to the reasons and steps that go into solving a problem. Give credit for the process. Ask for, and value, the explanations students give to tell how they reached a solution. Emphasize that there may be more than one way to arrive at a correct answer. Encourage discussions about the advantages (such as efficiency and generalizability) of various methods. For some students, it may be appropriate to point out that the easiest and most efficient way for a

human to solve a problem may not be the most efficient way for a computer program to solve that same problem.

How do I use these strategies with my students?

To use problem-solving strategies with your students, follow these guidelines.

1. Teach problem-solving strategies directly by using the Problem Solving Activity Worksheets and the Problem-Solving Strategies features in the text.
2. When you discuss the Examples in the Lesson, point out the step where you formulate a Plan for the problem.
3. Ask students to tell the class how they arrived at an answer.
4. Discuss with the class whether another strategy would have worked; compare different strategies that students may have used for a given problem.
5. Give recognition and credit for correct procedures as well as for correct answers; teach students that there may be only one correct answer but several correct ways to obtain that answer.
6. In reteaching students, ask them to explain their thoughts as they work the problem. This gives you a chance to suggest more effective problem-solving strategies.

How will problem solving help me implement the *Standards*?

The importance of problem solving to all education cannot be overestimated. To serve this goal effectively, the mathematics curriculum must provide many opportunities for all students to meet problems that interest and challenge them and that, with appropriate effort, they can solve. (p. 139, NCTM Standards) ■

Critical Thinking Questions

What are critical thinking questions?

Critical thinking questions ask the students to engage in mathematical thinking and to construct, symbolize, apply, and generalize mathematical ideas. This thinking can include investigations, constructing meanings from new situations, making and providing arguments for conjectures, working cooperatively, and creative and self-directed learning.

"Critical thinking" refers to types of thinking that are of a higher order in Bloom's taxonomy; these include application, analysis, synthesis, and evaluation.

Application involves applying concepts and ideas to new situations and is often signalled by these verbs: apply, build, choose, solve, plan, develop, construct, demonstrate, show.

Analysis means finding the underlying structure and breaking it down into stages or processes. This frequently involves looking for patterns and classifying examples and is often signalled by these verbs: relate, classify, compare, contrast, diagram, analyze, recognize.

Synthesis means bringing together data from various sources to come up with a new conclusion and is often signalled by these verbs: design, create, develop, make up, what happens if, invent, write a formula for.

Evaluation means judging the quality or worth of something, for example, which method works, or is more efficient. It is often signalled by these expressions: which is better, prove or disprove, evaluate, conclude, defend, choose, select, do you agree, judge, what do you think.

For example, in your geometry class you might ask the students to prove one triangle congruent to another, and then discuss alternative approaches. You could ask questions such as these—if a problem uses a coordinate approach, what might the transformational (rotate, reflect, etc.) approach be? How could a synthetic (through postulates and proofs) approach also be done with coordinates and an informal argument? Critical thinking also includes investigating and exploring to derive properties, generalizations, and procedures.

How do I use these questions with my students?

Critical thinking questions appear throughout *Geometry*. The Investigation that introduces each chapter and the Motivator that introduces each lesson (in the *Annotated Teacher's Edition*) often include such questions.

The Teaching Suggestions for each lesson include Critical Thinking questions pertinent to that lesson. Many of the *Activity Worksheets* end with critical thinking questions.

How will critical thinking help me implement the *Standards*?

The *Standards* call for "an environment that encourages students to explore, formulate and test conjectures, prove generalizations, and discuss and apply the results of their investigations." (p. 128, NCTM *Standards*)

Critical thinking questions are a vital part of such an environment. ■

Reading, Writing, and Discussing Math

What is the role of reading, writing, and discussing math?

The fourth general goal set forth by the *Standards* is "Learning to communicate mathematically." In the past, we have sometimes allowed students to spell mathematical terms incorrectly, or to read aloud "*x* two" instead of "*x*-squared" or "*x* to the power of two" because, as students hastened to explain, "This isn't English class." We now recognize that it is just as important that correct English be used in math class as it is that correct math be used in science class. These disciplines are inter-related and teach skills which will be integrated on the job. When we require students to use their newly-acquired skills only in a particular class, we participate in an artificial separation of knowledge into school courses.

We also now realize that precise and accurate communication about mathematics is closely connected to doing precise and accurate mathematics. For the classroom teacher, communication about mathematics brings an extra benefit. When a student writes or speaks about a problem, the teacher can often identify a mistaken or incomplete understanding that leads to an error in problem solving.

How do I encourage my students to communicate mathematically?

Ask a student to read an exercise (from the Classroom Exercises, for example) aloud and then answer the question. This helps you assess whether the student knows the vocabulary and the meaning of the various symbols.

Ask a student to put an exercise from the previous night's homework (in the Written Exercises) on the chalkboard and then explain the steps and process to the class. You could then ask if anyone else followed a different procedure to reach the same result. After the second student has explained the alternate method, ask the class to discuss the differences and the advantages and drawbacks of each method. This works particularly well with proofs, as there are often two equally valid ways to reach the same conclusion.

Ask students frequently to justify their conclusions (*before* you say whether or not the answer is correct). This encourages them to look beyond a correct answer to think about and defend their procedures and reasoning.

Encourage students to verbalize conjectures *before* they read in the text about certain facts and relationships. Discuss the closure questions that are given in *Geometry* at the end of Investigations that begin each chapter (in the *Annotated Teacher's Edition*) and the various *Activity Worksheets* (in the *Teacher's ResourceBank™*). Ask students to keep a math journal in which they record their questions, ideas, and reactions to the activities and discussions in class and a summary of what they read in the text.

Students learn to value those activities that they see you value. If you take time to listen to them as they learn to communicate mathematically, and if you allow their reading, writing, and discussing mathematics to contribute to your on-going evaluation of their achievements, they will know that these things really do count.

How will communicating mathematically help me meet the *Standards*?

The *Standards* (p. 6) put it this way:
  This is best accomplished in problem situations in which students have an opportunity to read, write, and discuss ideas in which the use of the language of mathematics becomes natural. As students communicate their ideas, they learn to clarify, refine, and consolidate their thinking. ■

Technology

What is the role of technology?

The NCTM *Standards* speak of "removing the 'computational gate' to the study of high school mathematics" (p. 130, NCTM *Standards*) and goes on to say this:

  By assigning computational algorithms to calculator or computer processing, this curriculum seeks not only to move students forward but to capture their interest. (p. 130, NCTM *Standards*)

It is particularly important to note that the non-college-intending student must have better preparation for the jobs of tomorrow.

  The ever-increasing role of technology in our society further argues for a curriculum that moves all students beyond computation. (p. 130, NCTM *Standards*)

Technology—the use of the calculator, computer, and graphing calculator—enables students to "study mathematics that is more interesting and useful and not characterized as remedial" and this in turn "will enhance students' self-concepts as well as their attitudes toward, and interest in, mathematics." (p. 131, NCTM *Standards*)

How do I use technology with my students?

Almost everyone today uses a computer at work. The use of computers on the job and at home continues to grow. Make sure that *all* your students have access to whatever technology is available. The emphasis in today's curriculum is not on writing programs, but on investigation and foreshadowing of mathematical ideas and applications by means of the computer.

The Computer Investigation pages in *Geometry* pose problems that can be solved by using an interactive computer

software package that constructs angles, triangles, and other figures, such as *The Geometric Supposer* from Sunburst Communications. The Technology Worksheets in the *Teacher's Resource-Bank™* provide students with problems that use a computer, a calculator, or a graphing calculator to explore various topics in geometry and related mathematics. Use the Technology Software disk to run the BASIC programs that are part of the Technology Worksheets. Those Technology Worksheets that use the graphing calculator (a Casio fx-7000G) assume no prior knowledge of the calculator; complete instructions help students to begin using this new technology.

In the *Geometry* text, a special icon, a calculator key marked with a C, tells you when problems are especially suited to the calculator. However, you may want to encourage students to use handheld calculators whenever they wish; this simulates mathematics as it is really used, both on the job and in scientific applications. Frequent use of calculators also helps students learn the importance of *estimating* before they calculate, and *checking* afterwards to be sure that answers are reasonable. "Appropriate use of calculators enhances children's understanding and mastery of arithmetic." (p. 47, *Everybody Counts, A Report to the Nation on the Future of Mathematics Education*, published by the National Academy of Sciences) The *Standards* also make this point:

  Contrary to the fears of many, the availability of calculators and computers has expanded students' capability of performing calculations. There is no evidence to suggest that the availability of calculators makes students dependent on them for simple calculations. (page 8, NCTM *Standards*)

Specific technology resources are listed for each chapter in *Geometry* within a Technology paragraph on the introductory interleaf pages. Read this

introduction to see how technology can be used in the following chapter.

How will technology help me implement the *Standards*?

The *Standards* say this:

> For example, students should first use an interactive computer software package that allows experimentation with figures and relations to observe [a relationship] across several trials... In the second phase, they would provide a deductive argument verifying their discovery. (page 159, NCTM *Standards*)

The Computer Investigation pages in *Geometry* suggest such experiments. ■

Summary

At the beginning of this section, we listed the five general goals for all students from the NCTM *Standards*.

1. Learning to value mathematics
2. Becoming confident in one's own ability
3. Becoming a mathematical problem solver
4. Learning to communicate mathematically
5. Learning to reason mathematically

These articles have explained how various instructional techniques used with the textbook, special features, and *Activity Worksheets* from *Geometry* can help you achieve those goals.

1. Cooperative learning groups help students value mathematics, as do the Applications worksheets and features in *Geometry*, and the Math Connections discussed in the *Annotated Teacher's Edition*.
2. Manipulatives and models, cooperative learning groups, problem-solving strategies, critical thinking questions, and technology used with *Geometry* and the accompanying worksheets all help students become confident in their own abilities.
3. Problem-solving strategies, and the opportunities to practice and increase them provided by the Problem Solving features in the textbook, as well as the Problem Solving Activity Worksheets, help students become mathematical problem solvers.
4. Cooperative learning groups, closure questions for each lesson in the

Annotated Teacher's Edition, the questions that end the Activity Worksheets, the features called Focus on Reading, and questions throughout the Written Exercises that say "Explain" and "Why," all help students to learn to communicate mathematically.

5. The entire *Geometry* text, with its accompanying materials, helps the students learn to reason mathematically. The traditional content on synthetic geometry, which teaches the students to write formal and informal proofs, strengthens their reasoning abilities and their skills in communicating that reasoning to others. The chapters on geometry from an algebraic perspective (Chapter 10: *Coordinate Geometry* and Chapter 15: *Transformations*) also broaden the students' ability to reason mathematically.

The *Standards* make this comment (in the context of program evaluation):

> The mathematics classroom envisioned in the Standards *is one in which calculators, computers, courseware, and manipulative materials are readily available and regularly used in instruction.* (p. 243, NCTM *Standards*)

Geometry provides you with materials, in the textbook, in the *Annotated Teacher's Edition*, and in the *Activity Worksheets*, to help you meet these goals and standards as you teach your students. ■

Symbol List

$\overleftrightarrow{MN}$	line MN
MN	distance between M and N
$\overline{MN}$	segment MN
$\overrightarrow{MN}$	ray MN
$\cong$	is congruent to
$\angle AOB$	angle AOB
m $\angle AOB$	degree measure of $\angle AOB$
$\therefore$	therefore
$\perp$	is perpendicular to
$\not\perp$	is not perpendicular to
$\sqrt{}$	square root
$\parallel$	is parallel to
$\not\parallel$	not parallel to
$\not\cong$	is not congruent to
$\triangle XYZ$	triangle XYZ
$\square ABCD$	parallelogram $ABCD$
$\sim$	is similar to
$\not\sim$	is not similar to
$\odot$	circle
$\overset{\frown}{AB}$	arc AB
$m\overset{\frown}{AB}$	degree measure of arc AB
$\wedge$	logical "and"
$\vee$	logical "or"
$\sim$	logical "not"
$\approx$	approximately equal to
$\neq$	not equal to
$\overset{?}{=}$	possibly equal to
$\rightarrow$	logical "If . . ., then . . ."

1 GEOMETRIC FIGURES

OVERVIEW

This chapter takes a close look at basic sets of points. Distance between two points is described in terms of the coordinates of the points. Students will identify sets of collinear and coplanar points. Angle concepts will include measurement, constructions, and supplementary and complementary addition properties. Logical thinking will be introduced with exercises involving conjunction and disjunction statements.

OBJECTIVES

- To identify the intersection of two geometric figures
- To identify and name collinear and coplanar points
- To find the distance between two points on a number line
- To find the coordinate of the midpoint of a given segment
- To apply the Midpoint Formula
- To name geometric figures
- To construct an angle congruent to a given angle, the bisector of a given angle, and the midpoint of a segment
- To apply the Angle Addition Postulate
- To find the measure of the complement and the supplement of an angle with a given measure, when possible
- To solve problems involving supplementary and complementary angles
- To write the conjunction and the disjunction of a given statement and to determine their truth values
- To graph conjunctions and disjunctions

PROBLEM SOLVING

Using the Segment Addition Postulate, students learn to solve problems involving the distance between points and lengths of given segments. Lesson 1.3 introduces the Midpoint Formula. The problem solving strategy Making a Model (See pages 11–13) can be used to locate midpoints of segments and construct congruent segments. The problem solving strategy Making a Diagram can be used with the Application in Lesson 1.5 to answer questions on bearing.

TECHNOLOGY

Computer: The Computer Investigation before Lesson 1.5 illustrates properties and characteristics of adjacent, supplementary, and complementary angles. This leads students to concepts of angle addition and angle bisectors, which are the topics of Lessons 1.5 and 1.6.

Calculator: You may want to encourage students to use a calculator for Exercises 24–26 on page 9 to facilitate computations.

SPECIAL FEATURES

Focus on Reading pp. 3, 18
Algebra Review p. 4
Mixed Review pp. 9, 14, 25, 31, 34, 37
Brainteaser p. 14
Midchapter Review p. 19
Computer Investigation p. 20
Application: Surveying p. 25
Application: Astronomical Units p. 31
Application: Electrical Circuits p. 35
Key Terms p. 38
Key Ideas and Review Exercises pp. 38–39
Chapter 1 Test p. 40
College Prep Test p. 41

PLANNING GUIDE

Lesson	Basic	Average	Above Average	Resources
1.1 pp. 3–4	FR all CE all WE 1–10 odd AR all	FR all CE all WE 1–14 odd AR all	FR all CE all WE 1–17 odd AR all	Reteaching p. 1 Practice p. 2
1.2 pp. 8–9	CE all WE 1–20 odd	CE all WE 1–26 odd	CE all WE 1–29 odd	Reteaching p. 3 Practice p. 4
1.3 pp. 13–14	CE all WE 1–11 odd Brainteaser	CE all WE 1–23 odd Brainteaser	CE all WE 1–26 odd Brainteaser	Reteaching p. 5 Practice p. 6
1.4 pp. 18–20	FR all CE all WE 1–15 odd Midchapter Review CI Activity 1	FR all CE all WE 1–21 odd Midchapter Review CI all	FR all CE all WE 1–24 odd Midchapter Review CI all	Reteaching p. 7 Practice p. 8
1.5 pp. 23–25	CE all WE 1–12 odd Application 1	CE all WE 1–20 odd Application all	CE all WE 1–22 odd Application all	Reteaching p. 9 Practice p. 10
1.6 pp. 29–31	CE all WE 1–19 odd Application 1	CE all WE 1–28 odd Application all	CE all WE 1–31 odd Application all	Reteaching p. 11 Practice p. 12
1.7 pp. 33–35	CE all WE 1–10 odd Application 1–2	CE all WE 1–14 odd Application all	CE all WE 1–17 odd Application all	Reteaching p. 13 Practice p. 14
1.8 p. 37	CE 1–2 WE 1–11 odd	CE all WE 1–17 odd	CE all WE 1–22 odd	Reteaching p. 15 Practice p. 16
Chapter 1 Review pp. 38–39	all	all	all	
Chapter 1 Test p. 40	1–8, 10, 12–18 odd	1–23 odd	all	
College Prep Test p. 41	all	all	all	

CE = Classroom Exercises   WE = Written Exercises   FR = Focus on Reading   AR = Algebra Review   CI = Computer Investigation
NOTE: For each level, all students should be assigned all Mixed Review exercises.

▰▰▰ INVESTIGATION

This activity can be used to reinforce the definition of angle bisector taught in Lesson 1.5. In order to do the construction, the students should know the definition of angle bisector. From that, they should figure out how to place the strips of tape in order to construct the angle bisector.

On a separate sheet of paper, have students use a straightedge to draw a large angle. Then, give each student two strips of transparent tape. Ask them to construct the angle bisector using the transparent tape.

The tape strips should be placed to intersect the rays of the angle in such a way that a parallelogram is formed within the interior of the angle. Then, a diagonal can be drawn from the vertex of the original angle to the vertex of the opposite angle of the parallelogram, forming the angle bisector.

Sonya Kovalevskaya (1850–1891)—As a child, this famous mathematician was fascinated by the mathematics on the temporary wallpaper in her room. Her father had covered the walls with the calculus notes he bought as a student.

More About the Mathematician

Sonya Kovalevskaya left Russia, where the universities were closed to women, to study mathematics in Germany. She was the first woman in history to be granted a doctorate in mathematics. She won the Prix Bordin of the French Academy of Sciences with her work on the motion of a rigid body about a point. The picture shows sketches of the various spinning tops she used to illustrate her conclusions, which have useful modern applications in studying the flow of fluids. She also published a paper on the shape of Saturn's rings.

1.1 Introduction to Geometric Figures

Teaching Resources

Problem Solving Worksheet 1
Project Worksheet 1
Quick Quizzes 1
Reteaching and Practice
  Worksheets pp. 1, 2

Objectives To identify and name points, lines, and planes
To name the intersection of two geometric figures

When you look at the sky on a clear night, you can see thousands of stars. They look so small that they appear to be dots. Each of these tiny dots suggests the simplest figure studied in geometry—a *point*. The three basic figures of geometry are the point, the line, and the plane.

A point is represented by a dot and named by a capital letter. A point has no size but is used only to indicate *position*. All geometric figures consist of points.

The points are named *A*, *B*, and *C* and are read as point *A*, point *B*, and point *C*.

A line is represented by a straight mark with an arrowhead at each end. A line consists of an endless, or infinite, number of points. It has no width or thickness and extends infinitely in two directions.

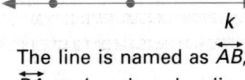

The line is named as $\overleftrightarrow{AB}$, $\overleftrightarrow{BA}$, or *k* and read as line *AB*, line *BA*, or line *k*.

A plane is represented by a four-sided figure and named by a capital letter in script. It is a flat surface, like the image of a picture projected onto a screen. It has length and width, but no thickness. A plane continues without end. There are an infinite number of points and lines in a plane.

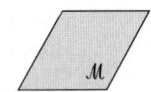

The plane is named $\mathcal{M}$ and is read as plane $\mathcal{M}$.

EXAMPLE 1 Give seven different names for the line represented at the right.

Solution A line may be named by any two points on the line or by a lowercase letter. Thus, the line may be named $\overleftrightarrow{FG}$, $\overleftrightarrow{GF}$, $\overleftrightarrow{FH}$, $\overleftrightarrow{HF}$, $\overleftrightarrow{GH}$, $\overleftrightarrow{HG}$, or *l*.

1.1 Introduction to Geometric Figures **1**

GETTING STARTED

Motivator

Draw a number line on the chalkboard with points labeled at 0 and 1. Ask these questions:

Which point is halfway between 0 and 1? $\frac{1}{2}$ Which point is halfway between $\frac{1}{2}$ and 0? $\frac{1}{4}$ How many points are there between 0 and 1? Infinitely many What does this suggest about the size of a point? No size

TEACHING SUGGESTIONS

Lesson Note

Students may have trouble identifying the abstract concepts of point, line, and plane with the physical models. To help them see that a plane has no thickness, use a transparency on an overhead projector. The projected picture is an example of a plane figure: it has no thickness and cannot be picked off the screen.

Math Connections

Astronomy: The vocabulary of this lesson can also be used to describe or draw the constellations—for example, The Big Dipper, Scorpius, and Orion.

Critical Thinking Questions

Application: Ask students these questions. Is the period at the end of a sentence a point? Explain your answer. No, the period has size; it is a physical model of a point. Is a piece of cardboard a plane? No, it too is a physical model.

Additional Example 1

Give seven names for the line represented below.

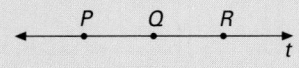

$\overleftrightarrow{PQ}$, $\overleftrightarrow{QP}$, $\overleftrightarrow{PR}$, $\overleftrightarrow{RP}$, $\overleftrightarrow{QR}$, $\overleftrightarrow{RQ}$, or *t*.

Lesson Note

To help students visualize figures in space, have them draw a cube or rectangular solid as follows.

(1) Draw two squares, one at the right and above the other.

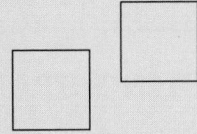

(2) Connect the corners, using dashed or light lines for the hidden lines. Label points as shown. The figure can be used to illustrate noncoplanar points such as A, F, G, and D, as well as intersecting planes.

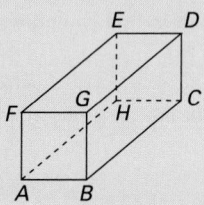

Use other physical examples in a similar way. In the classroom, a wall and the ceiling intersect in a line. The ceiling, front wall, and side wall intersect at a corner point.

Common Error Analysis

Error: Students tend to name the intersection of two planes as a *point* rather than as a *line*.

Use the following activity to help students see that the intersection of two planes is a *line*. Have the students put a dot in the middle of a piece of paper. Label the dot P. Fold the paper so that the crease contains the dot. Hold the paper with the fold so that it looks like a "V." The resulting planes intersect in the line of the crease, not just in the dot P.

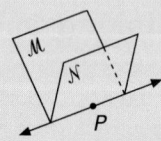

The undefined terms *point*, *line*, and *plane* can be used to define other terms.

Definition	**Space** is the set of all points.

Definition	Points are **collinear** if there exists a line that contains the points.

In the figure at the right, points A, B, and C are on the same line. These three points are collinear. However, there is no line that can contain points A, B, C, and D. These four points are **noncollinear**.

Definition	Points are **coplanar** if there exists a plane that contains the points.

The diagram at the right shows portions of three planes. (NOTE: A plane continues without end, so only *part* of a plane can be illustrated.) This figure might be interpreted as a double picture frame on a shelf. Points H, E, and F are in the same plane—the left frame. These points are coplanar. However, no plane can contain points H, E, F, and I. These four points are **noncoplanar**.

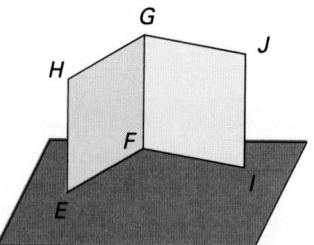

In all of the diagrams in this book, you should assume that points that *appear* to lie on a line are collinear. Similarly, assume that points that *appear* to lie on a plane are coplanar.

EXAMPLE 2   In the figure, find 3 collinear points, 3 noncollinear points, 4 coplanar points, and 4 noncoplanar points.

Solution   D, H, and F are collinear.
D, E, and F are noncollinear.
C, B, F, and G are coplanar.
D, A, B, and H are noncoplanar.
(Other answers are possible.)

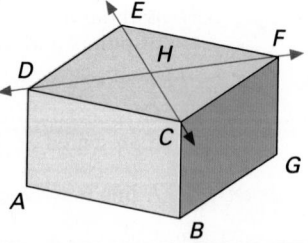

Additional Example 2

In the figure below, find 3 collinear points, 3 noncollinear points, 4 coplanar points, and 4 noncoplanar points.
U, X, and Q are collinear.
U, S, and P are noncollinear.
U, Q, P, and V are coplanar.
U, Q, P, and T are noncoplanar.
(Other answers are possible.)

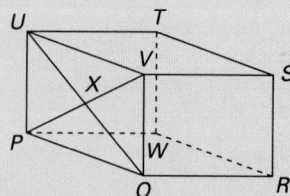

Following are some commonly used expressions describing various relationships among points, lines, and planes.

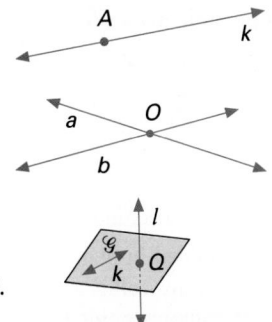

A is on *k*.
k passes through point *A*.
k contains point *A*.

Intersect means "cut" or "meet."
a intersects *b* at *O*.
The intersection of *a* and *b* is {*O*}.

Plane $\mathcal{G}$ contains line *k* and point *Q*.
k and *Q* lie on plane $\mathcal{G}$.
l intersects plane $\mathcal{G}$ at {*Q*}.
Q is the intersection of *l* and plane $\mathcal{G}$.

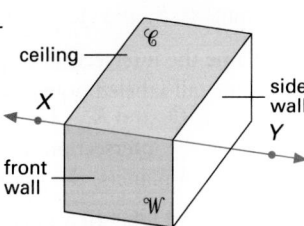

Definition

The **intersection** of two geometric figures is the set of points that are contained in both figures.

Think of the drawing at the right as a diagram of a classroom. The front wall and ceiling of the room are portions of two intersecting planes. They meet at a line.

Planes $\mathcal{C}$ and $\mathcal{W}$ contain $\overleftrightarrow{XY}$.
Planes $\mathcal{C}$ and $\mathcal{W}$ intersect at $\overleftrightarrow{XY}$.

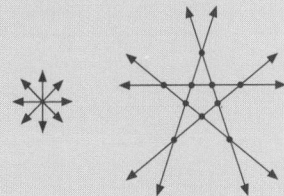
ceiling
side wall
front wall

Focus on Reading

Match the term on the right with its correct description on the left.

1. a flat surface   b   **a.** line
2. indicates position in space   d   **b.** plane
3. points that are on the same line   e   **c.** coplanar
4. extends in exactly two directions   a   **d.** point
5. the set of all points   f   **e.** collinear
6. points contained in the same plane   c   **f.** space

Classroom Exercises

Give a physical illustration of the set of points.

1. plane   **2.** point   **3.** line   **4.** noncollinear points
5. intersecting lines   **6.** intersecting planes   **7.** noncoplanar points

1.1 Introduction to Geometric Figures   **3**

 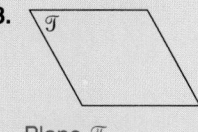
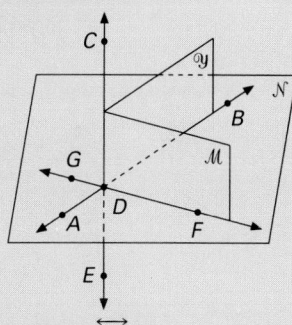
3

Additional Answers

Additional Answers

Classroom Exercises

1. Wall, ceiling, desktop.
2. Dot on the letter "i" on the chalkboard.
3. Top edge of wall.
4. Three corners of the chalkboard.
5. The joining of two blades of a pair of scissors.
6. The ceiling and a wall.
7. Three corners of the chalkboard and a point in the middle of the floor.

Written Exercises

15.

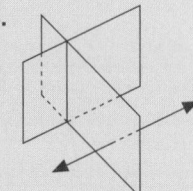

16.

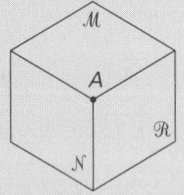

17.

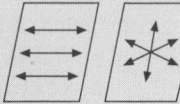

Exercise Note

Absolute value can also be defined as "the distance from zero without regard to direction."

Written Exercises

Write all possible names for the figure.

1.

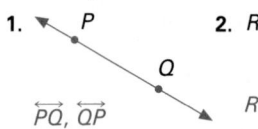

$\overleftrightarrow{PQ}, \overleftrightarrow{QP}$

2. R
   R

3.

$\mathcal{N}$

4.

$\overrightarrow{TU}, \overrightarrow{UT},$
$\overrightarrow{UV}, \overrightarrow{VU},$
$\overleftrightarrow{TV}, \overleftrightarrow{VT}, p$

True or false? If false, indicate why.

5. $\overleftrightarrow{AB}$ is contained in plane $\mathcal{M}$.   True
6. F, B, and A are collinear.   False. $\overleftrightarrow{FB}$ and $\overleftrightarrow{AB}$ are different lines.
7. Plane $\mathcal{M}$ contains point E.   False. E lies below $\mathcal{M}$.
8. Name a point that is collinear with points F and G.   B
9. What is the intersection of $\overleftrightarrow{DE}$ and plane $\mathcal{M}$?   B
10. Name two points that are coplanar with points F, B, and A.   C, G, or H

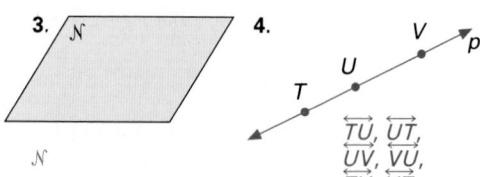

11. Name the intersection of planes $\mathcal{M}$ and $\mathcal{N}$. $\overleftrightarrow{SQ}$
12. Name all labeled points that are coplanar with S, Q, and R.   T, G, H
13. Name the intersection of $\overleftrightarrow{KL}$ and $\overleftrightarrow{GH}$.   S
14. In which plane(s) does point Q lie? $\mathcal{N}$, $\mathcal{M}$, $\mathcal{W}$

15. Draw and label two intersecting planes, with a line not contained in either of the planes.
16. Draw a diagram to show that three planes may intersect in a point.
17. Three lines may lie in the same plane in four ways such that there are 0, 1, 2, or 3 points of intersection. Draw all four cases.

Algebra Review   Prerequisite Review for Lesson 1.2

In geometry, absolute value can be of use in finding the distance between points. $|x|$ means "the **absolute value** of a number x." That is, $|x| = x$, if $x > 0$ and $|x| = -x$, if $x < 0$.

Example $|-8 + 3| = |-5| = -(-5) = 5$

Simplify.

1. $|-9 + 5|$ 4
2. $|-7 - 8|$ 15
3. $|-4 + 6 - 10|$ 8
4. $|-3 - 7 + 12|$ 2
5. $|5| - |-6|$ −1
6. $|4| + |7 - 8|$ 5
7. $-4 - |-8|$ −12
8. $|-8 + 0| - |-7|$ 1

1.2 Distance, Segments, and Rays

Objectives

To find the distance between two points on a number line
To apply the Ruler Postulate and the Segment Addition Postulate
To identify and name segments and rays

The length of the paper clip at the right can be given by different numbers of units depending upon the measurement system used.

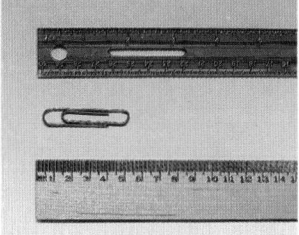

To find the distance between any two points on a line, it is necessary to agree upon a measuring device or ruler. Pick any two points *P* and *Q* on a line, with *Q* to the right of *P*. Assign the number 0 to *P* and the number 1 to *Q*.

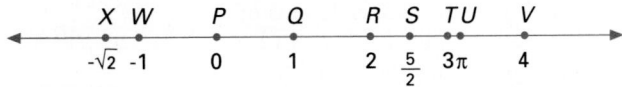

The distance from *P* to *Q* is 1. Write $PQ = 1$ or $QP = 1$, where the symbols *PQ* and *QP* mean "the distance between *P* and *Q*." Using *PQ* as a reference, the set of real numbers can now be associated with points on the line. When this is done, the line is called a **number line**.

Below, point *T* corresponds to 3, which is called the **coordinate** of *T*. The coordinate of a point on a number line is the number associated with that point.

On the number line above, the distance between the two points *W* and *R* with coordinates −1 and 2, respectively, is defined to be equal to the absolute value of the difference of their coordinates.

$$WR = |2 - (-1)| = |3| = 3, \text{ or } WR = |-1 - 2| = |-3| = 3$$

Definition

| The **distance** between any two points *A* and *B* with coordinates *m* and *n* is $|m - n|$ or $|n - m|$. | |

The ideas above are summarized in the Ruler Postulate. A **postulate** is a statement that is accepted without proof.

Teaching Resources

Application Worksheet 1
Quick Quizzes 2
Reteaching and Practice Worksheets, pp. 3, 4
Transparency 1

▰▰▰ GETTING STARTED

Prerequisite Quiz

1. Name the geometric figure shown.

$\overleftrightarrow{TW}$, or $\overleftrightarrow{WT}$

Simplify.

2. $8 - (-4)$ 12
3. $-4 - (-3)$ −1
4. $|-8|$ 8

5. $|7 - 11|$ 4
6. $|8 - (-1)|$ 9
7. $|-3 - (-5)|$ 2

Motivator

Ask students to name two ways to find the distance between the two points, −5 and 2, on a number line. Is this distance positive or negative? Why?   Positive; distance is always positive.

Draw a number line on the chalkboard with points −3, 1, and 5 labeled *A*, *B*, and *C* respectively. Then ask these questions.

What is *AB*?   4   *BC*?   4   *AC*?   8
What relationship exists between the three points? $AB + BC = AC$

▰▰▰ TEACHING SUGGESTIONS

Math Connections

Life Skills: Straight roads on a map can serve as models of rays, lines and segments.

Critical Thinking Questions

Application: Have students name some physical models of rays and segments that occur in nature. Ask: "Is the surface of the earth a plane?   No.   Is there a true physical model of a line?   No."   Use these questions for open-ended discussion.

Common Error Analysis

Error: Students sometimes cannot distinguish between $\overline{AB}$ and AB.

Make a helpful analogy by considering this inaccurate statement: John is 120 lb. Actually, John is a person; he *weighs* 120 lb. $\overline{AB}$ is a segment; it has length AB. The summary at the end of the lesson is helpful in clarifying the distinction between the four basic symbols: $\overleftrightarrow{AB}$, $\overrightarrow{AB}$, $\overline{AB}$, and AB.

Checkpoint

For the given coordinates of A and B, find AB.

1. A: -4; B: 7   11
2. A: $3\frac{7}{10}$; B: 9 $5\frac{3}{10}$
3. The points A, Y and F are collinear. F is between A and Y. Use this information and the Segment Addition Postulate to draw a diagram and write an equation.

$$\overset{A \quad F \quad Y}{\bullet\!\!-\!\!-\!\!\bullet\!\!-\!\!-\!\!\bullet} \text{ or } \overset{Y \quad F \quad A}{\bullet\!\!-\!\!-\!\!\bullet\!\!-\!\!-\!\!\bullet}$$
$$AF + FY = AY \qquad YF + FA = YA$$

Write a symbol for each of the following.

4. $\overset{P \qquad Q}{\longleftrightarrow}$ $\overleftrightarrow{PQ}$

5. $\overset{T \qquad\qquad W}{\bullet\!-\!-\!-\!-\!-\!\bullet}$ $\overline{TW}$

6. $\overset{U \qquad J}{\bullet\!-\!-\!-\!\longrightarrow}$ $\overrightarrow{UJ}$

7. distance between D and Y DY

Postulate 1

Ruler Postulate

1. Any two distinct points on a line can be assigned coordinates 0 and 1.
2. There is a one-to-one correspondence between the real numbers and all points on the line.
3. To every pair of points, there corresponds exactly one positive number called the distance between the two points.

EXAMPLE 1 Find the distance between each pair of points listed below.

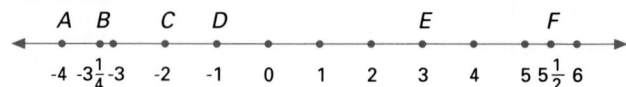

Points	Distance
D and C	$|-2 - (-1)| = |-2 + 1| = |-1| = 1$
B and F	$\left|5\frac{1}{2} - (-3\frac{1}{4})\right| = \left|5\frac{1}{2} + 3\frac{1}{4}\right| = \left|8\frac{3}{4}\right| = 8\frac{3}{4}$

In the figure, note that C is between A and B. There are an infinite number of points between A and B.

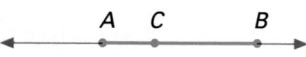

Definition

Segment AB ($\overline{AB}$) is the set of points consisting of points A, B, and all points between A and B. A and B are called the **endpoints** of the segment. Either $\overline{AB}$ or $\overline{BA}$ can be used to name the segment.

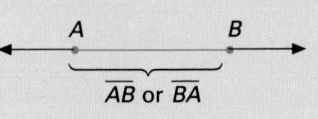

Keep in mind that:

(1) $\overline{AB}$ is a set of points.
(2) AB is the distance between points A and B, which is a number.
(3) The length of $\overline{AB}$ is AB—the distance between A and B.

In the figure, C is between A and B.

$$AC = 3 \quad CB = 6 \quad AB = 9$$
$$3 + 6 = 9$$

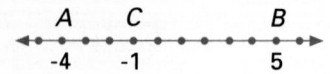

Therefore, $AC + CB = AB$

This suggests the Segment Addition Postulate stated on the next page.

6 Chapter 1 Geometric Figures

Additional Example 1

Find the distance between A and B on a number line for the given coordinates of A and B.

a. A: 6; B: -4 10
b. A: $3\frac{1}{6}$; B: $-2\frac{1}{3}$ $5\frac{1}{2}$
c. A: -4; B: 8   12
d. A: $-6\frac{1}{5}$; B: -2 $4\frac{1}{5}$

Postulate 2

Segment Addition Postulate

If C is between A and B, then
$$AC + CB = AB.$$

EXAMPLE 2   Points A, G, and R are collinear. Point R is between A and G.

Draw a diagram. Use the Segment Addition Postulate to write an equation.

Solution   Point A can be placed either to the left or to the right of point R.

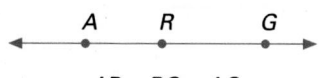

$$AR + RG = AG \qquad\qquad GR + RA = GA$$

Thus, the equation is $AR + RG = AG$, or $GR + RA = GA$.

EXAMPLE 3 G, R, and A are three collinear points such that A is between G and R. $GA = \frac{3}{5} AR$ and $GR = 24$. Find AR.

Plan   Let $x = AR$. Then $GA = \frac{3}{5}x$.

Solution
Write an equation. $GA + AR = GR$

Substitute. $\frac{3}{5}x + x = 24$

Multiply both sides of $3x + 5x = 120$
the equation by 5. $8x = 120$
$x = 15 \longrightarrow AR = 15$

Definition

Ray XY ($\overrightarrow{XY}$) consists of $\overline{XY}$ and all points P such that Y is between X and P. X is called the **endpoint** of the ray. The symbol $\overrightarrow{XY}$ is used to name the ray.

It is often possible to name a ray in more than one way. However, the first letter always names the endpoint of the ray, and the arrow above the two letters points to the right.

1.2 Distance, Segments, and Rays   **7**

Lesson Note

The Segment Addition Postulate is used extensively in proofs. In the next chapter, writing a proof requires writing an equation, given the diagram of a segment $\overline{AB}$ with a point C between A and B. Later, students are told that if a figure appears to be a segment, they need not state this in writing a proof.

The technique illustrated in Example 3 is very useful on standardized tests. Students can solve such problems without having to use equations with fractional coefficients. The equation $3x + 7x = 210$ is much easier to solve than $\frac{3}{7}x + x = 30$.

Closure

Have students answer these questions. How do you find the distance between two points on a number line?   Find the absolute value of their difference.   How can you draw a diagram indicating that points R, A, and F are collinear?   a line or segment containing all three points   You can use the Segment Addition Postulate to write an equation. What is the equation?   Given $\overleftrightarrow{AC}$ such that B is between A and C, $AB + BC = AC$.

Additional Example 2

Points K, T, and W are collinear. Point K is between T and W. Draw a diagram. Use the Segment Addition Postulate to write an equation.

$$\underset{TK + KW = TW}{\overset{T \quad K \quad W}{\bullet\!-\!\bullet\!-\!\bullet}} \text{ or } \underset{WK + KT = WT}{\overset{W \quad K \quad T}{\bullet\!-\!\bullet\!-\!\bullet}}$$

Additional Example 3

P, Q, and M are three collinear points such that M is between P and Q. $PM = \frac{3}{7} MQ$ and $PQ = 30$. Find PM.   9

Guided Practice

Classroom Exercises 1–3

Independent Practice

A Ex. 1–20, **B** Ex. 21–26, **C** Ex. 27–29

Basic: WE 1–20 odd
Average: WE 1–26 odd
Above Average: WE 1–29 odd

EXAMPLE 4 Write three names for the ray that has endpoint P and contains point T.

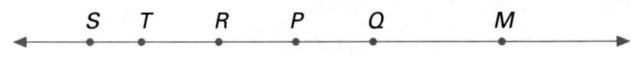

S   T   R   P   Q   M

Solution The ray begins at P and passes through T. Shade the drawing to show the ray.

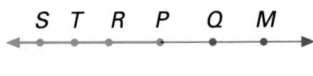

S   T   R   P   Q   M

Three names for the ray are $\overrightarrow{PR}$, $\overrightarrow{PT}$, and $\overrightarrow{PS}$.

Summary

line	ray	segment
A B	A B	A B
$\overleftrightarrow{AB}$	$\overrightarrow{AB}$	$\overline{AB}$

The distance between points C and D corresponds to a unique number, CD. CD is the length of $\overline{CD}$.

$$CD = |4-(-2)| = |6| = 6$$

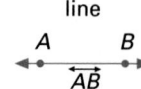

C ... D
-2 ... 4

$CD = 6$, or $DC = 6$

Classroom Exercises

Use the Segment Addition Postulate to write an equation for the segment.

1. A   T   Y
$AT + TY = AY$

2. W   U   O
$WU + UO = WO$

3. G   L   N
$GL + LN = GN$

Written Exercises

Find the indicated distance. Use the number line.

A   B   C   D   E   F   G   H
-12 -10 -8 -6 -4 -2   0   2   4   6   8   10

1. AC 6   **2.** CF 9   **3.** EB 7   **4.** GF 2   **5.** GB 14 **6.** HD 12 **7.** GA 17 **8.** DA 8

For the given coordinates of G and H, find GH.

9. $G: -7$; $H: 5$ 12

10. $G: 4\frac{5}{7}$; $H: 8$ $3\frac{2}{7}$

11. $G: -6.2$; $H: -2.5$ 3.7

Additional Example 4

Give three names for the ray with endpoint G and containing point R.

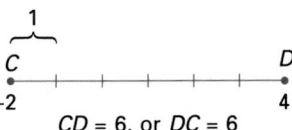

U   G   I   R   S

$\overrightarrow{GI}$, $\overrightarrow{GR}$, $\overrightarrow{GS}$

For the given set of collinear points, draw a diagram. Write an equation using the Segment Addition Postulate.

12. G, K, and T such that T is between G and K

13. R, A, and V such that R is between A and V

Give three names for the indicated ray.

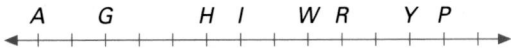

14. Endpoint I containing H
$\overrightarrow{IH}, \overrightarrow{IG}, \overrightarrow{IA}$

15. Endpoint W containing Y
$\overrightarrow{WR}, \overrightarrow{WY}, \overrightarrow{WP}$

16. Endpoint R containing I
$\overrightarrow{RW}, \overrightarrow{RI}, \overrightarrow{RH}, \overrightarrow{RA}, \overrightarrow{RG}$

Write a name for each of the following.

17.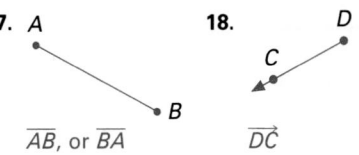
$\overline{AB}$, or $\overline{BA}$

18.
$\overrightarrow{DC}$

19. Distance between points H and W
HW, or WH

20.
$\overleftrightarrow{RS}$, or $\overleftrightarrow{SR}$

21. A, B, and C are three collinear points such that B is between A and C. $AB = \frac{3}{4}BC$ and $AC = 28$. Find AB.   12

22. P, Q, and R are three collinear points such that Q is between P and R. $PQ = \frac{4}{7}QR$ and $PR = 33$. Find QR.   21

23. For $\overline{AB}$, the coordinate of A is -6 and $AB = 7$. Find all possible coordinates of point B.   1, -13

Find the distance ST for the following coordinates.

24. $S{:}6.38$; $T{:}{-}7.91$ **25.** $S{:}19.07$; $T{:}4.63$ **26.** $S{:}{-}0.45$; $T{:}{-}8.54$

27. A, B, and C are three collinear points such that B is between A and C. $AB = \frac{3}{7}AC$ and $AB = 9$. Find BC.   12

28. Point T is on $\overrightarrow{MG}$, but T is not on $\overline{MG}$. $MT = \frac{5}{4}GT$ and $MT = 18$. Find MG. $3\frac{3}{5}$

29. For $\overline{GH}$, the coordinate of G is $2x - 6$ and the coordinate of H is $x - 5$. Find the coordinate of G if $GH = 6$.   8 or -16

Mixed Review   Prerequisite Review for Lesson 1.3

1. Name three collinear points.   *1.1* G, M, H
2. Name three noncollinear points.   *1.1* G, M, K; G, H, K; M, H, K
3. Name the intersection of $\overleftrightarrow{GH}$ and $\overleftrightarrow{MK}$ *1.1* M

Enrichment

Have the students draw a figure that meets all of the following conditions.

(1) $\overrightarrow{AC}$ and $\overrightarrow{BC}$ are drawn from the endpoints of $\overline{AB}$ such that A, B, and C are the vertices of a triangle; (2) D is a point such that $AC + CD = AD$; (3) B and D are the endpoints of a segment; (4) $\overline{BE}$ is a segment such that $BE + ED = BD$; (5) F is a point such that $AD + DF = AF$.

Now have the students examine the drawing to answer these questions.
1. Does E lie between A and F?   No   **2.** Which is greater, AF or CD?
3. Are B, E, and D collinear?   Yes AF

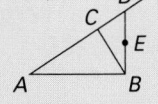

▰▰▰ GETTING STARTED

Prerequisite Quiz

Refer to the figure below.

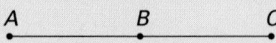

1. Write an equation using the Segment Addition Postulate. $AB + BC = AC$
2. Find AC for A: -3 and C: 8. 11
3. If $AB = \frac{4}{9} BC$ and $AC = 26$, find BC. 18

Motivator

Give each student two toothpicks and a small piece of modeling clay. Ask them if the toothpicks represent similar segments. Have them stick the toothpicks into the clay at different angles and positions. Explain how this shows that similar segments can occupy infinitely many positions in space. Ask these questions.

Are two segments having the same length actually the same segment? No, their positions may be different. Make a drawing to illustrate your answer.

▰▰▰ TEACHING SUGGESTIONS

Lesson Note

Stress the use of $\overline{AB} \cong \overline{CD}$ as interchangeable with $AB = CD$. This saves steps in proofs. It becomes cumbersome in a proof to have to say $\overline{AB} \cong \overline{CD}$ before you can say $AB = CD$. In a proof with P given as the midpoint of $\overline{AB}$, it is helpful for students to know that they can write either $AP = PB$ or $\overline{AP} \cong \overline{PB}$, by the definition of midpoint.

1.3 Congruent Segments and Constructions

Objectives

To identify and construct congruent segments
To locate the midpoint of a segment by construction
To apply the Midpoint Formula

In the diagram at the right, the two segments, $\overline{AB}$ and $\overline{CD}$, have the same length, 2 cm. The segments are said to be *congruent*. The symbol for congruence is $\cong$. The symbol $\not\cong$ is read "is not congruent to."

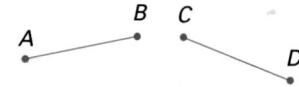

$AB = CD$: equal lengths
$\overline{AB} \cong \overline{CD}$: congruent segments

Definition

> **Congruent** ($\cong$) **segments** are segments that have the same length. $\overline{AB} \cong \overline{CD}$ is read "$\overline{AB}$ is congruent to $\overline{CD}$."

In the figure, markings (tick marks) are used to indicate that $\overline{AB}$ and $\overline{CD}$ are congruent. You should not assume congruence of segments unless they are indicated as congruent. By definition, *congruent segments* are segments that have the *same length*. Therefore, the statements $\overline{AB} \cong \overline{CD}$ and $AB = CD$ are equivalent.

In both figures below, R is between P and S. In the figure on the right, R divides $\overline{PS}$ into two congruent segments.

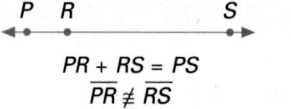
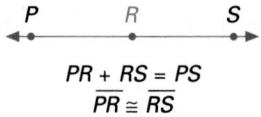

$PR + RS = PS$
$\overline{PR} \not\cong \overline{RS}$

$PR + RS = PS$
$\overline{PR} \cong \overline{RS}$

Definition

> M is the **midpoint** of $\overline{AB}$ if M lies on $\overline{AB}$ and $\overline{AM} \cong \overline{MB}$ ($AM = MB$).

Definition

> A **bisector of a segment** is a line, ray, segment, or plane that intersects the segment at its midpoint. A bisector divides the segment into two congruent segments.

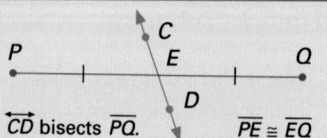

$\overrightarrow{CD}$ bisects $\overline{PQ}$. $\overline{PE} \cong \overline{EQ}$

10 Chapter 1 Geometric Figures

Postulate 3   Any segment has exactly one midpoint.

By agreement, a compass and straightedge are the only tools that may be used in the construction of a geometric figure. A straightedge is used for drawing segments. It need not have the coordinate markings of a ruler. A compass is used for drawing arcs, which are unbroken parts of a circle. A compass and straightedge can be used to construct a segment congruent to a given segment.

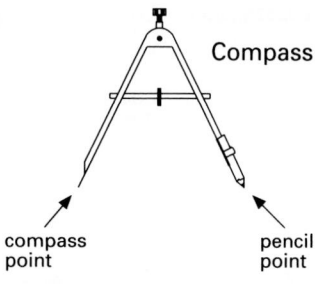

Compass

compass point   pencil point

Construction   Construct a segment congruent to a given segment.

P   Q

Construct $\overline{RS}$, where $\overline{RS} \cong \overline{PQ}$, on line k.

Use a straightedge to draw line k. Adjust the compass opening to correspond to the distance from P to Q.

Choose any point on line k and label it R. Set the compass point at R and draw an arc intersecting k. Label the point of intersection as point S.

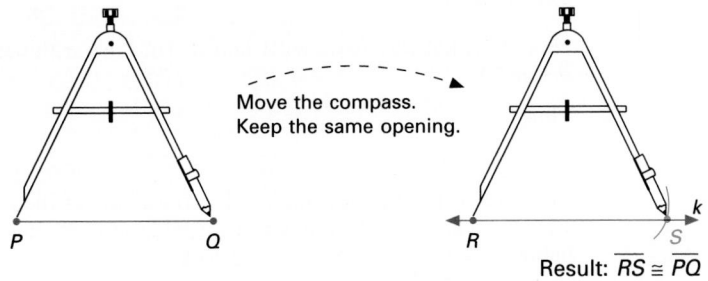

Move the compass. Keep the same opening.

P   Q   R   S   k

Result: $\overline{RS} \cong \overline{PQ}$

A compass and a straightedge can also be used to construct a point *equidistant* from two points. **Equidistant** means "the same distance."

Construction   Locate a point equidistant from two given points.

Let A and B be two given points. Open the compass wider than one-half of AB. Place the compass point on A. Draw an arc. Using the same compass opening,

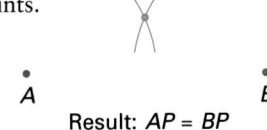

P

A   B

Result: $AP = BP$

1.3 Congruent Segments and Constructions   **11**

Math Connections

Building Trades: Constructions such as the ones in this lesson are often used by carpenters to make on-the-job marks on their pieces of wood. Students can experiment using string and a pencil instead of a metal compass.

Critical Thinking Questions

Analysis: Ask students if a line segment can have more than one bisector.   Yes. Have them make a drawing to justify the answer. Ask them what is special about the bisector that results from the construction on page 12.   Perpendicular bisector.

Common Error Analysis

Error: Sometimes students read hastily and become confused with an exercise or its instructions. In the figure below, $\overline{AB}$ bisects $\overline{CD}$.

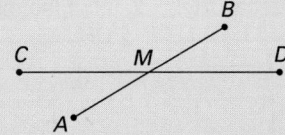

B

C   M   D

A

Careless readers may conclude incorrectly that $AM = MB$ since this may appear to be true in the figure. For this example, stress that when $\overline{AB}$ bisects $\overline{CD}$, $\overline{CD}$ is bisected, so $CM = MD$.

11

Finding the coordinate of the midpoint of a segment is easier to do by averaging the coordinates of the endpoints than by using the definition of midpoint. It is more difficult to find the distance between endpoints, take half of this, and then add the result to the coordinate of the first endpoint.

Constructions are used throughout this text to help students learn by discovery. Students do benefit by first learning to do constructions by manipulating the compass and straightedge. The constructions of this chapter cannot yet be proved. Note that Postulate 3 will be very important for indirect proofs later in the text.

Checkpoint

Write a statement of congruence and an equation for each of the following given statements.

1. $\overline{QU}$ bisects $\overline{JR}$.

$JT = TR$
$\overline{JT} \cong \overline{TR}$

2. M is the midpoint of $\overline{GH}$.

$\overline{GM} \cong \overline{MH}$
$GM = MH$

3. Find the coordinate of the midpoint M of $\overline{AB}$ whose endpoints have coordinates -6 and 4. -1

4. M is the midpoint of $\overline{AB}$. Find the coordinates of A and B. $A{:}4$; $B{:}18$

A M B
$(p-6)$ 11 $(p+8)$

---

place the compass point on B. Draw a second arc to intersect the first arc at point P. Point P is equidistant from the two given points, A and B.

Construction   Locate the midpoint of a given segment.
Locate M, the midpoint of $\overline{AB}$.

Locate a point P equidistant from A and B.	Locate a second point Q equidistant from A and B.	Draw $\overleftrightarrow{PQ}$ intersecting $\overline{AB}$ at point M.

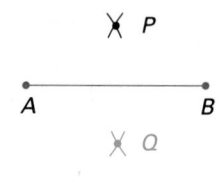
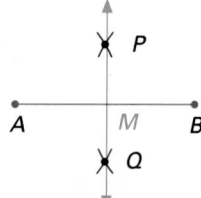

Result: M is the midpoint of $\overline{AB}$.

It is possible to find the coordinate of the midpoint of a segment when the coordinates of the endpoints are given. The procedure is the same as that used to find the arithmetic mean (average) of two numbers.

EXAMPLE 1 M is the midpoint of $\overline{AB}$. Find its coordinate, m.

A M B
-5 m 8

Solution   Point M is halfway between A and B. Find the arithmetic mean of -5 and 8.

Thus, $m = \dfrac{-5 + 8}{2} = \dfrac{3}{2}$, or $1\frac{1}{2}$.

The general formula for finding the coordinate of the midpoint of a segment when the coordinates of the endpoints are given is stated below.   Note: a derivation of the Midpoint Formula is required in Exercise 24.

Midpoint Formula

If the coordinates of the endpoints of $\overline{AB}$ are a and b, and m is the coordinate of the midpoint M, then $m = \dfrac{a + b}{2}$.

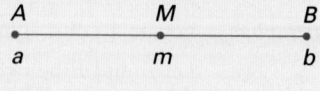

12   Chapter 1 Geometric Figures

Additional Example 1

Find the coordinate of the midpoint M of $\overline{AB}$ for $A{:}-7$ and $B{:}13$.   3

EXAMPLE 2   M is the midpoint of $\overline{AB}$. Find the coordinates of points A and B.

A ————— M ————— B
(p - 5)   6   (p + 2)

Solution

$$6 = \frac{(p - 5) + (p + 2)}{2}$$

$$6 = \frac{2p - 3}{2} \quad \longleftarrow \text{Multiply each side of the equation by 2 to eliminate fractions.}$$

$$12 = 2p - 3$$
$$15 = 2p$$
$$7\frac{1}{2} = p$$

The coordinate of A is $p - 5 = 7\frac{1}{2} - 5 = 2\frac{1}{2}$.

The coordinate of B is $p + 2 = 7\frac{1}{2} + 2 = 9\frac{1}{2}$.

Classroom Exercises

1. Tell how to construct a point equidistant from the endpoints of a segment.
2. Tell how to construct the midpoint of a segment.
3. What is a line called that intersects a segment at its midpoint?
4. According to our definition, what does it mean to say that two segments are congruent?
5. State the Midpoint Formula.

Written Exercises

Write a statement of congruence and an equation.

1. L is the midpoint of $\overline{TU}$. $TL \cong LU$, $TL = LU$

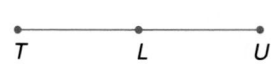

T ———— L ———— U

2. $\overleftrightarrow{RW}$ bisects $\overline{FG}$. $FK \cong KG$, $FK = KG$

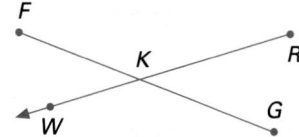

Draw a segment $\overline{PQ}$. (Exercises 3–5)

3. Construct a segment $\overline{GH}$ congruent to $\overline{PQ}$.
4. Locate a point T equidistant from the endpoints of $\overline{PQ}$.
5. Locate the midpoint of $\overline{PQ}$.

1.3 Congruent Segments and Constructions   **13**

Additional Example 2

M is the midpoint of $\overline{AB}$. Find the coordinates of points A and B.

A ———— M ———— B
(p − 7)   8   (p + 9)

A:0; B:16

Written Exercises

3.

4.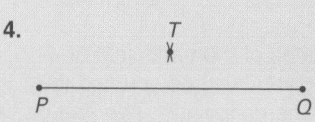

5.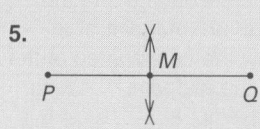

24.
$$AM = MB$$
$$m - a = b - m$$
$$2m = a + b$$
$$m = \frac{a + b}{2}$$

Mixed Review
1. The line containing X and Y.
2. The segment with endpts X and Y.
3. The ray with endpt X and containing Y.
4. The distance between X and Y.

Exercise Note
Make sure that some figures have no equal sides, no parallel sides, and no right angles.

Find the coordinate of the midpoint M of $\overline{AB}$ whose endpoints have the given coordinates.

6. 4 and 7 $5\frac{1}{2}$ **7.** -5 and 9   2   **8.** 0 and 13 $6\frac{1}{2}$

9. 5 and 12 $8\frac{1}{2}$ **10.** 3.2 and 7.6   5.4   **11.** -3.4 and -1.2 $-2.$

12. $3\frac{1}{2}$ and $7\frac{1}{2}$ $5\frac{1}{2}$ **13.** $-4\frac{1}{5}$ and $8\frac{3}{5}$ $2\frac{1}{5}$ **14.** $-2\frac{1}{2}$ and 7 $2\frac{1}{4}$

M is the midpoint of $\overline{AB}$. Find the coordinates of A, B, or M for the given data.

15. $A{:}p - 4$; $M{:}8$; $B{:}p + 12$ $A{:}0$, $B{:}16$ **16.** $A{:}3p - 1$; $M{:}10$; $B{:}2p + 6$ $A{:}8$, $B{:}12$

17. $A{:}p - 8$; $M{:}10$; $B{:}p + 6$ $A{:}3$, $B{:}17$ **18.** $A{:}2p - 2$; $M{:}14$; $B{:}4p + 6$ $A{:}6$, $B{:}22$

19. $A{:}-4$; $M{:}6 - p$; $B{:}2p - 4$ $M{:}1$, $B{:}6$ **20.** $A{:}12 - 2p$; $M{:}p - 6$; $B{:}16$ $A{:}-8$, $M{:}4$

Find the coordinate of the missing endpoint of $\overline{GH}$.

21. Midpoint $M{:}8$; $H{:}12$ 4   **22.** Midpoint $M{:}-5\frac{1}{2}$; $H{:}0$ -11

23. For $\overline{GH}$ above, if the coordinates of M and H are positive, is it reasonable that G must be positive?   No.

24. Let a and b be the coordinates of the endpoints of $\overline{AB}$. Let m be the coordinate of the midpoint M of $\overline{AB}$. Assume that $a < m < b$. Use the definition of midpoint to derive the Midpoint Formula: $m = \dfrac{a + b}{2}$.

25. The coordinates of the endpoints A and B of $\overline{AB}$ are 6 and 21, respectively. Find the coordinate of point C between A and B such that $AC = \frac{2}{3}(CB)$.   12

26. C, D, and E are the midpoints of $\overline{AB}$, $\overline{AC}$, and $\overline{CB}$, respectively. D and E have coordinates 6 and 13, respectively. Find the coordinates of A and B. $2\frac{1}{2}$, $16\frac{1}{2}$

Mixed Review

Identify the meaning of each symbol.   *1.2*

1. $\overleftrightarrow{XY}$ **2.** $\overline{XY}$ **3.** $\overrightarrow{XY}$ **4.** XY

▰ Brainteaser

Draw some four-sided, closed figures. Construct the midpoint of each side. Then connect the midpoints consecutively.

Is there a pattern in the shape of the figure formed by connecting the midpoints?   The resulting figure is always a parallelogram.

14   Chapter 1 Geometric Figures

Enrichment

Have the students use rulers and maps of the United States to answer these questions.

1. Name a city on the west coast of Florida that is collinear with Boston, Massachusetts, and Norfolk, Virginia.   Tampa or St. Petersburg

2. Think of San Francisco and New York City as the endpoints of a segment. Suppose you live at the midpoint. In which state do you live?   Nebraska

Now have each student make up three similar questions and challenge other students to answer them. If you wish, you can confine the activity to road maps of your state or local area.

1.4 Angle Measurement and Constructions

Objectives
To classify angles by their measures
To determine the measure of an angle by applying the Protractor Postulate
To construct an angle congruent to a given angle

The distance a football player can kick a football is determined by the force of the kick and the *angle* formed by the beginning path of the football and a horizontal ray on the ground.

Definition

An **angle** is a geometric figure consisting of two distinct rays with a common endpoint. The rays are the *sides* of the angle. The common endpoint is the *vertex* of the angle.

The symbol for angle is ∠.

In the diagram, the sides of the angle are $\overrightarrow{BA}$ and $\overrightarrow{BC}$.

The vertex is the common endpoint B.

The angle may be named in three ways: by three capital letters— ∠*CBA* or ∠*ABC*—where the middle letter names the vertex; by the vertex alone, when there is no possibility of confusion—∠*B*; or by a numeral placed between the rays—∠1.

EXAMPLE 1 Name each angle in the figure.

Solution Since there are three angles in the figure, it is unclear what is meant by ∠*Q*. None of these angles can be named by its vertex alone. Thus, the angles are ∠*PQR*, ∠1 (or ∠*SQR*), and ∠2 (or ∠*PQS*).

Additional Example 1

Name each angle in the figure.

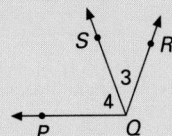

∠4, or ∠*PQS*, or ∠*SQP*; ∠3, or ∠*SQR*, or ∠*RQS*; ∠*PQR*, or ∠*RQP*

◼◼◼ GETTING STARTED

Prerequisite Quiz

Use the figure below. (Exercises 1–2)

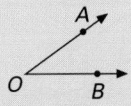

1. Name the common endpoint of the two rays. *O*
2. Name each ray in the figure. $\overrightarrow{OA}$, $\overrightarrow{OB}$
3. Name two rays with a common endpoint in the figure below. $\overrightarrow{SR}$, $\overrightarrow{ST}$

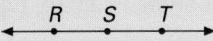

Motivator

Ask students these questions.
Why is it that the time read on a small wristwatch is the same as that read on a large wall clock (if both are accurate)? Same angle   How many degrees are there all the way around a circle?   360   What is a right angle?   angle of measure 90   What part of a rotation about a circle is a right angle? $\frac{1}{4}$

◼◼◼ TEACHING SUGGESTIONS

Lesson Note

In the figure for Example 1 students can identify an angle such as ∠1, or ∠*SQR*, by tracing the sides with a finger, moving from *S* to *Q*, then to *R*.

Math Connections

Trigonometry: Often in trigonometry it makes sense to consider rays that coincide and so form angles of measure 0 and 360, and to consider reflex angles with measures between 180 and 360. Here the discussion is limited to the set of angles given by the Protractor Postulate.

Critical Thinking Questions

Synthesis: As the hands of a clock move from midnight to the next midnight, how many times will they form a right angle?
44. The hands form a right angle twice every hour except for 3 o'clock and 9 o'clock, when a right angle occurs only once during the hour.

Checkpoint

1. Name each angle in as many ways as possible.

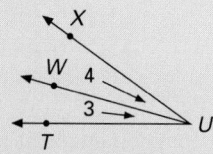

∠3, or ∠TUW, or ∠WUT; ∠4, or ∠WUX, or ∠XUW; ∠TUX, or ∠XUT

Find the measure of each angle. Tell whether the angle is acute, right, obtuse, or straight.

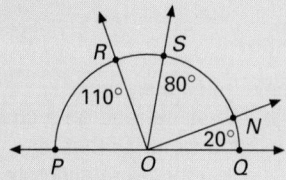

2. ∠RON   90 (right)
3. ∠POQ   180 (straight)
4. ∠ROQ   110 (obtuse)
5. ∠NOQ   20 (acute)
6. ∠ROS   30 (acute)

Lesson Note

To illustrate the four types of angles, open the board compass to various positions. Also, ask students to give clock times to illustrate the types. For example, 2:00, acute; 3:00, right; 5:00, obtuse; 6:00, straight.

Note the similarity between the Protractor and Ruler postulates. Restate the Protractor Postulate in less formal language, or try to lead the students to do this. Point out that later they will see that an angle can have a measure corresponding to any real number.

A **protractor** is used to measure an angle in **degrees** (symbol: °). In the figure at the right, the measure of angle *BOC* equals 70. This is written "m ∠*BOC* = 70."

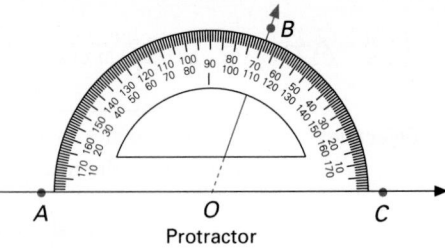

Protractor

In this book, you may assume that all angle measures are in degrees.

Using the protractor to measure an angle suggests that every angle has a measure, a basic assumption in geometry.

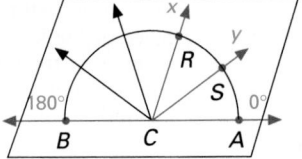

Postulate 4

Protractor Postulate

In a given plane, select any line $\overleftrightarrow{AB}$ and any point *C* between *A* and *B*. Also select any two points *R* and *S* on the same side of $\overleftrightarrow{AB}$ so that *S* is not on $\overrightarrow{CR}$. Then, there is a *pairing* of rays to real numbers from 0 to 180 in the following way:

1. $\overrightarrow{CA}$ is paired with 0 and $\overrightarrow{CB}$ is paired with 180.
2. If $\overrightarrow{CR}$ is paired with *x*, then $0 < x < 180$.
3. If $\overrightarrow{CR}$ is paired with *x* and $\overrightarrow{CS}$ is paired with *y*, then m ∠*RCS* = $|x - y|$.

In general, the Protractor Postulate says that to every angle there corresponds exactly one real number *n* such that $0 < n \leq 180$. The measure of the angle is *n*. We assume that the rays of an angle are distinct and cannot coincide. Therefore, an angle cannot have measure 0.

Angles may be classified into four categories by their measure.

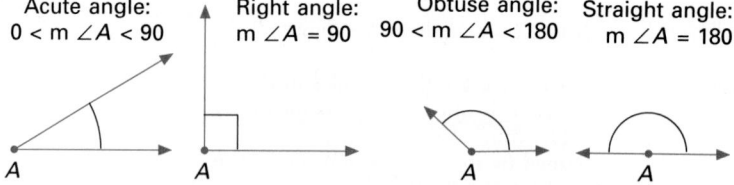

Acute angle: $0 < m \angle A < 90$ Right angle: $m \angle A = 90$ Obtuse angle: $90 < m \angle A < 180$ Straight angle: $m \angle A = 180$

Notice that the symbol ⌐ is used to indicate a right angle, and that a straight angle is a line with a labeled point indicating the vertex.

16   Chapter 1 Geometric Figures

EXAMPLE 2 Find the measure of ∠POR, ∠FOH, and ∠GOH. Classify each angle as acute, right, obtuse, or straight.

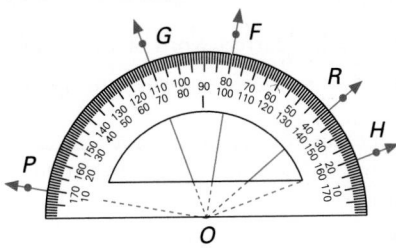

Solution

m ∠POR =
|170 − 40| =
130 (obtuse)

m ∠FOH =
|80 − 20| =
60 (acute)

m ∠GOH =
|110 − 20| =
90 (right)

Definition

Congruent (≅) angles are angles that have the same measure.
∠A ≅ ∠B means "∠A is congruent to ∠B."

In the figure, similar markings (arcs) are used to indicate that ∠A and ∠B are congruent. Unless marked as such, angles should not be assumed to be congruent. By definition, congruent angles are angles that have the same measure. Therefore, the statements ∠A ≅ ∠B and m ∠A = m ∠B are equivalent.

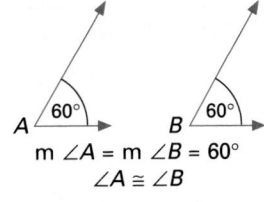

m ∠A = m ∠B = 60°
∠A ≅ ∠B

Construction Construct an angle congruent to a given angle.

Construct ∠B congruent to the given ∠A.

Using any compass opening, construct an arc with center A. Label points P and R on the line as shown. Draw line l; choose a point B. Using the same compass opening, construct an arc with center B, intersecting l at S.

Open the compass to the length PR. Construct an arc with center S, intersecting the other arc at point T. Draw BT⃗.

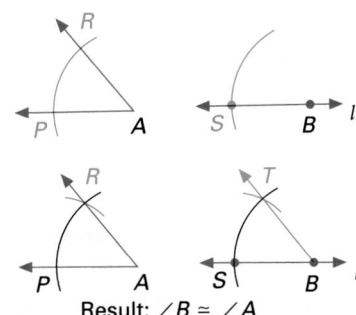

Result: ∠B ≅ ∠A

Additional Example 2

Find the measures of angles *BOA*, *DOB*, and *COB*. Classify each angle as acute, right, obtuse, or straight.

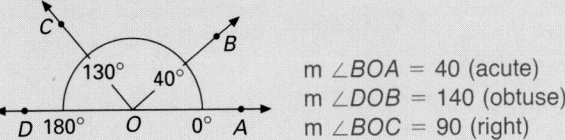

m ∠BOA = 40 (acute)
m ∠DOB = 140 (obtuse)
m ∠BOC = 90 (right)

Guided Practice

Classroom Exercises 1–6

Independent Practice

A Ex. 1–15, **B** Ex. 16–21, **C** Ex. 22–24

Basic: FR all, WE 1–15 odd, Midchapter Review all, CI activity 1

Average: FR all, WE 1–21 odd, Midchapter Review all, CI all

Above Average: FR all, WE 1–24 odd, Midchapter Review all, CI all

▬▬ *Focus on Reading*

Complete the sentence with a correct word or expression. (Exercises 1–3)

1. The equation m ∠A = m ∠B is equivalent to the statement _____. $\angle A \cong \angle B$
2. An angle with a measure of 180 is called a _____ _____.   straight angle
3. An _____ angle has a degree measure between 90 and 180.   obtuse
4. What is a protractor?   A tool used to measure angle size.
5. What is a degree?   A unit of angle measure.
6. What is a postulate?   A statement accepted without proof.

Classroom Exercises

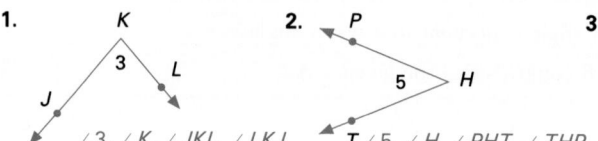

1. ∠W has sides _____ , _____. $\overrightarrow{WV}, \overrightarrow{WT}$
2. The vertex of ∠XYZ is _____.   Y
3. ∠W can also be called _____ or _____.   ∠TWV, ∠VWT
4. To measure an angle, use a _____.   protractor
5. Two angles are congruent if _____.   they have the same measure
6. To construct an angle congruent to ∠W, use a _____.   compass and straightedge

Written Exercises

Name the angle(s) in the figure in as many ways as possible.

1. ∠3, ∠K, ∠JKL, ∠LKJ

2. T ∠5, ∠H, ∠PHT, ∠THP

3. ∠1, ∠MFA, ∠AFM, ∠2;
∠WFM,
∠MFW,
∠AFW,
∠WFA

Use a protractor to draw an angle with the given measure. Then use a compass and straightedge to construct an angle congruent to that angle.
See Construction on page 17.

4. 25   **5.** 160   **6.** 95   **7.** 15

Find the measure of the angle. Tell whether the angle is acute, obtuse, right, or straight.

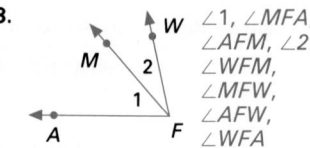

8. ∠MOX   **9.** ∠TOZ   **10.** ∠UOX   **11.** ∠TOM
70, acute   60, acute   100, obtuse   110, obtuse
12. ∠UOM   **13.** ∠ZOU   **14.** ∠ZOW   **15.** ∠TOX
30, acute   20, acute   90, right   180, straight

18   Chapter 1 Geometric Figures

Enrichment

Have the students imagine a distant planet on which an intelligent race of beings has divided a rotation through a circle into 50 "blips," abbreviated 50b. Thus, 50b = 360 degrees. Have the students make the following blip-degree conversions.

1. 25*b*   180
2. 15*b*   108
3. 5*b*   36
4. 90 degrees $12\frac{1}{2}b$
5. 60 degrees $8\frac{1}{3}b$
6. 1 degree $\frac{5}{36}b$

16. Name a right angle. ∠EAD
17. Name an acute angle with measure given in degrees. ∠FAB
18. Name an obtuse angle with measure given in degrees. ∠AFG
19. Find the measure of ∠GFE. 50
20. Find the measure of ∠AFB. 50
21. Name four straight angles.
    ∠AFE, ∠DGE, ∠CFB, ∠FGC

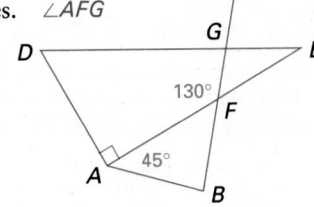

Refer to Figure 1 for Exercise 22. Use the Protractor Postulate to solve.

22. m ∠TOA = 3x + 50. m ∠SOA = x + 40.
    m ∠TOS = 50. Find m ∠TOA. 110

23. ∠TUV ≅ ∠XYZ, m ∠TUV = 3x − 20,
    and m ∠XYZ = $\frac{5x + 20}{3}$. Determine
    whether ∠TUV is a right angle. No, m ∠TUV = 40

24. A portion of ∠BAC is shown below.
    Construct an angle congruent to ∠BAC.
    Label the intersection of the lines point A.
    See Construction page 17.

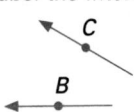

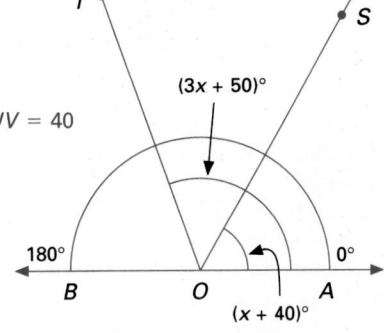

Figure 1

Midchapter Review

Points P, T, and Q are collinear points such that T is between P and Q.
(Exercises 1–3)

1. Use the Segment Addition Postulate to write an equation. 1.2
2. PT = $\frac{4}{5}$TQ, PQ = 18. Find PT. 1.2 8
3. Find the coordinate of the midpoint of $\overline{PQ}$ for coordinates P: −8
   and Q: 10. 1.3 m = 1

For Exercises 4–6, use A and B to represent the following with a
diagram and with symbols.

4. a line 1.1
5. a ray with endpoint A 1.2
6. a segment 1.2
7. Draw a segment. Locate its midpoint. 1.3

1. PT + TQ = PQ

4. $\overleftrightarrow{AB}$

5. $\overrightarrow{AB}$

6. $\overline{AB}$

7.

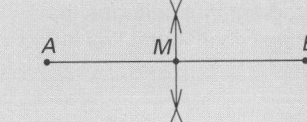

Computer Investigation

Supplementary and Complementary Angles

Use a computer software program that constructs angles of a given measure, labels and moves points, extends line segments, and measures line segments and angles.

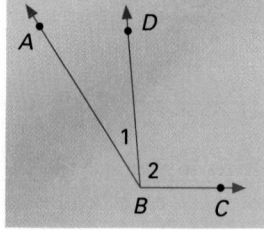

Notice in the figure at the right that there are three angles, $\angle 1$, $\angle 2$, and $\angle ABC$. $\angle 1$ and $\angle 2$ are called **adjacent angles**. Such angles will be formally defined in the next lesson. The activity below will help you discover an important property of adjacent angles.

Activity 1

Draw and label an angle of any measure as $\angle ABC$. Locate a point in the interior of $\angle ABC$. It will be labeled D. Draw $\overline{BD}$.

1. Find each of the following: m $\angle ABD$, m $\angle DBC$, and m $\angle ABC$.
2. Is there a numerical relationship between the measures of the three angles?
3. Construct a new angle. Then repeat Exercises 1–2 above for the new angle.
4. Repeat the process for a third angle.
5. Make a generalization about the relationship between two adjacent angles and the angle they form together.

Activity 2

In the lesson following this page, you will study two pairs of special types of adjacent angles. You will discover the property of one of these special types in the following activity.

Draw an angle with a measure of 150. The angle will be labeled $\angle CBA$. Extend side $\overline{CB}$ to a point D such that B is between C and D. Notice that $\angle CBA$ and $\angle DBA$ are adjacent angles. Although $\overrightarrow{BD}$ and $\overrightarrow{BC}$ may not actually appear as rays, they are called the outer rays of these adjacent angles.

1. Find m $\angle CBA$ and m $\angle DBA$.   150, 30
2. Find m $\angle CBD$.   180
3. Repeat the activity above using the steps of Exercises 1 and 2 for a different angle, say with measure 70.
4. Repeat the process for a third angle of measure 140.
5. Can you generalize the relationship between the measures of two adjacent angles whose **outer rays** form a **line**?   sum equals 180

20   Chapter 1 Computer Investigation

1.5 Adjacent Angles and Angle Bisectors

Objectives
To identify adjacent angles
To apply the Angle Addition Postulate
To apply the definition of angle bisector

An angle in a plane separates the plane into three sets of points: the *angle* itself, the *interior* of the angle, and the *exterior* of the angle. (This does not apply to a straight angle.) A point is an **interior point of an angle** if it is not on either side and is between two points of the angle, one on each side.

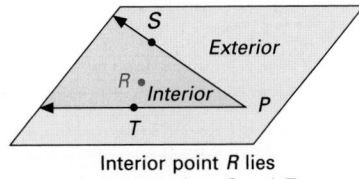

Interior point *R* lies *between* points *S* and *T*.

The figure at the right consists of three angles: ∠1, ∠2, and ∠CAB. ∠1 and ∠2 share vertex *A* and side $\overrightarrow{AD}$. No common interior points are shared by ∠1 and ∠2. ∠1 and ∠2 are called *adjacent angles*.

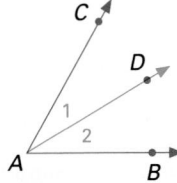

Definition

> **Adjacent angles** are two coplanar angles that have one common side and a common vertex, but no common interior points.

The figure at the right shows adjacent angles, ∠ABD and ∠DBC. The common side of the angles is $\overrightarrow{BD}$. The rays $\overrightarrow{BA}$ and $\overrightarrow{BC}$ are called the **outer rays** of the angles. Notice that m ∠ABC = 50 + 30, or 80. This suggests the following postulate.

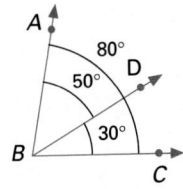

Postulate 5

> **Angle Addition Postulate**
> If *D* is in the interior of ∠ABC,
> then m ∠ABC = m ∠ABD + m ∠DBC

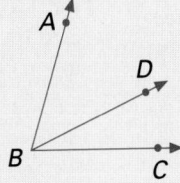

Manipulative Worksheet 2
Quick Quizzes 5
Reteaching and Practice
  Worksheets, pp. 9, 10

▰▰▰ GETTING STARTED

Prerequisite Quiz

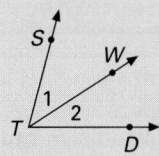

1. Name the sides of ∠1. $\overrightarrow{TS}$, $\overrightarrow{TW}$
2. Name the sides of ∠2. $\overrightarrow{TW}$, $\overrightarrow{TD}$
3. Name each angle.   ∠1, or ∠STW, or ∠WTS; ∠2, or ∠WTD, or ∠DTW; ∠STD, or ∠DTS

Motivator

On the blackboard, draw two adjacent angles *PQS* and *SQR* such that $\overrightarrow{QS}$ is in the interior of ∠PQR. Ask how many angles there are.   3   Ask your students to write an equation such as the one for segments in Lesson 1.2.   m ∠PQS + m ∠SQR = m ∠PQR

▰▰▰ TEACHING SUGGESTIONS

Lesson Note

Point out the similarity of the Angle Addition Postulate to the Segment Addition Postulate. Both are used to justify equations in proofs.

21

Lesson Note

Use numerical examples to show the relation between the measures of the *three* angles formed when two adjacent angles are given.

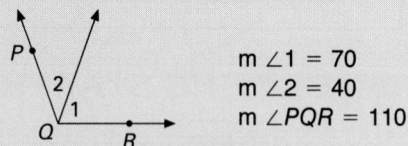

m $\angle 1 = 70$
m $\angle 2 = 40$
m $\angle PQR = 110$

Math Connections

Sports: A baseball field illustrates the Angle Addition Postulate if a line is drawn from home to second base. A line drive hit between second and third can be a model of a ray (except, of course, that it doesn't extend forever).

Critical Thinking Questions

Analysis: Ask students these questions. Into how many sets of points does a straight angle separate a plane? 2 sets. What could you name those sets? Are there points in the plane not included in these 2 sets? Yes. The straight angle itself is a set of points that is not in either of the 2 sets.

Common Error Analysis

Error: Students sometimes assume information from a given diagram, particularly in cases in which the common ray of two adjacent angles appears to be the angle bisector.

Students should understand that the Angle Addition Postulate does not imply that adjacent angles have equal measures. However, when it is stated that adjacent angles have equal measures, the Angle Addition Postulate also applies.

Make up a drill sheet with several exercises such as the one below. Can you write equations based on the definition of angle bisector and on the Angle Addition Postulate? Write the equation or equations that apply.

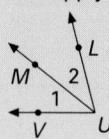

Given: $\overrightarrow{UM}$ bisects $\angle LUV$.
Yes; m $\angle LUM$ = m $\angle MUV$;
m $\angle LUM$ + m $\angle MUV$ = m $\angle LUV$

EXAMPLE 1 Given: m $\angle ABD = 130$, m $\angle 2 = 30$
Find m $\angle 1$.

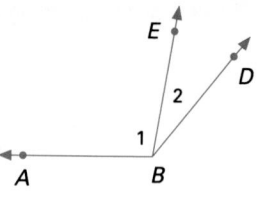

Plan Use the Angle Addition Postulate to write an equation.

Solution m $\angle 1$ + m $\angle 2$ = m $\angle ABD$

Solve for m $\angle 1$.   m $\angle 1$ + 30 = 130
m $\angle 1 = 100$

EXAMPLE 2 Given: m $\angle 1 = \frac{2}{3}$(m $\angle 2$), m $\angle ABC = 140$
Find m $\angle 1$ and m $\angle 2$.

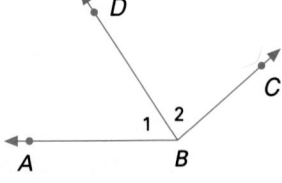

Plan Let x = m $\angle 2$. Then m $\angle 1 = \frac{2}{3}x$.

Solution Write an equation.   m $\angle 1$ + m $\angle 2$ = m $\angle ABC$
Substitute. $\frac{2}{3}x + x = 140$
Multiply both sides of $2x + 3x = 420$
the equation by 3. $5x = 420$
                           $x = 84 \longrightarrow$ m $\angle 2 = 84$
                           $\frac{2}{3}x = 56 \longrightarrow$ m $\angle 1 = 56$

In the figure at the right, $\angle 1$ and $\angle 2$ are adjacent angles. Suppose m $\angle 1$ = m $\angle 2$ = 30. Then $\overrightarrow{OC}$ divides $\angle AOB$ into two congruent angles. $\overrightarrow{OC}$ is called an **angle bisector**.

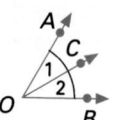

Definition $\overrightarrow{OC}$ bisects $\angle AOB$ if C is in the interior of $\angle AOB$ and $\angle AOC \cong \angle COB$ (m $\angle AOC$ = m $\angle COB$). $\overrightarrow{OC}$ is the **bisector** of $\angle AOB$.

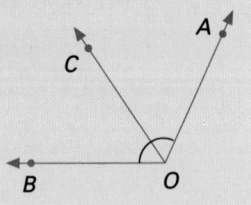

Additional Example 1

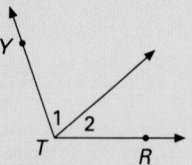

Given: m $\angle YTR = 100$, m $\angle 1 = 60$
Find m $\angle 2$. **40**

Additional Example 2

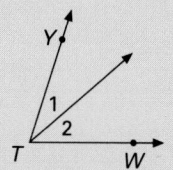

Given: m $\angle 1 = \frac{4}{5}$(m $\angle 2$), m $\angle YTW = 72$
Find m $\angle 1$. **32**

Construction Construct the bisector of a given angle.

Using O as the center, draw an arc, intersecting the sides of $\angle POQ$. Label the points of intersection G and H.

Next, use either the same or an enlarged compass opening to draw two arcs, one with center G, the other with center H.

Label the point where the two arcs intersect T. Draw $\overrightarrow{OT}$.

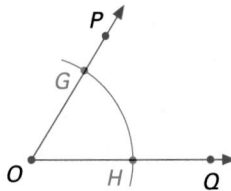 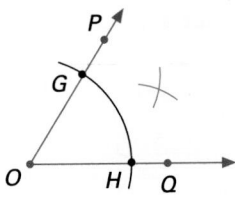 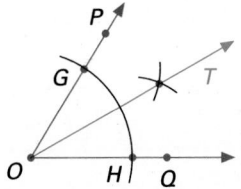

Result: $\overrightarrow{OT}$ is the bisector of $\angle POQ$.

Postulate 6 Every angle, except a straight angle, has exactly one bisector.

EXAMPLE 3 Assume that $\overrightarrow{PS}$ bisects $\angle RPQ$, m $\angle RPS = 5x + 4$, m $\angle SPQ = 7x - 10$. Find m $\angle RPS$.

Solution
m $\angle RPS = $ m $\angle SPQ$ ($\overrightarrow{PS}$ is a bisector.)
$$5x + 4 = 7x - 10$$
$$-2x + 4 = -10$$
$$-2x = -14$$
$$x = 7$$
Thus, m $\angle RPS = 5 \cdot 7 + 4 = 39.$ (m $\angle RPS = 5x + 4$)

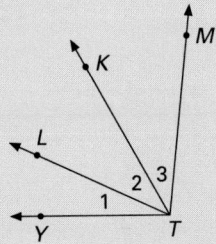

Classroom Exercises

1. m $\angle 1 = 35$, m $\angle 2 = 40$. Find m $\angle STU$.   75
2. m $\angle 1 = 20$, m $\angle STU = 75$. Find m $\angle 2$.   55
3. m $\angle 2 = 32$, m $\angle STU = 68$. Find m $\angle 1$.   36
4. Name two adjacent angles. $\angle 1$, $\angle 2$
5. Name two non-adjacent angles. $\angle STU$, $\angle 2$
   or $\angle STU$, $\angle 1$

Ex. 1–8

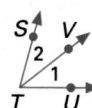

1.5 Adjacent Angles and Angle Bisectors   **23**

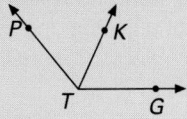

6. $\overrightarrow{TV}$ bisects $\angle STU$, m $\angle 1 = 37$. Find m $\angle STU$.   74

7. $\overrightarrow{TV}$ bisects $\angle STU$, m $\angle 2 = 37$. Find m $\angle STU$.   74

8. $\overrightarrow{TV}$ bisects $\angle STU$. Name two angles that are not congruent.
$\angle STU$, $\angle 2$ or $\angle STU$, $\angle 1$

Written Exercises

State whether $\angle 1$ and $\angle 2$ are adjacent. If not, tell why not.

1.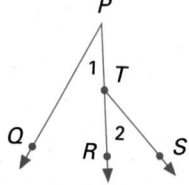

No, no common vertex

2.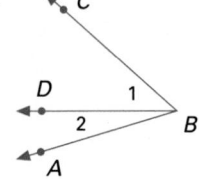

Yes

3.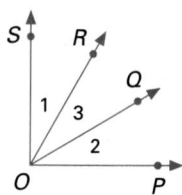

No, no common side

Find the indicated measure.

4. m $\angle POR = 80$,
m $\angle 2 = 25$
Find m $\angle 1$.   55

5. m $\angle QOS = 100$,
m $\angle 3 = 65$
Find m $\angle 2$.   35

6. m $\angle POR = 125$,
m $\angle 3 = 35$
Find m $\angle POS$.   160

7. m $\angle QOS = 49$,
m $\angle 1 = 23$
Find m $\angle POS$.   72

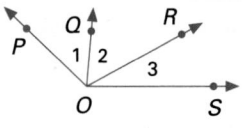

In the figure below, m $\angle FDB = 85$,
m $\angle DBA = 100$, m $\angle 2 = 30$, m $\angle 3 = 50$, m $\angle 4 = 25$,
and m $\angle 5 = 40$. Find each measure.

8. m $\angle EDC$ 65

9. m $\angle 6$ 45

10. m $\angle EBC$ 80

11. m $\angle 1$ 70

12. m $\angle FDC$ 110

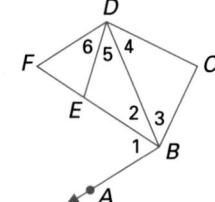

13. Draw an angle with a measure of 70. Construct an angle with $\frac{1}{2}$ of this measure.   See Construction on page 23.

14. Draw an angle with a measure of 130. Construct an angle with $\frac{1}{4}$ of this measure.
(HINT: $\frac{1}{4} = \frac{1}{2} \cdot \frac{1}{2}$)   Bisect twice. See Construction on page 23.

24   Chapter 1 Geometric Figures

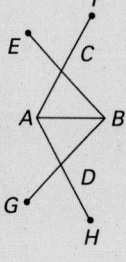

15. Given: m $\angle 1 = \frac{2}{5}$ (m $\angle 2$), m $\angle PQR = 49$. Find m $\angle 1$.   14

16. Given: m $\angle 2 = \frac{3}{5}$ (m $\angle 1$), m $\angle PQR = 64$. Find m $\angle 2$.   24

17. Given: m $\angle 1$ is twice m $\angle 2$, m $\angle PQR = 78$. Find m $\angle 1$.   52

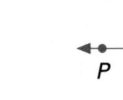

For Exercises 18–20, $\overrightarrow{QS}$ bisects $\angle PQR$.

18. Given: m $\angle 1 = 4x + 30$, m $\angle 2 = 2x + 40$. Find m $\angle 1$.   50

19. Given: m $\angle 1 = 42 - 2x$, m $\angle 2 = 30 + 4x$. Find m $\angle PQR$.   76

20. Given: m $\angle 1 = 6x + 18$, m $\angle 2 = 9x$. Find m $\angle 1$.   54

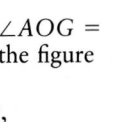

21. $\angle AOT$ and $\angle TOG$ are adjacent angles, m $\angle AOG = 100$, and m $\angle AOT = 3$(m $\angle TOG$). Draw the figure and find m $\angle TOG$.

22. In the figure at the right, $\overrightarrow{EB}$ bisects $\angle AED$, m $\angle AEB = 3x + 12$, m $\angle BED = x + 32$, and m $\angle BEC = \frac{1}{6}$m $\angle CED$. Find m $\angle AEC$.   48

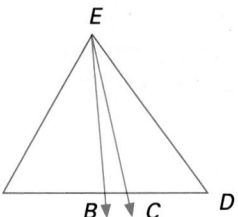

Mixed Review

Define the following.   *1.4, 1.5*

1. acute angle

2. right angle

3. obtuse angle

4. straight angle

5. congruent angles

6. adjacent angles

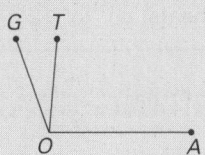

Application: *Surveying*

Surveyors provide accurate measurements of both distance and direction. *Bearing* is used to indicate direction. Bearing is stated as the number of degrees east or west of the north or south line. In the diagram, the bearing from Town A to Town B is N49°E, that is, 49° east of due north. The bearing from Town B to Town C is S20°E, or 20° east of due south.

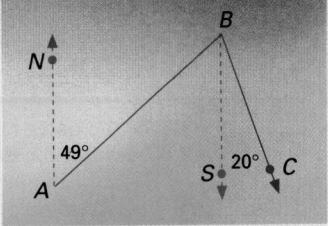

1. Using a protractor, find the bearing from Town B to Town A.   S49°W
2. Explain how to find the bearing from Town A to Town C, and then find it. Use a straightedge and protractor. Find m $\angle NAC$.   N86°E

▰▰▰ GETTING STARTED

Prerequisite Quiz

1. How are the sides of a straight angle related? They form a line.
2. What is the name of an angle whose measure is 90? Rt ∠
3. Use the figure below to write an equation for the diagram.
   m ∠3 + m ∠4 = m ∠YOP

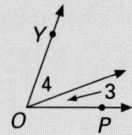

4. If m ∠3 = $\frac{2}{5}$(m ∠4) and m ∠YOP = 70, find m ∠3. 20

Motivator

Draw two adjacent angles on the chalkboard. Label them ∠1 and ∠2. Let m ∠1 = 20 and the measure of the large angle containing ∠1 and ∠2 equal 110. Ask your students if they can derive the measure of ∠2 from the information given. Yes; m ∠2 = 90.

▰▰▰ TEACHING SUGGESTIONS

Lesson Note

Stress the relationship of outer rays in the description of adjacent angles that are supplementary or complementary. Outer rays are opposite rays for supplementary angles; outer rays are perpendicular for complementary angles.

Students may ask why Postulate 7 is not a theorem as is Theorem 1.1. This is because the *interior* of a straight angle is undefined. Therefore, the Angle Addition Postulate cannot be applied.

1.6 Supplementary and Complementary Angles

Objectives
To find the measures of two supplementary angles
To find the measures of two complementary angles
To solve problems involving supplementary and complementary angles

A straight angle has a measure of 180. The sides are rays that together form a line. In the diagram, O is between A and B. Notice that the rays extend in *opposite* directions.

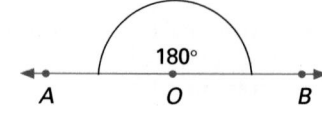

Definition

If point O is between points A and B on $\overleftrightarrow{AB}$, then $\overrightarrow{OA}$ and $\overrightarrow{OB}$ are called **opposite rays**. These rays are sides of ∠AOB, a straight angle.

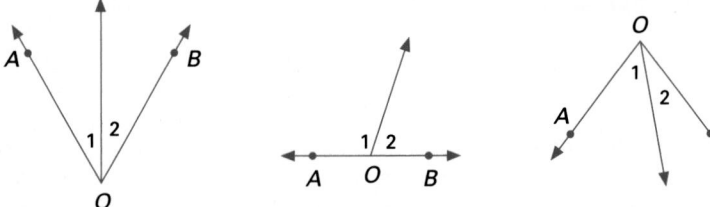

Each figure above shows a pair of adjacent angles, ∠1 and ∠2. In the second figure, the outer rays $\overrightarrow{OA}$ and $\overrightarrow{OB}$ form a line. In this case, ∠AOB is a straight angle and the outer rays are opposite rays.

Definition

Two adjacent angles whose outer rays are opposite rays are called a **linear pair** of angles.

Postulate 7

If the outer rays of two adjacent angles form a straight angle, then the sum of the measures of the angles is 180.

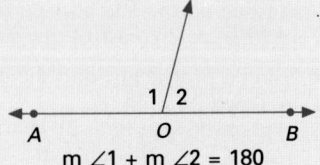

Definition	Two angles are **supplementary** if the sum of their measures is 180. Each angle is called a **supplement** of the other.	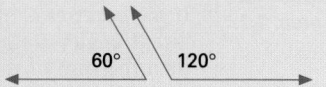 60° 120°

If supplementary angles are adjacent, then their outer rays form a straight angle. Postulate 7 may be restated as follows: If the outer rays of two adjacent angles form a straight angle, then the angles are supplementary. In this book, assume that angles that appear to be straight angles are straight angles.

EXAMPLE 1 Are the pairs of angles supplementary? Explain why or why not.

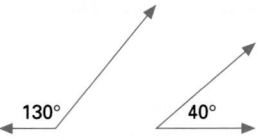
130° 40°

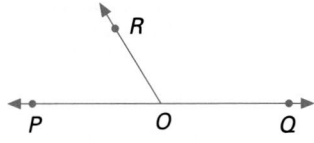
R
P   O   Q

Solution

No. 130 + 40 = 170. The sum of the measures is *not* 180.

Yes. The outer rays form a straight angle. The sum of the measures is 180 by Postulate 7.

Definition	Two lines are **perpendicular** if they intersect to form a right angle.

The symbol for perpendicular is ⊥. Thus, $\overrightarrow{OB} \perp \overrightarrow{OA}$ means "$\overrightarrow{OB}$ is perpendicular to $\overrightarrow{OA}$." Since segments and rays are parts of lines, then intersecting segments and rays are perpendicular if the lines that contain them are perpendicular.

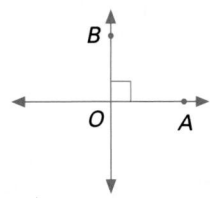
B
O   A

Sometimes, adjacent angles have their outer rays perpendicular. In the figure at the right, $\overrightarrow{OA} \perp \overrightarrow{OB}$ and ∠1 and ∠2 are acute angles. So ∠AOB is a right angle with m ∠AOB = 90. By the Angle Addition Postulate, m ∠1 + m ∠2 = m ∠AOB. Therefore, m ∠1 + m ∠2 = 90, by substituting 90 for m ∠AOB.

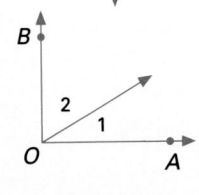
B
2   1
O   A

Consider the process of reasoning used above. Such a logical sequence of statements leading to a conclusion is called a **proof**. A statement that has been proved true, not just taken for granted, is called a **theorem**.

1.6 Supplementary and Complementary Angles **27**

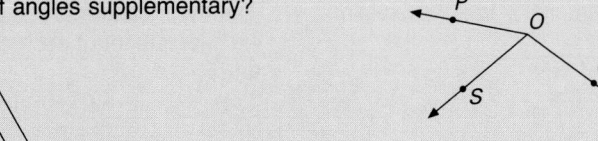

Lesson Note

Note that Theorem 1.1 is proven informally in paragraph form since the two-column proof is not introduced until the next chapter.

Checkpoint

Are the pairs of angles supplementary, complementary, or neither?

1.

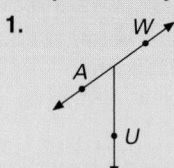

Supplementary

2.

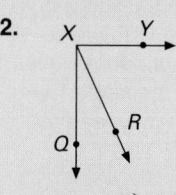

Given: $\overrightarrow{XQ} \perp \overrightarrow{XY}$ Complementary

Find the measure of a complement and a supplement, if possible, of each angle with the given measure.

3. 49   41; 131
4. 110   Not possible; 70
5. $3x - 20$ $110 - 3x$ if given angle is acute; $200 - 3x$ if given angle is not straight.
6. The measure of an angle is 10 less than 4 times the measure of its complement. Find the measure of each angle.   70, 20

Closure

Have students name at least two different ways which will guarantee that two angles will be (a) supplementary, and (b) complementary.   (1) Their sum is 180 or 90. (2) The difference between 180 or 90 and the measure of one angle is the measure of the other angle. (3) Their outer rays (if they are adjacent) form a line or a right angle.

Theorem 1.1

If the outer rays of two acute adjacent angles are *perpendicular*, then the sum of the measures of the angles is 90.

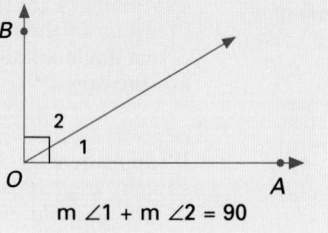

$m \angle 1 + m \angle 2 = 90$

Definition

Two angles are **complementary** if the sum of their measures is 90. Each angle is called a **complement** of the other.

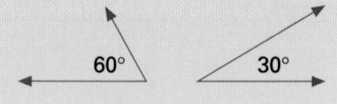

Theorem 1.1 may now be stated as follows:
If the outer rays of two acute adjacent angles are perpendicular, then the angles are *complementary*.

EXAMPLE 2   Is $\angle 1$ complementary to $\angle 2$? Explain why or why not.

Given: $m \angle 1 = 45$, $m \angle 2 = 55$ Given: $\overrightarrow{ED} \perp \overrightarrow{EF}$

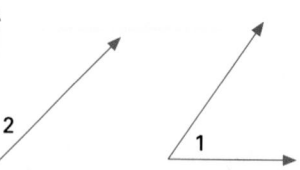   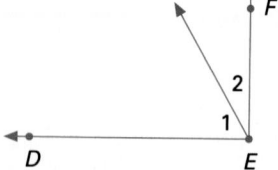

Solution   No. $45 + 55 = 100$. The sum of the measures is *not* 90.   Yes. The outer rays are perpendicular. The sum of the measures is 90 by Theorem 1.1.

EXAMPLE 3   Find the measure of a complement and supplement, if possible.

Angle	Measure	Measure of Complement	Measure of Supplement
A	35	55	145
B	97	Not possible	83
C	x	$90 - x$ for $\angle C$ acute	$180 - x$ for $\angle C$ not straight
D	$3x - 20$	$90 - (3x - 20) =$ $110 - 3x$ for $\angle D$ acute	$180 - (3x - 20) =$ $200 - 3x$ for $\angle D$ not straight

28   Chapter 1 Geometric Figures

Additional Example 2

Is $\angle 1$ complementary to $\angle 2$? Explain.

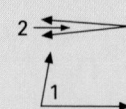

Given: $m \angle 1 = 80$, $m \angle 2 = 10$
Yes; $80 + 10 = 90$

Additional Example 3

Find the measure of the complement and the supplement of each angle, if possible.

a. 75   15, 105
b. y $90 - y$ if given angle is acute, $180 - y$, if given angle is not straight
c. 115   Not possible, 65
d. $7x + 30$ $60 - 7x$ if given angle is acute, $150 - 7x$, if given angle is not straight.

EXAMPLE 4   The measure of an angle is 60 less than twice the measure of its complement. Find the measure of the angles.

As you go about solving the problem, proceed in a step-by-step fashion. Determine exactly what you are looking for—the measures of two complementary angles. Then:

(1) Represent the *data*.   Let x = measure of complement.
$2x - 60$ = measure of the angle.
(60 less than twice x)

(2) Write the *equation*. $x + (2x - 60) = 90$
Sum of measures of complementary angles is 90.

(3) *Solve* the equation. $3x - 60 = 90$
$3x = 150$
$x = 50$ (measure of the complement)
Find the measure of the angle.
$2x - 60 = 2 \cdot 50 - 60$
$= 100 - 60 = 40$

(4) Check *solutions*.   Are the angles complementary?
$40 + 50 = 90$ ✔
Is angle measure 60 less than twice measure of complement?
$40 = 2 \cdot 50 - 60$
$40 = 100 - 60$
$40 = 40$ ✔

(5) Label the *answer*.   Thus, the measures of the angles are 40 and 50.

Classroom Exercises

Find the measure of a complement of an angle with the given measure.

1. 42   48   **2.** 65   25   **3.** 9   81   **4.** 25   65   **5.** 72   18   **6.** 16 74

Find the measure of a supplement of an angle whose measure is given.

7. 120   60   **8.** 80   100   **9.** 135   45   **10.** 70   110   **11.** 104   76   **12.** 139   41

Written Exercises

Are the indicated pairs of angles supplementary, complementary, or neither? Explain why or why not.

1. 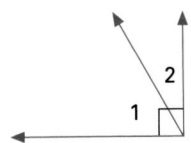   **2.** 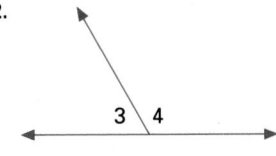   **3.**
75°
35°

1.6 Supplementary and Complementary Angles   **29**

Guided Practice

Classroom Exercises 1–12

Independent Practice

Ⓐ Ex. 1–19, Ⓑ Ex. 20–28, Ⓒ Ex. 29–31

Basic: WE 1–19 odd, Application 1
Average: WE 1–28 odd, Application all
Above Average: WE 1–31 odd, Application all

Additional Answers

Written Exercises

1. Complementary; the outer rays are perpendicular, therefore the sum of the measures of the angles is 90.
2. Supplementary; the sum of the measures of the angles is 180.
3. Neither; the outer rays form a straight angle, therefore the sum of the measures of the angles is neither 90 nor 180.

Additional Example 4

The measure of an angle is 20 more than 3 times the measure of its supplement. Find the measures of both angles.   40 and 140

10. If $0 < t < 90$, complement = $90 - t$; if $0 < t < 180$, supplement = $180 - t$
11. Complement = 58.3, supplement = 148.3
12. Complement = $23\frac{1}{2}$, supplement = $113\frac{1}{2}$
13. If $3 < m < 93$, complement = $93 - m$; if $3 < m < 183$, supplement = $183 - m$
14. If $-82 < x < 8$, complement = $82 + x$; if $-172 < x < 8$, supplement = $172 + x$
15. If $5 < x < 50$, complement = $100 - 2x$; if $5 < x < 95$, supplement = $190 - 2x$
16. ∠1 and ∠2 must be complementary because the sides of the frame must be perpendicular.
17. No. The supplement of an angle greater than 90 must be less than 90, that is, an acute angle.
18. No. If m ∠1 + m ∠2 = 90, then it is impossible for m ∠1 + m ∠2 = 180.
19. ∠BFA is supplementary to ∠BFE.
30. No. If this were possible, then $90 - x$ would equal $\frac{1}{2}(180 - x)$. But, if $90 - x = \frac{1}{2}(180 - x)$ then, $90 - x = 90 - \frac{1}{2}x$ so, $-x = -\frac{1}{2}x$ or, $x = \frac{1}{2}x$
This is impossible except when $x = 0$. Since no angle has measure 0, the complement of an angle cannot equal half its supplement.
31. Assume ∠ABC and ∠CBD are adjacent and complementary. By the Angle Addition Postulate we know that m ∠ABC + m ∠CBD = m ∠ABD. Since ∠ABC and ∠CBD are complementary, we know m ∠ABC + m ∠CBD = 90. So, m ∠ABD = 90. Thus, by the definition of perpendicular lines, $\overrightarrow{BA} \perp \overrightarrow{BD}$.

Find the measure of a complement and a supplement, if possible, of an angle with the given measure.

4. 32   58, 148
5. 74   16, 106
6. 19   71, 161
7. 99   None, 81
8. 136   None, 44
9. 180   None, none
10. t
11. 31.7
12. $66\frac{1}{2}$
13. $m - 3$
14. $8 - x$
15. $2x - 10$

16. The figure at the right shows pieces of a square-cornered picture frame. What must be the relationship between ∠1 and ∠2? Why?

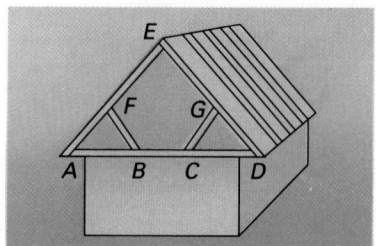

17. Is it reasonable to expect that a supplement of an obtuse angle can be a right angle? Why?

18. Is it reasonable to expect to find a pair of complementary angles that are also supplementary? Why?

19. The figure at the right shows rafters $\overline{AE}$ and $\overline{ED}$ of a roof with support braces $\overline{BF}$ and $\overline{CG}$. What relationship must exist between ∠BFA and ∠BFE?

20. The measure of an angle is 50 more than that of its complement. Find the measure of each angle.   70, 20
21. The measure of an angle is 70 more than that of its supplement. Find the measure of each angle.   125, 55
22. The measure of an angle is 3 times that of its supplement. Find the measure of each angle.   135, 45
23. The measure of an angle is equal to that of its supplement. Find the measure of each angle.   90, 90
24. The measure of an angle is 30 less than 5 times the measure of its complement. Find the measure of each angle.   70, 20
25. The measure of an angle is 20 more than 3 times the measure of its supplement. Find the measure of each angle.   40, 140
26. The measure of an angle is 26 less than 3 times the measure of its complement. Find the measure of each angle.   61, 29
27. The measure of an angle is $\frac{1}{5}$ the measure of its complement. Find the measure of each angle.   15, 75
28. The measure of an angle is $\frac{2}{3}$ the measure of its supplement. Find the measure of each angle.   72, 108
29. Four times the measure of the complement of an angle is 12 more than twice the difference of the measures of its supplement and its complement. Find the measure of each angle.   42, 48, 138
30. Can the measure of the complement of an angle be $\frac{1}{2}$ the measure of the supplement of the angle? Why or why not?
31. Prove: If complementary angles are adjacent, then their outer rays are perpendicular.

30   Chapter 1 Geometric Figures

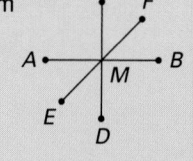

Mixed Review

1. Name the intersection of $\overleftrightarrow{FG}$ and plane $\mathcal{M}$. **1.1** *E*
2. Name a point that is collinear with points *C* and *E*. **1.1** *D*
3. Name 4 pairs of adjacent angles not contained in $\mathcal{M}$. **1.5** $\angle FEC, \angle GEC;$ $\angle FEB, \angle GEB;$ $\angle FED, \angle GED;$ $\angle FEA, \angle GEA$
4. Name 4 noncoplanar points. **1.1** *A, B, C, F;* other answers are possible.

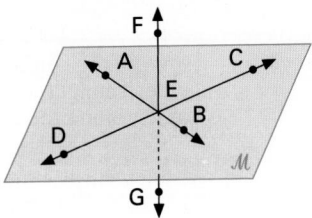

For the given coordinates of *A* and *B*, find *AB*. **1.2**

5. $A: -3; B = 8$ 11

6. $A: 2\frac{1}{2}; B: -\frac{3}{4}$ $3\frac{1}{4}$

Find the coordinate of the midpoint, *M* of $\overline{AB}$ whose endpoints have the given coordinates. **1.3**

7. 2 and 7 $4\frac{1}{2}$

8. $-3\frac{1}{2}$ and 6 $1\frac{1}{4}$

▰ Application: *Astronomical Units*

Astronomers measure distance within the solar system in astronomical units (A.U.). By the Ruler Postulate, the point 0 is assigned to the Sun and the point 1 is assigned to the Earth.

One astronomical unit, therefore, is the mean distance from the Sun to Earth. The mean distances of the other planets in astronomical units are shown in the table below.

Sun	0
Mercury	0.39
Venus	0.72
Earth	1
Mars	1.5
Jupiter	5.2
Saturn	9.5
Uranus	19.2
Neptune	30.1
Pluto	39.4

1. If one A.U. equals 92,960,000 mi, what is the distance in miles from the Sun to Mars?
139,440,000

2. At its closest, Venus is about 26 million miles from Earth. But can it also be as far as 159 million miles from Earth? Why?

2. Yes. When Venus is on the other side of the Sun, directly opposite from Earth, the distance between Venus and Earth is the sum of their distances from the Sun.

3. What is the distance from the Earth to Saturn?
ranges from 8.5 to 10.5 A.U.

Teaching Resources

Quick Quizzes 7
Reteaching and Practice Worksheets, pp. 13, 14

GETTING STARTED

Prerequisite Quiz

True or false?

1. A point has one dimension. F
2. A plane has no thickness. T
3. A segment contains only two points. F
4. $4 + 3 < 5$ F
5. $(-3)^2 = -9$ F

Motivator

Have students discuss the answers to these questions.

What two meanings does the word *or* have in everyday English? The word "or" can be exclusive as in "The light is on or off" or inclusive, as in "Do you want vanilla or chocolate?" The answer to the inclusive question could be "both." Math uses the inclusive "or." For example, what does it mean to say, "Let's go swimming *or* to the park"? Might you do both? Yes What does it mean to say "I will take the bus *or* the train"? Are you likely to do both? No When will a statement involving the word *and* be true? When both parts are true For example, is the following true? Why or why not? New Jersey is in the USA *and* California is a city. False because the second statement is false; California is a state, not a city.

TEACHING SUGGESTIONS

Lesson Note

This lesson is the first in a strand of formal logic that includes conditionals, converses, inverses, and biconditionals. The last two lessons in this chapter show the graphing of compound inequalities by using points, segments, rays, and lines, a topic that connects algebra with its geometric representations. The content of Chapter 2 is independent of the material in Lessons 1.7 and 1.8, although the logic strand is continued in Lessons 2.5 and 2.6.

1.7 Logic: Conjunction and Disjunction

Objectives
To write the conjunction (disjunction) of two statements
To determine if a conjunction (disjunction) is true
To determine the truth table of a conjunction (disjunction)

A statement is either true or false, but never both. The statement "$5 = 9 - 4$" is true. The statement "A line has an endpoint" is false. Two statements may be combined into a single statement connected by the word *and.*

Definition

If p and q are statements, the statement "*p and q*" is called the **conjunction** of p and q and written as $p \wedge q$. $p \wedge q$ is true when p and q are both true. $p \wedge q$ is false when at least one of the statements, p or q, is false.

Statement p: $5 = 9 - 4$ Statement q: A line has an endpoint.
Conjunction: $5 = 9 - 4$ *and* A line has an endpoint.
p $\wedge$ q

The conjunction "$5 = 9 - 4$ *and* A line has an endpoint" is false because one of the two statements is false ($p \wedge q$ is false).

The conjunction "Bread is a food *and* Blue is a color" is true because both parts are true. ($p \wedge q$ is true).

There are four different combinations of truth values for the two statements. The truth table below summarizes these combinations.

p	q	$p \wedge q$	
T	T	T	← Both statements are true.
T	F	F	← Only one statement is true.
F	T	F	
F	F	F	← Both statements are false.

EXAMPLE 1 Determine whether each of the following conjunctions is true.

a. A segment is a set of points (T) *and* $9 > 2$ (T)

Solution The conjunction is true since both parts are true.

b. $7 = 2 \cdot 3 + 1$ (T) *and* $-4 > 2$ (F)

Solution The conjunction is false since one part is false.

Additional Example 1

Determine whether each of the following conjunctions is true.

a. Space has only two dimensions *and* $(-5 < 3)$. F

b. A line is a set of points *and* *AB* is not a set of points. T

Two statements may also be combined into a single statement by the word *or*. In everyday language, this word is often used in the *exclusive* sense. Consider a parent saying to a child, "Tonight, you may go to the movies or you may go bowling." The parent means *either . . . or*: only one of the two choices, not both, may take place. In mathematics, however, the *inclusive* meaning of "or" is used. This means that if one or the other or *both* statements are true, then the single statement combining them by the word "or" is true.

Definition

> If p and q are statements, the statement "*p or q*" is called the **disjunction** of p and q and written as $p \lor q$. $p \lor q$ is true when at least one of the statements, p or q, is true. $p \lor q$ is false when p and q are both false.

The truth table for a disjunction is given below.

p	q	$p \lor q$	
T	T	T	← Both statements are true.
T	F	T	← At least one statement is true.
F	T	T	
F	F	F	← Both statements are false.

EXAMPLE 2 Determine if each of the following disjunctions is true.

a. A ray has two endpoints *or* A plane contains a finite number of points.

Solution (F) (F)

Thus, the disjunction is false since both of the statements are false.

b. Cows can fly *or* $2 + 3 = 5$.

Solution (F) (T)

Thus, the disjunction is true since at least one of the statements is true.

Classroom Exercises

Indicate whether the statement is true or false.

1. $3 = 1 + 2$ *or* $3 = 1 \cdot 2$ True
2. $3 = 1 + 2$ *and* $3 = 1 \cdot 2$ False
3. $6 = 4 \cdot 2$ *or* $12 = 7 \cdot 5$ False
4. $6 = 4 + 2$ *and* $12 = 7 + 5$ True
5. A disjunction is true if exactly one of its two parts is false.   True
6. A conjunction is true if one of its two parts is false.   False
7. If one part of a disjunction is true, then the disjunction is true.   True
8. A conjunction is false if one of its two parts is false.   True

Additional Example 2

Write a disjunction for the given pair of statements. Is the disjunction true?

A point has one dimension.
A plane is flat.
A point has one dimension $\lor$
A plane is flat.; T

Math Connections

Algebra: Conjunction and disjunction are used in writing solutions to quadratic inequalities.

Electronics: This notation is also used for logical circuits and switching diagrams as shown in the Application on page 35.

Critical Thinking Questions

Evaluation: Have students discuss whether this sentence is true or false: "I never tell the truth."   By the rules of formal logic, the truth or falsity of a statement that includes itself cannot be determined.

Checkpoint

True or false?

1. A ray has two endpoints *and* space is flat.   F
2. A ray has two endpoints *or* space is flat.   F
3. $(4 + 9 \le 16)$ *and* $(-1 \ge -2 + 1)$ T
4. $(4 + 9 \le 16)$ *or* $(-1 \ge -2 + 1)$ T
5. $[(-2)^2 < (-1)^2]$ *and* $(1 > 0.9)$ F
6. $[(-2)^2 < (-1)^2]$ *or* $(1 > 0.9)$ T

Closure

Ask students to answer these questions. What is an *and* statement called?
Conjunction

What is an *or* statement called?
Disjunction

When will a conjunction be true?   When both statements are true

When will a disjunction be true?   When at least one statement is true

▰▰▰ FOLLOW UP

Guided Practice

Classroom Exercises 1–8

Independent Practice

A Ex. 1–10, **B** Ex. 11–14, **C** Ex. 15–17

Basic: WE 1–10 odd, Application 1–2
Average: WE 1–14 odd, Application all
Above Average: WE 1–17 odd, Application all

▰▰▰GETTING STARTED

Prerequisite Quiz

Graph each inequality on a number line.

1. $x \geq 4$

2. $x \leq 6$

3. $x \geq -3$

4. $x \leq -2$

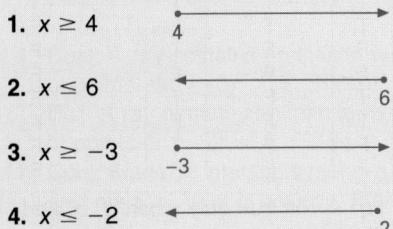

Motivator

Ask students these questions.

How do you graph $x \geq 4$?   Closed dot at 4, arrow pointing to the right.

How do you graph $x \leq 8$?   Closed dot at 8, arrow pointing to the left.

Graph these on separate number lines with their origins directly over each other. What points are graphed on *both* number lines? $4 \leq x \leq 8$

▰▰▰TEACHING SUGGESTIONS

Lesson Note

Since each graph in this lesson is a model of the geometric figures, point, segment, ray, or line, the symbols < and > are avoided. This is because the graph of $x > 3$ or the graph of $x < 3$ is not a ray since the endpoint is deleted and would be represented by an open circle.

Math Connections

Algebra: These graphs provide a way to represent sets of numbers that are the solutions to specified algebraic sentences.

1.8 Graphing Conjunctions and Disjunctions

Objective

To graph the conjunction or the disjunction of two inequalities

Inequalities can be graphed on a number line. The graph of $x \geq 5$ is shown below. A solid dot is placed at A. Then all points to the right of A are shaded.

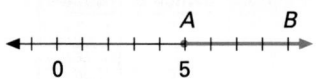

The graph is a ray: $\overrightarrow{AB}$

The graph of $x \leq 3$ is shown below. A solid dot is placed at C. Then all points to the left of C are shaded.

The graph is a ray: $\overrightarrow{CD}$

The graph of the conjunction (disjunction) of two inequalities can be drawn as follows.

First, graph each inequality on a separate number line. Then, on the third number line, locate either:
(1) the conjunction—the points *common to* the graphs of the two inequalities; or
(2) the disjunction—the points belonging to the graphs of *either* of the two inequalities, or both.

The resulting graph may be one of the four geometric figures that you have studied in this chapter—point, line, segment, or ray.

EXAMPLE 1

Graph the conjunction $x \geq 2$ *and* $x \leq 4$. Identify the resulting geometric figure, if possible.

Solution

The graph of $x \geq 2$ is the ray with endpoint P, extending to the right.

The graph of $x \leq 4$ is a ray with endpoint Q, extending to the left.

The graph of $x \geq 2$ *and* $x \leq 4$ is the set of points common to the graphs of both inequalities.

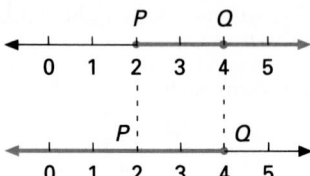

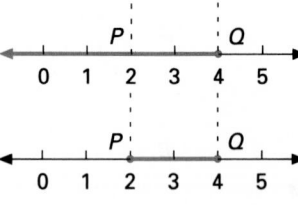

Thus, the graph of $x \geq 2$ *and* $x \leq 4$ is the segment $\overline{PQ}$.

Additional Example 1

Graph the conjunction $x \geq -4$ *and* $x \leq -1$. Identify the resulting geometric figure, if possible.

Segment; $\overline{ST}$

EXAMPLE 2   Graph the disjunction $x \leq -1$ or $x \leq 3$.

Identify the resulting geometric figure, if possible.

Solution   The graph of $x \leq -1$ is a ray with endpoint A, extending to the left.

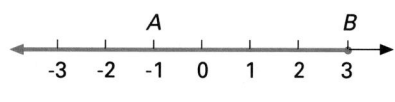

The graph of $x \leq 3$ is the ray with endpoint B, extending to the left.

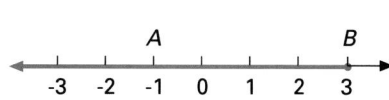

The graph of the disjunction is the set of points belonging to the graph of $x \leq -1$, or to the graph of $x \leq 3$, or to both.

Thus, the graph of $x \leq -1$ or $x \leq 3$ is the ray $\overrightarrow{BA}$.

Classroom Exercises

Describe the graph of each of the following.

1. $x \leq 5$ 2. $x \geq -6$ 3. $x \leq -7$ 4. $x \geq -2$ and $x \leq 2$

Written Exercises

Graph the inequality on a number line.

1. $x \geq 4$ 2. $x \leq 2$ 3. $x \geq -7$ 4. $x \leq 0$ 5. $x \geq -6$

Graph each of the following. Identify the resulting geometric figure(s).

6. $x \geq 1$ and $x \leq 7$ 7. $x \geq 4$ or $x \geq 8$ 8. $x \leq 5$ or $x \leq 12$
9. $x \leq -3$ and $x \geq -5$ 10. $x \geq -6$ or $x \leq 8$ 11. $x \geq -3$ and $x \geq -2$
12. $x \geq 3$ and $x \leq 3$ 13. $x \geq 3$ or $x \leq 3$ 14. $x \geq 1$ or $x \leq 6$
15. $x \leq 4$ or $x \geq 5$ 16. $x \leq -3$ or $x \geq 5$ 17. $x \geq -6$ and $x \leq -6$
18. $x \leq -1$ or $x \leq 3$ 19. $x \leq -1$ or $x \geq -1$ 20. $|x| \leq 7$
21. $[x \leq 3$ or $x \geq 5]$ and $[x \geq 3$ or $x \geq 5]$ 22. $x \leq -2$ and $x \geq 4$

Mixed Review

Is the conjunction (disjunction) true? If false, explain why.   *1.7*

1. A ray has two endpoints *or* a plane is flat.   True
2. A ray has two endpoints *and* a plane is flat.   False. A ray has only one endpoint.

Additional Example 2

Graph the disjunction $x \geq -4$ or $x \geq 2$. Identify the resulting geometric figure, if possible.

Ray; $\overrightarrow{TW}$

Enrichment

Have each student draw four figures similar to those shown below and then write a conjunction or a disjunction for each graph. Then have each student check the work of a classmate. Answers will vary.

Classroom Exercises

1. Ray with endpoint 5, extending left
2. Ray with endpoint −6, extending right
3. Ray with endpoint −7, extending left
4. Segment with endpoints 2 and −2

Written Exercises

1.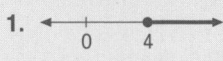

2.

3.

4.

5.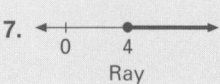

6.
   Segment

7.
   Ray

8.
   Ray

9.
   Segment

10.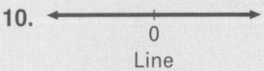
    Line

11.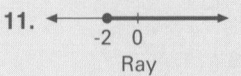
    Ray

12.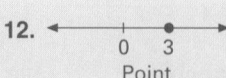
    Point

13.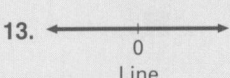
    Line

14.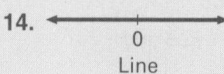
    Line

15.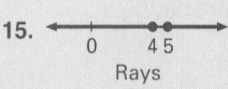
    Rays

16.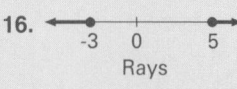
    Rays

17.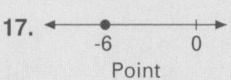
    Point

38

Chapter 1 Review

Key Terms

acute angle (p. 16)
adjacent angles (p. 21)
angle (p. 15)
angle bisector (p. 23)
bisector (p. 10)
collinear (p. 2)
complementary angles (p. 28)
congruent (p. 10)
conjunction (p. 32)
coplanar (p. 2)
coordinate (p. 5)
disjunction (p. 33)
distance (p. 5)
equidistant (p. 11)
intersection (p. 3)
line (p. 1)
linear pair (p. 26)

midpoint (p. 10)
noncollinear (p. 2)
noncoplanar (p. 2)
obtuse angle (p. 16)
perpendicular (p. 27)
plane (p. 1)
point (p. 1)
postulate (p. 5)
proof (p. 27)
ray (p. 7)
right angle (p. 16)
segment (p. 6)
space (p. 2)
straight angle (p. 16)
supplementary angles (p. 27)
theorem (p. 27)
vertex (p. 15)

Key Ideas and Review Exercises

1.1   A line can be named by any two points that lie on it.

1.2   To find the **distance** between two points on a number line, find the absolute value of the difference of their coordinates.

1. Give six possible names for the line shown. $\overleftrightarrow{AB}, \overleftrightarrow{AC}, \overleftrightarrow{BA}, \overleftrightarrow{BC}, \overleftrightarrow{CA}, \overleftrightarrow{CB}$

2. Find AB for $A:-7$ and $B:5$.   12

3. Use the Segment Addition Postulate to write an equation. $AB + BC = AC$

4. $AB = \frac{2}{5}BC$ and $AC = 21$. Find BC.   15

1.3   The **midpoint** of a segment divides the segment into two congruent segments.

To find the **coordinate of the midpoint** of a segment, find the arithmetic mean of the coordinates of the endpoints.

A **bisector** of a segment is a line, ray, segment, or plane that intersects the segment at its midpoint.

5. Find the coordinate of the midpoint of $\overline{AB}$ for $A:-8$ and $B:12$.   2

6. C is the midpoint of $\overline{AB}$. Find the coordinates of A and B for $A:(p - 8)$, $B:(3p + 4)$, and $C:6$. $A:-4, B:16$

38   Chapter 1 Review

Write a statement of congruence and an equation for the figure, using the Segment Addition Postulate.

7. Y is the midpoint of $\overline{XZ}$.

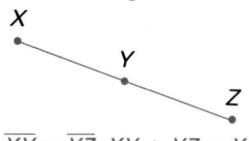

$\overline{XY} \cong \overline{YZ}$, $XY + YZ = XZ$

8. $\overleftrightarrow{SU}$ bisects $\overline{VW}$.

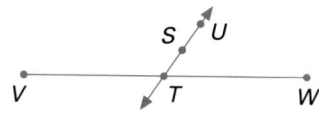

$\overline{VT} \cong \overline{TW}$, $VT + TW = VW$

9. Draw a segment. Locate its midpoint.   See Construction on page 12.

1.4, 1.5   To **classify angles,** refer to the following:

acute	right	obtuse	straight
$0 < m < 90$	$m = 90$	$90 < m < 180$	$m = 180$

Given: m $\angle PRS = 130$, m $\angle 2 = 20$, m $\angle URS = 40$

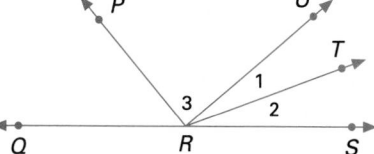

10. Find m $\angle 3$. Classify the angle. 90, right
11. Find m $\angle PRT$. Classify the angle. 110, obtuse
12. Name an angle bisector. $\overrightarrow{RT}$
13. Name a pair of adjacent angles. $\angle SRT$, $\angle TRU$
14. Draw an obtuse angle. Construct its bisector.   See Construction on page 23.

1.6   Two angles are **supplementary** if the sum of their measures is 180.
Two angles are **complementary** if the sum of their measures is 90.

Find the measure of a complement and a supplement, if possible, of the angle with the given measure.

15. 37 53, 143   **16.** 117 none, 63   **17.** 2 88, 178   **18.** $3x - 30$
19. The measure of an angle is 10 more than 7 times the measure of its complement. Find the measure of the angle. 80

1.7 $A \wedge B$ (**conjunction**) is true when A and B are both true. It is false otherwise.
$A \vee B$ (**disjunction**) is false when A and B are both false. It is true otherwise.

20. Write the conjunction and disjunction. Determine if each is true.
Collinear points are noncoplanar. Adjacent angles have points in common.

1.8   The **graph of the conjunction** of two inequalities is drawn by locating the points common to the graphs of the two inequalities.
The **graph of the disjunction** of two inequalities is drawn by locating the points belonging to the graphs of either inequality or both.

21. a. Graph the disjunction $x \leq 4$ *or* $x \leq 0$.
b. Graph the conjunction $x \geq -2$ *and* $x \leq 6$.

18. If $10 < x < 40$, comp $= 120 - 3x$; if $10 < x < 70$, supp $= 210 - 3x$
20. Conj: Collinear pts are noncoplanar and adj $\angle$s have pts in common. (False)
Disj: Collinear pts are noncoplanar or adj $\angle$s have pts in common. (True)
21. a.

b.

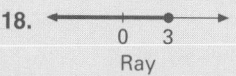

Additional Answers, page 37

18.

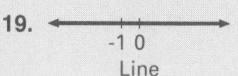

Ray

19.

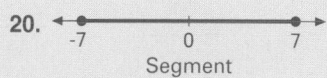

Line

20.

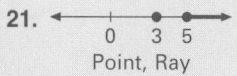

Segment

21.

Point, Ray

22. $\varnothing$

20. m ∠PTK = 80

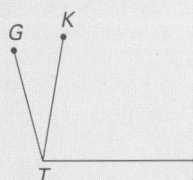

G K

T   P

21. Conj: pt has no length and the sum of
the meas of 2 comp ∠s = 180; false
Disj: pt has no length or the sum of the
meas of 2 comp ∠s = 180; true

22.

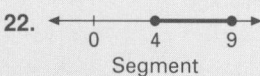

0   4   9

Segment

23.

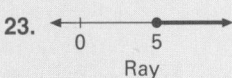

0   5

Ray

Chapter 1 Test

A—Level Ex.: 1–6, 9–17, 19, 21   B—Level
Ex.: 7–8, 18, 22–23   C—Level Ex.: starred

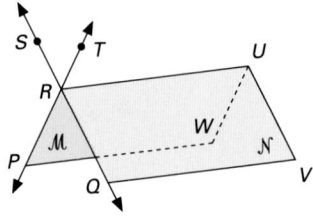

1. Identify the intersection of $\overleftrightarrow{PT}$ and $\overrightarrow{SQ}$.   *R*
2. Identify the intersection of planes $\mathcal{M}$ and $\mathcal{N}$. $\overleftrightarrow{RU}$
3. Name three collinear points.   *P, R, T*
4. Name four coplanar points.   *R, U, Q, V*
5. Find *AB* for *A*:−8 and *B*:6.   14
6. Find the coordinate of the midpoint of $\overline{AB}$ for
A:−4 and *B*:12.   4
7. $AC = \frac{4}{5}CB$ and $AB = 27$. Find *CB*.   15
8. *C* is the midpoint of $\overline{AB}$. Find the coordinates
of *A* and *B* for *A*:(16 − 2*p*), *B*:(*p* + 12), and *C*:12.
A:8, *B*:16

A   C   B

Draw an angle and label it ∠ABC. (Exercises 9–10)

9. Construct an angle congruent to ∠ABC.   See Construction on page 17.
10. Construct the bisector of ∠ABC.   See Construction on page 23.
11. Draw a segment. Locate the midpoint of the segment.   See Construction on page 12.

Use the diagram for Exercises 12–15.

12. By the Angle Addition Postulate,
m ∠2 + m ∠3 = _____   m∠SPQ

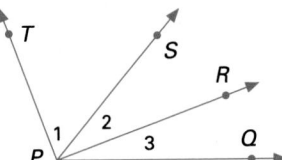

Given: m ∠TPR = 110, m ∠SPQ = 50, m ∠3 = 20

13. Find m ∠2. Classify the angle.   30, acute
14. Find m ∠1. Classify the angle.   80, acute
15. Given: $\overrightarrow{PR}$ bisects ∠SPQ, m ∠2 = 4*x* − 7, m ∠3 = 2*x* + 8
Find m ∠SPQ.   46

**Find the measure of a complement and a supplement, if possible, of
the angle with the given measure.**   **18.** 100 − 2*x* if 5 < *x* < 50, 190 − 2*x* if 5 < *x* < 95

16. 42   48, 138   **17.** 150   none, 30   **18.** 2*x* − 10

19. The measure of an angle is 10 less than 3 times the measure of its
complement. Find the measure of the angles.   25, 65
* **20.** ∠PTK and ∠KTG are adjacent angles.   m ∠PTK is 60 more than
$\frac{2}{3}$m ∠KTG.   m ∠PTG = 110. Draw the figure and find m ∠PTK.
21. Write the conjunction and the disjunction for statements (a) and (b).
Determine if each is true or false.
(a) A point has no length.
(b) The sum of the measures of two complementary angles is 180.

Graph the conjunction or disjunction. Identify the resulting geometric figure.

22. *x* ≥ 4 *and* *x* ≤ 9   **23.** *x* ≥ 5 *or* *x* ≥ 8

40   Chapter 1 Test

College Prep Test

Strategy for Achievement in Testing

In each item, you are to compare a quantity in Column 1 with a quantity in Column 2. Write the letter of the correct answer from these choices:

A—The quantity in Column 1 is greater than the quantity in Column 2.
B—The quantity in Column 2 is greater than the quantity in Column 1.
C—The quantity in Column 1 is equal to the quantity in Column 2.
D—The relationship cannot be determined from the given information.

(NOTE: Information centered above both columns refers to one or both of the quantities to be compared.)

Sample Question	Answer
Column 1 **Column 2**	m $\angle 1$ + m $\angle 2$ = 90
$\angle 1$ is complementary to $\angle 2$ m $\angle 1$ + m $\angle 2$ 80	Since 90 > 80, the answer is **A**.

	Column 1	Column 2		Column 1	Column 2

1.
C

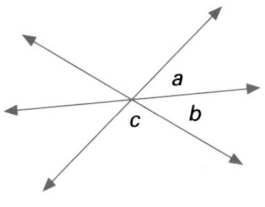

$1 - x$ 0

2. Three lines intersect in a point.
B

150        $c + a + b$

3. $\angle 1$ and $\angle 2$ are supplementary.
D

m $\angle 1$ m $\angle 2$

4.
C

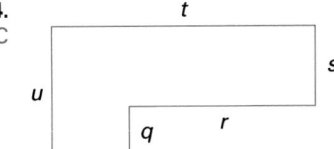

$p + q + r + s$ $t + u$

5. $x \neq 0$
A

$\frac{1}{x} \cdot x$ $x + (-x)$

6.   A   M   B
C

M is the midpoint of $\overline{AB}$.

AB $2 \cdot MB$

College Prep Test   **41**

2 PROOF

OVERVIEW

This chapter provides an introduction to the concept of proof. Students use the Segment Addition Postulate and Angle Addition Postulate to prove the equality of segment lengths and statements about complementary and supplementary angles. They use algebraic properties in multi-step geometric proofs, and identify deductive and inductive reasoning. The Vertical Angle Theorem is used to find angle measures and students apply the postulates that relate points, lines, and planes.

OBJECTIVES

- To recognize that two points determine exactly one line
- To draw conclusions from given statements and drawings
- To write statements in "if..., then..." form
- To prove theorems about lengths of segments and complementary and supplementary angles

PROBLEM SOLVING

The entire chapter teaches the strategy Using Definitions and Properties for determining the truth of a statement. Exercises 14–18 on pages 46–47 use the strategy Drawing a Diagram which is more fully explained in the Problem Solving Strategies Lesson on page 62. Example 3 on page 58 illustrates the Working Backwards strategy. Exercises 9–10 on page 66 help the students gain practice with the strategy of Marking the Parts. Written Exercises 12 through 15 on page 70 provide practice with the Drawing a Diagram and Looking for Patterns strategies. The strategy Making a Model helps clarify the concepts in Lesson 2.9.

TECHNOLOGY

Computer: The Computer Investigation on page 76 illustrates the construction and measurement of vertical angles. It serves as a preface to Lesson 2.8, which deals specifically with such angles. Students construct these angles on a computer, which helps them visualize and distinguish between adjacent and vertical angles.

SPECIAL FEATURES

Summary p. 45
Mixed Review pp. 47, 52, 56, 66, 70, 75, 81, 85
Algebra Review p. 47
Application: Tilt of the Earth p. 56
Midchapter Review p. 61
Problem Solving Strategies: Drawing a Diagram p. 62
Focus on Reading pp. 65, 73
Brainteaser p. 70
Computer Investigation: Vertical Angles p. 76
Key Terms p. 86
Key Ideas and Review Exercises pp. 86–87
Chapter 2 Test p. 88
College Prep Test p. 89
Cumulative Review (Chapters 1–2) pp. 90–91

PLANNING GUIDE

Lesson	Basic	Average	Above Average	Resources
2.1　pp. 45–47	CE all WE 1–9 odd AR all	CE all WE 1–15 odd AR all	CE all WE all AR all	Reteaching p. 17 Practice p. 18
2.2　pp. 51–52	CE all WE 1–9 odd	CE all WE 1–12 odd	CE all WE 1–16 odd	Reteaching p. 19 Practice p. 20
2.3　pp. 54–56	CE all WE 1–10 odd Application	CE all WE 1–14 odd Application	CE all WE all odd Application	Reteaching p. 21 Practice p. 22
2.4　pp. 59–62	CE all WE 1–9 odd Midchapter Review Problem Solving	CE all WE 1–15 odd Midchapter Review Problem Solving	CE all WE all odd Midchapter Review Problem Solving	Reteaching p. 23 Practice p. 24
2.5　pp. 65–66	FR all, CE all WE 1–8 odd	FR all, CE all WE 1–12 odd	FR all, CE all WE 1–14 odd	Reteaching p. 25 Practice p. 26
2.6　pp. 69–70	CE all WE 1–5 odd Brainteaser	CE all WE 1–11 odd Brainteaser	CE all WE 1–15 odd Brainteaser	Reteaching p. 27 Practice p. 28
2.7　pp. 73–76	FR 1–5 CE all WE 1–4, CI	FR 1–10 CE all WE 1–17 odd, CI	FR 1–15 CE all WE 1–20 odd, CI	Reteaching p. 29 Practice p. 30
2.8　pp. 79–81	CE all WE 1–10 odd	CE all WE 1–18 odd	CE all WE 1–20 odd	Reteaching p. 31 Practice p. 32
2.9　p. 85	CE all WE 1–6	CE all WE 1–9	CE all WE 1–11	Reteaching p. 33 Practice p. 34
Chapter 2 Review pp. 86–87	all	all	all	
Chapter 2 Test p. 88	all	all	all	
College Prep Test p. 89	all	all	all	
Cumulative Review pp. 90–91	1–12 odd	1–20 odd	all	

CE = Classroom Exercises　　WE = Written Exercises　　FR = Focus on Reading　　AR = Algebra Review　　CI = Computer Investigation
NOTE: For each level, all students should be assigned all Mixed Review exercises.

Project: This investigation helps students learn the meaning of "and" and "or" by sorting actual objects into a logic box. Use this project as a preview to Chapter 2.

Materials: Shoe box, tape, index cards, and a set of 18 attribute pieces: three colors (red, blue, yellow), three shapes (square, circle, triangle), and two sizes (small, large).

Have the students (in pairs) cut a door that will open and close in the end of a small box. The door should be labeled with an attribute such as "square" or "red." With tape, have them make a transparent pocket on the door to hold various labels (fig. 1) and insert a card with "red" printed on it.

Activity 1: Ask students to use the box to sort the attribute pieces. If an attribute card is red, it gets put in the box; if not, it stays out.

Activity 2: Have students add a partition to the box to make an "and" box (fig. 2). Ask them to sort the attribute pieces again but this time the attribute pieces placed in the second compartment must possess *both* attributes.

fig. 1 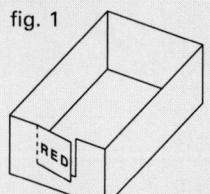 fig. 2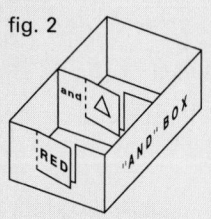

Change the label card and sort the attribute pieces again. An additional room can be added to the "and" box (fig. 3). Students can also make an "or" box (fig. 4). The attribute pieces in an "or" box will need to have only *one* of the attributes on the labels.

fig. 3 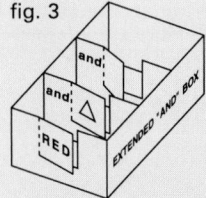 fig. 4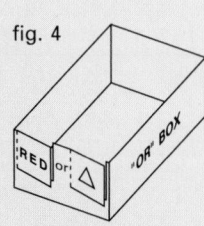

Have pairs of students record their findings using Project Worksheet 2. Then have the class as a whole discuss their conclusions.

2 PROOF

- ♥ *Babies are illogical*
- ♦ *Nobody is despised who can manage a crocodile*
- ♠ *Illogical persons are despised*
- ♣ *Conclusion?*

Lewis Carroll (1832–1898)—He used his real name, Charles Dodgson, at Oxford University in England where he taught mathematics for over 20 years. As Lewis Carroll, he put complicated logical arguments into his fantastic stories and humor into his mathematics publications.

More About the Mathematician

The Cheshire Cat and White Rabbit are characters from Carroll's *Alice in Wonderland* and *Through the Looking Glass.* The syllogism is from Dodgson's *Symbolic Logic.* One valid conclusion for this syllogism is that "Babies cannot manage crocodiles." Here is another of his syllogisms: All unripe fruit is unwholesome; all these apples are wholesome; no fruit, grown in the shade, is ripe. One conclusion is that "these apples were grown in the sun." Dodgson published many books, including *The Elements of Determinants (with Their Applications to Simultaneous Linear Equations and Algebraic Geometry)* and *Euclid and his Modern Rivals.*

2.1 Drawing Conclusions

Objectives
To draw conclusions from given statements
To state reasons for conclusions

Chapter 2 is an introduction to the methods used to prove geometric statements. When a lawyer presents his defense to a jury, he must be able to justify or give a convincing reason for every conclusion he draws. In geometry, you will be asked to draw conclusions and give reasons for them. A statement in geometry can be justified if you can provide a reason for the statement. Such reasons may include:

Definitions and geometric properties	Algebraic properties
Postulates	Arithmetic facts
Theorems that have been proved	Given or assumed information

The process of drawing conclusions and giving reasons is called deductive reasoning. Diagrams are provided for many examples and exercises in this text. The diagrams often supply the *given* information. In this lesson you will learn what may and what may not be assumed from a diagram.

EXAMPLE 1
What conclusion can be drawn from the diagram? Give a reason for the conclusion.

Plan
C is between A and B on $\overline{AB}$. Betweenness may be assumed from a diagram.

Solution
Conclusion: $AC + CB = AB$
Reason: Segment Addition Postulate

In Example 2, some information is given *in addition* to the diagram.

EXAMPLE 2
Given: F is the midpoint of $\overline{DE}$.
What conclusion can be drawn? Why?

Plan
More than one conclusion is possible. First, F is between D and E. Second, more specifically, a midpoint divides a segment into two congruent segments.

Solution
Conclusion: $DF + FE = DE$
Reason: Segment Addition Postulate
Conclusion: $\overline{DF} \cong \overline{FE}$ or $DF = FE$
Reason: Definition of midpoint

Teaching Resources

Manipulative Worksheet 3
Project Worksheet 2
Quick Quizzes 9
**Reteaching and Practice
Worksheets,** pp. 17, 18

▬▬ GETTING STARTED

Prerequisite Quiz

Use the given information to write an equation.

1. C is between A and B. $AC + CB = AB$
2. C is the midpoint of $\overline{AB}$. $AC = CB$
3. $\angle 1$ and $\angle 2$ are adjacent angles whose outer rays are perpendicular.
   $m \angle 1 + m \angle 2 = 90$
4. $\angle 1$ and $\angle 2$ are adjacent angles whose outer rays form a straight angle.
   $m \angle 1 + m \angle 2 = 180$

Motivator

You can draw conclusions from written statements but sometimes you can also assume information from diagrams. Ask your students to think of some geometric figures from which they can assume information that is not written down.
Segments containing 3 or more pts, adj ∠s, straight ∠s.

Additional Example 1

What conclusion can be drawn from the diagram? Give a reason for your answer.
$HR + RT = HT$, Seg Add Post

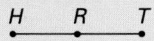

Additional Example 2

Refer to the figure in Additional Example 1. What conclusion can be drawn, given that R is the midpoint of $\overline{HT}$? Give a reason.
$\overline{HR} \cong \overline{RT}$, or $HR = RT$; (Def of midpt)

Lesson Note

This lesson paves the way for formal proofs in the next lesson. Students often find writing a conclusion from given data relatively easy but they do not easily justify that conclusion with a reason. At first, you might ask them to keep a list of all relevant reasons in front of them as they work. Once they learn to recognize the correct reason, they can move to citing the reason from memory, without the aid of the list. One primary objective of this lesson is to learn what information can acceptably be assumed from a diagram. The summary on page 45 lists this information for reference.

Math Connections

Design Engineering: Design engineers work in many fields—engineering, architecture, industrial design, and so on. Drawing accurate, carefully marked diagrams is essential to this type of work.

Critical Thinking Questions

Analysis: Ask students to distinguish between assuming information from a diagram and drawing conclusions from a diagram. Conclusions drawn from a diagram are based on information given explicitly or on assumptions allowed by general agreement. Assuming information from a diagram refers only to the assumptions allowed by general agreement.

In both diagrams below, you may assume that ∠1 and ∠2 are a pair of adjacent angles. Other information may or may not be assumed for each diagram as indicated.

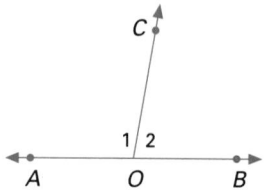 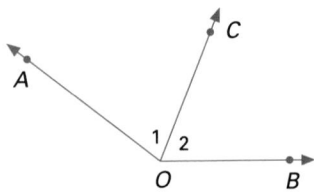

Assume: $\overrightarrow{OA}$ and $\overrightarrow{OB}$ are opposite rays forming a line.
∠AOB is a straight angle.
Do *not* assume: ∠1 ≅ ∠2, $\overrightarrow{OC} \perp \overleftrightarrow{AB}$, or m ∠1 > m ∠2.

Assume: $\overrightarrow{OA}$ and $\overrightarrow{OB}$ are not opposite rays.
C is in the interior of ∠AOB.
Do *not* assume: $\overrightarrow{OC}$ is an angle bisector, or ∠1 ≅ ∠2.

EXAMPLE 3   Given: $\overrightarrow{OC}$ bisects ∠AOB.
What conclusions can be drawn?
Give a reason for each conclusion.

Plan   C is in the interior of ∠AOB, so m ∠AOB = m ∠1 + m ∠2. More specifically, since $\overrightarrow{OC}$ is a bisector of ∠AOB, it divides the angle into two congruent angles.

Solution   Conclusion: m ∠AOB = m ∠1 + m ∠2
Reason: Angle Addition Postulate
Conclusion: ∠1 ≅ ∠2
Reason: Definition of angle bisector

EXAMPLE 4   Given: The figure as shown, with $\overrightarrow{SR} \perp \overrightarrow{ST}$.
Draw a conclusion and give a reason.

Plan   You can assume from the figure that ∠1 and ∠2 are acute adjacent angles. Use the fact that $\overrightarrow{SR} \perp \overrightarrow{ST}$ to draw a conclusion.

Solution   Conclusion: m ∠1 + m ∠2 = 90
Reason: If the outer rays of two acute adjacent angles are perpendicular, then the sum of the measures of the angles is 90 (Theorem 1.1).

Additional Example 3

Given: $\overrightarrow{OF}$ bisects ∠ROS.
What conclusions can be drawn?
Give a reason for each conclusion.

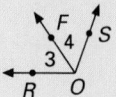

m ∠ROS = m ∠3 + m ∠4 (Angle Add Post); ∠3 ≅ ∠4 (Def of ∠ bis)

Additional Example 4

Given: The figure as shown
Draw a conclusion and give a reason.

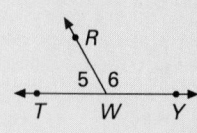

m ∠5 + m ∠6 = 180 (If outer rays of two adj ∠s form a st ∠, then the sum of their meas is 180.)

Summary

Information that may be assumed from a diagram:
- Angles are adjacent.
- Outer rays of two adjacent angles form a line or straight angle.
- A point is between two points.
- Points are collinear or coplanar.
- A point is in the interior of an angle.
- Geometric figures intersect.

Information that may *not* be assumed from a diagram:
- An angle is bisected.
- A segment is bisected.
- A point is the midpoint of a segment.
- Rays are perpendicular.
- Segments are congruent.
- Angles are congruent.
- An angle has a specific measure—for example, 60.

Classroom Exercises

In each exercise, indicate whether the stated conclusion may be drawn from the figure above it. Give a reason.

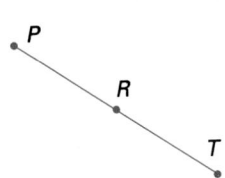
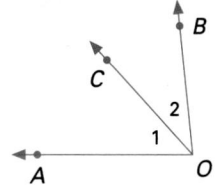
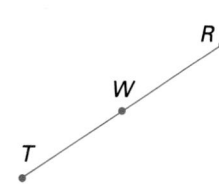

1. Conclusion: PR + RT = PT
2. Conclusion: R is between P and T

3. Conclusion: ∠1 ≅ ∠2
4. Conclusion: m ∠1 + m ∠2 = m ∠AOB

5. Conclusion: TW + WR = TR
6. Conclusion: TW = WR

Written Exercises

From the given information, draw a conclusion and express it as an equation. Give a reason for your answer.

1. Given: The figure below
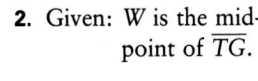

2. Given: W is the midpoint of $\overline{TG}$.
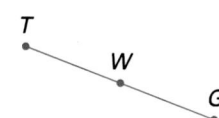

3. Given: $\overline{BD}$ bisects $\overline{AC}$.

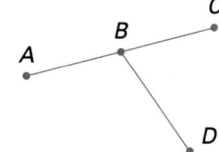

2.1 Drawing Conclusions **45**

Common Error Analysis

Error: Students may have difficulty drawing a correct conclusion when a given figure suggests three possible conclusions.

Comparison questions such as the following may help.

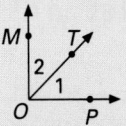

1. What conclusion can be drawn if no additional information is *given*? m ∠1 + m ∠2 = m ∠MOP
2. What must be stated as given to conclude m ∠1 = m ∠2? m ∠1 + m ∠2 = 90? $\overrightarrow{OT}$ is ∠ bis; $\overrightarrow{OM}$ ⊥ $\overrightarrow{OP}$

Checkpoint

For each of the following figures, draw a conclusion. Give a reason.

1.
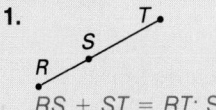
RS + ST = RT; Seg Add Post

2.
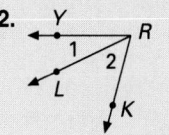
m ∠1 + m ∠2 = m ∠YRK (Angle Add Post)

3.
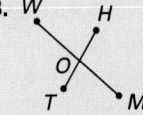
Given: $\overline{HT}$ bisects $\overline{WM}$. $\overline{WO}$ ≅ $\overline{OM}$, or WO = OM (Def of bis)

Closure

Ask your students what agreements we have about assumptions that you can make directly from a diagram. Adj angles, straight angle, point between two points, collinear or coplanar points, point in the angle interior, intersection. Have students make drawings to show 3 examples of information that you may *not* assume from a diagram. Bisection, midpoint, perpendicular, congruent, measure.

Enrichment

Challenge the students to solve the following problems. Ask them to state that no conclusion is possible, if that is what they decide.

1. C lies between A and B, and D lies between A and C. How is C related to D and B? C lies between D and B.

2. M is the midpoint of $\overline{AB}$, and A is the midpoint of $\overline{CM}$. What is the ratio $\frac{CA}{CB}$? $\frac{1}{3}$

3. A lies between C and B, and A lies between C and D. How is A related to B and D? A does not lie between B and D.

4. C is the midpoint of $\overline{AB}$ and B is the midpoint of $\overline{CD}$. How are AB and CD related? AB = CD

45

Guided Practice

Classroom Exercises 1–6

Independent Practice

A Ex. 1–9, **B** Ex. 10–15, **C** Ex. 16–18

Basic: WE 1–9 odd, AR all
Average: WE 1–15 odd, AR all
Above Average: WE all, AR all

Additional Answers

Classroom Exercises

1. Yes, Seg Add Post
2. Yes, between may be assumed for a diagram
3. No, cannot assume congruence
4. Yes, Angle Add Post
5. Yes, Segment Add Post
6. No, cannot assume congruence

Written Exercises

1. $HK + KL = HL$ (Seg Add Post)
2. $\overline{TW} \cong \overline{WG}$, $TW = WG$ (Def of midpt)
3. $AB = BC$, $\overline{AB} \cong \overline{BC}$ (Def of seg bis)
4. $m \angle 1 + m \angle 2 = 180$ (If the outer rays of 2 adjacent $\angle$s form a st $\angle$, then the sum of the meas of the $\angle$s is 180.)
5. $m \angle 1 + m \angle 2 = m \angle XYZ$ (Angle Add Post)
6. $\angle 1 \cong \angle 2$, $m \angle 1 = m \angle 2$ (Def of $\angle$ bis)
7. $m \angle 1 + m \angle 2 = 90$ (If the outer rays of 2 adj acute $\angle$s are $\perp$, then the sum of the meas of the $\angle$s is 90.)
8. $\overline{ME} \cong \overline{MP}$, $ME = MP$ (Def of congr seg)
9. $m \angle 1 + m \angle 2 = 180$ (If the outer rays of 2 adjacent $\angle$s form a st $\angle$, then the sum of the meas of the $\angle$s is 180.)
10. $m \angle CBD + m \angle CBE = 90$ (If the outer rays of 2 adj acute $\angle$s are $\perp$, then the sum of the meas of the $\angle$s is 90.)
11. $m \angle BEC = m \angle CED$, $\angle BEC \cong \angle CED$ (Def of $\angle$ bis)

4. Given: The figure below

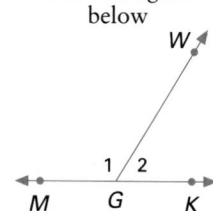

5. Given: The figure below

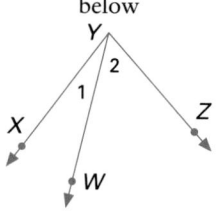

6. Given: $\overrightarrow{OR}$ bisects $\angle QOT$.

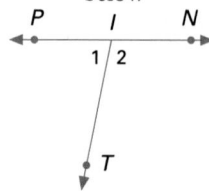

7. Given: $\overrightarrow{AG} \perp \overrightarrow{AT}$

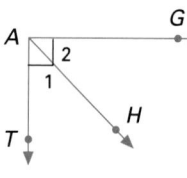

8. Given: The figure below

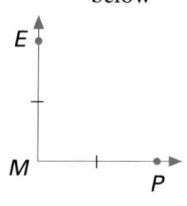

9. Given: The figure below

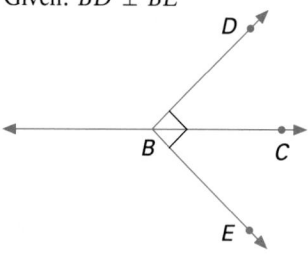

10. Given: $\overrightarrow{BD} \perp \overrightarrow{BE}$

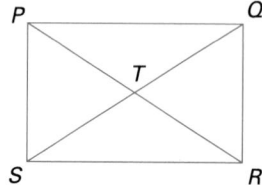

11. Given: $\overrightarrow{EC}$ bisects $\angle BED$.

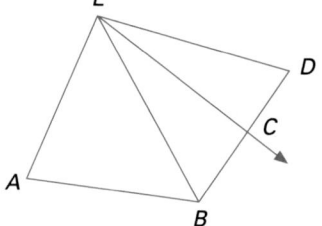

12. Given: $\overline{SQ}$ bisects $\overline{PR}$.

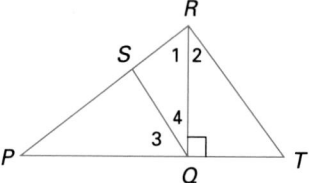

13. Given: $\overrightarrow{QP} \perp \overrightarrow{QR}$

Sketch a figure from the given information. Then determine whether the conclusion is correct. If it is, give the reason.

14. Given: $\angle GHK$, $\overrightarrow{HL}$ with L in the interior of $\angle GHK$, $\overrightarrow{HG} \perp \overrightarrow{HK}$
Conclusion: $\angle GHL \cong \angle LHK$

15. Given: $\angle APT$ and $\angle TPQ$ are adjacent angles, and $\overrightarrow{PA}$ and $\overrightarrow{PQ}$ are opposite rays.
Conclusion: $m \angle APT + m \angle TPQ = 180$

16. Given: ∠*CAB* with $\overline{BD}$ such that *D* is on $\overline{AC}$
Conclusion: $\overline{AC} \perp \overline{DB}$
17. Given: ∠*WXY*, *Z* on $\overline{XY}$, *V* on $\overline{XW}$, and $\overrightarrow{ZV}$ bisects ∠*XZW*.
Conclusion: m ∠*VZY* = m ∠*WZY* + m ∠*XZV*.
18. Given: ∠*PQR*, *T* on $\overline{QR}$, *S* is a point in the interior of ∠*PQR*;
and $\overline{PT}$ bisects $\overline{QS}$ at *U*.
Conclusion: *QS* = *QU* + *US*

Mixed Review

1. If m ∠*A* = 113, what is the measure of its supplement?   *1.8*   67
2. The measure of an angle is $\frac{4}{5}$ the measure of its complement.
Find the measure of each angle.   40, 50

The coordinates of the endpoints of a segment are −8 and 5.
(Exercises 3, 4)

3. Find the length of the segment described above.   *1.2*   13
4. Find the coordinate of the midpoint of the segment.   *1.5*   −1.5

Algebra Review

To solve linear equations:

(1) remove parentheses using the Distributive Property.
(2) put only those terms containing the variable on one side of the equation
by adding and/or subtracting.
(3) get the variable alone on one side by multiplying or dividing.

Example: Solve. $2a + 3(4 - 2a) = 2(a + 3)$
$$2a + 12 - 6a = 2a + 6$$
$$-4a + 12 = 2a + 6$$
$$12 - 6a = 6$$
$$-6a = -6$$
$$a = 1$$

Solve.

1. $a + 9 = 15$ 6   **2.** $s - 8 = 4$ 12   **3.** $-7 + n = -5$ 2
4. $6 + 2k = 10$ 2   **5.** $8d - 14 = 26$ 5   **6.** $17 - 3k = -10$ 9
7. $6 + 10c = 8c + 12$ 3   **8.** $7x - 11 = -10 + 8x$ −1   **9.** $-9 + 8x = x - 30$ −3
10. $a + 11 = 2(a + 3)$ 5   **11.** $-5(y + 2) = 20$ −6   **12.** $x + 2 = -3(x - 6)$ 4
13. $5(r - 1) = 2r + 4(r - 1)$ −1   **14.** $-2(3 - 4z) + 7z = 12z - (z + 4)$ $\frac{1}{2}$

12. $\overline{PT} \cong \overline{TR}$, *PT* = *TR* (Def of seg bis)
13. m ∠3 + m ∠4 = 90 (If the outer rays
of 2 adj acute ∠s are ⊥, then the sum
of the meas of the ∠s is 90.)

14.
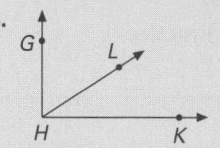
False.

15.
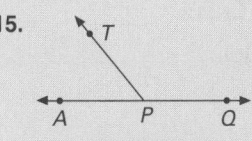
True. Postulate 7

16.
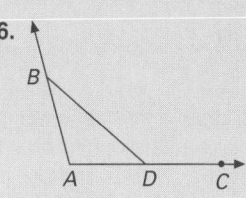
False.

17.
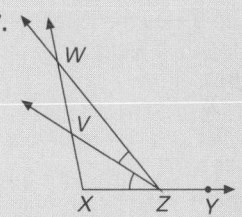
True. BY Angle Add Post, m ∠*VZY* =
m ∠*WZY* + m ∠*VZW*; $\overrightarrow{ZV}$ bis ∠*XZW*,
so m ∠*VZW* = m ∠*XZV*; ∴ by sub, m
∠*VZY* = m ∠*WZY* + m ∠*XZV*.

18.
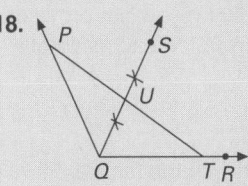
True. Seg Add Post

◢ GETTING STARTED

Prerequisite Quiz

Solve each equation.

1. $x + 5 = 9$ 4
2. $\frac{x}{3} = 5$ 15
3. $3x - 7 = 8$ 5
4. $9x - 8 = 5x$ 2

Motivator

On the chalkboard, draw a line segment with endpoint A and B and midpoint M. Ask your students to draw a conclusion from the information on the chalkboard. $AM = MB$ Ask them to prove their conclusion by writing statements in a statement column, giving the reason for each statement in a reason column.

◢ TEACHING SUGGESTIONS

Lesson Note

Algebraic properties are used frequently as reasons in proofs. For example, in the figure below, if $AB = CD$, then you can prove that $AC = BD$ by adding BC to each side of $AB = CD$.

It is helpful to begin with the familiar algebraic equations, instead of with geometric applications.

2.2 Introduction to Proof

Objectives   To identify properties of congruence and equality
To give missing reasons in proofs
To write proofs using properties of equality

Jane wants to prove that she qualifies to try out for J.V. volleyball. The following two-column format helps her to organize her thinking about her qualifications.

Qualifications for J.V.

(1) Team members can only be freshmen or sophomores.
(2) They must be physically fit.
(3) They must be passing all academic subjects.

Statement	Reason
1. I am a freshman or sophomore.	1. I am in the 9th grade.
2. I am physically fit.	2. I have a fitness statement signed by a doctor.
3. I am passing all subjects.	3. My average grade in every subject is B.
4. Therefore, I qualify to try out.	4. I meet all requirements.

An argument such as the one above is called a *proof.* You will learn in this lesson how to write a *proof* of a conclusion in mathematics.

Lengths of segments and measures of angles are real numbers. Therefore, the properties of real numbers may be applied to geometric situations. Four of these properties are listed below.

Addition Property of Equality: If $a = b$, then $a + c = b + c$ for all real numbers a, b, and c.

Subtraction Property of Equality: If $a = b$, then $a - c = b - c$ for all real numbers a, b, and c.

Multiplication Property of Equality: If $a = b$, then $ac = bc$ for all real numbers a, b, and c.

Division Property of Equality: If $a = b$, then $\frac{a}{c} = \frac{b}{c}$ for all real numbers a, b, and c ($c \neq 0$).

These properties can be used to justify steps in solving an equation.

48   Chapter 2 Proof

EXAMPLE 5   Given: $\overline{AB} \perp \overline{BC}$, m $\angle 1$ = m $\angle 2$
Prove: m $\angle 3$ + m $\angle 2$ = 90

Give the reason for Step 2 in the proof below.

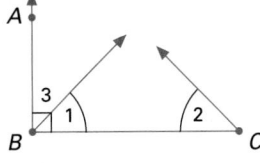

Statement	Reason
1. $\overline{AB} \perp \overline{BC}$	1. Given
2. m $\angle 1$ + m $\angle 3$ = 90	2. _____
3. m $\angle 1$ = m $\angle 2$	3. Given
4. $\therefore$ m $\angle 2$ + m $\angle 3$ = 90	4. Sub (Sub m $\angle 2$ for m $\angle 1$ in statement 2)

Solution   Use $\overline{AB} \perp \overline{BC}$ given in Step 1 to conclude that m $\angle 1$ + m $\angle 3$ = 90. The reason is: "If the outer rays of two adjacent acute angles are perpendicular, then the sum of the measures of the angles is 90."

Classroom Exercises

Give the missing reason.

1.
Statement	Reason
1. $x - 8 = 14$	1. Given
2. $x = 22$	2. _____

2.
Statement	Reason
1. $x = y$	1. Given
2. $3x = 3y$	2. _____

3.
Statement	Reason
1. $\overline{AB} \cong \overline{ST}$	1. Given
2. $\overline{ST} \cong \overline{YZ}$	2. Given
3. $\overline{AB} \cong \overline{YZ}$	3. _____

4.
Statement	Reason
1. m $\angle 1$ = m $\angle 2$	1. Given
2. m $\angle 3$ = m $\angle 4$	2. Given
3. m $\angle 1$ + m $\angle 3$ = m $\angle 2$ + m $\angle 4$	3. _____

Written Exercises

Which property does each statement illustrate?

1. $AB = AB$ Reflex Prop of Eq
2. If m $\angle 1$ = m $\angle 2$, then m $\angle 2$ = m $\angle 1$ Sym Prop of Eq
3. If $\overline{AB} \cong \overline{PQ}$ and $\overline{PQ} \cong \overline{TW}$, then $\overline{AB} \cong \overline{TW}$ Trans Prop of Cong
4. If $x = 5$ and $x + y = c$, then $5 + y = c$ Sub
5. If $x = y$ and $a = b$, then $x + a = y + b$ Equations may be added.

Write a proof for each of the following.

6. Given: $x - 7 = 4$
Prove: $x = 11$

7. Given: $7x = 28$
Prove: $x = 4$

8. Given: $3x + 7 = 22$
Prove: $x = 5$

Additional Example 5

Give the missing statement.

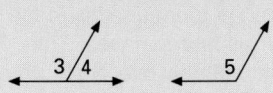

Statement	Reason
1. m $\angle 3$ = m $\angle 5$	1. Given
2. m $\angle 3$ + m $\angle 4$ = 180	2. Given
3. m $\angle 5$ + m $\angle 4$ = 180	3. Sub

11. (2) If the outer rays of 2 adj ∠s form a
   st ∠, then the sum of the meas of the
   ∠s is 180.
13. No; not reflexive and not symmetric
14. No; not reflexive and not symmetric
15. No; not reflexive and not transitive
16. No; not reflexive and not transitive

Give the missing reason(s) for each of the following proofs.

9. Given: $AT = BG$
   Prove: $AT - BT = BG - BT$

Statement	Reason
1. $AT = BG$	1. Given
2. $\therefore AT - BT = BG - BT$	2. Subt Prop of Eq

10. Given: $\overrightarrow{AB}$ bisects $\angle CAD$, $\angle 2 \cong \angle 1$.
   Prove: $\angle 3 \cong \angle 1$

Statement	Reason
1. $\overrightarrow{AB}$ bisects $\angle CAD$.	1. Given
2. $\angle 3 \cong \angle 2$	2. Def ang bis
3. $\angle 2 \cong \angle 1$	3. Given
4. $\therefore \angle 3 \cong \angle 1$	4. Trans Prop of Congr

11. Given: $m \angle 1 = m \angle 3$
   Prove: $m \angle 2 + m \angle 3 = 180$

Statement	Reason
1. $m \angle 1 = m \angle 3$	1. Given
2. $m \angle 1 + m \angle 2 = 180$	2. _____
3. $\therefore m \angle 3 + m \angle 2 = 180$	3. _____

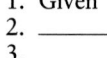

12. Given: M is the midpoint of $\overline{AB}$.
   Prove: $AM = \frac{1}{2}AB$

Statement	Reason
1. M is the midpoint of $\overline{AB}$.	1. Given
2. $AM = MB$	2. Def of midpt
3. $AM + MB = AB$	3. Seg Add Post
4. $AM + AM = AB$	4. Sub
5. $2AM = AB$	5. Add
6. $\therefore AM = \frac{1}{2}AB$	6. Div Prop of Eq

Equality (=) and congruence (≅) are examples of relations. A relation
that is reflexive, symmetric, and transitive is an equivalence relation.
Which of the following are equivalence relations? Why or why not?   None

13. >   **14.** is the brother of   **15.** ≠   **16.** is a complement of

Mixed Review

1. Find AB.   *1.2*   14
2. Find the coordinate of the midpoint of $\overline{AB}$.   *1.3*
3. $AC = \frac{3}{4}CB$. Find C.   *1.2*   −2

−1
-8   6
A C   B

Enrichment

The figure at the left is an aerial view of a water cascade. Water flows in at point A
and out at point B. The numbers give the height of each of the four surfaces in the
2×2 pattern. In the first arrangement, water will flow over all four surfaces. The
second arrangement is unsatisfactory because water would not flow over the surface
having a height of 4. Challenge the students to find as many arrangements as they
can for a 3×3 water cascade.
HINT: There are 42 possible arrangements,
two of which are shown at the right.

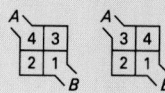

2.3 Writing Proofs in Geometry

Objective To write proofs using algebraic and geometric properties

In the last lesson you wrote algebraic proofs and supplied missing reasons in geometric proofs. Now you will form a plan of reasoning and write complete geometric proofs.

(1) What is given in writing? What can be assumed from the diagram?
(2) What conclusions can be made from each given statement?
(3) What reason can you give to justify each conclusion?
(4) In what order can you prove these conclusions to arrive at what you are asked to prove?

EXAMPLE 1 Given: $\overrightarrow{QP} \perp \overrightarrow{QR}$, m $\angle 1$ = m $\angle 3$
Prove: m $\angle 3$ + m $\angle 2$ = 90

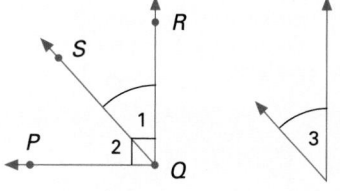

Proof

Statement	Reason
1. $\overrightarrow{QP} \perp \overrightarrow{QR}$	1. Given
2. m $\angle 1$ + m $\angle 2$ = 90	2. If outer rays of two adj acute $\angle$s are $\perp$, then the sum of the $\angle$ meas is 90.
3. m $\angle 1$ = m $\angle 3$	3. Given
4. m $\angle 3$ + m $\angle 2$ = 90	4. Sub

EXAMPLE 2 Given: $\overline{PQ}$ bisects $\overline{AB}$ at R; $\overline{AR} \cong \overline{RQ}$.
Prove: $\overline{RB} \cong \overline{RQ}$

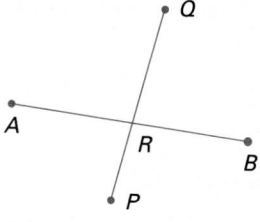

Plan Use the key word *bisects* to write an equation: $\overline{RB} \cong \overline{AR}$. Knowing $\overline{RB} \cong \overline{AR}$ and given $\overline{AR} \cong \overline{RQ}$, you may conclude $\overline{RB} \cong \overline{RQ}$, by the Transitive Property (or Substitution).

Proof

Statement	Reason
1. $\overline{PQ}$ bisects $\overline{AB}$ at R.	1. Given
2. $\overline{RB} \cong \overline{AR}$	2. A bis divides a seg into two $\cong$ segs.
3. $\overline{AR} \cong \overline{RQ}$	3. Given
4. $\therefore \overline{RB} \cong \overline{RQ}$	4. Trans Prop

GETTING STARTED

Prerequisite Quiz

Draw a conclusion. Give a reason.

1.

$$P \quad W \quad Y$$

PW + WY = PY (Seg Add Post)

2. Given: *W* is the midpoint of $\overline{PY}$. $\overline{PW} \cong \overline{WY}$, or PW = WY (Def of midpt)

Use the figure below.

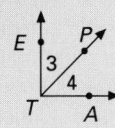

3. Given: $\overrightarrow{TE} \perp \overrightarrow{TA}$ m $\angle 3$ + m $\angle 4$ = 90 (If outer rays of two acute adj $\angle$s are $\perp$, sum of $\angle$ meas is 90.)
4. Given: $\overrightarrow{TP}$ bisects $\angle ETA$. m $\angle 3$ = m $\angle 4$, or $\angle 3 \cong \angle 4$ (Def $\angle$ bis)

Motivator

Ask your students what process they would use to solve an equation such as 2x + 8 = 12. Have them tell what they do in each step of the solution.

Additional Example 1

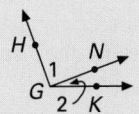

Prove that m $\angle 1$ + m $\angle 2$ = m $\angle HGK$.
Proof: N is in the interior of $\angle HGK$. (Given); m $\angle 1$ + m $\angle 2$ = m $\angle HGK$ (Angle Add Post)

Additional Example 2

Given $\overrightarrow{TW}$ bisects $\overline{LP}$. Draw a conclusion. Prove it.

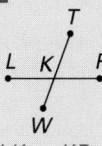

Conclusion: LK = KP or $\overline{LK} \cong \overline{KP}$
Proof: TW bisects $\overline{LP}$ (Given); LK = KP or $\overline{LK} \cong \overline{KP}$ (Def of seg bis)

Lesson Note

Students need guidance as they make the transition from drawing a conclusion and giving a reason to writing a formal two-column proof. This is why the first exercises require only completing the steps in a proof. Students thus have a chance to become familiar with the format.

Linking arrows are very helpful aids to clear thinking with longer proofs. With the arrows, the student shows which two or three steps lead to a particular conclusion.

Math Connections

Law: Lawyers often use logical arguments to establish a client's innocence. In many cases they can present a string of conditional statements and reasons to lead the jury to the desired conclusion.

Critical Thinking Questions

Synthesis: There is usually more than one way to write a proof. Write an example of one conclusion that follows from two or more different statements.   One answer might be this: With a diagram of the adjacent ∠s, the statement $\overrightarrow{BD}$ is an ∠ bis can be concluded from the statement ∠ABD ≅ ∠CBD, or from the statement m∠ABD = $\frac{1}{2}$m∠ABC or from the statement m∠ABC = 2m∠DBC.   Write a statement that can produce several different conclusions.   One possible answer is this: With a diagram of $\overline{AB}$ intersecting $\overline{CD}$ at E and the given statement that $\overline{CD}$ bisects $\overline{AB}$, the statement $\overline{AE} ≅ \overline{EB}$ or E is midpt of $\overline{AB}$ can be concluded.

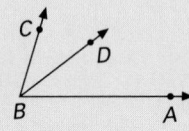

Classroom Exercises

Draw a conclusion from the diagram.

1.
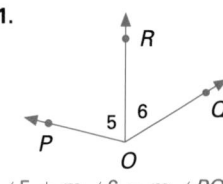

m ∠5 + m ∠6 = m ∠POQ

2.
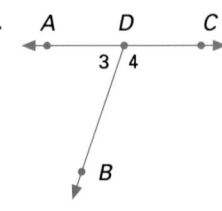

m ∠3 + m ∠4 = 180

3.
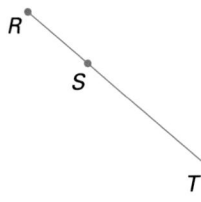

RS + ST = RT

Written Exercises

1. Given: $\overrightarrow{OB}$ bisects ∠TOY.
   Prove: m ∠1 = m ∠2

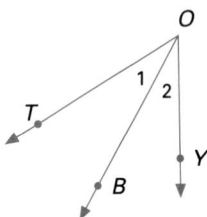

2. Given: $\overrightarrow{GK} \perp \overrightarrow{GL}$
   Prove: m ∠3 + m ∠4 = 90

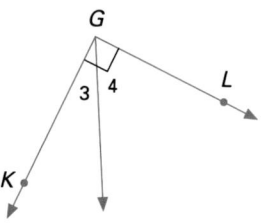

3. Given: $\overline{UR}$ bisects $\overline{YK}$.
   Prove: $\overline{YU} ≅ \overline{KU}$

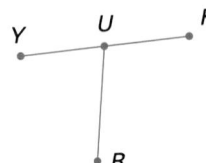

4. Given: $\overline{AB}$ and $\overline{CD}$ bisect each other.
   Prove: $\overline{AE} ≅ \overline{EB}$ and $\overline{EC} ≅ \overline{ED}$

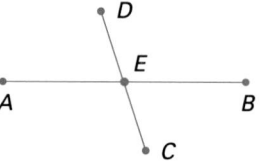

5. Given: m ∠3 = m ∠4
   Prove: m ∠2 + m ∠4 = 180

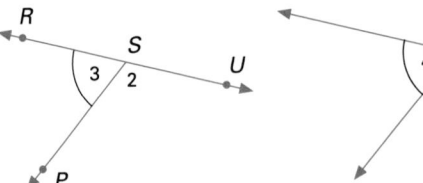

6. Given: $\overline{TY} \perp \overline{TW}$, m ∠5 = m ∠6
   Prove: m ∠4 + m ∠6 = 90

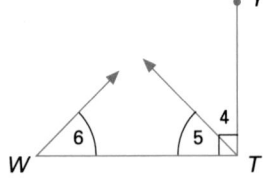

54   Chapter 2 Proof

Enrichment

Advertisements in magazines and newspapers can be considered as arguments aimed at convincing consumers to buy the advertiser's product. Have students bring to class examples of such advertisements. Then have them discuss in groups of 3 or 4 whether the arguments presented in the collected ads are convincing. Reasons should be given for their conclusions.

7. Given: $\overrightarrow{AB}$ bisects $\angle PAR$, and
$\angle 3 \cong \angle 2$
Prove: $\angle 1 \cong \angle 2$

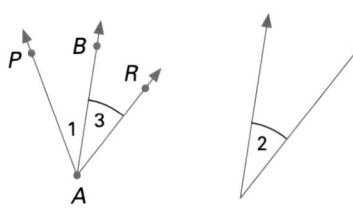

8. Given: $\angle 1 \cong \angle 2$, and $\angle 2 \cong \angle 3$
Prove: $\angle 1 \cong \angle 3$

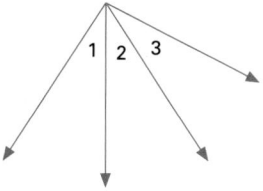

9. Given: $\overline{TL}$ bisects $\overline{BG}$; $\overline{TA} \cong \overline{BA}$.
Prove: $\overline{TA} \cong \overline{AG}$

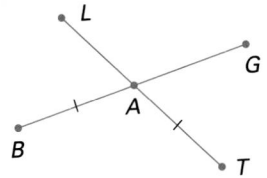

10. Given: $AD = AB$
Prove: $AD + BC = AC$

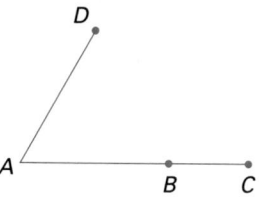

11. Given: B is the midpoint of $\overline{AC}$.
Prove: $AB + CD = BD$

12. Given: $\overrightarrow{OD}$ bisects $\angle AOC$.
Prove: $m \angle 4 + m \angle 5 = m \angle DOB$

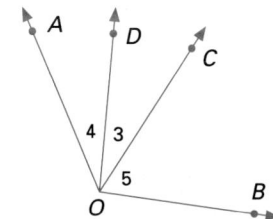

13. Given: $\overline{BA} \perp \overline{BD}$, and $\overline{BD}$ bisects
$\angle EBC$.
Prove: $m \angle 1 + m \angle 3 = 90$

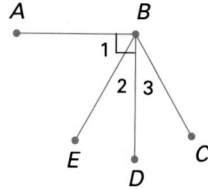

14. Given: $\overrightarrow{AB}$ bisects $\angle CAE$, $m \angle 2 = m \angle 4$.
Prove: $m \angle 1 + m \angle 3 = 180$

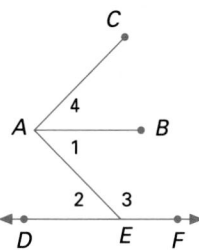

Checkpoint

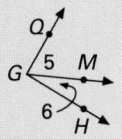

1. Given: $\overrightarrow{GH} \perp \overrightarrow{GQ}$
Prove: $m \angle 5 + m \angle 6 = 90$

Statement	Reason
1. $\angle 5$ and $\angle 6$ are adj $\angle$s	1. Def adj $\angle$s
2. $\overrightarrow{GH} \perp \overrightarrow{GQ}$	2. Given
3. $m \angle 5 + m \angle 6 = 90$	3. If outer rays of two adj acute $\angle$s are $\perp$, then sum of $\angle$ meas is 90.

2.

Prove: $RU + UP = RP$

Statement	Reason
1. U is between R and P	1. Given
2. $RU + UP = RP$	2. Seg Add Post

Closure

Ask students to name 4 steps to follow in formulating a plan for writing a proof.
(1) Find what is given.
(2) Ask what conclusions can be drawn from each given statement.
(3) Give a reason to justify each statement.
(4) Write the statements in an order that leads to the desired conclusion.

◤▬▬ FOLLOW UP

Guided Practice

Classroom Exercises 1–3

Independent Practice

A Ex. 1–10, **B** Ex. 11–14, **C** Ex. 15–16

Basic: WE 1–10 odd, Application
Average: WE 1–14 odd, Application
Above Average: WE 1–16 odd, Application

1.

Statement	Reason
1. $\overrightarrow{OB}$ bisects $\angle TOY$	1. Given
2. $\angle 1 \cong \angle 2$	2. Def of ∠ bisector

2.

Statement	Reason
1. $\overrightarrow{GK} \perp \overrightarrow{GL}$	1. Given
2. m $\angle 3$ + m $\angle 4$ = 90	2. If the outer rays of 2 adjacent ∠s are ⊥, then the sum of their meas is 90.

3.

Statement	Reason
1. $\overline{UR}$ bisects $\overline{YK}$	1. Given
2. $\overline{YU} \cong \overline{KU}$	2. Def of segment bis

4.

Statement	Reason
1. $\overline{AB}$ and $\overline{CD}$ bisect each other	1. Given
2. $\overline{AE} \cong \overline{EB}$ and $\overline{EC} \cong \overline{ED}$	2. Def of seg bis

5.

Statement	Reason
1. m $\angle 3$ = m $\angle 4$	1. Given
2. m $\angle 3$ + m $\angle 2$ = 180	2. If the outer rays of adj ∠s form a st ∠, then the sum of their meas is 180.
3. m $\angle 2$ + m $\angle 4$ = 180	3. Sub

6.

Statement	Reason
1. $\overline{TY} \perp \overline{TW}$	1. Given
2. m $\angle 5$ = m $\angle 6$	2. Given
3. m $\angle 4$ + m $\angle 5$ = 90	3. If the outer rays of two adj ∠s are ⊥ then the sum of their meas is 90.
4. m $\angle 4$ + m $\angle 6$ = 90	4. Sub

7.

Statement	Reason
1. $\overrightarrow{AB}$ bisects $\angle PAR$	1. Given
2. $\angle 3 = \angle 2$	2. Given
3. $\angle 1 = \angle 3$	3. Def of ∠ bis
4. $\angle 1 = \angle 2$	4. Trans Prop of Congr

8.

Statement	Reason
1. $\angle 1 \cong \angle 2$, $\angle 2 \cong \angle 3$	1. Given
2. $\angle 1 \cong \angle 3$	2. Trans Prop of Congr

Draw a figure for each of the following. Prove what is asked.

15. Given: $\overrightarrow{AC} \perp \overrightarrow{AB}$, D is a point in the interior of $\angle CAB$ such that m $\angle ADB$ = m $\angle CAD$.
    Prove: m $\angle ADB$ + m $\angle DAB$ = 90

16. Given: $\overleftrightarrow{GH}$ with P between G and H, and W between P and H, such that W is the midpoint of $\overline{PH}$.
    Prove: $GP + WH = GW$

Mixed Review

1. Name the intersection of planes $\mathcal{M}$ and $\mathcal{N}$.   **1.1** $\overleftrightarrow{GD}$
2. Name a point collinear with F and B.   **1.1** H
3. Name a point coplanar with F, G, and E.   **1.1** D
4. $\overline{AB} \cong \overline{BC}$. Is B the midpoint of $\overline{AC}$?   **1.3** No
5. Is $\angle AGF$ adjacent to $\angle FGD$?   **1.5** No

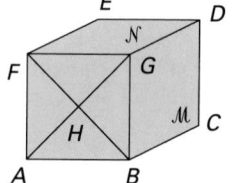

![Application icon] **Application:**   *Tilt of the Earth*

If we could watch the Earth from space as it revolves around the Sun, we would see that the planet's polar axis is tilted at a constant angle. The hemisphere (northern or southern) closer to the Sun at a given time receives sunlight longer during the 24-hour day. On the days of the solstices, the difference in the amount of light received is at a maximum. On the days of the equinoxes, the amount of light received is the same.

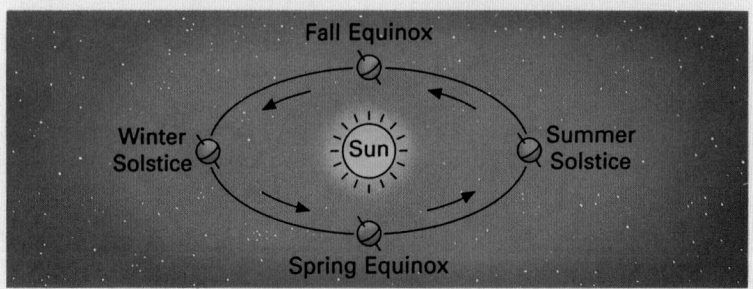

1. Explain why the North Pole has sunlight all day long on June 21, the day of the Summer Solstice.
2. Why are day and night the same length all over the world on March 21, the day of the Spring Equinox?
3. The diagram is labeled in relation to the seasons in the Northern Hemisphere. How would it need to be changed for the Southern Hemisphere?

56   Chapter 2 Proof

9.

Statement	Reason
1. $\overline{TL}$ bisects $\overline{BG}$ and $\overline{TA} \perp \overline{BA}$	1. Given
2. $\overline{BA} = \overline{AG}$	2. Def of seg bis
3. $\overline{TA} \cong \overline{AG}$	3. Trans Prop of Congr

10.

Statement	Reason
1. $AD = AB$	1. Given
2. $AB + BC = AC$	2. Seg Add Post
3. $AD + BC = AC$	3. Sub

See page 62 for answers to Ex. 11–16 and Application.

2.4 Proofs and More Complex Figures

Objectives
To apply alternate forms of the Segment Addition and Angle Addition Postulates

To use algebraic properties in multistep geometric proofs

The Subtraction Property of Equality can be used to obtain three equivalent statements of the Segment Addition Postulate. Similarly, the Angle Addition Postulate can be stated in three equivalent ways.

Segment Addition Postulate

Given: $\overline{AB}$, with C on $\overline{AB}$

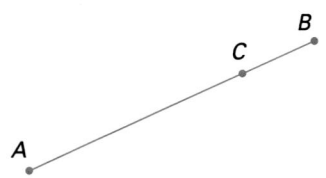

1. $AC + CB = AB$
2. $AC = AB - CB$
3. $CB = AB - AC$

Angle Addition Postulate

Given: $\angle ACB$, $\overrightarrow{CD}$ with D in the interior of $\angle ACB$

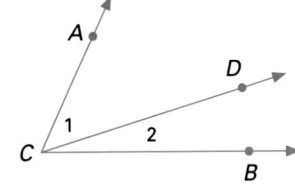

1. $m \angle 1 + m \angle 2 = m \angle ACB$
2. $m \angle 1 = m \angle ACB - m \angle 2$
3. $m \angle 2 = m \angle ACB - m \angle 1$

EXAMPLE 1

Given: $m \angle ABC = m \angle HGF$,
$\quad\quad m \angle 1 = m \angle 3$

Prove: $m \angle 2 = m \angle 4$

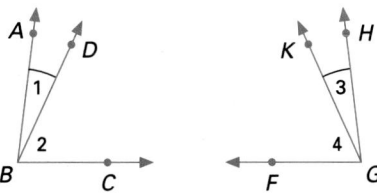

Plan
Decide what to do with $m \angle ABC$ and $m \angle 1$ in the *given* to get $m \angle 2$ in the *prove* part. Notice that $m \angle ABC - m \angle 1 = m \angle 2$ and $m \angle HGF - m \angle 3 = m\angle 4$.

Proof

Statement	Reason
1. $m \angle ABC = m \angle HGF$	1. Given
2. $m \angle 1 = m \angle 3$	2. Given
3. $m \angle ABC - m \angle 1 = m \angle 2$; $m \angle HGF - m \angle 3 = m \angle 4$	3. $\angle$ Add Post
4. $m \angle ABC - m \angle 1 = m \angle HGF - m \angle 3$	4. Equations may be subtracted.
5. $\therefore m \angle 2 = m \angle 4$	5. Sub

Teaching Resources

Quick Quizzes 12
Reteaching and Practice Worksheets, pp. 23, 24

■■■ GETTING STARTED

Prerequisite Quiz

Write an equation using the Angle Addition Postulate or the Segment Addition Postulate.

1.
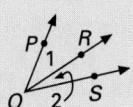

$PQ + QT = PT$

2.

$m \angle 1 + m \angle 2 = m \angle PQS$

3. Solve $AC + CB = AB$ for AC by applying the Subtraction Property of Equality. $AC = AB - CB$

4. Solve $m \angle 1 + m \angle 2 = m \angle ACB$ for $m \angle 2$. $m \angle 2 = m \angle ACB - m \angle 1$

Motivator

Ask students what equation results when they subtract y from each side of the equation $x + y = z$. $x = z - y$ Ask them what equation results when they subtract x from each side. $y = z - x$ Write this statement on the chalkboard: "If C is between A and B, then $AC + CB = AB$. Ask students what equation results when they subtract CB from each side of $AC + CB = AB$. $AC = AB - CB$

Additional Example 1

Given: $m \angle SOQ = m \angle POR$
Prove: $m \angle 6 = m \angle 8$

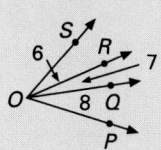

Proof: $m \angle SOQ = m \angle POR$ (Given); $m \angle SOQ - m \angle 7 = m \angle POR - m \angle 7$ (Subt Prop of Eq); $m \angle SOQ - m \angle 7 = m \angle 6$, $m \angle POR - m \angle 7 = m \angle 8$ (Angle Add Post); $m \angle 6 = m \angle 8$ (Sub)

Lesson Note

Tell the students that from now on, by agreement, statements which are given by a drawing, such as "*B* is between *A* and *C*" and "∠1 and ∠2 are adjacent angles," need not be stated in a proof before applying the Segment Addition Postulate or the Angle Addition Postulate. This keeps proofs from becoming unwieldy as they grow more complex.

Students can use the Segment Addition Postulate as a reason for any of the three forms of the equation, $AC + CB = AB$.

Math Connections

Problem Solving: Choosing the most appropriate form of the Segment or Angle Addition Postulate gives the students practice in planning the approach to solving a problem.

Critical Thinking Questions

Application: Given that $\overrightarrow{BD}$ and $\overrightarrow{BE}$ are in the interior of rt ∠*ABC*, what information is needed to conclude that m ∠*DBE* = $22\frac{1}{2}$? Both rays are ∠ bis. Show how this conclusion may be reached. Let $\overrightarrow{BD}$ bis rt ∠*ABC* and $\overrightarrow{BE}$ bis ∠*DBC* (draw a diagram). By the Angle Add Post and def of ∠ bis, m ∠*DBC* = 45 and m ∠*DBE* = $22\frac{1}{2}$.

EXAMPLE 2   Given: $\overline{AC} \cong \overline{BD}$
　　　　　　　 Prove: $\overline{AB} \cong \overline{CD}$

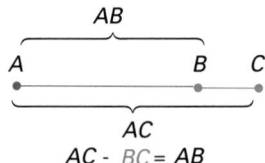

Plan   Look at the parts of the figure. Notice the relationships of the segments in the *given* and the *prove* statements.

AB

A　　　　　B　C

AC
$AC - BC = AB$

CD

B　C　　　　　D

BD
$BD - BC = CD$

This tells you to subtract BC from each side of the equation $AC = BD$.

Proof

Statement	Reason
1. $\overline{AC} \cong \overline{BD}$ ($AC = BD$)	1. Given
2. $AC - BC = AB$; $BD - BC = CD$	2. Seg Add Post
3. $AC - BC = BD - BC$	3. Subt Prop of Eq
4. ∴ $AB = CD$ ($\overline{AB} \cong \overline{CD}$)	4. Sub

EXAMPLE 3   Given: $\overrightarrow{BA} \perp \overrightarrow{BC}$, and $\overrightarrow{BC}$ bisects ∠*DBE*.
　　　　　　　 Prove: m ∠1 + m ∠3 = 90

Plan   It is often helpful to *work backwards* in developing proofs. In this example, you can show that m ∠1 + m ∠2 = 90. Therefore, if you can also show that m ∠3 = m ∠2, you can prove the conclusion—m ∠1 + m ∠3 = 90—by substitution.

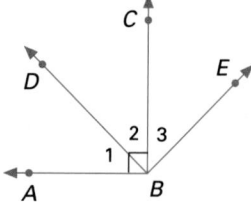

Notice that because $\overrightarrow{BC}$ bisects ∠*DBE*, m ∠2 = m ∠3.

Proof

Statement	Reason
1. $\overrightarrow{BA} \perp \overrightarrow{BC}$	1. Given
2. m ∠1 + m ∠2 = 90	2. If outer rays of acute adjacent angles are perpendicular, then the sum of the angle measures is 90.
3. $\overrightarrow{BC}$ bisects ∠*DBE*.	3. Given
4. m ∠3 = m ∠2	4. Def of ∠ bis
5. ∴ m ∠1 + m ∠3 = 90	5. Sub

Additional Example 2

Given: $XN = XW$,
　　　　$XM = XY$
Prove: $MN = YW$

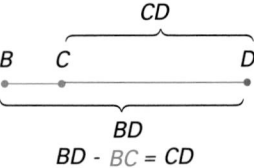

Proof: $XN = XW$ (Given); $XM = XY$ (Given); $XN - XM = XW - XY$ (Subt Prop of Eq); $XN - XM = MN$, $XW - XY = YW$ (Seg Add Post); $MN = YW$ (Sub)

Additional Example 3

Given: $\overrightarrow{OU}$ bisects ∠*TOS*; m ∠5 = m ∠7.

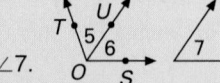

Prove: m ∠7 = m ∠6
Proof: $\overrightarrow{OU}$ bisects ∠*TOS* (Given); m ∠5 = m ∠7 (Given); m ∠5 = m ∠6 (Def of ∠ bis); m ∠7 = m ∠6 (Sub)

Classroom Exercises

Complete each equation from the information provided by the diagram above it.

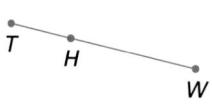

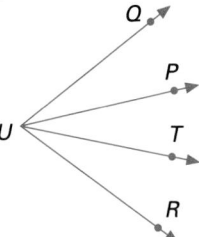

1. $TW = TH + \underline{HW}$

2. $TH = TW - \underline{HW}$

3. $\underline{HW} = TW - TH$

4. $m \angle QUR = m \angle QUT + \underline{m \angle TUR}$

5. $m \angle PUR - m \angle PUT = \underline{m \angle TUR}$

6. $m \angle RUP - \underline{m \angle PUT} = m \angle TUR$

Written Exercises

Copy the proof. Write the missing statements and reasons.

1. Given: $AP = AQ$, $AM = AN$
Prove: $MP = NQ$

Statement	Reason
1. $AP = AQ$	1. Given
2. $AM = AN$	2. Given
3. $AP - AM = AQ - AN$	3. Equations may be subtracted.
4. $AP - AM = \underline{MP}$, $AQ - AN = \underline{NQ}$	4. Seg Add Post
5. $\therefore MP = NQ$	5. Sub

2. Given: $PR = QS$
Prove: $PQ = RS$

Statement	Reason
1. $PR = QS$	1. Given
2. $PR - \underline{QR} = QS - \underline{QR}$	2. Subt Prop of Eq
3. $PR - QR = PQ$, $QS - QR = RS$	3. Seg Add Post
4. $\therefore PQ = RS$	4. Sub

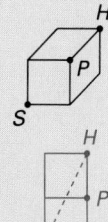
Checkpoint

1. Given: $m \angle 1 = m \angle 2$
Prove: $m \angle 3 + m \angle 2 = 180$

Statement	Reason
1. $m \angle 1 = m \angle 2$	1. Given
2. $\angle 1$ and $\angle 3$ are adj $\angle$s	2. Def of adj $\angle$s
3. $m \angle 1 + m \angle 3 = 180$	3. If outer rays of adj $\angle$s form a st $\angle$, then sum of meas = 180.
4. $m \angle 2 + m \angle 3 = 180$	4. Sub

2. Given: $PQ = AB$, $RQ = CB$
Prove: $RP = CA$

Statement	Reason
1. $PQ = AB$, $RQ = CB$	1. Given
2. $RQ - PQ = CB - AB$	2. Subt Prop of Eq
3. $RQ - PQ = RP$, $CB - AB = CA$	3. Seg Add Post
4. $RP = CA$	4. Sub

Closure

To summarize the lesson, ask students to write three forms of the Segment and Angle Addition Postulates. (Use art on p. 57.) $AC + CB = AB$, $AC = AB - CB$, $CB = AB - AC$; $m \angle 1 + m \angle 2 = m \angle ACB$, $m \angle 1 = m \angle ACB - m \angle 2$, $m \angle 2 = m \angle ACB - m \angle 1$

▆▆▆ FOLLOW UP

Guided Practice

Classroom Exercises 1–6

Independent Practice

A Ex. 1–9, **B** Ex. 10–15, **C** Ex. 16–17

Basic: WE 1–9 odd, Midchapter Review all, Problem Solving

Average: WE 1–15 odd, Midchapter Review all, Problem Solving

Above Average: WE 1–17 odd, Midchapter Review all, Problem Solving

Additional Answers

Written Exercises

4.

Statement	Reason
1. *SW* = *GY*, *TW* = *HY*	1. Given
2. *SW* − *TW* = *GY* − *HY*	2. Subt Prop of Eq
3. *SW* − *TW* = *ST*, *GY* − *HY* = *GH*	3. Seg Add Post
4. *ST* = *GH*	4. Sub

5.

Statement	Reason
1. m ∠5 = m ∠8, m ∠6 = m ∠7	1. Given
2. m ∠5 + m ∠6 = m ∠*ABC*, m ∠7 + m ∠8 = m ∠*EFG*	2. Angle Add Post
3. m ∠5 + m ∠6 = m ∠7 + m ∠8	3. Add Prop Eq
4. m ∠*ABC* = m ∠*EFG*	4. Sub

6.

Statement	Reason
1. *QS* bisects ∠*PQR*, ∠1 ≅ ∠3	1. Given
2. ∠1 ≅ ∠2	2. Def of ∠ bis
3. ∠2 ≅ ∠3	3. Trans Prop of Congr

7.

Statement	Reason
1. m ∠3 = m ∠5	1. Given
2. m ∠3 + m ∠4 = 180	2. If the outer rays of 2 adj ∠s form a st ∠, then the sum of the meas = 180.
3. m ∠4 + m ∠5 = 180	3. Sub

8.

Statement	Reason
1. ∠3 ≅ ∠5	1. Given
2. m ∠3 + m ∠4 = m ∠*ABE*, m ∠5 + m ∠4 = m ∠*DBC*	2. Angle Add Post
3. m ∠3 + m ∠4 = m ∠5 + m ∠4	3. Add Prop of Eq
4. m ∠*ABE* = m ∠*DBC*	4. Sub

9.

Statement	Reason
1. *AB* = *RS*	1. Given
2. *AB* + *BR* = *AR*, *RS* + *BR* = *BS*	2. Seg Add Post
3. *AB* + *BR* = *RS* + *BR*	3. Add Prop of Eq
4. *AR* = *BS*	4. Sub

---

3. A carpenter cuts off a 1-ft piece from each of two boards of the same length. Explain why the remaining board lengths must be equal.
Subt Prop of Eq

Write the proof. Give statements and reasons.

4. Given: *SW* = *GY*, *TW* = *HY*
Prove: *ST* = *GH*

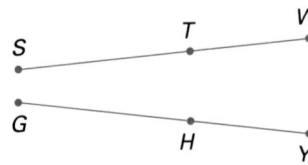

5. Given: m ∠5 = m ∠8, m ∠6 = m ∠7
Prove: m ∠*ABC* = m ∠*EFG*

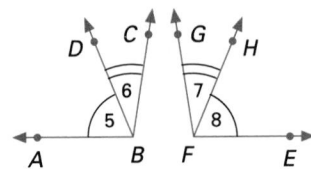

6. Given: $\overrightarrow{QS}$ bisects ∠*PQR*, and ∠1 ≅ ∠3
Prove: ∠2 ≅ ∠3

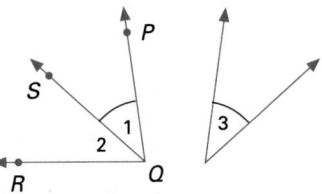

7. Given: m ∠3 = m ∠5
Prove: m ∠4 + m ∠5 = 180

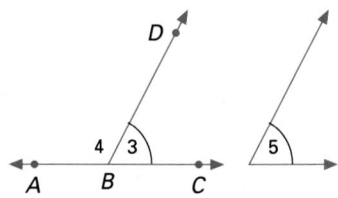

8. Given: ∠3 ≅ ∠5
Prove: ∠*ABE* ≅ ∠*DBC*

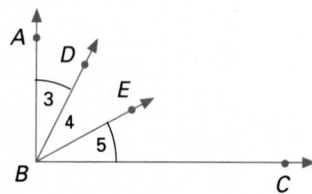

9. Given: *AB* = *RS*
Prove: *AR* = *BS*

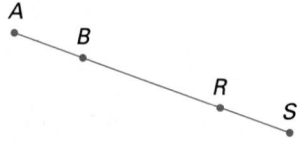

10. Given: *AB* = *BD*, *BC* = *BE*
Prove: *DE* = *AC*

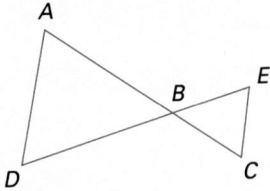

11. Given: $\overline{WX} \perp \overline{WY}$, m ∠5 = m ∠7
Prove: m ∠6 + m ∠7 = 90

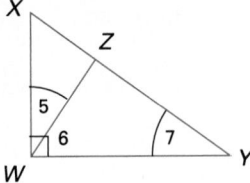

60 Chapter 2 Proof

10.

Statement	Reason
1. *AB* = *BD*, *BC* = *BE*	1. Given
2. *AB* + *BC* = *AC*, *DB* + *BE* = *DE*	2. Seg Add Post
3. *AB* + *BC* = *DB* + *BE*	3. Equations may be added.
4. *AC* = *DE*	4. Sub

11.

Statement	Reason
1. $\overline{WX} \perp \overline{WY}$, m ∠5 = m ∠7	1. Given
2. m ∠5 + m ∠6 = 90	2. If the outer rays of 2 acute adj ∠s are ⊥, then the sum of their meas is 90.
3. m ∠7 + m ∠6 = 90	3. Sub

12. Given: $AB = AE$, $AC = AD$
Prove: $BC = ED$

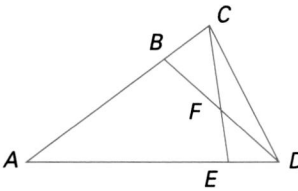

13. Given: $\overline{PQ} \perp \overline{PS}$, $\overline{PR} \perp \overline{PT}$
Prove: m $\angle 3$ = m $\angle 2$

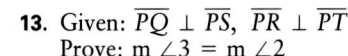
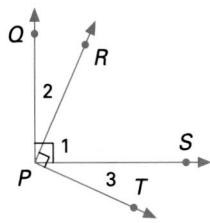

14. Given: $\overrightarrow{ST} \perp \overrightarrow{SY}$, m $\angle 4$ = m $\angle 5$
Prove: m $\angle 5$ + m $\angle 6$ = 90

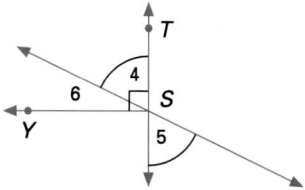

15. Given: $\overrightarrow{BE}$ bisects $\angle ABD$.
Prove: m $\angle 6$ + m $\angle 8$ = m $\angle EBC$

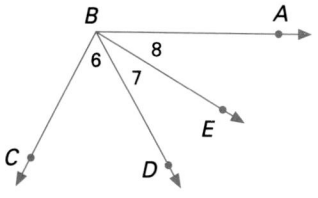

16. Given: $\overrightarrow{BF}$ bisects $\angle ABE$, and $\overline{BF} \perp \overline{BD}$.
Prove: m $\angle 1$ + m $\angle 3$ = 90

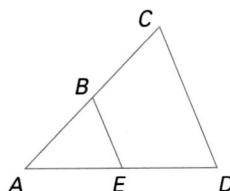

17. Given: B is the midpoint of $\overline{AC}$, $AC = AD$, $AB = BE$, and $BE = AE$.
Prove: $BE + ED = AC$

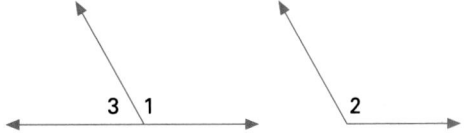

Midchapter Review

Which property is illustrated? (Exercises 1–2) *2.2*

1. $PQ = PQ$ Reflex Prop

2. If $x = y$, then $x + c = y + c$ Add Prop of Eq

3. Given: m $\angle 1$ = m $\angle 2$. Draw a conclusion about $\angle 2$ and $\angle 3$. Give a proof for it. *2.3, 2.4*

12.

Statement	Reason
1. $AB = AE$, $AC = AD$	1. Given
2. $AC - AB = BC$, $AD - AE = ED$	2. Seg Add Post
3. $AC - AB = AD - AE$	3. Equations may be subtracted.
4. $BC = ED$	4. Sub

13.

Statement	Reason
1. $\overline{PQ} \perp \overline{PS}$, $\overline{PR} \perp \overline{PT}$	1. Given
2. m $\angle 1$+ m $\angle 2$ = 90, m $\angle 1$ + m $\angle 3$ = 90	2. If the outer rays of 2 adj acute $\angle$s are $\perp$, then the sum of their meas is 90.
3. m $\angle 1$ + m $\angle 2$ = m $\angle 1$ + m $\angle 3$	3. Trans Prop of Eq
4. m $\angle 2$ = m $\angle 3$	4. Subt Prop of Eq

14.

Statement	Reason
1. $\overrightarrow{ST} \perp \overrightarrow{SY}$, m $\angle 4$ = m $\angle 5$	1. Given
2. m $\angle 4$ + m $\angle 6$ = 90	2. If the outer rays of two adj acute $\angle$s are $\perp$, then the sum of their meas is 90.
3. m $\angle 5$ + m $\angle 6$ = 90	3. Sub

15.

Statement	Reason
1. $\overrightarrow{BE}$ bisects $\angle ABD$	1. Given
2. m $\angle 7$ = m $\angle 8$	2. Def of $\angle$ bis
3. m $\angle 8$ + m $\angle 7$ = m $\angle EBC$	3. Angle Add Post
4. m $\angle 6$ + m $\angle 8$ = m $\angle EBC$	4. Sub

See page 70 for the answer to Ex. 16.

17.

Statement	Reason
1. $AC = AD$	1. Given
2. $AE + AD$	2. Seg Add Post
3. $BE = AE$	3. Given
4. $BE + ED = AC$	4. Sub

Midchapter Review

3.

Statement	Reason
1. m $\angle 1$ = m $\angle 2$	1. Given
2. m $\angle 1$ + m $\angle 3$ = 180	2. If the outer rays of 2 adj $\angle$s form a st $\angle$, then the sum of their meas is 180.
3. $\therefore$ m $\angle 2$ + m $\angle 3$ = 180	3. Sub

11.

Statement	Reason
1. *B* is the midpoint of $\overline{AC}$	1. Given
2. *AB* = *BC*	2. Def of midpt
3. *BC* + *CD* = *BD*	3. Seg Add Post
4. *AB* + *CD* = *BD*	4. Sub

12.

Statement	Reason
1. $\overrightarrow{OD}$ bisects $\angle AOC$	1. Given
2. m $\angle 3$ + m $\angle 5$ = m $\angle DOB$	2. Angle Add Post
3. m $\angle 3$ = m $\angle 4$	3. Def of $\angle$ bis
4. m $\angle 4$ + m $\angle 5$ = m $\angle DOB$	4. Sub

13.

Statement	Reason
1. $\overline{BA} \perp \overline{BD}$ and $\overline{BD}$ bisects $\angle EBC$	1. Given
2. m $\angle 2$ = m $\angle 3$	2. def of $\angle$ bis
3. m $\angle 1$ + m $\angle 2$ = 90	3. If the outer rays of two adj $\angle$s are $\perp$ then the sum of their meas is 90.
4. m $\angle 1$ + m $\angle 3$ = 90	4. Sub

14.

Statement	Reason
1. $\overline{AB}$ bisects $\angle CAE$ and m $\angle 2$ = m $\angle 4$	1. Given
2. m $\angle 4$ = m $\angle 1$	2. Def of $\angle$ bis
3. m $\angle 2$ = m $\angle 1$	3. Sub
4. m $\angle 2$ + m $\angle 3$ = 180	4. The sum the meas of a linear pair of angles is 180.
5. m $\angle 1$ + m $\angle 3$ = 180	5. Sub

15.

Statement	Reason
1. $\overline{AC} \perp \overline{AB}$ and m $\angle ADB$ = m $\angle CAD$	1. Given
2. m $\angle CAD$ + m $\angle DAB$ = 90	2. If the outer rays of two adj $\angle$s are $\perp$, then the sum of their meas is 90.
3. m $\angle ADB$ + m $\angle DAB$ = 90	3. Sub

Problem Solving Strategies

Drawing a Diagram

Objective   To solve problems by drawing a diagram

At times, you may need to draw a diagram according to given information. It is important that your diagram represent the given situation accurately. The problems that follow will give you practice in drawing diagrams.

Example   Two cities, *A* and *B*, are located 1,200 mi from each other. *B* is directly west of *A*. Bob starts at *B* and drives 300 mi south. He then drives 200 mi east. Adam lives 200 mi south of *A*. He drives 100 mi south and then 100 mi west. How far apart are the two men?

This is not a difficult problem when a correct diagram is drawn, but without a diagram it is easy to make mistakes. The figure to the right shows the distance and the answer, 900 mi.

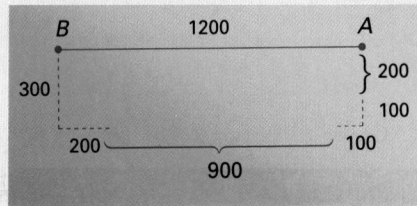

1. In a certain city the avenues run east/west and the streets run north/south. The avenues are named *A*, *B*, *C*, etc. from south to north. The streets are named 1st, 2nd, 3rd, etc. from west to east. Both the streets and avenues are spaced one-tenth of a mile apart. Alice lives at the intersection of *A* and 1st; Betty lives at the intersection of *F* and 5th. How far do they live from each other? No short cuts are possible because of buildings. $\frac{9}{10}$ mi

2. Towns *D*, *E*, and *F* are located along the same straight road. *E* is between *D* and *F*. *D* is 4 mi from *E* and 11 mi from *F*. The three towns are planning to build a movie theater. An equal number of people are expected to attend from each town. Where should they build the theater so that the total miles driven will be least? (Consider one person traveling from each town.)   *E*

3. A radio station sends out signals in all directions (circular) for 60 mi. Another radio station 100 mi away also sends out signals for 60 mi. If you are driving along a road that directly connects these two stations, what is the distance in miles through which you can hear both stations?   20 mi

62   Chapter 2 Proof

16.

Statement	Reason
1. *W* is midpoint of $\overline{PH}$	1. Given
2. *GP* + *PW* = *GW*	2. Seg Add Post
3. *PW* = *WH*	3. Def of midpt
4. *GP* + *WH* = *GW*	4. Sub

Application

1. On June 21, the North pole is tilted toward the sun all day.

2. No part of the earth is tilted toward the sun. So, no part of the earth receives sunlight for a longer time.

3. Exchange Summer with Winter and Fall with Spring.

2.5 Conditional Statements

Objectives To rewrite statements in "If . . . , then . . . " form
To determine whether conditional statements are true

Although theorems and problems in geometry are often written with separate *given* and *prove* (conclusion) statements, this format is not required. Consider the following example.

Given: M is the midpoint of $\overline{AB}$.
Conclusion: $AM = MB$

A   M   B

The given (*hypothesis*) and the conclusion can be written as a single conditional statement.

If M is the midpoint of $\overline{AB}$, then $AM = MB$.
    given or hypothesis   conclusion

Often letters such as p, q, r, s, and t are used to represent statements or parts of statements.

The statement "If p, then q" is written $p \rightarrow q$.

A statement that is in the form "If p, then q" is a **conditional** statement. The part of the statement following "if" is the **hypothesis**. The part of the statement following "then" is the **conclusion**.

Many statements that are not in "if . . . , then . . ." form can be written in this form.

Statement:	A right angle has a measure of 90.
In "If . . . , then . . ." form:	If an angle is a right angle, then its measure is 90.
Statement:	The midpoint of a segment divides the segment into two congruent segments.
In "If . . . , then . . ." form:	If a point is the midpoint of a segment, then it divides the segment into two congruent segments.

EXAMPLE 1 Write the statement in "If . . . , then . . . " form.
The angles formed when a ray bisects an angle are congruent.

Solution If a ray bisects an angle, then the angles formed are congruent.

Teaching Resources

Problem Solving Worksheet 2
Quick Quizzes 13
**Reteaching and Practice
Worksheets,** pp. 25, 26

▰▰▰ GETTING STARTED

Prerequisite Quiz

True or false?

1. Two angles are supplementary if the sum of their measures is 180.   T
2. $5 + 4 < 9$ F
3. The sides of a right angle are perpendicular to each other.   T
4. $\sqrt{16} = 8$ F

Motivator

Students have already used conditional statements. A conditional is a statement in the form used in each of the following statements.

If the outer rays of two adjacent angles form a straight angle, then the sum of their measures is 180.

If M is the midpoint of $\overline{AB}$, then $AM = MB$. If you use Hotdoor toothpaste, then you have fewer cavities.

Ask your students to give the necessary form or pattern of words for a statement to be called a conditional.   If ..., then ....

Additional Example 1

Write the statement in "If . . ., then . . ." form.

Congruent angles have equal measures.
If angles are congruent, then they have equal measures.

Lesson Note

Conditional statements are at the foundation of much of the work in geometry. Stress that every proof involves a hypothesis leading to a conclusion. Students understand proofs better when they write what they are proving in the form of a conditional. This should have higher priority than work with truth tables. "If ..., then ..." statements will be discussed again ... when the *converse* of a conditional is introduced (p. 109).

To explain the truth table, refer to the paragraphs about the salesman on page 64. A false value should be assigned to $p \rightarrow q$ only if the agreement is violated, that is, if he sells $10,000 worth of furniture, but does not receive the bonus. Cases of an unsatisfied or "false" hypothesis do not violate the agreement, so the conditional is assigned a value of "true." Geometry is usually concerned with true conditional statements.

Math Connections

Research: Determining the truth of conditional statements is similar to determining whether a correlation or causation relationship exists in research. When two events occur together, a scientist must carefully search for proof to discover whether one actually causes the other.

Critical Thinking Questions

Evaluation: Ask students to give a counter-example to disprove each statement.

1. Complementary angles are right angles.
2. The acute angles in a right triangle have equal measures.
3. All supplementary angles form a linear pair.   Answers may vary. Sample answers are given. 1. Angles with measures of 30 and 60 are complementary. 2. The acute angles may have measures of 35 and 55. 3. If supplementary angles are not adjacent, they do not form a linear pair.

The truth table at the right shows all possible truth values for p and q. A conditional is false only when the hypothesis p is true and the conclusion q is false.

p	q	$p \rightarrow q$
T	T	T
T	F	F
F	T	T
F	F	T

This can be illustrated by the following example: A salesman is told, "*If* you sell $10,000 worth of furniture, *then* we will give you a $500 bonus." The only situation that breaks this promise is the one in which the salesman sells $10,000 worth of furniture (making the hypothesis, or p, true), and he is *not* given a $500 bonus (making the conclusion, or q, false).

A case that proves the conditional false is a **counterexample**. Notice that if the salesman does *not* sell $10,000 worth of furniture and is *not* given $500, the promise is still kept. The conditional is true even though the hypothesis and conclusion are both false. This example can also be used to show that a conditional is true for the other two cases given in the truth table.

The conditional $p \rightarrow q$ can be read in any one of the following ways:

hypothesis   conclusion

(1) If p, then q (3) q if p
(2) p implies q (4) p only if q

EXAMPLE 2   Copy the statement. Underline the hypothesis once and the conclusion twice. Then rewrite the statement in two other equivalent forms.

Two angles are supplementary if the sum of the angle measures is 180.

Solution   The format as stated is "q if p."

Two angles are supplementary if the sum of the angle measures is 180.

If p, then q	If the sum of two angle measures is 180, then the angles are supplementary.
p implies q	The fact that the sum of two angle measures is 180 implies that the angles are supplementary.

Recall that in any proof, the *given* is the hypothesis, and the *proof* is the conclusion.

Additional Example 2

Copy the conditional. Underline the hypothesis once and the conclusion twice. Rewrite the conditional in two other equivalent forms.

Adjacent acute angles are complementary if the outer rays are perpendicular. *Adjacent acute angles are* complementary if *the outer rays are* perpendicular. (1) If *the outer rays are* perpendicular, then two adj acute ∠s are comp. (2) Outer rays are ⊥ implies that adj acute ∠s are comp.

EXAMPLE 3

Examine the proof below. State what is proved in "If . . . , then . . ." form.

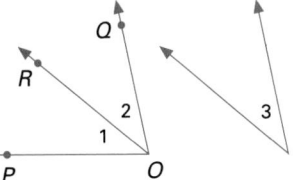

Statement	Reason
1. m ∠2 = m ∠3	1. Given
2. $\overrightarrow{OR}$ bisects ∠POQ.	2. Given
3. m ∠1 = m ∠2	3. Def of ∠ bis
4. ∴ m ∠1 = m ∠3	4. Sub

Plan There are two *given* statements. Hypothesis: m ∠2 = m ∠3 and $\overrightarrow{OR}$ bisects ∠POQ. The conclusion is the last statement: m ∠1 = m ∠3.

Solution Conditional: If m ∠2 = m ∠3 and $\overrightarrow{OR}$ bisects ∠POQ, then m ∠1 = m ∠3.

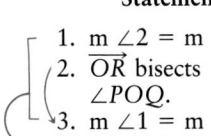
Focus on Reading

Copy and complete each statement.

1. A statement of the form "If . . . , then . . ." is called a <u>conditional</u>.
2. In $p \rightarrow q$, p is called the <u>hypothesis</u>.
3. In $p \rightarrow q$, q is called the <u>conclusion</u>.
4. $p \rightarrow q$ is false only if q is <u>false</u> and p is <u>true</u>.

Classroom Exercises

Name the hypothesis and the conclusion of the statement.

1. $r \rightarrow s$
2. If $x = 4$, then $5 > x$
3. If $a < b$ and $b < c$, then $a < c$
4. m ∠A < 90 if ∠A is acute
5. $AB = BC$ only if $BC = CD$

Written Exercises

Use the information to write a statement in "If . . ., then . . ." form.

1. Given: ∠A is obtuse.
   Conclusion: 90 < m ∠A < 180

2. Given: Two segments are congruent.
   Conclusion: The segments have the same length.

Copy the conditional. Underline the hypothesis once and the conclusion twice. State whether the conditional is true.

3. If $x^3 = -27$, then $x = -3$
4. If $x^2 + 1 = 1$, then $x = 2$
5. If $x < 4$, then $x < 5$
6. If $x < -4$, then $x < -5$

2.5 Conditional Statements **65**

Additional Example 3

Examine the proof below. Write the conclusion in "If..., then..." form.
Given: $3x - 4 = 11$
Prove: $x = 5$

Conclusion: If $3x - 4 = 11$, then $x = 5$.

Statement	Reason
1. $3x - 4 = 11$	1. Given
2. $3x = 15$	2. Add Prop of Eq
3. $x = 5$	3. Div prop of Eq

Common Error Analysis

Error: Students often have difficulty seeing that "p only if q" is in fact a correct rendering of $p \rightarrow q$. Many students think that the phrase would apply more appropriately to $q \rightarrow p$, since the word *if* comes before q in the phrase "p only if q".

Show the students that the phrase "p only if q" has the same truth table as "if p, then q", which is the usual rendering of the logical statement $p \rightarrow q$.

Checkpoint

Copy the conditional. Underline the hypothesis once and the conclusion twice. Is the conditional true?

1. If two angles are adjacent, <u>then they are complementary</u>.   F
2. If $17 = 6 \cdot 2 + 5$, then <u>$5^3 = 15$</u>.   F
3. A ray is an angle bisector only <u>if the ray divides the angle into two congruent angles</u>.   T

Closure

When is a conditional true? What is the hypothesis of a conditional? What is the conclusion of a conditional?

FOLLOW UP

Guided Practice

Classroom Exercises 1–5

Independent Practice

A Ex. 1–8, **B** Ex. 9–12, **C** Ex. 13–14

Basic   FR all, WE 1–8 odd
Average   FR all, WE 1–12 odd
Above Average   FR all, WE 1–14 odd

Additional Answers

Classroom Exercises

1. Hyp: r Concl: s
2. Hyp: $x = 4$ Concl: $5 > x$
3. Hyp: $a < b$ and $b < c$ Concl: $a < c$
4. Hyp: ∠A is acute Concl: m ∠A < 90 .
5. Hyp: $AB = BC$ Concl: $BC = CD$

Written Exercises

1. If $\angle A$ is obtuse, then $90 < m \angle A < 180$.
2. If 2 segs are $\cong$, then the segs have the same length.
3. Hyp: $x^3 = -27$, concl: $x = -3$ (True)
4. Hyp: $x^2 + 1 = 1$, concl: $x = 2$ (False)
5. Hyp: $x < 4$ concl: $x < 5$ (True)
6. Hyp: $x < -4$ concl: $x < -5$ (False)
7. If two angles are supplementary, then the sum of their measures is 180.
8. If a ray bisects an angle, then it divides the angle into two congruent angles.
9. <u>An angle is acute</u> if its measure is less than 90.
   a. If the measure of an angle is less than 90, then the angle is acute.
   b. The measure of an angle is less than 90 implies that the angle is acute.
10. The measure of an angle is 90 <u>only if the sides of the angle are perpendicular.</u>
    a. If the measure of an angle is 90, then the sides of the angle are perpendicular.
    b. The measure of an angle is 90 implies that the sides of the angle are perpendicular.

Mixed Review

1. A rt $\angle$ has $\perp$ sides *and* a plane consists of a finite number of pts. False; r is false; both must be true.
2. A rt $\angle$ has $\perp$ sides *or* a plane consists of a finite number of pts. True; p is true; at least one must be true.
3. A rt $\angle$ has $\perp$ sides *and* the sum of the meas of supp $\angle$s is 180. True; both p and q are true.
4. The sum of the meas of 2 supp $\angle$s is 180 or a plane consists of a finite number of points. True; q is true; one statement must be true.

Write the statement in "If . . . , then . . ." form.

7. Supplementary angles are two angles the sum of whose measures is 180.
8. A ray that bisects an angle divides the angle into two congruent angles.

Copy the statement. Underline the hypothesis once and the conclusion twice. Rewrite the statement in two other equivalent forms.

9. An angle is acute if its measure is less than 90.
10. The measure of an angle is 90 only if the sides of the angle are perpendicular.

Examine the proof. State what is proved in "If . . . , then . . ." form.

11.

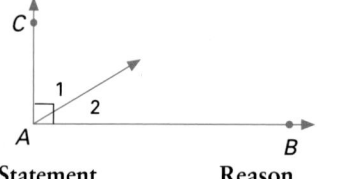

Statement	Reason
1. $\overrightarrow{AB} \perp \overrightarrow{AC}$	1. Given
2. $\therefore$ m $\angle 1$ + m $\angle 2 = 90$	2. If outer rays of two acute adjacent $\angle$s are $\perp$, then the sum of measures is 90.

If $\overrightarrow{AB} \perp \overrightarrow{AC}$, then m $\angle 1$ + m $\angle 2 = 90$.

12.

Statement	Reason
1. $\overline{CD}$ bisects $\overline{AB}$.	1. Given
2. $AD = BD$ ($\overline{AD} \cong \overline{BD}$)	2. Def of bis
3. $BD = TY$	3. Given
4. $\therefore AD = TY$	4. Trans Prop (or sub)

If $\overline{CD}$ bisects $\overline{AB}$ and $BD = TY$, then $AD = TY$.

Copy and complete the truth table.

13.

p	q	$p \rightarrow q$	$p \wedge (p \rightarrow q)$
T	T	T	T
T	F	F	F
F	T	T	F
F	F	T	F

14.

p	q	$p \rightarrow q$	$p \vee (p \rightarrow q)$	$[p \vee (p \rightarrow q)] \rightarrow q$
T	T	T	T	T
T	F	F	T	F
F	T	T	T	T
F	F	T	T	F

Mixed Review   1.1, 1.4, 1.6, 1.7

Use the statements in Exercises 1–4 below.

p: A right angle has perpendicular sides.
q: The sum of the measures of two supplementary angles is 180.
r: A plane consists of a finite number of points.

1. Write the conjunction $p \wedge r$. Is the conjunction true? Give a reason.
2. Write the disjunction $p \vee r$. Is the disjunction true? Give a reason.
3. Write the conjunction $p \wedge q$. Is the conjunction true? Give a reason.
4. Write the disjunction $q \vee r$. Is the disjunction true? Give a reason.

Enrichment

Tell students to consider each given statement as a promise. Ask them to explain under what conditions the promise is broken.

1. If it is cold, then I wear my coat.
2. I will use an umbrella only if it is raining.
3. Our team will go to the Rose Bowl if we win this game.

1. When it is cold and I do not wear my coat. 2. When it rains and I do not use an umbrella. 3. When our team wins the game and we do not go the Rose Bowl.

2.6 Deductive and Inductive Reasoning

Objectives   To draw conclusions from given statements
To determine whether a conclusion is reached deductively or inductively
To use inductive reasoning to draw conclusions

The proofs you have been writing in this chapter are examples of *deductive reasoning*. Using deductive reasoning, you draw conclusions that follow from other statements.

EXAMPLE 1   Draw a conclusion that follows from the given statements.

Given: If a person studies geometry, then he or she develops an appreciation for logic. Maria studies geometry.

Solution   Conclusion: Maria develops an appreciation for logic.

The given statements are often called **premises**. If the premises are true, then the conclusion drawn deductively is true, provided the argument used is valid. This example illustrates the importance of establishing a collection of statements in geometry that are accepted as true. Recall that such statements include undefined terms, definitions, and postulates.

EXAMPLE 2   Draw a conclusion from the given statement.

Given: $\overrightarrow{AB}$ bisects $\angle XAY$.

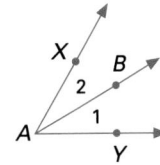

Solution   Conclusion: $\angle 1 \cong \angle 2$. The definition of angle bisector, an accepted statement since Lesson 1.5, applies in this situation.

Another kind of deductive argument is illustrated by the following.

Given: If you live in Chicago, then you live in Illinois. $(p \rightarrow q)$

If you live in Illinois, then you live in the U.S.A. $(q \rightarrow r)$

If you live in Chicago, then you live in the U.S.A. $(p \rightarrow r)$

In this way, conditionals can be linked together.

Given: $p \rightarrow q$ and $q \rightarrow r$ Conclusion: $p \rightarrow r$

▬▬ GETTING STARTED

Prerequisite Quiz

Draw a conclusion that follows from each statement.

1. m $\angle A$ = 35 and m $\angle B$ = 55
   m $\angle A$ + m $\angle B$ = 90, or $\angle A$ and $\angle B$ are comp $\angle$s.
2. B is the midpoint of $\overline{AC}$. $AB = BC$
3. A man worked for 2 h at a rate of $8/h. The man made $16.

Motivator

Have students draw two adjacent right angles and have them bisect each angle. Then have them measure the angle formed by the bisectors. Ask them to repeat the experiment several times. Then ask them to complete this statement: The bisectors of a pair of adjacent right angles form a _____ angle.   right   Then discuss with them whether the results of their drawings are acceptable as a proof of the statement. No, a deductive proof is needed.

Additional Example 1

Draw a conclusion that follows from the given statements.

Birds have wings. Eagles are birds.
Conclusion: Eagles have wings.

Additional Example 2

Draw a conclusion from the given statement.

Given: $\angle TRS$ is a right angle.
Conclusion:
m $\angle 1$ + m $\angle 2$ = 90

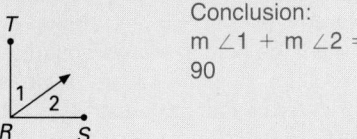

Lesson Note

Inductive reasoning is sometimes used as a discovery strategy in geometry. Students should be encouraged to use constructions, measurements, and test cases as ways to find and understand ideas. Be sure that students understand that in inductive reasoning, you arrive at a conclusion based on several observations. The greater the number of observations, the greater the likelihood that the conclusion is correct. Emphasize that inductive reasoning is not a method of proof. Rather, inductive reasoning can lead to the formulation of general statements that can be proved deductively. This table lists some important differences between deductive and inductive reasoning.

Deductive Reasoning	Inductive Reasoning
1. Conclusions are based on accepted statements (definitions, properties, postulates, theorems, and given information)	1. Conclusions are based on observations.
2. Conclusions must be true if the premises are true.	2. Conclusions may be true, but are not necessarily true.

Math Connection

Life Skills: People do not always realize that conclusions based on inductive reasoning are not necessarily true. Consumers especially are exposed to many advertisements that state conclusions based on inductive reasoning.

Another way of reaching conclusions is to use *inductive reasoning*. Scientists use this reasoning when, after repeating an observation or experiment, they make a generalization. For example, by repeated observations it is established that water freezes at 32°F or 0°C.

EXAMPLE 3 In each diagram of intersecting lines, use a protractor to find the measure of a pair of nonadjacent angles. Draw a conclusion inductively.

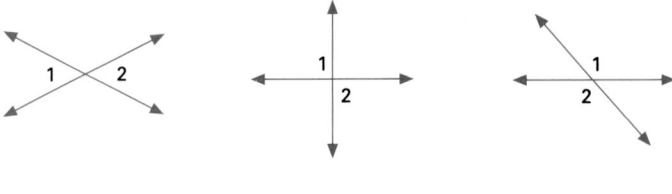

$$\text{m} \angle 1 = \text{m} \angle 2 \qquad \text{m} \angle 1 = \text{m} \angle 2 \qquad \text{m} \angle 1 = \text{m} \angle 2$$

Solution Conclusion: If two lines intersect, the pair of nonadjacent angles formed are equal in measure, or congruent.

Although the conclusions drawn inductively may *seem* true, they may not always *be* true. The ancient Egyptians farmed land in the shape of long narrow triangles. They observed that the formula

$$A = \frac{1}{2} \times \text{base length} \times \text{side length},$$

worked for finding the area of each of their triangular fields. They then concluded that this formula was true for all triangles. Actually, the correct formula is

$$A = \frac{1}{2} \times \text{base length} \times \text{height length}.$$

In the land the Egyptians measured, there was very little difference between the height length and true length of a side.

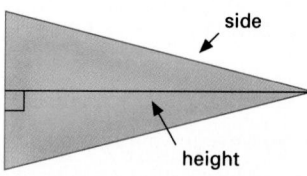

In geometry, inductive reasoning often serves to discover a new concept or theorem. However, to prove or to establish the truth of a statement, deductive reasoning is used.

Additional Example 3

Continue the pattern by adding odd numbers starting with the number 1. Then draw a conclusion inductively.
1 + 3 = 4
1 + 3 + 5 = 9
Conclusion: the sum is always equal to a perfect square. (It is the square of the number of terms that are added.)

Classroom Exercises

True or false. If false, give a reason.

1. Inductive reasoning is used to prove statements in geometry.
2. If $a \rightarrow b$ and $b \rightarrow c$, then it follows that $a \rightarrow c$. T

Draw a valid conclusion from the given statements.

3. If M is the midpoint of $\overline{AB}$, then $AM = MB$. G is the midpoint of $\overline{WY}$. *WG = GY*
4. If you go to the West Coast, you will see beautiful sunsets. If you see beautiful sunsets, you will feel exhilarated.

Written Exercises

Write a conclusion that follows from the given statements. (Exercises 1–5)

1. If a person brushes his or her teeth with No-Cav toothpaste, then he or she will not develop cavities. Mary brushes with No-Cav.
2. If today's date is December 31, then tomorrow is New Year's Day. Today's date is December 31.
3. If two angles are complementary, then they are congruent. $\angle A$ and $\angle B$ are complementary.
4. If you are under 16 years of age and live in New Jersey, then you cannot get a driver's license. If you cannot get a driver's license, then you cannot legally drive a car in New Jersey.
5. If you have a cold, then you will need rest. If you need rest, then you should not exercise strenuously.
6. Draw a triangle. Measure each of the angles. Find the sum of the measures of the three angles. Repeat this exercise with four more triangles. Inductively draw a conclusion.
7. In each of the following figures, $\overline{CD}$ bisects $\angle ACB$. Use a ruler to find AD and BD for the first three figures. Draw a conclusion. Measure $\overline{AD}$ and $\overline{BD}$ in figure **d**. What does the result illustrate about inductive reasoning?

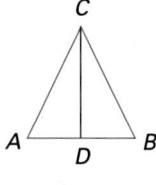
a.

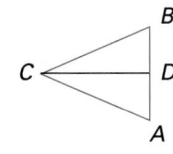
b.

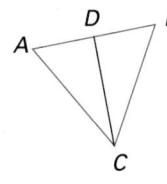
c.

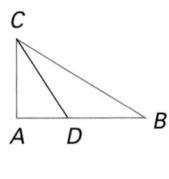
d.

Enrichment

Have students draw acute angles all the way around a common vertex and then measure each angle until they return to the starting segment.

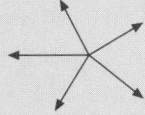

Ask students to repeat this procedure several times and to draw a conclusion inductively. Ask if they can prove deductively that the sum of measures of the angles around a point will always be equal to 360.   One way is to show that the sum of the measures is equal to four right angles.

Critical Thinking Questions

Analysis: Contrast the deductive reasoning of mathematics with the inductive reasoning that must often be used in science.   Use the topic for open-ended discussion. Mention some false theories that scientists once believed, such as a flat world or that the sun revolved around the earth.

Checkpoint

Write a conclusion that follows from each statement.

1. If 1,996 is divisible by 4, then there will be a presidential election in 1,996. 1,996 is divisible by 4.   Conclusion: There will be an election in 1996.
2. If 372 and 1,248 are even numbers, then their product will be an even number. 372 and 1,248 are even numbers. Conclusion: Their product is even.

Tell whether each conclusion is drawn deductively or inductively.

3. While keeping records for 50 consecutive days, an observer notes that every day the moon rises about 50 min later than it did the day before. The observer concludes that this will always happen. Inductive
4. A gallon of paint covers about 250 ft². A painter estimates that he will need 3 gallons to cover a room that contains 700 ft² of surface.   Deductive

Closure

Ask students to summarize the differences between inductive and deductive reasoning.

◼◼◼ FOLLOW UP

Guided Practice

Classroom Exercises 1–4

Independent Practice

🅐 Ex. 1–5, 🅑 Ex. 6–11, 🅒 Ex. 12–15

Basic: WE 1–5 odd, Brainteaser
Average: WE 1–11 odd, Brainteaser
Above Average: WE 1–15 odd, Brainteaser

Additional Answers

Classroom Exercises

1. False. Geometry uses deductive reasoning.
4. If you go to the West Coast, you will feel exhilarated.

Written Exercises

1. Mary will not develop cavities.
2. Tomorrow is New Year's Day.
3. $\angle A$ and $\angle B$ are congruent.
4. If you are under 16 years of age and live in New Jersey, then you cannot legally drive a car in New Jersey.
5. If you have a cold, then you should not exercise strenuously.
6. The sum of the measures of a triangle is 180 degrees.
7. Inductive reasoning does not always lead to true conclusions.

Mixed Review

1. none, 70
2. 55, 145
3. $(90 - x)$ for x acute; $(180 - x)$
4. $(60 - x)$ for $(x + 30)$ acute, $(150 - x)$
5.

   45 55 65 75

Additional Answer, page 61

16.

Statement	Reason
1. $\overrightarrow{BF}$ bisects $\angle ABE$, $\overrightarrow{BF} \perp \overrightarrow{BD}$	1. Given
2. m $\angle 1$ = m $\angle 4$	2. Def of $\angle$ bis
3. m $\angle ABC$ = m $\angle 1$ + m $\angle 2$ + m $\angle 3$ + m $\angle 4$	3. Angle Add Post
4. m $\angle 1$ + m $\angle 2$ = 90.	4. If the outer rays of 2 adj acute $\angle$s are $\perp$, then the sum of their measures is 90.
5. m $\angle 4$ − m $\angle 2$ = 90.	5. Sub
6. 90° + m $\angle 3$ + m $\angle 1$ = m $\angle ABC$	6. Sub
7. m $\angle ABC$ = 180.	7. Def of st $\angle$
8. 90 + m $\angle 3$ + m $\angle 1$ = 180.	8. Sub
9. m $\angle 3$ + m $\angle 1$ = 90.	9. Subt Prop of Eq

Tell whether the conclusion drawn is reached deductively or inductively. (Exercises 8–11)

8. A student finds the sum of the measures of the angles of 5 four-sided figures. In each figure, the sum is 360. The student concludes that in any four-sided figure, the sum of the measures of the angles is 360.   Inductively

9. A person reading a map observes that the distance between two Towns A and B is equal to the sum of the distances from A to C and from B to C. The person concluded that Town C is between A and B.   Deductively

10. A student has had good experiences with his high-school math teachers. He concludes that all high school math teachers are nice people.   Inductively

11. A wall 10 ft wide consists of 20 bricks across. A mason concludes that a wall 15 ft wide will consist of 30 bricks.   Deductively

There is a pattern for determining the number of segments that can be drawn between each of the points in a set of points, no three of which are collinear. Find the number of such segments for the given number of points, no three of which are collinear. To get started, draw figures like the ones shown. Then see if you can discover a pattern.

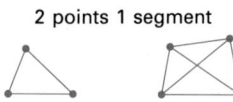

2 points 1 segment

12. 5 points   10
13. 6 points   15
14. 7 points   21
15. 20 points   190

Mixed Review

Find the measure of a complement and a supplement, if possible, of an angle with the given measure.   *1.6*

1. m $\angle A$ = 110
2. m $\angle B$ = 35
3. m $\angle A$ = x
4. m $\angle A$ = x + 30
5. Graph the conjunction: $15 < x - 30 < 45$.   *1.8*

Brainteaser

The "proof" below claims to show that 2 = 0. Find the error in reasoning.

1. Let $x = 1$ and $y = 1$.
2. $x = y$ Substitute.
3. $x^2 = y^2$ Square both sides.
4. $x^2 - y^2 = 0$ Subtract y^2 from both sides.
5. $(x + y)(x - y) = 0$ Factor.
6. $x + y = 0$ Divide each side by $(x - y)$.
7. $2 = 0$ Substitute.

Step 6 is invalid; cannot divide by 0. Thus, Step 7 is also invalid.

70 Chapter 2 Proof

2.7 Proving Theorems About Angles

Objective   To prove and apply theorems about supplementary angles and complementary angles

Two angles are supplementary if the sum of their measures is 180. Suppose that m $\angle A$ = 110 and m $\angle B$ = 110. The measure of a supplement of $\angle A$ is 70. Similarly, the measure of a supplement of $\angle B$ is 70.
The supplements of the two angles are congruent.

Theorem 2.1   If two angles are supplements of congruent angles, then they are congruent. (Supplements of congruent angles are congruent.)

Given: $\angle B \cong \angle C$, $\angle A$ is a
supplement of $\angle B$,
and $\angle D$ is a
supplement of $\angle C$.

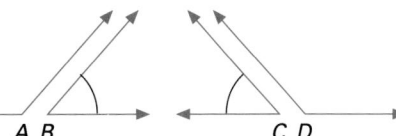

A B   C D

Prove: $\angle A \cong \angle D$

Plan   m $\angle A$ + m $\angle B$ = 180 and m $\angle D$ + m $\angle C$ = 180, so
m $\angle A$ + m $\angle B$ = m $\angle D$ + m $\angle C$. Use subtraction to get
m $\angle A$ = m $\angle D$.

Proof

Statement	Reason
1. $\angle A$ is a supplement of $\angle B$; $\angle D$ is a supplement of $\angle C$.	1. Given
2. m $\angle A$ + m $\angle B$ = 180, m $\angle D$ + m $\angle C$ = 180	2. Def of supp $\angle$s
3. m $\angle A$ + m $\angle B$ = m $\angle D$ + m $\angle C$	3. Sub
4. m $\angle B$ = m $\angle C$	4. Given
5. $\therefore$ m $\angle A$ = m $\angle D$ ($\angle A \cong \angle D$)	5. Equations may be subtracted.

A **corollary** of a theorem is a theorem whose proof follows from the
original theorem in a few steps. The statement below is a corollary of
Theorem 2.1.

Corollary   If two angles are supplements of the same angle, then they are congruent. (Supplements of the same angle are congruent.)

A theorem or a corollary may be used as a reason in a proof. In
Example 1, notice how Theorem 2.1 is used as a reason in Step 3.

Teaching Resources

Quick Quizzes 15
**Reteaching and Practice
Worksheets,** pp. 29, 30

GETTING STARTED

Prerequisite Quiz

1. Given: $\angle 7$ is supp to $\angle 8$. Write an
equation relating $\angle 7$ and $\angle 8$.
m $\angle 7$ + m $\angle 8$ = 180
2. Given: $\angle 5$ and $\angle 6$ are comp $\angle$s. Write
an equation relating $\angle 5$ and $\angle 6$.
m $\angle 5$ + m $\angle 6$ = 90
3. m $\angle A$ = 35. Find the measure of a
complement of A.   55
4. Find the measure of the supplement of
an angle with measure 140.   40

Motivator

Ask your students to recall the definition of
supplementary angles.   Sum of their meas
= 180   Complementary angles?   Sum of
their meas = 90

Suppose that two angles are congruent. Ask
students which angle will have the greater
supplement.   Neither   Why?   If $x = y$,
then $180 - x = 180 - y$

TEACHING SUGGESTIONS

Lesson Note

It is difficult to learn a new concept that is introduced only on the abstract level. Therefore, the ideas of this lesson are first introduced by concrete illustrations before the statements and proofs of Theorems 2.1 and 2.2. Give other examples such as these: If m ∠A = 60 and m ∠B = 60, find the supplement of each.

Ask these questions: Is there a pattern? Suppose two angles have the same measure. What would have to be true about their supplements? Why? Point out the advantages of using the two theorems and how they can save steps. For instance, in Example 1, it is not necessary to write equations and then use the Subtraction Property of Equality in order to prove ∠2 ≅ ∠3. Emphasize the shorter form of the theorems: Supp (comp) of ≅ ∠s, or the same ∠, are ≅.

Math Connections

Building Trades: Facts about complementary angles often help builders cut materials to a specified measure at the site of a construction.

Critical Thinking Questions

Analysis: Ask this question: If supplements/complements of the same angle are congruent, does it follow that the supplementary/complementary relationship is transitive? Explain. No. If the supp/comp relationship were trans, then the following would be true: If ∠A and ∠B are supp/comp and ∠B and ∠C are supp/comp, then ∠A and ∠C are supp/comp. This is false because it implies that ∠A ≅ ∠C, which is not always true.

EXAMPLE 1 Given: ∠1 ≅ ∠4
Prove: ∠2 ≅ ∠3

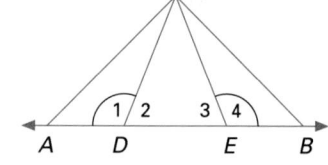

Proof

Statement	Reason
1. ∠1 ≅ ∠4	1. Given
2. ∠2 is a supplement of ∠1; ∠3 is a supplement of ∠4.	2. If outer rays of two adj ∠s form a st ∠, then the ∠s are supp.
3. ∠2 ≅ ∠3	3. Supp of ≅ ∠s are ≅.

You know that two angles are complementary if the sum of their measures is 90. A proof similar to that of Theorem 2.1 can be given to prove a theorem about complements of congruent angles.

Theorem 2.2 If two angles are complements of congruent angles, then they are congruent. (Complements of congruent angles are congruent.)

Given: ∠B ≅ ∠C, ∠A is a complement of ∠B, and ∠D is a complement of ∠C.

Prove: ∠A ≅ ∠D

You will be asked to prove this theorem in Exercise 15.

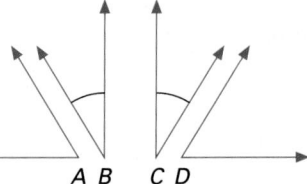

Corollary If two angles are complements of the same angle, then they are congruent. (Complements of the same angle are congruent.)

EXAMPLE 2 Given: m ∠UQT = 25, m ∠SQT = 25, $\overrightarrow{QT} \perp \overleftrightarrow{PR}$

What conclusion can be drawn about angle measures? Why?

Plan Since $\overrightarrow{QT} \perp \overleftrightarrow{PR}$, ∠PQU is a complement of ∠UQT. Similarly, ∠RQS is a complement of ∠SQT. Apply Theorem 2.2.

Solution ∠PQU ≅ ∠RQS since they are complements of congruent angles. Thus, m ∠PQU = 90 − 25 = 65, and m ∠RQS = 65.

Additional Example 1

Given: ∠3 is a supplement of ∠5.
Prove: ∠4 ≅ ∠5

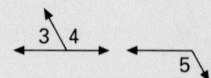

Proof: ∠3 supp ∠5 (Given); ∠3 and ∠4 are adj ∠s (Def of adj ∠s); ∠3 supp ∠4 (If outer rays of two adj ∠s form a st ∠, then ∠s are supp.); m ∠4 = m ∠5 (Supp of the same ∠ are ≅.)

| Theorem 2.3 | If two angles are right angles, then they are congruent. |

You will be asked to prove this theorem in Exercise 17.

Focus on Reading

Write the term for each abbreviation.

1. adj   **2.** bis   **3.** coll   **4.** comp   **5.** ⊥
6. cor   **7.** def   **8.** ∠   **9.** midpt   **10.** Post
11. prop   **12.** rt   **13.** st   **14.** supp   **15.** Thm

Classroom Exercises

Complete the statement of each conclusion. Give a reason for each conclusion.

1. Given: ∠5 ≅ ∠6
Conclusion: ∠3 ≅ _____

2. Given: ∠6 is a supplement of ∠8.
Conclusion: ∠8 ≅ _____

3. Given: ∠5 is a supplement of ∠7.
Conclusion: ∠3 ≅ _____

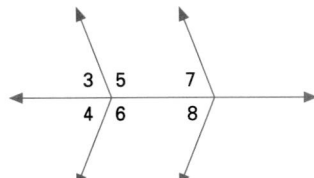

Written Exercises

State a conclusion that may be drawn. Give a reason.

1. Given: m ∠1 = 60, m ∠2 = 60

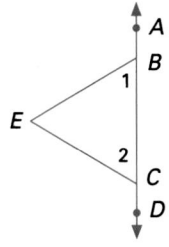

m ∠EBA = m ∠ECD. Supp of ≅ ∠s are ≅.

2. Given: m ∠ABF = 42, m ∠ECD = 42, $\overrightarrow{BA} \perp \overrightarrow{BC}$, $\overrightarrow{CD} \perp \overrightarrow{BC}$

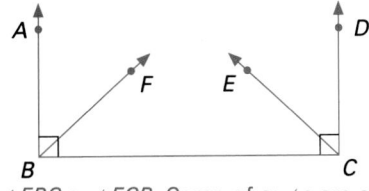

∠FBC ≅ ∠ECB. Comp of ≅ ∠s are ≅.

Additional Example 2

Given: $\overrightarrow{EA} \perp \overrightarrow{EC}$, $\overrightarrow{EB} \perp \overrightarrow{ED}$,
     m ∠BEC = 35

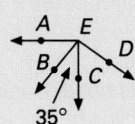

What conclusion can be drawn about angle measures? Why?
∠AEB ≅ ∠CED, comp of the same ∠ are ≅; m ∠AEB = m ∠CED = 90 − 35, or 55

Checkpoint

What conclusion can be drawn? Why?

1. Given: $\overrightarrow{BD} \perp \overleftrightarrow{AC}$

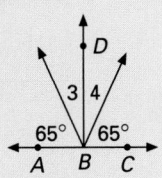

∠3 ≅ ∠4, comp of ≅ ∠s are ≅; m ∠3 = m ∠4 = 90 − 65 = 25

2.

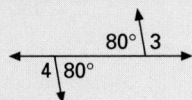

∠3 ≅ ∠4, supp of ≅ ∠s; m ∠3 = m ∠4 = 180 − 80 = 100

3. Given: ∠5 ≅ ∠6
Prove: ∠7 ≅ ∠8

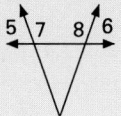

Statement	Reason
1. ∠5 ≅ ∠6	1. Given
2. ∠7 and ∠5 are adj ∠s; ∠8 and ∠6 are adj ∠s	2. Def of adj ∠s
3. ∠7 supp ∠5; ∠8 supp ∠6	3. If outer rays of adj ∠s form a st ∠, then ∠s are supp.
4. ∠7 ≅ ∠8	4. Supp of ≅ ∠s are ≅.

Closure

Ask students to summarize what they know about supplements and complements of congruent angles.

FOLLOW UP

Guided Practice

Classroom Exercises 1–3

Independent Practice

A Ex. 1–4, **B** Ex. 5–17, **C** Ex. 18–20

Basic: FR 1–5, WE 1–4, Computer Investigation

Average: FR 1–10, WE 1–17 odd, Computer Investigation

Above Average: FR 1–15, WE 1–20 odd, Computer Investigation

Additional Answers

Focus on Reading

1. Adjacent
2. Bisect
3. Collinear
4. Complementary
5. Perpendicular
6. Corollary
7. Definition
8. Angle
9. Midpoint
10. Postulate
11. Property
12. Right
13. Straight
14. Supplementary
15. Theorem

Classroom Exercises

1. ∠4; Supp ∠s of ≅ ∠s are ≅.
2. ∠4; Supp ∠s of the same ∠ are ≅.
3. ∠7; Supplements of the same ∠ are ≅.

Written Exercises

5.

Statement	Reason
1. m ∠1 = 60; m ∠4 = 60.	1. Given
2. m ∠1 + m ∠EBA = 180, m ∠4 + m ∠ECD = 180.	2. If the outer rays of 2 adj ∠s form a st ∠, then ∠s are supp.
3. m ∠EBA ≅ m ∠ECD	3. Supp of ≅ ∠s are ≅.

6.

Statement	Reason
1. m ∠ABF = 42, m ∠ECP = 42, $\overrightarrow{BA}$ ⊥ $\overline{BC}$, $\overline{CP}$ ⊥ $\overline{BC}$	1. Given
2. m ∠ABF + m ∠FBC = 90 and m ∠ECP + m ∠ECB = 90	2. If the outer rays of two adj ∠s are ⊥, the ∠s are comp.
3. m ∠FBC ≅ m ∠BCB	3. Comp of ≅ ∠s are ≅.

7.

Statement	Reason
1. $\overline{AB}$ ⊥ $\overline{AD}$, $\overline{BC}$ ⊥ $\overline{CD}$, ∠ACB ≅ DAC	1. Given
2. m ∠DAC + m ∠CAB = 90 and m ∠ACB + m ∠ACD = 90	2. If the outer rays of two adj ∠s are ⊥, the ∠s are comp.
3. m ∠DCA ≅ m ∠CAB	3. Comp of ≅ ∠s are ≅.

3. Given: $\overline{AB}$ ⊥ $\overline{AD}$, $\overline{BC}$ ⊥ $\overline{CD}$, and ∠DAC ≅ ∠ACB

m ∠DCA = m ∠CAB, comp of ≅ ∠s are ≅.

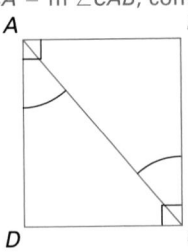

4. Given: ∠FBC ≅ ∠ECD
∠ABF ≅ ∠BCE. Supp of ≅ ∠s are ≅.

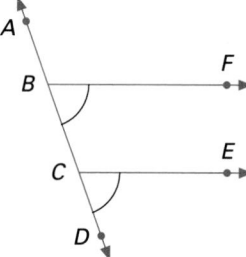

5–8. Write a proof of each conclusion in Exercises 1–4.

9. The figure at the right shows a roof-truss system. The beam from *A* to *B* is perpendicular to the crossbeam from *C* to *D*, and ∠ABE ≅ ∠ABF. Draw a conclusion about m ∠1. Prove it.

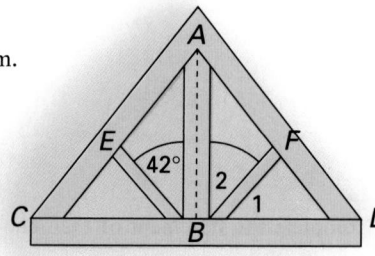

Write the proof. Give statements and reasons.

10. Given: ∠3 is a supplement of ∠2.
Prove: ∠2 ≅ ∠1

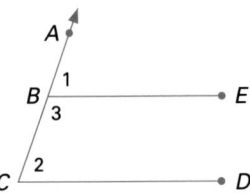

11. Given: $\overrightarrow{QP}$ ⊥ $\overline{QR}$, and ∠3 is a complement of ∠2.
Prove: ∠1 ≅ ∠3

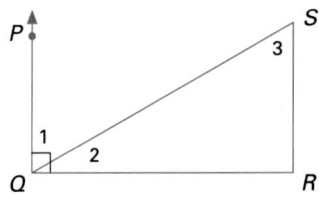

12. Given: $\overrightarrow{CA}$ bisects ∠ECF, and $\overrightarrow{CA}$ ⊥ $\overleftrightarrow{BD}$.
Prove: m ∠DCE = m ∠BCF

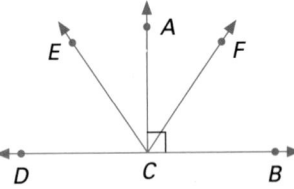

13. Given: m ∠2 = m ∠5, m ∠3 = m ∠5
Prove: m ∠4 = m ∠1

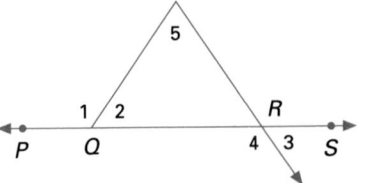

Enrichment

Explain that the methods of geometry can help dispel optical illusions. In this figure, for example, $\overline{EB}$ seems to be longer than $\overline{AE}$.

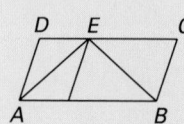

Have the students use a compass to determine that $\overline{EB}$ and $\overline{AE}$ have equal lengths.

Now, have each student draw one or two additional optical illusions to be used for a bulletin board display. This should include three-dimensional illusions. The students may find library resources useful in investigating this topic.

14. Prove the corollary to Theorem 2.1.
16. Prove the corollary to Theorem 2.2.

15. Prove Theorem 2.2.
17. Prove Theorem 2.3.

18. Given: $\overrightarrow{QP}$ bisects $\angle SQT$.
Prove: $\angle 2$ is a supplement of $\angle 3$.

19. Given: $\angle 1$ is a supplement of $\angle 4$.
Prove: $\angle 2$ is a supplement of $\angle 3$.

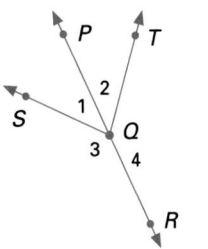

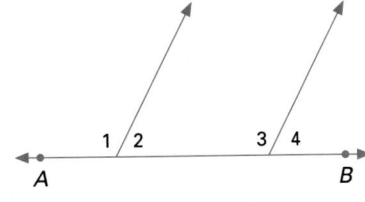

20. Given: $\overline{BA} \perp \overline{BC}$, $\overline{CD} \perp \overline{CB}$, $\angle 2 \cong \angle 3$
Prove: $\angle 1$ is a supplement of $\angle DCF$.

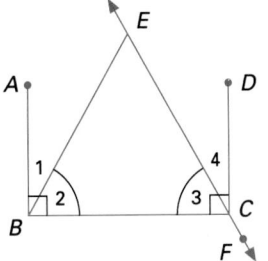

Mixed Review

Determine if the statement is true or false. *1.7, 2.5*

1. $2 \cdot 5 + 4 = 15$ *or* the measure of the complement of $\angle A$ is $180 - m \angle A$.   **False**

2. If $\overrightarrow{AB}$ is a ray, then a segment has one endpoint.   **True**

3. For what truth values of p and q is $p \wedge q$ true?   *p and q both true*

4. For what truth values of p and q is $p \vee q$ false?   *p and q both false*

5. For what truth values of p and q is $p \rightarrow q$ false?   *p, true; q, false*

6. Can you conclude from the figure that $GK + KL = GL$? Why or why not?   *2.1* Yes. Betweenness may be assumed.

7. Can you conclude that $GK = PT$? Why or why not?   *2.1* No. Congruence may not be assumed.

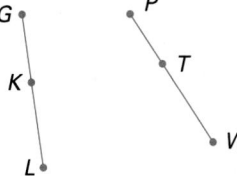

2.7 Proving Theorems About Angles   **75**

8.

Statement	Reason
1. $\angle FBC \cong \angle ECD$	1. Given
2. m $\angle ABF$ + m $\angle FBC$ = 180 and m $\angle ECD$ + m $\angle BCE$ = 180	2. If the outer rays of 2 adj $\angle$s form a st $\angle$, then the angles are supp.
3. m $\angle ABF$ = m $\angle BCE$	3. Supp of $\cong \angle$s are $\cong$.

9. Conclusion: m $\angle 1 = 48$; Proof:

Statement	Reason
1. $\overline{AB} \perp \overline{CD}$; $\angle ABE \cong \angle ABF$	1. Given
2. m $\angle 1$ + m $\angle 2 = 90$	2. If outer rays of 2 adj acute $\angle$s are $\perp$, the sum of the meas of the $\angle$s is 90.
3. m $\angle 2 = 42$	3. Def of congr $\angle$s
4. m$\angle 1$ = 48	4. Eq may be subt.

10.

Statement	Reason
1. $\angle 3$ and $\angle 2$ are supp	1. Given
2. $\angle 1$ and $\angle 3$ are supp	2. If the outer rays of 2 adj $\angle$s form a st $\angle$, then the $\angle$s are supp.
3. $\therefore \angle 2 \cong \angle 1$	3. Supp $\angle$s of the same $\angle$ are $\cong$.

11.

Statement	Reason
1. $\overrightarrow{QP} \perp \overline{QR}$, $\angle 3$ and $\angle 2$ are comp	1. Given
2. $\angle 1$ and $\angle 2$ are comp	2. If the outer rays of 2 adj $\angle$s are $\perp$, then the $\angle$s are comp.
3. $\therefore \angle 1 \cong \angle 3$	3. Comp $\angle$s of the same $\angle$ are $\cong$.

12.

Statement	Reason
1. $\overrightarrow{CA}$ bis $\angle ECF$; $\overrightarrow{CA} \perp \overleftrightarrow{BD}$	1. Given
2. $\angle ECA \cong \angle ACF$	2. Def of $\angle$ bis
3. $\angle DCE$ and $\angle ECA$ are comp; $\angle ACF$ and $\angle BCF$ are comp	3. If the outer rays of 2 adj acute $\angle$s are $\perp$, then the $\angle$s are comp.
4. $\therefore$ m $\angle DCE$ = m $\angle BCF$	4. Comp $\angle$s of $\cong \angle$s are $\cong$.

13.

Statement	Reason
1. m $\angle 2$ = m $\angle 5$, m $\angle 3$ = m $\angle 5$	1. Given
2. m $\angle 2$ = m $\angle 3$	2. Sub
3. $\angle 1$ and $\angle 2$ are supp; $\angle 4$ and $\angle 3$ are supp	3. If the outer rays of 2 adj $\angle$s form st $\angle$, then the $\angle$s are supp.
4. $\therefore \angle 1 \cong \angle 4$	4. Supp $\angle$s of $\cong \angle$s are $\cong$.

14. Using the diagram at the top of page 72:

Statement	Reason
1. $\angle 1$ and $\angle 2$ are supp; $\angle 1$ and $\angle 3$ are supp	1. Given
2. m $\angle 1$ + m $\angle 2$ = 180, m $\angle 1$ + m $\angle 3$ = 180	2. Def of supp $\angle$s
3. m $\angle 1$ + m $\angle 2$ = m $\angle 1$ + m $\angle 3$	3. Sub
4. $\therefore$ m $\angle 2$ = m $\angle 3$	4. Subt Prop of Eq

See page 76 for the answers to Ex. 15–18, and page 80 for Ex. 19–20.

15. Using the diagram in the middle of page 72:

Statement	Reason
1. ∠A and ∠B are comp; ∠C and ∠D are comp; m ∠B = m ∠C	1. Given
2. m ∠A + m ∠B = 90, m ∠C + m ∠D = 90	2. Def of comp ∠s
3. m ∠A + m ∠B = m ∠C + m ∠D	3. Trans Prop of Eq
4. m ∠A + m ∠B = m ∠B + m ∠D	4. Sub
5. ∴ m ∠A = m ∠D	5. Subt Prop of Eq

16. Using the diagram in the middle of page 72:

Statement	Reason
1. ∠A and ∠B are comp, ∠D and ∠B are comp	1. Given
2. m ∠A + m ∠B = 90, m ∠D + m ∠B = 90	2. Def of comp ∠s
3. m ∠A + m ∠B = m ∠D + m ∠B	3. Trans Prop of Eq
4. ∴ m ∠A = m ∠D	4. Subt Prop of Eq

17.

Statement	Reason
1. ∠1 and ∠2 are rt ∠s	1. Given
2. m ∠1 = 90, m ∠2 = 90	2. Def of rt ∠s
3. m ∠1 = m ∠2	3. Trans Prop of Eq

18.

Statement	Reason
1. QP bis ∠SQT	1. Given
2. m ∠2 = m ∠1	2. Def of ∠ bis
3. m ∠1 + m ∠3 = 180	3. If the outer rays of 2 adj ∠s form a st ∠, then the sum of their meas is 180.
4. m ∠2 + m ∠3 = 180	4. Sub
5. ∴ ∠2 and ∠3 are supp	5. Def of supp ∠s

Computer Investigation

Vertical Angles

Use a computer software program that constructs line segments, labels and moves points, extends line segments, and measures line segments and angles.

The figure to the right shows two segments, $\overline{AB}$ and $\overline{DC}$, intersecting at point E. The activities are designed to help you discover a relationship between certain pairs of angles formed by two intersecting lines. This relationship will be stated and proved in the next lesson.

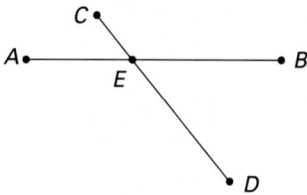

Activity 1

Draw a segment, $\overline{AB}$, of some length, say 8 units. Draw a point C not on $\overline{AB}$. Draw a second point, D, on the opposite side of $\overline{AB}$ from point C. Draw $\overline{CD}$. Label the point of intersection of the two segments E.

1. Find m ∠DEA, m ∠BEC, m ∠DEB, and m ∠AEC.
2. Clear the screen and draw a new pair of intersecting segments. Repeat the directions of Exercise 1 for this new drawing.
3. Repeat the process above for a third pair of intersecting segments. (Keep this pair of intersecting segments on the screen.)

Activity 2

Label a point F somewhere in the interior of ∠AED. Draw $\overline{EF}$.

4. Find m ∠FED and m ∠FEA.
5. Does either angle measure of Exercise 4 equal m ∠CEB? Can you guess why these angle measures are not equal when those of the previous activity were equal?

Summary

Try to state the generalization of the discovery activities of this page. Use the following questions to help make the generalization.

1. What are ∠AEC and ∠CEB called?   adjacent
2. Can you think of a description of ∠CEB and ∠AED to distinguish them from the type above?   vertical
3. Complete the following generalization:
   If two lines intersect, then ___vertical___ are congruent.

2.8 Vertical Angles

Objective
To identify vertical angles
To apply the Vertical Angles Theorem

As shown at the right, two intersecting lines form two pairs of
nonadjacent angles. ∠1 and ∠2
are nonadjacent. ∠3 and ∠4 are
nonadjacent.

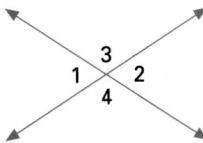

Definition

> **Vertical angles** are two nonadjacent angles formed by intersecting
> lines.

EXAMPLE 1 State whether the given angles are vertical. Give a reason.

∠XOW and ∠YOZ

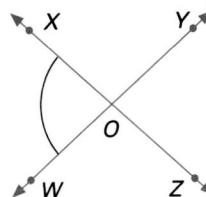

∠AOB and ∠DOC

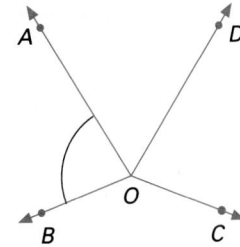

Solutions

∠XOW and ∠YOZ are vertical
angles. They are nonadjacent angles formed by two intersecting
lines.

∠AOB and ∠DOC are not vertical angles. They are not formed by
intersecting lines.

EXAMPLE 2
Name a second angle to form a
pair of vertical angles: ∠DOC
and _____.

Plan
Copy the diagram. Shade the
lines that intersect to form
∠DOC.

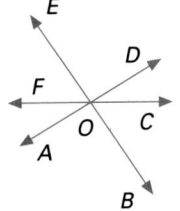

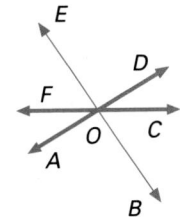

Solution
Thus, ∠DOC and ∠AOF are vertical angles.

2.8 Vertical Angles **77**

Teaching Resources

Quick Quizzes 16
**Reteaching and Practice
 Worksheets,** pp. 31, 32
Transparency 3

▰▰▰ GETTING STARTED

Prerequisite Quiz

1. Solve $3x + 40 = x + 70$.   15
2. Find the supplement of an angle with
   measure 135.   45
3. ∠1 and ∠2 are adjacent angles whose
   outer rays are perpendicular. If m ∠1 =
   50, find m ∠2.   40
4. If ∠A and ∠B are both supplements of
   ∠C, what relationship exists between ∠A
   and ∠B?   ∠A ≅ ∠B

Motivator

(Use the first diagram on p. 77.)
Given: m ∠1 = 40. Find m ∠2.   40
Given: m ∠1 = 50. Find m ∠2 and m ∠3.
50, 130

▰▰▰ TEACHING SUGGESTIONS

Lesson Note

When vertical angles are shown in a figure,
you can state in a proof that they are
vertical by definition. It is not necessary to
state that the lines are straight or that they
intersect.

Lead students to discover the Vertical
Angles Theorem. Have them determine
angle measures in figures such as the one
in Example 3.

Additional Example 1

State whether the given angles are
vertical. Give a reason for the answer.

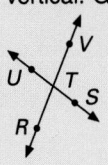

∠UTV and ∠RTS
Yes; they are non-
adjacent ∠s formed
by two intersecting lines.

Additional Example 2

Name a second angle to form a pair of
vertical angles.
∠QOR   ∠UOT

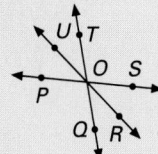

Math Connections

Design engineering: Many designs for home furnishings and decorative items are based on the Vertical Angle Theorem. Some examples are the direction pointer on a weather vane, ironing boards, and garden shears.

Critical Thinking Questions

Synthesis: Ask students to compare the angle relationships of complementary and supplementary angles with those of vertical angles. Comp and supp ∠s are not always ≅, but vertical ∠s are always ≅. Vertical ∠s are nonadj ∠s formed by intersecting lines supp and comp ∠s may be adjacent, or non-adjacent. What is another name for supplementary angles that are also vertical angles? Right angles

Common Error Analysis

Error: In complex drawings, students may identify pairs of angles as vertical, even when the angles are not formed by intersecting lines.

In the first figure below, ∠1 and ∠2 are not vertical. This can be seen by shading the sides of the angles thought to be vertical, as shown in the second figure. For vertical angles, the shadings form an "X." For complex drawings, have students shade sides of angles they think are vertical and look for the "X."

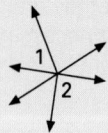 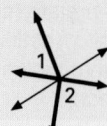

You may have noticed that vertical angles always appear to have the same measure. The following theorem proves this is true.

Theorem 2.4

> **Vertical Angles Theorem:** Vertical angles are congruent.

Given: A pair of vertical angles (∠1 and ∠2)

Prove: The angles are congruent (∠1 ≅ ∠2).

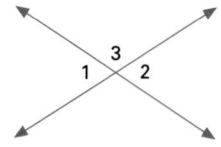

Proof

Statement	Reason
1. ∠1 and ∠2 are vert ∠s.	1. Given
2. ∠1 is a supp of ∠3; ∠2 is a supp of ∠3.	2. If the outer rays of two adj ∠s form a st ∠, then the ∠s are supp.
3. ∴ ∠1 ≅ ∠2	3. Supp of same ∠ are ≅.

Notice that to prove the theorem, you only need to show that ∠1 ≅ ∠2. Every statement in the proof about these angles is also true for any other pair of vertical angles.

EXAMPLE 3   Find the measure of each of the numbered angles.

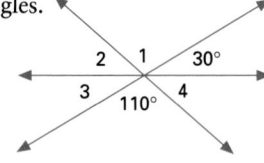

Solutions   Look for vertical angles. m ∠1 = 110, m ∠3 = 30 by the Vertical Angles Theorem. m ∠2 = 180 − (110 + 30) = 40. m ∠2 and m ∠4 are vertical angles. m ∠4 = m ∠2 = 40.

EXAMPLE 4   Find m ∠APB.

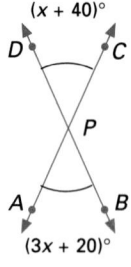

Solution   m ∠APB = m ∠DPC by Vertical Angles Theorem

$$3x + 20 = x + 40$$
$$2x = 20$$
$$x = 10$$

Thus, m ∠APB = 3x + 20
= 3 · 10 + 20 = 50.

78   Chapter 2 Proof

Additional Example 3

Find the measure of each of the numbered angles.

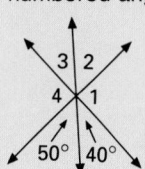

m ∠1 = 90,
m ∠2 = 50,
m ∠3 = 40,
m ∠4 = 90

Additional Example 4

Find m ∠APB.   115

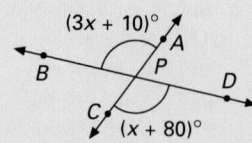

EXAMPLE 5   Given: $\overrightarrow{PQ} \perp \overrightarrow{PR}$

Prove: m $\angle 2$ + m $\angle 3$ = 90

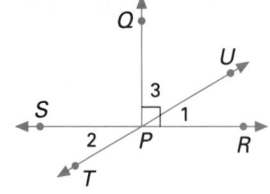

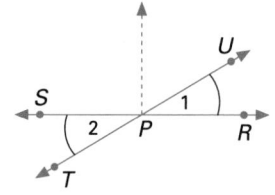

m $\angle 1$ = m $\angle 2$

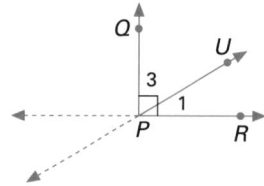

m $\angle 1$ + m $\angle 3$ = 90

Proof

	Statement	Reason
1.	$\overrightarrow{PQ} \perp \overrightarrow{PR}$	1. Given
2.	$\angle 1$ and $\angle 2$ are vert $\angle$s.	2. Def of vert $\angle$s
3.	m $\angle 1$ = m $\angle 2$ ($\angle 1 \cong \angle 2$)	3. Vert $\angle$s are $\cong$
4.	m $\angle 1$ + m $\angle 3$ = 90	4. If outer rays of two acute adj $\angle$s are $\perp$, then the sum of the measures is 90.
5.	m $\angle 2$ + m $\angle 3$ = 90	5. Sub

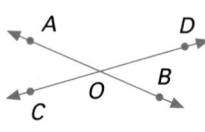

Corollary   If two lines are perpendicular, then four right angles are formed.

Classroom Exercises

State whether the given pair of angles is vertical. Give a reason for the answer.

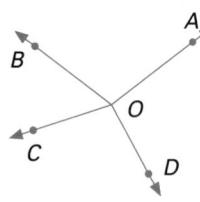

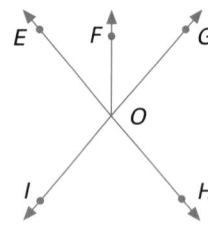

1. $\angle AOD$ and $\angle BOC$

2. $\angle AOC$ and $\angle COB$

3. $\angle AOB$ and $\angle DOC$

4. $\angle BOC$ and $\angle BOA$

5. $\angle GOH$ and $\angle HOI$

6. $\angle IOH$ and $\angle EOG$

2.8 Vertical Angles   **79**

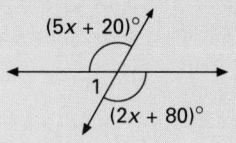
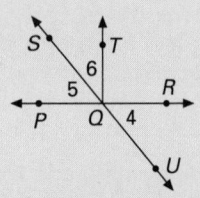

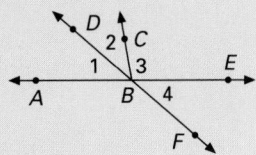
79

Classroom Exercises

1. Yes, they are non-adj ∠s formed by intersecting lines.
2. No, they are adj ∠s.
3. No, they are not formed by intersecting lines.
4. No, they are adj and not formed by intersecting lines.
5. No, they are adj ∠s.
6. Yes, they are non-adj ∠s formed by intersecting lines.

Additional Answers, page 75

19.
Statement	Reason
1. ∠1 and ∠4 are supp	1. Given
2. m ∠1 + m ∠2 = 180, m ∠3 + m ∠4 = 180	2. If the outer rays of 2 adj ∠s form a st ∠, then the sum of their meas = 180.
3. m ∠1 + m ∠4 = 180	3. Def of supp ∠s
4. m ∠1 + m ∠2 = m ∠1 + m ∠4	4. Trans Prop of Eq
5. m ∠2 = m ∠4	5. Subt Prop of Eq
6. m ∠3 + m ∠2 = 180	6. Sub
7. ∴ ∠2 and ∠3 are supp	7. Def of supp ∠s

20.
Statement	Reason
1. $\overline{BA} \perp \overline{BC}$, $\overline{CD} \perp \overline{CB}$, ∠2 ≅ ∠3	1. Given
2. ∠2 and ∠1 are comp; ∠3 and ∠4 are comp.	2. If the outer rays of 2 adj acute ∠s are ⊥, then the ∠s are comp.
3. m ∠1 = m ∠4	3. Complements of ≅ ∠s are ≅.
4. m ∠4 + m ∠DCF = 180.	4. If the outer rays of 2 adj ∠s form a st ∠, then the sum of their meas = 180.
5. m ∠1 + m ∠DCF = 180.	5. Sub
6. ∠1 and ∠DCF are supp ∠s.	6. Def of supp ∠s.

Find the measure of the angle.

7. m ∠1   50
8. m ∠2   130

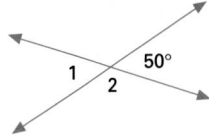

Written Exercises

Find the measure of the angle.

1. m ∠4   65
2. m ∠2   55
3. m ∠1   60
4. m ∠3   60

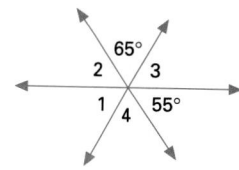

Find the indicated measure.

5. m ∠AOB   70

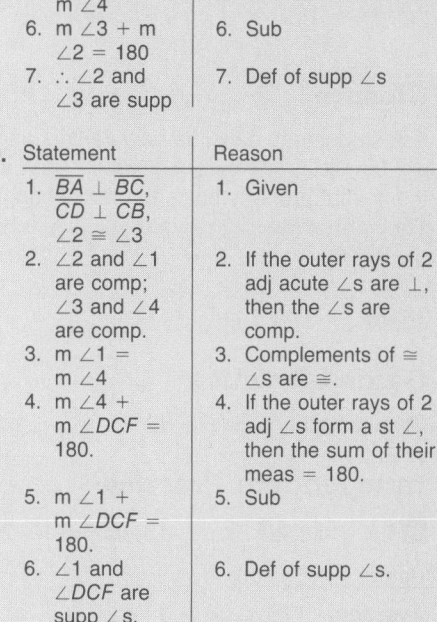

6. m ∠RST   34

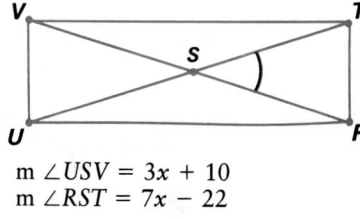

m ∠USV = 3x + 10
m ∠RST = 7x − 22

Use the figure at the right for Exercises 7–10.

7. Given: m ∠PQT = 2x − 20, m ∠SQR = x + 10
Find m ∠TQR.   140

8. Given: m ∠PQT = 3y − 20, m ∠SQR = y + 20
Find m ∠SQR.   40

9. Given: m ∠SQR = 5x − 10, m ∠TQP = x + 30
Find: m ∠PQS.   140

10. Given: m ∠PQT = 8y + 8, m ∠RQS = 10y − 12
Find: m ∠PQS.   92

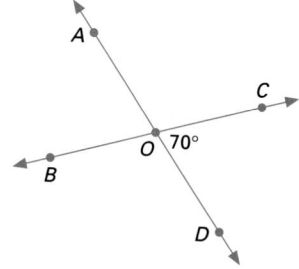

80   Chapter 2 Proof

Enrichment

Have students research home design and construction techniques to determine how vertical angles are used in these trades. Have the class make a list of the ways in which vertical angles are used in Lesson 2.9.

In the figure at the right, $\overleftrightarrow{US} \perp \overleftrightarrow{PR}$ and m $\angle VQU = 23$.
Find the indicated measure.

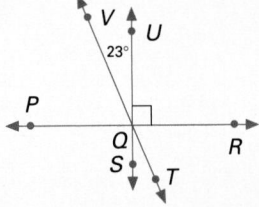

11. m $\angle SQT$ 23
12. m $\angle TQR$ 67
13. m $\angle PQT$ 113

14. Given: m $\angle 2 =$ m $\angle 3$
Prove: m $\angle 2 =$ m $\angle 1$

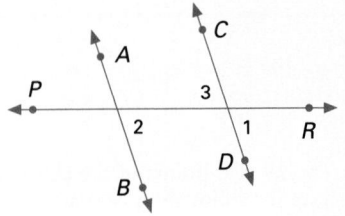

15. Given: The figure with $\overleftrightarrow{AC}$ intersecting $\overleftrightarrow{DE}$ at B
Prove: m $\angle 5 +$ m $\angle 4 =$ m $\angle DBF$

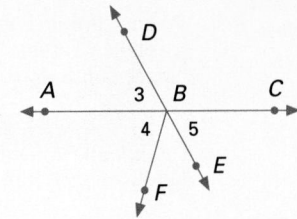

Given: $\overrightarrow{FB}$ bisects $\angle AFC$. Find each of the following values.

16. x 15
17. m $\angle DFE$ 30
18. m $\angle DFA$ 150

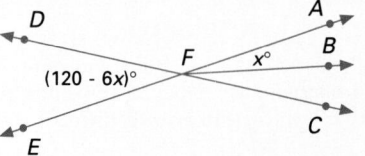

19. Write an explanation of how to identify vertical angles.
20. Prove the corollary to the Vertical Angles Theorem.

Mixed Review 1.1

1. What is the intersection of planes $\mathcal{M}$ and $\mathcal{P}$? $\overleftrightarrow{AB}$
2. Points A, B, G, and ___ are coplanar.   C or F
3. Points C, D, and ___ are collinear.   E
4. Which labeled planes contain point C? $\mathcal{M}, \mathcal{P}, \mathcal{Q}$
5. What is the intersection of planes $\mathcal{M}$ and $\mathcal{Q}$? $\overleftrightarrow{FG}$
6. Is B between F and E?   **1.2** No, not collinear
7. Name a pair of supplementary angles.   **1.6** $\angle DCF$ and $\angle FCE$

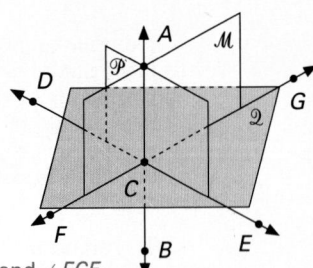

2.8 Vertical Angles   **81**

Written Exercises

14.

Statement	Reason
1. m $\angle 2 =$ m $\angle 3$	1. Given
2. m $\angle 3 =$ m $\angle 1$	2. Vert $\angle$s are $\cong$.
3. m $\angle 2 =$ m $\angle 1$	3. Trans Prop of Eq

15.

Statement	Reason
1. m $\angle 3 =$ m $\angle 5$	1. Vert $\angle$s are $\cong$.
2. m $\angle 3 +$ m $\angle 4 =$ m $\angle DBF$	2. Angle Add Post
3. m $\angle 5 +$ m $\angle 4 =$ m $\angle DBF$	3. Sub

19. Vertical $\angle$s are two non-adjacent $\angle$s with a common vertex. The sides of both angles must be contained in 2 lines, or segments, rays, which intersect at the common vertex of the angles.

20.

Statement	Reason
1. $\overleftrightarrow{AB}$ intersects $\overleftrightarrow{CD}$ at O, $\overleftrightarrow{AB} \perp \overleftrightarrow{CD}$	1. Given
2. $\angle AOC$ and $\angle AOD$ are rt $\angle$s	2. Def of $\perp$
3. m $\angle AOC =$ 90, m $\angle AOD = 90$	3. Def of rt $\angle$
4. $\angle AOC \cong \angle BOD$ (m $\angle AOC =$ m $\angle BOD$); $\angle AOD \cong \angle COB$ (m $\angle AOD =$ m $\angle COB$)	4. Vert $\angle$s are $\cong$.
5. m $\angle BOD =$ 90, m $\angle COB = 90$	5. Sub
6. $\angle BOD$ and $\angle COB$ are rt $\angle$s	6. Def of rt $\angle$

◼︎GETTING STARTED

Prerequisite Quiz

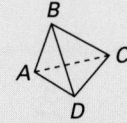

Refer to the four noncoplanar points, *A*, *B*, *C*, and *D*, in the figure of a pyramid.

1. Name all sets of three noncollinear points.
   {*A, B, C*}; {*A, B, D*}; {*A, C, D*}; {*B, C, D*}
   (NOTE: Students' answers need not use brackets.)
2. Name all sets, if any, of three collinear points.   None
3. Name all lines that contain edges of the pyramid. $\overleftrightarrow{AB}$, $\overleftrightarrow{AC}$, $\overleftrightarrow{AD}$, $\overleftrightarrow{BC}$, $\overleftrightarrow{BD}$, $\overleftrightarrow{CD}$
4. Name a line that does not intersect $\overleftrightarrow{BD}$. $\overleftrightarrow{AC}$

Motivator

On the chalkboard, draw two distinct points, *A* and *B*. Draw a straight line containing both *A* and *B*. Then, ask your students if they can draw another different line that contains the same two points.   No.   Ask them what conclusion this suggests about the number of lines that can be made to contain two different points.   One line.

2.9 Postulates and Theorems: Points, Lines, Planes

Objective   To apply postulates and theorems relating points, lines, and planes

In Chapter 1, *point*, *line*, and *plane* were accepted as undefined terms. *Space* was defined as the set of all points. Now that the idea of proof has been introduced, theorems that relate points, lines, and planes can be developed. First, it is necessary to state some postulates.

Postulate 8

A line contains at least two points.
A plane contains at least three noncollinear points.
Space contains at least four noncoplanar points.

If you walk in a straight line between any two points in the classroom, there is exactly one path. This suggests the following postulate.

Postulate 9

For any two points, there is exactly one line containing them.

The phrases "exactly one" and "one and only one" are used interchangeably in mathematics. Also, in this book phrases such as "two points" or "two lines" mean two distinct, or different, points or lines.

EXAMPLE   Can two lines intersect in more than one point? Give an informal argument to support the answer.

Solution   No, they cannot. The following *indirect* argument can be used. Suppose lines *l* and *m* had two points of intersection, *P* and *Q*. Then there would be two lines containing these two points. This contradicts Postulate 9, since only one line can contain any two points.

The argument in the Example is an **indirect proof**. This type of proof will be studied in greater detail in the next chapter.

Additional Example

Explain why a plane contains at least three lines.

By Postulate 8, a plane contains at least three noncollinear points. Call them *A*, *B*, and *C*. Then, by Postulate 9, each pair of points is contained in a line. So the plane contains the three lines, $\overleftrightarrow{AB}$, $\overleftrightarrow{BC}$, and $\overleftrightarrow{AC}$.

| **Theorem 2.5** | Two lines intersect at exactly one point. |

A ruler (line) placed on a desk top (plane) illustrates the basis for another postulate about lines and planes. If any two points of the ruler touch the plane of the desk top, then the entire ruler (without bending) must lie on the flat surface.

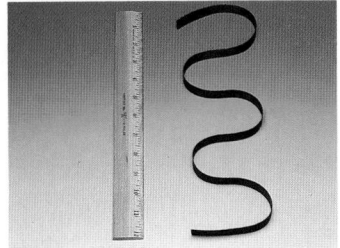

| **Postulate 10** | If two points of a line are in a given plane, then the line itself is in the plane. |

In the figure, $\overrightarrow{PQ}$ intersects plane $\mathcal{M}$ in point O. But the line is not contained in the plane. It has one and only one point in common with the plane. This result is stated below as a theorem. Proof is based on the kind of indirect argument used in Example 1.

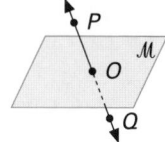

| **Theorem 2.6** | If a line intersects a plane but is not contained in the plane, then the intersection is exactly one point. |

Proof Suppose line l intersects plane $\mathcal{P}$ in two or more points. Then, according to Postulate 10, the line is contained in the plane. This contradicts the information given that the line is *not* contained in the plane. Therefore, the line intersects the plane in exactly one point.

| **Postulate 11** | If two planes intersect, then they intersect in exactly one line. |

You may have noticed that some four-legged chairs wobble a bit even when placed on a level floor. On the same floor, a three-legged stool does not wobble. This is because the ends of the legs of a three-legged stool always lie in exactly one plane, whereas the ends of the legs of a four-legged stool need not.

2.9 Postulates and Theorems: Points, Lines, Planes **83**

TEACHING SUGGESTIONS

Lesson Note

In this lesson, geometric relations between points, lines, and planes are introduced formally through postulates and theorems. These will be useful later in indirect proofs. Note that any statement about a "line" refers to a "straight line."

Math Connections

Life Skills: People use Postulate 9 in daily life. For instance, a seamstress marks points with straight pins that will determine a straight hem line. A construction crew marks two points in order to paint a straight line along the road that cars are not to cross. Ask students to name some other uses of Postulate 9 in daily life.

Critical Thinking Questions

Application: Discuss with the class why a door fastened to a frame by two hinges can move (to form many planes) but when the door is fastened at the latch, only one plane is possible.   3 points determine a plane

Enrichment

Discuss the difficulty of drawing three-dimensional figures on a two dimensional sheet of paper. Explain that success in doing this depends on the use of *perspective*. For example, this figure of a cube gives the appearance that all the edges are congruent. By checking with a compass, however, it can be seen that this is not so. Have the students use library resources to look up the word *perspective*. Then have them make a bulletin-board display on the use of perspective in geometry, using drawings of their own as well as illustrations taken from newspapers and magazines.

Checkpoint

Which postulate or theorem guarantees each of the following statements?

1. Given a point P not on $\overleftrightarrow{AB}$, exactly one plane contains point P and line $\overleftrightarrow{AB}$.
   Thm 2.7
2. If a plane contains points T and Q, then every point on $\overleftrightarrow{TQ}$ is in the plane.
   Post 10

Indicate whether the statement is *always true, sometimes true, or never true.*

3. Vertical angles are coplanar.   Always
4. A line intersects a plane in exactly two points.   Never
5. Two planes intersect in a line.   Always
6. Two nonintersecting lines lie in the same plane.   Sometimes

Closure

Ask your students to list the relationships that have been discovered in this lesson about points, lines, and planes.

Ask them what they now know about the number of lines containing two points? the intersection of two planes? the intersection of a line and a plane? the number of planes containing two points? the number of planes containing three noncollinear points?

Postulate 12   Three noncollinear points are contained in exactly one plane.

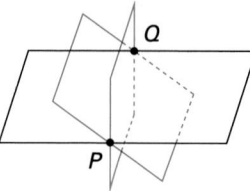

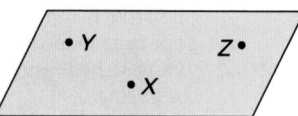

Two points: infinite number of planes

Three noncollinear points: exactly one plane

Postulates 8, 10, and 12 can be used to prove the following theorem.

Theorem 2.7   A line and a point not on the line are contained in exactly one plane.

Given: Line l and point P not on l

Prove: Exactly one plane contains P and l.

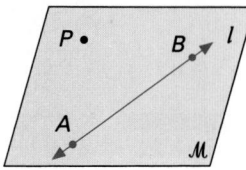

Plan   A line contains at least two points, say, A and B. Exactly one plane contains the three noncollinear points P, A, and B.

Proof

Statement	Reason
1. Line l and point P not on l	1. Given
2. l contains at least two points. Name these points A and B.	2. A line contains at least two points.
3. P, A, and B are noncollinear.	3. Def of noncoll points
4. Exactly one plane contains P, A, and B (plane $\mathcal{M}$).	4. Three noncoll points are contained in exactly one plane.
5. ∴ Exactly one plane contains P and l.	5. If two points of a line lie in a plane, then the line lies in the plane.

The following theorem is a direct consequence of Theorem 2.7.

Theorem 2.8   Two intersecting lines are contained in exactly one plane.

Classroom Exercises

Which postulate or theorem guarantees the truth of the statement?

1. $\overleftrightarrow{AB}$ and $\overleftrightarrow{CD}$, two different intersecting lines, have only one point in common. Theorem 2.5
2. Given two points R and S, there is one and only one line containing them. Postulate 9
3. If r and s are two different intersecting lines, then one and only one plane can contain both of them. Theorem 2.8

Written Exercises

1. Can a line that does not lie in a plane intersect the plane in more than one point? Give a reason for your answer.
2. Can two planes intersect in exactly one point? Give a reason for your answer.
3. How many different planes can contain the same line? Infinitely many
4. How many different planes can contain a given obtuse angle? Exactly one
5. Given any three points in space, is it possible that there is no plane that contains all three of the points? Give a reason for your answer.
6. Given any four points in space, is it possible that there is no plane that contains all four points? Give an example to support your answer, using points and planes of the classroom.

Indicate whether the statement is always true, sometimes true, or never true. (Exercises 7–9)

7. Three coplanar points are collinear. Sometimes
8. Three collinear points are coplanar. Always
9. Two lines are contained in exactly one plane. Sometimes
10. Draw a figure illustrating that it is possible for four planes to have only one point in common.
11. Prove Theorem 2.8.

Mixed Review

For each of the Exercises 1–3, draw a conclusion. Then prove it. *2.1, 2.2*

1. Given: $\angle AOB$
2. Given: $\overrightarrow{OC}$ bisects $\angle AOB$.
3. Given: $\overrightarrow{OB} \perp \overrightarrow{OA}$
4. Given: $m \angle 1 = m \angle 3$
   Prove: $m \angle 2 + m \angle 3 = m \angle AOB$

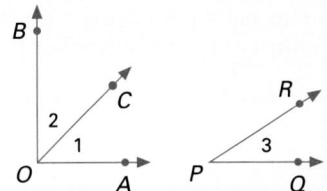

2.9 Postulates and Theorems: Points, Lines, Planes **85**

■▶ FOLLOW UP

Guided Practice
Classroom Exercises 1–3

Independent Practice
A Ex. 1–6, **B** Ex. 7–9, **C** Ex. 10–11

Basic: WE 1–6
Average: WE 1–9
Above Average: WE 1–11

Additional Answers
Written Exercises

1. No, Thm 2.6
2. No, Post 11
5. No, Post 12
6. Yes, three pts on the floor are contained by the plane containing the floor, but a pt on the ceiling is not contained by that plane.
10. One way of illustrating the intersection can be described as follows: Three planes may intersect in a line, like pages of a book that intersect at the binding. A fourth plane may intersect that line at a point. (Other configurations are possible.)

11.

Statement	Reason
1. Lines l and m intersect	1. Given
2. l and m intersect at a pt, call it A	2. Thm 2.5
3. l contains a pt B other than A, m has a pt C other than A	3. Post 8
4. A, B, and C are noncoll	4. Def of noncoll pts
5. Exactly one plane $\mathcal{N}$ contains A, B and C	5. Post 12
6. $\mathcal{N}$ contains l, $\mathcal{N}$ contains m	6. Post 10
7. ∴ Exactly one plane contains l and m	7. Steps 6 & 7

See page 86 for the answers to Mixed Review Ex. 1–4.

Chapter 2 Review

Key Terms

conditional (p. 63)
corollary (p. 71)
deductive reasoning (p. 67)
inductive reasoning (p. 68)
Reflexive Property (p. 49)

Substitution Property (p. 49)
Symmetric Property (p. 49)
Transitive Property (p. 49)
vertical angles (p. 77)

Key Ideas and Review Exercises

2.1, 2.2   To draw a conclusion from given data and to write a two-column proof

Information is *given* in two ways:
(1) in writing or   (2) as implied in a geometric figure.

Draw a conclusion and write a proof.

1. Given: The figure at the right
2. Given: T is the midpoint of $\overline{PW}$.

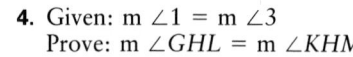

2.2, 2.3   To use algebraic properties as reasons in proofs
2.4   • If $a = b$, then $a + c = b + c$, $a - c = b - c$, $ac = bc$, and $\dfrac{a}{c} = \dfrac{b}{c}$ $(c \neq 0)$

• Equations may be added to or subtracted from each other.
• An equation may be substituted for its equal.

To use any of the three equivalent statements of the Segment Addition Postulate or the Angle Addition Postulate in a proof

3. Given: $QR = BV$, $PR = AV$
Prove: $PQ = AB$

4. Given: m $\angle 1$ = m $\angle 3$
Prove: m $\angle GHL$ = m $\angle KHM$

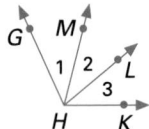

2.5   A conditional, "If p, then q," can be written as $p \rightarrow q$.
A conditional is false only when p is true and q is false.

Rewrite the given and the conclusion as a conditional. Is the conditional true?

5. Given: $\angle A$ and $\angle B$ are supplementary.
Conclusion: $\angle A$ and $\angle B$ are congruent.

2.6   Use deductive reasoning to draw conclusions.
                Use inductive reasoning to make generalizations.

6. $p \rightarrow q$ and $q \rightarrow r$ **7.** 101, 88, 75, 62, <u> 49 </u>

2.7   Complements of the same angle (or congruent angles) are congruent.
                Supplements of the same angle (or congruent angles) are congruent.

8.

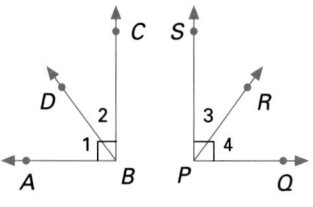

Given: $\overrightarrow{BA} \perp \overrightarrow{BC}$, $\overrightarrow{PS} \perp \overrightarrow{PQ}$,
        m $\angle 1$ = m $\angle 4$
Prove: m $\angle 2$ = m $\angle 3$

9.

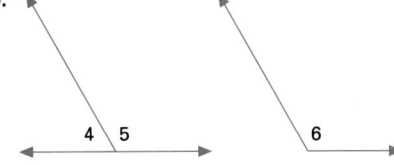

Given: $\angle 6$ is supplementary to $\angle 4$.
Prove: $\angle 5 \cong \angle 6$

2.8   Vertical angles are angles formed by two intersecting lines that are neither
                adjacent nor straight angles. Vertical angles are congruent.

10.

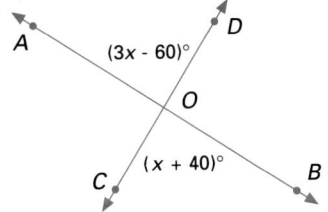

Find m $\angle BOD$.   90

11.

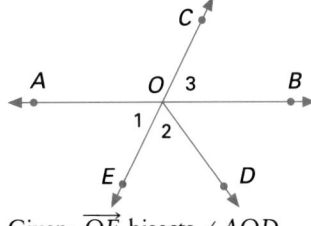

Given: $\overrightarrow{OE}$ bisects $\angle AOD$.
Prove: m $\angle 2$ = m $\angle 3$

2.9   The postulates and theorems that follow relate points, lines, and planes.
- For any two points, there is exactly one line containing them.
- Two lines intersect in at most one point.
- If a line intersects a plane but is not contained in the plane, then the intersection is exactly one point.
- If two planes intersect, then they intersect in exactly one line.
- Three noncollinear points are contained in exactly one plane.
- Two intersecting lines are contained in exactly one plane.
- A line and a point not on the line are contained in exactly one plane.

Indicate whether the statement is always true, sometimes true, or never true.

12. Two different lines can have several points in common.   Never

13. Three points are contained in exactly one plane.   Sometimes

14. Two planes have exactly two points in common.   Never

Chapter 2 Review   **87**

4.

Statement	Reason
1. m $\angle 1$ = m $\angle 3$	1. Given
2. m $\angle 1$ + m $\angle 2$ = m $\angle GHL$, m $\angle 3$ + m $\angle 2$ = m $\angle KHM$	2. Angle Add Post
3. m $\angle 1$ + m $\angle 2$ = m $\angle 3$ + m $\angle 2$	3. Add Prop of Eq
4. $\therefore$ m $\angle GHL$ = m $\angle KHM$	4. Sub

5. If $\angle A$ and $\angle B$ are supp, then $\angle A$ and $\angle B$ are $\cong$. (False)

6. $p \rightarrow r$

8.

Statement	Reason
1. $\overrightarrow{BA} \perp \overrightarrow{BC}$, $\overrightarrow{PS} \perp \overrightarrow{PQ}$, m $\angle 1$ = m $\angle 4$	1. Given
2. $\angle 1$ and $\angle 2$ are comp, $\angle 3$ and $\angle 4$ are comp.	2. If the outer rays of 2 adj acute $\angle$s are $\perp$, then the $\angle$s are comp.
3. $\therefore$ m $\angle 2$ = m $\angle 3$ ($\angle 2 \cong \angle 3$).	3. Comp $\angle$s of $\cong \angle$s are $\cong$.

9.

Statement	Reason
1. $\angle 6$ and $\angle 4$ are supp.	1. Given
2. $\angle 5$ and $\angle 4$ are supp.	2. If the outer rays of 2 adj $\angle$s form a st $\angle$, then the $\angle$s are supp.
3. $\therefore \angle 5 \cong \angle 6$	3. Supp $\angle$s of the same $\angle$ are $\cong$.

11.

Statement	Reason
1. $\overrightarrow{OE}$ bisects $\angle AOD$	1. Given
2. m $\angle 1$ = m $\angle 2$	2. Def of $\angle$ bis
3. m $\angle 1$ = m $\angle 3$ ($\angle 1 \cong \angle 3$)	3. Vert $\angle$s are $\cong$.
4. $\therefore$ m $\angle 2$ = m $\angle 3$	4. Sub

5. If ∠K and ∠A are supp, then ∠K and ∠A are adj. (False)

6. Concl: *RS* + *ST* = *RT* Proof: *S* is between *R* and *T* on $\overline{RT}$ (Given); ∴ *RS* + *ST* = *RT* (Seg Add Post)

7. Concl: ∠7 ≅ ∠8

Statement	Reason
1. $\overrightarrow{GH}$ bis ∠*MGR*	1. Given
2. ∴ ∠7 ≅ ∠8	2. Def of ∠ bis

8.
Statement	Reason
1. *PW* = *SY*, *PT* = *YU*	1. Given
2. *PW* − *PT* = *TW*, *SY* − *YU* = *US*	2. Seg Add Post
3. *PW* − *PT* = *SY* − *YU*	3. Equations may be subtracted.
4. ∴ *TW* = *US*	4. Sub

9.
Statement	Reason
1. ∠2 and ∠3 are supp.	1. Given
2. ∠3 and ∠4 are supp.	2. If the outer rays of 2 adj ∠s form a st ∠, then the ∠s are supp.
3. ∠2 ≅ ∠4	3. Supp ∠s of the same ∠ are ≅.
4. ∠1 ≅ ∠2	4. Vert ∠s are ≅.
5. ∴ ∠1 ≅ ∠4	5. Sub

10. If 2 acute adj ∠s are comp, then their outer rays are ⊥. Two acute adj ∠s that are comp implies that their outer rays are ⊥.

11. If ···, then ··· A conditional statement is true in all cases except when the hypothesis is true and the conclusion is false.

See page 89 for the answer to Ex. 12.

---

1. How many lines can be drawn that contain two points *P* and *Q*? Why?   One; exactly one line contains two points.

2.
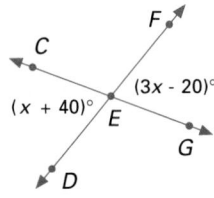
135°   1

Find m ∠ 1.   45

3.
25°   *B*
A   *O*   1

Given: $\overrightarrow{OA}$ ⊥ $\overrightarrow{OB}$
Find m ∠1.   65

4.
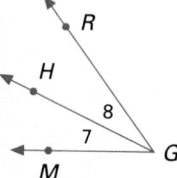
C   *F*
(3x − 20)°
(*x* + 40)°   *E*
D   *G*

Find m ∠*CEF*.   110

5. Write a statement in "If . . . , then . . ." form. Is the conditional true?
Given: ∠K and ∠A are supplementary.
Conclusion: ∠K and ∠A are adjacent.

For Items 6 and 7, draw a conclusion. Then prove it.

6. *R*   *S*   *T*
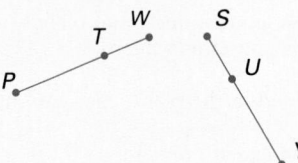

7. Given: $\overrightarrow{GH}$ bisects ∠*MGR*.
R
H
8
7
M   *G*

8. Given: *PW* = *SY*, *PT* = *YU*
Prove: *TW* = *US*
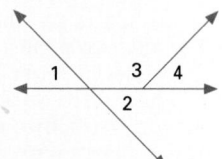
W   *S*
T   *U*
P
Y

9. Given: ∠2 is supplementary to ∠3.
Prove: ∠1 ≅ ∠4
1   3   4
2

10. Rewrite the statement in two equivalent forms. Two acute adjacent angles are complementary only if their outer rays are perpendicular.

11. Give the format of a conditional statement and explain how to determine if it is true.

* 12. Given: ∠*BAC* is an acute angle, $\overline{CB}$ ⊥ $\overline{CA}$, *D* lies on $\overrightarrow{AB}$ such that *B* is between *A* and *D*, and ∠*BCD* ≅ ∠*A*.
Prove: ∠*A* is complementary to the supplement of ∠*ACD*.

Strategy for Achievement in Testing

A helpful test-taking strategy is to discard any given information *that is not relevant*. In Item 1 below, the value $4x + 3y$ is not needed for the solution of the problem. Instead, use the fact that $\angle AQP$ and $\angle BQP$ are supplementary to find z. Then w is found using the congruence of vertical angles.

Choose the one best answer to each item.

1.

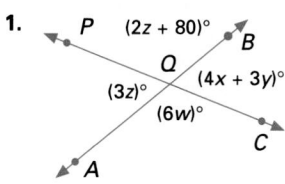

Find the value of w.   B

(A) 120   (B) 20   (C) 60   (D) 80
(E) none of these

2.

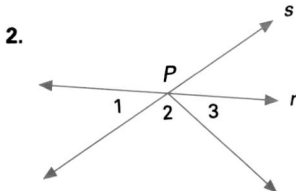

Lines r and s intersect at P; m $\angle 1 = 3x$, and $\angle 3 \cong \angle 1$.
Express m $\angle 2$ in terms of x.   D

(A) x (B) $180 - 3x$ (C) $6x$
(D) $180 - 6x$ (E) $3x$

3. Five bananas cost as much as three pears. If bananas cost 30 cents each, what is the cost of each pear?   E

(A) $1.50   (B) 18¢   (C) 10¢
(D) 30¢   (E) 50¢

4.

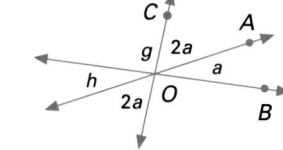

Find the value of $h + g$.   D

(A) a (B) $180 - \frac{1}{2}a$ (C) $2a$

(D) $180 - 2a$ (E) $\frac{1}{2}a$

5. The formula for the area of a rectangle is $A = lw$, where l is the length and w is the width. If the area of the rectangle below is 1, then $y =$ ____A____.

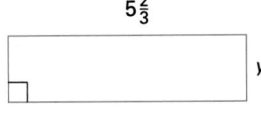

(A) $\frac{3}{17}$ (B) $\frac{17}{3}$ (C) 17   (D) 1
(E) None of these

6. In the figure below, all segments intersect at right angles. Then $x + y =$ ____E____.

(A) $w + r$
(B) $e + t$
(C) $t + r + w + e$
(D) $2t + 2w$
(E) None of these

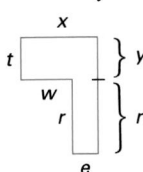

12.

Statement	Reason
1. $\angle BCD \cong \angle A$, $\overline{CB} \perp \overline{CA}$.	1. Given
2. m $\angle ACB = 90$.	2. Def of $\perp$
3. The meas of the supp of $\angle ACD = 180 - \angle ACD$.	3. Def of supp $\angle$s
4. m $\angle ACD =$ m $\angle ACB +$ m $\angle BCD$	4. Angle Add Post
5. m $\angle ACD = 90 +$ m $\angle BCD$	5. Sub
6. m $\angle ACD = 90 +$ m $\angle A$	6. Sub
7. The meas of the supp of $\angle ACD = 180 - (90 +$ m $\angle A)$.	7. Sub
8. The meas of the supp of $\angle ACD = 90 - \angle A$.	8. Distr Prop
9. The meas of the comp of $\angle A = 90 - \angle A$.	9. Def of comp $\angle$s
10. $\therefore$ The meas of the supp of $\angle ACD =$ the meas of the comp of $\angle A$.	10. Sub

89

6.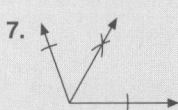

0   9

Ray

7.

13. If a ray bisects an ∠, then the two ∠s formed are equal in meas.

14.

Statement	Reason
1. $\overrightarrow{GH} \perp \overrightarrow{GL}$, m ∠1 = m ∠2	1. Given
2. m ∠1 + m ∠3 = 90	2. If the outer rays of 2 adj acute ∠s are ⊥, then the sum of their meas is 90.
3. ∴ m ∠2 + m ∠3 = 90	3. Sub

15.

Statement	Reason
1. TW = TZ, RW = YZ	1. Given
2. TW − RW = TR, TZ − YZ = TY	2. Seg Add Post
3. TW − RW = TZ − YZ	3. Equations may be subtracted.
4. ∴ TR = TY	4. Sub

16.

Statement	Reason
1. ∠2 and ∠3 are supp.	1. Given
2. m ∠1 + m ∠2 = 180.	2. If the outer rays of 2 adj ∠s form a st ∠, then the sum of the meas of the ∠s is 180.
3. ∠1 and ∠2 are supp.	3. Def of supp ∠s
4. ∴ ∠1 ≅ ∠3	4. Supp ∠s of the same ∠ are ≅.

17. When both p and q are false; when both p and q are true

Cumulative Review *(Chapters 1–2)*

Choose the best possible answer for each exercise. (Exercises 1–5)

1. What does $\overline{AB}$ represent?   C   *1.2*
   (A) plane   (B) line
   (C) segment   (D) distance
   (E) ray

2. Two distinct planes can intersect   *1.1*
   in a _____.   B
   (A) point   (B) line   (C) plane
   (D) ray   (E) segment

3. If m ∠A = x, then the measure   *1.8*
   of a supplement of ∠A is
   _____.   E
   (A) x (B) $180 + x$ (C) $90 - x$
   (D) $x - 180$ (E) $180 - x$

4. The coordinates of the endpoints   *1.5*
   of a segment are −6 and 10.
   Find the coordinate of the mid-
   point of the segment.   C
   (A) 4   (B) 6   (C) 2   (D) 8
   (E) none of these

5. A line that does not lie in a giv-   *2.8*
   en plane can intersect the plane
   in _____.   A
   (A) exactly one point
   (B) exactly two points
   (C) a line
   (D) two or more points
   (E) a ray

6. Graph the conjunction. Identify   *1.4*
   the resulting geometric figure.
   $x \geq 4$ *and* $x \geq 9$

7. Draw an obtuse angle. Con-   *1.6*
   struct the bisector of the angle.

8. The measures of the two angles   *1.7*
   formed by an angle bisector of
   an angle are $3x - 60$ and $x +$
   10. Find the measure of each
   angle.   45, 45, 90

9. The measure of an angle is   *1.8*
   twice the measure of its complement. Find the measure of
   each angle.   60, 30

10. Indicate whether the following   *2.5,*
   conditional is true or false: If   *2.7*
   two angles are vertical, then the
   angles are not congruent.
   False

Use the figure for Exercises 11 and 12.

11. Given: m ∠1 = 45,
       m ∠2 = 3x
   Find x.   15

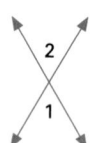

12. Given: m ∠1 = 4x − 30,
       m ∠2 = x + 30
   Find: m ∠2.   50

13. Write a statement in "If . . .,   *2.5*
then . . ." form.
Given: A ray bisects an angle.
Prove: The two angles formed are
equal in measure.

14. Given: $\overrightarrow{GH} \perp \overrightarrow{GL}$, m $\angle 1$ = m $\angle 2$
Prove: m $\angle 2$ + m $\angle 3$ = 90

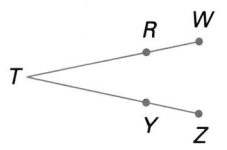

15. Given: $TW = TZ$, $RW = YZ$ *2.4*
Prove: $TR = TY$

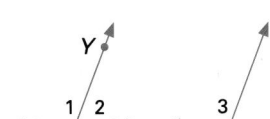

16. Given: $\angle 2$ is supplementary to   *2.6*
$\angle 3$.
Prove: $\angle 1 \cong \angle 3$

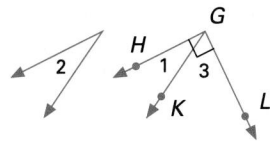

17. When is $p \lor q$ false?   *1.7*
When is $p \land q$ true?

18. Draw a conclusion for each   *2.1,*
figure below and express it as   *2.8*
an equation. Give a reason for
your answer.

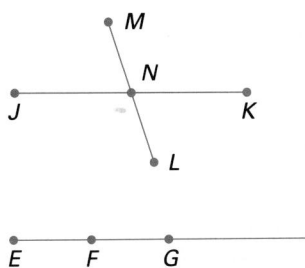

19. Given: $\overrightarrow{OU}$ bisects   *2.3*
$\angle ROT$.
Prove: m $\angle 4$ + m $\angle 5$
= m $\angle UOS$

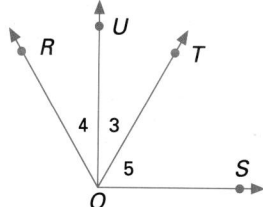

20. Given: m $\angle RSU$ = m $\angle VST$ *2.4*
Prove: m $\angle 5$ = m $\angle 6$

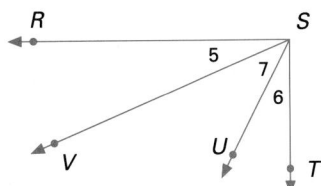

Cumulative Review   **91**

18. Each of the following conclusions can
be drawn:
First figure: $JN + NK = JK$, $MN + NL$
$= ML$ (Seg Add Post); m $\angle JNM$ + m
$\angle MNK$ = 180, m $\angle MNK$ + m $\angle KNL$ =
180, m $\angle KNL$ + m $\angle LNJ$ = 180, m
$\angle LNJ$ + m $\angle JNM$ = 180 (If the outer
rays of 2 adj $\angle$s form a st $\angle$, then the
sum of the meas of the $\angle$s is 180.); m
$\angle JNM$ = m $\angle KNL$, m $\angle MNK$ = m
$\angle JNL$ (Vert $\angle$s are $\cong$.)
Second figure: $EF + FG = EG$, $EF +$
$FH = EH$, $EG + GH = EH$, $FG + GH$
$= FH$, $EF + FG + GH = EH$ (Seg Add
Post)

19.

Statement	Reason
1. $\overrightarrow{OU}$ bis $\angle ROT$	1. Given
2. m $\angle 4$ = m $\angle 3$	2. Def $\angle$ bis
3. m $\angle 3$ + m $\angle 5$ = m $\angle UOS$	3. Angle Add Post
4. $\therefore$ m $\angle 4$ + m $\angle 5$ = m $\angle UOS$	4. Sub

20.

Statement	Reason
1. m $\angle RSU$ = m $\angle VST$	1. Given
2. m $\angle 5$ + m $\angle 7$ = m $\angle RSU$, m $\angle 6$ + m $\angle 7$ = m $\angle VST$	2. Angle Add Post
3. m $\angle 5$ + m $\angle 7$ = m $\angle 6$ + m $\angle 7$	3. Sub
4. $\therefore$ $\angle 5$ = m $\angle 6$	4. Add Prop of Eq

3 PARALLELISM

OVERVIEW

In this chapter the student explores and proves various relationships among lines, on a plane and in space, as well as the relationships of angles formed by parallel lines and transversals. Lesson 3.5 presents the parallel postulate and its converses. Students learn how to write indirect proofs. They solve problems using the theorem giving the sum of the measures of the angles in a triangle. This chapter discusses the negation, the contrapositive, and the inverse of a given statement.

OBJECTIVES

- To identify relationships between parallel and skew lines
- To apply the theorem giving the sum of the measures of the angles of a triangle
- To apply the theorem about angles formed by a transversal of parallel lines
- To construct a line parallel to a given line and passing through a given point not on the given line
- To write a negation, a biconditional, an inverse, and a contrapositive of a given statement
- To prove theorems about angles in a triangle

PROBLEM SOLVING

This chapter uses the strategy of Making a Model extensively as students explore lines and figures in space. Page 95 gives specific instructions for drawing a 3-dimensional picture. Encourage students to manipulate pencils and sheets of paper to represent lines and planes. The Written Exercises on page 96 use the strategy of Drawing a Diagram. For the Written Exercises on page 99, students may need to shade the angles in a drawing to see their relationships. Help students to develop a chart, or other systematic approach, for the Brainteaser on page 108. An effective strategy throughout this chapter is Writing a List of Known Facts. Page 133 discusses the strategy of Using an Alternate Approach.

TECHNOLOGY

Computer: The Computer Investigation on page 120 leads students to the discovery of different relationships between the interior and exterior angles of a triangle.

Calculator: Students may find a calculator helpful for the Mixed Review problems on page 96 and throughout Lesson 3.7 beginning on page 121.

SPECIAL FEATURES

Mixed Review pp. 96, 99, 105, 108, 119, 125, 129, 132
Brainteaser p. 108
Midchapter Review p. 114
Algebra Review p. 114
Computer Investigation: Angles of a Triangle p. 120
Focus on Reading p. 123
Application: Light Rays p. 125
Problem Solving Strategies: Using an Alternate Approach p. 133
Key Terms p. 134
Key Ideas and Review Exercises pp. 134–135
Chapter 3 Test p. 136
College Prep Test p.137

PLANNING GUIDE

Lesson	Basic	Average	Above Average	Resources
3.1 pp. 95–96	CE all WE 1–8	CE all WE 1–12	CE all WE all	Reteaching p. 35 Practice p. 36
3.2 pp. 99	CE all WE 1–12	CE all WE 1–16	CE all WE 1–19	Reteaching p. 37 Practice p. 38
3.3 pp. 104–105	CE all WE 1–13	CE all WE 1–22 odd	CE all WE 1–26 odd	Reteaching p. 39 Practice p. 40
3.4 pp. 107–108	CE all WE 1–6 Brainteaser	CE all WE 1–10 Brainteaser	CE all WE 1–12 Brainteaser	Reteaching p. 41 Practice p. 42
3.5 pp. 112–114	CE all WE 1–15 odd Midchapter Review AR all	CE all WE 1–23 odd Midchapter Review AR all	CE all WE 1–27 odd Midchapter Review AR all	Reteaching p. 43 Practice p. 44
3.6 pp. 117–120	CE all WE 1–10 CI 1	CE all WE 1–18 odd CI all	CE all WE 1–25 odd CI all	Reteaching p. 45 Practice p. 46
3.7 pp. 123–125	FR all CE all WE 1–11 Application	FR all CE all WE 1–21 odd Application	FR all CE all WE 1–24 odd Application	Reteaching p. 47 Practice p. 48
3.8 p. 129	CE all WE 1–7	CE all WE 1–10	CE all WE all	Reteaching p. 49 Practice p. 50
3.9 pp. 132–133	CE all WE 1–5 Problem Solving	CE all WE 1–7 Problem Solving	CE all WE 1–9 Problem Solving	Reteaching p. 51 Practice p. 52
Chapter 3 Review pp. 134–135	1–20 odd	all	all	
Chapter 3 Test p. 136	1–7, 9–11, 15–19 22–23	1–14, 21, 24 odd	all	
College Prep Test p. 137	all	all	all	

CE = Classroom Exercises   WE = Written Exercises   FR = Focus on Reading   AR = Algebra Review   CI = Computer Investigation
NOTE: For each level, all students should be assigned all Mixed Review exercises.

INVESTIGATION

Manipulative: This activity leads students to make discoveries about the angles that are formed when parallel lines are cut by a transversal. Use this investigation to introduce Lesson 3.2. To extend this investigation, use Manipulative Worksheet 5.

Materials: Each student will need a sheet of paper, a straight edge, transparent tape, a pen, a pencil, and a protractor.

Have each student place a piece of tape 4 to 5 inches long on a sheet of paper and trace the tape's edges with a pencil to represent a pair of parallel lines. With a straight edge, have each student draw a line across the tape, intersecting the parallel lines at an angle, as shown below.

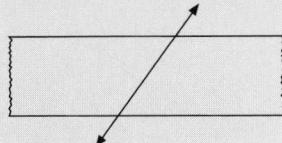

Tell students that the line they drew is called a *transversal* because it crosses the pair of parallel lines.

1. Ask students to count the number of angles they constructed by drawing the transversal.　8

Have students measure each of the eight angles with their protractors, write the respective angle measures, and label the angles as shown below.

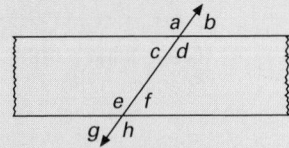

Ask students the following questions.

2. How many vertical angle pairs exist?　4
3. What term is used to describe angles *c*, *d*, *e*, and *f*?　Interior
4. Are angles *e* and *d* on the same or alternate sides of the transversal? Are they on the interior or the exterior of the parallel lines?　Alternate interior
5. Write several informal statements to generalize which pairs of angles are equal and which pairs of angles are supplementary.　Answers will vary.

2, 3, 5, 7, 11, 13, 17...

Euclid (about 300 B.C.)—Writing in Greek on rolls of Egyptian papyrus, Euclid produced the world's most enduring geometry text. This book, called *The Elements,* has been printed in more than 1,000 different editions worldwide. We know very little about Euclid's life.

More About the Mathematician

The Parallel Postulate (Postulate 14 on page 109) is one of the essential postulates of Euclidean geometry. The drawing shows many lines through point *P*, only one of which is parallel to $\overleftrightarrow{MN}$. Various non-Euclidean geometries, developed since 1800, replace this postulate with others. The triangle is a sketch for Euclid's proof that the base angles of an isosceles triangle are congruent. The first few prime numbers (2, 3, 5, 7, and 11, 13, 17. . .) are included in the picture because Euclid proved that the number of prime numbers is infinite (using the indirect method of proof).

3.1 Parallel and Skew Lines

Objective   To identify relationships among parallel lines and skew lines

---

Definition   Coplanar lines are lines that lie in the same plane. Coplanar lines either intersect or are parallel.

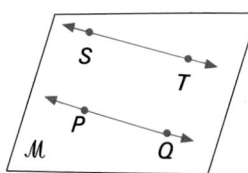

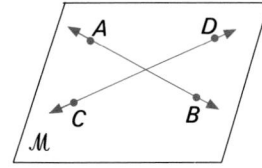

$\overleftrightarrow{PQ}$ and $\overleftrightarrow{ST}$ do not intersect. $\overleftrightarrow{PQ}$ is parallel to $\overleftrightarrow{ST}$ ($\overleftrightarrow{PQ} \parallel \overleftrightarrow{ST}$).

$\overleftrightarrow{AB}$ and $\overleftrightarrow{CD}$ intersect. $\overleftrightarrow{AB}$ is not parallel to $\overleftrightarrow{CD}$ ($\overleftrightarrow{AB} \nparallel \overleftrightarrow{CD}$).

Definition   Parallel lines are coplanar lines that do not intersect.

In space there exists a third possible relationship between two lines. $\overleftrightarrow{AB}$ and $\overleftrightarrow{CD}$ do not intersect, and they are not parallel since they are not coplanar. $\overleftrightarrow{AB}$ and $\overleftrightarrow{CD}$ are *skew lines*.

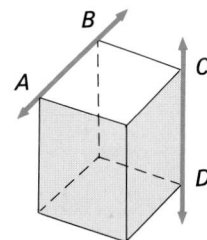

Definition   Skew lines are noncoplanar lines.

EXAMPLE 1   Identify the indicated pair of lines as parallel or skew.

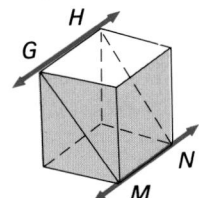

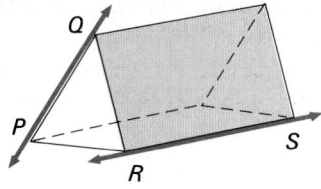

$\overleftrightarrow{GH}$ and $\overleftrightarrow{MN}$ are parallel.

$\overleftrightarrow{PQ}$ and $\overleftrightarrow{RS}$ are skew.

Teaching Resources

Manipulative Worksheet 5
Project Worksheet 3
Quick Quizzes 18
Reteaching and Practice
  Worksheets, pp. 35, 36
Transparency 4

GETTING STARTED

Prerequisite Quiz

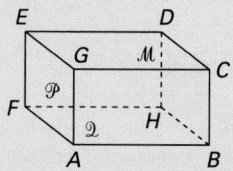

Plane $\mathcal{P}$ contains points *E*, *F*, and *A*; plane $\mathcal{M}$ contains points *E*, *G*, and *C*; and plane $\mathcal{Q}$ contains points *G*, *A*, and *B*.

1. Name the intersection of planes $\mathcal{P}$ and $\mathcal{Q}$. $\overleftrightarrow{AG}$
2. Is *B* coplanar with points *E*, *G*, and *D*? Why?   No. Points *E*, *G* and *D* are in plane $\mathcal{M}$. *B* is not.
3. $\overleftrightarrow{GC}$ is the intersection of what two planes? $\mathcal{M}$ and $\mathcal{Q}$
4. Name a plane that contains points *E*, *G* and *C*. $\mathcal{M}$

Motivator

Ask your students to give two physical models of lines in this room that
1. are parallel.   top and bottom of one wall
2. intersect.   top and side of one wall
3. are not parallel and do not intersect. the diagonal of the floor and a ceiling edge.

Additional Example 1

Identify the indicated pair of lines as parallel or skew.
$\overleftrightarrow{ED}$ and $\overleftrightarrow{BC}$ skew $\overleftrightarrow{AB}$ and $\overleftrightarrow{DC}$ $\parallel$

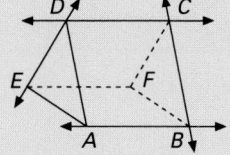

Lesson Note

Use physical examples from the classroom to illustrate parallel and skew lines. For example, the intersection of the front wall and the ceiling is skew to the line of intersection of the back wall and any side wall.

Drawing a box and using the steps described before Example 3 will help students to visualize skew and parallel lines.

Math Connections

Aviation: Air traffic controllers watch airplane flight paths and guide them away from paths that intersect. Nonintersecting paths can be represented by skew or parallel lines.

Critical Thinking Questions

Application: Airplanes heading north, south, east, and west are required to fly at different specified altitudes. Ask students to explain how this keeps air traffic flowing without crashing. The flight paths lie in four different planes and do not intersect.

Common Error Analysis

Error: For skew lines, segments, or rays, students may recall only that part of the definition which says they do not intersect, forgetting that to be skew, they cannot be coplanar. Emphasize that these figures are not coplanar if they are skew. For example, in the figure below, $\overleftrightarrow{AB}$ and $\overleftrightarrow{CD}$ do not intersect, but they are *not* skew since they *are* coplanar.

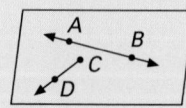

Segments or rays can also be parallel. In the figure, $\overline{GH} \parallel \overrightarrow{VW}$ since $\overleftrightarrow{GH} \parallel \overleftrightarrow{VW}$

    ↑   ↑
segment ‖ ray

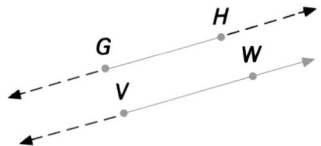

Definition | *Segments* or *rays* are *parallel* if the lines that contain them are parallel.

Similarly, *segments* or *rays* are *skew* if the lines that contain them are skew.

By definition, if two coplanar lines do not intersect, then they are parallel. This is not true for segments or rays, as seen below.

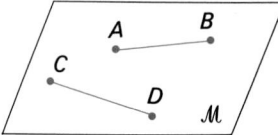   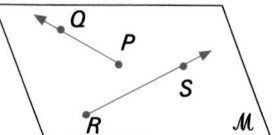

$\overline{AB}$ and $\overline{CD}$ are segments that do not intersect. But, $\overline{AB} \not\parallel \overline{CD}$.

$\overrightarrow{PQ}$ and $\overrightarrow{RS}$ are rays that do not intersect. But, $\overrightarrow{PQ} \not\parallel \overrightarrow{RS}$.

EXAMPLE 2 Which pairs of segments or rays appear to be parallel?

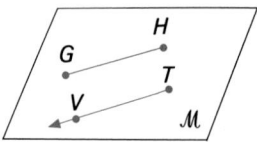

$\overline{GH} \parallel \overrightarrow{TV}$.

$\overrightarrow{CD}$ intersects $\overrightarrow{AB}$.
$\overrightarrow{CD} \not\parallel \overrightarrow{AB}$.

It is frequently helpful to draw a box to illustrate geometric relationships between lines in space. A diagram of a four-sided box with top and bottom can be drawn by following these steps.

Additional Example 2

Which pairs of segments or rays appear to be parallel?

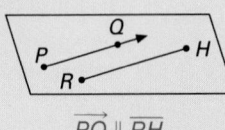

$\overrightarrow{PQ} \parallel \overline{RH}$ $\overline{TS} \not\parallel \overline{MN}$ ($\overrightarrow{TS}$ intersects $\overline{MN}$.)

Step 1
Draw the front of the box, as shown. Then draw the back slightly above and to the right of the front, using two dashed segments, as shown.

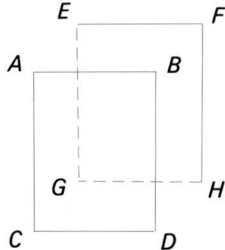

Step 2
Draw four segments to complete the box as shown. Use dashed segments for any segments that would be invisible.

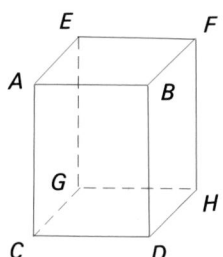

Practice drawing this figure. It is referred to in the next example. Exercises occasionally require that reasons be given why a statement is sometimes, always, or never true. In such cases, a geometric figure may be drawn to illustrate when the statement is true and when it is false.

EXAMPLE 3   Indicate whether the statement is always, sometimes, or never true. Give a reason for your answer.

| Parallel lines are skew. | Parallel segments do not intersect. | Nonparallel segments intersect. |

Solution

Never true. Reason: by definition of skew lines.

Always true. Reason: by definition of parallel lines and segments.

Sometimes true. Reason: Refer to the diagram above. $\overline{AB} \not\parallel \overline{BF}$. $\overline{AB}$ and $\overline{BF}$ intersect. $\overline{AB} \not\parallel \overline{HD}$. $\overline{AB}$ and $\overline{HD}$ are skew.

Classroom Exercises

Tell whether the pair of lines, rays, or segments appears to be parallel, intersecting, skew, or none of these.

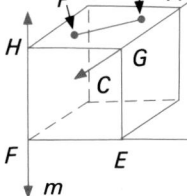

1. l and m
2. $\overline{BC}$ and m
3. $\overline{AB}$ and l
4. $\overrightarrow{QP}$ and $\overline{AB}$
5. $\overline{FE}$ and $\overline{AB}$
6. $\overline{BC}$ and l
7. $\overline{AD}$ and m
8. $\overline{FE}$ and $\overrightarrow{AG}$
9. $\overline{BH}$ and m

1. Skew
2. Parallel
3. Intersecting
4. None
5. Parallel
6. Skew
7. Parallel
8. Skew
9. Intersecting

Checkpoint

Draw a diagram of each of the following.

1. Two coplanar rays that do not intersect and are not parallel

2. Two skew segments

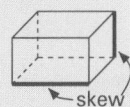

Use the figure below to identify the following.

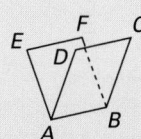

3. Segment ∥ to $\overline{EF}$ $\overline{DC}$ or $\overline{AB}$
4. Segment skew to $\overline{EA}$ $\overline{BC}$ or $\overline{CD}$
5. Segment $\not\parallel$ and not skew to $\overline{AB}$ $\overline{AD}$, $\overline{BC}$, $\overline{AE}$, or $\overline{BF}$

Closure

Draw a cube on the chalkboard and label its vertices. Have your students answer the following questions.

What two segments are
1. intersecting?
2. parallel?
3. skew?

◼◼◼FOLLOW UP

Guided Practice

Classroom Exercises 1–9

Independent Practice

🅐 Ex. 1–8, 🅑 Ex. 9–12, 🅒 Ex. 13–14

Basic: WE 1–8

Average: WE 1–12

Above Average: WE all

Additional Example 3

Indicate whether the statement is sometimes, always, or never true. Noninteresecting segments are parallel.
Sometimes; refer to the figures.

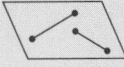 

Coplanar rays that do not intersect are parallel.
Sometimes; refer to the figures.

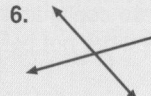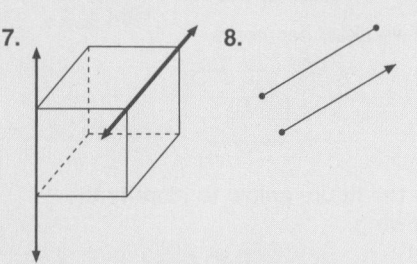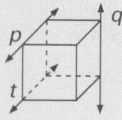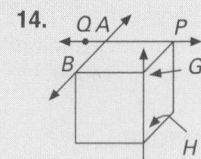
Written Exercises

For each of the following refer to labeled segments or lines in the figure of the highway and overpass shown at the right. Give an example.

1. parallel segments $\overline{ST}$ and $\overline{PQ}$
2. skew lines $\overleftrightarrow{GK}$ and $\overleftrightarrow{AB}$
3. intersecting segments $\overline{PQ}$ and $\overline{QR}$
4. coplanar segments which do not intersect and are not parallel $\overline{ST}$ and $\overline{QR}$

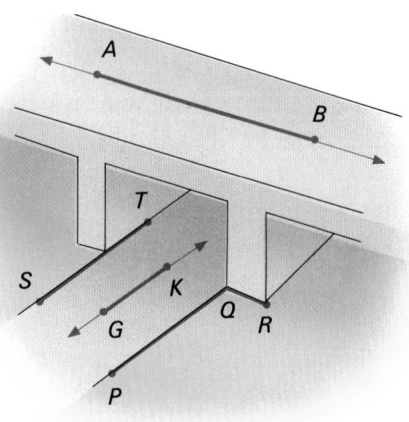

Draw a diagram to illustrate each of the following descriptions.

5. two nonintersecting segments that are not parallel
6. two intersecting lines
7. two skew lines
8. a segment that is parallel to a ray

Indicate whether the statement is always true, sometimes true, or never true. Give a reason for your answer.

9. Two coplanar lines that do not intersect are parallel.
10. Two lines that are not skew are parallel.
11. A segment in the plane of the ceiling of your classroom intersects a segment in the plane of the floor.
12. Two lines in space that do not intersect are parallel.

Draw the figure described.

13. Lines p and q are skew. Lines q and t are skew. But $p \parallel t$.
14. $\overleftrightarrow{AB}$ and $\overleftrightarrow{GH}$ are skew. $\overleftrightarrow{GH}$ and $\overleftrightarrow{PQ}$ are skew. $\overleftrightarrow{AB} \perp \overleftrightarrow{PQ}$.

Mixed Review

1. The measure of one of two complementary angles is $\frac{2}{3}$ the measure of the other. Find the measure of each angle. *1.7* 36, 54
2. The measure of one of a pair of vertical angles is $3x - 40$. The measure of the other is $x + 20$. Find the measure of each angle. *2.8* 50
3. The coordinates of the endpoints of $\overline{AB}$ are -8 and 12. Find the coordinate of the midpoint of $\overline{AB}$. *1.4* 2

96 Chapter 3 Parallelism

Enrichment

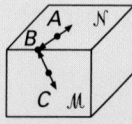

Draw this diagram and challenge the students to find a flaw in this reasoning.

1. Line $\overleftrightarrow{AB}$ lies in plane $\mathcal{N}$.
2. Line $\overleftrightarrow{BC}$ lies in plane $\mathcal{M}$.
3. Skew lines always lie in separate planes. Thus, $\overleftrightarrow{AB}$ and $\overleftrightarrow{BC}$ are skew.

4. But $\overleftrightarrow{AB}$ and $\overleftrightarrow{BC}$ intersect in point B.
5. Therefore, skew lines can intersect.

The flaw is that even though $\mathcal{M}$ and $\mathcal{N}$ are separate planes, there is actually a single plane that contains both $\overleftrightarrow{AB}$ and $\overleftrightarrow{BC}$; namely, the plane determined by points A, B, and C. This plane is not one of the 6 planes of the box.

3.2 Transversals and Special Angle Relationships

Objective To identify special pairs of angles formed by the intersection of two lines by a transversal

In the figure at the right, line *t* intersects line *m* at S and line *n* at W. Line *t* is called a *transversal* of *m* and *n*.

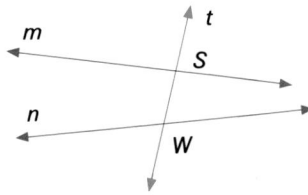

Definition A **transversal** is a line, ray, or segment that intersects two or more coplanar lines, rays, or segments, each at a different point.

EXAMPLE 1 Identify each transversal in the figure and the lines to which it is a transversal.

$\overrightarrow{PQ}$ is a transversal of $\overrightarrow{PR}$ and $\overrightarrow{QR}$.
$\overrightarrow{PR}$ is a transversal of $\overrightarrow{PQ}$ and $\overrightarrow{QR}$.
$\overrightarrow{QR}$ is a transversal of $\overrightarrow{PQ}$ and $\overrightarrow{PR}$.

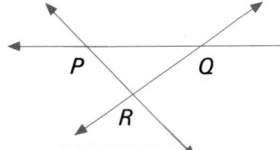

In the definitions of the special angles that follow, the *interior* and *exterior* of two lines cut by a transversal will be used. These regions are shown in the figure at the right. In each case, line *t* is a transversal of lines *m* and *n*.

When a transversal intersects two lines, special pairs of angles are formed. Such pairs will be identified by arcs in illustrations that follow.

∠3 and ∠5 are **alternate interior angles**. Two angles are alternate interior angles if:
(1) they lie on opposite sides of a transversal;
(2) they are both interior angles; and
(3) they are nonadjacent.
Another pair of alternate interior angles is ∠4 and ∠6.

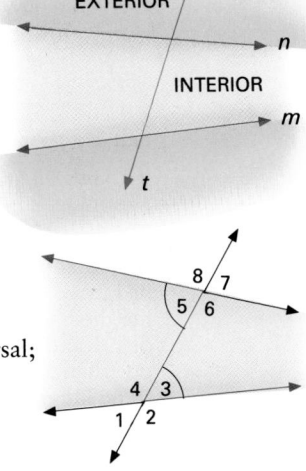

3.2 Transversals and Special Angle Relationships **97**

GETTING STARTED

Prerequisite Quiz
Refer to lines $\overleftrightarrow{DE}$, $\overleftrightarrow{FG}$, and $\overleftrightarrow{HI}$.

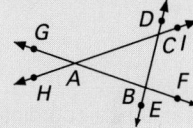

1. $\overleftrightarrow{GF}$ intersects which two lines, at what points? *HI at A; ED at B*
2. $\overleftrightarrow{HI}$ intersects which two lines, at what points? *GF at A; DE at C*
3. Point C is the intersection of which two lines? *HI and DE*

Motivator

Draw two parallel lines labeled *m* and *n*. Draw a line (transversal) that intersects line *m* at point *P* and line *n* at point *Q*. Have the students shade in the region which is in the **interior** of lines *m* and *n*.

Ask the students the following questions. Which pair of angles are on **alternate** sides of $\overleftrightarrow{PQ}$ and in the **interior** of lines *m* and *n*? Which pair of angles are in the same **corresponding** position with respect to *m*, *n*, and $\overleftrightarrow{PQ}$?

Additional Example 1

Identify each transversal in the figure.

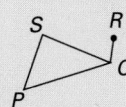

$\overline{PQ}$ is a transversal to $\overline{PS}$ and $\overline{QR}$, and a transversal to $\overline{PS}$ and $\overline{QS}$.

$\overline{SP}$ is a transversal to $\overline{SQ}$ and $\overline{PQ}$.
$\overline{SQ}$ is a transversal to $\overline{SP}$ and $\overline{QR}$.
$\overline{SQ}$ is a transversal to $\overline{SP}$ and $\overline{PQ}$.

Lesson Note

The following mnemonic devices help students recognize alternate interior angles and corresponding angles.

Alternate interior angles:
Think of a *Z* or a backward *Z*.

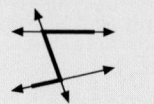 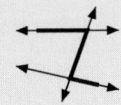

Corresponding angles:
Think of quadrants of a coordinate grid system: ∠1 and ∠2 are each in Quadrant III.

Note that diagrams in geometry often show segments of transversals rather than the transversals themselves.

Math Connections

Language: According to *Webster's New World Dictionary*, *trans* is a Latin-based prefix which means "over, across, or through." The word "transversal" uses this prefix to express the idea of going across. Ask students to list other words that use the prefix *trans* and to use them correctly in a sentence.

Critical Thinking Questions

Application: Instruct the students to write the alphabet in capital letters on a sheet of paper and use the following questions for open-ended discussion. Answers may vary according to how individuals form their letters; possible answers are given.

What are some capital letters whose shapes form alternate interior angles? A, H, M, N, W, Z. Corresponding angles? E, F Can you think of any capital letters whose shapes form alternate exterior angles? There are none. Why don't the letters L, T, and Y have any of the special angle relationships that are mentioned in the lesson? None have transversals.

∠4 and ∠8 are **corresponding angles.**
Two angles are corresponding angles if:
(1) they lie on the same side of a transversal;
(2) one angle is interior, one exterior; and
(3) they are nonadjacent.
Three other pairs of corresponding angles are: ∠1 and ∠5, ∠2 and ∠6, ∠3 and ∠7.

∠1 and ∠7 are **alternate exterior angles.**
Two angles are alternate exterior angles if:
(1) they lie on opposite sides of a transversal;
(2) they are both exterior angles; and
(3) they are nonadjacent.

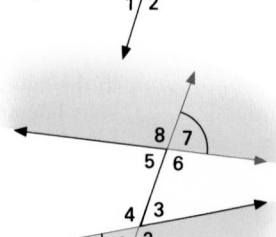

EXAMPLE 2 Identify the given pair of angles as alternate interior, alternate exterior, corresponding, or none of these:
∠4 and ∠6, ∠2 and ∠8, ∠4 and ∠8

Plan Redraw the figure to show each pair of angles.

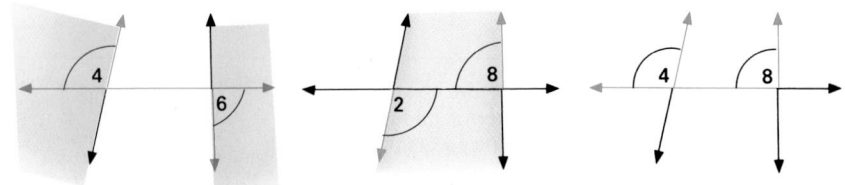

Solution ∠4 and ∠6 are alternate exterior angles. ∠2 and ∠8 are alternate interior angles. ∠4 and ∠8 are corresponding angles.

∠1 and ∠8 are none of these special pairs of angles.

A special pair of angles can be identified more easily by shading and extending their sides. By extending the sides of ∠1 and ∠3, their angle measures are not changed, but they are clearly seen to be alternate interior angles.

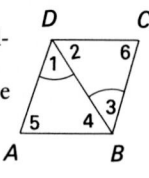 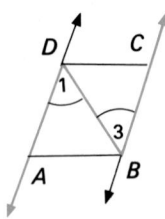

Additional Example 2

Identify each given pair of angles as alternate interior, alternate exterior, corresponding, or none of these.

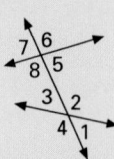

∠7 and ∠1 Alt ext ∠s
∠8 and ∠2 Alt int ∠s
∠6 and ∠3 None of these
∠5 and ∠1 Corr ∠s

Classroom Exercises

1. *l* is a transversal to _____.   *p* and *q*
2. *m* is a transversal to _____.   *p* and *q*

Identify the pair of angles as alternate interior, alternate exterior, corresponding, or none of these.

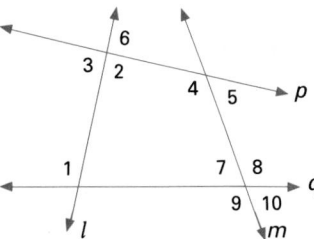

3. ∠1 and ∠2   Alt int
4. ∠3 and ∠4   Corr
5. ∠4 and ∠9   Corr
6. ∠1 and ∠10   Alt ext
7. ∠5 and ∠6   None
8. ∠4 and ∠8   Alt int

Written Exercises

Identify the pair of angles as alternate interior, alternate exterior, corresponding, or none of these.

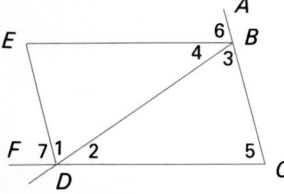

1. ∠6 and ∠10   Corr
2. ∠7 and ∠9   Alt int
3. ∠6 and ∠12   Alt ext
4. ∠5 and ∠10   None
5. ∠8 and ∠12   Corr
6. ∠5 and ∠7   None

7. ∠1 and ∠2   Corr
8. ∠2 and ∠5   None
9. ∠1 and ∠3   Alt int
10. ∠5 and ∠1   Corr
11. ∠2 and ∠4   Corr
12. ∠6 and ∠2   Alt ext

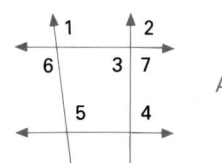

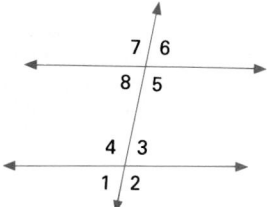

13. ∠1 and ∠3   Alternate interior
14. ∠ABD and ∠2   None of these
15. ∠E and ∠6   Alternate interior
16. ∠7 and ∠5   Corresponding
17. ∠4 and ∠FDB   None of these
18. ∠7 and ∠3   None of these
19. ∠DBA and ∠5   Corresponding

Mixed Review

1. Given: m ∠8 = m ∠3
   Prove: m ∠6 = m ∠3   *2.8*
2. Given: m ∠3 = m ∠8
   Prove: m ∠1 = m ∠6   *2.8*
3. Given: ∠3 is supplementary to ∠5.
   Prove: ∠6 ≅ ∠3   *2.7*
4. Given: ∠4 is supplementary to ∠8.
   Prove: ∠2 is supplementary to ∠8.   *2.8*

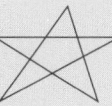

Checkpoint

Identify each pair of angles as alternate interior, alternate exterior, corresponding, or none of these.

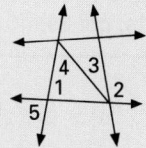

1. ∠1 and ∠2   Corr ∠s
2. ∠3 and ∠4   Alt int ∠s
3. ∠2 and ∠4   None of these
4. ∠2 and ∠5   Alt ext ∠s

Closure

Ask the students the definitions of alternate interior angles, alternate exterior angles, and corresponding angles. Have your students draw a figure, such as in Example 2, and label each of the alternate interior, alternate exterior, and corresponding angles.

▰▰▰ FOLLOW UP

Guided Practice

Classroom Exercises 1–8

Independent Practice

A Ex. 1–12, **B** Ex. 13–16, **C** Ex. 17–19

Basic: WE 1–12

Average: WE 1–16

Above Average: WE 1–19

Additional Answers

Mixed Review

Statement	Reason
1. m ∠8 = m ∠3	1. Given
2. m ∠8 = m ∠6	2. Vert ∠s are ≅.
3. m ∠3 = m ∠6	3. Sub

See page 103 for the answers to Mixed Review Ex. 2–4.

Enrichment

Challenge the students to begin at a point and draw a continuous figure that consists of 5 line segments, in which each segment is a transversal for exactly two other segments. For a hint, suggest that the figure is a familiar one.

▰▰▰ GETTING STARTED

Prerequisite Quiz

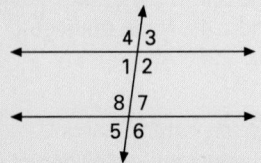

Identify each pair of angles as corresponding, alternate interior, or alternate exterior.

1. $\angle 1$ and $\angle 7$ Alt int $\angle$s
2. $\angle 6$ and $\angle 4$ Alt ext $\angle$s
3. $\angle 1$ and $\angle 5$ Corr $\angle$s
4. Given: m $\angle 4$ = m $\angle 6$
   Prove: m $\angle 8$ = m $\angle 4$

Statement	Reason
1. m $\angle 4$ = m $\angle 6$	1. Given
2. m $\angle 8$ = m $\angle 6$ or $\angle 8 \cong \angle 6$	2. Vert $\angle$s are $\cong$.
3. m $\angle 4$ = m $\angle 8$	3. Sub

Motivator

On the chalkboard, draw two parallel lines labeled *l* and *m*. Draw a transversal that intersects lines *l* and *m*.

Ask the students the following questions. Which angles are alternate interior angles? Do they seem to be congruent? Yes   Do lines *l* and *m* appear to be parallel or non parallel? Parallel   Do you think this would be true if the alternate interior angles were not congruent? No

3.3 Proving Lines Parallel

Objectives   To prove that under certain conditions lines are parallel
To apply the Alternate Interior Angle Postulate

You learned in Lesson 1.4 how to construct an angle that has the same measure as a given angle. A similar procedure can now be used to construct a congruent pair of alternate interior angles.

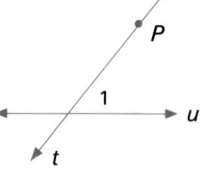

Construction   Given: Point *P* on line *t*, line *u* forming $\angle 1$ with *t*

Construct: A line *v* through point *P* forming $\angle 2$ such that $\angle 1 \cong \angle 2$, and $\angle 1$ and $\angle 2$ are alternate interior angles.

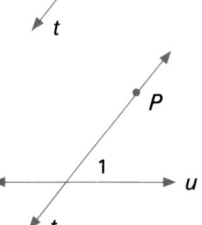

Construct $\angle 2$ congruent to $\angle 1$, with vertex *P*, on the left side of the transversal *t*. Extend the ray to form line *v*.

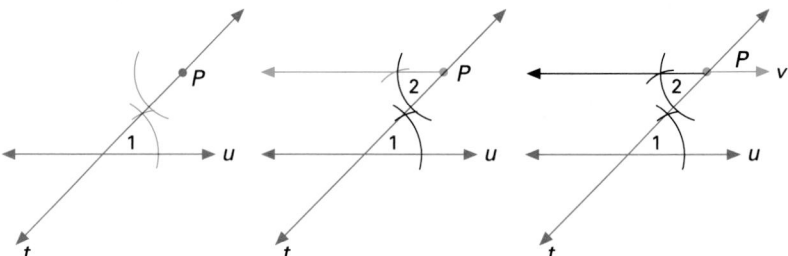

Result: $\angle 1$ and $\angle 2$ are congruent alternate interior angles.

Notice in the construction above that line *v* appears to be parallel to line *u*. This construction suggests the following postulate.

Postulate 13

The Alternate Interior Angles Postulate: If a transversal intersects two lines so that alternate interior angles are congruent (equal in measure), then the lines are parallel.

The Alternate Interior Angles Postulate can be used to show that lines are parallel if certain other pairs of angles are congruent.

In the figure, let m ∠1 = 130 and m ∠3 = 130. Now, ∠1 and ∠3 are correspond-ing angles, not alternate interior angles. But because ∠1 and ∠2 are vertical an-gles, m ∠2 = 130. Therefore ∠2 and ∠3 are *congruent* alternate interior angles (m ∠2 = m ∠3 = 130). Therefore, *m* ∥ *l*. This suggests the following theorem.

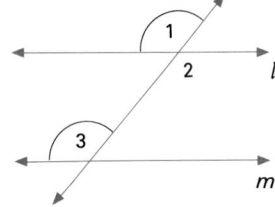

Theorem 3.1

If a transversal intersects two lines so that corresponding angles are congruent, then the lines are parallel.

Given: ∠1 ≅ ∠3
Prove: *m* ∥ *n*

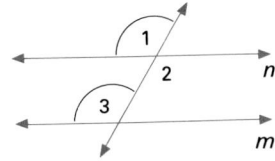

Proof

	Statement		Reason
1.	∠1 ≅ ∠3	1.	Given
2.	∠2 ≅ ∠1	2.	Vert ∠s are ≅.
3.	∠2 ≅ ∠3	3.	Trans Prop of ≅ (or Sub)
4.	∴ *m* ∥ *n*	4.	Alt Int ∠s Post

In the figure, ∠1 and ∠2 are **interior an-gles on the same side of the transversal.** If ∠2 is supplementary to ∠1, then it can be shown that *m* ∥ *n*. A numerical exam-ple of this is shown below. This relation-ship will then be proved as a theorem. Suppose that ∠2 is supplementary to ∠1, and m ∠1 = 120. The three sequential diagrams below illustrate that *m* ∥ *n*.

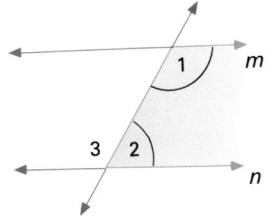

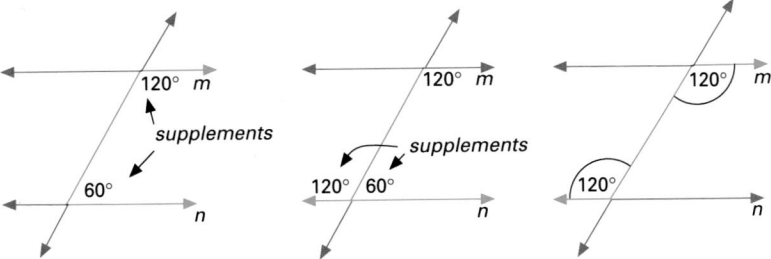

Alternate interior angles are ≅. Therefore *m* ∥ *n*.

Lesson Note

Emphasize that all the properties discussed in this lesson apply to segments and rays as well as to lines. In writing proofs, it would be cumbersome to restate a postulate or theorem about lines in terms of segments or rays.

Have the students list methods of proving lines parallel in their own words instead of simply reading or memorizing the summary at the end of the lesson.

Math Connections

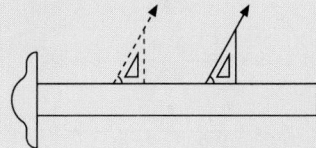

Technical Drawing: Use the following example to show students an application of how congruent corresponding angles formed by a transversal determine parallel lines. Draftsmen use a T-square and a plastic triangle to draw parallel lines. The T-square is used as a transversal. By sliding a triangle along the horizontal edge of the T-square to form congruent corresponding angles, the draftsman can draw many parallel lines.

Critical Thinking Question

Synthesis: Refer students to the Summary at the bottom of page 103. Ask them how to prove lines parallel, given that alternate exterior angles formed by a transversal are congruent. Plans for proof may be based on congruent alternate interior angles and their vertical angles. Several other answers are possible.

Common Error Analysis

Error: In complex figures (see Example 3) where two pairs of lines appear to be parallel, students may not recognize which lines can be *proved* parallel by using the angles in the figure.

Have the students shade the sides of the angles or trace them with a finger to help them identify the common side and the lines that are parallel. Have them look for the "**Z**" (or backward "**Z**") that is formed by alternate interior angles.

Checkpoint

Determine which lines are parallel. Give a reason for your answer.

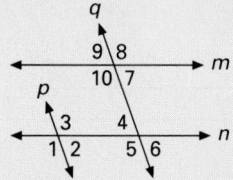

1. m ∠6 = 80, m ∠7 = 80   *m ∥ n*; Corr ∠s are ≅.
2. m ∠3 = m ∠5   *p ∥ q*;   Alternate interior ∠s are ≅.
3. m ∠4 = 60, m ∠10 = 120   *m ∥ n*; int ∠s are supp on the same side of transv.
4. m ∠4 = 55, m ∠7 = 55   *m ∥ n*; Alternate interior ∠s are ≅
5. ∠1 ≅ ∠5   *p ∥ q*; Corr ∠s are ≅

Closure

Have your students name four ways to prove lines are parallel.   Alternate interior angles are congruent. Corresponding angles are congruent. Interior angles on same side of transversal are supplementary. Two lines perpendicular to the same line are parallel.

Theorem 3.2   If two lines are intersected by a transversal so that interior angles on the same side of the transversal are supplementary, then the lines are parallel.

Given: ∠1 and ∠2 are supplementary.
Prove: $\overleftrightarrow{AB} \parallel \overleftrightarrow{CD}$

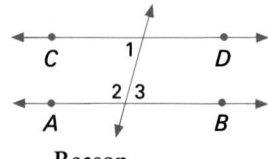

Proof

Statement	Reason
1. ∠1 and ∠2 are supplementary.	1. Given
2. ∠3 and ∠2 are supplementary.	2. If the outer rays of the two adj ∠s form a st ∠, then the ∠s are supp.
3. ∠1 ≅ ∠3	3. Supp of the same ∠ are ≅.
4. ∴ $\overleftrightarrow{AB} \parallel \overleftrightarrow{CD}$	4. Alt Int ∠s Post.

In the Figure, $\overleftrightarrow{AB} \perp \overleftrightarrow{PQ}$ and $\overleftrightarrow{CD} \perp \overleftrightarrow{PQ}$. The corresponding angles, ∠1 and ∠2, therefore have the same measure, 90. So, the two lines are parallel. This is stated below as a theorem.

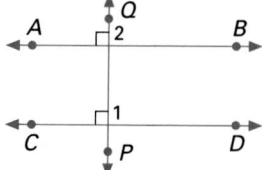

Theorem 3.3   In a plane, if two lines are perpendicular to the same line, then they are parallel.

EXAMPLE 1   Given: m ∠1 = m ∠3, m ∠2 = m ∠3
Prove: *l ∥ n*

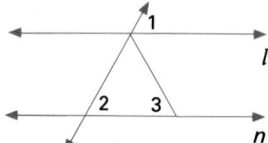

Plan   Look for a special pair of angles.
∠1 and ∠2 are *corresponding* angles.

Proof

Statement	Reason
1. m ∠1 = m ∠3	1. Given
2. m ∠2 = m ∠3	2. Given
3. m ∠1 = m ∠2 (∠1 ≅ ∠2)	3. Sub
4. ∴ *l ∥ n*	4. If corr ∠s ≅, then lines ∥.

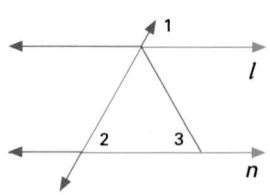

Additional Example 1

Given: $\overline{BD}$ bisects ∠CBA, m ∠2 = m ∠3
Prove: $\overline{EF} \parallel \overline{BD}$

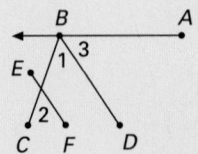

Proof: $\overline{BD}$ bisects ∠CBA (Given); m ∠1 = m ∠3 (def of ∠ bisects); m ∠2 = m ∠3 (Given); m ∠2 = m ∠1 or ∠2 ≅ ∠1 (Sub); $\overline{EF} \parallel \overline{BD}$ (If corr ∠s are ≅, then lines are ∥.)

All the properties of parallel lines in this lesson apply to segments and rays as well.

EXAMPLE 2 Find the value of x so that $m \parallel n$.

Solution The *sum* of measures of $\angle ADC$ and $\angle DAB$ must be 180.

$$(3x + 20) + (x + 40) = 180$$
$$4x + 60 = 180$$
$$4x = 120$$
$$x = 30$$

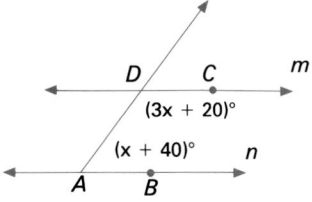

EXAMPLE 3 Given: $\overline{AB} \parallel \overline{ED}$. Find m $\angle BCF$.

Plan Through C, draw an auxiliary line segment $\overline{CG}$ that is parallel to $\overline{AB}$ and $\overline{ED}$.

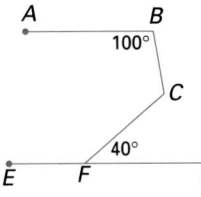

Solution

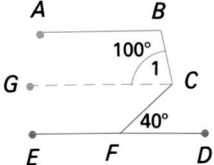 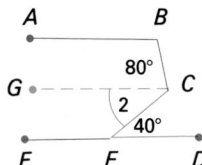 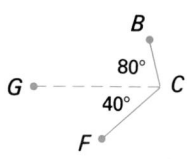

$\angle ABC$ and $\angle 1$ are supplementary if $\overline{AB} \parallel \overline{CG}$.

Therefore, m $\angle 1 = 180 - 100 = 80$.

m $\angle 2 = 40$ (alternate interior angles of parallels $\overline{CG}$ and $\overline{ED}$)

Then m $\angle BCF =$ m $\angle 1 +$ m $\angle 2 = 80 + 40 = 120$.

Thus, m $\angle BCF = 120$.

Summary

Ways to prove lines, rays, or segments parallel:

- Alternate interior angles congruent
- Corresponding angles congruent
- Interior angles on same side of transversal supplementary
- Two lines, segments, or rays perpendicular to the same transversal

3.3 Proving Lines Parallel   **103**

Additional Example 2

Find the value of x so that $k \parallel l$.

$x = 63$

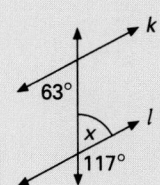

Additional Example 3

Find m $\angle QRS$ so that $\overrightarrow{PQ} \parallel \overline{TS}$.

105

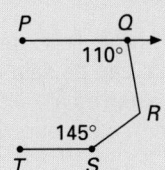

Guided Practice

Classroom Exercises 1–6

Independent Practice

A Ex. 1–13, **B** Ex. 14–22, **C** Ex. 23–26

Basic: WE 1–13
Average: WE 1–22 odd
Above Average: WE 1–26 odd

Additional Answers, page 99

2.

Statement	Reason
1. m $\angle 3$ = m $\angle 8$	1. Given
2. m $\angle 8$ = m $\angle 6$, m $\angle 3$ = m $\angle 1$	2. Vert $\angle$s are $\cong$.
3. m $\angle 1$ = m $\angle 6$	3. Sub

3.

Statement	Reason
1. $\angle 3$ and $\angle 5$ are supp.	1. Given
2. $\angle 6$ and $\angle 5$ are supp.	2. If the outer rays of 2 adj $\angle$s form a st $\angle$, then the $\angle$s are supp.
3. $\angle 3 \cong \angle 6$	3. Supp of the same $\angle$ are $\cong$.

4.

Statement	Reason
1. $\angle 4$ and $\angle 8$ are supp.	1. Given
2. m $\angle 4$ = m $\angle 2$	2. Vert $\angle$s are $\cong$.
3. m $\angle 4$ + m $\angle 8$ = 180.	3. Def of supp $\angle$s
4. m $\angle 2$ + m $\angle 8$ = 180	4. Sub
5. $\angle 2$ and $\angle 8$ are supp.	5. Def of supp $\angle$s

Additional Answers

Written Exercises

1. $\overline{PQ} \parallel \overline{RS}$; Corr ∠s are ≅.
2. $l \parallel m$; Alt int ∠s are ≅.
3. $\overline{RS} \parallel \overline{PQ}$; Int ∠s on the same side of the transv are supp.
4. $l \parallel m$; Corr ∠s are ≅.
5. $l \parallel m$; Alt int ∠s are ≅.
6. $l \parallel m$; Int ∠s on the same side of transv are supp.
7. $\overline{PQ} \parallel \overline{RS}$; Corr ∠s are ≅.
8. $l \parallel m$; Corr ∠s are ≅.
9. $l \parallel m$; Corr ∠s are ≅.
10. $\overline{PQ} \parallel \overline{RS}$; Int ∠s on the same side of transv are supp.
11. $\overline{PT}$ is a transversal to $\overline{PQ}$ and $\overline{RS}$.
12. Alternate int ∠s
13. $\overline{PT}$ is a transversal to the studs. ∠1 and ∠2 are corresponding ∠s.

21.

Statement	Reason
1. m ∠3 = m ∠2, $\overline{BC}$ bis ∠DBE	1. Given
2. m ∠1 = m ∠2	2. Def of ∠ bis
3. m ∠3 = m ∠1	3. Sub
4. ∴ $\overline{AD} \parallel \overline{BC}$	4. If alt int ∠s are ≅, then lines are ∥.

Classroom Exercises

State the reason why the lines, rays, or segments are parallel.

1.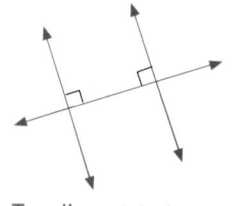

Corr ∠s ≅

2.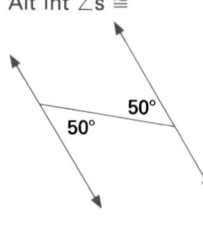

Two lines ⊥ to transversal

3. Alt int ∠s ≅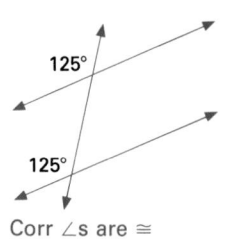

4.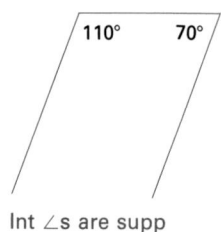

Int ∠s are supp

5.

Corr ∠s are ≅

6.

Int ∠s are supp

Written Exercises

Determine which lines or segments are parallel. Give a reason for your answer.

1. m ∠5 = 65, m ∠9 = 65
2. m ∠1 = 65, m ∠7 = 65
3. m ∠5 = 65, m ∠12 = 115
4. m ∠2 = 115, m ∠6 = 115
5. m ∠6 = 115, m ∠4 = 115
6. m ∠6 = 115, m ∠1 = 65
7. m ∠10 = 115, m ∠6 = 115
8. m ∠8 = 115, m ∠4 = 115
9. m ∠5 = 65, m ∠1 = 65
10. m ∠11 = 115, m ∠6 = 65

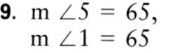

(Ex. 1–10)

11. In the drawing of the partial garage frame, the ceiling joist $\overline{PQ}$ and soleplate $\overline{RS}$ are parallel. How is the "let-in" corner brace $\overline{PT}$ related to $\overline{PQ}$ and $\overline{RS}$?
12. How are ∠RTP and ∠QPT related?
13. How is the corner brace $\overline{PT}$ related to the vertical studs it crosses? How are ∠1 and ∠2 related?

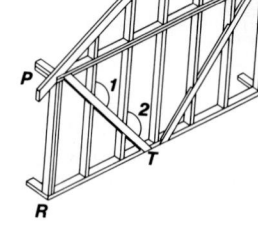

104 Chapter 3 Parallelism

Enrichment

Have the students answer this question by making drawings that satisfy the given conditions.

The sides of two angles lie in pairs of parallel lines. What can be said about the two angles?   The angles are either congruent (∠1 and ∠2) or supplementary (∠1 and ∠3).

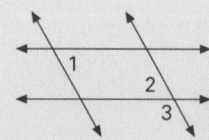

Find the value of *x* so that *m* ∥ *n*. (Exercises 14–16)

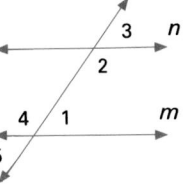

14. m ∠4 = 3*x* − 10, m ∠2 = *x* + 80   45
15. m ∠1 = 3*x* − 10, m ∠2 = 2*x* + 40   30
16. m ∠5 = 5*x* − 40, m ∠3 = 3*x*   20
17. m ∠1 = $\frac{4}{5}$ m ∠2. Find m ∠1, m ∠2.   80, 100

Find m ∠*ABC* so that $\overline{AE} \parallel \overline{CD}$. (Exercises 18–20)

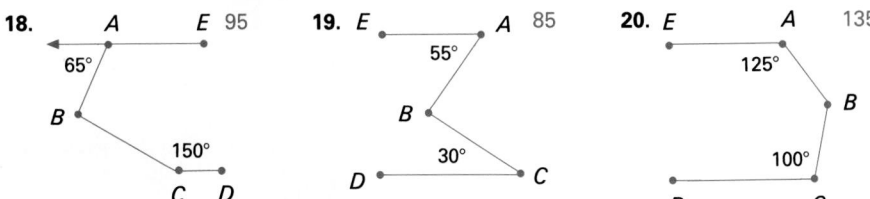

18. *A*   *E* 95 ... 65° ... *B* ... 150° ... *C*   *D*

19. *E* ... *A* 85 ... 55° ... *B* ... 30° ... *D* ... *C*

20. *E* ... *A* 135 ... 125° ... *B* ... 100° ... *D* ... *C*

21. Given: m ∠3 = m ∠2, $\overline{BC}$ bisects ∠*DBE*.
Prove: $\overline{AD} \parallel \overline{BC}$

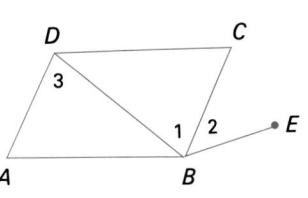

22. Given: m ∠1 = m ∠3, ∠2 is a supplement of ∠3.
Prove: $\overline{AB} \parallel \overline{CD}$

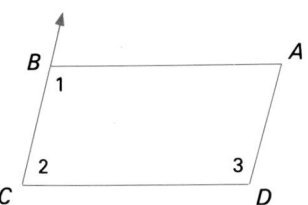

23. Prove Theorem 3.3.
24. Given: m ∠*VQT* = m ∠*WTQ*, m ∠1 = m ∠2
Prove: $\overline{PR} \parallel \overline{SU}$
25. m ∠3 = *x*², m ∠5 = 12*x* + 72
Find *x* so that $\overline{PR} \parallel \overline{SU}$.   6
26. m ∠3 = *x*, m ∠5 = 20*y* + 60, m ∠4 = 5*y* + 20
Find *x* and *y* so that $\overline{PR} \parallel \overline{SU}$.   40, 4

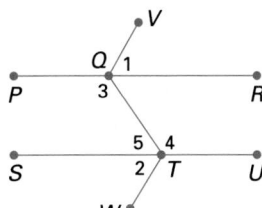

Mixed Review

Write the statement using logic symbols. (Exercises 1–3)

1. If *p*, then *q*   2.5 *p* → *q*   **2.** *p* and *q*   1.7 *p* ∧ *q*   **3.** *p* or *q*   1.7 *p* ∨ *q*
4. Identify the hypothesis and the conclusion of the conditional.
If a transversal cuts two lines so that a pair of corresponding angles
is congruent, then the lines are parallel.   2.5

22.

Statement	Reason
1. m ∠1 = m ∠3, ∠2 and ∠3 are supp	1. Given
2. m ∠2 + m ∠3 = 180	2. Def of supp ∠s
3. m ∠2 + m ∠1 = 180	3. Sub
4. ∴ $\overline{AB} \parallel \overline{CD}$	4. If int ∠s on the same side of transv are supp, then lines are ∥.

23.

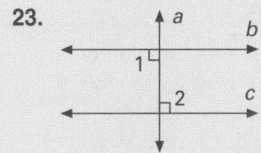

Statement	Reason
1. b ⊥ a, c ⊥ a	1. Given
2. m ∠1 = 90, m ∠2 = 90	2. Def of ⊥
3. m ∠1 = m ∠2	3. Sub
4. b ∥ c	4. If alt int ∠s are ≅, then lines are ∥.

24.

Statement	Reason
1. m ∠*VQT* = m ∠*WTQ*, m ∠1 = m ∠2	1. Given
2. m ∠*VQT* = m ∠1 + m ∠*RQT*, m ∠*WTQ* = m ∠2 + m ∠*STQ*	2. Angle Add Post
3. m ∠1 + m ∠*RQT* = m ∠2 + m ∠*STQ*	3. Sub
4. m ∠*RQT* = m ∠*STQ*	4. Subt Prop of Eq
5. ∴ $\overline{PR} \parallel \overline{SU}$	5. If alt int ∠s are ≅, then lines are ∥.

Mixed Review

4. Hyp: a transv cuts two lines so that a pair of corr ∠s are ≅. Concl: The lines are parallel.

◼️◼️ GETTING STARTED

Prerequisite Quiz

Indicate whether the conclusion is true. If false, give a reason.

1. Given: ∠1 and ∠2 are complementary.
   Conclusion: m ∠1 + m ∠2 = 180
   False. The sum is 90, not 180.
2. Given: ∠A and ∠B are vertical angles.
   Conclusion: ∠A ≅ ∠B   True
3. Given: *m* and *n* are skew lines.
   Conclusion: *m* and *n* intersect at point *P*.
   False. Skew lines cannot intersect.

Motivator

In an ancient Roman legend, there is the story of the man who was given the choice of opening two doors. Behind one was a ferocious lion who had not eaten in two days. Behind the other was a bag of gold. The man opens one door. There is the bag of gold. Ask the students what the man can conclude lies behind the other door even though he has not seen behind the door?
The ferocious lion

◼️◼️ TEACHING SUGGESTIONS

Lesson Note

Indirect proof is based on showing that every alternative conclusion is impossible. For example, suppose that the grades given in a school are *A, B, C, D,* and *E*. A teacher tells a student, "Your grade will not be an *A, C, D,* or *E*." The student *indirectly* concludes that the grade will be *B*.
In this lesson, indirect proofs are based on two alternatives only, the statement to be proved and its opposite. Emphasize the steps outlined in Example 1: (1) Assume the opposite of what is to be proved. (2) Reason directly until a contradiction is reached. (3) State that the assumption in the first step must be false. Therefore, the original conclusion is true.

3.4 Introduction to Indirect Proof

Objective   To write indirect proofs

Up until now, you have been writing direct proofs. Step by step, such proofs lead to a true conclusion. Another method of proving theorems is that of the *indirect* proof. The indirect proof shows that a conclusion cannot possibly be false. *Indirect* proof is frequently used by lawyers.

For example, suppose that Mr. Al Lee Bigh receives a summons for illegal parking on July 13. His lawyer argues *indirectly*. Mr. Bigh can prove that on July 13 his car was parked in another city 1,000 miles away. This leads to a *contradiction*. His car cannot be in two cities at the same time. The assumption of guilt must be false. Mr. Bigh, therefore, did not park illegally, and his alibi clears him.

Writing indirect proofs involves recognizing contradictory statements. For example:
  (1) ∠1 is complementary to ∠2.
  (2) m ∠1 + m ∠2 = 120

Statement 2 contradicts the definition of complementary angles. The sum of the measures of complementary angles is 90, not 120.

One kind of indirect proof makes use of the fact that there are only two possible truth values for any mathematical statement. A statement is either true or false, but not both.

To write an indirect proof, proceed as follows:
  (1) Accept the given information as true. Assume the *opposite* of what is to be proved.
  (2) Reason directly until there is a contradiction of the given or another known fact.
  (3) State that the assumption of the opposite of what was to be proved must be false. So, the original conclusion must be true because it is the only other possibility.

Sometimes it is difficult or impossible to find a direct method of proof. In such cases, indirect proof may be used. An example of indirect proof is illustrated on the next page.

106   Chapter 3 Parallelism

EXAMPLE   Write an indirect proof.

Given: m ∠1 ≠ m ∠2
Prove: m ∠1 ≠ m ∠3

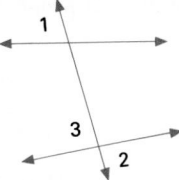

Proof
(Indirect)

1. Assume the opposite of what is to be proved.
2. Reason directly until a contradiction is reached.

Given: m ∠1 ≠ m ∠2
Assume: m ∠1 = m ∠3
m ∠2 = m ∠3 because vertical angles have the same measure. Then, by substitution, m ∠1 = m ∠2.

Now, at this point, m ∠1 ≠ m ∠2 is given, and m ∠1 = m ∠2 is concluded. This is a contradiction!

3. State that the assumption in the first step must be false. So, the original conclusion must be true.

So, the assumption that m ∠1 = m ∠3 must be *false*. Therefore, m ∠1 ≠ m ∠3.

Classroom Exercises

Explain why the two statements are contradictory.

1. *m* and *n* are skew. *m* intersects *n* at point *B*.

2. The sides of ∠1 are ⊥. m ∠1 = 60

3. ∠1 is a supplement of ∠2. m ∠1 + m ∠2 = 165

4. ∠1 and ∠2 are vertical angles. m ∠1 ≠ m ∠2

5. *m* ∥ *n*. The two lines *m* and *n* meet at point *P*.

6. ∠*A* is obtuse. m ∠*A* = 40

Written Exercises

For Exercises 1–2, write an indirect proof. (NOTE: ⊥̸ means "is not perpendicular to.")

1. Given: *P* is not the midpoint of $\overline{AB}$.
   Prove: *AP* ≠ *PB*

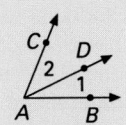

2. Given: m ∠1 ≠ m ∠2
   Prove: *r* ⊥̸ *s*

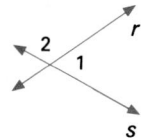

Machine Repair: Mechanics use indirect reasoning to identify problems in malfunctioning machines. The mechanic assumes a part to be faulty and tests the suspected part in a machine that works properly. If the machine continues to work properly during the test, then the assumption that the part is broken leads to a contradiction and to the conclusion that the part is not faulty. The faulty part may then be identified by the process of elimination.

Critical Thinking Questions

Logic: Let the students consider the following example of indirect reasoning in everyday life and ask them to think of other examples. Suppose that the inside light of your car does not turn on as usual when the car door is opened, but the radio, the headlights, the air conditioner and the windshield wipers are all still functioning normally. Explain the indirect reasoning that is involved in deciding that the light bulb must be burned out or the light socket is broken.   Assume the car battery is low. Then that contradicts the fact that other battery-operated devices are functioning properly. Therefore, the light bulb must have burned out or the light socket is broken.

Checkpoint

Explain why the statements are contradictory. (Ex. 1–2)

1. ∠*A* and ∠*B* are supplementary. m ∠*A* + m ∠*B* = 90   Sum of measure of supplementary ∠s is 180, not 90.

2. *m* and *n* are parallel lines. *m* and *n* intersect at point *T*.   Parallel lines do not have any points in common.

3. Give an indirect proof.
   Given: *a* ∦ *b*
   Prove: m ∠1 + m ∠3 ≠ 180

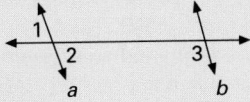

Proof: Given: *a* ∦ *b*. Assume m ∠1 + m ∠3 = 180; m ∠1 = m ∠2 (Vert ∠s are ≅.); m ∠2 + m ∠3 = 180 (Sub); *a* ∥ *b* (If int ∠s on same side of trans are supp, lines are ∥). Then *a* ∥ *b* and *a* ∦ *b*. But this is a contradiction. So, the assumption m ∠1 + m ∠3 = 180 must be false. Therefore, m ∠1 + m ∠3 ≠ 180.

Additional Example

Write an indirect proof.

Given: ∠1 ≠ ∠2
Prove: $\overrightarrow{AD}$ does not bisect ∠*CAB*.

Proof: Assume ∠1 ≠ ∠2 and $\overrightarrow{AD}$ bisects ∠*CAB*. Then ∠1 ≅ ∠2 and ∠1 ≠ ∠2 is a contradiction. So the assumption that $\overrightarrow{AD}$ bisects ∠*CAB* is false. Therefore, $\overrightarrow{AD}$ does not bisect ∠*CAB*.

107

Closure

Ask the students the following questions. Why is an **indirect proof** called indirect? Instead of directly proving a statement, you disprove the opposite. Have them give the three steps in writing an indirect proof. Assume the opposite of what is to be proved. Reason directly until there is a contradiction of the given. State that the opposite of what was to be proved must be false; so, the original conclusion must be true. Have the students give an example of a pair of contradictory statements. Answers may vary.

◤◤◤ FOLLOW UP

Guided Practice

Classroom Exercises 1–6

Independent Practice

◼A Ex. 1–6, ◼B Ex. 7–10, ◼C Ex. 11–12

Basic: WE 1–6, Brainteaser

Average: WE 1–10, Brainteaser

Above Average: WE 1–12, Brainteaser

Additional Answers

Classroom Exercises

1. Skew lines do not intersect.
2. If the sides of ∠1 are ⊥, then m ∠1 = 90.
3. If ∠1 and ∠2 are supp, m ∠1 + m ∠2 = 180.
4. Vertical angles are congruent ∴ m ∠1 = m ∠2
5. If lines m and n are ∥, they cannot intersect.
6. 90 < m ∠A < 180 by the definition of an obtuse angle.

Written Exercises

1. Given: P is not the midpt of $\overline{AB}$.
   Assume: $AP = PB$. Then $AP \cong PB$, by def of ≅ segs, and so P is a midpt of $\overline{AB}$. But P is not a midpt of $\overline{AB}$ and P is a midpt of $\overline{AB}$ is a contradiction. So the assumption that $AP = PB$ must be false. Therefore, $AP \neq PB$.

See pages 133–134 for the answers to Ex. 2–12 and Brainteaser.

For Exercises 3–8, write an indirect proof. Use the diagram below.

3. Given: $l \not\perp m$
   Prove: m ∠1 ≠ m ∠3

4. Given: $m \not\parallel n$
   Prove: ∠3 ≇ ∠2

5. Given: $l \perp m$, $m \not\parallel n$
   Prove: $l \not\perp n$

6. Given: $m \not\parallel n$
   Prove: m ∠4 + m ∠2 ≠ 180

7. Given: $m \not\parallel n$
   Prove: m ∠1 + m ∠2 ≠ 180

8. Given: m ∠3 ≠ m ∠2
   Prove: m ∠1 + m ∠2 ≠ 180

9. Prove indirectly that if m ∠1 ≠ m ∠2, then ∠1 and ∠2 are not vertical angles.

Essay.

10. Write an explanation of how an indirect proof is written. Indicate the three major steps.

11. Given: $\overleftrightarrow{AB}$ and $\overleftrightarrow{CD}$ are skew.
    Prove: $\overleftrightarrow{AC}$ and $\overleftrightarrow{BD}$ are skew.
    (HINT: If two lines are not skew, they either intersect or are parallel.)

12. The assertion that the given is true and the conclusion is false in an indirect proof may be stated as $p \wedge \sim q$. The contradiction of this assertion is $\sim(p \wedge \sim q)$. Use truth tables to show that $\sim(p \wedge \sim q)$ is logically equivalent to $p \rightarrow q$.

Mixed Review

Indicate whether each of the following is always true, sometimes true, or never true.

1. Three coplanar points are collinear.   *2.9* Sometimes true
2. Exactly one plane contains two lines.   *2.9* Sometimes true
3. Two distinct nonparallel planes intersect in one line.   *2.9* Always true
4. Two distinct lines intersect in several points.   *2.9* Never true
5. If m ∠A = m ∠B and m ∠B = m ∠C, then m ∠A = m ∠C.   *2.2* Always true

◤◤◤ Brainteaser

Ms. Brown, Ms. Green, and Ms. Blue live on the same street and wear coats of these three colors, but none wears a color matching her name. The one wearing the brown coat lives across the street from the other two. Ms. Brown lives next to the woman who drives to work. Ms. Blue takes the train to work. Match the women's names with the color of coat that each wears.

Enrichment

Have the students solve this puzzle indirectly by eliminating choices that conflict with the given information. At North High School, Jim, Betty, Carol, and Don are a freshman, a sophomore, a junior, and a senior, but not necessarily in that order. On the basis of the following information, determine the grade in which each student is enrolled.

1. Jim and Don are not eligible to graduate this year.
2. Betty was on the Honor Roll her freshman and sophomore years.
3. Carol went to her class dance with a boy who was one grade ahead of her.
4. Betty is one grade ahead of Jim.
5. Carol is two grades behind Jim.   Jim, junior; Betty, senior; Carol, freshman; Don, sophomore

3.5 Converses and the Parallel Postulate

Objectives
To form the converse of a conditional and to determine whether it is true

To prove and apply theorems about angles formed by a transversal of parallel lines

Interchanging the hypothesis and the conclusion of a conditional produces another conditional, called the converse. The converse of a true conditional may or may not be true. Consider this example:

> *Conditional*
> If you live in California, then you live in the United States. True
>
> *Converse*
> If you live in the United States, then you live in California. False

If a conditional is represented symbolically as $p \rightarrow q$, then $q \rightarrow p$ is the converse. $q \rightarrow p$ is false in the example above.

Definition

The **converse** of a conditional is the statement formed by interchanging the hypothesis and the conclusion.

EXAMPLE 1 A compact form of the Alternate Interior Angle Postulate is this: If alternate interior angles are congruent, then two lines are parallel. Write the converse of this conditional.

Solution If two lines are parallel, then alternate interior angles are congruent.

In the figure, line m contains point P and $m \parallel k$. It would seem that any other lines containing P, such as l or n, would intersect k. This suggests that m is the only line through P that is parallel to k.

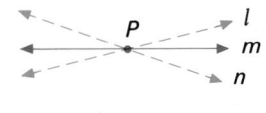

Postulate 14

The Parallel Postulate: Through a point not on a line, there is exactly one line parallel to the given line.

With the help of an *auxiliary line*, the converse of the Alternate Interior Angle Postulate can now be proved by using the Parallel Postulate. An auxiliary line is a line added in a diagram to help in a proof.

Teaching Resources

Problem Solving Worksheet 3
Quick Quizzes 22
Reteaching and Practice
  Worksheets, pp. 43, 44

▰▰▰ GETTING STARTED

Prerequisite Quiz

1. Identify the hypothesis and conclusion of the conditional:
   If angles are vertical angles, then the angles are congruent.   Hypothesis: Angles are vertical angles. Conclusion: Angles are congruent.

For the given data, state the reason why
$a \parallel b$.

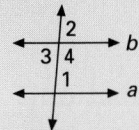

2. $\angle 1 \cong \angle 3$ Alt int $\angle$s are $\cong$.
3. $\angle 2 \cong \angle 1$ Corr $\angle$s are $\cong$.
4. $m \angle 1 + m \angle 4 = 180$.   Int $\angle$s on same side of trans are supp.

Motivator

For the conditional below:
Write the following statement on the board. If you live in California, then you live in the USA. Ask the students what the hypothesis is.   You live in California.   What the conclusion is?   You live in the USA.   Ask if the statement is true. Ask what the statement would be if the hypothesis and conclusion were exchanged.   If you live in the USA, then you live in California.   Have them determine if the new statement is true or false and why.   False. You live in the USA, yet do not live in California.

Additional Example 1

Write the converse of the conditional: If corresponding angles are congruent, then lines are parallel.   If lines are parallel, then corresponding angles are congruent.

Lesson Note

Some geometry textbooks postulate both the Alternate Interior Angle Postulate *and* its converse. Since indirect proof is an important part of geometry, we chose to teach this method in a previous lesson and then apply it in *proving* indirectly the converse of the Alternate Interior Angle Postulate.

Try to get students to discover the Parallel Postulate by using a real world model. For example, suppose a farmer has a barn. A post in the ground will be one of the support posts for building a fence parallel to the barn. How many different fences can be constructed containing this post? Only one. Thus, the Parallel Postulate is intuitively suggested.

Math Connections

Design: Architects use parallel lines in the design and structure of buildings. Discuss with the students the use of parallel lines in structural stability and aesthetic design. Ask them to name other examples of the use of parallel lines in everyday situations.

Critical Thinking Questions

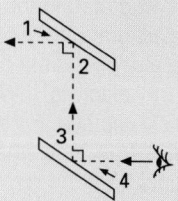

Application: Show students the diagram of a periscope that is made with two parallel mirrors which face each other. Tell them m ∠1 = 45 and ask the following questions. Identify a transversal.   The vertical path of vision between the mirrors.   Are ∠1 and ∠4 alternate exterior angles?   No; they are not formed by a common transversal.   How can you prove ∠1 ≅ ∠4?   Answers may vary. Sample plan for proof: Show m ∠2 = 45 by Subt Prop of Eq, because 45 + 90 + m ∠2 = 180. The mirrors are ∥, so m ∠3 = 45 because alt int ∠s are ≅. m ∠4 = 45 by Subt Prop of Eq, because 45 + 90 + m ∠4 = 180. Therefore, ∠1 ≅ ∠4 (Sub).

110

Theorem 3.4   If two parallel lines are intersected by a transversal, then alternate interior angles are congruent.

Given: $\overleftrightarrow{AB} \parallel \overleftrightarrow{CD}$
Prove: ∠DEF ≅ ∠EFB

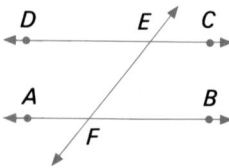

Proof (Indirect)

Given: $\overleftrightarrow{AB} \parallel \overleftrightarrow{CD}$

Assume that ∠DEF ≇ ∠EFB. At E, construct ∠1 such that ∠1 ≅ ∠EFB. Call the new line formed by this angle $\overleftrightarrow{GE}$. Then $\overleftrightarrow{GE} \parallel \overleftrightarrow{AB}$ by the Alternate Interior Angle Postulate. So, $\overleftrightarrow{DC}$ and $\overleftrightarrow{GE}$ both contain point E and are both parallel to $\overleftrightarrow{AB}$. This contradicts the Parallel Postulate. The assumption that ∠DEF ≇ EFB is false. Therefore, ∠DEF ≅ ∠EFB.

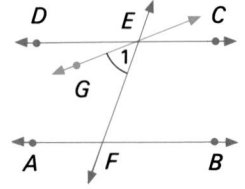

Theorem 3.4 can be used to prove the converse of Theorem 3.1.

Theorem 3.5   If two parallel lines are intersected by a transversal, then corresponding angles are congruent.

Given: $\overleftrightarrow{PQ} \parallel \overleftrightarrow{RS}$
Prove: ∠1 ≅ ∠3

Proof

Statement	Reason
1. $\overleftrightarrow{PQ} \parallel \overleftrightarrow{RS}$	1. Given
2. ∠2 ≅ ∠3	2. Alt int ∠s of ∥ lines are ≅.
3. ∠1 ≅ ∠2	3. Vert ∠s are ≅.
4. ∴ ∠1 ≅ ∠3	4. Trans Prop

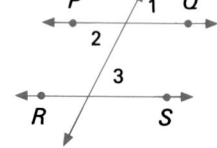

Recall that if interior angles on the same side of a transversal are supplementary, then the lines are parallel. The converse of this is stated as Theorem 3.6. You will be asked to prove this theorem in Exercise 24.

Theorem 3.6   If two parallel lines are intersected by a transversal, then interior angles on the same side of the transversal are supplementary.

110   Chapter 3 Parallelism

Additional Example 2

Given: m ∠1 = $\frac{4}{5}$ m ∠2, a ∥ b
Find m ∠2. 100

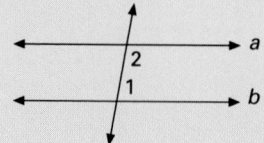

EXAMPLE 2   Given: $\overleftrightarrow{AB} \parallel \overleftrightarrow{CD}$, m $\angle 1 = \frac{2}{3}$ m $\angle 2$.
Find m $\angle 1$.

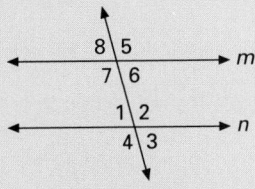

Plan   By Theorem 3.6, $\angle 1$ and $\angle 2$ are supplementary:
m $\angle 1 +$ m $\angle 2 = 180$.

Solution   Let m $\angle 2 = x$ and m $\angle 1 = \frac{2}{3}x$.   m $\angle 1 = \frac{2}{3}$ m $\angle 2$.

$$\frac{2}{3}x + x = 180$$
$$2x + 3x = 540$$
$$5x = 540$$
$$x = 108$$

Therefore, m $\angle 1 = \frac{2}{3} \cdot 108 = 72$

EXAMPLE 3   Given: $p \parallel q$, m $\angle 1 =$ m $\angle 3$
Prove: $r \parallel s$

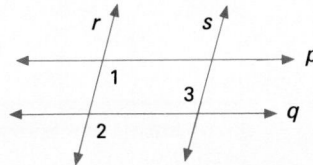

Plan r is parallel to s if m $\angle 2 =$ m $\angle 3$.

Proof

Statement	**Reason**
1. $p \parallel q$	1. Given
2. m $\angle 1 =$ m $\angle 2$	2. Corr $\angle$s of $\parallel$ lines are $\cong$ (have equal measure).
3. m $\angle 1 =$ m $\angle 3$	3. Given
4. m $\angle 2 =$ m $\angle 3$	4. Sub
5. $\therefore r \parallel s$	5. If alt int $\angle$s are $\cong$, then lines are $\parallel$.

Sometimes you have to decide whether you have enough information given to draw a conclusion. For example, before using a theorem about parallel lines to find an unknown measure, make sure that all of the conditions of that theorem are satisfied.

EXAMPLE 4   Based on the given information, find the measures of $\angle 1$ and $\angle 2$. If not enough information is given, state this fact.

Given: $\overline{AD} \parallel \overline{BC}$
Find m $\angle 1$ and m $\angle 2$.

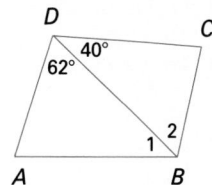

Additional Example 3

Given: $a \parallel b$, m $\angle 1 +$ m $\angle 2 = 180$
Prove: $c \parallel d$

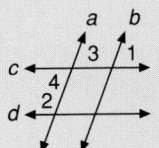

Proof: $a \parallel b$ (Given); m $\angle 1 +$ m $\angle 2 = 180$ (Given); m $\angle 3 =$ m $\angle 1$, or $\angle 3 \cong \angle 1$ (Corr $\angle$s of $\parallel$ lines are $\cong$.); m $\angle 4 =$ m $\angle 3$, or $\angle 4 \cong \angle 3$ (Vert. $\angle$s are $\cong$.); m $\angle 4 =$ m $\angle 1$ (Sub); m $\angle 4 +$ m $\angle 2 = 180$ (Sub); $c \parallel d$ (If int $\angle$s on the same side of trans are supp, lines are $\parallel$.)

Checkpoint

Given: $m \parallel n$. Find each indicated measure.

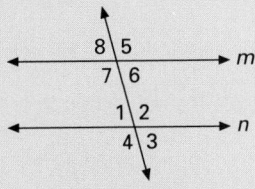

1. Given: m $\angle 7 = 39$. Find m $\angle 2$. 39
2. Given: m $\angle 1 = 105$. Find m $\angle 7$. 75
3. Given: m $\angle 8 = 65$. Find m $\angle 2$. 115
4. Given: m $\angle 1 = \frac{3}{7}$ m $\angle 7$. Find m $\angle 1$. 54
5. Given: m $\angle 6 = (7x - 30)$, m $\angle 3 = (-2x + 60)$. Find m $\angle 4$. 140

Closure

Ask the students to state the Parallel Postulate. Have them explain how to form the converse of a conditional.   Interchange the hypothesis and the conclusion.   Have the students describe what the theorems in this lesson conclude about angle pairs if two parallel lines are intersected by a transversal.   Alternate interior angles and corresponding angles are congruent. Interior angles on the same side of the transversal are supplementary.

▋▋▋▋FOLLOW UP

Guided Practice

Classroom Exercises 1–3

Independent Practice

A Ex. 1–15, **B** Ex. 16–23, **C** Ex. 24–27

Basic: WE 1–15 odd, Midchapter Review, AR all

Average: WE 1–23 odd, Midchapter Review, AR all

Above Average: WE 1–27 odd, Midchapter Review, AR all

Written Exercises

9. If the sum of the meas of 2 ∠s is 90, then the ∠s are comp. (True)

10. If 2 lines do not intersect, then they are ∥. (False)

11. If 2 lines are not ∥, then they are skew. (False)

12. m ∠3 = 60, not enough info to find m ∠4.

13. m ∠1 = 25, not enough info to find m ∠2.

14. m ∠ABC = 130, not enough info to find m ∠ADC.

15. m ∠3 = 56, m ∠ 5 = 75

16.

Statement	Reason
1. r ∥ s	1. Given
2. ∠1 and ∠5 are supp, ∠2 and ∠6 are supp.	2. If the outer rays of 2 adj ∠s form a st ∠, then the ∠s are supp.
3. m ∠6 = m ∠5	3. If lines are ∥, then corr ∠s are ≅.
4. ∴ ∠1 ≅ ∠2	4. Supp ∠s of ≅ ∠s are ≅.

17.

Statement	Reason
1. r ∥ s	1. Given
2. m ∠3 = m ∠6	2. Vert ∠s are ≅.
3. m ∠6 = m ∠5	3. If lines are ∥, then corr ∠s are ≅.
4. m ∠3 = m ∠5	4. Sub
5. m ∠1 + m ∠5 = 180	5. If the outer rays of 2 adj ∠s form a st ∠, then the ∠s are supp.
6. m ∠1 + m ∠3 = 180	6. Sub
7. ∴ ∠1 and ∠3 are supp.	7. Def of supp ∠s

18.

Statement	Reason
1. l ∥ m	1. Given
2. m ∠4 = m ∠5	2. If lines are ∥, then alt int ∠s are ≅.
3. m ∠5 + m ∠1 = 180	3. If the outer rays of 2 adj ∠ form a st ∠, then the sum of their meas is 180.
4. m ∠4 + m ∠1 = 180	4. Sub
5. ∠4 and ∠1 are supp.	5. Def of supp ∠s

Solution Shade the parallel segments $\overline{AD}$, $\overline{BC}$ and transversal $\overline{BD}$. ∠2 and ∠ADB are alternate interior angles of $\overline{AD}$ and $\overline{BC}$. So, m ∠2 = 62.

You do not know whether $\overline{AB}$ and $\overline{CD}$ are parallel. The measure of ∠1 cannot be found.

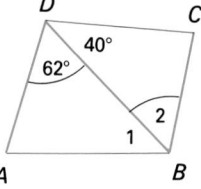

So, m ∠2 = 62, but not enough information is provided to determine m ∠1.

Theorem 3.7 If a transversal is perpendicular to one of two parallel lines, then it is perpendicular to the other.

Theorem 3.8 In a plane, if two lines are parallel to the same line, then they are parallel to each other.

Classroom Exercises

Given *l* ∥ *m*. Find the measures.

1. m ∠1 = _____ 75
2. m ∠2 = _____ 105
3. m ∠3 = _____ 105

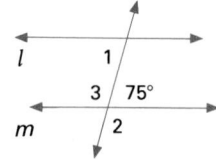

Written Exercises

If $\overleftrightarrow{AB}$ ∥ $\overleftrightarrow{CD}$, find the measure of the indicated angle.

1. Given: m ∠1 = 37
   Find m ∠8.   37

2. Given: m ∠3 = 40
   Find m ∠5.   140

3. Given: m ∠7 = 100
   Find m ∠4.   100

4. Given: m ∠8 = 30
   Find m ∠2.   150

5. Given: m ∠8 = 3x + 30, m ∠3 = x + 80
   Find m ∠8.   105

6. Given: m ∠4 = 4x + 20, m ∠8 = 3x + 90
   Find m ∠4.   60

7. Given: m ∠3 = $\frac{4}{5}$ m ∠5
   Find m ∠5.   100

8. Given: m ∠4 = $\frac{2}{7}$ m ∠8
   Find m ∠4.   40

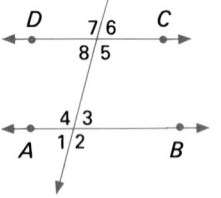

112 Chapter 3 Parallelism

Additional Example 4

Based on the given information, find the measures of ∠3 and ∠4. If not enough information is given, state this fact.

Given: $\overline{PS}$ ∥ $\overline{QR}$ m ∠3 = 70; m ∠4 cannot be found since $\overline{RS}$ is not given ∥ to $\overline{PQ}$.

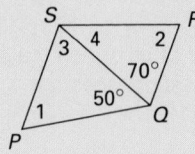

Write the converse of the conditional. Determine whether the converse is true. (Exercises 9–11)

9. If two angles are complementary, then the sum of the angle measures is 90.

10. If two lines are parallel, then the two lines do not intersect.

11. If two lines are skew, then they are not parallel.

Based on the given information, find the indicated angle measures. If not enough information is given, state this fact. (Exercises 12–15)

12. Given: $\overline{DC} \parallel \overline{AB}$, m $\angle 2 = 60$, m $\angle 1 = 40$
Find m $\angle 3$ and m $\angle 4$.

13. Given: $\overline{AD} \parallel \overline{BC}$, m $\angle 3 = 35$, m $\angle 4 = 25$
Find m $\angle 1$ and m $\angle 2$.

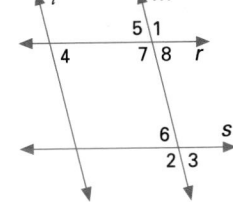

14. Given: $\overline{AD} \parallel \overline{BC}$, m $\angle 5 = 50$, m $\angle 3 = 20$
Find m $\angle ABC$ and m $\angle ADC$.

15. Given: $\overline{DC} \parallel \overline{AB}$, m $\angle 1 = 49$, m $\angle 2 = 56$
Find m $\angle 5$ and m $\angle 3$.

16. Given: $r \parallel s$
Prove: $\angle 1 \cong \angle 2$

17. Given: $r \parallel s$
Prove: $\angle 1$ and $\angle 3$ are supplementary.

18. Given: $l \parallel m$
Prove: $\angle 1$ and $\angle 4$ are supplementary.

19. Given: $l \parallel m$, m $\angle 4 =$ m $\angle 3$
Prove: $r \parallel s$

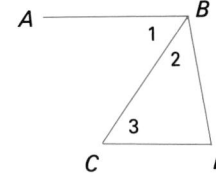

20. Given: $\overline{AB} \parallel \overline{CD}$
Prove: m $\angle 3$ + m $\angle 2 =$ m $\angle ABD$

21. Given: $\overline{BC}$ bisects $\angle ABD$; $\overline{AB} \parallel \overline{CD}$.
Prove: m $\angle 3 =$ m $\angle 2$

22. Given: $\overline{AB} \parallel \overline{CD}$, $\overline{BD} \perp \overline{CD}$
Prove: $\overline{BD} \perp \overline{AB}$
(Theorem 3.7)

23. Given: $\overline{AB} \parallel \overline{CD}$, $\overline{BD} \perp \overline{CD}$
Prove: $\angle 3$ and $\angle 2$ are complementary.

24. Prove Theorem 3.6.

25. Write a direct proof of Theorem 3.8. (HINT: Draw a transversal.)

26. Write an indirect proof of Theorem 3.8.

27. Prove: If two parallel lines are intersected by a transversal, then the bisectors of a pair of alternate interior angles are parallel.

19.

Statement	Reason
1. $l \parallel m$, m $\angle 4$ = m $\angle 3$	1. Given
2. m $\angle 4$ = m $\angle 5$	2. If lines are $\parallel$, then alt int $\angle$s are $\cong$.
3. m $\angle 6$ = m $\angle 3$	3. Vert $\angle$s are $\cong$.
4. m $\angle 6$ = m $\angle 5$	4. Sub
5. $r \parallel s$	5. If corr $\angle$s are $\cong$, then lines are $\parallel$.

20.

Statement	Reason
1. $\overline{AB} \parallel \overline{CD}$	1. Given
2. m $\angle 1$ + m $\angle 2$ = m $\angle ABD$	2. Angle Add Post
3. m $\angle 3$ = m $\angle 1$	3. If lines are $\parallel$, then alt int $\angle$s are $\cong$.
4. $\therefore$ m $\angle 3$ + m $\angle 2$ = m $\angle ABD$	4. Sub

21.

Statement	Reason
1. $\overline{BC}$ bis $\angle ABD$, $\overline{AB} \parallel \overline{CD}$	1. Given
2. m $\angle 1$ = m $\angle 2$	2. Def of $\angle$ bis
3. m $\angle 1$ = m $\angle 3$	3. If lines are $\parallel$, then alt int $\angle$s are $\cong$.
4. $\therefore$ m $\angle 2$ = m $\angle 3$	4. Sub

22.

Statement	Reason
1. $\overline{AB} \parallel \overline{CD}$, $\overline{BD} \perp \overline{CD}$	1. Given
2. m $\angle CDB$ = 90	2. Def of $\perp$
3. m $\angle ABD$ + m $\angle CDB$ = 180	3. If lines are $\parallel$, then int $\angle$s on same side of the transv are supp.
4. m $\angle ABD$ + 90 = 180	4. Sub
5. m $\angle ABD$ = 90	5. Subt Prop of Eq
6. $\therefore \overline{BD} \perp \overline{AB}$	6. Def of $\perp$

23.

Statement	Reason
1. $\overline{AB} \parallel \overline{CD}$, $\overline{BD} \perp \overline{CD}$	1. Given
2. $\overline{BD} \perp \overline{AB}$	2. If a transv is $\perp$ to one of 2 $\parallel$ lines, then it is $\perp$ to the other.
3. m $\angle 1$ + m $\angle 2$ = 90	3. If the outer rays of 2 adj acute $\angle$s are $\perp$, then the sum of their meas is 90.
4. m $\angle 1$ = m $\angle 3$	4. If lines are $\parallel$, then alt int $\angle$s are $\cong$.
5. m $\angle 2$ + m $\angle 3$ = 90	5. Sub
6. $\therefore \angle 2$ and $\angle 3$ are comp.	6. Def of comp $\angle$s

Enrichment

The Parallel Postulate of this lesson is Euclid's. However, since postulates are given without proof, other postulates are possible, providing systems of *non-Euclidean geometry*. Consider these: Through a point not on a line, (1) there can be more than one line parallel to the given line, (2) there is no line parallel to the given line. The first of these postulates was investigated by Janos Bolyai and Nikolai Lobachevsky. The resulting system is known as *hyperbolic geometry*. The second postulate was studied by Bernhard Riemann. The resulting system is known as *elliptic geometry*. Have the students do research and write reports on the subject of non-Euclidean geometry.

24.

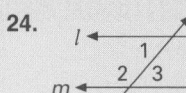

Statement	Reason
1. $l \parallel m$	1. Given
2. m $\angle 1$ = m $\angle 3$	2. If lines are ∥, then alt int $\angle$s are ≅.
3. m $\angle 2$ + m $\angle 3$ = 180	3. If the outer rays of 2 adj $\angle$s form a st $\angle$, then the sum of their meas is 180.
4. m $\angle 1$ + m $\angle 2$ = 180	4. Sub
5. ∴ $\angle 1$ and $\angle 2$ are supp.	5. Def of supp $\angle$s

25.

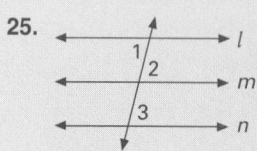

Statement	Reason
1. $l \parallel m$, $m \parallel n$	1. Given
2. m $\angle 1$ = m $\angle 2$	2. Alt int $\angle$s are ≅.
3. m $\angle 2$ = m $\angle 3$	3. If lines are ∥, then corr $\angle$s are ≅.
4. m $\angle 1$ = m $\angle 3$	4. Trans Prop of Eq
5. $l \parallel n$	5. If alt int $\angle$s are ≅, then lines are ∥.

27.

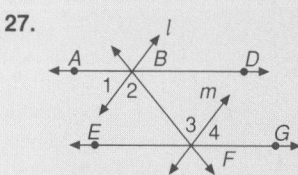

Statement	Reason
1. $\overleftrightarrow{AD} \parallel \overleftrightarrow{EG}$. l bis $\angle ABF$, m bis $\angle BFG$	1. Given
2. m $\angle ABF$ = m $\angle BFG$	2. If lines are ∥, then alt int $\angle$s are ≅.
3. m $\angle ABF$ = m $\angle 1$ + m $\angle 2$, m $\angle BFG$ = m $\angle 3$ + m $\angle 4$	3. Angle Add Post
4. m $\angle 1$ + m $\angle 2$ = m $\angle 3$ + m $\angle 4$	4. Sub
5. m $\angle 1$ = m $\angle 2$, m $\angle 3$ = m $\angle 4$	5. Def of $\angle$ bis
6. m $\angle 2$ + m $\angle 2$ = m $\angle 3$ + m $\angle 3$ (2 × m $\angle 2$ = 2 × m $\angle 3$	6. Sub
7. m $\angle 2$ = m $\angle 3$	7. Div Prop of Eq
8. ∴ $l \parallel m$	8. If alt int $\angle$s are ≅, then lines are ∥.

Midchapter Review

True or false? If false, draw a diagram to illustrate why. *3.1*

1. If two lines are not parallel, then they must intersect.
2. Two lines parallel to the same plane are parallel.
3. A line parallel to a plane must be parallel to any line in that plane.
4. Given: $\angle 4 \cong \angle 5$ *3.2, 3.3*
   Which segments are parallel? Why? $\overline{AE} \parallel \overline{FB}$, alt int $\angle$s
5. Given: $\overline{ED} \parallel \overline{AC}$, m $\angle 2 = \frac{4}{5}$m $\angle 8$.
   m $\angle 3 =$ _____ *3.5* 80
6. Given: $\overline{ED} \parallel \overline{AC}$, $\angle 6 \cong \angle 2$
   Prove: $\angle 6 \cong \angle 3$ *3.5*
7. Given: $\overline{ED} \not\parallel \overline{AC}$
   Prove indirectly: $\angle 1 \not\cong \angle 2$ *3.4*
8. Write the converse of the following conditional.
   If $\angle 6 \cong \angle 7$, then $\overline{EF} \parallel \overline{AB}$ *3.5*

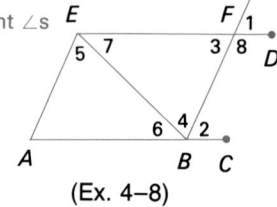

(Ex. 4–8)

Algebra Review

Objective: To solve fractional equations

To solve a fractional equation:
(1) Multiply each side by the LCD.
(2) Solve the resulting equation.

Example: Solve.

$$\frac{3x - 5}{4} = \frac{2x + 3}{3} \qquad \text{(LCD = 12)}$$

$$12 \cdot \frac{3x - 5}{4} = 12 \cdot \frac{2x + 3}{3} \quad \text{Multiply each side by 12.}$$

$$3(3x - 5) = 4(2x + 3) \quad \text{Solve.}$$

$$9x - 15 = 8x + 12$$

$$x = 27$$

Solve for x.

1. $\dfrac{4x - 2}{3} = \dfrac{5x + 1}{4}$ 11

2. $\dfrac{2y - 5}{3} = \dfrac{3y + 2}{4}$ -26

3. $\dfrac{z + 2}{3} = \dfrac{4z - 3}{6} - \dfrac{z}{2}$ -7

4. $\dfrac{2w - 3}{6} - \dfrac{w + 5}{9} = \dfrac{w - 1}{2}$ -2

5. $\dfrac{4}{t} + \dfrac{3}{5} = 3$ $\dfrac{5}{3}$

6. $\dfrac{3}{4v} = \dfrac{5}{6} - \dfrac{2}{3v}$ $\dfrac{17}{10}$

114   Chapter 3 Parallelism

26. Using the same diagram from Exercise 25: Assume that $l \parallel m$, $m \parallel n$ and $l \not\parallel n$. Then, because $\angle 1$ and $\angle 2$ are alt int $\angle$s of 2 ∥ lines, m $\angle 1$ = m $\angle 2$. Since $\angle 2$ and $\angle 3$ are corr $\angle$s of 2 ∥ lines, m $\angle 2$ = m $\angle 3$. By Sub, m $\angle 1$ = m $\angle 3$. Since $\angle 1$ and $\angle 3$ are ≅ alt int $\angle$s created by a trans through l and n, then $l \parallel n$. But $l \parallel n$ and $l \not\parallel n$ is a contradiction. So, the assumption that $l \not\parallel n$ is false. Therefore, $l \parallel n$.

See page 120 for the answers to Midchapter Review Ex. 1–3, 6–8.

3.6 The Angles of a Triangle

Objective

To apply the theorem about the sum of the measures of the angles of a triangle

Triangle ABC ($\triangle ABC$) is formed by three segments joining three noncollinear points, A, B, and C. Each of these points is a **vertex** (plural: **vertices**) of the triangle. The segments are the **sides** of the triangle.

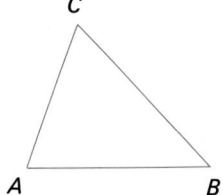

Sides of $\triangle ABC$: $\overline{AB}$, $\overline{BC}$, $\overline{CA}$
Vertices of $\triangle ABC$: A, B, C
Angles of $\triangle ABC$: $\angle A$, $\angle B$, $\angle C$

Definition

> A **triangle** is a figure formed by three segments joining three noncollinear points.

Recall that a straight angle has a measure of 180. In the figure, m $\angle AOC$ + m $\angle 3$ = 180, or by the Angle Addition Postulate, (m $\angle 1$ + m $\angle 2$) + m $\angle 3$ = 180.

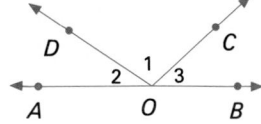

This concept of straight angle, together with properties of parallel lines, can be used to discover the sum of the measures of the angles of a triangle.

In the figure, $\overleftrightarrow{AB} \parallel \overleftrightarrow{CD}$, a side of $\triangle COD$.

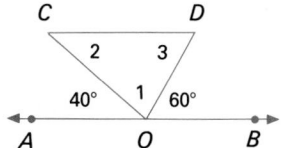

(1) Because $\angle AOB$ is a straight angle,
$$40 + m\ \angle 1 + 60 = 180$$
$$100 + m\ \angle 1 = 180$$
$$m\ \angle 1 = 80$$

(2) Since $\overleftrightarrow{AB} \parallel \overleftrightarrow{CD}$, m $\angle 2$ = 40 and m $\angle 3$ = 60 (alt int $\angle$s of $\parallel$ lines are $\cong$).

From (1) and (2) above, m $\angle 1$ + m $\angle 2$ + m $\angle 3$ becomes 80 + 40 + 60, or 180.

Therefore, m $\angle 1$ + m $\angle 2$ + m $\angle 3$ = 180.

If you were to perform this experiment with other pairs of sample values for m $\angle AOC$ and m $\angle BOD$, say 50 and 70, or 20 and 55, you would be able to *inductively* make a generalization about the sum of the measures of the angles of a triangle.

3.6 The Angles of a Triangle **115**

Teaching Resources

Manipulative Worksheet 6
Quick Quizzes 23
Reteaching and Practice Worksheets, pp. 45, 46
Technology Worksheet 3

▄▄▄GETTING STARTED

Prerequisite Quiz

Given $\overline{AB} \parallel \overline{CD}$.
Find each indicated angle measure.

1. m $\angle 1$ 55
2. m $\angle 2$ 50
3. m $\angle 3$ 75
4. m $\angle 4$ 130
5. m $\angle 5$ 105

Motivator

Ask your students to draw $\triangle ABC$ on a sheet of paper and cut out the triangle. Have them cut off the three corners or angles and place the three angles adjacent to each other. Ask the students what appears to be true about the sum of the measures of the three angles A, B, and C. The sum of the measures of the angles is 180.

▄▄▄TEACHING SUGGESTIONS

Lesson Note

Students have seen that if the outer rays of two adjacent angles form a line, then the sum of the angle measures is 180. Point out that, in the figure below, $\angle AOC$ is supplementary to $\angle 3$. Therefore, since m $\angle 1$ + m $\angle 2$ = m $\angle AOC$, m $\angle 1$ + m $\angle 2$ + m $\angle 3$ = 180. Thus, three angles can be consecutively adjacent angles with the outermost rays forming a line.

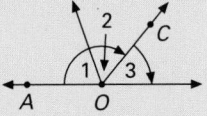

Math Connections

Polygons: Mention to students that the interior of a polygon can be readily divided into triangles. You can find the sum of the measures of the angles in any polygon if you know that the sum of the measures of the angles in a triangle is 180.

Critical Thinking Questions

Synthesis: Refer students to Theorem 3.9 at the top of page 116 for the following questions. Classify the angles in a right triangle.   Two acute ∠s and a rt ∠   Is it possible to have more than one obtuse angle in a triangle?   No; the sum would be greater than 180.

Checkpoint

Find the measure of each angle of △ ABC for the given data.

1. m ∠A = 2x, m ∠B = 3x, m ∠C = 5x
   m ∠A = 36, m ∠B = 54,
   m ∠C = 90
2. $\overline{CA}$ ⊥ $\overline{CB}$, m ∠A = 40   m ∠C = 90,
   m ∠B = 50
3. m ∠C = 144, m ∠B = $\frac{2}{7}$ m ∠A
   m ∠B = 8, m ∠A = 28
4. ∠C ≅ ∠A, m ∠B = 140   m ∠A =
   m ∠C = 20
5. m ∠A = 30, m ∠C is twice m ∠B.
   m ∠B = 50, m ∠C = 100

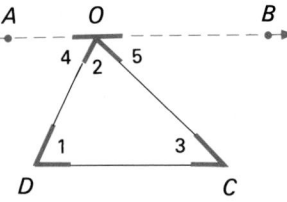

Theorem 3.9   The sum of the measures of the angles of a triangle is 180.

Given: △DOC
Prove: m ∠1 + m ∠2 + m ∠3 = 180

Proof

	Statement		Reason
1.	△DOC	1.	Given
2.	Through O, draw $\overleftrightarrow{AB} \parallel \overline{DC}$.	2.	Par Post
3.	m ∠1 = m ∠4, m ∠3 = m ∠5	3.	Alt int ∠s of ‖ lines have equal measure.
4.	m ∠4 + m ∠DOB = 180	4.	If outer rays of two adj ∠s form a st ∠, then the sum of their measures is 180.
5.	m ∠DOB = m ∠2 + m ∠5	5.	∠ Add Post
6.	m ∠4 + m ∠2 + m ∠5 = 180	6.	Sub
7.	∴ m ∠1 + m ∠2 + m ∠3 = 180	7.	Sub (Steps 3 and 6)

EXAMPLE 1   Given: m ∠A = 2x + 15, m ∠B = 3x − 5, m ∠C = 4x + 35

Find m ∠A, m ∠B, and m ∠C.

Solution   The sum of measures of the angles of a triangle is 180.

$$(2x + 15) + (3x − 5) + (4x + 35) = 180$$
$$9x + 45 = 180$$
$$9x = 135$$
$$x = 15$$

Check
m ∠A = 2x + 15 = 2 · 15 + 15 = 45
m ∠B = 3x − 5 = 3 · 15 − 5 = 40
m ∠C = 4x + 35 = 4 · 15 + 35 = 95

45 + 40 + 95 = 180

Recall that if the sides of an angle are perpendicular, the measure of the angle is 90. This idea is used in the next example.

Additional Example 1

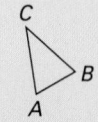

Given: △ABC with
   m ∠A = 5x + 20,
   m ∠B = 2x + 50,
   m ∠C = 5x − 10

Find the measure of each angle of the triangle.   m ∠A = 70, m ∠B = 70, m ∠C = 40

EXAMPLE 2   Given: $\triangle PQR$, with $\overline{PQ} \perp \overline{PR}$, m $\angle Q = \frac{2}{3}$ m $\angle R$
Find m $\angle R$.

Plan $\overline{PQ} \perp \overline{PR}$. So, m $\angle P = 90$.
Since m $\angle Q = \frac{2}{3}$ m $\angle R$, let m $\angle R = x$ and m $\angle Q = \frac{2}{3}x$.

Solution
$$m \angle P + m \angle Q + m \angle R = 180$$
$$90 + \tfrac{2}{3}x + x = 180$$
$$3(90 + \tfrac{2}{3}x + x) = 3 \cdot 180$$
$$270 + 2x + 3x = 540$$
$$5x = 270$$
$$x = 54 \Rightarrow m \angle R = 54$$

Classroom Exercises

Find the measure of the third angle of $\triangle ABC$.

1. m $\angle A = 60$, m $\angle B = 40$, m $\angle C =$ ___80___
2. m $\angle A = 70$, m $\angle B = 70$, m $\angle C =$ ___40___
3. m $\angle A = 90$, m $\angle B = 30$, m $\angle C =$ ___60___
4. m $\angle A = 45$, m $\angle B = 45$, m $\angle C =$ ___90___

Find the measure of each angle of $\triangle ABC$.

5. m $\angle A = x$, m $\angle B = 2x$, m $\angle C = 3x$ 30, 60, 90
6. m $\angle A = 40$, m $\angle B = x$, m $\angle C = x + 10$ 65, 75

Written Exercises

Find the measure of each angle of $\triangle ABC$ for the given data.

1. m $\angle A =$ m $\angle B =$ m $\angle C$ 60, 60, 60
2. $\overline{AC} \perp \overline{BC}$, m $\angle A = 20$ 70, 90
3. m $\angle A = x$, m $\angle B = x$, m $\angle C = 3x$ 36, 36, 108
4. m $\angle A = x + 2$, m $\angle B = 3x - 10$, m $\angle C = 4x - 4$
5. m $\angle A = x$, m $\angle B = x + 20$, m $\angle C = x + 40$
6. m $\angle A = 3x + 20$, m $\angle B = x - 10$, m $\angle C = 90 - 2x$
7. m $\angle A = \frac{3}{7}$ m $\angle B$, $\overline{AC} \perp \overline{BC}$ 27, 63, 90
8. $\overline{AC} \perp \overline{BC}$, m $\angle A = \frac{5}{4}$ m $\angle B$ 50, 40, 90
9. In $\triangle PQR$, m $\angle P = 70$. Angles Q and R have equal measures.
Find m $\angle Q$.   55
10. In $\triangle ABC$, m $\angle C = 150$. Find m $\angle A$ if m $\angle A = 2(m \angle B)$.   20

Additional Example 2

Given: $\overline{AB} \perp \overline{AC}$, m $\angle B = \frac{7}{2}$ m $\angle C$
Find m $\angle C$.   20

117

14.

Statement	Reason
1. $\overline{PS} \parallel \overline{QR}$, $\overline{PQ} \parallel \overline{RS}$	1. Given
2. m∠3 = m∠PQS, m∠2 = m∠QSP	2. If lines are ∥, then alt int ∠s are ≅.
3. m∠1 + m∠PQS + m∠QSP = 180	3. The sum of the meas of the ∠s of a △ is 180.
4. ∴ m∠1 + m∠2 + m∠3 = 180.	4. Sub

15.

Statement	Reason
1. $\overline{CD} \parallel \overline{AB}$	1. Given
2. m∠5 + m∠ACD = 180	2. If lines are ∥, then int ∠s on same side of the transv are supp.
3. m∠ACD = m∠2 + m∠3	3. Angle Add Post
4. m∠5 + m∠2 + m∠3 = 180	4. Sub
5. m∠2 = m∠4	5. If lines are ∥, then alt int ∠s are ≅.
6. ∴ m∠5 + m∠3 + m∠4 = 180.	6. Sub

In the figure of the frame of a house under construction, $\overline{EG} \perp \overline{AB}$, $\overline{FH} \perp \overline{AB}$, $\overline{CE} \perp \overline{AD}$, $\overline{CF} \perp \overline{DB}$, $\overline{CD} \perp \overline{AB}$. (Exercises 11–13)

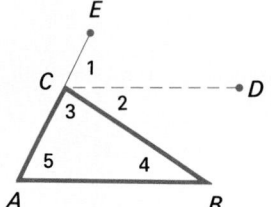

11. If the pitch of the roof (∠1) measures 44, find the measure of the angle between roof rafters, m∠ADB. (Assume m∠DAB = m∠DBA.) 92

12. If the pitch of the roof measures 30, find m∠4. (∠4 is the angle formed by brace $\overline{EG}$ and the roof rafter $\overline{AD}$.) 60

13. Find the measure of the angle between the roof rafters (m∠ADB) if m∠2 = 20. 140

14. Given: $\overline{PS} \parallel \overline{QR}$, $\overline{PQ} \parallel \overline{RS}$
Prove: m∠1 + m∠2 + m∠3 = 180

15. Use the figure below to prove Theorem 3.9 given that $\overline{CD}$ is parallel to $\overline{AB}$.

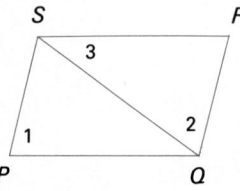

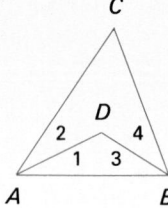

Find the measure of each numbered angle. (Exercises 16–18)

16. Given: $\overleftrightarrow{AB} \parallel \overleftrightarrow{CD}$

m∠1 = 50, m∠2 = 50

17. Given: $\overrightarrow{CD}$ bisects ∠ACB.

m∠1 = 51, m∠2 = 51, m∠3 = 97, m∠4 = 83

18. Given: $\overline{CD} \parallel \overline{AB}$

m∠1 = 34, m∠2 = 34, m∠3 = 40

In the figure below, $\overline{AD}$ bisects ∠CAB, and $\overline{BD}$ bisects ∠CBA.

19. Given: m∠C = 70, m∠1 = 20
Find m∠D. 125

20. Given: m∠C = 50, m∠1 = 30
Find m∠D. 115

21. Given: m∠1 = m∠3, m∠D = 100
Find m∠C. 20

Enrichment

Draw this figure on the chalkboard.

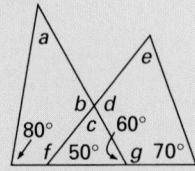

Have the students determine the measures of the angles labeled a–g.

a, 40; b, 110; c, 70; d, 110; e, 60; f, 130; g, 120 Next, have the students calculate the sum of the measures of the angles of the two quadrilaterals in the figure. 360 in each Discuss the possibility that the sum of the measures of the angles of every quadrilateral is 360. Challenge the students to design a proof that this is true. Answers will vary

In the figure, $\overline{ED} \parallel \overline{AB}$, $\overline{GF}$ bisects $\angle DGC$; $\overline{CF}$ bisects $\angle BCG$.

22. Given: m $\angle 2 = 40$. Find m $\angle F$.   90
23. Given: m $\angle GCB = 70$. Find m $\angle F$.   90
24. Given: m $\angle CGD = 100$. Find m $\angle F$.   90
25. The results of Exercises 22–24 inductively suggest the following property: If two lines are parallel, then the bisectors of interior angles on the same side of a transversal are _____.   ⊥

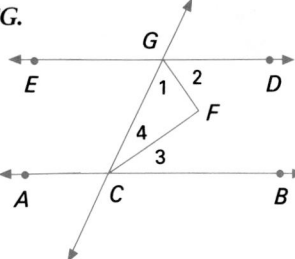

Mixed Review

1. Given: C is the midpoint of $\overline{BD}$; $AB = BC$.
   Prove: $AB = CD$ 1.3, 2.2, 2.4
2. Given: m $\angle 2 =$ m $\angle 4$, m $\angle 1 =$ m $\angle 3$
   Prove: m $\angle ABC =$ m $\angle EDC$ 2.2, 2.4
3. Given: $\overline{AB} \perp \overline{BD}$, $\overline{ED} \perp \overline{BD}$, m $\angle 1 =$ m $\angle 3$
   Prove: m $\angle 2 =$ m $\angle 4$ 1.6, 2.3, 2.4
4. Given: $\overline{BG}$ bisects $\angle ABC$; $\angle 2 \cong \angle 3$.
   Prove: $\angle 1 \cong \angle 3$ 1.5, 2.3, 2.4

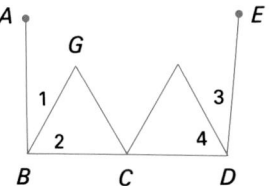

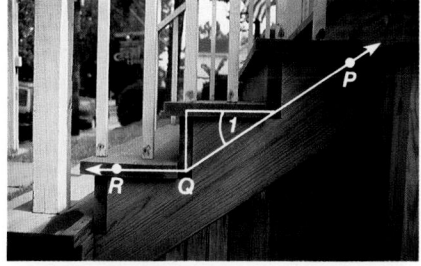

5. In the portion of the stairway shown, m $\angle PQR = 120$. Find m $\angle 1$. Explain why m $\angle 1$ can be found.   3.5

6. This photo shows parking stripes and concrete dividers in a parking lot. How are the positions of the stripes related? What name can be applied to a line drawn through the center of the row of dividers as it crosses the parking stripes? How can this photo be used to illustrate postulates or theorems you have studied in this chapter?   3.5

Mixed Review

1. Statement	Reason
1. C is the midpt of $\overline{BD}$, $AB = BC$	1. Given
2. $BC = CD$	2. Def of midpt
3. $\therefore AB = CD$	3. Sub

2. Statement	Reason
1. m $\angle 2 =$ m $\angle 4$, m $\angle 1 =$ m $\angle 3$	1. Given
2. m $\angle ABC =$ m $\angle 1 +$ m $\angle 2$, m $\angle EDC =$ m $\angle 3 +$ m $\angle 4$	2. Angle Add Post
3. m $\angle 1 +$ m $\angle 2 =$ m $\angle 3 +$ m $\angle 4$	3. Equations may be added.
4. $\therefore$ m $\angle ABC =$ m $\angle EDC$	4. Sub

3. Statement	Reason
1. $\overline{AB} \perp \overline{BD}$, $\overline{ED} \perp \overline{BD}$, m $\angle 1 =$ m $\angle 3$	1. Given
2. m $\angle ABC = 90$, m $\angle EDC = 90$	2. Def of ⊥
3. m $\angle ABC =$ m $\angle EDC$	3. Sub
4. m $\angle ABC -$ m $\angle 1 =$ m $\angle EDC -$ m $\angle 3$	4. Equations may be subtracted.
5. m $\angle 2 =$ m $\angle ABC -$ m $\angle 1$, m $\angle 4 =$ m $\angle EDC -$ m $\angle 3$	5. Angle Add Post
6. $\therefore$ m $\angle 2 =$ m $\angle 4$	6. Sub

4. Statement	Reason
1. $\overline{BG}$ bis $\angle ABC$, $\angle 2 \cong \angle 3$	1. Given
2. $\angle 1 \cong \angle 2$	2. Def of $\angle$ bis
3. $\therefore \angle 1 \cong \angle 3$	3. Trans Prop of Congr

5. m $\angle 1 = 60$. In this case, assume that the planes of the steps are $\parallel$ and that they form rt $\angle$s with the risers.
6. The stripes are parallel; transversal; Answers may vary.

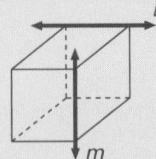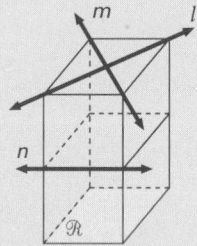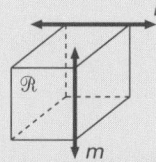
Computer Investigation

Angles of a Triangle

Use a computer software program that constructs random triangles by classification (acute, obtuse, etc.), labels points, extends segments, and measures angles.

The figure at the right shows a triangle with one side, $\overline{BC}$, extended to point *D* in the *exterior* of the triangle. ∠*ACD* is called an *exterior angle* of △*ABC*.

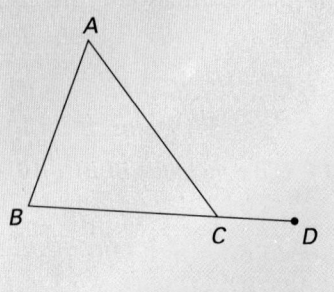

The activity below will help you discover some properties of the angles of a triangle and of an exterior angle of a triangle.

Activity 1

Draw and label an acute triangle *ABC*.

1. Find the measures of the three angles of the triangle.
2. Find m ∠*A* + m ∠*B* + m ∠*C*.
3. Draw a right triangle. Find the measures of the three angles. Find the sum of their measures.
4. Draw an obtuse triangle and find the sum of the measures of the three angles.
5. Draw an isosceles triangle. Find the sum of the measures of the three angles.
6. What general property of the angles of any triangle is suggested by the results of the five exercises above?   sum of ∠s = 180

Activity 2

Draw and label an obtuse triangle *ABC*. Extend $\overline{BC}$ to a point *D* so that *C* is between *B* and *D*. As mentioned above, ∠*ACD* is an exterior angle of △*ABC*.

7. Find m ∠*ACD* and the measure of each of the two angles of △*ABC* that are not adjacent to ∠*ACD*.
8. What numerical relationship exists between the three angle measures of Exercise 7?   m ∠*A* + m ∠*B* = m ∠*ACD*
9. Clear the screen and draw and label an acute triangle *ABC*, with an exterior angle. Then find the measure of the exterior angle and each of the two angles of △*ABC* that are not adjacent to the exterior angle. What relationship exists between the three angles?
m ext. ∠ = sum of int. ∠s not adjacent to ext. ∠.

3.7 Exterior and Remote Interi a Triangle

Teaching Resources

Quick Quizzes 24
**Reteaching and Practice
Worksheets,** pp. 47, 48

Objectives
To find the measures of exterior and remote interior angles of a triangle
To write proofs using the Exterior Angle Theorem

Side $\overline{AB}$ of $\triangle ABC$ is extended to point D. $\angle 1$ is adjacent and supplementary to $\angle 2$. $\angle 1$ is called an *exterior angle* of $\triangle ABC$. $\angle A$ and $\angle C$ of $\triangle ABC$ are called the *remote interior angles* of the exterior $\angle 1$.

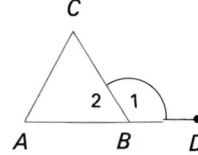

Definitions

An **exterior angle** of a triangle is an angle that is adjacent and supplementary to one of the angles of the triangle. The other two angles of the triangle are called the **remote interior angles** of that exterior angle.

Theorem 3.10

Exterior Angle Theorem: The measure of an exterior angle of a triangle is equal to the sum of the measures of its two remote interior angles.

Given: $\triangle ABC$ with exterior $\angle 1$
Prove: m $\angle 1$ = m $\angle A$ + m $\angle C$

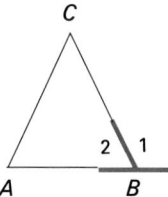

Proof

Statement	Reason
1. $\triangle ABC$ with exterior $\angle 1$	1. Given
2. $\angle 1$ is supplementary to $\angle 2$.	2. Def of ext $\angle$
3. m $\angle 1$ + m $\angle 2$ = 180	3. If outer rays of two adj $\angle$s form a st $\angle$, then sum of their meas is 180.
4. m $\angle A$ + m $\angle C$ + m $\angle 2$ = 180	4. Sum of measures of $\angle$s of a $\triangle$ = 180.
5. m $\angle 1$ + m $\angle 2$ = m $\angle A$ + m $\angle C$ + m $\angle 2$	5. Sub
6. $\therefore$ m $\angle 1$ = m $\angle A$ + m $\angle C$	6. Subt Prop of Eq

■■■**GETTING STARTED**

Prerequisite Quiz

1. In $\triangle ABC$ if m $\angle A$ = 40 and m $\angle B$ = 70; find m $\angle C$.   70
2. In $\triangle ABC$ if m $\angle A$ = x, m $\angle B$ = 3x, and m $\angle C$ = 6x, find m $\angle C$.   108
3. If $\overline{CB} \perp \overline{CA}$ and m $\angle A$ is twice m $\angle B$, find m $\angle B$.   30

Motivator

On the chalkboard, draw $\triangle ABC$. Extend $\overline{AB}$ beyond B to form $\angle CBD$. Is it interior or **exterior** to the triangle?   Exterior   Of the three $\angle$s in the interior of the $\triangle$, which two are the most removed or **remote** from the **exterior** $\angle CBD$? $\angle A$ and $\angle C$

Additional Example 1

Find m $\angle A$.   110

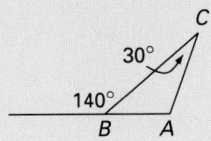

Additional Example 2

Given: m $\angle 1$ = x + 40, m $\angle 2$ = 2x + 5, m $\angle 3$ = 5x + 25
Find m $\angle 4$.   105

Lesson Note

Emphasize that an exterior angle of a triangle is at a vertex. Thus, in the figure below, ∠1 is not an exterior angle.

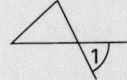

Help students discover the relation between the measures of an exterior angle and the two remote interior angles of a triangle. Draw a figure such as the one below.

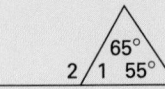

Ask the students to find m ∠2 by first finding m ∠1. Then have them compare m ∠2 with the sum of the measures of the two remote interior angles. m ∠1 = 60; Then m ∠2 = 180 − 60 = 120. Also, 65 + 55 = 120.

Math Connections

Exterior angles of a triangle may be compared to exterior angles of a polygon. Mention to students that the sum of measures of the exterior angles of a triangle, one at each vertex, is equal to the sum of measures of exterior angles in any other polygon. The sum is always 360.

Critical Thinking Questions

Synthesis: Ask students to prove that the sum of the exterior angles of a triangle, one at each vertex, equals 360.   Name the angles of the triangle ∠1, ∠2, and ∠3. By the Exterior Angle Theorem, the measures of the exterior angles are m ∠1 + m ∠2, m ∠1 + m ∠3, and m ∠2 + m ∠3. By Theorem 3.9, m ∠1 + m ∠2 + m ∠3 = 180. Therefore, m ∠1 + m ∠2 + m ∠1 + m ∠3 + m ∠2 + m ∠3 = 180 + 180, or 360 by Sub.

EXAMPLE 1   Find m ∠B.

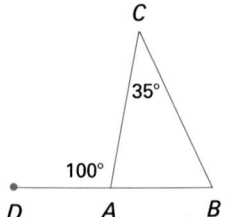

Solution   ∠DAC is an exterior angle.
m ∠DAC = m ∠C + m ∠B
100 = 35 + m ∠B
m ∠B = 65

EXAMPLE 2   Given: m ∠1 = 10x − 6
m ∠3 = 3x + 4
m ∠4 = 4x + 2
Find m ∠2.

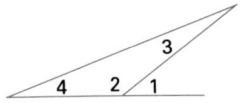

Plan   ∠1 is an exterior angle. Use Theorem 3.10 to write an equation. Solve for x to find m ∠1. Then use the fact that ∠2 is supplementary to ∠1 to find m ∠2.

Solution
$$m \angle 1 = m \angle 3 + m \angle 4$$
$$10x - 6 = (3x + 4) + (4x + 2)$$
$$10x - 6 = 7x + 6$$
$$3x = 12$$
$$x = 4$$
$$m \angle 1 = 10x - 6 = 10 \cdot 4 - 6 = 34$$
$$m \angle 1 = 34$$

Then, since ∠1 and ∠2 are supplementary, m ∠2 = 180 − 34 = 146.

EXAMPLE 3   Given: $\overline{DC} \parallel \overline{AB}$
Prove: m ∠1 = m ∠3 + m ∠2

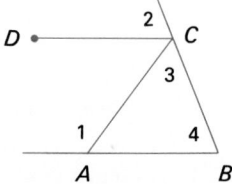

Proof

Statement	Reason
1. $\overline{DC} \parallel \overline{AB}$	1. Given
2. m ∠2 = m ∠4	2. Corr ∠s of ∥ lines have equal measures.
3. m ∠1 = m ∠3 + m ∠4	3. Ext ∠ Theorem
4. ∴ m ∠1 = m ∠3 + m ∠2	4. Sub

Additional Example 3

Given: $\overline{AC}$ bisects ∠DCE.
Prove: m ∠1 = m ∠4 + m ∠3

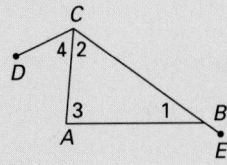

Proof: $\overline{AC}$ bisects ∠DCB (Given); m ∠4 = m ∠2 (Def of ∠ bis); m ∠1 = m ∠2 + m ∠3 (Ext Angle Thm); m ∠1 = m ∠4 + m ∠3 (Sub)

Focus on Reading

Indicate whether the statement is always true, sometimes true, or never true.

1. An exterior angle of a triangle is equal in measure to either of its remote interior angles.   Never true
2. An exterior angle of a triangle is supplementary to the angle of the triangle that is not one of its remote interior angles.   Always true
3. The angle that is adjacent to an exterior angle of a triangle is one of the remote interior angles.   Never true

Classroom Exercises

Find the missing angle measures.

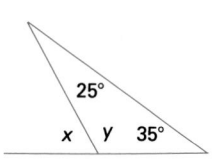
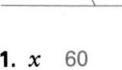
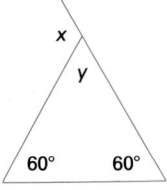
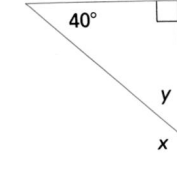

1. x   60
2. y   120
3. x   120
4. y   60
5. x   130
6. y   50

Written Exercises

Find the indicated angle measure. (Exercises 1–6)

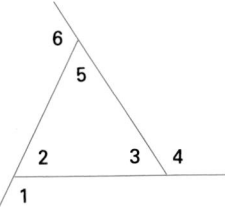

1. m ∠5 = 43, m ∠3 = 37. Find m ∠1.   80
2. m ∠3 = 60, m ∠2 = 40. Find m ∠6.   100
3. m ∠5 = 32, m ∠2 = 70. Find m ∠4.   102
4. m ∠1 = 160, m ∠3 = 100. Find m ∠5.   60
5. m ∠6 = 130, m ∠3 = 75. Find m ∠2.   55
6. m ∠4 = 60, m ∠2 = 39. Find m ∠5.   21

Find the indicated angle measure. (Exercises 7–16)

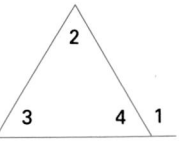

7. m ∠2 = 7x, m ∠3 = 3x, m ∠1 = 60
   Find m ∠2.   42
8. m ∠2 = 3x, m ∠3 = 3x, m ∠1 = 120
   Find m ∠2.   60
9. m ∠3 = 40, m ∠2 = 5x, m ∠1 = 7x
   Find m ∠2.   100
10. m ∠3 = x + 8, m ∠2 = 2x + 3, m ∠1 = 5x − 11. Find m ∠1.   44

3.7 Exterior and Remote Interior Angles of a Triangle   **123**

Common Error Analysis

Error: A typical error is to use the adjacent interior angle instead of one of the remote interior angles when solving problems related to the Exterior Angle Theorem. Emphasize the word "remote." If students are not familiar with this word, have them look up its meaning in a dictionary as part of their homework assignment. Provide practice in identifying the remote interior angles with chalkboard examples.

Checkpoint

Find each indicated angle measure.

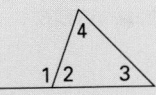

1. m ∠1 = 140, m ∠4 = 80. Find m ∠3.
   60
2. m ∠4 = 65, m ∠3 = 25. Find m ∠1.
   90
3. m ∠3 = 5x, m ∠4 = 7x, m ∠1 = (60).
   Find m ∠3.   25
4. m ∠1 = 5x − 25, m ∠3 = 2x − 5,
   m ∠4 = x + 10. Find m ∠2.   130

Closure

Ask your students how to form an exterior angle of a triangle.   Extend a side of the △.   Ask how they identify the two angles of the triangle that are remote interior angles of this exterior angle.   They are the angles that are not adjacent and supplementary to the exterior angle.   Have them explain the algebraic relationship between the measures of these three angles.   The measure of the exterior ∠ is equal to the sum of the measures of the two remote interior ∠s.

Independent Practice

A Ex. 1–11, **B** Ex. 12–21, **C** Ex. 22–24

Basic: FR all, WE 1–11, Application

Average: FR all, WE 1–21 odd, Application

Above Average: FR all, WE 1–24 odd, Application

Additional Answers

Written Exercises

17.

Statement	Reason
1. $\overleftrightarrow{PQ} \parallel \overleftrightarrow{ST}$	1. Given
2. m ∠3 = m ∠4	2. If lines are ∥, then alt int ∠s are ≅.
3. m ∠1 = m ∠2 + m ∠3	3. Meas of ext ∠ of a △ = sum of meas of remote int ∠s.
4. m ∠1 = m ∠2 + m ∠4	4. Sub

18.

Statement	Reason
1. $\overline{CE}$ bis ∠DCB, m ∠ECD = m ∠B	1. Given
2. m ∠ECD = m ∠ECB	2. Def of ∠ bis
3. m ∠B = m ∠ECB	3. Sub
4. $\overline{CE} \parallel \overline{AB}$	4. If alt int ∠s are ≅, lines are ∥.

11. m ∠1 = 10x − 4, m ∠2 = 2x + 6, m ∠3 = 4x + 2
Find m ∠2.   12

12. m ∠1 = 110, m ∠2 is 10 less than 3 times m ∠3.
Find m ∠3.   30

13. m ∠1 = 70, m ∠3 is 10 more than twice m ∠2.
Find m ∠3.   50

14. m ∠1 = 140, m ∠2 = $\frac{5}{9}$ m ∠3. Find m ∠3.   90

15. m ∠3 = 60, m ∠1 is twice as large as m ∠2.
Find m ∠1.   120

16. m ∠4 = 70, m ∠2 is three times as large as m ∠3.
Find m ∠2.   82.5

17. Given: $\overleftrightarrow{PQ} \parallel \overleftrightarrow{ST}$
Prove: m ∠1 = m ∠2 + m ∠4

18. Given: $\overline{CE}$ bisects ∠DCB; m ∠ECD = m ∠B
Prove: $\overline{CE} \parallel \overline{AB}$

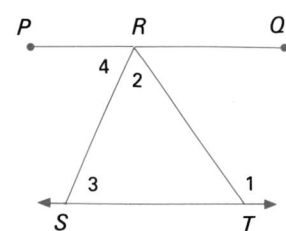

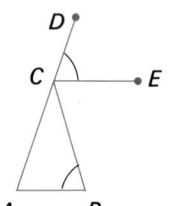

19.

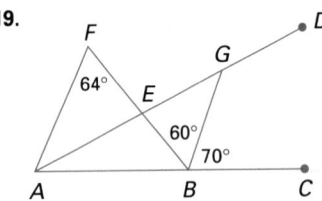

Given: $\overline{AD}$ bisects ∠FAC.
Find m ∠BGD.   143

20.

Given: m ∠GFB = 110, $\overline{GF} \parallel \overline{AD}$, $\overline{FB} \parallel \overline{ED}$, ∠1 ≅ ∠2
Find m ∠3.   140

21. In triangle PQR, m ∠P = $\frac{4}{5}$ m ∠Q and the measure of an exterior angle at R is 144. Find m ∠Q.   80

22. Find the measure of each angle of a triangle whose exterior angles, one at each vertex, have the measures 2x + 10, x + 20, and x + 30.   20, 85, 75

23. In △ABC, D is on $\overline{CB}$ such that m ∠CAD = m ∠CDA. m ∠CAB − m ∠ABC = 30. Find m ∠BAD.   15

24. Each exterior angle of △ABC has the same measure. Find the measure of each exterior angle.   120

Mixed Review

True or false? If false, explain why. *1.7*

1. A line has two endpoints *or* a plane is flat.   True
2. Vertical angles are congruent *and* parallel lines intersect.

Graph and name the resulting geometric figure. *1.8*

3. $x \geq 5$ or $x \leq 7$
4. $x \geq 5$ and $x \leq 7$

 # Application: *Light Rays*

The figure at the right shows a ray of light, called the incident ray, and its reflection, called the reflection ray. In physics, the *Law of Reflection of Light* is stated as below.

If an incident ray of light along $\overline{EC}$ is reflected from a mirror ($\overline{AB}$), the angle between the incident ray and the perpendicular or *normal* $\overline{CN}$ to the mirror is congruent to the angle between the reflected light ray along $\overline{CD}$ and the normal $\overline{CN}$.

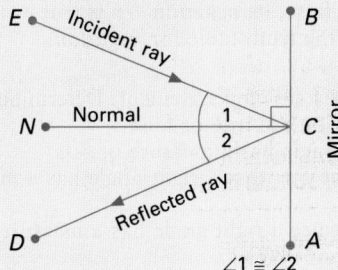

$\angle 1 \cong \angle 2$

Exercise

$\overline{SR}$ and $\overline{VR}$ represent mirrors which are at right angles. An incident light ray along $\overline{PQ}$ is reflected from $\overline{VR}$ at Q, then from $\overline{SR}$ at T, and then emerges along the path $\overline{TU}$. Prove $\overline{UT} \parallel \overline{PQ}$.

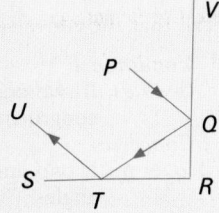

3.7 Exterior and Remote Interior Angles of a Triangle   **125**

Mixed Review

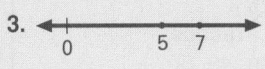
2. False; T $\wedge$ F is F.

3.
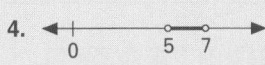
Line

4.
Segment

Application: Light Rays

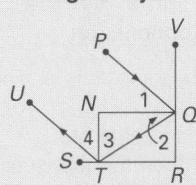

Statement	Reason
1. $\overline{SR} \perp \overline{VR}$	1. Given
2. $\overline{QN} \perp \overline{VR}$, $\overline{NT} \perp \overline{SR}$	2. Def of normal
3. $\overline{NT} \parallel \overline{QR}$, $\overline{NQ} \parallel \overline{TR}$	3. Two lines $\perp$ same line are $\parallel$.
4. m $\angle 2$ + m $\angle 3$ + m $\angle TNQ$ = 180	4. Sum of meas of the $\angle$s of a $\triangle$ = 180.
5. $\overline{NT} \perp \overline{NQ}$	5. If a transv is $\perp$ to one of 2 $\parallel$ lines, then it is $\perp$ to the other.
6. m $\angle TNQ$ = 90	6. Def of $\perp$
7. m $\angle 2$ + m $\angle 3$ = 90	7. Subt Prop of Eq
8. 2(m $\angle 2$ + m $\angle 3$) = 180	8. Mult Prop of Eq
9. m $\angle 2$ + m $\angle 2$ + m $\angle 3$ + m $\angle 3$ = 180	9. Distr Prop
10. m $\angle 2$ = m $\angle 1$, m $\angle 3$ = m $\angle 4$	10. Def of normal
11. m $\angle 1$ + m $\angle 2$ + m $\angle 3$ + m $\angle 4$ = 180	11. Sub
12. m $\angle 1$ + m $\angle 2$ = m $\angle PQT$, m $\angle 3$ + m $\angle 4$ = m $\angle QTU$	12. Angle Add Post
13. m $\angle PQT$ + m $\angle QTU$ = 180	13. Sub
14. $\therefore \overline{PQ} \parallel \overline{UT}$	14. If 2 lines are cut by a transv so that int $\angle$s on same side of transv are supp, the lines are $\parallel$.

GETTING STARTED

Prerequisite Quiz

For Exercises 1 and 2, give the hypothesis and conclusion.

1. $p \rightarrow q$ The hypothesis is p. The conclusion is q.
2. If the sides of an angle are perpendicular, then the measure of the angle is 90.   The hypothesis is: The sides of an angle are perpendicular. The conclusion is: The measure of the angle is 90.
3. Write the converse of the conditional in Exercise 2.   If the measure of an angle is 90, then the sides of the angle are perpendicular.
4. Write a true conditional whose converse is false.   One example is: If two angles are vertical, then the angles are congruent.

Motivator

Ask your students what the word *negation* means.   Negation means the opposite or negative of a statement.   Write the following statement on the board: It is true that if two angles are supplementary, then the sum of their angle measures is 180. Ask the students what the negation of the statement would be.   It is not true that two angles are supplementary, then the sum of the angle measures is 180.   Ask if the negation is true or false.   False

Objectives To write the negation, the contrapositive, and the inverse of a statement and to determine whether they are true

To write a biconditional and to determine whether it is true

The following logic statements have already been presented.
• conditional: $p \rightarrow q$
• converse of conditional $p \rightarrow q$: $q \rightarrow p$
• disjunction: $p \lor q$
• conjunction: $p \land q$

Another important statement in logic is the negation.

Definition

If p is a statement, then the statement "not p" is called the **negation** of p. The negation of p is written as $\sim p$.

Statement	Negation
p: l is parallel to m	$\sim p$: l is not parallel to m; or it is not true that l is parallel to m.

When a statement p is true, its negation $\sim p$ is false.
When a statement p is false, its negation $\sim p$ is true.
This is summarized in the truth table for *negation*.

p	$\sim p$
T	F
F	T

EXAMPLE 1 Write a negation of the following statement. Determine whether the statement and its negation are true or false.
Statement (p): A right angle has a measure of 30.
Negation ($\sim p$): It is not true that a right angle has a measure of 30.

Solution The statement is false since a right angle has a measure of 90.
The negation is therefore true.

Recall that the converse of a true conditional need not be true.

  Conditional
  ($p \rightarrow q$): If two angles are right angles, then the two angles are congruent. (True)
  Converse
  ($q \rightarrow p$): If two angles are congruent, then the two angles are right angles. (Not necessarily true)

126   Chapter 3 Parallelism

Additional Example 1

Write the negation of the following statement. Determine whether the statement and its negation are true or false.

Statement: Parallel lines are skew.
False

Negation: It is not true that parallel lines are skew.   True

Statements are said to be logically equivalent if they have the same truth values. A conditional and its converse are not logically equivalent, as shown in the truth table.

p	q	$p \rightarrow q$	$q \rightarrow p$
T	T	T	T
T	F	F	T
F	T	T	F
F	F	T	T

$\leftarrow$ $\begin{cases} \text{The conditional } (p \rightarrow q) \text{ and its} \\ \text{converse } (q \rightarrow p) \text{ have different truth} \\ \text{values.} \end{cases}$

A conditional and its converse can be written as one statement using the expression "if and only if." This is illustrated below.

Conditional: If two lines are parallel, then alternate interior angles are congruent.

Converse: If alternate interior angles are congruent, then lines are parallel.

Single Statement: Two lines are parallel if and only if alternate interior angles are congruent; *or*,
Alternate interior angles are congruent if and only if lines are parallel.

Definition

When a conditional statement and its converse are combined by "if and only if," the resulting statement is called a **biconditional**.

The biconditional "*p* if and only if *q*" is also written $p \leftrightarrow q$.

The biconditional $p \leftrightarrow q$ is true only when the conditional $p \rightarrow q$ and its converse $q \rightarrow p$ are both true.

EXAMPLE 2 Write the converse of the given statement. Then write a biconditional. Determine whether the biconditional is true and give a reason.

Statement: If an angle has a measure of 90, then the sides of the angle are perpendicular.

Solution

Converse: If the sides of an angle are perpendicular, then the angle has a measure of 90. (True)

Biconditional: An angle has a measure of 90 if and only if the sides of the angle are perpendicular (or: The sides of an angle are perpendicular if and only if the angle has a measure of 90).

The biconditional is true because both the statement and its converse are true.

TEACHING SUGGESTIONS

Lesson Note

Point out the similarity between the negation of a negation and the negative of a negative for signed numbers. Note the logical meaning of the following example of poor grammar: "I don't have no books." This double negation, understood logically, would mean "I do have books." In proving a theorem, emphasize that it is sometimes easier to prove the contrapositive of the given conditional than the conditional itself.

Math Connections

Life Skills: Discuss with students how knowing the truth values of the alternate forms of a conditional statement may help consumers evaluate advertising claims about a product. For example, it may be true that if you wash your clothes with No Spot laundry detergent, your clothes will be clean; however, the converse of that statement is not necessarily true.

Critical Thinking Questions

Analysis: Ask students to complete these exercises.

1. Under what conditions will the conditional $p \rightarrow q$ have a false converse?   When p is false and q is true.
2. Give an example of a conditional having a false converse.   Answers will vary.
3. Under what conditions will the conditional $p \rightarrow q$ have a true converse?   When p and q are both true or both false or when p is true and q is false.
4. Give an example of a conditional having a true converse.   Answers will vary.

Additional Example 2

Write the converse of the given statement. Then write a biconditional. Determine whether the biconditional is true and give a reason.

Statement: If M is the midpoint of $\overline{AB}$, then $AM = MB$
Converse: If $AM = MB$, then M is the midpoint of $\overline{AB}$. True

Biconditional: M is the midpoint of $\overline{AB}$ if and only if $AM = MB$; or $AM = MB$ if and only if M is the midpoint of $\overline{AB}$. True, since both the statement and its converse are true.

Common Error Analysis

Error: When writing the contrapositive of a conditional, students may forget to negate *both* parts, often writing the converse rather than the contrapositive.

To help students avoid this error, point out that *contra-* means *against* or *opposite*, so that contrapositive suggests negation. This helps students understand the opposite position of the two parts of a conditional.

Checkpoint

Write the indicated form of each statement and determine whether that form is true or false. (Ex. 1–3).

1. Negation of: The sides of a straight angle are perpendicular. It is not the case that the sides of a straight angle are perpendicular. True

2. Inverse of: In a plane, if two lines are not perpendicular to the same line, then they are not parallel. In a plane, if two lines are perpendicular to the same line, then they are parallel. True

3. Contrapositive of: If two lines are parallel, then the lines do not intersect. If two lines do intersect, then they are not parallel. True

4. Write the converse of the conditional of Exercise 3. Then form a biconditional. Is it true? Converse: If two lines do not intersect, then the two lines are parallel. Biconditional: Two lines are parallel if and only if they do not intersect. False, because the converse is false.

Closure

Ask your students how to form each of the following.

(1) the negation of a statement Write the negative or opposite of the statement.
(2) the inverse of a conditional Negate the hypothesis and the conclusion.
(3) the contrapositive of a conditional Interchange the hypothesis and conclusion and negate each of them.
(4) a biconditional of the two conditionals Combine a conditional statement and its converse using "if and only if."

Another statement that can be formed from a conditional is the contrapositive.

Definition

> The **contrapositive** of a conditional is the statement formed by interchanging the hypothesis and the conclusion and negating each of them. The contrapositive of $p \rightarrow q$ is $\sim q \rightarrow \sim p$.

The converse of a true conditional may not be true. However, the *contrapositive* of a true conditional is *always* true.

> *Statement*: If two angles are vertical, then the angles have equal measure. (True)
>
> *Contrapositive*: If two angles do *not* have equal measure, then they are *not* vertical angles.

A truth table can be used to show that whenever a conditional is true, so is its contrapositive. Likewise, whenever a conditional is false, so is its contrapositive. A conditional and its contrapositive are therefore *logically equivalent*. The significance of this relationship is that a conditional may be proved by proving its contrapositive instead.

Another statement formed from a conditional is the *inverse*. The inverse of a true conditional may or may not be true.

Definition

> The **inverse** of a conditional statement is the statement formed by negating the hypothesis and the conclusion. The inverse of $p \rightarrow q$ is $\sim p \rightarrow \sim q$.

EXAMPLE 3 Write the inverse of the following true statements. Is the inverse true or false?

a. *Statement*: If two angles are vertical, then the angles have equal measures.

Solution *Inverse*: If two angles are *not* vertical, then the angles do *not* have equal measure.

The inverse is false. Two angles may be non-vertical and yet have equal measure.

b. *Statement*: If two angles are congruent, then they have equal measures.

Solution *Inverse*: If two angles are *not* congruent, then they do *not* have equal measures.

The inverse is true.

Additional Example 3

Write the inverse of the following true statements. Is the inverse true or false?
(a) If alternate interior angles are congruent, then lines are parallel. Inverse: If alternate interior angles are not congruent, then lines are not parallel. True

(b) If two angles are adjacent, then the angles have a common vertex. Inverse: If two angles are not adjacent, then the angles do not have a common vertex. False, the angles may be non-adjacent and yet have a common vertex.

Classroom Exercises

For statements p and q and the conditional $p \rightarrow q$, match the term on the left with its symbolic representation on the right.

1. biconditional d
2. conjunction e
3. converse f
4. disjunction b
5. inverse c
6. negation a

a. $\sim p$
b. $p \vee q$
c. $\sim p \rightarrow \sim q$
d. $p \leftrightarrow q$
e. $p \wedge q$
f. $q \rightarrow p$

Written Exercises

Write the negation of the statement. Is the negation true or false?

1. Parallel lines intersect. Negation true
2. An obtuse angle has a measure less than 90. Negation true
3. The sides of a right angle are perpendicular. Negation false

Write the inverse and the contrapositive of the statement. Label each as true or false.

4. If the sum of the measures of two angles is 90, then the angles are complementary.
5. If two angles are adjacent and their outer rays form a straight angle, then the two angles are supplementary.
6. If an angle is a straight angle, then its measure is 90.
7. If an angle is a right angle, then its measure is not 20.

Write the converse of the statement. Then write a biconditional of the statement. Determine if the biconditional is true.

8. If M is the midpoint of $\overline{AB}$, then M divides $\overline{AB}$ into two congruent segments.
9. If corresponding angles are congruent, then lines are parallel.
10. If an angle is not acute, then the angle is obtuse.
11. Verify by truth table that $\sim(\sim p)$ is logically equivalent to p.
12. Verify by truth table that a conditional is logically equivalent to its contrapositive.

Mixed Review

Indicate whether the statement is always true, sometimes true, or never true.

1. Two coplanar segments that do not intersect are parallel. *3.1* Sometimes true
2. Two lines in a plane that do not intersect are parallel. *3.3* Always true
3. A ray and a segment that are coplanar intersect if they are not parallel. *3.3* Sometimes true

3.8 Negation, Contrapositive, Inverse, Biconditional **129**

Enrichment

Review the definition of conjunction and disjunction. Have the students verify that the following statements are equivalent by using a truth table.

$\sim (p \wedge q)$ and $\sim p \vee \sim q$

p	q	$p \wedge q$	$\sim (p \wedge q)$	$\sim p$	$\sim q$	$\sim p \vee \sim q$
T	T	T	F	F	F	F
T	F	F	T	F	T	T
F	T	F	T	T	F	T
F	F	F	T	T	T	T

FOLLOW UP

Guided Practice

Classroom Exercises 1–6

Independent Practice

A Ex. 1–7, **B** Ex. 8–10, **C** Ex. 11–12

Basic: WE 1–7
Average: WE 1–10
Above Average: WE all

Additional Answers

Written Exercises

1. ‖ lines do not intersect. (True)
2. An obtuse ∠ does not have a meas less than 90. (True)
3. The sides of a rt ∠ are not ⊥. (False)
4. Inv: If the sum of the meas of 2 ∠s is not 90, then the ∠s are not comp. (True) Contr: If 2 ∠s are not comp, then the sum of their meas is not 90. (True)
5. Inv: If the outer rays of 2 adj ∠s do not form a st ∠, then the ∠s are not supp. (True) Contr: If 2 adj ∠s are not supp, then their outer rays do not form a st ∠. (True)
6. Inv: If an ∠ is not a st ∠, then its meas is not 90. (False) Contr: If the meas of an ∠ is not 90, then the ∠ is not a st ∠. (False)
7. Inv: If an ∠ is not a rt ∠, then its meas is 20. (False) Contr: If the meas of an ∠ is 20, then it is not a rt ∠. (True)
8. Conv: If M divides $\overline{AB}$ into 2 ≅ seg, then M is the midpt of $\overline{AB}$. Bicond: M is the midpt of $\overline{AB}$ if and only if M divides $\overline{AB}$ into 2 ≅ seg. (True)
9. Conv: If lines are ‖, then corr ∠s are ≅. Bicond: Corr ∠s are ≅ if and only if lines are ‖. (True)
10. Conv: If an ∠ is obtuse, then it is not acute. Bicond: An ∠ is not acute if and only if it is obtuse. (False)

11.

p	$\sim p$	$\sim(\sim p)$
T	F	T
F	T	F

12.

p	q	$p \rightarrow q$	$\sim q$	$\sim p$	$\sim q \rightarrow \sim p$
T	T	T	F	F	T
T	F	F	T	F	F
F	T	T	F	T	T
F	F	T	T	T	T

3.9 Parallel Lines and Planes

Objectives   To determine whether statements about parallel lines and planes are
true
To prove statements about parallel lines and planes

In space, two lines can be parallel, intersecting, or skew. However, two
distinct planes either intersect or are parallel.

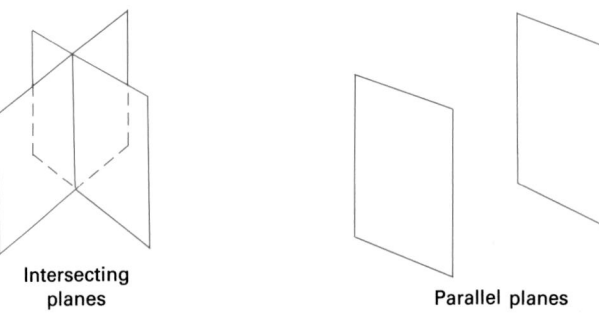

Intersecting
planes

Parallel planes

Definition | **Parallel planes** are planes that do not intersect.

There are three possible relationships between a line and a plane.

(1) The line can be
contained in the
plane. The inter-
section of the line
and the plane is
the line itself.

(2) The line can in-
tersect the plane
in exactly one
point.

(3) The line and
plane can have no
points in com-
mon. They do not
intersect.

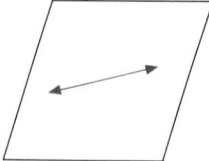 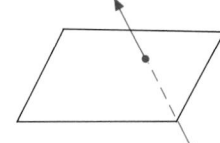 

Definition | A line and a plane are *parallel* if they do not intersect.

130   Chapter 3 Parallelism

Prerequisite Quiz

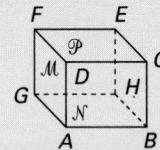

1. Name the intersection of planes $\mathcal{M}$ and
$\mathcal{N}$. $\overleftrightarrow{DA}$
2. $\overleftrightarrow{FD}$ intersects $\overleftrightarrow{DC}$ in what point? D
3. True or false? D is coplanar with points
G, A, and B.   False
4. Name the planes containing point D. $\mathcal{M}$,
$\mathcal{N}$, $\mathcal{P}$

Motivator

Have your students give a physical model of
two planes in the classroom that intersect.
The ceiling and the wall   Ask them to
describe this intersection geometrically.
A line.

Lesson Note

Emphasize that planes, in contrast to lines,
can be related in only *two* ways. Planes are
parallel or intersecting, whereas lines are
parallel, intersecting, or skew. Whenever
possible, find physical examples to help
students visualize in three dimensions. For
example, two planes parallel to the same
line may not be parallel to each other. To
illustrate this, use the front wall and a side
wall of the classroom as planes parallel to
the line which you represent as you stand in
the middle of the room. An example of
Theorem 3.11 is the intersection of the front
wall of the classroom with the ceiling and
floor. Emphasize that one counterexample
can show that a supposed property is false,
but one example cannot be used to prove a
statement true.

Theorem 3.11 | If two parallel planes are intersected by a third plane, then the lines of intersection are parallel.

Given: Plane $\mathscr{P} \parallel$ plane $\mathscr{Q}$, $\mathscr{P}$ intersects $\mathscr{R}$ in $\overleftrightarrow{AB}$;
  $\mathscr{Q}$ intersects $\mathscr{R}$ in $\overleftrightarrow{CD}$.
Prove: $\overleftrightarrow{AB} \parallel \overleftrightarrow{CD}$

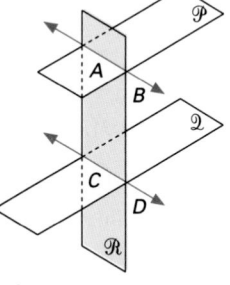

Indirect Proof
Given: $\mathscr{P} \parallel \mathscr{Q}$. Assume $\overleftrightarrow{AB} \not\parallel \overleftrightarrow{CD}$. $\overleftrightarrow{AB}$ and $\overleftrightarrow{CD}$ are coplanar (both in plane $\mathscr{R}$), and must therefore intersect at some point. Call this point T. Because T is on $\overleftrightarrow{AB}$, the intersection of planes $\mathscr{P}$ and $\mathscr{R}$, it is in $\mathscr{P}$. Because T is on $\overleftrightarrow{CD}$, the intersection of planes $\mathscr{Q}$ and $\mathscr{R}$, it is also in $\mathscr{Q}$.

Therefore, planes $\mathscr{P}$ and $\mathscr{Q}$ have a point in common, T. Yet it is given that $\mathscr{P}$ and $\mathscr{Q}$ are *parallel*. This is a contradiction because parallel planes have no points of intersection. The assumption that $\overleftrightarrow{AB} \not\parallel \overleftrightarrow{CD}$ must be false. Also, coplanar lines $\overleftrightarrow{AB}$ and $\overleftrightarrow{CD}$ cannot be skew.

Therefore, $\overleftrightarrow{AB} \parallel \overleftrightarrow{CD}$.

You have learned to prove a conditional by either a direct or indirect method of proof. Recall that a conditional is false only when the hypothesis is true and the conclusion is false. To prove a conditional false, you must find *one* example for which the conclusion is false when the hypothesis is true. Such an example is called a **counterexample**.

This counterexample can be a drawing which illustrates one case where the conditional is false.

EXAMPLE	True or false? Give a reason.
	Two lines parallel to the same plane are parallel.
Plan	Try to find a counterexample that illustrates that when the *hypothesis*, "two lines are parallel to the same plane," is true, the *conclusion*, "the lines are parallel," is false.
Solution	In the figure, $\overleftrightarrow{AC} \parallel$ plane $\mathscr{M}$ and $\overleftrightarrow{BD} \parallel$ plane $\mathscr{M}$. The *hypothesis* is true. But, $\overleftrightarrow{AC} \not\parallel \overleftrightarrow{BD}$. The *conclusion* is false. Therefore, the statement, "two lines parallel to the same plane are parallel," is false. The drawing is a counterexample.

Math Connections

Logic: Theorems of intersecting and parallel planes are proved using indirect proof, which is taught in Lesson 3.4. This offers a good review and application of indirect proof.

Critical Thinking Questions

Synthesis: Ask students to compare the extension of parallel lines to parallel planes. Compare the angle relationships between a transversal and parallel lines with the relationship between the angles formed by parallel planes and an intersecting plane. Do the special angle relationships of transversal lines hold true with intersecting planes of parallel planes? Use the question for open-ended discussion. Yes; the angle relationships also apply to planes. A good model is the cardboard dividers used in glassware packaging.

Common Error Analysis

Error: Students try to decide whether a relationship in space is true by thinking only in terms of two dimensions. Help students learn to visualize these relationships in three dimensions.

Drawing a box may help to visualize spatial relationships and find counterexamples, as shown below.

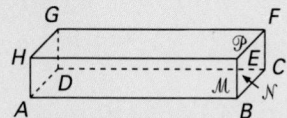

$\mathscr{M} \parallel \overleftrightarrow{GD}$ and $\mathscr{N} \parallel \overleftrightarrow{GD}$, but $\mathscr{M} \not\parallel \mathscr{N}$. $\overleftrightarrow{AB} \parallel \mathscr{P}$, but $\overleftrightarrow{AB} \not\parallel$ every line in $\mathscr{P}$; $\overleftrightarrow{AB} \not\parallel \overleftrightarrow{HG}$, which is in $\mathscr{P}$.

Additional Example

True or false. Give a reason.
If two lines are parallel and one of them intersects a plane, then the other intersects the plane. False. $\overleftrightarrow{AB} \parallel \overleftrightarrow{DC}$ and $\overleftrightarrow{AB}$ intersects $\mathscr{M}$, but $\overleftrightarrow{DC} \parallel \mathscr{M}$

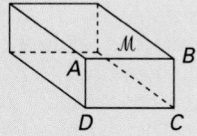

Checkpoint

Find a counterexample to the statement.

1. A line parallel to a given line is parallel to any plane that contains the given line.
2. Two planes perpendicular to the same plane are parallel.
3. Two lines perpendicular to the same line are parallel.

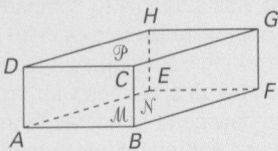

1. False; $\overleftrightarrow{DA} \parallel \overleftrightarrow{CB}$, but $\overleftrightarrow{DA} \not\parallel \mathcal{M}$ which contains $\overleftrightarrow{CB}$.
2. $\mathcal{M} \perp \mathcal{P}$ and $\mathcal{N} \perp \mathcal{P}$, but $\mathcal{M} \not\parallel \mathcal{N}$.
3. $\overleftrightarrow{DC} \perp \overleftrightarrow{CB}$ and $\overleftrightarrow{GC} \perp \overleftrightarrow{CB}$ but $\overleftrightarrow{DC} \not\parallel \overleftrightarrow{CG}$.

Closure

Have the students give the definition of counterexample and explain how it can be used to prove that a conditional is false. One example for which the conclusion is false is when the hypothesis is true. A conditional is false only when the hypothesis is true and the conclusion is false. Ask your students to recall the relationships that can exist between two distinct lines. Parallel, intersecting, or skew Two distinct planes. parallel or intersect Have them give examples of physical models in the classroom.

■■■FOLLOW UP

Guided Practice

Classroom Exercises 1–3

Independent Practice

A Ex. 1–5, **B** Ex. 6–7, **C** Ex. 8–9

Basic: WE 1–5, Problem Solving

Average: WE 1–7, Problem Solving

Above Average: WE 1–9, Problem Solving

Additional Answers

See page 137 for the answers to Ex. 3, 5, 6–9.

Classroom Exercises

1. What three possible relationships can exist between two distinct lines?   Parallel, intersecting, skew
2. What two possible relationships can exist between two distinct planes?   Intersecting, parallel
3. Explain how to prove a conditional false using a counterexample.   Find example where hypothesis is true, but conclusion is false.

Written Exercises

True or false? If false, draw a diagram to illustrate a counterexample.

1. Two lines parallel to the same line are parallel to each other.   True
2. Two planes parallel to the same plane are parallel to each other.   True
3. Two planes parallel to the same line are parallel to each other.
4. Two planes perpendicular to the same line are parallel.   True
5. If two lines are parallel, then every plane containing one of the lines is parallel to every plane containing the other line.
6. A line parallel to a plane is parallel to every line in that plane.
7. Given: planes $\mathcal{M}$ and $\mathcal{N}$ intersect in $\overleftrightarrow{AB}$; $\overleftrightarrow{CD} \parallel \mathcal{M}$, $\overleftrightarrow{CD} \parallel \mathcal{N}$, $\overleftrightarrow{CD}$ and $\overleftrightarrow{AB}$ are not skew.
   Prove: $\overleftrightarrow{CD} \parallel \overleftrightarrow{AB}$

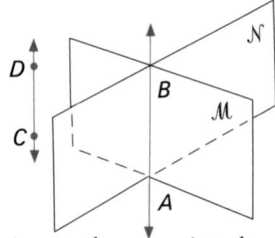

8. Prove: If a line intersects a plane in exactly one point, then it is not parallel to any line that is contained in the plane. (Use indirect proof.)
9. Assume the property that, through a point not on a plane, there is exactly one line and one plane parallel to the given plane. Prove: If a line intersects one of two parallel planes, then it also intersects the second plane.

Mixed Review

1. Given: m $\angle 6 = 75$, m $\angle 1 = 30$
   Find m $\angle 3$.   **3.7**   105
2. Given: m $\angle 2 = 70$, m $\angle 1 = \frac{2}{3}$ m $\angle 6$
   Find m $\angle 1$ and m $\angle 6$.   **3.6.**   44, 66
3. Given: $\overleftrightarrow{AB} \parallel \overleftrightarrow{CD}$, m $\angle 4 = 3x + 20$, m $\angle 1 = x + 40$
   Find m $\angle 1$.   **3.5.**   50

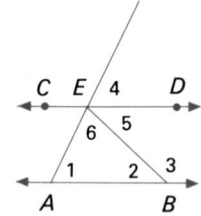

Enrichment

Discuss the fact that a single counterexample is sufficient to refute a statement, while any number of supporting examples are not sufficient to prove a statement true. This explains the need for formal, deductive proofs in geometry. In many settings, however, deductive proofs are impossible. For example, the only way to "prove" that it is the nature of water to boil at 212°F is by demonstration. This is an example of the scientific method, or inductive reasoning. Technology Worksheet 3 addresses this question as a computer investigation of different mathematical conjectures.

Problem Solving Strategies

Using an Alternate Approach

Many problems cannot be solved by a straightforward method. An alternate approach may have to be tried. For example, consider the drawing at the right. It is given that $p \parallel q$. You are asked to find m $\angle 1$ + m $\angle 2$.

A straightforward attempt might be trying to find the measure of both angles separately. This will not work since not enough information is given. This means that you have to look at the problem another way.

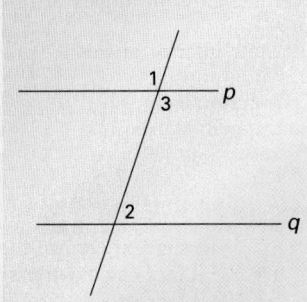

Example   Given: $p \parallel q$
Find m $\angle 1$ + m $\angle 2$.

Solution   m $\angle 1$ = m $\angle 3$ by the Vertical Angles Property.

Then m $\angle 2$ + m $\angle 3$ = 180 since these angles are on the same side of a transversal, interior angles of parallel lines are supplementary. By substitution, m $\angle 1$ + m $\angle 2$ = 180.

Thus, we found m $\angle 1$ + m $\angle 2$ *without* knowing the measure of each angle separately.

1. In a later lesson you will prove the formula for the area of a triangle: $A = \frac{1}{2}h \cdot b$ where h is the length of the altitude (a perpendicular to a side of the triangle) and b is the length of the side to which the perpendicular is drawn, the **base**. In the figure at the right, if the area of triangle ABC is 30, then find the area of triangle BDC.

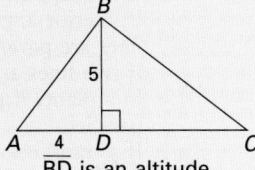

$\overline{BD}$ is an altitude.

2. If the triangle and rectangle at the right have equal areas, and if $\frac{pr}{2}$ = 60, then xy = ?

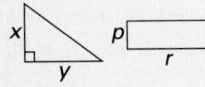

3. The formula for the volume of a cube is $V = s^3$, where s is the length of an edge. Without calculating any roots, find the length of the edge of a cube which has a volume = 238.328 cm³ and the area of a face = 38.44 cm². (HINT: Write the formula in terms of s and s^2.) $s = \dfrac{V}{s^2}$ = 6.2 cm

Problem Solving Strategies   **133**

9. Given: m$\angle 1 \neq$ m $\angle 2$. Assume: $\angle 1$ and $\angle 2$ are vert $\angle$s. Then $\angle 1 \cong \angle 2$ (m $\angle 1$ = m$\angle 2$). But m $\angle 1 \neq$ m $\angle 2$ and m $\angle 1$ = m $\angle 2$ is a contradiction. So the assumption that $\angle 1$ and $\angle 2$ are vert $\angle$s must be false. Therefore, $\angle 1$ and $\angle 2$ are not vertical $\angle$s.

10. Begin an indirect proof by accepting the given information as true and assuming that the conclusion is false (Step 1). Then reason until there is a contradiction of the given or another known fact (Step 2). The assumption that the conclusion is false is thereby shown to be untrue, and so the conclusion is true (Step 3).

Problem Solving Strategies
1. 20   2. 240

Additional Answers, page 108

2. Given: m $\angle 1 \neq$ m $\angle 2$. Assume: $r \perp s$. Then $\angle 1$ and $\angle 2$ are rt $\angle$s (m $\angle 1$ = 90 and m $\angle 2$ = 90), by def of $\perp$ lines; and so m $\angle 1$ = m $\angle 2$, by sub. But m $\angle 1 \neq$ m $\angle 2$ and m $\angle 1$ = m $\angle 2$ is a contradiction. So the assumption that $r \perp s$ must be false. Therefore, $r \not\perp s$.

3. Given: $l \not\perp m$. Assume: m $\angle 1$ = m $\angle 3$. Then m $\angle 1$ + m $\angle 3$ = 180 (if the outer rays of 2 adj $\angle$s form a st $\angle$, the sum of the meas of the $\angle$s is 180), and so m $\angle 1$ + m $\angle 1$ = 180 (subs). Thus 2(m $\angle 1$) = 180, and m $\angle 1$ = 90 (div prop of eq). Hence, $l \perp m$ (def of $\perp$ lines). But $l \perp m$ and $l \not\perp m$ is a contradiction. So the assumption that m $\angle 1$ = m $\angle 3$ must be false. Therefore, m $\angle 1 \neq$ m $\angle 3$.

4. Given: $m \not\parallel n$. Assume: $\angle 3 \cong \angle 2$. Then $\angle 2$ and $\angle 3$ are $\cong$ corr $\angle$s, so $m \parallel n$. But $m \parallel n$ and $m \not\parallel n$ is a contradiction. So the assumption that $\angle 3 \cong \angle 2$ is false. Therefore $\angle 3 \not\cong \angle 2$.

5. Given: $l \perp m$, $m \not\parallel n$. Assume: $l \perp n$. Then l is a transv $\perp$ to both m and n. So $m \parallel n$. But $m \not\parallel n$ and $m \parallel n$ is a contradiction. So the assumption that $l \perp n$ is false. Therefore, $l \not\perp n$.

6. Given: $m \not\parallel n$. Assume: m $\angle 4$ + m $\angle 2$ = 180. Then, $\angle 4$ and $\angle 2$ are supp int $\angle$s on the same side of the transv, $m \parallel n$. But $m \parallel n$ and $m \not\parallel n$ is a contradiction. So the assumption that m $\angle 4$ + m $\angle 2$ = 180 is false. Therefore, m $\angle 4$ + m $\angle 2 \neq$ 180.

7. Given: $m \not\parallel n$. Assume: m $\angle 1$ + m $\angle 2$ = 180. Then, since $\angle 1$ and $\angle 3$ form a st $\angle$, m $\angle 1$ + m $\angle 3$ = 180. Since $\angle 2$ and $\angle 3$ are both supp $\angle$s of $\angle 1$, they are $\cong$. Since they are $\cong$ corr $\angle$s, $m \parallel n$. But $m \parallel n$ and $m \not\parallel n$ is a contradiction. So the assumption that m $\angle 1$ + m $\angle 2$ = 180 is false. Therefore, m $\angle 1$ + m $\angle 2 \neq$ 180.

8. Given: m $\angle 3 \neq$ m $\angle 2$. Assume: m $\angle 1$ + m $\angle 2$ = 180. Then, $\angle 1$ and $\angle 2$ are supp by def of supp $\angle$s. Since $\angle 1$ and $\angle 3$ form a st $\angle$, $\angle 3$ and $\angle 1$ are supp. Then, since $\angle 3$ and $\angle 2$ are both supps of $\angle 1$, m $\angle 2$ = m $\angle 3$. But m $\angle 2$ = m $\angle 3$ and m $\angle 2 \neq$ m $\angle 3$ is a contradiction. So, the assumption that m $\angle 1$ + m $\angle 2$ = 180 is false. Therefore, m $\angle 1$ + m $\angle 2 \neq$ 180.

133

5. $m \parallel n$ alt int $\angle$s are $\cong$
6. $p \parallel q$ corr $\angle$s are $\cong$

Additional Answers, page 108

11. Given: $\overleftrightarrow{AB}$ and $\overleftrightarrow{CD}$ are skew. Assume: $\overleftrightarrow{AC}$ and $\overleftrightarrow{BD}$ are not skew. Since $\overleftrightarrow{AC}$ and $\overleftrightarrow{BD}$ are not skew, they either intersect or are $\parallel$. For $\overleftrightarrow{AC}$ and $\overleftrightarrow{BD}$ to either intersect or be $\parallel$, they must lie in the same plane, and therefore pts A, B, C, and D must also lie in the same plane. Since $\overleftrightarrow{AB}$ and $\overleftrightarrow{CD}$ are skew, A, B, C, and D cannot all lie in the same plane. But A, B, C and D lie in the same plane and A, B, C and D do not all lie in the same plane is a contradiction. So, the assumption that $\overleftrightarrow{AC}$ and $\overleftrightarrow{BD}$ are not skew is false. Therefore, $\overleftrightarrow{AC}$ and $\overleftrightarrow{BD}$ are skew.

12.

p	q	$p \rightarrow q$	p	$\sim q$	$p \wedge \sim q$	$\sim(p \wedge \sim q)$
T	T	T	T	F	F	T
T	F	F	T	T	T	F
F	T	T	F	F	F	T
F	F	T	F	T	F	T

Brainteaser
Ms. Blue wears the brown coat; Ms. Brown, the green coat; Ms. Green, the blue coat.

Key Terms

alternate exterior angles (p. 98)
alternate interior angles (p. 97)
biconditional (p. 127)
contrapositive (p. 128)
converse (p. 109)

corresponding angles (p. 98)
exterior (p. 97)
exterior angles (p. 121)
indirect proof (p. 106)
inverse (p. 128)
negation (p. 126)

parallel lines (p. 93)
parallel planes (p. 130)
remote interior angles (p. 121)
skew lines (p. 93)
transversal (p. 97)
triangle (p. 115)

Key Ideas and Review Exercises

3.1 In space there are three possible relationships between two lines:
  • Lines can be intersecting.
  • Lines can be parallel.
  • Lines can be skew, that is, noncoplanar.

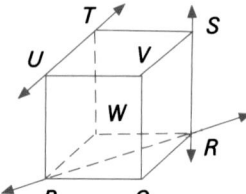

In the figure at the right:
1. Name a pair of lines that appear to be skew. $\overleftrightarrow{UT}, \overleftrightarrow{SR}$
2. Name a segment that appears to be parallel to $\overleftrightarrow{UT}$. $\overline{SV}$
3. Name a segment that appears to be skew to $\overleftrightarrow{PR}$. $\overline{QV}$
4. What line can be drawn to be parallel to $\overleftrightarrow{PR}$? $\overleftrightarrow{US}$

3.2, Properties of parallel lines are summarized by the theorems below.
3.3,   • If alternate interior angles are congruent, then lines are parallel.
3.5   • If corresponding angles are congruent, then lines are parallel.
  • If interior angles on the same side of a transversal are supplementary, then lines are parallel.
  • If two lines are parallel to the same line, then they are parallel.
  • In a plane, if two lines are perpendicular to the same line, then they are parallel.
  • If a transversal is perpendicular to one of two parallel lines, then it is perpendicular to the other.

Determine which lines are parallel. Give a reason.

5. m $\angle 1 = 105$, m $\angle 2 = 105$ $m \parallel n$
6. m $\angle 3 = 80$, m $\angle 8 = 80$ $p \parallel q$

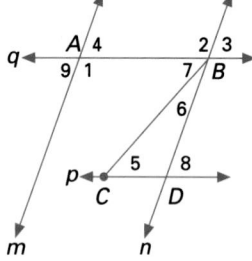

For Exercises 7–9, $m \parallel n$

7. Given: m $\angle 4 = 65$. Find m $\angle 2$.   **3.5**   115
8. Given: m $\angle 9 = 70$. Find m $\angle 3$.   70
9. Given: $\angle 4 \cong \angle 8$
  Prove: $p \parallel q$

3.4   To write an indirect proof:
  (1) Assume that the desired conclusion is false.
  (2) Reason directly until a contradiction is reached.
  (3) State that the assumption in the first step is false, so the original conclusion must be true.

10. Given: P is on $\overline{AB}$; P is not the midpoint of $\overline{AB}$.
  Prove indirectly that $\overline{AP} \neq \overline{PB}$.

3.6,   To solve problems about angle measures of triangles, use the following properties of triangles:
3.7
  • The sum of the measures of the angles of a triangle is 180.
  • The measure of an exterior angle of a triangle is equal to the sum of the measures of its two remote interior angles.

11. m $\angle 1 = 2x$, m $\angle 3 = 3x$, m $\angle 4 = 5x$. Find m $\angle 1$.   36
12. $\overline{AC} \perp \overline{AB}$, m $\angle 1 = \frac{4}{5}$ m $\angle 4$. Find m $\angle 1$ and m $\angle 4$.
13. m $\angle 3 = 4x + 20$, m $\angle 4 = x + 30$, m $\angle 2 =$ 40, 50
  $2x + 110$. Find m $\angle 1$.   30
14. Given: $\overline{AD} \parallel \overline{CE}$
  Prove: m $\angle 2 = $ m $\angle 4 + $ m $\angle 6$

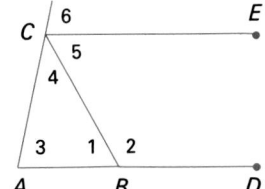

3.8   To identify and determine whether certain statements are true:
  • Negation of p: $\sim p$ (True when p is false)
  • Converse of $p \rightarrow q$: $q \rightarrow p$ (Not necessarily true, even if $p \rightarrow q$ is true)
  • Contrapositive of $p \rightarrow q$: $\sim q \rightarrow \sim p$ (True if $p \rightarrow q$ is true)
  • Inverse of $p \rightarrow q$: $\sim p \rightarrow \sim q$ (Not necessarily true, even if $p \rightarrow q$ is true)
  • Biconditional: $p \leftrightarrow q$ (True if $p \rightarrow q$ and $q \rightarrow p$ are both true)

Write the indicated expression. Determine if it is true. (Exercises 15–17)

15. Negation of "Parallel lines intersect."
16. Inverse of "If lines are parallel, then alternate interior angles are congruent."
17. Contrapositive of "If the sides of an angle are perpendicular, then the angle has a measure of 90."
18. Write the converse of "If lines are parallel, then corresponding angles are congruent." Then write a biconditional and determine if it is true.

3.9   **Determine whether the statements are true or false. Draw a diagram to support your answer. (Exercises 19–20)**

19. Two lines parallel to the same plane are parallel to each other.
20. A plane parallel to one of two skew lines contains the other.

9.

Statement	Reason
1. $\angle 4 \cong \angle 8$, $m \parallel n$	1. Given
2. $\angle 4 \cong \angle 3$	2. If lines are $\parallel$, then corr $\angle$s are $\cong$.
3. $\angle 3 \cong \angle 8$	3. Trans Prop of Congr
4. $\therefore p \parallel q$	4. If corr $\angle$s are $\cong$, then lines are $\parallel$.

10. Assume that P is on $\overline{AB}$, that P is not the midpt of $\overline{AB}$, and that $AP = PB$. Since $AP = PB$, then P is the midpt of $\overline{AB}$. But P is the midpt of $\overline{AB}$ and P is not the midpt of $\overline{AB}$ is a contradiction. So the assumption that $AP = PB$ is false. Therefore, $AP \neq PB$.

14.

Statement	Reason
1. $\overline{AD} \parallel \overline{CE}$	1. Given
2. m $\angle 3 = $ m $\angle 6$	2. If lines are $\parallel$, then corr $\angle$s are $\cong$.
3. m $\angle 2 = $ m $\angle 4 + $ m $\angle 3$	3. The meas of an ext $\angle$ of a $\triangle$ is equal to the sum of the meas of its 2 remote int $\angle$s.
4. $\therefore$ m $\angle 2 = $ m $\angle 4 + $ m $\angle 6$	4. Sub

15. $\parallel$ lines do not intersect. (True)
16. If lines are not $\parallel$, then alt int $\angle$s are not $\cong$. (True)
17. If an $\angle$ does not have a meas of 90, then the sides of the $\angle$ are not $\perp$. (True)
18. Conv: If corr $\angle$s are $\cong$, then lines are $\parallel$. (True)
  Bicond: Lines are $\parallel$ if and only if corr $\angle$s are $\cong$. (True)

19.

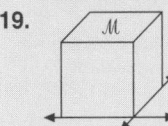

False. $l \parallel \mathcal{M}$, $n \parallel \mathcal{M}$, but $l \nparallel n$.

20.

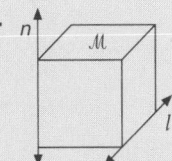

False. $\mathcal{M} \parallel l$, and n and l are skew; but $\mathcal{M}$ does not contain n.

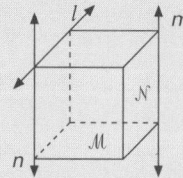

Exercises 1, 2, 4

1. False. *l* and *m* do not intersect, but *l* $\not\parallel$ *m*.
2. False. $\mathcal{M} \parallel l$, and $\mathcal{N} \parallel l$; but $\mathcal{M} \not\parallel \mathcal{N}$.
4. False. *l* and *m* are skew, and *n* $\parallel$ *m*; but *n* $\not\parallel$ *l*.
6. $\overline{RQ} \parallel \overline{SP}$

7.
Statement	Reason
1. $\overline{SP} \parallel \overline{RQ}$	1. Given
2. m $\angle 2 =$ m $\angle 4$	2. If lines are $\parallel$, then alt int $\angle$s are $\cong$.
3. m $\angle 3 +$ m $\angle 4 =$ m $\angle PSR$	3. Angle Add Post
4. ∴ m $\angle 2 +$ m $\angle 3 =$ m $\angle PSR$	4. Sub

8.
Statement	Reason
1. $\overline{SP} \parallel \overline{RQ}$, m $\angle PSR =$ m $\angle PQR$	1. Given
2. m $\angle 4 =$ m $\angle 2$	2. If lines are $\parallel$, then alt int $\angle$s are $\cong$.
3. m $\angle PSR =$ m $\angle 4 +$ m $\angle 3$, m $\angle PQR =$ m $\angle 2 +$ m $\angle 1$	3. Angle Add Post
4. m $\angle 4 +$ m $\angle 3 =$ m $\angle 2 +$ m $\angle 1$	4. Sub
5. m $\angle 3 =$ m $\angle 1$	5. Subt Prop of Eq
6. ∴ $\overline{PQ} \parallel \overline{RS}$	6. If alt int $\angle$s are $\cong$, then lines are $\parallel$.

15.

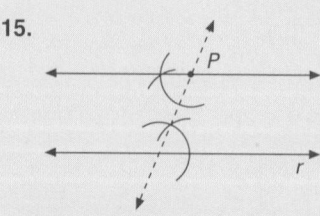

19. A rt $\angle$ does not have a meas of 80. (True)
20. 2 lines are $\parallel$ if and only if int $\angle$s on same side of a transv are supp. (True)

136

True or false? If false, draw a figure to show why. (Exercises 1–5)

1. Lines which do not intersect are parallel.
2. Two planes parallel to the same line are parallel.
3. Skew lines do not intersect.   True
4. A line parallel to one of two skew lines is parallel to the other.
5. The measure of an exterior angle of a triangle equals the sum of the measures of its two remote interior angles.   True

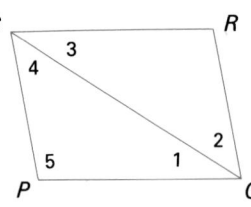

6. Given: m $\angle 4 = 40$, m $\angle 2 = 40$ Which segments are parallel?
7. Given: $\overline{SP} \parallel \overline{QR}$ Prove: m $\angle 2 +$ m $\angle 3 =$ m $\angle PSR$
*8. Given: $\overline{SP} \parallel \overline{RQ}$, m $\angle PSR =$ m $\angle PQR$ Prove: $\overline{PQ} \parallel \overline{RS}$
9. Given: $\angle 1 \cong \angle 4$, m $\angle 5 = 140$ Find m $\angle 1$.   20

For Exercises 10–14, $m \parallel n$.

10. Given: m $\angle 8 = 100$. Find m $\angle 4$.   100
11. Given: m $\angle 1 = 70$. Find m $\angle 3$.   70
12. Given: m $\angle 2 = \frac{2}{7}$ m $\angle 3$. Find m $\angle 1$. 140
13. Given: m $\angle 2$ is 60 less than twice m $\angle 3$. Find m $\angle 1$.   80
14. Given: m $\angle 10 = 3x + 30$, m $\angle 9 = x + 40$. Find m $\angle 9$.   45

15. Draw a line *r* and a point *P* not on *r*. Construct a line through *P* parallel to *r*.
16. Given: m $\angle A = 20$, $\angle 1 \cong \angle C$. Find m $\angle C$.   80
17. Given: m $\angle A = 2x + 10$, m $\angle 1 = 3x + 20$, m $\angle C = 15x - 50$. Find m $\angle C$.   100
18. Given: m $\angle A = 70$, m $\angle 2 = 120$. Find m $\angle C$.   50

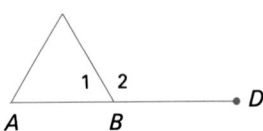

19. Write the negation of the statement "A right angle has a measure of 80." Is the negation true?

Use the following statement to answer Exercises 20–22.

"If two lines are parallel, then interior angles on the same side of a transversal are supplementary."

20. Write a biconditional for the statement and its converse. Is it true?
21. Write the inverse of the statement. Is it true?
22. Write the contrapositive of the statement. Is it true?

---

21. If 2 lines are not $\parallel$, then the int $\angle$s on same side of a transv are not supp. (True)
22. If int $\angle$s on same side of a transv are not supp, then lines are not $\parallel$. (True)

In some cases, it may not be necessary to know the measures of two angles of a triangle to find the measure of the third. For example:

Given: m ∠C = 50, $\overline{AD}$ and $\overline{BD}$ are angle bisectors. Find x.

You don't have to find m ∠1 and m ∠2 separately to solve the problem. Rather, find their sum and subtract it from 180.

Since m ∠C = 50, m ∠CAB + m ∠CBA = 180 − 50 = 130.

Then, from the definition of angle bisectors, m ∠1 + m ∠2 = $\frac{1}{2} \cdot 130 = 65$.

Now, x = 180 − (m ∠1 + m ∠2) = 180 − 65 = 115

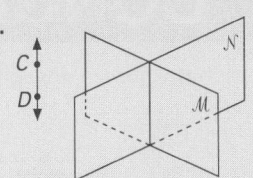

1. Find the sum of the degree measures
A of the numbered angles.

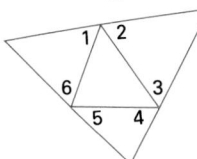

(A) 360   (B) 180   (C) 540   (D) 270
(E) It cannot be determined from the given information.

2.

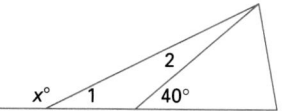

If ∠1 ≅ ∠2 then x = ___B___.
(A) 60   (B) 160   (C) 40
(D) 140   (E) 120

3.

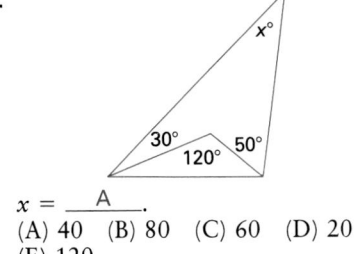

x = ___A___.
(A) 40   (B) 80   (C) 60   (D) 20
(E) 120

4.

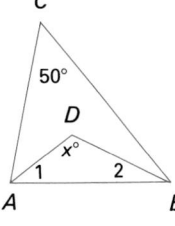

m ∠P − m ∠Q = ___D___.
(A) 115   (B) 65   (C) 40   (D) 0
(E) 50

5.

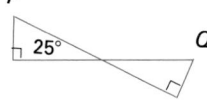

a = ___D___.
(A) 30   (B) 55   (C) 115
(D) 15   (E) 50

6. $\overline{PQ} \perp \overline{PR}$, m ∠1 = m ∠2

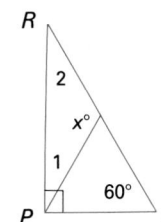

x = ___B___.
(A) 60   (B) 120   (C) 100   (D) 150
(E) None of these answers

9. Given: $\mathcal{P} \parallel \mathcal{Q}$, and line m intersects $\mathcal{P}$ at (just the one) point G of line m. Assume: m does not intersect $\mathcal{Q}$, i.e. m ∥ $\mathcal{Q}$. There is a plane $\mathcal{R}$ which contains m and is ∥ $\mathcal{Q}$. $\mathcal{R}$ contains m, so it contains the point G. But there is just one plane that contains the point G and is ∥ to $\mathcal{Q}$, so planes $\mathcal{R}$ and $\mathcal{P}$ must be the same plane. Thus $\mathcal{P}$ contains m and hence all points of m (infinitely many), which contradicts the assertion that m intersects $\mathcal{P}$ (at just one point). So the assumption that m does not intersect $\mathcal{Q}$ must be false. Therefore, m intersects $\mathcal{Q}$.

Additional Answers, page 132

3.

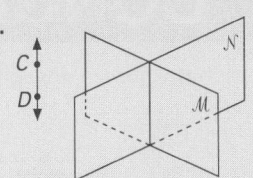

False. $\mathcal{M}$ and $\mathcal{N}$ are both ∥ $\overleftrightarrow{CD}$, but $\mathcal{M} \nparallel \mathcal{N}$.

5.

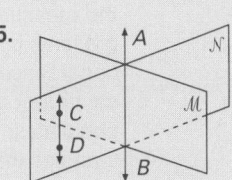

False. $\overleftrightarrow{CD} \parallel \overleftrightarrow{AB}$, but $\mathcal{M} \nparallel \mathcal{N}$.

6.

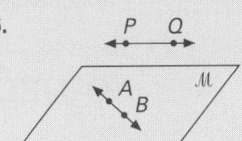

False. $\overrightarrow{PQ} \parallel \mathcal{M}$, but $PQ \nparallel \overleftrightarrow{AB}$.

7. Given: $\mathcal{M}$ and $\mathcal{N}$ intersect in $\overleftrightarrow{AB}$; $\overleftrightarrow{CD} \parallel \mathcal{N}$, $\overleftrightarrow{CD} \parallel \mathcal{M}$, $\overleftrightarrow{CD}$ is not skew to $\overleftrightarrow{AB}$. Assume: $\overleftrightarrow{CD} \nparallel \overleftrightarrow{AB}$. Since $\overleftrightarrow{CD}$ is not skew or ∥ to $\overleftrightarrow{AB}$, $\overleftrightarrow{CD}$ intersects $\overleftrightarrow{AB}$. Thus, $\overleftrightarrow{CD}$ shares at least one pt with both $\mathcal{N}$ and $\mathcal{M}$, since $\overleftrightarrow{AB}$ lies in each. But $\overleftrightarrow{CD}$ shares no pts with $\mathcal{N}$ or $\mathcal{M}$, since $\overleftrightarrow{CD} \parallel$ to both. But $\overleftrightarrow{CD}$ shares a pt with $\mathcal{N}$ and $\mathcal{M}$ and $\overleftrightarrow{CD}$ shares no pts with $\mathcal{N}$ or $\mathcal{M}$ is a contradiction. So, the assumption that $\overleftrightarrow{CD} \nparallel \overleftrightarrow{AB}$ is false, so $\overleftrightarrow{CD} \parallel \overleftrightarrow{AB}$.

8. Given: line l is in plane $\mathcal{N}$, line m intersects $\mathcal{N}$ at 1 pt. Assume: l ∥ m. Since l ∥ m, then l and m are contained by the same plane and do not intersect. Since m intersects $\mathcal{N}$ in just 1 pt, lines l and m are ∥ in another plane, plane $\mathcal{P}$. Since $\mathcal{P}$ contains every pt on l, $\mathcal{P}$ intersects $\mathcal{N}$ at line l. The pt of intersection of m and $\mathcal{N}$ is also in $\mathcal{P}$, so either l intersects m in plane $\mathcal{N}$ or $\mathcal{N}$ is $\mathcal{P}$. But if l intersects m then l ∦ m, and if $\mathcal{N}$ is $\mathcal{P}$, then only 1 pt of line m is in $\mathcal{P}$, so l ∦ m. But l ∦ m and l ∥ m is a contradiction, so the assumption that l ∥ m is false. Therefore, l ∦ m.

4 CONGRUENT TRIANGLES

OVERVIEW

This chapter classifies triangles according to the measures of their angles and sides and presents proofs dealing with triangle congruence. Students prove triangles congruent using the SSS, ASA, and SAS postulates. They prove the congruence of triangles that are parts of complex figures, the congruence of sides and angles, and prove segments parallel by means of congruent triangles.

OBJECTIVES

- To classify triangles according to the measures of angles and the lengths of sides
- To solve problems involving the measure of sides and angles in triangles
- To determine whether triangles are congruent, given their congruent parts
- To prove triangles congruent when they are parts of complex figures

PROBLEM SOLVING

The Application on page 152 provides students with an opportunity to use a chart. Students must read and interpret the information given in the chart in order to solve the problems. The Algebra Review on page 166 will refresh students' knowledge of solving systems of equations.

TECHNOLOGY

Computer: Students use a computer to construct different triangles by the SSS, ASA, and SAS Postulates in the Computer Investigation on page 146. The computer activities help students discover how the congruence of triangles can be established by showing the congruence of three pairs of parts.

SPECIAL FEATURES

Focus on Reading pp. 141, 164
Mixed Review pp. 142, 145, 152, 160, 166, 169
Computer Investigation: Constructing Triangles p. 146
Application: Angles in Sand Piles p. 152
Midchapter Review p. 157
Algebra Review p. 166
Key Terms p. 170
Key Ideas and Review Exercises pp. 170–171
Chapter 4 Test p. 172
College Prep Test p. 173
Cumulative Review (Chapters 1–4) pp. 174–175

PLANNING GUIDE

Lesson	Basic	Average	Above Average	Resources
4.1 pp. 141–142	FR all CE all WE 1–6	FR all CE all WE 1–11	FR all CE all WE 1–15	Reteaching p. 53 Practice p. 54
4.2 pp. 144–146	CE all WE 1–9 CI 1	CE all WE 1–14 CI all	CE all WE 1–16 CI all	Reteaching p. 55 Practice p. 56
4.3 pp. 149–152	CE all WE 1–10 odd Application	CE all WE 1–14 odd Application	CE all WE 1–17 odd Application	Reteaching p. 57 Practice p. 58
4.4 pp. 155–157	CE all WE 1–9 odd Midchapter Review	CE all WE 1–13 odd Midchapter Review	CE all WE 1–16 odd Midchapter Review	Reteaching p. 59 Practice p. 60
4.5 pp. 158–160	CE all WE 1–6 odd	CE all WE 1–10 odd	CE all WE 1–13 odd	Reteaching p. 61 Practice p. 62
4.6 pp. 164–166	FR all, CE all WE 1–12 odd AR all	FR all, CE all WE 1–16 odd AR all	FR all, CE all WE 1–18 odd AR all	Reteaching p. 63 Practice p. 64
4.7 pp. 168–169	CE all WE 1–4	CE all WE 1–8	CE all WE 1–10	Reteaching p. 65 Practice p. 66
Chapter 4 Review pp. 170–171	1–15 odd	all	all	
Chapter 4 Test p. 172	1–2, 4–10	1–11	all	
College Prep Test p. 173	all	all	all	
Cumulative Review pp. 174–175	1–18 odd	1–35 odd	all	

CE = Classroom Exercises   WE = Written Exercises   FR = Focus on Reading   AR = Algebra Review   CI = Computer Investigation
NOTE: For each level, all students should be assigned all Mixed Review exercises.

■ **INVESTIGATION**

Project: This activity helps students discover informally the SSS congruence by making a model of a triangle with straws. Use this project to prepare students for Chapter 4.

Materials: 3 straws (each a different length), glue, and a sheet of $8\frac{1}{2}$ x 11 plain white paper (cut in half).

Activity 1: With the students in pairs, have them investigate whether they can put the ends of the straws together to form a triangle. If so, ask them to rearrange the three straws to form as many different triangles as possible. After leading the students to conclude that there is, at most, one way to put the straws together to form a triangle, have them glue the straws down on a half sheet of paper to form a triangle and label it *ABC*.

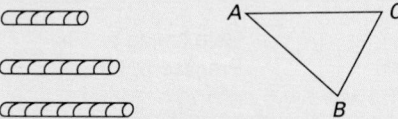

Activity 2: Instruct students to trace △*ABC* using the other half sheet of paper (or measure the outside side lengths of △*ABC* and draw one with the same dimensions). Label the new triangle *DEF*. Now have students compare the two triangles. Have students slide, turn, or flip one sheet of paper over as they compare △*ABC* and △*DEF*. Ask them questions such as these: "Are the sides of the two triangles the same length? Do the angles have the same measure?" Have students list their observations about the correspondence of the two triangles. For example:

Matching Angles	Matching Sides
∠*A* and ∠*D*	$\overline{AB}$ and $\overline{DE}$
∠*B* and ∠*E*	$\overline{BC}$ and $\overline{EF}$
∠*C* and ∠*F*	$\overline{CA}$ and $\overline{FD}$

Students' lists may vary.

Use Project Worksheet 4 to give students additional practice with congruent triangles.

To recognize his achievements in mathematics and science, President George Washington appointed Benjamin Banneker a member of the commission which surveyed and laid out the streets for the District of Columbia. Banneker also wrote an Almanac, for which he did all the intricate mathematical calculations himself.

More About the Mathematician

Benjamin Banneker was born in 1731 in Maryland. Even in his youth, his farming neighbors frequently brought mathematical puzzles and problems for him to solve. One neighbor, a mathematician and astronomer, shared books and scientific instruments with Banneker, who educated himself in astronomy. With this background, he accurately predicted a solar eclipse. He then extended his calculations to create, in 1792, an Almanac which was, at that time, a comprehensive collection of scientific information. Banneker continued to produce his Almanac for the next 10 years.

4.1 Triangle Classifications

Objectives To classify triangles according to measures of angles and lengths of sides
To find measures of angles and lengths of sides of a triangle

Triangles can be classified according to the measures of their angles. Recall the definition of acute and obtuse angles.

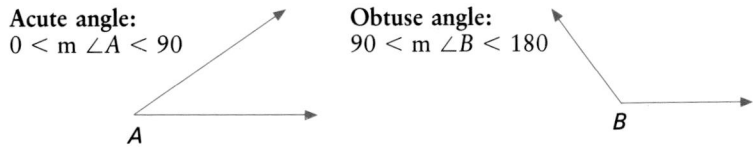

Acute angle:
$0 < m \angle A < 90$

Obtuse angle:
$90 < m \angle B < 180$

Definitions

An **acute triangle** is a triangle with three acute angles.
A **right triangle** is a triangle with one right angle.
An **obtuse triangle** is a triangle with one obtuse angle.
An **equiangular triangle** is a triangle with three congruent angles.

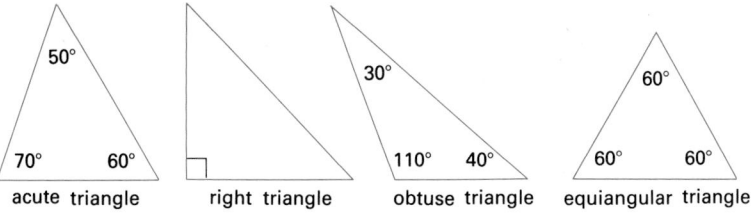

acute triangle   right triangle   obtuse triangle   equiangular triangle

Triangles can also be classified according to the lengths of their sides.

Definitions

An **equilateral triangle** is a triangle with three congruent sides.
An **isosceles triangle** is a triangle with at least two congruent sides.
A **scalene triangle** is a triangle with no congruent sides.

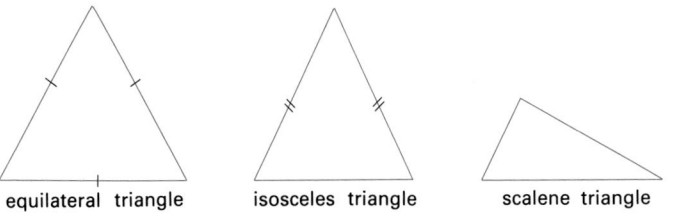

equilateral triangle   isosceles triangle   scalene triangle

4.1 Triangle Classifications   **139**

Additional Example 1

Classify each triangle as indicated.

By sides:

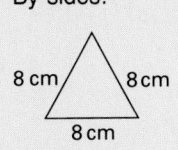

8 cm   8 cm
8 cm

Equilateral

By angles:

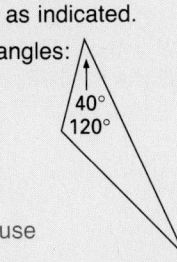

40°
120°

Obtuse

By angles and sides:

7′
110°
7′

Obtuse and isosceles

Teaching Resources

Project Worksheet 4
Quick Quizzes 27
Reteaching and Practicing
   Worksheets, pp. 53, 54
Technology Worksheet 4

▰▰▰GETTING STARTED

Prerequisite Quiz

Classify each angle as acute, right, obtuse, or straight.

1. m ∠A = 90   Right
2. m ∠A = 17   Acute
3. m ∠A = 180   Straight
4. m ∠A = 105   Obtuse
5. If the measures of two angles of a triangle are 40 and 80, what is the measure of the third angle?   60

Motivator

Ask your students to describe an acute angle.   An angle with a measure < 90.   an obtuse angle.   An angle with a measure > 90 but < 180.   Ask them if they think a triangle can contain two obtuse angles, and to explain why.   No. The sum of the angle measures of a triangle is 180.   Ask them if they know what two words are suggested by the word **equiangular**.   Equal angles

▰▰▰TEACHING SUGGESTIONS

Lesson Note

Triangles are classified in two ways, by angles and by sides. Point out that any triangle can be classified both ways. For example, a triangle with a right angle, and with the two sides that include the right angle congruent, is an *isosceles right triangle*.

Review the property for the sum of the measures of the angles of a triangle (Theorem 3.9, page 116). This property is used to prove that, in a right triangle, the two angles other than the right angle are complementary and acute.

139

Math Connections

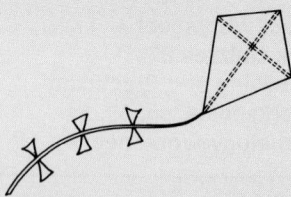

Kiting: Kites are the oldest form of aircraft. Although there are hundreds of different kinds of kites, the *two-stick kite* is the oldest type. Its frame consists of two rods that intersect at right angles to form two isosceles triangles which share the same base.

Critical Thinking Questions

Synthesis: Have students describe with an inequality statement the sum of the measures of the acute angles in an obtuse isosceles triangle. In $\triangle ABC$, if $90 < m \angle C < 180$, then $0 < m \angle A + m \angle B < 90$. Have them describe with an inequality statement the sum of any two angles of an acute triangle. In $\triangle ABC$, if $0 < m \angle C < 90$, then $90 < m \angle A + m \angle B < 180$.

Common Error Analysis

Error: Students often think that the base of a right triangle or an isosceles triangle must always be horizontal. This leads to errors in identifying the congruent sides of an isosceles triangle or the legs and hypotenuse of a right triangle. Draw several isosceles and right triangles on the chalkboard, as shown.

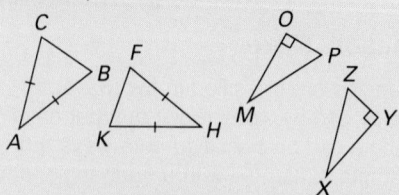

Have students name the congruent sides and the base of each isosceles triangle. Have them name the legs and the hypotenuse of each right triangle.

140

EXAMPLE 1   Classify each triangle as indicated.

By sides:

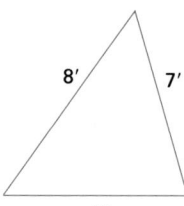

By angles:

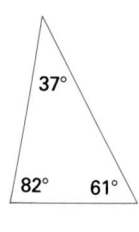

By sides and angles:

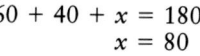

Solution   scalene   acute   isosceles, right

EXAMPLE 2   Classify the triangle by its angles.

Plan   Find the measure of the third angle.

Solution   Let x = measure of third angle.

$$60 + 40 + x = 180$$
$$x = 80$$

The measure of each angle of the triangle is less than 90. Thus, the triangle is acute.

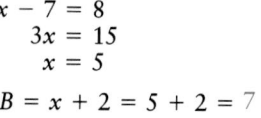

EXAMPLE 3 $\triangle ABC$ is isosceles with $AC = BC$. $AC = 5x - 7$, $BC = 2x + 8$, and $AB = x + 2$. Find AB.

Plan   Draw and label the figure.
Use $AC = BC$ to write an equation.

Solution
$$5x - 7 = 2x + 8$$
$$3x - 7 = 8$$
$$3x = 15$$
$$x = 5$$

$$AB = x + 2 = 5 + 2 = 7$$

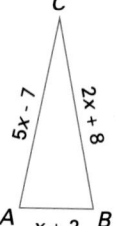

In right $\triangle ABC$, m $\angle C = 90$

$$m \angle A + m \angle B + m \angle C = 180$$
$$m \angle A + m \angle B + 90 = 180$$
$$m \angle A + m \angle B = 90$$

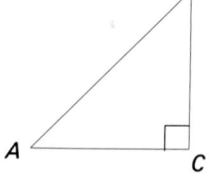

140   Chapter 4 Congruent Triangles

Additional Example 2

Classify the triangle by its angles.
Right triangle

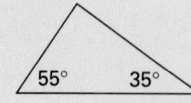

Additional Example 3

$\triangle PQR$ is isosceles with $QP = QR$. $QP = 6x - 4$, $QR = 3x + 11$, and $PR = x^2 - 7$. Find PR.   18

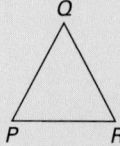

Therefore, $\angle A$ and $\angle B$ are complementary. Since the sum of the measures of $\angle A$ and $\angle B$ is 90, each angle is acute. This leads to Theorem 4.1.

Theorem 4.1

In a right triangle, the two angles other than the right angle are complementary and acute.

EXAMPLE 4 The measure of one acute angle of a right triangle is $\frac{2}{3}$ the measure of the other acute angle. Find the measure of the larger acute angle.

Plan Let x = larger angle
$\frac{2}{3}x$ = smaller angle

Solution
$$x + \frac{2}{3}x = 90 \quad \longleftarrow \text{ Sum of measures of complementary } \angle\text{s is } 90$$
$$3x + 2x = 270$$
$$5x = 270$$
$$x = 54$$

The measure of the larger acute angle is 54.

Focus on Reading

Determine whether the statement is always true, sometimes true, or never true.

1. An equilateral triangle is isosceles. Always true
2. An equiangular triangle is a right triangle. Never true
3. A right triangle is a triangle with three acute angles. Never true
4. An obtuse triangle is isosceles. Sometimes true
5. An obtuse triangle is a right triangle. Never true

Classroom Exercises

Classify the triangle for the lengths of sides given.

1. 5 cm, 4 cm, 6 cm   2. 4 in., 4 in., 4 in.   3. 6 ft, 2 ft, 6 ft   4. 7 in., 5 in., 6 in.
   Scalene   Equilateral   Isosceles   Scalene

Classify the triangle for the angle measures given.

5. 90, 40, 50   6. 130, 20, 30   7. 60, 60, 60   8. 50, 50, 80
   Right   Obtuse   Equiangular   Acute

Classify triangle *ABC* by its sides.

1. $AB = 7$, $BC = 6$, $AC = 8$ Scalene
2. $AC = 7$, $BC = 5$, $AB = 7$ Isosceles

Classify triangle *ABC* by its angles.

3. m $\angle A = 40$, m $\angle C = 60$ Acute
4. m $\angle B = 70$, m $\angle C = 10$ Obtuse
5. $\triangle ABC$ is isosceles with $AC = BC$, $AC = 4x - 1$, $BC = 2x + 11$, and $AB = 3x + 1$. Find AB. 19

Closure

Ask your students to define each of the following.

acute triangle A triangle with 3 acute angles.
right triangle A triangle with one right angle.
obtuse triangle A triangle with one obtuse angle.
equiangular triangle A triangle with 3 congruent angles.
equilateral triangle A triangle with 3 congruent sides.
isosceles triangle A triangle with 2 congruent sides.
scalene triangle A triangle with no congruent sides.

FOLLOW UP

Guided Practice

Classroom Exercises 1–8

Independent Practice

A Ex. 1–6, **B** Ex. 7–11, **C** Ex. 12–15

Basic: FR all, WE 1–6
Average: FR all, WE 1–11
Above Average: FR all, WE 1–15

Additional Example 4

In a right triangle, the measure of one acute angle is $\frac{5}{4}$ the measure of the other. Find the measure of the larger acute angle. 50

Additional Answers

12. If △ is isos, then $2n - 1 = n + 7$, or $n + 7 = 3n - 9$, or $2n - 1 = 3n - 9$. Solving for n in all 3 cases gives $n = 8$. Substituting 8 for n shows that all sides have length 15.

13.

Statement	Reason
1. △ABC, m ∠C = 90	1. Given
2. m ∠A + m ∠B + m ∠C = 180	2. Sum of meas of a △ is 180.
3. m ∠A + m ∠B = 90	3. Subt prop of eq.
4. ∠A and ∠B are comp	4. Def of comp ∠s
5. Also, m ∠A > 0 and m ∠B > 0, so m ∠A < 90 and m ∠B < 90 i.e. ∠A and ∠B are acute	5. Subt prop of eq

14.

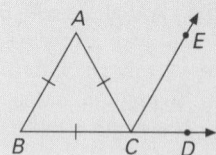

Statement	Reason
1. △ABC is equiangular, $\overrightarrow{CE}$ bis ∠ACD	1. Given
2. m ∠A = 60, m ∠B = 60	2. Def of equiangular △
3. m ∠ACD = 120	3. Meas of ext ∠ of △ = sum of the meas of the 2 remote int ∠s.
4. m ∠ACE = $\frac{1}{2}$(m ∠ACD)	4. Def of ∠ bis
5. m ∠ACE = $\frac{1}{2}$(120), or m ∠ACE = 60	5. Sub
6. m ∠ACE = m ∠A	6. Sub
7. $\overrightarrow{AB} \parallel \overrightarrow{CE}$	7. If alt int ∠s are ≅, then lines are ∥.

15. Assume a △ is obtuse and it has a rt ∠. Then, since it is an obtuse △, it has an angle with meas > 90. Then the sum of the meas of an obtuse ∠ and a rt ∠ is > 180. But the sum of the meas of the ∠s of a △ is always 180. So, the assumption that the obtuse △ has a rt ∠ is false. Therefore, an obtuse △ cannot have a rt ∠.

142

Written Exercises

Classify the triangle by its sides.

1. 8 m · 8 m · 8 m   Equilateral

2. 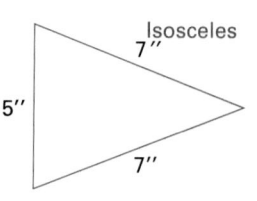 Isosceles 7″ · 5″ · 7″

3. 2′ · 4′ · 3′   Scalene

Classify the triangle by its angles.

4. The measures of two angles are 25 and 35.   Obtuse
5. The measures of two angles are 25 and 65.   Right
6. The measures of two angles are 75 and 45.   Acute
7. △ABC is isosceles with $AC = BC$. $AC = 6x - 5$, $BC = 4x + 7$, and $AB = 5x - 2$. Find AB.   28
8. △ABC is equilateral with $BC = 3x + 7$ and $AC = 5x + 1$. Find AB.   16
9. The measure of an acute angle of a right triangle is $\frac{4}{5}$ the measure of the other acute angle. Find the measure of the smaller acute angle.   40
10. In a right triangle, the measure of one acute angle is 30 less than twice the measure of the other acute angle. Find the measure of each angle.   40, 50
11. △PQR is isosceles with $PR = QR$. $PR = 10$, $QR = 3x - 11$, $PQ = x^2 - 25$. Find PQ.   24
12. The lengths of the sides of the triangle are given by $2n - 1$, $n + 7$, and $3n - 9$. Prove that if the triangle is isosceles, then the triangle is equilateral.
13. Prove Theorem 4.1.
14. Prove that the bisector of an exterior angle of an equiangular triangle is parallel to a side of a triangle.
15. Write an indirect proof that an obtuse triangle cannot be a right triangle.

Mixed Review

1. Given: The coordinates of the endpoints of a segment are −6 and 4. Find the coordinate of the midpoint.   **1.3** −1
2. The measures of a pair of alternate interior angles of two parallel lines are $3x - 20$ and $x + 40$. Find the measure of each angle.   **3.5** 70
3. If two adjacent angles are complementary, then their outer rays are ____.   **1.6** ⊥

Enrichment

Before class, prepare two figures like the ones below. Use small flat sticks with a single pin attaching them at each corner. Have the students observe that the triangle is a rigid figure, whereas the square is easily collapsed. Explain that this is the main reason why the triangle is a fundamental form in architecture and construction.

For a bulletin board display, encourage the students to cut out pictures which illustrate the practical uses of the triangle in architecture and construction.

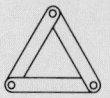
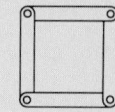

4.2 Congruence of Triangles

Teaching Resources

Application Worksheet 4
Manipulative Worksheet 7
Quick Quizzes 28
Reteaching and Practice
  Worksheets, pp. 55, 56

Objectives

To identify congruent triangles
To identify corresponding parts of congruent triangles
To determine missing measures in congruent triangles

Using the diagrams, trace Figure *A* and then place the tracing on top of Figure *B*. Since the two figures coincide, they are called congruent. Geometric figures are *congruent* if they have the same size and shape.

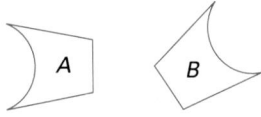

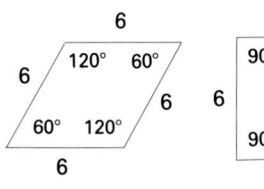 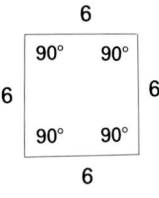 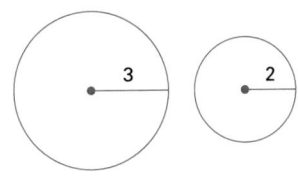

The lengths of the sides of the figures above are the same. But the figures differ in shape. The figures are *not* congruent.

The figures above have the same shape. But they differ in size. The figures are *not* congruent.

To determine whether two triangles are congruent, imagine overlaying one triangle on the other. The sides and angles of one triangle should be congruent to the corresponding sides and angles of the other triangle.

Definition

Two triangles are **congruent** ($\cong$) if corresponding angles are congruent and corresponding sides are congruent.

The triangles at the right are congruent. The sets of arcs and tick marks indicate that the corresponding angles are congruent (equal in measure) and the corresponding sides are congruent (equal in length). To indicate that triangle *ABC* is congruent to triangle *DEF*, write $\triangle ABC \cong \triangle DEF$. The order of the letters always indicates the corresponding vertices.

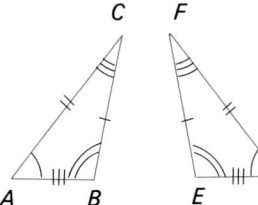

$$\triangle ABC \cong \triangle DEF$$

4.2 Congruence of Triangles   **143**

TEACHING SUGGESTIONS

Lesson Note

Emphasize that corresponding vertices of congruent triangles can be matched by using the corresponding order of letters. Thus, for △MYU ≅ △PER, the vertices correspond as follows: M to P, Y to E, U to R. The order of letters is also helpful in identifying corresponding congruent sides. For example, since △MYU ≅ △PER, $\overline{MU}$ ≅ $\overline{PR}$. The interpretation of congruence up to this point has been restricted to angles and segments. Congruence has meant equal in angle measure or equal in segment length. Now, congruence of geometric figures has a more general interpretation: same size and same shape, not merely same size.

Math Connections

Puzzles: When you put together a jigsaw puzzle, you apply congruence by placing the puzzle pieces in congruent spaces where they fit.

Critical Thinking Questions

Synthesis: Ask students to make a list of all the possible ways to state correctly that △ABC ≅ △XYZ. △ABC ≅ △XYZ; △BCA ≅ △YZX; △CAB ≅ △ZXY; △BAC ≅ △YXZ; △CBA ≅ △ZYX; △ACB ≅ △XZY Have them list all the ways to state correctly that △ABC ≅ △ZXY. △ABC ≅ △ZXY; △BCA ≅ △XYZ; △CAB ≅ △YZX; △BAC ≅ △XZY; △ACB ≅ △ZYX; △CBA ≅ △YXZ Ask students to draw a diagram of congruent triangles ABC and ZXY, where $\overline{AC}$ ≅ $\overline{BC}$, ∠A ≅ ∠B, and to mark the congruent sides and angles. Classify the triangles. Isos △

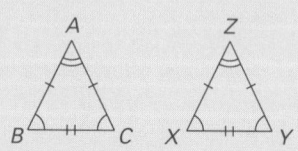

EXAMPLE 1 Using the markings shown, complete the congruence statement:
△PQR ≅ _____.

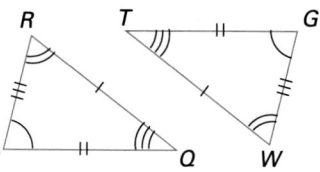

Solution From the markings, you know that

(1) ∠P ≅ ∠G,
(2) ∠Q ≅ ∠T, and
(3) ∠R ≅ ∠W.

So, △PQR ≅ △GTW.

EXAMPLE 2 Given: △PWG ≅ △SEM

Identify the corresponding sides and the corresponding angles. Draw the two triangles and mark the corresponding parts.

Solution Use the order of letters to identify the corresponding parts.

Corresponding sides:
△PWG ≅ △SEM
$\overline{PW}$ ≅ $\overline{SE}$
$\overline{WG}$ ≅ $\overline{EM}$
$\overline{PG}$ ≅ $\overline{SM}$

Corresponding angles:
△PWG ≅ △SEM
∠P ≅ ∠S
∠W ≅ ∠E
∠G ≅ ∠M

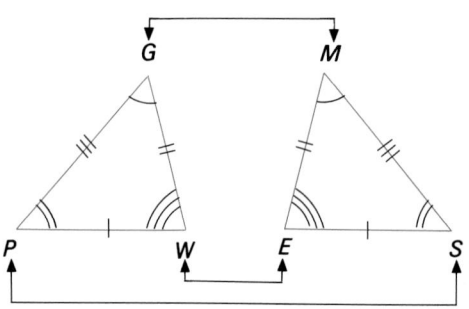

Classroom Exercises

Name the corresponding angles for the pair of congruent triangles.

1. △TYP ≅ △ERM
2. △QWZ ≅ △UTA
3. △PTC ≅ △DYH

Additional Example 1

Using the markings shown, complete the congruence statement: △TGF ≅ _____.
△YLK

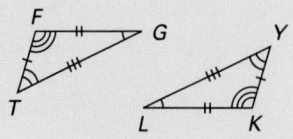

Additional Example 2

Given: △YUP ≅ △NAG
Identify the corresponding sides and the corresponding angles. Draw the two triangles and label the corresponding parts.
∠Y ≅ ∠N, ∠U ≅ ∠A, ∠P ≅ ∠G, $\overline{YU}$ ≅ $\overline{NA}$, $\overline{UP}$ ≅ $\overline{AG}$, $\overline{YP}$ ≅ $\overline{NG}$

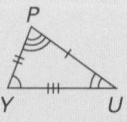

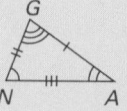

Name the corresponding sides for the pair of congruent triangles.

4. $\triangle GUM \cong \triangle PAT$ **5.** $\triangle RUT \cong \triangle WDG$ **6.** $\triangle SAD \cong \triangle LID$

Written Exercises

Use the markings on the figures to complete the congruence statement.

1.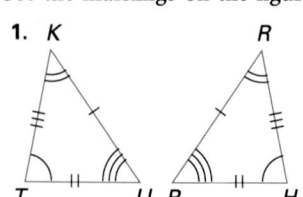

$\triangle TKU \cong \underline{\triangle HRP}$

2.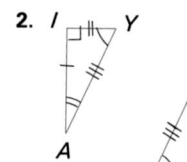

$\triangle LQP \cong \underline{\triangle IAY}$

3.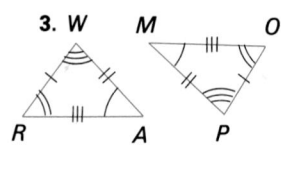

$\triangle RAW \cong \underline{\triangle OMP}$

Complete the congruence statement for the given pair of congruent triangles.

4. $\triangle RTY \cong \triangle PUM$
$\angle T \cong \underline{\hspace{1cm}} \angle U$

5. $\triangle STY \cong \triangle APQ$
$\angle Y \cong \underline{\hspace{1cm}} \angle Q$

6. $\triangle DFT \cong \triangle LKP$
$\underline{\hspace{1cm}} \cong \angle L \quad \angle D$

7. $\triangle AFK \cong \triangle YWE$
$\overline{AF} \cong \underline{\hspace{1cm}} \overline{YW}$

8. $\triangle KLB \cong \triangle PSM$
$\overline{KB} \cong \underline{\hspace{1cm}} \overline{PM}$

9. $\triangle FAD \cong \triangle UKT$
$\overline{AD} \cong \underline{\hspace{1cm}} \overline{KT}$

For Exercises 10–12, draw a diagram of the two congruent triangles.
Mark the corresponding parts and identify the corresponding sides and
the corresponding angles.

10. $\triangle TER \cong \triangle YUQ$ **11.** $\triangle EJS \cong \triangle IPU$ **12.** $\triangle DAS \cong \triangle HOP$

13. Given: $\triangle ABC \cong \triangle UKT$. $AB = 4$, $BC = 5$, $AC = 6$. Find UT.   6

14. Given: $\triangle GHT \cong \triangle MOW$. m $\angle G = 40$, m $\angle H = 70$. Find m $\angle M$
and m $\angle W$.   40, 70

15. Given: $\triangle ABC \cong \triangle FGH$. $AB = 4x + 5$, $FG = 2x + 13$, $AC = 3x - 1$. Find FH.   11

16. Given: $\triangle PQR \cong \triangle RQP$. Prove: $\triangle PQR$ is isosceles.

Mixed Review

1. The measure of the exterior angle of a triangle is 100. The measure
of one of its remote interior angles is 75. Find the measure of its
other remote interior angle.   *3.7*   25

2. Two angles are supplementary. The measure of one angle is twice
the measure of the other. Find the measure of each angle.   *1.6*   60, 120

3. One of the acute angles of a right triangle measures 42. Find the
measure of the other acute angle.   *3.6*   48

Enrichment

On the chalkboard, draw an equilateral
triangle with its three medians. However,
do *not* discuss these features of the
drawing. Instead, ask the students to
identify all pairs or groupings of triangles
that appear to be congruent.

$\triangle AOE \cong \triangle BOE$, $\triangle AOD \cong \triangle BOF$,
$\triangle COD \cong \triangle COF$, $\triangle AOC \cong \triangle BOC$,

$\triangle CEB \cong \triangle CEA \cong$
$\triangle AFB \cong \triangle AFC \cong$
$\triangle BDA \cong \triangle BDC$

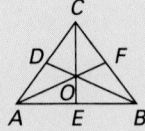

Repeat this activity with an isosceles
triangle that is not equilateral. The
students should find that fewer triangles
appear to be congruent.

Checkpoint

1. Use the markings shown to complete the
congruence statement.

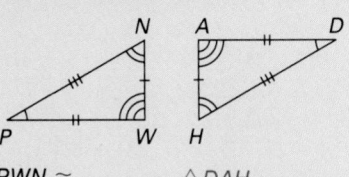

$\triangle PWN \cong \underline{\hspace{1cm}} \triangle DAH$

**For each pair of congruent triangles,
complete the congruence statement.**

2. $\triangle YUQ \cong \triangle ASP$; $\angle Q \cong \underline{\hspace{1cm}} \angle P$
3. $\triangle KJP \cong \triangle TME$; $\overline{KP} \cong \underline{\hspace{1cm}} \overline{TE}$
4. Draw a diagram of the two congruent
triangles, $\triangle UQM \cong \triangle VZR$. Mark the
corresponding parts and identify the
corresponding sides and the
corresponding angles.

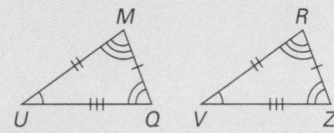

$\overline{UQ} \cong \overline{VZ}$, $\overline{UM} \cong \overline{VR}$, $\overline{MQ} \cong \overline{RZ}$,
$\angle U \cong V$, $\angle M \cong \angle R$, $\angle Q \cong \angle Z$

Closure

Ask the students how to identify the
corresponding sides (angles) of two
congruent triangles if they are given the
corresponding angles (sides).   By the order
of the letters.   Write $\triangle AGK \cong \triangle PTW$.
What are the corresponding sides and
angles.
$\overline{AG} \cong \overline{PT}$, $\overline{GK} \cong \overline{TW}$, $\overline{KA} \cong \overline{WP}$; $\angle A \cong \angle P$,
$\angle G \cong \angle T$, $\angle K \cong \angle W$

▰▰▰FOLLOW UP

Guided Practice

Classroom Exercises 1–6

Independent Practice

A Ex. 1–9, **B** Ex. 10–14, **C** Ex. 15–16

Basic: WE 1–9, Cl 1
Average: WE 1–14, Cl all
Above Average: WE 1–16, Cl all

Additional Answers

See page 146 for the answers to Classroom
Ex. 1–6 and Written Ex. 10–12, 16.

Additional Answers, page 145

Classroom Exercises

1. $\angle T \cong \angle E$, $\angle Y \cong \angle R$, $\angle P \cong \angle M$
2. $\angle Q \cong \angle U$, $\angle W \cong \angle T$, $\angle Z \cong \angle A$
3. $\angle P \cong \angle D$, $\angle T \cong \angle Y$, $\angle C \cong \angle H$
4. $\overline{GU} \cong \overline{PA}$. $\overline{UM} \cong \overline{AT}$, $\overline{GM} \cong \overline{PT}$
5. $\overline{RU} \cong \overline{WD}$, $\overline{UT} \cong \overline{DG}$, $\overline{RT} \cong \overline{WG}$
6. $\overline{SA} \cong \overline{LI}$, $\overline{AD} \cong \overline{ID}$, $\overline{SD} \cong \overline{LD}$

Written Exercises

10–12: Answers will vary.

10.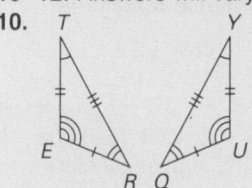

11.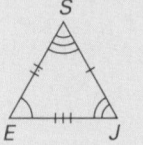

12.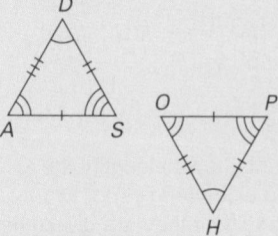

16.

Statement	Reason
1. $\triangle PQR \cong$ $\triangle RQP$	1. Given
2. $\overline{PQ} \cong \overline{RQ}$	2. Def of $\cong \triangle$s
3. $\overline{RQ} \cong \overline{QR}$	3. Reflex Prop of Congr
4. $\overline{PQ} \cong \overline{QR}$	4. Sub
5. $\triangle PQR$ is isos	5. Def of isos $\triangle$

 Computer Investigation

Constructing Triangles

Use a computer software program that constructs triangles by definition of SSS, SAS, or ASA, and measures line segments and angles.

You have seen that two triangles are congruent if three pairs of corresponding angles are congruent and if three pairs of corresponding sides are congruent. The computer activities below will help you discover that congruence of certain combinations of three pairs of parts is enough to establish congruence of the triangles.

Activity 1

$\triangle ABC$ has side lengths 5, 6, and 7. If you construct a triangle with two side lengths 5 and 6 and measure of angle between the 5 and 6 sides equal to m $\angle A$, do you think the new triangle will be congruent to $\triangle ABC$?

1. Construct a triangle with sides of lengths 5, 6, and 7.
2. Find the measures of the angles of this triangle.
3. Construct a new triangle with two sides of lengths 5 and 6, and a measure of the angle between the 5 and 6 sides equal to the measure of the corresponding angle of the original triangle.
4. Find the measures of the remaining side and angles of this new triangle. Is this new triangle congruent to the original triangle?
5. Construct a new triangle with sides of lengths 4, 7, and 9.
6. Repeat Exercises 2–4 referring to this new triangle.

Summary
The results of Activity 1 suggest that congruence of triangles can be established by showing a certain combination of three pairs of corresponding parts congruent. Generalize from the above activity.

Activity 2

$\triangle ABC$ has side lengths 6, 7, and 8. If you construct a triangle with a side of length 8 and angles of measures equal to measures of the angles on both ends of the corresponding side, do you think the new triangle will be congruent to $\triangle ABC$?

7. Draw $\triangle ABC$. Then construct the triangle of measurements given above. Determine if the triangles are congruent.
8. Try to generalize the suggested pattern.

4.3 The SAS Postulate

Objectives

To supply missing reasons in proofs of triangle congruence
To prove triangles congruent using the SAS Postulate
To construct congruent triangles using the SAS correspondence

Two triangles are congruent if three pairs of corresponding angles are congruent and three pairs of corresponding sides are congruent. However, it is not necessary to show congruence of all six pairs of parts in order to guarantee congruence of two triangles.

One or two pairs of congruent parts will *not* guarantee congruence.

Two pairs of congruent sides:
The triangles are *not* congruent.

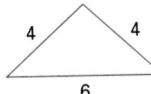 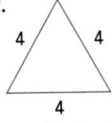

Congruence of triangles can be established by showing congruence of certain combinations of *three* pairs of parts. The combinations of pairs of parts that are required will be explained in this lesson and lessons that follow.

△DEF below was constructed using a compass and straightedge, so that $\overline{DE} \cong \overline{AB}$, $\angle E \cong \angle B$, and $\overline{EF} \cong \overline{BC}$.

Since both $\overline{AB}$ and $\overline{BC}$ are part of $\angle B$, $\angle B$ is the **included angle** for sides $\overline{AB}$ and $\overline{BC}$.

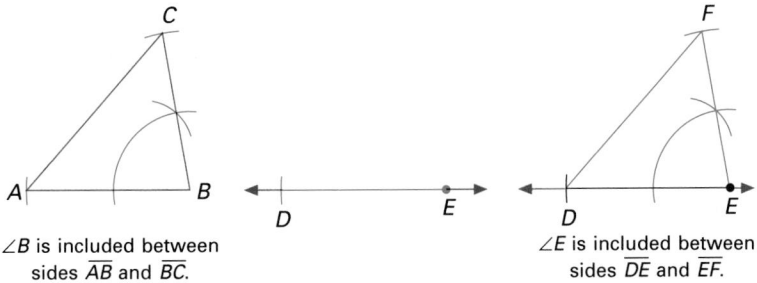

∠B is included between
sides $\overline{AB}$ and $\overline{BC}$.

∠E is included between
sides $\overline{DE}$ and $\overline{EF}$.

Now trace △DEF and fit it on top of △ABC. You can see that they have the same size and shape. Therefore, △ABC ≅ △DEF. This suggests the following postulate.

Teaching Resources

Quick Quizzes 29
Reteaching and Practice
Worksheets, pp. 57, 58

▰▰ GETTING STARTED

Prerequisite Quiz

Name the segments or angles that can be proved congruent from the given data.

1. Given: $\overline{PQ}$
bisects ∠RPS.

$\angle RPQ \cong \angle SPQ$

2.

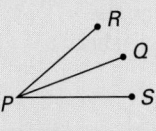

$\angle DOA \cong \angle COB$,
$\angle DOC \cong \angle AOB$

3. Given: $\overline{AB}$ bisects $\overline{CD}$.

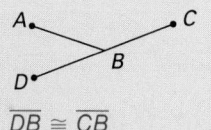

$\overline{DB} \cong \overline{CB}$

Motivator

Ask your students if they think that two triangles will be congruent if three pairs of corresponding angles are congruent. No Have them draw two such triangles that are not congruent.

Ask them if two triangles with two pairs of congruent sides and a pair of included congruent angles are congruent. Yes Demonstrate that this is true by constructing two such congruent triangles on the chalkboard.

Lesson Note

Construction is used as a tool for discovery. Once a triangle is constructed congruent to the triangle drawn, it might be helpful to have students cut out one of the triangles and see that it coincides with the other triangle.

Math Connections

Land Surveying: Surveyors can use the SAS Congruence Postulate to measure the distance across obstacles such as lakes and mountains by measuring the unobstructed corresponding side of a congruent triangle.

Critical Thinking Questions

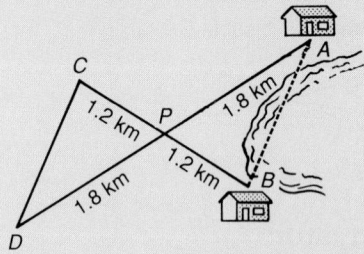

Application: Two cabins are located at points A and B on either side of a lake. To find the distance AB between the cabins, Eva measures the distances PA, PB, PC, and PD, as shown in the diagram. How will Eva use this information to find AB? Eva knows that since $\triangle PAB \cong \triangle PDC$ by SAS, then $AB = CD$ by def of $\cong \triangle$s. So, Eva measures the distance CD on land to find the distance AB between the cabins.

Common Error Analysis

Error: Students may forget to check their proofs to make sure that they include all three of the pairs of parts needed to prove two triangles congruent.

It is helpful to put a letter A or S (or an asterisk) by each pair of corresponding congruent parts. There must be *three* letters (or asterisks) preceding the statement claiming congruence of triangles. Another useful technique is to have the students write next to the SAS reason the number of each step that states the congruence of the corresponding parts.

Postulate 15   **SAS Postulate for Congruence of Triangles:** If two sides and the included angle of one triangle are congruent to the corresponding two sides and included angle of a second triangle, then the triangles are congruent.

EXAMPLE 1   Which pairs of triangles are congruent by the SAS Postulate?

Solution

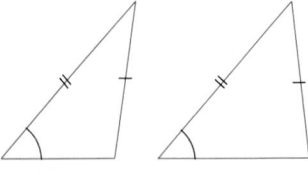

The angles marked congruent are not included between the sides marked congruent. The SAS Postulate does not apply. (In this case, the triangles are not congruent.)

The angles marked congruent are included between the sides marked congruent. The triangles are congruent.

EXAMPLE 2   To prove the triangles congruent by the SAS Postulate, determine which other pair of sides or angles needs to be congruent.

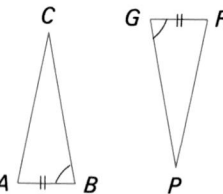

Solution   Copy the figure and mark the congruent side and angle.

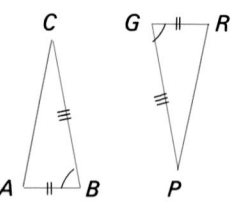

Needed: $\overline{BC} \cong \overline{GP}$

Copy the figure and mark the congruent sides.

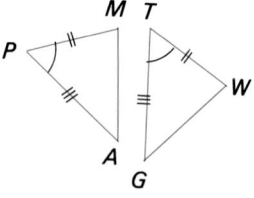

Needed: $\angle P \cong \angle T$

148   Chapter 4 Congruent Triangles

Additional Example 1

Which pairs of triangles are congruent by the SAS Postulate?

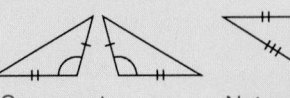

Congruent

Not congruent ($\cong \angle$s are not included between $\cong$ sides.)

Additional Example 2

In order to prove the triangles congruent by the SAS Postulate, which other pair of sides or angles must be congruent?

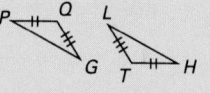
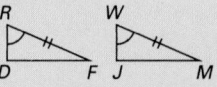

Needed: $\angle Q \cong \angle T$ Needed: $\overline{DR} \cong \overline{JW}$

EXAMPLE 3   Given: $\overline{TS}$ and $\overline{PR}$ bisect each other.
Prove: $\triangle TPQ \cong \triangle SRQ$

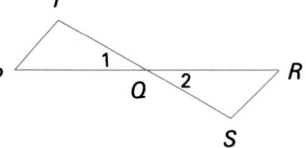

Plan   Since $\overline{TS}$ and $\overline{PR}$ bisect each other, $\overline{PQ} \cong \overline{RQ}$ and $\overline{TQ} \cong \overline{SQ}$. $\angle 1$ and $\angle 2$ are vertical angles.

Proof

Statement	Reason
1. $\overline{TS}$ and $\overline{PR}$ bisect each other.	1. Given
2. $\overline{PQ} \cong \overline{RQ}$	**(S)** 2. Def of bis
3. $\overline{TQ} \cong \overline{SQ}$	**(S)** 3. Def of bis
4. $\angle 1 \cong \angle 2$	**(A)** 4. Vert $\angle$s are $\cong$.
5. $\therefore \triangle TPQ \cong \triangle SRQ$	5. SAS

EXAMPLE 4   Given: $\overline{AD}$ bisects $\angle A$; $\overline{AC} \cong \overline{AB}$.
Prove: $\triangle ACD \cong \triangle ABD$

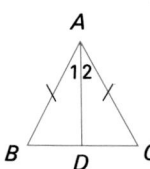

Plan   Since $\overline{AD}$ bisects $\angle A$, $\angle 1 \cong \angle 2$. $\overline{AD}$ forms a side of both triangles.

Proof

Statement	Reason
1. $\overline{AD}$ bisects $\angle A$.	1. Given
2. $\angle 1 \cong \angle 2$	**(A)** 2. Def of $\angle$ bis
3. $\overline{AC} \cong \overline{AB}$	**(S)** 3. Given
4. $\overline{AD} \cong \overline{AD}$	**(S)** 4. Reflex Prop
5. $\therefore \triangle ACD \cong \triangle ABD$	5. SAS

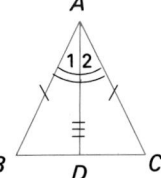

Notice that the Reflexive Property (Step 4) allows you to find a pair of congruent sides. This property is commonly used in geometric proofs.

Classroom Exercises

For each pair of sides of the triangle, identify the included angle.

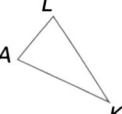

1. $\overline{PW}$ and $\overline{WQ}$ $\angle W$
2. $\overline{PQ}$ and $\overline{PW}$ $\angle P$
3. $\overline{QP}$ and $\overline{WQ}$ $\angle Q$
4. $\overline{AK}$ and $\overline{AL}$ $\angle A$
5. $\overline{KA}$ and $\overline{LK}$ $\angle K$
6. $\overline{AL}$ and $\overline{LK}$ $\angle L$

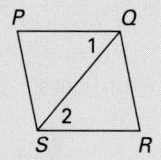
Checkpoint

To prove the two triangles congruent by the SAS Postulate, which other pair of sides or angles must be proved congruent?

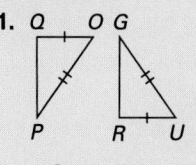

1. $\angle O \cong \angle U$

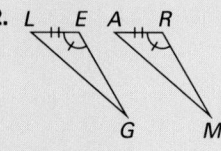

2. $\overline{EG} \cong \overline{RM}$

3. Given: $\overline{QS}$ bisects $\angle PSR$, $\overline{PS} \cong \overline{RS}$
Prove: $\triangle SPQ \cong \triangle SRQ$

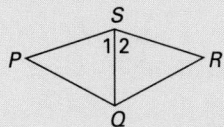

Statement	Reason
1. $\overline{QS}$ bis. $\angle PSR$	1. Given
2. $\angle 1 \cong \angle 2$	2. Def of $\angle$ bis
3. $\overline{SQ} \cong \overline{SQ}$	3. Reflex Prop of Congr
4. $\overline{PS} \cong \overline{RS}$	4. Given
5. $\triangle PSQ \cong \triangle RSQ$	5. SAS

5.

Statement	Reason
1. $\overline{QP} \perp \overline{QR}$, $\overline{TS} \perp \overline{TU}$, $\overline{QR} \cong \overline{TU}$, $\overline{QP} \cong \overline{TS}$	1. Given
2. m $\angle Q$ = 90, m $\angle T$ = 90	2. Def of $\perp$
3. m $\angle Q$ = m $\angle T$	3. Sub
4. $\triangle PQR \cong \triangle STU$	4. SAS

6.

Statement	Reason
1. $\overline{AD} \cong \overline{CD}$, $\angle 1 \cong \angle 2$	1. Given
2. $\overline{DB} \cong \overline{DB}$	2. Reflex Prop of Congr
3. $\triangle ADB \cong \triangle CDB$	3. SAS

7.

Statement	Reason
1. $\angle 1 \cong \angle 2$, $\overline{EC} \cong \overline{ED}$, E is the midpt of $\overline{AB}$	1. Given
2. $\overline{AE} \cong \overline{BE}$	2. Def of midpt
3. $\triangle ACE \cong \triangle BDE$	3. SAS

8.

Statement	Reason
1. $\overline{PO} \cong \overline{QO}$, $\overline{RO} \cong \overline{SO}$	1. Given
2. $\angle POR \cong \angle QOS$	2. Vert $\angle$s are $\cong$.
3. $\triangle POR \cong \triangle QOS$	3. SAS

Use the markings to determine if there is enough information to show the two triangles are congruent. (Congruent or not enough information)

7.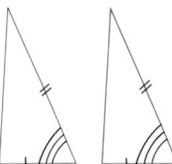

Congruent

8.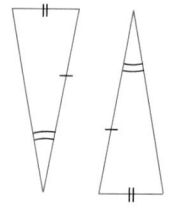

Not enough information

9.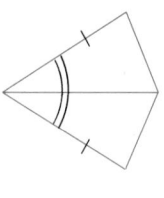

Congruent

10.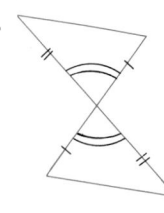

Congruent

Written Exercises

To prove the two triangles congruent by the SAS Postulate, which additional pair of sides or angles must be congruent?

1.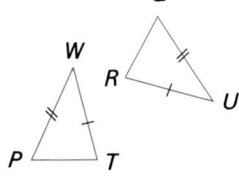

$\angle W \cong \angle U$

2.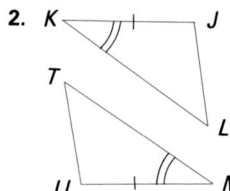

$\overline{KL} \cong \overline{MT}$

3.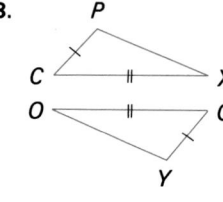

$\angle C \cong \angle G$

Supply the missing statements and reasons in the proof.

4. Given: B is the midpoint of $\overline{AC}$; $\overline{AE} \perp \overline{AC}$, $\overline{CD} \perp \overline{CA}$, $\overline{EA} \cong \overline{DC}$.
Prove: $\triangle EAB \cong \triangle DCB$

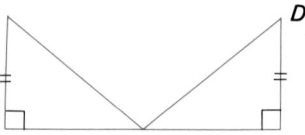

Statement	Reason
1. $\overline{AE} \perp \overline{AC}$, $\overline{CD} \perp \overline{CA}$	1. Given
2. m $\angle A$ = 90, m $\angle C$ = 90	2. Def of $\perp$
3. m $\angle A$ = m $\angle C$ ($\angle A \cong \angle C$)	(A) 3. Sub
4. B is the midpoint of $\overline{AC}$.	4. Given
5. $\overline{AB} \cong \overline{CB}$	(S) 5. Def of midpt
6. $\overline{EA} \cong \overline{DC}$	(S) 6. Given
7. $\therefore \triangle EAB \cong \triangle DCB$	7. SAS

Additional Example 4

Given: M bisects $\overline{GK}$, $\overline{LG} \perp \overline{GK}$, $\overline{PK} \perp \overline{GK}$, $\overline{LG} \cong \overline{PK}$
Prove: $\triangle LGM \cong \triangle PKM$

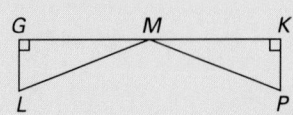

Proof: M bisects $\overline{GK}$ (Given); $\overline{GM} \cong \overline{KM}$ (def of bis); $\overline{LG} \perp \overline{GK}$, $\overline{PK} \perp \overline{GK}$ (Given); m $\angle G$ = 90, m $\angle K$ = 90 (def of $\perp$); m $\angle G$ = m $\angle K$, or $\angle G \cong \angle K$ (Sub); $\overline{LG} \cong \overline{PK}$ (Given); $\triangle LGM \cong \triangle PKM$ (SAS)

5. Given: $\overline{QP} \perp \overline{QR}$, $\overline{TS} \perp \overline{TU}$, $\overline{QR} \cong$ $\overline{TU}$, $\overline{QP} \cong \overline{TS}$
Prove: $\triangle PQR \cong \triangle STU$

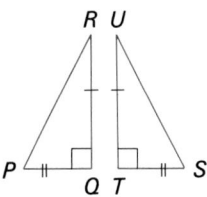

6. Given: $\overline{AD} \cong \overline{CD}$, $\angle 1 \cong \angle 2$
Prove: $\triangle ADB \cong \triangle CDB$

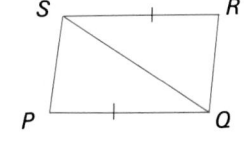

7. Given: $\angle 1 \cong \angle 2$, $\overline{EC} \cong \overline{ED}$, E is the midpoint of $\overline{AB}$.
Prove: $\triangle ACE \cong \triangle BDE$

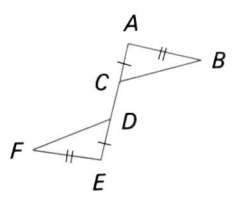

8. Given: $\overline{PO} \cong \overline{QO}$, $\overline{RO} \cong \overline{SO}$
Prove: $\triangle POR \cong \triangle QOS$

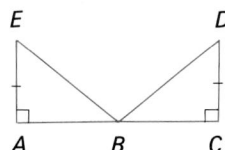

9. Given: $\overline{AB} \parallel \overline{FE}$, $\overline{AB} \cong \overline{EF}$, $\overline{AC} \cong \overline{ED}$
Prove: $\triangle ABC \cong \triangle EFD$

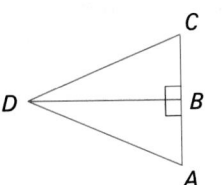

10. Given: $\overline{PQ} \parallel \overline{RS}$, $\overline{PQ} \cong \overline{RS}$
Prove: $\triangle PQS \cong \triangle RSQ$

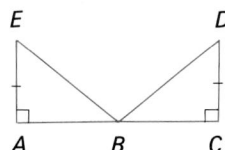

11. Given: $\overline{DB} \perp \overline{AC}$, $\overline{DB}$ bisects $\overline{AC}$.
Prove: $\triangle ABD \cong \triangle CBD$

12. Given: $\overline{EA} \perp \overline{AB}$, $\overline{DC} \perp \overline{CB}$, $\overline{EA} \cong$ $\overline{DC}$, B is the midpoint of $\overline{AC}$.
Prove: $\triangle EAB \cong \triangle DCB$

13. Given: $\triangle ABC$ with $\overline{BA} \cong \overline{CA}$, point D on $\overline{BC}$ such that $\overline{AD}$ bisects $\angle A$.
Prove: $\triangle BAD \cong \triangle CAD$

14. Draw an acute triangle. Construct a triangle congruent to it using SAS.

9. Statement	Reason
1. $\overline{AB} \cong \overline{EF}$, $\overline{AC} \cong \overline{ED}$, $\overline{AB} \parallel \overline{FE}$	1. Given
2. $\angle A \cong \angle E$	2. If lines are ∥, then alt int ∠s are ≅.
3. $\triangle ABC \cong \triangle EFD$	3. SAS

10. Statement	Reason
1. $\overline{RS} \parallel \overline{PQ}$, $\overline{PQ} \cong \overline{RS}$	1. Given
2. $\angle PQS \cong \angle RSQ$	2. If lines are ∥, then alt int ∠s are ≅.
3. $\overline{SQ} \cong \overline{SQ}$	3. Reflex Prop of Congr
4. $\triangle PQS \cong \triangle RSQ$	4. SAS

11. Statement	Reason
1. $\overline{DB} \perp \overline{AC}$, $\overline{DB}$ bis $\overline{AC}$	1. Given
2. m $\angle DBC = 90$, m $\angle DBA = 90$	2. Def of ⊥
3. m $\angle DBC = $ m $\angle DBA$	3. Sub
4. $\overline{DB} \cong \overline{DB}$	4. Reflex Prop of Congr
5. $\overline{BC} \cong \overline{BA}$	5. Def of seg bis
6. $\triangle ABD \cong \triangle CBD$	6. SAS

12. Statement	Reason
1. $\overline{EA} \perp \overline{AB}$, $\overline{DC} \perp \overline{CB}$, $\overline{EA} \cong \overline{DC}$, B is midpt of $\overline{AC}$	1. Given
2. m $\angle EAB = 90$, m $\angle DCB = 90$	2. Def of ⊥
3. m $\angle EAB = $ m $\angle DCB$	3. Sub
4. $\overline{AB} \cong \overline{CB}$	4. Def of midpt
5. $\triangle EAB \cong \triangle DCB$	5. SAS

13. Statement	Reason
1. $\overline{BA} \cong \overline{CA}$, $\overline{AD}$ bis $\angle A$	1. Given
2. $\angle BAD \cong \angle CAD$	2. Def of ∠ bis
3. $\overline{AD} \cong \overline{AD}$	3. Reflex Prop of Congr
4. $\triangle BAD \cong \triangle CAD$	4. SAS

14.

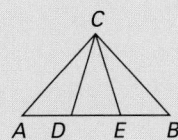

15. If the ∠ included by the ≅ sides of an isosc △ is bis, then the resulting △s are ≅.

16.

Statement	Reason
1. $\overline{AD} \parallel \overline{BC}$, ∠1 ≅ ∠3, $\overline{AC} \cong \overline{FB}$	1. Given
2. ∠1 ≅ ∠2	2. If lines are ∥, then alt int ∠s are ≅.
3. ∠2 ≅ ∠3	3. Trans Prop of Congr
4. $\overline{BC} \cong \overline{BC}$	4. Reflex Prop of Congr
5. △ABC ≅ △FCB	5. SAS

17.

Statement	Reason
1. $\overline{CF} \parallel \overline{BE}$, ∠3 ≅ ∠4, $\overline{BC} \cong \overline{BE}$	1. Given
2. ∠4 ≅ ∠5	2. If lines are ∥, then alt int ∠s are ≅.
3. ∠3 ≅ ∠5	3. Trans Prop of Congr
4. $\overline{BF} \cong \overline{BF}$	4. Reflex Prop of Congr
5. △CBF ≅ △EBF	5. SAS

15. State the result of Example 4 as a theorem about isosceles triangles.
16. Given: $\overline{AD} \parallel \overline{BC}$, ∠1 ≅ ∠3, $\overline{AC} \cong \overline{FB}$
Prove: △ABC ≅ △FCB
17. Given: $\overline{CF} \parallel \overline{BE}$, ∠3 ≅ ∠4, $\overline{BC} \cong \overline{BE}$
Prove: △CBF ≅ △EBF

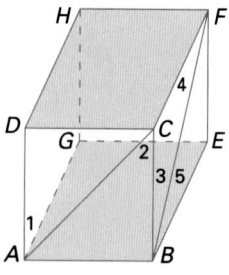

Mixed Review

In the figure at the right, $\overline{DE} \parallel \overline{AF}$.

1. Given: m ∠2 = 40. Find m ∠4. *3.5* 40
2. Given: m ∠2 = 65. Find m ∠DCB. *3.5* 115
3. Given: m ∠1 = 60, m ∠3 is twice m ∠2. Find m ∠2. *3.6*
4. Given: m ∠6 = 120, m ∠1 = 40 Find m ∠3. *3.7* 80

■ **Application:** *Angles in Sand Piles*

Materials such as sand, gravel, and coal are often stored in supply yards in loose piles. The angle made between the side of the pile and the (horizontal) ground is called the *angle of repose*. This angle varies depending on the characteristics of the material. For example, wet sand has a greater angle of repose than an equal volume of dry sand. The angle of repose can be used to estimate the relative sizes of piles, cross-sectional area, volume of a pile, ground area needed for storing given quantities, minimum amount of sand needed to bank a dam.

Material	Angle of repose (approximate)
Sand (dry)	30
Sand (wet)	38
Gravel	35
Salt	36
Coal	37
Cement	40
Crushed Stone	35

1. Which would form a higher pile—an equal volume of dry sand or wet sand? Wet sand
2. If a pile of salt and a pile of coal had equal heights, which pile would have the greater volume? Salt
3. Which would require a greater area for storage, a given volume of cement or of gravel? Gravel

152 Chapter 4 Congruent Triangles

4.4 SSS and ASA Congruence Proofs

Objectives   To supply missing reasons in proofs of triangle congruence
To construct congruent triangles using the SSS and ASA patterns
To prove triangles congruent using the SSS, ASA, and SAS postulates

Triangle congruence properties can be applied in various ways. Using triangles you can indirectly find the distance across a lake (see Exercise 15). Two new congruence postulates are developed in this lesson. △*DEF*, below, was constructed so that $\overline{DE} \cong \overline{AB}$, $\overline{EF} \cong \overline{BC}$, and $\overline{FD} \cong \overline{CA}$.

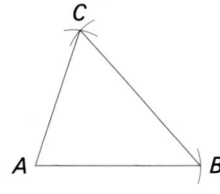 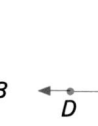 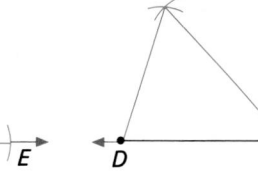

Trace △*DEF* and fit it on top of △*ABC*. You can see that they have the same size and shape. Therefore, △*ABC* ≅ △*DEF*. This suggests the following postulate.

Postulate 16   **SSS Postulate for Congruence of Triangles:** If the three sides of one triangle are congruent to the corresponding three sides of a second triangle, then the triangles are congruent.

A triangle can be constructed congruent to another triangle by an *angle-side-angle* method. Construct △*DEF* so that $\overline{DE} \cong \overline{AB}$, ∠*D* ≅ ∠*A*, and ∠*E* ≅ ∠*B*.

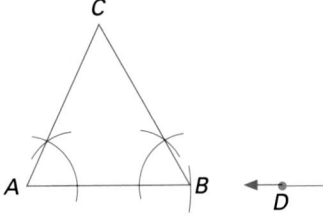  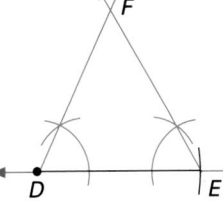

Additional Example 1

For each diagram, indicate whether the two triangles are congruent and write a statement of congruence. Justify your statement.

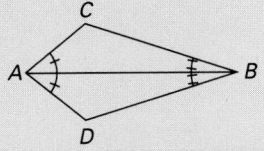

Yes, △*ACB* ≅ △*ADB*. Reason: ASA
($\overline{AB} \cong \overline{AB}$ by the Reflex Prop of Congr)

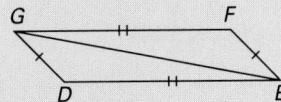

Yes, △*GDE* ≅ △*EFG*. Reason: SSS
($\overline{EG} \cong \overline{EG}$ by the Reflex Prop of Congr)

Teaching Resources

Problem Solving Worksheet 4
Quick Quizzes 30
Reteaching and Practice Worksheets, pp. 59, 60

▰ GETTING STARTED

Prerequisite Quiz

1. Draw an angle. Construct an angle congruent to it.

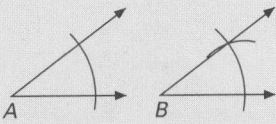

2. Draw a segment. Construct a segment congruent to it.

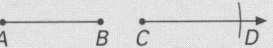

3. Name a pair of congruent parts in the triangles shown below. $\overline{TY} \cong \overline{TY}$

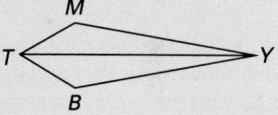

Motivator

You have seen that it is not necessary to show six pairs of corresponding congruent parts to establish congruency of two triangles. The SAS Postulate of the last lesson demonstrated that the correct combination of three pairs of corresponding congruent parts is enough. Ask the students if they can think of other combinations of three pairs of corresponding congruent parts that establish congruency.   SSS, ASA

▰ TEACHING SUGGESTIONS

Lesson Note

The SSS and ASA Postulates are introduced in this lesson. The SAS Postulate of the last lesson is also applied to give students experience with all three postulates and practice in choosing the appropriate one.

Math Connections

Carpentry: The triangular braces that carpenters use to support shelves and cabinets must be congruent so that the shelves will be level and parallel.

Critical Thinking Questions

Application: Refer students to the diagram of a young tree supported by three pieces of rope.

Each piece of rope is 5 ft long and is tied to the trunk of the tree 4 ft above the ground. If each rope is tied to a separate stake in the ground that is 3 ft away from the tree, then the young tree stands perpendicular to the ground. What congruence patterns allow you to prove that the three triangles formed by the ropes with the ground and tree are congruent? SAS or SSS

Common Error Analysis

Error: After using the SSS Postulate, students sometimes attempt to prove two triangles congruent by an angle-angle-angle method, which is invalid.

Have the students draw two triangles with the same angle measures, but having sides of different lengths. Discuss the fact that, although the angles of the triangles are congruent, the triangles themselves are not congruent.

Postulate 17

ASA Postulate for Congruence of Triangles: If two angles and the included side of one triangle are congruent to the corresponding two angles and included side of a second triangle, then the triangles are congruent.

The *included side* mentioned in Postulate 17 is the side whose endpoints are the vertices of the two given angles. You now have three methods of establishing congruence of triangles: SSS, SAS, and ASA.

EXAMPLE 1 For each diagram, indicate whether the two triangles are congruent and, if so, write a statement of congruence. Justify your statement.

Solution

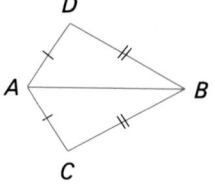
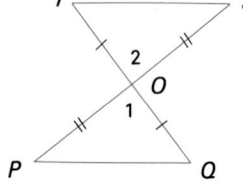
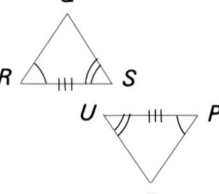

Yes. △ADB ≅ △ACB
Reason: SSS

Yes. △TOS ≅ △QOP
Reason: SAS

Yes. △QRS ≅ △TPU
Reason: ASA

EXAMPLE 2 Given: $\overline{AB}$ bisects ∠CAD, ∠1 ≅ ∠2.
Prove: △CAB ≅ △DAB

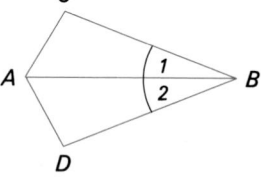

Proof

Statement		Reason
1. $\overline{AB}$ bisects ∠CAD.		1. Given
2. ∠CAB ≅ ∠DAB	(A)	2. Def of ∠ bis
3. ∠1 ≅ ∠2	(A)	3. Given
4. $\overline{AB}$ ≅ $\overline{AB}$	(S)	4. Reflex Prop
5. ∴ △CAB ≅ △DAB		5. ASA

EXAMPLE 3 Given: $\overline{CD}$ bisects $\overline{AB}$, $\overline{AC}$ ≅ $\overline{BC}$.
Prove: △ADC ≅ △BDC

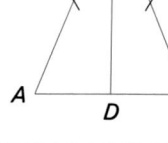

Proof

Statement		Reason
1. $\overline{CD}$ bisects $\overline{AB}$.		1. Given
2. $\overline{AD}$ ≅ $\overline{BD}$	(S)	2. Def of bis
3. $\overline{CD}$ ≅ $\overline{CD}$	(S)	3. Reflex Prop
4. $\overline{AC}$ ≅ $\overline{BC}$	(S)	4. Given
5. ∴ △ADC ≅ △BDC		5. SSS

Additional Example 2

Given: *T* bisects $\overline{PW}$, $\overline{PS}$ ≅ $\overline{WU}$,
$\overline{ST}$ ≅ $\overline{UT}$
Prove: △SPT ≅ △UWT

Proof: *T* bisects $\overline{PW}$. (Given); $\overline{PT}$ ≅ $\overline{WT}$ (def of bis); $\overline{PS}$ ≅ $\overline{WU}$ (Given); $\overline{ST}$ ≅ $\overline{UT}$ (Given); △SPT ≅ △UWT (SSS)

Classroom Exercises

For each pair of angles of the triangle, identify the included side.

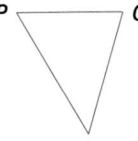

1. $\angle P$ and $\angle Q$ $\overline{PQ}$
2. $\angle P$ and $\angle R$ $\overline{PR}$
3. $\angle Q$ and $\angle R$ $\overline{QR}$

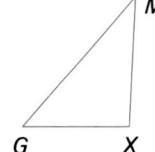

4. $\angle G$ and $\angle M$ $\overline{GM}$
5. $\angle G$ and $\angle X$ $\overline{GX}$
6. $\angle M$ and $\angle X$ $\overline{MX}$

Indicate whether the pair of triangles is congruent, and, if so, state why.

7.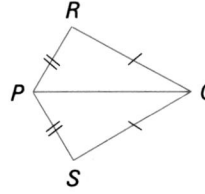

Congruent; SSS

8.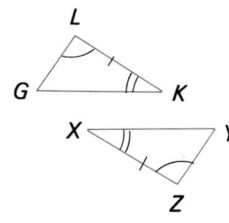

Congruent; ASA

9.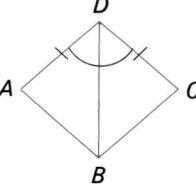

Congruent; SAS

Written Exercises

Complete the statement.

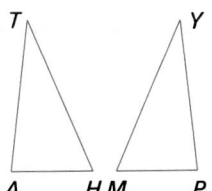

1. To prove $\triangle TAH \cong \triangle YPM$ by ASA, show $\angle A \cong \underline{\angle P}$, $\angle H \cong \underline{\angle M}$, and $\underline{\overline{AH}} \cong \underline{\overline{PM}}$.

2. To prove $\triangle TAH \cong \triangle YPM$ by SAS, show $\overline{TH} \cong \underline{\overline{YM}}$ $\overline{HA} \cong \underline{\overline{MP}}$, and $\underline{\angle H} \cong \underline{\angle M}$.

Supply the missing statements and reasons for the proof.

3. Given: $\overline{TS} \cong \overline{QR}$, $\overline{TU} \cong \overline{QP}$,
$\overline{TU} \parallel \overline{QP}$
Prove: $\triangle TSU \cong \triangle QRP$

Statement		Reason
1. $\overline{TS} \cong \overline{QR}$	(S)	1. Given
2. $\overline{TU} \cong \overline{QP}$	(S)	2. Given
3. $\overline{TU} \parallel \overline{QP}$		3. Given
4. $\angle T \cong \angle Q$	(A)	4. Alt int $\angle$s
5. $\therefore \triangle TSU \cong \triangle QRP$		5. SAS

4.4 SSS and ASA Congruence Proofs   **155**

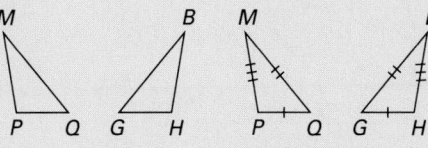
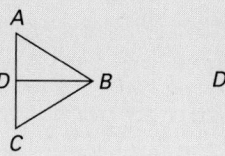
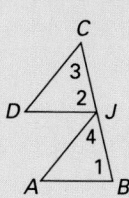

Written Exercises

4.

Statement	Reason
1. $\overline{PR}$ and $\overline{MN}$ bis each other	1. Given
2. $NP = MP$, $TP = RP$	2. Def of seg bis
3. $\angle NPT \cong \angle MPR$	3. Vert $\angle$s are $\cong$.
4. $\triangle NTP \cong \triangle MRP$	4. SAS

5.

Statement	Reason
1. $\overline{CD}$ bis $\angle C$, $\angle 1 \cong \angle 2$	1. Given
2. $m \angle ACD = m \angle BCD$	2. Def of $\angle$ bis
3. $\overline{CD} \cong \overline{CD}$	3. Reflex Prop of Congr
4. $\triangle CDB \cong \triangle CDA$	4. ASA

6.

Statement	Reason
1. $\overline{AB} \parallel \overline{CD}$, $\angle B \cong \angle D$, $\overline{AB} \cong \overline{CD}$	1. Given
2. $\angle BAF \cong \angle DCE$	2. Alt int $\angle$s of 2 $\parallel$ lines are $\cong$.
3. $\triangle ABF \cong \triangle CDE$	3. ASA

7.

Statement	Reason
1. $\overline{PG} \cong \overline{SG}$, $\overline{TP} \cong \overline{TS}$	1. Given
2. $\overline{TG} \cong \overline{TG}$	2. Reflex Prop of Eq
3. $\triangle TPG \cong \triangle TSG$	3. SSS

8.

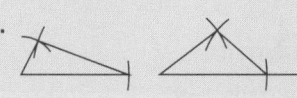

9.

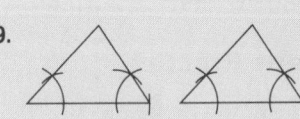

10.

Statement	Reason
1. $\overline{OE} \perp \overline{MP}$, OE bis $\angle MOP$	1. Given
2. $m \angle OEM = 90$, $m \angle OEP = 90$	2. Def of $\perp$
3. $m \angle OEM = m \angle OEP$	3. Sub
4. $\overline{OE} \cong \overline{OE}$	4. Reflex Prop of Congr
5. $m \angle MOE = m \angle POE$	5. Def of bis
6. $\triangle MOE \cong \triangle POE$	6. ASA

---

4. Given: $\overline{TR}$ and $\overline{MN}$ bisect each other.
Prove: $\triangle NTP \cong \triangle MRP$

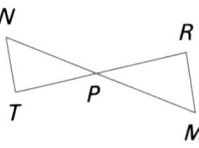

5. Given: $\overline{CD}$ bisects $\angle ACB$; $\angle 1 \cong \angle 2$.
Prove: $\triangle CDA \cong \triangle CDB$

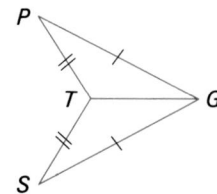

6. Given: $\overline{AB} \parallel \overline{CD}$, $\angle B \cong \angle D$, $\overline{AB} \cong \overline{CD}$
Prove: $\triangle ABF \cong \triangle CDE$

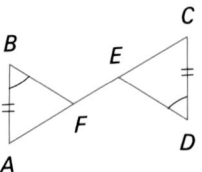

7. Given: $\overline{PG} \cong \overline{SG}$, $\overline{TP} \cong \overline{TS}$
Prove: $\triangle TPG \cong \triangle TSG$

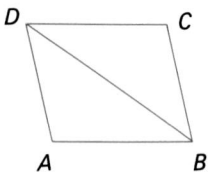

8. Draw an obtuse triangle. Construct a triangle congruent to it by SSS.

9. Draw a scalene triangle. Construct a triangle congruent to it by ASA.

10. Given: $\overline{OE} \perp \overline{MP}$, $\overline{OE}$ bisects $\angle MOP$.
Prove: $\triangle MOE \cong \triangle POE$

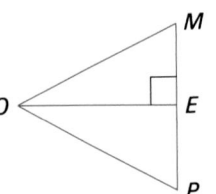

11. Given: $\overline{AD} \parallel \overline{BC}$, $\overline{DC} \parallel \overline{BA}$
Prove: $\triangle ADB \cong \triangle CBD$

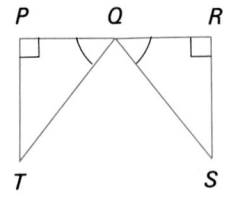

12. Given: Q is the midpoint of $\overline{PR}$; $\overline{TP} \perp \overline{PR}$, $\overline{SR} \perp \overline{RP}$, $\angle TQP \cong \angle SQR$.
Prove: $\triangle TPQ \cong \triangle SRQ$

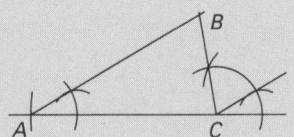

Ex. 12

13. It is given that $\overline{RS}$ bisects $\angle PRQ$ of $\triangle PRQ$. Also, $\overline{RS} \perp \overline{PQ}$ at point S. Prove that the two triangles formed are congruent.

Enrichment

Have the students draw two acute angles, $\angle A$ and $\angle B$, and segments labeled $\overline{AB}$, $\overline{AC}$, and $\overline{BD}$.

Now have the students construct the following triangles. Suggest that they plan each construction by first marking the given parts on a rough sketch of the triangle.

Construct $\triangle ABC$ using $\angle A$, $\angle B$, and $\overline{AC}$. Plan: (1) Construct $\angle C$ so that $m\angle C = 180 - (m \angle A + m \angle B)$. (2) Construct $\triangle ABC$ by ASA using $\angle C$, $\angle A$, and $\overline{AC}$.

14. A carpenter has to make a triangular brace for the gate at the right. He cuts two strips of wood of equal length, $\overline{PQ}$ and $\overline{RS}$. They are nailed to the gate so that $\overline{PQ} \parallel \overline{RS}$. Prove that the triangles formed are congruent. He checks his construction before driving the nails all the way in by measuring $\overline{PR}$ and $\overline{QS}$ to make sure they are equal. Why?

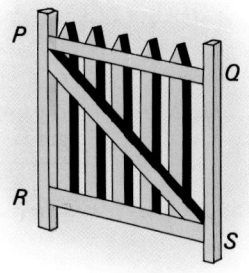

15. Jane wants to find the distance from her camp tent at G to a tree at M on the opposite side of a lake. She sets a stake at L so that $\overline{GM} \perp \overline{GL}$. She places a second stake at H in line with $\overline{GL}$, making $GL = HL$. She then stretches a tape from H, making $\overline{HN} \perp \overline{HL}$, and placing a stake at K, the intersection of $\overrightarrow{ML}$ with $\overrightarrow{HN}$. Prove $\triangle GLM \cong \triangle HLK$. Why will measuring $\overline{HK}$ now give the distance across the lake?

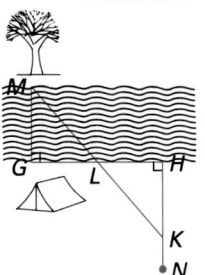

16. For what special case can you say that $\triangle ABC \cong \triangle BAC$?

Midchapter Review

Classify each triangle by its sides. *4.1*

1. 3 cm, 3cm, 3cm
Equilateral

2. 5 in., 7 in., 6 in.
Scalene

3. 6 ft, 4 ft, 6 ft   Isosceles

Classify each triangle by its angles. *4.1*

4. The measures of two angles are 22 and 68.   Right
5. The measures of two angles are 28 and 32.   Obtuse

Given: $\triangle ABC \cong \triangle XYZ$, $AC = 8$, $BC = 6$, m $\angle A = 37$, m $\angle B = 53$ *4.2*

6. Find the length of $\overline{YZ}$.   6   **7.** Find the measure of $\angle Z$.   90
8. Given $\overline{AB}$ and $\overline{CD}$ bisect each other, complete the proof.   *4.3*

Statement	Reason
1. $\overline{AB}$ and $\overline{CD}$ bisect each other.	1. Given
2. $AE = BE$, $CE = DE$	2. Def of bis
3. m $\angle 1 = $ m $\angle 2$	3. Vert $\angle$s are $\cong$.
4. $\triangle AEC \cong \triangle BED$	4. SAS

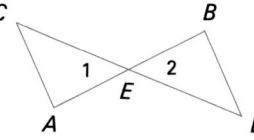

11.

Statement	Reason
1. $\overline{AD} \parallel \overline{BC}$, $\overline{DC} \parallel \overline{BA}$	1. Given
2. $\angle ADB \cong \angle CBD$, $\angle CDB \cong \angle ABD$	2. If lines are $\parallel$, then alt int $\angle$s are $\cong$.
3. $\overline{DB} \cong \overline{DB}$	3. Reflex Prop of Congr
4. $\triangle ADB \cong \triangle CBD$	4. ASA

12.

Statement	Reason
1. Q is the midpt of $\overline{PR}$, $\angle TQP \cong \angle SQR$, $\overline{TP} \perp \overline{PR}$, $\overline{SR} \perp \overline{RP}$	1. Given
2. $\overline{PQ} \cong \overline{RQ}$	2. Def of seg bis
3. m $\angle TPQ = 90$, m $\angle SRQ = 90$	3. Def of $\perp$
4. m $\angle TPQ = $ m $\angle SRQ$	4. Sub
5. $\triangle TPQ \cong \triangle SRQ$	5. ASA

13.

Statement	Reason
1. $\overline{RS}$ bis $\angle PRQ$, $\overline{RS} \perp \overline{PQ}$	1. Given
2. $\angle PRS \cong \angle QRS$	2. Def $\angle$ bis
3. $\overline{RS} \cong \overline{RS}$	3. Reflex Prop of Congr
4. m $\angle PSR = 90$, m $\angle QSR = 90$	4. Def of $\perp$
5. m $\angle PSR = $ m $\angle QSR$	5. Sub
6. $\triangle PRS \cong \triangle QRS$	6. ASA

14.

Statement	Reason
1. $\overline{PQ} \cong \overline{RS}$, $\overline{PQ} \parallel \overline{RS}$	1. Given
2. $\overline{PS} \cong \overline{PS}$	2. Reflex Prop of Congr
3. $\angle QPS \cong \angle RSP$	3. If lines are $\parallel$, then alt int $\angle$s are $\cong$.
4. $\triangle QPS \cong \triangle RSP$	4. SAS

To make sure that $\overline{PQ}$ is still $\parallel$ to $\overline{RS}$.

15.

Statement	Reason
1. $GL = LH$, $\overline{GM} \perp \overline{GL}$, $\overline{HN} \perp \overline{HL}$	1. Given
2. m $\angle G = 90$, m $\angle H = 90$	2. Def of $\perp$
3. m $\angle G = $ m $\angle H$	3. Sub
4. $\angle MLG \cong \angle KLH$	4. Vert $\angle$s are $\cong$.
5. $\triangle GLM \cong \triangle HLK$	5. ASA

They are corr parts of $\cong \triangle$s.

16. Isos $\triangle$ such that $AC = BC$ and $\angle A \cong \angle B$.

▰▰ GETTING STARTED

Prerequisite Quiz

Name the segments or angles that can be proved congruent from the given data. Give a reason.

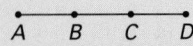

1. Given: $AB = CD$ $\overline{AC} \cong \overline{BD}$; adding BC to each side of $AB = CD$ and using Seg Add Post, $AC = BD$, or $\overline{AC} \cong \overline{BD}$.

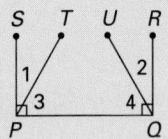

2. Given: $\overline{PS} \perp \overline{PQ}$, $\overline{QR} \perp \overline{QP}$, $\angle 1 \cong \angle 2$
   $\angle 3 \cong \angle 4$; comp of $\cong \angle$s are $\cong$.

Motivator

Explain to the students that sometimes parts of geometric figures are congruent that are not actually sides or angles of triangles they want to prove congruent. Draw the figure in Example 1 on the chalkboard with $\angle 1 \cong \angle 2$. Have students answer the following questions. Are these angles of the two triangles?   No   Are they related to angles of the triangles?   Yes   How can you use this relationship to prove a pair of angles of the triangles congruent?   Since the angles are supplementary and one pair of angles is congruent then the other pair of angles is congruent.

▰▰ TEACHING SUGGESTIONS

Lesson Note

In this lesson, the students first apply the techniques of Chapter 2 to prove segments or angles congruent so that they can then prove triangles congruent.

4.5 Congruence in Complex Figures

Objective To prove triangles are congruent when they are parts of complex figures

EXAMPLE Given: $\angle 1 \cong \angle 2$, $\overline{EA} \cong \overline{DC}$,
   B is the midpoint of $\overline{AC}$.
   Prove: $\triangle EAB \cong \triangle DCB$

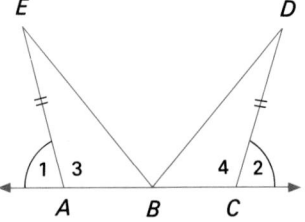

Proof	Statement	Reason
	1. $\angle 1 \cong \angle 2$	1. Given
	2. $\angle 1$ and $\angle 3$ are supplementary. $\angle 2$ and $\angle 4$ are supplementary.	2. If the outer rays of two adj $\angle$s form a st $\angle$, then the $\angle$s are supp.
	3. $\angle 3 \cong \angle 4$	(A) 3. Supp of $\cong \angle$s are $\cong$.
	4. B is the midpoint of $\overline{AC}$.	4. Given
	5. $\overline{AB} \cong \overline{CB}$	(S) 5. Def of midpt
	6. $\overline{EA} \cong \overline{DC}$	(S) 6. Given
	7. $\therefore \triangle EAB \cong \triangle DCB$	7. SAS

Classroom Exercises

Supply the missing statements and reasons in the proof.

Given: $\overline{XZ} \cong \overline{WY}$, $\angle 1 \cong \angle 2$, $\angle X \cong \angle W$
Prove: $\triangle MXY \cong \triangle NWZ$

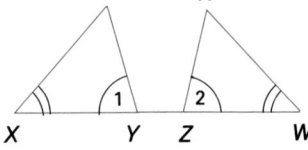

Statement	Reason
1. $\angle 1 \cong \angle 2$	(A) 1. ___Given___
2. $\angle X \cong \angle W$	(A) 2. ___Given___
3. $\underline{\overline{XZ} \cong \overline{WY}}$	3. Given
4. $XZ - \underline{YZ} = WY - \underline{YZ}$	4. Subt Prop of Eq
5. $XZ - YZ = XY$, $WY - YZ = ZW$	5. ___Seg Add Post___
6. $\underline{XY = WZ}$	(S) 6. Sub
7. $\therefore \triangle MXY \cong \triangle NWZ$	7. ___ASA___

Additional Example

Given: $\overline{AG} \perp \overline{AC}$, $\overline{CD} \perp \overline{CA}$, B bisects $\overline{AC}$, $\angle 1 \cong \angle 2$, $\angle 5 \cong \angle 6$
Prove: $\triangle ABF \cong \triangle CBE$

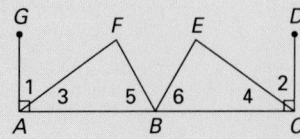

Proof: $\overline{AG} \perp \overline{AC}$, $\overline{CD} \perp \overline{CA}$ (Given); $\angle 1 \cong \angle 2$ (Given); $\angle 3$ comp. $\angle 1$, $\angle 4$ comp. $\angle 2$ (If outer rays of adj $\angle$s are $\perp$, the $\angle$s are comp); $\angle 3 \cong \angle 4$ (Comp of $\cong \angle$s are $\cong$.); B bisects $\overline{AC}$ (Given); $\overline{AB} \cong \overline{CB}$ (def of bis); $\angle 5 \cong \angle 6$ (Given); $\triangle ABF \cong \triangle CBE$ (ASA)

Written Exercises

1. Given: $\overline{AC} \cong \overline{CE}$, $\overline{AB} \cong \overline{DE}$, $\overline{BG} \cong \overline{DF}$, $\angle 1 \cong \angle 2$
Prove: $\triangle BGC \cong \triangle DFC$

2. Given: $\angle 3 \cong \angle 4$, $\angle 5 \cong \angle 6$, C is the midpoint of $\overline{BD}$.
Prove: $\triangle BGC \cong \triangle DFC$

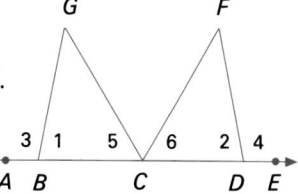

3. Given: $\overline{AB} \cong \overline{CD}$, $\angle 3 \cong \angle 2$, $\overline{EC} \cong \overline{FB}$
Prove: $\triangle AEC \cong \triangle DFB$

4. Given: $\angle 1 \cong \angle 4$, $\overline{EC} \cong \overline{FB}$, $\angle E \cong \angle F$
Prove: $\triangle AEC \cong \triangle DFB$

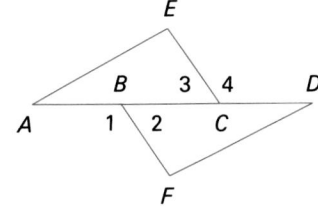

5. Given: $\overline{AG} \cong \overline{DG}$, $\overline{FG} \cong \overline{EG}$, $\overline{AB} \cong \overline{DC}$, $\angle A \cong \angle D$
Prove: $\triangle FAB \cong \triangle EDC$

6. Given: $\angle 1 \cong \angle 2$, $\overline{FB} \perp \overline{AD}$, $\overline{EC} \perp \overline{AD}$, $\overline{FB} \cong \overline{EC}$
Prove: $\triangle FAB \cong \triangle EDC$

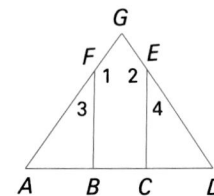

7. Given: $\overline{RP} \cong \overline{US}$, $\overline{RQ} \cong \overline{UT}$, $\overline{RP} \parallel \overline{US}$, $\overline{QW} \cong \overline{TV}$
Prove: $\triangle PQW \cong \triangle STV$

8. Given: $\angle 1 \cong \angle 2$, $\overline{AC} \cong \overline{DB}$, $\overline{GB} \cong \overline{FC}$
Prove: $\triangle BAG \cong \triangle CDF$

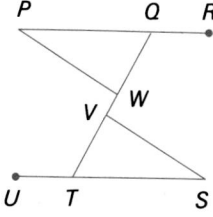

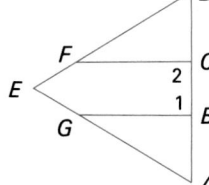

Math Connections

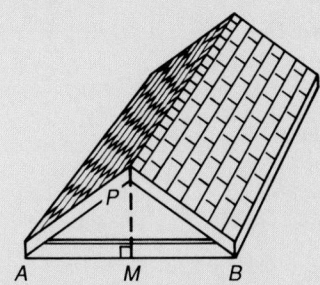

Carpentry: To build a gable roof, a carpenter joins the rafters $\overline{PA}$ and $\overline{PB}$ so that the end of the ridgepole, P, is directly above the midpoint M of the tiebeam $\overline{AB}$. $\triangle APB$, formed by the rafters and the tiebeam, is isosceles.

Critical Thinking Questions

Analysis: Ask students to draw unequal segments $\overline{AC}$ and $\overline{BD}$ perpendicular to each other at their common midpoint E. Join the endpoints of the segments to form a four-sided figure. Then have students answer these questions.

1. Write a congruence relation for four triangles and name the congruence pattern. $\triangle AED \cong \triangle CED \cong \triangle CEB \cong \triangle AEB$; SAS

2. How can you use your answer to Question 1 to conclude that triangles ADC and ABC are isosceles? Since $\triangle AED \cong \triangle CED \cong \triangle CEB \cong \triangle AEB$, then $\overline{AD} \cong \overline{CD} \cong \overline{AB} \cong \overline{CB}$ by def of $\cong$ $\triangle$s. Therefore, $\triangle ADC$ and $\triangle ABC$ are isos $\triangle$s by def.

Common Error Analysis

Error: In proving two triangles congruent, students tend to use any given congruent angles or segments as corresponding parts even when they are not parts of the relevant triangles.

Emphasize that corresponding parts must be sides or angles of the two triangles. In the example $\angle 1$ and $\angle 2$ are *exterior angles*, not corresponding parts, of the triangles. For this reason there is no "(A)" by Step 1 of the proof. Use $\angle 1$ and $\angle 2$ to prove $\angle 3 \cong \angle 4$, where $\angle 3$ and $\angle 4$ *are* angles of the triangles.

Enrichment

Have each student draw an equilateral triangle 6 inches on a side. Then challenge the students to partition the triangle into 9 congruent equilateral triangles by drawing 6 segments.

As an alternative, give the students the following figure and challenge them to draw 3 segments which will produce 9 congruent triangles. (The 3 segments are dashed in the answer below.)

Checkpoint

1. Given: ∠1 ≅ ∠2,
$\overline{AB} \cong \overline{CB}$
Prove: △ABE ≅
△CBE

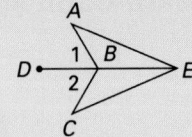

Statement	Reason
1. ∠1 ≅ ∠2	1. Given
2. ∠ABE supp ∠1, ∠CBE supp ∠2.	2. If the outer rays of two adj ∠s form a st ∠, the ∠s are supp.
3. $\overline{AB} \cong \overline{CB}$	3. Given
4. $\overline{BE} \cong \overline{BE}$	4. Reflex Prop of Congr
5. △ABE ≅ △CBE	5. SAS

2. Given: $\overline{WT} \perp \overline{WR}$,
$\overline{RS} \perp \overline{RW}$,
∠1 ≅ ∠2,
$\overline{PM}$ bisects $\overline{WR}$.
Prove: △PWQ ≅ △MRQ

Statement	Reason
1. $\overline{PM}$ bis $\overline{WR}$	1. Given
2. $\overline{WQ} \cong \overline{RQ}$	2. Def of bis
3. ∠5 ≅ ∠6	3. Vert ∠s are ≅.
4. $\overline{WT} \perp \overline{WR}$, $\overline{RS} \perp \overline{RW}$	4. Given
5. ∠3 comp ∠1, ∠4 comp ∠2	5. If outer rays of two adj ∠s are ⊥, ∠s are comp.
6. ∠1 ≅ ∠2	6. Given
7. ∠3 ≅ ∠4	7. Comp ∠s of ≅ ∠s are ≅.
8. △PWQ ≅ △MRQ	8. ASA

Closure

Ask the students to explain how to prove corresponding parts of triangles congruent using given congruency of sides or angles that are not sides or angles of the triangles. Use relationships between the parts.

▰▰▰▰ FOLLOW UP

Guided Practice

Classroom Exercises 1–7

Independent Practice

Ⓐ Ex. 1–6, **Ⓑ** Ex. 8–10 **Ⓒ** Ex. 11–13

Basic: WE 1–6 odd

Average: WE 1–10 odd

Above Average: WE 1–13 odd

9. Given: $\overline{AD} \cong \overline{CB}$, $\overline{CE} \parallel \overline{BF}$, ∠1 and ∠2 are supplementary.
Prove: △CED ≅ △BFA

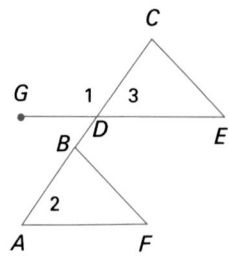

10. Given: $\overline{RV} \perp \overline{PT}$, ∠1 ≅ ∠2, ∠3 ≅ ∠4, R is the midpoint of $\overline{QS}$.
Prove: △QRW ≅ △SRU

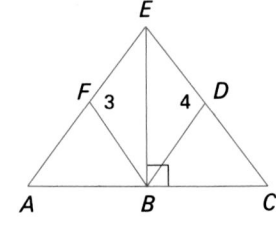

11. Given: $\overline{PQ} \cong \overline{RS}$, ∠1 ≅ ∠2, ∠3 ≅ ∠4, $\overline{WP} \perp \overline{PS}$, $\overline{SX} \perp \overline{PS}$
Prove: △PUR ≅ △STQ

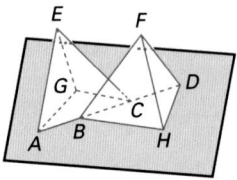

12. Given: $\overline{EB}$ bisects ∠FBD; ∠3 ≅ ∠4, $\overline{FB} \cong \overline{DB}$, $\overline{EB} \perp \overline{AC}$.
Prove: △FAB ≅ △DCB

13. Given: $\overline{AB} \cong \overline{DC}$, $\overline{AE} \cong \overline{DF}$, and $\overline{CE} \cong \overline{BF}$
Prove: △ACE ≅ △DBF

Mixed Review

1. Write the contrapositive of $p \to q$ *3.8*   ~q → ~p
2. The endpoints of $\overline{AB}$ have coordinates −4 and 6. Find the coordinate of M, the midpoint of $\overline{AB}$.   *1.3*   1
3. Graph and name the resulting geometric figure: $x \geq 5$ and $x \leq 12$ *1.8*   Line segment
4. On a number line, what is the distance between −2.5 and 1.6?   *1.2*   4.1

Additional Answers

See pages 649–650 for the answers to Ex. 1–13.

Mixed Review

3. Line segment

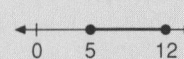

4.6 The Third Angle and AAS Theorems

Objectives   To apply the Third Angle Theorem
To apply the AAS Theorem
To recognize patterns for proving triangles congruent

Recall that the sum of the measures of the
three angles of a triangle is 180. In the figure, the measures of two angles of one triangle are equal to the measures of two
angles of the second triangle. Find the measures of the third angles. Repeat the process
with another pair of triangles. You should
notice a pattern.

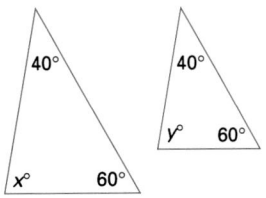

Theorem 4.2

Third Angle Theorem: If two angles of one triangle are congruent to
two angles of a second triangle, then the third angles of the triangles
are congruent.

Given: $\angle A \cong \angle D$, $\angle B \cong \angle E$
Prove: $\angle C \cong \angle F$

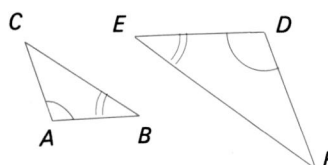

Proof

Statement	Reason
1. m $\angle A$ = m $\angle D$, m $\angle B$ = m $\angle E$	1. Given
2. m $\angle A$ + m $\angle B$ + m $\angle C$ = 180	2. Sum of meas of $\angle$s of a $\triangle$ = 180
3. m $\angle D$ + m $\angle E$ + m $\angle F$ = 180	3. Sum of meas of $\angle$s of a $\triangle$ = 180
4. m $\angle A$ + m $\angle B$ + m $\angle C$ = m $\angle D$ + m $\angle E$ + m $\angle F$	4. Sub
5. m $\angle A$ + m $\angle B$ + m $\angle C$ = m $\angle A$ + m $\angle B$ + m $\angle F$	5. Sub
6. $\therefore$ m $\angle C$ = m $\angle F$	6. Subt Prop of Eq
7. $\angle C \cong \angle F$	

EXAMPLE 1   Given: m $\angle A$ = m $\angle P$, m $\angle B$ = m $\angle Q$,
m $\angle C$ = $4x - 30$, m $\angle R$ = $2x + 60$
Find m $\angle C$.

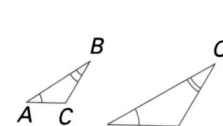

Additional Example 1

Given: m $\angle A$ = m $\angle D$, m $\angle B$ = m $\angle E$,
m $\angle C$ = $(7x - 10)$, m $\angle F$ = $(2x + 20)$
Find m $\angle C$.   32

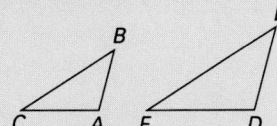

▰▰▰ GETTING STARTED

Prerequisite Quiz

**Find the measure of the third angle of
each triangle, given the measures of two
of its angles.**

1. 50 and 75   55
2. 100 and 30   50
3. x and y $180 - (x + y)$, or $180 - x - y$

**Use the markings shown to determine
why each pair of triangles is congruent.**

4. 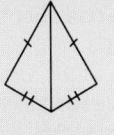   5.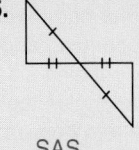

SSS   SAS

Motivator

Have your students tell you what they know
about the sum of the measures of the three
angles of a triangle.   Their sum equals 180.
On the chalkboard, draw two triangles with
angle measures of 40, 60, x and 40, 60, y
respectively. Have the students determine
the value of x and the value of y. $x = y = $
80   Ask them what this suggests.   If two
pairs of corresponding angles of a pair of
triangles are congruent then the third pair of
angles are congruent.

▰▰▰ TEACHING
SUGGESTIONS

Lesson Note

The Third Angle Theorem can be introduced
following the proof that the sum of the
measures of the angles of a triangle is 180.
However, it seems more appropriate to
prove the Third Angle Theorem in this
chapter when it is applied in the proof of the
AAS Theorem for congruence.

Emphasize the Summary of ways to prove
triangles congruent in this lesson.

Math Connections

Logic: The use of a counterexample to disprove a statement can be compared to an indirect proof. Indirect proof requires finding a contradiction in order to disprove the assumption. Counterexamples also present a contradiction to an assumption.

Critical Thinking Questions

Analysis: Use the following experiment to help students explore the relationship between triangles having at least two congruent corresponding angles. Have students compare two different pairs of these triangles, one pair having two *non-corresponding* sides congruent, the other pair having two *corresponding* sides congruent.

Draw any two triangles in which two angles and a side of one triangle are congruent to two corresponding angles and a non-corresponding side of the other. Draw two more triangles that fit this description. Is there a pattern? How do the triangles appear to be related? Are they congruent? The triangles will have the same shape, but not necessarily the same size. Therefore, the triangles are not necessarily congruent. Compare the pairs of triangles that you drew with non-corresponding sides congruent to the pair of triangles *PQR* and *ABC* with corresponding angles congruent and $\overline{QR} \cong \overline{BC}$. Explain why these triangles with congruent corresponding sides are congruent, but those with congruent non-corresponding sides are not necessarily congruent. By def of congruent triangles, congruent sides must be corresponding.

Common Error Analysis

Error: Students often do not look critically at the given data. For example, given the figures below, they may be tempted to conclude congruence by SAS.

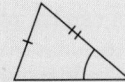 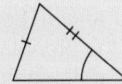

Emphasize the need to check that the angle is included by the two corresponding congruent sides since SSA is not a congruence pattern.

Solution

$$m \angle C = m \angle R \text{ by the Third Angle Theorem}$$
$$4x - 30 = 2x + 60$$
$$2x - 30 = 60 \qquad\qquad m \angle C = 4x - 30$$
$$2x = 90 \qquad\qquad\qquad\quad = 4 \cdot 45 - 30$$
$$x = 45 \qquad\qquad\qquad\quad = 150$$

So far, only three congruence patterns have been used to prove triangles congruent: SSS, SAS, and ASA. Other possibilities of side and angle combinations should be considered. Recall that a statement can be disproved by finding a counterexample.

Is AAA a congruence pattern? Consider the angles and sides in the figure. $\angle A \cong \angle X$, $\angle B \cong \angle Y$, and $\angle C \cong \angle Z$. Yet, corresponding sides are not congruent. The triangles have the same shape, but not the same size. Therefore, the AAA pattern *does not* guarantee triangle congruence.

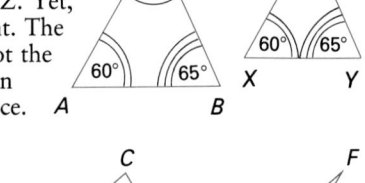

Is SSA a congruence pattern? Once again, a counterexample can be used to show that this property is not always true. For the two triangles at the right, $\angle A \cong \angle D$, $\overline{AC} \cong \overline{DF}$, and $\overline{CB} \cong \overline{FE}$. Yet, the triangles have different shapes. Therefore, the triangles are not congruent. The SSA pattern *does not* guarantee congruence either.

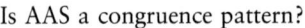

Is AAS a congruence pattern?

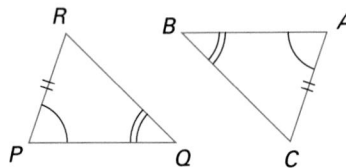

Since $\angle P \cong \angle A$ and $\angle Q \cong \angle B$, $\angle R \cong \angle C$ by the Third Angle Theorem. So the triangles *are* congruent, by ASA. Thus, the AAS pattern guarantees congruence, as stated in the next theorem.

Theorem 4.3

> **AAS Theorem:** If two angles and a non-included side of one triangle are congruent to the corresponding two angles and side of a second triangle, then the triangles are congruent.

You will be asked to prove this theorem in Exercise 17. Use the figure above for reference.

EXAMPLE 2   Given: ∠1 ≅ ∠2, $\overline{BD}$ bisects ∠D.
Prove: △ABD ≅ △CBD

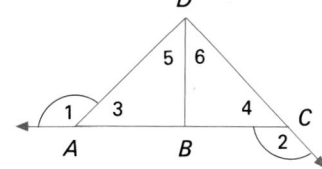

Proof

Statement	Reason
1. $\overline{BD}$ bisects ∠D.	1. Given
2. ∠5 ≅ ∠6	**(A)** 2. Def of ∠ bis
3. $\overline{BD} ≅ \overline{BD}$	**(S)** 3. Reflex Prop
4. ∠1 ≅ ∠2	4. Given
5. ∠1 and ∠3 are supplementary; ∠2 and ∠4 are supplementary.	5. If outer rays of two adj ∠s form a st ∠, then the ∠s are supp.
6. ∠3 ≅ ∠4	**(A)** 6. Supp of ≅ ∠s are ≅.
7. ∴ △ABD ≅ △CBD	7. AAS

 Summary

You *can* prove triangles are congruent by establishing these congruences.

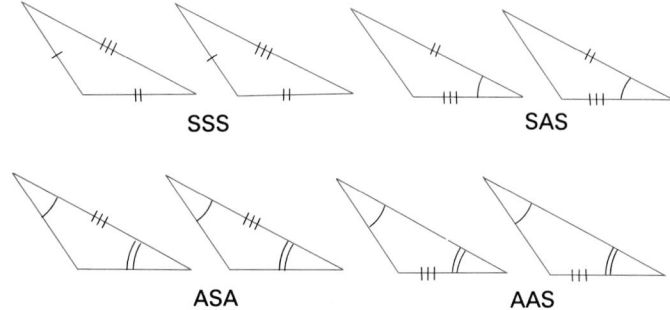

SSS   SAS

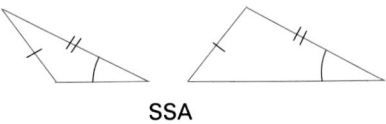 

ASA   AAS

You *cannot* prove triangles congruent by establishing these congruences.

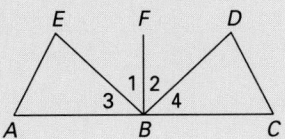

SSA

AAA

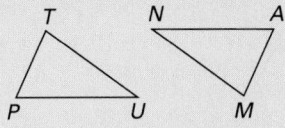

Additional Answers

Written Exercises

10.

Statement	Reason
1. $\angle A \cong \angle C$, $\overline{AC}$ bis $\overline{DE}$	1. Given
2. $\overline{DB} \cong \overline{EB}$	2. Def of seg bis
3. $\angle DBA \cong$ $\angle EBC$	3. Vert $\angle$s are $\cong$.
4. $\triangle ADB \cong$ $\triangle CEB$	4. AAS

11.

Statement	Reason
1. $\overline{PQ} \perp \overline{RS}$, $\angle R \cong \angle S$	1. Given
2. $m\angle RQP = 90$, m $\angle SQP = 90$	2. Def of $\perp$
3. m $\angle RQP =$ m $\angle SQP$	3. Sub
4. $\overline{PQ} \cong \overline{PQ}$	4. Reflex Prop of Congr
5. $\triangle PQR \cong$ $\triangle PQS$	5. AAS

12.

Statement	Reason
1. $\overline{AB} \parallel \overline{DC}$, $\overline{AD} \parallel \overline{CB}$	1. Given
2. $\angle CDB \cong$ $\angle ABD$, $\angle ADB \cong$ $\angle CBD$	2. If lines are $\parallel$, then alt int $\angle$s are $\cong$.
3. $\overline{DB} \cong \overline{DB}$	3. Reflex Prop of Congr
4. $\triangle ADB \cong$ $\triangle CBD$	4. ASA

13.

Statement	Reason
1. $\angle P \cong \angle S$, $\overline{PU} \cong \overline{SR}$, $\overline{PR} \parallel \overline{US}$, $\overline{QU} \parallel \overline{TR}$	1. Given
2. $\angle 2 \cong \angle 1$	2. If lines are $\parallel$, then alt int $\angle$s are $\cong$.
3. $\angle 2 \cong \angle 3$	3. If lines are $\parallel$, then corr $\angle$s are $\cong$.
4. $\angle 1 \cong \angle 3$	4. Trans Prop of Congr
5. $\triangle PUQ \cong$ $\triangle SRT$	5. AAS

14.

Statement	Reason
1. $\overline{PR} \cong \overline{US}$, $\overline{QR} \cong \overline{UT}$, $\angle 3 \cong \angle 4$, $\overline{RT} \parallel \overline{QU}$, $\angle P \cong \angle S$	1. Given
2. $PR - QR =$ PQ, $US -$ $UT = ST$	2. Seg Add Post
3. $PR - QR =$ $US - UT$	3. Equations may be subtracted.
4. $PQ = ST$	4. Sub
5. $\angle 1 \cong \angle 4$	5. If lines are $\parallel$, then corr $\angle$s are $\cong$.
6. $\angle 1 \cong \angle 3$	6. Trans Prop of Congr
7. $\triangle PUQ \cong$ $\triangle SRT$	7. ASA

Focus on Reading

Indicate whether or not the set of given data guarantees congruence of two triangles.

1. congruence of three pairs of sides and three pairs of angles   Yes
2. congruence of three pairs of sides   Yes
3. congruence of three pairs of angles   No
4. congruence of two pairs of angles and any one pair of sides   Yes
5. congruence of two pairs of sides and any pair of non-included angles   No

Classroom Exercises

Based on the markings shown, determine whether the two triangles are congruent. State why or why not.

1.    **2.**    **3.**

Yes; ASA   No; SSA not sufficient   No; AAA not sufficient

Written Exercises

Based on the given information, determine whether the triangles are congruent. State why or why not. (Exercises 1–4)

1. $\overline{AC} \cong \overline{DF}$, $\overline{BC} \cong \overline{EF}$, $\angle C \cong \angle F$ Yes; SAS
2. $\overline{FE} \cong \overline{CB}$, $\angle A \cong \angle D$, $\angle E \cong \angle B$ Yes; AAS
3. $\angle A \cong \angle D$, $\angle B \cong \angle E$, $\angle C \cong \angle F$ No; AAA not sufficient
4. $\angle A \cong \angle D$, $\angle B \cong \angle E$, $\overline{AB} \cong \overline{DE}$ Yes; ASA

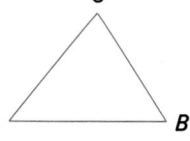

Find the indicated angle measure. (Exercises 5–7)

5. Given: $\angle A \cong \angle D$, $\angle B \cong \angle E$, m $\angle C = 4x - 20$, m $\angle F = 2x + 10$. Find m $\angle C$.   40
6. Given: m $\angle A =$ m $\angle D$, m $\angle C =$ m $\angle F$, m $\angle B = 4y + 10$, m $\angle E = y + 70$. Find m $\angle E$.   90
7. Given: m $\angle A =$ m $\angle D$, m $\angle B =$ m $\angle E$, m $\angle C = 2x - 30$, m $\angle F = x + 5$. Find m $\angle C$.   40

164   Chapter 4 Congruent Triangles

Enrichment

Challenge the students to draw a triangle and then draw 5 additional triangles inside the first, so that all 6 triangles have an AAA relationship.

Each triangle has angles of 30, 30, and 120.

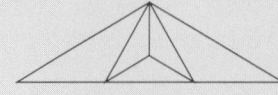

Complete the proof. (Exercises 8–9)

8. Given: $\overline{AD} \parallel \overline{CB}$,
  $\angle A \cong \angle C$
Prove: $\triangle ADB \cong \triangle CBD$

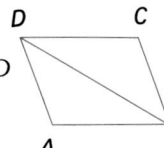

9. Given: $\angle R \cong \angle S$,
  $\overline{PQ}$ bisects $\overline{RS}$;
  $\overline{PQ}$ bisects $\angle RPS$.
  Prove: $\triangle PQR \cong \triangle PQS$

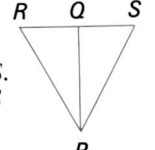

Statement	Reason
1. $\angle A \cong \angle C$	**(A)** 1. Given
2. $\overline{AD} \parallel \overline{CB}$	2. Given
3. $\angle ADB \cong \angle CBD$	**(A)** 3. Alt int $\angle$s of $\parallel$ lines are $\cong$.
4. $\overline{DB} \cong \overline{DB}$	**(S)** 4. <u>Reflex Prop</u>
5. $\triangle ADB \cong \triangle CBD$	5. <u>AAS</u>

Statement	Reason
1. $\angle R \cong \angle S$	**(A)** 1. Given
2. $\overline{PQ}$ bisects $\overline{RS}$.	2. Given
3. $\overline{RQ} \cong \overline{SQ}$	**(S)** 3. <u>Def of seg bis</u>
4. $\overline{PQ}$ bisects $\angle RPS$.	4. Given
5. $\angle RPQ \cong \angle SPQ$	**(A)** 5. <u>Def of $\angle$ bis</u>
6. $\triangle PQR \cong \triangle PQS$	6. <u>AAS</u>

10. Given: $\overline{AC}$ bisects $\overline{DE}$;
  $\angle A \cong \angle C$.
Prove: $\triangle ADB \cong \triangle CEB$

11. Given: $\overline{PQ} \perp \overline{RS}$, $\angle R \cong \angle S$
Prove: $\triangle PQR \cong \triangle PQS$

12. Given: $\overline{AB} \parallel \overline{DC}$, $\overline{AD} \parallel \overline{CB}$
Prove: $\triangle ADB \cong \triangle CBD$

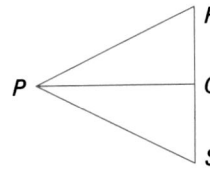

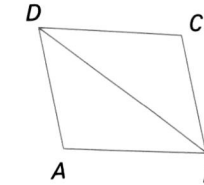

13. Given: $\angle P \cong \angle S$
  $\overline{PU} \cong \overline{SR}$,
  $\overline{PR} \parallel \overline{US}$,
  $\overline{QU} \parallel \overline{TR}$
Prove: $\triangle PUQ \cong \triangle SRT$

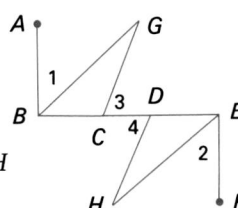

14. Given: $\overline{PR} \cong \overline{US}$,
  $\overline{QR} \cong \overline{UT}$,
  $\angle 3 \cong \angle 4$,
  $\overline{RT} \parallel \overline{QU}$,
  $\angle P \cong \angle S$
Prove: $\triangle PUQ \cong \triangle SRT$

15. Given: $\overline{AB} \perp \overline{BE}$,
  $\overline{EF} \perp \overline{BE}$,
  $\angle 1 \cong \angle 2$,
  $\angle 3 \cong \angle 4$,
  $\overline{BD} \cong \overline{EC}$
Prove: $\triangle BCG \cong \triangle EDH$

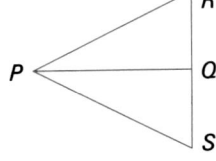

16. Given: $\overline{CG} \parallel \overline{DH}$,
  $\overline{AB} \parallel \overline{FE}$,
  $\angle 1 \cong \angle 2$,
  $\overline{HE} \cong \overline{BG}$
Prove: $\triangle BCG \cong \triangle EDH$

15.

Statement	Reason
1. $\overline{AB} \perp \overline{BE}$, $\overline{EF} \perp \overline{BE}$, $\angle 1 \cong \angle 2$, $\angle 3 \cong \angle 4$, $\overline{BD} \cong \overline{EC}$	1. Given
2. $\angle GBC$ and $\angle 1$ are comp, $\angle HED$ and $\angle 2$ are comp	2. If the outer rays of 2 adj acute $\angle$s are $\perp$, then the $\angle$s are comp.
3. $\angle GBC \cong \angle HED$	3. Comp $\angle$s of $\cong \angle$s are $\cong$.
4. $\angle GCB$ and $\angle 3$ are supp, $\angle HDE$ and $\angle 4$ are supp	4. If the outer rays of 2 adj $\angle$s form a st $\angle$, then the $\angle$s are supp.
5. $\angle GCB \cong \angle HDE$	5. Supp $\angle$s of $\cong \angle$s are $\cong$.
6. $BD - CD = EC - CD$	6. Subt Prop of Eq
7. $BD - CD = BC$, $EC - CD = ED$	7. Seg Add Post
8. $BC = ED$	8. Sub
9. $\triangle BCG \cong \triangle EDH$	9. ASA

16.

Statement	Reason
1. $\overline{CG} \parallel \overline{DH}$, $\overline{AB} \parallel \overline{FE}$, $\angle 1 \cong \angle 2$, $\overline{HE} \cong \overline{BG}$	1. Given
2. $\angle 3 \cong \angle 4$	2. If lines are $\parallel$, then alt int $\angle$s are $\cong$.
3. $\angle BCG$ and $\angle 3$ are supp, $\angle EDH$ and $\angle 4$ are supp	3. If the outer rays of 2 adj $\angle$s form a st $\angle$, then the $\angle$s are supp.
4. $\angle BCG \cong \angle EDH$	4. Supp $\angle$s of $\cong \angle$s are $\cong$.
5. $\angle ABC \cong \angle FED$	5. If lines are $\parallel$, then alt int $\angle$s are $\cong$.
6. $m\angle ABC = m\angle 1 + m\angle CBG$, $m\angle FED = m\angle 2 + m\angle HED$	6. Angle Add Post
7. $m\angle 1 + m\angle CBG = m\angle 2 + m\angle HED$	7. Sub
8. $m\angle CBG = m\angle HED$	8. Equations may be subtracted.
9. $\triangle BCG \cong \triangle EDH$	9. AAS

17. Statement	Reason
1. Using the ≅ △s *ABC* and *PQR* on page 162: ∠*P* ≅ ∠*A*, ∠*Q* ≅ ∠*B*, $\overline{PR}$ ≅ $\overline{AC}$	1. Given
2. ∠*R* ≅ ∠*C*	2. If 2 ∠s of ≅ △s are ≅, then the third ∠s of the △s are ≅.
3. △*PQR* ≅ △*ABC*	3. ASA

18. Statement	Reason
1. ∠*CDE* ≅ ∠*CED*, ∠*ACD* ≅ ∠*BCE*, $\overline{AE}$ ≅ $\overline{BD}$	1. Given
2. ∠*CEB* and ∠*CED* are supp, ∠*CDA* and ∠*CDE* are supp	2. If the outer rays of 2 adj ∠s form a st ∠, then the ∠s are supp.
3. ∠*CEB* ≅ ∠*CDA*	3. Supp ∠s of ≅ ∠s are ≅.
4. *AE* − *ED* = *AD*, *BD* − *ED* = *BE*	4. Seg Add Post
5. *AE* − *ED* = *BD* − *ED*	5. Subt Prop of Eq
6. *AD* = *BE*	6. Sub
7. △*ADC* ≅ △*BEC*	7. AAS

Mixed Review

1. See Construction on page 23.

17. Prove Theorem 4.3, the AAS Theorem.
18. Given: △*ABC* with points *D* and *E* on $\overline{AB}$ such that *D* is between *A* and *E*, $\overline{DC}$ and $\overline{EC}$ drawn such that ∠*CDE* ≅ ∠*CED*, $\overline{AE}$ ≅ $\overline{BD}$, ∠*ACD* ≅ ∠*BCE*.
    Prove: △*ADC* ≅ △*BEC*

Mixed Review

1. Draw an angle. Construct the bisector of this angle. *1.5*
2. The measures of two angles formed by the bisector of a given angle are $4x - 20$ and $x + 40$. Find the measures of the angles formed. *1.5* 60
3. The measure of an angle is three times the measure of its complement. Find the measures of the two angles. *1.6* 67.5, 22.5
4. The measure of one acute angle of a right triangle is $\frac{2}{3}$ the measure of the other acute angle. Find the measure of each acute angle. *4.1* 36, 54

Algebra Review

To solve a system of two equations with two variables by substitution:
(1) Solve for one variable in one equation.
(2) Substitute the expression for that variable in the second equation.
(3) Solve the resulting equation.
(4) Substitute the result from (3) in either one of the original equations and solve for the remaining variable.
(5) Check by substituting the values into both original equations.

Example: Solve and check. $a - 2b = 11$
$5a + 4b = 27$

(1)	$a - 2b = 11$	Solve for *a*.
	$a = 2b + 11$	
(2)	$5a + 4b = 27$	Substitute $(2b + 11)$ for *a*.
	$5(2b + 11) + 4b = 27$	
(3)	$10b + 55 + 4b = 27$	Solve for *b*.
	$b = -2$	
(4)	$a - 2(-2) = 11$	Substitute (-2) for *b*.
	$a = 7$	Solve for *a*.

(5) Check. $7 - (2)(-2) \overset{?}{=} 11$ $5(7) + 4(-2) \overset{?}{=} 27$
                      $7 + 4 = 11$ ✔ $35 - 8 = 27$ ✔

Solve and check.

1. $2a + b = 12$
$a + 2b = 9$
$a = 5, b = 2$

2. $-3p - q = 13$
$2p + 3q = -4$
$p = -5, q = 2$

3. $x - 2y = 8$
$2y = 3x - (-16)$
$x = -12, y = -10$

166 Chapter 4 Congruent Triangles

4.7 Applying Congruence: Corresponding Parts

Objectives To prove congruence of sides and angles by first proving congruence of triangles

To prove that segments are parallel by first proving that triangles are congruent

Recall from the definition of triangle congruence that Corresponding Parts of Congruent Triangles are Congruent. This is abbreviated as CPCTC. If two triangles can be proved congruent, then any pair of corresponding sides or pair of corresponding angles are congruent.

EXAMPLE 1 Given: $\overline{EF}$ bisects $\angle BEC$; $\overline{BE} \cong \overline{CE}$.
Prove: $\angle 1 \cong \angle 2$

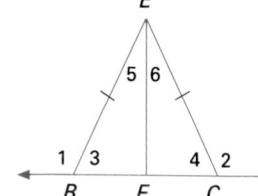

Plan $\angle 1$ and $\angle 2$ are supplements, respectively, of $\angle 3$ and $\angle 4$. Use SAS to prove the triangles congruent. Then, $\angle 3 \cong \angle 4$ by CPCTC. Finally, show that $\angle 1 \cong \angle 2$.

Proof

Statement	Reason
1. $\overline{BE} \cong \overline{CE}$	(S) 1. Given
2. $\overline{EF}$ bisects $\angle BEC$.	2. Given
3. $\angle 5 \cong \angle 6$	(A) 3. Def of $\angle$ bis
4. $\overline{EF} \cong \overline{EF}$	(S) 4. Reflex Prop
5. $\triangle BFE \cong \triangle CFE$	5. SAS
6. $\angle 3 \cong \angle 4$	6. CPCTC
7. $\angle 1$ and $\angle 3$ are supplementary; $\angle 2$ and $\angle 4$ are supplementary.	7. If the outer rays of two adj $\angle$s form a st $\angle$, then the $\angle$s are supp.
8. $\therefore \angle 1 \cong \angle 2$	8. Supp of $\cong \angle$s are $\cong$.

Lines can be proved parallel by showing that a pair of alternate interior angles is congruent. Sometimes this can be done by showing that the angles are corresponding parts of congruent triangles.

Additional Example 1

Given: M bisects $\overline{AP}$, $\angle S \cong \angle T$,
$\angle 4 \cong \angle 1$
$\overline{AR} \perp \overline{AP}$, $\overline{PQ} \perp \overline{PA}$
Prove: $\overline{SM} \cong \overline{TM}$

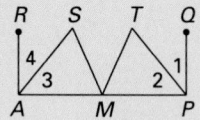

Proof: M bis $\overline{AP}$ (Given); $\overline{AM} = \overline{PM}$ (Def of bis); $\overline{AR} \perp \overline{AP}$, $\overline{PQ} \perp \overline{PA}$ (Given); $\angle 3$ comp. $\angle 4$, $\angle 2$ comp. $\angle 1$ (If outer rays of two adj $\angle$s are $\perp$, then $\angle$s are comp); $\angle 4 \cong \angle 1$ (Given); $\angle 3 \cong \angle 2$ (Comp of $\cong$ $\angle$s are $\cong$); $\angle S \cong \angle T$ (Given); $\triangle SAM \cong \triangle TPM$ (AAS); $\overline{SM} \cong \overline{TM}$ (CPCTC)

Teaching Resources

Quick Quizzes 33
Reteaching and Practice
  Worksheets, pp. 65, 66
Transparency 5

▰▰ GETTING STARTED

Prerequisite Quiz

Indicate whether the two triangles are congruent. Give a reason.

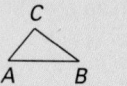 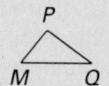

1. Given: $\overline{AB} \cong \overline{MQ}$, $\angle C \cong \angle P$, $\angle B \cong \angle Q$
   Yes; AAS
2. Given: $\overline{AC} \cong \overline{MP}$, $\overline{AB} \cong \overline{MQ}$, $\overline{BC} \cong \overline{QP}$
   Yes; SSS
3. Given: $\overline{AC} \cong \overline{MP}$, $\overline{AB} \cong \overline{MQ}$, $\angle B \cong \angle Q$
   No; SSA is not a congruence pattern
4. Given: $\angle C \cong \angle P$, $\angle B \cong \angle Q$, $\overline{CB} \cong \overline{PQ}$
   Yes; ASA

Motivator

Ask your students the following questions. If two triangles are congruent, what is true about pairs of corresponding parts? They are congruent. What are some ways to prove lines parallel? Show alternate interior angles and corresponding angles are congruent. How can you prove that a point M is the midpoint of $\overline{AB}$ or that $\overline{CD}$ bisects $\angle ACB$? Show $\overline{AM} \cong \overline{MB}$ or $\angle ACD \cong \angle BCD$.

▰▰ TEACHING SUGGESTIONS

Lesson Note

This lesson is an extension of the previous ones on congruence. Emphasize that once triangles are proved congruent, all pairs of corresponding parts are known to be congruent.

Math Connections

Automobile Industry: Cars are usually mass-produced on an assembly line. The same model of car manufactured in the same year can be considered congruent to every such model. When a part to a car needs to be replaced, the concept of *corresponding parts of congruent figures* can be applied. You can replace the broken part with a corresponding part from another car of the same model year.

Critical Thinking Questions

Synthesis: If two triangles are congruent, then are the corresponding sides necessarily parallel? Why or why not?   No; congruent triangles may change positions and still be congruent.   If a pair of corresponding sides of congruent triangles are parallel, are all corresponding sides necessarily parallel? Why or why not? Sometimes, but not necessarily. One counterexample is a pair of congruent triangles formed by intersecting lines of different lengths.

Checkpoint

1. Given: $\overline{PS} \cong \overline{PQ}$.
   $\overline{SR} \cong \overline{QR}$
   Prove: $\angle SPR \cong$
   $\angle QPR$

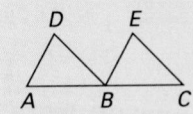

Statement	Reason
1. $\overline{PS} \cong \overline{PQ}$, $\overline{SR} \cong \overline{QR}$	1. Given
2. $\overline{PR} \cong \overline{PR}$	2. Reflex Prop of Congr
3. $\triangle PSR \cong \triangle PQR$	3. SSS
4. $\angle SPR \cong \angle QPR$	4. CPCTC

2. Given: $\overline{AB} \cong \overline{BC}$,
   $\overline{AD} = \overline{BE}$,
   $\overline{AD} \parallel \overline{BE}$
   Prove: $\overline{BD} \parallel \overline{CE}$

Statement	Reason
1. $\overline{AD} \parallel \overline{BE}$	1. Given
2. $\angle DAB \cong \angle EBC$	2. If lines are $\parallel$, corr $\angle$s are $\cong$
3. $\overline{AB} \cong \overline{BC}$, $\overline{AD} \cong \overline{BE}$	3. Given
4. $\triangle ABD \cong \triangle BCE$	4. SAS
5. $\angle ABD \cong \angle BCE$	5. CPCTC
6. $\overline{BD} \parallel \overline{CE}$	6. If corr $\angle$s are $\cong$, lines are $\parallel$

EXAMPLE 2   Given: $\overline{AD} \parallel \overline{CB}$, $\overline{AD} \cong \overline{CB}$
   Prove: $\overline{CD} \parallel \overline{AB}$

Proof

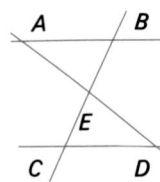

	Statement		Reason
	1. $\overline{AD} \cong \overline{CB}$	(S)	1. Given
	2. $\overline{AD} \parallel \overline{CB}$		2. Given
	3. $\angle 1 \cong \angle 2$	(A)	3. Alt int $\angle$s of $\parallel$ lines are $\cong$.
	4. $\overline{DB} \cong \overline{BD}$	(S)	4. Reflex Prop
	5. $\triangle DAB \cong \triangle BCD$		5. SAS
	6. $\angle 3 \cong \angle 4$		6. CPCTC
	7. $\therefore \overline{CD} \parallel \overline{AB}$		7. If alt int $\angle$s are $\cong$, then lines are $\parallel$.

Classroom Exercises

Provide a reason for each step in the proof below.
Given: $\overline{AB} \parallel \overline{CD}$, $\overline{AB} \cong \overline{CD}$
Prove: E is the midpoint of $\overline{AD}$.

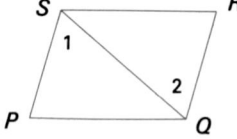

	Statement		Reason
1. $\overline{AB} \parallel \overline{CD}$		1.	Given
2. $\angle BAE \cong \angle CDE$	(A)	2.	Alt int $\angle$s
3. $\overline{AB} \cong \overline{CD}$	(S)	3.	Given
4. $\angle ABE \cong \angle DCE$	(A)	4.	Alt int $\angle$s
5. $\triangle ABE \cong \triangle DCE$		5.	ASA
6. $\overline{AE} \cong \overline{DE}$		6.	CPCTC
7. E is the midpoint of $\overline{AD}$.		7.	Def of midpt

Written Exercises

1. Given: $\overline{PS} \cong \overline{RQ}$, $\overline{SR} \cong \overline{QP}$
   Prove: $\angle 1 \cong \angle 2$
2. Given: $\overline{PS} \parallel \overline{RQ}$, $\overline{PQ} \parallel \overline{RS}$
   Prove: $\angle P \cong \angle R$

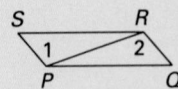

Additional Example 2

Given: $\overline{QR} \cong \overline{SP}$, $\overline{QP} \cong \overline{SR}$,
Prove: $\overline{SP} \parallel \overline{QR}$

Proof: $\overline{QR} \cong \overline{SP}$, $\overline{QP} \cong \overline{SR}$ (Given); $\overline{PR} \cong \overline{RP}$ (Reflex Prop of Congr); $\triangle SPR \cong \triangle QRP$ (SSS); $\angle 1 \cong \angle 2$ (CPCTC); $\overline{SP} \parallel \overline{QR}$ (If alt int $\angle$s are $\cong$, then lines are $\parallel$)

3. Given: $\overline{AC}$ and $\overline{DE}$ bisect each other.
Prove: $\overline{AD} \parallel \overline{CE}$
4. Given: $\overline{AD} \parallel \overline{CE}$, B is the midpoint of $\overline{AC}$.
Prove: B is the midpoint of $\overline{DE}$.

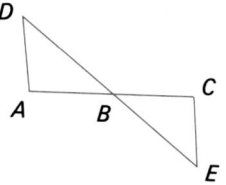

5. Given: $\overline{FS} \cong \overline{UQ}$,
$\overline{FP} \cong \overline{UR}$,
$\overline{FP} \parallel \overline{UR}$
Prove: $\overline{PQ} \parallel \overline{RS}$

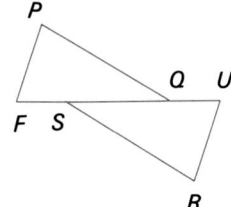

6. Given: $\overline{FP} \parallel \overline{UR}$,
$\overline{PQ} \parallel \overline{RS}$,
$\overline{PQ} \cong \overline{RS}$
Prove: $\overline{FQ} \cong \overline{US}$

7. Given: $\overline{TQ} \perp \overline{PR}$
$\overline{TQ} \perp \overline{US}$,
$\angle 1 \cong \angle 2$
Prove: $\overline{QU} \cong \overline{QS}$

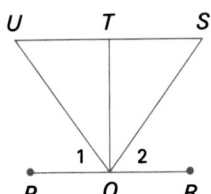

8. Given: $\overline{TQ}$ bisects $\angle UQS$;
$\overline{UQ} \cong \overline{SQ}$.
Prove: $\overline{TQ} \perp \overline{US}$

9. Given: $\overline{CB} \perp$ every line in plane $\mathcal{M}$
that passes through B,
$\angle 1 \cong \angle 2$, $\overline{DB} \cong \overline{IB}$
Prove: $\angle 5 \cong \angle 6$
10. Given: $\overline{BC} \perp$ every line in plane $\mathcal{N}$
that passes through C,
$\angle 3 \cong \angle 4$, $\overline{ED} \cong \overline{GI}$, $\overline{EC} \cong \overline{GC}$.
Prove: $\overline{BD} \cong \overline{BI}$

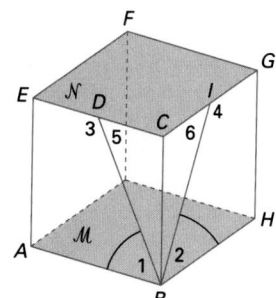

Mixed Review

In the figure, $\overline{QT} \parallel \overline{RS}$

1. Given: m $\angle 3 = 60$. Find m $\angle 5$. **3.5** 60
2. Given: m $\angle 2 = 70$. Find m $\angle QRS$. **3.5** 110
3. Given: m $\angle 3 = 75$, m $\angle 1 = 140$. Find m $\angle 4$. **3.6** 65
4. Given: m $\angle 2 = 40$, m $\angle 3$ is 4 times m $\angle 4$.
Find m $\angle 3$. **3.7** 112

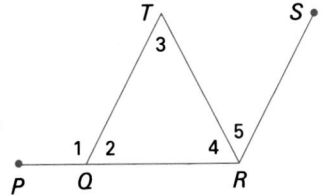

Enrichment

Discuss the fact that the parallel lines on a piece of paper can be used to divide a segment into a desired number of congruent segments. On the left, $\overline{AB}$ is divided into 4 congruent segments, and on the right,

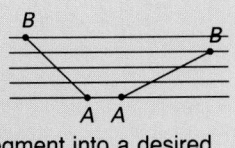

AAS;
CPCTC

$\overline{AB}$ is divided into 3 congruent segments. Have the students justify the validity of this device. For a hint, suggest that they draw right triangles.

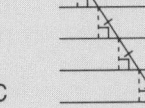

1. Scalene
2. Equil
3. Isos

12.

Statement	Reason
1. $\angle S \cong \angle R$, $\overline{PQ}$ bis $\angle SQR$	1. Given
2. $\angle SQP \cong \angle RQP$	2. Def of $\angle$ bis
3. $\overline{PQ} \cong \overline{PQ}$	3. Reflex Prop of Congr
4. $\triangle SPQ \cong \triangle RPQ$	4. AAS

Chapter 4 Review

Key Terms

acute triangle   (p. 139)
congruent triangles   (p. 143)
corresponding parts of
  congruent angles   (p. 143)
equiangular triangle   (p. 139)
equilateral triangle   (p. 139)

included angle   (p. 147)
included side   (p. 154)
isosceles triangle   (p. 139)
obtuse triangle   (p. 139)
right triangle   (p. 139)
scalene triangle   (p. 139)

Key Ideas and Review

4.1   To classify triangles by angles, remember the following.

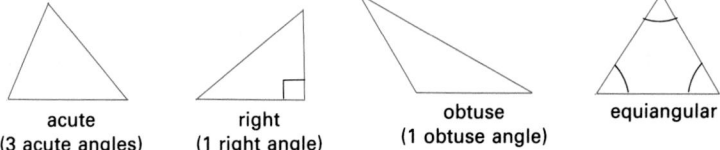

acute
(3 acute angles)

right
(1 right angle)

obtuse
(1 obtuse angle)

equiangular

To classify triangles by sides, remember the following.

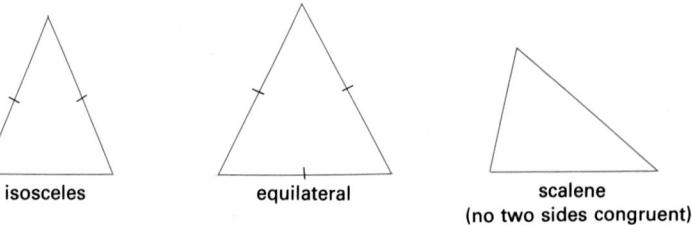

isosceles

equilateral

scalene
(no two sides congruent)

Classify each triangle by its sides.

1. 3 cm, 2 cm, 2.3 cm   **2.** 3 ft, 3 ft, 3 ft   **3.** 3 cm, 2 cm, 3 cm

Classify the triangle by its angles. (Exercises 4–5)

4. The measures of two angles are 35 and 55. Right
5. The measures of two angles are 70 and 50. Acute
6. The measure of one acute angle of a right triangle is $\frac{3}{2}$ the measure of the other acute angle. Find the measure of each acute angle. 54, 36
7. $\triangle ABC$ is isosceles with $AC = AB$. $AC = 4x - 2$, $AB = 2x + 6$, $BC = 3x - 1$. Find BC. 11

4.2, 4.6 To prove triangles congruent, use one of the following four congruence patterns.

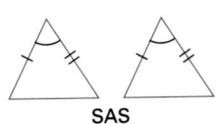

SAS

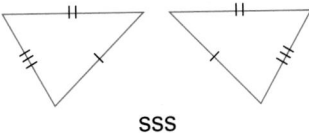

SSS

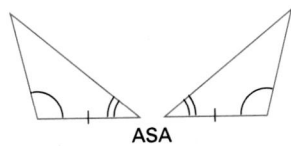

ASA

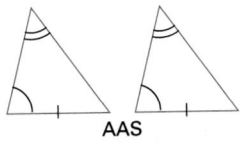

AAS

4.7 To prove corresponding parts of triangles congruent, first show that the triangles are congruent and then use CPCTC.

Based on the given information, determine whether the two triangles are congruent. State the reason. (Exercises 8–10) 8. No; AAA not sufficient

8. $\angle Y \cong \angle L$, $\angle P \cong \angle H$, $\angle T \cong \angle G$
9. $\overline{YT} \cong \overline{LG}$, $\overline{PT} \cong \overline{HG}$, $\angle T \cong \angle G$ Yes; SAS
10. $\angle P \cong \angle H$, $\angle T \cong \angle G$, $\overline{PT} \cong \overline{HG}$ Yes; ASA
11. Given: $\angle Y \cong \angle L$, $\angle T \cong \angle G$, m $\angle P = 3x - 40$, m $\angle H = x + 20$
    Find m $\angle P$.   m $\angle P = 50$

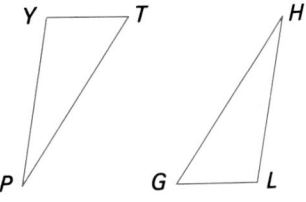

12. Given: $\angle S \cong \angle R$, $\overline{PQ}$ bisects $\angle SQR$.
    Prove: $\triangle SPQ \cong \triangle RPQ$

13. Given: $\overline{TU} \cong \overline{GY}$, $\overline{KY} \parallel \overline{HU}$, $\overline{KT} \perp \overline{TG}$, $\overline{HG} \perp \overline{TG}$
    Prove: $\angle K \cong \angle H$

14. Given: $\overline{MQ} \parallel \overline{WL}$, $\overline{MQ} \cong \overline{WL}$
    Prove: $\overline{ML} \parallel \overline{WQ}$

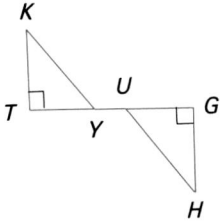

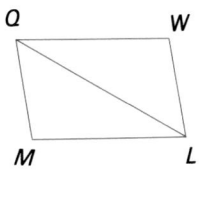

4.3, 4.4 To construct a triangle congruent to a given triangle, see pages 147 and 153.

15. Draw an acute triangle. Construct a triangle congruent to it, using the SSS pattern.

13.

Statement	Reason
1. $\overline{TU} \cong \overline{GY}$, $\overline{KY} \parallel \overline{HU}$, $\overline{KT} \perp \overline{TG}$, $\overline{HG} \perp \overline{TG}$	1. Given
2. $TU - YU = GY - YU$	2. Equations may be subtracted.
3. $TU - YU = TY$, $GY - YU = GU$	3. Seg Add Post
4. $TY = GU$	4. Sub
5. $\angle KYU \cong \angle HUY$	5. If lines are $\parallel$, then alt int $\angle$s are $\cong$.
6. $\angle KYT$ and $\angle KYU$ are supp, $\angle HUY$ and $\angle HUG$ are supp	6. If the outer rays of 2 adj $\angle$s form a st $\angle$, then the $\angle$s are supp.
7. $\angle KYT \cong \angle HUG$	7. Supp $\angle$s of $\cong \angle$s are $\cong$.
8. m $\angle KTY = 90$, m $\angle HGU = 90$	8. Def of $\perp$
9. m $\angle KTY = $ m $\angle HGU$	9. Sub
10. $\triangle KTY \cong \triangle HGU$	10. ASA
11. $\angle K \cong \angle H$	11. CPCTC

14.

Statement	Reason
1. $\overline{MQ} \parallel \overline{WL}$, $\overline{MQ} \cong \overline{WL}$	1. Given
2. $\angle MQL \cong \angle WLQ$	2. If lines are $\parallel$, then alt int $\angle$s are $\cong$.
3. $\overline{QL} \cong \overline{QL}$	3. Reflex prop of Congr
4. $\triangle QML \cong \triangle LWQ$	4. SAS
5. $\angle QLM \cong \angle LQW$	5. CPCTC
6. $\overline{ML} \parallel \overline{WQ}$	6. If alt int $\angle$s are $\cong$, then lines are $\parallel$.

15.

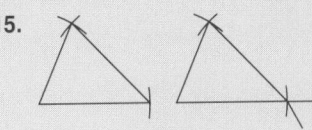

6.

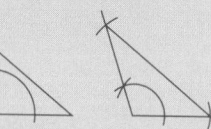

See Construction on page 147.

10.

Statement	Reason
1. $\overline{PQ}$ bis ∠MPT, $\overline{PQ}$ bis ∠MQT	1. Given
2. ∠MPQ ≅ ∠TPQ, ∠MQP ≅ ∠TQP	2. Def of ∠ bis
3. $\overline{PQ}$ ≅ $\overline{PQ}$	3. Reflex Prop of Congr
4. △PMQ ≅ △PTQ	4. ASA

11.

Statement	Reason
1. $\overline{PT}$ ≅ $\overline{RT}$, $\overline{ST}$ bis $\overline{PR}$	1. Given
2. $\overline{SP}$ ≅ $\overline{SR}$	2. Def of seg bis
3. $\overline{ST}$ ≅ $\overline{ST}$	3. Reflex Prop of Congr
4. △STP ≅ △STR	4. SSS
5. ∠PTS ≅ ∠RTS	5. CPCTC
6. $\overline{ST}$ bis ∠PTR	6. Def of ∠ bis

12.

Statement	Reason
1. $\overline{TW}$ ∥ $\overline{NM}$, $\overline{WY}$ ≅ $\overline{MX}$, ∠1 ≅ ∠2	1. Given
2. ∠TWX ≅ ∠NMY	2. If lines are ∥, then alt int ∠s are ≅.
3. WY − YX = MX − YX	3. Subt Prop of Eq
4. WY − YX = WX, MX − YX = MY	4. Seg Add Post
5. WX = MY	5. Sub
6. ∠WXT and ∠2 are supp, ∠MYN and ∠1 are supp	6. If the outer rays of 2 adj ∠s form a st ∠, then the ∠s are supp.
7. ∠WXT ≅ ∠MYN	7. Supp ∠s of ≅ ∠s are ≅.
8. △TWX ≅ △NMY	8. ASA
9. ∠T ≅ ∠N	9. CPCTC

Chapter 4 Test

A-Level Ex.: 1–2, 4–9   B-Level Ex.: 3, 10–11   C-Level Ex.: starred

Classify the triangle by its sides and by its angles.

1.

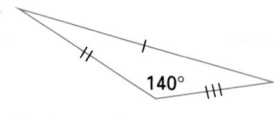

140°

Scalene; obtuse

2.

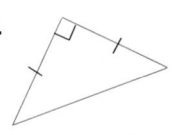

Isosceles; right

3. The measure of one acute angle of a right triangle is $\frac{3}{7}$ of the measure of the other acute angle. Find the measure of each acute angle.   27, 63

4. Given: △ABC is isosceles with AB = BC, AB = 12, BC = 3x + 6, AC = x² + 3x. Find AC.   AC = 10

5. Given: ∠A ≅ ∠D, ∠B ≅ ∠E, m ∠C = 4y + 20, m ∠F = y + 80. Find m ∠C.   100

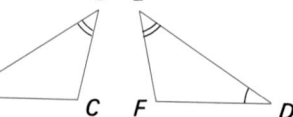

6. Draw an obtuse triangle. Construct a triangle congruent to this triangle, using an SAS pattern.

Based on the markings shown, determine whether the triangles are congruent. Provide a reason.

7.

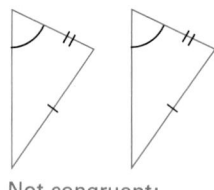

Not congruent; SSA not sufficient

8.

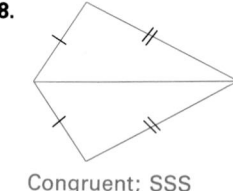

Congruent; SSS

9.

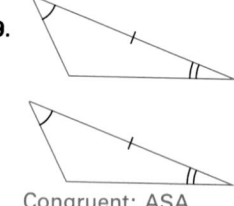

Congruent; ASA

10. Given: $\overline{PQ}$ bisects ∠MPT; $\overline{PQ}$ bisects ∠MQT.
Prove: △PMQ ≅ △PTQ

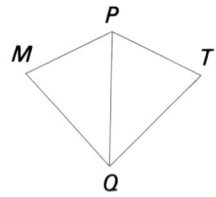

11. Given: $\overline{PT}$ ≅ $\overline{RT}$; $\overline{ST}$ bisects $\overline{PR}$.
Prove: $\overline{ST}$ bisects ∠PTR.

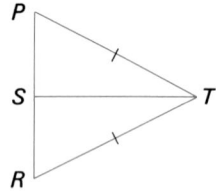

*12. Given: $\overline{TW}$ ∥ $\overline{NM}$; ∠1 ≅ ∠2; $\overline{WY}$ ≅ $\overline{MX}$
Prove: ∠T ≅ ∠N

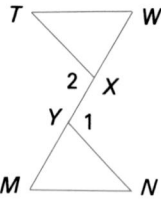

College Prep Test

In each item you are to compare a quantity in Column 1 with a quantity in Column 2. Write the letter of the correct answer from these choices:

A—The quantity in Column 1 is greater than the quantity in Column 2.
B—The quantity in Column 2 is greater than the quantity in Column 1.
C—The quantity in Column 1 is equal to the quantity in Column 2.
D—The relationship cannot be determined from the given information.

Column 1	Column 2

1. B

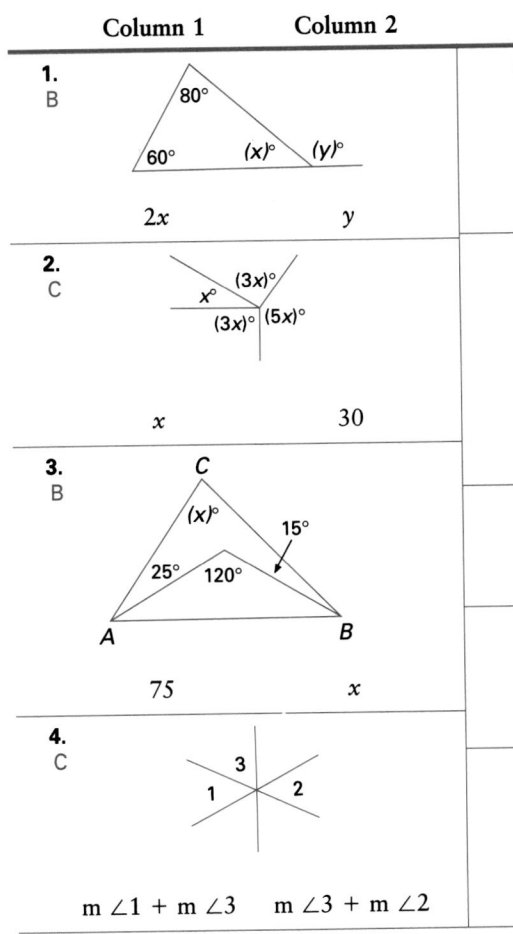

$2x$ y

2. C

x 30

3. B

75   x

4. C

$m \angle 1 + m \angle 3$ $m \angle 3 + m \angle 2$

Column 1	Column 2

5. D

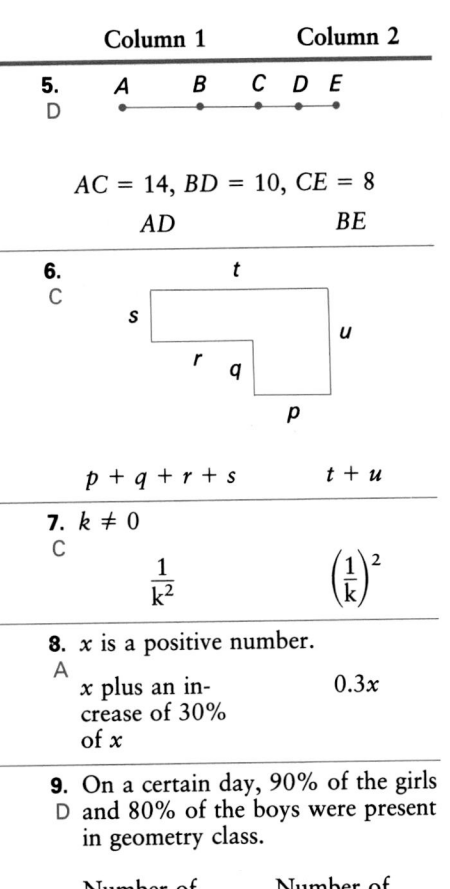

$AC = 14$, $BD = 10$, $CE = 8$

AD BE

6. C

$p + q + r + s$ $t + u$

7. $k \neq 0$

C

$\dfrac{1}{k^2}$ $\left(\dfrac{1}{k}\right)^2$

8. x is a positive number.

A

x plus an increase of 30% of x $0.3x$

9. On a certain day, 90% of the girls and 80% of the boys were present in geometry class.

D

Number of boys absent   Number of girls absent

5. Line seg

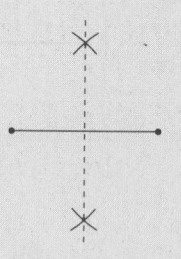

6.

7. $AB = CD$, $BP = DQ$ (Given); $AB + BP = CD + DQ$ (Equations may be added.); $AB + BP = AP$, $CD + DQ = CQ$ (Seg Add Post); ∴ $AP = CQ$ (Sub)

8. $AB = CD$ (Given); $AB + BP = AP$ (Seg Add Post); ∴ $CD + BP = AP$ (Sub)

11. Parallel lines are not skew. (True)

12. If 2 ∠s are ≅, then they are vert. (False)

18.

Statement	Reason
1. $\overline{DA}$ bis ∠CDB, $\overline{CD}$ ‖ $\overline{AB}$	1. Given
2. ∠1 ≅ ∠2	2. Def of ∠ bis
3. ∠1 ≅ ∠3	3. If lines are ‖, then alt int ∠s are ≅.
4. ∠2 ≅ ∠3	4. Trans Prop of Congr

20.

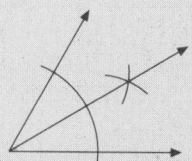

21. Conj: Vert ∠s are ≅ and the sum of the meas of 2 supp ∠s is 90. (False) Disj: Vert ∠s are ≅ or the sum of the meas of 2 supp ∠s is 90. (True)

25.

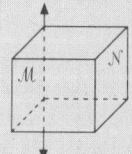

False. Planes $\mathcal{M}$ and $\mathcal{N}$ are 2 nonparallel planes that are both parallel to the dashed line.

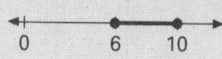
Cumulative Review (Chapters 1–4)

Choose the best possible answer from A–E. (Exercises 1–4)

1. Given: Coordinates of the end-points of a segment are −8 and 12. Find the coordinate of the midpoint. A   *1.3*
   (A) 2   (B) 20   (C) 4
   (D) −20   (E) None of these

2. If ∠A is acute and m ∠A = 2x, then the measure of its complement is ____. C   *1.6*
   (A) 180   (B) 90
   (C) 90 − 2x   (D) 180 − 2x
   (E) 2x − 90

3. Given: lines m and n are co-planar. Which of the following relationships is not possible? D   *3.1*
   (A) m ‖ n   (B) m and n intersect
   (C) m ⊥ n   (D) m and n skew
   (E) m and n form vertical angles.

4. The contrapositive of $p \rightarrow q$ is: *3.8*
   (A) $q \rightarrow p$ (B) $\sim p \rightarrow q$
   (C) $p \rightarrow \sim q$ (D) $\sim q \rightarrow \sim p$
   (E) None of these   D

5. Graph and name the resulting geometric measure. Line segment $x \geq 6$ and $x \leq 10$ *1.8*

6. Draw a segment. Construct its midpoint.   *1.3*

7. Given: $AB = CD$, $BP = DQ$
   Prove: $AP = CQ$ *2.3*

8. Given: $AB = CD$
   Prove: $CD + BP = AP$

9. Given: m ∠1 = 5x + 35, m ∠3 = 70
   Find x. 15   *2.8*

10. Given: m ∠1 = 7x − 10, m ∠2 = 4x + 50
    Find m ∠2. 130

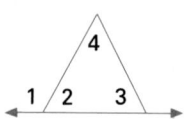

11. Write the negation of the statement "Parallel lines are skew." Is the negation true?   *3.8*

12. Write the converse of the statement "If two angles are vertical, then the angles are congruent." Is this converse true?   *3.5*

Use the figure below for Exercises 13–15.

13. Given: m ∠1 = 140, m ∠3 = 85 Find m ∠4. 55   *3.7*

14. Given: m ∠2 = 3x, m ∠3 = 4x, m ∠4 = 5x Find m ∠1. 135   *3.6*

15. m ∠4 = 3x + 20, m ∠3 = x + 10, m ∠1 = 110 Find m ∠3. 30   *3.7*

For Exercises 16–18, use the diagram below and assume $\overleftrightarrow{CD} \parallel \overleftrightarrow{BA}$.

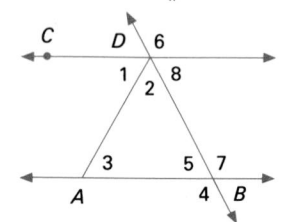

16. Given: m ∠4 = 100. Find m ∠6. 100   *3.5*

17. Given: m ∠8 = $\frac{4}{5}$ m ∠7. Find m ∠8. 80   *3.5*

18. Given: $\overline{DA}$ bisects ∠CDB. Prove: ∠3 ≅ ∠2   *3.5*

19. $AB = \frac{2}{3}BC$
and $AC = 25$.
Find AB. $AB = 10$ *1.2*

A •———• B •———• C

20. Draw an angle. Construct the   *1.5*
bisector of this angle.

21. Write the conjunction and dis-   *1.7*
junction of the statements and
determine if each is true or
false.
a. Vertical angles are
congruent.
b. The sum of the measures of
two supplementary angles is
90.

22. The measure of an angle is 3   *1.6*
times the measure of its sup-
plement. Find the measure of
each angle.   135, 45

23. How many different planes   *3.1*
can contain one line?   An infinite number

24. How many different planes can
contain two intersecting lines?   One

State whether the statement is true or
false. If false, draw a figure to show
why. (Exercises 25–26)

25. Two planes parallel to the   *3.9*
same line are parallel.   False

26. Two planes intersected by a
third plane are parallel.   False

27. Write the converse of "If two   *3.8*
lines are parallel, then alter-
nate interior angles are congruent." Then write a biconditional and determine if it is
true.

28. Given: $m \not\parallel n$ *3.4*
Prove: $\angle 1 \not\cong \angle 2$ (Use an indirect proof.)

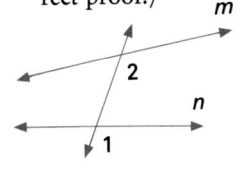

29. The measure of one acute angle   *4.1*
of a right triangle is $\frac{1}{2}$ the measure of the other. Find the
measure of each acute angle.   30, 60

Determine from the markings   *4.3*
whether the two triangles are con-   *4.6*
gruent. State why or why not.

30.   **31.**

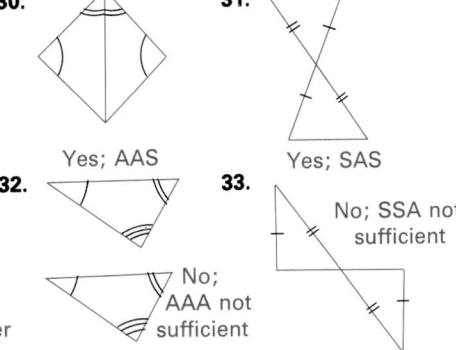

Yes; AAS   Yes; SAS

32.   **33.**

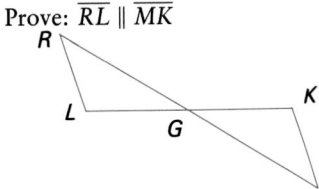

No; AAA not
sufficient

No; SSA not
sufficient

34. Given: $\overline{RM}$ and $\overline{LK}$ bisect each   *4.7*
other.
Prove: $\overline{RL} \parallel \overline{MK}$

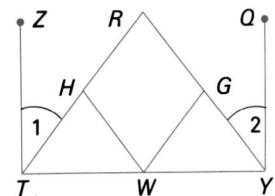

35. Given: $\angle 1 \cong \angle 2$, $\overline{ZT} \perp \overline{TY}$,   *4.7*
$\overline{QY} \perp \overline{TY}$, $\overline{TR} \cong \overline{YR}$,
$\overline{HR} \cong \overline{GR}$, W is the
midpoint of $\overline{TY}$.
Prove: $\overline{WH} \cong \overline{WG}$

26.

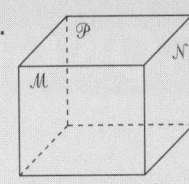

False. Plane $\mathcal{P}$ (top) intersects both
planes $\mathcal{M}$ and $\mathcal{N}$ (sides), but the planes
are not parallel.

27. Conv: If alt int $\angle$s are $\cong$, then the 2
lines are $\parallel$. (True)
Bicond: 2 lines are $\parallel$ if and only if alt int
$\angle$s are $\cong$. (True)

28. Assume that $m \not\parallel n$ and that $\angle 1 \cong \angle 2$.
Since $\angle 2$ and $\angle 1$ are corr $\angle$s and $\cong$,
then $m \parallel n$. But $m \parallel n$ and $m \not\parallel n$ is a
contradiction. So the assumption that
$\angle 1 \cong \angle 2$ is false. Therefore, $\angle 1 \not\cong \angle 2$.

34.

Statement	Reason
1. $\overline{RM}$ and $\overline{LR}$ bis each other	1. Given
2. $\overline{RG} \cong \overline{MG}$, $\overline{LG} \cong \overline{KG}$	2. Def of seg bis
3. $\angle RGL \cong \angle MGK$	3. Vert $\angle$s are $\cong$.
4. $\triangle RLG \cong \triangle MKG$	4. SAS
5. $\angle R \cong \angle M$	5. CPCTC
6. $\overline{RL} \parallel \overline{MK}$	6. If alt int $\angle$s are $\cong$, then lines are $\parallel$.

35.

Statement	Reason
1. $\angle 1 \cong \angle 2$, $\overline{ZT} \perp \overline{TY}$, $\overline{QY} \perp \overline{TY}$, $\overline{TR} \cong \overline{YR}$, $\overline{HR} \cong \overline{GR}$, W is the midpt of $\overline{TY}$	1. Given
2. $\angle WTH$ and $\angle 1$ are comp, $\angle WYG$ and $\angle 2$ are comp	2. If the outer rays of 2 adj acute $\angle$s are $\perp$, then the $\angle$s are comp.
3. $\angle WTH \cong \angle WYG$	3. Comp $\angle$s of $\cong \angle$s are $\cong$.
4. $\overline{WT} \cong \overline{WY}$	4. Def of midpt
5. $TR - HR = TH$, $YR - GR = YG$	5. Seg Add Post
6. $TR - HR = YR - GR$	6. Equations may be subtracted.
7. $TH = YG$	7. Sub
8. $\triangle WTH \cong \triangle WYG$	8. SAS
9. $\overline{WH} \cong \overline{WG}$	9. CPCTC

5 CONGRUENCE

OVERVIEW

In this chapter, students prove and apply theorems about isosceles triangles. They write proofs for overlapping triangles and identify parts of right triangles. They use the Hypotenuse-Leg Theorem, identify altitudes and medians of triangles, and prove theorems about them. They prove and apply theorems about perpendicular bisectors and construct perpendicular segments. Students prove and apply the Triangle Inequality Theorem, and, given the lengths of 3 sides decide whether a triangle can be constructed. They are also asked to write paragraph proofs.

OBJECTIVES

- To solve problems involving isosceles triangles
- To draw conclusions from diagrams with labeled data
- To construct a line perpendicular to a given line and passing through a given point on the line
- To determine whether a triangle can be constructed, given lengths of three sides
- To relate the measures of sides and angles of triangles with the symbol =, >, or <
- To prove relationships between segments and angles in complex figures
- To identify the hypotenuse, legs, altitudes, and median in a right triangle

PROBLEM SOLVING

Throughout Chapter 5, the problem solving strategy of Making a Model can be used to solve problems. Concrete models can help answer questions involving overlapping triangles (Lesson 5.2) by showing each triangle separately. Mathematical models can be used to discover theorems such as the Triangle Inequality Theorem (Lesson 5.8). Lessons 5.1 and 5.7 provide exercises that give students an opportunity to use algebraic equations to solve problems.

TECHNOLOGY

Computer: A Computer Investigation on page 193 enables students to discover properties of altitudes and medians of triangles. This activity can also be extended to introduce Lesson 5.6 on perpendicular bisectors.

Calculator: You may want to encourage students to use a calculator in Lesson 5.8, Exercises 21 and 22, to facilitate computations.

SPECIAL FEATURES

Mixed Review pp. 181, 183, 187, 198, 204, 209, 214, 219
Brainteaser pp. 187, 214
Midchapter Review p. 192
Computer Investigation: Altitudes and Medians of a Triangle p. 193
Focus on Reading pp. 202, 217
Application: Congruent Relationships p. 204
Algebra Review pp. 209, 219
Key Terms p. 220
Key Ideas and Review Exercises pp. 220, 221
Chapter 5 Test p. 222
College Prep Test p. 223

PLANNING GUIDE

Lesson	Basic	Average	Above Average	Resources
5.1 pp. 179–181	CE all WE 1–17 odd	CE all WE 1–24 odd	CE all WE 1–27 odd	Reteaching p. 67 Practice p. 68
5.2 pp. 183	CE all WE 1–6	CE all WE 1–21 odd	CE all WE 1–21	Reteaching p. 69 Practice p. 70
5.3 pp. 186–187	CE all WE 1–9 Brainteaser	CE all WE 1–17 odd Brainteaser	CE all WE 1–19 odd Brainteaser	Reteaching p. 71 Practice p. 72
5.4 pp. 190–193	CE all WE 1–8 Midchapter Review CI all	CE all WE 1–18 odd Midchapter Review CI all	CE all WE 1–21 odd Midchapter Review CI all	Reteaching p. 73 Practice p. 74
5.5 pp. 197–198	CE all WE 1–13 odd	CE all WE 1–18 odd	CE all WE 1–21 odd	Reteaching p. 75 Practice p. 76
5.6 pp. 202–204	FR all CE all WE 1–5 Application all	FR all CE all WE 1–8 Application all	FR all CE all WE 1–10 Application all	Reteaching p. 77 Practice p. 78
5.7 pp. 208–209	CE all WE 1–9 AR all	CE all WE 1–12 AR all	CE all WE 1–17 AR all	Reteaching p. 79 Practice p. 80
5.8 pp. 212–214	CE all WE 1–13 Brainteaser	CE all WE 1–20 odd Brainteaser	CE all WE 1–26 odd Brainteaser	Reteaching p. 81 Practice p. 82
5.9 pp. 217–219	FR all CE all WE 1–9 AR all	FR all CE all WE 1–13 AR all	FR all CE all WE 1–16 AR all	Reteaching p. 83 Practice p. 84
Chapter Review pp. 220–221	all	all	all	
Chapter Test p. 222	1–11, 13, 14	1–11	all	
College Prep Test	all	all	all	

CE = Classroom Exercises   WE = Written Exercises   FR = Focus on Reading   AR = Algebra Review   CI = Computer Investigation

NOTE: For each level, all students should be assigned all Mixed Review exercises.

INVESTIGATION

Manipulatives: Each pair of students makes one proportional divider using two strips of posterboard of equal length (8 inches by $\frac{1}{2}$ inch is a good size). Punch a hole in each strip two inches from the end and join the two strips with a metal fastener or brad.

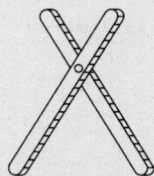

Working with partners, open the strips to form an angle. Place the divider on a sheet of paper and trace the angle formed by the longer legs onto the paper. With a straightedge, complete the drawing of the triangle on the paper. Repeat this by changing the opening of the dividers and drawing several different triangles. Measure all three angles of each triangle. This activity can help the students to discover Theorem 5.1.

Cut out three of the triangles. On one triangle, fold the triangles to find the midpoints of the sides, and then draw the medians.

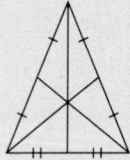 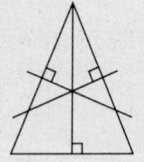 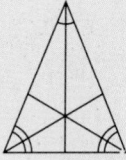

| medians | perpendicular bisectors | angle bisectors |

On another triangle, fold to find the midpoints of the sides and then use a protractor to draw the perpendicular bisectors of all three sides. On the third triangle, fold to find the angle bisectors. These investigations lead to the relationships of Lesson 5.6.

Albert Einstein (1879–1955)—Although he failed algebra, Einstein developed the Theory of Relativity that forms the foundation for modern physics and space exploration. He played the violin to relax.

More About the Mathematician

While in school, Einstein acquired the questioning attitude towards assumptions that helped him develop his general theory of relativity. This theory says that the speed of light is absolute while space and time intervals are relative. Einstein used the example of a moving train, relative to a passenger and an observer outside the train, to explain his theory. He was a self-taught mathematical genius whose statement that the energy content (E) of a body can be measured by its mass (m) times the square of the speed of light (c) is the most famous equation of the 20th century. He also developed the idea of the existence of a curved universe and of time as the fourth dimension.

5.1 Isosceles Triangles

Objectives
To prove and apply theorems about isosceles triangles
To solve problems involving isosceles triangles

Recall that an isosceles triangle has
at least two congruent sides.
Each of these sides is called a leg.
The angle formed by the legs is
called the **vertex angle**. The side opposite the vertex angle is called the
base. The angles opposite the legs
are called the **base angles**.

It can be proved that base angles are congruent. The drawing of an
auxiliary segment within an isosceles triangle allows the proof to be
easily completed.

Theorem 5.1

If two sides of a triangle are congruent, then the angles opposite
these sides are congruent. (The base angles of an isosceles triangle are
congruent.)

Given: $\triangle ABC$, $\overline{AC} \cong \overline{BC}$
Prove: $\angle A \cong \angle B$

Plan
Construct $\overline{CD}$, the angle bisector of $\angle ACB$.
Then prove that $\triangle ADC \cong \triangle BDC$.

Proof

Statement	Reason
1. $\overline{AC} \cong \overline{BC}$	(S) 1. Given
2. Draw $\overline{CD}$, the bisector of $\angle ACB$, intersecting $\overline{AB}$ at D.	2. Every $\angle$, except a st $\angle$, has exactly one bis.
3. $\angle 1 \cong \angle 2$	(A) 3. Def of $\angle$ bis
4. $\overline{CD} \cong \overline{CD}$	(S) 4. Reflex Prop
5. $\triangle ADC \cong \triangle BDC$	5. SAS
6. $\therefore \angle A \cong \angle B$	6. CPCTC

Corollary

If a triangle is equilateral, then it is also equiangular, and the measure of each angle is 60.

5.1 Isosceles Triangles **177**

Teaching Resources

Manipulative Worksheet 9
Problem Solving Worksheet 5
Project Worksheet 5
Quick Quizzes 34
**Reteaching and Practice
Worksheets,** pp. 67, 68
Technology 5
Transparencies 6A, 6B, 6C

GETTING STARTED

Prerequisite Quiz

Solve each equation.

1. $2x + 6 = 20$ 7
2. $x + (2x - 10) + (2x + 10) = 180$ 36
3. The measures of two angles of a triangle
   are 50 and 50. Find the measure of the
   third angle.   80
4. The three angles of a triangle have equal
   measures. Find the measure of each
   angle.   60
5. The measure of one angle of a triangle
   is 40. Find the measure of each of the
   other two angles if they are congruent.
   70

Motivator

Have your students construct triangle ABC
with $AB = AC = 6$ cm and $BC = 4$ cm.

Ask them what appears to be true about the
measures of $\angle B$ and $\angle C$. Then have them
verify by measuring these angles with a
protractor.

Have them complete the statement of the
following theorem suggested by this
construction: If two sides of a triangle are
congruent, then _____.   the angles
opposite these sides are congruent.

Lesson Note

Review the concept of "converse," which is used in the theorems of this lesson.
Use the Motivator to lead students to discover Theorem 5.1. It may be necessary to review the solution of linear equations.

Math Connections

Algebra: Lesson 5.1 introduces the properties of isosceles triangles and connects them with algebra by using algebraic equations to solve for the length of a missing side.

Critical Thinking Questions

Logic: What can you conclude about a given triangle $\triangle ABC$, if $\angle A \cong \angle B$?
$\triangle ABC$ is isosceles and $\overline{AC} \cong \overline{BC}$.

Common Error Analysis

Error: Students soon discover that if the measure of the vertex angle of an isosceles triangle is known, they can find the measure of each base angle without using algebra. They subtract the given measure from 180 and then take half of the result. However, they sometimes forget the last step.

Suggest that they add the three angles of the triangle to verify that the sum is 180, before continuing to the next exercise.

EXAMPLE 1 The measure of a base angle of an isosceles triangle is 10 less than twice the measure of the vertex angle. Find the measure of a base angle.

Plan Draw a diagram. Let x = measure of vertex angle. Then $2x - 10$ = measure of a base angle.

Solution

$$\mathrm{m} \angle A + \mathrm{m} \angle B + \mathrm{m} \angle C = 180$$
$$x + (2x - 10) + (2x - 10) = 180$$
$$5x - 20 = 180$$
$$5x = 200$$
$$x = 40$$
$$2x - 10 = (2 \cdot 40) - 10 = 70$$

Thus, the measure of a base angle is 70.

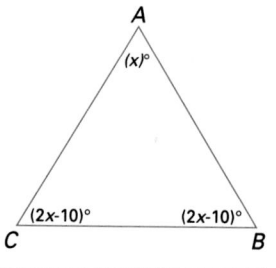

The converse of Theorem 5.1 is given below as Theorem 5.2.

Theorem 5.2 If two angles of a triangle are congruent, then the sides opposite these angles are congruent. (A triangle with two congruent sides is isosceles.)

The following corollary can be proved by using Theorem 5.2.

Corollary If a triangle is equiangular, then it is also equilateral.

The **perimeter** of a triangle is the sum of the lengths of its three sides. Example 2 applies the properties of isosceles triangles to a perimeter problem.

EXAMPLE 2 In $\triangle ABC$, $AC = BC$, $AB = 6$, and the perimeter is 20. Find AC.

Plan Draw $\triangle ABC$. Let x = length of $\overline{AC}$ and of $\overline{BC}$. Use the given perimeter to write an equation.

Solution

$$AC + BC + AB = 20$$
$$x + x + 6 = 20$$
$$2x + 6 = 20$$
$$x = 7$$

Thus, $AC = 7$.

Additional Example 1

The measure of the vertex angle of an isosceles triangle is 20. Find the measure of each base angle. 80

Additional Example 2

In $\triangle PQR$, $\angle P \cong \angle R$, $PR = 16$ and the perimeter is 36. Find QP. 10

EXAMPLE 3   Given: $\angle 3 \cong \angle 4$, D is the midpoint of $\overline{EC}$; $\angle 1 \cong \angle 2$
Prove: $\overline{EA} \cong \overline{CB}$

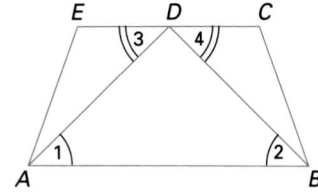

Plan   Show $\overline{EA} \cong \overline{CB}$ by proving $\triangle EDA \cong \triangle CDB$. $\angle 1$ and $\angle 2$ are *not* angles of those triangles; however, since $\angle 1 \cong \angle 2$, $\overline{AD} \cong \overline{BD}$.

Proof

Statement		Reason
1. $\angle 3 \cong \angle 4$	(A)	1. Given
2. D is midpt of $\overline{EC}$.		2. Given
3. $\overline{ED} \cong \overline{CD}$	(S)	3. Def of midpt
4. $\angle 1 \cong \angle 2$		4. Given
5. $\overline{AD} \cong \overline{BD}$	(S)	5. Sides opp $\cong$ $\angle$s of a $\triangle$ are $\cong$.
6. $\triangle EDA \cong \triangle CDB$		6. SAS
7. $\therefore \overline{EA} \cong \overline{CB}$		7. CPCTC

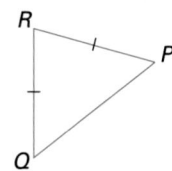

Classroom Exercises

1. Name the vertex angle, base, legs, and base angles of the isosceles triangle.   V: $\angle R$; B: $\overline{QP}$; L: $\overline{RQ}$, $\overline{RP}$; B $\angle$s: $\angle Q$, $\angle P$

Complete.

2. If $\overline{FG} \cong \overline{CG}$, then $\underline{\angle 1} \cong \underline{\angle 4}$.

3. If $\overline{EG} \cong \overline{ED}$, then $\underline{\angle 7} \cong \underline{\angle 3}$.

4. If $\angle 6 \cong \angle 5$, then $\underline{AG} \cong \underline{BG}$.

5. If $\angle 2 \cong \angle 7$, then $\underline{DG} \cong \underline{DE}$.

6. If $\overline{EG} \cong \overline{DG}$, then $\underline{\angle 2} \cong \underline{\angle 3}$.

7. If $\overline{AG} \cong \overline{BG}$, then $\underline{\angle 6} \cong \underline{\angle 5}$.

8. If $\angle 1 \cong \angle 4$, then $\underline{FG} \cong \underline{CG}$.

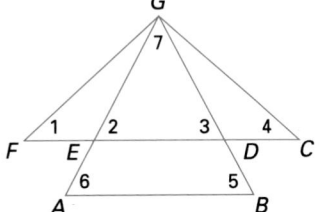

Refer to the figure below.

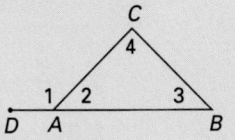

1. Given: $\overline{AC} \cong \overline{BC}$, m $\angle 4 = 100$. Find m $\angle 2$.   40
2. Given: $\overline{AC} \cong \overline{BC}$, $AB = 4$, and the perimeter of $\triangle ABC$ is 24. Find BC.   10
3. Given: $\overline{AC} \cong \overline{BC}$, m $\angle 1 = 150$. Find m $\angle 4$.   120
4. Given: $\overline{AC} \cong \overline{BC}$, m $\angle 4$ is three times m $\angle 3$. Find m $\angle 3$.   36
5. Given: $\angle 3$ is a supplement of $\angle 1$. Prove: $\overline{AC} \cong \overline{BC}$

Statement	Reason
1. $\angle 3$ supp $\angle 1$	1. Given
2. $\angle 2$ supp $\angle 1$	2. If outer rays of 2 adj $\angle$s form a straight $\angle$, the $\angle$s are supp.
3. $\angle 3 \cong \angle 2$	3. Supp of the same $\angle$ are $\cong$.
4. $\overline{AC} \cong \overline{BC}$	4. Sides opp $\cong$ $\angle$s of a $\triangle$ are $\cong$.

Closure

Ask your students to explain which angles of an isosceles triangle are the base angles and which sides of an isosceles triangle are the legs. Draw a triangle PYT given $PY = YT$ and ask them what conclusion they can give based upon the theorem in this lesson. $\angle P \cong \angle T$ Have them explain how they would find m $\angle P$ if m $\angle P = 2x - 10$ and m $\angle T = x + 30$.

▬▬ FOLLOW UP

Guided Practice

Classroom Exercises 1–8

Independent Practice

A Ex. 1–17, **B** Ex. 18–24, **C** Ex. 25–27

Basic: WE 1–17 odd
Average: WE 1–24 odd
Above Average: WE 1–27 odd

Additional Example 3

Given: $\overline{RT} \cong \overline{QT}$
Prove: $\angle 1 \cong \angle 2$

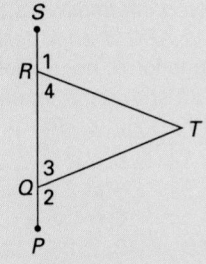

Proof: $\overline{RT} \cong \overline{QT}$ (Given); $\angle 3 \cong \angle 4$ ($\angle$s opp $\cong$ sides of a $\triangle$ are $\cong$); $\angle 1$ supp $\angle 4$, $\angle 3$ supp $\angle 2$ (if outer rays of 2 adj $\angle$s form a st $\angle$, the $\angle$s are supp); $\angle 1 \cong \angle 2$ (supp of $\cong$ $\angle$s are $\cong$)

Additional Answers

Written Exercises

14.

Statement	Reason
1. $\overline{AB} \cong \overline{BC}$	1. Given
2. $\angle 3 \cong \angle 1$	2. $\angle$s opp $\cong$ sides are $\cong$.
3. $\angle 2 \cong \angle 1$	3. Vert $\angle$s are $\cong$.
4. $\angle 3 \cong \angle 2$	4. Trans Prop of Congr

15.

Statement	Reason
1. $\angle 4 \cong \angle 2$	1. Given
2. $\angle 1 \cong \angle 2$	2. Vert $\angle$s are $\cong$.
3. $\angle 4 \cong \angle 1$	3. Trans Prop of Congr
4. $\overline{AB} \cong \overline{AC}$	4. Sides opp $\cong$ $\angle$s are $\cong$.

18.

Statement	Reason
1. $\overline{BA} \parallel \overline{CD}$, $\overline{CB} \cong \overline{DB}$	1. Given
2. $\angle C \cong \angle D$	2. $\angle$s opp $\cong$ sides are $\cong$.
3. $\angle 1 \cong \angle C$	3. If lines are $\parallel$, then corr $\angle$s are $\cong$.
4. $\angle 2 \cong \angle D$	4. If lines are $\parallel$, then alt int $\angle$s are $\cong$.
5. $\angle 1 \cong \angle 2$	5. Trans Prop of Congr

19.

Statement	Reason
1. $\overline{QV} \cong \overline{RW}$, $\overline{PQ} \cong \overline{PR}$, $\overline{PS} \cong \overline{PT}$	1. Given
2. $\angle Q \cong \angle R$	2. $\angle$s opp $\cong$ sides are $\cong$.
3. $PQ - PS = PR - PT$	3. Equations may be subtracted.
4. $PQ - PS = SQ, PR - PT = TR$	4. Seg Add Post
5. $SQ = TR$	5. Sub
6. $\triangle SQV \cong \triangle TRW$	6. SAS
7. $\angle 1 \cong \angle 2$	7. CPCTC

20.

Statement	Reason
1. $\overline{DA} \cong \overline{DB}$, $\angle 1 \cong \angle 2$	1. Given
2. m $\angle DAB =$ m $\angle DBA$	2. $\angle$s opp $\cong$ sides are $\cong$.
3. m $\angle 1 +$ m $\angle DAB =$ m $\angle 2 +$ m $\angle DBA$	3. Equations may be added.
4. m $\angle 1 +$ m $\angle DAB =$ m $\angle CAB$, m $\angle 2 +$ m $\angle DBA =$ m $\angle CBA$	4. Angle Add Post
5. m $\angle CAB =$ m $\angle CBA$	5. Sub
6. $\overline{AC} \cong \overline{BC}$	6. Sides opp $\cong$ $\angle$s are $\cong$.

180

Written Exercises

Find the measure of each angle of the triangle.

1.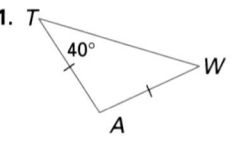

m $\angle W = 40$, m $\angle A = 100$

2.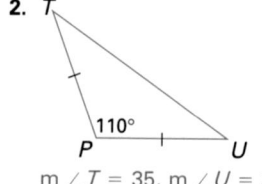

m $\angle T = 35$, m $\angle U = 35$

3.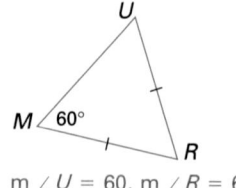

m $\angle U = 60$, m $\angle R = 60$

In $\triangle ABC$, $AC = BC$. Find the measure of each angle of the triangle. (Exercises 4–5)

4. m $\angle C = x$, m $\angle A = 2x - 20$ 44, 68, 68

5. m $\angle C = 2x + 20$, m $\angle A = 6x + 10$ 40, 70, 70

6. The measure of the vertex angle of an isosceles triangle is twice the measure of a base angle. Find the measure of a base angle.   45

7. The measure of the vertex angle of an isosceles triangle is 20 less than twice the measure of a base angle. Find the measure of the vertex angle.   80

8. Given: $\overline{BF} \cong \overline{CF}$, m $\angle 5 = 130$. Find m $\angle 4$.   25

9. Given: $\overline{BF} \perp \overline{FC}$, $\overline{BF} \cong \overline{FC}$. Find m $\angle 2$.   45

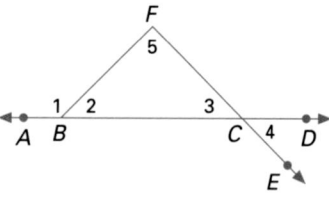

10. Given: $\overline{BC} \cong \overline{FC}$, m $\angle 4 = 30$. Find m $\angle 1$.   105

11. Given: $\overline{FB} \cong \overline{FC}$, $FB = 10$, $BC = 15$. Find the perimeter of $\triangle FBC$.   35

12. Given: $\angle 2 \cong \angle 3$, $BF = 7\frac{1}{2}$, $BC = 11$. Find the perimeter of $\triangle FBC$.   26

13. Given: $\overline{BF} \cong \overline{CF}$, $BC = 20$, the perimeter of $\triangle BFC$ is 48. Find FC.   14

14. Given: $\overline{AB} \cong \overline{BC}$
Prove: $\angle 3 \cong \angle 2$

15. Given: $\angle 4 \cong \angle 2$
Prove: $\overline{AB} \cong \overline{AC}$

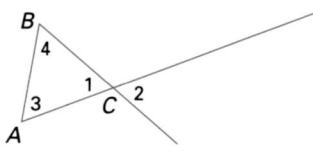

16. The perimeter of an equiangular triangle is 45 cm. Find the measure of each side.   15 cm

17. $\triangle XYZ$ is equiangular and XY is 27 cm. Find the perimeter of $\triangle XYZ$.   81 cm

---

Enrichment

Draw this figure on the chalkboard.

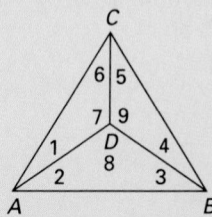

Have the students tell whether each set of given information is sufficient to prove that $\triangle ABC$ is equilateral. (If the information is not sufficient, the students should show counterexamples.)

1. $\overline{CD} \cong \overline{DA} \cong \overline{DB}$ ns
2. $\angle 1 \cong \angle 6$, $\angle 2 \cong \angle 3$, $\angle 4 \cong \angle 5$ ns
3. $\angle 7 \cong \angle 8 \cong \angle 9$ ns
4. $\angle 1 \cong \angle 2 \cong \angle 3 \cong \angle 4 \cong \angle 5 \cong \angle 6$ s
5. $\overline{AD} \cong \overline{DB}$, $\angle 1 \cong \angle 2$, $\angle 4 \cong \angle 3$ ns

18. Given: $\overline{BA} \parallel \overline{CD}$,
$\overline{CB} \cong \overline{DB}$
Prove: $\angle 1 \cong \angle 2$

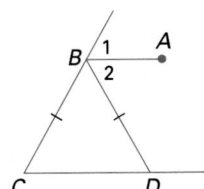

19. Given: $\overline{PQ} \cong \overline{PR}$,
$\overline{PS} \cong \overline{PT}$,
$\overline{QV} \cong \overline{RW}$
Prove: $\angle 1 \cong \angle 2$

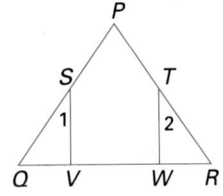

20. Given: $\overline{DA} \cong \overline{DB}$,
$\angle 1 \cong \angle 2$
Prove: $\overline{AC} \cong \overline{BC}$

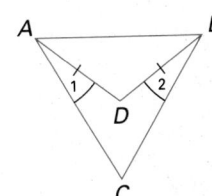

21. Prove the corollary to Theorem 5.1.
22. Prove Theorem 5.2. (HINT: Construct an auxiliary segment.)
23. $\triangle GHK$ is isosceles, with $\overline{GH}$ as the base. Point L is on $\overline{GH}$ such that $\overline{KL}$ bisects $\angle K$. m $\angle G = 40$. Find m $\angle GKL$ and m $\angle GLK$.   m $\angle GKL = 50$, m $\angle GLK = 90$
24. Prove the corollary to Theorem 5.2.
25. Prove: In an isosceles triangle, the bisector of an exterior angle of the vertex angle is parallel to the base.
26. Given: $\overline{BA}$ is perpendicular to every line in plane $\mathcal{M}$ passing through A; $\angle ACB \cong \angle ADB$
Prove: $\triangle BCD$ is isosceles.
27. Given: $\overline{AE} \perp \overline{CD}$, E is the midpoint of $\overline{CD}$; $\overline{BA}$ is perpendicular to every line in plane $\mathcal{M}$ passing through A.
Prove: $BC = BD$

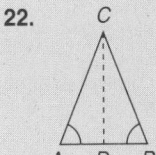

Mixed Review

1. The measure of an angle is twice the measure of its complement. Find the measure of the angle.   *1.6* 60
2. $\angle A$ and $\angle B$ are vertical angles. m $\angle A = 4x - 40$, m $\angle B = 2x + 60$. Find m $\angle A$.   *2.7* 160
3. The measure of an exterior angle of a triangle is 150. The measure of one of the two remote interior angles is $\frac{2}{3}$ the measure of the other. Find the measure of each remote interior angle.   *3.7* 60, 90

5.1 Isosceles Triangles   **181**

21.

Statement	Reason
1. $\triangle ABC$ is equilateral.	1. Given
2. $\overline{AB} \cong \overline{AC}$, $\overline{CA} \cong \overline{CB}$	2. Def of equil $\triangle$
3. $\angle C \cong \angle B$, $\angle B \cong \angle A$	3. $\angle$s opp $\cong$ sides are $\cong$.
4. $\angle C \cong \angle A$	4. Trans Prop of Congr
5. m $\angle C$ + m $\angle B$ + m $\angle A = 180$	5. Sum of meas of $\angle$s of $\triangle = 180$.
6. m $\angle B$ + m $\angle B$ + m $\angle B = 180$, or $3 \times$ m $\angle B = 180$	6. Sub
7. m $\angle B = 60$	7. Div Prop of Eq
8. m $\angle A = 60$, m $\angle C = 60$	8. Sub

22.

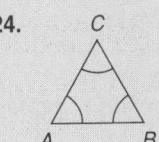

Statement	Reason
1. m $\angle A$ = m $\angle B$	1. Given
2. Draw $\overline{CD}$, the bis of $\angle ACB$, intersecting $\overline{AB}$ at D.	2. Every $\angle$, except a st $\angle$, has exactly one bis.
3. m $\angle ACD \cong \angle BCD$	3. Def $\angle$ bis
4. $\overline{CD} \cong \overline{CD}$	4. Reflex Prop of Eq
5. $\triangle ACD \cong \triangle BCD$	5. AAS
6. $\overline{AC} \cong \overline{BC}$	6. CPCTC

24.

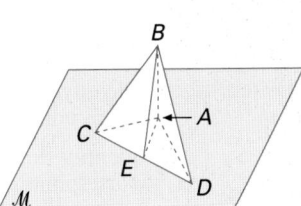

Statement	Reason
1. $\triangle ABC$ is equiangular.	1. Given
2. m $\angle A$ = m $\angle B$, m $\angle A$ = m $\angle C$	2. Def equiangular $\triangle$
3. $CA = CB$, $CB = AB$	3. Sides opp $\cong$ $\angle$s are $\cong$.
4. $CA = AB$	4. Sub
5. $\triangle ABC$ is equil	5. Def of equil $\triangle$

See page 651 for the answers to Ex. 25–27.

181

Prerequisite Quiz

Given: △*ADF* ≅ △*PMG*
Draw the two triangles and complete each of the following.

1. $\overline{AF}$ ≅ _____ $\overline{PG}$
2. ∠*P* ≅ _____ ∠*A*
3. $\overline{GM}$ ≅ _____ $\overline{FD}$
4. $\overline{MP}$ ≅ _____ $\overline{DA}$
5. ∠*F* ≅ _____ ∠*G*
6. ∠*M* ≅ _____ ∠*D*

Motivator

Draw the figure in the example on the chalkboard. Have your students copy the drawing on paper and then shade △*QSU*. Ask them which △ appears ≅ to this triangle. △*RTU* Have them recopy the figure, and shade △*QST*. Ask them which △ appears ≅ to this triangle. △*RTS*

━━━━**TEACHING**
      SUGGESTIONS

Lesson Note

Emphasize the importance of shading and of drawing triangles separately. A useful teaching technique is to show overhead projections with pairs of congruent triangles in different colors.

Math Connections

Quilting: Lap quilting is the technique of joining three layers of material together to form square blocks which are then sewn together to make a quilt. Each block is made up of smaller shapes that are pieced together to form a pattern. Although there are a variety of designs that you can create, the most common consist of geometric figures which often have overlapping triangles.

5.2 Introduction to Overlapping Triangles

Objective To identify and name overlapping triangles

When triangles overlap, redrawing them separately may help identify those that are required in a problem or proof.

EXAMPLE Identify the triangles that contain ∠1 and ∠2 as corresponding angles.

Identify the triangles that contain $\overline{QS}$ and $\overline{RT}$ as corresponding sides.

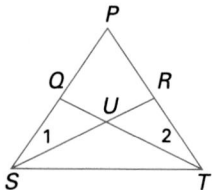

Solution Draw the three pairs of triangles separately.

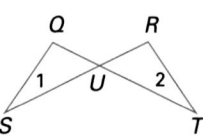 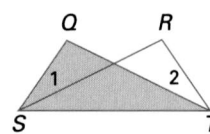 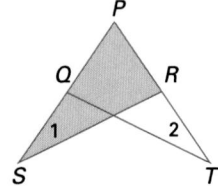

△*QSU* and △*RTU* contain $\overline{QS}$ and $\overline{RT}$, *and* ∠1 and ∠2.

△*QST* and △*RTS* contain $\overline{QS}$ and $\overline{TR}$. (∠1 is *not* an angle of △*QST*. ∠2 is *not* an angle of △*RTS*.)

△*PSR* and △*PTQ* contain ∠1 and ∠2. ($\overline{QS}$ is *not* a side of △*PSR*. $\overline{RT}$ is *not* a side of △*PTQ*.)

182   Chapter 5 Congruence

Additional Example

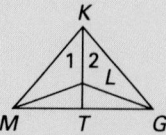

a. Identify the triangles that contain ∠1 and ∠2 as corresponding angles.
   △*MKL* and △*GKL*; △*MKT* and △*GKT*
b. Identify the triangles that contain $\overline{ML}$ and $\overline{GL}$ as corresponding sides.
   △*MKL* and △*GKL*; △*MLT* and △*GLT*

Classroom Exercises

Identify a triangle in the figure that appears to be congruent to the given triangle.

1. △ATB   △BRA
2. △SAR   △SBT
3. △TAO   △RBO
4. △BAR
5. △ORB
4. △ABT
5. △OTA

6. △YTW   △YUM
7. △YTM   △YUW
8. △MUY   △WTY
9. △WYU
10. △MTY
9. △MYT
10. △WUY

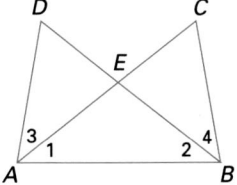

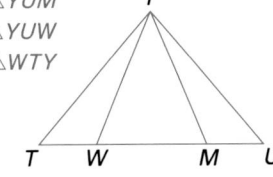

Written Exercises

Name each pair of triangles of which the given sides or angles are corresponding parts. Draw each pair of triangles separately.

1. ∠D and ∠C
2. $\overline{AD}$ and $\overline{BC}$
3. $\overline{ED}$ and $\overline{EC}$
4. ∠1 and ∠2
5. ∠3 and ∠4

6. ∠1 and ∠2
7. ∠5 and ∠6
8. $\overline{PQ}$ and $\overline{RQ}$
9. ∠3 and ∠4
10. $\overline{UV}$ and $\overline{SV}$
11. $\overline{VP}$ and $\overline{VR}$
12. $\overline{PS}$ and $\overline{RU}$
13. ∠UPR and ∠SRP

14. $\overline{IB}$ and $\overline{JA}$
15. $\overline{ID}$ and $\overline{JD}$
16. $\overline{EA}$ and $\overline{CB}$
17. $\overline{EB}$ and $\overline{CA}$
18. ∠1 and ∠2
19. $\overline{EH}$ and $\overline{CH}$
20. ∠AEC and ∠BCE
21. ∠DAJ and ∠DBI

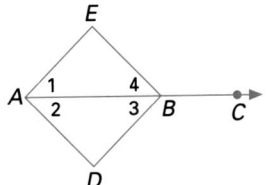

Mixed Review

1. Given: $\overline{AE} \cong \overline{AD}$, $\overline{AB}$ bisects ∠EAD.
   Prove: ∠E ≅ ∠D   4.3
2. Given: ∠EBC ≅ ∠DBC, ∠1 ≅ ∠2
   Prove: $\overline{AE} \cong \overline{AD}$ 4.6
3. Given: $\overline{AE} \parallel \overline{BD}$, $\overline{AE} \cong \overline{BD}$
   Prove: $\overline{AD} \cong \overline{BE}$ 4.3

Enrichment

Have the students draw regular hexagon *ABCDEF* with its nine diagonals. Then challenge the students to name all the triangles whose vertices are also vertices of the hexagon.

ABC, ABD, ABE, ABF, ACD, ACE, ACF, ADE, ADF, AEF, BCD, BCE, BCF, BDE, BDF, BEF, CDE, CDF, CEF, DEF

Critical Thinking Questions

Application: Show students the following block of patchwork for a quilt and have them answer the following question.

How many triangles are contained within this one block? Begin by counting the smallest triangles. Each quarter has 4 △s, so 4 · 4 = 16; the diagonals together form 4 more △s; each of the 2 diagonals divides the square into 2 △s, so 2 · 2 = 4. Thus, there are 24 triangles.

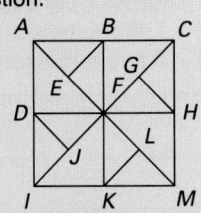

Checkpoint

Name each pair of triangles for which the given sides or angles are corresponding parts.

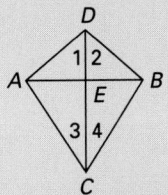

1. $\overline{AE}$, $\overline{BE}$ △AED, △BED; △AEC, △BEC
2. $\overline{AD}$, $\overline{BD}$ △ADE, △BDE; △ADC, △BDC
3. ∠1, ∠2   △ADE, △BDE, △ADC, △BDC

Closure

Ask your students to describe two methods for identifying corresponding parts of congruent triangles in figures involving overlapping triangles. Shade pairs of congruent triangles in different colors. Draw the pairs of triangles separately.

▇▇▇FOLLOW UP

Guided Practice

Classroom Exercises 1–10

Independent Practice

A Ex. 1–5, **B** Ex. 6–21

Basic: WE 1–5
Average: WE 1–21 odd
Above Average: WE 1–21 odd

Additional Answers

See page 193 for the answers to Written Ex. 1–21, and Mixed Review Ex. 1–3.

▰▰▰GETTING STARTED

Prerequisite Quiz

Name the pairs of triangles for which the given pairs of sides or angles are corresponding parts.

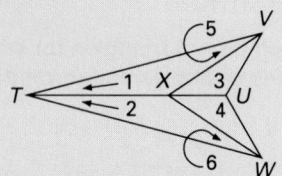

1. ∠1, ∠2   △TVX, △TWX; △TVU, △TWU
2. $\overline{TV}$, $\overline{TW}$ △TVX, △TWX; △TVU, △TWU
3. $\overline{XV}$, $\overline{XW}$ △TXV, △TXW; △XVU, △XWU
4. ∠5, ∠6   △TVX, △TWX

Motivator

Draw the figure in Example 1 on the chalkboard and have the students copy the figure. Tell them that *DA = DE*, and *AC = EC*, and have them shade the triangles containing these pairs of corresponding sides. Ask them to explain how they could prove that these triangles are congruent.
SSS

▰▰▰TEACHING SUGGESTIONS

Lesson Note

Overhead projections will help the students to see the three separate pairs of triangles in Example 2. Again, have students draw separate figures for triangles that overlap.

5.3 Congruence and Overlapping Triangles

Objective   To write proofs involving congruent overlapping triangles

EXAMPLE 1   Given: $\overline{BD}$ bisects ∠ADE; ∠1 ≅ ∠2
Prove: $\overline{AD}$ ≅ $\overline{ED}$

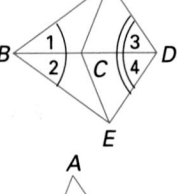

Plan   Consider the three pairs of triangles separately. Mark congruent parts.

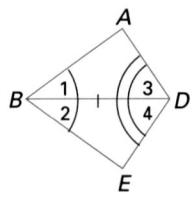 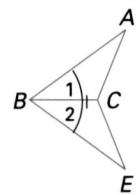 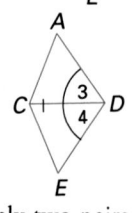

Three pairs of parts are marked. The triangles are congruent by ASA.

Only two pairs of parts are marked. This does not establish triangle congruence.

Only two pairs of parts are marked. This does not establish triangle congruence.

So, use △BAD ≅ △BED to prove $\overline{AD}$ ≅ $\overline{ED}$.

Proof

Statement	Reason
1. ∠1 ≅ ∠2	(A) 1. Given
2. $\overline{BD}$ bisects ∠ADE.	2. Given
3. ∠3 ≅ ∠4	(A) 3. Def of ∠ bis
4. $\overline{BD}$ ≅ $\overline{BD}$	(S) 4. Reflex Prop
5. △BAD ≅ △BED	5. ASA
6. ∴ $\overline{AD}$ ≅ $\overline{ED}$	6. CPCTC

The Reflexive Property of Congruence can be used to state that angles common to two triangles are congruent. In the figure below, ∠A is common to △ACE and △AFB.

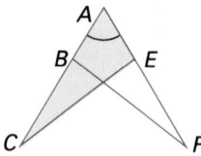 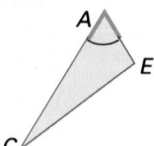 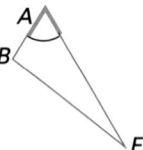

184   Chapter 5 Congruence

Additional Example 1

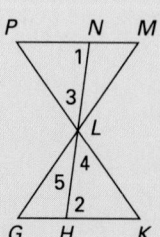

Given: *L* is the midpoint of $\overline{PK}$, $\overline{LN}$ ≅ $\overline{LH}$
Prove: ∠1 ≅ ∠2

Proof: *L* is the midpt of $\overline{PK}$ (Given); $\overline{LP}$ ≅ $\overline{LK}$ (def of midpt); $\overline{LN}$ ≅ $\overline{LH}$ (Given); ∠3 ≅ ∠4 (vert ∠s are ≅); △PLN ≅ △KLH (SAS); ∠1 ≅ ∠2 (CPCTC)

Sometimes it may be necessary to prove a pair of triangles congruent so that their corresponding parts can be used to prove a *second* pair of triangles congruent. These procedures are illustrated in the next example.

EXAMPLE 2 Given: $\overline{DB}$ bisects $\angle ADC$; $\angle 3 \cong \angle 4$
Prove: $\angle 5 \cong \angle 6$

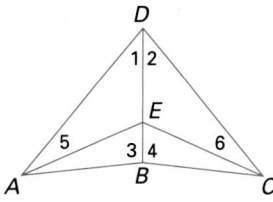

Plan Consider the pairs of triangles separately. Determine which pairs of triangles must be proved congruent to prove that $\angle 5 \cong \angle 6$.

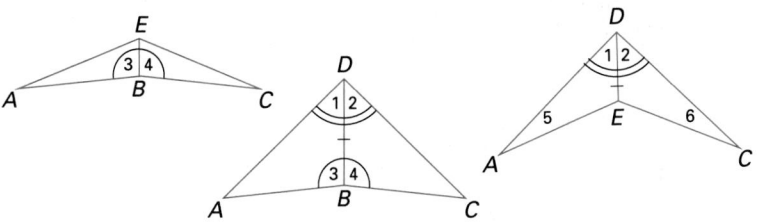

First prove $\triangle ADB \cong \triangle CDB$. Use this result to prove $\triangle ADE \cong \triangle CDE$. Then, it follows that $\angle 5 \cong \angle 6$ by CPCTC.

Proof

Statement	Reason
1. $\overline{DB}$ bisects $\angle ADC$.	1. Given
2. $\angle 1 \cong \angle 2$	**(A)** 2. Def of $\angle$ bis
3. $\angle 3 \cong \angle 4$	**(A)** 3. Given
4. $\overline{DB} \cong \overline{DB}$	**(S)** 4. Reflex Prop
5. $\triangle ADB \cong \triangle CDB$	5. ASA
6. $\overline{AD} \cong \overline{CD}$	**(S)** 6. CPCTC
7. $\overline{DE} \cong \overline{DE}$	**(S)** 7. Reflex Prop
8. $\angle 1 \cong \angle 2$	**(A)** 8. Given
9. $\triangle ADE \cong \triangle CDE$	9. SAS
10. $\angle 5 \cong \angle 6$	10. CPCTC

In more complex proofs such as the one above, it is sometimes helpful to draw a flow diagram before actually writing out the steps.

Flow Diagram

$\overline{DB}$ bisects $\angle ADC$. $\rightarrow \angle 1 \cong \angle 2$
$\angle 3 \cong \angle 4 \rightarrow \triangle ADB \cong \triangle CDB \rightarrow \overline{AD} \cong \overline{CD}$
$\overline{DB} \cong \overline{DB}$ $\angle 1 \cong \angle 2 \rightarrow \triangle ADE \cong \triangle CDE \rightarrow \angle 5 \cong \angle 6$
                    $\overline{DE} \cong \overline{DE}$

Math Connections

Problem Solving: Proving overlapping triangles congruent can be compared to writing a computer program, in which identifying intermediate goals can simplify the task. Flow diagrams for proofs and flow charts for programs both present an overview of a plan from which intermediate goals can be identified.

Critical Thinking Questions

Analysis: Refer students to Example 1 on page 184 and ask the following questions. Which triangles must be congruent in order to prove $\overline{AC} \cong \overline{EC}$? $\triangle ACB \cong \triangle ECB$ or $\triangle ACD \cong \triangle ECD$ In Example 1, $\triangle BAD$ is proved congruent to $\triangle BED$. Explain how to prove congruence in the two pairs of triangles contained within these triangles. $\overline{AB} \cong \overline{EB}$ by CPCTC and $\triangle ABC \cong \triangle EBC$ by SAS; $\overline{AD} \cong \overline{ED}$ by CPCTC and $\triangle ACD \cong \triangle ECD$ by SAS.

Common Error Analysis

Error: Students may fail to see that corresponding parts which are used to prove one pair of triangles congruent can be used *again* as corresponding parts of a second pair of triangles.

Some students find it helpful to repeat a step for the second pair of triangles. For example, in Additional Example 2, the step $\overline{CB} \cong \overline{AB}$ could be repeated before stating that $\triangle CEB \cong \triangle AEB$.

Additional Example 2

Given: $\overline{CD} \cong \overline{AD}$, $\overline{CB} \cong \overline{AB}$
Prove: $\angle 1 \cong \angle 2$

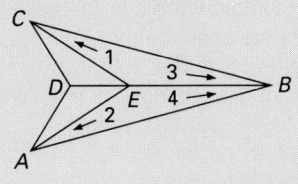

Proof: $\overline{CD} \cong \overline{AD}$ (Given); $\overline{CB} \cong \overline{AB}$ (Given); $\overline{DB} \cong \overline{DB}$ (Reflex Prop of Congr); $\triangle CDB \cong \triangle ADB$ (SSS); $\angle 3 \cong \angle 4$ (CPCTC); $\overline{ED} = \overline{EB}$ (Reflex Prop of Eq); $\triangle CEB \cong \triangle AEB$ (SAS); $\angle 1 \cong \angle 2$ (CPCTC)

Checkpoint

Use the figure below.

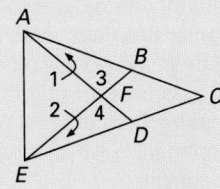

1. Given: $\angle 1 \cong \angle 2$, $\overline{AF} \cong \overline{EF}$
Prove: $\overline{FB} \cong \overline{FD}$

Statement	Reason
1. $\angle 1 \cong \angle 2$, $\overline{AF} \cong \overline{EF}$	1. Given
2. $\angle 3 \cong \angle 4$	2. vert $\angle$s $\cong$
3. $\triangle AFB \cong \triangle EFD$	3. ASA
4. $\overline{FB} \cong \overline{FD}$	4. CPCTC

2. Given: $\overline{AC} \cong \overline{EC}$, $\overline{DC} \cong \overline{BC}$
Prove: $\overline{AD} \cong \overline{EB}$

Statement	Reason
1. $\overline{AC} \cong \overline{EC}$, $\overline{DC} \cong \overline{BC}$	1. Given
2. $\angle C \cong \angle C$	2. Reflex Prop of Congr
3. $\triangle ACD \cong \triangle ECB$	3. SAS
4. $\overline{AD} \cong \overline{EB}$	4. CPCTC

Closure

Explain why it is sometimes necessary to prove two pairs of triangles congruent to prove a required pair of angles congruent. What are the steps in such a procedure? Have students draw a figure in which two angles are congruent by the Reflexive Property of Equality.

▰▰FOLLOW UP

Guided Practice

Classroom Exercises 1–7

Independent Practice

A Ex. 1–10, **B** Ex. 11–17, **C** Ex. 18–19

Basic: WE 1–10 odd, Brainteaser
Average: WE 1–17 odd, Brainteaser
Above Average: WE 1–19 odd, Brainteaser

Classroom Exercises

Name pairs of triangles that can be proved congruent from the given data and state the congruence pattern used as a reason.

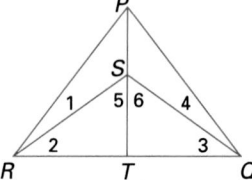

1. **Given:** $\overline{PR} \cong \overline{PS}$, $\angle 1 \cong \angle 2$
2. **Given:** $\angle 3 \cong \angle 4$, $\angle 5 \cong \angle 6$ $\triangle RTQ \cong \triangle STQ$ (ASA)
3. **Given:** $\angle 7 \cong \angle 8$, $\angle 5 \cong \angle 6$ $\triangle RTQ \cong \triangle STQ$ (AAS)
4. **Given:** $\overline{PR} \cong \overline{PS}$, $\overline{TR} \cong \overline{TS}$ $\triangle PRT \cong \triangle PST$ (SSS)
5. **Given:** $\overline{TR} \cong \overline{TS}$, $\overline{QR} \cong \overline{QS}$ $\triangle RTQ \cong \triangle STQ$ (SSS)
6. **Given:** $\angle 9 \cong \angle 10$, $\overline{PR} \cong \overline{PS}$, $\overline{TR} \cong \overline{TS}$ $\triangle PRT \cong \triangle PST$ (SAS)
7. **Given:** $\overline{TR} \cong \overline{TS}$, $\angle PTR \cong \angle PTS$ $\triangle PRT \cong \triangle PST$ (SAS)
   1. $\triangle PRT \cong \triangle PST$ (SAS), $\triangle PRQ \cong \triangle PSQ$ (SAS)

Written Exercises

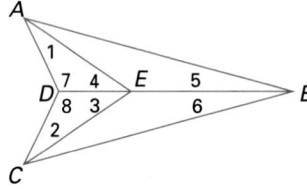

1. **Given:** $\angle 3 \cong \angle 4$, $\overline{AE} \cong \overline{CE}$
   Prove: $\overline{AD} \cong \overline{CD}$
2. **Given:** $\angle 7 \cong \angle 8$, $\overline{BD}$ bisects $\angle ABC$.
   Prove: $\angle DAB \cong \angle DCB$

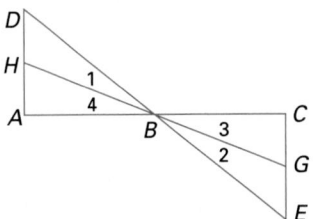

3. **Given:** $\overline{PS}$ bisects $\angle RPQ$; $\overline{PR} \cong \overline{PQ}$.
   Prove: $\overline{RS} \cong \overline{QS}$
4. **Given:** $\overline{PT} \perp \overline{RQ}$, $\overline{PT}$ bisects $\overline{RQ}$.
   Prove: $\overline{PT}$ bisects $\angle RPQ$.

5. **Given:** $\overline{AD} \parallel \overline{CE}$, $\overline{HA} \cong \overline{GC}$
   Prove: $\overline{HB} \cong \overline{GB}$
6. **Given:** $\overline{HG}$ and $\overline{DE}$ bisect each other.
   Prove: $\overline{HD} \cong \overline{GE}$
7. **Given:** B is the midpoint of $\overline{AC}$; $\overline{AD} \perp \overline{AC}$, $\overline{EC} \perp \overline{AC}$
   Prove: B is the midpoint of $\overline{DE}$.

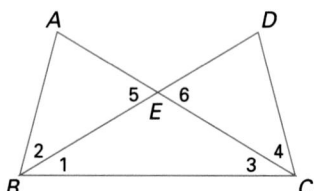

8. **Given:** $\angle 1 \cong \angle 3$, $\angle ABC \cong \angle DCB$
   Prove: $\overline{AB} \cong \overline{DC}$
9. **Given:** $\angle A \cong \angle D$, $\overline{AE} \cong \overline{DE}$
   Prove: $\overline{AB} \cong \overline{DC}$
10. **Given:** $\overline{AB} \cong \overline{DC}$, $\angle 1 \cong \angle 3$, $\angle 2 \cong \angle 4$
    Prove: $\overline{AC} \cong \overline{DB}$

Enrichment

Have the students draw an equilateral triangle 8 cm on a side. Then have them connect the midpoints of the sides of the triangle to form a second triangle inside the first. Next, have them draw a third triangle by connecting the midpoints of the sides of the second triangle.
Ask the students: (1) What type of triangles are the second and third triangles? Equilateral (2) What is the ratio of the perimeters of the first, second, and third triangles? 24:12:6, or 4:2:1 Ask them how the sum of the perimeters of the second, third, fourth, fifth..., triangles compares with the perimeter of the first triangle? The sum of the perimeters approaches the perimeter of the first triangle (24 cm).

11. Given: $\overline{PR}$ bisects $\angle SPT$; $\angle 3 \cong \angle 4$
    Prove: $\overline{QS} \cong \overline{QT}$
12. Given: $\overline{PR}$ bisects $\angle SRT$; $\overline{SR} \cong \overline{TR}$
    Prove: $\angle 1 \cong \angle 6$
13. Given: $\overline{PS} \cong \overline{PT}$, $\overline{SR} \cong \overline{TR}$
    Prove: $\angle 2 \cong \angle 5$

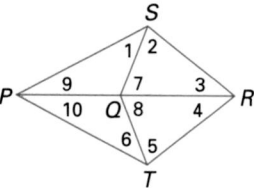

14. Given: $\overline{AD}$ and $\overline{CF}$ bisect each other.
    Prove: $\angle 5 \cong \angle 6$
15. Given: $\overline{AD}$ bisects $\overline{BE}$; $\angle 5 \cong \angle 6$
    Prove: $\overline{AC} \cong \overline{DF}$
16. Given: $\overline{AC} \parallel \overline{FD}$, $\overline{BC} \cong \overline{EF}$
    Draw a flow diagram to prove $\overline{AC} \cong \overline{FD}$.
17. Write the proof for the flow diagram of Exercise 16.

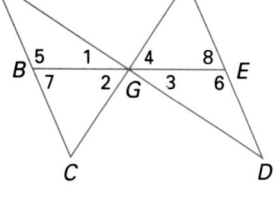

18. Given: $\triangle AED$ and $\triangle BCD$ are equilateral; D is the midpoint of $\overline{EC}$.
    Prove: $XE = XC$
19. Given: $AX = BX$, $CX = EX$
    Prove: $\overline{AB} \parallel \overline{EC}$

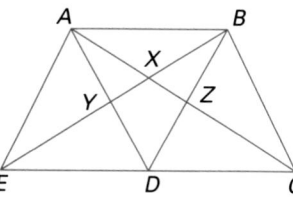

Mixed Review

Assume $\overline{PR} \parallel \overline{SU}$. *3.5, 3.7*

1. Given: m $\angle 2 = 40$
   Find m $\angle 3$.   40
2. Given: m $\angle 2 = 60$, m $\angle 5 = 50$
   Find m $\angle 4$.   110
3. Given: m $\angle 5 = 7x$, m $\angle 2 = 3x$, m $\angle 6 = 2x$
   Find m $\angle 2$.   45

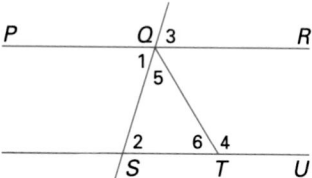

Brainteaser

$\triangle ABC$ is equilateral. M, N, and P are the midpoints of the sides of $\triangle ABC$. X, Y, and Z are the midpoints of the sides of $\triangle MNP$. D, E, and F are the midpoints of the sides of $\triangle XYZ$. What can you conclude about $\triangle DEF$? Explain your reasoning.

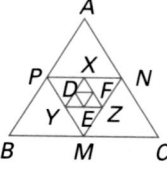

▰▰GETTING STARTED

Prerequisite Quiz

Give the congruence pattern for showing
△PQR ≅ △GHK. (Ex. 1-3)

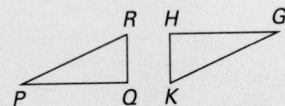

1. Given: $\overline{PQ} \perp \overline{RQ}$, $\overline{GH} \perp \overline{KH}$, $\overline{PQ} \cong \overline{GH}$, $\overline{RQ} \cong \overline{KH}$ SAS
2. Given: m ∠Q = 90, m ∠H = 90, m ∠R = m ∠K, $\overline{PR} = \overline{GK}$ AAS
3. Given: ∠Q is a right angle, ∠H is a right angle, ∠P ≅ ∠G, $\overline{PQ} \cong \overline{GH}$ ASA
4. Which of the following cannot, in general, be used to prove triangles congruent? C
  (A) SSS (B) AAS (C) SSA (D) SAS

Motivator

Draw a right triangle on the chalkboard. Ask your students if they can recall from previous math courses what the side **opposite** the right angle of a right triangle is called. Hypotenuse Ask them if they can recall what the other two sides are called.
Legs

5.4 Congruence of Right Triangles

Objectives   To identify the parts of a right triangle
         To write proofs involving congruence of right triangles

Definition   The sides forming the right angle of a right triangle are called **legs**. The side opposite the right angle is the **hypotenuse**.

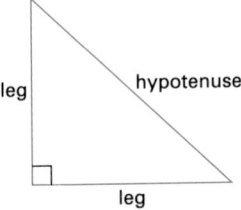

Right triangles can be proved congruent by any of the methods developed for other triangles.

Recall that the SAS congruency pattern requires that the corresponding angle be *included between* the sides. SSA does not, in general, guarantee the congruence of two triangles, as we have seen. However, the SSA pattern *can* be used for the special case of right triangles.

Theorem 5.3   **Hypotenuse-Leg (HL) Theorem:** Two right triangles are congruent if the hypotenuse and a leg of one are congruent, respectively, to the hypotenuse and corresponding leg of the other.

Given: Rt △s ABC and DEF with
     rt ∠s B and E, $\overline{AC} \cong \overline{DF}$,
     $\overline{BC} \cong \overline{EF}$
Prove: △ABC ≅ △DEF

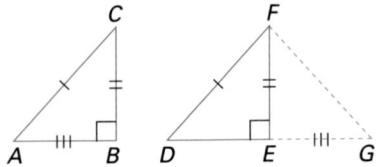

Plan   Extend $\overline{DE}$ to point G so that $\overline{AB} \cong \overline{EG}$. Prove △ABC ≅ △GEF. Then by CPCTC, (1) ∠G ≅ ∠A and (2) $\overline{AC} \cong \overline{GF}$. Show that △DGF is isosceles. Then, ∠D ≅ ∠G. From this and (1), ∠D ≅ ∠A. Then, △ABC ≅ △DEF by SAS.

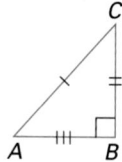
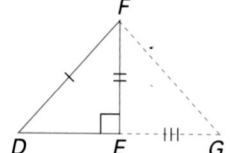

Proof

	Statement		Reason
	1. Rt △s *ABC* and *DEF* with rt ∠s *B* and *E*		1. Given
	2. Extend $\overline{DE}$ so that $\overline{EG} \cong \overline{BA}$.	(S)	2. Construction
	3. Draw $\overline{FG}$.		3. Any two points determine a line.
	4. $\overline{FE} \perp \overline{DG}$		4. Def of rt ∠
	5. ∠*FEG* is a rt ∠.		5. Def of rt ∠
	6. ∠*ABC* ≅ ∠*GEF*	(A)	6. All rt ∠s are ≅.
	7. $\overline{EF} \cong \overline{BC}$	(S)	7. Given
	8. △*ABC* ≅ △*GEF*		8. SAS
	9. $\overline{AC} \cong \overline{GF}$		9. CPCTC
	10. $\overline{AC} \cong \overline{DF}$	(S)	10. Given
	11. $\overline{GF} \cong \overline{DF}$		11. Trans Prop Cong
	12. ∠*D* ≅ ∠*G*		12. If 2 sides of a △ are ≅, then ∠s opp ≅ sides are ≅.
	13. ∠*A* ≅ ∠*G*		13. CPCTC
	14. ∠*A* ≅ ∠*D*	(A)	14. Trans Prop Cong
	15. ∠*B* ≅ ∠*DEF*	(A)	15. All rt ∠s are ≅.
	16. ∴ △*ABC* ≅ △*DEF*		16. AAS

EXAMPLE 1   Given: $\overline{PS} \perp \overline{QS}$, $\overline{PR} \perp \overline{QR}$, $\overline{PS} \cong \overline{QR}$
Prove: $\overline{PR} \cong \overline{QS}$

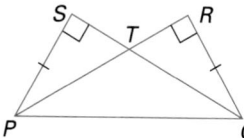

Proof

	Statement		Reason
	1. $\overline{PS} \perp \overline{QS}$, $\overline{PR} \perp \overline{QR}$		1. Given
	2. ∠*S* and ∠*R* are rt ∠s.		2. ⊥s form rt ∠s.
	3. △*PSQ* and △*QRP* are rt △s.		3. Def of rt △
	4. $\overline{PQ} \cong \overline{PQ}$	(H)	4. Reflex Prop
	5. $\overline{PS} \cong \overline{QR}$	(L)	5. Given
	6. △*PSQ* ≅ △*QRP*		6. HL
	7. ∴ $\overline{PR} \cong \overline{QS}$		7. CPCTC

Additional Example 1

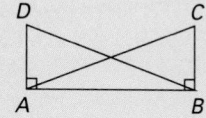

Given: $\overline{DA} \perp \overline{AB}$, $\overline{CB} \perp \overline{BA}$, $\overline{AC} \cong \overline{BD}$
Prove: ∠*D* ≅ ∠*C*

Proof: $\overline{DA} \perp \overline{AB}$, $\overline{CB} \perp \overline{BA}$ (Given);
∠*DAB* and ∠*CBA* are rt ∠s (def of ⊥);
△*DAB* and △*CBA* are rt △s (def of rt △); $\overline{AC} \cong \overline{BD}$ (Given); $\overline{AB} \cong \overline{BA}$ (Reflex Prop of Congr); △*DAB* ≅ △*CBA* (HL);
∠*D* ≅ ∠*C* (CPCTC)

■■■ **TEACHING SUGGESTIONS**

Lesson Note

Some teachers prefer to teach all special cases of right triangle congruence patterns, HL, HA, LL, and LA. In this book only the unique case, HL, is presented. For the other cases, methods that apply to any pairs of triangles are used: AAS (instead of HA), SAS (instead of LL), ASA or AAS (instead of LA).

Math Connections

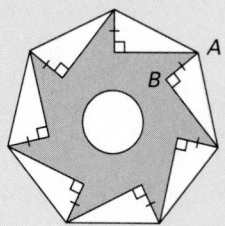

Saw Mills: A circular saw blade is made by cutting congruent right triangles from a 7-sided figure whose sides are all the same length. A right angle is important in keeping the blades from breaking off because the 90°-angle can withstand the pressure of cutting against an object.

Critical Thinking Questions

Application: Refer students to the diagram of the circular saw blade above and ask the following questions.

If $\overline{AB}$ is cut the same length for each tooth, explain why this guarantees that all the triangles removed will be congruent.   The congruent sides of the figure from which the saw is cut are the hypotenuse segments of the congruent right triangles that are removed. $\overline{AB}$ is one of the corresponding legs of the rt △s. Therefore, by the HL Thm, the △s cut away are all ≅.

Checkpoint

Use the figure below for Exercises 1 and 2.

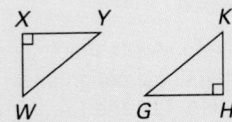

1. Given: $\overline{XY} \perp \overline{XW}$, $\overline{HG} \perp \overline{HK}$, $\overline{XY} \cong \overline{HG}$, $\overline{XW} \cong \overline{HK}$

Prove: $\angle W \cong \angle K$

Statement	Reason
1. $\overline{XY} \perp \overline{XW}$, $\overline{HG} \perp \overline{HK}$	1. Given
2. $\angle X$ and $\angle H$ are rt $\angle$s	2. Def of $\perp$
3. $\angle X \cong \angle H$	3. All rt $\angle$s are $\cong$
4. $\overline{XY} \cong \overline{HG}$, $\overline{XW} \cong \overline{HK}$	4. Given
5. $\triangle WXY \cong \triangle KHG$	5. SAS
6. $\angle W \cong \angle K$	6. CPCTC

2. Given: $\overline{XY} \perp \overline{XW}$, $\overline{HG} \perp \overline{HK}$, $\overline{WX} \cong \overline{KH}$, $\overline{WY} \cong \overline{KG}$

Prove: $\overline{WX} \cong \overline{KH}$

Statement	Reason
1. $\overline{XY} \perp \overline{XW}$, $\overline{HG} \perp \overline{HK}$	1. Given
2. $\angle X$ and $\angle Y$ are rt $\angle$s	2. Def of $\perp$
3. $\triangle WXY$ and $\triangle KHG$ are rt $\triangle$s	3. Def of rt $\triangle$
4. $\overline{WX} \cong \overline{KH}$, $\overline{WY} \cong \overline{KG}$	4. Given
5. $\triangle WXY \cong \triangle KHG$	5. HL
6. $\overline{WX} \cong \overline{KH}$	6. CPCTC

Closure

Ask your students to distinguish between the hypotenuse and the legs of a right triangle. The hypotenuse is the side opposite the right angle and the legs are the other two sides. Ask them to explain the special case in which the congruency pattern SSA can be used to prove right triangles congruent. By the Hypotenuse-Leg Theorem, which is a special case of SSA, two right triangles are congruent if the hypotenuse and leg of one right triangle are congruent to the hypotenuse and corresponding leg of the other right triangle.

EXAMPLE 2 Given: $\overline{VP} \perp \overline{VQ}$, $\overline{TR} \perp \overline{TS}$, $\overline{PR} \cong \overline{SQ}$, $\overline{QV} \cong \overline{RT}$

Prove: $\angle P \cong \angle S$

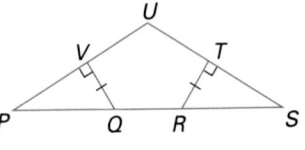

Plan Show $\overline{PQ} \cong \overline{SR}$. Then prove rt $\triangle PQV$ congruent to rt $\triangle SRT$.

Proof

Statement	Reason
1. $\overline{VP} \perp \overline{VQ}$, $\overline{TR} \perp \overline{TS}$	1. Given
2. $\overline{QV} \cong \overline{RT}$	(L) 2. Given
3. $\angle PVQ$ and $\angle STR$ are rt $\angle$s.	3. $\perp$s form rt $\angle$s.
4. $\triangle PVQ$ and $\triangle STR$ are rt $\triangle$s.	4. Def of rt $\triangle$
5. $PR = SQ$ ($\overline{PR} \cong \overline{SQ}$)	5. Given
6. $PR - QR = SQ - QR$	6. Subt Prop of Eq
7. $PR - QR = PQ$, $SQ - QR = SR$	7. Seg Add Post
8. $PQ = SR$ ($\overline{PQ} \cong \overline{SR}$)	(H) 8. Sub
9. $\triangle PVQ \cong \triangle STR$	9. HL for $\cong$ rt $\triangle$s
10. $\therefore \angle P \cong \angle S$	10. CPCTC

Classroom Exercises

Based on the markings, state whether the pair of right triangles can be proved congruent. If congruent, state one of the triangle-congruence theorems as a reason.

1.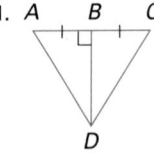

Yes; SAS

2.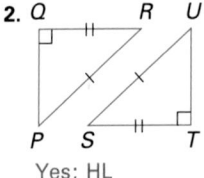

Yes; HL

3.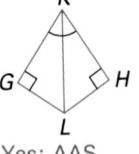

Yes; AAS

Written Exercises

1. Given: $\overline{QR} \perp \overline{QP}$, $\overline{HF} \perp \overline{HG}$, $\overline{QR} \cong \overline{HF}$, $\overline{QP} \cong \overline{HG}$

Prove: $\triangle RQP \cong \triangle FHG$

2. Given: $\overline{QR} \perp \overline{QP}$, $\overline{HF} \perp \overline{HG}$, $\overline{RQ} \cong \overline{FH}$, $\overline{RP} \cong \overline{FG}$

Prove: $\angle P \cong \angle G$

3. Given: $\overline{QR} \perp \overline{QP}$, $\overline{HF} \perp \overline{HG}$, $\angle R \cong \angle F$, $\overline{PR} \cong \overline{GF}$

Prove: $\overline{RQ} \cong \overline{FH}$

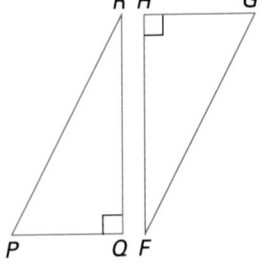

Additional Example 2

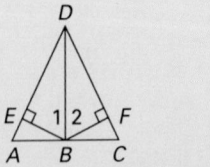

Given: $\overline{BE} \perp \overline{AD}$, $\overline{BF} \perp \overline{CD}$, $\overline{AE} \cong \overline{CF}$, $\overline{AD} \cong \overline{CD}$

Prove: $\angle 1 \cong \angle 2$

Proof: $\overline{BE} \perp \overline{AD}$, $\overline{BF} \perp \overline{CD}$ (Given); $\angle DEB$ and $\angle DFB$ are rt $\angle$s (Def of $\perp$); $\triangle DEB$ and $\triangle DFB$ are rt $\triangle$s (Def of rt $\triangle$); $\overline{AE} \cong \overline{CF}$ or $AE = CF$ (Given); $\overline{AD} \cong \overline{CD}$ or $AD = CD$ (Given); $AD - AE = CD - CF$ (Subtr Prop of Eq); $ED = FD$ or $\overline{ED} \cong \overline{FD}$ (Seg Add Post); $\overline{BD} \cong \overline{BD}$ (Reflex Prop of Congr); $\triangle DEB \cong \triangle DFB$ (HL); $\angle 1 \cong \angle 2$ (CPCTC)

4. Given: $\overline{AF} \perp \overline{AD}$, $\overline{DE} \perp \overline{AD}$,
$\angle F \cong \angle E$, $\overline{FB} \cong \overline{EC}$
Prove: $\triangle ABF \cong \triangle DCE$

5. Given: $\overline{AF} \perp \overline{AD}$, $\overline{DE} \perp \overline{AD}$,
$\overline{FA} \cong \overline{ED}$, $\overline{FB} \parallel \overline{EC}$
Prove: $\triangle ABF \cong \triangle DCE$

6. Given: $\overline{AF} \perp \overline{AD}$, $\overline{DE} \perp \overline{AD}$, B is the
midpoint of $\overline{AC}$; C is the mid-
point of $\overline{BD}$; $\overline{FB} \cong \overline{EC}$.
Prove: $\overline{AF} \cong \overline{DE}$

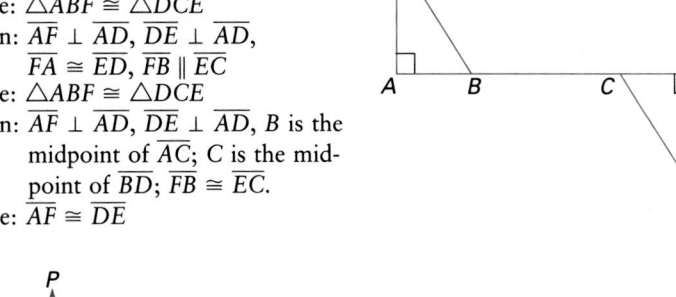

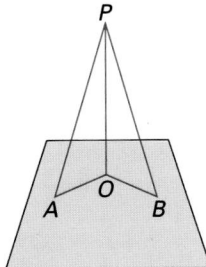

7. $\overline{OP}$ represents a TV antenna pole perpendicular to the plane of a roof (perpendicular to every line in the roof
passing through O). If guy wires $\overline{AP}$
and $\overline{BP}$ are congruent, prove that A
and B must be the same distance
from O.

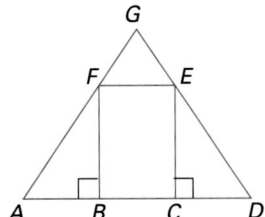

8. Rafters $\overline{AG}$ and $\overline{DG}$ are congruent.
At points F and E, equidistant from
G, braces $\overline{FB}$ and $\overline{EC}$ are constructed
perpendicular to $\overline{AD}$. Prove that the
braces must be congruent.

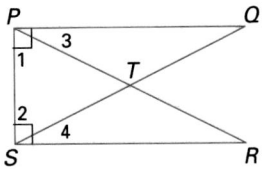

9. Given: $\overline{PQ} \perp \overline{PS}$, $\overline{SR} \perp \overline{SP}$, $\overline{PT} \cong \overline{ST}$
Prove: $\angle Q \cong \angle R$
10. Given: $\overline{PQ} \perp \overline{PS}$, $\overline{SR} \perp \overline{SP}$, $\overline{PR} \cong \overline{SQ}$
Prove: $\overline{PT} \cong \overline{ST}$
11. Given: $\overline{PQ} \perp \overline{PS}$, $\overline{SR} \perp \overline{SP}$, $\angle 3 \cong \angle 4$
Prove: $\angle Q \cong \angle R$

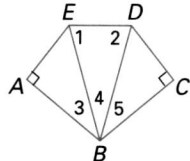

12. Given: $\overline{AB} \perp \overline{AE}$, $\overline{CB} \perp \overline{CD}$,
$\angle 1 \cong \angle 2$, $\overline{AB} \cong \overline{CB}$
Prove: $\overline{AE} \cong \overline{CD}$
13. Given: $\overline{AB} \perp \overline{AE}$, $\overline{CB} \perp \overline{CD}$,
$\angle ABD \cong \angle EBC$,
$\overline{AB} \cong \overline{CB}$
Prove: $\triangle EBD$ is isosceles.

5.4 Congruence of Right Triangles **191**

Enrichment

Have the students draw a 30-60-90 right
triangle. Then, without using a ruler, have
them prove that the side opposite the 30°
angle has a measure one-half that of the
hypotenuse. Tell them that they may use
an auxiliary line. If you wish, suggest that
they draw $\overline{CP}$, as shown below, such that
m $\angle 1 = 30$ and m $\angle 2 = 60$.

Since m $\angle 1 = 30$,
$\triangle APC$ is isosceles
with $AP = PC$.
Also, $\triangle PBC$ is
equilateral, with $PC = PB = BC$. Thus,
$AB = AP + PB = BC + BC$
$AB = 2 BC$, or $BC = \frac{1}{2}AB$

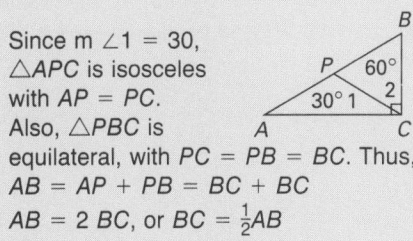

191

4.

Statement	Reason
1. $\overline{AF} \perp \overline{AD}$, $\overline{DE} \perp \overline{AD}$, $\angle F \cong \angle E$, $\overline{FB} \cong \overline{EC}$	1. Given
2. m $\angle FAB =$ 90, m $\angle EDC = 90$	2. Def of $\perp$
3. m $\angle FAB =$ m $\angle EDC$	3. Sub
4. $\triangle ABF \cong$ $\triangle DCE$	4. AAS

5.

Statement	Reason
1. $\overline{AF} \perp \overline{AD}$, $\overline{DE} \perp \overline{AD}$, $\overline{FA} \cong \overline{ED}$, $\overline{FB} \parallel \overline{EC}$	1. Given
2. m $\angle FAB =$ 90, m $\angle EDC = 90$	2. Def of $\perp$
3. m $\angle FAB =$ m $\angle EDC$	3. Sub
4. $\angle FBA \cong$ $\angle ECD$	4. If lines are $\parallel$, then alt ext $\angle$s are $\cong$.
5. $\triangle ABF \cong$ $\triangle DCE$	5. SAS

6.

Statement	Reason
1. $\overline{AF} \perp \overline{AD}$, $\overline{DE} \perp \overline{AD}$, $\overline{FB} \cong \overline{EC}$, B is midpt of $\overline{AC}$, C is midpt of $\overline{BD}$.	1. Given
2. $\triangle FAB$ and $\triangle EDC$ are rt $\triangle$s.	2. Def of rt $\triangle$
3. $\overline{AB} \cong \overline{BC}$, $\overline{BC} \cong \overline{CD}$	3. Def of midpt
4. $\overline{AB} \cong \overline{CD}$	4. Trans Prop of Congr
5. $\triangle FAB \cong$ $\triangle EDC$	5. HL
6. $\overline{AF} \cong \overline{DE}$	6. CPCTC

7.

Statement	Reason
1. $\overline{AP} \cong \overline{BP}$, $\overline{PO} \perp \overline{AO}$, $\overline{PO} \perp \overline{BO}$	1. Given
2. $\overline{PO} \cong \overline{PO}$	2. Reflex Prop of Congr
3. $\triangle POA \cong$ $\triangle POB$	3. HL
4. $\overline{AO} \cong \overline{BO}$	4. CPCTC

8.

Statement	Reason
1. $\overline{AG} \cong \overline{DG}$, $\overline{GF} \cong \overline{GE}$, $\overline{FB} \perp \overline{AD}$, $\overline{EC} \perp \overline{AD}$	1. Given
2. $AG - GF =$ $DG - GE$	2. Equations may be subtracted.
3. $AG - GF =$ AF, $DG - GE = DE$	3. Seg Add Post
4. $AF = DE$	4. Sub
5. $\angle A \cong \angle D$	5. $\angle$s opp $\cong$ sides are $\cong$.
6. $\triangle AFB \cong$ $\triangle DEC$	6. AAS
7. $\overline{FB} \cong \overline{EC}$	7. CPCTC

192

The methods stated below are used to prove right triangles congruent. Prove each one. (Exercises 14–16)

14. Hypotenuse-Acute Angle Method (HA): Two right triangles are congruent if the hypotenuse and an acute angle of one are congruent to the hypotenuse and the corresponding acute angle of the other.

15. Leg-Leg Method (LL): Two right triangles are congruent if the two legs of one are congruent to the corresponding legs of the other.

16. Based on the pattern of Exercises 14–15, write the meaning of the LA method for proving right triangles congruent. Prove it.

17. Draw a flow diagram for the proof in Exercise 18.

18. Given: $\overline{SP} \cong \overline{SQ}$, $\overline{PR} \perp \overline{SQ}$, $\overline{QT} \perp \overline{SP}$
Prove: $\triangle UPQ$ is isosceles.

19. Draw a flow diagram for the proof in Exercise 20.

20. Given: $\angle 3 \cong \angle 4$, $\overline{PR} \perp \overline{SQ}$, $\overline{QT} \perp \overline{SP}$
Prove: $\triangle SPQ$ is isosceles.

21. Given: $\overline{PR} \perp \overline{SQ}$, $\overline{QT} \perp \overline{SP}$, $\overline{UT} \cong \overline{UR}$
Prove: $\overline{SR} \cong \overline{ST}$

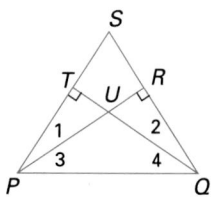

Midchapter Review

1. Given: $\overline{PR} \cong \overline{QR}$, m $\angle R = 50$
Find m $\angle Q$. **5.1** 65

2. Given: $\angle P \cong \angle Q$, $PR = 4x - 2$, $QR = 2x + 8$, $PQ = 12$. Find the perimeter of $\triangle PQR$. **5.1** 48

3. Given: $\angle P \cong \angle Q$, m $\angle R$ is 3 times m $\angle P$.
Find m $\angle Q$. **5.1** 36

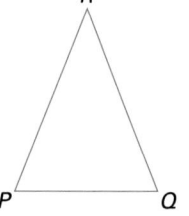

4. Name the triangles in the second figure on the right. **5.2** $\triangle QST$, $\triangle QRT$, $\triangle STP$, $\triangle RTP$, $\triangle QSP$, $\triangle QRP$

5. Given: $\overline{QS} \perp \overline{PS}$, $\overline{QR} \perp \overline{PR}$, $\angle 3 \cong \angle 4$
Prove: $\triangle QSP \cong \triangle QRP$ **5.3, 5.4**

6. Given: $\overline{SP} \cong \overline{RP}$, m $\angle QSP = 90$, m $\angle QRP = 90$
Prove: $\angle 3 \cong \angle 4$ **5.3, 5.4**

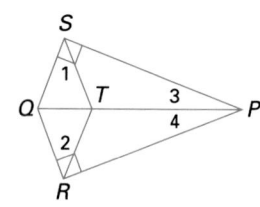

7. Given: $\overline{QP}$ bisects $\angle SQR$; $\angle 3 \cong \angle 4$, $\overline{ST} \cong \overline{RT}$
Prove: $\angle 1 \cong \angle 2$ **5.2, 5.3**

See pages 653–654 for Written Ex. 9–21 and Midchapter Review Ex. 5–7.

Computer Investigation

Altitudes and Medians of a Triangle

Use a computer to construct altitudes and medians of triangles by classification, and to measure the resulting figures.

In △*ABC*, $\overline{CM}$ is a segment from a vertex *C* to *M*, the midpoint of the opposite side $\overline{AB}$. $\overline{CM}$ is a **median** of △*ABC*.

In △*GHK*, $\overline{GP}$ is a segment from a vertex *G*, perpendicular to the opposite side $\overline{HK}$. $\overline{GP}$ is an **altitude** of △*GHK*.

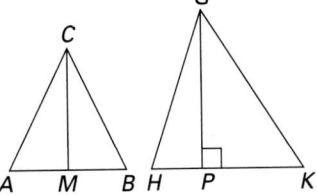

Activity 1

Draw an acute triangle.

1. Draw a median from each of the vertices.
2. Are the medians in the interior of the triangle? Do they intersect?
3. Are the medians congruent?
4. Determine if the median bisects the vertex angle from which it is drawn.

Answer questions 1–4 above for the following types of triangles which you are to draw using the computer.

5. obtuse
6. isosceles
7. equilateral

Activity 2

Draw an acute triangle.

8. Draw an altitude from each of the vertices.
9. Are the altitudes in the interior of the triangle? Do they intersect?
10. Is any altitude also a median?
11. Are the altitudes congruent?

Answer questions 8–11 above for the following types of triangles.

12. obtuse
13. isosceles
14. equilateral

Summary

Make some generalizations from the exercises above.

15. Are medians always in the interior of a triangle?
16. When do medians bisect the vertex angles from which they are drawn?
17. When will a median and an altitude be the same?

Additional Answers, page 183

Written Exercises

1. △*DAE*, △*CBE*; △*DBA*, △*CAB*
2. △*DAE*, △*CBE*; △*DBA*, △*CAB*
3. △*DAE*, △*CBE*
4. △*DBA*, △*CAB*
5. △*DAE*, △*CBE*
6. △*TUV*, △*TSV*; △*TPV*, △*TRV*; △*TQP*, △*TQR*
7. △*PUV*, △*RSV*
8. △*PVQ*, △*RVQ*; △*QTR*, △*QTP*
9. △*TUR*, △*TSP*; △*TUV*, △*TSV*
10. △*PUV*, △*RSV*; △*TUV*, △*TSV*
11. △*PVQ*, △*RVQ*; △*PVT*, △*RVT*; △*PVU*, △*RVS*
12. △*PSR*, △*RUP*; △*PST*, △*RUT*
13. △*TPR*, △*TRP*; △*UPR*, △*SRP*; △*TPQ*, △*TRQ*
14. △*IAB*, △*JBA*; △*DIB*, △*DJA*
15. △*IDE*, △*JDC*; △*IDB*, △*JDA*
16. △*EAB*, △*CBA*; △*EAI*, △*CBJ*; △*EAH*, △*CBH*; △*EAC*, △*CBE*; △*EAD*, △*CBD*
17. △*EBA*, △*CAB*; △*EBC*, △*CAE*; △*EBD*, △*CAD*; △*EBG*, △*FAC*
18. △*AEI*, △*BCJ*; △*AEH*, △*BCH*; △*AEB*, △*BCA*
19. △*EHA*, △*CHB*; △*EHC*, △*CHE*
20. △*AEC*, △*BCE*; △*AEF*, △*BCG*
21. △*IAH*, △*JBH*; △*DAJ*, △*DBI*; △*AFC*, △*BGE*; △*DAC*, △*DBE*

Mixed Review

1.

Statement	Reason
1. $\overline{AE} \cong \overline{AD}$, $\overline{AB}$ bis ∠*EAD*	1. Given
2. ∠1 ≅ ∠2	2. Def of ∠ bis
3. $\overline{AB} \cong \overline{AB}$	3. Reflex Prop of Congr
4. △*AEB* ≅ △*ADB*	4. SAS
5. ∠*E* ≅ ∠*D*	5. CPCTC

2.

Statement	Reason
1. ∠*EBC* ≅ ∠*DBC*, ∠1 ≅ ∠2	1. Given
2. ∠4 and ∠*EBC* are supp, ∠3 and ∠*DBC* are supp	2. If the outer rays of 2 adj ∠s form a st ∠, then the ∠s are supp.
3. ∠4 ≅ ∠3	3. supp ∠s of ≅ ∠s are ≅.
4. $\overline{AB} \cong \overline{AB}$	4. Reflex Prop of Congr
5. △*AEB* ≅ △*ADB*	5. ASA
6. $\overline{AE} \cong \overline{AD}$	6. CPCTC

3.

Statement	Reason
1. $\overline{AE} \parallel \overline{BD}$, $\overline{AE} \cong \overline{BD}$	1. Given
2. $\overline{AB} \cong \overline{AB}$	2. Reflex Prop of Congr
3. ∠1 ≅ ∠3	3. If lines are ∥, then alt int ∠s are ≅.
4. △*AEB* ≅ △*BDA*	4. SAS
5. $\overline{AD} \cong \overline{BE}$	5. CPCTC

193

▰▰▰GETTING STARTED

Prerequisite Quiz

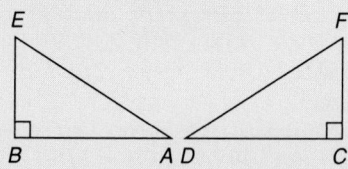

1. Given: $\overline{EB} \perp \overline{AB}$, $\overline{FC} \perp \overline{CD}$, $AB = DC$,
   $EB = FC$
   Prove: $\triangle ABE \cong \triangle DCF$ SAS
2. Given: $\overline{EB} \perp \overline{AB}$, $\overline{FC} \perp \overline{CD}$, m $\angle A =$
   m $\angle D$, $AB = DC$
   Prove: m $\angle E =$ m $\angle F$ ASA
3. Given: $\overline{EB} \perp \overline{AB}$, $\overline{FC} \perp \overline{CD}$, m $\angle E =$
   m $\angle F$, $AE = DF$
   Prove: $AB = DC$ AAS, CPCTC

Motivator

Ask the students the meaning of the word
median.   Middle   Ask students to
construct isosceles triangle ABC with $CA =$
CB. Then have them construct a segment
from vertex C to the midpoint P on $\overline{AB}$. Ask
them if $\overline{CP}$ seems to be perpendicular to
$\overline{AB}$.   Yes

▰▰▰TEACHING
  SUGGESTIONS

Lesson Note

Stress that all the ways for proving triangles
congruent still apply to proving right
triangles congruent. When you summarize,
point out that the only new method is HL.

5.5 Altitudes and Medians of Triangles

Objectives   To identify altitudes and medians of triangles
 To prove and apply theorems about altitudes and medians of triangles

Definition   A **median** of a triangle is a segment whose endpoints are a vertex of
 the triangle and the midpoint of the opposite side.

Every triangle has three medians,
one from each vertex. As shown
in the illustration, each median
bisects the opposite side.

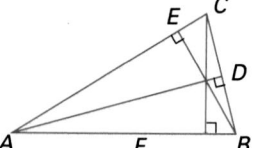

Notice that each median, except
for its endpoints, is in the interior
of the triangle. For convenience, this fact may be stated as "a median is
in the interior of a triangle."

Definition   An **altitude** of a triangle is a segment from a vertex of the triangle
 perpendicular to the line containing the opposite side.

Every triangle has three altitudes, one from
each vertex. In acute triangle ABC, as
shown at the right, all three altitudes are in
the interior of the triangle.

Unlike medians, altitudes are not necessarily in the interior of a
triangle. The cases for right and obtuse triangles are seen below.

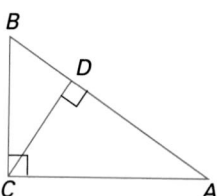   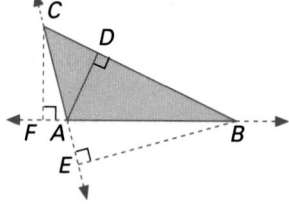

Right Triangle
Altitudes $\overline{BC}$ and $\overline{CA}$ are
legs of the triangle.

Obtuse Triangle
Altitudes $\overline{CE}$ and $\overline{BE}$ are in
the exterior of the triangle.

EXAMPLE 1   Which segments are altitudes of △*ABC*? Which are medians?

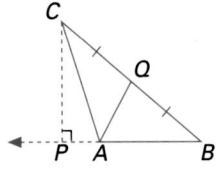

Solutions $\overline{CP}$ is an altitude. $\overline{BR}$ is a median.
$\overline{AQ}$ is a median. $\overline{SB}$ is an altitude.

Sometimes an altitude can also be a median. Such is the case for the altitude from the vertex angle of an isosceles triangle. This is demonstrated by the next theorem. Notice that the proof is presented in *paragraph form*. In a paragraph proof, reasons that are expected to be clear to the reader may be omitted for the sake of brevity.

Theorem 5.4

The altitude from the vertex angle to the base of an isosceles triangle is a median (the altitude bisects the base).

Given: △*ABC* is isosceles, with $\overline{AC} \cong \overline{BC}$;
$\overline{CD}$ is an altitude.
Prove: $\overline{CD}$ is a median.

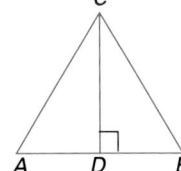

Proof (Paragraph Form)

Since $\overline{CD}$ is an altitude, $\overline{CD} \perp \overline{AB}$ and two right triangles are formed. $\overline{AC} \cong \overline{BC}$ because △*ABC* is isosceles. Also, $\overline{CD} \cong \overline{CD}$ by the Reflexive Property. Therefore, △*ADC* ≅ △*BDC* by HL. Then $\overline{AD} \cong \overline{BD}$ by CPCTC. So, *D* is the midpoint of $\overline{AB}$, and altitude $\overline{CD}$ is also a median.

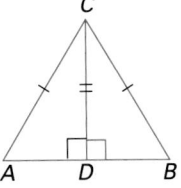

Medians that are drawn to corresponding sides of congruent triangles are called **corresponding medians**.

5.5 Altitudes and Medians of Triangles   **195**

Additional Example 1

Name an altitude.
Name a median.

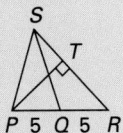

median: $\overline{SQ}$; altitude: $\overline{PT}$

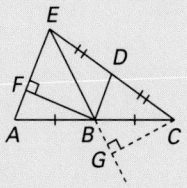

Closure

Ask your students the following questions. **What is a median of a triangle?** A segment whose endpoints are a vertex of the triangle and the midpoint of the opposite side. **What is an altitude of a triangle?** A segment from a vertex of the triangle perpendicular to the line containing the opposite side. **When can a median also be an altitude?** When it is perpendicular to the line containing the opposite side. **What do you know about an altitude from the vertex angle to the base of an isosceles triangle?** It is also a median.

 **FOLLOW UP**

Guided Practice

Classroom Exercises 1–8

Independent Practice

A Ex. 1–13, **B** Ex. 14–18, **C** Ex. 19–21

Basic: WE 1–13 odd
Average: WE 1–18 odd
Above Average: WE 1–21 odd

Additional Answers

Classroom Exercises

7. Yes. Endpts of a median are a vertex and the midpt of the opp side.
8. No. In an obtuse △, 2 altitudes lie in the ext of the △. In a rt △, 2 altitudes are legs.

Written Exercises

6.
Statement	Reason
1. $\overline{PR} \cong \overline{QR}$, $\overline{RS}$ is an alt.	1. Given
2. $\overline{PS} \cong \overline{QS}$	2. In an isos △, the alt from the vertex is a median.
3. $\overline{RS} \cong \overline{RS}$	3. Reflex Prop of Congr
4. $\triangle PSR \cong \triangle QSR$	4. SSS

7.
Statement	Reason
1. $\overline{RS}$ is a median, $\overline{PR} \cong \overline{QR}$	1. Given
2. $\overline{PS} \cong \overline{QS}$	2. Def of median
3. $\overline{RS} \cong \overline{RS}$	3. Reflex Prop of Congr
4. $\triangle PSR \cong \triangle QSR$	4. SSS

Theorem 5.5 | Corresponding medians of congruent triangles are congruent.

Given: $\triangle ABC \cong \triangle EDF$, $\overline{CP}$ is a median of $\triangle ABC$; $\overline{FQ}$ is a median of $\triangle EDF$.
Prove: $\overline{CP} \cong \overline{FQ}$

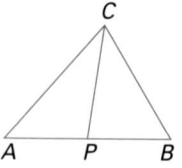 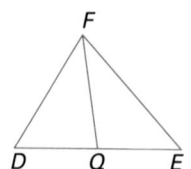

Proof

Statement	Reason
1. $\triangle ABC \cong \triangle EDF$, $\overline{CP}$ and $\overline{FQ}$ are medians.	1. Given
2. $AC = EF$ ($\overline{AC} \cong \overline{EF}$)	**(S)** 2. CPCTC
3. $\angle A \cong \angle E$	**(A)** 3. CPCTC
4. $AB = ED$ ($\overline{AB} \cong \overline{ED}$)	4. CPCTC
5. $AP = \frac{1}{2}AB$, $EQ = \frac{1}{2}ED$	5. Def of median
6. $\frac{1}{2}AB = \frac{1}{2}ED$	6. Mult Prop of Eq
7. $AP = EQ$	**(S)** 7. Sub
8. $\triangle APC \cong \triangle EQF$	8. SAS
9. $\therefore \overline{CP} \cong \overline{FQ}$	9. CPCTC

EXAMPLE 2 Given: $\triangle ABC \cong \triangle PQR$, $\overline{CM}$ and $\overline{RN}$ are corresponding medians; $CM = 9a - 2$, $RN = 4(a + 2)$
Find RN.

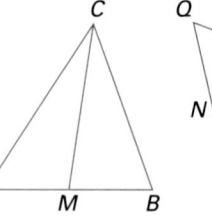 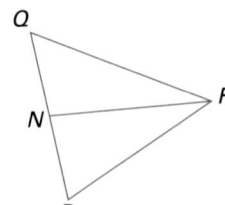

Solution $CM = RN$ by (Theorem 5.5)

$$9a - 2 = 4(a + 2)$$
$$9a - 2 = 4a + 8$$
$$5a - 2 = 8$$
$$5a = 10$$
$$a = 2$$

$RN = 4(a + 2) = 4(2 + 2) = 16$

Theorem 5.6 | Corresponding altitudes of congruent triangles are congruent.

196 Chapter 5 Congruence

Additional Example 2

Given: $\overline{LH}$ and $\overline{QS}$ are corresponding medians of congruent triangles, $LH = x^2$, $QS = 5x - 4$. Find QS.
16 or 1

Classroom Exercises

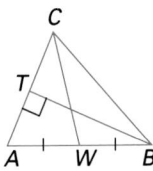
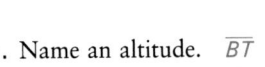
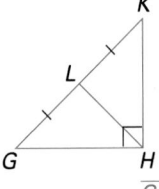
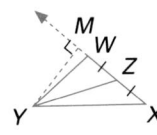

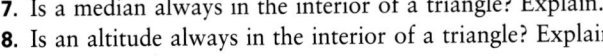
GH or KH

1. Name an altitude. BT
2. Name a median. CW
3. Name an altitude.
4. Name a median. HL
5. Name an altitude. YM
6. Name a median. YZ

7. Is a median always in the interior of a triangle? Explain.
8. Is an altitude always in the interior of a triangle? Explain.

Written Exercises

Complete the statement by referring to the diagram at right.

1. $\overline{DF}$ is a median of $\triangle ADG$.
2. $\triangle ADG$ has altitude AB .
3. $\triangle ABG$ has altitudes AB and GB .
4. $\triangle ABD$ has altitudes AB and DB .
5. The median to $\overline{AD}$ in $\triangle ADG$ is GC

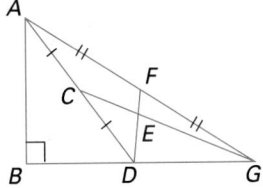

6. Given: $\overline{PR} \cong \overline{QR}$, $\overline{RS}$ is an altitude.
   Prove: $\triangle PSR \cong \triangle QSR$
7. Given: $\overline{RS}$ is a median; $\overline{PR} \cong \overline{QR}$.
   Prove: $\triangle PSR \cong \triangle QSR$
8. Given: $\overline{RS}$ is a median and an altitude.
   Prove: $\triangle PRQ$ is isosceles.

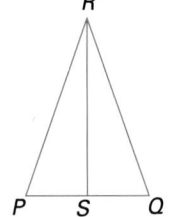

Find the indicated length. (Exercises 9–10)

9. Given: $\triangle TVS \cong \triangle XZW$,
   $\overline{SU}$ and $\overline{WY}$ are
   corresponding me-
   dians; $SU = 4x - 3$
   and $WY = 2x + 7$.
   Find SU. 17

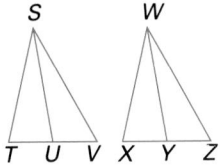

5.5 Altitudes and Medians of Triangles   **197**

8.

Statement	Reason
1. $\overline{RS}$ is a median and an alt.	1. Given
2. $\overline{PS} \cong \overline{QS}$	2. Def of median
3. $\angle PSR$ and $\angle QSR$ are rt $\angle$s.	3. Def of alt
4. $\angle PSR \cong \angle QSR$	4. All rt $\angle$s are $\cong$.
5. $\overline{RS} \cong \overline{RS}$	5. Reflex Prop of Congr
6. $\triangle PSR \cong \triangle QSR$	6. SAS
7. $\overline{PR} \cong \overline{QR}$	7. CPCTC
8. $\triangle PRQ$ is isos.	8. Def of isos $\triangle$

11.

Statement	Reason
1. $\overline{PR}$ is an alt of $\triangle PQS$.	1. Given
2. m $\angle 1$ + m $\angle 3$ + m $\angle PRQ$ = 180, m $\angle 2$ + m $\angle 4$ + m $\angle PRS$ = 180	2. Sum of meas of $\angle$s of $\triangle$ = 180.
3. m $\angle 2$ + m $\angle 4$ + m $\angle PRS$ = m $\angle 1$ + m $\angle 3$ + m $\angle PRQ$	3. Sub
4. $\angle PRS$ and $\angle PRQ$ are rt $\angle$s.	4. Def of alt
5. m $\angle PRS$ = m $\angle PRQ$	5. All rt $\angle$s are =.
6. m $\angle 2$ + m $\angle 4$ + m $\angle PRS$ = m $\angle 1$ + m $\angle 3$ + m $\angle PRS$	6. Sub
7. m $\angle 2$ + m $\angle 4$ = m $\angle 1$ + m $\angle 3$	7. Subt Prop of Eq

12.

Statement	Reason
1. $\overline{PQ} \perp \overline{PS}$, m $\angle 1$ = m $\angle 4$	1. Given
2. m $\angle 1$ + m $\angle 2$ = 90	2. Def of $\perp$
3. m $\angle 4$ + m $\angle 2$ = 90	3. Sub
4. m $\angle 4$ + m $\angle 2$ + m $\angle PRS$ = 180	4. Sum of meas of $\angle$s of $\triangle$ = 180.
5. 90 + m $\angle PRS$ = 180	5. Sub
6. m $\angle PRS$ = 90	6. Subt Prop of Eq
7. $\overline{PR}$ is an alt of $\triangle PQS$.	7. Def of alt

197

197

13.

Statement	Reason
1. $\overline{PR}$ is a median, $\angle 3 \cong \angle 4$	1. Given
2. $QR = SR$	2. Def of median
3. $\overline{PQ} \cong \overline{PS}$	3. Sides opp $\cong$ $\angle$s are $\cong$.
4. $\triangle PQR \cong \triangle PSR$	4. SAS
5. $\angle 1 \cong \angle 2$	5. CPCTC
6. $\overline{PR}$ bis $\angle QPS$	6. Def of $\angle$ bis

14.

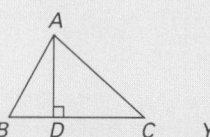 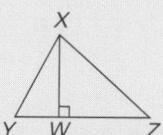

Proof is given here for the case of obtuse triangles.

Statement	Reason
1. $\triangle ABC \cong \triangle XYZ$, $\overline{AD} \perp \overline{BC}$, $\overline{XW} \perp \overline{YZ}$	1. Given
2. $\angle ADB$ and $\angle XWY$ are rt $\angle$s.	2. Def of $\perp$
3. $\angle ADB \cong \angle XWY$	3. All rt $\angle$s are $\cong$.
4. $\overline{AB} \cong \overline{XY}$	4. CPCTC
5. $\angle ABC \cong \angle XYZ$	5. CPCTC
6. $\triangle ADB \cong \triangle XWY$	6. AAS
7. $\overline{AD} \cong \overline{XW}$	7. CPCTC

15.

Statement	Reason
1. $\overline{DB}$ bis $\angle ADC$; $\overline{BE}$ and $\overline{BF}$ are altitudes of $\triangle$s ABD and CBD, respectively.	1. Given
2. $\angle EDB \cong \angle FDB$	2. Def of $\angle$ bis
3. $\angle BED$ and $\angle BFD$ are rt $\angle$s.	3. Def of alt
4. $\angle BED \cong \angle BFD$	4. All rt $\angle$s are $\cong$.
5. $\overline{BD} \cong \overline{BD}$	5. Reflex Prop of Congr
6. $\triangle BED \cong \triangle BFD$	6. AAS
7. $\overline{BE} \cong \overline{BF}$	7. CPCTC

See pages 654–655 for the answers to Ex. 16–21.

10. Given: $\triangle RPQ \cong \triangle UST$, $\overline{QV}$ and $\overline{TW}$ are corresponding altitudes; $QV = 2(t + 1)$ and $TW = 4(2t - 1)$. Find TW.   4

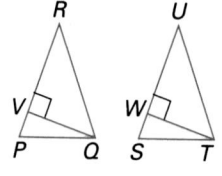

11. Given: $\overline{PR}$ is an altitude of $\triangle PQS$.
Prove: m $\angle 4$ + m $\angle 2$ = m $\angle 1$ + m $\angle 3$

12. Given: $\overline{PQ} \perp \overline{PS}$, $\angle 4 \cong \angle 1$
Prove: $\overline{PR}$ is an altitude of $\triangle PQS$.

13. Given: $\angle 3 \cong \angle 4$, $\overline{PR}$ is a median.
Prove: $\overline{PR}$ bisects $\angle QPS$.

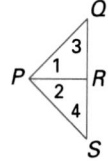

14. Prove Theorem 5.6.

15. Given: $\triangle ACD$, $\overline{DB}$ bisects $\angle ADC$; $\overline{BE}$ and $\overline{BF}$ are altitudes of $\triangle$s ABD and CBD, respectively.
Prove: $\overline{BE} \cong \overline{BF}$

16. Given: $\overline{DB}$ is an altitude from the vertex angle of isosceles $\triangle ADC$; $\overline{BE}$ and $\overline{BF}$ are medians of $\triangle$s ABD and CBD, respectively.
Prove: $\overline{BE} \cong \overline{BF}$

17. Given: $\overline{DB}$ is a median from the vertex of isosceles $\triangle ADC$; $\angle 1 \cong \angle 2$. Draw a flow diagram showing that $\overline{BE} \cong \overline{BF}$.

18. Write a proof for Exercise 17 in two-column form.

Prove or disprove the given statement.

19. If an altitude of a triangle bisects an angle of the triangle, then the triangle is isosceles.

20. The two medians drawn to the congruent sides of an isosceles triangle are congruent.

21. The angle bisector of a base angle of an isosceles triangle is also an altitude.

Mixed Review

1. Draw an angle. Construct the bisector of the angle.   *1.5*

2. The measure of one of two complementary angles is $\frac{2}{3}$ the measure of the other. Find the measure of each angle.   *1.6*   36, 54

3. In $\triangle ABC$, $\overline{AC} \cong \overline{AB}$ and m $\angle A = 30$. Find m $\angle B$.   *5.1*   75

Mixed Review

1.

5.6 Perpendicular Bisectors

Objectives
To prove and apply theorems about perpendicular bisectors
To construct perpendicular lines and segments

There are two segments and one line that are of special importance in the study of triangles. They are shown in the figures below.

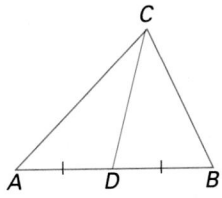 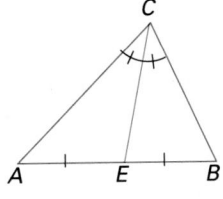 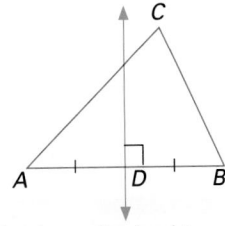

Recall that a *median* is a segment that has a vertex and the midpoint of the opposite side as endpoints.

An *angle bisector* bisects an angle of the triangle.

A *perpendicular bisector* of a side is perpendicular to that side at its midpoint.

The median, the angle bisector, and the perpendicular bisector are not necessarily contained in the same line. However, if a triangle is isosceles, then the bisector of the vertex angle is both a median and a perpendicular bisector of the base.

Theorem 5.7

The bisector of the vertex angle of an isosceles triangle is the perpendicular bisector of the base.

Given: $\overline{AC} \cong \overline{BC}$, $\overline{CD}$ bisects $\angle ACB$.
Prove: $\overline{CD}$ bisects $\overline{AB}$ and $\overline{CD} \perp \overline{AB}$.

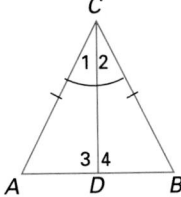

Plan
Prove $\triangle ADC \cong \triangle BDC$. Then $\overline{AD} \cong \overline{BD}$ and $\angle 3 \cong \angle 4$. Note then that m $\angle 3$ + m $\angle 4$ = 180. So, m $\angle 3$ = m $\angle 4$ = 90.

5.6 Perpendicular Bisectors **199**

Teaching Resources

Quick Quizzes 39
Reteaching and Practice Worksheets, pp. 77, 78

▰▰▰ GETTING STARTED

Prerequisite Quiz

1. Draw a line segment. Then construct its midpoint.   See Lesson 1.3.
2. Draw an angle. Then construct its bisector.   See Lesson 1.5.
3. Given: $\overline{AC} \cong \overline{BC}$, $\overline{CD}$ bisects $\angle ACB$
   Prove: $\overline{AD} \cong \overline{BD}$

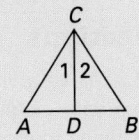

Statement	Reason
1. $\overline{AC} \cong \overline{BC}$, $\overline{CD}$ bis $\angle ACB$	1. Given
2. $\angle 1 \cong \angle 2$	2. Def of ∠ bis
3. $\overline{CD} \cong \overline{CD}$	3. Reflex Prop of Congr
4. $\triangle ACD \cong \triangle BCD$	4. SAS
5. $\overline{AD} \cong \overline{BD}$	5. CPCTC

Motivator

Ask the students to construct isosceles $\triangle ABC$ with $CA = CB$. Then have them draw the bisector of vertex angle C. Ask them what observations they can make about the intersection of the angle bisector and base, $\overline{AB}$.   The angle bisector is perpendicular to the base and bisects the base.   Ask them what is meant by the word equidistant.   Equal distance

Lesson Note

Once Theorem 5.8 has been proved, emphasize that it can be applied without first proving triangles congruent. Students often have difficulty with the terminology "points equidistant from the endpoints of a segment." Suggestion: Draw a segment on the chalkboard and ask a student to mark a point that appears to be equidistant from the endpoints. Show that the midpoint is a special case, and that other points, not on the segment, are also equidistant from its endpoints.

Math Connections

Decorating: To achieve a visual balance, interior decorators can use perpendicular bisectors in floor plans to locate the place on a wall to hang pictures or place murals so that they are equidistant from two major objects in a room.

Corollary | The bisector of the vertex angle of an isosceles triangle is also a median and an altitude of the triangle.

Construction Construct the perpendicular bisector of a segment.

Follow the same steps as for constructing its midpoint (see Lesson 1.3). Notice in this construction that $PA = PB$—that is, P is equidistant from A and B. Also, $QA = QB$—that is, Q is equidistant from A and B. The next theorem states that the line containing P and Q is the perpendicular bisector of $\overline{AB}$.

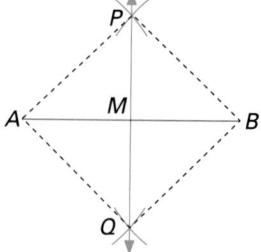

Theorem 5.8 | A line containing two points, each equidistant from the endpoints of a given segment, is the perpendicular bisector of the segment.

Given: $\overline{PA} \cong \overline{PB}$, $\overline{QA} \cong \overline{QB}$
Prove: $\overleftrightarrow{PQ}$ is the perpendicular bisector of $\overline{AB}$.

Plan Notice that $\triangle APB$ is isosceles. Prove $\triangle APQ \cong \triangle BPQ$ to get $\angle 1 \cong \angle 2$. Thus, $\overrightarrow{PR}$ is the bisector of the vertex angle of an isosceles triangle. Apply Theorem 5.7.

In the figure used for Theorem 5.8, points P and Q were placed on opposite sides of segment $\overline{AB}$. Points P and Q can also be placed on the same side of $\overline{AB}$ in the proof of this theorem.

The important idea in Theorem 5.8 is that points equidistant from the endpoints of the segment are on the perpendicular bisector of the segment. This idea is useful for constructing a perpendicular to a line through any point.

Construction A line perpendicular to a given line through a point not on the line.

Given: Line l and point P not on l
Construct: $\overleftrightarrow{PQ} \perp l$

• P

(See following page for construction.)

$l \longleftrightarrow$

200 Chapter 5 Congruence

Draw an arc with center P, intersecting l at two points. Label the points A and B.

Using A and B as centers, swing equal arcs. Label the point of intersection Q.

Draw $\overrightarrow{PQ}$.

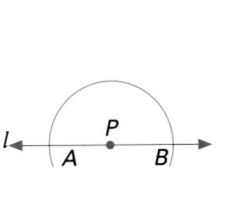

PA = PB

QA = QB

Points P and Q are equidistant from A and B. Result: $\overleftrightarrow{PQ} \perp l$ by Theorem 5.8.

Construction A line perpendicular to a given line through a point on the line.

Given: Line l and point P on l
Construct: $\overleftrightarrow{QP} \perp l$

Draw an arc with center P intersecting l at two points. Label the points A and B.

Using A and B as centers, swing equal arcs. Label their intersection Q.

Draw $\overleftrightarrow{QP}$.

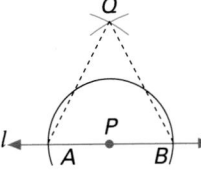

PA = PB

QA = QB

Result: $\overleftrightarrow{QP} \perp l$

You will be asked to prove the result of this construction in Exercise 13.

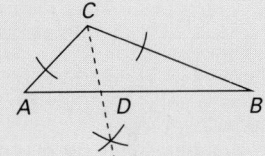

Checkpoint

Draw a conclusion based on the given information. State which theorem from this lesson is applied. (Ex. 1–3)

1. Given: $\overline{HM}$ bisects $\overline{GK}$, $\overline{GM} \cong \overline{KM}$

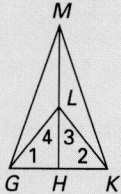

2. Given: $\angle 1 \cong \angle 2$, $\angle 3 \cong \angle 4$
3. $\overline{LH} \perp \overline{GK}$, H is the midpoint of $\overline{GK}$.
4. Prove the conclusion of Exercise 2 above.
   1. $\overline{MH}$ is $\perp$ bis of $\overline{GK}$ (Thm 5.8)
   2. $\overline{LH}$ is $\perp$ bis of $\overline{GK}$ (Thm 5.7)
   3. $\overline{GM} \cong \overline{KM}$ and $\overline{GL} \cong \overline{KL}$ (Thm 5.9)

4. Statement	Reason
1. $\angle 1 \cong \angle 2$	1. Given
2. $\overline{GL} \cong \overline{KL}$	2. Sides opp $\cong \angle$s of a $\triangle$ are $\cong$.
3. $\triangle GKL$ is isos	3. Def of isos $\triangle$
4. $\angle 3 \cong \angle 4$	4. Given
5. $\overline{LH}$ bis $\angle GLK$	5. Def of $\angle$ bis
6. $\overline{LH}$ is $\perp$ bis of $\overline{GK}$	6. Bis of vertex $\angle$ of isos $\triangle$ is $\perp$ bis of base.

Closure

Ask the students what kind of a triangle has an angle bisector, a median, and a perpendicular bisector contained in the same line.   Isosceles triangle   Ask them what it means to say that two points are equidistant from the ends of a segment. The points are the same distance from the ends of the segment.   Have them explain the relationship of the line joining two such points to the original segment.   It is the perpendicular bisector of the segment.

Theorem 5.9   Any point on the perpendicular bisector of a segment is equidistant from the endpoints of the segment.

Given: l is the perpendicular bisector of $\overline{AB}$; P is any point on l.
Prove: P is equidistant from the endpoints of $\overline{AB}$.

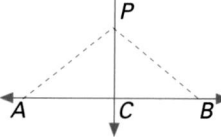

Plan   Draw $\overline{PA}$ and $\overline{PB}$, and prove $\triangle PAC \cong \triangle PBC$.

EXAMPLE 1   Given: $\overline{CD} \perp \overline{AB}$, $\overline{AD} \cong \overline{BD}$
Prove: $\angle A \cong \angle B$

Plan $\overline{CD}$ is the $\perp$ bisector of $\overline{AB}$. Thus, C is equidistant from A and B, that is, $AC = BC$.

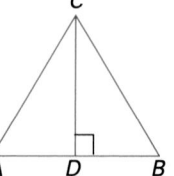

Proof

Statement	Reason
1. $\overline{CD} \perp \overline{AB}$, $\overline{AD} \cong \overline{BD}$	1. Given
2. D is the midpoint of $\overline{AB}$.	2. Def of midpt
3. $\overline{CD}$ is the $\perp$ bisector of $\overline{AB}$.	3. Def of $\perp$ bis
4. $AC = BC$ ($\overline{AC} \cong \overline{BC}$)	4. Any point on $\perp$ bis of a seg is equidistant from seg endpts.
5. $\therefore \angle A \cong \angle B$	5. $\angle$s opp $\cong$ sides of a $\triangle$ are $\cong$.

Focus on Reading

Use the figure and its markings to complete the sentence.

1. $\overline{DB}$ is the _____ of $\overline{AC}$.   Perpendicular bisector
2. Point D is _equidistant_ from the _endpoints_ of $\overline{AC}$.
3. $\angle ADC$ is the _vertex angle_ of the isosceles triangle.
4. ___$\overline{DB}$___ is a median of $\triangle DAC$.
5. Any point on $\overline{DB}$ is equidistant from points ___A___ and ___C___.

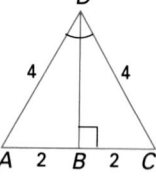

Additional Example 1

Given: $\overline{PQ} \perp \overline{SR}$, T is midpoint of $\overline{SR}$
Prove: $\overline{SQ} \cong \overline{QR}$

Proof: $\overline{PQ} \perp \overline{SR}$ (Given); T is midpoint of $\overline{SR}$ (Given); $\overline{PQ}$ is $\perp$ bis of $\overline{SR}$ (def of $\perp$ bis); $QS = QR$ or $\overline{QS} \cong \overline{QR}$ (any point on $\perp$ bis of seg is equid from seg endpoints)

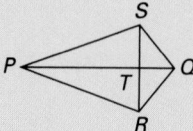

Classroom Exercises

Based on the markings in the figure, state which segment is a perpendicular bisector of another segment. Justify your answer.

1.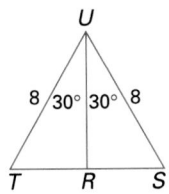

$\overline{UR} \perp$ bis of $\overline{TS}$
(Thm 5.7)

2.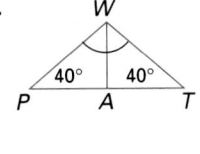

$\overline{WA} \perp$ bis of $\overline{PT}$
(Thm 5.7)

3.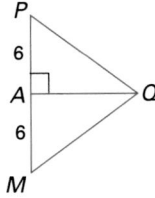

$\overline{QA} \perp$ bis of $\overline{MP}$
(Def of $\perp$ bis.)

4.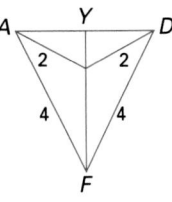

$\overline{YF} \perp$ bis of $\overline{AD}$
(Thm 5.8)

Written Exercises

1. Draw a line and a point not on the line. Construct the perpendicular to the line from the point.

2. Draw a line. Mark a point on the line. Construct the perpendicular to the line at the point.

Write the proof without showing triangles congruent. (Exercises 3–8)

3. Given: $\overline{TQ}$ bisects $\angle PTR$; $\overline{TP} \cong \overline{TR}$.
Prove: $\overline{TQ}$ is $\perp$ bis of $\overline{PR}$.

4. Given: $\overline{TQ} \perp \overline{PR}$, $\overline{PQ} \cong \overline{RQ}$
Prove: $\angle 5 \cong \angle 6$

5. Given: $\angle 3 \cong \angle 4$, $\overline{TQ}$ bisects $\angle PTR$.
Prove: $\overline{PQ} \cong \overline{RQ}$

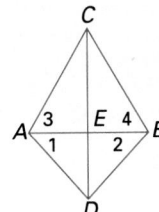

6. Given: $\angle 3 \cong \angle 4$, $\angle 1 \cong \angle 2$
Prove: $\overline{CD}$ is $\perp$ bis of $\overline{AB}$.

7. Given: $\overline{CD}$ is $\perp$ bis of $\overline{AB}$.
Prove: $\angle 1 \cong \angle 2$

8. Given: $\angle 3 \cong \angle 4$, $\angle CAD \cong \angle CBD$
Prove: $\overline{CD}$ is $\perp$ bis of $\overline{AB}$.

9. Prove that there is exactly one perpendicular to a line from a point not on that line. (HINT: Use an indirect proof.)

10. Prove that there is exactly one perpendicular to a line at a point on that line. (HINT: Use an indirect proof.)

 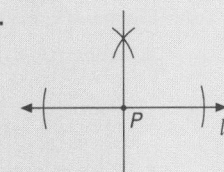
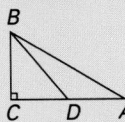

6.

Statement	Reason
1. $\angle 3 \cong \angle 4$, $\angle 1 \cong \angle 2$	1. Given
2. $\overline{AC} \cong \overline{BC}$, $\overline{AD} \cong \overline{BD}$	2. Sides opp $\cong \angle$s are $\cong$.
3. $\overline{CD}$ is $\perp$ bis of $\overline{AB}$	3. A line with 2 pts, each equidist from endpts of seg, is $\perp$ bis of the seg.

7.

Statement	Reason
1. $\overline{CD}$ is $\perp$ bis of $\overline{AB}$	1. Given
2. $\overline{AD} \cong \overline{BD}$	2. Any pt on $\perp$ bis of a seg is equidist from endpts of seg.
3. $\angle 1 \cong \angle 2$	3. $\angle$s opp $\cong$ sides are $\cong$.

8.

Statement	Reason
1. $\angle 3 \cong \angle 4$, $\angle CAD \cong \angle CBD$	1. Given
2. $\overline{AC} \cong \overline{BC}$	2. Sides opp $\cong \angle$s are $\cong$.
3. C is equidist from A and B.	3. Def of equidist
4. m $\angle CAD -$ m $\angle 3 =$ m $\angle CBD -$ m $\angle 4$	4. Equations may be subtracted.
5. m $\angle CAD -$ m $\angle 3 =$ m $\angle 1$, m $\angle CBD -$ m $\angle 4 =$ m $\angle 2$	5. Angle Add Post
6. m $\angle 1 =$ m $\angle 2$	6. Sub
7. $\overline{AD} \cong \overline{BD}$	7. Sides opp $\cong \angle$s are $\cong$.
8. D is equidist from A and B	8. Def of equidist
9. $\overline{CD}$ is $\perp$ bis of $\overline{AB}$.	9. A line with 2 pts, each equidist from endpts of seg is $\perp$ bis of seg.

9. Assume there are two distinct line segments from a point P perpendicular to a line at points A and B. Then a $\triangle APB$ is formed, with m $\angle PAB =$ m $\angle PBA = 90$. But m $\angle APB > 0$, since $\overline{PA}$ and $\overline{PB}$ are distinct. Thus, the sum of the measures of the angles of the $\triangle$ will be greater than 180, which contradicts the theorem that the sum of the measures of the angles is 180. Therefore, the assumption must be false; we conclude that there can be only one perpendicular.

See page 221 for the answers to Mixed Review Ex. 2 and Application Ex. 1–3.

204

1. List the four congruence patterns for triangles. *4.3–4.6* SSS, SAS, ASA, AAS
2. Draw an angle. Construct a second angle congruent to it. *1.4*

Find the complement and supplement of angle A, where possible. *1.6*

3. m $\angle A = 40$
C: 50, S: 140

4. m $\angle A = 140$
C: None, S: 40

5. m $\angle A = x$
C: $(90 - x)$
S: $(180 - x)$

6. m $\angle A = 3x - 10$
C: $(100 - 3x)$
S: $(190 - 3x)$

Application: *Congruent Relationships*

The congruent relationships in an isosceles triangle and its altitude, a perpendicular bisector, are used often in construction. Following are different examples of how these properties can be applied.

1. Electric wires are to be strung between two poles across a ravine. The power company needs to know how much wire is needed, but it is impossible to measure directly across the ravine. How could this distance be found using the given diagram? Why does this work?

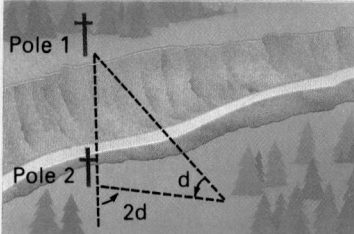

2. A homemade carpenter's level has an isosceles triangular frame, with a plumb bob hanging from the vertex. The plumb bob should always hang vertically. How can this be used to show whether or not a surface is horizontal? Why does this work?

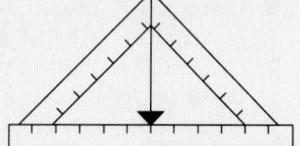

3. An artist is painting a large design in which an angle must be bisected. A carpenter's square is the only measuring device available. How can the square be used to bisect the angle? Why does this work?

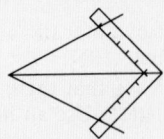

204   Chapter 5 Congruence

10. Assume that there are two distinct segments $\overline{PA}$ and $\overline{PB}$ perpendicular to a point P on a line. Then m $\angle 1 =$ m $\angle 2 = 90$. But m $\angle APB > 0$, so m $\angle 1 +$ m $\angle 2 +$ m $\angle APB > 180$, which contradicts the result that the sum must be 180. Thus, the assumption that there can be two distinct segments perpendicular to the line at a point on the line must be false. We conclude there can be only one such segment.

5.7 Inequalities in a Triangle

Objective   To prove and apply inequality relationships in a triangle

Algebraic inequalities are frequently used in geometric proofs. The formal definitions of ">" and "<" are suggested by the following:
$8 > 5$ since there exists a positive number, 3, such that $5 + 3 = 8$.

Definition

> $a > b$ if there exists a positive number c such that $b + c = a$.
> $a < b$ if there exists a positive number c such that $a + c = b$.

The definition of ">" can be used in proving the following theorem.

Theorem 5.10

Exterior Angle Inequality Theorem: The measure of an exterior angle of a triangle is greater than the measure of either of its remote interior angles.

Given: $\triangle ABC$ with exterior $\angle 1$
Prove: m $\angle 1 >$ m $\angle 3$ and m $\angle 1 >$ m $\angle 2$

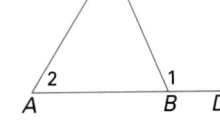

Proof

Statement	Reason
1. $\triangle ABC$ with exterior $\angle 1$	1. Given
2. m $\angle 1 =$ m $\angle 2 +$ m $\angle 3$	2. Ext $\angle$ Thm
3. $\therefore$ m $\angle 1 >$ m $\angle 3$, m $\angle 1 >$ m $\angle 2$	3. Def of $>$

The following are properties of inequalities:

(1) If $a > b$ and $b > c$, then $a > c$ (Trans Prop of Ineq)
(2) If $a > b$ and $c = d$, then $a + c > b + d$ (Add Prop of Ineq)
(3) If $a > b$ and $c > d$, then $a + c > b + d$
(4) If $a > b$ and $b = c$, then $a > c$ (Sub Prop of Ineq)

Any triangle with sides of unequal length also has angles of unequal measure. The orders of inequality for the sides and the angles are related. In the figure, $AB > AC$.

It appears that m $\angle C >$ m $\angle B$. The larger angle appears to be opposite the longer side. This is proved in the following theorem.

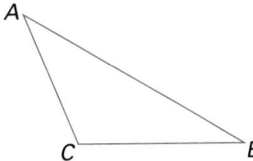

Teaching Resources

Quick Quizzes 40
Reteaching and Practice Worksheets, pp. 79, 80

▬ GETTING STARTED

Prerequisite Quiz

Insert the correct inequality symbol to make each statement true.

1. 5 _____ 8 $<$
2. 7 + 4 _____ 10 $>$

Find the measure of the largest angle in triangle *ABC*.

3. m $\angle A = (x - 30)$, m $\angle B = (2x + 10)$, m $\angle C = x$ 110
4. m $\angle A = (2x - 5)$, m $\angle B = (x + 10)$, m $\angle C = (x + 15)$ 75
5. The measure of an acute angle of a right triangle is 40. Find the measure of the other acute angle.   50

Motivator

Prepare a sheet of paper with three different scalene triangles for each student. Ask the students to measure each side and each angle of the first triangle. Have them name the longest side and the angle with the greatest measure. In the second triangle, have them measure each side to find the longest. Have them predict the angle with the greatest measure, and verify with a protractor. Have them repeat the above steps for the third triangle. Ask them if they can generalize.   The angle with the greatest measure is opposite the longest side.

▬ TEACHING SUGGESTIONS

Lesson Note

Try to lead students to discover intuitively the properties of inequalities listed on this page. Most students are familiar with algebraic inequalities, but few have seen the formal definition $a < b$. To make the definition more meaningful, it is helpful to ask questions such as: How do we know that $-5 < 2$? What positive number can you add to -5 to get 2?

Math Connections

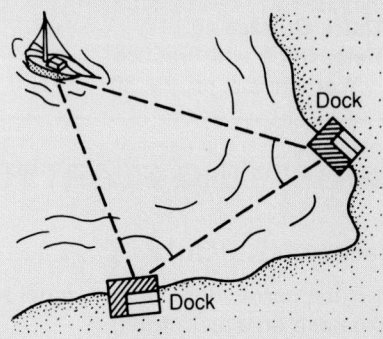

Dock

Dock

Navigation: Ships can use the inequalities of a triangle to determine their course. If the captain of an inbound ship wants to find out which of two ports is closer, she can have people at each dock find the measure of the angles between the ship and the other dock and send her the information over the radio. From this information, the captain can use Theorem 5.12 from page 206 to determine which port is closer.

Critical Thinking Questions

Application: The elasticity of rubber bands is being tested with two 7-inch metal prongs hinged together at one end with the rubber band stretching between the two other ends of the prongs. When stretched to a maximum length, the majority of rubber bands allow the prongs to open 60°. What is a good estimate for the maximum length of most rubber bands? Explain.   Both prongs are 7 in. long, so the △ is isos and the base ∠s are ≅. Thus, the △ is equiangular and equilat; therefore, the max length of the rubber band is 7 in.   What can you conclude about the maximum length of a rubber band that allows the prongs to open more than 60°?   The max length of the rubber band is greater than 7 in.

Common Error Analysis

Error: In problems such as Additional Example 2, students tend to think that the longest segment is opposite the angle with the greatest measure in the entire diagram. Thus, they might conclude that $\overline{QS}$ is the longest side.

Emphasize the need to first look at each triangle separately as in Example 2.

Theorem 5.11   If one side of a triangle is longer than another side, then the measure of the angle opposite the longer side is greater than the measure of the angle opposite the shorter side.

Given: △RPQ with RQ > RP
Prove: m ∠RPQ > m ∠Q

Plan   Locate point S on $\overline{RQ}$ so that $\overline{RP} \cong \overline{RS}$. Then △RPS is isosceles.

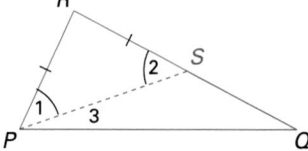

Proof

Statement	Reason
1. △RPQ with RQ > RP	1. Given
2. On $\overline{RQ}$, choose S so that RS = RP.	2. Ruler Post
3. Draw $\overline{PS}$.	3. Two points determine a line.
4. △RPS is isosceles.	4. Def of isos △
5. ∠1 ≅ ∠2 (m ∠1 = m ∠2)	5. Base ∠s of isos △ are ≅.
6. m ∠RPQ = m ∠1 + m ∠3	6. ∠ Add Post
7. m ∠RPQ > m ∠1	7. Def of >
8. m ∠RPQ > m ∠2	8. Sub Prop of Ineq
9. m ∠2 > m ∠Q	9. Ext ∠ Ineq Thm
10. ∴ m ∠RPQ > m ∠Q	10. Trans Prop of Ineq

The converse of Theorem 5.11 can be proved by using the following property in an indirect proof.

Trichotomy Property
For any two real numbers a and b, only one of three possible relationships exists: $a < b$ or $a = b$ or $a > b$.

Theorem 5.12   If one angle of a triangle has a greater measure than a second angle, then the side opposite the greater angle is longer than the side opposite the smaller angle.

Given: △RPQ with m ∠P > m ∠Q
Prove: RQ > RP

Plan   According to the Trichotomy Property, one of three cases holds: RQ < RP, RQ = RP, or RQ > RP. Prove that each of the first two cases leads to a contradiction. Then the third case must hold.

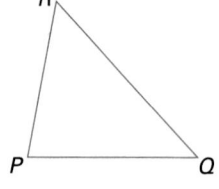

206   Chapter 5 Congruence

Additional Example 1

Name the longest side of the given triangle.

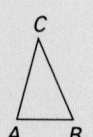

m ∠A = (x + 50)
m ∠B = (3x − 10)
m ∠C = (2x − 10)
$\overline{BC}$

| **Theorem 5.13** | In a scalene triangle, the longest side is opposite the largest angle and the largest angle is opposite the longest side. |

You will be asked to prove Theorem 5.13 in Exercise 9.

EXAMPLE 1   Name the longest side of the given triangle.

Solution

$$m \angle A + m \angle B + m \angle C = 180$$
$$(x - 30) + (2x + 10) + x = 180$$
$$4x - 20 = 180$$
$$4x = 200$$
$$x = 50$$

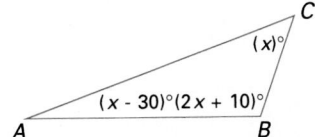

$$m \angle A = x - 30 \qquad m \angle B = 2x + 10 \qquad m \angle C = x$$
$$= 50 - 30 = 20 \qquad = 2 \cdot 50 + 10 = 110 \qquad = 50$$

Thus, the longest side is $\overline{AC}$.

EXAMPLE 2   Name the longest segment in the figure.

Solution   In $\triangle ABD$, $\overline{DB}$ is the longest side since it is opposite the largest angle. In $\triangle BCD$, $\overline{DC}$ is longer than any other side, including $\overline{DB}$. Therefore, $\overline{DC}$ is the longest segment in the figure.

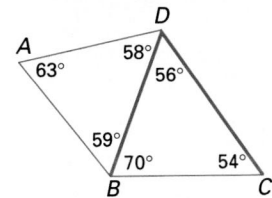

EXAMPLE 3   Given: m $\angle Q <$ m $\angle P$, $RP > RS$
Prove: m $\angle Q <$ m $\angle 1$

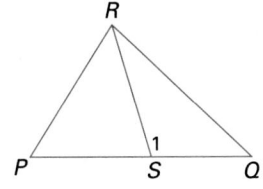

Proof

Statement	Reason
1. m $\angle Q <$ m $\angle P$, or m $\angle P >$ m $\angle Q$	1. Given
2. $RQ > RP$	2. The longer side is opp the larger $\angle$ in a $\triangle$.
3. $RP > RS$	3. Given
4. $RQ > RS$	4. Tran Prop of Ineq
5. $\therefore$ m $\angle 1 >$ m $\angle Q$, or m $\angle Q <$ m $\angle 1$	5. The larger $\angle$ is opp the longer side in a $\triangle$.

Checkpoint

Complete each statement.

1. For $\triangle PQR$, if $PQ > PR$, then _____ > _____ $\angle R$, $\angle Q$

2. For $\triangle GHK$, if $\angle K < \angle G$, then _____ < _____ GH, HK

Determine the longest segment. (Ex. 3–4)

3. 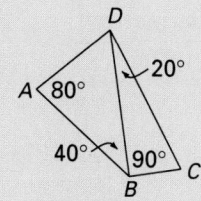 $\overline{DC}$

4. $\overline{AB}$ (Since m $\angle A = 50$, m $\angle B = 45$, m $\angle C = 85$)

Closure

Ask your students to explain the Trichotomy Property.   Only one of three possible relationships exists for any two real numbers, a and b: $a < b$, $a = b$, $a > b$. Ask them what inequality relationships exist for triangles with respect to side lengths corresponding to angle measure.   Theorem 5.11, Theorem 5.12

◣◣◣ FOLLOW UP

Guided Practice

Classroom Exercises 1–3

Independent Practice

A Ex. 1–9, **B** Ex. 10–12, **C** Ex. 14–17

Basic: WE 1–9, AR all

Average: WE 1–12, AR all

Above Average: WE 1–17, AR all

Additional Answers

Classroom Exercises

1. $\angle P$, $\angle R$, $\angle T$ 2. $\angle W$, $\angle G$, $\angle A$
3. $\angle T$, $\angle B$, $\angle U$

Written Exercises

1. $\overline{CB}$ 2. $\overline{AR}$ 3. $\overline{BK}$

Additional Example 2

Name the longest segment in the figure. $\overline{QR}$

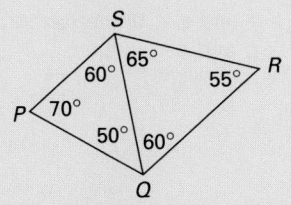

Additional Example 3

Given: $\overline{DC} \parallel \overline{AB}$,
m $\angle 3 >$ m $\angle 1$
Prove: $BD > AD$

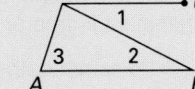

Proof: $\overline{DC} \parallel \overline{AB}$ (Given); m $\angle 1 =$ m $\angle 2$ (Alt int $\angle$s of $\parallel$ lines have = meas); m $\angle 3 >$ m $\angle 1$ (Given); m $\angle 3 >$ m $\angle 2$ (Sub); $BD > AD$ (Longer side is opp larger $\angle$ in a $\triangle$)

207

4.

Statement	Reason
1. $AD = BD$	1. Given
2. $\angle 1 \cong \angle 2$	2. $\angle$s opp $\cong$ sides are $\cong$.
3. m $\angle 2 >$ m $\angle 3$	3. Ext $\angle$ Ineq Thm
4. m $\angle 1 >$ m $\angle 3$	4. Sub Prop of Ineq
5. $DC > AD$	5. Side opp larger $\angle$ is longer.

5.

Statement	Reason
1. m $\angle 2 >$ m $\angle P$	1. Given
2. m $\angle 2 =$ m $\angle 1$	2. Vert $\angle$s are $\cong$.
3. m $\angle 1 >$ m $\angle P$	3. Sub Prop of Ineq
4. $PT > QT$	4. Side opp larger $\angle$ is longer.

6.

Statement	Reason
1. $XZ > YZ$, m $\angle X >$ m $\angle Z$	1. Given
2. m $\angle Y >$ m $\angle X$	2. $\angle$ opp longer side is larger.
3. m $\angle Y >$ m $\angle Z$	3. Trans Prop of Ineq
4. $XZ > XY$	4. Side opp larger $\angle$ is longer.

7.

Statement	Reason
1. $\overline{SQ}$ bis $\angle PSR$	1. Given
2. $\angle 1 \cong \angle 2$	2. Def of $\angle$ bis
3. m $\angle 3 >$ m $\angle 2$	3. Ext $\angle$ Ineq Thm
4. m $\angle 3 >$ m $\angle 1$	4. Sub Prop of Ineq
5. $RS > RQ$	5. Side opp larger $\angle$ is longer.

8.

Statement	Reason
1. $\overline{AB} \parallel \overline{CD}$, m $\angle 1 >$ m $\angle 3$	1. Given
2. m $\angle 1 =$ m $\angle 2$	2. If lines are $\parallel$, then alt int $\angle$s are $\cong$.
3. m $\angle 2 >$ m $\angle 3$	3. Sub
4. $ED > EC$	4. Side opp larger $\angle$ is longer.

9.

Statement	Reason
1. Scalene $\triangle ABC$, m $\angle B >$ m $\angle C$, m $\angle C >$ m $\angle A$	1. Given
2. $AC > AB$, $AB > BC$	2. Side opp larger $\angle$ is longer.
3. $AC > BC$	3. Trans Prop of Ineq
4. $\overline{AC}$ is longest side and is opp the largest $\angle$.	4. $\overline{AC}$ is opp $\angle B$

208

Classroom Exercises

Name the angles in order from smallest to largest.

1.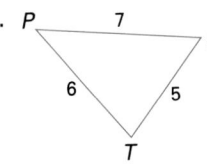

2.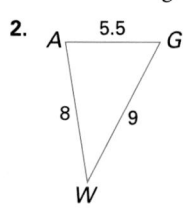

3.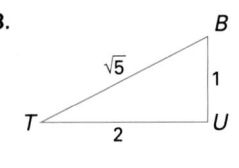

Written Exercises

In Exercises 1–3, determine the longest segment.

1.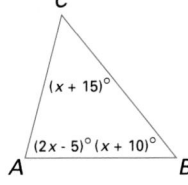

2.

3.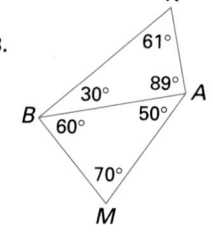

4. Given: $AD = BD$
Prove: $DC > AD$

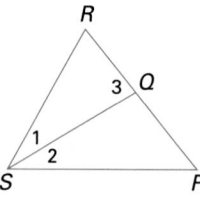

5. Given: m $\angle 2 >$ m $\angle P$
Prove: $PT > QT$

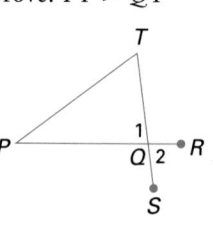

6. Given: $XZ > YZ$, m $\angle X >$ m $\angle Z$
Prove: $XZ > XY$

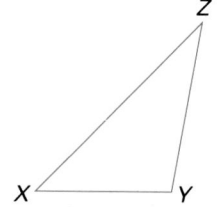

7. Given: $\overline{SQ}$ bisects $\angle PSR$.
Prove: $RS > RQ$

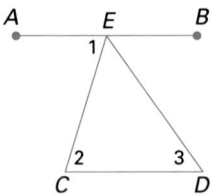

8. Given: $\overline{AB} \parallel \overline{CD}$, m $\angle 1 >$ m $\angle 3$
Prove: $ED > EC$

9. Prove Theorem 5.13.

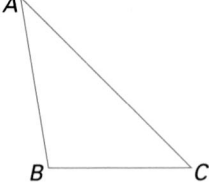

208   Chapter 5 Congruence

Enrichment

Have the students decide whether each statement is true or false.

1. $>$ is commutative.   F
2. $\neq$ is commutative.   T
3. $<$ is transitive.   T
4. If $a < b$ and $x < y$, then $a + x < b + y$ T
5. If $a < b$ and $x > y$, then $a + x > b + y$ F
6. If $x < y$ and a is a real number, then $ax < ay$ F
7. If $x > y$ and $a < 0$, then $ax < ay$ T
8. If $a = b$ and $x < y$, then $a - x < b - y$ F

Complete each statement for △ABC.

10. If $AB > BC$, then $\underline{m \angle C} > \underline{m \angle A}$ **11.** If $AC < AB$, then $\underline{m \angle B} < \underline{m \angle C}$

12. If $m \angle A < m \angle B$, then $\underline{BC} < \underline{AC}$ **13.** If $m \angle C > m \angle B$, then $\underline{AC} < \underline{AB}$

14. Prove Theorem 5.12.

15. Given: $SQ > SP$,
          $SR = SQ$
   Prove: $m \angle P + m \angle R$
          $> m \angle PQR$

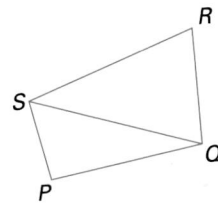

16. Given: $YW = YZ$,
          $YZ > YX$
   Prove: $m \angle XWZ <$
          $m \angle X + m \angle Z$

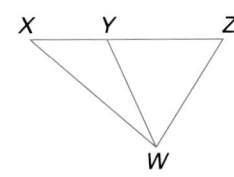

17. Given: $AB = BC$
   Prove: $AB + BC > AC$

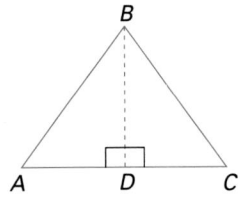

Mixed Review

1. The coordinates of the endpoints of a segment are -8 and 12. Find the coordinate of the midpoint of the segment.   *1.3*   2

2. The measures of two angles formed by an angle bisector of an angle are $3x - 20$ and $x + 40$. Find the measure of each angle.   *1.5*   70

3. The measure of one acute angle of a right triangle is 4 times the measure of the other. Find the measure of the larger acute angle.   *4.1*   72

4. Graph and name the resulting geometric figure: $x \geq 5$ *and* $x \geq 7$.   *1.8*

▰▰ / Algebra Review

Solve an inequality as you would solve an equation. However, when you multiply or divide either side of an inequality by a negative number, reverse the inequality symbol.

Example: Solve. $-3x + 14 < 8$
                             $-3x < -6$
                             $x > 2$

The solution consists of all numbers greater than 2.

Solve.

1. $-15 < 9y - 6$ $-1 < y$ **2.** $3a + 8 < -19$ $a < -9$ **3.** $-4x + 12 > -16$ $x < 7$

4. $4x + 13 \leq 3x - 10$ **5.** $\frac{2a}{3} + 9 \leq 11$ $a \leq 3$ **6.** $-4 + \frac{3m}{2} < m$ $m < 8$
     $x \leq -23$

---

14. In △RPQ, $m \angle P > m \angle Q$ (Given). By the Trichotomy Property, either $RQ < RP$ or $RQ = RP$ or $RQ > RP$.
(Case 1) Assume $RQ < RP$. By Thm 5.11, $m \angle P < m \angle Q$. But this contradicts the given information, so the assum is false.
(Case 2) Assume $RQ = RP$. Then △RPQ is isos, with $m \angle P = m \angle Q$. But this contradicts the given information, so the assum is false.
(Case 3) $RQ > RP$. Since Cases 1 and 2 are both false, this is the only remaining poss. it must be true.

15.

Statement	Reason
1. $SQ > SP$, $SR = SQ$	1. Given
2. $m \angle P > m \angle SQP$	2. ∠ opp longer side is larger.
3. $m \angle R = m \angle SQR$	3. ∠ opp ≅ sides are ≅.
4. $m \angle P + m \angle R > m \angle SQP + m \angle SQR$	4. Add Prop of Ineq
5. $m \angle SQP + m \angle SQR = m \angle PQR$	5. Angle Add Post
6. $m \angle P + m \angle R > m \angle PQR$	6. Sub Prop of Ineq

16.

Statement	Reason
1. $YW = ZY$, $YZ > YX$	1. Given
2. $m \angle YWZ = m \angle Z$	2. ∠s opp ≅ sides are ≅.
3. $YX < YW$	3. Sub Prop of Ineq
4. $m \angle XWY < m \angle X$	4. ∠ opp longer side is larger.
5. $m \angle XWY + m \angle YWZ < m \angle Z + m \angle X$	5. Add Prop of Ineq
6. $m \angle XWY + m \angle YWZ = m \angle XWZ$	6. Angle Add Post
7. $m \angle XWZ < m \angle Z + m \angle X$	7. Sub Prop of Ineq

Mixed Review

4. Ray

See page 223 for the answer to Ex. 17.

5.8 The Triangle Inequality Theorem

Objectives

To apply the Triangle Inequality Theorem
To determine whether a triangle can be constructed, given lengths for three sides
To write paragraph proofs

▰▰GETTING STARTED

Prerequisite Quiz

In △*ABC*, $\overline{CB} \perp \overline{CA}$, and m ∠*A* = 40.
Insert the symbol < or > to make each inequality true.

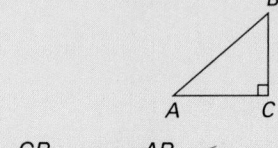

1. *CB* _____ *AB*   <
2. *CA* _____ *AB*   <
3. *AC* _____ *BC*   >

Motivator

Ask the students to draw segments of lengths 3 cm, 5 cm, and 10 cm. Then ask them to construct a triangle with sides of these lengths. Ask them to explain why this cannot be done.   The 3 cm and 5 cm segments are too short and cannot meet to form a third vertex.   Have them try again using an 8 cm segment in place of the 10 cm segment.   The 3 cm and 5 cm segments meet on the 8 cm, so the third vertex lies on the 8 cm segment.   Have them guess how the original 10 cm segment could be changed to make the triangle construction possible.   The length of the segment should be less than 8 cm, which is the sum of the other two segment lengths.

The first figure below shows a segment from point *P* perpendicular to line *l*. The second figure shows several segments drawn from *P* to *l*.

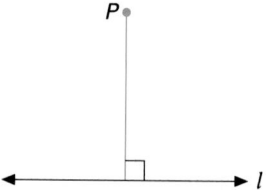 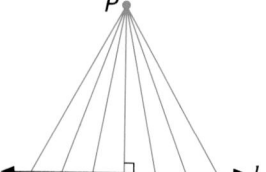

Notice that of all the segments drawn from point *P* to line *l*, the shortest one appears to be the perpendicular segment. This leads to the following theorem. The proof of the theorem is presented in paragraph form.

▰▰
Theorem 5.14

The perpendicular segment from a point to a line is the shortest segment from the point to the line.

Given: $\overline{PA} \perp l$
Prove: $\overline{PA}$ is the shortest segment from *P* to *l*.

Proof
(Paragraph
Form)

Draw any segment, $\overline{PQ}$ from *P* to *l*, along with $\overline{PA} \perp l$. Then show that *PQ* > *PA*, as follows:

△*PAQ* is a right triangle. Thus, ∠1 is a right angle. This makes ∠2 an acute angle and m ∠1 > m ∠2. Therefore, *PQ* > *PA*, since in a triangle the longer side is opposite the larger angle.

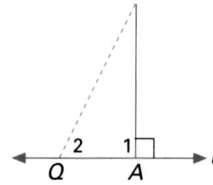

▰▰
Corollary

The longest side of a right triangle is the hypotenuse.

Given: Rt △*ABC* with hypotenuse $\overline{AB}$
Prove: $\overline{AB}$ is the longest side of the triangle.

EXAMPLE 1   If possible, draw a triangle with the given lengths for the three sides.

2 cm, 3 cm, 6 cm   2 cm, 4 cm, 6 cm

Solution

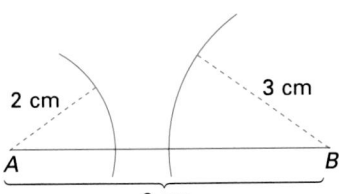   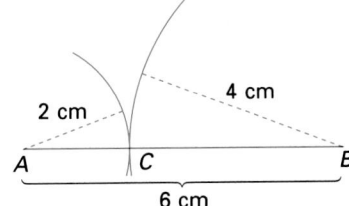

The 2-cm and 3-cm segments will not meet to form a third vertex, C. No triangle is formed.

$2 + 3 < 6$

The 2-cm and 4-cm segments meet on $\overline{AB}$. So, the third vertex lies on $\overline{AB}$. No triangle is formed.

$2 + 4 = 6$

Example 1 suggests that certain restrictions exist for the lengths of the sides of a triangle.

Theorem 5.15   **The Triangle Inequality Theorem:** The sum of the lengths of any two sides of a triangle is greater than the length of the third side.

Given: $\triangle ABC$
Prove: 1. $AB + BC > AC$
     2. $AC + AB > BC$
     3. $AC + BC > AB$

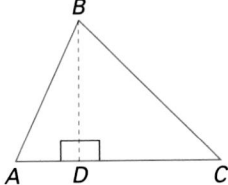

Proof

Statement	Reason
1. From B, construct $\overline{BD} \perp \overline{AC}$.	1. From a pt not on a line, exactly one $\perp$ seg can be constructed to that line.
2. $AB > AD$; $BC > DC$	2. The longest side of a rt $\triangle$ is the hyp.
3. $AB + BC > AD + DC$	3. Add Prop of Ineq
4. $AC = AD + DC$	4. Seg Add Post
5. $\therefore AB + BC > AC$	5. Sub

The proofs of parts 2 and 3 are done similarly.

5.8 The Triangle Inequality Theorem   **211**

Additional Example 1

Refer to the two constructions in Example 1. Which construction, the first or the second, illustrates why the triangle cannot be constructed?

a. $\triangle FGH$, with $FG = 3$ cm, $GH = 5$ cm and $FH = 2$ cm   Second

b. $\triangle RST$, with $RS = 4$ in., $ST = 12$ in., and $RT = 5$ in.   First

Additional Example 2

Can a triangle be constructed with sides of the given lengths?

a. 7 m, 6 m, 10m   Yes

b. 4 ft, 8 ft, 3 ft   No, $4 + 3 \not> 8$

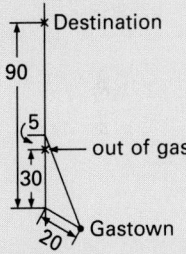

Common Error Analysis

Error: When students are given the lengths of two sides of a triangle, and asked to find the restrictions on the third side, they tend to give only *one* restriction.

Thus, for Example 3, students might conclude only that $x < 10$ and omit the other restriction that x must also be greater than 2. Ask them to try to draw a triangle with sides 4, 6, and 1.

Checkpoint

Determine whether a triangle could have sides of the given lengths. If not, indicate why not.

1. 4 cm, 6 cm, 5 cm   Yes
2. 5 yd, 8 yd, 3 yd   No, $5 + 3 \not> 8$
3. 14 ft, 22 ft, 7 ft   No, $14 + 7 \not> 22$
4. Find the shortest segment from T to $\overline{UV}$. Why is it the shortest?
   $\overline{TW}$. By Thm 5.14, since m $\angle TWU = 90$ so that $\overline{TW} \perp \overline{UV}$.

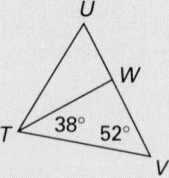

Closure

Have students state the Triangle Inequality Theorem.   The sum of the lengths of any two sides of a triangle is greater than the length of the third side.   Have the students give three possible inequalities for the construction of a triangle if the lengths of two sides of the triangle are 5 and 7.
$7 + 5 > x, 7 + x > 5, 5 + x > 7$

EXAMPLE 2   Can a triangle be constructed with sides of the given lengths?

4 m, 8 m, 3 m   7 in, 2 in, 6 in

Solution   Check the sum of the two smallest lengths.
$3 + 4 = 7$, and $7 < 8$. $2 + 6 = 8$, and $8 > 7$.
A triangle cannot be constructed.   A triangle can be constructed.

EXAMPLE 3   Find the restrictions on x such that $\triangle ABC$ can be constructed.

Solution   Use Theorem 5.15 to write three inequalities.
Sum of two lengths > third length
   (1) $4 + 6 > x$
   (2) $4 + x > 6$
   (3) $x + 6 > 4$

The third inequality is true in any case because $6 > 4$ and x is positive.

Solve the first two inequalities.
   $4 + 6 > x$ and $4 + x > 6$
   $10 > x$ and $x > 2$
So, $x > 2$ and $x < 10$, or $2 < x < 10$

Therefore, x can be any number between 2 and 10.

EXAMPLE 4   Given: m $\angle 1$ = m $\angle 2$
Prove: $BD + DC > AC$

Plan $AD + DC > AC$.

Substitute BD for AD since m $\angle 1$ = m $\angle 2$.

Proof

Statement	Reason
1. m $\angle 1$ = m $\angle 2$ ($\angle 1 \cong \angle 2$)	1. Given
2. $AD = BD$ ($\overline{AD} \cong \overline{BD}$)	2. Sides opp $\cong \angle$s of a $\triangle$ are $\cong$.
3. $AD + DC > AC$	3. $\triangle$ Ineq Thm
4. $\therefore BD + DC > AC$	4. Sub

Classroom Exercises

Tell whether or not it is possible to form a triangle with these sides.

1. 3, 5, 7   Yes
2. 2, 6, 3   No
3. 7, 5, 9   Yes
4. 4 m, 6 m, 1 m   No
5. 4.3, 6, 0.9   No
6. 3, 10, 13   No
7. $1\frac{1}{2}$, $2\frac{1}{2}$, $4\frac{1}{2}$ No
8. 5.1 ft, 7 ft, 2.3 ft   Yes

Additional Example 3

Find the restrictions on x such that $\triangle ABC$ can be constructed.
$2 < x < 12$

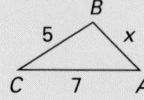

Additional Example 4

Given: m $\angle 1$ = m $\angle 2$
Prove: $QT + TS + PR > SR$
Proof: m $\angle 1$ = m $\angle 2$ or $\angle 1 \cong \angle 2$ (Given); $\overline{PT} \cong \overline{QT}$ or $PT = QT$ (Sides opp $\cong \angle$s of a $\triangle$ are $\cong$); $PT + TS = PS$ (Seg Add Post); $PS + PR > SR$ ($\triangle$ Ineq. Thm); $PT + TS + PR > SR$ (Sub); $QT + TS + PR > SR$ (Sub)

Written Exercises

Determine whether a triangle can be formed having the given lengths for sides. If not, indicate why.

1. 4 cm, 5 cm, 2 cm   Yes
2. 4 in, 7 in, 3 in   No; 4 + 3 ≯ 7
3. 9 cm, 3 cm, 7 cm   Yes

Complete each statement.

4. The longest side of a right triangle is the ___hypotenuse___.
5. The shortest segment from a point to a line is the ___perpendicular___ segment
6. Determine the shortest segment from P to $\overline{RS}$. Explain why it is shortest.
   $\overline{PQ}$, since $\overline{PQ} \perp \overline{RS}$

7. From P, construct the shortest segment to l.

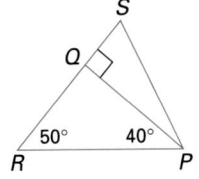

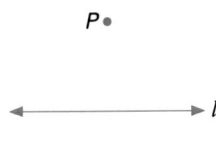

Find the restrictions on x for constructing the triangle.

8.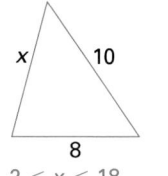
   2 < x < 18

9.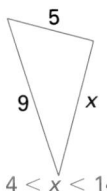
   4 < x < 14

10.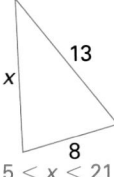
   5 < x < 21

11.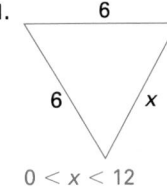
   0 < x < 12

12. Two sides of a triangle have lengths 20 and 30. The length of the third side can be any number between ___ and ___.   10, 50

13. The lengths of two sides of a triangle are 15 and 8. Use an inequality to express the range of the length of the third side.   7 < x < 23

14. The length of the base of an isosceles triangle is 12. What can be said about the lengths of the legs of the triangle?   Legs are longer than 6.

The map at the right shows air distances from New York to Chicago, Dallas, and Miami. Use the Triangle Inequality Theorem to find the range of distances between:

15. Chicago and Dallas.   661 < CD < 2,087
16. Dallas and Miami.   282 < DM < 2,466
17. Chicago and Miami.   379 < CM < 1,805

5.8 The Triangle Inequality Theorem   **213**

FOLLOW UP

Guided Practice

Classroom Exercises 1–8

Independent Practice

A Ex. 1–13, **B** Ex. 14–20, **C** Ex. 21–26

Basic: WE 1–13, Brainteaser
Average: WE 1–20 odd, Brainteaser
Above Average: WE 1–26 odd, Brainteaser

Additional Answers

Written Exercises

7.

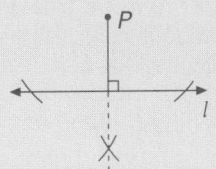

18.

Statement	Reason
1. ∠A ≅ ∠C	1. Given
2. $\overline{AD} \cong \overline{DC}$	2. Sides opp ≅ ∠s are ≅.
3. AD + BD > AB	3. △ Ineq Thm
4. BD + DC > AB	4. Sub

19. Rt △ABC, hyp = $\overline{AB}$ (Given), ∠C is rt ∠, so ∠A and ∠B are acute ∠s. m ∠A < m ∠C and m ∠B < m ∠C. Since m ∠A < m ∠C then BC < AB, since for two sides of a △, the longer side is opp the larger ∠. Also, since m ∠B < m ∠C, AC < AB. Therefore, the hyp $\overline{AB}$, is the longest side.

Enrichment

Draw this figure on the chalkboard. Challenge the students to find a point P on l such that AP + BP is as small as possible. If after a time they cannot find a valid construction, explain the following construction at the board, and challenge the students to prove that it is valid.

Construction: (1) Construct $\overline{AC} \perp l$.
(2) Plot Q on $\overrightarrow{AC}$ such that AC = CQ.
(3) Draw $\overline{QB}$ intersecting l at P. P is the point on l such that AP + BP is as small as possible.

Plan of proof:
(1) Show that BQ = AP + BP.
(2) For any other point, X, on l, BX + XQ > BQ. Therefore, BX + XQ > AP + BP.

Use with Enrichment

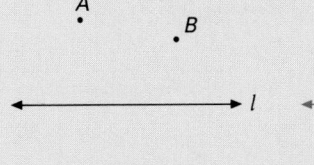

20. Part One: Rt △ABC, (hyp = $\overline{AC}$) (Given); Construct $\overline{DB}$ so that $\overline{DB} \perp \overline{AC}$ (From a pt, a ⊥ can be constructed to a line.); AB > AD, BC > DC (Longest side of rt △ is hyp.); AB + BC > AD + DC (Add Prop of Ineq); AC = AD + DC (Seg Add Post); AB + BC > AC (Sub Prop of Ineq)

Part Two: AC > AB (Hyp is longest side of rt △.); BC > 0 (Length is always pos); AC + BC > AB (Add Prop of Ineq)

Part Three: AC > BC (Hyp is longest side rt △.); AB > 0 (Length is always pos.); AC + AB > BC (Add Prop of Ineq.)

23. Part One: Obtuse △ABC, ∠B is obtuse, ∠A and ∠C are acute (Given); Construct $\overline{BD} \perp \overline{AC}$ (From a pt, a ⊥ can be constructed to a line.); AB > AD, BC > DC (longest side of rt △ is hyp.); AB + BC > AD + DC (Add Prop of Ineq); AD + DC = AC (Seg Add Post); AB + BC > AC (Sub Prop of Ineq)

Part Two: AC > AB (Side opp larger ∠ is longer.); BC > 0 (Length is pos.); AC + BC > AB (Add Prop of Ineq)

Part Three: AC > BC (Side opp larger ∠ is longer.); AB > 0 (Length is pos.); AC + AB > BC (Add Prop of Ineq)

24.

Statement	Reason
1. Q is on $\overline{XY}$	1. Given
2. XZ + XQ > ZQ, ZY + QY > ZQ	2. △ Ineq Thm
3. XZ + ZY + XQ + QY > 2(ZQ)	3. Add Prop of Ineq
4. XQ + QY = XY	4. Seg Add Post
5. XZ + ZY + XY > 2(ZQ)	5. Sub Prop of Ineq

25.

Statement	Reason
1. RQ + RS > SQ, SP + ST > PT	1. △ Ineq Thm
2. RQ + RS + SP + ST > SQ + PT	2. Add Prop of Ineq
3. RS + SP = RP, ST + TQ = SQ	3. Seg Add Post
4. RQ + RP + ST > ST + TQ + PT	4. Sub Prop of Ineq
5. RQ + RP > PT + TQ	5. Subt Prop of Ineq

---

18. In the figure at the right, ∠A ≅ ∠C.
Prove: BD + DC > AB

19. Write a paragraph proof of the Corollary to Theorem 5.14.

20. Prove Theorem 5.15 for a right triangle.

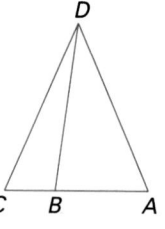

In △ABC, AB and BC are given, find the range of values for AC.

37.48 < AC < 92.7
21. AB = 27.61, BC = 65.09   **22.** AB = 94.83, BC = 101.48

23. Prove Theorem 5.15 for an obtuse triangle.   6.65 < AC < 196.31

24. Prove that if Q is a point on $\overline{XY}$ of △XYZ, then 2 · ZQ < XZ + ZY + XY.

25. Prove: PR + RQ > PT + TQ   **26.** Prove: AC + BD > AD + BC

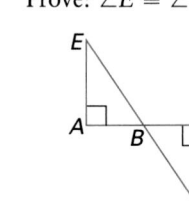   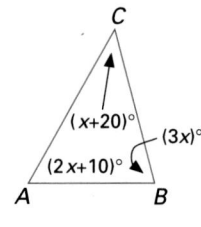

Mixed Review

1. The measure of the vertex angle of an isosceles triangle is 110. Find the measure of each base angle. **5.1** 35

2. Given: $\overline{QS} \cong \overline{QW}$, $\overline{PS} \cong \overline{PW}$
Prove: ∠1 ≅ ∠2 **5.3**

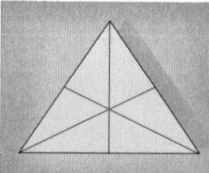

3. Given: $\overline{AE} \perp \overline{AC}$, $\overline{CD} \perp \overline{AC}$, $\overline{AC}$ bisects $\overline{ED}$.
Prove: ∠E ≅ ∠D **4.6**

4. Determine the longest segment. **5.7** $\overline{AC}$

Brainteaser

The figure to the right shows a triangle with three medians. How many triangles in all are there in the figure? **16**

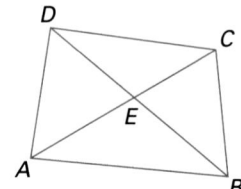

214   Chapter 5 Congruence

26.

Statement	Reason
1. BC < CE + EB, AD < DE + AE	1. △ Ineq Thm
2. BC + AD < CE + EB + DE + AE	2. Add Prop of Ineq
3. CE + AE = AC, EB + DE = BD	3. Seg Add Post
4. BC + AD < AC + BD	4. Sub Prop of Ineq

See page 222 for the answers to Mixed Review Ex. 2–3.

5.9 Inequalities for Two Triangles

Objective To apply the SAS and SSS Inequality Theorems

Think of two sticks hinged together and connected with an elastic band. As the angle between the sticks becomes greater, the length of the band increases. This suggests Theorem 5.16.

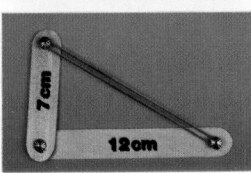

Theorem 5.16

The SAS Inequality Theorem: If two sides of one triangle are congruent, respectively, to two sides of a second triangle, and the included angle of the first triangle has a greater measure than the included angle of the second triangle, then the third side of the first triangle is longer than the third side of the second triangle.

Given: $\triangle ABC$ and $\triangle DEF$, with $\overline{CA} \cong \overline{FD}$, $\overline{CB} \cong \overline{FE}$, and m $\angle C >$ m $\angle F$
Prove: $AB > DE$

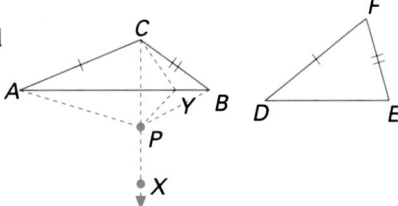

Proof

Statement	Reason
1. $\overline{CA} \cong \overline{FD}$	**(S)** 1. Given
2. Construct $\overrightarrow{CX}$ such that m $\angle ACX =$ m $\angle F$.	**(A)** 2. Protractor Post
3. Locate point P on $\overrightarrow{CX}$ such that $CP = FE$ ($\overline{CP} \cong \overline{FE}$).	**(S)** 3. Ruler Post
4. $\triangle APC \cong \triangle DEF$	4. SAS
5. Construct bis of $\angle PCB$.	5. Protractor Post
6. Locate point Y, the intersection of $\overline{AB}$ and the bis of $\angle PCB$.	6. Ruler Post

Teaching Resources

Quick Quizzes 42
Reteaching and Practice
  Worksheets, pp. 83, 84
Transparencies 11, 12

▬ GETTING STARTED

Prerequisite Quiz

Complete each statement.

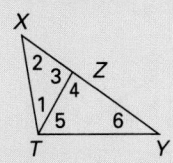

1. If $XT > XZ$, then m $\angle$ _____ $<$ m $\angle$ _____. 1, 3
2. If m $\angle 4 <$ m $\angle 6$, then _____ $>$ _____. TZ, TY
3. If m $\angle 1 = 30$ and m $\angle 2 = 50$, then _____ $>$ _____. TZ, XZ
4. If m $\angle 1 = 20$, m $\angle 3 = 80$, then _____ $>$ _____. XT, XZ
5. Given: $\overline{CY}$ bisects $\angle PCB$,
   $\overline{CP} \cong \overline{CB}$
   Prove: $\overline{PY} \cong \overline{BY}$

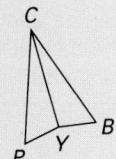

Statement	Reason
1. $\overline{CY}$ bisects $\angle PCB$	1. Given
2. $\angle PCY \cong \angle BCY$	2. Def of $\angle$ bis
3. $\overline{YC} \cong \overline{YC}$	3. Reflex Prop of Congr
4. $\overline{CP} \cong \overline{CB}$	4. Given
5. $\triangle PCY \cong \triangle BCY$	5. SAS
6. $\overline{PY} \cong \overline{BY}$	6. CPCTC

Motivator

Have the students draw a triangle with sides of lengths 5 cm and 6 cm and an included angle with a measure of 40. Then have them draw a second triangle with sides of lengths 5 cm and 6 cm and an included angle with a measure of 50. Have them predict which triangle will have the longer side opposite the given included angle and then have them verify by measuring the sides. The second triangle has the longer side.

TEACHING SUGGESTIONS

Lesson Note

Ask students to use the figures at the top of page 215 to try to verbalize an SAS inequality theorem. The figures suggest why this is sometimes called the Hinge Theorem.

Math Connections

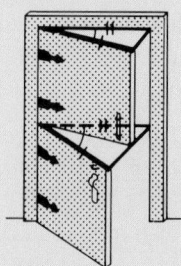

Dutch Doors: Dutch doors are divided horizontally so that the upper and lower parts may be opened separately, as shown above. The triangles formed by opening the doors have congruent corresponding sides as marked. The SAS Inequality Theorem can be applied to show that the lengths of the third sides of the triangles vary with the measurements of the angles included by the congruent corresponding sides.

Critical Thinking Questions

Analysis: Ask the students to compare the SAS Inequality Theorem with the SSS Inequality Theorem. Given 2 △s with 2 pair of corr sides ≅, the SAS Ineq Thm is the converse of SSS Ineq Thm. How can these two triangle inequality theorems be stated as one biconditional? Is the biconditional true? Why or why not? If 2 △s have 2 pair of corr sides ≅, then the meas of the incl ∠ of one △ is greater than the meas of the incl ∠ of the other △ if and only if the side opp the incl ∠ in the first △ is greater than the side opp the incl ∠ of the other △. True; because both conditionals are true.

---

7. $\overline{CY} \cong \overline{CY}$	(S) 7.	Reflex Prop
8. $\angle PCY \cong \angle BCY$	(A) 8.	Def of ∠ bis
9. $FE = CB$ ($\overline{FE} \cong \overline{CB}$)	9.	Given
10. $CP = CB$ ($\overline{CP} \cong \overline{CB}$)	(S) 10.	Trans Prop ·
11. $\triangle PCY \cong \triangle BCY$	11.	SAS
12. $PY = BY$ ($\overline{PY} \cong \overline{BY}$)	12.	CPCTC
13. $AY + PY > AP$	13.	△ Ineq Thm
14. $AY + BY > AP$	14.	Sub
15. $AY + BY = AB$	15.	Seg Add Post
16. $AP = DE$	16.	CPCTC
17. $\therefore AB > DE$	17.	Sub Prop of Ineq

The SSS Inequality Theorem can be proved indirectly by using the Trichotomy Property.

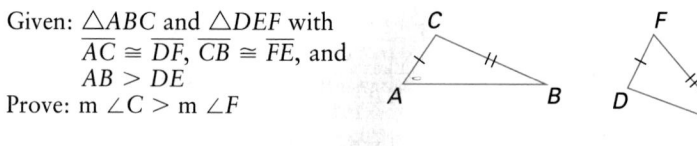

Theorem 5.17

The SSS Inequality Theorem: If two sides of one triangle are congruent, respectively, to two sides of a second triangle, and the length of the third side of the first triangle is greater than the length of the third side of the second triangle, then the angle opposite the third side of the first triangle has a greater measure than the angle opposite the third side of the second triangle.

Given: $\triangle ABC$ and $\triangle DEF$ with $\overline{AC} \cong \overline{DF}$, $\overline{CB} \cong \overline{FE}$, and $AB > DE$
Prove: m $\angle C >$ m $\angle F$

Plan

Use an indirect proof. By the Trichotomy Property, m $\angle C <$ m $\angle F$, or m $\angle C =$ m $\angle F$, or m $\angle C >$ m $\angle F$. Show that the first two assumptions lead to a contradiction.

EXAMPLE 1

Given: $\overline{AD} \cong \overline{DC}$, $AB > BC$
Prove: m $\angle 1 >$ m $\angle 2$

Proof

Statement	Reason
1. $\overline{AD} \cong \overline{DC}$	1. Given
2. $\overline{DB} \cong \overline{DB}$	2. Reflex Prop
3. $AB > BC$	3. Given
4. $\therefore$ m $\angle 1 >$ m $\angle 2$	4. SSS Ineq Thm

Additional Example 1

Insert the correct inequality symbol and give a reason: YU _____ XT

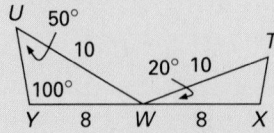

$YU > XT$ by the SAS Inequality Theorem, since m $\angle YWU = 30$ and m $\angle TWX = 20$.

EXAMPLE 2 Given: $\overline{DB}$ bisects $\overline{AC}$; m ∠1 > m ∠2
Prove: m ∠A > m ∠C

Plan Use the SAS Inequality Theorem to prove
$DC > DA$. Then, in △ACD, m ∠A > m ∠C.

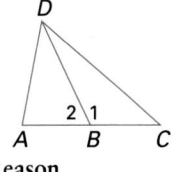

Proof

Statement	Reason
1. $\overline{DB}$ bisects $\overline{AC}$.	1. Given
2. $\overline{AB} \cong \overline{BC}$	2. Def of seg bis
3. $\overline{DB} \cong \overline{DB}$	3. Reflex Prop
4. m ∠1 > m ∠2	4. Given
5. $DC > DA$	5. SAS Ineq Thm
6. ∴ m ∠A > m ∠C	6. In a △, larger ∠ is opp longer side.

Focus on Reading

Indicate whether the statement is sometimes true, always true, or never true.

1. The lengths of two sides of a triangle are equal, respectively, to the lengths of two sides of a second triangle, and the third side of the first triangle is longer than the third side of the second triangle.   Sometimes true

2. The lengths of two sides of a triangle are equal, respectively, to the lengths of two sides of a second triangle, the included angle of the first triangle has a greater measure than the included angle of the second triangle, and the third sides are equal in length.   Never true

3. The legs of a right triangle are congruent, respectively, to the legs of a second right triangle, and the hypotenuse of the first right triangle is longer than the hypotenuse of the second right triangle.   Never true

Classroom Exercises

Write an inequality comparing the indicated side lengths or the indicated angle measures.

1.
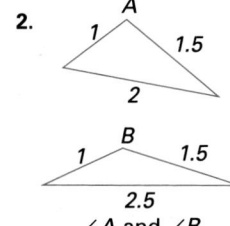
$\overline{AB}$ and $\overline{BC}$
$AB > BC$

2.
4 cm 40° 4 cm 30°

∠A and ∠B
m ∠B > m ∠A

3.
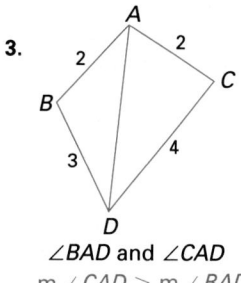
∠BAD and ∠CAD
m ∠CAD > m ∠BAD

5.9 Inequalities for Two Triangles **217**

Additional Example 2

Given: $\overline{SQ}$ bisects $\overline{PR}$, $PS > RS$.
Prove: m ∠1 > m ∠2

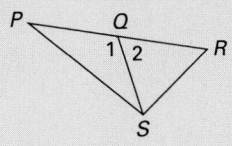

Proof: $\overline{SQ}$ bisects $\overline{PR}$ (Given); $\overline{PQ} \cong \overline{RQ}$ (Def of bis); $\overline{QS} \cong \overline{SQ}$ (Reflex Prop of Congr); $PS > RS$ (Given); m ∠1 > m∠2 (SSS Ineq Thm)

Checkpoint

Insert the correct inequality symbol.

1. $PQ = 4$, $QR = 4$, m ∠1 = 125. PS _____ RS >
2. m ∠P = m ∠R, $PQ = 10$, $QR = 7$. m ∠4 _____ m ∠3 <
3. Given: $\overline{SQ}$ bisects $\overline{PR}$, $SR < SP$. Prove: m ∠1 > m ∠2

Statement	Reason
1. $\overline{SQ}$ bisects $\overline{PR}$.	1. Given
2. $\overline{QP} = \overline{QR}$	2. Def of bis
3. $\overline{QS} = \overline{QS}$	3. Reflex Prop of Congr
4. $SR < SP$ or $SP > SR$	4. Given
5. m ∠1 > m ∠2	5. SSS Ineq Thm

Closure

Ask the students to explain what the SSS Inequality Theorem states.   If two sides of one △ are congruent, respectively, to two sides of a second △, and the length of the third side of the first △ is greater than the third side of the second △, then the angle measure opposite the third side of the first △ is greater than the angle measure opposite the third side of the second △.
Ask them to recall the SAS Inequality Theorem.   If two sides of one △ are congruent, respectively, to two sides of a second △, and the included angle of the first △ has a greater measure than the included angle of the second △, then the third side of the first △ is longer than the third side of the second △.

FOLLOW UP

Guided Practice

Classroom Exercises 1–3

Independent Practice

A Ex. 1–8, **B** Ex. 9–13, **C** Ex. 14–16

Basic: FR all, WE 1–8, AR all
Average: FR all, WE 1–13, AR all
Above Average: FR all, WE 1–16, AR all

Additional Answers

4.

Statement	Reason
1. $\overline{SQ}$ bis $\overline{PR}$, $SR > SP$	1. Given
2. $\overline{PQ} \cong \overline{QR}$	2. Def of seg bis
3. $\overline{SQ} \cong \overline{SQ}$	3. Reflex Prop of Congr
4. m $\angle 2 >$ m $\angle 1$	4. SAS Ineq Thm

5.

Statement	Reason
1. m $\angle P >$ m $\angle R$, $PQ = QR$	1. Given
2. $SR > SP$	2. Side opp larger $\angle$ is longer.
3. $\overline{SQ} \cong \overline{SQ}$	3. Reflex Prop of Congr
4. m $\angle 2 >$ m $\angle 1$	4. SSS Ineq Thm

6.

Statement	Reason
1. m $\angle P =$ m $\angle R$, m $\angle 3 >$ m $\angle 4$	1. Given
2. $\overline{SP} \cong \overline{SR}$	2. Sides opp $\cong \angle$s of $\triangle$ are $\cong$.
3. $\overline{SQ} \cong \overline{SQ}$	3. Reflex Prop of Congr
4. $PQ > QR$	4. SAS Ineq Thm

7.

Statement	Reason
1. $QR = PS$, $QS < PR$	1. Given
2. $\overline{PQ} \cong \overline{PQ}$	2. Reflex Prop of Congr
3. m $\angle PQR >$ m $\angle SPQ$	3. SSS Ineq Thm

8.

Statement	Reason
1. T is midpt of $\overline{QS}$, m $\angle PTQ >$ m $\angle PTS$	1. Given
2. $TS = TQ$	2. Def of midpt
3. $\overline{TP} \cong \overline{TP}$	3. Reflex Prop of Congr
4. $PQ > PS$	4. SAS Ineq Thm

9.

Statement	Reason
1. $TP > TQ$, $SQ = PR$	1. Given
2. $\overline{PQ} \cong \overline{PQ}$	2. Reflex Prop of Congr
3. m $\angle 1 >$ m $\angle 4$	3. $\angle$ opp longer side is larger.
4. $PS > QR$	4. SAS Ineq Thm

Written Exercises

Insert the correct inequality symbol. (Exercises 1–3)

1. $AB = BC$, m $\angle 1 = 30$, m $\angle 2 = 20$
$AD \underline{\leq} DC$.

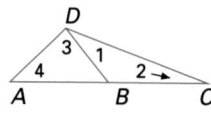

2. $DA = CB$, m $\angle DAB = 115$, $\overline{DA} \parallel \overline{CB}$
$DB \underline{>} AC$.

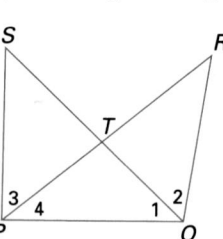

3. m $\angle 1 \underline{>}$ m $\angle 2$.

3 cm
3 cm 3 cm
4 cm
2.5 cm

4. Given: $\overline{SQ}$ bisects $\overline{PR}$; $SR > SP$
Prove: m $\angle 2 >$ m $\angle 1$

5. Given: m $\angle P >$ m $\angle R$, $PQ = QR$
Prove: m $\angle 2 >$ m $\angle 1$

6. Given: m $\angle P =$ m $\angle R$, m $\angle 3 >$ m $\angle 4$
Prove: $PQ > QR$

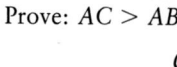

7. Given: $QR = PS$, $QS < PR$
Prove: m $\angle PQR >$ m $\angle SPQ$

8. Given: T is the midpoint of $\overline{QS}$; m $\angle PTQ >$ m $\angle PTS$
Prove: $PQ > PS$

9. Given: $TP > TQ$, $SQ = PR$
Prove: $PS > QR$

10. Given: m $\angle 4 >$ m $\angle 1$, m $\angle 3 >$ m $\angle 2$, $PS = QR$
Prove: $QS > PR$

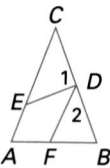

Essay

11. Write an explanation of why Theorem 5.16 may be referred to as the Hinge Theorem.

12. Given: $QS = TR$, $SR > TQ$
Prove: $PR > PQ$

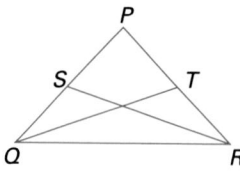

13. Given: $\overline{ED} \cong \overline{DF}$, m $\angle 1 >$ m $\angle 2$,
D is midpoint of $\overline{CB}$;
$\overline{AE} \cong \overline{AF}$
Prove: $AC > AB$

Enrichment

Have the students draw any triangle and label it ABC. Next have them place a point P anywhere in the interior of the triangle. Then ask them to prove that $AP + BP + CP$ is greater than one-half the perimeter of the triangle.

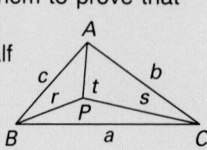

Proof: $r + s > a$, $s + t > b$, $t + r > c$ (Triangle Ineq Thm); $2r + 2s + 2t > a + b + c$ (Add Prop of Ineq); $2(r + s + t) >$ perimeter of $\triangle ABC$ (def of perimeter); $r + s + t > \frac{1}{2}$ perimeter of $\triangle ABC$ (Mult Prop of Ineq)

14. Given: m ∠DBC = m ∠DCB,
 m ∠ADB < m ∠ADC
 Prove: m ∠ACB < m ∠ABC

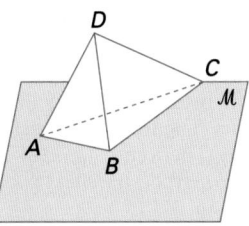

15. Given: AE = AB = ED = DC,
 EB < EC
 Prove: m ∠AEB > m ∠DCE

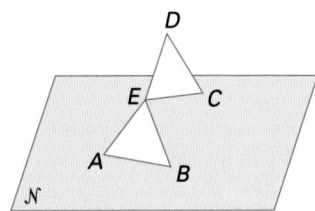

16. Prove the SSS Inequality Theorem.

Mixed Review

In the figure, $\overline{AB} \parallel \overline{EC}$ 3.5, 3.7

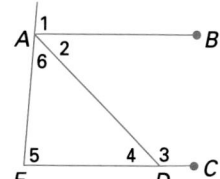

1. Find m ∠1 if m ∠5 = 40. **3.5** 40
2. Find m ∠5 if m ∠EAB = 150. **3.5** 30
3. Given: m ∠5 = 70, m ∠6 = 40. Find m ∠3. **3.7** 110
4. Given: m ∠4 = 30, m ∠6 is twice m ∠5. Find m ∠5. **3.7** 50

Algebra Review

To multiply or divide fractions:

Use $\dfrac{a}{b} \cdot \dfrac{c}{d} = \dfrac{a \cdot c}{b \cdot d}$ or $\dfrac{a}{b} \div \dfrac{c}{d} = \dfrac{a \cdot d}{b \cdot c}$.

Example: Divide $\dfrac{x^2 - 10x + 16}{x^2 - 49} \div \dfrac{3x - 24}{x - 7}$.

$$\dfrac{x^2 - 10x + 16}{x^2 - 49} \div \dfrac{3x - 24}{x - 7} = \dfrac{(x^2 - 10x + 16)(x - 7)}{(x^2 - 49)(3x - 24)}$$

$$= \dfrac{\cancel{(x - 8)}(x - 2)\cancel{(x - 7)}}{\cancel{(x - 7)}(x + 7)(3)\cancel{(x - 8)}} = \dfrac{x - 2}{3(x + 7)}$$

Multiply or divide.

1. $\dfrac{x^2 - 4}{4x + 12} \cdot \dfrac{2x + 6}{4x + 8}$
 $\dfrac{(x - 2)}{8}$

2. $\dfrac{x^2 - 2x - 8}{x^2 - 25} \div \dfrac{3x - 9}{x - 5}$
 $\dfrac{x^2 - 2x - 8}{3x^2 + 6x - 45}$

3. $\dfrac{m^2 - 4}{a + 1} \cdot \dfrac{7a + 7}{2m + 4}$
 $\dfrac{7m - 14}{2}$

5.9 Inequalities for Two Triangles **219**

16. Given: △ABC and △DEF on p. 216
with $\overline{AC} \cong \overline{DF}$, $\overline{CB} \cong \overline{FE}$, AB > DE.
Case 1: Assume that m ∠C < m ∠F.
Then by SAS Ineq Thm AB < DE,
which contradicts the given information,
so m ∠C ≮ m ∠F.
Case 2: Assume that m ∠C = m ∠F.
Then △ACB ≅ △DFE by SAS, so AB
= DE by CPCTC. This contradicts the

given information, so m ∠C ≠ m ∠F.
Therefore, m ∠C > m ∠F by the Trich
Prop.

10.

Statement	Reason
1. m ∠4 > m ∠1, m ∠3 > m ∠2, PS = QR	1. Given
2. $\overline{PQ} \cong \overline{PQ}$	2. Reflex Prop of Congr
3. m ∠4 + m ∠3 > m ∠1 + m ∠2	3. Add Prop of Ineq
4. m ∠4 + m ∠3 = m ∠SPQ, m ∠1 + m ∠2 = m ∠RQP	4. Angle Add Post
5. m ∠SPQ > m ∠RQP	5. Sub
6. QS > PR	6. SAS Ineq Thm

11. As the hinge opens wider, the third side becomes longer. The two ≅ sides are hinged together. SAS refers to the 2 ≅ sides and the included ∠.

12.

Statement	Reason
1. QS = TR, SR > TQ	1. Given
2. $\overline{QR} \cong \overline{QR}$	2. Reflex Prop of Congr
3. m ∠SQR > m ∠TRQ	3. SSS Ineq Thm
4. PR > PQ	4. Side opp larger ∠ is longer.

13.

Statement	Reason
1. $\overline{ED} \cong \overline{DF}$, m ∠1 > m ∠2, D is midpt of CB, $\overline{AE} \cong \overline{AF}$	1. Given
2. $\overline{DB} \cong \overline{DC}$	2. Def of Midpt
3. EC > FB	3. SAS Ineq Thm
4. EC + AE > FB + AE	4. Add Prop of Ineq
5. EC + AE > FB + AF	5. Sub
6. EC + AE = AC, FB + AF = AB	6. Seg Add Post
7. AC > AB	7. Sub Prop of Ineq

14.

Statement	Reason
1. m ∠DBC = m ∠DCB, m ∠ADB < m ∠ADC	1. Given
2. $\overline{DB} \cong \overline{DC}$	2. Sides opp ≅ ∠s are ≅.
3. $\overline{AD} \cong \overline{AD}$	3. Reflex Prop of Congr
4. AB < AC	4. SAS Ineq Thm
5. m ∠ACB < m ∠ABC	5. ∠ opp longer side is larger.

See page 223 for the answer to Ex. 15.

219

3.

Statement	Reason
1. $\angle 4 \cong \angle 1$	1. Given
2. $\angle 3$ and $\angle 4$ are supp, $\angle 2$ and $\angle 1$ are supp.	2. If the outer rays of 2 adj $\angle$s form a st $\angle$, then the $\angle$s are supp.
3. $\angle 3 \cong \angle 2$	3. Supp $\angle$s of $\cong \angle$s are $\cong$.
4. $\overline{PS} \cong \overline{RS}$	4. Sides opp $\cong \angle$s are $\cong$.

4. $\overline{SQ}$ is $\perp$ bis of $\overline{PR}$. If a seg bisects the vertex $\angle$ of an isos $\triangle$, then it is $\perp$ bis of base.

5.

Statement	Reason
1. $\overline{BC} \cong \overline{BA}$, $\overline{CD} \cong \overline{AD}$	1. Given
2. $\overline{BD} \cong \overline{BD}$	2. Reflex Prop of Congr
3. $\triangle BCD \cong \triangle BAD$	3. SSS
4. $\angle 5 \cong \angle 6$	4. CPCTC
5. $\overline{ED} \cong \overline{ED}$	5. Reflex Prop of Congr
6. $\triangle CDE \cong \triangle ADE$	6. SAS
7. $\angle 3 \cong \angle 4$	7. CPCTC

6.

Statement	Reason
1. $\overline{CA} \perp \overline{BD}$ $\angle 1 \cong \angle 2$	1. Given
2. m $\angle DBA = 90$, m $\angle DBC = 90$	2. Def of $\perp$
3. m $\angle DBA = $ m $\angle DBC$	3. Sub
4. $\overline{CE} \cong \overline{AE}$	4. Sides opp $\cong \angle$s are $\cong$.
5. $\triangle CBE \cong \triangle ABE$	5. HL
6. $\overline{CB} \cong \overline{AB}$	6. CPCTC
7. $\overline{BD} \cong \overline{BD}$	7. Reflex Prop of Congr
8. $\triangle DBC \cong \triangle DBA$	8. SAS
9. $\angle 5 \cong \angle 6$	9. CPCTC

Key Terms

altitude   (p. 194)
base   (p. 177)
base angles   (p. 177)
equiangular triangle   (p. 177)
equilateral triangle   (p. 177)
hypotenuse   (p. 188)
leg of isosceles triangle   (p. 177)
leg of right triangle   (p. 188)

median   (p. 194)
perimeter of equilateral triangle   (p. 178)
perpendicular bisector   (p. 199)
SAS Inequality Theorem   (p. 215)
SSS Inequality Theorem   (p. 216)
trichotomy property   (p. 206)
vertex angle   (p. 177)

Key Ideas and Review Exercises

5.1   To write proofs and solve problems about isosceles triangles, use these properties of an isosceles triangle:

- Two sides (legs) are $\cong$: $\overline{AB} \cong \overline{AC}$
- $\angle$s opposite $\cong$ legs are $\cong$: $\angle B \cong \angle C$
- Bis of vertex $\angle$ is $\perp$ bis of base.

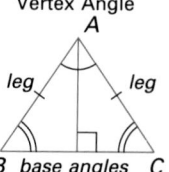

Vertex Angle

1. Given: $PS = RS$, m $\angle 3 = 80$. Find m $\angle PSR$.   20
2. Given: m $\angle PSR = 2x$, m $\angle P = 3x - 10$, $PS = RS$. Find m $\angle PSR$.   50
3. Given: $\angle 4 \cong \angle 1$
   Prove: $\overline{PS} \cong \overline{RS}$
4. Given: $\overline{PS} \cong \overline{RS}$, $\overline{QS}$ bisects $\angle PSR$. What conclusion about $\overline{SQ}$ and $\overline{PR}$ can be drawn? Why?

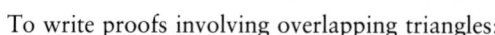

5.3, 5.4   To prove right triangles congruent, use the Hypotenuse-Leg congruence pattern or any of the patterns established for congruence of triangles.

To write proofs involving overlapping triangles:

(1) Separate the pairs of triangles.
(2) Determine which pair of triangles you must prove congruent. If necessary, prove a pair of triangles congruent so that their corresponding sides or angles can be used to prove a second pair of triangles congruent.

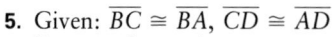

5. Given: $\overline{BC} \cong \overline{BA}$, $\overline{CD} \cong \overline{AD}$
   Prove: $\angle 3 \cong \angle 4$
6. Given: $\overline{CA} \perp \overline{BD}$, $\angle 1 \cong \angle 2$
   Prove: $\angle 5 \cong \angle 6$

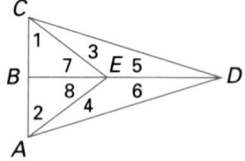

5.5   To work with altitudes and medians, use the following theorems.

- Corr alt of ≅ △s are ≅.
- Corr medians of ≅ △s are ≅.
- The alt to the base of an isos △ is a median.

7. Prove that the altitude to the base of an isosceles triangle bisects the vertex angle.

5.6   To apply properties of perpendicular bisectors

(1) If *P* and *Q* are equidistant (*PA* = *PB*, *QA* = *QB*) from the endpoints of a segment (*A* and *B*), then $\overline{PQ}$ is the perpendicular bisector of $\overline{AB}$.

(2) Any point *W* on the perpendicular bisector of a segment $\overline{AB}$ is equidistant (*WA* = *WB*) from the endpoints of the segment.

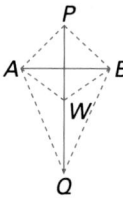

Based on the given information, what conclusions can be drawn?

8. Given: $\overline{AD} \cong \overline{AC}$, $\overline{BD} \cong \overline{BC}$.
9. Given: $\overline{AB}$ is the perpendicular bisector of $\overline{DC}$.

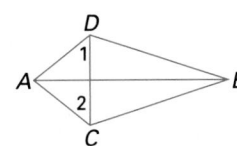

5.7–
5.9   To work with inequalities in triangles, use

Triangle Inequality Theorem.

- *BA* + *AC* > *BC*
- If *AB* > *AC*, then m ∠*C* > m ∠*B*
- If m ∠*A* > m ∠*B*, then *CB* > *CA*

SAS Inequality Theorem

- If *AB* = *DE*, *AC* = *DF*, m ∠*A* > m ∠*D*, then *BC* > *EF*

SSS Inequality Theorem

- If *AB* = *DE*, *AC* = *DF*, *BC* > *EF*, then m ∠*A* > m ∠*D*

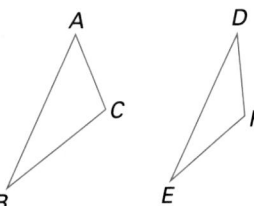

10. Name the longest side.
$\overline{AB}$

11. Write an inequality for the restrictions on *x*.
1 < *x* < 7

12. Given: ∠1 ≅ ∠2,
     PQ > *PR*
     Prove: *PQ* > *PS*

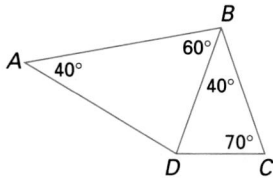

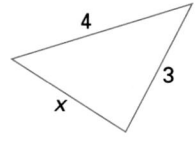

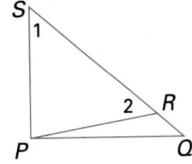

Chapter 5 Review   **221**

Mixed Review

2.

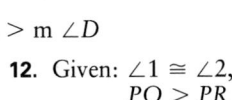

3. Locate two points on the angle an equal distance from the vertex. Then place the square over the angle such that the distances from these two points to the vertex of the right angle of the square are equal. Draw a line from the vertex of the square to the vertex of the angle to be bisected. This line will be the bisector of the angle in question, because the △s formed are congruent (SSS).

7.

Statement	Reason
1. △*ABC* with alt $\overline{AD}$ and *AB* = *AC*	1. Given
2. $\overline{AD}$ is a median.	2. Alt to base of isos△ is a median.
3. *BD* = *CD*	3. Def of median
4. $\overline{AD} \cong \overline{AD}$	4. Reflex Prop of Congr
5. △*ADB* ≅ △*ADC*	5. SSS
6. ∠*CAD* ≅ ∠*BAD*	6. CPCTC
7. $\overline{AD}$ bis ∠*BAC*, the vertex ∠	7. Def of ∠ bis

8. $\overline{AB}$ ⊥ bis of $\overline{DC}$; $\overline{AB}$ is bis of ∠s *CAD* and *CBD*.
9. $\overline{AD} \cong \overline{AC}$, $\overline{DB} \cong \overline{CB}$

12.

Statement	Reason
1. ∠1 ≅ ∠2, *PQ* > *PR*	1. Given
2. $\overline{PR} \cong \overline{PS}$ (*PR* = *PS*)	2. Sides opp ≅ ∠s are ≅.
3. *PQ* > *PS*	3. Subt Prop of Ineq

Additional Answers, page 204

Application

1. Find a vantage point such that the angle (labelled measure *d*) formed by the lines of sight to the poles has one-half the measure of the angle (labelled measure 2*d*) exterior to the triangle. It is easy to show (using the exterior angle theorem) that the △ is isosceles. The distance from the vantage point to Pole 2 (which can be measured) is equal to the distance between the two poles.

2. Place the level upright on the surface to be judged, with the isosceles triangle resting on its base. The surface is horizontal if the plumb bob points to the center mark of the base. This works because the altitude of an isosceles triangle bisects the base; and since there is only one perpendicular from a point to a line (see Problem 9, page 203), a line from the vertex of an isosceles triangle perpendicular to the base must be an altitude. Thus the angle formed by the vertical plumb line and the base are right angles, so the base is horizontal.

8.

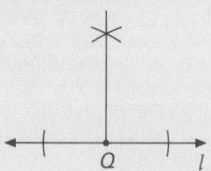

12.

Statement	Reason
1. $\angle 1 \cong \angle 2$, $\overline{AB} \cong \overline{CD}$, $\angle 3 \cong \angle 4$	1. Given
2. $\overline{EB} \cong \overline{EC}$	2. Sides opp $\cong \angle$s are $\cong$.
3. $\angle ABE$ and $\angle 1$ are supp, $\angle DCE$ and $\angle 2$ are supp.	3. If the outer rays of 2 adj $\angle$s form a st $\angle$, then the $\angle$s are supp.
4. $\angle ABE \cong \angle DCE$	4. Supp $\angle$s of $\cong \angle$s are $\cong$.
5. $\triangle ABE \cong \triangle DCE$	5. SAS
6. $\overline{EA} \cong \overline{ED}$ $\angle A \cong \angle D$	6. CPCTC
7. $\triangle ABF \cong \triangle DCG$	7. AAS
8. $\overline{FA} \cong \overline{GD}$	8. CPCTC
9. $EA = FA + FE$, $ED = GD + GE$	9. Seg Add Post
10. $FA + FE = GD + GE$	10. Sub
11. $\overline{FE} \cong \overline{GE}$	11. Subt Prop of Eq

Additional Answer, page 214

2.

Statement	Reason
1. $\overline{QS} \cong \overline{QW}$, $\overline{PS} \cong \overline{PW}$	1. Given
2. $\overline{QP} \cong \overline{QP}$	2. Reflex Prop
3. $\triangle QSP \cong \triangle QWP$	3. SSS
4. $\angle SQP \cong \angle WQP$	4. CPCTC
5. $\overline{QR} \cong \overline{QR}$	5. Reflex Prop
6. $\triangle QSR \cong \triangle QWR$	6. SAS
7. $\angle 1 \cong \angle 2$	7. CPCTC

Chapter 5 Test

A—Level Ex.: 1–4, 13–14   B—Level Ex.: 5–11
C—Level Ex.: starred

1. The measure of a base angle of an isosceles triangle is 80. Find the measure of the vertex angle.   **20**

Draw a conclusion from the diagram. (Exercises 2–3)
Name the longest side. (Exercise 4)   **2.** $\overline{TR}$ is $\perp$ bis of $\overline{PQ}$.   **3.** $\overline{WU}$ is $\perp$ bis of $\overline{TV}$.

2.

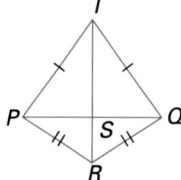

3.

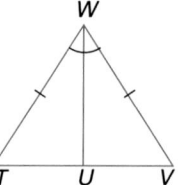

4.

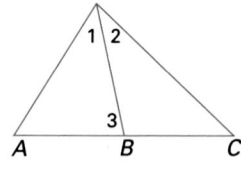

5. In $\triangle ABC$, m $\angle A = 2x - 10$, m $\angle B = x + 40$, m $\angle C = x + 30$ Name the longest side. $\overline{AC}$
6. In $\triangle ABC$, $AB = 6$, $BC = 7$, $AC = 5$. Name the largest angle. $\angle A$
7. In $\triangle ABC$, $AB = 4$, $BC = 6$. What are the restrictions on the length of $\overline{AC}$? $2 < AC < 10$
8. Draw a line l with point Q on l. Construct a perpendicular to l at Q.

Complete the statement with the symbol =, >, or <.

9. B is the midpoint of $\overline{AC}$. m $\angle 1 = 40$, m $\angle 2 = 70$. $AB \underline{=} BC$
10. $AB = BC = 6$, m $\angle 1 = 50$, m $\angle A = 70$. $AD \underline{\le} CD$
11. $AD = BD$, m $\angle C <$ m $\angle 3$ $AD \underline{\le} DC$

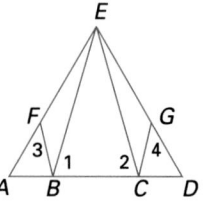

*12. Given: $\angle 1 \cong \angle 2$, $\overline{AB} \cong \overline{CD}$, $\angle 3 \cong \angle 4$
Prove: $\overline{FE} \cong \overline{GE}$

13. Identify the hypotenuse and legs of $\triangle RST$.   H: $\overline{RS}$; L: $\overline{RT}$ and $\overline{ST}$
14. Identify an altitude and a median of $\triangle RST$.   A: $\overline{RT}$ or $\overline{ST}$; M: $\overline{TU}$

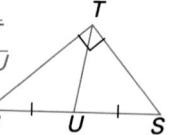

3.

Statement	Reason
1. $\overline{AE} \perp \overline{AC}$, $CD \perp AC$, $\overline{AC}$ bis $\overline{ED}$.	1. Given
2. $\angle$s A and C are rt. $\angle$s.	2. $\perp$ lines form rt $\angle$s.
3. $\angle A \cong \angle C$	3. All rt $\angle$s are $\cong$.
4. $\overline{EB} \cong \overline{BD}$	4. Def of bis
5. $\angle EBA \cong \angle DBC$	5. Vert $\angle$s are $\cong$.
6. $\triangle EBA \cong \triangle DBC$	6. AAS
7. $\angle E \cong \angle D$	7. CPCTC.

Indicate the one correct answer for each question.

1.
E
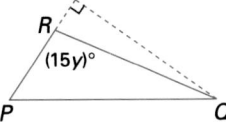

In △PQR above, which of the following could be a value of y?

(A) 4 (B) 12 (C) 2 (D) 6
(E) 8

2.
C
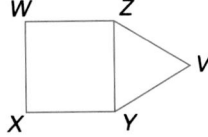

In the figure above, XYZW is a square (all sides congruent and four right angles). Also, $\overline{VY} \cong \overline{VZ} \cong \overline{XY}$
Then, m ∠V = ___.

(A) 15 (B) 45 (C) 60 (D) 90
(E) 30

3.
D
$$\begin{array}{r} 6\square2 \\ \times\ 8 \\ \hline 5,1\triangle6 \end{array}$$

In the correctly calculated product above, if □ and △ are replaced with different digits, then □ = ___.

(A) 6 (B) 2 (C) 5 (D) 4
(E) 7

4. In a senior class there are 400 boys
E and 500 girls. If 60% of the boys and 50% of the girls bought class rings, how many seniors did not buy class rings?

(A) 490 (B) 240 (C) 250
(D) 310 (E) 410

5.
A
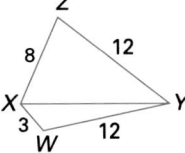

In the figure above, the perimeter of △XYZ is how much greater than the perimeter of △XYW?

(A) 5 (B) 4 (C) 15 (D) 7
(E) 20

6.
B
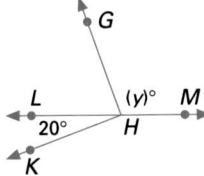

In the figure above, $\overline{GH} \perp \overline{KH}$
Find y.

(A) 70 (B) 110 (C) 200
(D) 160 (E) 100

7.
A
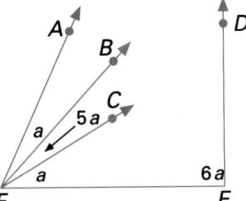

If the figure above were redrawn to scale so that a = 15, how many points of intersection would there be in addition to points E and F?

(A) 1 (B) 2 (C) 3 (D) 4
(E) 5

Additional Answer, page 209

17.

Statement	Reason
1. AB = BC	1. Given
2. Draw $\overline{BD}$ ⊥ $\overline{AC}$	2. Only one seg from pt not on line ⊥ the line.
3. △ABD is rt △.	3. Def of ⊥
4. m ∠ABD < 90	4. Non-rt ∠s of a rt △ are acute.
5. m ∠ADB = 90	5. Def of ⊥
6. m ∠ABD < m ∠ADB	6. Sub
7. AB > AD	7. Side opp larger ∠ is larger
8. △CBD is rt △.	8. Def of ⊥
9. m ∠CBD < 90	9. Non-rt ∠s of a rt △ are acute.
10. m ∠CDB = 90	10. Def of ⊥
11. m ∠CBD < m ∠CDB	11. Sub
12. BC > DC	12. Side opp larger angle is longer.
13. AB + BC > AD + DC	13. Inequalities may be added.
14. AD + DC = AC	14. Seg Add Post
15. AB + BC > AC	15. Sub

Additional Answer, page 219

15.

Statement	Reason
1. AE = AB = ED = DC, EB < EC	1. Given
2. m ∠A + m ∠B + m ∠AEB = 180; m ∠C + m ∠D = m ∠CED = 180	2. Sum of meas of ∠s of △ = 180
3. m ∠A + m ∠B + m ∠AEB = m ∠C + m ∠D + m ∠CED	3. Sub
4. m ∠B = m ∠AEB, m ∠C = m ∠CED	4. ∠s opp ≅ sides of △ are ≅.
5. m ∠A + 2(m ∠AEB) = m ∠D + 2(m ∠C)	5. Sub
6. m ∠A < m ∠D	6. SAS Ineq Thm
7. −1(m ∠A) > −1(m ∠D)	7. Mult Prop of Ineq
8. 2(m ∠AEB) > 2(m ∠DCE); m ∠AEB > m ∠DCE	8. Mult Prop of Ineq

6 POLYGONS

OVERVIEW

In this chapter, students identify, classify, and illustrate polygons. Students determine measures of interior and exterior angles of regular polygons, and investigate properties of quadrilaterals. They apply and prove theorems about parallelograms and use the Midsegment Theorem to solve problems. They also apply theorems about a transversal and three or more parallel lines.

OBJECTIVES

- To identify and name polygons
- To identify concave polygons
- To find the sum of angle measures, given the number of sides of a polygon
- To find the measure of each interior and exterior angle of a given regular polygon
- To determine the number of sides of a regular polygon, given the measure of its exterior angle
- To apply theorems about parallelograms
- To divide a given segment into a given number of congruent segments

PROBLEM SOLVING

The Problem Solving Strategies lesson on page 256 teaches students how Making a Table can help them see patterns in order to solve problems. Exercises 17–24 in Lesson 6.1 ask the student to make a table to discover the pattern in determining the number of diagonals from one vertex for a polygon with n sides.

TECHNOLOGY

Computer: The Computer Investigation on page 240 helps students discover the properties of parallelograms given in Theorems 6.3, 6.4, and 6.5 (Lesson 6.4). This activity can also be extended so that students can determine if a given quadrilateral is a parallelogram (Lesson 6.5).

Calculator: You may want to encourage students to use a calculator for Exercises 34–36 on page 233.

SPECIAL FEATURES

Mixed Review pp. 229, 234, 239, 251, 255, 261
Algebra Review p. 235
Application: Tiling and Soccer Ball p. 235
Application: Mirrors p. 239
Computer Investigation: Properties of a Parallelogram p. 240
Midchapter Review p. 246
Brainteaser p. 246
Problem Solving Strategies: Making a Table p. 256
Key Terms p. 262
Key Ideas and Review Exercises pp. 262–263
Chapter 6 Test p. 264
College Prep Test p. 265
Cumulative Review (Chapters 1–6) pp. 266–267

PLANNING GUIDE

Lesson	Basic	Average	Above Average	Resources
6.1 pp. 228–229	CE all WE 1–10	CE all WE 1–23 odd	CE all WE 1–26 odd	Reteaching p. 85 Practice p. 86
6.2 pp. 232–235	CE all WE 1–24 odd Application AR all	CE all WE 1–42 odd Application AR all	CE all WE 1–49 odd Application AR all	Reteaching p. 87 Practice p. 88
6.3 pp. 238–240	CE all WE 1–15 odd Application CI 1–2	CE all WE 1–18 odd Application CI all	CE all WE 1–21 odd Application CI all	Reteaching p. 89 Practice p. 90
6.4 pp. 244–246	CE all WE 1–18 odd Midchapter Review Brainteaser	CE all WE 1–27 odd Midchapter Review Brainteaser	CE all WE 1–30 odd Midchapter Review Brainteaser	Reteaching p. 91 Practice p. 92
6.5 pp. 250–251	CE all WE 1–13 odd	CE all WE 1–20 odd	CE all WE 1–25 odd	Reteaching p. 93 Practice p. 94
6.6 pp. 254–256	CE all WE 1–13 Problem Solving	CE all WE 1–20 odd Problem Solving	CE all WE 1–23 odd Problem Solving	Reteaching p. 95 Practice p. 96
6.7 pp. 260–261	CE all WE 1–10	CE all WE 1–16	CE all WE 1–20	Reteaching p. 97 Practice p. 98
Chapter 6 Review pp. 262–263	1–15 odd	all	all	
Chapter 6 Test p. 264	1–8, 10, 12–19	1–19	all	
College Prep Test p. 265	all	all	all	
Cumulative Review pp. 266–267	1–26 odd	all	all	

CE = Classroom Exercises   WE = Written Exercises   FR = Focus on Reading   AR = Algebra Review   CI = Computer Investigation
NOTE: For each level, all students should be assigned all Mixed Review exercises.

■ INVESTIGATION

Project: This project directs students in constructing a tangram puzzle. This will give them practice in following directions as well as in exploring the sizes and shapes of various polygons. Use this activity as an introduction to Chapter 6.

Tell students that the seven-piece tangram puzzle originated in China, where it gained popularity in the early 1800s. According to legend, a man named Tan dropped a beautiful, square ceramic tile on the floor and it broke into seven pieces. He never could put the pieces back together, but he succeeded in using the pieces to create different geometric shapes and patterns. (See Project Worksheet 6 for a picture of the puzzle).

Materials: Sheet of plain white paper, pencil, scissors, ruler.

Have students draw a figure (square) 4 × 4 in. and label the corners *A*, *B*, *C*, and *D*. Use the following directions to construct a tangram puzzle on the chalkboard as the students construct their own on their paper.

1. Draw diagonal $\overline{AC}$ in the square.
2. Label *E* and *F* the midpoints of $\overline{AB}$ and $\overline{BC}$, respectively. Draw $\overline{EF}$.
3. Label *G* the midpoint of $\overline{EF}$. Draw $\overline{GD}$.
4. Construct a line segment perpendicular to $\overline{AC}$ from point *E*.
5. Construct a line segment from point *G* to $\overline{AC}$, parallel to $\overline{BC}$.

Check students' constructions or have them exchange with another student to check for accuracy. Have students cut out their finished puzzles and experiment with making various shapes, or direct them to complete Project Worksheet 6.

Fig. 1

Fig. 2

Lewis Latimer (1848–1928)—By inventing a durable carbon filament for the electric light bulb, Latimer solved the problem of changing electric current into light. As an associate of Alexander Graham Bell, he also drew plans for the first telephone patent.

More About the Mathematician

Lewis Latimer, son of a fugitive slave, was an outstanding inventor and draftsman. He also wrote the first textbook on the electrical lighting system as well as poems and essays. He was a charter member of the Edison Pioneers, a group of distinguished scientists and inventors who worked with Thomas Edison. He helped create the electric industry and supervised the installation of electric lights for the cities of New York, Philadelphia, and London, England. Electric lamp sockets that he designed and made are in the Smithsonian Institution in Washington, D.C.

6.1 Introducing Polygons

Objectives
To identify and name polygons and their parts
To identify and draw convex and concave polygons
To identify regular polygons

Some geometric figures have curved parts. Others, including polygons, consist entirely of segments.

Definition

A **polygon** is the union of three or more coplanar segments such that:
1. each endpoint is shared by exactly two segments;
2. segments intersect only at their endpoints; and
3. intersecting segments are noncollinear.

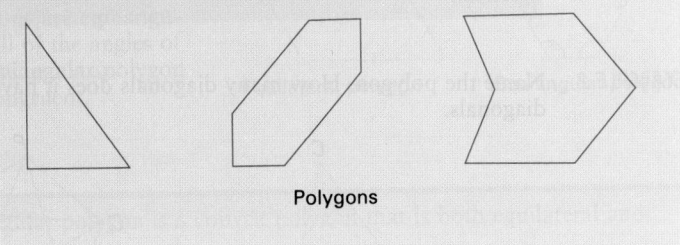

Polygons

Each segment of a polygon is called a **side**. Each endpoint of a side is called a **vertex** of the polygon. For each polygon, the number of sides is equal to the number of vertices. A polygon is usually named by listing its vertices in order.

Adjacent sides of a polygon intersect at a vertex. *Nonadjacent* sides do not intersect. The vertices of *consecutive angles* are endpoints of the same side of the polygon. If two angles are not consecutive, then they are called **nonconsecutive**.

EXAMPLE 1 Name two sides of polygon *ABCDE* that are adjacent to $\overline{AB}$. Name two angles that are nonconsecutive with ∠C. Name two angles that are consecutive with ∠C.

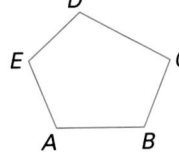

Solution $\overline{BC}$ and $\overline{EA}$ are adjacent to $\overline{AB}$.
∠E and ∠A are nonconsecutive with ∠C.
∠B and ∠D are consecutive with ∠C.

6.1 Introducing Polygons **225**

Additional Example 1

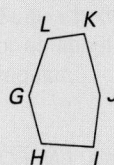

Name two angles of polygon *GHIJKL* that are consecutive with ∠K. ∠L, ∠J

Name three sides that are not adjacent to $\overline{HI}$.
$\overline{JK}, \overline{KL}, \overline{LG}$

Teaching Resources

Application Worksheet 6
Project Worksheet 6
Quick Quizzes 43
Reteaching and Practice Worksheets, pp. 85, 86

GETTING STARTED

Prerequisite Quiz

1. Draw △*XYZ*. Identify the sides and vertices of this triangle. $\overline{XY}, \overline{YZ}, \overline{XZ}$, ∠X, ∠Y, ∠Z
2. Given △*ABC* with *AB* = *BC* = 8 cm and m ∠C = 60. Find m ∠A, m ∠B, and *AC*. 60, 60, 8 cm

Motivator

Have students suggest words beginning with the prefixes "tri−, quad−, penta−, and deca−." Have them connect the meaning of the words with each prefix. You may wish to have students work in groups, and assign one prefix to each group.

TEACHING SUGGESTIONS

Lesson Note

Have the students recall familiar uses of polygons. The stop sign is an octagon. The Pentagon in Washington, D.C is a five-sided building. Note that a concave polygon can be thought of as hollowed out like a cave, or as "bashed in." Advanced students can compare properties of concave and convex polygons.

Students can find a formula for the number of diagonals of a convex polygon. From each of *n* vertices there are *n* − 3 diagonals. Since the number *n*(*n* − 3) counts each diagonal twice, the formula is $\frac{n(n-3)}{2}$.

FOLLOW UP

Guided Practice

Classroom Exercises 1–10

Independent Practice

A Ex. 1–10, **B** Ex. 11–23, **C** Ex. 24–26

Basic: WE 1–10
Average: WE 1–23 odd
Above Average: WE 1–26 odd

Additional Answers

Classroom Exercises

1. No, segments intersect at points other than endpoints.
3. No, not every endpoint is shared by 2 segments.
5. No, not every side is a segment.
6. No, segments are not coplanar.

Classroom Exercises

State whether the figure is a polygon. If it is not, explain why.

1.

2.  Yes

3.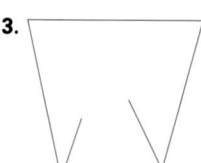

4. 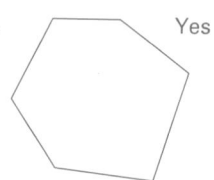 Yes

5.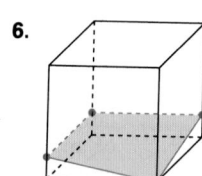

6.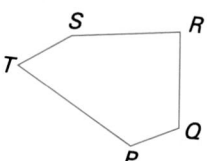

Complete the statement for polygon *PQRST*.

7. $\overline{RS}$ and ___$\overline{ST, RQ}$___ are adjacent sides.
8. $\overline{TP}$ and ___$\overline{SR, RQ}$___ are nonadjacent sides.
9. $\angle S$ and ___$\angle T, \angle R$___ are consecutive angles.
10. $\angle P$ and ___$\angle S, \angle R$___ are nonconsecutive angles.

Written Exercises

Name the polygon. State whether it is convex or concave.

1.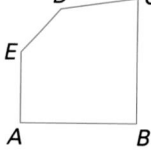
Pentagon *ABCDE*; convex

2.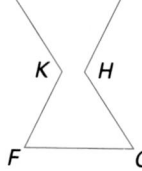
Hexagon *FGHIJK*; concave

3.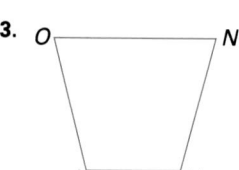
Quadrilateral *LMNO*; convex

Determine whether the polygon is equilateral, equiangular, or regular.

4.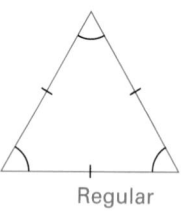
Regular

5.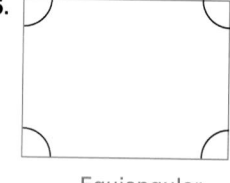
Equiangular

6.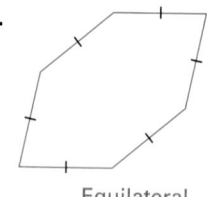
Equilateral

228 Chapter 6 Polygons

Enrichment

Have each student draw a quadrilateral and a convex pentagon, showing all possible diagonals, and then count the number of nonoverlapping regions in each figure. The quadrilateral will contain 4 such regions and the pentagon will contain 11.
Now have the students determine the number of nonoverlapping regions formed by the diagonals of a convex hexagon and of a convex heptagon.

25 and 50, or if three diagonals go through the same point, 24 and 49.
As an additional activity, the pentagon and hexagon drawings can be drawn on light cardboard and cut along the lines to make jigsaw puzzles.

Draw the figure.

7. a convex pentagon

8. a concave hexagon

9. a concave pentagon

10. a convex octagon

11. an equiangular quadrilateral

12. an equilateral hexagon

13. a regular triangle

14. a regular quadrilateral

15. a regular pentagon

16. a regular octagon

Complete the table.

	Number of sides of polygon	Number of diagonals from one vertex	Number of diagonals
17.	3	0	0
18.	4	1	2
19.	5	2	5
20.	6	3	9
21.	7	4	14
22.	8	5	20
23.	50	47	1,175
24.	n	$n-3$	$\dfrac{n(n-3)}{2}$

25. Draw an equilateral concave polygon.

26. The edges of the opening of a wrench are parallel. Standard nuts are regular polygons in shape. Will the wrench fit nuts with any given number of sides? Explain.

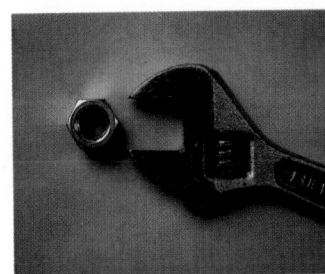

Mixed Review

1. Write the converse of "If a polygon is regular, then it is equiangular." Is the converse true or false? **3.5** If a polygon is equiangular, then it is regular; false

2. Write the inverse of "If a polygon is regular, then it is equilateral." Is the inverse true or false? **3.8** If a polygon is not regular, then it is not equilateral; false

3. Write the contrapositive of "If a polygon is regular, then it is equilateral." Is the contrapositive true or false? **3.8** If a polygon is not equilateral, then it is not regular; true

4. Write the inverse if "If a polygon is convex, then it is not concave." Is the inverse true or false? **3.8** If a polygon is not convex, then it is concave; true

6.1 Introducing Polygons **229**

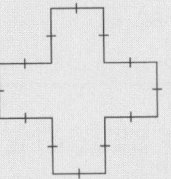

Teaching Resources

Manipulative Worksheet 11
Quick Quizzes 44
Reteaching and Practice
   Worksheets, pp. 87, 88
Technology 6

▰▰▰ GETTING STARTED

Prerequisite Quiz

1. What is the sum of the measures of the angles of a triangle?   180

How many sides does each of the following polygons have?

2. quadrilateral   4
3. heptagon   7
4. octagon   8
5. decagon   10

How many diagonals does each of the following polygons have?

6. triangle   0
7. quadrilateral   2
8. pentagon   5
9. octagon   20

Motivator

On the chalkboard, draw the three polygons on page 230. Explain to the students that when they are having difficulty in visualizing the separate parts or polygons contained in a complicated figure, they can apply the strategy of breaking the more complicated figure down into several smaller figures. One way to do this for polygons is to draw all possible diagonals from one of the vertices. Ask the students how many diagonals can be drawn that way in a quadrilateral.   1   In a pentagon.   2   In a hexagon.   3   Then have them tell you how many triangles are formed in the quadrilateral.   2   In the pentagon.   3   In the hexagon.   4   Ask them to give the sum of the angle measures in a triangle.   180 Ask them how they can find the sum of the angle measures of polygons by forming triangles.   The sum of the angle measures is the product of the number of triangles formed, which is two less than the number of sides, and 180.

6.2 Interior Angles of Polygons

Objectives
To find the sum of the angle measures of a convex polygon
To find angle measures and the number of sides of polygons
To find angle measures and the number of sides of regular polygons

To store honey, bees build honey-combs with hundreds of *hexagonal* compartments. Why do you think a hexagon is better for this purpose than a circle, a square, or a triangle? (The computer storage shown in the photo is a modern industrial "honey-comb".)

Quadrilateral

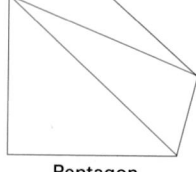
Pentagon

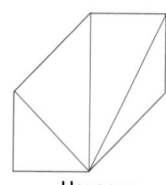
Hexagon

In each figure above the diagonals *from one vertex* to each of the other vertices form triangles. In each case, the total number of triangles formed is 2 less than the number of sides of the polygon. If a polygon has n sides, then $n - 2$ triangles are formed. The sum of the measures of the interior angles of the polygon is the sum of the measures of the angles of these triangles.

Polygon	Number of Sides	Number of Triangles	Sum of Angle Measures
quadrilateral	4	2	$2 \cdot 180 = 360$
pentagon	5	3	$3 \cdot 180 = 540$
hexagon	6	4	$4 \cdot 180 = 720$
heptagon	7	5	$5 \cdot 180 = 900$
octagon	8	6	$6 \cdot 180 = 1,080$

▰▰▰ Theorem 6.1   The sum of the measures of the interior angles of a convex polygon with n sides is $(n - 2)180$.

Additional Example 1

Find the sum of the measures of the angles of a dodecagon. $(12 - 2) \cdot 180$ or 1800

Corollary 1   The sum of the measures of the interior angles of a convex quadrilateral is 360.

In this book, the *interior angles* of a polygon will simply be called the **angles of the polygon.**

EXAMPLE 1   Find the sum of the measures of the angles of a 22-gon.

Solution $n = 22$
Sum $= (22 - 2)180 = 20 \cdot 180 = 3{,}600$

EXAMPLE 2   In a hexagon, the measure of one angle is twice that of a second angle. The remaining angles are congruent, each with a measure three times that of the second angle. Find the measure of each angle.

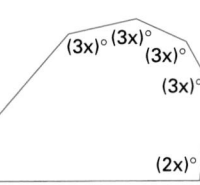

Plan   By Theorem 6.1, the sum of the measures of the angles is $(6 - 2)180 = 720$. Let $x =$ measure of the second angle.

Solution
$$x + 2x + 3x + 3x + 3x + 3x = 720$$
$$15x = 720$$
$$x = 48$$

The measure of the second angle is 48.

The measure of the original angle is $2 \cdot 48 = 96$.

The measure of each of the remaining angles is $3 \cdot 48 = 144$.

If you know the sum of the angle measures of a polygon, you can use the formula $S = (n - 2)180$ to find the number of sides of the polygon.

EXAMPLE 3   The sum of the measures of the angles of a convex polygon is 3,960. Find the number of sides of the polygon.

Solution
$$(n - 2)180 = 3{,}960$$
$$n - 2 = \frac{3{,}960}{180} \quad \longleftarrow \quad \text{It is easier to divide by 180 than}$$
$$n - 2 = 22 \qquad\qquad \text{to expand the left-hand side first.}$$
$$n = 24$$

The polygon has 24 sides.

Additional Example 2

The measures of four angles of a pentagon are 90, 108, 108 and 114. Find the measure of the fifth angle.   120

Additional Example 3

The sum of the measures of the angles of a convex polygon is 2,520. Find the number of sides of the polygon.   16

TEACHING SUGGESTIONS

Lesson Note

Have the students draw a convex quadrilateral, pentagon, and hexagon, and measure the angles with a protractor. Then have them compare the sums of the angle measures of these polygons. Most students will quickly see a pattern emerge. Theorem 6.1 is applied only to convex polygons because of the restriction that angle measures are less than or equal to 180.

Math Connections

Snowflakes: Snowflakes vary in size, from over an inch in diameter to tiny invisible specks. They consist of many crystals that are formed by freezing water. The crystals that make up a snowflake are always arranged around a center, forming either 6–pointed stars or thin plates of hexagonal shape. From the center of each snowflake, branches spread out in six directions, each branch being identical to the others. However, because of their complexity no two snowflakes are ever alike.

Critical Thinking Questions

Synthesis: Explain to students how some regular polygonal shapes, such as squares and triangles, can be used like tiles to completely cover an area without leaving gaps and without overlapping. This is called a tessellation. Ask them what the sum of the measures of the adjacent angles around a single vertex must be in a tessellation? 360   What regular polygon will combine with a regular hexagon to form a tessellation? Explain why and describe its pattern.   A reg △. Int ∠s of a regular hexagon and a regular △ meas 120 and 60, respectively. 2 hexagons and 2 △s alternate around a single vertex: 120 + 60 + 120 + 60 = 360.

Common Error Analysis

Error: When students find the number of sides of a polygon given the sum of the interior angle measures, they may divide the sum by 180 and then forget to add 2.

Help them to remember the formula, $(n - 2)180 =$ sum of measures of interior angles, where 180 is multiplied by the number of *triangles*, $(n - 2)$, and not by n.

Checkpoint

1. Find the sum of the measures of the angles of a nonagon.   1,260
2. Find the measure of the fourth angle of a quadrilateral if the measures of three of the angles are 78, 82, and 60.   140
3. If the sum of the measures of the angles of a polygon is 1,620, how many sides does the polygon have?   11
4. Find the measure of each angle of a regular octagon.   135

Closure

Ask the students how to find the sum of the measures of the interior angles of a convex polygon.   Subtract two from the number of sides and multiply by 180.   Ask them to explain how they can find the sum of the measures of the angles in a polygon if they were to forget the formula.   First determine the number of triangles formed by drawing diagonals from one vertex. Then multiply that number by 180.

◾FOLLOW UP

Guided Practice

Classroom Exercises 1–8

Independent Practice

A Ex. 1–24, **B** Ex. 25–42, **C** Ex. 43–49

Basic: WE 1–24 odd, Application, AR all

Average: WE 1–42 odd, Application, AR all

Above Average: WE 1–49 odd, Application, AR all

A regular polygon was defined to be an equilateral and equiangular convex polygon. Using the definition and Theorem 6.1, you can find the measure of each angle of a regular polygon if the number of sides is known.

EXAMPLE 4   Find the measure of each angle of a regular hexagon.

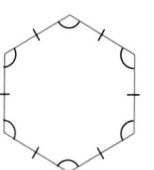

Solution   The sum of the measures is $(6 - 2)180 = 720$. Each angle has the same measure.

Therefore, the measure of each angle is $\frac{720}{6} = 120$.

Example 4 can be generalized as a corollary to Theorem 6.1.

Corollary 2   The measure of an angle of a regular polygon with n sides is $\frac{(n - 2)180}{n}$.

EXAMPLE 5   The measure of each angle of a regular polygon is 150. How many sides does the polygon have?

$$150 = \frac{(n - 2)180}{n}$$
$$150n = (n - 2)180$$
$$150n = 180n - 360 \quad \longleftarrow \quad \text{Here, it is easier to expand first.}$$
$$-30n = -360$$
$$n = 12$$

The polygon has 12 sides.

Classroom Exercises

How many sides does a polygon have if the measures of its angles have the given sum?

1. 180   3
2. 3 · 180   5
3. 6 · 180   8
4. 720   6
5. 360   4
6. 1,800   12
7. 18,000   102
8. 180,000   1,002

Written Exercises

Find the sum of the measures of the angles of the convex polygon.

1. a pentagon   540
2. a hexagon   720
3. a decagon   1,440
4. a 30-gon   5,040
5. a 62-gon   10,800
6. a 100-gon   17,640

232   Chapter 6 Polygons

Additional Example 4

Find the measure of each angle of a regular 18-gon.   160

Additional Example 5

The measure of each angle of a regular polygon is 162. How many sides does the polygon have?   20

Can the three angle measures given belong to a convex quadrilateral? If so, find the measure of the fourth angle.

7. 80, 50, 90   140   **8.** 100, 70, 120   70   Not possible
9. 60, 60, 60
10. 95, 97, 83   85   **11.** 132, 112, 145   Not possible   **12.** 38, 43, 57
                                                                            Not possible

The sum of the measures of the angles of a convex polygon is given. Find the number of sides of the polygon.

13. 900   7   **14.** 1,260   9   **15.** 1,980   13
16. 3,600   22   **17.** 4,500   27   **18.** 7,560   44

Find the measure of an angle of the regular polygon.

                                                                                        144
19. a square   90   **20.** a pentagon   108   **21.** a decagon
22. a 20-gon   162   **23.** a 30-gon   168   **24.** a 100-gon
                                                                                              176.4

Each angle of a regular polygon has the given measure. How many sides does the polygon have?

25. 60   3   **26.** 135   8   **27.** 108   5

Find the measure of each angle of quadrilateral *ABCD*.

28. m $\angle A = 10x$, m $\angle B = 6x + 10$, m $\angle C = 12x - 10$, m $\angle D = 8x$ 100, 70, 110, 80
29. m $\angle A = 8x + 5$, m $\angle B = 10x + 5$, m $\angle C = 10x - 8$,
     m $\angle D = 13x - 11$ 77, 95, 82, 106
30. The sum of the measures of four angles of a pentagon is 498. What is the measure of the unknown angle?   42
31. The sum of the measures of nine angles of a decagon is 1,320. What is the measure of the unknown angle?   120
32. Three angles of a hexagon are congruent. The other three angles are also congruent. Each of the first three angles has a measure twice that of one of the second three angles. What is the measure of each angle of the hexagon?   80, 80, 80, 160, 160, 160
33. In a pentagon, the measure of one angle is twice that of a second angle. The remaining angles are congruent, each having a measure of three times that of the second angle. What is the measure of each angle of the pentagon?   90, 45, 135, 135, 135

Can the number be the sum of the angle measures of a polygon?

💾 **34.** 10,180   No   💾 **35.** 15,660   Yes   💾 **36.** 18,180
                                                                                                     Yes

Can the number be the measure of an angle of a regular polygon? If the measure is possible, how many sides does the polygon have?

37. 90   Yes; 4   **38.** 100   No   **39.** 120   Yes; 6
40. 140   Yes; 9   **41.** 160   Yes; 18   **42.** 175   Yes; 72

Additional Answers

Written Exercises

43.

Statement	Reason
1. Quad *ABCD* is convex.	1. Given
2. m $\angle DAB$ = m $\angle DAC$ + m $\angle CAB$, m $\angle DCB$ = m $\angle DCA$ + m $\angle BCA$	2. Angle Add Post
3. m $\angle D$ + m $\angle DAC$ + m $\angle DCA$ = 180, m $\angle B$ + m $\angle BCA$ + m $\angle BAC$ = 180	3. Sum of the meas of the $\angle$s of a $\triangle$ = 180.
4. m $\angle D$ + m $\angle DAC$ + m $\angle DCA$ + m $\angle B$ + m $\angle BCA$ + m $\angle BAC$ = 360	4. Equations may be added.
5. m $\angle D$ + m $\angle DAB$ + m $\angle B$ + m $\angle DCB$ = 360	5. Sub

44.

Statement	Reason
1. Pentagon *EFGHI* is convex.	1. Given
2. m $\angle IEF$ = m $\angle IEH$ + m $\angle FEH$, m $\angle EFG$ = m $\angle EFH$ + m $\angle GFH$, m $\angle GHI$ = m $\angle GHF$ + m $\angle FHE$ + m $\angle EHI$	2. Angle Add Post
3. m $\angle I$ + m $\angle IEH$ + m $\angle EHI$ = 180, m $\angle EHF$ + m $\angle HEF$ + m $\angle EFH$ = 180, m $\angle G$ + m $\angle GFH$ + m $\angle GHF$ = 180	3. Sum of the meas of the $\angle$s of a $\triangle$ = 180.
4. m $\angle I$ + m $\angle IEH$ + m $\angle EHI$ + m $\angle EHF$ + m $\angle HEF$ + m $\angle EFH$ + m $\angle G$ + m $\angle GFH$ + m $\angle GHF$ = 540	4. Equations may be added.
5. m $\angle I$ + m $\angle IEF$ + m $\angle EFG$ + m $\angle G$ + m $\angle GHI$ = 540	5. Sub

---

Enrichment

Have each student draw a figure similar to the concave quadrilateral below. Then have them use protractors to measure each of the angles. By comparing results with each other, ask the students if they can formulate a pattern or rule.

m $\angle D$ = m $\angle A$ + m $\angle B$ + m $\angle C$

Next, have the students follow the same procedure with a figure similar to this.

m $\angle H$ = m $\angle E$ + m $\angle F$ + m $\angle G$ + m $\angle I$ − 180

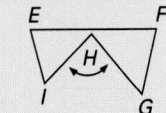

45.

Statement	Reason
1. Quad *ABCD* is convex.	1. Given
2. m ∠*EAB* + m ∠*AEB* + m ∠*EBA* = 180, m ∠*BEC* + m ∠*ECB* + m ∠*CBE* = 180, m ∠*ECD* + m ∠*CED* + m ∠*EDC* = 180, m ∠*DEA* + m ∠*DAE* + m ∠*ADE* = 180	2. Sum of the meas of the ∠s of a △ = 180.
3. m ∠*EAB* + m ∠*EBA* + m ∠*AEB* + m ∠*BEC* + m ∠*ECB* + m ∠*CBE* + m ∠*ECD* + m ∠*CED* + m ∠*EDC* + m ∠*DEA* + m ∠*DAE* + m ∠*ADE* = 720	3. Equations may be added.
4. m ∠*DEC* + m ∠*CEB* + m ∠*BEA* + m ∠*AED* = 360	4. Sum of the meas of the ∠s around a pt = 360.
5. m ∠*EAB* + m ∠*EBA* + m ∠*ECB* + m ∠*EBC* + m ∠*ECD* + m ∠*EDC* + m ∠*DAE* + m ∠*ADE* = 360	5. Equations may be subtracted.
6. m ∠*DAB* = m ∠*DAE* + m ∠*BAE*, m ∠*ABC* = m ∠*ABE* + m ∠*CBE*, m ∠*BCD* = m ∠*BCE* + m ∠*DCE*, m ∠*CDA* = m ∠*CDE* + m ∠*ADE*	6. Angle Add Post
7. m ∠*ABC* + m ∠*BCD* + m ∠*CDA* + m ∠*DAB* = 360	7. Sub

See page 235 for the answer to Ex. 46 and 240 for the answers to Ex. 47–48.

Use the figures at the right for Exercises 43–44.

43. Prove Theorem 6.1 for convex quadrilaterals (Corollary 1).

44. Prove Theorem 6.1 for pentagons.

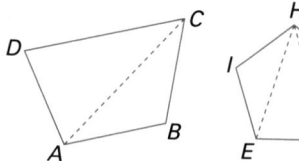

Another approach can be used to prove Theorem 6.1. Draw triangles with a common vertex in the interior of the polygon, as shown below. Any interior point may be chosen for the vertex. Then use the fact that the sum of the measures of the angles with the common vertex is 360.

45. Prove Theorem 6.1 for convex quadrilaterals.

46. Prove Theorem 6.1 for convex hexagons.

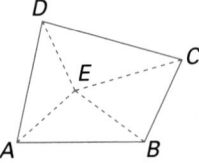 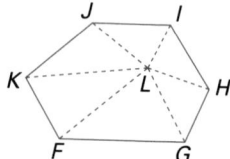

47. Prove Corollary 2.

48. Prove: Opposite sides of a regular hexagon are parallel. (HINT: Draw a transversal through vertices of the sides to be proven parallel such that it is ⊥ to one side.)

49. Using a protractor, measure each angle of the given concave quadrilateral. Does the formula of Theorem 6.1 seem to apply?   No

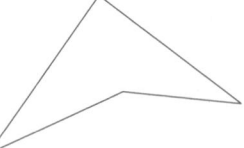

Mixed Review

Tell whether the statement is true or false.

1. Alternate interior angles of parallel lines are congruent.   *3.5*   True

2. Exterior angles of parallel lines on the same side of a transversal are supplementary.   *3.7*   True

3. If corresponding angles are supplementary, then lines are parallel.   *3.3*   False

4. Any point on the perpendicular bisector of a segment is equidistant from the endpoints of the segment.   *5.6*   True

Application: *Tiling and Soccer Ball*

Different tile patterns can be arranged by fitting together various types of regular polygons. For example, octagons and squares can be put together, as shown below. Can the following combinations of tiles be fitted together? If so, make paper cutouts or drawings to demonstrate.

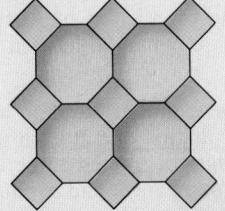

1. squares and hexagons   No
2. pentagons and triangles   No
3. pentagons and squares   No
4. hexagons and triangles   Yes
5. squares and triangles   Yes
6. hexagons and pentagons   No
7. octagons and triangles   No

8. The cover of a soccer ball is made of two types of regular polygons sewn together. What are these polygons? Why was this combination not possible when tile patterns were considered in Exercises 1–7 above?   Pentagons and hexagons; a ball's surface is not a plane.

Algebra Review

To solve a fractional equation of the form $\dfrac{ax + b}{dx} = c$ for x,

where $d \neq 0$, $x \neq 0$:

(1) Multiply both sides of the equation by dx.
(2) Solve for x.

Example: Solve $\dfrac{5x - 16}{3x} = 3$.

Multiply by $3x$. $5x - 16 = 9x$
$-4x = 16$
$x = -4$

Solve the equation.

1. $\dfrac{2x - 5}{3x} = 1$ -5

2. $\dfrac{3x + 4}{2x} = 2$ 4

3. $\dfrac{5x - 28}{3x} = 3$ -7

4. $\dfrac{4x - 10}{2x} = 4$ -2.5

5. $\dfrac{6x - 3}{5x} = 3$ $-\dfrac{1}{3}$

6. $\dfrac{10x + 2}{2x} = 7$ $\dfrac{1}{2}$

Additional Answer, page 234

46.

Statement	Reason
1. Hexagon *FGHIJK* is convex.	1. Given
2. m ∠*FLG* + m ∠*LGF* + m ∠*GFL* = 180, m ∠*GLH* + m ∠*HGL* + m ∠*GHL* = 180, m ∠*HLI* + m ∠*LIH* + m ∠*IHL* = 180, m ∠*ILJ* + m ∠*LJI* + m ∠*JIL* = 180, m ∠*JLK* + m ∠*LKJ* + m ∠*KJL* = 180, m ∠*KLF* + m ∠*LFK* + m ∠*FKL* = 180	2. Sum of the meas of the ∠s of a △ = 180.
3. m ∠*FLG* + m ∠*LGF* + m ∠*GFL* + m ∠*GLH* + m ∠*HGL* + m ∠*GHL* + m ∠*HLI* + m ∠*LIH* + m ∠*IHL* + m ∠*ILJ* + m ∠*LJI* + m ∠*JIL* + m ∠*JLK* + m ∠*LKJ* + m ∠*KJL* + m ∠*KLF* + m ∠*LFK* + m ∠*FKL* = 1,080	3. Equations may be added.
4. m ∠*FLG* + m ∠*GLH* + m ∠*HLI* + m ∠*ILJ* + m ∠*JLK* + m ∠*KLF* = 360	4. Sum of the meas of the ∠s around a pt = 360.
5. m ∠*LGF* + m ∠*GFL* + m ∠*HGL* + m ∠*GHL* + m ∠*LIH* + m ∠*IHL* + m ∠*LJI* + m ∠*JIL* + m ∠*LKJ* + m ∠*KJL* + m ∠*LFK* + m ∠*FKL* = 720	5. Equations may be subtracted.
6. m ∠*KFL* + m ∠*GFL* = m ∠*KFG*, m ∠*FGL* + m ∠*HGL* = m ∠*FGH*, m ∠*GHL* + m ∠*IHL* = m ∠*GHI*, m ∠*HIL* + m ∠*JIL* = m∠*HIJ*, m ∠*IJL* + m ∠*KJL* = m ∠*IJK*, m ∠*JKL* + m ∠*FKL* = m ∠*JKF*	6. Angle Add Post
7. m ∠*KFG* + m ∠*FGH* + m ∠*GHI* + m ∠*HIJ* + m ∠*IJK* + m ∠*JKF* = 720	7. Sub

6.3 Exterior Angles of Polygons

Objectives   To identify the exterior angles of a polygon
To find the measures of exterior angles of polygons
To explain why the exterior angles of a regular polygon are congruent

▰▰GETTING STARTED

Prerequisite Quiz

1. If the measure of an angle is 78, what is the measure of its supplement? 102
2. The measure of an exterior angle of a triangle is (12*x* + 14). The measures of the two remote interior angles are (5*x* + 7) and (9*x* − 7). What is the measure of the exterior angle? 98
3. Find the sum of the measures of the angles of a hexagon. 720
4. Find the measure of each angle of a regular octagon. 135

Motivator

Have students imagine walking around a large triangle with their right arm stretched out in front of them. Each time they turn a corner as they walk, their arm will move from one side of an angle to the second side. Explain to them that each of these angles formed is an exterior angle of the triangle. Ask them how many circles they completed as they walked around the triangle. One Have them tell you how many degrees. 360 Have them explain what would happen if they walked around a quadrilateral, a pentagon, or some other polygon. The same thing Ask them what conclusion they can draw from this. Theorem 6.2 page 236

▰▰TEACHING SUGGESTIONS

Lesson Note

Have the students draw several polygons with one exterior angle at each vertex. Guide them to discovering Theorem 6.2 about the sum of the exterior angle measures by having them measure the angles and find their sums.

A convex polygon has two exterior angles at each vertex. A pentagon has ten exterior angles.

Trace the pentagon at the right as well as the exterior angles, one at each vertex. Cut out the wedges of tracing paper determined by each exterior angle and arrange them to fit around a point. This suggests the sum of these angle measures is 360.

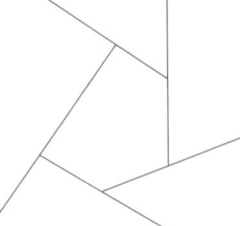

Theorem 6.2   The sum of the measures of the exterior angles, one at each vertex, of any convex polygon is 360.

Plan   Sketch a polygon from which to generalize. Use Theorem 6.1, which concerns the sum of the measures of *interior* angles.

Proof (Paragraph Form)   At each vertex of a polygon, the sum of the measures of the interior angle and an exterior angle is 180. For example, in the figure at right, m ∠1 + m ∠6 = 180.

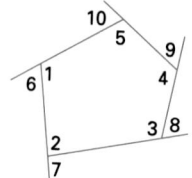

In a polygon of *n* sides, there are *n* vertices. Therefore, the sum of the measures of all pairs of interior angles and exterior angles is 180*n*. It was stated in Theorem 6.1 that the sum of the measures of the interior angles of a polygon with *n* sides is (*n* − 2)180. So, the sum of the measures of the exterior angles of a polygon with *n* sides is the difference of the sum of interior angles from 180.

$$180n - (n - 2)180 = 180n - (180n - 360)$$
$$= 180n - 180n + 360$$
$$= 360$$

236   Chapter 6 Polygons

Additional Example 1

What is the sum of the measures of the exterior angles, one at each vertex, of a convex heptagon? 360

EXAMPLE 1   What is the sum of the measures of the exterior angles, one at each vertex, of a convex dodecagon?

Solution   Using Theorem 6.2, the sum is 360.

EXAMPLE 2   The measures of the exterior angles, one at each vertex, of a quadrilateral are m $\angle 1 = t + 4$, m $\angle 2 = t + 6$, m $\angle 3 = t + 8$, and m $\angle 4 = t + 10$. Find the measure of each exterior angle.

Solution

$$m \angle 1 + m \angle 2 + m \angle 3 + m \angle 4 = 360$$
$$(t + 4) + (t + 6) + (t + 8) + (t + 10) = 360$$
$$4t + 28 = 360$$
$$4t = 332$$
$$t = 83$$

Therefore, m $\angle 1 = 83 + 4 = 87$, m $\angle 2 = 83 + 6 = 89$, m $\angle 3 = 83 + 8 = 91$, and m $\angle 4 = 83 + 10 = 93$.

All exterior angles of a regular polygon are congruent since they are supplements of congruent angles. The following corollary to Theorem 6.2 is based on this fact.

Corollary   The measure of an exterior angle of a regular polygon with n sides is $\frac{360}{n}$.

EXAMPLE 3   The measure of an exterior angle of a regular polygon is $3x + 1$. The measure of a second exterior angle of the same polygon is $4x - 12$. Identify the polygon.

Solution   Exterior angles of a regular polygon have the same measure.

$$4x - 12 = 3x + 1$$
$$x = 13$$
$$3x + 1 = 3 \cdot 13 + 1 = 40$$

Each exterior angle has a measure of 40. By the corollary to Theorem 6.2,

$$40 = \frac{360}{n}$$
$$40n = 360$$
$$n = 9$$

Therefore, the polygon is 9-sided (a nonagon).

Math Connections

Hallways: Use the following example to discuss with students how the interior angles determine a regular polygon. When visiting his dad's office building, Jeremy notices that, although the outside of the building appears circular, the walkway inside is polygonal. Although he cannot see the shape of the floor, he figures it out by walking down the hallway counting his steps and estimating the angle of each turn. He begins at a turn and, after every 16 steps, he turns right at an angle of approximately 120. After 5 turns, Jeremy returns to his starting point and concludes that the hallway is shaped like a regular hexagon.

Critical Thinking Questions

Synthesis: Ask students to classify the triangles that are formed by extending, until they intersect, two sides of a regular polygon that has more than 4 sides. Base $\angle$s are $\cong$ because supp $\angle$s of $\cong$ $\angle$s are $\cong$; therefore, they are isos $\triangle$s. Have students extend the angles of a regular nonagon, as described above, to form a 9–pointed star. Ask them what the measure of the angle at each point is. Since each base $\angle$ of isos $\triangle$s formed is an ext $\angle$ with meas $\frac{360}{9} = 40$, the meas of the vertex $\angle$ (point of the star) is $180 - 2(40) = 100$.

Additional Example 2

The measures of the exterior angles, one at each vertex, of a pentagon are m $\angle A$ = $(6s + 8)$, m $\angle B$ = $(7s + 2)$, m $\angle C$ = $(8s - 2)$, m $\angle D$ = $(6s - 8)$, and m $\angle E$ = $(8s + 10)$. Find the measure of each exterior angle. m $\angle A = 68$, m $\angle B = 72$, m $\angle C = 78$, m $\angle D = 52$, and m $\angle E = 90$.

Additional Example 3

The measure of an exterior angle of a regular polygon is $(7x + 2)$. The measure of a second exterior angle of the same polygon is $(9x - 6)$. Identify the polygon. Dodecagon

Common Error Analysis

Error: Students tend to miss the significance of the words "one at each vertex" in the statement of Theorem 6.2 on page 236.

Have them draw a pentagon like the one at the bottom of page 236. Then have them extend each side of the polygon in the opposite direction from that shown in the diagram. Emphasize that these exterior angles are different from the ones shown in the student text and ask them to explain why each exterior angle they drew has the same measure as that shown in the text. Supplements of the same angle have the same measure.

Checkpoint

1. What is the sum of the measures of the exterior angles of a dodecagon? 360
2. The measures of the exterior angles of a quadrilateral, one at each vertex, are $m\angle 1 = (9x + 8)$, $m\angle 2 = (10x + 5)$, $m\angle 3 = (11x + 7)$, and $m\angle 4 = (13x - 4)$. Find the measure of each exterior angle. 80, 85, 95, 100
3. Find the measure of each exterior angle of a regular octagon. 45
4. Find the number of sides of a regular polygon if the measure of an exterior angle is 20. 18

Closure

Have the students explain which angles of a polygon are exterior angles. They are angles in the exterior of a polygon which are formed when one side of the polygon is extended past the vertex. Ask them what they know about the measures of exterior angles of convex polygons. Theorem 6.2 page 236

◼◼◼FOLLOW UP

Guided Practice

Classroom Exercises 1–14

Independent Practice

🅐 Ex. 1–15, 🅑 Ex. 16–18, 🅒 Ex. 19–21

Basic: WE 1–15 odd, Application, CI 1–2
Average: WE 1–18 odd, Application, CI all
Above Average: WE 1–21 odd, Application, CI all

Classroom Exercises

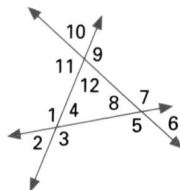

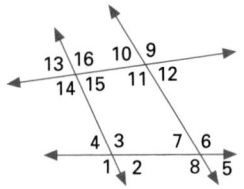

1. Name the interior angles. 4, 8, 12
2. Name the exterior angles.
3. Find $m\angle 1 + m\angle 9 + m\angle 5$. 360
4. Find $m\angle 4 + m\angle 12 + m\angle 8$. 180
5. Find $m\angle 2 + m\angle 6 + m\angle 10$. 180
6. Name the exterior angles.
7. Find $m\angle 6 + m\angle 12 + m\angle 16 + m\angle 4$.
8. Find $m\angle 15 + m\angle 11 + m\angle 7 + m\angle 3$.
9. Find $m\angle 13 + m\angle 9 + m\angle 5 + m\angle 1$.
10. Find $m\angle 14 + m\angle 16 + m\angle 10 + m\angle 12 + m\angle 6 + m\angle 8 + m\angle 2 + m\angle 4$. 720

Find the measure of each exterior angle of the given regular polygon.

11. a pentagon
12. an octagon
13. a decagon
14. a 15-gon

Written Exercises

Copy the polygon. Draw an exterior angle at each vertex. Use a protractor to measure each of these exterior angles. What is the sum of the measures? Does this sum agree with Theorem 6.2?

1. 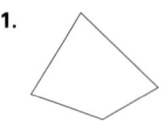 360; yes; sum of ext ∠s = 360

2. 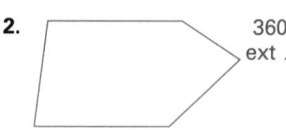 360; yes; sum of ext ∠s = 360

3. The measures of the exterior angles of a quadrilateral, one at each vertex, are x, $2x$, $3x$ and $4x$. Find the measure of each exterior angle. 36, 72, 108, 144
4. The measures of the exterior angles of a hexagon, one at each vertex, are $5x + 8$, $3x + 13$, $5x - 3$, $6x + 10$, $4x - 5$, and $5x + 1$. Find x. 12

Find the measure of each exterior angle of the given regular polygon.

5. triangle 120
6. quadrilateral 90
7. hexagon 60
8. dodecagon 30
9. 20-gon 18
10. 100-gon 3.6

238 Chapter 6 Polygons

Enrichment

Challenge the students with this problem. Let I be the measure of an interior angle of a regular polygon, and E the measure of its adjacent exterior angle. Confirm that for a regular hexagon, $I = (2)E$. The multiplier, m, is 2. For a square, $I = (1)E$, the multiplier being 1. Now for any polygon of n sides, discover a formula for the multiplier m in terms of n. $m = \frac{n-2}{2}$

Using this formula and the relationship $E + I = 180$, find the measure of each interior and exterior angle of a regular decagon and 100-gon. Compare your results with values obtained using other formulas studied in this chapter.
Decagon: $I = 144$, $E = 36$;
100-gon: $I = 176.4$, $E = 3.6$

The measure of an exterior angle of a regular polygon is given. Find the number of sides of the polygon.

11. 60   6   **12.** 45   8   **13.** 30   12   **14.** 24
                                                                                       15

Essay.

15. The exterior angles of a regular polygon are congruent. Explain why.

16. Prove the corollary to Theorem 6.2.

17. The measure of an exterior angle of a regular polygon is $2x + 21$. The measure of a second exterior angle of the same polygon is $4x - 3$. Identify the polygon.   Octagon

18. The measure of an exterior angle of a regular polygon is $12x - 6$. The measure of a second exterior angle of the same polygon is $8x + 6$. Identify the polygon.   Dodecagon

19. A regular polygon has obtuse exterior angles. How many sides does the polygon have?   3

20. Of the regular polygons with obtuse interior angles, which has the largest exterior angle?   Pentagon

21. Is there a regular polygon with an interior angle congruent to an exterior angle? If so, draw a figure to illustrate.   Yes; a square

Mixed Review

Refer to the figure to complete the statement.

1. $\angle 3$ and $\angle$___5___ are alternate interior angles.   *3.2*

2. $\angle 1$ and $\angle$___5___ are corresponding angles.   *3.2*

3. $\angle 2$ and $\angle$___8___ are alternate exterior angles.   *3.2*

4. $\angle 4$ and $\angle$___5___ are interior angles on the same side of the transversal.   *3.2*

5. Given $\angle 8 \cong \angle 4$, prove $m \parallel l$.   *3.3*   If 2 lines are intersected by a transversal so that corr $\angle$s are $\cong$, the lines are $\parallel$.

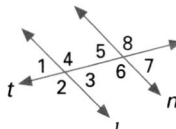

Application: *Mirrors*

Two mirrors can be arranged so that a regular polygon is shown. Draw a segment on a sheet of paper and arrange the two mirrors as shown. Move the mirrors until a regular hexagon is formed. Measure the angle formed by the mirrors. Move the mirrors to form other regular polygons. What is the relationship between the measure of the angle formed by the mirrors and the number of sides of the of the regular polygon?
As the angle decreases, the number of sides of the polygon increases.

6.3 Exterior Angles of Polygons   **239**

47.

Statement	Reason
1. Reg polygon with *n* sides	1. Given
2. Polygon has *n* ∠s.	2. Number of sides of polygon = number of ∠s
3. measure of each ∠ is equal, call it *x*.	3. Def of reg polygon
4. *nx* = (*n* − 2)180	4. Sum of meas of int ∠s of convex polygon = (*n* − 2)180.
5. $x = \frac{(n-2) \cdot 180}{n}$	5. Div Prop of Eq

48.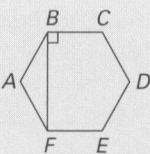

Statement	Reason
1. Reg hexagon *ABCDEF*	1. Given
2. Draw $\overline{BF} \perp \overline{BC}$.	2. Only 1 seg exists from pt not on line ⊥ to line.
3. m ∠*CBF* = 90	3. Def of ⊥
4. m ∠*C* = 120, m ∠*D* = 120, m ∠*E* = 120	4. Meas of ∠s of reg polygon = $\frac{(n-2)180}{n}$
5. m ∠*C* + m ∠*D* + m ∠*E* + m ∠*CBF* + m ∠*EFB* = 540	5. Sum of meas of int ∠s of convex polygon = (*n* − 2)180.
6. m ∠*EFB* = 90	6. Subt Prop of Eq
7. $\overline{BC} \parallel \overline{FE}$	7. If int ∠s on same side of a transv are supp, then lines are ∥.

 # Computer Investigation

Properties of a Parallelogram

Use a computer software program that draws random parallelograms, labels points, and measures line segments or angles.

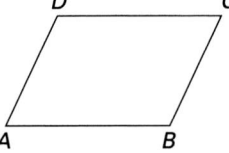

Quadrilateral *ABCD* has two pairs of opposite sides parallel: $\overline{CB} \parallel \overline{DA}$, $\overline{CD} \parallel \overline{BA}$. Such a figure is called a **parallelogram**. Opposite angles are angles *A* and *C* and angles *B* and *D*. The following activities are designed to help you discover some properties of parallelograms formally proved in the next lesson.

Activity 1

Draw a random parallelogram.

1. Find the lengths of all four sides.

2. Draw a new random parallelogram. Measure its sides.

3. Repeat the activity of Exercise 2 for a third random parallelogram.

Activity 2

What relationship seems to exist between opposite angles? ≅

4. Use the measuring tool of the program to test your hypothesis for at least three different parallelograms.

Activity 3

What relationship seems to exist between consecutive angles? Supp

5. Use the measuring tool of the program to test your hypothesis for at least three different parallelograms.

Activity 4

Draw a random parallelogram and label it *ABCD*. Draw the two diagonals $\overline{DB}$ and $\overline{AC}$. Label their point of intersection *E*.

6. Are the diagonals congruent? Measure to verify. No

7. Do the diagonals bisect each other? Measure to verify. Yes

8. Repeat the activities of Exercises 6 and 7 for two other random parallelograms.

Summary

What generalizations appear to be true of a parallelogram for: opposite sides? opposite angles? consecutive angles? diagonals?
∥ and ≅; ≅; supp; bis each other

240 Chapter 6 Computer Investigation

6.4 Quadrilaterals and Parallelograms

Objectives   To identify parts and investigate properties of quadrilaterals
To prove and apply theorems about parallelograms

The figure to the right shows
two patterns for building a
structure such as a bridge,
tower, or scaffold. Which do
you think gives greater strength
and why?

A **quadrilateral** is a polygon with four sides. $ABCD$ is a quadrilateral.

$\overline{AB}$ and $\overline{BC}$ are adjacent sides.
$\overline{AB}$ and $\overline{CD}$ are opposite sides.
$\overline{AC}$ and $\overline{BD}$ are diagonals.
$\angle ABC$ and $\angle BAD$ are consecutive angles.
$\angle ABC$ and $\angle ADC$ are opposite angles.

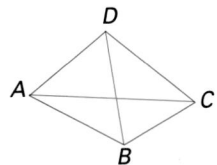

Definition   A **parallelogram** is a quadrilateral with both pairs of opposite sides
parallel. ($\square EFGH$ means parallelogram $EFGH$.)

Theorem 6.3   A diagonal of a parallelogram forms two congruent triangles.

Given: $\square ABCD$ with diagonal $\overline{AC}$
Prove: $\triangle ABC \cong \triangle CDA$

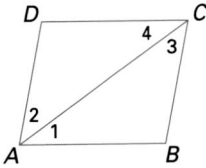

Plan   Use ASA. $\angle 1$ and $\angle 4$, and $\angle 2$ and $\angle 3$
are alternate interior angles.

Proof

Statement	Reason
1. $\square ABCD$ with diag $\overline{AC}$	1. Given
2. $\overline{AB} \parallel \overline{DC}$, $\overline{AD} \parallel \overline{BC}$	2. Def of $\square$
3. $\angle 1 \cong \angle 4$, $\angle 3 \cong \angle 2$ **(A, A)**	3. Alt int $\angle$s of $\parallel$ lines are $\cong$.
4. $\overline{AC} \cong \overline{CA}$ **(S)**	4. Reflex Prop
5. $\therefore \triangle ABC \cong \triangle CDA$	5. ASA

Note that in the proof of this theorem, it would be incorrect to state
that $\triangle ABC \cong \triangle DCA$.

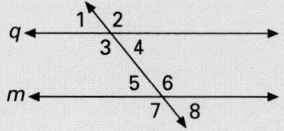
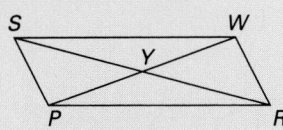

Lesson Note

As you introduce properties of parallelograms, have students verify them with drawings and measurements. If you use an overhead projector, a simple way to create a variety of parallelograms is to draw several sets of parallel lines various distances apart. By crossing and moving two of them, different parallelograms will be formed.

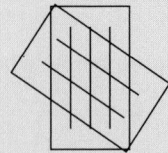

Sufficient conditions for proving a quadrilateral to be a parallelogram are introduced in the next lesson. The theorems in this lesson refer to the properties or necessary conditions of a parallelogram, although this term is not used until Chapter 7.

Math Connections

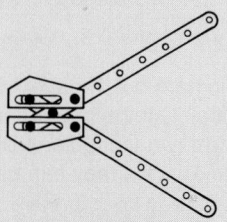

Pliers: The "pliers" shown above are made so that the jaws are always parallel. The brads connecting the cardboard jaws to the plastic handles form the vertices of a parallelogram.

Critical Thinking Questions

Analysis: Ask students to draw a diagram and answer the following question. In $\square ABCD$ with diagonals $\overline{AC}$ and $\overline{BD}$ intersecting at E, is $\triangle DEC \cong \triangle AEB$? Why or why not?   No. By def of $\cong$ polygons, $\triangle DEC \cong \triangle AEB$ states that $\angle EDC \cong \angle EAB$ and $\angle ECD \cong \angle EBA$, but these $\angle$s are supp, not $\cong$.

EXAMPLE 1   *WXYZ* is a parallelogram. Which of the following pairs of triangles are congruent?

$\triangle XYZ$ and $\triangle YXW$ $\triangle XYZ$ and $\triangle ZWX$

Solution $\triangle XYZ$ and $\triangle YXW$ are not necessarily congruent. $\triangle XYZ$ and $\triangle ZWX$ are congruent by Theorem 6.3.

In the proof of Theorem 6.3, you saw that $\triangle ABC \cong \triangle CDA$. Therefore, the opposite sides of $\square ABCD$ are congruent (CPCTC). The corresponding angles, B and D, are also congruent. By using diagonal $\overline{BD}$ you can show that $\angle DAB \cong \angle DCB$. Therefore, the opposite angles of $\square ABCD$ are congruent. This suggests the following corollaries.

Corollary 1   Opposite sides of a parallelogram are congruent.

Corollary 2   Opposite angles of a parallelogram are congruent.

EXAMPLE 2   In $\square ABCD$, $AB = 3x + 15$, $CD = 5x - 3$
Find AB.

Plan   Since opposite sides of a parallelogram are congruent, $AB = CD$.

Solution
$$AB = CD$$
$$3x + 15 = 5x - 3$$
$$x = 9$$
$$AB = 3(9) + 15 = 42$$

If two parallel lines are intersected by a transversal, then the interior angles on the same side of the transversal are supplementary. This can be used to prove Theorem 6.4.

Theorem 6.4   Consecutive angles of a parallelogram are supplementary.

Additional Example 2

In $\square$ *EFGH*, m $\angle E = (9x - 3)$, m $\angle G = (8x + 6)$. Find m $\angle E$.   78

EXAMPLE 3 In ▱PQRS, PQ = 23 and m ∠Q = 81
Find SR and m ∠R.

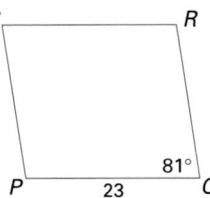

Solution Since PQ = 23, SR = 23
m ∠Q + m ∠R = 180
m ∠R = 180 − m ∠Q
m ∠R = 180 − 81 = 99

EXAMPLE 4 Given: ▱ABCD with AB = AC
Prove: ∠ACB ≅ ∠CDA

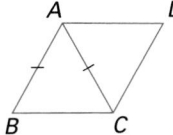

Proof

Statement	Reason
1. ▱ABCD with AB = AC	1. Given
2. ∠ACB ≅ ∠ABC	2. In a △, ∠s opp ≅ sides are ≅.
3. ∠ABC ≅ ∠CDA	3. Opp ∠s of a ▱ are ≅.
4. ∴ ∠ACB ≅ ∠CDA	4. Trans Prop

You can use corresponding parts of congruent triangles to prove the following theorem about the diagonals of a parallelogram.

Theorem 6.5

The diagonals of a parallelogram bisect each other.

Given: ▱ABCD with diagonals AC and BD
intersecting at O
Prove: AC and BD bisect each other.

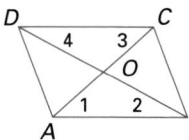

Proof

Statement	Reason
1. ▱ABCD with diags AC and BD intersecting at O	1. Given
2. AB ∥ CD	2. Opp sides of a ▱ are ∥.
(A, A) 3. ∠1 ≅ ∠3, ∠2 ≅ ∠4	3. Alt int ∠s of ∥ lines are ≅.
(S) 4. AB ≅ CD	4. Opp sides of a ▱ are ≅.
5. △ABO ≅ △CDO	5. ASA
6. AO ≅ CO, BO ≅ DO	6. CPCTC
7. ∴ AC and BD bisect each other.	7. Def of a bis

6.4 Quadrilaterals and Parallelograms **243**

Additional Example 3

In ▱MTRA, MA = 7 and m ∠M = 65.
Find TR and m ∠T. TR = 7;
m ∠T = 115

Additional Example 4

Given: ▱ABCD with AE ≅ CF
Prove: DE ≅ BF

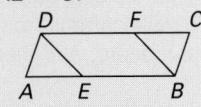

Proof: ▱ABCD
with AE ≅ CF
(Given); AD ≅ CB (Opp sides of a ▱ are
≅); ∠A ≅ ∠C (Opp ∠s of a ▱ are ≅);
△ADE ≅ △CBF (SAS); DE ≅ BF
(CPCTC)

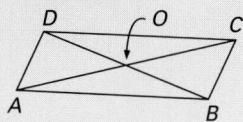
243

Guided Practice

Classroom Exercises 1–8

Independent Practice

🅐 Ex. 1–18, 🅑 Ex. 19–27, 🅒 Ex. 28–30

Basic: WE 1–18 odd, Midchapter Review, Brainteaser

Average: WE 1–27 odd, Midchapter Review, Brainteaser

Above Average: WE 1–30 odd, Midchapter Review, Brainteaser

Additional Answers

Written Exercises

17.

Statement	Reason
1. □ABCD, ∠A ≅ ∠DBA	1. Given
2. DA = DB	2. Sides opp ≅ ∠s are ≅.
3. DA = BC	3. Opp sides of □ are ≅.
4. DB = BC	4. Sub

18.

Statement	Reason
1. □RSTU, $\overline{RT}$ bis ∠URS	1. Given
2. $\overline{RU} \parallel \overline{ST}$	2. Def of □
3. ∠STR ≅ ∠URT	3. If lines are ‖, then alt int ∠s are ≅.
4. ∠URT ≅ ∠TRS	4. Def of ∠ bis
5. ∠TRS ≅ ∠STR	5. Trans Prop of Congr
6. $\overline{RS} ≅ \overline{ST}$	6. Sides opp ≅ ∠s are ≅.
7. △RST is isos.	7. Def of isos △

22.

Statement	Reason
1. □ABCD with diag $\overline{AC}$	1. Given
2. △ADC ≅ △CBA	2. Diag of □ forms 2 ≅ △s.
3. $\overline{AD} ≅ \overline{BC}$, $\overline{CD} ≅ \overline{BA}$	3. CPCTC

23.

Statement	Reason
1. □ABCD with diags $\overline{AC}$ and $\overline{BD}$	1. Given
2. △ADC ≅ △CBA	2. Diag of □ forms 2 ≅ △s.
3. ∠CBA ≅ ∠CDA	3. CPCTC
4. △ADB ≅ △CBD	4. Diag of □ forms 2 ≅ △s.
5. ∠BAD ≅ ∠DCB	5. CPCTC

■ **Summary**

Properties of a Parallelogram

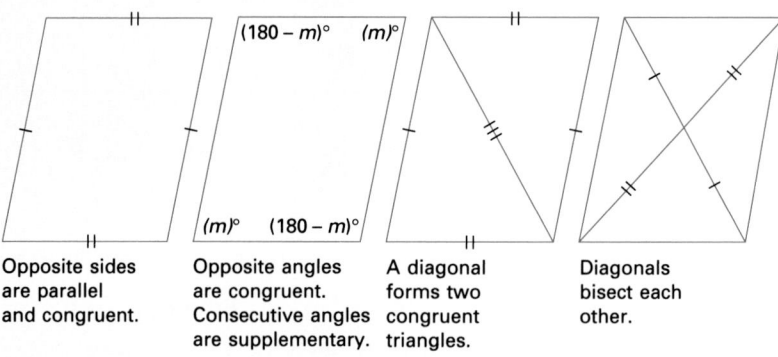

Opposite sides are parallel and congruent.

Opposite angles are congruent. Consecutive angles are supplementary.

A diagonal forms two congruent triangles.

Diagonals bisect each other.

Classroom Exercises

Complete the statement for parallelogram *TUVW*.

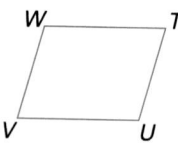

1. $\overline{WV}$ is opposite ___$\overline{TU}$___.
2. ∠U and ___∠T or ∠V___ are supplementary.
3. ∠T ≅ ___∠V___
4. $\overline{TU}$ and ___$\overline{UV}, \overline{WT}$___ are adjacent.
5. ∠W is opposite ___∠U___.
6. ∠T and ___∠W, ∠U___ are consecutive angles.
7. Which sides of □ *TUVW* are parallel? ___$\overline{TU} \parallel \overline{VW}, \overline{UV} \parallel \overline{WT}$___
8. Which sides of □ *TUVW* are congruent? ___$\overline{TU} ≅ \overline{VW}, \overline{UV} ≅ \overline{WT}$___

Written Exercises

***ABCD* is a parallelogram. State whether the two triangles are congruent. If so, explain why.**

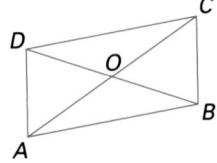

1. △ABC and △BCD   No
2. △ABC and △CDA   Yes; diag forms 2 ≅ △s.
3. △CDB and △ABD   Yes; diag forms 2 ≅ △s.
4. △CDB and △BAD   No

Enrichment

Before class, cut out several small parallelograms from a sheet of light cardboard, and then cut them along the

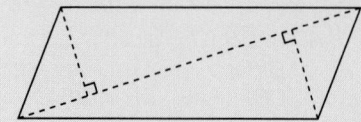

lines shown in this drawing. Use each group of pieces as a puzzle. Have pairs of students work on a puzzle; the object is to use all four triangles to fit into the shape of a parallelogram.

After each pair has succeeded, have them design similar puzzles of their own and challenge each other with them.

EFGH is a parallelogram. Find the unknown measure.

5. EF = 17, GH = <u>17</u>

6. m ∠EFG = 67, m ∠GHE = <u>67</u>

7. m ∠HEF = 119, m ∠GHE = <u>61</u>

8. m ∠HGF = 125, m ∠EHG = <u>55</u>

9. EF = 12x, GH = 10x + 12, GH = <u>72</u>

10. EF = 5x − 7, GH = 3x + 1, EF = <u>13</u>

11. EH = 2x + 2, FG = 3x − 5, FG = <u>16</u>

12. EO = 3x + 2, GO = 5x − 8, EO = <u>17</u>

13. FO = 4x + 13, HO = 5x + 1, FH = <u>122</u>

14. m ∠EFG = 6x + 6, m ∠FGH = 3x + 3, m ∠FGH = <u>60</u>

15. m ∠EFG = 12x − 24, m ∠GHE = 9x + 12, m ∠EFG = <u>120</u>

16. HO = EO, m ∠HOE = 100, m ∠OHE = <u>40</u>

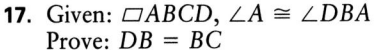

17. Given: ▱ABCD, ∠A ≅ ∠DBA
Prove: DB = BC

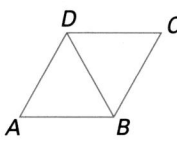

18. Given: ▱RSTU, $\overline{RT}$ bisects ∠URS.
Prove: △RST is isosceles.

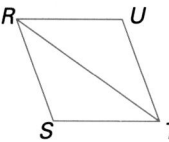

Explain why the figure cannot be a parallelogram.

19.

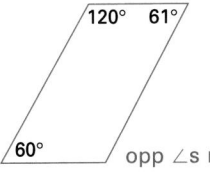

opp ∠s not ≅

20.

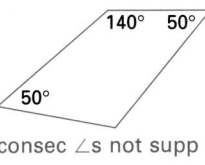

consec ∠s not supp

21.

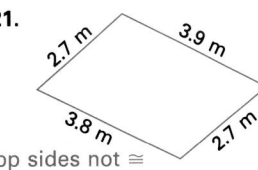

opp sides not ≅

22. Prove Corollary 1.   **23.** Prove Corollary 2.   **24.** Prove Theorem 6.4.

25. Given: ▱ACDF, ▱BCEF
Prove: △ABF ≅ △DEC

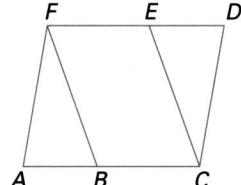

26. Given: ▱ABCD, diagonals $\overline{AC}$ and $\overline{BD}$ intersect at O.
Prove: O is the midpoint of $\overline{EF}$.

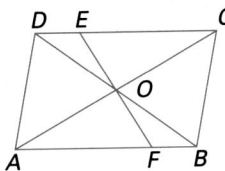

24.

Statement	Reason
1. ▱ABCD	1. Given
2. $\overline{BC}$ ∥ $\overline{AD}$	2. Def of ▱
3. ∠A and ∠B are supp. ∠C and ∠D are supp.	3. If lines are ∥, then int ∠s on same side of transv are supp.
4. $\overline{BA}$ ∥ $\overline{CD}$	4. Def of ▱
5. ∠B and ∠C are supp. ∠A and ∠D are supp.	5. If lines are ∥, then int ∠s on same side of transv are supp.

25.

Statement	Reason
1. ▱ACDF, ▱BCEF	1. Given
2. FD = AC, AF = DC, BF = CE, BC = FE	2. Opp sides of ▱ are ≅.
3. FD − FE = AC − BC	3. Equations may be subtracted.
4. FD − FE = ED, AC − BC = AB	4. Seg Add Post
5. ED = AB	5. Sub
6. △ABF ≅ △DEC	6. SSS

26.

Statement	Reason
1. ▱ABCD, diags $\overline{AC}$ and $\overline{BD}$ intersect at O.	1. Given
2. $\overline{DO}$ ≅ $\overline{BO}$	2. Diags of ▱ bis each other.
3. $\overline{DC}$ ∥ $\overline{BA}$	3. Def of ▱
4. ∠DEO ≅ ∠BFO	4. If lines are ∥, then alt int ∠s are ≅.
5. ∠DOE ≅ ∠BOF	5. Vert ∠s are ≅.
6. △BOF ≅ △DOE	6. AAS
7. $\overline{EO}$ ≅ $\overline{FO}$	7. CPCTC
8. O is midpt of $\overline{EF}$.	8. Def of midpt

27.

Statement	Reason
1. □JLMP, K is midpt of $\overline{JL}$, N is midpt of $\overline{PM}$.	1. Given
2. $NM = \frac{1}{2}PM$, $JK = \frac{1}{2}JL$	2. Def of midpt
3. $PM = JL$	3. Opp sides □ are ≅.
4. $\frac{1}{2}PM = \frac{1}{2}JL$	4. Mult Prop of Eq
5. $NM = JK$	5. Sub
6. $PM \parallel JL$	6. Def of □
7. $\angle NMO \cong \angle KJO$	7. If lines are ∥, then alt int ∠s are ≅.
8. $\angle JOK \cong \angle MON$	8. Vert ∠s are ≅.
9. $\triangle MON \cong \triangle JOK$	9. AAS
10. $\overline{JO} \cong \overline{MO}$	10. CPCTC
11. O is midpt of $\overline{JM}$.	11. Def of midpt

29.

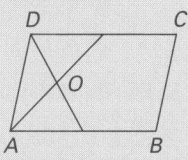

Statement	Reason
1. □ABCD, $\overline{DO}$ bis ∠ADC, $\overline{AO}$ bis ∠DAB	1. Given
2. m∠DAB + m∠ADC = 180	2. Consec ∠s of □ are supp.
3. m∠OAD = $\frac{1}{2}$ × m∠DAB, m∠ODA = $\frac{1}{2}$ × m∠ADC	3. Def of ∠ bis
4. m∠OAD + m∠ODA = $\frac{1}{2}$(m∠ADC + m∠DAB)	4. Equations may be added.
5. m∠OAD + m∠ODA = 90	5. Sub
6. m∠OAD + m∠ODA + m∠DOA = 180	6. Sum of meas of ∠ of △ = 180
7. m∠DOA = 90	7. Equations may be subtracted.
8. $\overline{DO} \perp \overline{AO}$	8. Def of ⊥

See page 256 for the answer to Ex. 30.

246

---

27. Given: □JLMP, K is the midpoint of $\overline{JL}$; N is the midpoint of $\overline{PM}$.
Prove: O is the midpoint of JM.

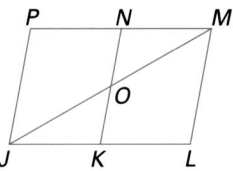

28. In □XYZW, the sum of the measures of ∠X and ∠Z is 30 greater than the measure of ∠Y. Find the measure of ∠X.   70

29. Prove: The bisectors of consecutive angles of a parallelogram are perpendicular.

30. Prove: The bisectors of opposite angles of a parallelogram are parallel or concurrent.

Midchapter Review

Name each polygon; tell whether it is concave or convex; then tell whether it is equilateral, equiangular, or regular.   *6.1*

1. 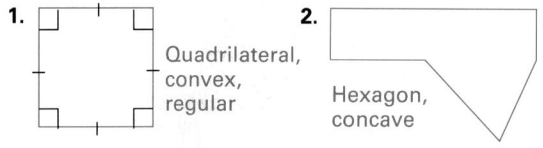 Quadrilateral, convex, regular

2. Hexagon, concave

3. 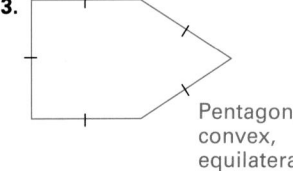 Pentagon, convex, equilateral

4. Find the sum of the measures of the angles of a pentagon.   *6.2*   540

5. What is the measure of each angle of a convex regular octagon?   *6.2*   135

6. What is the measure of each exterior angle of a regular pentagon?   *6.3*   72

7. The measures of two exterior angles of a regular polygon are $2x + 10$ and $3x - 15$. Identify the polygon.   *6.3*   Hexagon

Brainteaser

The design shown was made from a set of wooden tiles. Only three different kinds of tiles were used. Each tile is a parallelogram with all sides congruent. Without measuring, find the angle measures of the three different parallelograms.

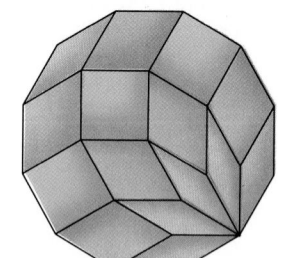

Blue 90, 90, 90, 90 Orange 30, 150, 30, 150
Green 120, 60, 120, 60

246   Chapter 6 Polygons

6.5 Quadrilaterals That Are Parallelograms

Teaching Resources

Problem Solving Worksheet 6
Quick Quizzes 47
Reteaching and Practice Worksheets, pp. 93, 94

Objectives
To determine if a given quadrilateral is a parallelogram
To prove that certain quadrilaterals are parallelograms

Experimenting with quadrilaterals suggests that there may be other ways to prove that a quadrilateral is a parallelogram. Fasten four sticks, two of one length and two of another length, at their ends to form a quadrilateral with opposite sides of equal length. The quadrilateral formed appears to be a parallelogram. If the sticks are moved so that the measures of the angles between them are changed, then the quadrilateral remains a parallelogram.

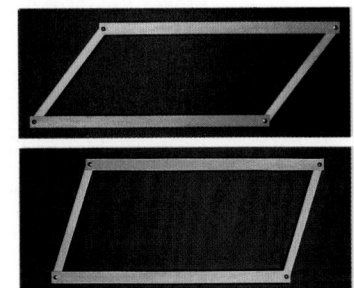

Theorem 6.6
If both pairs of opposite sides of a quadrilateral are congruent, then the quadrilateral is a parallelogram.

Given: Quad *ABCD* with $\overline{AB} \cong \overline{CD}$,
   $\overline{CB} \cong \overline{AD}$
Prove: *ABCD* is a parallelogram.

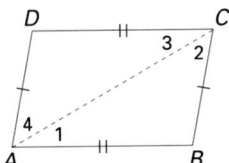

Plan
Draw $\overline{AC}$ and prove $\triangle ABC \cong \triangle CDA$. Use congruent pairs of alternate interior angles to prove opposite sides parallel.

Proof

Statement		Reason
1. Quad *ABCD* with $\overline{AB} \cong \overline{CD}$, $\overline{CB} \cong \overline{AD}$	**(S, S)**	1. Given
2. Draw $\overline{AC}$.		2. Two points determine a line.
3. $\overline{AC} \cong \overline{AC}$	**(S)**	3. Reflex Prop
4. $\triangle ABC \cong \triangle CDA$		4. SSS
5. $\angle 1 \cong \angle 3, \angle 2 \cong \angle 4$		5. CPCTC
6. $\overline{AB} \parallel \overline{CD}, \overline{CB} \parallel \overline{AD}$		6. If alt int $\angle$s are $\cong$, then lines are $\parallel$.
7. $\therefore$ *ABCD* is a $\square$.		7. Def of a parallelogram

6.5 Quadrilaterals That Are Parallelograms   **247**

▬▬▬**GETTING STARTED**

Prerequisite Quiz

Refer to the figure below.

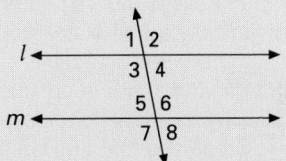

1. If $\angle 4 = $ _____ , then $l \parallel m$. $\angle 5$ or $\angle 8$
2. If m $\angle 3 +$ _____ $= 180$, then $l \parallel m$.
   $\angle 5$ or $\angle 8$

True or false?

3. Two triangles can be proved congruent by an SSS correspondence.   T
4. Two triangles can be proved congruent by an SSA correspondence.   F
5. Two triangles can be proved congruent by an SAS correspondence.   T

Motivator

Perform a demonstration for the students by fastening two sticks together at their centers with opposite sides of equal length. Stretch a large rubber band or elastic cord around their ends. Ask them what shape is formed. Quadrilateral-parallelogram   Ask them whether the figure will still be a parallelogram if the angle measure between the sticks is changed.   Yes   Ask them what conclusions they can draw from this demonstration.   Theorem 6.6 page 247

TEACHING SUGGESTIONS

Lesson Note

Example 3 illustrates the fact that the converse of a theorem is not necessarily true. A diagonal of a parallelogram does form two congruent triangles with the sides of the parallelogram. However, the converse is not true.

Students must understand that a statement is considered true only if it is always true. If it is true for some cases, but not others, then it is not logically a true statement. The lesson summary lists *sufficient conditions* for proving that a quadrilateral is a parallelogram. *Necessary conditions*, which refer to properties that a given figure must have, are formally identified in Chapter 7. Conditions that are both necessary and sufficient can be described with a biconditional such as the following:
A quadrilateral is a parallelogram if and only if the diagonals bisect each other.
A quadrilateral is a parallelogram if and only if the opposite sides are parallel.
Since a definition is reversible, it can always be expressed in "If and only if. . .!" form. That is, the conditions given in a definition are both necessary and sufficient.

Math Connections

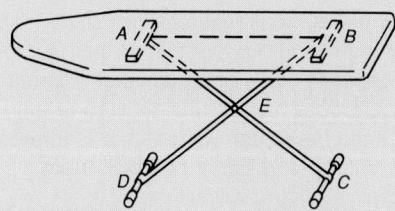

Ironing Board: The folding legs of an ironing board are designed in such a way that the board remains parallel to the floor. In the figure above, $\overline{AB}$ lies in the plane of the board. The points D and C lie in the plane of the floor, because the legs rest on the floor. By joining these four points, quadrilateral $ABCD$ can be formed. Since the legs, $\overline{AC}$ and $\overline{DB}$ in the figure, bisect each other at E, $ABCD$ is a parallelogram and the opposite sides, $\overline{AB}$ and $\overline{DC}$ are always parallel.

Theorem 6.7   If the diagonals of a quadrilateral bisect each other, then the quadrilateral is a parallelogram.

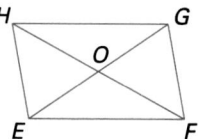

Given: Diagonals $\overline{EG}$ and $\overline{FH}$ of quad $EFGH$ bisect each other.
Prove: $EFGH$ is a parallelogram.

Plan   Use SAS to prove $\triangle EOF \cong \triangle GOH$ and $\triangle GOF \cong \triangle EOH$. Then opposite sides of $EFGH$ are congruent.

EXAMPLE 1   Is the quadrilateral a parallelogram? Explain why.

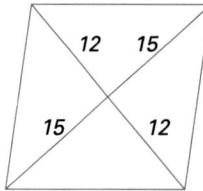

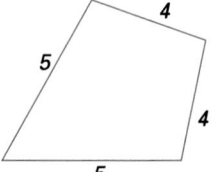

Solution   It is a parallelogram, since diagonals bisect each other.   It is a parallelogram, since both pairs of opposite sides are congruent.   The opposite sides are not congruent. The figure is not a parallelogram.

The two following theorems are also useful when proving that a quadrilateral is a parallelogram.

Theorem 6.8   If two sides of a quadrilateral are parallel and congruent, then the quadrilateral is a parallelogram.

Given: Quad $ABCD$ with
$\overline{AB} \parallel \overline{CD}, \overline{AB} \cong \overline{CD}$
Prove: $ABCD$ is a parallelogram.

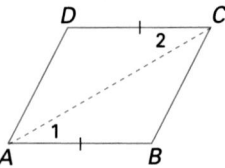

Plan   Draw $\overline{AC}$. Use SAS to prove $\triangle ABC \cong \triangle CDA$.

248   Chapter 6 Polygons

Additional Example 1

Is the quadrilateral a parallelogram? Explain why.

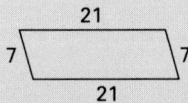

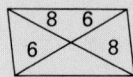

Yes; diagonals bisect each other.

Yes; opposite sides are congruent.

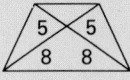

No; diagonals do not bisect each other.

Proof	Statement	Reason

Proof

Statement
1. Quad *ABCD* with $\overline{AB} \cong \overline{CD}$
2. Draw $\overline{AC}$.
3. $\overline{AB} \parallel \overline{CD}$
4. $\angle 1 \cong \angle 2$
5. $\overline{AC} \cong \overline{CA}$
6. $\triangle ABC \cong \triangle CDA$
7. $\overline{BC} \cong \overline{DA}$
8. ∴ *ABCD* is a parallelogram.

Reason
(S) 1. Given
2. Two points determine a line.
3. Given
(A) 4. Alt int ∠s of ‖ lines are ≅.
(S) 5. Reflex Prop
6. SAS
7. CPCTC
8. If both pairs of opp sides are ≅, then quad is a ▱.

Theorem 6.9 | If both pairs of opposite angles of a quadrilateral are congruent, then the quadrilateral is a parallelogram.

EXAMPLE 2

Given: Quad *ABCD* with $\angle A \cong \angle C$, $\overline{AB} \parallel \overline{DC}$
Prove: *ABCD* is a parallelogram.

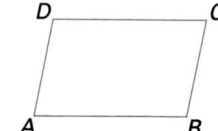

Proof

Statement
1. Quad *ABCD* with $\overline{AB} \parallel \overline{DC}$
2. $\angle D$ and $\angle A$ are supp; $\angle B$ and $\angle C$ are supp.
3. $\angle A \cong \angle C$
4. $\angle D \cong \angle B$
5. ∴ *ABCD* is a parallelogram.

Reason
1. Given
2. If lines are ‖, then int ∠s on same side of transv are supp.
3. Given
4. Supp of ≅ ∠s are ≅.
5. If a quad has both pairs of opp ∠s ≅, then it is a ▱.

EXAMPLE 3

True or false? If a diagonal of a quadrilateral forms two congruent triangles, then the quadrilateral is a parallelogram.

Solution

False. *EFGH* is a parallelogram, but *ABCD* is not.

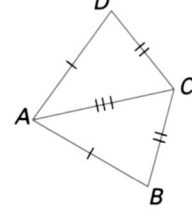
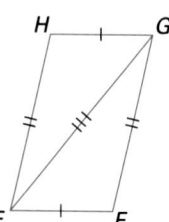

6.5 Quadrilaterals That Are Parallelograms **249**

Critical Thinking Questions

Synthesis: Ask students to experiment with drawing pairs of segments of equal and unequal lengths that bisect each other. Then ask them what other properties must the legs of the ironing board described in Math Connections have in order for the board to remain stable and parallel to the floor regardless of the height to which the board is adjusted. The legs must be ≅.

Checkpoint

Using only the given information, determine whether *EFGH* is a parallelogram. Explain why. (Ex. 1–5)

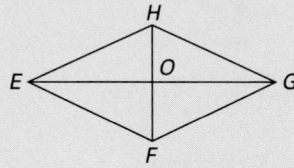

1. $\overline{HG} \parallel \overline{EF}$ No; insufficient information
2. $\angle HEF \cong \angle HGF$, $\angle EHG \cong \angle EFG$ Yes; Thm 6.9
3. $\overline{HO} \cong \overline{FO}$, $\overline{EO} \cong \overline{GO}$ Yes; Thm 6.7
4. $\overline{EF} \cong \overline{HG}$, $\overline{EF} \parallel \overline{HG}$ Yes; Thm 6.8
5. $\overline{EG} \cong \overline{HF}$ No; not a sufficient condition
6. Given: $\angle A \cong \angle C$, $\angle 1 \cong \angle 2$
   Prove: *ABCD* is a parallelogram.

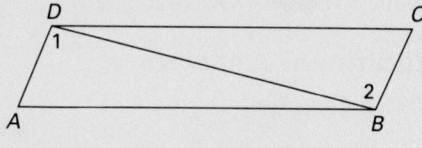

Statement	Reason
1. $\angle A \cong \angle C$, $\angle 1 \cong \angle 2$	1. Given
2. $\overline{DB} \cong \overline{BD}$	2. Reflex Prop of Cong
3. $\triangle ABD \cong \triangle CDB$	3. AAS
4. $\overline{AD} \cong \overline{CB}$ $\overline{AB} \cong \overline{CD}$	4. CPCTC
5. *ABCD* is a ▱.	5. If opp sides of a quad are ≅, then quad is a ▱.

Additional Example 2

Given: Quad *ABCD* with $\overline{AE} \cong \overline{CF}$, $\overline{DE} \cong \overline{BF}$, $\angle 1 \cong \angle 2$
Prove: *ABCD* is a parallelogram.

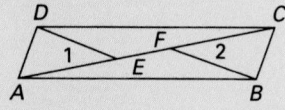

Proof: $\overline{AE} \cong \overline{CF}$, $\overline{DE} \cong \overline{BF}$, $\angle 1 \cong \angle 2$ (Given); $\triangle ADE \cong \triangle CBF$ (SAS); $\angle DAE \cong \angle BCF$ (CPCTC); $\overline{AD} \parallel \overline{CB}$ (If alt int ∠s are ≅, then lines are ‖.); $\overline{AD} \cong \overline{CB}$ (CPCTC); *ABCD* is a ▱. (If two sides of a quad are ‖ and ≅, then quad is a ▱.)

Closure

Write the list of properties 1–7 on the chalkboard and have the students go down the list adding one property at a time to what they already know, until they know for certain that they have a parallelogram.
(1) It has four sides.
(2) One diagonal is bisected by the other diagonal.
(3) It has at least one pair of adjacent supplementary angles.
(4) It has at least one pair of congruent angles.
(5) It has at least one pair of parallel sides.
(6) These parallel sides are congruent.
(7) Both diagonals bisect each other.
properties 1–6

 **FOLLOW UP**

Guided Practice

Classroom Exercises 1–8

Independent Practice

A Ex. 1–13, **B** Ex. 14–20, **C** Ex. 21–25

Basic: WE 1–13 odd

Average: WE 1–20 odd

Above Average: WE 1–25 odd

Additional Answers

Written Exercises

14.

Statement	Reason
1. Quad *ABCD* with diags $\overline{AC}$ and $\overline{BD}$, $\overline{AC}$ and $\overline{BD}$ bis each other at *O*	1. Given
2. $\overline{AO} \cong \overline{OC}$, $\overline{DO} \cong \overline{OB}$	2. Def of seg bis
3. ∠*DOC* ≅ ∠*BOA*, ∠*BOC* ≅ ∠*DOA*	3. Vert ∠s are ≅.
4. △*COD* ≅ △*AOB*, △*COB* ≅ △*AOD*	4. SAS
5. $\overline{DC} \cong \overline{BA}$, $\overline{DA} \cong \overline{BC}$	5. CPCTC
6. *ABCD* is a ▱.	6. If 2 pair opp sides are ≅, then quad is ▱.

 **Summary**

Sufficient Conditions for Proving a Quadrilateral is a Parallelogram

When proving that a quadrilateral is a parallelogram, it is enough to show that the quadrilateral satisfies any *one* of the five conditions below. Each is a *sufficient* condition.

1. Both pairs of opposite sides are parallel. (Definition)
2. Both pairs of opposite sides are congruent.
3. Diagonals bisect each other.
4. Two sides are parallel and congruent.
5. Both pairs of opposite angles are congruent.

Classroom Exercises

Form sufficient conditions for proving that *ABCD* is a parallelogram by completing the statement.

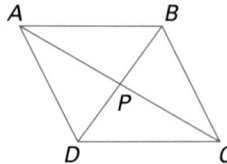

∠*BCD*, ∠*ABC*

1. $\overline{AB} \cong$ _____ and $\overline{AD} \cong$ _____ $\overline{DC}, \overline{BC}$
2. ∠*BAD* ≅ _____ and ∠*ADC* ≅ _____
3. ∠*ABD* ≅ _____ and ∠*DAC* ≅ _____
4. $\overline{AP} \cong$ _____ and $\overline{DP} \cong$ _____ $\overline{PC}, \overline{BP}$
5. $\overline{AB} \parallel$ _____ and $\overline{BC} \parallel$ _____ $\overline{DC}, \overline{AD}$
6. $\overline{AB} \cong$ _____ and $\overline{AB} \parallel$ _____ $\overline{DC}, \overline{DC}$
7. $\overline{AD} \cong$ _____ and ∠*ADC* is supplementary to _____. $\overline{BC}$, ∠*DCB*
8. △*ABD* ≅ _____ △*CDB*

   3. ∠*BDC*, ∠*BCA*

Written Exercises

Using only the given information, determine whether *ABCD* is a parallelogram. Give a reason for your answer.

1. $\overline{AB} \parallel \overline{DC}$ and $\overline{BC} \parallel \overline{AD}$ Yes (Def of ▱)
2. $\overline{AB} \cong \overline{BC} \cong \overline{CD}$ No (Not sufficient)
3. $\overline{AB} \cong \overline{DC}$ and $\overline{BC} \cong \overline{AD}$ Yes (Thm 6.6)
4. $\overline{AB} \cong \overline{DC}$ and $\overline{AB} \parallel \overline{DC}$ Yes (Thm 6.9)
5. $\overline{AB} \cong \overline{DC}$ and $\overline{BC} \parallel \overline{AD}$ No (Not sufficient)
6. $\overline{AC} \cong \overline{DB}$ No (Not sufficient)
7. ∠*DAB* ≅ ∠*BCD* No (Not sufficient)
8. $\overline{AO} \cong \overline{CO}$ and $\overline{BO} \cong \overline{DO}$ Yes (Thm 6.7)
9. △*ABC* ≅ △*ADC* No (Not sufficient)
10. $\overline{AO} \cong \overline{BO}$ and $\overline{CO} \cong \overline{DO}$ No (Not sufficient)

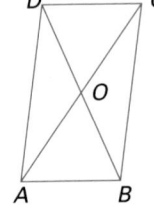

Additional Example 3

True or false? If the diagonals of a quadrilateral are congruent, then the quadrilateral is a parallelogram.

False. *TQRS* is a parallelogram but *PWXY* is not.

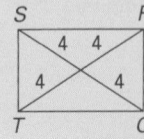

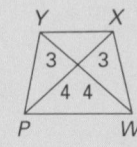

11. m $\angle DAB = 42$,
m $\angle ABC = 138$, **Yes**
m $\angle BCD = 42$ (Thm 6.9)

12. m $\angle ABD = 37$,
m $\angle CDB = 37$, **Yes**
$AB = CD = 12$ (Thm 6.8)

13. m $\angle ABD = 37$,
m $\angle CDB = 37$,
$AB = BC$ **No**
(Not sufficient)

14. Prove Theorem 6.7.

15. Prove Theorem 6.9.

16. Given: $\overline{ED} \cong \overline{BA}$,
$\triangle AEF \cong \triangle DBC$
Prove: $ABDE$ is a parallelogram.

17. Given: $\square ACDF$,
$\overline{FE} \cong \overline{CB}$
Prove: $ABDE$ is a parallelogram.

(Ex. 16–19)

18. Given: $\square ABDE$,
$\overline{FE} \cong \overline{CB}$
Prove: $\triangle AEF \cong \triangle DBC$

19. Given: $\overline{AF} \parallel \overline{CD}$,
$\triangle AFE \cong \triangle DCB$
Prove: $ABDE$ is a parallelogram.

20. Given: Equilateral $\triangle$s GHJ and IHJ
Prove: $GHIJ$ is a parallelogram.

True or false? Give a proof or a counterexample to defend your answer.

21. If consecutive angles of a quadrilateral are supplementary, then the quadrilateral is a parallelogram.

22. If the diagonals of a quadrilateral are perpendicular, then the quadrilateral is a parallelogram.

23. If one pair of opposite sides and one pair of opposite angles of a quadrilateral are congruent, then the quadrilateral is a parallelogram.

24. If the sum of the lengths of any two adjacent sides of a quadrilateral is constant, then the quadrilateral is a parallelogram.

25. If the sum of the measures of any two consecutive angles of a quadrilateral is constant, then the quadrilateral is a parallelogram.

Mixed Review

Find the measure of each angle of $\triangle ABC$.

1. m $\angle A = 36$, $\angle B$ is a right angle. *3.6* B: 90, C: 54

2. $\angle A$ and $\angle C$ are congruent; $\angle B$ is a right angle. *3.6* A: 45, B: 90, C: 45

3. $\angle A$ and $\angle C$ are complementary; m $\angle A = 2(\text{m} \angle C)$. *3.6* A: 60, B: 90, C: 30

4. $\overline{AB} \cong \overline{BC}$ and m $\angle A = 27$ *5.1* B: 126, C: 27

5. m $\angle A = 90$, the exterior angle at B has measure 140. *3.7* B: 40, C: 50

6. The exterior angles at A and B each have measure 120. *3.6* A: 60, B: 60, C: 60

15.

Statement	Reason
1. Quad $ABCD$, m $\angle A$ = m $\angle C$	1. Given
2. m $\angle B$ = m $\angle D$ ($\angle B \cong \angle D$)	2. Given
3. m $\angle A$ + m $\angle B$ + m $\angle C$ + m $\angle D = 360$	3. Sum of meas of int $\angle$s of a convex quad = 360.
4. m $\angle A$ + m $\angle B$ + m $\angle A$ + m $\angle B = 360$ ($2 \times$ m $\angle A$ + $2 \times$ m$\angle B$ = 360)	4. Sub
5. m $\angle A$ + m $\angle B = 180$	5. Div Prop of Eq
6. m $\angle C$ + m $\angle B = 180$	6. Sub
7. $\overline{AB} \parallel \overline{CD}$, $\overline{BC} \parallel \overline{AD}$	7. If int $\angle$s on same side of a transv are supp, then lines are $\parallel$.
8. $ABCD$ is a $\square$	8. Def of $\square$

16.

Statement	Reason
1. $\overline{ED} \cong \overline{BA}$, $\triangle AEF \cong \triangle DBC$	1. Given
2. $\overline{AE} \cong \overline{DB}$	2. CPCTC
3. $ABDE$ is a $\square$.	3. If 2 pair opp sides are $\cong$, then quad is $\square$.

17.

Statement	Reason
1. $\square ACDF$, $\overline{FE} \cong \overline{CB}$ ($FE = CB$)	1. Given
2. $\overline{FD} \parallel \overline{AC}$	2. Def of $\square$
3. $FD = AC$	3. Opp sides of $\square$ are $\cong$.
4. $FD - FE = AC - BC$	4. Equations may be subtracted.
5. $FD - FE = ED$, $AC - BC = AB$	5. Seg Add Post
6. $ED = AB$	6. Sub
7. $ABDE$ is $\square$	7. If one pair opp sides are $\parallel$ and $\cong$, then quad is $\square$.

See page 655 for the answers to Ex. 18–25.

▰▰ GETTING STARTED

Prerequisite Quiz

1. If M is the midpoint of $\overline{AB}$, what two segments are known to be congruent?
   $\overline{AM} = \overline{MB}$
2. If $\overline{PM} \cong \overline{MQ}$, is M the midpoint of $\overline{PQ}$?
   Not necessarily

Which of the following prove that *JKLM* is a parallelogram?

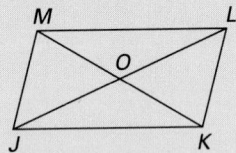

3. $\overline{JK} \parallel \overline{LM}, \overline{JM} \parallel \overline{LK}$ Yes
4. $\overline{JK} \cong \overline{LM}, \overline{JK} \parallel \overline{LM}$ Yes
5. $\overline{JO} \cong \overline{OL}, \overline{MO} \cong \overline{OK}$ Yes

Motivator

On the chalkboard draw a triangle like the one on page 252. Ask your students the following questions. How wide is the triangle?   It depends on where you measure.   Is it wider at the top or at the bottom?   Bottom   Is it half as wide in the middle as it is at the base?   Yes   How could you check your answer?   By measuring

▰▰ TEACHING
SUGGESTIONS

Lesson Note

Flow diagrams are used as an aid in planning proofs. Some teachers allow students to use flow diagrams as an alternative to traditional two-column proofs. Note that the proof arrows shown in earlier chapters actually form the skeleton of a flow diagram.

6.6 The Midsegment Theorem

Objective   To apply the Midsegment Theorem

A **triangle midsegment** joins the midpoints of two sides of a triangle. Draw several triangles, and construct midsegments. Measure the midsegment and the third side of each triangle.

Theorem 6.10   **Midsegment Theorem:** The segment joining the midpoints of two sides of a triangle is parallel to the third side, and its length is half the length of the third side.

Given: $\triangle ABC, \overline{CD} \cong \overline{AD}, \overline{CE} \cong \overline{BE}$
Prove: $\overline{DE} \parallel \overline{AB}, DE = \frac{1}{2}AB$

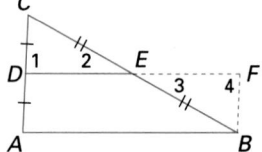

Proof

Statement	Reason
1. $\triangle ABC, \overline{CD} \cong \overline{AD}, \overline{CE} \cong \overline{BE}$	(S) 1. Given
2. Extend $\overline{DE}$ to point F such that $\overline{FE} \cong \overline{DE}$.	(S) 2. Ruler Post
3. E is the midpt of $\overline{DF}$.	3. Def of midpt
4. Draw $\overline{BF}$.	4. Two points determine a line.
5. $\angle 2 \cong \angle 3$	(A) 5. Vert $\angle$s are $\cong$.
6. $\triangle CED \cong \triangle BEF$	6. SAS
7. $\angle 1 \cong \angle 4$	7. CPCTC
8. $\overline{AD} \parallel \overline{BF}$	8. If alt int $\angle$s are $\cong$, then lines are $\parallel$.
9. $\overline{CD} \cong \overline{BF}$	9. CPCTC
10. $\overline{AD} \cong \overline{BF}$	10. Sub (Steps 1 and 9)
11. $ABFD$ is a $\square$.	11. If one pair of opp sides is $\parallel$ and $\cong$, then quad is a $\square$.
12. $\overline{DF} \cong \overline{AB}$ ($DF = AB$)	12. Opp sides of a $\square$ are $\cong$.
13. $DE = \frac{1}{2}DF$	13. Def of midpt
14. $\therefore DE = \frac{1}{2}AB$	14. Sub
15. $\therefore \overline{DE} \parallel \overline{AB}$	15. Opp sides of a $\square$ are $\parallel$.

Additional Example 1

Find the length of $\overline{UT}$ if X is the midpoint of $\overline{TV}$, W is the midpoint of $\overline{UV}$, and $\overline{XW}$ = 13.   26

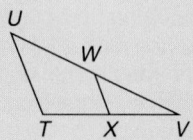

EXAMPLE 1 Find RQ if S is the midpoint of $\overline{PR}$, T is the midpoint of $\overline{PQ}$, and $ST = 7$.

Solution By Theorem 6.10, $ST = \frac{1}{2}RQ$.

Therefore, $RQ = 14$.

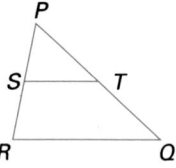

EXAMPLE 2 Prove that the segments joining the midpoints of adjacent sides of a convex quadrilateral form a parallelogram.

Given: Quad $ABCD$ with E, F, G, and H the midpts of $\overline{AB}$, $\overline{BC}$, $\overline{CD}$, and $\overline{DA}$, respectively

Prove: $EFGH$ is a parallelogram.

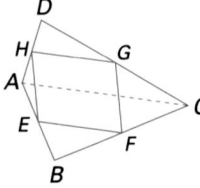

Plan Make a flow diagram.

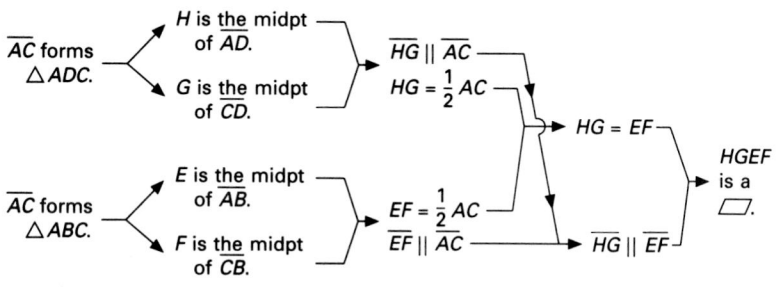

Proof

Statement	Reason
1. Draw $\overline{AC}$ to form $\triangle ABC$ and $\triangle ADC$.	1. Two pts determine a line; def of $\triangle$.
2. E is midpt of $\overline{AB}$; F is midpt of $\overline{BC}$.	2. Given
3. $EF = \frac{1}{2}AC$ and $\overline{EF} \parallel \overline{AC}$	3. Midseg Thm
4. G is midpt of $\overline{CD}$; H is midpt of $\overline{DA}$.	4. Given
5. $HG = \frac{1}{2}AC$ and $\overline{HG} \parallel \overline{AC}$	5. Midseg Thm
6. $EF = HG$ ($\overline{EF} \cong \overline{HG}$)	6. Sub
7. $\overline{EF} \parallel \overline{HG}$	7. Lines $\parallel$ to same lines are $\parallel$.
8. $\therefore$ $EFGH$ is a $\square$.	8. If one pair of opp sides is $\parallel$ and $\cong$, then quad is a $\square$.

6.6 The Midsegment Theorem **253**

Additional Example 2

Prove that the segments joining the midpoints of the sides of an equilateral triangle form another equilateral triangle.

Given: $\triangle RST$ with $RS = ST = RT$; W, X, and Y are midpoints of $\overline{RS}$, $\overline{ST}$, and $\overline{RT}$ resp.

Prove: $XY = YW = WX$, $RS = ST = RT$, and W, X, and Y are midpoints of $\overline{RS}$, $\overline{ST}$, and $\overline{RT}$, resp. (Given); $XY = \frac{1}{2}RS$, $YW = \frac{1}{2}ST$, $WX = \frac{1}{2}RT$ (Midseg Thm); $\frac{1}{2}RS = \frac{1}{2}ST = \frac{1}{2}RT$ (Div Prop); $XY = YW = WX$ (Sub)

Math Connections

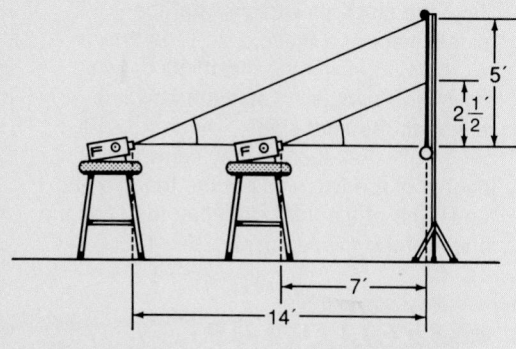

Image Projection: Refer students to the diagram above and discuss the following application of proportional triangles. Tasha has just returned from a vacation and wants to show her friends slides of the places she visited. As Tasha sets up the screen and the slide projector, she notices that the image only uses half of the screen. To enlarge the image to twice the size, Tasha moves the projector from 7 ft to 14 ft away from the screen without adjusting the angle of projection. The image becomes twice as large.

Critical Thinking Questions

Application: Ask students to draw a diagram and answer the following questions. Milan and Terry say goodbye and start walking in different, but not opposite, directions at 6 ft/sec and 4 ft/sec, respectively. After 5 seconds, they are 24 ft apart. If they continue at the same speeds and in the same directions, how long will it be until they are 48 ft apart?   5 sec more

Checkpoint

In $\triangle XYZ$, U, V, and W are midpoints of $\overline{XZ}$, $\overline{ZY}$, and $\overline{XY}$, respectively. Use this information to complete each statement.

1. $\overline{WV} \parallel$ _____ XZ
2. If $UW = 6$, then $ZY = $ _____ 12
3. If $WV = 8$, then _____ $= 16$ XZ
4. If $UV = 8t - 27$ and $XY = 4t + 6$, then $UV = $ _____ 13
5. If $UW = 3x + 2$ and $ZY = 7x - 5$, then $ZY = $ _____ 58

Have the students explain what the midsegment of a triangle is.   It is the segment that joins the midpoints of two sides of a triangle.   Ask them how it is related to the third side.   Theorem 6.10 page 252   Ask them if they know the lengths of the two sides of the triangle and the length of the third side, can they find the length of the midsegment.   Yes

▬▬▬FOLLOW UP

Guided Practice

Classroom Exercises 1–6

Independent Practice

A Ex. 1–13, **B** Ex. 14–20, **C** Ex. 21–23

Basic: WE 1–13, Problem Solving

Average: WE 1–20 odd, Problem Solving

Above Average: WE 1–23 odd, Problem Solving

Additional Answers

Written Exercises

15.

Statement	Reason
1. O is midpt of $\overline{LM}$, P is midpt of $\overline{MN}$, Q is midpt of $\overline{NL}$.	1. Given
2. $\overline{OP} \parallel \overline{NL}$, $\overline{OQ} \parallel \overline{MN}$	2. Midseg Thm
3. $\overline{OP} \parallel \overline{NQ}$, $\overline{OQ} \parallel \overline{NP}$	3. If seg is $\parallel$ to line, it is also $\parallel$ to any seg of the line.
4. $OPNQ$ is $\square$.	4. Def of $\square$

16.

Statement	Reason
1. O is midpt of $\overline{LM}$, P is midpt of $\overline{MN}$, Q is midpt of $\overline{NL}$.	1. Given
2. $\overline{NL} \parallel \overline{OP}$, $OP = \frac{1}{2}NL$	2. Midseg Thm
3. $QL = \frac{1}{2}NL$	3. Def of midpt
4. $\overline{OP} \cong \overline{QL}$	4. Sub
5. $\angle MOP \cong \angle OLQ$	5. If lines are $\parallel$, then corr $\angle$s are $\cong$.
6. $\overline{LO} \cong \overline{OM}$	6. Def of midpt
7. $\triangle LOQ \cong \triangle OMP$	7. SAS

EXAMPLE 3   Given: D and E are the midpoints of $\overline{AC}$ and $\overline{BC}$, respectively; $DE = x + 8$, $AB = 3x - 4$. Find DE and AB.

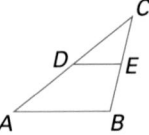

Solution

$$DE = \tfrac{1}{2}AB \text{ by the Midsegment Theorem}$$
$$x + 8 = \tfrac{1}{2}(3x - 4)$$
$$2x + 16 = 3x - 4$$
$$-x = -20$$
$$x = 20$$

$$AB = 3x - 4 = 3 \cdot 20 - 4 = 56$$
$$DE = \tfrac{1}{2}AB = \tfrac{1}{2} \cdot 56 = 28$$

Classroom Exercises

Find the indicated length if D is the midpoint of $\overline{AC}$ and E is the midpoint of $\overline{BC}$.

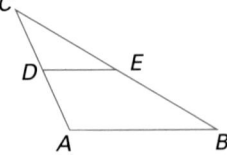

1. If $AB = 12$, then $DE = \underline{6}$

2. If $AB = 15$, then $DE = \underline{7.5}$

3. If $AB = 3.7$, then $DE = \underline{1.85}$

4. If $DE = 12$, then $AB = \underline{24}$

5. If $DE = 15$, then $AB = \underline{30}$

6. If $DE = 3.7$, then $AB = \underline{7.4}$

Written Exercises

In $\triangle FGH$, I, J, and K are the midpoints of $\overline{FG}$, $\overline{GH}$, and $\overline{HF}$, respectively. Using this information, complete the statements.

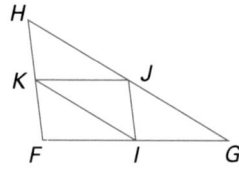

1. $\overline{KJ} \parallel \underline{\overline{FG}}$

2. $\overline{FH} \parallel \underline{\overline{JI}}$

3. If $HG = 12$, then $KI = \underline{6}$

4. If $FH = 18$, then $\underline{JI} = 9$

5. If $KJ = 3$, then $FG = \underline{6}$

6. If $IJ = 7$, then $\underline{FH} = 14$

7. If $FG = 6x - 4$ and $KJ = 2x + 1$, then $FG = \underline{14}$

8. If $GH = 3x - 8$ and $KI = x + 4$, then $KI = \underline{20}$

Additional Example 3

Given: S and T are midpoints of $\overline{PR}$ and $\overline{QR}$, resp. $ST = 3x - 4$, $PQ = 5x + 2$ Find PQ.   52

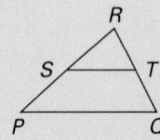

In the △s shown, P, S, Q, T, G, and E are the midpoints of $\overline{CA}$, $\overline{AT}$, $\overline{CT}$, $\overline{AB}$, $\overline{AD}$, and $\overline{BD}$, respectively. Complete the statements.

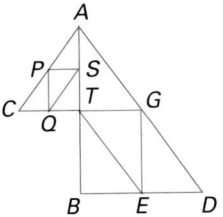

9. $\overline{TE} \parallel$ _$\overline{AD}$_

10. $\overline{CA} \parallel$ _$\overline{QS}$_

11. If $QS = 5$, then $CA =$ _10_

12. If $AG = 7.5$, then $TE =$ _7.5_

13. If $AB = 12$, then $AS =$ _3_

14. If the perimeter of △ACT is 24, then the perimeter of △PSQ is _12_.

O, P, and Q are the midpoints of $\overline{LM}$, $\overline{MN}$, and $\overline{NL}$, respectively.

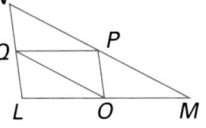

15. Prove: OPNQ is a parallelogram.

16. Prove: △LOQ ≅ △OMP

17. Prove: △LOQ ≅ △PQO

18. Prove: △LOQ ≅ △QPN

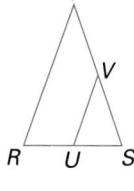

19. Given: U and V are the midpoints of $\overline{RS}$ and $\overline{ST}$, respectively. △RST is isosceles with $\overline{RT} \cong \overline{ST}$.
    Prove: △USV is isosceles.

20. Given: U and V are the midpoints of $\overline{RS}$ and $\overline{ST}$, respectively. △USV is isosceles with $\overline{UV} \cong \overline{SV}$.
    Prove: △RST is isosceles.

21. Prove: The segments joining the midpoints of the opposite sides of a quadrilateral bisect each other.

22. Given: H, I, and J are the midpoints of $\overline{DF}$, $\overline{EF}$, and $\overline{GF}$, respectively.
    Prove: ∠JHI ≅ ∠KDL

23. Given: H, I, J, K, and L are the midpoints of $\overline{DF}$, $\overline{EF}$, $\overline{GF}$, $\overline{DG}$, and $\overline{DE}$, respectively.
    Prove: △JHI ≅ △KDL

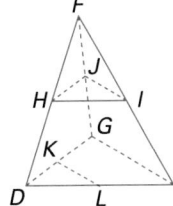

Mixed Review

Referring to the figure at the right, indicate whether the conditional statement is true or false.

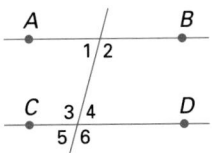

1. If $\overline{AB} \parallel \overline{CD}$, then ∠1 ≅ ∠4. **3.5** True

2. If ∠2 ≅ ∠4, then $\overline{AB} \parallel \overline{CD}$. False

3. If ∠2 ≅ ∠3, then $\overline{AB} \parallel \overline{CD}$. True

4. If $\overline{AB} \parallel \overline{CD}$, then ∠2 ≅ ∠6. **3.5** True

17.

Statement	Reason
1. O is midpt of $\overline{LM}$, P is midpt of $\overline{MN}$, Q is midpt of $\overline{NL}$.	1. Given
2. $\overline{QL} \parallel \overline{PO}$, $\overline{PQ} \parallel \overline{LO}$	2. Midseg Thm
3. OPQL is ▱.	3. Def of ▱
4. △LOQ ≅ △PQO	4. Diag of ▱ forms 2 ≅ △s.

18.

Statement	Reason
1. O is midpt of $\overline{LM}$, P is midpt of $\overline{MN}$, Q is midpt of $\overline{NL}$.	1. Given
2. $NQ = QL$, $NP = \frac{1}{2}NM$	2. Def of midpt
3. $PO = \frac{1}{2}NL$, $QO = \frac{1}{2}NM$	3. Midseg Thm
4. $QO = NP$	4. Sub
5. $QP = \frac{1}{2}LM$	5. Midseg Thm
6. $LO = \frac{1}{2}LM$	6. Def of midpt.
7. $QP = LO$	7. Sub
8. △LOQ ≅ △QPN	8. SSS

19.

Statement	Reason
1. U is midpt of $\overline{RS}$, V is midpt of $\overline{ST}$, $\overline{RT} \cong \overline{ST}$	1. Given
2. $VU = \frac{1}{2}RT$	2. Midseg Thm
3. $VS = \frac{1}{2}ST$	3. Def of midpt
4. $VS = \frac{1}{2}RT$	4. Sub
5. $VS = VU$	5. Sub
6. △USV is isos	6. Def of isos △

See pages 655–656 for the answers to Ex. 20–23.

Enrichment

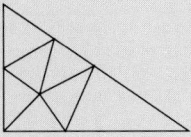

Challenge the students to partition a right triangle into regions such that each region is an acute triangle.
Then challenge them to find, by experimentation, the least number of acute triangles into which a right triangle can be partitioned. Have them submit a drawing to support the answer. 7

Problem Solving

1. Check students' drawings.

Additional Answer, page 246

30.

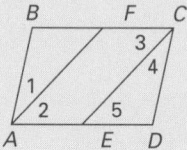

Statement	Reason
1. $\square ABCD$, $\overline{AF}$ bis $\angle BAD$, $\overline{CE}$ bis $\angle BCD$	1. Given
2. m $\angle 3$ = m $\angle 4$, m $\angle 1$ = m $\angle 2$	2. Def of $\angle$ bis
3. m $\angle BAD$ = m $\angle BCD$	3. Opp $\angle$s of $\square$ are $\cong$.
4. m $\angle BAD$ = m $\angle 1$ + m $\angle 2$, m $\angle BCD$ = m $\angle 3$ + m $\angle 4$	4. Angle Add Post
5. m $\angle 1$ + m $\angle 2$ = m $\angle 3$ + m $\angle 4$	5. Sub
6. m $\angle 2$ + m $\angle 2$ = m $\angle 3$ + m $\angle 3$ (2 × m $\angle 2$ = 2 × m $\angle 3$)	6. Sub
7. m $\angle 2$ = m $\angle 3$	7. Div Prop of Eq
8. $\overline{BC} \parallel \overline{AD}$	8. Def of $\square$
9. m $\angle 3$ = m $\angle 5$	9. If lines are $\parallel$, then alt int $\angle$s are $\cong$.
10. m $\angle 5$ = m $\angle 2$	10. Sub
11. $\overline{AF} \parallel \overline{CE}$	11. If corr $\angle$s are $\cong$, then lines are $\parallel$.

Problem Solving Strategies

Making a Table

A frequently used strategy for solving problems is to make a table of known data, and then look for a pattern. The technique is illustrated in the example below.

Example Five points are marked on a circle. All possible segments are drawn, joining each pair of points. How many segments are drawn?

Solution Draw three circles. Experiment with 2, 3, and 4 points. Look for a pattern. Make a table showing the number of segments drawn in each case.

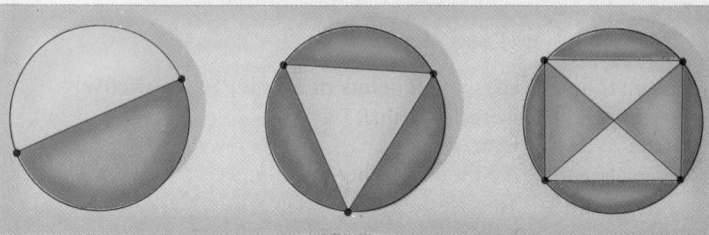

Number of points	2	3	4	5
Number of segments	1	3	6	?

The pattern appears to be that each time a point is added, the number of *new* segments added is one less than the total number of points. If this pattern holds, then for 5 points there should be 6 + (5 − 1) = 10 segments.

Exercises

1. Draw a circle with 6 points on it. Show that 15 segments are formed when all pairs of points are joined.

2. Extend the table above to predict the number of segments for 8 points. 7 pts, 21 segments; 8 pts, 28 segments

3. Develop a formula for predicting the number of segments for any number of points on a circle for the example above. Then use the formula to predict the number of segments for 20 points.

Number of segments for k points = $\dfrac{k(k-1)}{2}$; 190

256 Problem Solving Strategies

6.7 Lines Parallel to Many Lines

Objectives To apply theorems about transversals to three or more parallel lines
To find the distance between two parallel lines

In Chapter 5, it was proven that the shortest segment from a point to a line is the segment perpendicular to the line. Therefore, **the distance from a point to a line** is defined as the length of the perpendicular segment from the point to the line.

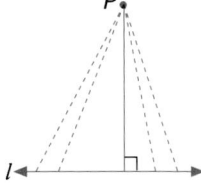

Recall also that the distance between two points is measured along a line. In Chapter 5, the triangle inequality allowed you to conclude that $AC < AB + BC$. In other words, the shortest path along line segments from point A to point C is the direct path $\overline{AC}$.

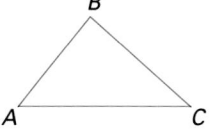

This notion of distance being measured along shortest paths can now be extended to a discussion of parallel lines. First, the following theorem assures that there is a unique distance between parallel lines.

Theorem 6.11 If two lines are parallel, then all points of each line are equidistant from the other line.

Given: $l \parallel m$, P and Q are on l; $\overline{PA} \perp m$,
$\overline{QB} \perp m$.
Prove: $PA = QB$

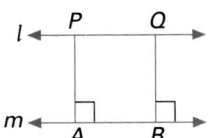

Plan Make a flow diagram.

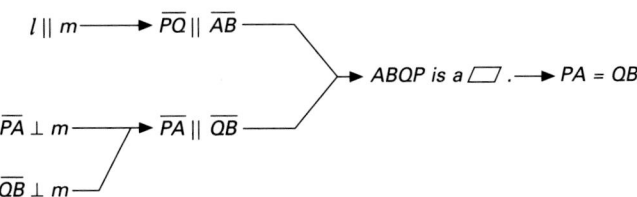

Teaching Resources

Quick Quizzes 49
**Reteaching and Practice
Worksheets,** pp. 97, 98

■■■ GETTING STARTED

Prerequisite Quiz

If *ABCD* is a parallelogram, which of the following are true? (Ex. 1–4)

1. $\overline{AB} \cong \overline{AD}$ Not necessarily
2. $\overline{AB} \parallel \overline{DC}$ Yes
3. $\overline{AB} \perp \overline{BC}$ Not necessarily
4. $\overline{AD} \cong \overline{BC}$ Yes
5. If $\overline{AC} \cong \overline{DF}$, $\angle A \cong \angle D$, and $\angle B \cong \angle E$, is $\triangle ABC \cong \triangle DEF$? If so, why?
   Yes, AAS

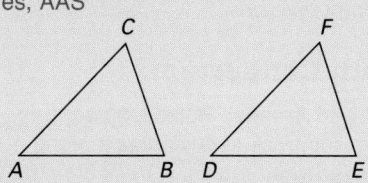

Motivator

Ask the students to estimate in feet how far away they are from the teacher's desk. Answers will vary.   Ask them how they would find the shortest distance between where they are and a wall in the room. They would measure the distance (segment) perpendicular from where they are to the wall.   Have the students locate two parallel lines in the classroom. Ask them what observation they can make about the distance between the points on the lines. The distance is the same.

Lesson Note

It may seem obvious to students that a perpendicular segment is the shortest distance between a point and a line. But a discussion of distances can be useful. For example, the air distance from Miami, Florida to New Orleans, Louisiana is 669 miles; the road distance is 892 miles. A look at the map will show why this is so. It is important, therefore, that we agree on definitions for distance.

An overhead projector transparency of a sheet of ruled notebook paper and a transparency of a ruler can be used to introduce Theorem 6.12. The transparency of the ruler can be moved around to show different divisions.

Math Connections

Railroad Tracks: Railroads use a two-railed track to guide trains along predetermined routes. A track is made up of two parallel steel rails fastened to a series of parallel wooden cross ties, forming many adjacent rectangles between the rails of the track. Since all of the cross ties have the same length and the distance between the steel rails is always the same, this example can be used to represent Theorem 6.11 from page 257 and the corollary to Theorem 6.12 on page 259.

Definition — The **distance between two parallel lines** is the distance from a point on one line to the other line.

A typical ruler is not always helpful for dividing certain distances equally. For example, it is difficult to divide a two-inch segment into thirds with a standard ruler. As shown at right, a sheet of ruled notebook paper can be used to do this. The equally spaced lines on the paper help to locate equally spaced marks on the two-inch segment. This suggests the following theorem.

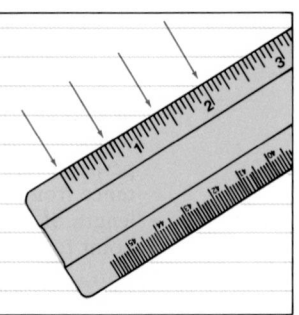

Theorem 6.12 — If three parallel lines cut off congruent segments on one transversal, then they cut off congruent segments on every transversal.

Given: $l \parallel m \parallel n$ with transversals $\overleftrightarrow{AE}$ and $\overleftrightarrow{BF}$, $\overline{AC} \cong \overline{CE}$
Prove: $\overline{BD} \cong \overline{DF}$

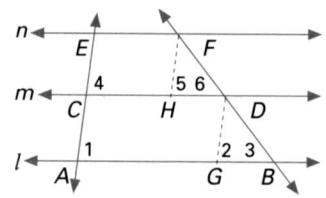

Plan — Draw $\overline{FH} \parallel \overline{EC}$ and $\overline{DG} \parallel \overline{AC}$. Prove that $AGDC$ and $CHFE$ are parallelograms and that $\triangle GBD \cong \triangle HDF$.

Proof

Statement	Reason
1. $l \parallel m \parallel n$ with transv $\overleftrightarrow{AE}$ and $\overleftrightarrow{BF}$	1. Given
2. Draw $\overline{DG} \parallel \overline{CA}$ and $\overline{FH} \parallel \overline{EC}$.	2. Through a given point, there is exactly one line $\parallel$ to a given line.
3. $AGDC$ and $CHFE$ are $\square$s.	3. If 2 pairs of opp sides of a quad are $\parallel$, then it is a $\square$.
4. $\overline{CE} \cong \overline{HF}$, $\overline{GD} \cong \overline{AC}$	4. Opp sides of $\square$ are $\cong$.
5. $\overline{AC} \cong \overline{CE}$	5. Given
6. $\overline{GD} \cong \overline{HF}$	(S) 6. Trans Prop Cong
7. $\angle 1 \cong \angle 2$, $\angle 4 \cong \angle 5$	7. Corr $\angle$s of $\parallel$ lines are $\cong$.
8. $\angle 1 \cong \angle 4$	8. Corr $\angle$s of $\parallel$ lines are $\cong$.
9. $\angle 2 \cong \angle 5$	(A) 9. Trans Prop Cong
10. $\angle 3 \cong \angle 6$	(A) 10. Corr $\angle$s of $\parallel$ lines are $\cong$.
11. $\triangle GBD \cong \triangle HDF$	11. AAS
12. $\therefore \overline{BD} \cong \overline{DF}$	12. CPCTC

258 Chapter 6 Polygons

Additional Example 1

$\overline{AB} \parallel \overline{CD} \parallel \overline{EF} \parallel \overline{GH}$; $\overline{AC} \cong \overline{CE} \cong \overline{EG}$; $BD = 10$
Find BH.   30

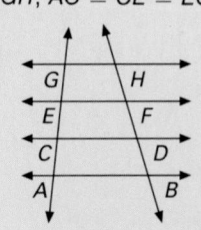

Additional Example 2

$\overline{XP} \parallel \overline{YQ} \parallel \overline{ZR}$, $XY = 9$, $YZ = 9$, $PQ = 3t + 2$, $PR = 7t + 1$
Find QR.
$QR = 3 \cdot 3 + 2$ or 11

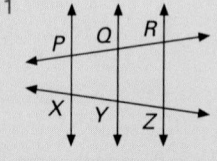

Corollary — If any number of parallel lines cut off congruent segments on one transversal, then they cut off congruent segments on every transversal.

EXAMPLE 1 $l \parallel m \parallel n \parallel o$, $AB = BC = CD$, $EF = 5$. Find FH.

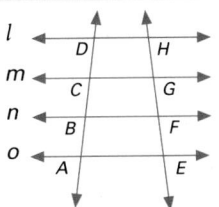

Solution $EF = FG = GH = 5$, $FG + GH = FH$
$FH = 5 + 5 = 10$

EXAMPLE 2 $\overleftrightarrow{AF} \parallel \overleftrightarrow{BE} \parallel \overleftrightarrow{CD}$, $FE = 7$, $ED = 7$, $AB = 7x - 5$,
$BC = 3x + 3$. Find AB.

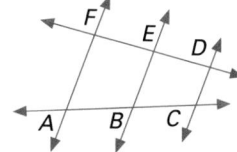

Solution $AB = BC$ by Theorem 6.12
$7x - 5 = 3x + 3$
$4x = 8$
$x = 2$

Thus, $AB = 7x - 5 = 7 \cdot 2 - 5 = 9$.

EXAMPLE 3 Given: $\angle 1 \cong \angle 2$, $\angle 2 \cong \angle 3$, $\overline{BD} \cong \overline{DF}$
Prove: $\overline{AC} \cong \overline{CE}$

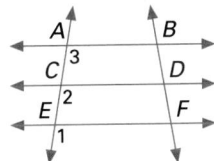

Proof

Statement	Reason
1. $\angle 1 \cong \angle 2$, $\angle 2 \cong \angle 3$	1. Given
2. $\overline{AB} \parallel \overline{CD}$, $\overline{CD} \parallel \overline{EF}$	2. If corr $\angle$s are $\cong$, lines are $\parallel$.
3. $\overline{AB} \parallel \overline{CD} \parallel \overline{EF}$	3. Lines $\parallel$ to the same line are $\parallel$.
4. $\overline{BD} \cong \overline{DF}$	4. Given
5. $\therefore \overline{AC} \cong \overline{CE}$	5. If $\parallel$ lines cut off $\cong$ segs on one transv, they cut off $\cong$ segs on every transv.

Theorem 6.13 — If a segment is parallel to one side of a triangle and contains the midpoint of a second side, then this segment bisects the third side.

Using Theorem 6.12 and its corollary, you can divide a segment into any number of congruent segments.

Additional Example 3

Given: $\angle 1 \cong \angle 2$, $\angle 2 \cong \angle 3$, $\angle 3 \cong \angle 4$,
$\overline{ST} \cong \overline{TW} \cong \overline{WX}$
Prove: $\overline{GH} \cong \overline{TK}$

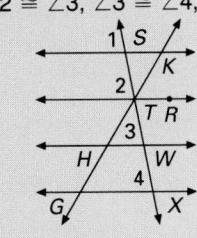

Proof: $\angle 1 \cong \angle 2$, $\angle 2 \cong \angle 3$, $\angle 3 \cong \angle 4$
(Given); $\overline{SK} \parallel \overline{TR}$, $\overline{TR} \parallel \overline{TW}$, $\overline{TW} \parallel \overline{GX}$ (If
corr $\angle$s are $\cong$, lines are $\parallel$.); $\overline{SK} \parallel \overline{TR} \parallel$
$\overline{TW} \parallel \overline{GX}$ (Lines $\parallel$ to the same line are $\parallel$.);
$\overline{ST} \cong \overline{TW} \cong \overline{WX}$ (Given); $\overline{GH} \cong \overline{TK}$ (If $\parallel$
lines cut off $\cong$ seg on 1 trans, they cut
off $\cong$ seg on every trans.)

Critical Thinking Questions

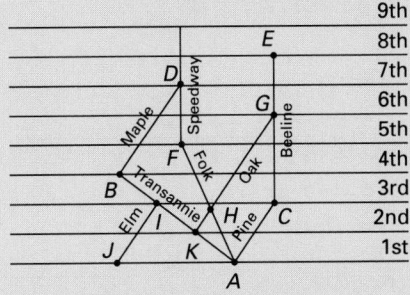

The numbered streets run east-west and are 135 m apart. Speedway runs perpendicular to 8th Street. Beeline is parallel to Speedway. All streets that are named after trees are parallel. Elm is perpendicular to Transannie. Point A is 555 m from point B and 300 m from point C. What is the shortest route from point J to point E and how long is that route? Explain.
The shortest route is J to I to K to G to E. $JI < JA$ and $IK = AK$, so $JI + IK < JA + AK$. The dist of the shortest route is 1335 m, or about 1.3 km. Since the numbered streets are $\parallel$ and all are 135 m apart, they also cut $\cong$ segs on all other streets. If $AB = 555$ m, then each seg that is cut along Transannie is 185 m long. If $AC = 300$ m, then each seg that is cut along Pine, and along streets that are $\parallel$ to Pine, are 150 m long. Therefore, the sum of the lengths of the segs along the route is $2(150) + 185 + 4(150) + 2(135) = 1355$ m.

Common Error Analysis

Error: Students sometimes assume congruence between segments on different transversals intersected by three or more parallel lines. Have students refer to the diagram to determine whether each statement is true or false and to explain each answer.

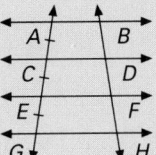

Given: $\overleftrightarrow{AB} \parallel \overleftrightarrow{CD} \parallel \overleftrightarrow{EF} \parallel \overleftrightarrow{GH}$, $\overline{AC} \cong \overline{CE} \cong \overline{EG}$

1. $\overline{BD} \cong \overline{FH}$ True by the Corollary to Theorem 6.12.
2. $\overline{CE} \cong \overline{DF}$ False. The segments are not cut off by the same transversal.
3. $\overline{BF} \cong \overline{DH}$ True; by the Corollary to Theorem 6.12 and the Seg Add Post.

Checkpoint

$\overline{AD} \parallel \overline{BE} \parallel \overline{CF}, \overline{AB} \cong \overline{BC}$

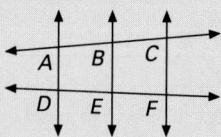

1. $DE = 3x, EF = 4x - 3$
   Find DE. 9
2. $DE = 3x - 2, EF = x + 8$
   Find DF. 26
3. $DF = 2x + 6, EF = 3x - 7$
   Find DE. 8

Closure

Have the students explain how to find the distance between two parallel lines.   The distance between two parallel lines is the perpendicular distance from a point on one line to the other line.   Ask them to explain the relationship between parallel lines and transversals.   Theorem 6.12 page 258 and Corollary page 259

◼◼◼ FOLLOW UP

Guided Practice

Classroom Exercises 1–10

Independent Practice

🅐 Ex. 1–10, 🅑 Ex. 11–16, 🅒 Ex. 17–20

Basic: WE 1–10

Average: WE 1–16

Above Average: WE 1–20

Additional Answers

Written Exercises

7.–8. See Construction on page 260.

9.

Statement	Reason
1. $\angle 1 \cong \angle 5$, $\angle 5 \cong \angle 9$, $\overline{GH} \cong \overline{HI}$	1. Given
2. $\overline{LG} \parallel \overline{KH}$, $\overline{KH} \parallel \overline{JI}$	2. If corr $\angle$s are $\cong$, then lines are $\parallel$.
3. $\overline{LG} \parallel \overline{JI}$	3. If 2 lines are $\parallel$ to same line, then they are $\parallel$.
4. $\overline{LK} \cong \overline{KJ}$	4. If 3 $\parallel$ lines make $\cong$ segs on 1 transv, then they make $\cong$ segs on every transv.

260

Construction   Divide a segment into three congruent segments.

Given: $\overline{AB}$
Choose any point not on $\overline{AB}$. Label it C. Draw $\overrightarrow{AC}$.

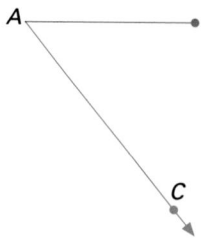

Use a compass with any convenient opening to construct $\overline{AX} \cong \overline{XY} \cong \overline{YZ}$.

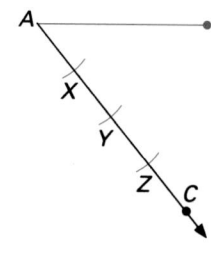

Draw $\overline{BZ}$. Construct $\overline{XP} \parallel \overline{BZ}$ and $\overline{YQ} \parallel \overline{BZ}$.

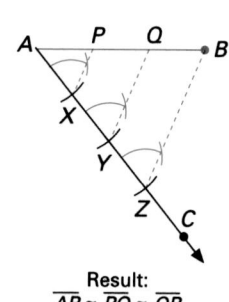

Result:
$\overline{AP} \cong \overline{PQ} \cong \overline{QB}$

Classroom Exercises

Given: $l \parallel m \parallel n \parallel o$; $\overline{AD} \cong \overline{DG} \cong \overline{GJ}$. Using this information, complete the statement.

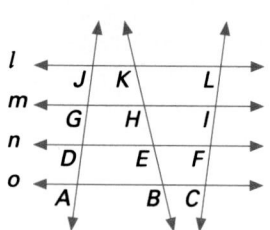

1. If $KH = 7$, then $HE = \underline{7}$
2. If $LI = 9$, then $FC = \underline{9}$
3. If $KE = 14$, then $KH = \underline{7}$
4. If $LC = 12$, then $IF = \underline{4}$
5. If $KE = 15$, then $HB = \underline{15}$
6. If $IC = 20$, then $LF = \underline{20}$
7. If $EB = 1$, then $KB = \underline{3}$
8. If $LF = 16$, then $LC = \underline{24}$
9. If $HE = 4$, then $KE = \underline{8}$
10. If $IF = 9$, then $LC = \underline{27}$

Written Exercises

Given: $\overleftrightarrow{AB} \parallel \overleftrightarrow{CD} \parallel \overleftrightarrow{EF}$, $\overline{AC} \cong \overline{CE}$. Using this information, complete the statement.

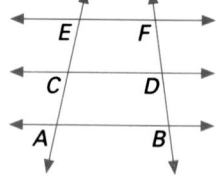

1. $BD = 5x + 1, DF = 2x + 13, BD = \underline{21}$
2. $BD = 2x + 1, DF = 4x - 7, DF = \underline{9}$
3. $BD = 3x + 4, DF = 4x - 5, DF = \underline{31}$
4. $BD = 3x + 4, BF = 7x - 2, DF = \underline{34}$
5. $DF = 2x + 5, BF = 5x - 2, BD = \underline{29}$
6. $DF = 3x + 2, BF = 12x + 1, BD = \underline{3.5}$

260   Chapter 6 Polygons

Enrichment

Challenge the students to prove that opposite sides of a regular hexagon are parallel.
Given: Regular hexagon $ABCDEF$
Prove: $\overline{AB} \parallel \overline{ED}$

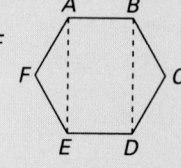

Proof: $ABCDEF$ is a reg hexagon. (Given); Draw $\overline{AE}$ and $\overline{BD}$. (2 pts are contained in exactly 1 line); $\overline{AF} \cong \overline{BC}$, $\overline{FE} \cong \overline{CD}$, $\angle F \cong \angle C$ (Def of reg polygon); $\triangle AFE \cong \triangle BCD$ (SAS); $\overline{AE} \cong \overline{BD}$ (CPCTC); $\overline{AB} \cong \overline{ED}$ (Def of reg polygon); $ABDE$ is a $\square$ (Quad with opp sides $\cong$ is a $\square$.); $\therefore \overline{AB} \parallel \overline{ED}$ (Def of $\square$)
Other proofs are possible.

7. Draw a segment 15 cm long. Divide it into three congruent segments.

8. Draw a segment 17 cm long. Divide it into five congruent segments.

9. Given: $\angle 1 \cong \angle 5$, $\angle 5 \cong \angle 9$, $\overline{GH} \cong \overline{HI}$
Prove: $\overline{LK} \cong \overline{KJ}$

10. Given: $\angle 3 \cong \angle 6$, $\angle 8 \cong \angle 9$, $\overline{LK} \cong \overline{KJ}$
Prove: $\overline{GH} \cong \overline{HI}$

11. Given: $\angle 3 \cong \angle 10$, $\overline{KH} \parallel \overline{JI}$, $\overline{GH} \cong \overline{HI}$
Prove: $\overline{LK} \cong \overline{KJ}$

12. Given: $\overline{LG} \parallel \overline{KH}$, $\angle 5 \cong \angle 9$, $\overline{LK} \cong \overline{KJ}$
Prove: $\overline{GH} \cong \overline{HI}$

13. Construct a pair of parallel lines. Find the distance between the lines.

14. Prove Theorem 6.11.

15. Prepare a flow diagram for the proof of Theorem 6.12.

16. Prove the corollary to Theorem 6.12 for four parallel lines.

17. Prove Theorem 6.13.

Provide a counterexample to show that the conditional is false.

18. If three lines cut off congruent segments on each of two transversals, then the lines are parallel.

19. If a segment joining a point on each of two sides of a triangle is half as long as the third side, then it is parallel to the third side.

20. The yard lines of a football field must be marked off for the first game of the season. A permanent marker is at the end of each goal line. There are no other markers on the field. How can the maintenance crew lay off the yard lines so that they are parallel to the goal line and parallel to each other?

Mixed Review

Complete the statement, using $\square MNOP$. **6.5**

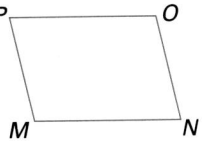

1. If $MN = 6$, then $PO = \underline{\quad 6 \quad}$

2. If $PM = 8.7$, then $ON = \underline{\quad 8.7 \quad}$

3. If $MN + NO = 37$, then $OP + PM = \underline{\quad 37 \quad}$

4. If $PM + MN = 4.9$, then $MN + NO = \underline{\quad 4.9 \quad}$

5. If m $\angle M = 78$, then m $\angle O = \underline{\quad 78 \quad}$

6. If m $\angle N = 83$, then m $\angle O = \underline{\quad 97 \quad}$

10.

Statement	Reason
1. $\angle 3 \cong \angle 6$, $\angle 8 \cong \angle 9$, $\overline{LK} \cong \overline{KJ}$	1. Given
2. $\overline{LG} \parallel \overline{KH}$, $\overline{KH} \parallel \overline{JI}$	2. If alt int $\angle$s are $\cong$, then lines are $\parallel$.
3. $\overline{LG} \parallel \overline{JI}$	3. If 2 lines are $\parallel$ to the same line, then they are $\parallel$.
4. $\overline{GH} \cong \overline{HI}$	4. If 3 $\parallel$ lines make $\cong$ segs on 1 transv, then they make $\cong$ segs on every transv.

11.

Statement	Reason
1. $\angle 3 \cong \angle 10$, $\overline{KH} \parallel \overline{JI}$, $\overline{GH} \cong \overline{HI}$	1. Given
2. $\overline{JI} \parallel \overline{LG}$	2. If alt int $\angle$s are $\cong$, then lines are $\parallel$.
3. $\overline{KH} \parallel \overline{LG}$	3. If 2 lines are $\parallel$ to the same line, then they are $\parallel$.
4. $\overline{LK} \cong \overline{KJ}$	4. If 3 $\parallel$ lines make $\cong$ segs on 1 transv, then they make $\cong$ segs on every transv.

12.

Statement	Reason
1. $\overline{LG} \parallel \overline{KH}$, $\angle 5 \cong \angle 9$, $\overline{LK} \cong \overline{KJ}$	1. Given
2. $\overline{KH} \parallel \overline{JI}$	2. If corr $\angle$s are $\cong$, then lines are $\parallel$.
3. $\overline{LG} \parallel \overline{JI}$	3. If 2 lines are $\parallel$ to the same line, then they are $\parallel$.
4. $\overline{GH} \cong \overline{HI}$	4. If 3 $\parallel$ lines make $\cong$ segs on 1 transv, then they make $\cong$ segs on every transv.

13.

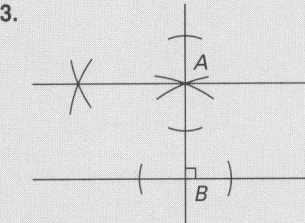

Construct $\perp$ from 1 line to the other.

See page 262 for the answers to Ex. 14–20.

14.

Statement	Reason
1. $l \parallel m$, $\overline{PA}$ $\perp m$, $\overline{QB}$ $\perp m$	1. Given
2. $\overline{PA} \parallel \overline{QB}$	2. Lines $\perp$ to the same line are $\parallel$.
3. $ABQP$ is a $\square$.	3. Def of $\square$
4. $\overline{PA} \cong \overline{QB}$	4. Opp sides of $\square$ are $\cong$.

15. Student flowcharts may vary. See page 667 for flowchart.

16.

Statement	Reason
1. $l \parallel m \parallel n \parallel p$, with transv $\overleftrightarrow{AD}$ and $\overleftrightarrow{EH}$, $\overline{AB} \cong \overline{BC}$ $\cong \overline{CD}$	1. Given
2. $\overline{FG} \cong \overline{GH}$, $\overline{FG} \cong \overline{FE}$	2. If 3 $\parallel$ lines make segs on 1 transv, then they make $\cong$ segs on every transv.
3. $\overline{GH} \cong \overline{FE}$	3. Trans Prop of Congr

17.

Statement	Reason
1. $\overline{AD} \cong \overline{BD}$, $\overline{DE} \parallel \overline{BC}$	1. Given
2. Draw $l \parallel$ $\overline{DE}$ through pt A.	2. Through a given pt, there exists exactly 1 line $\parallel$ to given line.
3. $l \parallel \overline{BC}$	3. If 2 lines are $\parallel$ to the same line, then they are $\parallel$.
4. $AE \cong \overline{EC}$	4. If 3 $\parallel$ lines make $\cong$ segs on 1 transv, then they make $\cong$ segs on every transv.

18.

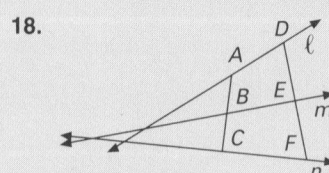

$AB = BC$, $DE = EF$; l, m and n are $\parallel\!\!\!\backslash$.

19.

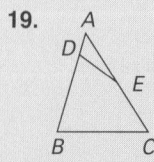

$DE = \frac{1}{2}BC$; $\overline{DE} \parallel\!\!\!\backslash \overline{BC}$.

20. Sidelines are laid $\perp$ to the goal lines, then the yd lines are laid $\perp$ to the sidelines. Yard lines and goal lines are $\parallel$ because they are all $\perp$ to the sidelines. Measure diags to verify rect.

262

Key Terms

adjacent sides   (p. 225)
concave   (p. 227)
consecutive angles   (p. 225)
convex   (p. 227)
diagonal   (p. 226)
distance between parallel lines   (p. 258)
equiangular polygon   (p. 227)
equilateral polygon   (p. 227)

opposite angles   (p. 241)
opposite sides   (p. 241)
parallelogram   (p. 241)
polygon   (p. 225)
quadrilateral   (p. 241)
regular polygon   (p. 227)
triangle midsegment   (p. 252)

Key Ideas and Review Exercises

6.1   To identify convex and concave polygons, determine if a segment connecting any two points is in the interior of the polygon. If so, the polygon is convex. If not, it is concave.

To identify a regular polygon, determine if all sides and all angles of the polygon are congruent. If angles are congruent, it is equiangular. If sides are congruent, it is equilateral.

1. Name the polygon. Is it convex or concave?   Concave; pentagon

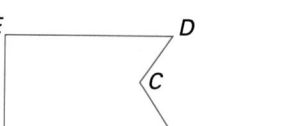

2. Is the given polygon equilateral, equiangular, or regular?   Equiangular

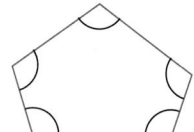

6.2   To find the sum of the angle measures of a polygon with n sides, use:
Sum $= (n - 2)180$.

To find the measure of each angle of a regular polygon with n sides, use:
Measure $= \dfrac{(n - 2)180}{n}$.

3. Find the sum of the measures of the angles of a dodecagon.   1,800

4. Find the measure of each angle of a regular hexagon.   120

6.3   The sum of the measures of the exterior angles, one at each vertex, of a polygon is 360.

5. The measures of three exterior angles of a quadrilateral are 37, 58, and 92. Find the measure of the exterior angle at the fourth vertex.   173

6. Find the measure of an exterior angle of a regular octagon.   45

6.4   To prove theorems or to determine properties of parallelograms, use one or more of the following:
- (1) Opposite sides of a parallelogram are parallel.
- (2) Opposite sides of a parallelogram are congruent.
- (3) Opposite angles of a parallelogram are congruent.
- (4) Consecutive angles of a parallelogram are supplementary.
- (5) The diagonals of a parallelogram bisect each other.

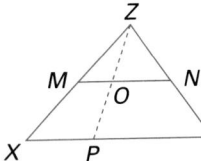

7. $ABCD$ is a parallelogram. m $\angle A = 79$, m $\angle B =$ _____   101

8. $ABCD$ is a parallelogram. $AO = 2x + 1$, $AC = 5x - 5$, $AO =$ _____   15

9. Given: $\square ABCD$
Prove: $\triangle AOD \cong \triangle COB$

6.5   To prove that a quadrilateral is a parallelogram, use one of the following:
- (1) Both pairs of opposite sides are parallel.
- (2) Both pairs of opposite sides are congruent.
- (3) The diagonals bisect each other.
- (4) One pair of opposite sides is parallel and congruent.
- (5) Both pairs of opposite angles are congruent.

10. $\overline{XY} \parallel \overline{WZ}$, $\overline{XY} \cong \overline{WZ}$. Is $WXYZ$ a parallelogram? Why?   Yes, one pair of opp sides $\parallel$ and $\cong$.

11. Given: $\triangle ZYX \cong \triangle XWZ$
Prove: $XWZY$ is a parallelogram.

6.6   Midsegment Theorem: The segment joining the midpoints of two sides of a triangle is parallel to the third side, and its length is half the length of the third side.

12. M and N are the midpoints of $\overline{XZ}$ and $\overline{YZ}$, respectively; $MN = 12$. Find XY.   24

13. Given: M and N are the midpoints of $\overline{XZ}$ and $\overline{YZ}$, respectively.
Prove: m $\angle OPY = 180 -$ m $\angle MOZ$

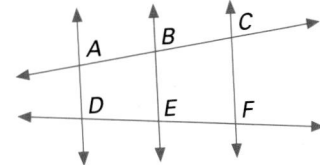

6.7   Three or more parallel lines that cut off congruent segments on one transversal cut off congruent segments on every transversal they intersect.

14. $\overleftrightarrow{AD} \parallel \overleftrightarrow{BE} \parallel \overleftrightarrow{CF}$, $AB = 14$, $BC = 14$, $DE = 3x + 2$, $DF = 7x - 1$. Find EF.   17

15. Draw a segment 12 cm long. Using a straightedge and compass, divide it into five congruent segments.

9.

Statement	Reason
1. $\square ABCD$	1. Given
2. $\overline{DO} \cong \overline{BO}$, $\overline{AO} \cong \overline{CO}$	2. Diags of $\square$ bis each other.
3. $\angle AOD \cong \angle COB$	3. Vert $\angle$s are $\cong$.
4. $\triangle AOD \cong \triangle COB$	4. SAS

11.

Statement	Reason
1. $\triangle ZYX \cong \triangle XWZ$	1. Given
2. $\overline{ZY} \cong \overline{XW}$, $\overline{YX} \cong \overline{WZ}$	2. CPCTC
3. $XWZY$ is $\square$.	3. If 2 pair opp sides are $\cong$, then quad is $\square$.

13.

Statement	Reason
1. M is midpt of $\overline{XZ}$, N is midpt of $\overline{YZ}$.	1. Given
2. $\overline{MN} \parallel \overline{XY}$	2. Midseg Thm
3. m $\angle MOZ =$ m $\angle XPO$	3. If lines are $\parallel$, then corr $\angle$s are $\cong$.
4. m $\angle OPY +$ m $\angle XPO = 180$.	4. If outer rays of 2 adj $\angle$s form a st $\angle$, then sum of their meas $= 180$.
5. m $\angle OPY = 180 -$ m $\angle XPO$.	5. Subt Prop of Eq
6. m $\angle OPY = 180 -$ m $\angle MOZ$.	6. Sub

15. See Construction on page 260.

9. m ∠A = 75, m ∠B = 85, m ∠C = 95,
m ∠D = 105

19. See Construction on page 260.

20.

Statement	Reason
1. □ABCD, BM = CN	1. Given
2. $\overline{DC} \parallel \overline{AB}$	2. Def of □
3. $\overline{DN} \parallel \overline{AM}$	3. If lines are ∥, then segs contained by them are ∥.
4. $\overline{DC} \cong \overline{AB}$	4. Opp sides of □ are ≅.
5. DC − CN = AB − BM	5. Equations may be subtracted.
6. DC − CN = DN, AB − BM = AM	6. Seg Add Post
7. DN = AM	7. Sub
8. AMND is □.	8. If 1 pair sides are ∥ and ≅, then quad is □.

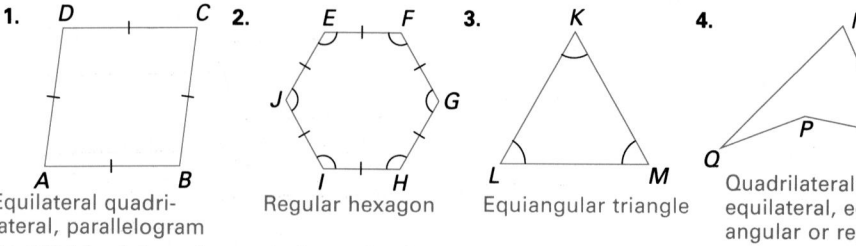

Chapter 6 Test

A-Level Ex.: 1—8, 10, 12—19   B-Level Ex.: 9, 11
C-Level Ex.: starred

Name the polygon. Is the polygon equilateral, equiangular, or regular?

1. Equilateral quadrilateral, parallelogram

2. Regular hexagon

3. Equiangular triangle

4. Quadrilateral; not equilateral, equiangular or regular

5. Which of the polygons in Items 1—4 are concave?   4

6. Find the sum of the measures of the angles of a 100-gon.   17,640

7. Find the measure of each angle of a regular octagon.   135

8. Find the measure of each exterior angle of a regular pentagon.   72

9. In quad ABCD, m ∠A = x − 10, m ∠B = x, m ∠C = x + 10 and m ∠D = x + 20. Find the measure of each angle.

10. The measures of the exterior angles of a pentagon, one at each vertex, are 45, 86, 135, and 54. Find the measure of the fifth exterior angle.   40

11. Find the number of sides in a regular polygon if the measure of each exterior angle is 12.   30

12. Name a pair of consecutive angles of quad RSTU.   ∠STU, ∠TUR; or ∠TUR, ∠URS; or ∠URS, ∠RST; or ∠RST, ∠STU

True or false?

13. Adjacent sides of a parallelogram are congruent.   False

14. Opposite angles of a parallelogram are supplementary.   False

15. If each pair of opposite sides is congruent in a quadrilateral, then it is a parallelogram.   True

16. Diagonals of a parallelogram are perpendicular.   False

Given: WXYZ is a parallelogram.

17. m ∠XWZ + m ∠YXW = _____   180

18. WA = x + 5, YA = 2x − 7. Find WA.   17

19. Draw a segment four inches long. Using a straightedge and a compass, divide it into five congruent segments.

*20. Given: □ABCD, BM = CN
Prove: AMND is a parallelogram.

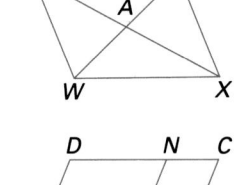

Strategy for Achievement in Testing

Questions on standardized achievement tests often involve geometric figures. If a diagram is not given, you may want to draw one. Try to make a reasonably accurate diagram, but it is important to avoid special cases. For example, if a question refers to a quadrilateral, do not draw a parallelogram. A good diagram uses all of the given information, and only that information.

Choose the best answer to each question or problem.

1. Which of the following is true of a
D line, but not true of a segment?
   (A) has exactly one endpoint
   (B) contains an infinite number of points
   (C) is named by two points
   (D) has no endpoints
   (E) has a midpoint

2. Which number is closest to 3?
B (A) $\frac{10}{3}$ (B) 2.9   (C) 11
   (D) $\sqrt{8}$ (E) $\frac{11}{4}$

3. An acute angle can have a measure of:
D I. 89.999   II. 0.0001   III. 90.0001
   (A) I only   (B) II only   (C) III only
   (D) I and II only   (E) I and III only

4. If $l \parallel m$, then
E I. $\angle 3$ and $\angle 5$ are supplementary.
   II. m $\angle 7$ = m $\angle 8$
   III. m $\angle 3$ = m $\angle 7$ + m $\angle 6$
   (A) I only   (B) II only   (C) III only
   (D) I and II only   (E) I, II, and III

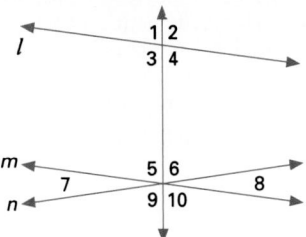

5. m $\angle A$ = 68 and m $\angle C$ = 73
A

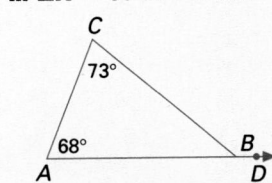

Which of these is not possible?
   (A) m $\angle CBD$ is acute.
   (B) m $\angle ABC$ is acute.
   (C) m $\angle CBD$ = 141
   (D) m $\angle ABC$ < 141
   (E) None of these

6. If $\overline{AB} \cong \overline{TP}$, $\overline{BC} \cong \overline{PS}$, and $\overline{CA} \cong \overline{ST}$,
E then:
   (A) $\triangle ABC \cong \triangle TSP$
   (B) $\triangle ABC \cong \triangle PTS$
   (C) $\triangle ABC \cong \triangle SPT$
   (D) $\triangle ABC \cong \triangle STP$
   (E) None of these

7. Each angle of a regular polygon with
D r sides has a measure of _____.
   (A) $\frac{360}{r}$ (B) $(r - 2)180$
   (C) 360   (D) $\frac{(r - 2)180}{r}$ (E) 60

8. To construct an angle of 150, you
E should first construct a(n) _____.
   (A) angle of 45   (B) angle bisector
   (C) isosceles triangle
   (D) pair of parallel lines
   (E) equilateral triangle

3. Triangles have three sides *and* triangles have three angles. Triangles have three sides *or* triangles have three angles.

4.

6. See Construction on page 17.
7. See Construction on page 23.
8. m ∠1 = 40, m ∠2 = 30, m ∠3 = 30, m ∠4 = 80, m ∠5 = 40, m ∠6 = 30, m ∠7 = 110
10. ∠1 and ∠6, ∠1 and ∠7, ∠2 and ∠5, ∠2 and ∠8, ∠3 and ∠5, ∠3 and ∠8, ∠4 and ∠6, ∠4 and ∠7
13. Conv: If pts lie in the same plane, then they are coplanar. (True)
Inv: If points are not coplanar, then they do not lie in the same plane. (True)
Contrapos: If pts do not lie in the same plane, then they are not coplanar. (True)

Cumulative Review *(Chapters 1–6)*

1. If C is the midpoint of AB, find the coordinates of A and B for $A:(p - 6)$, $B:(4p + 2)$, and $C:(8)$. $A(-2)$, $B(18)$ *1.3*

2. The measure of the supplement of an angle is $2\frac{1}{2}$ times the measure of its complement. Find the angle. 30 *1.6*

3. Write the conjunction and the disjunction of the two statements: Triangles have three sides. Triangles have three angles. *1.7*

4. Graph $x \le 8$ *and* $x \ge 3$ on a number line. *1.8*

5. Are ∠AOB and ∠CBO adjacent? If not, why not? No, ∠s have different vertices. *1.5*

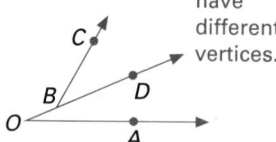

6. Draw an obtuse angle and label it ∠WXY. Construct ∠ABC congruent to ∠WXY. *1.4*

7. Construct the bisector of ∠ABC in Exercise 5. *1.5*

8. Solve for x below and find all the angles. *2.7*

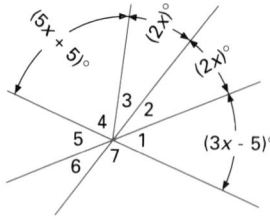

Assume $\overrightarrow{BP}$ bisects ∠ABC.

9. Given: Adjacent angles ABP and PBC which are ≇. To write an indirect proof that $\overrightarrow{BP}$ is not the bisector of ∠ABC, what assumption must be made? *3.4*

10. Two parallel lines are intersected by a transversal. Which nonadjacent pairs of angles are supplementary? *3.5*

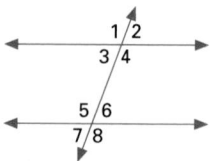

11. Given: Isosceles triangle ABC with $\overline{AC} \cong \overline{BC}$, and exterior ∠DCB. If m ∠A = 72, what is m ∠DCB? 144 *3.6*

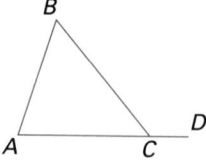

12. In △ABC, m ∠A = 2x, m ∠B = 3x − 60, and m ∠C = x. Find m ∠B. 60 *3.6*

13. Write the converse, inverse, and contrapositive of the following statement and tell whether each is true or false: If points are coplanar, then they lie in the same plane. *3.5, 3.8*

14. Given: △ABC and △XYZ, $\overline{AC} \cong \overline{XZ}$, $\overline{CB} \cong \overline{ZY}$, ∠B ≅ ∠Y. Is enough information given to prove △ABC ≅ △XYZ? Why or why not? No, angle must be included between 2 ≅ sides. *4.6*

15. The measure of one acute angle of a right triangle is $\frac{2}{3}$ the measure of the other. Find the measures of the two acute angles. *4.1* 36, 54

16. Given: Isosceles triangle ABC with $AC = BC$. If $AC = x + 8$, $BC = 3x - 2$, and $AB = 2x + 1$, find AB. *5.1* 11

17. Given: Isosceles triangle ABC with $AC = BC$, and $\overline{AD}$ bisecting $\angle CAB$, and $\overline{BD}$ bisecting $\angle CBA$. If m $\angle C = 40$, find m $\angle ADB$. *5.2* 110

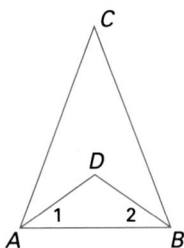

18. Given: P and Q are equidistant from X and Y. What conclusion can be drawn? *5.6* $\overline{PQ}$ is $\perp$ bis of $\overline{XY}$.

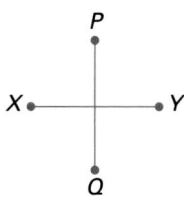

19. In $\triangle DEF$, $\overline{DE} \perp \overline{EF}$. Which side of the triangle is the hypotenuse? *4.1* $\overline{FD}$

20. In $\triangle ABC$, if $AB = 5$, and $BC = 10$, what restrictions are there on the length of $\overline{AC}$? *5.7* $5 < AC < 15$

21. If the measure of each interior angle of a polygon is 135, how many sides does the polygon have? *6.2* 8

22. Is Quad $ABCD$ a $\square$ if *6.5*
 (a) $\overline{AB} \parallel \overline{CD}$ and $AD = BC$? No
 (b) $\overline{AB} \parallel \overline{CD}$ and $\overline{AD} \parallel \overline{BC}$? Yes
 (c) $\overline{AM} \cong \overline{MC}$ and $\overline{DM} \cong \overline{BM}$? Yes
 (d) m $\angle DAB = $ m $\angle BCD$ and $\angle DAB$ is supplement of $\angle ABC$?
 Yes

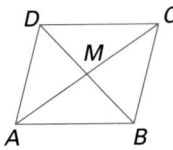

23. Given: $\overline{AB}$ is the perpendicular bisector of $\overline{CD}$; $\overline{AC} \cong \overline{BD}$. *5.6*
 Prove: $\overline{AM} \cong \overline{BM}$

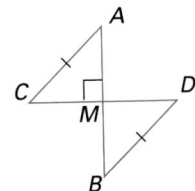

24. What is the sum of the measures of the angles of a decagon? *6.2* 1,440

25. Given: $m \parallel n$ and angle measures as indicated. Find x. *2.7* 84

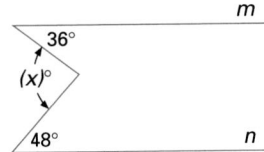

26. Given: the solid figure as shown *3.1*
 (a) Name two skew segments. n, q
 (b) Name two coplanar segments. m, n

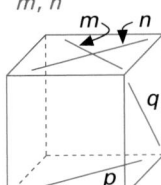

23.

Statement	Reason
1. $\overline{AB}$ is $\perp$ bis of $\overline{CD}$, $\overline{AC} \cong \overline{BD}$	1. Given
2. $\overline{CM} \cong \overline{DM}$	2. Def of $\perp$ bis
3. $\angle AMC$ and $\angle BMD$ are rt $\angle$s	3. $\perp$ lines form rt $\angle$s.
4. $\triangle AMC$ and $\triangle BMD$ are rt $\triangle$s	4. Def of rt $\triangle$
5. $\triangle AMC \cong \triangle BMD$	5. HL
6. $\overline{AM} \cong \overline{BM}$	6. CPCTC

7 SPECIAL QUADRILATERALS

OVERVIEW

In this chapter, students prove congruence of quadrilaterals and identify their corresponding parts. They also prove and apply theorems about rectangles, rhombuses, squares, and about medians and altitudes of general and isosceles trapezoids. They identify necessary and sufficient conditions for special quadrilaterals and are required to construct quadrilaterals with specific, given properties.

OBJECTIVES

- To match congruent parts of congruent quadrilaterals
- To write congruence statements about quadrilaterals
- To identify rectangles, rhombuses, and squares
- To identify sufficient conditions for special quadrilaterals
- To apply theorems about medians and altitudes of trapezoids
- To construct a rectangle given two adjacent sides

PROBLEM SOLVING

The problem solving strategies Drawing a Diagram and Finding a Counterexample help students understand the concept of sufficient conditions. Exercises 1–24 on page 281 ask students to draw a diagram or find a counterexample to determine whether certain stated conditions are sufficient for a quadrilateral to be a rectangle, rhombus, or square.

TECHNOLOGY

Computer: The Computer Investigation on page 284 enables students to discover properties of the *bases, legs,* and *diagonals* of trapezoids. This activity can be used to explore the definitions and theorems in Lessons 7.4 and 7.5.

Calculator: A calculator can be helpful in Exercises 16–18 on page 289 to simplify computations.

SPECIAL FEATURES

Mixed Review pp. 272, 277, 289, 294, 297
Algebra Review p. 272
Focus on Reading p. 275
Midchapter Review p. 282
Application: Conditions for Quadrilaterals p. 283
Computer Investigation: Necessary Conditions p. 284
Brainteaser p. 297
Key Terms p. 298
Key Ideas and Review Exercises pp. 298–299
Chapter 7 Test p. 300
College Prep Test p. 301

PLANNING GUIDE

Lesson	Basic	Average	Above Average	Resources
7.1 pp. 271–272	CE all WE 1–4 AR all	CE all WE 1–9 odd AR all	CE all WE 1–12 odd AR all	Reteaching p. 99 Practice p. 100
7.2 pp. 275–277	FR all CE all WE 1–18 odd	FR all CE all WE 1–28 odd	FR all CE all WE 1–31 odd	Reteaching p. 101 Practice p. 102
7.3 pp. 280–284	CE all WE 1–12 Midchapter Review Application 1–2 CI odd	CE all WE 1–30 odd Midchapter Review Application all CI all	CE all WE 1–34 odd Midchapter Review Application all CI all	Reteaching p. 103 Practice p. 104
7.4 pp. 288–289	CE all WE 1–15 odd	CE all WE 1–21 odd	CE all WE 1–25 odd	Reteaching p. 105 Practice p. 106
7.5 pp. 292–294	CE all WE 1–20 odd	CE all WE 1–33 odd	CE all WE 1–35 odd	Reteaching p. 107 Practice p. 108
7.6 pp. 296–297	CE all WE 1–5 Brainteaser	CE all WE 1–9 Brainteaser	CE all WE 1–12 Brainteaser	Reteaching p. 109 Practice p. 110
Chapter 7 Review pp. 298–299	all	all	all	
Chapter 7 Test p. 300	1–13, 17	1–17	all	
College Prep Test p. 301	all	all	all	

CE = Classroom Exercises   WE = Written Exercises   FR = Focus on Reading   AR = Algebra Review   CI = Computer Investigation

NOTE: For each level, all students should be assigned all Mixed Review exercises.

Investigation

Project: As an introduction to quadrilaterals, introduce the students to the game, "Quadrilateral Concentration." In this game, students explore common properties of quadrilaterals and use their own words to describe the properties.

Materials: 1 sheet stiff paper, glue, 20 quadrilateral cards and a scorecard (from Project Worksheet 7).

Procedure:

1. Have students glue the uncut sheet of gamecards onto a sheet of manila or construction paper. Have them cut out one set of cards for each group of 2–4 students and spread them out face down.

2. For each turn, a player gets to turn up 2 cards. The object is to name as many common properties as possible for the two figures on the cards within 30 seconds. Group members decide what constitutes "similar characteristics" for the group. The following are some characteristics that would be acceptable: opposite sides congruent, two sides congruent, opposite angles congruent, or two sides parallel. (Players can assume congruence from the appearance of the figures alone.)

3. Each time a player names properties, he or she writes down the number of similarities named on his or her scorecard. The player then turns the cards face down again.

4. The next player then takes a turn. The cards should always remain in the same position so that players can benefit from remembering where the figures are.

5. The game ends when each player has taken the same number of turns. Each player totals the numbers on his or her scorecard. The winner is the player with the highest score.

Stephen Hawking (Jan. 8, 1942)—Hawking developed a new model for black holes (stars that have collapsed). By means of mathematical calculations, he demonstrated that there can be "mini" black holes too small to see that weigh as much as Mt. Everest.

More About the Mathematician

As a college student, Hawking was diagnosed as having Lou Gehrig's disease and told that he had less than two years to live. Since that time he has published many books, won numerous prizes and revolutionized our thinking about the history of the universe. Hawking uses a model of a pebble causing ripples in a pond to explain the structure of a black hole. He can talk and move only by means of electronic equipment but with help from his many friends and his wife and three children he continues to perform his thought experiments. He explores and mathematically maps the universe, extending his ideas not only through space but also through time.

7.1 Proving Quadrilaterals Congruent

Objectives   To identify and name congruent quadrilaterals and their parts
To prove quadrilaterals congruent

Congruent quadrilaterals have the same size and shape. In this lesson, postulates and theorems for congruent triangles will be used to prove theorems concerning congruence of convex quadrilaterals.

Definition   Congruent quadrilaterals are quadrilaterals whose corresponding sides are congruent and whose corresponding angles are congruent.

Quad $ABCD \cong$ quad $EFGH$

$\angle A \cong \angle E$	$\overline{AB} \cong \overline{EF}$
$\angle B \cong \angle F$	$\overline{BC} \cong \overline{FG}$
$\angle C \cong \angle G$	$\overline{CD} \cong \overline{GH}$
$\angle D \cong \angle H$	$\overline{DA} \cong \overline{HE}$

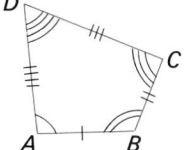
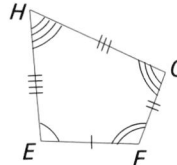

EXAMPLE 1   Using the markings below, write a congruence statement.

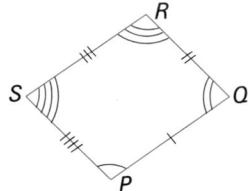
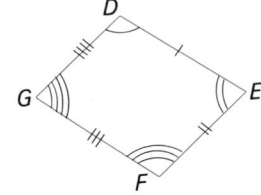

Solution   Quad $PQRS \cong$ quad $DEFG$

In the case of triangles, three pairs of congruent parts were needed to establish congruent triangles. It does *not* follow, however, that four pairs of congruent parts can establish congruent quadrilaterals.

$\overline{AB} \cong \overline{EF}$
$\overline{BC} \cong \overline{FG}$
$\overline{CD} \cong \overline{GH}$
$\overline{DA} \cong \overline{HE}$

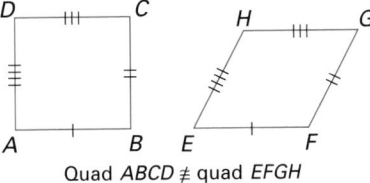

Quad $ABCD \not\cong$ quad $EFGH$

7.1 Proving Quadrilaterals Congruent   **269**

Additional Example 1

According to the markings on the quadrilaterals, what is a correct congruence statement?
Quad $LMNO \cong$ quad $ZYXW$

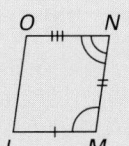
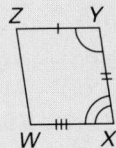

Teaching Resources

Project Worksheet 7
Quick Quizzes 50
**Reteaching and Practice
Worksheets,** pp. 99, 100

▐▐▐ GETTING STARTED

Prerequisite Quiz

Complete the congruence statements.

1. If $\triangle BCD \cong \triangle JKL$, then $\overline{DC} \cong$ _____.
   $\underline{LK}$
2. If $\overline{AB} \cong \overline{FH}$, $\overline{BC} \cong \overline{HG}$, and $\overline{CA} \cong \overline{GF}$,
   then $\triangle ABC \cong \triangle$ _____.   FHG

Motivator

Use cardboard or wooden strips fastened together with small bolts or brads to show a quadrilateral shape. Ask your students whether they can change the shape of the quadrilateral if they do not change the lengths of the sides of the quadrilateral. Yes.   Ask them to guess what condition is necessary to keep the shape rigid and what conclusion they can make about two congruent quadrilaterals.   At least one angle must be fixed. Congruent quadrilaterals have five pairs of congruent parts.

▐▐▐ TEACHING SUGGESTIONS

Lesson Note

In order to prove quadrilaterals (or other polygons) congruent, the students must draw auxiliary lines that form pairs of triangles whose congruent parts provide the congruences necessary to prove quadrilaterals congruent. Relate SASAS and ASASA for quadrilaterals to SAS and ASA for triangles. Remind students that a single counterexample can be used to disprove a statement, but no number of examples can be used to prove something. As an illustration from arithmetic, there are an infinite number of examples in which the sum and product of two numbers are equal. Some examples are: 0, 0; 2, 2; 3, $\frac{3}{2}$; 4, $\frac{4}{3}$; n, $\frac{n}{n-1}$. However, the sum and product of two numbers are not always equal.

269

Math Connections

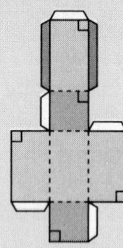

Boxes: Moving companies store many packing boxes at a time, keeping them as flat, unfolded pieces of cardboard until they are ready to be used. In the diagram above, the quadrilaterals that must be congruent in order for the box to fit together properly are shaded to match each other. Discuss with students which of the congruence patterns for quadrilaterals can be used to prove them congruent.

Critical Thinking Questions

Application: A construction crew leader gives four pieces of wood to each crew member and instructs them to build a 4–sided frame to be used as a mold for concrete. Each crew member received pieces of wood of the same size. When the leader returns, he finds that all of the frames are different shapes. Why? SSSS is not a congruence pattern for quadrilaterals. What are the simplest instructions the crew leader could have given to ensure that the frames would all be the same size and shape? He could have given the meas of two incl ∠s at which to connect 3 of the pieces (SASAS).

Common Error Analysis

Error: Students do not use included sides or included angles when applying the congruence theorems for quadrilaterals. Have students draw quadrilaterals having three corresponding congruent sides and two corresponding angles, only one of which is an included angle as illustrated below.

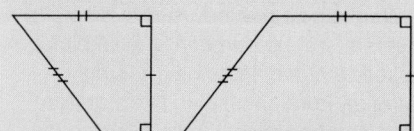

Have them find the measures of the remaining angles and sides and conclude whether the quadrilaterals are congruent.

Four pairs of congruent sides do not establish congruence of quadrilaterals because a quadrilateral, unlike a triangle, is not rigid. In the previous figure, the sides of quadrilateral *ABCD* collapse inward at angles *B* and *D* to form quadrilateral *EFGH*. In general, it will be necessary to establish *five* pairs of congruences to prove quadrilaterals congruent.

Theorem 7.1

Two quadrilaterals are congruent if any three sides and the included angles of one are congruent, respectively, to three sides and the included angles of the other. (SASAS for congruent quadrilaterals)

Given: $\overline{AB} \cong \overline{EF}$, $\overline{BC} \cong \overline{FG}$, $\overline{CD} \cong \overline{GH}$, $\angle B \cong \angle F$, $\angle C \cong \angle G$
Prove: Quad *ABCD* ≅ quad *EFGH*

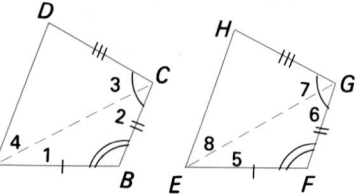

Plan

Draw $\overline{AC}$ and $\overline{EG}$. $\triangle ABC \cong \triangle EFG$ by SAS. Using corresponding parts of these triangles, prove that $\triangle ACD \cong \triangle EGH$. Use CPCTC to complete the proof.

EXAMPLE 2

Given: ▱*ABCD* and ▱*EFGH*, $\overline{AB} \cong \overline{EF}$, $\overline{AD} \cong \overline{EH}$, $\angle A \cong \angle E$
Prove: ▱*ABCD* ≅ ▱*EFGH*

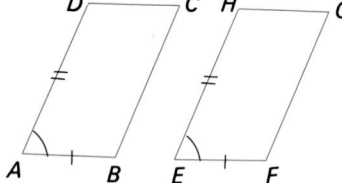

Proof

Statement	Reason
1. $\overline{AB} \cong \overline{EF}$	(S) 1. Given
2. $\angle A \cong \angle E$	(A) 2. Given
3. $\overline{AD} \cong \overline{EH}$	(S) 3. Given
4. ▱*ABCD*, ▱*EFGH*	4. Given
5. $\angle B$ and $\angle A$ are supp; $\angle F$ and $\angle E$ are supp.	5. Adj ∠s of a ▱ are supp.
6. $\angle B \cong \angle F$	(A) 6. Supp of ≅ ∠s are ≅.
7. $\overline{BC} \cong \overline{AD}$, $\overline{FG} \cong \overline{EH}$	7. Opp sides of a ▱ are ≅.
8. $\overline{BC} \cong \overline{FG}$	(S) 8. Sub (Steps 3 and 7)
9. ∴ ▱*ABCD* ≅ ▱*EFGH*	9. SASAS

An angle-side-angle-side-angle congruence pattern can also be proved.

Additional Example 2

Given: ▱*IJKL* and ▱*MNOP*, $\overline{IJ} \cong \overline{MN}$, $\angle J \cong \angle N$, $\overline{JK} \cong \overline{NO}$
Prove: ▱*IJKL* ≅ ▱*MNOP* using ASASA.

Proof: $\overline{IJ} \cong \overline{MN}$, $\angle J \cong \angle N$, $\overline{JK} \cong \overline{NO}$, *IJKL* and *MNOP* are ▱s (Given); $\angle I$ supp. $\angle J$, $\angle M$ supp. $\angle N$, $\angle K$ supp. $\angle J$, $\angle O$ supp. $\angle N$ (Consec ∠s of a ▱ are supp); $\angle I \cong \angle M$, $\angle K \cong \angle O$ (Supps of ≅ ∠s are ≅.); ▱*IJKL* ≅ ▱*MNOP* (ASASA)

Theorem 7.2

Two quadrilaterals are congruent if any three angles and the included sides of one are congruent, respectively, to three angles and the included sides of the other. (ASASA for congruent quadrilaterals)

Given: $\angle F \cong \angle J$, $\angle G \cong \angle K$,
$\angle H \cong \angle L$, $\overline{FG} \cong \overline{JK}$,
$\overline{GH} \cong \overline{KL}$
Prove: Quad $EFGH \cong$ quad $IJKL$

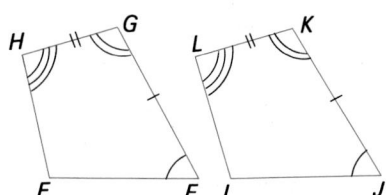

Classroom Exercises

Based on the markings shown, write a correct congruence statement.

1.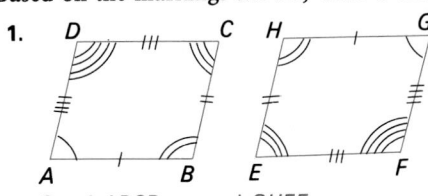

Quad $ABCD \cong$ quad $GHEF$

2.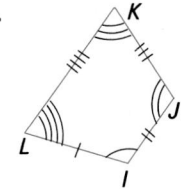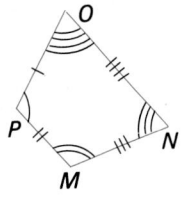

Quad $IJKL \cong$ quad $PMNO$

Complete the congruence statement.

Given: Quad $QRST \cong$ quad $XWVU$

3. $\angle R \cong$ $\underline{\angle W}$
4. $\overline{TS} \cong$ $\underline{\overline{UV}}$
5. $\angle W \cong$ $\underline{\angle R}$

Given: Quad $YZAB \cong$ quad $CDEF$

6. $\overline{FE} \cong$ $\underline{\overline{BA}}$
7. $\angle E \cong$ $\underline{\angle A}$
8. $\overline{DE} \cong$ $\underline{\overline{ZA}}$

Written Exercises

Name the pairs of congruent sides and the pairs of congruent angles for the congruence statement.

1. Quad $ABCD \cong$ quad $WXYZ$

2. Quad $LMVQ \cong$ quad $XCRN$

3. Given: $\overline{EB}$ is the $\perp$ bisector of $\overline{AC}$; $\overline{AF} \cong \overline{CD}$, $\angle A \cong \angle C$.
   Prove: Quad $ABEF \cong$ quad $CBED$

4. Given: $\overline{EB}$ is the $\perp$ bisector of $\overline{AC}$; $\overline{BE}$ bisects $\angle FED$;
   $\overline{FE} \cong \overline{DE}$.
   Prove: Quad $ABEF \cong$ quad $CBED$

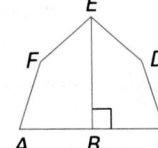

Checkpoint

1. If quad $KLMN \cong$ quad $WVUT$, $\angle M \cong$ $\angle$_____.   U
2. If quad $KLMN \cong$ quad $WVUT$, $\overline{NK} \cong$ _____. $\overline{TW}$
3. Based on the markings shown, give a correct congruence statement.   quad $RSTU \cong$ quad $PONM$

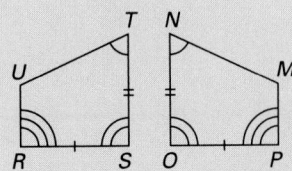

4. Which congruence theorem could be used to show that the two quadrilaterals above are congruent?   ASASA

Closure

On the chalkboard, draw two quadrilaterals, $ABCD$ and $EFGH$, that look congruent. Next to the figures, write the following list of congruent parts. Have the students go down the list one item at a time adding it to what is already known. Have them tell you when they know for certain that the two quadrilaterals are congruent.

1. $\overline{AB} \cong \overline{EF}$
2. $\angle A \cong \angle E$
3. $\overline{DA} \cong \overline{HE}$
4. $\angle B \cong \angle F$
5. $\overline{CD} \cong \overline{GH}$
6. $\overline{BC} \cong \overline{FG}$
7. $\angle C \cong \angle G$

Numbers 1–6

◼◼◼ FOLLOW UP

Guided Practice

Classroom Exercises 1–8

Independent Practice

Ⓐ Ex. 1–4, **Ⓑ** Ex. 5–9, **Ⓒ** Ex. 10–12

Basic: WE 1–4, AR all
Average: WE 1–9 odd, AR all
Above Average: WE 1–12 odd, AR all

Written Exercises

1. $\overline{AB} \cong \overline{WX}$, $\overline{BC} \cong \overline{XY}$, $\overline{CD} \cong \overline{YZ}$, $\overline{DA} \cong \overline{ZW}$; $\angle A \cong \angle W$, $\angle B \cong \angle X$, $\angle C \cong \angle Y$, $\angle D \cong \angle Z$

2. $\overline{LM} \cong \overline{XC}$, $\overline{MV} \cong \overline{CR}$, $\overline{VQ} \cong \overline{RN}$, $\overline{QL} \cong \overline{NX}$; $\angle L \cong \angle X$, $\angle M \cong \angle C$, $\angle V \cong \angle R$, $\angle Q \cong \angle N$

3.

Statement	Reason
1. $\overline{EB}$ is ⊥ bis of $\overline{AC}$, $\overline{AF} \cong \overline{CD}$, $\angle A \cong \angle C$	1. Given
2. $\angle ABE$, $\angle CBE$ are rt ∠s.	2. Def of ⊥
3. $\angle ABE \cong \angle CBE$	3. Rt ∠s are ≅.
4. $\overline{AB} \cong \overline{BC}$	4. Def of bis
5. $\overline{BE} \cong \overline{BE}$	5. Reflex Prop of Congr
6. Quad $ABEF \cong$ quad $CBED$.	6. SASAS

4.

Statement	Reason
1. $\overline{EB}$ is ⊥ bis of $\overline{AC}$, $\overline{BE}$ bis $\angle FED$, $\overline{FE} \cong \overline{DE}$	1. Given
2. $\angle FEB \cong \angle DEB$	2. Def of ∠ bis
3. $\overline{BE} \cong \overline{BE}$	3. Reflex Prop of Congr
4. $\angle ABE$, $\angle CBE$ are rt ∠s.	4. Def of ⊥
5. $\angle ABE \cong \angle CBE$	5. Rt ∠s are ≅.
6. $\overline{AB} \cong \overline{CB}$	6. Def of bis
7. Quad $ABEF \cong$ quad $CBED$.	7. SASAS

5.

Statement	Reason
1. $\overline{AF} \cong \overline{BD}$, $\overline{AG} \cong \overline{BC}$, $\angle A \cong \angle B$	1. Given
2. $\overline{AB} \cong \overline{AB}$	2. Reflex Prop of Congr
3. Quad $ABDG \cong$ quad $BAFC$.	3. SASAS

See pages 656–657 for the answers to Ex. 6–12.

272

---

5. Given: $\overline{AF} \cong \overline{BD}$, $\overline{AG} \cong \overline{BC}$, $\angle A \cong \angle B$
   Prove: Quad $ABDG \cong$ quad $BAFC$

6. Given: $\triangle GEF \cong \triangle CED$, $\overline{AF} \cong \overline{BD}$
   Prove: Quad $ABDG \cong$ quad $BAFC$

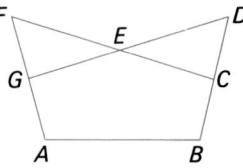

7. Given: $\overline{TX} \cong \overline{WU}$, $\triangle TZY \cong \triangle WZV$
   Prove: Quad $TUVZ \cong$ quad $WXYZ$

8. Given: $\triangle TUW \cong \triangle WXT$, $\overline{TY} \cong \overline{XY}$, $\overline{UV} \cong \overline{WV}$, $\overline{TZ} \cong \overline{WZ}$
   Prove: Quad $TUVZ \cong$ quad $WXYZ$

9. Given: $\square TUWX$, $\overline{YV}$ and $\overline{TW}$ bisect each other at Z.
   Prove: Quad $TUVZ \cong$ quad $WXYZ$

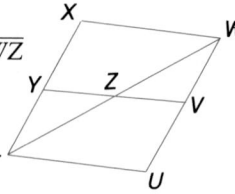

10. Prove Theorem 7.1.

11. Prove Theorem 7.2.

12. There are several patterns to be considered for proving convex quadrilaterals congruent. Consider these patterns: SSAAS, SSASS, AASAA, AASSA, and ASSAS. Choose one. Prove that it establishes congruence or show by counterexample that it does not.

Mixed Review

Determine whether the statement is true or false.

1. Opposite sides of a parallelogram are congruent. **6.4** True

2. If both pairs of opposite sides of a quadrilateral are congruent, then the quadrilateral is a parallelogram. **6.5** True

3. Diagonals of a parallelogram are congruent. **6.4** False

4. If the diagonals of a quadrilateral are congruent, then the quadrilateral is a parallelogram. **6.5** False

Algebra Review

To divide a polynomial by a monomial, divide each term of the polynomial by the monomial.

Example   Divide $18p^5 + 9p^3 - 12p^2 \div 3p^2$.

$$\frac{18p^5 + 9p^3 - 12p^2}{3p^2} = \frac{18p^5}{3p^2} + \frac{9p^3}{3p^2} - \frac{12p^2}{3p^2} = 6p^3 + 3p - 4$$

Divide.

1. $(t^9 + t^7 - t^5) \div t^2$
2. $(m^6 - m^4 + m^2) \div m^2$
3. $(25x^6 - 15x^4 + 10x^2) \div 5x^2$
4. $(14y^7 + 28y^6 - 35y^5) \div 7y^3$

1. $t^7 + t^5 - t^3$ 2. $m^4 - m^2 + 1$ 3. $5x^4 - 3x^2 + 2$ 4. $2y^4 + 4y^3 - 5y^2$

7.2 Necessary Conditions: Rectangles, Rhombuses, and Squares

Objectives
To identify rectangles, rhombuses, and squares
To prove and apply properties of rectangles, rhombuses, and squares
To identify necessary conditions for special quadrilaterals

A parallelogram is a quadrilateral with both pairs of opposite sides parallel. The pairs of opposite sides and opposite angles of a parallelogram are congruent. Some special parallelograms and their properties will be considered in this lesson. (The overlapping of classes of parallelograms is shown by the diagram on the right.)

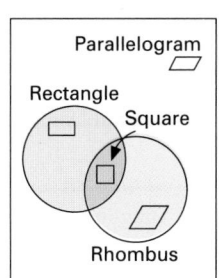

Definitions

A **rectangle** is a parallelogram with four right angles.
A **rhombus** is a parallelogram with four congruent sides.
A **square** is a rectangle with four congruent sides.

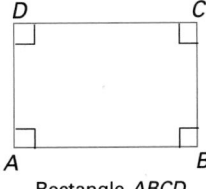
Rectangle *ABCD*

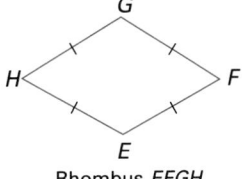
Rhombus *EFGH*

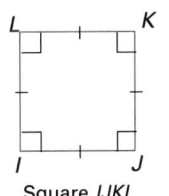
Square *IJKL*

From these definitions, you can see that figures may be classified in more than one way:

> All rectangles, rhombuses, and squares are parallelograms.
> All squares are rectangles.
> All squares are rhombuses.

A **necessary condition** is a property or characteristic that must be satisfied in order to achieve a desired result. For example, for a given quadrilateral to be a parallelogram, it must satisfy the necessary condition that the opposite sides be parallel. If the opposite sides are not parallel, then the quadrilateral is not a parallelogram.

Teaching Resources

Application Worksheet 7
Quick Quizzes 51
Reteaching and Practice Worksheets, pp. 101, 102
Transparency 15

◢ GETTING STARTED

Prerequisite Quiz

Identify each statement as always true, sometimes true, or never true for parallelograms.

1. Opposite sides are parallel.   Always
2. Consecutive angles are supplements.   Always
3. Opposite sides are congruent.   Always
4. Diagonals are congruent.   Sometimes
5. Opposite angles are congruent.   Always

Motivator

Have the students look around the room and find surfaces that are rectangles. Have them explain what makes them rectangles. They have four right angles.   Next ask them if there are any surfaces that are squares. Ask them if the square surfaces are also rectangles.   Yes   Have them explain how the surfaces are different. Have them look for "diamond" shaped surfaces and explain to them that these are called "rhombuses." Have them explain what is special about these surfaces.   The four sides are congruent.

Lesson Note

This lesson deals with two different concepts. It introduces properties of the rectangle, rhombus, and square. It also introduces the idea of necessary conditions as properties that a given figure must have. For example, the angles of a square must be right angles. Any figure with a different kind of angle is not a square.

You may wish to introduce the idea of necessary conditions using nonmathematical examples. For example, have students discuss what is necessary for a person to make a good basketball or football player, a rock drummer, a salesperson, or a student. There may be some disagreement on these qualifications. However, discussion should be based on a recognition of properties that such people must have. Note that "rhombuses," rather than Greek "rhombi," is used as the plural of "rhombus" in this text.

Math Connections

Flow Charts: An operations flow chart is a diagram that shows the steps involved in a job. Various quadrilaterals and other shapes are commonly used to classify each step by the action that is involved. For example, a rhombus often indicates a decision point at which the steps that follow may vary.

Critical Thinking Questions

Analysis: Ask students to define a square in two words.   Answers may vary. Reg ▱
Ask students to draw a diagram and answer the following question.

Quad *ABCD* is a parallelogram with diagonals $\overline{AC}$ and $\overline{BD}$, and with $AC > BD$. In order to make this parallelogram a rectangle, which two angles need to decrease in measure? $\angle ADC$ and $\angle ABC$

EXAMPLE 1   Which of the following are necessary conditions for a parallelogram to be a square? Why?

   a. The parallelogram has four congruent sides.
   b. The parallelogram has at least two right angles.

Solution   Both conditions are necessary.
   a. Since a square is a rhombus, it must have four congruent sides.
   b. Since a square is also a rectangle, it must have four right angles.

From Lesson 6.4, you know certain properties of a parallelogram. These properties are necessary conditions for a quadrilateral to be a parallelogram. Necessary conditions for rhombuses and rectangles will now be established.

Theorem 7.3   The diagonals of a rhombus are perpendicular.

Given: Rhom *ABCD* with diagonals $\overline{AC}$
   and $\overline{BD}$
Prove: $\overline{AC} \perp \overline{BD}$

Plan   All sides of a rhombus are congruent, so $\overline{AD} \cong \overline{AB}$ and $\overline{CD} \cong \overline{CB}$. Points *A* and *C* are each equidistant from the endpoints of $\overline{BD}$.

Proof

Statement	Reason
1. Rhomb *ABCD* with diags $\overline{AC}$ and $\overline{BD}$	1. Given
2. $\overline{AD} \cong \overline{AB}$, $\overline{CD} \cong \overline{CB}$	2. Def of rhom
3. $\therefore \overline{AC} \perp \overline{BD}$	3. A line containing 2 pts, each equidist from endpts of a seg, is $\perp$ bis of the seg (Thm 5.8).

Theorem 7.4   The diagonals of a rectangle are congruent.

Given: Rect *EFGH* with diagonals $\overline{HF}$ and $\overline{GE}$
Prove: $\overline{HF} \cong \overline{GE}$

Plan   Use the definition of a rectangle and properties of a parallelogram to prove $\triangle HEF \cong \triangle GFE$.

274   Chapter 7 Special Quadrilaterals

Additional Example 1

Which of the following are necessary conditions for a parallelogram to be a rhombus? Why?
(1) The parallelogram has two congruent adjacent sides.   Necessary; a rhombus is a ▱ with all sides ≅.

(2) The parallelogram has congruent diagonals.
Not necessary; a rhombus does not need to have ≅ diags.

Theorem 7.5   Each diagonal of a rhombus bisects two angles of the rhombus.

Given: Rhom IJKL, with diagonals $\overline{IK}$ and $\overline{JL}$
Prove: $\overline{IK}$ bisects $\angle JIL$ and $\angle JKL$; $\overline{JL}$ bisects $\angle ILK$ and $\angle IJK$.

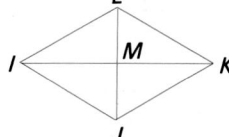

Plan   Use the fact that diagonals of a rhombus are perpendicular and that diagonals of a parallelogram bisect each other.

EXAMPLE 2   Given: Rect IJKL, $JL = 5x - 1$, $IK = 3x + 5$
Find JL.

Solution
$$JL = IK$$
$$5x - 1 = 3x + 5$$
$$2x = 6$$
$$x = 3$$
$$JL = 5 \cdot 3 - 1 = 14$$

Summary

The table below shows necessary conditions for the given figures.

Condition	Rectangle	Rhombus	Square
All sides are ≅.		X	X
4 right ∠s	X		X
≅ diagonals	X		X
⊥ diagonals		X	X
Diagonals bisect ∠s.		X	X

Focus on Reading

Suppose you were proving each statement. State the hypothesis and the conclusion.

1. If a parallelogram has four congruent sides, then it is a rhombus.
2. If a parallelogram is a rectangle, then the diagonals are congruent.
3. The angles of a rectangle are right angles.
4. The adjacent sides of a rhombus are congruent.

Additional Example 2

Given: Rhombus ABCD,
  $AE = 7x - 5$,
  $EC = 3x + 3$
Find AC.   18

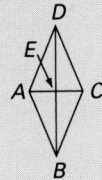

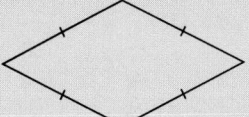
275

Guided Practice

Classroom Exercises 1–3

Independent Practice

A Ex. 1–18, **B** Ex. 19–28, **C** Ex. 29–31

Basic: FR all, WE 1–18 odd

Average: FR all, WE 1–28 odd

Above Average: FR all, WE 1–31 odd

Additional Answers

Focus on Reading

1. Hypoth: A parallelogram has four congruent sides. Concl: The parallelogram is a rhombus.
2. Hypoth: A parallelogram is a rectangle. Concl: Its diagonals are congruent.
3. Hypoth: A figure is a rectangle. Concl: It has four right angles.
4. Hypoth: A figure is a rhombus. Concl: Its adjacent sides are congruent.

Written Exercises

2. False

6. False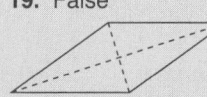

9. False 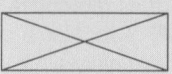   19. False

20. True. A square is a rect. The diags of a rect are ≅.
21. True A square is a rhom. The diags of a rhom are ⊥.
22. False

27.

Statement	Reason
1. Rect *EFGH*, $\overline{HF}$ and $\overline{GE}$ are diags.	1. Given
2. ∠*HEF* and ∠*GFE* are rt ∠s.	2. Def of rect
3. ∠*HEF* ≅ ∠*GFE*	3. Rt ∠s are ≅.
4. $\overline{HE}$ ≅ $\overline{GF}$	4. Opp sides of ▱ are ≅.
5. $\overline{EF}$ ≅ $\overline{EF}$	5. Reflex Prop of Congr
6. △*HEF* ≅ △*GFE*	6. SAS
7. $\overline{HF}$ ≅ $\overline{GE}$	7. CPCTC

Classroom Exercises

Based only on the markings shown, identify the parallelogram as a rectangle, a rhombus, a square, or none of these. A figure may be correctly identified in more than one way.

1.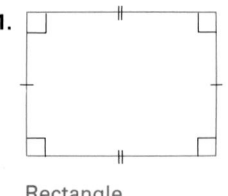

Rectangle

2.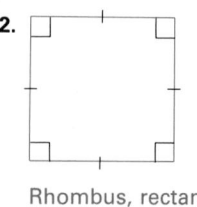

Rhombus, rectangle, square

3.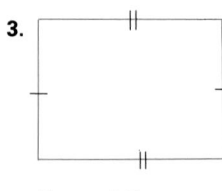

None of these

Written Exercises

True or false? If false, draw a picture to defend your answer.

1. All squares are rhombuses.   True
2. All rhombuses are rectangles.
3. Some rhombuses are rectangles.   True
4. Some parallelograms are squares.   True
5. Some parallelograms are rectangles.   True
6. No rectangles are squares.
7. All squares are rectangles.   True
8. All rhombuses are parallelograms.   True
9. All rectangles are squares.
10. Some rhombuses are squares.   True

Is the statement a necessary condition for a quadrilateral to be a rectangle?

11. Opposite angles are congruent.   Yes
12. Adjacent angles are congruent.   Yes
13. Diagonals are congruent.   Yes
14. Diagonals are perpendicular.   No

Find the indicated measure.

15. Given: Square *ABCD*, $AB = 23$
    Find *BC*.   23

16. Given: Rect *ABCD*, $BD = 13$
    Find *AC*.   13

17. Given: Rhom *ABCD*
    Find m ∠*AEB*.   90

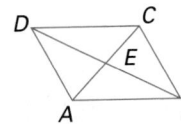

18. Given: Rect *ABCD*, $AC = 7x + 5$, $BD = 14x - 2$
    Find *AC*.   12

If the statement is true, explain why. If false, draw a counterexample.

19. The diagonals of a rhombus are congruent.
20. The diagonals of a square are congruent.   True
21. The diagonals of a square are perpendicular.   True
22. A diagonal of a rectangle bisects opposite angles.

276   Chapter 7 Special Quadrilaterals

Enrichment

On the chalkboard, draw a quadrilateral in the form of a kite. Stress the fact that a kite is not a parallelogram. Then have each student list as many necessary conditions for a kite as he or she can.

(1) Two pairs of adj sides are ≅.
(2) One pair of opp ∠s are ≅.
(3) Diags are ⊥.
(4) The longer diag bisects the shorter one.
(5) The longer diag bisects a pair of opp ∠s.

Other necessary conditions can be listed.

Find the indicated measure.

23. Given: $\square ABCD$, $AC = 9x + 1$,
      $AE = 5x - 1$
   Find CE.   14

24. Given: $\square ABCD$, $BD = 4x + 4$,
      $ED = 3x - 2$
   Find BE.   10

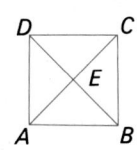

25. Given: Square $ABCD$
   Find m $\angle EAB$.   45

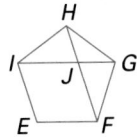

26. Given: Rhom $XYZW$,
      m $\angle XYW = 3x + 10$,
      m $\angle ZYW = 4x - 5$
   Find m $\angle XYW$.   55

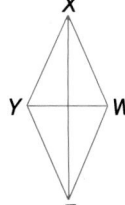

27. Prove Theorem 7.4.

28. Prove Theorem 7.5.

29. Prove: The midpoint of the hypotenuse of a right triangle is equidistant from the three vertices of the triangle. (HINT: Construct a rectangle.)

30. Given: Regular pentagon $EFGHI$,
      diagonals $\overline{FH}$ and $\overline{GI}$
      intersect at J.
   Prove: $EFJI$ is a parallelogram.

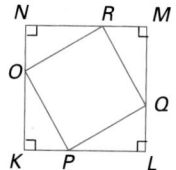

31. Given: Square $KLMN$,
      $\overline{NO} \cong \overline{KP} \cong \overline{LQ} \cong \overline{MR}$
   Prove: $OPQR$ is a square.

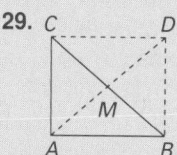

Mixed Review

Is the statement true or false?

1. Adjacent sides of a parallelogram are congruent.   *6.4* False

2. If exactly two adjacent sides of a quadrilateral are congruent, then the quadrilateral is a parallelogram.   *6.5* False

3. If all consecutive angles of a quadrilateral are congruent, then the quadrilateral is a parallelogram.   *6.5* True

4. Consecutive angles of a parallelogram are congruent.   *6.4* False

5. Diagonals of a parallelogram bisect each other.   *6.4* True

6. If the diagonals of a quadrilateral bisect each other, then the quadrilateral is a parallelogram.   *6.5* True

7.2 Necessary Conditions: Rectangles, Rhombuses, and Squares   **277**

28.

Statement	Reason
1. Rhom $IJKL$, $\overline{IK}$ and $\overline{JL}$ are diags.	1. Given
2. $\overline{IK} \perp \overline{JL}$	2. Diags of rhom are $\perp$.
3. $\angle LMK$, $\angle JMK$, $\angle IML$, and $\angle IMJ$ are rt $\angle$s.	3. Def of $\perp$
4. $\angle LMK \cong \angle JMK$; $\angle IML \cong \angle IMJ$	4. Rt $\angle$s are $\cong$.
5. $\overline{LM} \cong \overline{JM}$	5. Diags of $\square$ bis each other.
6. $\overline{MK} \cong \overline{MK}$; $\overline{IM} \cong \overline{IM}$	6. Reflex Prop of Congr
7. $\triangle LMK \cong \triangle JMK$; $\triangle LMI \cong \triangle JMI$	7. SAS
8. $\angle LMK \cong \angle JMK$; $\angle LMI \cong \angle JMI$	8. CPCTC
9. $\overline{IK}$ bis $\angle JKL$; $\overline{IK}$ bis $\angle JIL$.	9. Def of $\angle$ bis

29.

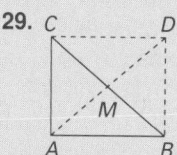

Statement	Reason
1. $\triangle ABC$, m $\angle CAB = 90$, M is midpt of $\overline{BC}$.	1. Given
2. Draw $\overline{BD} \perp \overline{AB}$, and draw $\overline{CD} \perp \overline{AC}$.	2. At given pt on line, $\perp$ may be constructed.
3. $\overline{CA} \parallel \overline{BD}$, $\overline{CD} \parallel \overline{BA}$	3. In plane, 2 lines $\perp$ to same line are $\parallel$.
4. $ABCD$ is $\square$.	4. Def of $\square$
5. $\angle CDB$ is rt $\angle$.	5. Opp $\angle$s in $\square$ are $\cong$.
6. $ABCD$ is rect.	6. Def of rect
7. Draw $\overline{AD}$	7. 2 pts determine a line.
8. $\overline{AD}$ contains M.	8. Diags of $\square$ bis each other; a line has 1 midpt.
9. $\overline{AD} \cong \overline{CB}$	9. Diags of rect are $\cong$.
10. $AM = \frac{1}{2}AD$, $BM = \frac{1}{2}CB$	10. Def of bis
11. $\overline{AM} \cong \overline{CM} \cong \overline{BM}$	11. Sub

See pages 283–284 for the answers to Ex. 30–31.

■■■GETTING STARTED

Prerequisite Quiz

If a quadrilateral has each given property, will it be a parallelogram?

1. Opposite sides are parallel. Yes
2. Opposite sides are congruent. Yes
3. The sum of the measures of the four angles is 360. No
4. One pair of opposite sides is congruent. No
5. Diagonals are congruent. No

Motivator

Ask the students to explain what it means if someone tells them that it is necessary for them to do something. They must do it. Ask them what it means if someone tells them that what they have done is sufficient. They have done enough and they do not need to do anymore.

■■■TEACHING SUGGESTIONS

Lesson Note

This lesson deals with conditions that are sufficient to prove that a quadrilateral is a rectangle, a rhombus, or a square. Sufficient conditions guarantee a result. For example, a quadrilateral must be a rectangle if it has four right angles, or if it is a parallelogram with one angle. However, one right angle in a quadrilateral is not a sufficient condition for the quadrilateral to be a rectangle. You can use models to illustrate sufficient conditions. Take four sticks of equal length to make a quadrilateral. It will always be a rhombus. Having four congruent sides is a sufficient condition for a quadrilateral to be a rhombus.

7.3 Sufficient Conditions: Rectangles, Rhombuses, and Squares

Objectives To identify sufficient conditions for special quadrilaterals
To prove that a figure is a rectangle, rhombus, or square

Think about the following statements.
(1) If a quadrilateral has right angles, then it is a rectangle.
(2) If a quadrilateral has four right angles, then it is a rectangle.
(3) If a quadrilateral is a rhombus, then it is a rectangle.
Statement 1 is false. Having a right angle is a *necessary* condition for a rectangle, but it does not assure that a quadrilateral is a rectangle. Statement 2 is true. Every quadrilateral with four right angles is a rectangle. This is a **sufficient** condition for a rectangle because it guarantees that the figure is a rectangle. Statement 3 is false. Being a rhombus is *neither* necessary nor sufficient to be a rectangle.

Theorem 7.6 A parallelogram with one right angle is a rectangle.

Theorem 7.7 A parallelogram with two adjacent congruent sides is a rhombus.

Theorem 7.8 A parallelogram with perpendicular diagonals is a rhombus.

Given: $\square ABCD$, $\overline{AC} \perp \overline{BD}$
Prove: $ABCD$ is a rhombus.

Plan Show that $\overline{AD}$ and $\overline{AB}$ are $\cong$ corresponding sides of $\cong$ $\triangle$s AED and AEB.

Proof

Statement	Reason
1. $ABCD$ is a parallelogram.	1. Given
2. $\overline{DE} \cong \overline{BE}$	(S) 2. Diags of a $\square$ bis each other.
3. $\overline{AC} \perp \overline{BD}$	3. Given
4. $\angle DEA \cong \angle BEA$	(A) 4. $\perp$ lines form $\cong$ rt $\angle$s.
5. $\overline{AE} \cong \overline{AE}$	(S) 5. Reflex Prop
6. $\triangle AED \cong \triangle AEB$	6. SAS
7. $\overline{AD} \cong \overline{AB}$	7. CPCTC
8. $\therefore ABCD$ is a rhombus.	8. A $\square$ with two adj $\cong$ sides is a rhom.

Additional Example 1

Consider a quadrilateral with four right angles. Is this property a sufficient condition for the quadrilateral to be a square?

The quadrilateral shown has four right angles, but does not have four congruent sides. Having four right angles is not a sufficient condition for a quadrilateral to be a square.

EXAMPLE 1 Consider a quadrilateral with perpendicular diagonals. Is this property a sufficient condition for the quadrilateral to be a rhombus? If it is not sufficient, provide a counterexample.

Solution The quadrilateral at the right has perpendicular diagonals, but it does not have four congruent sides. Having perpendicular diagonals is *not* a sufficient condition for a quadrilateral to be a rhombus.

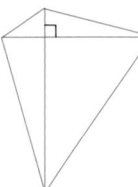

Two other sufficient conditions for a parallelogram to be a rectangle or a rhombus are given as theorems below.

Theorem 7.9 A parallelogram with congruent diagonals is a rectangle.

Theorem 7.10 A parallelogram with a diagonal that bisects opposite angles is a rhombus.

There are sufficient conditions for proving that a quadrilateral not specifically identified as a parallelogram is a rectangle, a rhombus, or a square.

Theorem 7.11 A quadrilateral with four congruent sides is a rhombus.

EXAMPLE 2 True or false? A quadrilateral with congruent diagonals is a rectangle. Draw a figure to support your answer.

Solution False. It is possible for a quadrilateral to have congruent diagonals and not be a rectangle. Congruent diagonals are necessary, but *not* sufficient for proving that a quadrilateral is a rectangle.

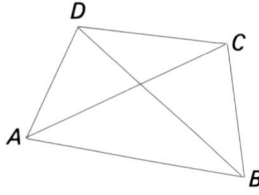

$$\overline{AC} \cong \overline{BD}$$

Gates: Some folding elevator gates are made in such a way that many rows of rhombuses are formed. The vertical sides of the gate remain parallel as it opens and closes because, although they are not a physical part of the gate, the diagonals of a rhombus remain perpendicular as they get longer or shorter.

Critical Thinking Questions

Application: Some necessary conditions for an equilateral triangle are: has 3 sides, has 3 angles, is a triangle, has 3 congruent angles, has 3 congruent sides, the measure of each angle is 60. Ask students which are sufficient conditions for an equilateral triangle.   Is a △ and has 3 ≅ sides *or* has 3 ≅ ∠s *or* the meas of each ∠ is 60.   Ask students to list necessary and sufficient conditions for an isosceles triangle. Necessary: has 3 sides, has 3 ∠s, is a △, has 2 ≅ sides, has 2 ≅ ∠s, the bis of the vert ∠ is ⊥ to the base at its midpt, the alt from the vert is a median, the alt to the base bis the vert ∠, the median to the base bis the vert ∠s Sufficient: is a △ and has 2 ≅ sides *or* has 2 ≅ ∠s.

Additional Example 2

True or false? A quadrilateral with perpendicular diagonals is a rhombus. Draw a picture to defend your answer.
False

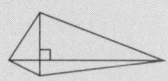

279

Checkpoint

Which of the following are sufficient conditions for a rectangle to be a square? (Ex. 1–2)

1. The diagonals are congruent.   Not sufficient
2. The diagonals are perpendicular.   Sufficient

Based only on the markings shown, identify each parallelogram as a rectangle, a rhombus, a square, or none of these.

3. Rect

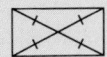

4. Sq, rh, rect

Closure

Draw a quadrilateral *ABCD* that looks like a square on the chalkboard. Next to the figure write the following list of properties. Have the students go down this list adding one property at a time to what they already know, until they know for certain that the quadrilateral is a square.

1. $\overline{AB} \cong \overline{BC}$
2. *ABCD* is a parallelogram.
3. $\angle C$ is a right angle.
4. $\overline{BC} \cong \overline{CD}$
5. $\angle D$ is a right angle.
6. $\angle A \cong \angle B \cong \angle C \cong \angle D$
7. $\overline{AB} \cong \overline{BC} \cong \overline{CD} \cong \overline{DA}$ Properties 1–3

◼◼FOLLOW UP

Guided Practice

Classroom Exercises 1–6

Independent Practice

A Ex. 1–12, **B** Ex. 13–30, **C** Ex. 31–34

Basic: WE 1–12, Midchapter Review, Application 1–2, CI odd

Average: WE 1–30 odd, Midchapter Review, Application, CI all

Above Average: WE 1–34 odd, Midchapter Review, Application, CI all

---

EXAMPLE 3   Is the following a sufficient condition for a quadrilateral to be a square?

One pair of opposite angles are right angles.

Solution   This is *not* a sufficient condition for a quadrilateral to be a square. A counterexample can be devised by drawing two noncongruent right triangles with a common hypotenuse.

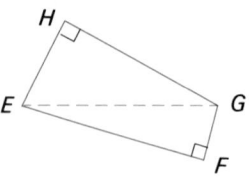

◼◼ Summary

The following chart shows sufficient conditions for a *parallelogram* to be a rectangle, rhombus, or square.

Condition	Rectangle	Rhombus	Square
2 adj ≅ sides		X	
1 right ∠	X		
≅ diagonals	X		
⊥ diagonals		X	
1 right ∠ and 2 adj ≅ sides			X
diags bis opp ∠s.		X	

Classroom Exercises

Based only on the markings, identify the parallelogram as a rectangle, a rhombus, a square, or none of these. A figure may be correctly identified in more than one way.

1.
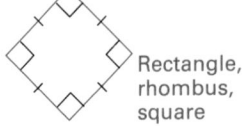
Rectangle, rhombus, square

2.
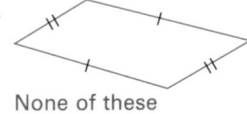
None of these

3.

Rectangle

4.
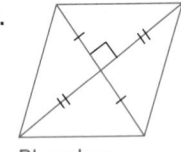
Rhombus

5.
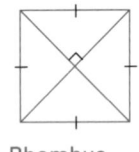
Rhombus

6.
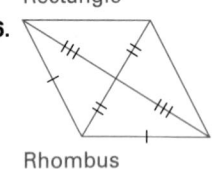
Rhombus

Additional Example 3

Is the following a sufficient condition for a quadrilateral to be a rectangle? Two consecutive angles are right angles.   No

Written Exercises

True or false? Draw a picture to defend your answer.

1. A quadrilateral with two consecutive right angles is a rectangle.
2. A parallelogram with two consecutive right angles is a rectangle.
3. A rhombus with two consecutive right angles is a square.
4. A quadrilateral with perpendicular diagonals is a rhombus.
5. A parallelogram with perpendicular diagonals is a rhombus.
6. A quadrilateral with four congruent sides is a rhombus.

Determine whether the stated condition is sufficient for the figure to be a square. If not sufficient, draw a diagram of a counterexample.

7. a quadrilateral with at least one right angle
8. a quadrilateral with at least two right angles
9. a quadrilateral with at least three right angles
10. a quadrilateral with four right angles
11. a parallelogram with at least one right angle
12. a rhombus with at least one right angle   Sufficient
13. Given: $\overline{AB} \cong \overline{DC}$, $\overline{AB} \parallel \overline{DC}$, $\overline{AC} \cong \overline{BD}$
    Prove: $ABCD$ is a rectangle.
14. Given: $\overline{AE} \cong \overline{BE} \cong \overline{CE} \cong \overline{DE}$
    Prove: $ABCD$ is a rectangle.

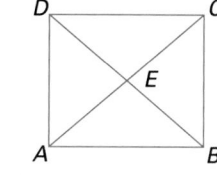

Determine whether the condition is sufficient for the figure to be a rectangle. Defend your answer with a proof or a counterexample.

15. a quadrilateral with congruent diagonals
16. a quadrilateral with at least one right angle
17. a quadrilateral with congruent diagonals that bisect each other
18. a parallelogram with opposite angles supplementary

Determine whether the condition is sufficient for the figure to be a rhombus. Provide a proof or a counterexample.

19. a quadrilateral with diagonals that are perpendicular bisectors of each other
20. a quadrilateral with two pairs of adjacent congruent sides

Determine whether the condition is sufficient for the figure to be a square. Provide a proof or a counterexample.

21. a parallelogram with one right angle
22. a rectangle with two consecutive congruent angles
23. a rectangle with perpendicular diagonals
24. a rhombus with congruent diagonals

7.3 Sufficient Conditions: Rectangles, Rhombuses, and Squares   **281**

 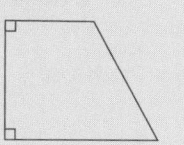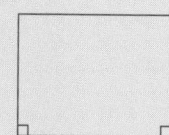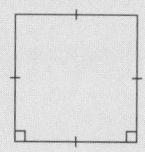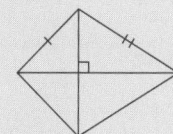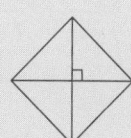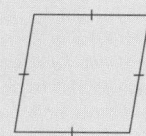

Enrichment

Challenge the students to prove the following statement: The segments joining the midpoints of adjacent sides of a rectangle form a rhombus.

Given: Rect $ABCD$ with E, F, G, H the midpoints of $\overline{AB}$, $\overline{BC}$, $\overline{CD}$, and $\overline{DA}$, respectively.

Prove: $EFGH$ is a rhombus.

Proof: $ABCD$ is a rect, E, F, G, H are midpts of $\overline{AB}$, $\overline{BC}$, $\overline{CD}$, $\overline{DA}$, resp. (Given); Draw $\overline{AC}$ and $\overline{BD}$. (2 pts are contained in exactly 1 line); $\overline{AC} \cong \overline{BD}$ or $AC = BD$ (Diag of rect. are $\cong$.); $EF = \frac{1}{2}AC$, $FG = \frac{1}{2}BD$, $GH = \frac{1}{2}AC$, $HE = \frac{1}{2}BD$ (Midseg Thm); $EF = FG = GH = HE$ or $\overline{EF} \cong \overline{FG} \cong \overline{GH} \cong \overline{HE}$ (Sub); $EFGH$ is a rhombus. (A quad with 4 $\cong$ sides is a rhombus.)

14.

Statement	Reason
1. $\overline{AE} \cong \overline{BE} \cong$ $\overline{CE} \cong \overline{DE}$	1. Given
2. $ABCD$ is a ▱.	2. If diags bis each other, quad is a ▱.
3. $AE + CE =$ $DE + BE$	3. Equations may be added.
4. $AE + CE =$ AC; $DE +$ $BE = BD$	4. Seg Add Post
5. $AC = BD$	5. Sub
6. $ABCD$ is a rect.	6. ▱ with ≅ diags is a rect.

15. Not sufficient **16.** Not sufficient

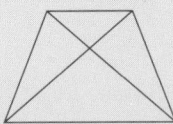 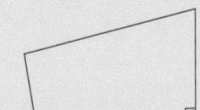

17. Sufficient.

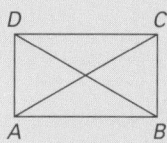

Statement	Reason
1. Quad $ABCD$ with $\overline{BD} \cong$ $\overline{AC}$, BD and AC bis each other	1. Given
2. $ABCD$ is a ▱.	2. If diags bis each other, quad is a ▱.
3. $ABCD$ is a rect.	3. If diags of ▱ are ≅, ▱ is a rect.

18. Sufficient.

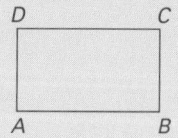

Statement	Reason
1. ▱ $ABCD$ with $\angle A$ and $\angle C$ opp and supp	1. Given
2. $\angle A \cong \angle C$	2. Opp ∠s of a ▱ are ≅.
3. $m \angle A + m$ $\angle C = 180$	3. Def of supp ∠s
4. $2 \times m \angle A$ $= 90$	4. Sub
5. $m \angle A = 90$	5. Div Prop Eq
6. $ABCD$ is a rect.	6. ▱ with 1 rt ∠ is a rect.

See pages 657–659 for the answers to Written Ex. 19–34 and Midchapter Review Ex. 1.

Essay.

25. Explain in your own words the difference between a necessary condition for a quadrilateral to be a rectangle and a sufficient condition for a quadrilateral to be a rectangle.

26. Prove Theorem 7.6.   **27.** Prove Theorem 7.7.

28. Prove Theorem 7.9.   **29.** Prove Theorem 7.10.

30. Prove Theorem 7.11.

Is the condition sufficient for the two given figures to be congruent? Give a proof or a counterexample.

31. two squares with a diagonal of the first congruent to a diagonal of the second

32. two rhombuses with a diagonal of the first congruent to a diagonal of the second

33. two rectangles with two adjacent sides of the first congruent, respectively, to two adjacent sides of the second

34. two rectangles with a side and a diagonal of the first congruent to a side and a diagonal of the second

Midchapter Review

1. Given: Rectangles $ABCD$ and $WXYZ$, $AB = WX$
Prove: Quad $ABCD \cong$ quad $WXYZ$

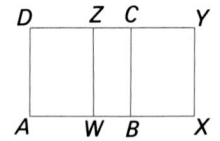

Given: Rhom $ABCD$, $DB = 10$, m $\angle ABC = 120$

2. Find XB. $XB = 5$ **3.** Find m $\angle AXB$.   90

4. Find m $\angle ABX$.   60

True or false? If false, give a counterexample.

5. A rhombus with one right angle is a square.   True

6. A parallelogram with two congruent sides is a rhombus.   False, rectangle

7. Every square is a rhombus.   True

Determine whether the given condition is only necessary or sufficient for the figure to be a rectangle.

8. a quadrilateral with one right angle   Necessary, not sufficient

9. a parallelogram with one right angle   Necessary, sufficient

Application: *Conditions for Quadrilaterals*

We use necessary and sufficient conditions in everyday life to minimize our efforts. Determine the conditions for each of the following situations.

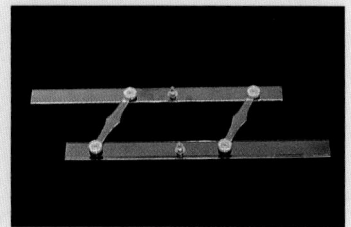

1. Parallel rulers are used to draw parallel lines. Consider the photo at the left. How are the parallel rulers made? What conditions will insure that the rulers are parallel in all positions?

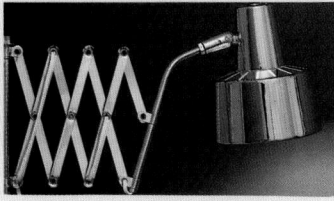

2. What type of quadrilateral is used in making the lamp shown above? What insures that the lamp moves to and from the wall in a straight line?

3. The lamp shown above can be pivoted into many different positions. The light itself can be raised or lowered while pointing in the same direction. Why is this so?

4. A shortwave radio antenna is stretched between a pole on one side of a house and a tree on the other. Its length cannot be measured directly. What kind of geometric figure could be used on the ground in front of the house to determine the distance? What would be necessary and sufficient conditions for that figure to achieve the desired result?

7.3 Sufficient Conditions: Rectangles, Rhombuses, and Squares   **283**

Application

1. Each side of the device is ≅ to the side opp. The ▱ thus formed will have opp sides ‖, no matter what the ∠s formed by the sides are.

2. Rhombus. The assembly is held in place such that one diag is always vert. The other diag must be horiz.

3. The arm of the lamp consists of 2 long connected ▱s. Each ▱ has one short side fixed, so that the side opp is always ‖ to a fixed line as it moves.

4. Several answers are possible. A rect can be formed. The length of the side opp the antenna wire will be = to the length of the wire. A necessary and sufficient condition that the figure be a rect is that it be a quad with 3 rt ∠s.

Additional Answer, page 277

30.

Statement	Reason
1. Reg pentagon *EFGHI* with diags $\overline{FH}$ and $\overline{GI}$ intersect at *J*	1. Given
2. $\overline{EF} \cong \overline{FG} \cong \overline{GH}$	2. Def of reg polygon
3. m ∠E = m ∠EFG = m ∠FGH = m ∠GHI	3. Def of reg polygon
4. m ∠E = m ∠EFG = m ∠FGH = m ∠GHI = 108.	4. Each int ∠ of reg polygon = $\frac{(n-2)180}{n}$
5. m ∠FGH + m ∠GHF + m ∠HFG = 180.	5. Sum of meas ∠s in △ = 180.
6. ∠GHF ≅ ∠HFG	6. ∠s opp ≅ sides of △ are ≅.
7. 108 + m ∠GHF + m ∠HFG = 180	7. Sub
8. m ∠GHF + m ∠HFG = 72	8. Subt Prop of Eq
9. 2 · ∠GHF = 72	9. Sub
10. m ∠GHF = m ∠HFG = 36	10. Div Prop of Eq
11. m ∠EFG = m ∠EFH + m ∠HFG	11. Angle Add Post
12. 108 = m ∠EFH + 36	12. Sub
13. m ∠EFH = 72	13. Subtr Prop of Eq
14. m ∠E + m ∠EFH = 108 + 72 = 180	14. Eq may be added.
15. $\overline{IE} \parallel \overline{JF}$	15. If 2 int ∠s on same side of transv are supp, then lines are ‖.
16. By a similar process $\overline{IJ} \parallel \overline{EF}$; ∴ *EFJI* is a ▱.	16. Def of ▱

31.

Statement	Reason
1. Sq *KLMN*, $\overline{NO} \cong$ $\overline{KP} \cong \overline{LQ} \cong \overline{MR}$	1. Given
2. $\overline{NK} \cong \overline{KL} \cong \overline{LM} \cong$ $\overline{MN}$	2. Def of sq
3. $NK - NO = OK$, $KL - KP = PL$, $LM - LQ = QM$, $MN - MR = RN$	3. Seq Add Post
4. $NK - NO = KL -$ $KP = LM - LQ =$ $MN - MP$	4. Eq may be subt.
5. $OK = PL = QM =$ RN	5. Sub
6. *KLMN* is a rect.	6. Def of sq
7. $\angle K, \angle L, \angle M$, and $\angle N$ are rt $\angle$s.	7. Def of rect.
8. $\angle K \cong \angle L \cong \angle M \cong$ $\angle N$	8. Rt $\angle$s are $\cong$.
9. $\triangle OKP \cong \triangle PLQ \cong$ $\triangle QMR \cong \triangle RNO$	9. SAS
10. $\overline{OP} \cong \overline{PQ} \cong \overline{QR} \cong$ $\overline{RO}$	10. CPCTC
11. Quad *OPQR* is a $\square$.	11. If opp sides of a quad are $\cong$, the quad is a $\square$.
12. m $\angle OPL$ = m $\angle KOP$ + m $\angle K$	12. Ext $\angle$ Thm
13. m $\angle OPL$ = m $\angle OPQ$ + m $\angle QPL$	13. $\angle$ Add Post
14. m $\angle K$ = 90	14. Def of rt $\angle$.
15. m $\angle KOP$ + 90 = m $\angle OPQ$ + m $\angle QPL$	15. Sub
16. $\angle KOP \cong \angle QPL$	16. CPCTC
17. 90 = m $\angle OPQ$	17. Eq may be subt.
18. $\angle ORQ \cong \angle OPQ$	18. Opp $\angle$s of $\square$ are $\cong$.
19. m $\angle ORQ$ = 90	19. Sub
20. $\angle OPQ$ and $\angle PQR$ are supp, $\angle ROP$ and $\angle OPQ$ are supp.	20. Consec. $\angle$s of $\square$ are supp.
21. m $\angle OPQ$ + m $\angle PQR$ = 180, m $\angle ROP$ + m $\angle OPQ$ = 180	21. Def of supp $\angle$s
22. m $\angle PQR$ = 90, m $\angle ROP$ = 90	22. Eq may be subt.
23. $\angle OPQ, \angle PQR,$ $\angle QRP,$ and $\angle ROP$ are rt $\angle$s.	23. Def of rt $\angle$
24. Quad *OPQR* is a rectangle.	24. Def of rectangle
25. Rect *OPQR* is a square	25. Def of square

Computer Investigation

Necessary Conditions

Use a computer software program that draws random trapezoids, labels points, constructs segments between two given points, and measures line segments.

The quadrilateral *ABCD* at the right has only one pair of parallel sides: $\overline{AB} \parallel \overline{CD}$. Such a quadrilateral is defined formally in the next lesson as a **trapezoid**. The computer activity below will help you discover some of the properties of trapezoids.

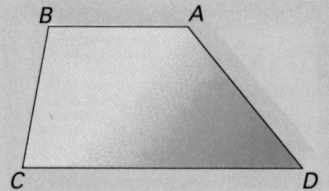

Activity 1

Draw a random trapezoid. The nonparallel sides $\overline{BC}$ and $\overline{AD}$ do not have to be congruent.

1. Find *CD* and *BA*.
2. Do you think that the parallel sides of a trapezoid could ever be congruent? Draw three more random trapezoids, measuring the lengths of the parallel sides of each. Are the parallel sides of a trapezoid ever congruent?   No
3. Recall that if one pair of sides of a quadrilateral are both *parallel* and *congruent*, the quadrilateral is a **parallelogram** and both pairs of sides are parallel. How does this explain the conclusion of Exercise 2? $\parallel$ sides of trap $\neq$.

Activity 2

Draw a random trapezoid *ABCD*. Draw the diagonals $\overline{AC}$ and $\overline{BD}$.

4. Are the diagonals congruent?   No
5. Do the diagonals bisect each other?   No
6. Repeat Exercises 4–5 for two more random trapezoids.
7. Can the diagonals of a trapezoid ever bisect each other? HINT: Recall that if the diagonals of a quadrilateral bisect each other, then the quadrilateral must be a _____.   parallelogram, no
8. Can you draw a trapezoid with congruent diagonals?   Yes, isos trap

Summary

Can a trapezoid have two pairs of congruent sides?   No
Can the diagonals of a trapezoid bisect each other?   No
Can the diagonals of a trapezoid ever be congruent? When?   Yes; isos trap

284   Computer Investigation

7.4 Trapezoids

Objectives To identify bases, legs, medians, and altitudes of trapezoids
To apply theorems about medians and altitudes of trapezoids

A parallelogram has both pairs of opposite sides parallel. Another special quadrilateral, the *trapezoid*, has only one set of parallel sides.

The photo shows a table in the shape of a trapezoid. What are some advantages of this shape? How might a number of such tables be arranged for a conference?

Definition A **trapezoid** is a quadrilateral with exactly one pair of parallel sides.

The sides of a trapezoid are given specific names. The *bases* are the two parallel sides. The *legs* are the two nonparallel sides.

EXAMPLE 1 Which sides of each trapezoid are the bases? Which sides are the legs?

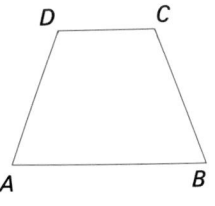

$\overline{AB} \parallel \overline{DC}$

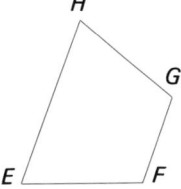

$\overline{FG} \parallel \overline{EH}$

Solution $\overline{AB}$ and $\overline{DC}$ are the bases.
$\overline{AD}$ and $\overline{BC}$ are the legs.

$\overline{FG}$ and $\overline{EH}$ are the bases.
$\overline{EF}$ and $\overline{HG}$ are the legs.

EXAMPLE 2 Determine whether the diagonals of a trapezoid bisect each other. Use a drawing to illustrate your answer.

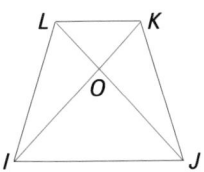

Solution Since $IO \neq KO$ and $JO \neq LO$, the diagonals $\overline{IK}$ and $\overline{JL}$ do not bisect each other.

7.4 Trapezoids **285**

Teaching Resources

Manipulative Worksheet 13
Quick Quizzes 53
**Reteaching and Practice
  Worksheets,** pp. 105, 106
Transparencies 17A, 17B

▆▆▆GETTING STARTED

Prerequisite Quiz

Refer to △*ABC*. (Ex. 1-3)

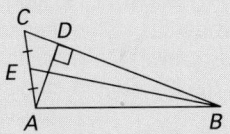

1. Identify an altitude. $\overline{AD}$
2. Identify the base to which the altitude is drawn. $\overline{BC}$
3. Identify a median. $\overline{BE}$
4. Given ▱*FGHI*, which sides are parallel?
   $\overline{FG}$, $\overline{HI}$; $\overline{GH}$, $\overline{FI}$

Motivator

Ask the students if a quadrilateral must have parallel sides. No Ask them what a quadrilateral with two pairs of parallel sides is called. Parallelogram Have them tell you what a quadrilateral with exactly one pair of parallel sides is called. Then have them find surfaces in the room with that shape. Trapezoid

Additional Example 1

Which sides of each trapezoid are the bases? the legs? ($\overline{QT} \parallel \overline{RS}$; $\overline{WX} \parallel \overline{ZY}$)

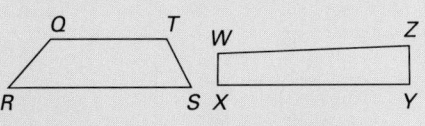

Bases: $\overline{QT}$, $\overline{RS}$ Bases: $\overline{WX}$, $\overline{ZY}$
Legs: $\overline{QR}$, $\overline{TS}$ Legs: $\overline{WZ}$, $\overline{XY}$

Additional Example 2

Make drawings to determine whether or not the diagonals of a trapezoid must be congruent. No

285

Lesson Note

A trapezoid has sometimes been defined as a quadrilateral with at least two parallel sides. This makes a parallelogram a special trapezoid, so that any theorem proved for trapezoids is true for parallelograms. However, in this text, a trapezoid is defined as a quadrilateral with exactly two parallel sides. This definition makes it easier for students to classify quadrilaterals. Before the proof of Theorem 7.12, students can discover properties of the median of a trapezoid by drawing trapezoids, then measuring the bases and median.

Math Connections

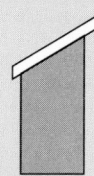

Structure: Many common objects with straight edges have only two parallel sides. Consequently, trapezoids can be found in many objects that you see every day, such as the side of a speaker's podium, as shown above. Ask students to think of other common objects (or parts of them) that have a trapezoidal shape. The side view of the bristles of a used toothbrush, the front view of some rooftops, the front view of a dam.

Critical Thinking Questions

Analysis: A trapezoid has a pair of parallel sides. Is this a necessary condition, a sufficient condition, or neither? It is necessary, but not sufficient that a trapezoid have a pair of parallel sides.

Definition

An **altitude** of a trapezoid is a perpendicular segment from any point on one base to the line containing the other base.

A triangle has exactly three altitudes. A trapezoid, however, has infinitely many altitudes.

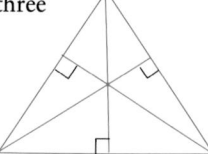 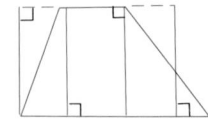

Theorem 7.12

All altitudes of a trapezoid are congruent.

Given: Trap $ABCD$ with $\overline{DC} \parallel \overline{AB}$ and altitudes $\overline{EH}$ and $\overline{FG}$
Prove: $\overline{EH} \cong \overline{FG}$

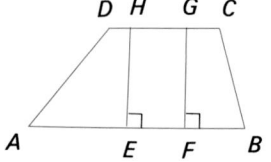

Proof

Statement	Reason
1. Trap $ABCD$, alt $\overline{EH}$ and $\overline{FG}$	1. Given
2. $\overline{AB} \parallel \overline{DC}$	2. Given
3. $\overline{EH} \perp \overline{AB}, \overline{FG} \perp \overline{AB}$	3. Def of alt of trap
4. $\therefore \overline{EH} \cong \overline{FG}$ ($EH = FG$)	4. If 2 lines are $\parallel$, all pts of each are equidist from other line.

Another special segment in a trapezoid is the median.

Definition

The **median** of a trapezoid is the segment that joins the midpoints of the legs.

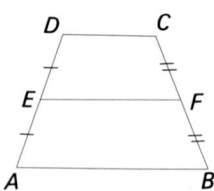
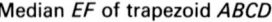

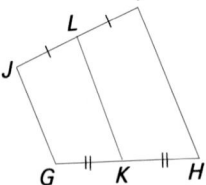

Median $\overline{EF}$ of trapezoid $ABCD$ Median $\overline{LK}$ of trapezpoid $GHIJ$

| **Theorem 7.13** | The median of a trapezoid is parallel to its bases. Its length is half the sum of the lengths of the two bases. |

Given: Trap $ABCD$ with median $\overline{XY}$, $\overline{AB} \parallel \overline{DC}$

Prove: $\overline{DC} \parallel \overline{XY} \parallel \overline{AB}$,
$XY = \frac{1}{2}(DC + AB)$

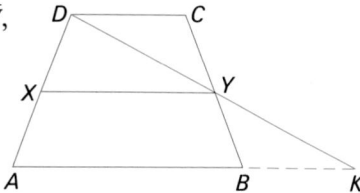

Proof

Statement	Reason
1. Trap $ABCD$ with median $\overline{XY}$, $\overline{AB} \parallel \overline{DC}$	1. Given
2. Draw $\overleftrightarrow{DY}$ intersecting $\overleftrightarrow{AB}$ at K.	2. Two pts determine a line.
3. $\angle DYC \cong \angle KYB$	**(A)** 3. Vert $\angle$s are $\cong$.
4. $CY = BY$	**(S)** 4. Def of median
5. $\angle DCB \cong \angle KBC$	**(A)** 5. Alt int $\angle$s of $\parallel$ lines are $\cong$.
6. $\triangle DCY \cong \triangle KBY$	6. ASA
7. $DC = KB$	7. CPCTC
8. $AB + KB = AK$	8. Seg Add Post
9. $AB + DC = AK$	9. Sub
10. $DY = KY$	10. CPCTC
11. $DX = AX$	11. Def of median
12. $\overline{XY} \parallel \overline{AB}$, $XY = \frac{1}{2}AK$	12. $\triangle$ Midseg Thm
13. $\therefore \overline{XY} \parallel \overline{DC} \parallel \overline{AB}$	13. 2 lines $\parallel$ to same line are $\parallel$.
14. $\therefore XY = \frac{1}{2}(AB + DC)$	14. Sub

EXAMPLE 3 Given: Trapezoid $ABCD$, median $\overline{EF}$, $AB = 20$, $EF = 15$
Find DC.

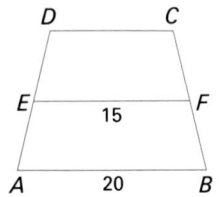

Solution

$15 = \frac{1}{2}(20 + DC)$
$30 = 20 + DC$
$DC = 10$

Both sides of the equation were multiplied by 2 to simplify computations.

Additional Example 3

Given: Trapezoid $JKLM$, median $\overline{NO}$,
$JK = 13$, $NO = 19$
Find ML.   25

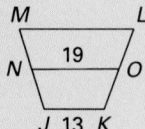

Common Error Analysis

Error: Students often misinterpret the definition of median of a trapezoid to mean that all segments formed by the medians are congruent to each other.

Have students draw a trapezoid such as $ABCD$ shown below, find the midpoints K and P of the legs, and draw median $\overline{PK}$. Then have them use a ruler to measure $\overline{CK}$, $\overline{KB}$, $\overline{DP}$, and $\overline{PA}$ and compare the measures.

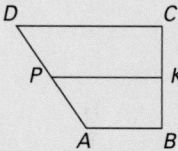

Checkpoint

1. Make a drawing to determine if it is possible for a trapezoid to have two right angles.   Yes

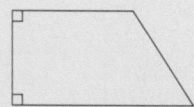

2. Find the length of the median of the trapezoid.   21

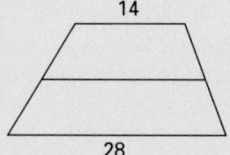

3. $ABCD$ is a trapezoid with median $\overline{EF}$. What is the length of the median if the lengths of bases $\overline{AB}$ and $\overline{DC}$ are 21 and 7?   14

Closure

Have the students explain the following parts of a trapezoid.

1. The bases.   The two parallel sides.
2. The legs.   The two non-parallel sides.
3. The median.   The segment that joins the midpoints of the legs.
4. An altitude.   Definition page 286.

Ask the students to explain the relationship of all altitudes of a trapezoid.   They are congruent.   Have them explain how the median and bases of a trapezoid are related.   Theorem 7.13 page 287

Guided Practice

Classroom Exercises 1–6

Independent Practice

A Ex. 1–15, **B** Ex. 16–21, **C** Ex. 22–25

Basic: WE 1–15 odd
Average: WE 1–21 odd
Above Average: WE 1–25 odd

Additional Answers

Written Exercises

8. Possible   **9.** Possible

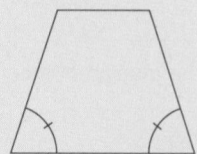

10. Not possible; figure would be rhomb with 2 pairs of ∥ sides.

11. Possible

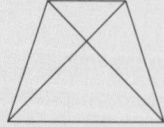

12. Not possible. Two of the ∠s would be int ∠s on the same side of a transversal with a sum of 180, so each would have meas = 90. The 3rd ≅ ∠ would then have meas 90, so the 4th would have meas 90 also. Hence, the figure would be rect.

13. Not possible; figure would be rect.

14. Possible   **15.** Possible

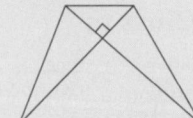

EXAMPLE 4 Given: Trap $GHIJ$, median $\overline{KL}$,
   $JI = 4x - 4$, $KL = 3x + 6$,
   $GH = 5x - 5$
Find KL.

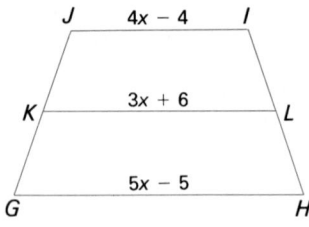

Solution
$$3x + 6 = \tfrac{1}{2}[(4x - 4) + (5x - 5)]$$
$$6x + 12 = (4x - 4) + (5x - 5)$$
$$21 = 3x$$
$$7 = x$$
$$KL = 3 \cdot 7 + 6 = 27$$

Classroom Exercises

Use the diagram to name the parts of the figure.

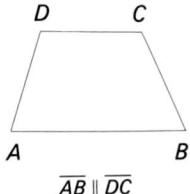
$\overline{AB} \parallel \overline{DC}$

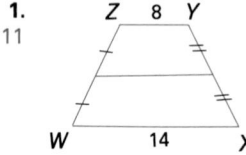
$\overline{MN} \parallel \overline{PO}$

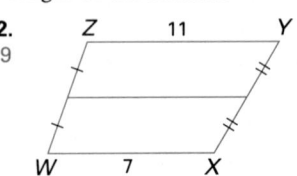
$\overline{AD} \parallel \overline{BC}$

1. The bases are $\underline{\overline{AB}, \overline{DC}}$.   **3.** $\underline{\overline{TR}}$ is an altitude.   **5.** $\underline{\overline{EG}}$ is an altitude.
2. The legs are $\underline{\overline{AD}, \overline{BC}}$.   **4.** $\underline{\overline{QS}}$ is a median.   **6.** $\underline{\overline{FH}}$ is a median.

Written Exercises

$WXYZ$ is a trapezoid with $\overline{WX} \parallel \overline{ZY}$. Find the length of the median.

1.
11
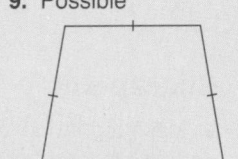

2.
9
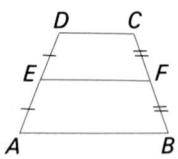

$ABCD$ is a trapezoid, with median $\overline{EF}$. Find the indicated length.

3. $AB = 15$, $DC = 5$, $EF = \underline{10}$
4. $AB = 28$, $DC = 22$, $EF = \underline{25}$
5. $AB = 24$, $DC = \underline{16}$, $EF = 20$
6. $AB = 31$, $DC = \underline{21}$, $EF = 26$
7. $AB = \underline{19}$, $DC = 3$, $EF = 11$

Additional Example 4

Given: Trapezoid $PQRS$, median $\overline{TU}$,
   $PQ = 4x - 13$, $TU = 2x + 6$, $SR = 3x + 4$
Find PQ. $x = 7$; $PQ = 15$

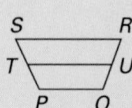

Determine whether the trapezoid described is possible. If it is possible, use a drawing to illustrate your answer.

8. a trapezoid with two congruent sides
9. a trapezoid with three congruent sides
10. a trapezoid with four congruent sides
11. a trapezoid with two congruent angles
12. a trapezoid with three congruent angles
13. a trapezoid with four congruent angles
14. a trapezoid with congruent diagonals
15. a trapezoid with diagonals that are perpendicular

GHIJ **is a trapezoid, with median** $\overline{KL}$. **Find the indicated length.**

16. $GH = 3x - 5$, $JI = x + 3$, $KL = x + 7$
Find KL.　15
17. $GH = 4x + 3$, $JI = 2x - 1$, $KL = 3x + 1$
Find GH.　Depends on x.
18. $GH = 4x - 3$, $JI = x + 5$, $KL = 3x$
Find JI.　7

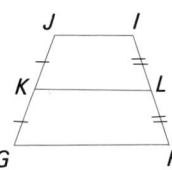

Construct a counterexample for each statement.

19. Consecutive angles of a trapezoid are congruent.
20. Consecutive angles of a trapezoid are supplementary.
21. The legs of a trapezoid are congruent.

Prove the conclusion, or disprove it by giving a counterexample.

Given: Trap *ABCD*, with median $\overline{GF}$; $\overline{AC}$ intersects $\overline{GF}$ at H; $\overline{DE} \perp \overline{AC}$ at H.

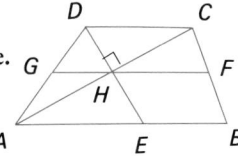

22. Conclusion: *EBCD* is a trapezoid.
24. Conclusion: $\triangle AHG$ is isosceles.
23. Conclusion: $\triangle AED$ is isosceles.
25. Conclusion: $\triangle HFC$ is isosceles.

Mixed Review

1. Prove: The base angles of an isosceles triangle are congruent.　5.1
2. Prove: Opposite angles of a parallelogram are congruent.　6.4

Determine whether the stated condition is necessary or not, and whether it is sufficient or not for a quadrilateral to be a parallelogram.　7.2, 7.3

3. Two opposite angles are congruent.　Necessary, not sufficient
4. Four angles are right angles.　Sufficient, not necessary
5. Both pairs of opposite sides are congruent.　Sufficient and necessary

19. False. $\angle A \not\equiv \angle B$.

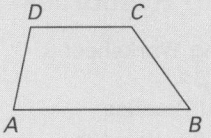

20. False. $\angle E$ and $\angle F$ are not supp.

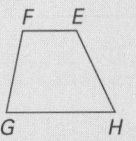

21. $\overline{AD} \not\equiv \overline{BC}$.

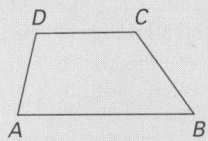

22. Concl cannot be proved or disproved. *EBCD* is either trap or ▱.

23.

Statement	Reason
1. Trap *ABCD* with median $\overline{GF}$, $\overline{AC}$ intersects $\overline{GF}$ at H, $\overline{DE} \perp \overline{AC}$ at H	1. Given
2. G is midpt $\overline{AD}$.	2. Def of median
3. $\overline{AG} \cong \overline{GD}$	3. Def of midpt.
4. $\overline{DC} \parallel \overline{GF} \parallel \overline{AB}$	4. Median of trap is $\parallel$ to bases.
5. $\overline{HD} \cong \overline{HE}$	5. If 3 $\parallel$ lines cut $\cong$ segs on 1 transv, they cut $\cong$ segs on every transv.
6. $\angle DHA$ and $\angle EHA$ are rt $\angle$s.	6. Def of $\perp$
7. $\angle DHA \cong \angle EHA$	7. Rt $\angle$s are $\cong$.
8. $\overline{AH} \cong \overline{AH}$	8. Reflex Prop of Congr
9. $\triangle AHD \cong \triangle AHE$	9. SAS
10. $\overline{AE} \cong \overline{AD}$	10. CPCTC
11. $\triangle AED$ is isos.	11. Def of isos $\triangle$

See page 659 for the answers to Written Ex. 24–25 and Mixed Review Ex. 1–2.

Enrichment

The drawing shows nine points, labeled *A* through *I*, in a square array. The nine points determine three horizontal and three vertical lines.

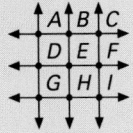

Using any combination of four points as the vertices of a trapezoid, name all the trapezoids in which the parallel sides are neither horizontal nor vertical lines.
AIFB, AIHD, GCBD, GCFH

Prerequisite Quiz

1. How does an isosceles triangle differ
   from a non-isosceles triangle?   Two
   sides are ≅.
2. In △ABC, AC = 8, BC = 8, m ∠A = 64
   Find m ∠B.   64
3. In trap ABCD, $\overline{AB}$ and $\overline{DC}$ are the bases,
   m ∠A = 72. Find ∠D.   108

Motivator

Ask the students, "What is an isosceles
triangle?"   A triangle with two congruent
sides.   Then ask them what they think an
isosceles trapezoid would be.   A trapezoid
with two congruent legs   Have them explain
what they know about the base angles of an
isosceles triangle.   The angles are
congruent.   Ask them if they think this
would be true for an isosceles trapezoid.
Yes

7.5. Isosceles Trapezoids

Objectives   To identify isosceles trapezoids
To prove and apply theorems about isosceles trapezoids
To identify necessary and sufficient conditions for isosceles trapezoids

Definition   An **isosceles trapezoid** is a trapezoid with congruent legs.

It was proved earlier that the base angles of an isosceles triangle are
congruent. A similar theorem for isosceles trapezoids can now be
proved.

Theorem 7.14   The base angles of an isosceles trapezoid are congruent.

Given: Trap ABCD, $\overline{AB} \parallel \overline{DC}$, $\overline{AD} \cong \overline{BC}$
Prove: ∠A ≅ ∠B and ∠ADC ≅ ∠BCD

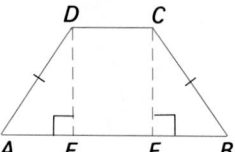

Plan   Draw perpendiculars $\overline{DE}$ and $\overline{CF}$ to form congruent right triangles
AED and BFC. ∠A and ∠B are corresponding parts of these triangles.

Proof

Statement	Reason
1. Trap ABCD, $\overline{AB} \parallel \overline{DC}$	1. Given
2. Draw $\overline{DE} \perp \overline{AB}$, $\overline{CF} \perp \overline{AB}$.	2. There is exactly one ⊥ from a pt to a line.
3. ∠DEA and ∠CFB are rt ∠s.	3. Def of ⊥
4. △AED and △BFC are rt △s.	4. Def of rt △
5. $\overline{AD} \cong \overline{BC}$	(H) 5. Given
6. $\overline{DE} \cong \overline{CF}$ (DE = CF)	(L) 6. ∥ lines are equidistant.
7. △AED ≅ △BFC	7. HL
8. ∠A ≅ ∠B	8. CPCTC
9. ∠A and ∠ADC are supp; ∠B and ∠BCD are supp.	9. ∥ lines form supp int ∠s on same side of transv.
10. ∴ ∠ADC ≅ ∠BCD	10. Supp of ≅ ∠s are ≅.

The next theorem is the converse of Theorem 7.14.

Theorem 7.15

If the base angles of a trapezoid are congruent, then the trapezoid is isosceles.

EXAMPLE 1   Find m ∠W and m ∠Y in isosceles trapezoid WXYZ.

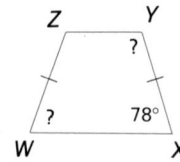

Plan   m ∠W = m ∠X by Theorem 7.14. ∠X and ∠Y are supplementary since they are interior angles on the same side of transversal $\overline{XY}$, and $\overline{WX} \parallel \overline{ZY}$.

Solution
m ∠W = m ∠X	m ∠X + m ∠Y = 180
m ∠W = 78	m ∠Y = 180 − m ∠X
	m ∠Y = 180 − 78 = 102

EXAMPLE 2   Given: Trap MNOP, $\overline{MN} \parallel \overline{PO}$, ∠M ≅ ∠N,
MP = 5x − 12, NO = x + 8
Find NO.

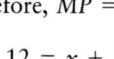

Plan   MNOP is an isosceles trapezoid by Theorem 7.15. Therefore, MP = NO.

Solution
$$5x − 12 = x + 8$$
$$4x = 20$$
$$x = 5$$
$$NO = 5 + 8, \text{ or } 13$$

Theorem 7.16

The diagonals of an isosceles trapezoid are congruent.

Given: Isosceles trap ABCD, $\overline{AD} \cong \overline{BC}$, $\overline{AB} \parallel \overline{DC}$
Prove: $\overline{AC} \cong \overline{BD}$

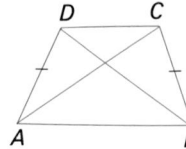

Some quadrilaterals with congruent diagonals are isosceles trapezoids, while others are not. However, every trapezoid with congruent diagonals is isosceles, as stated in Theorem 7.17.

7.5 Isosceles Trapezoids   **291**

Additional Example 1

Find m ∠F and m ∠H in isosceles trapezoid EFGH.   m ∠F = 119;
m ∠H = 61

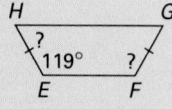

Additional Example 2

Given: Trap IJKL $\overline{LK} \parallel \overline{IJ}$, ∠I ≅ ∠J,
IL = 3x + 5, JK = 6x − 7
Find IL.   x = 4; IL = 17

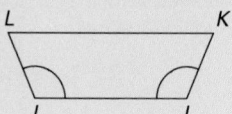

291

Critical Thinking Questions

Ask students how to prove Theorem 7.15 on page 291.

Trap *ABCD* with base ∠s *A* and *B* ≅ (Given). Draw $\overline{DE} \perp \overline{AB}$ and $\overline{CF} \perp \overline{AB}$. By def of ⊥, m ∠*DEA* = 90 and m ∠*CFB* = 90. m ∠*DEA* = m ∠*CFB* by sub. $\overline{DE}$ and $\overline{CF}$ are ∥ by def of trap. *DE* = *CF* because ∥ lines are equidist. Thus, △*DEA* ≅ △*CFB* by AAS, and $\overline{DA} \cong \overline{CB}$ by CPCTC. Therefore, trap *ABCD* is isos by def of isos trap.

Checkpoint

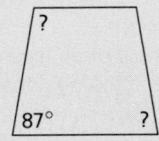

1. Find the indicated angle measures in the isosceles trapezoid.   87, 93
2. Given: Isos trap *MNOP*, $\overline{PO} \parallel \overline{MN}$, m ∠*M* = 5*x* + 3, m ∠*P* = 10*x* − 3. Find m ∠*M*.   *x* = 12; m ∠*M* = 63

Closure

Have the students explain the difference between an isosceles trapezoid and a non-isosceles trapezoid.   An isosceles trapezoid has two congruent legs and a non-isosceles trapezoid does not.   Have them tell you what they know about the base angles and diagonals of an isosceles trapezoid.   They are congruent.

▰▰▰ FOLLOW UP

Guided Practice

Classroom Exercises 1–9

Independent Practice

A Ex. 1–20, **B** Ex. 21–33, **C** Ex. 34–35

Basic: WE 1–20 odd

Average: WE 1–33 odd

Above Average: WE 1–35 odd

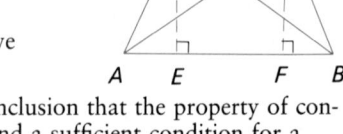

Theorem 7.17   If the diagonals of a trapezoid are congruent, then the trapezoid is isosceles.

Given: Trap *ABCD*, $\overline{DC} \parallel \overline{AB}$, $\overline{AC} \cong \overline{BD}$
Prove: $\overline{BC} \cong \overline{AD}$

Plan   Draw $\overline{DE} \perp \overline{AB}$ and $\overline{CF} \perp \overline{AB}$.
Prove △*AFC* ≅ △*BED* and then prove
△*BAC* ≅ △*ABD*.

Theorems 7.16 and 7.17 allow the conclusion that the property of congruent diagonals is *both* a necessary and a sufficient condition for a trapezoid to be isosceles. Every isosceles trapezoid has congruent diagonals, and every trapezoid with congruent diagonals is isosceles.

EXAMPLE 3   Is the property of one pair of congruent opposite sides a necessary condition for a quadrilateral to be an isosceles trapezoid? Is it a sufficient condition? Why?

Solution   It is a necessary condition. By definition, an isosceles trapezoid *must* have congruent legs.   It is *not* a sufficient condition. A quadrilateral can have a pair of congruent opposite sides and not be an isosceles trapezoid.

Classroom Exercises

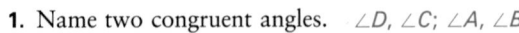

Refer to the trapezoids for Exercises 1–9.

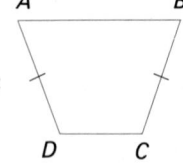

1. Name two congruent angles.   ∠*D*, ∠*C*; ∠*A*, ∠*B*
2. Name two pairs of supplementary angles.   ∠*A*, ∠*D*; ∠*B*, ∠*C*
3. Name two parallel sides. $\overline{DC} \parallel \overline{AB}$
4. Name two nonparallel sides. $\overline{AD}$ and $\overline{BC}$

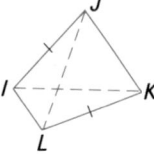

5. If trapezoid *IJKL* is isosceles, then ∠*IJK* ≅ _____∠*LKJ*_____.
6. If trapezoid *IJKL* is isosceles, then $\overline{JL} \cong$ _____$\overline{IK}$_____.

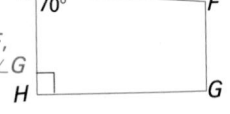

7. Name two supplementary angles.   ∠*H* and ∠*G*, ∠*E* and ∠*F*, ∠*H* and ∠*F*,
8. Name two nonsupplementary angles.   ∠*E* and ∠*G*
9. Why is trapezoid *EFGH* not isosceles?   Base ∠s not ≅

Additional Example 3

Is the property "two consecutive angles of a quadrilateral are congruent" a necessary condition for the quadrilateral to be an isosceles trapezoid? Is it a sufficient condition? Why?

Necessary; base ∠s of isos trap have been proved ≅.

Not Sufficient; quad can have 2 consec ∠s ≅ and not be an isos trap.

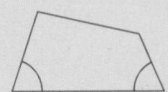

Written Exercises

Find the indicated angle measure in the isosceles trapezoid.

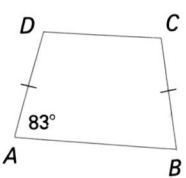
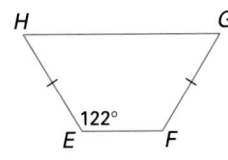
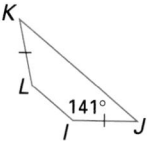

1. m ∠B = ___83___   **3.** m ∠H = ___58___   **5.** m ∠K = ___39___
2. m ∠C = ___97___   **4.** m ∠G = ___58___   **6.** m ∠J = ___39___

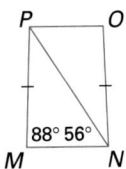
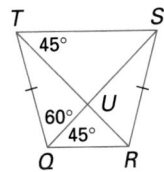
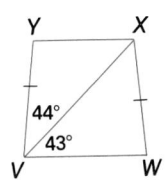

7. m ∠MPN = ___36___   **9.** m ∠QTU = ___30___   **11.** m ∠W = ___87___
8. m ∠PNO = ___32___   **10.** m ∠TSU = ___45___   **12.** m ∠VXW = ___50___

13. Given: $\overline{AC} \cong \overline{BC}$, $\overline{DE} \parallel \overline{AB}$
Prove: *ABED* is an isosceles trapezoid.
14. Given: Isosceles trap *ABED*, $\overline{DE} \parallel \overline{AB}$
Prove: △*ABC* is isosceles.

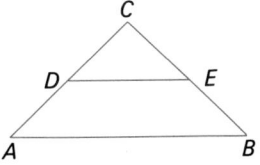

Use isosceles trapezoid *XYZW* with $\overline{WZ} \parallel \overline{XY}$
for Exercises 15–18.

15. Given: m ∠W = 9t + 20,
m ∠Z = 12t − 10
Find m ∠W. 110

16. Given: m ∠X = 12t,
m ∠Y = 11t + 4
Find m ∠X. 48

17. Given: m ∠W = 5m + 6,
m ∠X = 2m + 6
Find m ∠X. 54

18. Given: m ∠Z = 7n − 6,
m ∠Y = 4n − 12
Find m ∠Z. 120

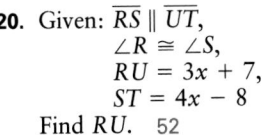

19. Given: $\overline{RS} \parallel \overline{UT}$,
∠R ≅ ∠S,
RU = 4x − 6,
ST = x + 15
Find *ST*. 22

20. Given: $\overline{RS} \parallel \overline{UT}$,
∠R ≅ ∠S,
RU = 3x + 7,
ST = 4x − 8
Find *RU*. 52

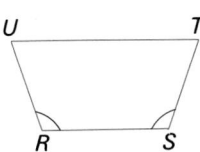

Enrichment

Discuss the symmetry of the isosceles trapezoid. Point out that, like the isosceles triangle, the isosceles trapezoid is used extensively in the construction of buildings, monuments, bridges, and other forms of architecture.

Have the students create a bulletin board display of pictures illustrating the practical as well as the artistic importance of the isosceles trapezoid.

Additional Answers

Written Exercises

13.

Statement	Reason
1. $\overline{AC} \cong \overline{BC}$, $\overline{DE} \parallel \overline{AB}$	1. Given
2. *ABED* is a trap.	2. Def of trap
3. ∠A ≅ ∠B	3. ∠s opp ≅ sides of △ are ≅.
4. *ABED* is isos trap.	4. If base ∠s of trap ≅, trap is isos.

14.

Statement	Reason
1. Isos trap *ABED*, $\overline{DE} \parallel \overline{AB}$	1. Given
2. ∠A ≅ ∠B	2. Base ∠s of isos trap are ≅.
3. $\overline{AC} \cong \overline{BC}$	3. Sides opp ≅ ∠s of △ are ≅.
4. △ABC is isos.	4. Def of isos △

25.
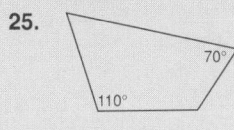

26.
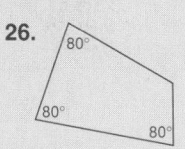

27.
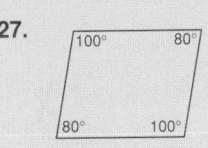

28.
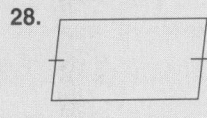

29.

Statement	Reason
1. Trap *ABCD*, ∠A ≅ ∠B	1. Given
2. Draw $\overline{DE} \perp \overline{AB}$, and draw $\overline{CF} \perp \overline{AB}$.	2. There is exactly 1 ⊥ from pt to line.
3. ∠AED and ∠BFC are rt ∠s	3. Def of ⊥
4. $\overline{AED} \cong \overline{BFC}$	4. Rt ∠s are ≅.
5. △AED ≅ △BFC	5. AAS
6. $\overline{AD} \cong \overline{BC}$	6. CPCTC
7. *ABCD* is isos trap.	7. Def of isos trap

30.

Statement	Reason
1. Isos trap *ABCD*, $\overline{AD} \cong \overline{BC}$, $\overline{AB} \parallel \overline{DC}$	1. Given
2. m ∠ABC = m ∠BAD	2. Base ∠s of isos trap are ≅.
3. $\overline{AB} \cong \overline{AB}$	3. Reflex Prop of Congr
4. △ABC ≅ △BAD	4. SAS
5. $\overline{AC} \cong \overline{BD}$	5. CPCTC

31.

Statement	Reason
1. Trap $ABCD$, $\overline{DC}$ ‖ $\overline{AB}$, $\overline{AC}$ ≅ $\overline{BD}$	1. Given
2. Draw $\overline{CF}$ ⊥ $\overline{AB}$ and $\overline{DE}$ ⊥ $\overline{AB}$	2. There is 1 ⊥ from pt to line.
3. $\angle AFC$ and $\angle BED$ are rt ∠s.	3. Def of rt ∠
4. $\triangle AFC$ and $\triangle BED$ are rt △s.	4. Def of rt △
5. $\overline{CF}$ ≅ $\overline{DE}$	5. Alts of trap are ≅.
6. $\triangle AFC$ ≅ $\triangle BED$	6. HL
7. $\angle BAC$ ≅ $\angle ABD$	7. CPCTC
8. $\overline{BA}$ ≅ $\overline{AB}$	8. Reflex Prop of Congr
9. $\triangle BAC$ ≅ $\triangle ABD$	9. SAS
10. $\overline{BC}$ ≅ $\overline{AD}$	10. CPCTC

32.

Statement	Reason
1. Isos trap $ABCD$, $\overline{DC}$ ‖ $\overline{AB}$	1. Given
2. $\overline{AD}$ ≅ $\overline{BC}$	2. Def of isos trap
3. $\angle DAB$ ≅ $\angle CBA$	3. Base ∠s of isos trap are ≅.
4. $\overline{AB}$ ≅ $\overline{AB}$	4. Reflex Prop of Congr
5. $\triangle ABD$ ≅ $\triangle BAC$	5. SAS

33.

Statement	Reason
1. Isos trap $ABCD$, $\overline{DC}$ ‖ $\overline{AB}$	1. Given
2. $\overline{DA}$ ≅ $\overline{CB}$	2. Def of isos trap
3. $\overline{AC}$ ≅ $\overline{BD}$	3. Diags of isos trap are ≅.
4. $\overline{DC}$ ≅ $\overline{DC}$	4. Reflex Prop of Congr
5. $\triangle ACD$ ≅ $\triangle BDC$	5. SSS
6. $\angle DAC$ ≅ $\angle CBD$	6. CPCTC
7. $\angle DEA$ ≅ $\angle CEB$	7. Vert ∠s are ≅.
8. $\triangle AED$ ≅ $\triangle BEC$	8. AAS

See page 298 for the answers to Ex. 34–35.

294

Is the property a necessary condition for a trapezoid to be an isosceles trapezoid? Is it a sufficient condition?

21. Diagonals are perpendicular to each other. Not necessary, not sufficient

22. One pair of opposite sides is congruent. Necessary and sufficient

23. Opposite angles are supplementary. Necessary and sufficient

24. Two pairs of consecutive angles are supplementary. Necessary, not sufficient

Is the property a necessary condition for a quadrilateral to be an isosceles trapezoid? Is it a sufficient condition? If not sufficient, give a counterexample.

25. Opposite angles are supplementary. Necessary, not sufficient

26. Two pairs of consecutive angles are supplementary. Necessary, not sufficient

27. Two pairs of consecutive angles are congruent. Necessary, not sufficient

28. One pair of opposite sides is congruent. Necessary, not sufficient

29. Prove Theorem 7.15.

30. Prove Theorem 7.16.

31. Prove Theorem 7.17.

32. Given: Isosceles trap $ABCD$, $\overline{DC}$ ‖ $\overline{AB}$
Prove: $\triangle ABD$ ≅ $\triangle BAC$

33. Given: Isosceles trap $ABCD$, $\overline{DC}$ ‖ $\overline{AB}$
Prove: $\triangle AED$ ≅ $\triangle BEC$

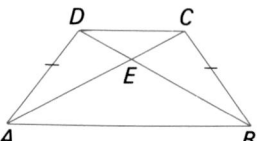

34. Given: $\overline{FB}$ ⊥ $\overline{AB}$, $\overline{FB}$ ⊥ $\overline{BC}$, $\overline{AB}$ ≅ $\overline{BC}$, M is the midpoint of $\overline{AF}$; N is the midpoint of $\overline{FC}$.
Prove: $ACNM$ is an isosceles trapezoid.

35. Given: M is the midpoint of $\overline{AF}$; N is the midpoint of $\overline{FC}$; $\triangle AFE$ ≅ $\triangle CFG$.
Prove: $ACNM$ is an isosceles trapezoid.

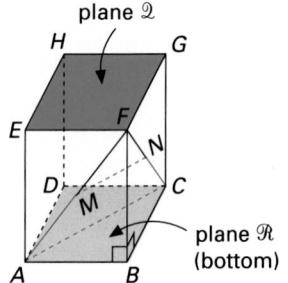

Mixed Review

Is the statement true or false?

1. Two quadrilaterals can be proved congruent by a side-side-side-side pattern. *7.1* False

2. A rhombus has congruent diagonals. *7.2* False

3. Draw an acute angle. Construct an angle congruent to it. *1.4* See Lesson 1.4

4. Draw an obtuse triangle. Construct a triangle congruent to it. *4.3* See Lesson 4.3. The construction for an obtuse triangle is similar.

7.6 Constructing Quadrilaterals

Objective To construct quadrilaterals with given properties

Special quadrilaterals can be constructed by using their unique properties. To do so efficiently, make use of the *sufficient conditions* developed in the preceding lessons.

EXAMPLE 1 Construct a rectangle with adjacent sides congruent to two given segments.

_____ a   _____ b

Plan A parallelogram with one right angle is a rectangle.
Construct one right angle. Then construct a quadrilateral with opposite sides congruent.

Solution Construct a perpendicular at *E*. Construct $\overline{EF}$ and $\overline{EH}$ so that $EF = a$ and $EH = b$.

Swing arcs of length *a* from *H* and length *b* from *F*. Label the point of intersection *G* and draw $\overline{HG}$ and $\overline{FG}$.

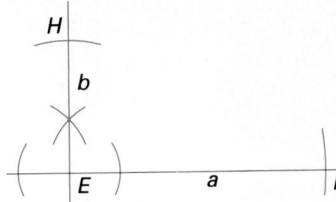
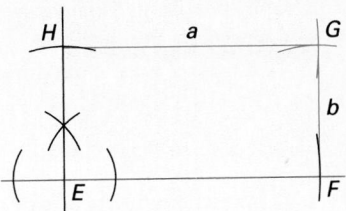

Because the opposite sides in the construction are congruent, the result is a parallelogram. Then, since ∠*E* is a right angle, *EFGH* is the required rectangle.

EXAMPLE 2 Construct a rhombus with diagonals congruent to two given segments.

_____ a   _____ b

Plan A parallelogram with diagonals that are perpendicular is a rhombus.
Construct diagonals that are perpendicular bisectors of each other.

7.6 Constructing Quadrilaterals   **295**

Teaching Resources

Quick Quizzes 55
Reteaching and Practice
Worksheets, pp. 109, 110

■ GETTING STARTED

Prerequisite Quiz

1. Draw a segment approximately two inches long. Construct a segment congruent to this segment. (see Lesson 1.3)
2. Draw an acute angle. Construct an angle congruent to this angle. (see Lesson 1.4)
3. Draw a line and a point on the line. Construct the perpendicular to the line at that point. (see Lesson 5.6)

Motivator

Have the students give sufficient conditions for a parallelogram to be the following quadrilaterals.

1. A rectangle.   It has one right angle and congruent diagonals.
2. A rhombus.   It has two adjacent congruent sides, perpendicular diagonals, and diagonals that bisect opposite angles.
3. A square.   It has one right angle and two adjacent congruent sides.

Ask them if they could construct the above quadrilaterals if they were given two segments and they know the sufficient conditions for each special parallelogram. Yes   Ask them how they could construct a rectangle given two segments with different lengths.   Construct one right angle with the given segments forming adjacent sides of the rectangle. Then construct a quadrilateral with congruent opposite sides.

Additional Example 1

Draw a segment. Construct a square with sides congruent to the given segment.
The construction is similar to that for Example 1.

TEACHING SUGGESTIONS

Lesson Note

Review basic ruler and compass constructions. Emphasize that to construct a rectangle, a rhombus, or a square, the construction must be based on sufficient conditions for a quadrilateral to be that figure. For example, to construct a rhombus, one sufficient condition is that the quadrilateral have four congruent sides. A construction can be based on this.

Math Connections

Architecture: An architect is sometimes asked to restructure parts of a building that is already built. Whether the change consists of adding another room or dividing one room into two, the architect must work with pre-existing conditions. The additional structure must have specific properties in order to be combined with the existing structure.

Critical Thinking Questions

Analysis: Ask students to identify the resulting quadrilateral and give reasons for their answers.

Construct congruent alternate interior angles. Construct the perpendicular bisector of the transversal. Extend the noncommon sides of the angles and the perpendicular bisector until they intersect on both sides of the transversal. Connect all four vertices to form a quadrilateral.   Rhom; ⊥ diags, one diag bis the other, one pair of sides is ∥.

Checkpoint

Draw a segment and an angle. Construct a rhombus with sides congruent to this segment and an angle congruent to this angle.   Given segment and angle:

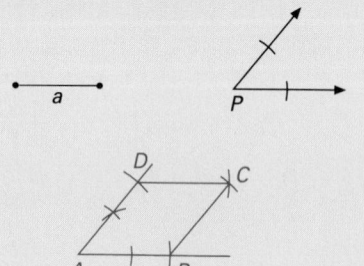

ABCD is the required rhombus.

296

Solution   Construct $\overline{AC}$ with $AC = a$. Construct the perpendicular bisector of $\overline{AC}$.

Construct a segment of length b. Bisect the segment to find the length of $\frac{1}{2}b$.

On the perpendicular bisector of $\overline{AC}$, mark off segments of length $\frac{1}{2}b$ to determine B and D. Connect points A, B, C, and D.

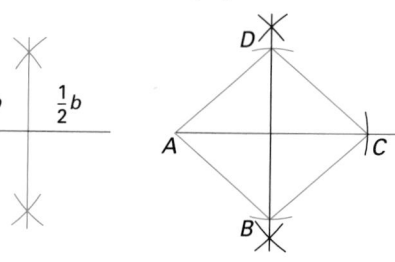

Classroom Exercises

Is the construction sufficient or not sufficient to result in a rhombus?

1. Construct a quadrilateral with two adjacent congruent sides.   Not sufficient
2. Construct a quadrilateral with four congruent sides.   Sufficient
3. Construct a quadrilateral with perpendicular diagonals.   Not sufficient
4. Construct a parallelogram with perpendicular diagonals.   Sufficient
5. Construct a parallelogram with two adjacent congruent sides.   Sufficient

Written Exercises

1. Draw two segments of two different lengths. Construct a rectangle with sides congruent to these segments.
2. Draw a segment. Construct a square with sides congruent to this segment.
3. Construct a parallelogram with adjacent sides of lengths a and b and an angle congruent to the given angle.
4. Construct a rhombus with sides of length c and an angle congruent to the given angle.

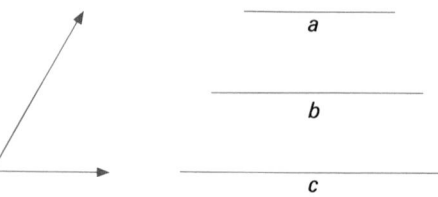

5. Draw two segments. Construct a rhombus with diagonals congruent to these two segments.

Additional Example 2

Draw a segment. Construct a square with diagonals congruent to the given segment.   The construction is similar to that for Example 2.

6. Draw a segment. Construct a parallelogram with one side congruent to this segment, another side half as long, and an angle of measure 60.

7. Construct a quadrilateral congruent to quad *ABCD* using an SASAS pattern.

8. Construct a quadrilateral congruent to quad *ABCD* using an ASASA pattern.

9. Is it possible to construct a quadrilateral congruent to quad *ABCD* using an SSSS pattern?   Not possible

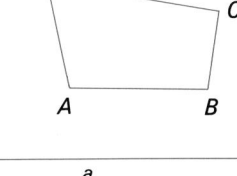

10. Construct a parallelogram with diagonals of lengths *a* and *b* and one side of length *c*.

a

b

c

11. Construct an isosceles trapezoid with three sides of length *s* and a diagonal of length *d*.

s   d

12. Construct a trapezoid with bases of lengths *a* and *b*, a leg of length *c*, and a diagonal of length *d*.

a   c

b   d

Mixed Review

Is the statement true or false?   *5.6, 7.4*

1. All altitudes of a triangle are congruent.   False
2. All altitudes of a trapezoid are congruent.   True

State whether the congruence pattern is sufficient to establish congruence of two triangles.   *4.3, 4.4, 4.6*

3. SAS
Sufficient

4. SSA
Not sufficient

5. AAA
Not sufficient

6. SSS
Sufficient

7. AAS
Sufficient

◢ Brainteaser

The figure to the right is a square with an equilateral triangle on top of it. Can you find the value of *x*?   15

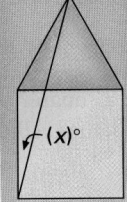

(x)°

7.6 Constructing Quadrilaterals   **297**

Chapter Review

1. $\overline{LM} \cong \overline{ZY}$, $\overline{MN} \cong \overline{YX}$, $\overline{NO} \cong \overline{XW}$, $\overline{OL} \cong \overline{WZ}$; $\angle L \cong \angle Z$, $\angle M \cong \angle Y$, $\angle N \cong \angle X$, $\angle O \cong \angle W$.

Additional Answers, page 294

34.

Statement	Reason
1. $\overline{FB} \perp \overline{AB}$, $\overline{FB} \perp \overline{BC}$, $\overline{AB} \cong \overline{BC}$, M is midpt of $\overline{AF}$, N is midpt of $\overline{FC}$.	1. Given
2. $\angle FBA$ and $\angle FBC$ are rt $\angle$s.	2. Def of $\perp$
3. $\angle FBA \cong \angle FBC$	3. Rt $\angle$s are $\cong$.
4. $\overline{FB} \cong \overline{FB}$	4. Reflex Prop of Congr
5. $\triangle ABF \cong \triangle CBF$	5. SAS
6. $\overline{AF} \cong \overline{FC}$	6. CPCTC
7. $\frac{1}{2}AF = \frac{1}{2}FC$	7. Mult Prop of Eq
8. $AM = \frac{1}{2}AF$, $NC = \frac{1}{2}FC$	8. Def of midpt
9. $AM = NC$	9. Sub
10. $\overline{MN} \parallel \overline{AC}$	10. Midseg Thm
11. $ACNM$ is isos trap.	11. Def of isos trap

35.

Statement	Reason
1. M is midpt of $\overline{AF}$, N is midpt of $\overline{FC}$, $\triangle AFE \cong \triangle CFG$	1. Given
2. $\overline{AF} \cong \overline{CF}$	2. CPCTC
3. $\frac{1}{2}AF = \frac{1}{2}CF$	3. Mult Prop of Eq
4. $AM = \frac{1}{2}AF$, $NC = \frac{1}{2}FC$	4. Def of midpt
5. $AM = NC$	5. Sub
6. $\overline{MN} \parallel \overline{AC}$	6. Midseg Thm
7. $ACNM$ is isos trap	7. Def of isos trap

Chapter 7 Review

Key Terms

altitude of a trapezoid (p. 286)
congruent quadrilaterals (p. 269)
isosceles trapezoid (p. 290)
median of a trapezoid (p. 286)
necessary condition (p. 273)
rectangle (p. 273)
rhombus (p. 273)
square (p. 273)
sufficient condition (p. 278)
trapezoid (p. 285)

Key Ideas and Review Exercises

7.1 To identify and name congruent quadrilaterals, determine corresponding sides and angles.

1. If quad $LMNO \cong$ quad $ZYXW$, name the pairs of congruent sides and the pairs of congruent angles.

2. Based on the marks shown in the figure, complete the congruence statement: Quad $JKLM \cong$ _____.
quad $QPON$

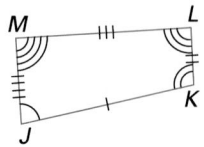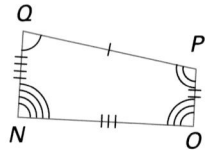

3. Identify two congruence patterns that can be used to prove that two given quadrilaterals are congruent. SASAS; ASASA

7.2 To prove and apply properties of rectangles, rhombuses, and squares, and to identify necessary conditions for them, use one or more of the following:
• A rectangle is a parallelogram with four right angles.
• A rhombus is a parallelogram with four congruent sides.
• A square is a rectangle with four congruent sides.
• The diagonals of a rhombus are perpendicular.
• The diagonals of a rectangle are congruent.
• Each diagonal of a rhombus bisects two opposite angles of the rhombus.

4. Is the property that diagonals are congruent a necessary condition for a parallelogram to be a square? Yes

5. In rhombus $ABCD$, $AB = 5x + 3$, $BC = 9x - 5$. Find AB. 13

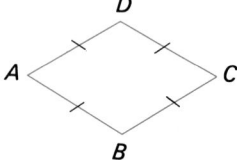

Chapter 7 Review

6. Given: Rect $XYZW$, M is the midpoint of $\overline{XY}$.
Prove: $WM = ZM$

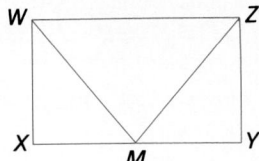

7.3 To prove that a figure is a rectangle, a rhombus, or a square, and to identify sufficient conditions for them, use one or more of the following:
- A parallelogram with perpendicular diagonals is a rhombus.
- A parallelogram with at least one right angle is a rectangle.
- A parallelogram with congruent diagonals is a rectangle.
- A parallelogram with a diagonal that bisects its opposite angles is a rhombus.
- A quadrilateral with four congruent sides is a rhombus.
- A parallelogram with two adjacent congruent sides is a rhombus.
- A parallelogram with two adjacent congruent sides and one right angle is a square.

7. Prove: A parallelogram with two consecutive congruent angles is a rectangle.

8. Prove: A rectangle with two adjacent congruent sides is a square.

7.4 A trapezoid is a quadrilateral with exactly one pair of parallel sides. The median of a trapezoid has length one-half the sum of the lengths of its bases. The median is parallel to the bases.

9. Can a trapezoid have exactly one right angle? Provide a drawing to support your answer.

10. The lengths of the two bases of a trapezoid are 9 and 17. What is the length of the median?   13

7.5 The base angles of an isosceles trapezoid are congruent.
If the base angles of a trapezoid are congruent, then the trapezoid is isosceles.
The diagonals of an isosceles trapezoid are congruent.
If the diagonals of a trapezoid are congruent, then the trapezoid is isosceles.

11. In quadrilateral $PQRS$, if $\overline{PQ} \parallel \overline{RS}$, and $\angle P \cong \angle Q$, is $PQRS$ an isosceles trapezoid? Why or why not?   Only if $\angle P$ and $\angle Q$ are not rt $\angle$s.

12. In isosceles trapezoid $LMNO$, $\overline{LO} \cong \overline{MN}$, m $\angle L = 5x + 10$, and m $\angle M = 6x$. Find m $\angle L$.   60

7.6 To construct a special quadrilateral (rhombus, rectangle, square, or trapezoid), use one or more of the unique properties of that quadrilateral.

13. Draw two segments. Construct a rectangle with sides congruent to the segments.

14. Draw two segments. Construct a rhombus with diagonals congruent to the segments.

Chapter 7 Review **299**

6.

Statement	Reason
1. Rect $XYZW$, M is midpt of $\overline{XY}$.	1. Given
2. $XYZW$ is a ▱.	2. Def of rect
3. $\overline{WX} \cong \overline{ZY}$	3. Opp sides of ▱ are $\cong$.
4. m $\angle X = 90$, m $\angle Y = 90$	4. Def of rect
5. m $\angle X =$ m $\angle Y$	5. Sub
6. $XM = YM$	6. Def of midpt
7. $\triangle WXM \cong \triangle ZYM$	7. SAS
8. $\overline{WM} \cong \overline{ZM}$	8. CPCTC

7.

Statement	Reason
1. ▱$ABCD$, $\angle A \cong \angle B$	1. Given
2. $\overline{AD} \parallel \overline{BC}$	2. Def of ▱
3. $\angle A$ and $\angle B$ are supp.	3. If 2 ∥ lines are intersected by transv, int $\angle$s on same side of transv are supp.
4. m $\angle A$ + m $\angle B = 180$.	4. Def of supp $\angle$s
5. $2 \times$ m $\angle A = 180$.	5. Sub
6. m $\angle A = 90$	6. Div Prop of Eq
7. ▱$ABCD$ is a rect.	7. ▱ with 1 rt $\angle$ is a rect.

8.

Statement	Reason
1. Rect $EFGH$, $\overline{HE} \cong \overline{EF}$	1. Given
2. $EFGH$ is ▱	2. Def of rect
3. $\overline{HE} \cong \overline{FG}$, $\overline{EF} \cong \overline{HG}$	3. Opp sides of ▱ are $\cong$.
4. $\overline{HE} \cong \overline{EF} \cong \overline{HG} \cong \overline{FG}$	4. Sub
5. $EFGH$ is a sq.	5. Def of sq

9. No

13. Follow steps of Example 1, p. 295.
14. Follow steps of Example 2, p. 295.

14. False.   **15. False.**

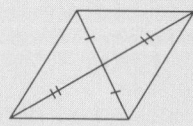 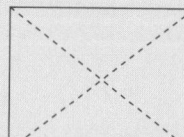

17.

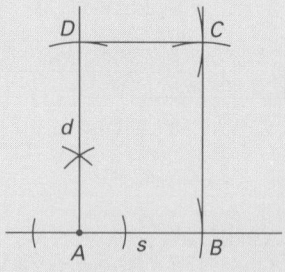

Construct $AB = s$. Construct line $\perp$ to $\overline{AB}$ at A. With A as ctr and d as rad, draw arc intersecting $\perp$ at D. With D as ctr and s as rad, and B as ctr and d as rad, swing arcs inters at C. ABCD is required rect.

*18.

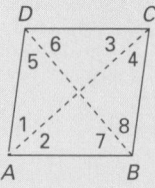

Statement	Reason
1. Rhom ABCD with diags $\overline{AC}$ and $\overline{BD}$	1. Given
2. $\overline{AD} \cong \overline{AB}$, $\overline{CD} \cong \overline{CB}$	2. Def of rhom
3. $\overline{AC} \cong \overline{AC}$	3. Reflex Prop of Congr
4. $\triangle ADC \cong \triangle ABC$	4. SSS
5. $\angle 1 \cong \angle 2$, $\angle 3 \cong \angle 4$	5. CPCTC
6. $\overline{AD} \cong \overline{CD}$, $\overline{AB} \cong \overline{CB}$	6. Def of rhom
7. $\overline{DB} \cong \overline{DB}$	7. Reflex Prop of Congr
8. $\triangle DAB \cong \triangle DCB$	8. SSS
9. $\angle 5 \cong \angle 6$, $\angle 7 \cong \angle 8$	9. CPCTC
10. $\overline{AC}$ and $\overline{DB}$ bis $\angle$s DAB, DCB and $\angle$s ADC, ABC respectively.	10. Def of $\angle$ bis

See page 301 for the answer to Ex. 19.

300

If quad $ABCD \cong$ quad $PQRS$, which angle or side is congruent to the given angle or side?

1. $\overline{BC} \cong$ $\underline{\overline{QR}}$ **2.** $\overline{SP} \cong$ $\underline{\overline{DA}}$ **3.** $\angle B \cong$ $\underline{\angle Q}$ **4.** $\angle S \cong$ $\underline{\angle D}$

5. Based on the markings shown, write a correct congruence statement. Which congruence pattern can be used to prove the two quadrilaterals congruent?
Quad $ABCD \cong$ quad $XWZY$; SASAS

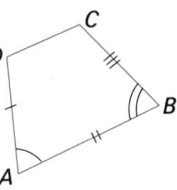 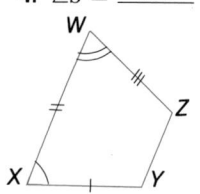

Based only on the markings shown, identify the parallelogram as a rectangle, rhombus, square, or none of these.

6.   **7.**   **8.**   **9**

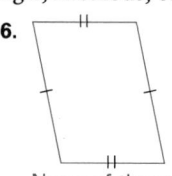 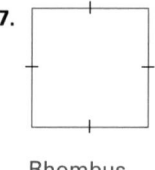 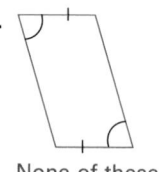 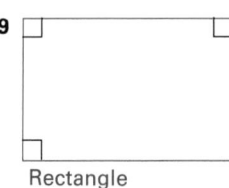

None of these   Rhombus   None of these   Rectangle

Is the stated condition *necessary* for a quadrilateral to be a rhombus?

10. a quadrilateral with perpendicular diagonals   Yes

11. a quadrilateral with four congruent sides   Yes

Is the stated condition *sufficient* for a quadrilateral to be a rectangle?

12. a quadrilateral with three right angles   Yes

13. a quadrilateral with congruent diagonals   No

True or false? Draw a picture to defend your answer.

14. A parallelogram with diagonals that bisect each other is a rhombus.

15. A quadrilateral with congruent diagonals is an isosceles trapezoid.

16. Given: Trap $ABCD$, median $\overline{EF}$, $AB = x + 2$, $DC = 2x + 1$, $EF = 2x - 3$. Find EF.   15

17. Construct a rectangle with sides of lengths d and s.

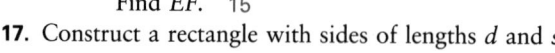

* **18.** Prove: The diagonals of a rhombus bisect the opposite angles.

* **19.** Prove: The midpoint of the hypotenuse of a right triangle is equidistant from the three vertices of the triangle.

College Prep Test

In each item you are to compare a quantity in Column 1 with a quantity in Column 2. Write the letter of the correct answer from these choices.

A—The quantity in Column 1 is greater than the quantity in Column 2.
B—The quantity in Column 2 is greater than the quantity in Column 1.
C—The quantity in Column 1 is equal to the quantity in Column 2.
D—The relationship cannot be determined from the given information.

	Column 1	Column 2
1. A	A prime factor of 77 is 11.	
	11	the other prime factor
2. C	$x = 5$	
	$5x + 5$	$7x - 5$
3. D	$\angle A$ and $\angle B$ are complementary. $\angle A$ is acute.	
	m $\angle A$	m $\angle B$
4. C	$\angle C$ and $\angle D$ are vertical angles. $\angle C$ is obtuse.	
	m $\angle C$	m $\angle D$
5. A	m $\angle G <$ m $\angle F <$ m $\angle E$	
	m $\angle E$	m $\angle G$
6. B	$\angle A$ and $\angle B$ are the base angles of isosceles trapezoid $ABCD$; m $\angle A = 75$.	
	m $\angle B$	m $\angle C$
7. A	$\angle C$ is an obtuse angle in $\triangle ABC$.	
	AB	BC

	Column 1	Column 2
8. A	In $\triangle XYZ$, $XY = 9$, $YZ = 10$, and $ZX = 11$	
	m $\angle X$	m $\angle Z$
9. A	$ABCDEF$ is a regular hexagon. $\angle 1$ is an angle of $ABCDEF$. $\angle 2$ is an exterior angle of $ABCDEF$.	
	m $\angle 1$	m $\angle 2$
10. A	$\angle A$ is the vertex angle of isosceles $\triangle ABC$. m $\angle A = 74$	
	m $\angle A$	m $\angle B$
11. D	$\angle A$ is a base angle of isosceles $\triangle ABC$. m $\angle A = 74$.	
	AB	BC
12. C or D	$ABCDE$ is an equilateral pentagon.	
	m $\angle C$	m $\angle E$
13. C	$x = 0$	
	$3x$	$5x$
14. B	Quad $ABCD \cong$ quad $EFGH$. $AB = 12$ and $FG = 5$.	
	BC	EF

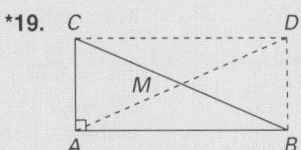

---

12. In the case of a regular pentagon, all sides and all angles are $\cong$. Thus, the answer is C. In the case of an equilateral pentagon that is not equiangular, the answer is D.

Additional Answer, page 300

*19.

Statement	Reason
1. Rt $\triangle ABC$, M is midpt of $\overline{BD}$	1. Given
2. Draw $\overline{AD}$ so that $AM = MD$; Draw $\overline{CD}$ and $\overline{BD}$.	2. 2 pts determine line.
3. $MC = MB$	3. Def of midpt
4. $ABCD$ is a $\square$.	4. If diags of quad bis each other, quad is $\square$.
5. $\angle A$ is rt $\angle$	5. Def of rt $\triangle$
6. $ABCD$ is a rect.	6. $\square$ with 1 rt $\angle$ is a rect.
7. $AD = CB$	7. Diags of rect are $\cong$.
8. $\frac{1}{2}AD = \frac{1}{2}CB$	8. Mult Prop of Eq
9. $AM = \frac{1}{2}AD$, $CM = MB = \frac{1}{2}CB$	9. Def of bis
10. $AM = CM = MB$	10. Sub

Additional Answers, page 297

8. On a line, construct $\overline{A'B'} = AB$. Construct $\angle A' \cong \angle A$, $\angle B' \cong \angle B$. On side $\angle B'$, construct $B'C' = BC$. At C, construct $\angle C' \cong \angle C$. Extend sides of $\angle A'$ and $\angle C'$ to intersect at D'. $A'B'C'D' \cong ABCD$.

10. On a line construct $PQ = b$ with midpt M. On a separate line construct $AC = a$ with midpt R. Draw 2 arcs with ctr A and rad c, and ctr R and rad $\overline{MQ}$, intersecting at B. Extend $\overline{BR}$. With B as ctr and rad b, draw arc intersecting $\overline{BR}$ at D. $ABCD$ is required $\square$.

---

11. Construct $AB = s$. With A as ctr and s as rad, and B as ctr and d as rad, draw two arcs intersecting at D. With B as ctr and s as rad, and A as ctr and d as rad, draw two arcs intersecting at C on the same side of $\overleftrightarrow{AB}$ as D. $ABCD$ is required trap.

12. Construct $AB = a$. With A as ctr and c as rad, and B as ctr and d as rad, draw arcs intersecting at D. Construct line $m \parallel \overleftrightarrow{AB}$ through D. On m construct $DC = b$, with C on same side $\overleftrightarrow{AD}$ as B. Draw $\overline{BC}$. $ABCD$ is required trap.

8 SIMILARITY

OVERVIEW

In this chapter, students use and solve proportions to find the geometric mean of two numbers. They apply the AA Similarity Postulate for triangles and prove and apply the SAS and SSS Similarity Theorems. Students use these theorems to solve problems dealing with proportional parts in similar triangles. They also construct segments with lengths of a given ratio.

OBJECTIVES

- To solve proportions
- To find the geometric mean of two numbers
- To name corresponding congruent angles and corresponding proportional sides, given two similar polygons
- To use proportions relating to similar triangles
- To apply the AA, SAS, and SSS Similarity Theorems for triangles
- To apply theorems concerning proportional parts in triangles
- To derive proportions from a given proportion

PROBLEM SOLVING

For the Application on page 307, it may help students to use the problem solving strategy, Drawing a Diagram, to visualize and solve the exercise. Drawing a Diagram can also help students with the Application on page 317 as they try to find the height of a tree. Students explore the strategy, Using a Formula, to solve the exercises for the Problem Solving Strategies lesson on page 312.

TECHNOLOGY

Computer: The Computer Investigation on page 322 enables students to construct and measure triangles and encourages them to make generalizations that lead to the discovery of the SAS and SSS Similarity Theorems.

Calculator: A calculator will be helpful in Exercises 2 and 3 in the Problem Solving Strategies lesson on page 312. Encourage students to use a calculator to solve proportions to the nearest tenth for Exercises 22–24 on page 321.

SPECIAL FEATURES

Mixed Review pp. 307, 311, 321, 327, 331
Application: Pitch of a Roof p. 307
Problem Solving Strategies: Using a Formula to Find the Golden Ratio p. 312
Midchapter Review p. 316
Application: Finding the Height of a Tree p. 317
Computer Investigation: Similar Triangles p. 322
Focus on Reading p. 325
Algebra Review p. 327
Key Terms p. 332
Key Ideas and Review Exercises pp. 332–333
Chapter 8 Test p. 334
College Prep Test p. 335
Cumulative Review (Chapters 1–8) pp. 336–337

PLANNING GUIDE

Lesson	Basic	Average	Above Average	Resources
8.1 pp. 305–307	CE all WE 1–28 odd Application	CE all WE 1–41 odd Application	CE all WE 1–51 odd Application	Reteaching p. 111 Practice p. 112
8.2 pp. 310–312	CE all WE 1–4 Problem Solving	CE all WE 1–15 Problem Solving	CE all WE 1–18 Problem Solving	Reteaching p. 113 Practice p. 114
8.3 pp. 315–317	CE all WE 1–12 odd Midchapter Review Application	CE all WE 1–20 odd Midchapter Review Application	CE all WE 1–25 odd Midchapter Review Application	Reteaching p. 115 Practice p. 116
8.4 pp. 320–322	CE all WE 1–17 odd CI 1	CE all WE 1–24 odd CI all	CE all WE 1–28 odd CI all	Reteaching p. 117 Practice p. 118
8.5 pp. 325–327	FR all CE all WE 1–8 AR all	FR all CE all WE 1–16 odd AR all	FR all CE all WE 1–22 odd AR all	Reteaching p. 119 Practice p. 120
8.6 pp. 330–331	CE all WE 1–14	CE all WE 1–22	CE all WE 1–27 odd	Reteaching p. 121 Practice p. 122
Chapter 8 Review pp. 332–333	all odd	all	all	
Chapter 8 Test p. 334	1–13	all	all	
College Prep Test p. 335	all	all	all	
Cumulative Review pp. 336–337	1–25 odd	1–33 odd	all	

CE = Classroom Exercises   WE = Written Exercises   FR = Focus on Reading   AR = Algebra Review   CI = Computer Investigation

NOTE: For each level, all students should be assigned all Mixed Review exercises.

INVESTIGATION

Project: The mathematicians of the school of Pythagoras were fascinated by the five-pointed star and the regular pentagon. These figures display many interesting characteristics and ratios. In this investigation, students learn to construct a pentagon and use that construction to draw a five-pointed star.

Materials: Paper, pencil, compass, and ruler

Use the following directions to construct a pentagon and five-pointed star at the chalkboard as the students construct their own on their paper.

1. Open the compass to approximately $\frac{3}{4}$ of an inch. Draw a circle with center O. Draw a diameter $\overline{AB}$. Construct $\overline{CO}$, the $\perp$ bisector of $\overline{AB}$. Construct D, the midpoint of $\overline{OB}$. (fig. 1)

2. With D as a center and $\overline{CD}$ as a radius, draw arc $\widehat{CE}$. Draw $\overline{CE}$. (fig. 2)

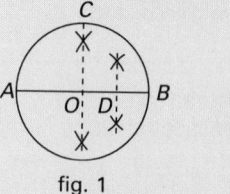

fig. 1

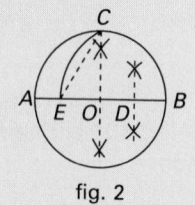

fig. 2

3. Open compass to length CE. Starting at C, draw a series of arcs on the circle. Draw the segments. (fig. 3)

4. Draw diagonals to form a five-pointed star. (fig. 4)

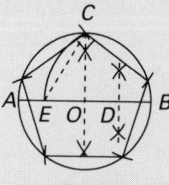

fig. 3

fig. 4

Students will use their constructions to draw a figure called the lute of Pythagoras in Project Worksheet 8.

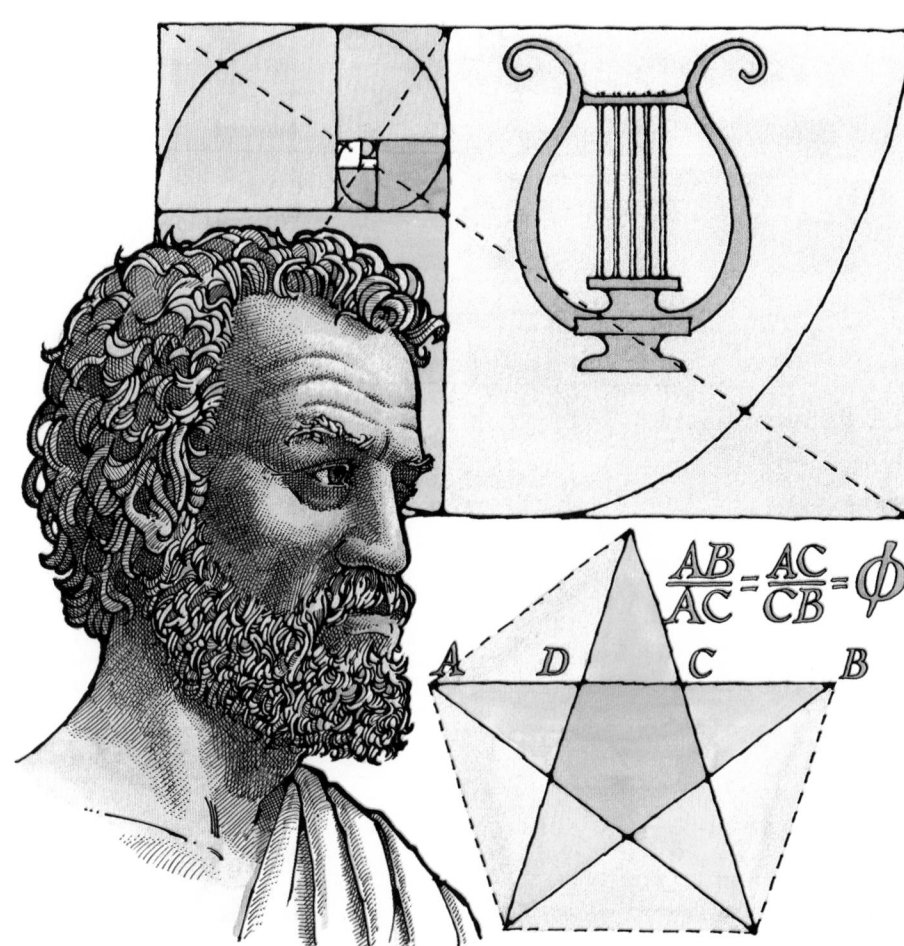

$$\frac{AB}{AC} = \frac{AC}{CB} = \phi$$

Pythagoras (about 569 B.C.)—Pythagoras founded a secret society of mathematicians. One secret the members were sworn not to reveal was the newly-discovered fact that the square root of two cannot be expressed as a fraction.

More About the Mathematician

Pythagoras explored the relationship between music and mathematics on instruments such as the lyre (the small harp in the picture). He found that a note sounded by plucking a string, and the note from a string that is half as long, are a musical octave apart. He concluded that all beauty and nature can be expressed by whole-number relationships. Therefore, the Pythagoreans called the square root of two "irrational." Members of the Pythagorean school signed their manuscripts with the five-pointed star. The Golden Ratio, which is approximately 1.618, occurs in many ways in this figure. This ratio $[\frac{1 + \sqrt{5}}{2}]$ also occurs in the curve of many seashells.

8.1 Ratio and Proportion

Objectives
To solve proportions
To find the geometric mean of two numbers

When a slide is projected, the screen image has the same shape as the image on the slide, but its size differs. An important question then arises: How do measures of the slide and of its image compare?

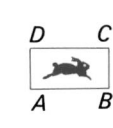
Slide

Image

Definition

A **ratio** is a comparison of two numbers by division. The ratio of a to b ($b \neq 0$) may be written: a to b, $\frac{a}{b}$, or $a:b$.

As with fractions, it often is best to write a ratio in simplest form. For example, $\frac{14}{56}$ in simpler form is $\frac{1}{4}$.

Look at the illustration. Measure the sides. Note that $\dfrac{AB}{BC} = \dfrac{A'B'}{B'C'}$.

Definition

A **proportion** is an equation of the form $\frac{a}{b} = \frac{c}{d}$ ($b \neq 0$, $d \neq 0$). a and d are the **extremes** of the proportion; b and c are the **means**.

Theorem 8.1

In a proportion the product of the extremes equals the product of the means.

Plan

Prove that $\frac{a}{b} = \frac{c}{d}$ ($b \neq 0$, $d \neq 0$) implies $ad = bc$. This is called the **cross product**.

Proof

Statement	Reason
1. $\frac{a}{b} = \frac{c}{d}$ ($b \neq 0$, $d \neq 0$)	1. Given
2. $\frac{a}{b}(bd) = \frac{c}{d}(bd)$	2. Mult Prop of Eq
3. $a \cdot \frac{1}{b}(bd) = c \cdot \frac{1}{d}(bd)$	3. Def of division
4. $a \cdot (bd)\frac{1}{b} = c \cdot (bd)\frac{1}{d}$	4. Comm Prop of Mult

8.1 Ratio and Proportion **303**

GETTING STARTED

Prerequisite Quiz

1. Classify each of the following as true or false.
   a. $\frac{3}{4} = \frac{6}{8}$ True
   b. $\frac{2}{3} = \frac{4}{7}$ False
2. Solve for x.
   a. $\frac{5}{8} = \frac{x}{32}$ 20
   b. $\frac{2}{3} = \frac{7}{x}$ 10.5

Motivator

Ask the students how they can numerically compare two things that have different sizes, for example a board one foot long and another board two feet long. One board is one foot longer than the other board. Ask them if this is a comparison by difference. Yes Ask them to compare the boards in a different way. One board is twice as long as the other. Can they tell you what type of comparison this is? It is a ratio with the ratio being two to one. Have them explain what a ratio is. It is a comparison of two numbers by division.

Lesson Note

This lesson contains a review of algebraic skills and concepts. Have the students give other examples that illustrate ratio and proportion. The students can construct a pantograph to make enlargements or reductions of drawings thus giving illustrations of proportional segments. Point out that ratios are written in the form $a:b$ and proportions in the form $a:b = c:d$. The study of the Golden Ratio can be an enrichment topic.

Emphasize that the definition of geometric mean relates to positive numbers only. Emphasize also, that in writing a proportion involving the geometric mean of two numbers, the means of the proportion are the *same* number.

Math Connections

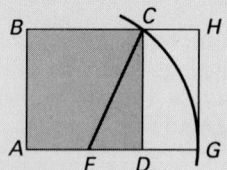

Golden Rectangle: The Golden Rectangle, discovered by the Greeks in the fifth century, B.C., is considered an especially pleasing geometric shape. Examples of the Golden Rectangle appear in ancient and modern paintings, art, and architecture. To construct a golden rectangle, first construct a square $ADCB$. Then, find the midpoint, F, of $\overline{AD}$, and draw $\overline{FC}$ as shown. Extend $\overline{AD}$ to G such that $FC = FG$. Complete the rectangle by constructing $\overline{GH}$ perpendicular to $\overline{AG}$. An unusual property of Golden Rectangle $AGHB$ is that when the original square $ADCB$ is removed, the remaining rectangle, $CDGH$, is also a Golden Rectangle.

Critical Thinking Questions

Application: Ask students to write the ratio of a Golden Rectangle's width to its length (called the Golden Ratio) in terms of x, if the length of the side of the original square is $2x$ and $CF = x\sqrt{5}$. $2:1+\sqrt{5}$

---

5. $a \cdot (db) \frac{1}{b} = (cb) d \cdot \frac{1}{d}$ 5. Comm and Assoc Prop of Mult

6. $(ad) b \cdot \frac{1}{b} = (bc) d \cdot \frac{1}{d}$ 6. Assoc and Comm Prop of Mult

7. $b \cdot \frac{1}{b} = 1, d \cdot \frac{1}{d} = 1$ 7. Prop of Mult Inverses

8. $ad \cdot 1 = bc \cdot 1$ 8. Sub

9. $\therefore ad = bc$ 9. Identity Prop of Mult

Corollary

If the product of the extremes equals the product of the means, then a proportion exists.

In other words, $ad = bc$ implies $\frac{a}{b} = \frac{c}{d}$ ($b \neq 0$, $d \neq 0$).

EXAMPLE 1   Solve the proportion $\frac{6}{x} = \frac{9}{12}$.

Solution
$$\frac{6}{x} = \frac{9}{12}$$
$$6 \cdot 12 = 9 \cdot x$$
$$72 = 9x$$
$$8 = x$$

Theorem 8.2

If $\frac{a}{b} = \frac{c}{d}$ (assume no denominator equals zero), then:

(1) $\frac{b}{a} = \frac{d}{c}$ (4) $\frac{a - b}{b} = \frac{c - d}{d}$

(2) $\frac{a}{c} = \frac{b}{d}$ (5) $\frac{a}{b} = \frac{a + c}{b + d}$

(3) $\frac{a + b}{b} = \frac{c + d}{d}$

Plan

To prove Part 3, show that $(a + b)d = b(c + d)$. If the product of the extremes equals the product of the means, then a proportion exists.

EXAMPLE 2   State three proportions that follow when Theorem 8.2 is applied to $\frac{2}{3} = \frac{6}{9}$. Check by using Theorem 8.1.

Solution

(1) $\frac{3}{2} = \frac{9}{6}$ True, since $3 \cdot 6 = 9 \cdot 2 = 18$

(2) $\frac{2}{6} = \frac{3}{9}$ True, since $2 \cdot 9 = 3 \cdot 6 = 18$

(3) $\frac{5}{3} = \frac{15}{9}$ $\frac{2 + 3}{3} = \frac{6 + 9}{9}$, true; since $5 \cdot 9 = 3 \cdot 15 = 45$

304   Chapter 8 Similarity

Additional Example 1

Solve the proportion.

a. $\frac{7}{x} = \frac{21}{56}$ $x = 18\frac{2}{3}$

b. $\frac{x}{5} = \frac{4}{3}$ $x = 6\frac{2}{3}$

Additional Example 2

a. If $\frac{a}{b} = \frac{3}{4}$, then $\frac{a - b}{b} = \frac{3 - x}{4}$, what is the value of x?   4

b. If $\frac{a}{3} = \frac{5}{b}$, then $\frac{a}{x} = \frac{3}{b}$. What is the value of x?   5

c. If $\frac{4}{5} = \frac{a}{7}$, then $\frac{x + 5}{5} = \frac{a + 7}{7}$. What is the value of x?   4

One special proportion has the form $\frac{a}{m} = \frac{m}{b}$. For example, if $\frac{3}{m} = \frac{m}{27}$, then $81 = m^2$ and $m = 9$ or $m = -9$.

The number 9 is called the **geometric mean** of 3 and 27.

Definition

The **geometric mean** of two positive numbers, a and b, is the positive number m, such that $\frac{a}{m} = \frac{m}{b}$.

EXAMPLE 3 **a.** Find the geometric mean m of 6 and 15.   **b.** Find the geometric mean m of a and $9a$ $(a > 0)$.

Solutions

$\dfrac{6}{m} = \dfrac{m}{15}$

$m^2 = 90$

$m = \sqrt{90}$

$\quad = \sqrt{9} \cdot \sqrt{10}$

$\quad = \pm 3\sqrt{10}$

$3\sqrt{10}$ is the geometric mean.

$\dfrac{a}{m} = \dfrac{m}{9a}$

$m^2 = 9a^2$

$m = \pm 3a$

$3a$ is the geometric mean.

EXAMPLE 4 The measures of three angles of a triangle are in the ratio 2:3:5. Find the measure of the largest angle.

Solution Draw a diagram. Let the angle measures be $2x$, $3x$, and $5x$ (ratio 2:3:5). Write an equation using the fact that the sum of the measures of the angles of a triangle is 180.

$2x + 3x + 5x = 180$

$10x = 180$

$x = 18$

$5x = 5 \cdot 18 = 90$

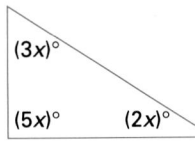

Therefore, 90 is the measure of the largest angle.

Classroom Exercises

Write the ratio in simplest form.

1. $\frac{16}{64}$ $\frac{1}{4}$ **2.** 18 to 54 $\frac{1}{3}$ **3.** 28:52 $\frac{7}{13}$ **4.** 20 to 85 $\frac{4}{17}$ **5.** 1.5:2 $\frac{3}{4}$

Common Error Analysis

Error: Students fail to note the distinction between the solutions of these problems: "Solve $\frac{2}{x} = \frac{x}{8}$ for x" and "Find the geometric mean of 2 and 8."

The first problem has two numbers in the solution set, $\{4, -4\}$. The second has the unique solution, 4, because, by definition, the geometric mean of two positive numbers is a positive number.

Checkpoint

1. Solve the proportion for x. $\frac{3}{x} = \frac{45}{30}$ 2
2. Identify the means and extremes for the proportion $\frac{2}{4} = \frac{5}{10}$ m: 4, 5; ext: 2, 10
3. Find the geometric mean of 6 and 15.
   $3\sqrt{10}$
4. Find the ratio of x to y when $7x = 5y$.
   5:7
5. A 45-inch board is to be cut into two pieces in the ratio of 2 to 13. What will be the length of each piece? 6 in, 39 in

Closure

Have your students explain what a ratio is and have them give an example. Definition of ratio page 303. Have them explain what a proportion is. Definition page 303. Ask them to explain the relationship between the extremes and means in a proportion. The product of the extremes equals the product of the means. Ask them to give an example of the geometric mean of two positive numbers. Answers may vary but should be of the form $\frac{a}{m} = \frac{m}{b}$ where a, b, and m are positive.

Additional Example 3

Find the geometric mean, x, of each of the following pairs of numbers.
a. 3 and 12 6
b. 2 and 75 $5\sqrt{6}$
c. $\sqrt{3}$ and $\sqrt{27}$ 9
d. 4 and $x + 3$ 6

Additional Example 4

Two complementary angles have measures in the ratio 4 to 5. What is the measure of each angle? 40, 50

Guided Practice

Classroom Exercises 1–5

Independent Practice

Ⓐ Ex. 1–28, Ⓑ Ex. 29–41, Ⓒ Ex. 42–51

Basic: WE 1–28 odd, Application
Average: WE 1–41 odd, Application
Above Average: WE 1–51 odd, Application

Additional Answers

Written Exercises

17. $\frac{4}{1} = \frac{8}{2}; \frac{1}{2} = \frac{4}{8}; \frac{1+4}{4} = \frac{2+8}{8}; \frac{1-4}{4} = \frac{2-8}{8}; \frac{1}{4} = \frac{1+2}{4+8}$

18. $\frac{12}{6} = \frac{6}{3}; \frac{6}{3} = \frac{12}{6}; \frac{6+12}{12} = \frac{3+6}{6}; \frac{6-12}{12} = \frac{3-6}{6}; \frac{6}{12} = \frac{6+3}{12+6}$

42. False. $\frac{3}{4} = \frac{6}{8}$, but $\frac{3+1}{4} \neq \frac{6+1}{8}$

43. False. $\frac{4}{5} = \frac{8}{10}$, but $\frac{4+2}{5+2} \neq \frac{8+2}{10+2}$

44. True.

Statement	Reason
1. $\frac{x}{y} = \frac{r}{s}$	1. Given
2. $xs = yr$	2. If proport, then prod of means = prod of extremes.
3. $xr + xs = xr + yr$	3. Add Prop of Eq
4. $x(r + s) = r(x + y)$	4. Distr Prop
5. $\frac{x}{y+x} = \frac{r}{s+r}$	5. If prod of means = prod of extremes, then a proport exists.

Written Exercises

Is the proportion true or false?

1. $\frac{3}{2} = \frac{18}{12}$ T **2.** $\frac{5}{6} = \frac{6}{7}$ F **3.** $\frac{5}{7} = \frac{25}{49}$ F **4.** $\frac{7}{11} = \frac{77}{121}$ T

5. $\frac{2}{3} = \frac{2+2}{3+3}$ T **6.** $\frac{19}{23} = \frac{19^2}{23^2}$ F **7.** $\frac{6}{11} = \frac{6 \cdot 13}{11 \cdot 13}$ T **8.** $\frac{12}{18} = \frac{12-9}{18-9}$ F

Solve the proportion for x.

9. $\frac{3}{4} = \frac{x}{24}$ 18 **10.** $\frac{x}{7} = \frac{35}{49}$ 5 **11.** $\frac{5}{9} = \frac{x}{108}$ 60 **12.** $\frac{6}{15} = \frac{3}{x}$ 7.5

13. $\frac{1}{x} = \frac{x}{4}$ ±2 **14.** $\frac{2}{x} = \frac{x}{8}$ ±4 **15.** $\frac{x}{4} = \frac{64}{x}$ ±16 **16.** $\frac{x}{1} = \frac{25}{x}$ ±5

Use Theorem 8.2 to state five proportions that follow from the given proportion. Use Theorem 8.1 to check.

17. $\frac{1}{4} = \frac{2}{8}$ **18.** $\frac{6}{12} = \frac{3}{6}$

Find the value of x in the proportion.

19. If $\frac{a}{b} = \frac{3}{4}$, then $\frac{a+b}{b} = \frac{3+x}{4}$ 4 **20.** If $\frac{p}{q} = \frac{9}{13}$, then $\frac{p-q}{q} = \frac{9-x}{13}$ 13

Find the geometric mean m of the pair of positive numbers.

21. 1 and 9 3 **22.** 5 and 20 10 **23.** 12 and 3 6 **24.** 6 and 7 $\sqrt{42}$

25. 9 and 5 $3\sqrt{5}$ **26.** p and q $\sqrt{pq}$ **27.** p^2 and q^2 pq **28.** c and $9c$ $3c$

Is the statement true for all values of the variables? (Assume each denominator ≠ 0.) (Exercises 29–32)

29. $\frac{cx}{cy} = \frac{dx}{dy}$ T **30.** $\frac{c+x}{c+y} = \frac{d+x}{d+y}$ F **31.** $\frac{ab}{bc} = \frac{ad}{dc}$ T **32.** $\frac{p-q}{r-q} = \frac{p+q}{r+q}$ F

33. The ratio of the measures of two complementary angles is 4 to 5. What are the measures of the angles? 40, 50

34. The ratio of the measures of two supplementary angles is 3 to 6. What are the measures of the angles? 60, 120

35. The measures of three angles of a triangle are in the ratio 5:6:9. What are the measures of the angles? 45, 54, 81

36. The measures of three angles of a triangle are in the ratio 2:4:9. What are the measures of the angles? 24, 48, 108

37. The perimeter of a rectangle is 60 cm. The ratio of the length of a base to the length of a corresponding altitude of the rectangle is 5 to 7. Find the lengths of the base and altitude. $b = 12.5$, $a = 17.5$

38. The perimeter of an isosceles triangle is 99 cm. The ratio of the length of the base to the length of a leg is 3:4. What are the lengths of the base and leg of the triangle? $b = 27$, $l = 36$

39. If two boxes of cereal cost $1.38, what will seven boxes cost? $4.83

Enrichment

In a **geometric sequence**, each term after the first is obtained by multiplying the previous term by a number called the **common ratio**. For example, in the sequence 2, 6, 18, 54, each term after 2 is obtained by multiplying the previous term by 3. Notice that the sequence can be written as 2, 2(3), 2(3²), 2(3³). This suggests a method of finding the common ratio of a sequence beginning with 2 whose fourth term is 54.

$$2x^3 = 54; x = 3$$

1. The fifth term of a geometric sequence that begins with 1 is 81. Find the common ratio. 3
2. The fourth term of a geometric sequence that begins with 4 is 256. Find the common ratio. 4

40. A baseball player has a batting average of .320. If he has been at bat 225 times, how many hits does he have?   72

41. Pat answered correctly 80% of the questions on a test. If there were 15 questions, how many were answered correctly?   12

Tell whether the proportion is true for all values of the variables. If it is true, give a proof. If false, give a numerical counterexample. (Assume each denominator is not equal to zero.)

42. If $\frac{x}{y} = \frac{r}{s}$, then $\frac{x + c}{y} = \frac{r + c}{s}$

43. If $\frac{x}{y} = \frac{r}{s}$, then $\frac{x + a}{y + a} = \frac{r + a}{s + a}$

44. If $\frac{x}{y} = \frac{r}{s}$, then $\frac{x}{x + y} = \frac{r}{r + s}$

45. If $\frac{x}{y} = \frac{r}{s} = \frac{m}{n}$, then $\frac{x}{y} = \frac{x + r + m}{y + s + n}$

46. Prove the corollary to Theorem 8.1.
47. Prove Property 1 of Theorem 8.2.
48. Prove Property 2 of Theorem 8.2.
49. Prove Property 3 of Theorem 8.2.
50. Prove Property 4 of Theorem 8.2.
51. Prove Property 5 of Theorem 8.2.

Mixed Review

Assume $l \parallel m$. Determine whether the two angles are congruent.   *3.5*
(Exercises 1–4)

1. $\angle 3$ and $\angle 5$
2. $\angle 1$ and $\angle 8$
3. $\angle 2$ and $\angle 7$
4. $\angle 1$ and $\angle 5$

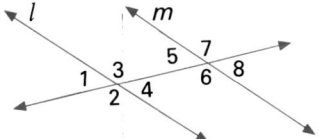

1. Not necessarily
2. Yes
3. Yes
4. Yes

5. If the ratio of the measures of $\angle 4$ and $\angle 6$ is 2 to 3, what are the measures of the angles?   *8.1*   72, 108

Application: *Pitch of a Roof*

The pitch of a roof is the ratio of the rise of the roof to the run of the roof. For example a 4-in-12 roof rises 4 inches for every 12 inches of run.

If the building is 24 ft wide, what is the total rise for a 4-in-12 roof? For a 6-in-12 roof?   4 ft, 6 ft

8.1 Ratio and Proportion   **307**

Statement	Reason
1. $\frac{x}{y} = \frac{r}{s} = \frac{m}{n}$	1. Given
2. $xs = ry$	2. If proport, then prod of means = prod of extremes.
3. $xy + xs = xy + ry$	3. Add Prop of Eq
4. $x(y + s) = y(x + r)$	4. Distr Prop
5. $\frac{x}{y} = \frac{x + r}{y + s}$	5. If prod of means = prod of extremes, then a proport exists.
6. $\frac{x + r}{y + s} = \frac{m}{n}$	6. Sub
7. $n(x + r) = m(y + s)$	7. If proport, then prod of means = prod of extremes.
8. $n(x + r) + (x + r)(y + s) = m(y + s) + (x + r)(y + s)$	8. Add Prop of Eq
9. $(x + r)[n + (y + s)] = (y + s)[m + (x + r)]$	9. Distr Prop
10. $(x + r)(y + s + n) = (y + s)(x + r + m)$	10. Comm Prop, Assoc Prop
11. $\frac{x + r}{y + s} = \frac{x + r + m}{y + s + n}$	11. If prod of means = prod of extremes, then a proport exists.
12. $\frac{x}{y} = \frac{x + r + m}{y + s + n}$	12. Sub

46.

Statement	Reason
1. $ad = bc$, $b \neq 0$, $d \neq 0$	1. Given
2. $ad \cdot 1 = bc \cdot 1$	2. Mult Prop of Eq
3. $\frac{b}{b} = 1$, $\frac{d}{d} = 1$	3. A nonzero number div by itself = 1.
4. $ad \cdot \frac{b}{b} = bc \cdot \frac{d}{d}$	4. Sub
5. $\frac{a}{b} \cdot db = \frac{c}{d} \cdot db$	5. Comm and Assoc Props for Mult
6. $\frac{a}{b} = \frac{c}{d}$	6. Div Prop of Eq

47.

Statement	Reason
1. $\frac{a}{b} = \frac{c}{d}$, $b \neq 0$, $d \neq 0$	1. Given
2. $bc = ad$	2. If proport, then prod of means = prod of extremes.
3. $\frac{b}{a} = \frac{d}{c}$	3. If prod of means = prod of extremes, then proportion exists.

See page 312 for the answers to Ex. 48–51.

▰▰▰GETTING STARTED

Prerequisite Quiz

Which of these figures appear to be similar?

1, 6; 2, 4; 3, 5

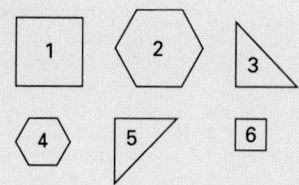

Motivator

Ask the students what they mean when they say that two people are similar. They are alike. Ask them if this means that they are completely alike. No Have the students guess how much alike two geometric figures must be to be called similar. They must have congruent corresponding angles and equal ratios of the lengths of corresponding sides.

▰▰▰TEACHING SUGGESTIONS

Lesson Note

The general notation and symbolism for similarity are introduced via similar polygons. Point out to students that similarity involves two conditions: congruent corresponding angles and proportional corresponding sides.

Some students may need help in writing similarity statements correctly and in "matching up" corresponding sides in proportions. For enrichment, students could develop and prove a theorem about the perimeters of similar polygons.

8.2 Similar Polygons

Objectives
To identify similar polygons
To find the measures of parts of similar polygons
To explain the difference between similar and congruent polygons

The figures below are not the same size, but they have the same shape. Such figures are called similar polygons.

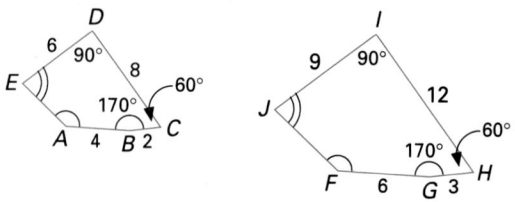

Polygon *ABCDE* is similar to polygon *FGHIJ*.

Definition
Two **polygons are similar** (~) if corresponding angles are congruent and the ratios of the lengths of corresponding sides are equal.

Polygon *ABCDE* ~ polygon *FGHIJ* means that "polygon *ABCDE* is similar to polygon *FGHIJ*." Note that the corresponding vertices must be named in the same order. The symbol ≁ means *is not similar to*.

EXAMPLE 1 Is square *KLMN* similar to rectangle *OPQR*? Why or why not?

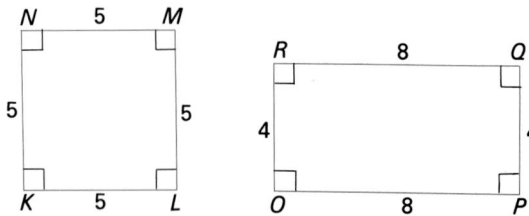

Solution Corresponding angles are congruent, since all the angles are right.

Corresponding sides are not proportional: $\frac{KL}{OP} = \frac{5}{8}$ and $\frac{LM}{PQ} = \frac{5}{4}$.

Therefore, square *KLMN* ≁ rectangle *OPQR*.

308 Chapter 8 Similarity

Additional Example 1

Is square *ABCD* similar to square *WXYZ*? Yes

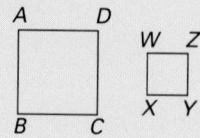

Having congruent corresponding angles is not a sufficient condition to prove the similarity of two polygons. Nor is having corresponding proportional sides sufficient. Both congruent corresponding angles *and* corresponding proportional sides must be true. In the two quadrilaterals,

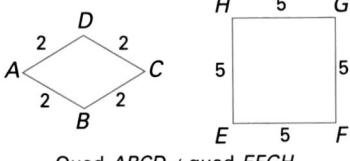

Quad *ABCD* ≁ quad *EFGH*.

$$\frac{AB}{EF} = \frac{BC}{FG} = \frac{CD}{GH} = \frac{DA}{HE} = \frac{2}{5}.$$

Given that the lengths of corresponding sides of similar polygons are proportional, use a proportion to find the unknown lengths of sides.

EXAMPLE 2 Polygon *ABCDE* ~ polygon *FGHIJ*. *AB* = 16, *BC* = 20, *CD* = 14, *DE* = 15, *EA* = 25, and *FG* = 24. Find *GH*.

Solution

$$\frac{AB}{FG} = \frac{BC}{GH}$$
$$\frac{16}{24} = \frac{20}{GH}$$
$$16(GH) = 20 \cdot 24 = 480$$
$$GH = 30$$

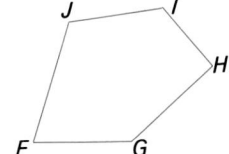

Similarity provides a basis for important applications of geometry.

EXAMPLE 3 A blueprint for a house was drawn on a scale in which $\frac{1}{8}$ in. on the plan represents 1 ft of actual length. Part of the blueprint is shown. What is the actual length of the living room?

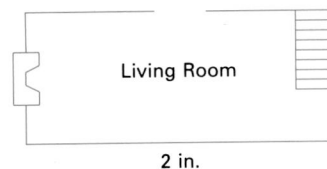

Living Room

2 in.

Plan Write a proportion using the ratio $\frac{\text{scale length}}{\text{actual length}}$.

Solution The length of the room on the blueprint is 2 in.

$$\frac{\frac{1}{8} \text{ in.}}{1 \text{ ft}} = \frac{2 \text{ in.}}{x \text{ ft}}$$
$$\frac{1}{8}x = 2, \text{ so } x = 16$$

Thus, the living room is 16 ft long.

Math Connections

Projections: An overhead projector shows an image on the screen that is similar to the image that is on a transparency. Discuss with students other examples of projected images that are similar to an original image: enlarged photographs, reduced photocopies, slides projected onto a screen, and so on.

Critical Thinking Questions

Analysis: Ask students which of the following types of polygons are similar to all other polygons of their classification.

1. **Regular octogons** Yes. The meas of each int ∠ = 135 and the length of all sides of a reg octogon are ≅, so the ratio of lengths of corr sides of any 2 reg octogons is proport.
2. **Isosceles triangles** No. The meas of int ∠s may differ.
3. **Rhombuses** No. All sides of a rhom are ≅, so the ratio of lengths of corr sides of any 2 rhoms is proport, but the meas of int ∠s may vary among rhoms.
4. **Rectangles** No. The meas of each int ∠ = 90, but length of sides may vary among rects.
5. **Squares** Yes. The meas of each int ∠ = 90 and the sides of a sq are all ≅, so the ratio of lengths of corr sides of any 2 squares is proport.
6. **Regular polygons** No. The number of sides and the number of angles may vary; therefore, their shapes may differ.

Common Error Analysis

Error: Students do not identify corresponding sides of similar figures correctly.

Emphasize that corresponding sides lie opposite corresponding angles of similar polygons. Have students mark one pair of corresponding angles of similar polygons and immediately identify the corresponding opposite sides. Have them continue this procedure for all angles of the polygons.

Checkpoint

1. Which of the following pairs of polygons are similar?   c

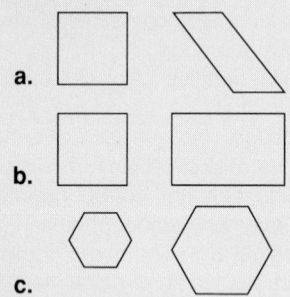

a.

b.

c.

2. For each pair of similar polygons, find the unknown length. (a) *ABCDE ~ MNOPQ; BC = 15, NO = 17, DE = 20, PQ = _____.* $\frac{68}{3}$ (b) *ABCD ~ XYZW; AB = 16, CD = 20, ZW = 27, XY = _____.* $\frac{108}{5}$

Closure

Ask the students what they know about the measures of the sides and angles of two congruent polygons.   The measures of corresponding sides and corresponding angles are congruent.   Ask them what they know about the measures of the sides and angles of two similar polygons.   Definition page 308   Have them explain what congruent polygons and similar polygons have in common.   Corresponding angles are congruent.   Then ask them how congruent polygons and similar polygons are different.   The relationship between corresponding sides

---

Classroom Exercises

Determine whether the polygons are similar on the basis of the marking alone. Explain your answer.

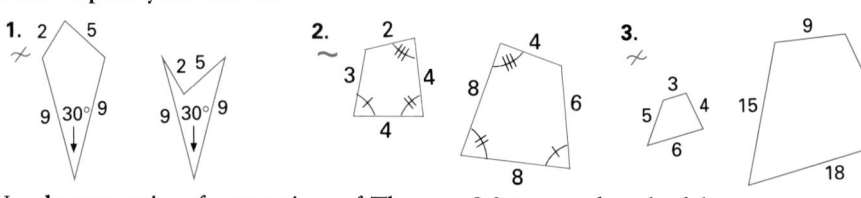

Use the properties of proportions of Theorem 8.2 to complete the following statements. Quad *RSTU ~* quad *JKLM.*

4. $\frac{RS}{ST} = \frac{?}{KL}$ *JK*

5. $LM \cdot RS = TU \cdot \underline{\qquad}$ *JK*

6. $\frac{RS + JK}{JK} = \frac{TU + ?}{LM}$ *LM*

Written Exercises

Find the unknown lengths for the pair of similar polygons.

1. *ABCD ~ EFGH*

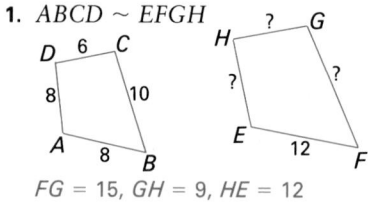

FG = 15, GH = 9, HE = 12

2. Square *IJKL ~* square *MNOP*

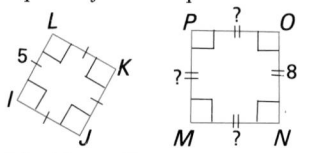

All sides = 8

3. *▱QRST ~ ▱UVWX*

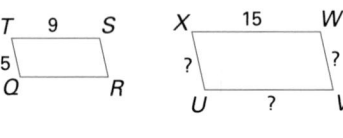

UV = 15, XU = VW = $8\frac{1}{3}$

4. *ABCDE ~ FGHIJ*

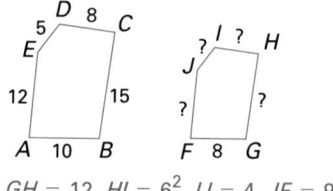

GH = 12, HI = $6\frac{2}{5}$, IJ = 4, JF = $9\frac{3}{5}$

For Exercises 5–8, find the missing measures.

5. Given: Rhomb *ABCD ~* rhomb *EFGH, AB = 7, EF = 12,* m ∠*A = 55*   12, 125
   Find *FG* and m ∠*F.*

6. Given: Pentagon *IJKLM ~* pentagon *NOPQR, IJ = 24, JK = 20, KL = 28, NO = 32*
   Find *OP* and *PQ.* $26\frac{2}{3}, 37\frac{1}{3}$

7. Given: Polygon *STUVW ~* polygon *XYZAB, XY = 32, YZ = 36, ST = 4c + 2, TU = 5c + 2*
   Find *ST* and *TU.*   4, $4\frac{1}{2}$

8. Given: Polygon *CDEFG ~* polygon *HIJKL, DE = 13, EF = 15, IJ = 5x + 2, JK = 7x − 2*
   Find *IJ* and *JK.* $19\frac{1}{2}, 22\frac{1}{2}$

Enrichment

Have students draw, to scale, the floor plan of a room—either at home or at school. The following steps should be followed.

(1) Choose an appropriate scale. Suggest one-quarter inch or one-half inch to one foot as possible scales.
(2) Do a sketch of the outline of the room, including any furniture that is to be in the final drawing.
(3) Measure all lengths that are needed and write them on the sketch.
(4) Use a proportion to convert each length to the appropriate scale length, e.g. 9 ft = 4.5 in.
(5) Draw a careful diagram using the scale lengths.

9. Given: Square $ABCD$, square $XYZW$
Prove: $ABCD \sim XYZW$

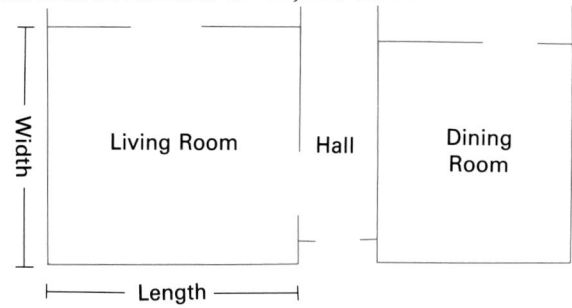

10. If two polygons are similar, are they necessarily congruent? Explain.   No, sides not always ≅ but ∠s are ≅.

11. If two polygons are congruent, are they necessarily similar? Explain.   Yes, ∠s are ≅ and sides corr in $\frac{1}{1}$ ratio

A portion of a blueprint for a house is shown. The scale used is $\frac{1}{8}$ of an inch to the foot. In Exercises 12–15, find the indicated measure.

| | | |
| Width | Living Room | Hall | Dining Room |

Length

12. How wide is the living room?   12 ft
13. How long is the living room?   13 ft
14. How wide is the dining room?   10 ft
15. How wide is the hall?   4 ft
16. Draw two equilateral pentagons that are not similar.
17. Draw two equiangular pentagons that are not similar.
18. Refer to an encyclopedia or book of art history to find uses of the **golden ratio** in painting, architecture, or photography. Look also under Golden Section, Divine Proportion, and Golden Rectangle.

Mixed Review

1. List four necessary conditions for a quadrilateral to be a parallelogram.   *6.6*
2. List four sufficient conditions for a quadrilateral to be a parallelogram.   *6.6*
3. If a parallelogram has congruent diagonals, is the parallelogram necessarily a square? If not, show a counterexample.   *7.3* No, rectangle

8.2 Similar Polygons   **311**

Mixed Review

1. Both pairs of opp sides ∥; one pair of opp ∠s ≅; one pair of opp sides ≅; diag bis each other.
2. Both pairs of opp sides ∥; both pairs of opp sides ≅; one pair of opp sides ≅ and ∥; diag bis each other.

■ **FOLLOW UP**

Guided Practice

Classroom Exercises 1–6

Independent Practice

🅰 Ex. 1–4, 🅱 Ex. 5–15, 🅲 Ex. 16–18

Basic: WE 1–4, Problem Solving
Average: WE 1–15, Problem Solving
Above Average: WE 1–18, Problem Solving

Additional Answers

Classroom Exercises

1. Not similar: corr ∠s not ≅
2. Similar: sides proport and ∠s ≅
3. Not similar: sides proport but corr ∠s not necessarily ≅

Written Exercises

9.

Statement	Reason
1. Sq $ABCD$, sq $XYZW$	1. Given
2. $ABCD$, $XYZW$ are rects with $AB = BC = CD = DA$ and $XY = YZ = ZW = WX$.	2. Def of sq
3. $ABCD$, $XYZW$ each have 4 rt ∠s.	3. Def of rect
4. $\angle A \cong \angle X$, $\angle B \cong \angle Y$, $\angle C \cong \angle Z$, $\angle D \cong \angle W$	4. All rt ∠s are ≅.
5. $\frac{AB}{XY} = \frac{AB}{XY}$	5. Reflex Prop of Eq
6. $\frac{AB}{XY} = \frac{BC}{YZ} = \frac{CD}{ZW} = \frac{DA}{WX}$	6. Sub
7. $ABCD \sim XYZW$	7. Def of ∼ polygons

16. 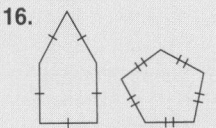   **17.**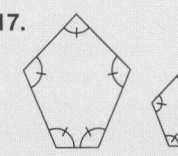

18. The floor plan of the Parthenon is an example.

311

48.

Statement	Reason
1. $\frac{a}{b} = \frac{c}{d}$, $b \neq 0$, $d \neq 0$	1. Given
2. $\frac{a}{b} \cdot \frac{b}{c} = \frac{c}{d} \cdot \frac{b}{c}$	2. Mult Prop of Eq
3. $\frac{a}{c} \cdot \frac{b}{b} = \frac{b}{d} \cdot \frac{c}{c}$	3. Assoc and Comm Props for Mult
4. $\frac{b}{b} = 1$, $\frac{c}{c} = 1$	4. A nonzero number div by itself = 1.
5. $\frac{a}{c} \cdot 1 = \frac{b}{d} \cdot 1$	5. Sub
6. $\frac{a}{c} = \frac{b}{d}$	6. Id Prop of Mult

49.

Statement	Reason
1. $\frac{a}{b} = \frac{c}{d}$, $b \neq 0$, $d \neq 0$	1. Given
2. $ad = bc$	2. If proport, then prod of means = prod of extremes.
3. $ad + bd = bc + bd$	3. Add Prop of Eq
4. $(a + b)d = (c + d)b$	4. Distr Prop
5. $\frac{a + b}{b} = \frac{c + d}{d}$	5. If prod of means = prod of extremes, then a proport exists.

50.

Statement	Reason
1. $\frac{a}{b} = \frac{c}{d}$, $b \neq 0$, $d \neq 0$	1. Given
2. $ad = bc$	2. If proport, then prod of means = prod of extremes.
3. $ad - bd = bc - bd$	3. Subt Prop of Eq
4. $(a - b)d = (c - d)b$	4. Distr Prop
5. $\frac{a - b}{b} = \frac{c - d}{d}$	5. If prod of means = prod of extremes, then a proportion exists.

51.

Statement	Reason
1. $\frac{a}{b} = \frac{c}{d}$, $b \neq 0$, $d \neq 0$	1. Given
2. $ad = bc$	2. If proport, then prod of means = prod of extremes.
3. $ab + ad = ab + bc$	3. Add Prop of Eq
4. $a(b + d) = b(a + c)$	4. Distr Prop
5. $\frac{a}{b} = \frac{a + c}{b + d}$	5. If prod of means = prod of extremes, then a proport exists.

Problem Solving Strategies

Using a Formula to Find the Golden Ratio

Look at the three rectangles. Which do you think is most pleasing to the eye? Most people choose the one in the middle. Why? We don't really know, but this shape has been used throughout history to create works of art and beautiful buildings such as the Parthenon.

In the golden rectangle the long side is the geometric mean between the width and the long side plus the width. To solve this proportion you must use the quadratic formula.

$$\frac{w}{l} = \frac{l}{w + l}$$

Let $l = 1$ and cross multiply.
$$w(w + 1) = 1$$
$$w^2 + w - 1 = 0$$
$$w = \frac{-1 \pm \sqrt{5}}{2} \approx 0.618$$

The **golden ratio** in mathematical literature is the reciprocal $\frac{1}{w}$. See Exercise 3.

A golden rectangle can be constructed as follows: Draw square $ABCD$; find the midpoint of $\overline{AB}$ and draw segment $\overline{TC}$ as shown. Extend the segment $\overline{AB}$ to R by using the length of the segment $\overline{TC}$ from T. Complete the rectangle by drawing $\overline{RS} \perp$ to $\overline{AR}$.

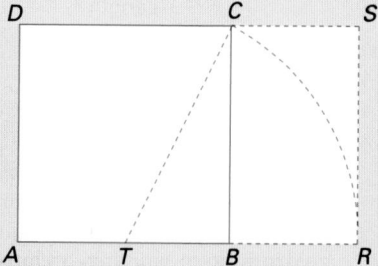

Exercises

1. Since all golden rectangles are similar, we can choose any convenient lengths. Let a side of the square used for the construction above have length 2. Use the Pythagorean Theorem to find the length of $\overline{TC}$. Now set up the proportion used to define the golden ratio and prove that it is a true proportion.

2. Using a calculator, find the ratio $\frac{AB}{AR}$ to the nearest thousandth.   0.618

3. Using a calculator, find the ratio $\frac{AR}{AB}$ to the nearest thousandth.   1.618

Problem Solving Strategies

1. $\frac{AB}{AR} = \frac{AR}{AB + AR}$;

$TC^2 = 1^2 + 2^2$
$TC^2 = 1 + 4$
$TC^2 = 5$
$TC = \sqrt{5}$
$AR = 1 + \sqrt{5}$

$\frac{2}{1 + \sqrt{5}} = \frac{1 + \sqrt{5}}{2 + 1 + \sqrt{5}}$
$2(3 + \sqrt{5}) = (1 + \sqrt{5})^2$
$6 + 2\sqrt{5} = 1 + 2\sqrt{5} + 5$
$6 + 2\sqrt{5} = 6 + 2\sqrt{5}$

8.3 Similar Triangles

Objectives
To write and use proportions relating to similar triangles
To apply the AA Similarity Postulate

A printer can quickly tell if two rectangular designs are similar by placing one on top of the other so that two corners coincide. If the diagonal of the larger rectangle contains the diagonal of the smaller, forming similar triangles, then the two rectangles are similar. Many other practical applications of geometry depend upon properties of similar triangles.

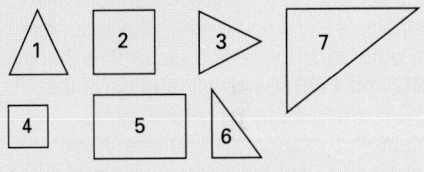

similar

not similar

Definition

Two **triangles are similar** if corresponding angles are congruent and corresponding sides are proportional.

EXAMPLE 1

Given: $\triangle XYZ \sim \triangle TUV$, $XY = 12$, $YZ = 15$, $ZX = 18$, $TU = 16$
Find UV and VT.

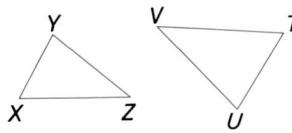

Solution

$$\frac{XY}{TU} = \frac{YZ}{UV} = \frac{ZX}{VT}$$

$$\frac{XY}{TU} = \frac{12}{16} = \frac{3}{4} \qquad \text{So, } \frac{15}{UV} = \frac{3}{4} \qquad \text{and} \qquad \frac{18}{VT} = \frac{3}{4}$$

$$3 \cdot UV = 60 \qquad\qquad 3 \cdot VT = 72$$
$$UV = 20 \qquad\qquad\quad VT = 24$$

Theorem 8.3

Congruent triangles are similar.

Theorem 8.4

Transitive Property of Triangle Similarity: If $\triangle ABC \sim \triangle DEF$ and $\triangle DEF \sim \triangle GHI$, then $\triangle ABC \sim \triangle GHI$.

Additional Example 1

If $\triangle ABC \sim \triangle XYZ$, find the unknown measures.

a. $AB = 12$, $XY = 16$, $BC = 8$, $YZ =$ _____ $10\frac{2}{3}$

b. $m \angle A = 14$, $m \angle B = 82$, $m \angle Y =$ 82, $m \angle X = 14$, $m \angle C =$ _____, $m \angle Z =$ _____   84; 84

▰ GETTING STARTED

Prerequisite Quiz

Which of the following geometric figures have the "same shape?"

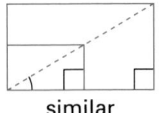

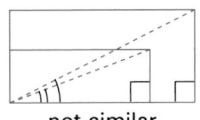

1, 3; 2, 4; 6, 7

Motivator

Ask the students what parts of a triangle determine its size.   The lengths of the sides Then ask them what they think determines its shape.   Angle measures   Ask them what they think are sufficient conditions for two triangles to have the same shape.
Congruent corresponding angles

▰ TEACHING SUGGESTIONS

Lesson Note

Begin the section by giving the measures of three angles such as: 90, 45, and 45. Ask the students to construct a triangle having these angles. This should produce a variety of triangles and lead to a discussion of similarity.

The definition of similar triangles contains both the condition that corresponding angles are congruent and that corresponding sides are proportional. Both conditions are not necessary for similar triangles. But both are needed for other similar polygons.
Review the theorems for congruent triangles. As with congruent triangles, it is important to "match" corresponding vertices and sides when writing relationships about similar figures. Students may need to be reminded that corresponding angles are congruent. Corresponding sides are opposite congruent angles.

Math Connections

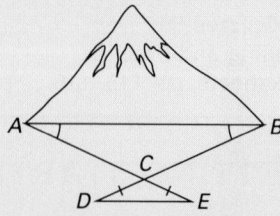

Surveying: Surveyors sometimes use similar triangles to measure inaccessible distances. For example, as shown in the diagram above, a surveyor can find the distance AB from one side of a mountain to the other side by setting up similar triangles ABC and EDC. Since all lengths of the similar triangles except AB may be obtained, the surveyor can use these measures to set up a proportion and solve for AB.

Critical Thinking Questions

Application: Refer students to the diagram of the surveyor's similar triangles above and ask them how the surveyor knows that the triangles are similar and what proportion can be used to find AB.   Base $\angle$s of $\triangle ABC$ are $\cong$ and the legs of $\triangle EDC$ are $\cong$. Both $\triangle$s are isos with $\cong$ vert $\angle$s, so corr $\angle$s are $\cong$. The $\triangle$s are $\sim$ by AA$\sim$. Proports: $\frac{AB}{ED} = \frac{AC}{EC}$ or $\frac{AB}{ED} = \frac{BC}{DC}$

Checkpoint

1. If $\triangle$s ABC and XYZ are similar and $\angle A \cong \angle Y$, $\angle B \cong \angle Z$, and $\angle C \cong \angle X$, then $\triangle ABC \sim \triangle$ _____   YZX

2. If $\triangle PQR \sim \triangle XYZ$, then $\angle P \cong \angle$ _____ and $\frac{PQ}{XY} = \frac{?}{YZ}$ X; QR

3. Given the two triangles ABC and XYZ. If m $\angle A$ = 30, m $\angle Z$ = 30, m $\angle C$ = 45, m $\angle X$ = 45, then $\triangle ABC \sim \triangle$ _____   ZYX

4. Are these triangles similar?   yes

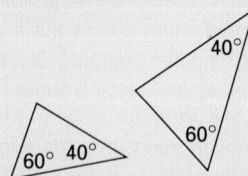

5. If two triangles are congruent, are they similar?   Yes

6. If two triangles are similar are they necessarily congruent?   No

---

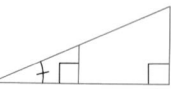

EXAMPLE 2   Write a correct similarity statement for the triangles if m $\angle X$ = 32, m $\angle Z$ = 87, m $\angle R$ = 87, and m $\angle Q$ = 32.

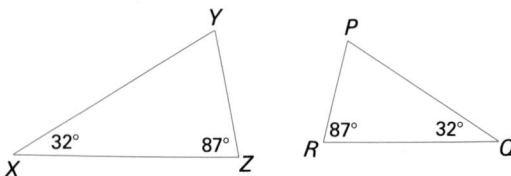

Solution   By the AA Similarity Postulate, $\triangle XZY \sim \triangle QRP$.

EXAMPLE 3   Given: $\overleftrightarrow{RS} \parallel \overline{XY}$
Prove: $\triangle ZRS \sim \triangle ZXY$

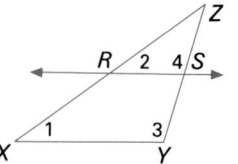

Proof

Statement	Reason
1. $\overleftrightarrow{RS} \parallel \overline{XY}$	1. Given
2. $\angle 1 \cong \angle 2$, $\angle 3 \cong \angle 4$	2. Corr $\angle$s of $\parallel$ lines are $\cong$.
3. $\therefore \triangle ZRS \sim \triangle ZXY$	3. AA $\sim$

Every proportion has a related pair of **cross products**: the product of the extremes and the product of the means. For example, in $\frac{AB}{CD} = \frac{EF}{GH}$, $AB \cdot GH$ and $CD \cdot EF$ are cross products. Given the cross products, you can write the corresponding proportion(s).

EXAMPLE 4   Write two proportions from $ZY \cdot AB = CD \cdot XW$.

Plan   Choose the factors of one of the cross products to be the means of the proportion; the factors of the other cross product are the extremes.

Solution $\frac{ZY}{CD} = \frac{XW}{AB}$,   or $\frac{XW}{ZY} = \frac{AB}{CD}$ (Other solutions are possible.)

Additional Example 2

For triangles LMN and FGH write a similarity statement.
m $\angle L$ = 48, m $\angle N$ = 61, m $\angle G$ = 48, m $\angle F$ = 61 $\triangle LNM \sim \triangle GFH$

Additional Example 3

Given: $\overline{AB} \parallel \overline{DE}$
Prove: $\triangle ACB \sim \triangle DCE$

$\overline{AB} \parallel \overline{DE}$ (Given);
$\angle ACB \cong \angle DCE$ (vert $\angle$s);
$\angle BAC \cong \angle EDC$ (Alt int $\angle$s);
$\triangle ACB \sim \triangle DCE$ (AA Similarity)

EXAMPLE 5 Given: $\square XYZW$
Prove: $XQ \cdot PQ = ZQ \cdot RQ$

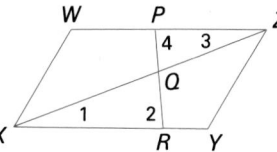

Plan Draw a flow diagram.

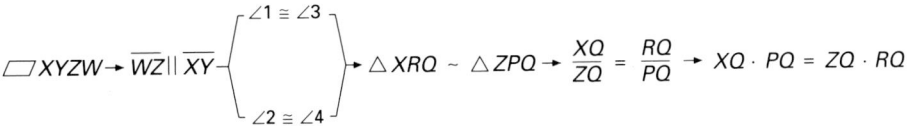

Classroom Exercises

State whether the two triangles are similar.

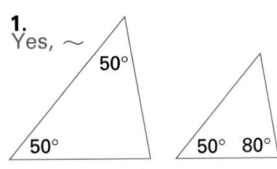

1. Yes, $\sim$

2. Yes, $\sim$

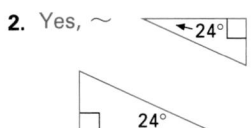

3. No, $\not\sim$

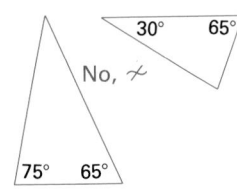

Complete the statement. $\triangle ABC \sim \triangle DEF$

4. $\angle A = \angle \underline{\quad}$, $\angle B = \angle \underline{\quad}$,
$\angle C = \angle \underline{\quad}$. *D, E, F*

5. $\dfrac{AB}{?} = \dfrac{BC}{?} = \dfrac{CA}{?}$

DE, EF, FD

Written Exercises

For Exercises 1–6, assume $\triangle ABC \sim \triangle DEF$. Find the missing measure.

1. $AB = 8$, $BC = 10$, $DE = 12$,
$EF = \underline{\quad}$ 15

2. $AB = 12$, $CA = 16$, $DE = 9$,
$FD = \underline{\quad}$ 12

3. $m \angle A = 80$, $m \angle B = 40$, $m \angle D = \underline{\quad}$, $m \angle E = \underline{\quad}$ 80, 40

4. $m \angle B = 65$, $m \angle C = 35$, $m \angle E = \underline{\quad}$, $m \angle F = \underline{\quad}$ 65, 35

5. $m \angle A = 40$, $m \angle B = 60$, $m \angle D = \underline{\quad}$, $m \angle F = \underline{\quad}$ 40, 80

6. $m \angle B = 35$, $m \angle C = 85$, $m \angle D = \underline{\quad}$, $m \angle E = \underline{\quad}$ 60, 35

For Exercises 7–10, use the diagram at right.

7. Given: $\overline{CD} \parallel \overline{AB}$
Prove: $\triangle ABE \sim \triangle DCE$

8. Given: $\angle A \cong \angle D$
Prove: $\triangle ABE \sim \triangle DCE$

9. Given: $\overline{CD} \parallel \overline{AB}$
Prove: $\dfrac{AB}{DC} = \dfrac{AE}{DE}$

10. Given: $\overline{CD} \parallel \overline{AB}$
Prove: $AE \cdot CE = BE \cdot DE$

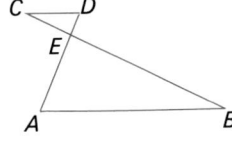

■■■FOLLOW UP

Guided Practice

Classroom Exercises 1–5

Independent Practice

A Ex. 1–12, **B** Ex. 13–20, **C** Ex. 21–25

Basic: WE 1–12 odd, Midchapter Review, Application

Average: WE 1–20 odd, Midchapter Review, Application

Above Average: WE 1–25 odd, Midchapter Review, Application

Additional Answers

Written Exercises

7.

Statement	Reason
1. $\overline{CD} \parallel \overline{AB}$	1. Given
2. $\angle D \cong \angle A$, $\angle C \cong \angle B$	2. If 2 lines are $\parallel$, then alt int $\angle$s are $\cong$.
3. $\triangle ABE \sim \triangle DCE$	3. AA$\sim$

8.

Statement	Reason
1. $\angle A \cong \angle D$	1. Given
2. $\angle AEB \cong \angle DEC$	2. Vert $\angle$s are $\cong$.
3. $\triangle ABE \sim \triangle DCE$	3. AA$\sim$

9.

Statement	Reason
1. $\overline{CD} \parallel \overline{AB}$	1. Given
2. $\angle D \cong \angle A$, $\angle C \cong \angle B$,	2. If 2 lines are $\parallel$, then alt int $\angle$s are $\cong$.
3. $\triangle ABE \sim \triangle DCE$	3. AA$\sim$
4. $\frac{AB}{DC} = \frac{AE}{DE}$	4. Def of $\sim \triangle$s

10.

Statement	Reason
1. $\overline{CD} \parallel \overline{AB}$	1. Given
2. $\angle D \cong \angle A$, $\angle C \cong \angle B$	2. If 2 lines are $\parallel$, then alt int $\angle$s are $\cong$.
3. $\triangle ABE \sim \triangle DCE$	3. AA$\sim$
4. $\frac{AE}{DE} = \frac{BE}{CE}$	4. Def of $\sim \triangle$s
5. $AE \cdot CE = BE \cdot DE$	5. If proport, then prod of means = prod of extremes.

Additional Example 4

Write two proportions that give the product $MN \cdot JK = OP \cdot GH$.

$\dfrac{MN}{OP} = \dfrac{GH}{JK}$, $\dfrac{JK}{OP} = \dfrac{GH}{MN}$
Other solutions are possible.

13.

Statement	Reason
1. $\overline{IJ} \perp \overline{FG}$, $\overline{HG} \perp \overline{FG}$	1. Given
2. $\angle IJF$ and $\angle G$ are rt $\angle$s.	2. Def of $\perp$
3. $\angle IJF \cong \angle G$	3. Rt $\angle$s are $\cong$.
4. $\angle F \cong \angle F$	4. Reflex Prop of Cong
5. $\triangle FJI \sim \triangle FGH$	5. AA~

14.

Statement	Reason
1. $\overline{IJ} \perp \overline{FG}$, $\triangle FJI \sim \triangle FGH$	1. Given
2. m $\angle IJF =$ m $\angle G$	2. Def of $\sim \triangle$s
3. $\angle IJF$ is a rt $\angle$.	3. Def of $\perp$
4. m $\angle IJF = 90$	4. Def of rt $\angle$
5. m $\angle G = 90$	5. Sub
6. $\overline{HG} \perp \overline{FG}$	6. Def of $\perp$

15.

Statement	Reason
1. $\overline{EB} \perp \overline{AC}$, $\overline{CF} \perp \overline{AE}$	1. Given
2. $\angle ABE$ and $\angle AFC$ are rt $\angle$s.	2. Def of $\perp$
3. $\angle ABE \cong \angle AFC$	3. Rt $\angle$s are $\cong$.
4. $\angle A \cong \angle A$	4. Reflex Prop of Congr
5. $\triangle ABE \sim \triangle AFC$	5. AA~

16.

Statement	Reason
1. $\overline{EB} \perp \overline{AC}$, $\overline{CF} \perp \overline{AE}$	1. Given
2. $\angle CBE$ and $\angle EFC$ are rt $\angle$s.	2. Def of $\perp$
3. $\angle CBE \cong \angle EFC$	3. Rt $\angle$s are $\cong$.
4. $\angle FDE \cong \angle CDB$	4. Vert $\angle$s are $\cong$.
5. $\triangle BCD \sim \triangle FED$	5. AA~

17.

Statement	Reason
1. $GHIJ$ is a trap.	1. Given
2. $\overline{JI} \parallel \overline{GH}$	2. Def of trap
3. $\angle JIG \cong \angle HGI$, $\angle IJH \cong \angle GHJ$	3. If 2 lines are $\parallel$, then alt int $\angle$s are $\cong$.
4. $\triangle JIK \sim \triangle HGK$	4. AA~
5. $\dfrac{GH}{IJ} = \dfrac{HK}{JK}$	5. Def of $\sim \triangle$s
6. $GH \cdot JK = IJ \cdot HK$	6. If proport, then prod of means = prod of extremes.

Based on the given information, is $\triangle ABC \sim \triangle XYZ$?

11. m $\angle A = 60$, m $\angle B = 50$, m $\angle X = 60$, m $\angle Z = 70$ Yes

12. m $\angle A = 55$, m $\angle C = 45$, m $\angle Y = 80$, m $\angle Z = 45$ Yes

13. Given: $\overline{IJ} \perp \overline{FG}$, $\overline{HG} \perp \overline{FG}$
Prove: $\triangle FJI \sim \triangle FGH$

14. Given: $\overline{IJ} \perp \overline{FG}$, $\triangle FJI \sim \triangle FGH$
Prove: $\overline{HG} \perp \overline{FG}$

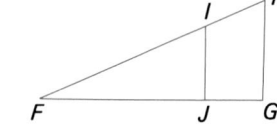

15. Given: $\overline{EB} \perp \overline{AC}$, $\overline{CF} \perp \overline{AE}$
Prove: $\triangle ABE \sim \triangle AFC$

16. Given: $\overline{EB} \perp \overline{AC}$, $\overline{CF} \perp \overline{AE}$
Prove: $\triangle BCD \sim \triangle FED$

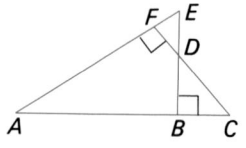

17. Given: Trap $GHIJ$
Prove: $GH \cdot JK = IJ \cdot HK$

18. Given: Trap $GHIJ$
Prove: $GK \cdot JK = IK \cdot HK$

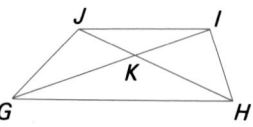

19. Given: $\triangle LMN \sim \triangle PQR$, altitudes $\overline{NO}$ and $\overline{RS}$, respectively
Prove: $\dfrac{NL}{RP} = \dfrac{NO}{RS}$

20. Given: $\triangle LMN \sim \triangle PQR$, altitudes $\overline{NO}$ and $\overline{RS}$, respectively
Prove: $ON \cdot SQ = OM \cdot SR$

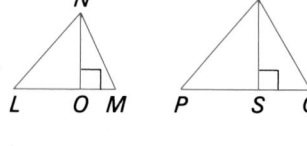

21. Given: Rt $\triangle ABC$, $\overline{ED} \perp \overline{AB}$, $\overline{BE}$ bis $\angle ABC$.
Prove: $\dfrac{AE}{AC} = \dfrac{BD}{BC}$

22. Given: Rt $\triangle ABC$, $\overline{ED} \perp \overline{AB}$, $\angle DEB \cong \angle EBA$
Prove: $\dfrac{1}{BC} + \dfrac{1}{AB} = \dfrac{1}{ED}$

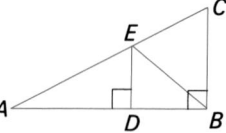

23. Prove Theorem 8.3.   **24.** Prove Theorem 8.4.

25. Prove Example 5 using a two-column proof.

Midchapter Review

Is the proportion true or false? *8.1*

1. $\dfrac{2}{6} = \dfrac{9}{27}$ True

2. $\dfrac{3}{4} = \dfrac{3+4}{4+3}$ False

3. $\dfrac{5}{9} = \dfrac{2 \cdot 5}{2 \cdot 9}$ True

Solve the proportion for x. *8.1*

4. $\dfrac{3}{x} = \dfrac{6}{12}$ $x = 6$

5. $\dfrac{1}{3} = \dfrac{x}{21}$ $x = 7$

6. $\dfrac{2}{x} = \dfrac{3}{20}$ $x = 13\frac{1}{3}$

316 Chapter 8 Similarity

Additional Example 5

Given: Rect $ABCD$
Prove: $AB \cdot CR = CD \cdot AR$

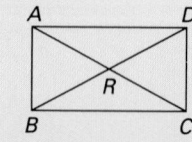

Quad $ABCD$ is rect (Given);
$\overline{AB} \cong \overline{CD}$ (Opp sides of a $\square$ are $\cong$);
$\overline{BR} \cong \overline{DR}$ $\overline{AR} \cong \overline{CR}$ (Diags of a $\square$ bis

each other.); $\triangle ABR \cong \triangle CDR$ (SSS);
$\triangle ABR \sim \triangle CDR$ ($\cong \triangle$s are $\sim$);
$\dfrac{AB}{CD} = \dfrac{AR}{CR}$ (Def of $\sim \triangle$s);
$AB \cdot CR = CD \cdot AR$ (In a proportion, prod of means = prod of extremes.)

Find the geometric mean of the pair of positive numbers. *8.1*

7. 1 and 16   4

8. 5 and 6 $\sqrt{30}$

9. p and q $\sqrt{p \cdot q}$

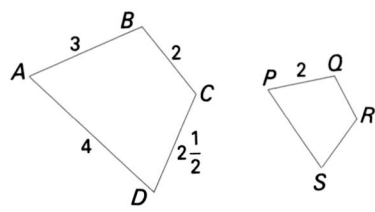

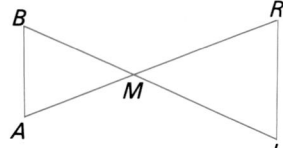

10. Find the missing lengths for the pair of similar quadrilaterals above.

$PS = 2\frac{2}{3}$, $RS = 1\frac{2}{3}$, $QR = 1\frac{1}{3}$

11. Write a similarity statement for the triangles above.   *8.3* $\triangle ABC \sim \triangle XYZ$
m $\angle B = 100$, m $\angle Y = 100$,
m $\angle A = 60$, m $\angle X = 60$

12. Given: $\overline{AB} \parallel \overline{RL}$. Write a similarity statement. Prove triangle similarity.   *8.3*
$\angle A \cong \angle R$ (If lines $\parallel$, alt int $\angle$s are $\cong$.);
$\triangle AMB \sim \triangle RML$; AA Post for $\triangle$s.

Application: *Finding the Height of a Tree*

One method of estimating the height of a tall tree involves properties of similar triangles. Place a mirror on level ground, as shown in the diagram, and stand where you can see the top of the tree in the mirror. Since the light bounces off a mirror at an angle equal to that at which it strikes the mirror, $\angle HRF \cong \angle TRB$. Assuming that the tree and you are both perpendicular to the ground, $\angle TBR \cong \angle HFR$; therefore, $\triangle TBR \sim \triangle HFR$ by AA Similarity. Measurements can then be made and the proportion solved for the height of the tree.

1. Suppose a person 6 ft tall places the mirror so that the top of the tree is visible. The mirror sits 20 ft from the tree and 1 ft from the person. How tall is the tree?   120 ft

2. Suppose the person repeats the procedure with another tree and finds $BR = 20$, $RF = 2$. How tall is this tree?   60 ft

3. Using a person's height is actually a simplification. What precisely should $\overline{HF}$ represent in the diagram? $\perp$ distance from eyes to ground

8.3 Similar Triangles   **317**

18.

Statement	Reason
1. *GHIJ* is a trap.	1. Given
2. $\overline{IJ} \parallel \overline{GH}$	2. Def of trap
3. $\angle JIG \cong \angle HGI$, $\angle IJH \cong \angle GHJ$	3. If 2 lines are $\parallel$, then alt int $\angle$s are $\cong$.
4. $\triangle JIK \sim \triangle HGK$	4. AA~
5. $\frac{JK}{HK} = \frac{IK}{GK}$	5. Def of $\sim$ $\triangle$s
6. $GK \cdot JK = IK \cdot HK$	6. If proport, then prod of means = prod of extremes.

19.

Statement	Reason
1. $\triangle LMN \sim \triangle PQR$, alts $\overline{NO}$ and $\overline{RS}$, respectively.	1. Given
2. $\angle L \cong \angle P$	2. Def of $\sim$ $\triangle$s
3. $\overline{NO} \perp \overline{LM}$, $\overline{RS} \perp \overline{PQ}$	3. Def of alt
4. $\angle NOL$ and $\angle RSP$ are rt $\angle$s.	4. Def of $\perp$
5. $\angle NOL \cong \angle RSP$	5. Rt $\angle$s are $\cong$.
6. $\triangle LON \sim \triangle PSR$	6. AA~
7. $\frac{NL}{RP} = \frac{NO}{RS}$	7. Def of $\sim$ $\triangle$s

20.

Statement	Reason
1. $\triangle LMN \sim \triangle PQR$, alts $\overline{NO}$ and $\overline{RS}$, respectively.	1. Given
2. $\angle M \cong \angle Q$	2. Def of $\sim$ $\triangle$s
3. $\overline{NO} \perp \overline{LM}$, $\overline{RS} \perp \overline{PQ}$	3. Def of alt
4. $\angle NOM$ and $\angle RSQ$ are rt $\angle$s.	4. Def of $\perp$
5. $\angle NOM \cong \angle RSQ$	5. Rt $\angle$s are $\cong$.
6. $\triangle NOM \sim \triangle RSQ$	6. AA~
7. $\frac{ON}{SR} = \frac{OM}{SQ}$	7. Def of $\sim$ $\triangle$s
8. $ON \cdot SQ = OM \cdot SR$	8. If proport, then prod of means = prod of extremes.

See pages 659–660 for the answers to Ex. 21–25.

Enrichment

Have students consider the given conditions for the triangles below. (The figures are not drawn proportionally.) Ask them to find conditions for the angles of the four smaller triangles that will make the two larger triangles similar if m $\angle 5$ is 35.

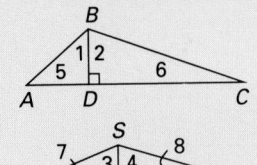

In $\triangle$s *ABD* and *RSQ*: m $\angle 5$ = m $\angle 7$ = 35, m $\angle 1$ = m $\angle 3$ = 55. In $\triangle$s *BDC* and *TQS*: m $\angle 2$ = m $\angle 8$, m $\angle 4$ = m $\angle 6$, m $\angle 2$ = m $\angle 4$, m $\angle 6$ = m $\angle 8$, m $\angle$s 2, 4, 6, 8 = 45. Given: $\triangle ABD \sim \triangle RSQ$ $\triangle BDC \sim \triangle TQS$

Teaching Resources

Manipulative Worksheet 15
Quick Quizzes 59
Reteaching and Practice
  Worksheets, pp. 117, 118
Transparency 18

▆▆▆▆GETTING STARTED

Prerequisite Quiz

Solve the proportion.

1. $\frac{2.5}{x} = \frac{5}{10}$ 5
2. $\frac{2}{3} = \frac{x}{8}$ $5\frac{1}{3}$

Motivator

Ask students how many points divide a
segment in half.   One   Have the students
draw a rectangle on a sheet of paper. Have
them mark the midpoints of two opposite
sides of the rectangle and connect them.
Ask them if this segment divides the
rectangular shape in half.   Yes   Next have
them draw a triangle on the paper and mark
the midpoints of two sides of the triangle.
Have them draw a segment connecting
these points and ask them what they notice
about this segment.   It is parallel to the
third side of the triangle and it bisects the
other two sides.

▆▆▆▆TEACHING SUGGESTIONS

Lesson Note

Distribute a page with several similar
triangles of various sizes and shapes for
students to measure. Have students draw a
line parallel to one side of each triangle and
intersecting the other two sides. Then have
them measure the segments cut off on the
two sides of each triangle and compare the
ratios. Emphasize the importance of writing
the ratios in the same order. Students may
not have observed that the definition of
similarity allows only for proportions such as
$\frac{AC}{CD} = \frac{BC}{EC}$ (See the figure on page 318.) With
Theorem 8.5 the students now are able to
use the proportion $\frac{CD}{DA} = \frac{CE}{EB}$. Using Theorem
8.2 it is possible to have other proportions
related to the triangle of Theorem 8.5. For
example, using the corollary to Theorem 8.1
results in $\frac{CD}{CE} = \frac{CA}{CB}$.

318

8.4 The Triangle Proportionality Theorem

Objectives   To prove theorems involving triangle proportionality
To apply the Triangle Proportionality Theorem and its converse

Two segments are divided proportionally when the lengths of the parts
of one segment form a proportion with the lengths of the parts of the
other segment.

Points P and Q divide the two segments below proportionally, since
$\frac{2}{3} = \frac{4}{6}$. $\overline{AB}$ and $\overline{CD}$ are divided proportionally if $\frac{AP}{PB} = \frac{CQ}{QD}$.

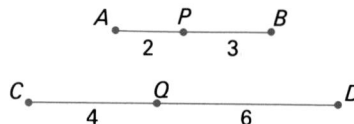

Theorem 8.5   **Triangle Proportionality Theorem:** If a line is parallel to one side of
a triangle and intersects the other two sides, then it divides the two
sides proportionally.

Given: $\triangle ABC$, with $\overleftrightarrow{DE} \parallel \overline{AB}$
Prove: $\frac{AD}{DC} = \frac{BE}{EC}$

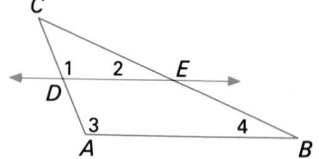

Proof

Statement	Reason
1. $\triangle ABC$ with $\overleftrightarrow{DE} \parallel \overline{AB}$	1. Given
2. $\angle 1 \cong \angle 3$, and $\angle 2 \cong \angle 4$	2. Corr $\angle$s of $\parallel$ lines are $\cong$.
3. $\triangle ABC \sim \triangle DEC$	3. AA $\sim$
4. $\frac{AC}{DC} = \frac{BC}{EC}$	4. Def of $\sim$ $\triangle$s
5. $\frac{AC - DC}{DC} = \frac{BC - EC}{EC}$	5. If $\frac{a}{b} = \frac{c}{d}$, then $\frac{a - b}{b} = \frac{c - d}{d}$
6. $AC - DC = AD$, $BC - EC = BE$	6. Seg Add Post
7. $\therefore \frac{AD}{DC} = \frac{BE}{EC}$	7. Sub

Theorem 8.2 leads to the following additional properties for $\triangle ABC$.

$$\frac{DC}{AC} = \frac{EC}{BC} \qquad \frac{DC}{AD} = \frac{EC}{BE} \qquad \frac{DC}{EC} = \frac{AD}{BE}$$

318   Chapter 8 Similarity

Additional Example 1

a. Find the missing measure. $\overline{AC} \parallel \overline{DE}$
4

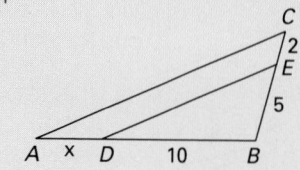

b. Find the missing measure. $\overline{AB} \parallel \overline{YZ}$
$6\frac{2}{3}$

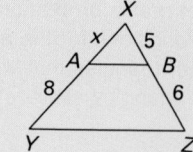

EXAMPLE 1 If $\overline{ST} \parallel \overline{RQ}$, which of the following are true?

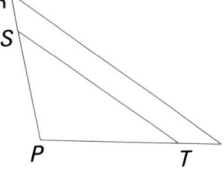

a. $\dfrac{PS}{SR} = \dfrac{PT}{TQ}$ b. $\dfrac{PS}{PR} = \dfrac{TQ}{PQ}$

Solution a. True, by Triangle Proportionality Theorem   b. False (unless $\overline{ST}$ is midsegment)

EXAMPLE 2 $\overline{ED} \parallel \overline{AB}$, $AE = 16$, $EC = 12$

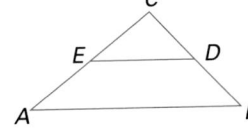

a. If $DC = 15$, find BD.   b. If $BD = 9$, find BC.

Solutions

$$\dfrac{AE}{EC} = \dfrac{BD}{DC}$$
$$\dfrac{16}{12} = \dfrac{BD}{15}$$
$$12 \cdot BD = 16 \cdot 15$$
$$BD = 20$$

$$\dfrac{AE}{AC} = \dfrac{BD}{BC}$$
$$\dfrac{16}{28} = \dfrac{9}{BC}$$
$$16 \cdot BC = 28 \cdot 9$$
$$BC = 15\tfrac{3}{4}$$

The converse of the Triangle Proportionality Theorem can be proved by using an auxiliary line.

Theorem 8.6 If a line divides two sides of a triangle proportionally, then the line is parallel to the third side of the triangle.

Given: $\triangle ABC$ with D between A and C, and E between B and C, $\dfrac{AD}{DC} = \dfrac{BE}{EC}$
Prove: $\overleftrightarrow{DE} \parallel \overline{AB}$

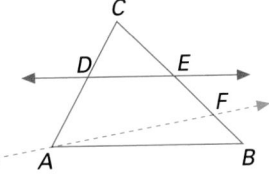

Plan Through point A, draw a line $\overleftrightarrow{AF}$ assumed parallel to $\overleftrightarrow{DE}$. Show that $BE = FE$.

EXAMPLE 3 Based on the information given, is $\overline{JK} \parallel \overline{HI}$?

Solution If $\overline{JK} \parallel \overline{HI}$, then $\dfrac{JH}{JG} = \dfrac{KI}{KG}$.
But this is not the case: $\dfrac{3}{5} \ne \dfrac{4}{6}$.
Therefore, $\overline{JK} \not\parallel \overline{HI}$.

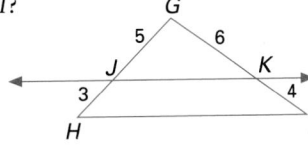

8.4 The Triangle Proportionality Theorem   **319**

Math Connections

Carpentry: Discuss with students the following application of the Triangle Proportionality Theorem.

Suppose a carpenter has a rectangular piece of wood that is $8\tfrac{1}{4}$ inches wide and he needs to divide it into 5 equally wide pieces. The carpenter discovers that the quotient of $8\tfrac{1}{4} \div 5$ involves a fraction that does not appear on the ruler. However, applying the Triangle Proportionality Theorem, the carpenter places the ruler diagonally, from 0 to 10 inches, across the width of the piece of wood, as shown in the diagram below. He then marks on the wood every 2 inches along the ruler, draws lines parallel to the side of the piece of wood and cuts 5 congruent pieces.

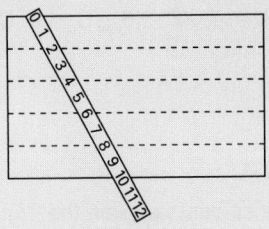

Critical Thinking Questions

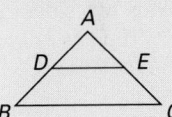

Application: Refer students to the figure above, where $\overline{DE} \parallel \overline{BC}$, and ask them which part of Theorem 8.2 on page 304, guarantees that $\dfrac{AE}{AD} = \dfrac{AC}{AB}$. By the $\triangle$ Proport Thm, $\dfrac{AE}{AC} = \dfrac{AD}{AB}$. Therefore, by Part (2) of Thm 8.2, $\dfrac{AE}{AD} = \dfrac{AC}{AB}$.

Common Error Analysis

Error: Students may have difficulty in writing the terms of the proportion in the correct order.

Emphasize that the letters designating corresponding lengths are in "corresponding locations" on the figures.

Additional Example 2

If $\overline{AB} \parallel \overline{YZ}$, which proportions are always true?

a. $\dfrac{AX}{XC} = \dfrac{BY}{YC}$ Yes
b. $\dfrac{CY}{AX} = \dfrac{CX}{BY}$ No

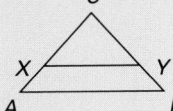

319

Checkpoint

1. The line intersecting the two sides of each triangle is parallel to the third side. Find the missing measurement. 36

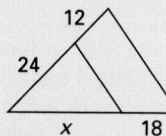

2. If $XP = 14$, $PZ = 10$, $YQ = 21$, $QZ = 15$, is $PQ \parallel XY$? Yes

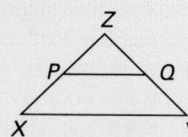

3. Find the unknown length. (Use figure for Exercise 2.)
   a. $PX = 4$, $XZ = 6$, $ZQ = 9$, $QY =$ _____ 18
   b. $XZ = 14$, $PZ = 5$, $YQ = 6$, $QZ =$ _____ $3\frac{1}{3}$

Closure

Have the students explain the Triangle Proportionality Theorem. Theorem 8.5 page 318. Ask them what they know about a line that divides two sides of a triangle proportionally. The line is parallel to the third side of the triangle.

▰▰▰ FOLLOW UP

Guided Practice

Classroom Exercises 1–4

Independent Practice

Ⓐ Ex. 1–17, Ⓑ Ex. 18–24, Ⓒ Ex. 25–28

Basic: WE 1–17 odd, Cl 1
Average: WE 1–24 odd, Cl all
Above Average: WE 1–27 odd, Cl all

Classroom Exercises

For Exercises 1–4, $\overline{DE} \parallel \overline{AB}$. Complete the proportion.

1. $\dfrac{CD}{CA} = \dfrac{CE}{?}$ CB

2. $\dfrac{CE}{CB} = \dfrac{?}{CA}$ CD

3. $\dfrac{?}{DC} = \dfrac{BC}{EC}$ AC

4. $\dfrac{CD}{?} = \dfrac{CE}{EB}$ DA

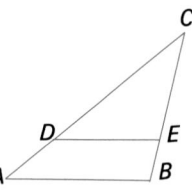

Written Exercises

Assume that the line intersecting the two sides of the triangle is parallel to the third side. Find the missing measure.

1. 8
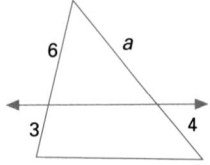

2. 5 b
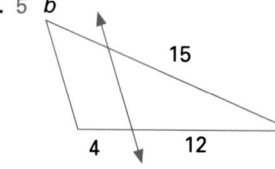

3. 6
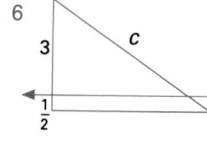

4. 3
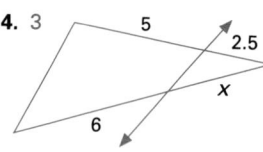

5. 4
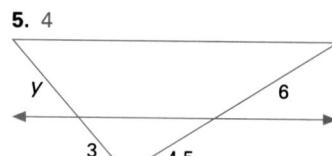

6. $2\frac{1}{12}$
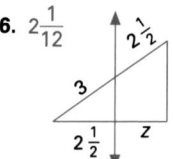

Based on the information given, is $\overline{IJ} \parallel \overline{FG}$?

7. $FI = 12$, $IH = 8$, $GJ = 15$, $JH = 10$ Yes
8. $FI = 15$, $IH = 20$, $GJ = 18$, $JH = 25$ No
9. $FH = 20$, $FI = 12$, $GH = 30$, $GJ = 18$ Yes
10. $FH = 18$, $FI = 10$, $GH = 27$, $JH = 15$ No

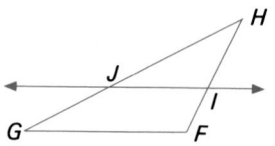

Determine whether the line(s) intersecting two of the sides of the triangle is(are) parallel to the third side.

11. Yes
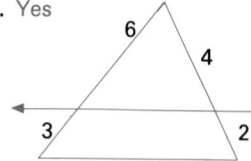

12. No
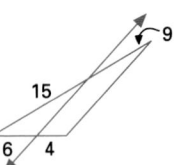

13. Vert, yes; horiz, no
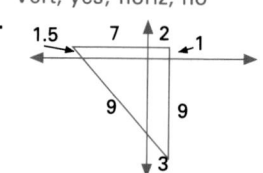

320   Chapter 8 Similarity

Additional Example 3

Based on the given information, is $\overline{MN} \parallel \overline{YA}$? No

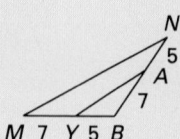

$\overline{IJ} \parallel \overline{FG}$. Find the indicated length.

14. $HJ = 4$, $HG = 6$, $HI = 6$, $HF = \underline{\quad 9 \quad}$
15. $HJ = 12$, $HG = 15$, $HF = 20$, $HI = \underline{\quad 16 \quad}$
16. $HJ = 12$, $JG = 24$, $HI = 8$, $IF = \underline{\quad 16 \quad}$
17. $HI = 15$, $IF = 10$, $JG = 12$, $HJ = \underline{\quad 18 \quad}$

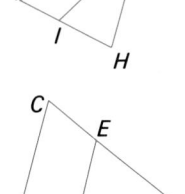

18. Given: $\dfrac{AD}{DB} = \dfrac{CE}{EB}$, $\angle BDE \cong \angle BED$
Prove: $\angle A \cong \angle C$
19. Given: $\angle A \cong \angle C$, $\angle BDE \cong \angle BED$
Prove: $\dfrac{AD}{DB} = \dfrac{CE}{EB}$

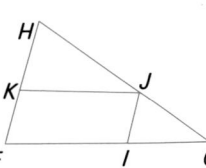

20. Given: $\overline{KJ} \parallel \overline{FG}$, $\overline{IJ} \parallel \overline{FH}$
Prove: $\dfrac{FK}{FH} = \dfrac{IG}{FG}$

21. Complete the proof of Theorem 8.6 in a two-column form.

Solve the following proportions, correct to the nearest tenth, using a calculator.

22. $\dfrac{25}{x} = \dfrac{49}{127}$ 64.8

23. $\dfrac{17}{55} = \dfrac{x}{9}$ 2.8

24. $\dfrac{101}{36} = \dfrac{74}{x}$ 26.4

25. Prove: If three parallel lines intersect two transversals, then they divide the transversals proportionally.

Prove or disprove with a counterexample.

26. Given: $\overline{DE} \parallel \overline{AB}$, $\overline{FE} \parallel \overline{CB}$
Prove or disprove: $\overline{FD} \parallel \overline{CA}$
27. Given: $\overline{FD} \parallel \overline{CA}$, $\overline{DE} \parallel \overline{AB}$
Prove or disprove: $\dfrac{GF}{FC} = \dfrac{GE}{EB}$

28. Give a proof of the Midsegment Theorem based on Theorem 8.6, the converse of the Triangle Proportionality Theorem.

Mixed Review

1. State the Symmetric Property of Equality.   *2.3*
2. State the Transitive Property of Congruence.   *2.3*
3. Draw a concave pentagon.   *6.1*
4. Draw a convex polygon.   *6.1*

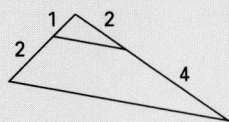
Additional Answers

Written Exercises

18.

Statement	Reason
1. $\dfrac{AD}{DB} = \dfrac{CE}{EB}$, $\angle BDE \cong \angle BED$	1. Given
2. $\overline{DE} \parallel \overline{AC}$	2. Midseg Thm
3. $\angle BDE \cong \angle A$, $\angle BED \cong \angle C$	3. If lines are $\parallel$, then corr $\angle$s are $\cong$.
4. $\angle A \cong \angle C$	4. Trans Prop of Congr

19.

Statement	Reason
1. $\angle A \cong \angle C$, $\angle BDE \cong \angle BED$	1. Given
2. m$\angle A$ + m$\angle C$ + m$\angle B$ = 180, m$\angle BDE$ + m$\angle BED$ + m$\angle B$ = 180.	2. Sum of meas of $\angle$s of $\triangle$ = 180.
3. m$\angle A$ + m$\angle C$ + m$\angle B$ = m$\angle BDE$ + m$\angle BED$ + m$\angle B$	3. Sub
4. m$\angle B$ = m$\angle B$	4. Reflex Prop of Eq
5. m$\angle A$ + m$\angle C$ = m$\angle BDE$ + m$\angle BED$	5. Subt Prop of Eq
6. 2 $\times$ m$\angle A$ = 2 $\times$ m$\angle BDE$	6. Sub
7. m$\angle A$ = m$\angle BDE$	7. Div Prop of Eq
8. $\overline{DE} \parallel \overline{AC}$	8. If 2 lines are inters by a transv so that corr $\angle$s are $\cong$, then lines are $\parallel$.
9. $\dfrac{AD}{DB} = \dfrac{CE}{EB}$	9. Triangle Prop Thm

20.

Statement	Reason
1. $\overline{KJ} \parallel \overline{FG}$, $\overline{IJ} \parallel \overline{FH}$	1. Given
2. $\dfrac{KH}{FK} = \dfrac{JH}{GJ}$, $\dfrac{JH}{GJ} = \dfrac{IE}{IG}$	2. Triangle Proport Thm
3. $\dfrac{KH}{FK} = \dfrac{IF}{IG}$	3. Trans Prop of Congr
4. $\dfrac{KH + FK}{FK} = \dfrac{IF + IG}{IG}$	4. If $\dfrac{a}{b} = \dfrac{c}{d}$; then $\dfrac{a+b}{b} = \dfrac{c+d}{d}$
5. $KH + FK = FH$; $IF + IG = FG$	5. Seg Add Post
6. $\dfrac{FH}{FK} = \dfrac{FG}{IG}$	6. Sub
7. $\dfrac{FK}{FH} = \dfrac{IG}{FG}$	7. If $\dfrac{a}{b} = \dfrac{c}{d}$; then $\dfrac{b}{a} = \dfrac{d}{c}$

See pages 660–661 for the answers to Written Ex. 21, 25–28 and Mixed Review Ex. 1–4.

321

Computer Investigation

Similar Triangles

Use a computer software program that constructs triangles by definition of SAS and SSS, labels points, and measures line segments and angles.

Activity 1

Draw the following triangles.

1. Construct and label $\triangle ABC$ with $AC = 2$, $AB = 1$, and m $\angle A = 40$.
2. Use the measuring tool of the software program to find CB, m $\angle B$, and m $\angle C$.
3. Construct a new triangle and label it $A'B'C'$ with $A'C' = 4$, $A'B' = 2$, and m $\angle A = 40$.
4. Find $C'B'$, m $\angle B'$, and m $\angle C'$.
5. Find $C'B' : CB$.
6. Why are the triangles now similar?

Activity 2

7. Construct $\triangle ABC$ with $AC = 3$, $AB = 2$, and m $\angle A = 30$.
8. Find the measure of each of the remaining parts.
9. Construct a new triangle and label it $\triangle A'B'C'$ with $A'B' = 6$, $A'C' = 9$, and m $\angle A = 30$.
10. Find the measures of each of the remaining parts.
11. Why are the two triangles now similar?

Activity 3

Lower case letters refer to the sides of the triangle opposite the angle of the same name; the side can also be named by the adjacent vertices. (Side a is opposite $\angle A$, also referred to as $\overline{BC}$.)

12. Construct and label $\triangle ABC$ with sides: $a = 2$, $b = 4$, and $c = 3$.
13. Find the measure of each angle.
14. Construct and label a new $\triangle A'B'C'$ with each side measuring twice the corresponding side lengths of $\triangle ABC$.
15. Find the measure of each angle of $\triangle A'B'C'$.
16. Why are the triangles now similar?
17. Repeat the steps of Exercises 13–17 for two new triangles with sides: $a = 3$, $b = 4$, $c = 5$; and $a' = 6$, $b' = 8$, $c' = 10$.
18. Try to generalize a new way to show triangles similar.

8.5 SAS and SSS Similarity Theorems

Objective   To apply the SAS and SSS Similarity Theorems

Given $\triangle ABC$, you can construct $\triangle XYZ$, so that $\angle X \cong \angle A$ and the sides $\overline{XY}$ and $\overline{XZ}$ are twice as long as sides $\overline{AB}$ and $\overline{AC}$, respectively. The two triangles will appear to be similar. This suggests the following theorem.

Theorem 8.7

SAS Similarity Theorem: If an angle of one triangle is congruent to an angle of another triangle and the corresponding sides that include these angles are proportional, then the triangles are similar.

Given: $\angle C \cong \angle F$, $\dfrac{AC}{DF} = \dfrac{BC}{EF}$
Prove: $\triangle ABC \sim \triangle DEF$

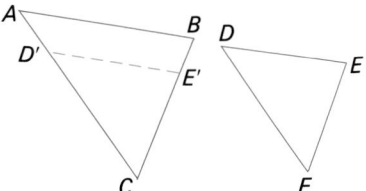

Proof

Statement	Reason
1. $\angle C \cong \angle F$	(A) 1. Given
2. Choose D' on $\overline{AC}$ so that $\overline{D'C} \cong \overline{DF}$ ($D'C = DF$)	(S) 2. Ruler Post
3. Choose E' on $\overline{BC}$ so that $\overline{E'C} \cong \overline{EF}$ ($E'C = EF$)	(S) 3. Ruler Post
4. $\triangle D'E'C \cong \triangle DEF$	4. SAS
5. $\triangle D'E'C \sim \triangle DEF$	5. $\cong \triangle$s are similar.
6. $\dfrac{AC}{DF} = \dfrac{BC}{EF}$	6. Given
7. $\dfrac{AC}{D'C} = \dfrac{BC}{E'C}$	7. Sub
8. $\overline{D'E'} \parallel \overline{AB}$	8. If line divides two sides of $\triangle$ proport, it is $\parallel$ to third side.
9. $\angle CD'E' \cong \angle A$ and $\angle CE'D' \cong \angle B$	9. Corr $\angle$s of $\parallel$ lines are $\cong$.
10. $\triangle ABC \sim \triangle D'E'C$	10. AA $\sim$
11. $\therefore \triangle ABC \sim \triangle DEF$	11. Trans Prop of $\triangle \sim$

This SAS similarity pattern is suggested by the SAS congruence theorem.

Another similarity pattern suggested by congruence is an SSS pattern.

▰▰ GETTING STARTED

Prerequisite Quiz

What method would you use to prove congruence for each of the following?

1. SAS   **2.** SSS

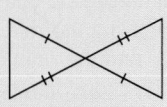   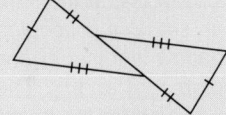

Motivator

Have the students define similar triangles. They have congruent corresponding angles and proportional corresponding sides. Ask them how the resulting triangle and the original triangle would be related if they were to triple the lengths of all three sides of the original triangle. The resulting triangle would be similar to the original, but its sides would be three times as long.

▰▰ TEACHING SUGGESTIONS

Lesson Note

Compare the theorems for similar triangles with the SAS and SSS theorems for congruent triangles. Students may need to be reminded about the importance of correct notation for similar triangles and proportions. The steps to prove line segments proportional are as follows:
(1) Find two triangles, each of which has two of the four segments as sides.
(2) Prove these triangles similar.
(3) Form a proportion using the corresponding sides.
(4) Transform the proportion as needed.

Math Connections

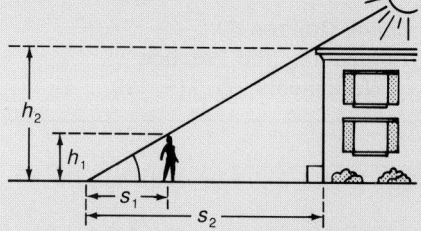

Shadows: The diagram above illustrates the application of the SAS Similarity Theorem in finding the height of a building. By measuring the length of a person's shadow, s_1, and the length of the building's shadow, s_2, at the same time of day, you can set up a proportion involving the ratio of shadow lengths, s_1 and s_2, and the ratio of heights, h_1 and h_2. Then the proportion can be solved for the height of the building, h_2.

Critical Thinking Questions

Application: Ask students what assumption is made in the application of the SAS Similarity Theorem in the Math Connections above. The ∠ the person forms with the ground is a rt ∠. If not, then by the △ Proport Thm, the sides of the △, shown in the diagram, will not be proport.

Checkpoint

For the following pairs of triangles:

1. Determine if they are similar.
2. Give the theorem that you would use to prove the triangles similar.
3. Write the similarity statement.
   No

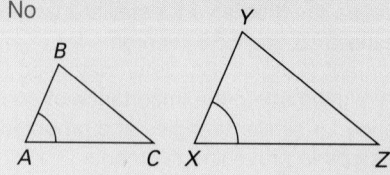

Yes, AA ~; △ABC ~ △DEF

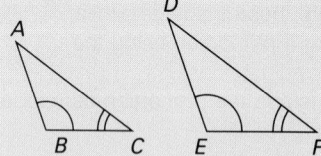

Closure

Have the students explain the two similarity theorems they learned in this lesson. SAS Similarity Theorem page 323 and SSS Similarity Theorem page 324

324

Theorem 8.8

SSS Similarity Theorem: If all three pairs of corresponding sides of two triangles are proportional, then the two triangles are similar.

Given: $\dfrac{CA}{FD} = \dfrac{AB}{DE} = \dfrac{CB}{FE}$

Prove: $\triangle ABC \sim \triangle DEF$

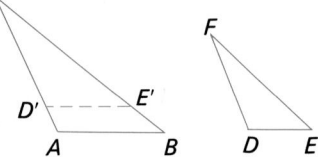

Plan

Assume $CA > FD$. Locate D' on $\overline{CA}$ and E' on $\overline{CB}$ so that $\overline{CD'} \cong \overline{FD}$ and $\overline{CE'} \cong \overline{FE}$. Use SAS Similarity to prove $\triangle ABC \sim \triangle D'E'C$. Prove $D'E' \cong DE$ by using these similar triangles and the given. Then $\triangle D'E'C \cong \triangle DEF$ by SSS. Since congruent triangles are similar, the Transitive Property for Similar Triangles completes the proof.

You now have available three methods for proving triangles similar: AA Similarity, SAS Similarity, SSS Similarity.

EXAMPLE 1 Is $\triangle PQR \sim \triangle XYZ$? Why or why not?

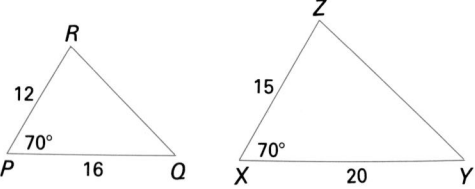

Solution m $\angle P = 70$ and m $\angle X = 70$, so $\angle P \cong \angle X$

$\dfrac{12}{15} = \dfrac{4}{5}$ and $\dfrac{16}{20} = \dfrac{4}{5}$, so $\dfrac{PR}{XZ} = \dfrac{PQ}{XY}$

Thus, $\triangle PQR \sim \triangle XYZ$ by SAS Similarity.

Some special triangles are easily shown to be similar. All angles of equiangular triangles have measure of 60. Therefore, all equiangular triangles are similar by the AA Similarity Postulate. Isosceles right triangles can be proved similar by the SAS Similarity Theorem.

EXAMPLE 2 Prove all equilateral triangles are similar.
Given: Equilateral △s ABC and DEF
Prove: $\triangle ABC \sim \triangle DEF$

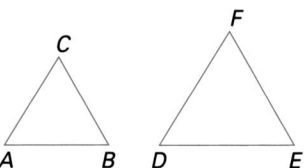

Additional Example 1

Is $\triangle ABC \sim \triangle CDE$? No

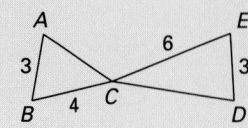

Additional Example 2

Prove: All isosceles right triangles are similar.
$AB = BC$,
$DE = EF$,
m $\angle E = 90$ (Given)
$\dfrac{AB}{DE} = \dfrac{BC}{EF}$ (Div Prop of Eq)
$\triangle ABC \sim \triangle DEF$ (SAS ~)

Proof	Statement	Reason

Statement	Reason
1. $\triangle ABC$ and $\triangle DEF$ are equilateral.	1. Given
2. $AB = BC = CA$, $DE = EF = FD$	2. Def of equil $\triangle$
3. $\dfrac{AB}{DE} = \dfrac{BC}{DE} = \dfrac{CA}{DE}$	3. Div Prop of Eq
4. $\dfrac{AB}{DE} = \dfrac{BC}{EF} = \dfrac{CA}{FD}$	4. Sub
5. $\therefore \triangle ABC \sim \triangle DEF$	5. SSS $\sim$

EXAMPLE 3 Given: $\dfrac{KI}{HI} = \dfrac{JI}{GI}$

Prove: $KJ \cdot HI = HG \cdot KI$

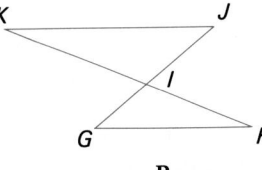

Proof		
Statement		Reason
1. $\dfrac{KI}{HI} = \dfrac{JI}{GI}$		1. Given
2. $\angle KIJ \cong \angle HIG$		2. Vert $\angle$s are $\cong$.
3. $\triangle KIJ \sim \triangle HIG$		3. SAS $\sim$
4. $\dfrac{KJ}{HG} = \dfrac{KI}{HI}$		4. Def of similar $\triangle$s
5. $\therefore KJ \cdot HI = HG \cdot KI$		5. In a proportion, the prod of extremes = the prod of means.

Focus On Reading

Select the correct spelling of each word.

1. allitude   altatude   altitude ✔

2. congruant   congruent ✔   congerant

3. coresponding   coressponding   corresponding ✔

4. extremes ✔   extreems   extreemes

5. similiar   simular   similar ✔

Classroom Exercises

Determine whether the two triangles are similar. Explain why or why not.

1.

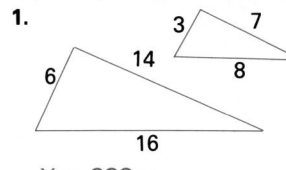

Yes, SSS$\sim$

2.

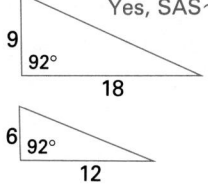

Yes, SAS$\sim$

3.

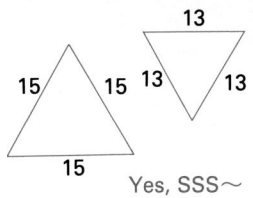

Yes, SSS$\sim$

Additional Example 3

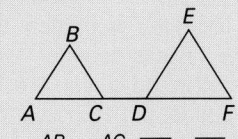

Given: $\dfrac{AB}{DE} = \dfrac{AC}{DF}$, $\overline{AB} \parallel \overline{DE}$

Prove: $AB \cdot EF = BC \cdot DE$

$\angle A \cong \angle D$ (corr $\angle$s of $\parallel$ lines)

$\triangle ABC \sim \triangle DEF$ (SAS Sim Thm)

$\dfrac{AB}{DE} = \dfrac{BC}{EF}$ (Def of sim $\triangle$s)

$AB \cdot EF = BC \cdot DE$ (If proport, the prod of means = prod of extremes.)

Guided Practice

Classroom Exercises 1–6

Independent Practice

A Ex. 1–8, **B** Ex. 9–24, **C** Ex. 25–28

Basic: FR all, WE 1–8, AR all

Average: FR all, WE 1–16 odd, AR all

Above Average: FR all, WE 1–22 odd, AR-all

Additional Answers

Written Exercises

9.

Statement	Reason
1. Equiangular $\triangle$s ABC and DEF	1. Given
2. m $\angle A =$ m $\angle B =$ m $\angle D =$ m $\angle E = 60$	2. Meas of $\angle$ of reg polygon with n sides $= \dfrac{(n-2)180}{n}$
3. $\triangle ABC \sim \triangle DEF$	3. AA$\sim$

10.

Statement	Reason
1. Isos rt $\triangle$s ABC and DEF	1. Given
2. m $\angle A = 90$, m $\angle D = 90$	2. Def of rt $\triangle$
3. m $\angle A =$ m $\angle D$	3. Sub
4. $AC = AB$, $DF = DE$	4. Def of isos $\triangle$
5. $\dfrac{AC}{DF} = \dfrac{AC}{DF}$	5. Reflex Prop of Eq
6. $\dfrac{AC}{DF} = \dfrac{AB}{DE}$	6. Sub
7. $\triangle ABC \sim \triangle DEF$	7. SAS$\sim$

11.

Statement	Reason
1. $\square GKLJ$	1. Given
2. $\overline{JL} \parallel \overline{GK}$, $\overline{JG} \parallel \overline{LK}$	2. Def of $\square$
3. $\angle JLI \cong \angle KHL$, $\angle IJL \cong \angle JGK$, $\angle JGK \cong \angle LKH$	3. If 2 $\parallel$ lines are inters by transv, then corr $\angle$s are $\cong$.
4. $\angle IJL \cong \angle LKH$	4. Trans Prop of Congr
5. $\triangle JLI \sim \triangle KHL$	5. AA$\sim$

12.

Statement	Reason
1. $\triangle KHL \sim \triangle JLI$	1. Given
2. $\angle HLK \cong \angle I$	2. Def of $\sim$ $\triangle$s
3. $\angle H \cong \angle H$	3. Reflex Prop of Congr
4. $\triangle KHL \sim \triangle GHI$	4. AA$\sim$

13.

Statement	Reason
1. $\frac{NQ}{NO} = \frac{NP}{NM}$	1. Given
2. $\angle N \cong \angle N$	2. Reflex Prop of Congr
3. $\triangle NPQ \sim \triangle NMO$	3. SAS~
4. $\frac{NQ}{NO} = \frac{QP}{OM}$	4. Def of $\sim \triangle$s

14.

Statement	Reason
1. $\frac{NP}{NM} = \frac{QN}{ON}$	1. Given
2. $\angle N \cong \angle N$	2. Reflex Prop of Congr
3. $\triangle PQN \sim \triangle MON$	3. SAS~
4. $\frac{QP}{OM} = \frac{NP}{NM}$	4. Def of $\sim \triangle$s
5. $QP \cdot NM = OM \cdot NP$	5. If proport, then prod of means = prod of extremes.

15.

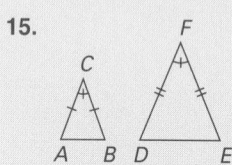

Statement	Reason
1. Isos $\triangle$s ABC and DEF with $\cong$ vert $\angle$s C and F.	1. Given
2. $AC = BC$, $DF = EF$	2. Def of isos $\triangle$
3. $\frac{AC}{AC} = \frac{DF}{DF} = 1$	3. A number div by itself = 1.
4. $\frac{AC}{BC} = \frac{DF}{EF}$	4. Sub
5. $\frac{AC}{DF} = \frac{BC}{EF}$	5. If $\frac{a}{b} = \frac{c}{d}$, then $\frac{a}{c} = \frac{b}{d}$.
6. $\triangle ABC \sim \triangle DEF$	6. SAS~

16.

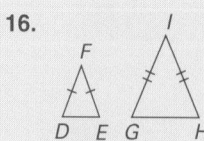

Statement	Reason
1. Isos $\triangle$s DEF and GHI with vert $\angle$s F and I, $\frac{DE}{GH} = \frac{EF}{HI}$.	1. Given
2. $EF = FD$, $HI = IG$	2. Def of isos $\triangle$s
3. $\frac{DE}{GH} = \frac{FD}{IG}$	3. Sub
4. $\frac{EF}{HI} = \frac{FD}{IG}$	4. Sub
5. $\triangle DEF \sim \triangle GHF$	5. SSS~

4.

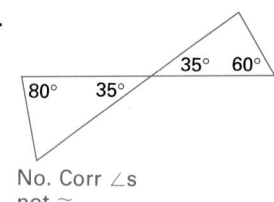

No. Corr $\angle$s not $\cong$

5.

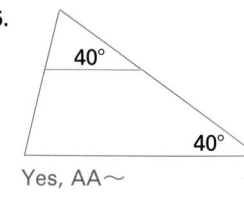

Yes, AA~

6.

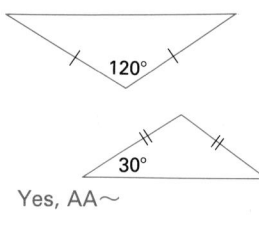

Yes, AA~

Written Exercises

On the basis of the given information, is $\triangle ABC \sim \triangle DEF$? Explain. (Exercises 1–8)

1. $\angle A \cong \angle D$, $\angle B \cong \angle E$ Yes, AA~

2. $\frac{AB}{DE} = \frac{BC}{EF} = \frac{CA}{FD}$ Yes, SSS~

3. $\angle A \cong \angle D$ $\frac{AB}{DE} = \frac{BC}{EF}$ No

4. $\frac{BC}{EF} = \frac{CA}{FD}$ No

5. m $\angle A$ = 40, m $\angle B$ = 40, m $\angle D$ = 40, m $\angle F$ = 100   Yes, AA~

6. $\frac{AC}{DF} = \frac{AB}{DE}$, $\angle A \cong \angle D$ Yes, SAS~

7. $\frac{AC}{DE} = \frac{AB}{DF} = \frac{AB}{EF}$ No

8. $\triangle ABC$ is equilateral, $\angle D \cong \angle F$ No

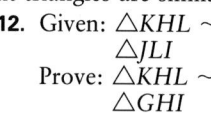

9. Prove that all equiangular triangles are similar.

10. Prove that all isosceles right triangles are similar.

11. Given: $\square GKLJ$
Prove: $\triangle JLI \sim \triangle KHL$

12. Given: $\triangle KHL \sim \triangle JLI$
Prove: $\triangle KHL \sim \triangle GHI$

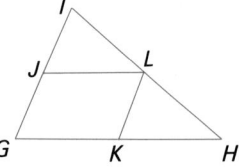

13. Given: $\frac{NQ}{NO} = \frac{NP}{NM}$
Prove: $\frac{NQ}{NO} = \frac{QP}{OM}$

14. Given: $\frac{NP}{NM} = \frac{QN}{ON}$
Prove: $QP \cdot NM = OM \cdot NP$

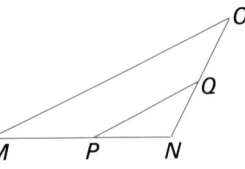

15. Prove that two isosceles triangles are similar if the vertex angle of one is congruent to the vertex angle of the other.

16. Prove that two isosceles triangles are similar if the base and a leg of one are proportional to the base and a leg of the other.

17. Prove Theorem 8.8 for $CA > FD$.

18. Prove or disprove: The triangle formed by connecting the midpoints of the sides of a triangle is similar to the given triangle.

19. Prove or disprove: The triangle formed by perpendiculars to the sides of a triangle is similar to the given triangle $\triangle ABC \sim \triangle EDF$.

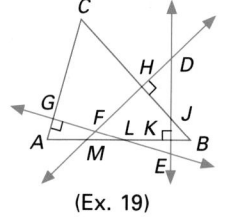

(Ex. 19)

Enrichment

The five segments within the right triangle in the figure are equally spaced and divide $\overline{AB}$ into 6 equal parts. If $BC = 10$, ask students to find the sum of the measures of the five segments.

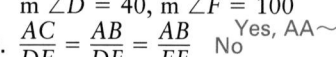

The segments form six similar triangles. The length of the segment next to $\overline{BC}$ can be found using the proportion: $5:6 = x:10$. $x = 50/6$. If this is done for each segment, we obtain the sum $150/6 = 25$. Another way is to form a square. If the segments are extended, each will equal 10, making a sum of 50. But the diagonal cuts the square in half so the sum of the segments is 25.

For which of the following patterns can two quadrilaterals be proved similar? Explain your reasoning step-by-step.

20. SASAS Similarity   **21.** ASASA Similarity

22. The steepness of a road is measured by its *grade*. A three-percent grade means that the road rises 3 ft for every 100 ft of horizontal length. If you drive up a three-percent grade for 2 mi, how many ft higher are you than when you started (1 mi = 5,280 ft)?   316.8 ft

Mixed Review

True or false?

1. A median of a triangle, except for its endpoints, is always in the interior of a triangle.   **5.5**   True

2. An altitude of a triangle, except for its endpoints, is always in the interior of a triangle.   **5.5**   False

3. An angle bisector of a triangle cannot be an altitude of the triangle.   **5.5**   False

4. A median of a triangle cannot be an altitude of the triangle.   **5.5**   False

5. A triangle cannot be constructed with sides 2, 3, and 5.   **5.5**   True

Algebra Review

To graph the solution set of an inequality, locate the points on the number line that correspond to the solutions.

Example: Graph the solution set of $x < 2$ *and* $x \geq -1$.

         Graph each part separately, then determine the intersection.

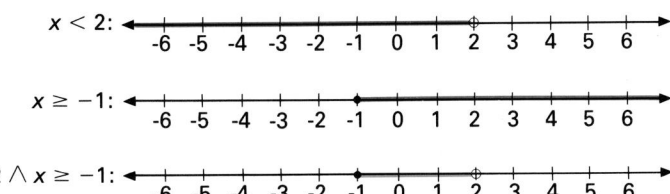

Graph the solution set of the inequality.

1. $x > 3$ **2.** $x < 5$ **3.** $x > -5$ **4.** $x < -3$

5. $a < 6$ **6.** $t > 0$ **7.** $s > -1$ **8.** $m < -3$

9. $x > -2$ *and* $x < 1$ **10.** $x < 0$ *and* $x > -2$ **11.** $x < 6$ *and* $x > 1$

12. $x > 0$ *and* $x < 4$ **13.** $x < -2$ *and* $x < 3$ **14.** $x > -2$ *and* $x > 0$

8.5 SAS and SSS Similarity Theorems   **327**

17.

Statement	Reason
1. △s ABC and DEF, $\frac{CA}{FD} = \frac{AB}{DE} = \frac{CB}{FE}$	1. Given
2. Locate D' on $\overline{CA}$ and E' on $\overline{CB}$ such that $\overline{CD'} \cong \overline{FD}$ and $\overline{CE'} \cong \overline{FE}$.	2. Ruler Post
3. Draw $\overline{DE}$	3. 2 pts determine a line.
4. $\frac{CA}{CD'} = \frac{CB}{CE'}$	4. Sub
5. $\angle C \cong \angle C$	5. Reflex Prop of Congr
6. △$ABC \sim$ △$D'E'C$	6. SAS~
7. $\frac{AB}{D'E'} = \frac{CA}{CD'}$	7. Def of ~ △s
8. $\frac{AB}{D'E'} = \frac{CA}{FD}$	8. Sub
9. $\frac{AB}{D'E'} = \frac{AB}{DE}$	9. Sub
10. $\frac{D'E'}{AB} = \frac{DE}{AB}$	10. If $\frac{a}{b} = \frac{c}{d}$, then $\frac{b}{a} = \frac{d}{c}$.
11. $D'E' = DE$	11. Mult Prop of Eq
12. △$D'E'C \cong$ △DEF	12. SSS
13. △$D'E'C \sim$ △DEF	13. $\cong$ △s are ~.
14. △$ABC \sim$ △DEF	14. Trans Prop for ~ △

18.

Statement	Reason
1. △ABC with midpts L, M, and N of $\overline{AB}$, $\overline{BC}$ and $\overline{CA}$, respectively.	1. Given
2. $\overline{MN}$, $\overline{LN}$, and $\overline{ML}$ are midsegs.	2. Def of midseg
3. $MN = \frac{1}{2}AB$, $LN = \frac{1}{2}BC$, $ML = \frac{1}{2}CA$	3. Midseg Thm
4. $\frac{MN}{AB} = \frac{1}{2}, \frac{LN}{BC} = \frac{1}{2}, \frac{ML}{CA} = \frac{1}{2}$	4. Div Prop of Eq
5. $\frac{MN}{AB} = \frac{LN}{BC} = \frac{ML}{CA}$	5. Sub
6. △$MNL \sim$ △ABC	6. SSS~

See pages 661–662 for the answers to Written Ex. 19–21 and Algebra Review Ex. 1–14.

Prerequisite Quiz

1. Write a proportion that results from the similarity $\triangle FGH \sim \triangle NLM$. $\frac{FG}{NL} = \frac{GH}{LM}$
2. In the following diagrams, name the altitudes. $\overline{AD}$ and $\overline{BD}$; $\overline{AB}$ and $\overline{BC}$

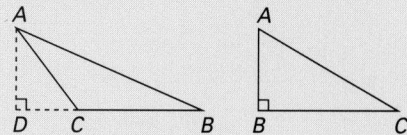

3. In the following diagrams, name the medians. $\overline{BD}$; $\overline{BD}$

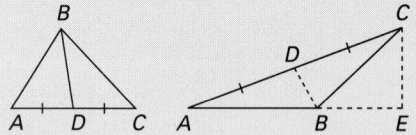

4. Draw an angle. Construct the bisector of this angle.   See construction, p. 23.

Motivator

Ask the students what would happen to the lengths of the altitudes if the lengths of the sides of a triangle were to double.   They would double.   Then ask them what would happen to the medians.   They would double.   Ask them why they think this happens.   The altitudes and medians are proportional to the corresponding sides.

▰▰▰ TEACHING SUGGESTIONS

Lesson Note

Start this chapter with a review of altitudes, medians, and angle bisectors. The material of this lesson is not as basic as that contained in the previous lessons. This topic can lead to interesting activities and the kinds of problems that are found in contests.

8.6 Segments in Similar Triangles

Objectives	To apply theorems concerning proportional parts in triangles To construct segments with lengths in a given ratio

Theorem 8.9   Corresponding altitudes of similar triangles are proportional to corresponding sides.

Given: $\triangle ABC \sim \triangle DEF$, altitudes $\overline{CG}$ and $\overline{FH}$
Prove: $\dfrac{CG}{FH} = \dfrac{AC}{DF}$

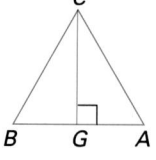
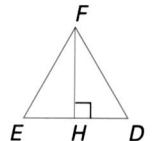

Proof

Statement	Reason
1. $\triangle ABC \sim \triangle DEF$, altitudes $\overline{CG}$ and $\overline{FH}$	1. Given
2. $\angle CGA$ and $\angle FHD$ are rt $\angle$s.	2. Def of alt
3. $\angle CGA \cong \angle FHD$	(A) 3. All rt $\angle$s are $\cong$.
4. $\angle A \cong \angle D$	(A) 4. Def of similar $\triangle$s
5. $\triangle CGA \sim \triangle FHD$	5. AA $\sim$
6. $\therefore \dfrac{CG}{FH} = \dfrac{AC}{DF}$	6. Def of similar $\triangle$s

Theorem 8.10   Corresponding medians of similar triangles are proportional to corresponding sides.

EXAMPLE 1 $\triangle LMN \sim \triangle PQR$, $\overline{NW}$ and $\overline{RO}$ are medians. Find NW and RO.

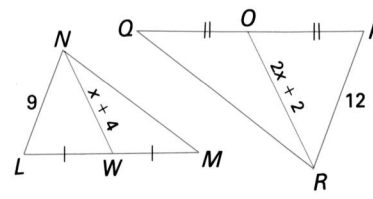

Solution

$$\frac{9}{12} = \frac{x + 4}{2x + 2}$$
$$9(2x + 2) = 12(x + 4)$$
$$18x + 18 = 12x + 48$$
$$6x = 30$$
$$x = 5 \text{ (So, } x + 4 = 9 \text{ and } 2x + 2 = 12)$$

Thus, $NW = 9$ and $RO = 12$.

Additional Example 1

If $\triangle ABC \sim \triangle XYZ$, and $\overline{BD}$ and $\overline{YW}$ are medians, find BD and YW. $BD = \frac{5}{7}$; $YW = \frac{8}{7}$

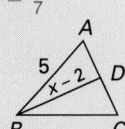
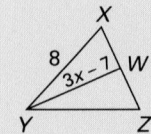

Theorem 8.11 The bisector of an angle of a triangle divides the opposite side of the triangle into segments proportional to the other two sides.

Given: $\triangle ABC$, $\overline{AD}$ bisects $\angle CAB$.

Prove: $\dfrac{DC}{DB} = \dfrac{AC}{AB}$

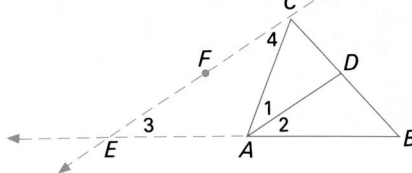

Plan

$\overline{AD}$ bisects $\angle CAB$. ⟶ $\angle 1 \cong \angle 2$

Draw $\overrightarrow{CF} \parallel \overline{DA}$. — $\angle 1 \cong \angle 4$ ⟶ $\angle 3 \cong \angle 4$ → $\overline{AE} \cong \overline{AC}$

$\angle 3 \cong \angle 2$

$\dfrac{CD}{DB} = \dfrac{AE}{AB}$ ⟶ $\dfrac{CD}{DB} = \dfrac{AC}{AB}$

Draw $\overrightarrow{BA}$ through $\overrightarrow{CF}$ at E.

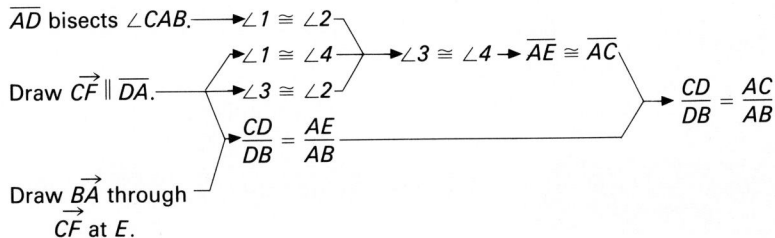

EXAMPLE 2 Given: $\overline{CD}$ bisects $\angle ACB$.
$AC = 12$, $BC = 28$, $AB = 20$
Find AD and DB.

Plan Let $x = AD$. Then $20 - x = DB$

Solution

$\dfrac{AD}{DB} = \dfrac{12}{28}$

$\dfrac{x}{20 - x} = \dfrac{12}{28}$

$28x = 12(20 - x)$

$28x = 240 - 12x$

$40x = 240$

$x = 6$ Therefore, $AD = 6$ and $DB = 14$.

Construction To divide a segment into two segments whose lengths have a ratio $\frac{2}{3}$.

Given: $\overline{AB}$

Construct P on $\overline{AB}$ so that $\dfrac{AP}{PB} = \dfrac{2}{3}$.

Carpentry: As mentioned in Math Connections for Lesson 8.4, a carpenter can use alternative methods of measuring when the unit of measurement that is needed is not marked on the measuring tool that is available. The carpenter could have also used the method given in the Construction of this lesson on pages 329-330 to divide the piece of wood into 5 congruent pieces.

Critical Thinking Questions

Analysis: Refer students to the $8\frac{1}{4}$-in piece of wood from the Math Connections of Lesson 8.4 and ask them what ratio the carpenter must use to perform the division by the method given in the Construction on pages 329–330. To cut 5 $\cong$ pieces, the resulting construction must mark the board in a ratio $\frac{1}{5} \cdot \frac{4}{5}$, or 1:4.

Checkpoint

1. Given $\triangle ABC \sim \triangle XYZ$, $\overline{BD}$, $\overline{YW}$ are altitudes. $AB = 25$, $XY = 30$, $BD = 20$. Find YZ. 24

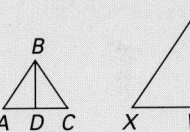

2. Given: $\triangle FGH \sim \triangle LMN$, $\overline{GA}$ and $\overline{MP}$ are medians. $GH = 14$, $MN = 21$, $GA = 16$. Find MP. 24

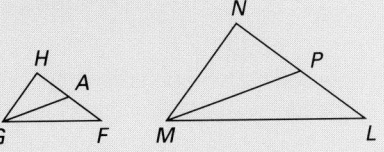

3. Given: $\triangle ABC$ with angle bisector $\overline{AD}$, $AB = 15$, $AC = 12$, $BD = 20$. Find CD. 16

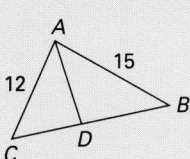

Additional Example 2

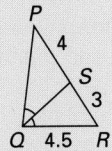

In $\triangle PQR$, $\overline{SQ}$ bisects $\angle Q$, $PS = 4$, $SR = 3$, $QR = 4.5$. Find PQ. 6

329

Closure

Ask students what would happen to the lengths of the altitudes and the lengths of the medians if they were to quadruple the lengths of the sides of similar triangles. The lengths of both would quadruple. Ask them what would happen to the lengths of the altitudes and the lengths of the medians if they were to shorten the lengths of the sides of similar triangles by one-third. The lengths of both would shorten by one-third.

■■■FOLLOW UP

Guided Practice

Classroom Exercises 1–2

Independent Practice

A Ex. 1–14, **B** Ex. 15–22, **C** Ex. 23–27

Basic: WE 1–14

Average: WE 1–22

Above Average: WE 1–27 odd

Additional Answers

Written Exercises

21.

Statement	Reason
1. $\overline{AE}$ bis $\angle DAB$, $\overline{CE}$ bis $\angle DCB$	1. Given
2. $\frac{AD}{AB} = \frac{DE}{BE}$, $\frac{DE}{BE} = \frac{DC}{CB}$	2. Bis of $\angle$ div opp side of $\triangle$ proport to other 2 sides
3. $\frac{AD}{AB} = \frac{DC}{CB}$	3. Sub

22. See Construction on page 329 using segs of length $3(XY)$ and $4(XY)$.

Draw any segment $\overline{XY}$.
Construct segments of lengths $2(XY)$ and $3(XY)$.

Construct $\triangle ABC$ so that $AC = 2(XY)$ and $BC = 3(XY)$. Bisect $\angle ACB$. Draw the angle bisector intersecting $\overline{AB}$ at P.

Result: $\dfrac{AP}{PB} = \dfrac{2}{3}$

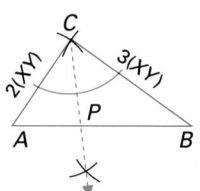

Classroom Exercises

Solve for x.

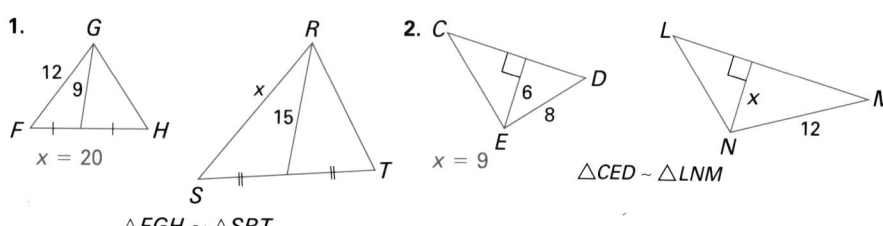

1. $x = 20$ $\triangle FGH \sim \triangle SRT$

2. $x = 9$ $\triangle CED \sim \triangle LNM$

Written Exercises

$\triangle ABC \sim \triangle EFG$, $\overline{CD}$ and $\overline{GH}$ are altitudes. Find the indicated measure.

1. $AC = 16$, $EG = 12$, $CD = 8$, $GH = $ ___6___

2. $AC = 18$, $EG = 21$, $CD = $ ___12___, $GH = 14$

3. $AC = 14$, $EG = 21$, $CD = 3x + 2$, $GH = 7x - 2$, $CD = $ ___8___

4. $AC = 10$, $EG = 15$, $CD = x + 3$, $GH = 2x + 2$, $GH = $ ___12___

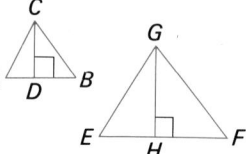

$\triangle IJK \sim \triangle MNO$, $\overline{KL}$ and $\overline{OP}$ are medians. Find the indicated measure.

5. $KI = 25$, $OM = 30$, $KL = 20$, $OP = $ ___24___

6. $KI = 25$, $OM = 35$, $OP = 14$, $KL = $ ___10___

7. $KI = 24$, $OM = 30$, $KL = 3x + 4$, $OP = 6x - 4$, $KL = $ ___16___

8. $KI = 6$, $OM = 9$, $KL = x + 3$, $OP = 2x + 1$, $OP = $ ___15___

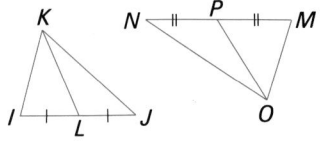

Enrichment

The diagram below shows a square inscribed in a right triangle. If the square has sides equal to 3, $y = 5$, and the perimeter of the lower triangle is 12, find the perimeter of the upper triangle.

After finding that $x = 4$, this problem can be done in two ways. (1) Using the properties of similarity, a student can find the values of a and b and add them to 3 to find the perimeter.
(2) A student might have the insight that the ratios of the perimeters will be equal to the ratio of sides and thus come directly to the conclusion that the perimeter is 9.

Given: $\overline{ST}$ bisects $\angle RSQ$ of $\triangle QRS$. Find the indicated measure.

9. $QS = 6$, $RS = 9$, $QT = 2$, $TR = $ ___3___
10. $QS = 8$, $RS = 10$, $TR = 5$, $QT = $ ___4___
11. $QS = 8$, $RS = 12$, $QR = 10$, $TR = $ ___6___
12. $QS = 12$, $RS = 16$, $QR = 14$, $QT = $ ___6___
13. $QS = 3$, $RS = 5$, $QT = 2x + 1$, $TR = 4x - 5$, $QT = $ ___21___
14. $QS = 16$, $RS = 20$, $QT = x + 3$, $TR = 2x - 3$, $TR = $ ___15___
15. Perimeter of $\triangle QRS = 27$, $QT = 4$, $TR = 5$, $QS = $ ___8___
16. Perimeter of $\triangle QRS = 80$, $QT = 14$, $TR = 18$, $SR = $ ___27___

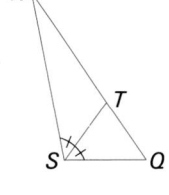

Given: $\triangle EFG \sim \triangle KLM$, $\overline{GH}$ and $\overline{MN}$ are altitudes; $\overline{EI}$ and $\overline{KO}$ are medians; and $\overline{FJ}$ and $\overline{LP}$ are angle bisectors. Find the indicated measure.

17. $KL = 3.6$, $EF = 2.4$, $KO = 2.1$, $EI = $ ___1.4___
18. $KM = 4$, $GE = 3.2$, $MN = $ ___3___, $GH = 2.4$
19. $MN = 2$, $GH = $ ___1.5___, $KO = 2.8$, $EI = 2.1$
20. $PL = 4.9$, $JF = 5.6$, $KO = $ ___2.1___, $EI = 2.4$

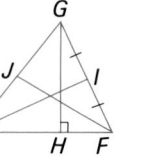 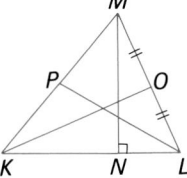

21. Given: $\overline{AE}$ bisects $\angle DAB$;
$\overline{CE}$ bisects $\angle DCB$.
Prove: $\dfrac{AD}{AB} = \dfrac{DC}{CB}$

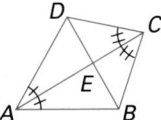

22. Draw a segment $\overline{XY}$. Construct P so that $\dfrac{XP}{PY} = \dfrac{3}{4}$.

23. Prove Theorem 8.10.   **24.** Prove Theorem 8.11.

25. Prove: If two triangles are similar, then corresponding altitudes are proportional to corresponding medians.

26. Prove: Diagonals of a trapezoid divide each other into proportional segments.

27. Prove or disprove the converse of Theorem 8.9.

Mixed Review

1. What is the geometric mean of 8 and 50? *8.1* 20
2. Give an example to show that the AAA pattern is insufficient for congruence of triangles. *4.6*
3. State the Triangle Inequality Theorem. *5.7*
4. How many diagonals does a convex hexagon have? *6.1* 9

8.6 Segments in Similar Triangles **331**

23.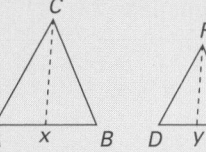

Statement	Reason
1. $\triangle ABC \sim$ $\triangle DEF$ with medians $\overline{CX}$ and $\overline{FY}$, respectively.	1. Given
2. $AX = BX$, $DY = EY$	2. Def of median
3. $AX + BX =$ AB, $DY +$ $EY = DE$	3. Seg Add Post
4. $2AX = AB$, $2DY = DE$	4. Sub
5. $\dfrac{CA}{FD} = \dfrac{AB}{DE}$; $\angle A \cong \angle D$	5. Def of $\sim \triangle$s
6. $\dfrac{CA}{FD} = \dfrac{2AX}{2DY}$	6. Sub
7. $\dfrac{CA}{FD} = \dfrac{AX}{DY}$	7. Id Prop for Mult
8. $\triangle CAX \sim$ $\triangle FDY$	8. SAS~
9. $\dfrac{CA}{FD} = \dfrac{CX}{FY}$	9. Def of $\sim \triangle$s

24.

Statement	Reason
1. $\triangle ABC$, $\overline{AD}$ bis $\angle CAB$	1. Given
2. Draw $\overleftrightarrow{CF}$ parallel to $\overline{AD}$.	2. Parallel Post
3. Extend $\overrightarrow{BA}$ to inters $\overleftrightarrow{CF}$ at E.	3. 2 lines inters in 1 pt.
4. In $\triangle CBE$, $\dfrac{DC}{DB} = \dfrac{AE}{AB}$.	4. Triangle Proport Thm
5. $\angle 3 \cong \angle 2$	5. If ∥ lines are inters by transv, then corr $\angle$s are $\cong$.
6. $\angle 4 \cong \angle 1$	6. If ∥ lines are inters by transv, then alt int $\angle$s are $\cong$.
7. $\angle 1 \cong \angle 2$	7. Def of $\angle$ bis
8. $\angle 3 \cong \angle 4$	8. Trans Prop of Congr
9. $\overline{AE} \cong \overline{AC}$	9. Sides opp $\cong \angle$s of $\triangle$ are $\cong$.
10. $\dfrac{DC}{DB} = \dfrac{AC}{AB}$	10. Sub

See page 332 for the answers to Written Ex. 25–27 and Mixed Review Ex. 2–3.

25.

Statement	Reason
1. $\triangle ABC \sim$ $\triangle DEF$ with alts $\overline{CW}$ and $\overline{FY}$ and medians $\overline{CX}$ and $\overline{FZ}$, respectively.	1. Given
2. $\frac{CW}{FY} = \frac{CB}{FE}$	2. Corr alts of $\sim \triangle$s are proport to corr sides.
3. $\frac{CB}{FE} = \frac{CX}{FZ}$	3. Corr medians of $\sim$ $\triangle$s are proport to corr sides.
4. $\frac{CW}{FY} = \frac{CX}{FZ}$	4. Sub

26.

Statement	Reason
1. Trap $ABCD$ with diags $\overline{DB}$ and $\overline{CA}$ inters at E.	1. Given
2. $\overline{DC} \parallel \overline{AB}$	2. Def of trap
3. $\angle DCE \cong$ $\angle EAB$, $\angle CDE \cong$ $\angle EBA$	3. If $\parallel$ lines are inters by transv, then alt int $\angle$s are $\cong$.
4. $\triangle DCE \cong$ $\triangle BAE$	4. AA$\sim$
5. $\frac{DE}{BE} = \frac{CE}{AE}$	5. Def of $\sim \triangle$s

27. See illustration for Thm 8.9.

Statement	Reason
1. $\frac{CG}{FH} = \frac{AB}{DE}, \frac{CG}{FH}$ $= \frac{BC}{EF}, \frac{CG}{FH} =$ $\frac{CA}{FD}$	1. Given
2. $\frac{AB}{DE} = \frac{BC}{EF} =$ $\frac{CA}{FD}$	2. Sub
3. $\triangle ABC \sim$ $\triangle DEF$	3. SSS$\sim$

Mixed Review

2. $\triangle ABC \not\cong \triangle A'B'C'$

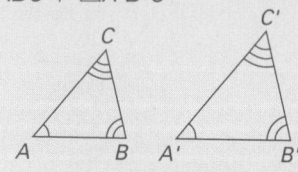

3. The sum of the lengths of any two sides of a triangle is greater than the length of the third side.

Key Terms

AA Similarity   p. 314
cross products   p. 314
extremes   p. 303
geometric mean   p. 305
means   p. 303
proportion   p. 303
ratio   p. 303

SAS Similarity   p. 323
SSS Similarity   p. 324
similar polygons   p. 308
similar triangles   p. 313
Transitive Property of Triangle Similarity   p. 313
Triangle Proportionality Theorem   p. 318

Key Ideas and Review Exercises

8.1   To solve a proportion, equate the product of the extremes and the product of the means.

To find the geometric mean of two positive numbers a and b, solve the proportion $\frac{a}{m} = \frac{m}{b}$ for the positive number m.

1. Solve the proportion: $\frac{8}{15} = \frac{x}{20}$. $10\frac{2}{3}$

2. Find the geometric mean of 12 and 75.   30

3. The ratio of the measures of two complementary angles is 2 to 7. What are the measures of the angles?   20, 70

8.2   To show that two polygons are similar, show that corresponding angles are congruent and corresponding sides are proportional.

4. Based on the information given, is quad $ABCD \sim$ quad $EFGH$?   No

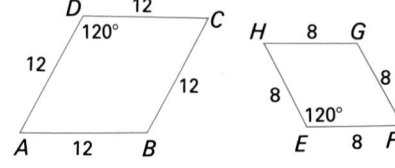

5. Quad $IJKL \sim$ quad $MNOP$. Find the missing lengths. $PM = 45$, $MN = 45$, $NO = 30$

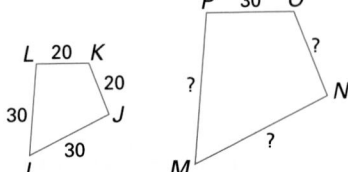

8.3   Two triangles are similar if two angles of one triangle are congruent to two angles of the other.

6. If m $\angle X = 65$, m $\angle Y = 55$, m $\angle L = 65$, and m $\angle N = 60$, is $\triangle XYZ \sim \triangle LMN$?   Yes

7. For similar triangles XYZ and LMN, write two proportions that give $YZ \cdot LN = MN \cdot XZ$. $\frac{YZ}{MN} = \frac{XZ}{LN}, \frac{YZ}{XZ} = \frac{MN}{LN}$

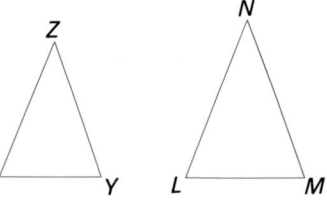

8.4 The Triangle Proportionality Theorem states that if a line is parallel to one side of a triangle and intersects the other two sides, then it divides the two sides proportionally.

The *converse* states that if a line divides two sides of a triangle proportionally, then the line is parallel to the third side of the triangle.

8. If $\overline{TV} \parallel \overline{XY}$, is $\dfrac{XT}{TZ} = \dfrac{ZV}{VY}$? Yes, only if $\overline{TV}$ is a midsegment

9. $\overline{TV} \parallel \overline{XY}$, $XT = 9$, $TZ = 12$, $ZV = 15$. Find VY. $11\frac{1}{4}$

10. If $XT = 54$, $TZ = 42$, $YV = 63$, and $VZ = 48$, is $\overline{TV} \parallel \overline{XY}$? No

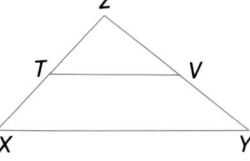

8.5 To prove two triangles similar, the SAS Similarity and the SSS Similarity theorems can be used. Not necessarily. SS is not sufficient.

11. If $\dfrac{GH}{IH} = \dfrac{JK}{LK}$, is $\triangle GHI \sim \triangle JKL$? Why?

12. Given: $\overline{GI} \cong \overline{IH}$, $\overline{JL} \cong \overline{KL}$, $\angle I \cong \angle L$
Prove: $\triangle GHI \sim \triangle JKL$

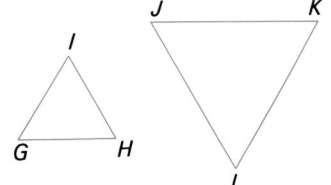

8.6 Corresponding altitudes (and medians) of similar triangles are proportional to corresponding sides.

The bisector of an angle divides the opposite side of a triangle into segments proportional to the other two sides.

Proportional segments can be constructed using an angle bisector.

13. Explain how to determine whether the medians of similar triangles are corresponding medians. Corr medians have endpts at corr vertices.

14. $\triangle XYZ \sim \triangle PQR$, $\overline{ZA}$ and $\overline{RB}$ are altitudes; $XY = 40$, $PQ = 60$, $ZA = 18$, $RB = \underline{\ 27\ }$.

15. $\overline{AD}$ bisects $\angle CAB$; $AC = 24$, $AB = 32$, $CD = 18$, $BD = \underline{\ 24\ }$.

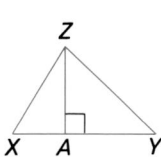
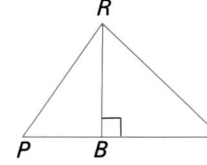
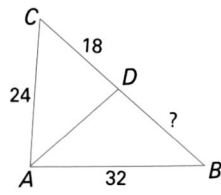

16. Draw a segment $\overline{AB}$. Construct C so that $\dfrac{AC}{CB} = \dfrac{3}{5}$.

12.

Statement	Reason
1. $\overline{GI} \cong \overline{IH}$, $\overline{JL}$ $\cong \overline{KL}$, $\angle I \cong$ $\angle L$	1. Given
2. $\dfrac{GI}{GI} = \dfrac{JL}{JL} = 1$	2. A number div by itself = 1.
3. $\dfrac{GI}{IH} = \dfrac{JL}{KL}$	3. Sub
4. $\dfrac{GI}{JL} = \dfrac{IH}{KL}$	4. If $\dfrac{a}{b} = \dfrac{c}{d}$, then $\dfrac{a}{c} = \dfrac{b}{d}$.
5. $\triangle GHI \sim$ $\triangle JKL$	5. SAS~

16. From A swing an arc of length 3, from B swing an arc of length 5, arcs intersecting at P. Bisect $\angle APB$. Label the intersection of $\overline{AB}$ and the angle bisector, C. Then, $\dfrac{AC}{CB} = \dfrac{3}{5}$.

14.

Statement	Reason
1. △KLM, alts $\overline{MN}$ and $\overline{KO}$	1. Given
2. m ∠MOP = 90, m ∠MNL = 90	2. Def of alt
3. m ∠MOP = m ∠MNL	3. Sub
4. m ∠OMP = m ∠OMN	4. Reflex Prop of Eq
5. △MOP ~ △MNL	5. AA~

15.

Statement	Reason
1. $\frac{m}{n} = \frac{p}{q}$	1. Given
2. $\frac{m}{n} - 1 = \frac{p}{q} - 1$	2. Subt Prop of Eq
3. $\frac{n}{n} = 1, \frac{q}{q} = 1$	3. A number div by itself = 1.
4. $\frac{m}{n} - \frac{n}{n} = \frac{p}{q} - \frac{q}{q}$	4. Sub
5. $\frac{1}{n}(m - n) = \frac{1}{q}(p - q)$	5. Distr Prop
6. $\frac{m - n}{n} = \frac{p - q}{q}$	6. Notation

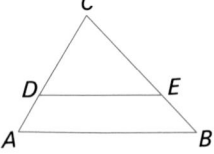

Chapter 8 Test

A—Level Ex.: 1–4, 6–13 B—Level Ex.: 5, 14 C—Level Ex.: starred

Solve the proportion for x. (Exercises 1–2)

1. $\frac{x}{15} = \frac{32}{40}$ 12

2. $\frac{21}{27} = \frac{x}{45}$ 35

3. Find the geometric mean of 8 and 2.
  4

$\overline{DE} \parallel \overline{AB}$. **Find the indicated length.**

4. DA = 12, CE = 12, EB = 15, CD = ___9.6___

5. CA = 35, CE = 25, EB = 10, AD = ___10___

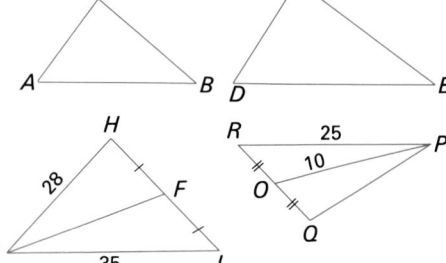

Based on the information given, is △ABC ~ △DEF? If so, why?

6. ∠A ≅ ∠D, ∠C ≅ ∠F   Yes, AA~

7. ∠B ≅∠E, $\frac{AC}{DF} = \frac{AB}{DE}$ No

8. $\frac{AB}{DE} = \frac{BC}{EF} = \frac{CA}{FD}$ Yes, SSS~

9. △GHI ~ △PQR. Find GF.   14

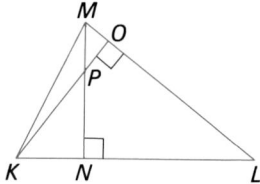

True or false? (Exercises 10–12)

10. If two triangles are similar, then their corresponding altitudes are congruent.   False

11. Congruent triangles are similar.   True

12. Two quadrilaterals are similar if their corresponding angles are congruent.   False

13. Given: $\overline{IJ}$ bisects ∠GIH; IG = 45, IH = 25, GH = 42
Find GJ.   27

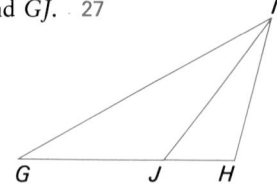

14. Given: $\overline{MN}$ and $\overline{KO}$ are altitudes of △KLM.
Prove: △MOP ~ △MNL

*15. Prove: if $\frac{m}{n} = \frac{p}{q}$, then $\frac{m - n}{n} = \frac{p - q}{q}$

Choose the one best answer to each question or problem.

1. In $\triangle ABC$, $\overline{DE} \parallel \overline{BC}$, $AE = 5$,
A $EB = 3$, and $AD = 4$. Find DC.

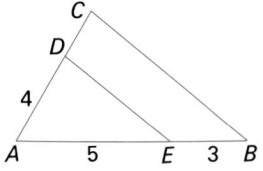

(A) 2.4   (B) 3.75   (C) 4
(D) 6.4   (E) Cannot be determined

2. In $\triangle FGH$, $FG = 6$, $GH = 7$, and
E $HF = 8$. Find IG.

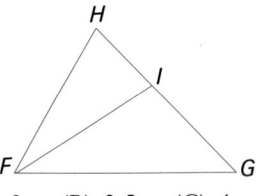

(A) 3   (B) 3.5   (C) 4   (D) 6
(E) Cannot be determined

3. If one dozen pencils cost one dollar,
B how many dollars will n pencils cost?

(A) $\dfrac{12n}{1.00}$ (B) $\dfrac{1.00n}{12}$ (C) $\dfrac{12}{1.00n}$

(D) $\dfrac{1.00}{12n}$ (E) Cannot be determined

4. Given: Trap $JKLM$, median $\overline{NO}$,
C $JK = 6x + 1$, $NO = 7x - 3$,
$ML = 4x + 3$. Find NO.

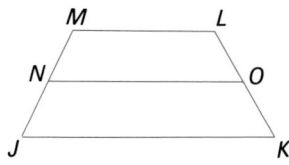

(A) 6.5   (B) 13   (C) 14.5
(D) 16   (E) Cannot be determined

5. If it takes Sam 15 min to type two
A pages, how many hours will it take
him to type 50 pages?
(A) 6.25   (B) 12.5
(C) 15   (D) 25
(E) Cannot be determined

6. On a map drawn to scale 0.125 in. =
C 10 mi, what is the actual distance in
miles between two cities which are
2.5 in. apart on the map?
(A) 12.5   (B) 25
(C) 200   (D) 1,250
(E) Cannot be determined

7. In the figure below, rect $PQRS$ is
B composed of five congruent rectangles; $SP = 75$. Find TU.

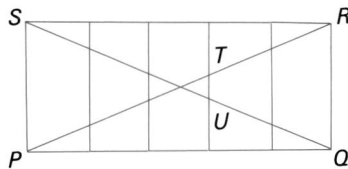

(A) 7.5   (B) 15
(C) 20   (D) 30
(E) Cannot be determined

8. A five foot-tall person casts a shadow
A 8 ft long. At the same time, a tree
casts a shadow 64 ft long. How many
feet tall is the tree?
(A) 40   (B) 51.2
(C) 80   (D) 102.4
(E) Cannot be determined

1. a. Collinear   b. Coplanar but
noncollinear c. Coplanar but
noncollinear d. Noncoplanar
e. Coplanar but noncollinear
f. Coplanar but noncollinear
2. See Construction on page 23.
3. ∠AOP ≅ ∠POB
4. If an angle is right, then its sides are ⊥.

6. Statement	Reason
1. △ABC	1. Given
2. Extend $\overline{AB}$ through a pt D.	2. 2 pts determine a line.
3. Construct $\overrightarrow{BE}$ such that $\overrightarrow{BE}$ ∥ to $\overline{AC}$.	3. Parallel Post
4. m ∠DBE + m ∠EBC = m ∠DBC	4. Angle Add Post
5. m ∠DBC + m ∠CBA = 180.	5. If outer rays of 2 adj ∠s form st ∠, then sum of ∠ meas = 180.
6. m ∠DBE + m ∠EBC + m ∠CBA = 180.	6. Sub
7. m ∠A = m ∠DBE	7. If 2 ∥ lines are inters by transv, then corr ∠s are ≅.
8. m ∠C = m ∠EBC	8. If 2 ∥ lines are inters by transv, then alt int ∠s are ≅.
9. m ∠A + m ∠C + m ∠CBA = 180.	9. Sub

15. See Construction on page 200.

---

1. Identify the following points as   *1.1*
collinear, coplanar but noncollinear, or noncoplanar.
(a) E, B   (b) A, B, C
(c) A, F, D   (d) A, B, C, D
(e) A, D, C, E   (f) A, B, C, F

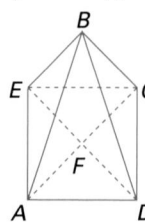

2. Draw an acute angle. Con-   *1.5*
struct the bisector of the angle.

3. State a conclusion from this   *2.1*
statement: $\overrightarrow{OP}$ bisects ∠AOB.

4. Write the following statement
in "If . . ., then . . ." form: A
right angle has perpendicular
sides.

5. True or false? Skew lines are   *3.1*
noncoplanar lines.   True

6. Give a proof of the theorem: *3.6*
The sum of the measures of the
angles of a triangle is 180.

7. Given: Parallel lines *l* and *m*   *3.5*
with transversal *t*. $\overline{BC}$ bisects
∠ABD and $\overline{DC}$ bisects ∠EDB.
What is the measure of ∠DCB?   90

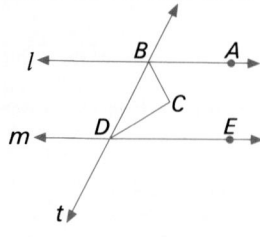

8. True or false? If two planes in-   *3.1*
tersect, then they intersect in
exactly one line.   True

9. What is the measure of the   *1.6*
angle formed by the bisectors
of two adjacent complementary
angles?   45

10. Find the measure of all the an-   *3.6*
gles of △ABC if the m ∠CBD =
150, and m ∠A is one-half the
m ∠C.   m ∠A = 50,
m ∠B = 30,
m ∠C = 100

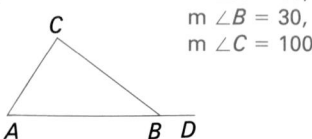

11. △ABC is isosceles with ∠A ≅   *5.1*
∠B; AC = x^2 − 10, AB = x
+ 2, CB = 3x. Find AB.   7

12. True or false? If the measures   *4.1*
of two angles of a triangle are
30 and 40, then the triangle is
obtuse.   True

13. Given: ∠A ≅ ∠E, $\overline{AC}$ ≅ $\overline{EF}$,   *4.4*
$\overline{CB}$ ≅ $\overline{FD}$. Is enough informa-
tion given to prove △ABC ≅   No, ASS
△EDF? Why or why not?   insufficient

14. The measure of one of the base   *5.1*
angles of an isosceles triangle is
20 less than the measure of the
vertex angle. Find the measure
of the vertex angle. $73\frac{1}{3}$

15. Draw a segment approximately   *5.6*
5 cm long. Construct the perpendicular bisector of the
segment.

16. Is there a regular polygon in   *6.3*
which an exterior angle has a
measure of 20?   Yes, 18-gon

17. Is the property, a quadrilateral has congruent opposite angles, a necessary condition for a quadrilateral to be a parallelogram?   Yes   *6.4, 7.2*

18. Is the property, consecutive angles are supplementary, a sufficient condition for a quadrilateral to be a parallelogram?   Yes   *6.5, 7.3*

19. Given: D and E are the midpoints of $\overline{AC}$ and $\overline{BC}$, respectively. $DE = 3x - 2$, $AB = 2x + 10$. Find AB.   17   *6.6*

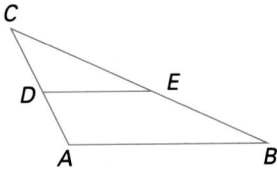

20. Is the property, a parallelogram has congruent diagonals, a necessary condition for a parallelogram to be a rhombus?   No   *7.2*

21. Is it possible to prove $\triangle ABC \cong \triangle DEF$ if $\angle A \cong \angle D$, $\overline{AB} \cong \overline{DE}$, and $\overline{BC} \cong \overline{EF}$?   No   *4.5*

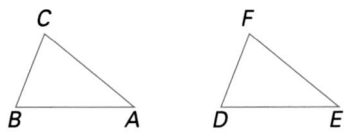

22. Construct an angle of 30° using straight edge and compass.   *1.5, 4.1*

23. Point D is the midpoint of $\overline{AB}$ and point P is located so that $PA = PB$. What is m $\angle PDA$?   90   *5.4*

24. Find the values of x and y in $\square ABCD$ if $AB = 3x + 2$, $BC = y + 12$, $CD = 5x - 8$, and $DA = 4y + 3$. $x = 5, y = 3$ *6.4*

25. Given: $\triangle ABC$ with M the midpoint of $\overline{AC}$, and N the midpoint of $\overline{BC}$. If $MN = 2x + 5$, and $AB = 2x + 15$. Find MN.   10   *6.6*

26. A regular polygon has each interior angle half as large as each exterior angle. How many sides does the polygon have?   3   *6.3*

27. If $4x = 2y$, what is the ratio of x to y?   1:2   *8.1*

28. Is the property, a parallelogram has at least one right angle, a sufficient condition for a parallelogram to be a rectangle?   Yes   *7.3*

29. Is the property, a trapezoid has congruent diagonals, a necessary condition for a trapezoid to be isosceles?   Yes   *7.5*

30. Find the geometric mean of 2 and 18.   6   *8.1*

31. Are any two regular pentagons similar? If so, prove it. If not, provide a counterexample.   *8.2*

32. True or false? If $\triangle LMN \sim \triangle TUV$, and $\triangle TUV \sim \triangle XYZ$, then $\dfrac{MN}{YZ} = \dfrac{NL}{ZX}$.   True   *8.3*

33. In $\triangle ABC$, $AB = 3$, $BC = 6$, and m $\angle B = 60$. In $\triangle EFG$, $EF = 8$, $FG = 16$, and m $\angle F = 60$. Are $\triangle ABC$ and $\triangle EFG$ similar? Why or why not?   Yes, SAS $\sim$ *8.5*

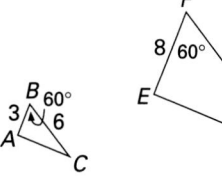

22. Construct an equilaleral $\triangle$ and bis one of its $\angle$s.

31. Yes.

Statement	Reason
1. Reg pentagons $ABCDE$ and $A'B'C'D'E'$	1. Given
2. m $\angle A$ = m $\angle B$ = m $\angle C$ = m $\angle D$ = m $\angle E$ = 108, m $\angle A'$ = m $\angle B'$ = m $\angle C'$ = m $\angle D'$ = m $\angle E'$ = 108	2. Meas of $\angle$s of regular polygon with n sides = $\dfrac{(n-2)180}{n}$.
3. m $\angle A$ = m $\angle A'$, m $\angle B$ = m $\angle B'$, m $\angle C$ = m $\angle C'$, m $\angle D$ = m $\angle D'$, m $\angle E$ = m $\angle E'$	3. Sub
4. $AB = BC = CD = DE = EA$, $A'B' = B'C' = C'D' = D'E' = E'A'$	4. Def of reg polygon
5. $\dfrac{AB}{A'B'} = \dfrac{AB}{A'B'}$	5. Reflex Prop of Eq
6. $\dfrac{AB}{A'B'} = \dfrac{BC}{B'C'} = \dfrac{CD}{C'D'} = \dfrac{DE}{D'E'} = \dfrac{EA}{E'A'}$	6. Sub
7. $ABCDE \sim A'B'C'D'E'$	7. Def of similar polygons

9 RIGHT TRIANGLES

OVERVIEW

Students find missing lengths of legs in right triangles using similarity properties and the Pythagorean Theorem. They also use the converse of the Pythagorean Theorem to determine whether a triangle is acute, obtuse, or right. Students will apply the properties of special right triangles to solve word problems involving right triangles.

OBJECTIVES

- To find the length of one side of a right triangle, given the lengths of the other two sides
- To find the missing lengths of sides in a right triangle, given measures of other parts
- To determine whether a triangle with given lengths of sides is acute, obtuse, or right
- To find the missing lengths of parts of isosceles triangles, equilateral triangles, and rhombuses, given lengths of other parts
- To solve problems using properties of right triangles

PROBLEM SOLVING

The problem solving strategy, Using a Formula, can be used throughout Chapter 9. All exercises in Lesson 9.2 ask students to use the Pythagorean Theorem to solve problems. The use of the Pythagorean Theorem is extended in Lesson 9.3 to include using the formula to determine whether triangles are acute, right, or obtuse. The strategy, Solving a Simpler Problem, can be used to help solve problems that involve using the Pythagorean Theorem with complex figures, as in Exercise 33 on page 348.

TECHNOLOGY

Computer: The Computer Investigation on page 359 enables students to explore the sine ratio in order to calculate lengths of sides or angle measures of given right triangles.

Calculator: The Using the Calculator lesson on page 365 familiarizes students with using calculators to compute trigonometric ratios. Encourage students to use a calculator throughout Chapter 9 to facilitate calculations.

SPECIAL FEATURES

Mixed Review pp. 343, 348, 353, 364, 369, 373
Application: Baseball "Diamond" p. 349
Algebra Review p. 353
Focus on Reading pp. 357, 367
Computer Investigation: A Special Ratio in Right Triangles p. 359
Midchapter Review p. 358
Using the Calculator p. 365
Brainteaser p. 369
Key Terms p. 374
Key Ideas and Review Exercises pp. 374, 375
Chapter 9 Test p. 376
College Prep Test p. 377

PLANNING GUIDE

Lesson	Basic	Average	Above Average	Resources
9.1 pp. 342–343	CE all WE 1–20 odd	CE all WE 1–27 odd	CE all WE 1–31 odd	Reteaching p. 123 Practice p. 124
9.2 pp. 346–349	CE all WE 1–21 Application 1–3	CE all WE 1–32 odd Application all	CE all WE 1–36 odd Application all	Reteaching p. 125 Practice p. 126
9.3 pp. 352–353	CE all WE 1–12 AR all	CE all WE 1–23 odd AR all	CE all WE 1–29 odd AR all	Reteaching p. 127 Practice p. 128
9.4 pp. 357–359	FR all CE all WE 1–12 odd CI 1–2	FR all CE all WE 1–25 odd CI all	FR all CE all WE 1–28 odd CI all	Reteaching p. 129 Practice p. 130
9.5 pp. 362–365	CE all WE 1–19 odd Calculator	CE all WE 1–28 odd Calculator	CE all WE 1–32 odd Calculator	Reteaching p. 131 Practice p. 132
9.6 pp. 367–369	FR all CE all WE 1–22 odd Brainteaser	FR all CE all WE 1–30 odd Brainteaser	FR all CE all WE 1–32 odd Brainteaser	Reteaching p. 133 Practice p. 134
9.7 pp. 371–373	CE all WE 1–7	CE all WE 1–13	CE all WE 1–18	Reteaching p. 135 Practice p. 136
Chapter 9 Review pp. 374–375	1–23 odd	all	all	
Chapter 9 Test p. 376	1–14, 19–22	1–22	all	
College Prep Test p. 377	all	all	all	

CE = Classroom Exercises   WE = Written Exercises   FR = Focus on Reading   AR = Algebra Review   CI = Computer Investigation

NOTE: For each level, all students should be assigned all Mixed Review exercises.

Technology: This activity allows students to use a calculator to explore the Golden Ratio and to experience the use of trigonometric functions in a simple formula. It can be used in connection with Lesson 9.5.

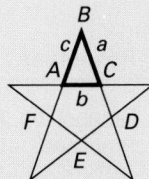

Draw a five-pointed "Star of Pythagoras" on the blackboard as shown, or refer the students to Technology Worksheet 9. Challenge the students to find the measures of the angles of △ABC. (The measure of each interior angle of the regular pentagon ACDEF is 108, so m ∠A = m ∠C = 72, and m ∠B = 36.) Explain to the students that this isosceles triangle is known as a *golden triangle* because the ratio of the length of a leg to the length of the base is 1.61803..., the Golden Ratio. Have the students verify this ratio numerically by using the the formula

$$\frac{a}{b} = \frac{\sin A}{\sin B}.$$

The students can divide sin 72 by sin 36 on their calculators to find $\frac{a}{b}$. They should compare this value with the value of the expression

$$\frac{1 + \sqrt{5}}{2},$$

which is the Golden Ratio (page 312.) Use Technology Worksheet 9 for further exploration of the Golden Ratio and the Star of Pythagoras.

$$xy^2 = a^2(a-x)$$

Maria Agnesi (1718–1799)—Maria Agnesi, the eldest of 21 children, could speak five languages before age 15. She is said to have written complete solutions to puzzling math problems while walking in her sleep.

More About the Mathematician

In 1748, Maria Agnesi published a two-volume work called *Analytical Institutions*, a systematic text that integrated and explained algebra, analysis, and recent work on the calculus. The book included many of her own methods and generalizations and contained the curve (pictured here) named after her. In 1801, an Englishman (who had learned Italian just so he could translate this book) mistranslated the name of the curve as "witch." Consequently, this curve is now widely known as the "Witch of Agnesi." Maria Agnesi's name was added to the faculty roll at the University of Bologna in Italy, but she never accepted an invitation to lecture there.

9.1 Right Triangle Similarity Properties

Objective To find the lengths of sides and altitudes of right triangles

In right triangle PRQ, the altitude $\overline{RS}$ from R to the hypotenuse forms two right triangles, $\triangle PSR$ and $\triangle QSR$. Suppose that m $\angle 1 = 50$. Then the measure of each of the other numbered angles can be found by using Theorem 4.1, which states that the acute angles of a right triangle are complementary.

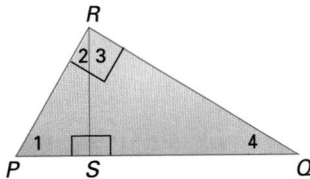

Note that in each of the three triangles above ($\triangle PSR$, $\triangle QSR$, $\triangle RPQ$), the measures of the angles are the same: 40, 50, and 90. Thus, by AA Similarity, the triangles are similar. The general case for this result is stated in the following theorem.

Theorem 9.1

In a right triangle, the altitude to the hypotenuse forms two similar right triangles, each of which is also similar to the original triangle.

Given: Right $\triangle PRQ$, with right $\angle PRQ$ and altitude $\overline{RS}$
Prove: $\triangle PRQ \sim \triangle PSR$, $\triangle PRQ \sim \triangle RSQ$, $\triangle PRS \sim \triangle RSQ$

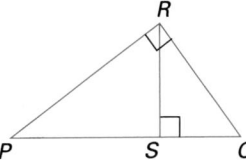

Proof

Statement	Reason
1. Rt $\triangle PRQ$ with rt $\angle PRQ$, altitude $\overline{RS}$	1. Given
2. $\angle PSR$ and $\angle QSR$ are rt $\angle$s.	2. Def of alt
3. $\angle PRQ \cong \angle PSR$	(A) 3. All rt $\angle$s are $\cong$.
4. $\angle P \cong \angle P$	(A) 4. Reflex Prop
5. $\therefore \triangle PSR \sim \triangle PRQ$	5. AA $\sim$
6. $\angle PRQ \cong \angle RSQ$	(A) 6. All rt $\angle$s are $\cong$.
7. $\angle Q \cong \angle Q$	(A) 7. Reflex Prop
8. $\therefore \triangle PRQ \sim \triangle RSQ$	8. AA $\sim$
9. $\therefore \triangle PSR \sim \triangle RSQ$	9. Trans Prop of Similarity

Teaching Resources

Application Worksheet 9
Manipulative Worksheet 16
Problem Solving Worksheet 10
Quick Quizzes 62
**Reteaching and Practice
Worksheets,** pp. 123, 124
Transparencies 20, 21

▰▰ GETTING STARTED

Prerequisite Quiz

Simplify.

1. $\sqrt{18}$ $3\sqrt{2}$
2. $\sqrt{12}$ $2\sqrt{3}$
3. $\sqrt{40}$ $2\sqrt{10}$
4. $\sqrt{32}$ $4\sqrt{2}$

Solve.

5. $x^2 = 45$ $\pm 3\sqrt{5}$
6. $x^2 = 12$ $\pm 2\sqrt{3}$
7. $64 = x^2 + 12x$ $4, -16$
8. $36 = x^2 + 5x$ $4, -9$

Motivator

Ask the students what they know about corresponding sides of two triangles if two angles of one triangle are congruent to two angles of the second triangle.   They are proportional.

▰▰ TEACHING SUGGESTIONS

Lesson Note

It will probably be necessary to review the AA Similarity Postulate and the properties of similar triangles.

Diagrams and labels such as those below help students to understand and remember Corollaries 1 and 2.

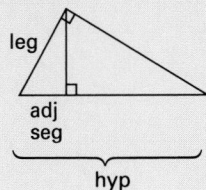
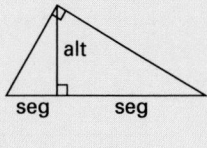

Math Connections

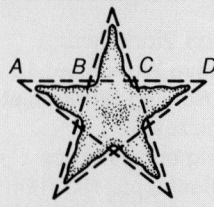

Nature: One of the characteristics of a starfish can be described with a proportion that is based on properties of a 5–pointed star formed from a regular pentagon. In the diagram, $\frac{AC}{AB} = \frac{AB}{BC}$, so AB is the geometric mean of AC and BC, and $(AB)^2 = (AC)(BC)$. These ratios are equal to the Golden Ratio, which is discussed in the Problem Solving Strategies lesson on page 312.

Critical Thinking Questions

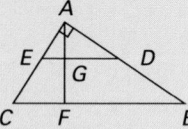

Application: Refer students to the diagram and ask the following question.

If $\overline{ED} \parallel \overline{CB}$ and $\overline{ED} \perp \overline{AF}$, explain how to prove that AB is the geometric mean of CB and FB. Since $\overline{ED} \parallel \overline{CB}$, then $\overline{AF} \perp \overline{CB}$ and $\overline{AF}$ is an alt in rt $\triangle ABC$. Therefore, by Corollary 2 on page 341 and the def of geometric mean, AB is the geometric mean of CB and FB.

Common Error Analysis

Error: Students tend to use the wrong segments when applying Corollary 2.

An abbreviated statement such as the following may be helpful.
Leg^2 = product of hypotenuse and segment nearer to or touching that leg.

In $\triangle PRQ$ at the right, h, m, and n represent the lengths of sides of the triangles formed by the altitude to the hypotenuse. Notice that since $\triangle PRS \sim \triangle RQS$ by Theorem 9.1, their corresponding sides are proportional.

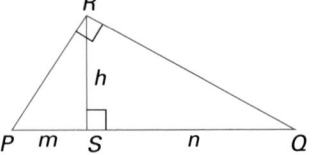

$$\frac{m}{h} = \frac{h}{n}, \text{ or } h^2 = mn$$

Corollary 1   In a right triangle, the square of the length of the altitude to the hypotenuse equals the product of the lengths of the segments formed on the hypotenuse.

Corollary 1 is often stated as follows:

In a right triangle, the length of the altitude to the hypotenuse is the geometric mean of the lengths of the segments formed on the hypotenuse.

EXAMPLE 1   Given: Right $\triangle ABC$ with altitude $\overline{BD}$ to hypotenuse $\overline{AC}$, $BD = 12$, $AD = 8$
   Find CD.

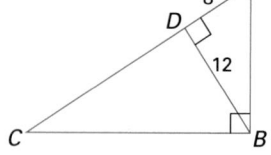

Solution
$(BD)^2 = AD \cdot CD$ (Corollary 1)
$12^2 = 8 \cdot CD$
$\frac{144}{8} = CD$
$CD = 18$

EXAMPLE 2   Given: Right $\triangle KHG$ with altitude $\overline{HT}$ to hypotenuse $\overline{KG}$, $KG = 14$, $KT = 6$
   Find HT.

Plan   Find the length of the other segment of the hypotenuse, TG.

$TG = KG - KT$
$= 14 - 6$
$= 8$

Then apply Corollary 1.

Solution   (Length of altitude)^2 = product of lengths of segments of hypotenuse

$(HT)^2 = 8 \cdot 6 = 48$
$HT = \sqrt{48} = \sqrt{16} \cdot \sqrt{3} = 4\sqrt{3}$

Additional Example 1

Given: Right $\triangle PQR$ with altitude $\overline{QS}$ to hypotenuse $\overline{PR}$, $PS = 6$, $SR = 8$
Find $\overline{QS}$. $4\sqrt{3}$

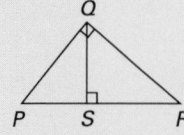

Additional Example 2

Given: Right $\triangle MNY$ with altitude $\overline{XN}$ to hypotenuse $\overline{MY}$, $MY = 12$, $XY = 4$
Find XN. $4\sqrt{2}$

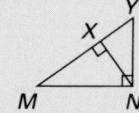

EXAMPLE 3   Right △ABC has its right ∠ at C. The altitude from C to the hypotenuse meets the hypotenuse at point D. CD = 4 and AD = 2. Find DB.

Plan   First draw a right triangle. Label the vertex of the right angle C. Draw the altitude from that vertex to $\overline{AB}$. The segments of the hypotenuse are then $\overline{AD}$ and $\overline{DB}$. AD = 2 and CD = 4. Let DB = x.

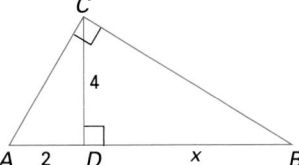

Solution   (Length of altitude)^2 = product of lengths of segments of hypotenuse

$$4^2 = 2 \cdot x$$
$$16 = 2x$$
$$8 = x$$

So, DB = 8

Corollary 2

If the altitude is drawn to the hypotenuse of a right triangle, then the square of the length of either leg equals the product of the lengths of the hypotenuse and the segment of the hypotenuse adjacent to that leg.

Given: Right △RTS with right ∠RTS
and altitude $\overline{TU}$
Prove: $s^2 = mt$ and $r^2 = nt$

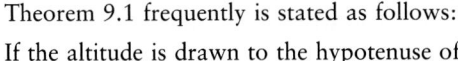

Plan   △RTS ~ △RUT, and △RTS ~ △TUS. Use corresponding sides of similar triangles to write proportions.

Theorem 9.1 frequently is stated as follows:

If the altitude is drawn to the hypotenuse of a right triangle, then the length of either leg is the geometric mean of the length of the hypotenuse and the length of the segment of the hypotenuse adjacent to that leg.

EXAMPLE 4   Given: Right △FDE, altitude $\overline{DG}$, FG = 5, GE = 4 Find FD.

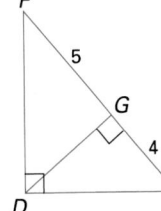

Solution   (length of leg)^2 = (length of hyp) · (length of adj seg)
$$(FD)^2 = 9 \cdot 5 = 45$$
$$FD = \sqrt{45} = \sqrt{9} \cdot \sqrt{5} = 3\sqrt{5}$$

Checkpoint

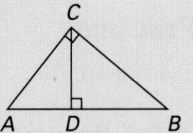

In right △**ABC**, $\overline{CD}$ is an altitude to hypotenuse $\overline{AB}$. Find the indicated lengths.

1. CD = 4, AD = 2, DB = _____   .8
2. AD = 8, DB = 4, CD = _____ $4\sqrt{2}$
3. AC = 6, AD = 2, DB = _____   16
4. AD = 7, DB = 5, CB = _____ $2\sqrt{15}$
5. AD = DB, CD = 4, AD = _____   4
6. AC = $2\sqrt{3}$, DB = 4, AD = _____   2

Closure

Ask the students to write two equations for a right triangle with altitude h as taught in this lesson.   Corollary 1 page 340, Corollary 2 page 341

Additional Example 3

Right △PQR has its right angle at Q. The altitude from Q to the hypotenuse meets the hypotenuse at point T, PR = 10 and PT = 8. Find QT.   4

Additional Example 4

Given: Right △TWX with altitude $\overline{WY}$, XW = $4\sqrt{3}$, XT = 8
Find XY.   6

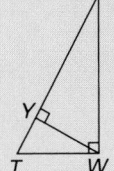

Guided Practice

Classroom Exercises 1–10

Independent Practice

A Ex. 1–20, **B** Ex. 21–27, **C** Ex. 28–31

Basic: WE 1–20 odd

Average: WE 1–27 odd

Above Average: WE 1–31 odd

Additional Answers

Written Exercises

24. See illustration on page 340.

Statement	Reason
1. Rt $\triangle PQR$ with rt $\angle PRQ$, alt $\overline{RS}$	1. Given
2. $\triangle RSQ \sim \triangle PSR$	2. Alt to hyp in rt $\triangle$ forms 2 $\sim$ rt $\triangle$s.
3. $\frac{h}{n} = \frac{m}{h}$	3. Def of $\sim \triangle$s
4. $h^2 = nm$	4. If proport, then prod of means = prod of extremes.

25. See illustration on page 341.

Statement	Reason
1. Rt $\triangle RST$ with rt $\angle RTS$ and alt $\overline{TU}$	1. Given
2. $\triangle RST \sim \triangle RTU$	2. Alt to hyp in rt $\triangle$ forms 2 $\sim \triangle$s, each $\sim$ to orig $\triangle$.
3. $\frac{s}{t} = \frac{m}{s}$	3. Def of $\sim \triangle$s
4. $s^2 = mt$	4. If proport, then prod of means = prod of extremes.

EXAMPLE 5   Given: Right $\triangle DEF$, $\overline{FG} \perp \overline{DE}$, $DF = 8$, $EG = 12$. Find DG.

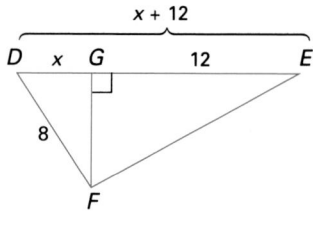

Plan   Let $DG = x$. [Length of segment adjacent to leg $\overline{DF}$]

Then $DE = x + 12$. [length of hypotenuse]

Solution
$$(DF)^2 = DE \cdot DG$$
$$8^2 = (x + 12)x$$
$$64 = x^2 + 12x$$
$$0 = x^2 + 12x - 64$$
$$0 = (x + 16)(x - 4)$$
$$x + 16 = 0 \quad or \quad x - 4 = 0$$
$$x = -16 \quad or \quad x = 4$$

A segment cannot have a negative length, so discard the solution -16. Thus, $DG = 4$.

Classroom Exercises

Identify the following using the given figure.

1. hypotenuse of $\triangle TZY$ $\overline{TY}$

2. hypotenuse of $\triangle WZY$ $\overline{YZ}$

3. legs of $\triangle TYZ$ $\overline{TZ}$, $\overline{YZ}$

4. legs of $\triangle TZW$ $\overline{TW}$, $\overline{ZW}$

5. legs of $\triangle WZY$ $\overline{WZ}$, $\overline{WY}$

6. altitude to hypotenuse of $\triangle TYZ$ $\overline{ZW}$

7. segment of hypotenuse of $\triangle TYZ$ adjacent to $\overline{ZY}$ $\overline{WY}$

8. segment of hypotenuse of $\triangle TYZ$ adjacent to $\overline{ZT}$ $\overline{TW}$

9. Use Corollary 1 to Theorem 9.1 to write an equation. $(WZ)^2 = TW \cdot WY$

10. Use Corollary 2 to Theorem 9.1 to write an equation.
$(TZ)^2 = TW \cdot TY$, $(ZY)^2 = WY \cdot TY$

Written Exercises

In right $\triangle SVU$, $\angle SVU$ is a right angle and $\overline{VT}$ is an altitude. Find the length.

1. $TV = 4$, $ST = 2$, $TU = \underline{\quad 8 \quad}$

2. $ST = 12$, $TU = 3$, $TV = \underline{\quad 6 \quad}$

3. $ST = 6$, $TU = 4$, $VT = \underline{\quad 2\sqrt{6} \quad}$

4. $TV = 4$, $TU = 6$, $ST = \underline{\qquad}$ $2\frac{2}{3}$

5. $ST = 8$, $SU = 10$, $TV = \underline{\quad 4 \quad}$

6. $SU = 12$, $TU = 8$, $TV = \underline{\quad 4\sqrt{2} \quad}$

7. $ST = 2$, $SU = 8$, $SV = \underline{\quad 4 \quad}$

8. $TU = 3$, $SU = 8$, $UV = \underline{\quad 2\sqrt{6} \quad}$

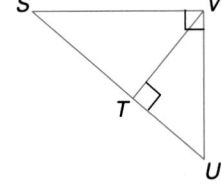

Additional Example 5

Given: Right $\triangle VTU$ with altitude $\overline{TW}$,
       $VT = 4$, $WU = 6$

Find VW.   2

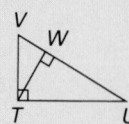

In right $\triangle QMP$, $\angle QMP$ is a right angle and $\overline{MN}$ is an altitude. Find the length.

9. $QM = 6$, $QN = 2$,
$QP = \underline{\quad 18 \quad}$

10. $QN = 8$, $NP = 4$,
$MP = \underline{\quad 4\sqrt{3} \quad}$

11. $MP = 8$, $QP = 16$,
$NP = \underline{\quad 4 \quad}$

12. $QN = 4$, $NP = 8$,
$MP = \underline{\quad 4\sqrt{6} \quad}$

13. $QN = 12$, $NP = 2$,
$MP = \underline{\quad 2\sqrt{7} \quad}$

14. $QN = 3$, $QP = 12$,
$MN = \underline{\quad 3\sqrt{3} \quad}$

15. $QM = 6$, $PN = 5$,
$NQ = \underline{\quad 4 \quad}$

16. $QN = 6$, $PM = 4$,
$QP = \underline{\quad 8 \quad}$

17. $QP = 10$, $MN = 4$,
$QN = \underline{\quad 8 \text{ or } 2 \quad}$

18. $PQ = 13$, $MN = 6$,
$NP = \underline{\quad 9 \text{ or } 4 \quad}$

19. $QN = NP$, $QM = 4$,
$QN = \underline{\qquad} \quad 2\sqrt{2}$

20. $QP = 20$, $MN = 8$,
$NP = \underline{\quad 4 \text{ or } 16 \quad}$

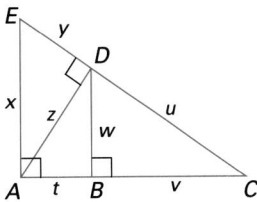

21. Right triangle RST has its right angle at S. The altitude from S meets the hypotenuse at U. $RU = 4$ and $UT = 6$. Find RS, SU, and ST. $2\sqrt{10}$, $2\sqrt{6}$, $2\sqrt{15}$

22. The right angle of right triangle KGA is at A. The altitude from A meets the hypotenuse at P. $PK = 32$ and $GA = 12$. Find GP and PA. $4, 8\sqrt{2}$

23. If the altitude to the hypotenuse of a right triangle bisects the hypotenuse and the length of the hypotenuse is 6, find the length of each leg. $3\sqrt{2}$

24. Prove Corollary 1 to Theorem 9.1.   **25.** Prove Corollary 2 to Theorem 9.1.

26. Prove: In a right triangle, the product of the lengths of the hypotenuse and the altitude to the hypotenuse equals the product of the lengths of the legs.

27. The lengths of the two legs and hypotenuse of a right triangle are a, b, and c respectively. Find the length of the altitude to the hypotenuse in terms of a, b, and c.

In the figure, m $\angle EAC = 90$, $\overline{DB} \perp \overline{AC}$, $\overline{AD} \perp \overline{EC}$

28. Given: $x = 4$, $u = 6$. Find y. 2

29. Prove: $w(u + y) = xu$

30. Given: $t = 9$, $v = 4$. Find x. 19.5

31. Given: $u = 6$, $t = 5$. Find x. $\frac{9}{2}\sqrt{5}$

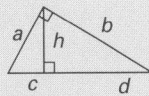

Mixed Review

1. The measures of the acute angles of a right triangle are in the ratio 3:7. Find the measure of each acute angle.   **8.1**   27, 63

2. Find the sum of the measures of the interior angles of a pentagon.   **6.2**   540

3. The measure of one of the angles of a parallelogram is 140. Find the measure of each of the other three angles.   **6.4**   40, 140, 40

9.1 Right Triangle Similarity Properties   **343**

26.

Statement	Reason
1. Rt $\triangle ABC$ with rt $\angle ABC$ and alt $\overline{BD}$	1. Given
2. $\triangle ABC \sim \triangle BDC$	2. Alt to hyp in rt $\triangle$ forms 2 $\sim$ $\triangle$s, each $\sim$ to orig $\triangle$.
3. $\frac{AC}{AB} = \frac{BC}{BD}$	3. Def of $\sim$ $\triangle$s
4. $AC \cdot BD =$ $AB \cdot BC$	4. If proport, then prod of means = prod of extremes.

27. Let alt $= h$. Then $\frac{a}{c} = \frac{h}{b}$; $h = \frac{ab}{c}$

29.

Statement	Reason
1. Rt $\triangle ACE$ with rt $\angle EAC$, $\overline{DB}$ $\perp$ $\overline{AC}$	1. Given
2. $\angle DBC$ is a rt $\angle$.	2. Def of $\perp$
3. $\angle EAC \cong$ $\angle DBC$	3. All rt $\angle$s are $\cong$.
4. $\angle C \cong \angle C$	4. Reflex Prop of Congr
5. $\triangle EAC \sim$ $\triangle DBC$	5. AA$\sim$
6. $\frac{x}{y + u} = \frac{w}{u}$	6. Def of $\sim$ $\triangle$s
7. $w(y + u) =$ xu	7. If proport, then prod of means = prod of extremes.

Teaching Resources

Project Worksheet 9
Quick Quizzes 63
Reteaching and Practice
  Worksheets, pp. 125, 126
Transparencies 22, 23

◤◤◤GETTING STARTED

Prerequisite Quiz

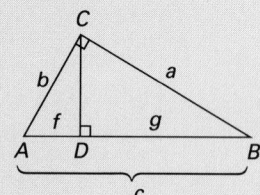

1. Write a formula for finding b in terms of f and c. $b^2 = fc$
2. Write a formula for finding a in terms of g and c. $a^2 = gc$

Solve.

3. $c^2 = 20$ $\pm 2\sqrt{5}$
4. $16 + b^2 = 64$ $\pm 4\sqrt{3}$

Motivator

Explain to the students that early Greek mathematicians discovered an important relationship between the lengths of the legs of a right triangle and the length of the hypotenuse. Have students find a^2, b^2, and c^2, for each set of data below. Then have them write an equation relating the three in each case. $a^2 + b^2 = c^2$

$a = 12$, $b = 5$, $c = 13$ 144, 25, 169
$a = 6$, $b = 8$, $c = 10$ 36, 64, 100
$a = 15$, $b = 8$, $c = 17$ 225, 64, 289

◤◤◤TEACHING SUGGESTIONS

Lesson Note

Students will be better able to handle the applications of the Pythagorean Theorem if the following properties are reviewed.

1. A diagonal divides a rectangle into two congruent *right* triangles.
2. The diagonals of a rhombus are perpendicular bisectors of each other.
3. The bisector of the vertex angle of an isosceles triangle is the perpendicular bisector of the base.

344

9.2 The Pythagorean Theorem

Objective   To apply the Pythagorean Theorem

In this lesson, you will explore a method to find the length of the hypotenuse of a right triangle, given the lengths of the two legs.

Theorem 9.2
Pythagorean Theorem: In any right triangle, the sum of the squares of the lengths of the legs is equal to the square of the length of the hypotenuse.

Given: Right $\triangle ACB$ with right $\angle ACB$
Prove: $a^2 + b^2 = c^2$

Plan   Draw $\overline{CD} \perp \overline{AB}$. $\overline{CD}$ is the altitude to hypotenuse $\overline{AB}$. Write two equations using corollary 2 to Theorem 9.1.

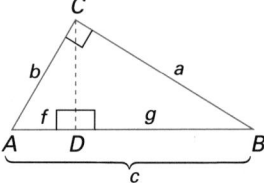

Proof

	Statement		Reason
1.	Rt $\triangle ACB$ with rt $\angle ACB$	1.	Given
2.	Draw $\overline{CD} \perp$ to $\overline{AB}$.	2.	From a pt not on a line, there is exactly one line $\perp$ to the line.
3.	$\overline{CD}$ is an altitude.	3.	Def of alt
4.	$b^2 = fc$, $a^2 = gc$	4.	Square of length of leg = prod of lengths of hyp and adj seg.
5.	$a^2 + b^2 = fc + gc$	5.	Add Prop of Eq
6.	$a^2 + b^2 = (f + g)c$	6.	Distr Prop
7.	$f + g = c$	7.	Seg Add Post
8.	$\therefore a^2 + b^2 = c \cdot c$ or $a^2 + b^2 = c^2$	8.	Sub

EXAMPLE 1   Find c.   Find b.

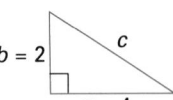   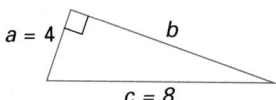

Solutions
$a^2 + b^2 = c^2$ $a^2 + b^2 = c^2$
$4^2 + 2^2 = c^2$ $4^2 + b^2 = 8^2$
$16 + 4 = c^2$ $16 + b^2 = 64$
$20 = c^2$ $b^2 = 48$
$c = \sqrt{20}$, or $2\sqrt{5}$ $b = \sqrt{48}$, or $4\sqrt{3}$

Additional Example 1

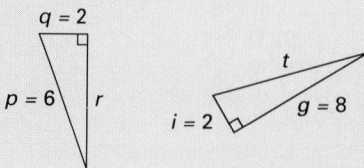

Find r. $4\sqrt{2}$ Find t. $2\sqrt{17}$

Additional Example 2

The length of a side of a square is 6. Find the length of a diagonal. $6\sqrt{2}$

EXAMPLE 2   The length of a diagonal of a square is 6. Find the length of a side.

Plan   Draw a square with a diagonal. The diagonal is the hypotenuse of each right triangle formed.

Let s = length of a side.

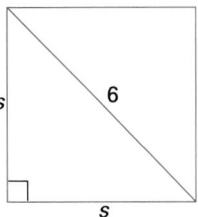

Solution
$$s^2 + s^2 = 6^2$$
$$2s^2 = 36$$
$$s^2 = 18$$
$$s = \sqrt{18} = \sqrt{9} \cdot \sqrt{2}, \text{ or } 3\sqrt{2}$$

Thus, the length of a side of the square is $3\sqrt{2}$.

EXAMPLE 3   The length of a side of a rhombus is 13 cm, and the length of one diagonal is 10 cm. Find the length of the other diagonal.

Plan   Draw a rhombus with its diagonals. Each side is a hypotenuse of a right triangle. The diagonals bisect each other.

Let $a = \frac{1}{2}$(length of the other diagonal).

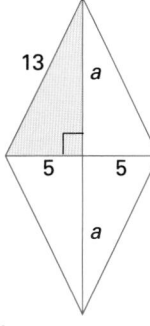

Solution
$$13^2 = a^2 + 5^2$$
$$169 = a^2 + 25$$
$$144 = a^2$$
$$12 = a$$

Thus, the length of the other diagonal is $2 \cdot 12$, or 24 cm.

EXAMPLE 4   The length of each leg of an isosceles triangle is 6 in. The length of the base is 4 in. Find the length of the altitude from the vertex angle.

Plan   Draw an isosceles triangle with an altitude from the vertex angle. Since by Theorem 5.7 such an altitude bisects the base, $CD = \frac{1}{2} \cdot 4$, or 2. Let h equal the length of the altitude. By the Pythagorean Theorem:

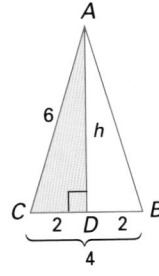

Solution
$$2^2 + h^2 = 6^2$$
$$4 + h^2 = 36$$
$$h = \sqrt{32}, = \sqrt{16} \cdot \sqrt{2}, \text{ or } 4\sqrt{2}$$

The length of the altitude is $4\sqrt{2}$ in.

Math Connections

History: The Pythagorean Theorem, one of the most familiar relationships in geometry, was used as early as 2000 B.C. by the ancient Egyptians. There have also been records of the relationship being used long ago by the Babylonians and the Chinese. About 500 B.C., the Greek mathematician Pythagoras became well-known for proving the theorem and teaching many of its applications. Thereafter, it was named in honor of Pythagoras.

Critical Thinking Questions

Application: Refer students to the diagram of three stacked pipes. Ask them to find the height h of the stack if the diameter of the outside of each pipe is 8 inches.

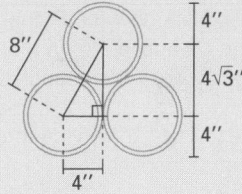

$h = 4 + \sqrt{8^2 - 4^2} + 4 = 8 + 4\sqrt{3} \approx$ 14.93 in

Common Error Analysis

Error: Students may think of the Pythagorean Theorem as $a^2 + b^2 = c^2$ rather than $(leg_1)^2 + (leg_2)^2 = hypotenuse^2$. This sometimes leads to the incorrect identification of a or b with the hypotenuse. Emphasize the verbal approach rather than the formula $a^2 + b^2 = c^2$.

Additional Example 3

The lengths of the diagonals of a rhombus are 6 cm and 8 cm. Find the length of a side of the rhombus.   5 cm

Additional Example 4

The length of the base of an isosceles triangle is 4 cm. The length of the altitude from the vertex angle is 6 cm. Find the length of each leg of the triangle. $2\sqrt{10}$ cm

Checkpoint

For Exercises 1–4, find the indicated length using the figure below.

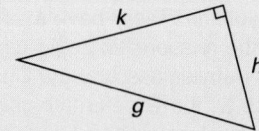

1. If $h = 4$, $k = 4$, find g. $4\sqrt{2}$
2. If $g = 13$, $k = 12$, find h. 5
3. If $g = 8$, $h = 4$, find k. $4\sqrt{3}$
4. If $k = 3$, $h = 6$, find g. $3\sqrt{5}$
5. Find the length of a diagonal of a square if the length of a side is 4 cm. $4\sqrt{2}$ cm
6. The lengths of the legs of an isosceles triangle are 8 ft, and the length of the altitude from the vertex angle to the base is 4 ft. Find the length of the base. $8\sqrt{3}$ ft

Closure

Have the students explain the Pythagorean Theorem in terms of the legs and the hypotenuse.   Theorem 9.2 page 344
Have them explain how the Pythagorean Theorem can be applied to find lengths of segments of squares or rectangles.   The diagonals of squares and rectangles form right triangles. Thus, the Pythagorean Theorem can be applied.

◢◣◢◣ FOLLOW UP

Guided Practice

Classroom Exercises 1–8

Independent Practice

A Ex. 1–21, **B** Ex. 22–32, **C** Ex. 33–36

Basic: WE 1–21, Application 1–3
Average: WE 1–32 odd, Application all
Above Average: WE 1–36 odd, Application all

EXAMPLE 5   Find the length of the altitude to the hypotenuse of a right triangle with legs of lengths 6 and 8.

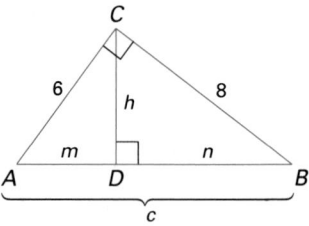

Solution   Use the Pythagorean Theorem to find c.

$$c^2 = 6^2 + 8^2$$
$$c^2 = 36 + 64$$
$$c^2 = 100$$
$$c = 10$$

According to Theorem 9.1, $\triangle CBD \sim \triangle ABC$

$$\frac{CD}{AC} = \frac{CB}{AB}$$
$$\frac{h}{6} = \frac{8}{10}$$
$$h = \frac{48}{10}, \text{ or } 4.8$$

Thus, the length of the altitude is 4.8.

Classroom Exercises

For Exercises 1–8, write an equation that can be solved to find a, b, or c.

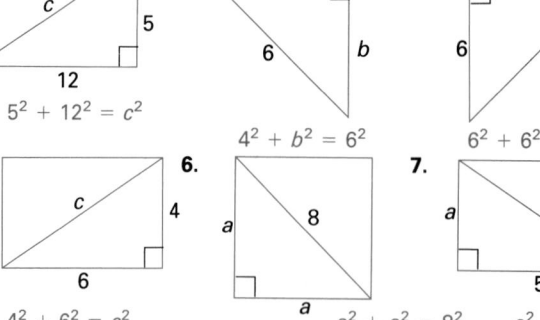

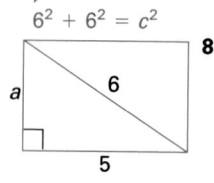

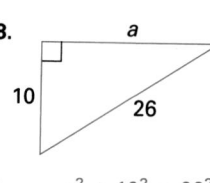

1. $5^2 + 12^2 = c^2$

2. $4^2 + b^2 = 6^2$

3. $6^2 + 6^2 = c^2$

4. $2^2 + b^2 = 4^2$

5. $4^2 + 6^2 = c^2$

6. $a^2 + a^2 = 8^2$

7. $a^2 + 5^2 = 6^2$

8. $a^2 + 10^2 = 26^2$

346   Chapter 9 Right Triangles

Additional Example 5

Find the length of the altitude to the hypotenuse of a right triangle with legs of lengths 4 and 2. $\frac{4\sqrt{5}}{5}$

Written Exercises

For Exercises 1–8, find *a*, *b*, or *c* for Classroom Exercises 1–8.
For Exercises 9–14, *p* and *q* are lengths of legs, and *r* is the length of
the hypotenuse of a right triangle.

9. If $p = 4$, $q = 2$, find r. $2\sqrt{5}$
10. If $r = 8$, $p = 2$, find q. $2\sqrt{15}$
11. If $r = 8$, $p = 6$, find q. $2\sqrt{7}$
12. If $q = 15$, $r = 17$, find p.   8
13. If $p = 10$, $q = 10$, find r. $10\sqrt{2}$
14. If $r = 10$, $p = 5$, find q. $5\sqrt{3}$

Find the indicated length to the nearest tenth.

15. $a = 12$, $b = 14$,
$c = \underline{18.4}$

16. $a = 18$, $b = 23$,
$c = \underline{29.2}$

17. $a = 18$, $c = 32$,
$b = \underline{26.5}$

18. $c = 37$, $b = 14$,
$a = \underline{34.2}$

19. $a = 13.6$, $b = 11.8$,
$c = \underline{18.0}$

20. $c = 23.5$, $b = 11.3$,
$a = \underline{20.6}$

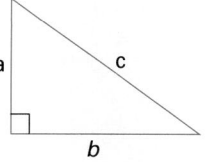

21. The figure at the right shows a shelf 12″ wide supported by 20″ braces like $\overline{AB}$. Find the distance $\overline{AC}$ from shelf to the brace.   16 in.

22. The lengths of two adjacent sides of a rectangle are 8 and 6. Find the length of a diagonal of the rectangle.   10

23. The length of each side of a rhombus is 8 ft. The length of one of its diagonals is 4 ft. Find the length of the other diagonal. $4\sqrt{15}$ ft

24. The lengths of the diagonals of a rhombus are 16 km and 8 km. Find the length of a side of the rhombus. $4\sqrt{5}$ km

25. The lengths of the legs of an isosceles triangle are each 6 in. The length of the base is 8 in. Find the length of an altitude from the vertex angle to the base. $2\sqrt{5}$ in

26. Find the length of the altitude to the hypotenuse of a right triangle if the lengths of the legs are 5 ft and 12 ft. $4\frac{8}{13}$ ft

27. Find the length of the altitude to the hypotenuse of a right triangle if the lengths of the legs are 15 cm and 8 cm. $7\frac{1}{17}$ cm

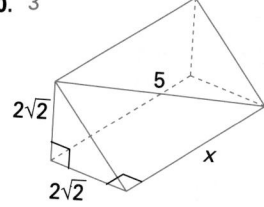

Find the value of *x* in each figure.

28.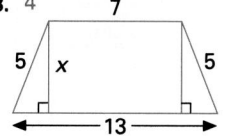

29. 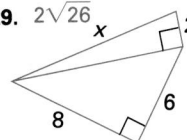 $2\sqrt{26}$

30.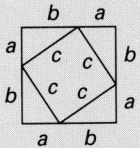

Additional Answers

Written Exercises

1. $c = 13$
2. $b = 2\sqrt{5}$
3. $c = 6\sqrt{2}$
4. $b = 2\sqrt{3}$
5. $c = 2\sqrt{13}$
6. $a = 4\sqrt{2}$
7. $a = \sqrt{11}$
8. $a = 24$

Enrichment

Draw this figure on the chalkboard. Challenge the students to prove the Pythagorean Theorem by using areas of the squares and the triangles in the figure.

area of large square = area of small square + areas of four congruent triangles:

$$(a + b)^2 = c^2 + 4(\tfrac{1}{2}ab)$$
$$a^2 + 2ab + b^2 = c^2 + 2ab$$
$$a^2 + b^2 = c^2$$

Have the students do individual research to find other proofs of the Pythagorean Theorem.

31. The base of an aerial fire truck ladder is 5 ft above the ground and 25 ft from the bottom of the wall of a burning building. How long must a ladder be to reach the roof, which is known to be 65 ft above the ground? (Fire companies find this out in advance for tall buildings.)
≈ 64.8 ft

32. Find d, the length of the diagonal of a rectangular solid with dimensions 2, 4, and 6. (HINT: Use separate triangles, as shown.) $2\sqrt{14}$

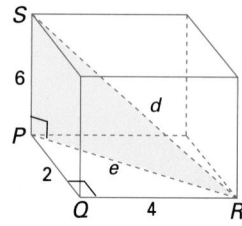 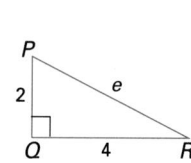 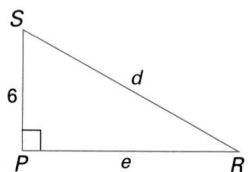

Find a formula for the length of a diagonal of a rectangular solid with sides of the given dimensions. (Exercises 33–35)

33. ℓ, w, and h
$\sqrt{\ell^2 + w^2 + h^2}$

34. e, e, e (a cube) $e\sqrt{3}$

35. x, $2\sqrt{3x}$, 6 $x + 6$

36. In $\triangle ABC$, $AB = 9$, $BC = 11$, $AC = 10$. Find the length of the altitude $\overline{BN}$. $\triangle ABC$ is *not* a right triangle. (HINT: Apply the Pythagorean Theorem twice, using the diagram at the right.) $6\sqrt{2}$

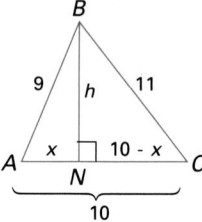

Mixed Review

1. Find the coordinate of the midpoint of a segment if the coordinates of its endpoints are −4 and 10.   *1.3* 3

2. The measure of the vertex angle of an isosceles triangle is 40. Find the measure of each base angle.   *4.1* 70

3. Given $\triangle ABC$ with m $\angle A = 40$, m $\angle B = 60$. Name the longest side of the triangle.   *5.8* $\overline{AB}$

Application: *Baseball "Diamond"*

The so-called baseball "diamond" is actually a square with sides of length 90 ft. The pitcher's mound is located near the center of the square, with the pitching rubber 60 ft 6 in. from home plate. (In the example and problems that follow, assume that the pitching rubber lies on the plane of the playing field. In fact, the rubber is slightly elevated by the mound.)

Example

Show that the pitching rubber is situated closer to home plate than to second base.

Solution

According to the Pythagorean Theorem,
$$(HS)^2 = (HF)^2 + (FS)^2$$
$$(HS)^2 = 90^2 + 90^2$$
$$(HS)^2 = 2 \cdot 90^2$$
$$HS = 90\sqrt{2} \approx 127.3 \text{ ft}$$

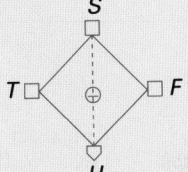

Since the rubber is 60.5 ft from home plate, it must be approximately 127.3 − 60.5, or 66.8 ft from second base.

Exercises

1. A third baseman fields a ground ball, steps on third to force a runner, and throws to first to get the batter. How far is his throw to first base? $90\sqrt{2} \approx 127.3$ ft

It sometimes surprises baseball fans when a fast runner is unable to beat a slow pitcher to first base on a ball hit to the first baseman. In the next four Exercises compute the distances related to this problem.

2. Compute the distance from the pitching rubber to the center of the diamond. ≈ 3.14

3. Compute the distance from the center of the "diamond" to first base. ≈ 63.64

4. Using the results of Exercises 2 and 3, compute the distance from the pitching rubber to first base. ≈ 63.72

5. How much further is it from home plate to first base than from the pitching rubber to first base? ≈ 26.28

GETTING STARTED

Prerequisite Quiz

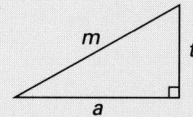

Find the indicated length. (Ex. 1–3)

1. If $t = 6$, $a = 8$, find m.   10
2. If $m = 6$, $a = 2$, find t. $4\sqrt{2}$
3. If $m = 8$, $t = 2$, find a. $2\sqrt{15}$

Simplify.

4. $\sqrt{7} \cdot \sqrt{7}$ 7
5. $2\sqrt{3} \cdot 2\sqrt{3}$ 12
6. $(\sqrt{5})^2$ 5

Motivator

Ask the students how they form the converse of a conditional.   Interchange the hypothesis and the conclusion.   Have them state the Pythagorean Theorem.   Theorem 9.2 page 344   Ask them to form the converse of this theorem.   Theorem 9.3 page 350

TEACHING SUGGESTIONS

Lesson Note

Have students draw a right triangle, an obtuse triangle, and an acute triangle. Have them measure the lengths of the sides of each triangle, and using a calculator or a table, compare the square of the longest side of each triangle to the sum of the squares of the other two sides.

Then ask students to formulate a conjecture that relates their findings to classifying triangles as right, acute, or obtuse.

9.3 Converse of the Pythagorean Theorem

Objective   To determine whether a triangle is right, acute, or obtuse

A person bought a triangular piece of riverfront property measuring 150' by 200' by 250' and wondered if it formed a right triangle. This lesson will show you how to answer this question.

Theorem 9.3   **Converse of the Pythagorean Theorem:** If the sum of the squares of the lengths of two sides of a triangle is equal to the square of the length of the third side, then the triangle is a right triangle.

Given: $\triangle PQR$ with $p^2 + q^2 = r^2$
Prove: $\triangle PQR$ is a right triangle.

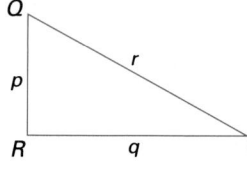

**Proof
(Paragraph
form)**   Construct $\triangle GHK$, with rt $\angle K$, $HK = p$, $GK = q$. By the Pythagorean Theorem,

$$(GH)^2 = p^2 + q^2$$

It is given that $r^2 = p^2 + q^2$.
Then $(GH)^2 = r^2$ and $GH = r$.
Because of the way $\triangle GHK$ was constructed and the fact that $GH = r$, $\triangle PQR \cong \triangle GHK$ by SSS. Then, by CPCTC, $\angle R$ is a right angle. Therefore, $\triangle PQR$ is a right triangle.

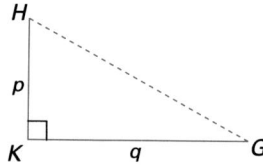

EXAMPLE 1   Given: $AB = 18$, $BC = 82$, $AC = 80$
Determine whether the triangle is a right triangle. If so, name the right angle.

Solution
$$(AB)^2 + (AC)^2 \overset{?}{=} (BC)^2$$
$$18^2 + 80^2 \overset{?}{=} 82^2$$
$$324 + 6{,}400 \overset{?}{=} 6{,}724$$
$$6{,}724 = 6{,}724 \checkmark$$

The triangle is a right triangle since $(AB)^2 + (AC)^2 = (BC)^2$.
The right angle is $\angle A$, the angle opposite the longest side $\overline{BC}$.

Additional Example 1

In $\triangle XYZ$, $XZ = 15$, $XY = 16$, and $YZ = 17$. Determine whether the triangle is a right triangle.
$15^2 + 16^2 \neq 17^2$; not a rt $\triangle$.

Additional Example 2

Determine whether the triangle is acute, obtuse, or right for three sides of the given lengths.

a. 15, 12, 9   Right
b. 3, 4, 6   Obtuse
c. 5, 7, 5   Acute

Theorem 9.4

If the square of the longest side of a triangle is greater (less) than the sum of the squares of the lengths of the other two sides, then the triangle is obtuse (acute).

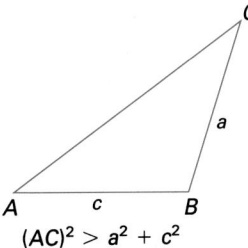

$(AC)^2 > a^2 + c^2$

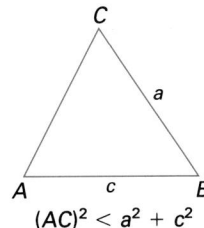

$(AC)^2 < a^2 + c^2$

Theorems 9.3 and 9.4 can be used to determine whether a triangle is acute, obtuse, or right. It may be necessary to use the algebraic property that $(\sqrt{a})^2 = \sqrt{a} \cdot \sqrt{a} = a$, for all $a \geq 0$.

EXAMPLE 2

Determine whether the triangle is acute, obtuse, or right for three sides of the given lengths.

a. 10, 2, 9   b. 15, 36, 39   c. $\sqrt{6}, \sqrt{3}, \sqrt{5}$

Solutions

Compare the square of the greatest length with the sum of the squares of the other two lengths.

a. $10^2 \overset{?}{=} 2^2 + 9^2$ b. $39^2 \overset{?}{=} 15^2 + 36^2$ c. $(\sqrt{6})^2 \overset{?}{=} (\sqrt{3})^2 + (\sqrt{5})^2$
   $100 \overset{?}{=} 4 + 81$ $1{,}521 \overset{?}{=} 225 + 1{,}296$ $6 \overset{?}{=} 3 + 5$
   $100 > 85$ $1{,}521 = 1{,}521$ $6 < 8$

The triangle is an obtuse triangle.   The triangle is a right triangle.   The triangle is an acute triangle.

EXAMPLE 3

Sketch parallelogram $ABCD$ with diagonals intersecting at O, $DC = 10$, $AC = 16$, and $BD = 12$. Determine whether $ABCD$ is a rhombus.

Solution

$OD = 6$, $OC = 8$ since diagonals of a $\square$ bisect each other.

$(OC)^2 + (OD)^2 \overset{?}{=} (DC)^2$
$8^2 + 6^2 \overset{?}{=} 10^2$
$64 + 36 \overset{?}{=} 100$
$100 = 100$ ✔

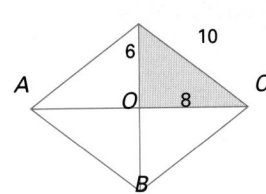

Therefore, $\triangle DOC$ is a right triangle, and the diagonals of $\square ABCD$ are perpendicular to each other.

Thus, $\square ABCD$ is a rhombus.

Math Connections

Egyptians: Egyptian land surveyors were named *rope stretchers* after their methods of surveying land, which applied the converse of the Pythagorean Theorem. Since a triangle with sides of lengths 3, 4, and 5 forms a right triangle, they used a rope with 11 knots, which divided the rope into 12 congruent units (including the outer segments without knots), to determine right angles. Discuss with students how variations in the length of a unit have no effect on forming a right triangle.

Critical Thinking Questions

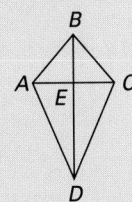

Analysis: The frame of a kite is made of two pieces of wood, one 16 inches long and one 2 feet long. In the diagram, if triangles ABE and CBE are isosceles right triangles, what is the perimeter of the kite? $16\sqrt{2} + 16\sqrt{5} \approx 58.4$ in   If the pieces of wood were connected at some point other than E, would the perimeter change?   Yes

Common Error Analysis

Error: Students incorrectly evaluate expressions such as $(3\sqrt{2})^2$ as $9\sqrt{2}$.

Emphasize that $(3\sqrt{2})^2$ means $(3\sqrt{2})(3\sqrt{2}) = (3 \cdot 3 \cdot \sqrt{2} \cdot \sqrt{2}) = 9 \cdot 2 = 18$. Have them evaluate several similar expressions.

Additional Example 3

Sketch parallelogram $PQRS$ with diagonals intersecting at M. $PQ = 12$, $PR = 8$, and $QS = 18$. Determine whether $PQRS$ is a rhombus.

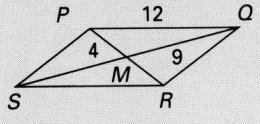

$4^2 + 9^2 \neq 12^2$. Therefore, $\triangle PQM$ is not a right triangle. So, $\square PQRS$ is not a rhombus.

Checkpoint

The lengths of the three sides of △*ABC* are given. Determine whether it is a right, an acute, or an obtuse triangle.

1. $A = 8$, $B = 7$, $C = 6$ Acute
2. $A = 7$, $B = 6$, $C = 12$ Obtuse
3. $A = 5$, $B = 12$, $C = 13$ Right

Determine whether parallelogram *TUVW* is a rhombus.

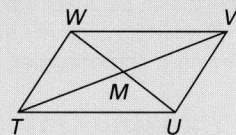

4. $WU = 18$, $TV = 24$, $WV = 15$ Yes
5. $WU = 4\sqrt{3}$, $TV = 2\sqrt{5}$, $WT = \sqrt{17}$ Yes
6. $TM = 5$, $WU = 12$, $WT = 7$ No

Closure

Ask the students how they can determine if a triangle is a right triangle given the lengths of its three sides.   Theorem 9.3 page 350
Then ask them how they can determine if a triangle is obtuse or acute from the lengths of its three sides.   Theorem 9.4 page 351

▰▰▰FOLLOW UP

Guided Practice

Classroom Exercises 1–9

Independent Practice

A Ex. 1–12, **B** Ex. 13–23, **C** Ex. 24–29

Basic: WE 1–12, AR all
Average: WE 1–23 odd, AR all
Above Average: WE 1–29 odd, AR all

Additional Answers

Written Exercises

23. If sq of longest side = sum of sqs of other 2 sides, then △ is rt. If sq of longest side > (or <) sum of sqs of other 2 sides, then △ is obtuse (or acute).

Classroom Exercises

Simplify.

1. $(\sqrt{7})^2$ 7   **2.** $(\sqrt{3})^2$ 3   **3.** $(\sqrt{y})^2$ y   **4.** $(\sqrt{2y})^2$ 2y   **5.** $(3\sqrt{2})^2$ 18

Indicate whether △*ABC* is a right triangle for the given lengths of *a*, *b*, and *c*. Explain why.

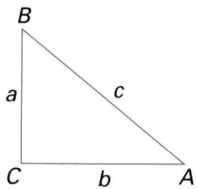

6. $a = 3$, $b = 4$, $c = 5$ Yes, $3^2 + 4^2 = 5^2$
7. $a = 5$, $b = 2$, $c = 6$ No, $5^2 + 2^2 < 6^2$
8. $a = 2$, $b = 1$, $c = \sqrt{5}$ Yes, $2^2 + 1^2 = (\sqrt{5})^2$
9. $a = \sqrt{3}$, $b = 2$, $c = \sqrt{7}$ Yes, $(\sqrt{3})^2 + 2^2 = (\sqrt{7})^2$

Written Exercises

The lengths of three sides of △*ABC* are given. Determine whether △*ABC* is a right triangle. If so, name the right angle.

1. $AC = 18$, $BC = 24$, $AB = 30$ Yes, $\angle C$ **2.** $BC = 30$, $AB = 72$, $AC = 78$ Yes, $\angle B$
3. $AC = 17$, $AB = 16$, $BC = 14$ No   **4.** $AC = 1.5$, $AB = 2$, $BC = 2.5$ Yes, $\angle A$

Determine whether the triangle formed with given lengths of the three sides is acute, obtuse, or right.

5. 24, 32, 50 Obtuse **6.** 9, 16, 19 Obtuse **7.** 18, 24, 28 Acute **8.** 0.6, 0.6, 1 Obtuse
9. $\frac{5}{13}$, 1, $\frac{12}{13}$ Right **10.** 8, 5, 5 Obtuse **11.** 7, 6, 9 Acute **12.** 7, 3, 5 Obtuse
13. $8x$, $17x$, $15x$ Right **14.** a, a, a Acute **15.** $2\sqrt{3}$, $\sqrt{30}$, $3\sqrt{2}$ Right **16.** a, a, $a\sqrt{2}$ Right

Determine whether the parallelogram *ABCD* is a rhombus, using the given data.

17. $AD = 10$, $AC = 16$, $DB = 12$ Yes
18. $DC = 4$, $DB = 4\sqrt{3}$, $AE = 2$ Yes
19. $DB = 2\sqrt{3}$, $AC = 2\sqrt{2}$, $AB = \sqrt{5}$ Yes
20. $AC = 2\sqrt{3}$, $DB = 2$, $AD = 2$ Yes

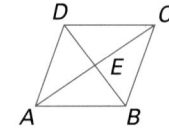

21. Determine whether △*BCD* is a right triangle.   Yes

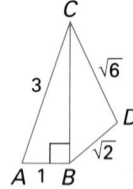

22. Given: $\overline{PQ} \perp$ plane $\mathcal{M}$. Determine whether △*RQS* is acute, obtuse, or right.   Acute

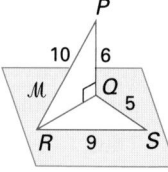

Enrichment

Explain that a *Pythagorean triple* is an ordered triple of positive integers (*a*, *b*, *c*) such that $a^2 + b^2 = c^2$. Two such triples are (3, 4, 5) and (5, 12, 13). Ask the students to find a few more such triples on their own. Then show them the following way of finding Pythagorean triples.

Let x and y be two positive integers such that $x < y$. Then $a = 2xy$, $b = y^2 - x^2$, and $c = y^2 + x^2$. For example, let $x = 2$ and $y = 6$. Then $a = 2xy = 24$, $b = y^2 - x^2 = 32$, and $c = y^2 + x^2 = 40$. Have the students check that $24^2 + 32^2 = 40^2$.   576 + 1,024 = 1,600
Then have the students use this method to find other triples.

Essay.

23. Describe in a short paragraph how to determine whether a triangle is right, acute, or obtuse, given the lengths of the three sides.

24. Given: $\overline{AD}$ is the altitude to $\overline{BC}$; $(AD)^2 = (DC) \cdot (BD)$
Prove: $\angle BAC$ is a right angle.

25. Given: $\overline{AD}$ is the altitude to $\overline{BC}$; $(AC)^2 = (BC) \cdot (DC)$
Prove: $\angle BAC$ is a right angle.

26. Prove Theorem 9.4 for an obtuse triangle.

27. The medians to the legs of a right triangle have lengths $2\sqrt{13}$ and $\sqrt{73}$. Find the lengths of the three sides of the triangle.   *6, 8, 10*

28. Given: x and y are the lengths of the legs of a right triangle with hypotenuse of length z. Show that the triangle with sides of lengths $x + 2$, $y + 2$, and $z + 2$ is an acute triangle.

29. In $\triangle ABC$, $\overline{AB}$ is the longest side. AC is 2 more than twice CB, and AC is 1 less than AB. Find the length of CB to guarantee that $\triangle ABC$ will be obtuse.   *CB < 5*

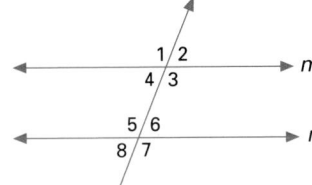

Mixed Review

In the figure at the right, $m \parallel n$.

1. Find m $\angle 7$ if m $\angle 1 = 105$.   *3.5*   105
2. m $\angle 4 = \frac{2}{3}$m $\angle 5$. Find m $\angle 5$.   *3.5*   108
3. m $\angle 4 = 3x + 20$, m $\angle 6 = x + 60$.
Find m $\angle 4$.   *3.5*   80
4. What is the corresponding angle to $\angle 2$?   *3.1* $\angle 6$

◢◢ Algebra Review

Example: To simplify an expression with a radical in the denominator:

Rationalize the denominator.

$$\frac{4}{\sqrt{2}} = \frac{4 \cdot \sqrt{2}}{\sqrt{2} \cdot \sqrt{2}} = \frac{4\sqrt{2}}{2} = 2\sqrt{2}$$

Simplify the expression by rationalizing the denominator.

1. $\frac{1}{\sqrt{2}}$ $\frac{\sqrt{2}}{2}$ **2.** $\frac{6}{\sqrt{3}}$ $2\sqrt{3}$ **3.** $\frac{4}{\sqrt{2}}$ $2\sqrt{2}$ **4.** $\frac{28}{\sqrt{7}}$ $4\sqrt{7}$ **5.** $\frac{10}{\sqrt{5}}$ $2\sqrt{5}$ **6.** $\frac{4}{\sqrt{6}}$ $\frac{2}{3}\sqrt{6}$

24.

Statement	Reason
1. Alt $\overline{AD}$ to $\overline{BC}$ in $\triangle ABC$, $(AD)^2 = DC \cdot BD$	1. Given
2. $\frac{AD}{BD} = \frac{DC}{AD}$	2. If $ad = bc$, then $\frac{a}{b} = \frac{c}{d}$
3. $\angle BDA \cong \angle ADC$	3. Rt $\angle$s are $\cong$.
4. $\triangle BDA \sim \triangle ADC$	4. SAS~
5. $\angle DAC \cong \angle B$, $\angle DAB \cong \angle C$	5. Def of ~ $\triangle$s
6. m $\angle ADB = 90$	6. Def of alt
7. m $\angle ADB = $ m $\angle DAC + $ m $\angle C$	7. Ext Angle Thm
8. m $\angle ADB = $ m $\angle DAC + $ m $\angle DAB$	8. Sub
9. m $\angle ADB = $ m $\angle BAC$	9. Angle Add Post
10. m $\angle BAC = 90$	10. Sub

25.

Statement	Reason
1. Alt $\overline{AD}$ to $\overline{BC}$, $(AC)^2 = BC \cdot DC$	1. Given
2. $\frac{AC}{BC} = \frac{DC}{AC}$	2. If $ad = bc$, then $\frac{a}{b} = \frac{c}{d}$
3. $\angle C \cong \angle C$	3. Reflex Prop of Congr
4. $\triangle ABC \sim \triangle ADC$	4. SAS~
5. m $\angle BAC = $ m $\angle ADC$	5. Def ~ $\triangle$s
6. $\overline{AD} \perp \overline{BC}$	6. Def of alt
7. $\angle ADC$ is a rt $\angle$.	7. Def of $\perp$
8. m $\angle ADC = 90$	8. Def of rt $\angle$
9. m $\angle BAC = 90$	9. Sub

See page 359 for the answers to Ex. 26, 28.

GETTING STARTED

Prerequisite Quiz

Simplify.

1. $\sqrt{8}$ $2\sqrt{2}$
2. $\sqrt{12}$ $2\sqrt{3}$
3. $\frac{4}{\sqrt{2}}$ $2\sqrt{2}$

Find the indicated lengths.

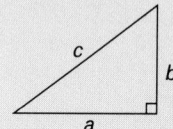

4. $a = 4, b = 4, c =$ _____ $4\sqrt{2}$
5. $c = 8, a = 4, b =$ _____ $4\sqrt{3}$
6. $c = 4, a = b =$ _____ $2\sqrt{2}$

Motivator

Have the students draw a right triangle
having an acute angle with a measure of
45. Have them find the measure of the other
acute angle.   45

Ask the students to classify the triangle.
An isosceles right triangle   Ask them what
they know about the legs of this right
triangle.   They are congruent.

TEACHING
SUGGESTIONS

Lesson Note

There are two approaches to solving
problems involving special right triangles.
1. Memorize the pattern for the lengths of
   the three sides, as shown in the lesson
   summary.
2. Use the Pythagorean Theorem if the legs
   are congruent.

For a 30–60–90 triangle, memorize only
that the length of the leg opposite the angle
of measure 30 is one-half the hypotenuse.
Then use the Pythagorean Theorem.

9.4 Two Special Types of Right Triangles

Objectives   To apply the properties of the 45-45-90 triangle
To apply the properties of the 30-60-90 triangle

The isosceles right triangle, shown at the
right, occurs frequently in the study of
mathematics. Notice that the measure of
each base angle is 45. This special triangle is
often called a *45-45-90 triangle.*

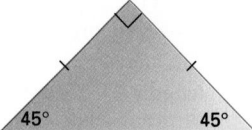

EXAMPLE 1   Find the length of the hypotenuse of the right triangle.

Solution
$h^2 = 4^2 + 4^2$
$h^2 = 16 + 16$
$h^2 = 32$
$h = \sqrt{32}$, or $\sqrt{16} \cdot \sqrt{2}$
$h = 4\sqrt{2}$

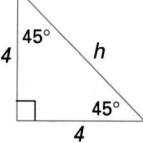

Notice a pattern for the 45-45-90 triangle.

Theorem 9.5   In a 45-45-90 triangle, the hypotenuse is $\sqrt{2}$ times
as long as a leg.

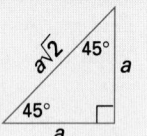

EXAMPLE 2 $\triangle XYZ$ is a 45-45-90 triangle. Find the length of a leg if the length of
the hypotenuse is 12.

Solution
$12 = a\sqrt{2}$
$a = \frac{12}{\sqrt{2}}$
$a = \frac{12 \cdot \sqrt{2}}{\sqrt{2} \cdot \sqrt{2}}$
$= \frac{12\sqrt{2}}{2}$, or $6\sqrt{2}$

Thus, the length of a leg of $\triangle XYZ$ is $6\sqrt{2}$.

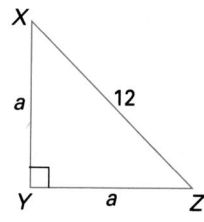

354   Chapter 9 Right Triangles

Additional Example 1

Find the length of the hypotenuse of
these right triangles. $10\sqrt{2}$; $12\sqrt{2}$

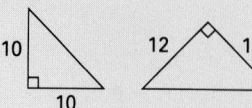

Additional Example 2

Find the length of a leg of 45–45–90
$\triangle ABC$ if the length of the hypotenuse is
6. $3\sqrt{2}$

Another special type of right triangle is the 30-60-90 triangle.

Theorem 9.6

In a 30-60-90 triangle:
(1) the hypotenuse is twice as long as the leg opposite the angle of measure 30; (2) the leg opposite the angle of measure 60 is $\sqrt{3}$ times as long as the leg opposite the angle of measure 30.

Given: Right $\triangle PQR$ with right $\angle PRQ$, m $\angle Q = 60$, m $\angle P = 30$
Prove: $QP = 2 \cdot QR$, $PR = \sqrt{3} \cdot QR$

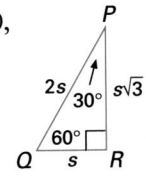

Plan

Extend $\overline{QR}$ to T so that $RT = RQ$. Draw $\overline{PT}$. Prove that $\triangle PRQ \cong \triangle PRT$. Show that m $\angle Q =$ m $\angle T =$ m $\angle QPT = 60$. Let $QR = s$. Then QP must equal $2s$. Use the Pythagorean Theorem to find PR in terms of s.

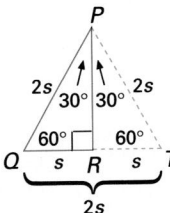

EXAMPLE 3

Find the lengths of the legs of a 30-60-90 triangle if the length of the hypotenuse is 8.

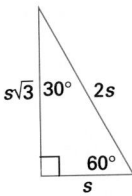

Solution

$$2s = 8$$
$$s = 4$$
Then, $s\sqrt{3} = 4\sqrt{3}$

Thus, the lengths of the legs are 4 and $4\sqrt{3}$.

EXAMPLE 4

$\triangle ABC$ is a 30-60-90 triangle. $AC = 12$. Find AB and BC.

Solution

$$AC = s\sqrt{3} = 12$$
$$s = \frac{12}{\sqrt{3}}$$
$$s = \frac{12 \cdot \sqrt{3}}{\sqrt{3} \cdot \sqrt{3}}$$
$$s = \frac{12\sqrt{3}}{3}$$
$$AB = s = 4\sqrt{3}$$
$$BC = 2s = 8\sqrt{3}$$

Thus, $AB = 4\sqrt{3}$ and $BC = 8\sqrt{3}$.

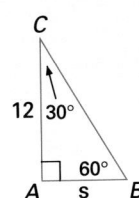

9.4 Two Special Types of Right Triangles **355**

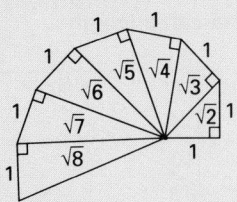
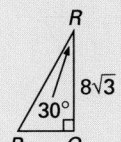

Checkpoint

Find the indicated lengths.

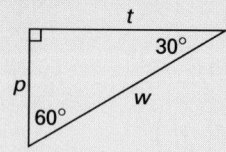

1. $w = 6$, $p = $ _____, $t = $ _____
   3; 3√3
2. $p = 8$, $w = $ _____, $t = $ _____
   16; 8√3
3. $t = 6$, $p = $ _____, $w = $ _____
   2√3; 4√3

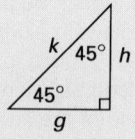

4. $k = 8$, $h = $ _____, $g = $ _____
   4√2; 4√2
5. $h = 5$, $g = $ _____, $k = $ _____
   5; 5√2
6. $g = 3\sqrt{2}$, $h = $ _____, $k = $ _____
   3√2; 6

Closure

Ask the students what relationship exists between the length of the hypotenuse and the length of each leg of a 45–45–90 triangle.   Theorem 9.5 page 354   Ask them which leg length of a 30–60–90 triangle is ½ the length of the hypotenuse. The leg opposite the angle measuring 30. Ask them how they find the length of the other leg.   Theorem 9.6 page 355

EXAMPLE 5   The length of an altitude of an equilateral triangle is $4\sqrt{3}$. Find the length of a side of the triangle.

Plan   Draw an equilateral triangle with an altitude. Two 30-60-90 triangles are formed. Label the sides of one of the triangles as s, $2s$, and $s\sqrt{3}$.

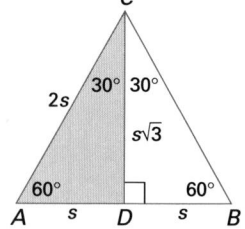

Solution
$$CD = s\sqrt{3} = 4\sqrt{3}$$
$$s = 4$$
$$AB = 2s, \text{ or } 8$$

Thus, the length of a side is 8.

EXAMPLE 6   Given: ▱$ABCD$, altitude $\overline{DE}$, $AD = 6$, m $\angle B = 120$
Find the length of the altitude, $\overline{DE}$.

Plan   Since $ABCD$ is a parallelogram, the consecutive angles A and B are supplementary. m $\angle A = 180 - 120 = 60$. Use the 30-60-90 triangle.

Solution   Use the 30-60-90 triangle at the right.

$$AD = 2s = 6$$
$$s = 3$$

Thus, $DE = s\sqrt{3} = 3\sqrt{3}$.

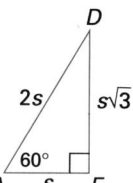

▰ Summary

Special Right Triangle Properties

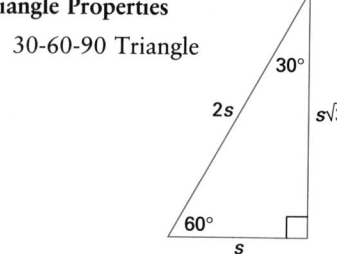

45-45-90 Triangle   30-60-90 Triangle

Additional Example 5

The length of a side of an equilateral triangle is 10. Find the length of an altitude of the triangle.   5√3

Additional Example 6

Given: ▱$ABCD$, altitude $\overline{DE}$, $AD = 4$, m $\angle B = 135$
Find the length of the altitude, $\overline{DE}$.   2√2

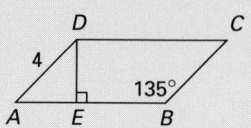

Focus on Reading

Copy and complete the sentence to make a true statement.

1. In a 30-60-90 triangle, the length of the leg opposite the angle measuring _____ is $\frac{1}{2}$ the length of the hypotenuse.   30
2. An altitude of an equilateral triangle divides the triangle into two _____-_____-_____ triangles.   30-60-90
3. In a 45-45-90 triangle, the length of the hypotenuse is _____. $\sqrt{2} \cdot$ (leg)
4. For a 30-60-90 triangle, the lengths of the three sides can be represented as s, $2s$ and _____. $s\sqrt{3}$

◼◼◼ FOLLOW UP

Guided Practice
Classroom Exercises 1–8

Independent Practice
A Ex. 1–12, **B** Ex. 13–25, **C** Ex. 26–28

Basic: FR all, WE 1–12 odd, CI 1–2
Average: FR all, WE 1–25 odd, CI all
Above Average: FR all, WE 1–28 odd, CI all

Classroom Exercises

Find the indicated lengths.

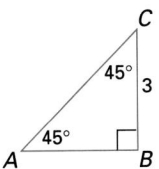 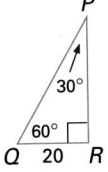 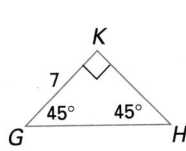 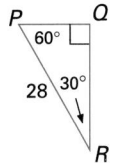

1. $AB =$ ___3___   **3.** $QP =$ ___40___   **5.** $KH =$ ___7___   **7.** $RQ =$ ___$14\sqrt{3}$___
2. $AC =$ ___$3\sqrt{2}$___   **4.** $PR =$ ___$20\sqrt{3}$___   **6.** $GH =$ ___$7\sqrt{2}$___   **8.** $PQ =$ ___14___

Written Exercises

For the given length, find each of the remaining two lengths.

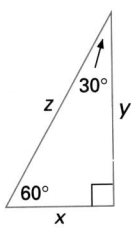 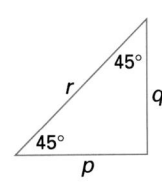 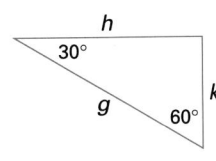

1. $x = 6$ $y = 6\sqrt{3}, z = 12$ **5.** $p = 6$ $q = 6, r = 6\sqrt{2}$ **9.** $k = 3.4$
2. $y = 6$ $x = 2\sqrt{3}, z = 4\sqrt{3}$ **6.** $r = 6$ $p = 3\sqrt{2}, q = 3\sqrt{2}$ **10.** $k = 6\sqrt{3}$
3. $z = 14$ $x = 7, y = 7\sqrt{3}$ **7.** $q = 4\sqrt{2}$ $p = 4\sqrt{2}, r = 8$ **11.** $g = 17$
4. $y = 4\sqrt{3}$ $x = 4, z = 8$ **8.** $r = 10$ $p = 5\sqrt{2}, q = 5\sqrt{2}$ **12.** $h = 2\sqrt{3}$
9. $g = 6.8, h = 3.4\sqrt{3}$ **10.** $g = 12\sqrt{3}, h = 18$ **11.** $h = \frac{17}{2}\sqrt{3}, k = \frac{17}{2}$ **12.** $g = 4, k = 2$

9.4 Two Special Types of Right Triangles   **357**

Enrichment

Draw this figure on the chalkboard.

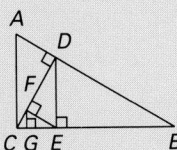

In right triangle ABC, m $\angle B = 30$, and

$CG = 1$ cm. Have the students find these measures.

1. $CF =$ _____   2 cm
2. $CE =$ _____   4 cm
3. $CD =$ _____   8 cm
4. $CB =$ _____   16 cm
5. $FG =$ _____ $\sqrt{3}$ cm
6. $FE =$ _____ $2\sqrt{3}$ cm
7. $DE =$ _____ $4\sqrt{3}$ cm
8. $DB =$ _____ $8\sqrt{3}$ cm

Written Exercises

24.

Statement	Reason
1. Rt $\triangle ABC$ with m $\angle B$ = 90, m $\angle C$ = 45, and m $\angle A$ = 45	1. Given
2. $AB = CB$ $(\overline{AB} \cong \overline{CB})$	2. Sides opp $\cong$ $\angle$s of $\triangle$ are $\cong$.
3. $(AB)^2 + (CB)^2 = (AC)^2$	3. Pythag Thm
4. $(AB)^2 + (AB)^2 = (AC)^2$	4. Sub
5. $2(AB)^2 = (AC)^2$	5. Distr Prop of Eq
6. $\sqrt{2} \cdot AB = AC$	6. $\sqrt{a^2} = \sqrt{a} \cdot \sqrt{a} = a$ for $a > 0$

25.

Statement	Reason
1. Rt $\triangle PQR$ with rt $\angle PRQ$, m $\angle Q = 60$, m $\angle QPR = 30$	1. Given
2. Extend $\overline{QR}$ to T so that $RT = RQ$.	2. 2 pts determine a line.
3. $\overline{PR} \cong \overline{PR}$	3. Reflex Prop of Congr
4. $\angle PRQ \cong \angle PRT$	4. Rt $\angle$s are $\cong$.
5. $\triangle PQR \cong \triangle PTR$	5. SAS
6. m $\angle T = $ m $\angle Q = 60$, m $\angle TPR = $ m $\angle QPR = 30$	6. CPCTC
7. m $\angle TPR + $ m $\angle QPR = $ m $\angle QPT$	7. Angle Add Post
8. m $\angle QPT = 60$	8. Sub
9. $\triangle QPT$ is equiang.	9. Def of equiang $\triangle$
10. $\triangle QPT$ is equil.	10. If $\triangle$ is equiang, then $\triangle$ is equil
11. Let $RQ = RT = s$	11. Notation
12. $QT = 2s$	12. Seg Add Post
13. $QP = QT$	13. Def of equil $\triangle$
14. $QP = 2s$	14. Sub
15. $s^2 + (PR)^2 = (2s)^2$	15. Pythag Thm
16. $(PR)^2 = 3s^2$	16. Subt Prop of Eq
17. $PR = s\sqrt{3}$	17. $\sqrt{a^2} = \sqrt{a} \cdot \sqrt{a} = a$ for $a > 0$

Find the indicated lengths using the given information.

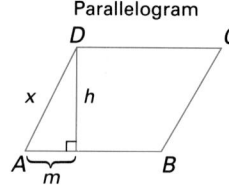

Parallelogram

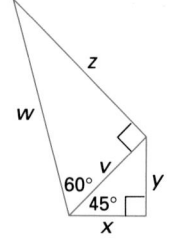

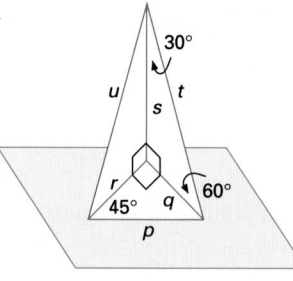

13. m $\angle B = 120$, $x = 18$, $h = \underline{9\sqrt{3}}$

14. m $\angle D = 150$, $x = 4$, $h = \underline{2}$

15. m $\angle B = 135$, $x = 6$, $m = \underline{3\sqrt{2}}$

16. $x = 4$, $w = \underline{8\sqrt{2}}$

17. $y = \sqrt{3}$, $z = \underline{3\sqrt{2}}$

18. $w = 6$, $x = \underline{}$

19. $z = 4\sqrt{3}$, $y = \underline{2\sqrt{2}}$

18. $\frac{3}{2}\sqrt{2}$

20. $t = 10$, $q = \underline{5}$

21. $s = 6\sqrt{3}$, $q = \underline{6}$

22. $p = 10$, $q = \underline{5\sqrt{2}}$

23. $t = 8$, $p = \underline{4\sqrt{2}}$

24. Prove Theorem 9.5

25. Prove Theorem 9.6

26. $\triangle ABC$ is a right triangle with a right angle at A. $\overline{AD}$ is an altitude $AB = 8$, m $\angle B = 30$. Find BD, AD, AC, and DC. $\quad 4\sqrt{3}, 4, \frac{8}{3}\sqrt{3}, \frac{4}{3}\sqrt{3}$

27. The length of each side of a square is 8. The four corners are cut off to form a regular octagon. Find its perimeter. $\quad 64\sqrt{2} - 64$

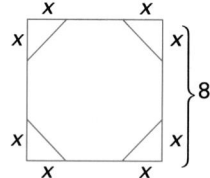

28. Find the perimeter of trapezoid $ABCD$. (HINT: Draw altitudes from C and D to the base $\overline{AB}$.) $39 + 5(\sqrt{3} + \sqrt{2})$

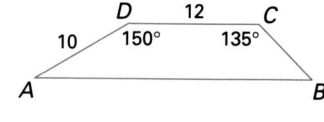

Midchapter Review

Find the indicated lengths using the given information. *9.1, 9.2, 9.4*

1. $p = 4$, $QP = 8$, $m = \underline{2}$

2. $m = 6$, $r = 4$, $h = \underline{2\sqrt{6}}$

3. $m = 4$, $r = 5$, $p = \underline{6}$

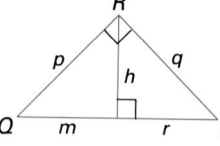

4. m $\angle Q = 60$, $p = 6$, $q = \underline{6\sqrt{3}}$

5. $p = 4$, $q = 4\sqrt{2}$, $QP = \underline{4\sqrt{3}}$

6. m $\angle P = 45$, $p = 4\sqrt{2}$, $m + r = \underline{8}$

Determine whether the triangle with given side lengths is acute, obtuse, or right. If the triangle is right, indicate which side is the hypotenuse. *9.3*

7. 4, 6, 7   Acute

8. 3, 7, 9   Obtuse

9. $\sqrt{6}, \sqrt{2}, 2\sqrt{2}$ Right, $2\sqrt{2}$

Computer Investigation

A Special Ratio in Right Triangles

Use a computer software program that draws angles of a given measure, constructs perpendicular segments, extends segments, labels points, and measures line segments and angles. In right $\triangle ACB$ $\overline{AB}$ is the hypotenuse. $\overline{BC}$ is the leg of the right $\triangle$, *opposite* the angle with measure 30. Notice the ratio

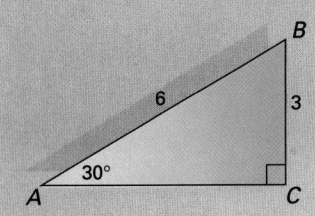

$$\frac{\text{length of leg opposite } \angle \text{ of } 30}{\text{length of hyp}} = \frac{3}{6}, \text{ or } \frac{1}{2}.$$

This ratio for a right triangle is very important in mathematics and will be formally defined in the next lesson.

Activity 1

Draw a right triangle as directed below.

1. Draw $\angle CBA$ with measure 30.
2. From C draw a segment $\overline{CD}$ perpendicular to $\overline{BA}$ intersecting $\overline{BA}$ at E; $\triangle BEC$ is a right triangle.
3. Measure to find CB (the length of the hypotenuse) and CE (the length of the leg opposite $\angle B$).
4. Find the ratio $\dfrac{CE}{CB}$ to the nearest hundredth of a unit.

Activity 2

Repeat the steps above for a new angle of measure 30. Draw the angle with its sides extended so that the new right triangle will not be necessarily congruent to the first triangle of Activity 1.

5. How does the ratio $\dfrac{CE}{CB}$ compare with the ratio of Activity 1?

Activity 3

Repeat the steps of Activities 1 and 2 above for right triangles each containing an acute angle with the given measure and the given name.

6. m $\angle CBA = 25$ 7. m $\angle CBA = 45$

Summary

Generalize the results of Activities 1–3 above.

In a right triangle with an acute angle of a given measure, what ratio is always the same, regardless of the lengths of the sides?

26.

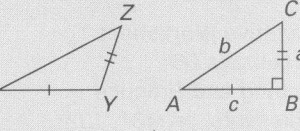

Statement	Reason
1. $\triangle XYZ$, $(XZ)^2 > (XY)^2 + (ZY)^2$	1. Given
2. Construct $\triangle ABC$ with rt $\angle B$ such that $XY = AB$ and $YZ = BC$	2. Protract Post and Ruler Post
3. $(AB)^2 + (BC)^2 = (AC)^2$	3. Pythag Thm
4. $(XZ)^2 > (AC)^2$	4. Sub Prop of Ineq: if $a > b$ and $b = c$, then $a > c$.
5. m $\angle Y >$ m $\angle B$	5. Converse SSS Ineq Thm
6. m $\angle B = 90$	6. Def of rt $\angle$
7. m $\angle Y > 90$	7. Sub
8. $\angle B$ is obtuse.	8. Def of obtuse $\angle$

28.

Statement	Reason
1. $x^2 + y^2 = z^2$	1. Given
2. $x + y > z$	2. Triang Ineq Thm
3. $x + y + 1 > z$	3. Add Prop of Ineq: if $a > b$, then $a + 1 > b$.
4. $4x + 4y + 4 > 4z$	4. Mult Prop of Ineq
5. $4x + 4y + 4 + 4 > 4z + 4$	5. Add Prop of Ineq
6. $x^2 + 4x + 4 + y^2 + 4y + 4 > z^2 + 4z + 4$	6. Add Prop of Ineq: if $a > b$ and $c = d$, then $a + c > b + d$.
7. $(x + 2)^2 + (y + 2)^2 > (z + 2)^2$	7. Factorization
8. $\triangle$ with sides $x + 2$, $y + 2$, and $z + 2$ is acute by Thm 9.4.	8. Thm 9.4.

359

Manipulative Worksheet 17
Quick Quizzes 66
Reteaching and Practice
  Worksheets, pp. 131, 132
Technology Worksheet 9
Transparency 24

▰▰▰GETTING STARTED

Prerequisite Quiz

Find a to the nearest tenth.

1. $0.4226 = \frac{a}{7}$ 3.0
2. $a = \frac{12}{0.9397}$ 12.8
3. $\frac{25}{a} = 0.4562$ 54.8

In △*ABC*, *AB* = *AC* and $\overline{AD}$ is an altitude.

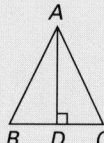

4. If m ∠*CAB* = 40, find m ∠*C*. 70
5. If m ∠*C* = 30, find m ∠*DAC*. 60

Motivator

Draw the three triangles on page 360 on the chalkboard. Label the triangles *ABC* with angle *A* having a measure of 60. Ask the students what the ratio of the length of the side opposite ∠*A* to the length of the hypotenuse is for each triangle. $\frac{\sqrt{3}}{2}$.

▰▰▰TEACHING SUGGESTIONS

Lesson Note

Only one trigonometric ratio is discussed in this lesson. Using just one ratio, the students learn more easily how to find a trigonometric ratio given an angle and, conversely, how to find an angle given the ratio. While students must know how to use a table, once this technique is mastered, they should be encouraged to use a calculator.

9.5 The Sine Ratio

Objective   To find side lengths and angle measures in right triangles using the sine ratio

Notice a pattern that applies to the three triangles below.

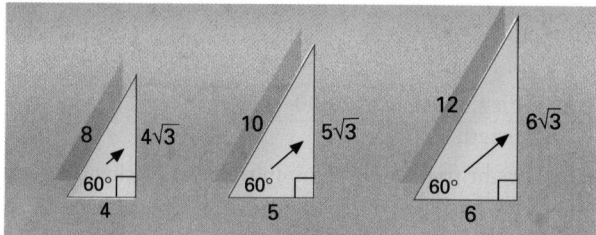

The ratio of the length of the side opposite the angle measuring 60 to the length of the hypotenuse is the same for each triangle.

$$\frac{\text{length of side opposite angle of } 60}{\text{length of hypotenuse}} = \frac{4\sqrt{3}}{8} = \frac{5\sqrt{3}}{10} = \frac{6\sqrt{3}}{12} = \frac{\sqrt{3}}{2}$$

Using a calculator or square root table, the ratio $\frac{\sqrt{3}}{2} \approx \frac{1.732}{2}$, or 0.8660. By the properties of similar triangles, it can be shown that this ratio is the same for any angle measuring 60, regardless of the lengths of the sides. This ratio is called *sine* of the angle measuring 60. So, the sine of 60 ≈ 0.8660. This is written and abbreviated as sin 60 ≈ 0.8660. For convenience, we agree to use = rather than the approximation symbol, ≈.

The sine of an angle is an example of a trigonometric ratio. The word *trigonometry* is derived from two Greek words meaning *triangle measurement.*

Definition

For right triangle *ABC* with angle *A* as shown,

sine of angle $A = \dfrac{\text{length of opposite leg}}{\text{length of hypotenuse}}$;

$\sin A = \dfrac{a}{c}$.

You can use the trigonometric table on page 643 to find decimal approximations for the sines of angles from 0 to 90.

Additional Example 1

Use the table of ratios to find sin 38 and sin 75.
sin 38 = 0.6157; sin 75 = 0.9659

EXAMPLE 1   Use the table of ratios to find sin 16 and sin 60.

Solution   Locate 16 and 60 in the angle measure column. Then read across to the sine column.

Angle Measure	Sin	Cos	Tan	Angle Measure	Sin	Cos	Tan
12°	.2079	.9781	.2126	58°	.8480	.5299	1.600
13°	.2250	.9744	.2309	59°	.8572	.5150	1.664
14°	.2419	.9703	.2493	→60° →	.8660	.5000	1.732
15°	.2588	.9659	.2679	61°	.8746	.4848	1.804
→16° →	.2756	.9613	.2867	62°	.8829	.4695	1.881
17°	.2924	.9563	.3057	63°	.8910	.4540	1.963

Thus, sin 16 = 0.2756 and sin 60 = 0.8660.

EXAMPLE 2   Find m ∠A to the nearest degree if sin A = 0.9394.

Solution   Read down the sine column. The value closest to 0.9394 is 0.9397. From this number read across to the angle column.

Angle Measure	Sin	Cos	Tan	Angle Measure	Sin	Cos	Tan
22°	.3746	.9272	.4040	68°	.9272	.3746	2.475
23°	.3907	.9205	.4245	69°	.9336	.3584	2.605
24°	.4067	.9135	.4452	70° ←	.9397	.3420	2.747
25°	.4226	.9063	.4663	71°	.9455	.3256	2.904
26°	.4384	.8988	.4877	72°	.9511	.3090	3.077

So, m ∠A = 70 to the nearest degree.

Equations can now be formed with the sine ratio in order to calculate side lengths or angle measures of given right triangles.

EXAMPLE 3   If m ∠A = 25 and c = 7, find a to the nearest tenth.

Solution   Write an equation using the sine ratio.

$$\sin A = \frac{\text{length of leg opposite } \angle A}{\text{length of hypotenuse}} = \frac{a}{c}$$

$\sin 25 = \dfrac{a}{7}$ [Use the table in Example 2 to find sin 25.]

$0.4226 = \dfrac{a}{7}$

$0.4226(7) = \left(\dfrac{a}{7}\right)7$

$2.9582 = a$ Thus, rounded to the nearest tenth, a is 3.0.

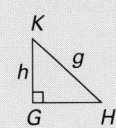

Additional Example 2

Find m ∠A to the nearest degree if sin A = 0.3446.   20

Additional Example 3

If m ∠H = 42 and g = 24, find h to the nearest tenth.   16.1

Math Connections

Language: The word *trigonometry* comes from two Greek words, *trigonon* and *metria*, meaning "triangle" and "measure," respectively. Trigonometry is a method of finding unknown measures of parts of a triangle by using the relationships that exist between the sides and angles of any triangle.

Critical Thinking Questions

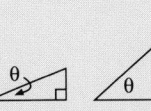 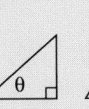 

Analysis: The value of a sine ratio depends on the measure of the angle whose sine ratio is found. Refer students to a series of right triangles, each with an acute angle of increasing measure, as shown above. Ask students to generalize (by observation or by using increasing values for θ) the effects on the value of a sine ratio as the angle's measure increases from 0 to 90.   As the meas of θ increases from 0 to 90, so does its sine ratio.

Checkpoint

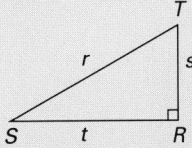

Find the indicated measure to the nearest tenth (sides) or to the nearest degree (angles).

1. m ∠S = 54, r = 23, s = _____   18.6
2. m ∠T = 28, r = 35, t = _____   16.4
3. r = 15, t = 13, m ∠T = _____   60
4. r = 46, s = 26, m ∠S = _____   34
5. m ∠S = 35, s = 8, r = _____   13.9
6. Triangle ABC has a right angle at C, AB = 18, and m ∠A = 72. Find CB to the nearest tenth.   17.1

Closure

Ask the students to give the definition of the sine of an acute angle of a right triangle. Definition on page 360 Ask them how they can find the measure of an angle of a right triangle if they know the lengths of the hypotenuse and a leg. Use the sine ratio. Then ask them how they can find the length of a leg if they know the measure of an acute angle and the length of the hypotenuse of a right triangle. Use the sine ratio.

◢FOLLOW UP

Guided Practice

Classroom Exercises 1–8

Independent Practice

Ⓐ Ex. 1–19, Ⓑ Ex. 20–28, Ⓒ Ex. 29–32

Basic: WE 1–19 odd, Calculator
Average: WE 1–28 odd, Calculator
Above Average: WE 1–32 odd, Calculator

EXAMPLE 4 The length of the hypotenuse of a right triangle is 9. The length of a leg is 6. Find the measure of each acute angle to the nearest degree.

Solution First sketch a right triangle. Write an equation using the sine ratio.

$$\sin M = \frac{\text{length of leg opposite angle}}{\text{length of hypotenuse}}$$

$$\sin M = \frac{y}{r} = \frac{6}{9}, \text{ or } \frac{2}{3}$$

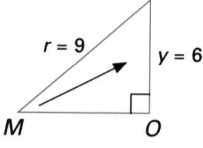

Divide to 5 decimal places: $\sin M = 0.66666$
Round to 4 decimal places: $\sin M = 0.6667$
Find closest value in sine column: m $\angle M = 42$ to the nearest degree.
m $\angle N = 90 - 42$, or 48.

Thus, the measures of the acute angles are approximately 42 and 48.

EXAMPLE 5 The measure of the vertex angle of an isosceles triangle is 40. The length of the altitude to the base is 12. Find the length of a leg to the nearest tenth.

Plan Draw the triangle. The altitude bisects the vertex angle. Find m $\angle B$.

$$\text{m } \angle B = 90 - \frac{1}{2} \cdot 40 = 90 - 20 = 70$$

Write an equation using the sine ratio for right $\triangle BDA$.

Solution
$$\sin 70 = \frac{12}{c}$$
$$c(\sin 70) = 12$$
$$c = \frac{12}{\sin 70} = \frac{12}{0.9397} = 12.77$$

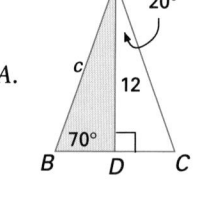

Thus, the length of a leg is 12.8, rounded to the nearest tenth.

Classroom Exercises

For the right triangle, give sin A and sin B as a ratio of lengths of sides.

1.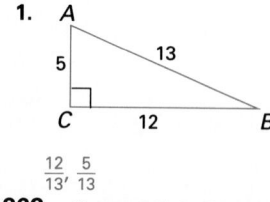

$\frac{12}{13}, \frac{5}{13}$

2.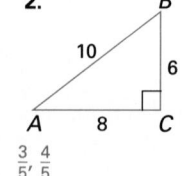

$\frac{3}{5}, \frac{4}{5}$

3.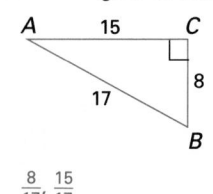

$\frac{8}{17}, \frac{15}{17}$

4.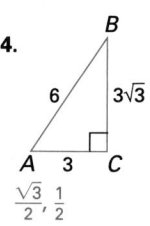

$\frac{\sqrt{3}}{2}, \frac{1}{2}$

Additional Example 4

The length of the hypotenuse of a right triangle is 7. The length of a leg is 4. Find the measure of each acute angle to the nearest degree. 35; 55

Additional Example 5

The measure of the vertex angle of an isosceles triangle is 50. The length of the base is 14. Find the length of a leg to the nearest tenth. 16.6

Write an equation that can be used to find the indicated side length or angle measure.

5.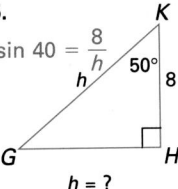
$\sin 40 = \dfrac{8}{h}$
$h = ?$

6.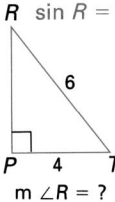
$\sin R = \dfrac{2}{3}$
m ∠R = ?

7.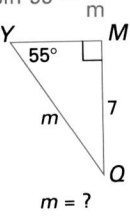
$\sin 55 = \dfrac{7}{m}$
$m = ?$

8.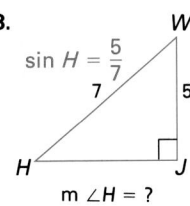
$\sin H = \dfrac{5}{7}$
m ∠H = ?

Written Exercises

Use the table on page 643 to find the indicated value.

1. sin 49   0.7547
2. sin 79   0.9816
3. sin 19   0.3256
4. sin 84   0.9945
5. sin 62   0.8829
6. sin 88   0.9994

Use the table on page 643 to find m ∠A to the nearest degree.

7. sin A = 0.6293   39
8. sin A = 0.9976   86
9. sin A = 0.5446   33
10. sin A = 0.6020   37
11. sin A = 0.9562   73
12. sin A = 0.3018   18

Find the indicated measure to the nearest tenth (sides) or to the nearest degree (angles).

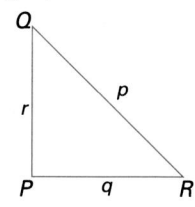

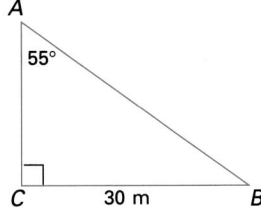

13. m ∠R = 42, p = 18, r = ___12.0___
14. m ∠Q = 65, p = 19, q = ___17.2___
15. q = 10, p = 12, m ∠Q = ___56___
16. q = 15, p = 20, m ∠Q = ___49___
20. A jet plane is 10 km above the ground in the figure below. Find the distance d from the runway.   191 km

17. m ∠ACB = 50, AB = 20, AC = ___23.7___
18. m ∠ACB = 40, AC = 16, AD = ___5.5___
19. AC = 30, AB = 20, m ∠A = ___71___
21. Find the property boundary from A to B.   36.6 m

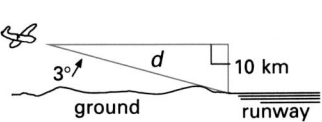

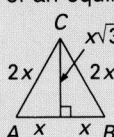

9.5 The Sine Ratio   **363**

32. Because in a rt △, side opp an acute ∠ is shorter than the hyp, the sine < 1. Because the side opp must have length > 0, the acute ∠ has sine > 0.

Mixed Review

1.

Statement	Reason
1. ∠3 ≅ ∠4, $\overline{DC} ≅ \overline{EC}$	1. Given
2. $\overline{BC} ≅ \overline{BC}$	2. Reflex Prop of Congr
3. △DCB ≅ △ECB	3. SAS
4. ∠1 ≅ ∠2	4. CPCTC

2.

Statement	Reason
1. ∠ABD ≅ ∠ABE, $\overline{BC}$ bis ∠DCE	1. Given
2. ∠1 is supp to ∠ABD, ∠2 is supp to ∠ABE.	2. If outer rays of adj ∠s form a st ∠, then sum of ∠ meas is 180.
3. ∠1 ≅ ∠2	3. Supp ∠s of ≅ ∠s are ≅.
4. $\overline{BC} ≅ \overline{BC}$	4. Reflex Prop of Congr
5. ∠3 ≅ ∠4	5. Def of ∠ bis
6. △BDC ≅ △BEC	6. ASA
7. ∠D ≅ ∠E	7. CPCTC

In Exercises 22–26, round answers to the nearest tenth (sides) or nearest degree (angles).

22. The length of the hypotenuse of a right triangle is 8. The length of a leg is 6. Find the measures of each acute angle. **49, 41**

23. Triangle *ABC* has a right angle at *A*. *AB* = 5, m ∠C = 40. Find *BC*. **7.8**

24. The measure of an acute angle of a right triangle is 20. The length of the leg opposite this angle is 10. Find the length of the hypotenuse. **29.2**

25. The measure of the vertex angle of an isosceles triangle is 32. The length of each leg is 16. Find the length of the altitude drawn to the base. **15.4**

26. The length of a leg of an isosceles triangle is 14, and the length of the base is 12. Find the measure of the vertex angle. **50**

27. An airplane takes off with an airspeed of 290 ft/s and climbs at an angle that measures 9 with the horizontal. How long will it take the plane to reach a height of 6,000 ft? **2 min 12 s**

28. A student solved the following problem. Triangle *ABC* has a right angle at *C*. *AB* = 49.4 and m ∠A = 72. Find *CB*. The student's answer was 64.3. Was this a reasonable answer? Why? No. $\overline{AB}$ is the hypotenuse, which is the longest side.

In Exercises 29–31, find *y* to the nearest tenth or degree.

29. **30.** **31.**

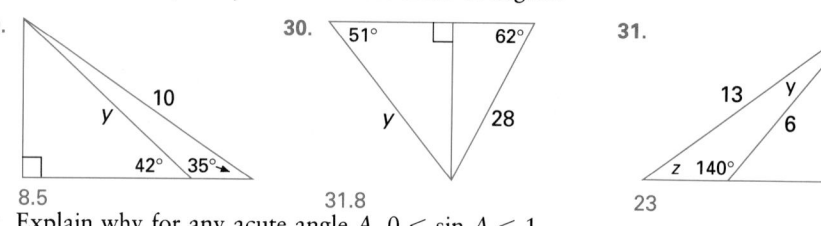

8.5   31.8   23

32. Explain why for any acute angle *A*, 0 < sin *A* < 1.

Mixed Review

1. Given: ∠3 ≅ ∠4, $\overline{DC} ≅ \overline{EC}$
Prove: ∠1 ≅ ∠2 **4.7**

2. Given: ∠ABD ≅ ∠ABE, $\overline{BC}$ bisects ∠DCE
Prove: ∠D ≅ ∠E **4.7**

3. Given: m ∠D = 60, m ∠3 = 20
Find m ∠ABD. **3.7 80**

4. Given: m ∠1: m ∠D: m ∠3 = 3:4:5.
Find m ∠ABD. **8.1 135**

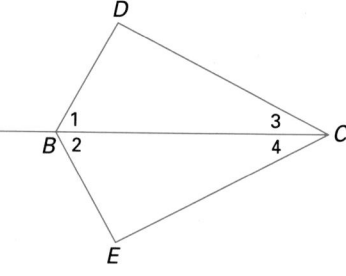

Using the Calculator

A scientific calculator with keys for computing trigonometric ratios can be used to find the sine of a given angle or the acute angle with a given sine ratio.

To find sin 65 to four decimal places:

(1) Make sure the calculator is in the degree mode. DEG must appear in the display window.
(2) Enter the angle measure 65.
(3) Press the ⬚sin⬚ key.
(4) The calculator will display 0.906307787.
(5) Round to four decimal places, 0.9063.
The calculator steps for finding sin 65 are as follows.

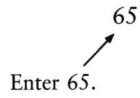

Enter 65.   Press ⬚sin⬚ key.

Example 1	Using the figure at the right, find a to the nearest tenth.

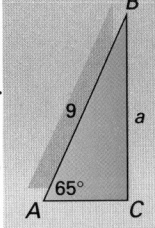

Solution

$$\sin 65 = \frac{a}{9}$$

$$9 \sin 65 = a$$

Calculator Steps:
$9 \times 65 \sin = 8.156770083$

Thus, $a = 8.2$ to the nearest tenth.

Example 2 Suppose you are given sin A = 0.8774. Find m ∠A.

Solution Finding m ∠A is the *inverse* of finding the sine ∠A. To do this, the ⬚INV⬚ (inverse) key or the ⬚2nd f⬚ (second-function) key is used. For example, if sin A = 0.8774, you can find m ∠A to the nearest degree by using the calculator as follows.

Calculator Steps:
0.8774 ⬚INV⬚ ⬚sin⬚   or   0.8774 ⬚2nd f⬚ ⬚sin⬚
Display: 61.33030001   Display: 61.33030001

Thus, m ∠A = 61, to the nearest degree.

Exercises

Do Written Exercises 1–12 of lesson 9.5 using a calculator.

GETTING STARTED

Prerequisite Quiz

Find the indicated measures to the nearest tenth (sides) or to the nearest degree (angles).

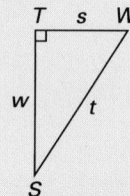

1. m $\angle W$ = 38, t = 22, w = _____   13.5
2. m $\angle S$ = 65, t = 19, s = _____   17.2
3. w = 43, t = 54, m $\angle W$ = _____   53
4. s = 13, t = 27, m $\angle S$ = _____   29

Motivator

On the chalkboard, draw the right triangle *ABC* shown in the definition on page 366. Have the students identify each of the following parts:
Hypotenuse   *c*
Leg opposite $\angle A$ *a*
Leg opposite $\angle B$ *b*
Leg adjacent to $\angle A$ *b*
Leg adjacent to $\angle B$ *a*
sin A $\frac{a}{c}$
sin B $\frac{b}{c}$

TEACHING SUGGESTIONS

Lesson Note

Since students already know how to use a table of trigonometric ratios (Lesson 9.5), less emphasis on this technique is needed now. After illustrating the use of the table, have students use a calculator. In real-life applications, calculators are used.

9.6 Other Trigonometric Ratios

Objective   To find side lengths and angle measures using sine, cosine, and tangent ratios

In this lesson, you will study two more ratios that will help you find the lengths of sides and the measures of angles.

Definition

In right triangle *ABC* with acute angle *A*:

cosine of angle $A = \dfrac{\text{length of adjacent leg}}{\text{length of hypotenuse}}$

$\cos A = \dfrac{b}{c}$

tangent of angle $A = \dfrac{\text{length of opposite leg}}{\text{length of adjacent leg}}$

$\tan A = \dfrac{a}{b}$

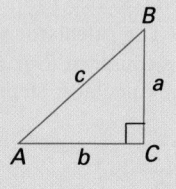

When solving problems involving right triangles, you must choose which trigonometric ratio to use.

EXAMPLE 1   For each triangle, find *x* to the nearest tenth.

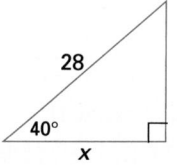
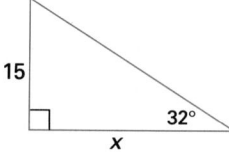

Plan   28 = length of hypotenuse
     x = length of adjacent leg
     Use the cosine ratio.

     15 = length of opposite leg
     x = length of adjacent leg
     Use the tangent ratio.

Solution

$\cos 40 = \dfrac{x}{28} = \dfrac{\text{adj}}{\text{hyp}}$

$0.7660 = \dfrac{x}{28}$

$0.7660(28) = x$

$21.4480 = x$

Thus, $x = 21.4$.

$\tan 32 = \dfrac{15}{x} = \dfrac{\text{opp}}{\text{adj}}$

$0.6249 = \dfrac{15}{x}$

$0.6249x = 15$

$x = \dfrac{15}{0.6249} = 24.00384$

Thus, $x = 24.0$.

Additional Example 1

For each triangle, find the indicated length to the nearest tenth.

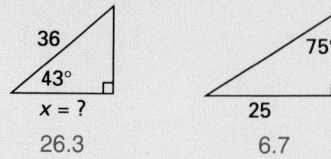

     26.3   6.7

EXAMPLE 2 Find x to the nearest degree.

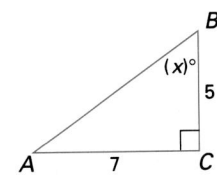 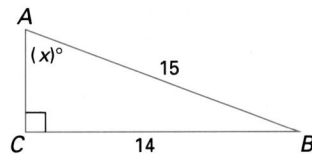

Plan

Given: lengths of both legs
Use the tangent ratio.

Given: length of opp leg and hypotenuse
Use the sine ratio.

Solutions

$\tan B = \dfrac{7}{5} = \dfrac{\text{opp}}{\text{adj}}$

$\tan B = 1.4000$

$m \angle B = 54$ [1.4000 is closer to 1.3764 than 1.4281.]

$\sin A = \dfrac{14}{15} = \dfrac{\text{opp}}{\text{hyp}}$

$\sin A = 0.93333$ ← Divide to 5 decimal places.

$\sin A = 0.9333$ ← Round to 4 decimal places.

$m \angle A = 69$

EXAMPLE 3 The length of the altitude to the base of an isosceles triangle is 16. The measure of a base angle is 55. Find the length of the base to the nearest tenth.

Solution

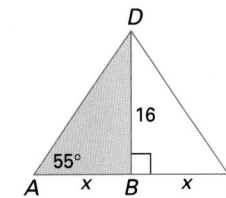

$\tan 55 = \dfrac{16}{x}$

$1.428 = \dfrac{16}{x}$

$1.428x = 16$

$x = 11.204$

$AC = 2x = 2(11.204) = 22.408$

Thus, the length of the base is 22.4 to the nearest tenth.

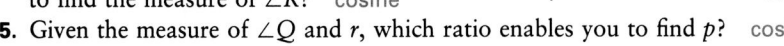

Focus On Reading

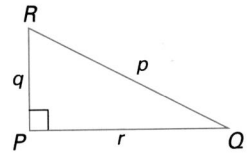

1. Name the hypotenuse. $\overline{RQ}$ or p
2. Name the leg adjacent to $\angle R$. $\overline{RP}$ or q
3. Name the leg opposite $\angle Q$. $\overline{RP}$ or q
4. Given p and q, which ratio enables you to find the measure of $\angle R$? cosine
5. Given the measure of $\angle Q$ and r, which ratio enables you to find p? cosine

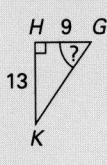 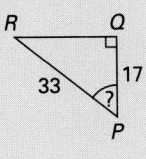
Math Connections

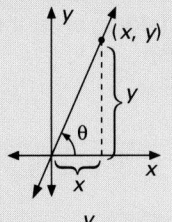

$\text{slope} = \dfrac{y}{x} = \tan \theta$

Algebra: The tangent ratio can be compared to the slope of a line. As shown above, the slope of a line that passes from left to right in an upward direction through the origin has a slope that is equal to the tangent ratio of the angle θ formed by the line with the horizontal axis.

Critical Thinking Questions

Analysis: Ask students to prove that the ratio of the sine of an angle to the cosine of that angle is equal to the tangent of the angle.

$\dfrac{\sin \theta}{\cos \theta} = \dfrac{\frac{opp}{hyp}}{\frac{adj}{hyp}} = \dfrac{opp}{adj} = \tan \theta$

Common Error Analysis

Error: Some students may find it difficult to identify the leg opposite, or adjacent to, an angle. Students may also have trouble determining which trigonometric ratio to use, when one side of the angle involved is not along a horizontal leg.

Give students practice with exercises such as the following,

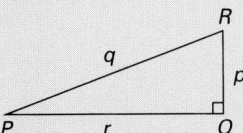

1. Name the leg adjacent to $\angle R$. p
2. Name the leg opposite $\angle R$. r
3. Name the leg adjacent to $\angle P$. r
4. $\tan R = \dfrac{?}{?}$ $\dfrac{r}{p}$
5. $\sin R = \dfrac{?}{?}$ $\dfrac{r}{q}$
6. $\cos R = \dfrac{?}{?}$ $\dfrac{p}{q}$

Checkpoint

Find the indicated measure to the nearest tenth (sides) or to the nearest degree (angles).

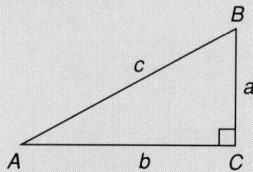

1. $c = 17$, m $\angle B = 36$, $a =$ _____   10.0
2. m $\angle A = 42$, $b = 16$, $a =$ _____   14.4
3. $b = 12$, m $\angle A = 25$, $c =$ _____   13.2
4. $b = 18$, $c = 33$, m $\angle A =$ _____   57
5. $b = 43$, $a = 25$, m $\angle B =$ _____   60
6. The length of the altitude to the base of an isosceles triangle is 18. The measure of a base angle is 32. Find the length of the base.   58

Closure

On the chalkboard, draw right triangle *ABC* shown in the definition on page 366. Have the students name each of the following ratios:

sin A $\frac{a}{c}$

cos A $\frac{b}{c}$

tan A $\frac{a}{b}$

sin B $\frac{b}{c}$

cos B $\frac{a}{c}$

tan B $\frac{b}{a}$

Classroom Exercises

Write an equation using the appropriate trigonometric ratio for finding *a*.

1.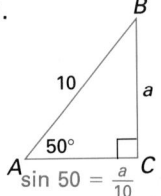
$\sin 50 = \frac{a}{10}$

2.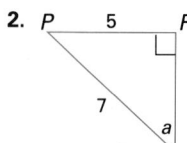
$\sin a = \frac{5}{7}$

3.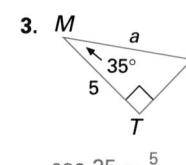
$\cos 35 = \frac{5}{a}$

4.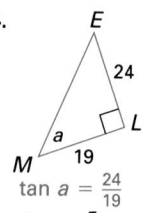
$\tan a = \frac{24}{19}$

5.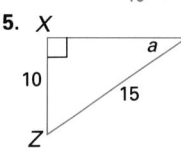
$\sin a = \frac{10}{15}$

6.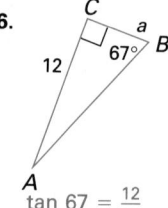
$\tan 67 = \frac{12}{a}$

7.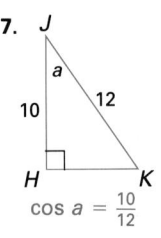
$\cos a = \frac{10}{12}$

8.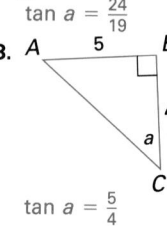
$\tan a = \frac{5}{4}$

Written Exercises

For Exercises 1–8, find the indicated measures for Classroom Exercises 1–8 above. (Give lengths to the nearest tenth of a unit and angle measures to the nearest degree.)

For Exercises 9–22, find the indicated measure (sides to the nearest tenth, angles to the nearest degree). Refer to the figure at the right.

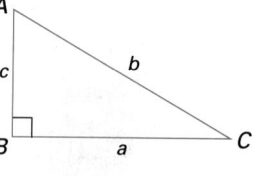

9. $b = 19$, m $\angle C = 48$, $a =$ __12.7__
10. m $\angle A = 29$, $b = 14$, $c =$ __12.2__
11. m $\angle A = 49$, $a = 19$, $c =$ __16.5__
12. $c = 13$, m $\angle C = 26$, $b =$ __29.7__
13. $c = 32$, m $\angle A = 54$, $b =$ __54.4__
14. $b = 35$, m $\angle C = 75$, $a =$ __9.1__
15. $c = 12$, $b = 16$, m $\angle C =$ __49__
16. $b = 18$, $a = 8$, m $\angle C =$ __64__
17. $a = 14$, $b = 16$, m $\angle A =$ __61__
18. $c = 20$, $a = 18$, m $\angle A =$ __42__
19. $c = 40$, m $\angle A = 70$, $b =$ __117.0__
20. $b = 30$, m $\angle C = 85$, $c =$ __29.9__
21. $c = 124$, $a = 275$, m $\angle A =$ __66__
22. m $\angle A = 65$, $a = 2,345$, $c =$ __1,093.5__

23. The length of the altitude to the base of an isosceles triangle is 14. The measure of a base angle is 62. Find the length of the base to the nearest unit.   15

24. The length of the altitude to the base of an isosceles triangle is 17. The length of the base is 18. Find the measure of a base angle to the nearest degree.   62

Enrichment

Explain that the sine and cosine ratios are called cofunctions. Draw this figure on the board and point out that $\frac{a}{c} =$ sin A = cos B.

In light of this observation, have the students study the table and derive a statement expressing a sine-cosine relationship.

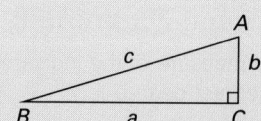

The sine of an an acute angle is equal to the cosine of the angle's complement.

25. Right triangle ABC has right angle at C, m $\angle A = 42$, and $AC = 24$. Find the length of the hypotenuse to the nearest unit. 32

26. Find, to the nearest unit, the length of the longer leg of a right triangle if the length of the hypotenuse is 8, and the measure of an acute angle is 22. 7

27. The lengths of the sides of a rectangle are 12 and 22. Find, to the nearest degree, the measure of the angle formed by a diagonal and the longer side. 29

28. The measure of the vertex angle of an isosceles triangle is 110. The length of a leg is 18. Find the length, to the nearest unit, of the altitude to the base. 10

29. The length of each side of a rhombus is 12. The measure of an angle between two adjacent sides is 40. Find, to the nearest unit, the length of each diagonal. 23, 8

30. Right triangle ABC has right angle at B. $\overline{BD}$ is an altitude to the hypotenuse. $DC = 5$ and $AD = 16$. Find m $\angle ABD$ to the nearest degree. 61

31. Given: $\overline{EB} \perp \overline{AD}$, $\overline{AC} \perp \overline{CD}$, $AB = BD$, m $\angle ABE = 62$, $CD = 8$. Find EB to the nearest unit. 5

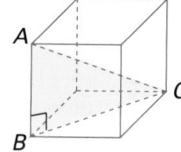

32. Given: length of each side of the cube is 10. Find m $\angle ACB$ to the nearest degree. 35

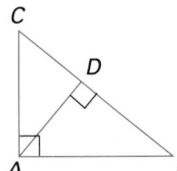

Mixed Review

In the figure at the right, $\overline{AC} \perp \overline{AB}$ and $\overline{AD} \perp \overline{CB}$. Find the indicated measure.

1. m $\angle B = 60$, $BC = 8$, $AC = \underline{\ 4\sqrt{3}\ }$ 9.4
2. m $\angle C = 45$, $BC = 6$, $AC = \underline{\ 3\sqrt{2}\ }$ 9.4
3. m $\angle B = 30$, $AB = 4$, $BC = \underline{\qquad}$ 9.4 $\frac{8}{3}\sqrt{3}$
4. $CD = 4$, $CB = 9$, $AC = \underline{\ 6\ }$ 9.1
5. $CD = 6$, $BD = 8$, $AD = \underline{\ 4\sqrt{3}\ }$ 9.1

Brainteaser

The practice field at a high school is a rectangle whose length is twice as long as its width. The shorter ends have semicircles on them. The distance around the inside of the track that encloses the field is 100 m. Find the area of the practice field to the nearest meter. 392 m²

Guided Practice

Classroom Exercises 1–8

Independent Practice

A. Ex. 1–22, **B** Ex. 23–30, **C** Ex. 31–32

Basic: FR all, WE 1–22 odd, Brainteaser

Average: FR all, WE 1–30 odd, Brainteaser

Above Average: FR all, WE 1–32 odd, Brainteaser

Additional Answers

Written Exercises

1. 7.7
2. 46
3. 6.1
4. 52
5. 42
6. 5.1
7. 34
8. 51

9.7 Applying Trigonometric Ratios

▰▰▰ GETTING STARTED

Prerequisite Quiz

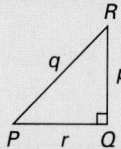

Find the indicated measure to the
nearest tenth (sides) or to the nearest
degree (angles).

1. m $\angle P$ = 40, q = 18, p = _____   11.6
2. q = 24, m $\angle R$ = 70, p = _____   8.2
3. m $\angle R$ = 37, p = 126, r = _____   94.9
4. q = 17, r = 13, m $\angle P$ = _____   40
5. q = 112, p = 79, m $\angle R$ = _____   45
6. p = 240, r = 78, m $\angle R$ = _____   18

Motivator

Ask the students if they would elevate or
depress their line of vision if they are
standing on the ground and look up to see
an airplane flying directly overhead.
Elevate. Ask them if the pilot of the
airplane would have to elevate or depress
his/her line of vision to see the person on
the ground. Depress

▰▰▰ TEACHING SUGGESTIONS

Lesson Note

Students are very much interested in real
applications. Try to borrow a surveyor's
transit. Discuss how a transit is used to
measure various angles involved in the
applications of this lesson.

Objective   To solve word problems using trigonometric ratios

Surveyors, pilots, and navigators frequently apply trigonometric ratios
to problems involving *angles of elevation* or *depression*.

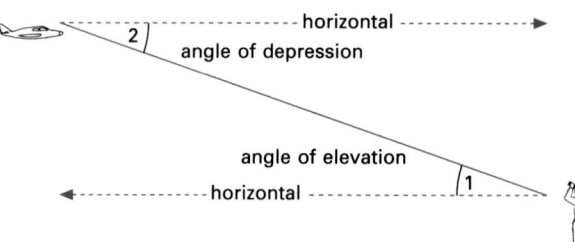

In the diagram above, $\angle 1$ indicates the *angle of elevation*. It is the angle by which the ground observer's line of vision must be raised, or elevated, with respect to the horizontal, to sight the plane.

$\angle 2$ is called the *angle of depression*. It is the angle by which the pilot's
line of vision must be lowered, or depressed, with respect to the horizontal, to sight the ground observer.

EXAMPLE 1   A plane is flying over level ground at an altitude of 900 m. When the
pilot sights a landing field, the measure of the angle of depression is 27.
Find the distance, to the nearest meter, from the point on the ground
directly under the pilot to the landing field.

This is known as the *ground distance*.

Plan   Draw a diagram with the plane at
P flying to the field at F. Then use
the tangent ratio to find GF.

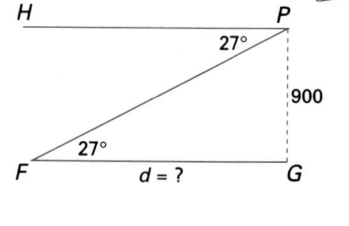

Solution

$$\tan 27 = \frac{PG}{GF}$$

$$0.5095 = \frac{900}{d}$$

$$0.5095d = 900$$

$$d = 1{,}766.4$$

So the ground distance, to the nearest meter, is 1,766 m.

Additional Example 1

A plane is flying over level ground at an
altitude of 7,000 ft. When the pilot sights
a landing field, the measure of the angle
of depression is 32. Find the ground
distance to the nearest foot. 11,202 ft

EXAMPLE 2   A tree 50 ft high casts a 35 ft shadow. Find, to the nearest degree, the measure of the angle of elevation of the sun.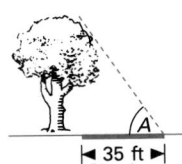

Solution   Draw a sketch.

$$\tan A = \frac{50}{35}$$

$$\tan A = 1.4285$$

$$m \angle A \approx 55$$

Thus, the measure of the angle of elevation is 55 to the nearest degree.

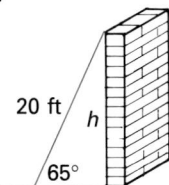

◄ 35 ft ►
Shadow

EXAMPLE 3   A 20-ft ladder is leaning against a wall. The foot of the ladder forms an angle of measure 65 with the ground. How far, to the nearest foot, is the top of the ladder from the ground?

Solution   Draw a sketch.

$$\sin 65 = \frac{h}{20}$$

$$0.9063 = \frac{h}{20}$$

$$h = 20(0.9063) = 18.126$$

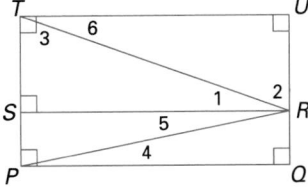

20 ft / h

65°

Thus, the top of the ladder is about 18 ft above the ground.

Classroom Exercises

Give the number of the angle and tell whether it is an angle of elevation or depression if a person at:

1. R sights the point T ∠1, elevation
2. P sights the point R ∠4, elevation
3. T sights the point R ∠6, depression
4. R sights the point P ∠5, depression

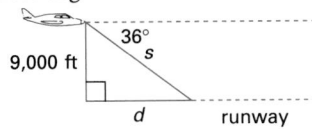

Written Exercises

Find the indicated measure to the nearest unit or the nearest degree.

1. The measure of the angle of depression is 36. The altitude of the plane is 9,000 ft. Find the ground distance d from the plane to the runway.   12,388 ft
2. Find the flight distance s from the plane to the runway.   15,311 ft

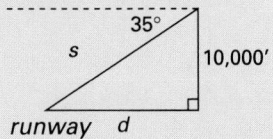

36°
9,000 ft   s

d   runway

Math Connections

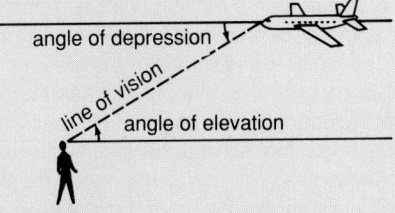

angle of depression
line of vision
angle of elevation

Parallel Lines: The angles formed by lines of vision are alternate interior angles. The angle of depression is congruent to the angle of elevation because both angles are formed with respect to two horizontal lines that are parallel.

Critical Thinking Questions

Analysis: As you move farther away from a building, does the angle of elevation to its roof increase or decrease?   It decreases. If you are looking from the window of a tall building at a car moving on the road, does the angle of depression increase or decrease as the car moves farther away? It decreases.

Checkpoint

Find the indicated measure to the nearest unit (lengths) or to the nearest degree (angles). (Ex. 1-2)

35°
s   10,000'

runway   d

1. The altitude of a plane is 10,000 ft. The measure of the angle of depression is 35. (a) Find the ground distance d from the plane to the runway.   14,281 ft   (b) Find the flight distance s from the plane to the runway.   17,434 ft
2. A building 360 ft tall casts a 180-ft shadow. Find the measure of the angle of elevation of the sun.   63

Additional Example 2

A building 275 ft tall casts a 160-ft shadow. Find, to the nearest degree, the measure of the angle of elevation of the top of the building.   60

Additional Example 3

A 40-ft ladder is leaning against a building. The ladder forms a 70° angle with the ground. How far, to the nearest foot, is the bottom of the ladder from the bottom of the building?   14 ft

Have the students explain what is meant by the angle of depression.   It is the angle formed when a line (of vision) is lowered with respect to the horizontal.   Ask them what is meant by the angle of elevation.   It is the angle formed when a line is raised with respect to the horizontal.   Ask them what is meant by the ground distance of an airplane from an object on the ground.   It is the distance from the point on the ground directly under the pilot to the object.

◼◼◼**FOLLOW UP**

Guided Practice

Classroom Exercises 1–4

Independent Practice

🅰 Ex. 1–7, 🅱 Ex. 8–13, 🅲 Ex. 14–18

Basic: WE 1–7
Average: WE 1–13
Above Average: WE 1–18

---

3. A 45-ft ladder makes an angle of measure 55 with the ground. How high up the wall does the ladder reach?   37 ft

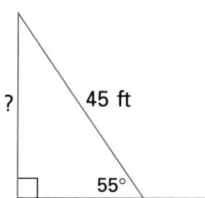

4. A building 200 ft tall casts a 155-ft shadow. Find the angle of elevation of the sun.   52

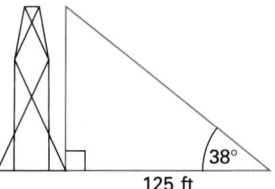

5. At a point 125 ft from the base of a tower, the angle of elevation of the top of the tower has a measure of 38. How high is the tower?   98 ft

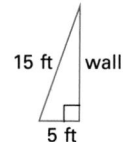

6. A 15-ft ladder leans against a wall. The foot of the ladder is 5 ft from the bottom of the wall. At what angle is the ladder to the ground?   71

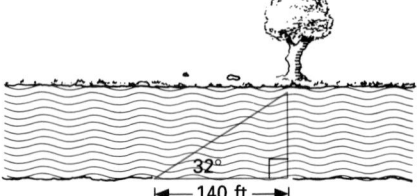

7. To estimate the width of a lake, Bob stood directly opposite a large tree. He then walked 140 ft along the bank and approximated the measure of the angle between his line of sight to the tree and the lake's edge as 32. Find the approximate width of the lake.   87 ft

8. A tree casts a 60-ft shadow. Find the height of the tree if the angle of elevation of the sun has a measure of 51.   74 ft

9. The top of a lighthouse is 250 ft above sea level. From the top, the measure of the angle of depression to a boat is 48. How far is the boat from the bottom of the lighthouse?   225 ft

10. The height of a cloud cover (called the "ceiling") at an airport was found the following way. A searchlight was pointed straight up. From a point on the ground 1,500 ft from the searchlight, the measure of the angle of elevation to the illumination of the clouds was 72. Find the height of the cloud cover.   4,617 ft

372   Chapter 9 Right Triangles

Enrichment

Show this drawing on the chalkboard.

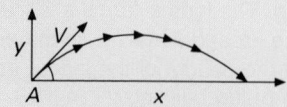

When a projectile is fired with velocity V at angle A, its initial velocity can be divided into two *components*, the horizontal velocity x, (downrange), and the vertical velocity y, (upward).
Have the students derive formulas for x and y in terms of V and A.
$x = V \cos A$; $y = V \sin A$
Given: $V = 100$ m/s; m $\angle A = 20$
Find: values of x and y
$x = $ _____ 100 cos 20 = 94, m/sec
$y = $ _____ 100 sin 20 = 34 m/sec

11. A pilot flying over level ground at an altitude of 2,400 ft sights a building. The angle of depression measures 6. Find the ground distance between the building and the point directly below the pilot. 22,835 ft

12. The surface of a ramp is 475 m long. It rises a vertical distance of 28 m. Find the measure of the angle of elevation. 3

13. A man in a sailboat is directly opposite the base of a 750-m high cliff. The angle of elevation measures 31. How far from land is the sailor? 1,248 m

Use the diagram to answer Exercises 14–16.

14. A navigator at *A* observes the measure of the angle of elevation to the top of the cliff to be 15. This measure changes to 26 at *B*. If the height of the cliff is known to be 750 m, how far has the boat moved to get from point *A* to point *B*? 1,261 m

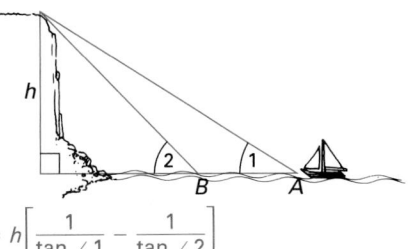

15. Write the formula for *AB* in terms of m ∠1, m ∠2, and *h*.

$$AB = h\left[\frac{1}{\tan \angle 1} - \frac{1}{\tan \angle 2}\right]$$

16. If the distance *AB* = 1,500 m, m ∠1 = 19, and m ∠2 = 28, how high is the cliff? 1,466 m

17. A 60-ft-high lighthouse stands on top of a cliff. The angles of elevation to the bottom and top of the lighthouse measure 39 and 43, respectively. Find the distance of the boat from the base of the cliff. 489 ft

18. A surveyor wants to measure the distances from points *A* and *B* to point *C* on the opposite side of a stream. Point *C* can be sighted from both *A* and *B*. She measures $\overline{AB}$ and angles *A* and *B* with the results indicated. Find *AC* and *BC* to nearest meter.
194 m, 175 m

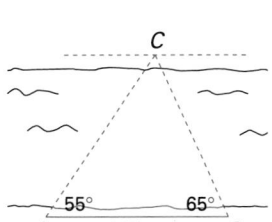

Mixed Review

1. Find the sum of the measures of the interior angles of a regular decagon. **6.2** 1,440

2. Points *A* and *B* on a number line have coordinates −8 and 6. Find the coordinate of *M*, the midpoint of $\overline{AB}$. **1.3** −1

3. Graph the conjunction $x \geq 3$ and $x \leq 8$ and describe the geometric figure. **1.8**

4. The ratio of the measures of the two acute angles of a right triangle is 2:7. Find the measure of each angle. **8.1** 20, 70

3.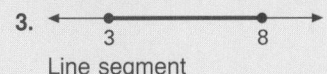
Line segment

Key Terms

angle of depression   (p. 370)
angle of elevation   (p. 370)
cosine   (p. 366)
Pythagorean Theorem   (p. 344)

sine   (p. 360)
tangent   (p. 366)
trigonometric ratio   (p. 360)

Key Ideas and Review Exercises

9.1   If the altitude is drawn to the hypotenuse of a right triangle, then
  • the square of the length of the altitude equals the product of the lengths of the segments of the hypotenuse: $h^2 = mn$,
  • and the square of the length of either leg equals the product of the lengths of the hypotenuse and the segment of the hypotenuse adjacent to that leg:
   $r^2 = nt$ and $s^2 = mt$.

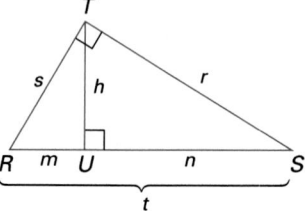

Find the indicated length, using the figure above.

1. $TU = 4$, $RU = 8$, $US =$ _____2_____

2. $RU = 2$, $US = 8$, $RT =$ _____$2\sqrt{5}$_____

3. $TS = 4$, $RS = 8$, $US =$ _____2_____

4. $RU = 12$, $ST = 8$, $SU =$ _____4_____

9.2   In any right triangle, the sum of the squares of the lengths of the legs equals the square of the length of the hypotenuse: $a^2 + b^2 = c^2$.

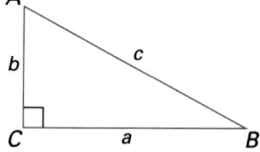

Find the indicated length, using the figure at the right.

5. $a = 4$, $b = 6$, $c =$ _____$2\sqrt{13}$_____

6. $a = 2$, $c = 7$, $b =$ _____$3\sqrt{5}$_____

7. Find the diagonal length of a square if the length of a side is 6 m. $6\sqrt{2}$ m

8. The length of each leg of an isosceles triangle is 4 ft. The length of the base is 6 ft. Find the length of the altitude from the vertex angle to the base. $\sqrt{7}$ ft

9.3 $\triangle XYZ$, with $\overline{XY}$ the longest side, is:
a *right* triangle if $z^2 = x^2 + y^2$,
an *obtuse* triangle if $z^2 > x^2 + y^2$, or
an *acute* triangle if $z^2 < x^2 + y^2$.

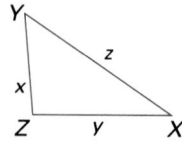

Determine whether the triangle formed with the given lengths of the three sides is acute, obtuse, or right. If the triangle is right, indicate which side is the hypotenuse.

9. 4, 3, 6   Obtuse **10.** 5, 6, 7   Acute **11.** $\sqrt{2}$, 1, $\sqrt{3}$ Right; $\sqrt{3}$ **12.** a, a, $a\sqrt{3}$ Obtuse

9.4   Properties of Special Right Triangles

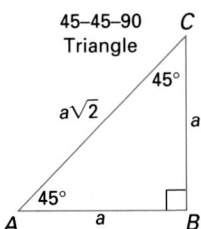

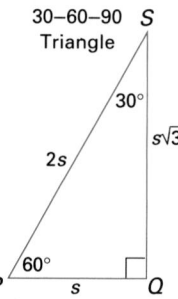

Use the figures above to find the indicated lengths.

13. $AC = 8$, $AB = \underline{4\sqrt{2}}$, $BC = \underline{4\sqrt{2}}$ **15.** $PS = 12$, $PQ = \underline{\;6\;}$, $SQ = \underline{6\sqrt{3}}$

14. $BC = 3\sqrt{2}$, $AB = \underline{3\sqrt{2}}$, $AC = \underline{\;6\;}$ **16.** $SQ = 6$, $PQ = \underline{2\sqrt{3}}$, $PS = \underline{4\sqrt{3}}$

Find the indicated length for an equilateral triangle with given lengths.

17. side: 8; altitude: $\underline{4\sqrt{3}}$ **18.** altitude: 4, side: $\underline{\quad\quad}$ $\frac{8}{3}\sqrt{3}$

9.5,   Trigonometric Ratios for a Right Triangle

9.6

$$\sin A = \frac{a}{c} \quad \sin B = \frac{b}{c} \quad \longleftarrow \frac{\text{opp}}{\text{hyp}}$$

$$\cos A = \frac{b}{c} \quad \cos B = \frac{a}{c} \quad \longleftarrow \frac{\text{adj}}{\text{hyp}}$$

$$\tan A = \frac{a}{b} \quad \tan B = \frac{b}{a} \quad \longleftarrow \frac{\text{opp}}{\text{adj}}$$

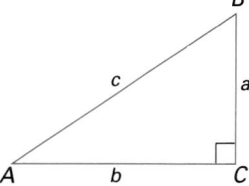

Find the indicated measure (sides to the nearest unit, angles to the nearest degree). Use the figure above.

19. $b = 8$, $a = 4$, m $\angle A = \underline{\;27\;}$ **20.** m $\angle A = 40$, $a = 5$, $c = \underline{\;8\;}$

21. $a = 20$, m $\angle B = 25$, $c = \underline{\;22\;}$ **22.** $b = 5$, $c = 15$, m $\angle A = \underline{\;71\;}$

9.7   Trigonometric ratios can be applied to word problems involving angles of elevation and angles of depression.

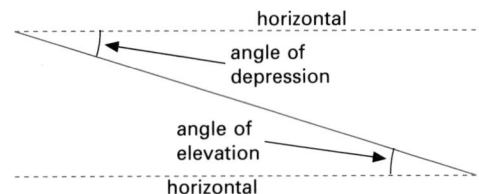

Find the required measure to the nearest unit.

23. A tree casts a 150-ft shadow. Find the height of the tree if the angle of elevation of the sun has a measure of 49.   173 ft

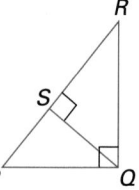

Use the right triangle at the right to find the indicated length.

1. $PR = 25$, $SP = 16$, $PQ =$ __20__
2. $SQ = 4$, $SP = 2$, $SR =$ __8__
3. $SR = 5$, $PQ = 6$, $SP =$ __4__
4. $SR = 6$, $SP = 2$, $SQ =$ __2√3__

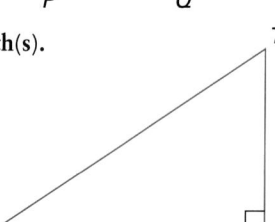

Use the right triangle at the right to find the indicated length(s).

5. $RS = 6$, $ST = 4$, $RT =$ __2√13__
6. $RT = 17$, $RS = 15$, $ST =$ __8__
7. m $\angle R = 45$, $RT = 6$, $ST =$ __3√2__
8. m $\angle R = 60$, $RS = 6$, $RT =$ ___, $ST =$ ___   12, 6√3
9. m $\angle T = 60$, $TS = 4$, $RT =$ ___, $RS =$ ___   8, 4√3
10. m $\angle T = 45$, $RS = 8$, $ST =$ ___, $RT =$ ___   8, 8√2

13. Right; hyp = 6

Determine whether the triangle with the given lengths of sides is acute, obtuse, or right. If the triangle is right, tell which side is the hypotenuse.   Obtuse

11. 5, 7, 9   Obtuse
12. 7, 8, 5   Acute
13. $4\sqrt{2}$, 2, 6
14. a, $a\sqrt{5}$, $a\sqrt{7}$

15. Find the length of the diagonal of a square with a side of length 4. $4\sqrt{2}$
16. Find the length of the altitude to the hypotenuse of a right triangle with legs of lengths 2 and 4. $\frac{4}{5}\sqrt{5}$
17. The lengths of the diagonals of a rhombus are 6 and 12. Find the length of a side of the rhombus. $3\sqrt{5}$
18. Find the length of a side of an equilateral triangle with altitude of length 6. $4\sqrt{3}$

Use the right triangle at the right to find the indicated measure (sides to the nearest unit, angles to the nearest degree).

19. $h = 6$, $g = 8$, m $\angle H =$ __49__
20. $h = 12$, m $\angle K = 62$, $k =$ __23__
21. $g = 15$, m $\angle H = 40$, $h =$ __10__

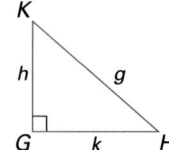

22. A tree casts a shadow of 140 ft. Find the height of the tree to the nearest foot if the angle of elevation of the sun measures 6.   15 ft
* 23. An angle of a rhombus measures 118. The length of the diagonal opposite that angle is 16 ft. Find the length of the other diagonal to the nearest foot.   10 ft

Indicate the one correct answer for each question.

1. In the figure below, $x + y =$ _____
C (A) 130   (B) 50   (C) 80
(D) 0   (E) 40

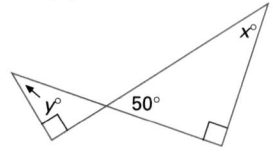

2. If the measure of one acute angle of a
C right triangle is 25, find the measure
of the supplement of the other.
(A) 65   (B) 155   (C) 115
(D) 35   (E) None of these

3. In the diagram below, $\angle RSQ$, $\angle QTS$,
E and $\angle QTP$ are right angles. Find QR.

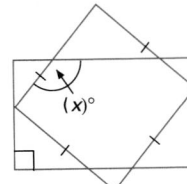

(A) 12   (B) 13   (C) 15
(D) 25   (E) 17

4. In the figure of the square and rectan-
E gle below, the value of x is.
(A) 120   (B) 100   (C) 150
(D) 135   (E) Not enough informa-
tion is given to answer the question.

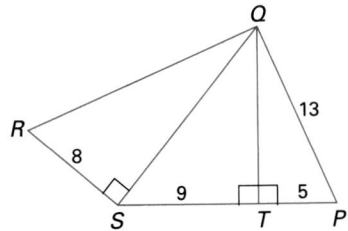

5.
A

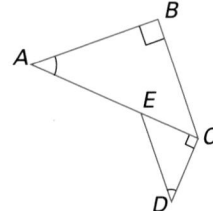

In the figure above, $DE = 5$,
$DC = 4$, $AE = 7$. $\angle A \cong \angle D$.
Find AB.
(A) 8   (B) 6   (C) 7   (D) 5
(E) 10

6. For the three squares below, find the
C ratio $\dfrac{PQ}{RS}$.

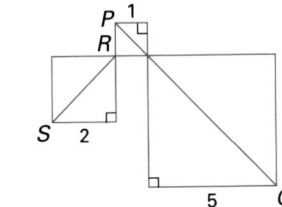

(A) $\frac{1}{2}$ (B) $\frac{3\sqrt{2}}{2}$ (C) $\frac{3}{1}$ (D) $\frac{3}{2}$

(E) $\dfrac{18}{\sqrt{2}}$

7. In the figure below, $AC = 4\sqrt{2}$,
B $AD = 2\sqrt{2}$, m $\angle DAB = 105$.
Find BA.

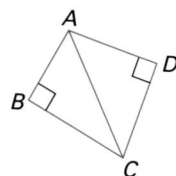

(A) 8   (B) 4   (C) $2\sqrt{2}$
(D) $\sqrt{2}$ (E) None of these

OVERVIEW

In this chapter, students use a coordinate system to graph ordered pairs of numbers and lines that meet given conditions. They will assign coordinates to vertices of polygons, determine the distance between two points and find the coordinates of the midpoint of a segment. Students will also determine perpendicularity and parallelism of two lines, as well as the slope of a line that is parallel or perpendicular to a given line.

OBJECTIVES

- To write coordinates of points graphed on a coordinate system
- To find the length of a segment, given the coordinates of its endpoints
- To find the coordinates of the midpoint of a segment, given the coordinates of the endpoints of the segment
- To find the coordinates of the endpoint of a segment, given the coordinates of the midpoint and of the other endpoint
- To place a quadrilateral in a coordinate system, so that the simplest coordinates are assigned to the vertices
- To find the slope of a segment, given the coordinates of its endpoints
- To draw a line with a given slope passing through a given point
- To write equations of lines meeting given conditions
- To graph lines with given equations
- To prove theorems using coordinate geometry

PROBLEM SOLVING

The Application on page 402 illustrates the problem solving strategy, Finding More Than One Way. Students are asked to solve the exercises in Lesson 10.5 using vector methods. This shows students that there are different approaches to solving problems.

TECHNOLOGY

Computer: The Computer Investigation on page 392 enables students to graph equations of the form $y = mx + b$ and discover the changes in the graph for different values of m and b.

Calculator: Encourage students to use a calculator to solve Exercises 27 and 28 on page 387.

SPECIAL FEATURES

Focus on Reading p. 381
Mixed Review pp. 383, 387, 396, 401, 407
Algebra Review pp. 387, 407
Midchapter Review p. 391
Computer Investigation: Equation of a Line p. 392
Application: Vector Methods p. 402
Key Terms p. 408
Key Ideas and Review Exercises pp. 408–409
Chapter 10 Test p. 410
College Prep Test p. 411
Cumulative Review (Chapters 1–10) pp. 412–413

PLANNING GUIDE

Lesson		Basic	Average	Above Average	Resources
10.1	pp. 381–383	FR all CE all WE 1–24 odd	FR all CE all WE 1–37 odd	FR all CE all WE 1–43 odd	Reteaching p. 137 Practice p. 138
10.2	pp. 386–387	CE all WE 1–22 odd AR all	CE all WE 1–28 odd AR all	CE all WE 1–33 odd AR all	Reteaching p. 139 Practice p. 140
10.3	pp. 390–392	CE all WE 1–19 odd Midchapter Review CI Activity 1	CE all WE 1–31 odd Midchapter Review CI all	CE all WE 1–34 odd Midchapter Review CI all	Reteaching p. 141 Practice p. 142
10.4	pp. 395–396	CE all WE 1–21 odd	CE all WE 1–39 odd	CE all WE 1–44 odd	Reteaching p. 143 Practice p. 144
10.5	pp. 400–402	CE all WE 1–26 odd Application	CE all WE 1–34 odd Application	CE all WE 1–37 odd Application	Reteaching p. 145 Practice p. 146
10.6	pp. 405–407	CE all WE 1–12 AR all	CE all WE 1–20 AR all	CE all WE 1–25 odd AR all	Reteaching p. 147 Practice p. 148
Chapter 10 Review pp. 408–409		1–16 odd	all	all	
Chapter 10 Test p. 410		1–6, 9–15, 17–19	1–20	all	
College Prep Test p. 411		all	all	all	
Cumulative Review pp. 412–413		1–33 odd	all	all	

CE = Classroom Exercises   WE = Written Exercises   FR = Focus on Reading   AR = Algebra Review   CI = Computer Investigation
NOTE: For each level, all students should be assigned all Mixed Review exercises.

◼◼◼ INVESTIGATION

In this investigation, students use manipulatives to develop an intuitive sense for which slopes are positive and which are negative, and for when the slope of one line is greater than or less than the slope of another line.

This investigation uses informal language without formal definitions to help students discover basic relationships before they learn the precise language of slope in Lesson 10.3. This activity will probably take the students less than half an hour to do. Students can do this investigation either alone or in groups of two or three. Each student needs a copy of Manipulative Worksheet 19 and each group needs to have several pencils, each about four inches long.

After students perform the experiments on the Worksheet, and draw their conclusions, have the whole class discuss the generalizations they have discovered and the conjectures they have made.

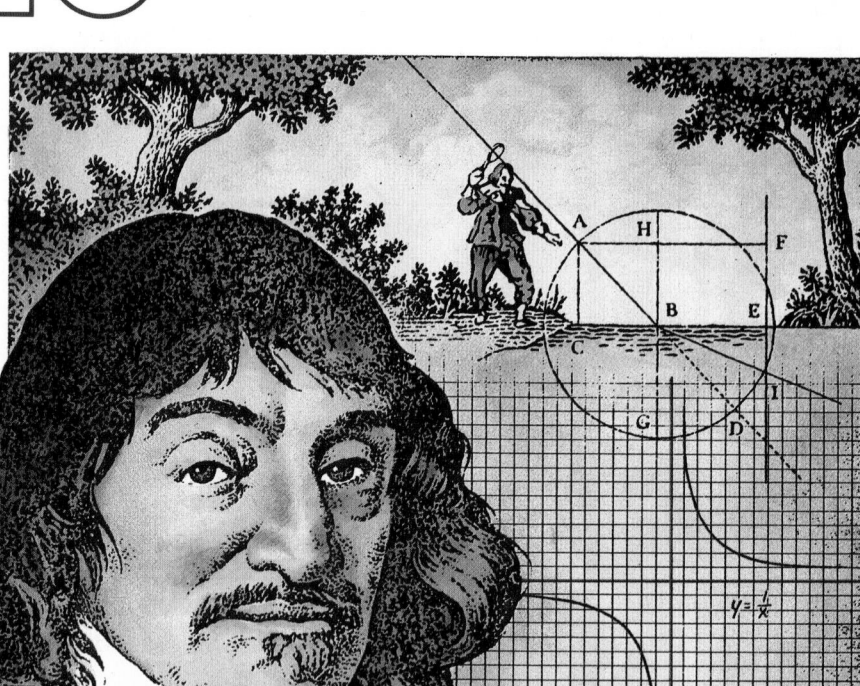

"*I think, therefore I am*"
(*cogito, ergo sum*)

René Descartes (1596–1650)—Descartes developed the idea that any point on a plane can be named by a pair of numbers. Today this idea is used in street maps, lines of longitude and latitude on a globe, and graph paper.

More About the Mathematician

Descartes thought that all knowledge should be worked out by chains of statements and reasons such as those used in geometry. The assertion "I think, therefore I am" was the basic axiom upon which his philosophy was based. Examples in his *Discourse on the Method of Reasoning Well* (published in 1637) included the behavior of lenses (an optics experiment is shown in the picture) and of shooting stars. His *Discourse* included a 106-page appendix on geometry. The hyperbola in the picture illustrates his assertion that every equation can be represented by a curve and every curve can be written as an equation—the basis for analytic geometry.

10.1 Coordinate Systems and Distance

Objectives   To graph ordered pairs of numbers
To find the distance between two points in a coordinate plane

Every point on a number line can be named
by exactly one real number.

To name points in a plane, a system based
on two perpendicular number lines is often
used.

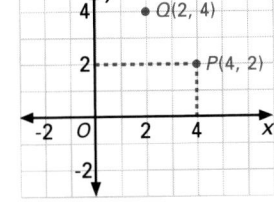

The horizontal line is called the **x-axis**; the
vertical line, the **y-axis**. The **origin** is the
point of intersection of the two axes.

For every point in a plane, there is a
corresponding ordered pair of numbers
(x, y) called the **coordinates** of the point.
For every ordered pair of numbers, there is a corresponding point,
called the **graph** of the ordered pair. The graphs of points $P(4, 2)$ and
$Q(2, 4)$ are shown above. Observe that the order of the coordinates is
important. The x-coordinate, or **abscissa**, is the horizontal distance
from the point to the y-axis. The y-coordinate, or **ordinate**, is the vertical distance from the point to the x-axis. The abscissa of P is 4 and the
ordinate of P is 2.

A plane where points are associated with ordered pairs in this way is
called a **coordinate plane**.

The axes separate the plane into four quadrants, numbered I, II, III, and IV in a counterclockwise manner.
Both coordinates of a point in Quadrant I are positive. Abscissas are negative in Quadrants II and III,
while ordinates are negative in Quadrants III and IV.
Points on the axes are not contained in any quadrant.

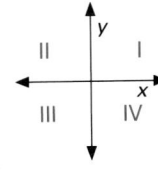

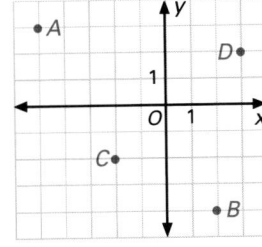

EXAMPLE 1   Write the coordinates of points A, B,
C, and D. Tell which quadrant con-
tains each point.

Solution $A(-5, 3)$ Quadrant II
$B(2, -4)$ Quadrant IV
$C(-2, -2)$ Quadrant III
$D(3, 2)$ Quadrant I

Additional Example 1

Write the coordinates of each of the
points and the quadrant in which it
is located.

$A (-2, 2)$ Quad II
$B (1, 5)$ Quad I
$C (4, -2)$ Quad IV
$D (5, 0)$ x-axis
$E (-5, -3)$ Quad III

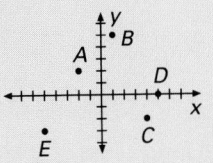

Teaching Resources

Manipulative Worksheet 18
Problem Solving Worksheet 11
Project Worksheet 10
Quick Quizzes 69
**Reteaching and Practice
Worksheets,** pp. 137, 138
Transparencies 25, 26A, 26B

◤ GETTING STARTED

Prerequisite Quiz

1. Draw a coordinate plane and plot the
following points. $(-3,4)$; $(2,0)$; $(-1,-2)$
Check students' graphs.
2. Find the distance between the two points
A and B on the given number line.   7, 7

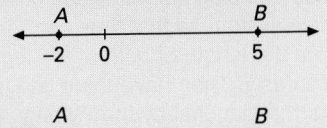

3. The point with coordinates $(0,0)$ is called
the _____.   Origin
4. Another name for the x-coordinate of a
point is the _____.   Abscissa

Motivator

Ask the students what kind of instructions
they would give someone who was trying to
get to their house. Ask them how they
would tell the person "how far." Ask them if
they would use an agreed-upon starting
point or direction to do this.   Yes

379

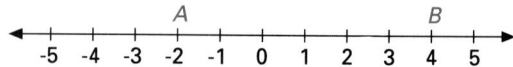
Lesson Note

Emphasize basic terminology. Point out that a point on the axes is not in a quadrant. Introduce the terms "Cartesian coordinate system" or "Cartesian plane," with a reference to René Descartes. As an optional assignment, students might do a project on Descartes. This is an excellent chapter in which to use calculators.

As a manipulative approach, make several paper right triangles with integral unit sides and a coordinate system with the same units. Ask the students to estimate the length of the sides of the triangles in terms of the given units.

Have students place the triangles on the coordinate system so that they can determine the length of both of the legs simultaneously. Then have them find the length of the diagonal by the Pythagorean Theorem, and also by placing it along the x-axis.

Math Connections

Cartography: One of the first coordinate systems ever developed is used to describe locations on a map. On a world atlas, *latitude* measures distance in degrees north or south of the equator. *Longitude* measures distance in degrees east or west of the prime meridian, which is assigned a longitude of 0°.

A city map is also a type of coordinate system in which areas are identified, usually by a letter and a number, such as B−4. Point out the difference between a Cartesian coordinate plane, in which specific points are located, and a city map, in which a coordinate system is used to specify areas of a square in which many objects are located.

The distance AB between the two points A and B on a number line is $|m - n|$ or $|n - m|$ where m and n are the coordinates of A and B, respectively. For example, AB below is $|4 - (-2)| = 6$.

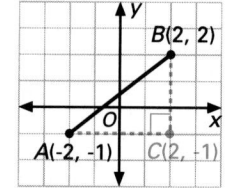

To find the distance between two points in a coordinate plane, the Pythagorean Theorem can be used.

EXAMPLE 2 Locate the points $A(-2,-1)$ and $B(2, 2)$ in a coordinate plane. Draw a right triangle which has the two points as vertices of the acute angles. Find the lengths of the legs of this triangle. Using the Pythagorean Theorem, find the distance between the two points.

$AC = |2 - (-2)| = 4$ ← difference of abscissas
$BC = |2 - (-1)| = 3$ ← difference of ordinates
$AB = \sqrt{(AC)^2 + (BC)^2}$
$AB = \sqrt{4^2 + 3^2} = 5$

Example 2 suggests a general formula for finding the distance between any two points on a plane.

Theorem 10.1

The Distance Formula: The distance d between $P_1(x_1,y_1)$ and $P_2(x_2,y_2)$ is given by the formula $d = \sqrt{(x_2 - x_1)^2 + (y_2 - y_1)^2}$.

Given: $P_1(x_1, y_1)$, $P_2(x_2, y_2)$
Prove: $d = P_1P_2$
   $= \sqrt{(x_2 - x_1)^2 + (y_2 - y_1)^2}$

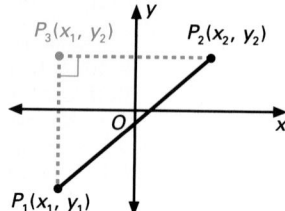

Outline of Proof

(1) Locate $P_3(x_1, y_2)$ and draw $\triangle P_1P_2P_3$.
(2) $\overline{P_1P_3}$ is vertical. (Why?) Therefore, $P_1P_3 = |y_2 - y_1|$
(3) $\overline{P_2P_3}$ is horizontal. (Why?) Therefore, $P_2P_3 = |x_2 - x_1|$
(4) $\triangle P_1P_2P_3$ is a right triangle. (Why?) So, by the Pythagorean Theorem,

$$d^2 = (P_1P_3)^2 + (P_2P_3)^2$$
$$d^2 = |y_2 - y_1|^2 + |x_2 - x_1|^2$$
$$d^2 = (y_2 - y_1)^2 + (x_2 - x_1)^2$$
$$d = \sqrt{(x_2 - x_1)^2 + (y_2 - y_1)^2}$$

380 Chapter 10 Coordinate Geometry

Additional Example 2

Locate the points $C(3,-5)$ and $D(-1,-3)$ in a coordinate plane. Draw a right triangle which has the two points as vertices of the acute angles with the legs parallel to the x and y axes. Find the lengths of the legs of this triangle. 2,4
Using the Pythagorean Theorem, find the distance between the two points. $2\sqrt{5}$

EXAMPLE 3   Use the Distance Formula to find the distance between $A(-1, -2)$ and $B(3, 3)$.

Solution   If $A = P_1$, and $B = P_2$, then $x_1 = -1$, $y_1 = -2$, $x_2 = 3$, and $y_2 = 3$
$$d = \sqrt{(x_2 - x_1)^2 + (y_2 - y_1)^2}$$
$$= \sqrt{[3 - (-1)]^2 + [3 - (-2)]^2}$$
$$= \sqrt{4^2 + 5^2}$$
$$= \sqrt{16 + 25}, \text{ or } \sqrt{41}$$

The Distance Formula can sometimes be used to prove properties of geometric figures.

EXAMPLE 4 $\triangle ABC$ has vertices $A(-2, 6)$, $B(6, 4)$, and $C(0, -2)$. Is $\triangle ABC$ scalene, isosceles, or equilateral?

Plan   Find the lengths of the three sides of $\triangle ABC$.

Solution
$$AB = \sqrt{[6 - (-2)]^2 + (4 - 6)^2}$$
$$= \sqrt{8^2 + (-2)^2} = \sqrt{64 + 4}$$
$$= \sqrt{68} = 2\sqrt{17}$$
$$AC = \sqrt{[0 - (-2)]^2 + (-2 - 6)^2}$$
$$= \sqrt{2^2 + (-8)^2}$$
$$= \sqrt{4 + 64} = \sqrt{68} = 2\sqrt{17}$$
$$BC = \sqrt{(0 - 6)^2 + (-2 - 4)^2}$$
$$= \sqrt{(-6)^2 + (-6)^2}$$
$$= \sqrt{36 + 36} = \sqrt{72} = 6\sqrt{2}$$

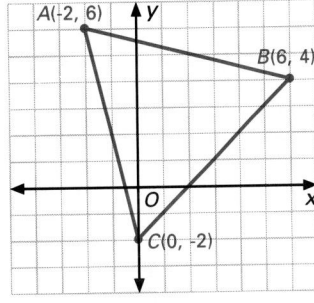

Since $AB = AC$, $\triangle ABC$ is isosceles.

▰ Focus on Reading

Translate each sentence into symbols.

1. The length of segment $\overline{PQ}$ is 7. $PQ = 7$
2. The distance between two points P and Q on a number line is the absolute value of the difference of the coordinates of the two points. $PQ = |x_2 - x_1|$
3. The distance between the two points A and B with coordinates (x, y) and (u, v) is 5. $5 = \sqrt{(x - u)^2 + (y - v)^2}$
4. The line containing the points D and E is perpendicular to the line containing points F and G. $\overleftrightarrow{DE} \perp \overleftrightarrow{FG}$

10.1 Coordinate Systems and Distance   **381**

381

Closure

Ask the students how they measure distances on a number line.   Measure the distances between two points.   Ask them what formula they use. $|m - n|$ where m and n are coordinates of the points.   Have them tell you the formula they use to measure distances on a plane.   The Distance Formula, page 380.   Have the students explain how they graph ordered pairs of numbers.   The first number tells which direction and how far to move on the horizontal x-axis and the second number tells which direction and how far to move along the vertical y-axis.

◼️ FOLLOW UP

Guided Practice

Classroom Exercises 1–8

Independent Practice

Ⓐ Ex. 1–24, Ⓑ Ex. 25–37, Ⓒ Ex. 38–42

Basic: FR all, WE 1–24 odd
Average: FR all, WE 1–37 odd
Above Average: FR all, WE 1–43 odd

Additional Answers

Written Exercises

7-10.

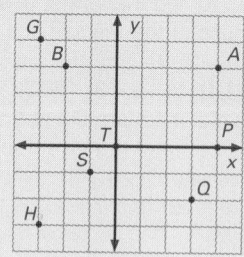

43. (5,0), (−5,0), (0,5), (0,−5), (3,4), (4,3), (−3,4), (−4, 3), (−3,−4), (−4,−3), (3,−4), (4,−3)

Mixed Review

1. See Construction on page 200.

Classroom Exercises

The figure at the right shows a street map. In some ways it is like the coordinate axes we have been studying. North is to the top and east is to the right. (Exercises 1–5)

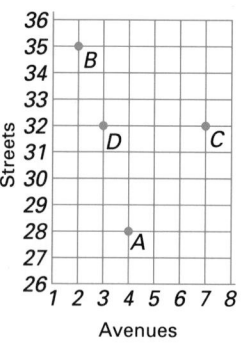

1. What would be your direction and distance in walking from A to B?   2 blocks W, 7 blocks N; 9 blocks
2. How would you walk from B to C?   5 blocks E, 3 blocks S
3. How would you walk from D to A?   1 block E, 4 blocks S
4. What is the address of point C?   7th Ave. and 32nd St.
5. What is the address of point D?   3rd Ave. and 32nd St.

6. In what quadrants is the abscissa positive?   I, IV
7. In what quadrants is the ordinate negative?   III, IV
8. In what quadrant is the point $P(3,-2)$?   IV

Written Exercises

Name the coordinates and quadrant of the following points in the diagram.

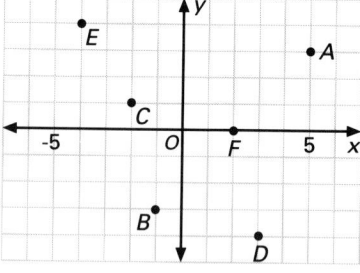

1. A (5,3) I   **2.** B (−1,−3) III   **3.** C (−2, 1) II
4. D (3,−4) IV   **5.** E (−4, 4) II   **6.** F (2,0)Pos x-axis

Draw a graph and locate the following points.

7. $A(4, 3)$ and $B(-2, 3)$
8. $P(4, 0)$ and $Q(3, -2)$
9. $S(-1, -1)$ and $T(0, 0)$
10. $G(-3, 4)$ and $H(-3, -3)$

Locate the pair of given points in a coordinate plane. Draw a right triangle that has the two points as vertices of the acute angles. Find the lengths of the legs and the length of the hypotenuse of the triangle.

11. $A(2, 1)$, $B(4, 5)$ 2, 4, $2\sqrt{5}$
12. $C(3, -2)$, $D(0, 4)$ 3, 6, $3\sqrt{5}$
13. $E(-1, -3)$, $F(4, 2)$ 5, 5, $5\sqrt{2}$
14. $G(0, -4)$, $H(-3, 2)$ 3, 6, $3\sqrt{5}$

Use the Distance Formula to find the distance between the given points.

15. $P_1(0, 0)$, $P_2(4, 3)$ 5
16. $P_1(1, 1)$, $P_2(4, 5)$ 5
17. $P_1(-1, 2)$, $P_2(11, 7)$ 13
18. $P_1(2, -3)$, $P_2(7, 9)$ 13
19. $P_1(1, 2)$, $P_2(5, 6)$ $4\sqrt{2}$
20. $P_1(6, 2)$, $P_2(1, 7)$ $5\sqrt{2}$

Enrichment

Plot the following points: $A(-1,1)$, $B(2,1)$, $C(5,0)$, $D(0,-3)$, $E(-3,-1)$, $F(5,-2)$, $G(3,5)$, $H(-3,5)$, $J(0,2)$.

Consider each point as a location at which a truck must make a delivery. The driver wishes to make the total distance as small as possible. The driver is to visit each point only once and does not have to end at the same point at which he started. Each student is to find two such routes the delivery man may take. For each route, find the distance between each stop, and then find the total distance for each route to the nearest hundredth. Use a calculator. Discuss the distances for the different routes.

Find the length of $\overline{AB}$.

21. $A(0, 9), B(-5, -3)$ 13

22. $A(-1, 2), B(2, -2)$ 5

23. $A(2, -3), B(-3, 1)$ $\sqrt{41}$

24. $A(-2, -3), B(3, 1)$ $\sqrt{41}$

ABC is a triangle. Use the lengths AB, BC, and AC to tell whether or not $\triangle ABC$ is isosceles.

25. $A(-2, 0), B(2, 0), C(0, 5)$ Yes

26. $A(0, 3), B(5, 1), C(0, -1)$ Yes

27. $A(0, 5), B(5, 0), C(7, 7)$ Yes

28. $A(1, 5), B(-4, 3), C(-1, 0)$ Yes

29. $A(4, 4), B(-6, 2), C(5, -1)$ No, scalene

30. $A(1, -6), B(-3, -3), C(-2, -2)$ Yes

ABC is a triangle. Use the lengths AB, BC, and AC to tell whether or not $\triangle ABC$ is a right triangle.

31. $A(0, 0), B(2, 4), C(4, -2)$ Yes

32. $A(1, -2), B(-5, 4), C(5, 2)$ Yes

Using the Distance Formula to find the lengths of the sides and tell whether or not $ABCD$ is a parallelogram.

33. $A(0, -2), B(5, -3), C(7, 1), D(2, 2)$ Yes **34.** $A(-4, -2), B(2, 0), C(3, 3), D(-2, 1)$

35. $A(-2, -1), B(3, -2), C(4, 1), D(-1, 2)$ Yes No

$ABCD$ is a parallelogram. Using Theorem 7.4, tell whether or not it is a rectangle. (Exercises 36–37)

36. $A(1, -2), B(7, 0), C(5, 3), D(-1, 1)$
No

37. $A(-4, -2), B(-2, -6), C(8, -1),$
$D(6, 3)$ Yes

38. $A(3, y)$ is a distance of 10 units from $B(-3, -1)$. Find all possible values of y. 7, −9

39. $C(-1, 2)$ is a distance of $\sqrt{17}$ units from $D(x, -2)$. Find all possible values of x. 0, −2

40. $(-2, -2)$ and $(2, -1)$ are the coordinates of opposite vertices of a square. What are the coordinates of the other vertices? $(-\frac{1}{2}, \frac{1}{2}), (\frac{1}{2}, -3\frac{1}{2})$

41. $(-1, -3), (2, -2),$ and $(1, 1)$ are the coordinates of three vertices of a square. What are the coordinates of the fourth vertex? $(-2, 0)$

42. $(-4, -1), (-3, 1),$ and $(1, -1)$ are the coordinates of three vertices of a rectangle. What are the coordinates of the fourth vertex? $(0, -3)$

43. Find the coordinates of all points with integer coordinates that are a distance of 5 units from the origin.

Mixed Review

1. Draw a segment 5 cm long. Construct the perpendicular bisector of the segment. **5.6**

Find the coordinate of the midpoint of each segment. **1.3**

2. $\overline{AG}$ −3 **3.** $\overline{EK}$ 1 **4.** $\overline{DJ}$ 0

5. $\overline{LF}$ 2 **6.** $\overline{KG}$ 2 **7.** $\overline{JB}$ −1

```
      A   B   C   D   E   F   G   H   I   J   K   L   M
  ◄───┼───┼───┼───┼───┼───┼───┼───┼───┼───┼───┼───┼───►
     -6  -5  -4  -3  -2  -1   0   1   2   3   4   5   6
```

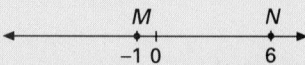
GETTING STARTED

Prerequisite Quiz

1. Find the midpoint of the segment $\overline{MN}$ on the given number line. $2\frac{1}{2}$

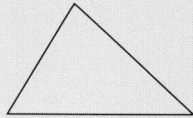

2. Locate the midpoint of each side of the given triangle. Check students' constructions.

Motivator

Ask the students what they mean when they say they are "halfway there." They have gone half the distance. Have them describe the location of the midpoint of two points. The point in the middle of two points, or the point halfway between the two points. Ask them how they would determine their "average" in their math class. Divide the sum of their grades by the number of grades.

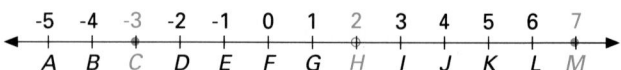
TEACHING SUGGESTIONS

Lesson Note

Teaching the lesson, you might want to begin by having students do the Algebra Review on page 387. Students will be asked to solve fractional equations throughout this lesson. Thus, the Algebra Review will allow them to practice the algebraic skills needed.

10.2 The Midpoint Formula

Objective   To find the coordinates of the midpoint of a segment

The coordinate x_m of the midpoint of a segment on a number line is the average of the coordinates of the endpoints. On the number line below, the midpoint of $\overline{CM}$ is 2, since $\frac{-3 + 7}{2} = 2$.

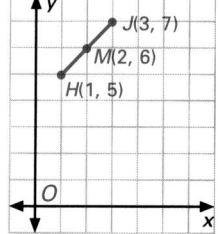

The coordinates of the midpoint of a segment in a coordinate plane can be found in a similar way.

$\overline{HJ}$, with endpoints $H(1, 5)$ and $J(3, 7)$, has the midpoint $M(2, 6)$. The abscissa 2 is the average of the abscissas 1 and 3. The ordinate 6 is the average of the ordinates 5 and 7.

This result is stated in the following theorem.

Theorem 10.2

The Midpoint Formula: Given $P(x_1, y_1)$ and $Q(x_2, y_2)$, the coordinates (x_m, y_m) of M, the midpoint of $\overline{PQ}$, are $\left(\frac{x_1 + x_2}{2}, \frac{y_1 + y_2}{2}\right)$.

Given: $P(x_1, y_1)$, $Q(x_2, y_2)$
Prove: $x_m = \dfrac{x_1 + x_2}{2}$, $y_m = \dfrac{y_1 + y_2}{2}$

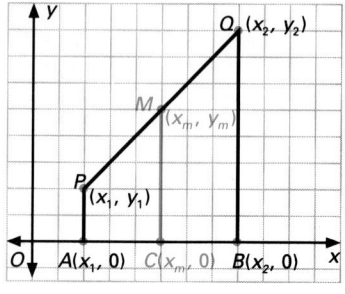

Proof   Draw $\overline{PA}$, $\overline{MC}$, and $\overline{QB}$ all $\perp$ to the x-axis. These are all $\parallel$ to the y-axis and therefore vertical. A, B, and C have the same x-coordinates as P, Q, and M, respectively. C is the midpoint of $\overline{AB}$. By Theorem 6.12, if three parallel lines cut congruent segments on one transversal, they cut congruent segments on all transversals. But the coordinates of the midpoint of two points on a number line is the average.

Therefore, $x_m = \dfrac{x_1 + x_2}{2}$ for all points on $\overline{MC}$.

Similarly, it can be shown that $y_m = \dfrac{y_1 + y_2}{2}$.

384   Chapter 10 Coordinate Geometry

Additional Example 1

Given $A(-1, -5)$ and $B(-6, 3)$, find the midpoint of $\overline{AB}$. $(-3.5, -1)$

EXAMPLE 1   Given $P(3, 2)$ and $Q(-5, -8)$, find the midpoint of $\overline{PQ}$.

Solution
$$x_m = \frac{x_1 + x_2}{2} \qquad y_m = \frac{y_1 + y_2}{2}$$
$$= \frac{3 + (-5)}{2} = -1 \qquad = \frac{2 + (-8)}{2} = -3$$
Therefore, the coordinates of the midpoint are $(-1, -3)$.

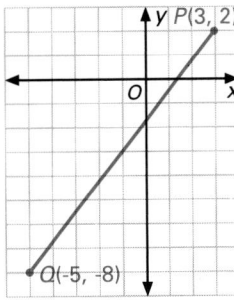

EXAMPLE 2   Given: $M(1, 2)$ is the midpoint of $\overline{AB}$. The coordinates of A are $(-3, 8)$. Find the co-ordinates of B.

Solution   Let A have the coordinates (x_1, y_1) and B have the coordinates (x_2, y_2).
$$x_m = \frac{x_1 + x_2}{2} \qquad y_m = \frac{y_1 + y_2}{2}$$
$$1 = \frac{-3 + x_2}{2} \qquad 2 = \frac{8 + y_2}{2}$$
$$2 = -3 + x_2 \qquad 4 = 8 + y_2$$
$$x_2 = 5 \qquad y_2 = -4$$
Therefore, the coordinates of B are $(5, -4)$.

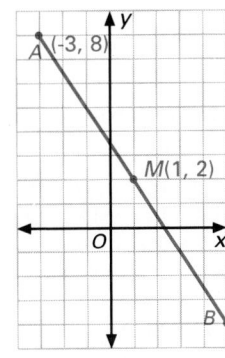

EXAMPLE 3 $ABCD$ is a quadrilateral with vertices $A(-1, -3)$, $B(9, -3)$, $C(11, 5)$, and $D(1, 5)$. Show that $ABCD$ is a parallelogram using the midpoint formula.

Plan   Show that diagonals $\overline{AC}$ and $\overline{BD}$ bisect each other by showing that they have the same midpoint.

Solution   For diagonal $\overline{AC}$,
$$x_m = \frac{-1 + 11}{2}, \, y_m = \frac{-3 + 5}{2}$$
$$x_m = 5, \, y_m = 1$$

For diagonal $\overline{BD}$,
$$x_m = \frac{9 + 1}{2}, \, y_m = \frac{-3 + 5}{2}$$
$$x_m = 5, \, y_m = 1$$

Since the diagonals have the same midpoint, they bisect each other, and the quadrilateral is a parallelogram.

10.2 The Midpoint Formula   **385**

Math Connections

Interior Decoration: On a scale drawing of a rectangular room that is 25 × 15 feet, an interior decorator assigns to the vertices the coordinates (0,0), (0,15), (25,0) and (25,15). The decorator can use the Midpoint Formula to find the center of the room where a light fixture will be hung on the ceiling. Since the room is rectangular, the diagonals are congruent and bisect each other. Therefore, the midpoint of either diagonal pinpoints the location of the center of the room.

Critical Thinking Questions

Analysis: Ask students to compare the Midpoint Formula with the formula for finding an average.   The formula for finding an average is based on a number line, but the Midpoint Formula is based on a coordinate plane. However, both formulas involve the sum of two coordinates divided by 2.

Common Error Analysis

Error: Students replace the "+" signs in the Midpoint Formula with "−" signs. Illustrate that the number obtained by taking half the difference of two numbers does not lie between the two given numbers, and that the average of two numbers does.

Additional Example 2

Given: $M(1, 4)$ is the midpoint of $\overline{AB}$. The coordinates of A are $(3, 2)$. Find the coordinates of B. $(-1, 6)$

Additional Example 3

$ABCD$ is a quadrilateral with vertices $A(0, -2)$, $B(7, 0)$, $C(5, 4)$, $D(-2, 2)$. Show that $ABCD$ is a parallelogram using the Midpoint Formula.
Midpoint of $\overline{AC}$ is $(\frac{5}{2}, 1)$, midpoint of $\overline{BD}$ is $(\frac{5}{2}, 1)$. Diagonals bisect each other.

Checkpoint

1. For each pair of points X and Y, find the coordinates of the midpoint of $\overline{XY}$.
   (a) X(7,3), Y(−2,−6) (2.5,−1.5)
   (b) X(15,−32), Y(−12,0) (1.5,−16)
2. M is the midpoint of $\overline{XY}$. Find the coordinates of Y.
   (a) M(−3,−1), X(−6,4) (0,−6)
   (b) M(2,5), X(4,−3) (0,13)
3. ABCD is a quadrilateral with the given vertices. Using the fact that a quadrilateral is a parallelogram if its diagonals bisect each other, decide whether ABCD is a parallelogram. A(0,0), B(3,1), C(6,5), D(3,4) Diag. midpts $(3,\frac{5}{2})$, ABCD is a □.
4. Given △ABC, with A(−1,1), B(2,1), C(1,4). Find the length of the median $\overline{AM}$. M(1.5,2.5), AM = $\sqrt{8.5}$

Closure

Ask the students how they can find the midpoint of a segment on a number line. Find the average of the coordinates of the endpoints. Then ask them how they can find the midpoint of a segment in a plane. Midpoint Formula, page 384 Ask them if it matters which quadrant the endpoints of the segment are in when they use the Midpoint Formula. No

Classroom Exercises

Find the coordinate of the midpoint of each given segment.

1. $\overline{GK}$ 2   2. $\overline{AG}$ −3   3. $\overline{DJ}$ 0   4. $\overline{AI}$ −2   5. $\overline{DL}$ 1   6. $\overline{FL}$ 2

For each pair of points A and B, find the coordinates of the midpoint of $\overline{AB}$.

7. A(0, 0), B(8, 8) (4,4)   8. A(−6, 6), B(0, 0) (−3,3)   9. A(0, 0), B(−4, 10) (−2,5)

Written Exercises

For each pair of points A and B, find the coordinates of the midpoint of $\overline{AB}$.

1. A(0, 0), B(4, 4) (2, 2)   2. A(3, 5), B(1, 7) (2, 6)   3. A(−4, 3), B(6,−5) (1,−1)
4. A(7,−6), B(−1, 2) (3,−2) 5. A(2, 9), B(−8,−1) (−3, 4) 6. A(−1, 0), B(−9,−4)
7. A(7, 5), B(0, 0) $(3\frac{1}{2}, 2\frac{1}{2})$ 8. A(2, 5), B(1, 2) $(1\frac{1}{2}, 3\frac{1}{2})$ 9. A(−1, 3), B(6,−2) $(2\frac{1}{2}, \frac{1}{2})$
10. A(3,−4), B(−8, 1)   11. A(2,−7), B(−9, 1)   12. A(−5,−4), B(−1,−9)

M is the midpoint of $\overline{AB}$. Find the coordinates of B.
6. (−5, −2)

13. A(0, 0), M(3, 3) (6, 6)   14. A(1, 2), M(3, 5) (5, 8)   10. $(-2\frac{1}{2},-1\frac{1}{2})$
15. A(4, 2), M(1,−2) (−2,−6)   16. A(1, 5), M(2,−2) (3,−9)   11. $(-3\frac{1}{2},-3)$
17. A(−3, 3), M(2,−4) (7,−11) 18. A(−3,−1), M(−1,−4) (1,−7) 12. $(-3,-6\frac{1}{2})$

ABCD is a quadrilateral with the given vertices. Using the fact that a quadrilateral is a parallelogram if its diagonals bisect each other, show that ABCD is a parallelogram. 19. Midpt of both: (3,3) 20. Midpt of both: $(1,\frac{1}{2})$

19. A(0, 1), B(6, 1), C(6, 5), D(0, 5)   20. A(−3, 0), B(−1,−3), C(5, 1), D(3, 4)
21. A(−2,−1), B(1,−1), C(4, 4), D(1, 4)   22. A(−3, 1), B(6,−1), C(7, 1), D(−2, 3)
23. △XYZ has vertices X(−3,−5), Y(1, 7), and Z(0, 4). Find the length of median $\overline{ZM}$. $\sqrt{10}$ 21. Midpt of both: $(1,1\frac{1}{2})$ 22. Midpt of both: (2, 1)
24. △RST has vertices R(−5, 0), S(5, 0), and T(0, 6). Find the length of median $\overline{TM}$. 6

ABC is a triangle with given vertices. Find the coordinates of M, the midpoint of $\overline{AB}$, and N, the midpoint of $\overline{AC}$. How does MN relate to BC?

25. A(−3, 6), B(−3, 0), C(1, 4)   26. A(4,−3), B(2, 1), C(−4,−5)

Enrichment

Have students experiment by drawing quadrilaterals within a coordinate system and connecting the midpoints of the sides. The result is always a parallelogram. Encourage students to try to prove the result. The first step is to generalize the naming of the four points, for example:

A(x₁, y₁), B(x₂, y₂), C(x₃, y₃), D(x₄, y₄).
$A(x_1, y_1), B(x_2, y_2), C(x_3, y_3), D(x_4, y_4)$.
Finding the midpoints is straightforward algebra. The midpoints of the diagonals of the new quadrilateral are then found to be identical. In general, the coordinates of the midpoint are

$$\frac{x_1 + x_2 + x_3 + x_4}{4},$$
$$\frac{y_1 + y_2 + y_3 + y_4}{4}$$

Find the coordinates of the midpoint M of $\overline{CD}$, correct to the nearest hundredth.

C 27. $C(7.3, 4.2)$, $D(5.8, -1.9)$
$M(6.55, 1.15)$

C 28. $C(-1.1, 5.6)$, $D(-3.8, -8.7)$
$M(-2.45, -1.55)$

29. Find the coordinates of two points that trisect the segment with endpoints $A(-3, 8)$ and $B(9, -1)$. $(1, 5)$, $(5, 2)$

30. Find the coordinates of the point on the segment with endpoints $A(0, 0)$ and $B(3, 6)$ that is twice as far from A as from B. $(2, 4)$

31. Find the coordinates of the point on the segment with endpoints $P_1(x_1, y_1)$ and $P_2(x_2, y_2)$ that is twice as far from P_1 as from P_2. $\left(\dfrac{2x_2 + x_1}{3}, \dfrac{2y_2 + y_1}{3} \right)$

32. Develop a formula for finding the coordinates of a point on the segment with endpoints $P_1(x_1, y_1)$ and $P_2(x_2, y_2)$ that is n times as far from P_1 as from P_2.

33. Use the formula from Exercise 32 to find the coordinates of a point on the segment with endpoints $A(-6, -3)$ and $B(9, 7)$ that is four times as far from A as from B. $(6, 5)$

32. $x = \dfrac{nx_2 + x_1}{n + 1}$, $y = \dfrac{ny_2 + y_1}{n + 1}$

Mixed Review

1. Draw a line. Through a given point not on the line, construct a line parallel to the given line. **3.3**

2. Draw a line. Through a given point not on the line, construct a line perpendicular to the given line. **5.6**

3. Draw a line. Through a given point on the line, construct a line perpendicular to the given line. **5.6**

Algebra Review

To solve a fractional equation of the form $\dfrac{ax + b}{c} = \dfrac{dx + e}{f}$, where $c \neq 0$, $f \neq 0$:
(1) Multiply each side of the equation by cf.
(2) Solve the resulting equation.

Example   Solve. $\dfrac{4x + 2}{3} = \dfrac{3x - 5}{2}$

$2(4x + 2) = 3(3x - 5)$ ◄—— result of multiplying both sides by $3 \cdot 2$
$8x + 4 = 9x - 15$
$x = 19$

Solve each equation.

1. $\dfrac{x - 2}{2} = \dfrac{x + 1}{3}$ 8

2. $\dfrac{x + 4}{3} = \dfrac{x - 3}{4}$ -25

3. $\dfrac{3x + 1}{4} = \dfrac{2x - 2}{3}$ -11

4. $\dfrac{5x + 4}{4} = \dfrac{4x - 2}{3}$ 20

10.2 The Midpoint Formula   **387**

▰▰▰GETTING STARTED

Prerequisite Quiz

1. Consider the points plotted below. What is the length of the leg labeled "run?" 6 What is the length of the leg labeled "rise?" 6

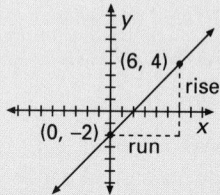

2. Consider the points $A(4,7)$ and $C(6,10)$. What is the change in x from A to C? 2 What is the corresponding change in y? 3

3. What is the slope of the segment $\overline{XY}$ with endpoints $X(4,7)$ and $Y(2,1)$? 3

Motivator

Ask the students if they would rather run up a steep or a not-so-steep hill. Ask them what is meant by "steep." The slope Then ask them what is meant by the pitch of a roof? The slope of the ground? The slant of a board? Ask them how these descriptions are related. They refer to the slope of something.

▰▰▰TEACHING SUGGESTIONS

Lesson Note

In this lesson slope is presented in two ways: first, as the ratio of lengths of legs in a right triangle with the positive or negative direction assigned by inspection; second, using the formula for the change in y over change in x. These two methods reinforce each other and students should always do a quick sketch to check the sign (positive or negative) that they obtain using the formula.

10.3 Slope of a Line

Objectives
To find the slope of a segment or a line
To draw a line given the slope and a point

In the photograph at the right, the roads are steeper at certain places than at others. The steepest parts of a road have the greatest *slope*. In mathematics, slope is the measure of the steepness of a segment or a line.

A segment determines a special right triangle having a horizontal and a vertical leg. The length of the vertical leg, sometimes called the *rise*, is $|y_2 - y_1|$. The length of the horizontal leg, sometimes called the *run*, is $|x_2 - x_1|$. The slope is the ratio of the lengths of these two legs, or the ratio of rise to run. It is necessary, however, to distinguish between a segment which rises from left to right, from one which descends from left to right. If a segment rises from left to right, then it will be assigned a positive slope. If it descends, then it will be assigned a negative slope. The formula in the following definition does this.

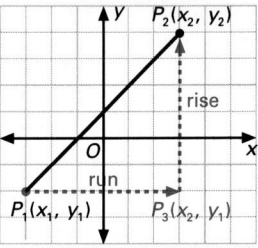

Definition

The **slope** of a segment $\overline{P_1P_2}$, with endpoints $P_1(x_1, y_1)$ and $P_2(x_2, y_2)$, $x_2 \neq x_1$, is the ratio $\dfrac{y_2 - y_1}{x_2 - x_1}$.

EXAMPLE 1 Find the slope m of $\overline{AB}$ for $A(-1,-2)$ and $B(3, 5)$. Draw the graph of $\overline{AB}$ to check.

Solution
$$m = \frac{y_2 - y_1}{x_2 - x_1}$$
$$= \frac{5 - (-2)}{3 - (-1)} = \frac{7}{4}$$

Therefore, the slope of $\overline{AB}$ is $\dfrac{7}{4}$.

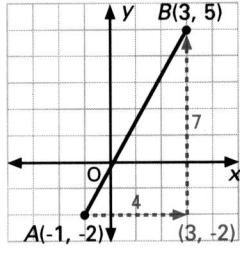

388 Chapter 10 Coordinate Geometry

Additional Example 1

Find the slope of $\overline{XY}$ for $X(-6,5)$ and $Y(3,-8)$. $-\dfrac{13}{9}$

A horizontal segment neither rises nor descends. Since the y-coordinates of all points on a horizontal segment are equal, the slope of a horizontal segment is 0. For example, the slope of $\overleftrightarrow{PQ}$ is $\frac{4-4}{4-1} = 0$. The x-coordinates of a vertical segment are all equal. Therefore, the slope of a vertical segment is undefined, because division by 0 is undefined. The slope of $\overleftrightarrow{MN}$, for example, is $\frac{-1-(-6)}{-2-(-2)} = \frac{5}{0}$, which is undefined.

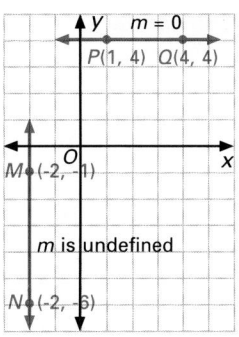

EXAMPLE 2   Is the segment with the given endpoints horizontal, vertical, or neither?
$P_1 = (4, -7)$, $P_2 = (4, 6)$

Solution   The slope is $\frac{-7-6}{4-4}$, or $\frac{-13}{0}$, which is undefined.
Therefore, the segment is vertical.

Theorem 10.3   All segments of a non-vertical line have equal slopes.

Given: $\overline{AB}$ and $\overline{DE}$ on line l
Prove: Slope of $\overline{AB}$ = slope of $\overline{DE}$

Outline of Proof:

1. Assign coordinates as follows:
$A(x_1, y_1)$, $B(x_2, y_2)$, $D(x_3, y_3)$, $E(x_4, y_4)$.
2. Locate $C(x_2, y_1)$ and $F(x_4, y_3)$, and draw $\triangle ABC$ and $\triangle DEF$.
3. $\overline{AC}$ and $\overline{DF}$ are horizontal and therefore parallel.
4. $\angle BAC \cong \angle EDF$ (Corr $\angle$s of $\parallel$ lines are $\cong$.)
5. $\angle C \cong \angle F$, since each is a right angle formed by the horizontal and vertical segments.
6. $\triangle ABC \sim \triangle DEF$ (AA~)
7. $\frac{BC}{AC} = \frac{EF}{DF}$ ($\sim\triangle$s and properties of proportion)
8. If the line rises to the right, slope of $\overline{AB} = \frac{BC}{AC}$, slope of $\overline{DE} = \frac{EF}{DF}$.

   If the line descends to the right, slope of $\overline{AB} = -\frac{BC}{AC}$, slope of $\overline{DE} = -\frac{EF}{DF}$.
9. Slope of $\overline{AB}$ = slope of $\overline{DE}$ (Substitution)

10.3 Slope of a Line   **389**

Math Connections

Construction: Measures of steepness are often expressed as a ratio of the difference in vertical change to the difference in horizontal change. However, these measures of steepness are expressed differently, depending on the object to which it refers. For example, the steepness of a road, called the *gradient*, is expressed as a percent. The *pitch* of a roof is expressed as a ratio, but it is different from the slope ratio. In the diagram below, the slope of $\overline{AB}$ is $\frac{BD}{AD}$, the measure of the rise to the run, but the pitch of the symmetrical roof is $\frac{BD}{AC}$, the measure of the rise to the span.

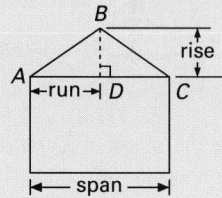

Critical Thinking Questions

Synthesis: Ask students to graph two intersecting lines in the same coordinate plane, one with a slope of 2 and the other with a slope of $-\frac{1}{2}$. In another coordinate plane, ask students to graph two intersecting lines, one with a slope of $\frac{3}{2}$ and the other with a slope of $-\frac{2}{3}$. Ask them to generalize about the relationship between the slopes of a pair of lines and their positions in a coordinate plane. Lines with slopes that are neg reciprocals are $\perp$. In the same coordinate plane, challenge students to plot a line that is perpendicular to $y = 2x + 3$ at any point on the line. Check student constructions.

Common Error Analysis

Error: Students write the slope formula as $\frac{x_2 - x_1}{y_2 - y_1}$.

Have students express slope as $\frac{\text{rise}}{\text{run}}$ or as $\frac{\text{vertical change}}{\text{horizontal change}}$ before writing the formula.

Checkpoint

1. Find the slope of the segment with the given endpoints.
   (a) $B(-3,7)$, $C(4,-5)$ $-\frac{12}{7}$
   (b) $P(0,5)$, $Q(-6,3)$ $\frac{1}{3}$
   (c) $R(2.3,-4.5)$, $S(1.7,5.2)$ $-\frac{97}{6}$
2. Find the slope of the line containing the given points.
   (a) $A(-3,-4)$, $B(4,3)$ 1
   (b) $A(-3,5)$, $B(-3,-4)$ Undefined
   (c) $A(9,2)$, $B(-3,7)$ $-\frac{5}{12}$
3. Given: $X(-2,-7)$, $Y(1,2)$, $Z(0,-1)$. Determine the slope of $\overline{XY}$ and $\overline{XZ}$. Use this information to determine if X, Y, and Z are collinear. $m = 3$, yes

Closure

Ask the students to explain what is meant by the slope of a line.   It is the ratio of the legs of a right triangle.   Ask them what formula they use to find the slope. $m = \frac{y_2 - y_1}{x_2 - x_1}$.   Have them explain the difference between a positive slope and a negative slope.   If a segment rises from left to right, then the slope is positive. If it descends from left to right, then the slope is negative. Have them explain what the slope of a horizontal line is.   Zero   of a vertical line. Undefined

---

You can draw a line if you are given its slope and the coordinates of a point on the line.

EXAMPLE 3   Draw a line with slope $= \frac{1}{2}$, containing the point $(-3,-4)$.

Solution   Graph the point $(-3,-4)$.
From $(-3,-4)$, move up 1 unit.
From that point, move to the right 2 units.
Mark that point.
Draw the line through the two points.

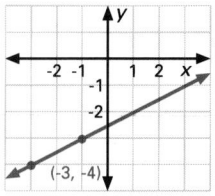

Classroom Exercises

Tell whether the slope of the line is positive, negative, zero, or undefined.

1. 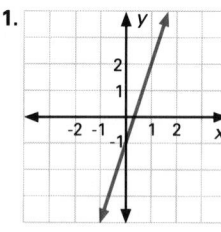 $m = 3$, positive

2. 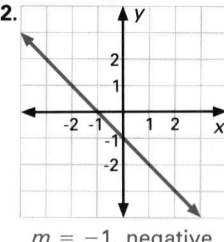 $m = -1$, negative

3. 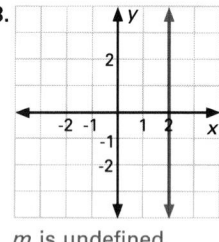 m is undefined.

Is the segment with the given endpoints horizontal, vertical, or neither?

4. $A(0, 1)$, $B(0,-3)$ Vert
5. $C(-3, 5)$, $D(2,-4)$ Neither
6. $E(5,-3)$, $F(-2,-3)$ Horiz
7. $G(5,-1)$, $H(4, 1)$ Neither
8. $I(-2, a)$, $J(5, a)$ Horiz
9. $K(b, 4)$, $L(b,-1)$ Vert

Written Exercises

Find the slope of the segment with the given endpoints.

1. $A(3, 5)$, $B(6, 9)$ $\frac{4}{3}$
2. $C(-3, 5)$, $D(6, 5)$ 0
3. $E(0, 0)$, $F(-3,-5)$ $\frac{5}{3}$
4. $G(-1, 4)$, $H(-1,-4)$
5. $I(3,-3)$, $J(2,-1)$ -2
6. $K(4,-4)$, $L(3,-5)$ 1
7. $M(5,-2)$, $N(-3, 4)$ $-\frac{3}{4}$
8. $O(-6, 1)$, $P(-7, 5)$ -4
9. $Q(-2,-1)$, $R(3,-4)$ $-\frac{3}{5}$

4. Undef

Additional Example 3

Draw a line with slope $\frac{2}{3}$ containing the point $(-2,-1)$.

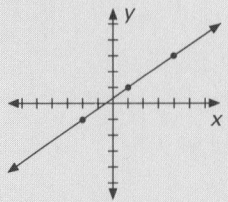

Find the slope of the line containing the given points.

10. $P_1(4, 2), P_2(6, 4)$ 1 **11.** $P_1(9, 2), P_2(3, 0)$ $\frac{1}{3}$ **12.** $P_1(-2, 4), P_2(-8, 2)$ $\frac{1}{3}$

13. $P_1(-4, 5), P_2(-2, 5)$ 0 **14.** $P_1(3, 6), P_2(-3, -3)$ $\frac{3}{2}$ **15.** $P_1(0, 2), P_2(-2, 1)$ $\frac{1}{2}$

Draw a line with the given slope containing the given point.

16. $m = \frac{2}{3}, (1, -3)$ **17.** $m = \frac{1}{2}, (-2, 4)$

18. $m = \frac{-3}{2}, (-3, 1)$ **19.** $m = \frac{-2}{5}, (1, 4)$

Use slopes to show whether the three given points are in the same line.

20. $A(0, 0), B(2, 4), C(6, 12)$ **21.** $D(-3, -2), E(0, 0), F(6, 4)$

22. $G(2, -5), H(-5, -5), I(0, -5)$ **23.** $J(0, 7), K(4, 1), L(6, 5)$

Find the unknown coordinate so that $\overline{PQ}$ will have the given slope.

24. $P(3, 5), Q(x, 1), m = 4$ 2 **25.** $P(4, y), Q(9, 2)$ $m = -5$ 27

26. $P(8, 8), Q(3, y), m = 0$ 8 **27.** $P(6, 9), Q(x, -2),$ m is undefined. 6

28. $P(4, 7), Q(-2, y), m = \frac{2}{3}$ 3 **29.** $P(x, -1), Q(-1, 3), m = \frac{-4}{5}$ 4

30. A carpenter must know the "pitch" or "slope" of a roof when building it. This is the number of inches of vertical rise in 12 in. of horizontal run. A slope of 3-in-12 means a vertical rise of 3 in. in a horizontal run of 12 in. How many inches would the roof rise in a run of 15 ft if it had a slope of 3-in-12? 45 in

31. If a roof rises 3 ft in a run of 15 ft, what is the slope of the roof? $\frac{1}{5}$

32. $\triangle ABC$ has vertices $A(1, -2), B(4, -5),$ and $C(2, 6).$ What is the slope of the median to side $\overline{BC}$? $\frac{5}{4}$

33. The coordinates of the midpoints of the sides of a triangle are $(1, 3), (4, -2),$ and $(5, 1).$ Find the coordinates of the vertices. (2, 6), (8, -4), (0, 0)

34. Discuss the relationship between the slope of a line and the trigonometric ratios of the angle that the line makes with the x-axis. $m = \dfrac{y_2 - y_1}{x_2 - x_1} = \tan A$

Midchapter Review

Name the coordinates and the quadrant for each of the points.

1. A $(-2, -3)$; III **2.** B $(3, -1)$; IV **3.** C $(-3, 3)$; II No

4. Is the triangle formed by points $A, B,$ and C isosceles? **10.1**

5. Find the coordinate of the midpoint of $\overline{AB}$. $(\frac{1}{2}, -2)$

6. Find the slope of $\overleftrightarrow{AB}$. $\frac{2}{5}$

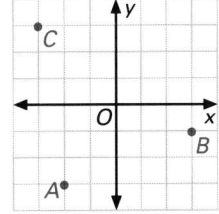

FOLLOW UP

Guided Practice

Classroom Exercises 1–9

Independent Practice

A Ex. 1–19, **B** Ex. 20–31, **C** Ex. 32–34

Basic: WE 1–19 odd, Midchapter Review, CI 1

Average: WE 1–31 odd, Midchapter Review, CI all

Above Average: WE 1–34 odd, Midchapter Review, CI all

Additional Answers

Written Exercises

16. 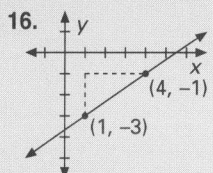 **17.**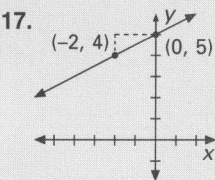

18. 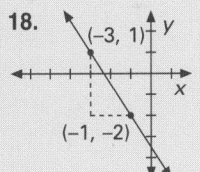 **19.**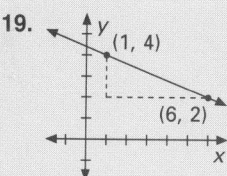

20. Yes; $m_{\overline{AB}} = 2, m_{\overline{BC}} = 2, m_{\overline{AC}} = 2$

21. Yes; $m_{\overline{DE}} = \frac{2}{3}, m_{\overline{EF}} = \frac{2}{3}, m_{\overline{DF}} = \frac{2}{3}$

22. Yes; $m_{\overline{GH}} = 0, m_{\overline{HI}} = 0, m_{\overline{IG}} = 0$

23. No; $m_{\overline{JK}} = -\frac{3}{2}, m_{\overline{KL}} = 2, m_{\overline{LJ}} = -\frac{1}{3}$

Enrichment

Refresh students' memories of the relationships between sides in 30-60-90 and 45-45-90 triangles. Then have them find tan 30, tan 45, and tan 60 in the first quadrant. You can now discuss the meaning of the tangent ratio for angles in the second quadrant. What might be a reasonable definition of tan 120 if it is to have the same meaning as slope? Draw a line at an angle of 120 degrees from the origin and lead students to the recognition that they can assign coordinates based on the 30-60-90 right triangle. This will lead to the discovery that tan 120 = −1.732. Have students find tan 135 and tan 150. Finally, ask what they think tan 90 and tan 180 should equal. Undefined; 0

Computer Investigation

Equation of a Line

Use a computer software program that graphs equations.

Activity 1

Draw the graphs below. If necessary, adjust the boundaries of the grid to accommodate the given lines.

1. Draw the graph of $y = \frac{2}{3}x$. NOTE: Be sure to graph $y = \frac{2}{3} \cdot x$, not $y = \frac{2}{3x}$.

2. Use the graph to find the slope of this line. Start wherever the line crosses the y-axis: $(0, 0)$. Count 3 blocks to the right along the x-axis. Then, count up the number of blocks until you are on the line: 2. The slope of the line is $\frac{\text{rise}}{\text{run}} = \frac{2}{3}$.

3. Draw the graph of $y = 3x + 1$. Use the graph to find the slope of the line. Start wherever the line crosses the y-axis: $(0,1)$. Move 1 block to the right along the x-axis. Then, count up the number of blocks until you are on the line: 3. The slope of this line is $\frac{\text{rise}}{\text{run}} = \frac{3}{1}$, or 3.

4. Predict the slope of the line $y = \frac{4}{5}x - 2$ before graphing it. Then verify your conclusion by graphing the line.

Activity 2

Draw the graph of each equation below on the same set of axes.

5. $y = \frac{3}{4}x$ 6. $y = \frac{3}{4}x + 2$ 7. $y = \frac{3}{4}x + 6$ 8. $y = \frac{3}{4}x - 4$

Use the graphs of Exercises 5–8 above to answer Exercises 9–11 below.

9. What is the slope of each line above? $\frac{3}{4}$

10. How are all the lines related to each other?   Parallel

11. For each line of Exercises 5–8, where does the line cross the y-axis?
    $(0, 0), (0, 2), (0, 6), (0, -4)$

Activity 3

Predict the slope and the point of crossing the y-axis for each line whose equation is given below. Verify by graphing.

12. $y = 4x - 7$ $4, (0, -7)$ 13. $y = \frac{3}{5}x - 8$ $\frac{3}{5}, (0, -8)$ 14. $y = -\frac{1}{5}x + 3$

$-\frac{1}{5}, (0, 3)$

Summary

The graph of an equation of the form $y = mx + b$ has a slope of
_____ and crosses the y-axis at _____. m, b

10.4 Equation of a Line

Objectives   To draw the graph of an equation of a line
To write an equation of a line in three forms

A **graph of an equation** is the set of points whose coordinates satisfy the given equation. An equation whose graph is a line is called a **linear equation**.

All points of a *vertical* line have the same x-coordinate or abscissa. All points with this x-coordinate lie on the given vertical line. The equation $x = a$, where a is any real number, is an equation of the vertical line which intersects the x-axis at $(a, 0)$. Similar reasoning leads to the conclusion that an equation of a *horizontal* line which intersects the y-axis at $(0, b)$ is $y = b$.

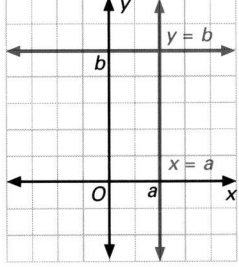

The formula for the slope of a line can be used to obtain an equation of a line through a given point with a given slope. If $P_1(x_1, y_1)$ is a specific point on a line and $P(x, y)$ is any other point on the line, then $m = \dfrac{y - y_1}{x - x_1}$ is an equation of the given line with slope m and containing $P_1(x_1, y_1)$. Using the Multiplication Property of Equality with this equation essentially proves the following theorem.

Theorem 10.4

An equation of a line with slope m containing the point $P_1(x_1, y_1)$ is $y - y_1 = m(x - x_1)$.

This is the **point-slope** form of the equation of a line.

EXAMPLE 1   Write an equation in point-slope form of the line with slope 2 containing the point $(3, -4)$.

Solution   Let $x_1 = 3$, $y_1 = -4$, and $m = 2$.
$$y - (-4) = 2(x - 3)$$

The **standard form** of an equation of a line is an equation of the form $ax + by = c$, where a and b are not both zero, and where $a \geq 0$. (In some texts $ax + by + c = 0$ is considered to be the standard form.)

10.4 Equation of a Line   **393**

Lesson Note

Discuss the relationship between the three forms of the equations of a line given in this lesson. That is, if two points are given, you can determine the slope and write an equation of the line either in *point-slope form* or in *standard form*. Or, if you know the slope and *y*-intercept of a line, you can write its equation in *slope-intercept form*. Be sure that students know how to read the slope and *y*-intercept from this third form of a linear equation. Emphasize how useful this is in graphing (see Example 3). Point out that when the *x*- and *y*-intercepts of the graph of a line are known, students can use these points to find the slope of a line and to write its equation in any of the three forms.

Point out that these methods are based on the definition of slope. Be sure that the students know the names of the methods. These will be standard terminology in future courses. Review the slopes of vertical and horizontal lines.

Remind students to note the form of the equation requested for each problem and not to simplify beyond that form.

Math Connections

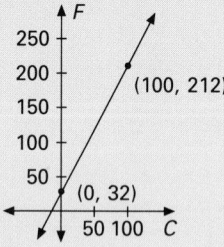

Meteorology: The formula that is used to convert temperature measurement from Celsius to Fahrenheit is $F = \frac{9}{5}C + 32$. From the graph of this equation, you can see that this is a linear relationship expressed in slope-intercept form; that is, the *y*-intercept is 32 and the slope is $\frac{9}{5}$.

EXAMPLE 2 Write an equation in standard form of the line containing points $(2, 1)$ and $(4, -4)$.

Plan Find the slope. Use the point-slope form of the equation of a line.

Solution
$$m = \frac{-4 - 1}{4 - 2} = \frac{-5}{2}$$

$$(y - 1) = \frac{-5}{2}(x - 2) \quad or \quad [y - (-4)] = \frac{-5}{2}(x - 4)$$
$$2(y - 1) = -5(x - 2) \qquad\qquad 2(y + 4) = -5(x - 4)$$
$$2y - 2 = -5x + 10 \qquad\qquad 2y + 8 = -5x + 20$$
$$5x + 2y = 12 \qquad\qquad 5x + 2y = 12$$

The **y-intercept** of a line is the *y*-coordinate of the point at which the line intersects the *y*-axis. If the *y*-intercept is 6, for example, then it intersects the *y*-axis at the point $(0, 6)$. In general, if the *y*-intercept is *b*, then the line crosses the *y*-axis at $(0, b)$.

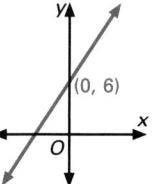

If the slope *m* and the *y*-intercept *b* of a line are known, then the equation is $(y - b) = m(x - 0)$, which simplifies to $y = mx + b$. This proves the following theorem.

Theorem 10.5 If a line has slope *m* and *y*-intercept *b*, then an equation of the line is $y = mx + b$.

$y = mx + b$ is the **slope-intercept** form of the equation of a line.

EXAMPLE 3 Draw the graph of $y = 3x - 2$.

Plan The slope *m* is 3. The *y*-intercept *b* is -2.

Solution Plot the point $(0, -2)$. From that point, go up three units and to the right one unit. Draw a point at $(1, 1)$. Draw a line through the two points.

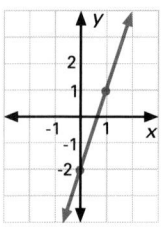

EXAMPLE 4 Write an equation in slope-intercept form of the line pictured below.

Solution The *y*-intercept *b* is -3. Since the line passes through $(0, -3)$ and $(4, 0)$, the slope *m* is
$$\frac{0 - (-3)}{4 - 0} = \frac{3}{4}.$$
Therefore, $y = \frac{3}{4}x - 3$ is the equation of the line.

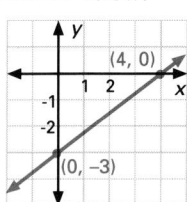

Additional Example 2

Write an equation in standard form of the line containing the points $(3,5)$ and $(-1,2)$. $3x - 4y = -11$

Additional Example 3

Draw the graph of $y = -2x - 4$.

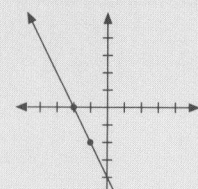

Classroom Exercises

Find the equation of the line with the given slope and *y*-intercept.

1. $m = 5, b = 2$
$y = 5x + 2$

2. $m = -1, b = 4$
$y = -x + 4$

3. $m = \frac{4}{3}, b = -2$
$y = \frac{4}{3}x - 2$

Find the slope and *y*-intercept of the line. Then draw the graph.

4. $y = 4x + 3$

5. $y = \frac{-2}{3}x - 2$

6. $y = -x + 1$

Written Exercises

What is an equation of a horizontal line that contains the given point?

1. $(3, 4)$ $y = 4$

2. $(-4, 3)$ $y = 3$

3. $(5, -2)$ $y = -2$

What is an equation of a vertical line that contains the given point?

4. $(3, 4)$ $x = 3$

5. $(-4, 3)$ $x = -4$

6. $(5, -2)$ $x = 5$

Write an equation in slope-intercept form of the line with the given slope and the given *y*-intercept.

7. $m = -3, b = 2$

8. $m = \frac{5}{4}, b = -6$

9. $m = \frac{1}{3}, b = -\frac{4}{3}$

10. $m = 0, b = 5$

11. $m = -2, b = 0$

12. $m = 1, b = 1$

13. Write an equation in point-slope form of the line having slope -3 and containing $(1, 2)$. $y - 2 = -3(x - 1)$

14. Write an equation in standard form of the line in Exercise 13. $3x + y = 5$

15. Find the *y*-intercept of the line with equation $2x - y = 3$. -3

Write an equation in point-slope form of the line with the given slope containing the given point.

16. $m = 4, (2, 3)$

17. $m = 3, (-1, 4)$

18. $m = -2, (4, 1)$

19. $m = -1, (-3, -1)$

20. $m = \frac{-2}{5}, (-2, -1)$

21. $m = \frac{-4}{3}, (4, 0)$

Write an equation in standard form of the line containing the given points.

22. $(2, 0), (4, 4)$ $2x - y = 4$

23. $(1, 3), (5, 2)$ $x + 4y = 13$

24. $(-1, 3), (5, -2)$ $5x + 6y = 13$

25. $(-2, 0), (0, -5)$ $5x + 2y = -10$

26. $(-2, -7), (3, -1)$ $6x - 5y = 23$

27. $(4, -1), (-6, -2)$ $x - 10y = 14$

Draw a graph of each equation.

28. $y = 2x + 1$

29. $y = 3x - 2$

30. $y = -4x - 4$

31. $y = -x - 2$

32. $y = \frac{-3}{2}x + 2$

33. $y = \frac{-3}{4}x - 3$

10.4 Equation of a Line   **395**

Critical Thinking Questions

Application: pH is a symbol for the degree of acidity or alkalinity of a solution. The pH level in Lake Tiki is measured every 2 years. The data is shown in the chart below. If the decline in the pH level is constant, what will the pH level be in the year 2036?
6.855

Year	pH
1985	7.62
1987	7.59
1989	7.56

Checkpoint

1. Given the point P(3,7):
   (a) What is the equation of a vertical line containing the point? $x = 3$
   (b) What is the equation of a horizontal line containing the point? $y = 7$
2. Write an equation in point-slope form of the line with the given slope containing the given point. $m = -\frac{5}{3}, (-3, -1)$.
   $y = -\frac{5}{3}x - 6$
3. Write an equation in standard form of the line containing the points $(-2, 4)$ and $(3, -1)$. $x + y = 2$
4. Draw the graph of $y = 3x - 5$.
   Check student drawings.

Additional Example 4

Write an equation in slope-intercept form of the line pictured below. $y = x - 3$

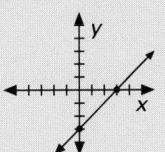

Closure

Ask the students to give the slope of the equation, $y = 3x - 2$.   **3**   Then ask them to give the y-intercept.   **−2**   Have them explain what each of these numbers describes about the graph of the line with the given equation.   The slope of the line describes the ratio of the vertical change to the horizontal change of the line. The y-intercept describes where the line intersects the y-axis.   Ask the students to give the three general forms of the equation for a line.
$y - y_1, = m(x - x_1)$; $ax + by = c$, and $y = mx + b$.

◼◼◼FOLLOW UP

Guided Practice

Classroom Exercises 1–6

Independent Practice

Ⓐ Ex. 1–21, Ⓑ Ex. 22–39, Ⓒ Ex. 40–44

Basic: WE 1–21 odd

Average: WE 1–39 odd

Above Average: WE 1–44 odd

Additional Answers

Classroom Exercises

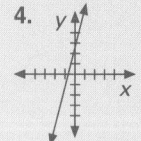 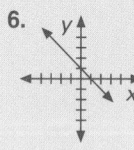

$m = 4,$ $m = -\frac{2}{3},$ $m = -1,$
$b = 3$ $b = -2$ $b = 1$

Written Exercises

7. $y = -3x + 2$

8. $y = \frac{5}{4}x - 6$

9. $y = \frac{1}{3}x - \frac{4}{3}$

10. $y = 0x + 5$

11. $y = -2x + 0$

12. $y = x + 1$

16. $y - 3 = 4(x - 2)$

17. $y - 4 = 3(x + 1)$

18. $y - 1 = -2(x - 4)$

19. $y + 1 = -1(x + 3)$

20. $y + 1 = -\frac{2}{5}(x + 2)$

See page 401 for the answers to Written Ex. 21, 28–33 and Mixed Review Ex. 1–6.

Write an equation for each line.

34.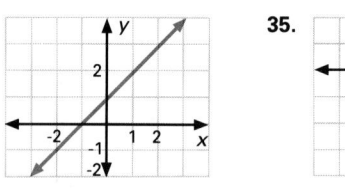

$y = x + 1$

35.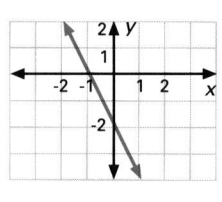

$y = -2x - 2$

36.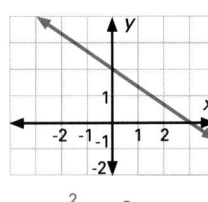

$y = -\frac{2}{3}x + 2$

37.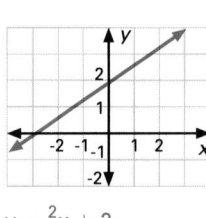

$y = \frac{2}{3}x + 2$

38.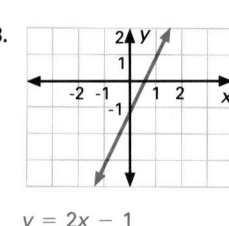

$y = 2x - 1$

39.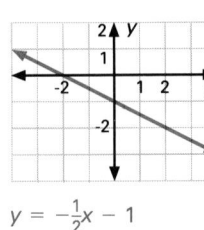

$y = -\frac{1}{2}x - 1$

40. What are the possible slopes of lines that make an angle of measure 30 with the x-axis? $m = \pm\frac{\sqrt{3}}{3}$

41. What are the possible slopes of lines that make an angle of measure 45 with the x-axis? $m = \pm 1$

42. What are the possible slopes of lines that make an angle of measure 60 with the x-axis? $m = \pm\sqrt{3}$

43. Write an equation of a line that makes an angle of measure 30 with the x-axis and which contains the point $(2, 1)$.

44. Write an equation of a line that makes an angle of measure 45 with the x-axis and which contains the point $(-3, 4)$. $y - 4 = 1(x + 3)$

43. $y - 1 = \frac{1}{\sqrt{3}}(x - 2)$ or $y - 1 = \frac{\sqrt{3}}{3}(x - 2)$

Mixed Review

1. Write the converse of "If two angles are congruent, then they are equal in measure."   **3.5**

2. Write the converse of "All right angles are congruent."   **3.5**

3. Write the inverse of "If two lines are parallel, then they are coplanar."   **3.8**

4. Write the inverse of "Two congruent triangles have equal areas."   **3.8**

5. Write the contrapositive of "If a triangle is scalene, then its sides are of different lengths."   **3.8**

6. Write the contrapositive of "An obtuse triangle has an obtuse angle."   **3.8**

Enrichment

Tell students that the graph at the right represents a bicycle trip. Each unit on the x-axis represents an hour; each unit on the y-axis represents five miles. Ask the following: How long did the entire trip take?   5 hours.   What was the average speed for the first two hours?   5 mi/h   What happened during the third hour?   The rider stopped.   What was the average speed for the entire trip?   4 mi/h

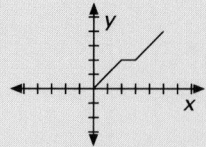

10.5 Parallel or Perpendicular Lines

Objectives
To determine if lines are parallel or perpendicular
To find the slope of a line parallel or perpendicular to a given line
To write an equation of a line parallel or perpendicular to a given line

The roofs of the buildings on the left have the same pitch or slope. It appears that the corresponding lines on the roofs are parallel.

In the figure, perpendicular segments have been constructed from points on lines l_1 and l_2 to the x-axis. If $\triangle ABP \sim \triangle CDQ$, then $\frac{PB}{AB} = \frac{QD}{CD}$. These ratios are the slopes of the lines l_1 and l_2. Since $\triangle ABP \sim \triangle CDQ$, $\angle PAB \cong \angle QCD$, and $\overline{AP} \parallel \overline{CQ}$. This suggests the following two theorems, which are given without proof.

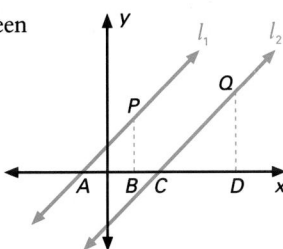

Theorem 10.6

If two non-vertical lines have equal slopes, then they are parallel.

Theorem 10.7

If two non-vertical lines are parallel, then they have equal slopes.

Theorems 10.6 and 10.7 are converses of each other. Remember that the proof of a theorem does not guarantee that its converse is true. The combination of Theorems 10.6 and 10.7 allows us to state that parallelism and equal slopes are necessary and sufficient conditions for each other.

Notice in Example 1 that it is not necessary to graph the lines before determining that they are parallel.

Additional Example 1

Do the following equations describe parallel lines? $3x - 2y = 0$ and $6x - 4y = 12$.   Yes

Additional Example 2

Write an equation in slope-intercept form of the line containing $P(-6,2)$ parallel to the line with an equation of $y = \frac{1}{2}x + 5$.
$y = \frac{1}{2}x + 5$

▰▰▰GETTING STARTED

Prerequisite Quiz

1. What is the slope of the line $y = 5x + 7$?   5
2. What is the slope of the line containing the points
   (a) $(-2,5)$ and $(7,2)$ $-\frac{1}{3}$
   (b) $(4,6)$ $(4,-2)$ Undef
   (c) $(9,1)$, $(3,1)$ 0
3. Write an equation of a line with a slope of 3 and containing the point $(4,1)$.
   $y = 3x - 11$

Motivator

Ask the students to define parallel lines. Coplanar lines that never intersect each other.   Ask them how the equation of a line represents the direction of the line.   The sign of the slope   Have the students guess what relationship exists between the slopes of non-intersecting lines.   Their slopes are equal.

▰▰▰TEACHING SUGGESTIONS

Lesson Note

As a memory aid, tell students that parallel lines are always the same distance apart and have the same slope. After the conditions for parallel and perpendicular lines are determined, the methods of Lesson 10.4 are applied to write the equations. Ask students to give the slope of vertical lines. The slope is undefined. They are still parallel because they are all perpendicular to the x-axis. But the formulas cannot be used in writing their equations because they involve division by 0 which is undefined.

Math Connections

Computer Programming: To draw a line in BASIC, you use the *Line* statement. The *Line* statement asks you to specify two points on a coordinate plane between which the computer will draw a line. If you wish to draw a line parallel to your first line, you can use the slope of the line to determine the plotting points of the second line.

Critical Thinking Questions

Application: In a coordinate plane, an equation of the line containing one side of a rectangle is $y = \frac{5}{3}x - 11$. If the adjacent side contains the point $(-2, -1)$, what is an equation for the adjacent side of the rectangle? $y + 1 = -\frac{3}{5}(x + 2)$; $y = -\frac{3}{5}x - \frac{11}{5}$

Checkpoint

1. Indicate whether the following pairs of lines are parallel, perpendicular or neither.
   (a) $2x - 4y = -3$, $8y - 4x = 8$
   Parallel
   (b) $2x - 3y = -3$, $3x - 2y = 5$ Neither
2. Given the line $y = 4x - 5$ and the point $(3, -6)$:
   (a) Write an equation in slope-intercept form of the line containing the given point and parallel to the given line.
   $y = 4x - 18$
   (b) Write an equation in slope-intercept form of the line containing the given point and perpendicular to the given line.
   $y = -\frac{1}{4}x - \frac{21}{4}$
3. Determine whether $\overline{AB}$ and $\overline{CD}$ are parallel, perpendicular or neither.
   $A(-3,5)$, $B(1,4)$, $C(-3,-1)$, $D(5,-2)$.
   $m_{\overline{AB}} = -\frac{1}{4}$, $m_{\overline{CD}} = -\frac{1}{8}$; Neither

---

EXAMPLE 1   Do the given equations describe parallel lines?

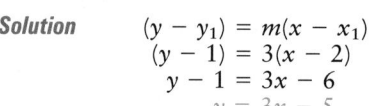
$$2x - 3y = -12 \qquad 9y = 6x - 9$$

Plan   Rewrite the equations in slope-intercept form. Then compare the slopes.

Solution
$$2x - 3y = -12$$
$$-3y = -2x - 12$$
$$y = \frac{2}{3}x + 4 \longrightarrow m = \frac{2}{3}$$

$$9y = 6x - 9$$
$$y = \frac{2}{3}x - 1 \longrightarrow m = \frac{2}{3}$$

Since the slopes are equal, the lines are parallel.

EXAMPLE 2   Write an equation in slope-intercept form of a line containing $P(2, 1)$ parallel to the line whose equation is $y = 3x - 1$.

Plan   Since the parallel lines have the same slope, use the slope of 3 and the point $(2, 1)$ in the point-slope form.

Solution
$$(y - y_1) = m(x - x_1)$$
$$(y - 1) = 3(x - 2)$$
$$y - 1 = 3x - 6$$
$$y = 3x - 5$$

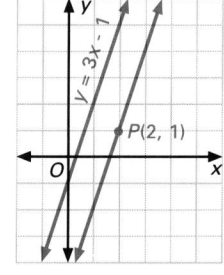

EXAMPLE 3   Determine whether $ABCD$ with vertices $A(-2, 2)$, $B(3,-1)$, $C(0,-6)$, and $D(-5,-3)$ is a parallelogram.

Plan   Show that opposite sides are parallel.

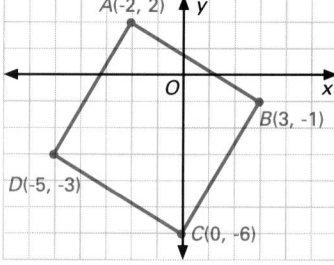

Solution

Slope of $\overline{AB} = \dfrac{2 - (-1)}{-2 - 3}$
$$= \frac{3}{-5} = -\frac{3}{5}$$

Slope of $\overline{CD} = \dfrac{-3 - (-6)}{-5 - 0}$
$$= \frac{3}{-5} = -\frac{3}{5}$$

Slope of $\overline{CB} = \dfrac{-6 - (-1)}{0 - 3}$
$$= \frac{-5}{-3} = \frac{5}{3}$$

Slope of $\overline{DA} = \dfrac{-3 - 2}{-5 - (-2)}$
$$= \frac{-5}{-3} = \frac{5}{3}$$

Since their slopes are equal, $\overline{AB} \parallel \overline{CD}$ and $\overline{CB} \parallel \overline{DA}$.
Therefore, $ABCD$ is a parallelogram.

Additional Example 3

Determine whether $ABCD$ with vertices $A(-3,1)$, $B(2,-2)$, $C(4,1)$, $D(-1,4)$ is a parallelogram.   Yes

Additional Example 4

If $A(-1,5)$, $B(1,1)$, $C(4,2)$, $D(2,-2)$, determine whether the lines $\overleftrightarrow{AB}$ and $\overleftrightarrow{CD}$ are parallel, perpendicular, or neither.
slope of $\overline{AB}$: -2; slope of $\overline{CD}$: 2; Neither

The relationship between perpendicularity and slope is not as obvious as that between parallelism and slope. Two perpendicular lines are shown in the diagram. The equation of one of the lines is $y = \frac{3}{4}x - 2$. The equation of the other is $y = -\frac{4}{3}x + 3$. Notice that the slopes of the lines are negative reciprocals of each other. In other words, the product $(\frac{3}{4})(-\frac{4}{3})$ of the two slopes is -1. This property is true for all non-vertical perpendicular lines.

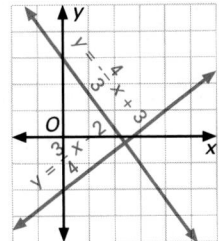

Theorem 10.8 | If the product of the slopes of two non-vertical lines is -1, then the lines are perpendicular.

EXAMPLE 4 For the points $A(0, 0)$, $B(5, 5)$, $C(-1, 1)$, and $D(4, 6)$ determine whether $\overleftrightarrow{AB}$ and $\overleftrightarrow{CD}$ are parallel, perpendicular, or neither of these.

Solution Slope of $\overleftrightarrow{AB} = \dfrac{5 - 0}{5 - 0} = \dfrac{5}{5} = 1$ Slope of $\overleftrightarrow{CD} = \dfrac{6 - 1}{4 - (-1)} = \dfrac{5}{5} = 1$
Since the slopes are equal, $\overleftrightarrow{AB} \parallel \overleftrightarrow{CD}$.

EXAMPLE 5 Write an equation in slope-intercept form of a line containing $(-2, 0)$, perpendicular to the line whose equation is $y = -\frac{2}{3}x + 3$.

Solution
$$mn = -1 \quad \longleftarrow \text{Let } m \text{ and } n \text{ be the slopes.}$$
$$-\frac{2}{3}n = -1 \quad \longleftarrow m = \text{slope of given line}$$
$$n = \frac{3}{2}$$
$$(y - 0) = \frac{3}{2}[x - (-2)] \quad \longleftarrow \text{line containing } (-2, 0)$$
$$y = \frac{3}{2}x + 2$$

Theorem 10.9 | The product of the slopes of two non-vertical perpendicular lines is -1.

EXAMPLE 6 Determine whether $ABCD$ from Example 3 is a rectangle.

Solution $ABCD$ is a parallelogram as shown in the solution of Example 3.
Since slope of $\overline{AB} \cdot$ slope of $\overline{BC} = -\frac{3}{5} \cdot \frac{5}{3} = -1$, $\overline{AB} \perp \overline{BC}$.
Therefore, $ABCD$ is a rectangle.

10.5 Parallel or Perpendicular Lines **399**

Additional Example 5

Write an equation in slope-intercept form of the line containing (4,1), and perpendicular to the line whose equation is $y = \frac{1}{3}x - 4$. $y = -3x + 13$

Additional Example 6

Determine whether $ABCD$ in Additional Example 3 is a rectangle. No

Guided Practice

Classroom Exercises 1–12

Independent Practice

🅰 Ex. 1–26, 🅱 Ex. 27–34, 🅲 Ex. 35–37

Basic: WE 1–26 odd, Application
Average: WE 1–34 odd, Application
Above Average: WE 1–37 odd, Application

Classroom Exercises

Indicate whether the pair of equations describes parallel lines, perpendicular lines, or neither.

1. $y = 3x + 2$ and $y = -3x - 2$ Neither
2. $y = 2x + 1$ and $y = 2x - 1$ ‖
3. $y = 4x + 5$ and $y = 4x - 3$ ‖
4. $y = 2x + 1$ and $y = -\frac{1}{2}x - 1$ ⊥
5. $y = -x + 1$ and $y = x - 1$ ⊥
6. $y = \frac{2}{3}x - 5$ and $y = \frac{2}{3}x - 1$ parallel
7. $y = 6$ and $x = 4$ ⊥
8. $y = 7$ and $y = -7$ ‖

Give the slope of a line parallel to the line with the given equation. Then give the slope of a line perpendicular to the line with the given equation.

9. $y = 2x + 2$
$2; -\frac{1}{2}$

10. $y = -3x - 2$
$-3; \frac{1}{3}$

11. $y = \frac{1}{2}x - 3$
$\frac{1}{2}; -2$

12. $y = \frac{-2}{3}x + 5$
$-\frac{2}{3}; \frac{3}{2}$

Written Exercises

Indicate whether the pair of equations describes parallel lines, perpendicular lines, or neither.

1. $2x + 3y = 7$, $3x + 2y = 9$ Neither
2. $3x - 4y = -2$, $3x + 4y = 6$ Neither
3. $-4x + 3y = 6$, $-12x + 9y = 9$ ‖
4. $2x + 5y = -4$, $-5x + 2y = 3$ ⊥

Write an equation in slope-intercept form of the line containing the given point and parallel to the line whose equation is given.

5. $y = 2x + 1$, $(0, 0)$ $y = 2x$
6. $y = -3x - 5$, $(1, 1)$ $y = -3x + 4$
7. $y = 5x - 2$, $(3, -4)$ $y = 5x - 19$
8. $y = -4x + 6$, $(-2, 1)$ $y = -4x - 7$
9. $y = \frac{3}{4}x$, $(4, -2)$ $y = \frac{3}{4}x - 5$
10. $y = \frac{3}{2}x + 1$, $(-2, -5)$ $y = \frac{3}{2}x - 2$

Write an equation in slope-intercept form of the line containing the given point and perpendicular to the line whose equation is given.

11. $y = 2x$, $(0, 0)$ $y = -\frac{1}{2}x$
12. $y = -3x + 1$, $(1, 1)$ $y = \frac{1}{3}x + \frac{2}{3}$
13. $y = x - 3$, $(-3, 2)$ $y = -x - 1$
14. $y = -x + 7$, $(-1, -1)$ $y = x$
15. $y = \frac{3}{4}x$, $(-1, 3)$ $y = -\frac{4}{3}x + \frac{5}{3}$
16. $y = -\frac{5}{3}x - 2$, $(2, -2)$ $y = \frac{3}{5}x - \frac{16}{5}$

For the given points A, B, C, and D, determine whether $\overleftrightarrow{AB}$ and $\overleftrightarrow{CD}$ are parallel, perpendicular, or neither of these.

17. $A(0, 0)$, $B(3, 3)$, $C(-1, 1)$, $D(2, -2)$ ⊥
18. $A(-3, 2)$, $B(-2, 5)$, $C(-4, -1)$, $D(2, 1)$ Neither
19. $A(3, 9)$, $B(-2, 4)$, $C(-1, 0)$, $D(4, 5)$ ‖
20. $A(6, -1)$, $B(-2, -2)$, $C(-6, 3)$, $D(0, 4)$ Neither

Determine whether or not $ABCD$ is a parallelogram.

21. $A(-3, 2)$, $B(-1, 4)$, $C(0, -3)$, $D(-2, -5)$ Yes
22. $A(2, -7)$, $B(7, -2)$, $C(6, 1)$, $D(1, -3)$ No

Enrichment

Use the vector approach shown in the Application on page 402 to determine whether $ABCD$ with vertices $A(3, 2)$, $B(-2, -1)$, $C(1, -6)$, and $D(6, -3)$ is a parallelogram.   *ABCD is a parallelogram.*

Use the vector approach to determine if ▱$ABCD$ is a rectangle.   ▱$ABCD$ is a rectangle.

Determine whether or not *ABCD* is a rectangle.

23. $A(-6,-3)$, $B(-1, 2)$, $C(1, 0)$, $D(-4,-5)$ Yes

24. $A(-1,-4)$, $B(-7, 0)$, $C(-9,-3)$, $D(-3,-7)$ Yes

Determine whether or not the diagonals of *ABCD* are perpendicular.

25. $A(0, 5)$, $B(4, 2)$, $C(0,-2)$, $D(-4,-2)$ No

26. $A(5, 4)$, $B(-1, 3)$, $C(0,-3)$, $D(6,-2)$ Yes

27. The slope of a line is 3. A line parallel to this given line contains the point $(-2, 5)$. The abscissa of a second point on this line is 0. Find the ordinate of this point.   11

28. The slope of a line is $-\frac{3}{4}$. A line perpendicular to this line contains the point $(2,-7)$. The ordinate of a second point on this line is 1. Find the abscissa of this point.   8

29. The slope of a line is -2. A line perpendicular to this line contains the point $(-1,-3)$. The abscissa of a second point on this line is 7. Find the ordinate of this point.   1

30. The slope of a line is $\frac{4}{5}$. A line perpendicular to this line contains the point $(5, 2)$. The ordinate of a second point on this line is -3. Find the abscissa of this point.   9

Classify the quadrilateral in all appropriate categories: rhombus, trapezoid, parallelogram, rectangle, square, and "kite". A kite has exactly one of its diagonals as a perpendicular bisector of the other.

31. $A(-2, 1)$, $B(-1, 4)$, $C(2, 3)$, $D(3,-4)$ Kite

32. $E(-1, 0)$, $F(2, 3)$, $G(6, 1)$, $H(3,-2)$ Parallelogram

33. $M(-4, 0)$, $N(-1, 2)$, $O(1,-1)$, $P(-2,-3)$ Parallelogram, Rectangle, Square

34. $R(-1, 1)$, $S(2, 2)$, $T(6, 1)$, $V(-3, 2)$ Trapezoid

Triangle *TRI* has vertices $T(3, 2)$, $R(-3,-1)$, and $I(1,-4)$.

35. Find an equation of the line containing the median from T to $\overline{RI}$.

36. Find an equation of the line containing the altitude from T to $\overline{RI}$.

37. Find an equation of the line containing the midsegment parallel to $\overline{RI}$. $3x + 4y = 2$ **35.** $9x - 8y = 11$ **36.** $4x - 3y = 6$

Mixed Review

1. Isos trap, rect, square, $\square$
3. Rect, rhomb, square, $\square$

Which property is a property of a trapezoid, an isosceles trapezoid, a parallelogram, a rectangle, a rhombus, or a square?   **6.5, 6.6**

1. Diagonals are congruent.

2. Diagonals are perpendicular.   Rhomb, square

3. Opposite angles are congruent.

4. Opposite sides are congruent. $\square$, rect, rhomb, square

5. At least one pair of consecutive angles is congruent.   Isos trap, rect, square

6. Diagonals bisect each other. $\square$, rect, rhomb, square

21. $y - 0 = -\frac{4}{3}(x - 4)$

28. 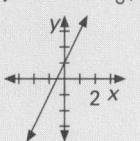   **29.**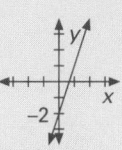

30. 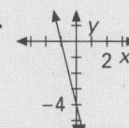   **31.**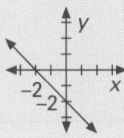

32. 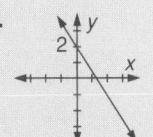   **33.**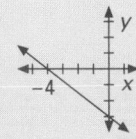

Mixed Review

1. If two angles are equal in measure, then they are congruent.

2. If angles are congruent, then they are right angles.

3. If two lines are not parallel, then they are not coplanar.

4. If two triangles are not congruent, then they do not have equal areas.

5. If the sides of a triangle do not have different lengths, then it is not scalene.

6. If a triangle does not have an obtuse angle, then it is not an obtuse triangle.

Application: *Vector Methods*

Anything that has both *magnitude* and *direction* can be represented by a vector. Vectors are drawn as arrows, because arrows have both magnitude (length) and direction.

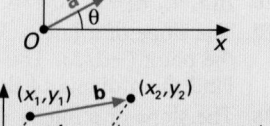

Vector **a** is in **standard position**, since it begins at the origin. Only one distinct vector can be drawn from the origin to (x, y), so we may refer to **a** simply as (x, y).

Any **free vector** such as **b**, with its **tail** at (x_1, y_1) and its **head** at (x_2, y_2), can be standardized: Subtract $(x_2, y_2) - (x_1, y_1) = (x_2 - x_1, y_2 - y_1)$ and relocate as shown. Vector **b'** has the same length and direction as **b**. (Why?)

We can multiply a vector in standard position by a real number c: $c(x, y) = (cx, cy)$. When c does not equal 0 or 1, the result is a new vector with a different length, but the same *or opposite* direction. Vectors that have the same or opposite direction, even collinear vectors, are called **parallel**.

We can use vector methods to determine if line segments are perpendicular or parallel. In the following examples we will think of line segments as vectors. (It won't matter which endpoints we call heads or tails.)

EXAMPLE 1   Given segment $\overline{AC}$ with endpoints $(-5, 1)$ and $(-1, 3)$ and segment $\overline{BD}$ with endpoints $(2, 1)$ and $(8, 4)$, determine whether the two are parallel.

Plan   Regard $\overline{AC}$ and $\overline{BD}$ as free vectors. Standardize them as vectors **a** and **b**. If **a** ∥ **b**, then $\overline{AC}$ ∥ $\overline{BD}$. Vectors **a** and **b** are parallel if **a** times some real number c equals **b**: $c\mathbf{a} = \mathbf{b}$.

Solution $\mathbf{a} = (-5, 1) - (-1, 3) = (-4, -2)$ $\mathbf{b} = (8, 4) - (2, 1) = (6, 3)$
$-1.5(-4, -2) = (6, 3)$, so **a** ∥ **b**. We conclude that $\overline{AC}$ ∥ $\overline{BD}$.

EXAMPLE 2   Given segment $\overline{AC}$ with endpoints $(3, 7)$ and $(4, 9)$ and segment $\overline{BD}$ with endpoints $(1, 3)$ and $(-1, 4)$, determine whether the two are perpendicular.

Plan   Regard $\overline{AC}$ and $\overline{BD}$ as free factors. Standardize them as vectors **a** and **b**. If **a** ⊥ **b**, then $\overline{AC}$ ⊥ $\overline{BD}$. Vectors **a** and **b** are perpendicular if their **dot product** $\mathbf{a} \cdot \mathbf{b} = 0$. For vectors $\mathbf{a} = (x_1, y_1)$ and $\mathbf{b} = (x_2, y_2)$, $\mathbf{a} \cdot \mathbf{b}$ is defined as $x_1 x_2 + y_1 y_2$. (Note: this is a real number, not a new vector.)

Solution $\mathbf{a} = (4, 9) - (3, 7) = (1, 2)$ $\mathbf{b} = (-1, 4) - (1, 3) = (-2, 1)$
$\mathbf{a} \cdot \mathbf{b} = (1 \cdot -2 + 2 \cdot 1) = 0$, so **a** ⊥ **b**. We conclude that $\overline{AC}$ ⊥ $\overline{BD}$.

Try using a vector approach to solve problems 17–20, page 400.

402   Application: Vectors

10.6 Proofs with Coordinates

Objectives
To assign coordinates to the vertices of a polygon in a plane
To use coordinate geometry to prove statements

Coordinate geometry can be used to prove many theorems. In all such cases the coordinate proof is not the only proof, but it is often the simpler proof. The following theorem was proved in this book as Theorem 6.10. Compare this coordinate proof to the earlier proof.

Theorem 10.10

Midsegment Theorem: The segment joining the midpoints of two sides of a triangle is parallel to the third side, and its length is half the length of the third side.

Given: $\triangle RST$, with
$R(0, 0)$, $S(2a, 0)$,
and $T(2b, 2c)$,
M is the midpoint
of $\overline{RT}$; N is the
midpoint of $\overline{ST}$.
Prove: $\overline{MN} \parallel \overline{RS}$, $MN = \frac{1}{2}(RS)$

Outline of Proof:

1. By the Midpoint Formula, the coordinates of M are (b, c) and of N are $(a + b, c)$.
2. The slope of $\overline{MN}$ = the slope of $\overline{RS}$, which is zero ($\overline{MN} \parallel \overline{RS}$).
3. $MN = |a + b - b| = a$
4. $RS = 2a$
5. $MN = \frac{1}{2}(RS)$

A first step in setting up a coordinate proof involving a polygon is to place the figure in a coordinate plane.
(1) If possible, place the polygon so that a side is on one of the axes with a vertex at the origin.
(2) If a polygon contains a right angle, place the polygon so that two sides lie on the x- and y-axes.

10.6 Proofs with Coordinates **403**

Teaching Resources

Quick Quizzes 74
Reteaching and Practice
  Worksheets, pp. 147, 148

▰▰▰ **GETTING STARTED**

Prerequisite Quiz

1. Given the points $X(-4,9)$ and $Y(6,-1)$:
(a) Find the length of $\overline{XY}$. $10\sqrt{2}$
(b) Find the midpoint of $\overline{XY}$. $(1,4)$
(c) Find the slope of the line $\overline{XY}$. -1
(d) Write an equation in slope-intercept form of the line $\overleftrightarrow{XY}$. $y = -x + 5$
(e) Write an equation in slope-intercept form of the line parallel to $\overline{XY}$ and containing the point $(-2,1)$. $y = -x - 1$
(f) Write an equation in slope-intercept form of the line perpendicular to $\overline{XY}$ and containing the point $(2,1)$. $y = x + 1$

Motivator

Ask the students what it means to "prove" something. It means to verify or show that something is true. Ask them to give different ways to prove geometric theorems. A direct proof and an indirect proof. Ask them how an indirect proof differs from a direct proof. An indirect proof shows that a conclusion cannot possibly be false. A direct proof shows step-by-step that a conclusion is true.

403

Lesson Note

This is another type of proof that is used extensively outside of geometry classes. A key to ease in using coordinates to write a proof is the placing of the figure on the axes. This should be done to keep the coordinates of the crucial points as simple as possible.

Math Connections

Problem Solving: Coordinate geometry is an important part of analytic geometry because it connects algebraic and geometric concepts and methods. The interrelationships between algebra and geometry provide a variety of powerful problem solving techniques.

Critical Thinking Questions

GRAPH 1   GRAPH 2

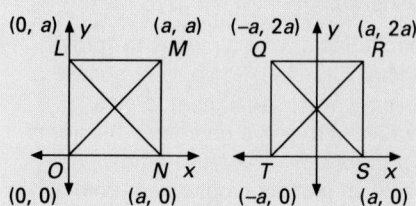

Analysis: Refer students to the diagram above and ask them what needs to be shown for each graph in order to conclude that the diagonals are perpendicular.

GRAPH 1: $\left(\frac{a-0}{0-a}\right)\left(\frac{a-0}{a-0}\right) = -1$

GRAPH 2: $\left(\frac{2a-0}{-a-a}\right)\left(\frac{2a-0}{a-(-a)}\right) = -1$

Discuss with students which placement makes this task easier.   Answers may vary.

(3) If the proof involves the Midpoint Formula, use coordinates such as $(2a, 2b)$ to avoid fractions in the proof.
(4) If the proof involves an isosceles triangle or trapezoid, it is sometimes convenient to place the figure with its axis of symmetry on the y-axis.
(5) If the proof involves perpendicular diagonals of a square or rhombus, it is sometimes convenient to place the figure so that the diagonals coincide with the x-axis and y-axis.

As long as you do not assume more than is given in the hypothesis of a theorem, the placement of the figure does not affect the validity of the proof. It may, however, simplify the computation necessary to complete the proof.

EXAMPLE 1   Which placement of the figure is preferable to prove that the median to the base of an isosceles right triangle is perpendicular to the base? Why?

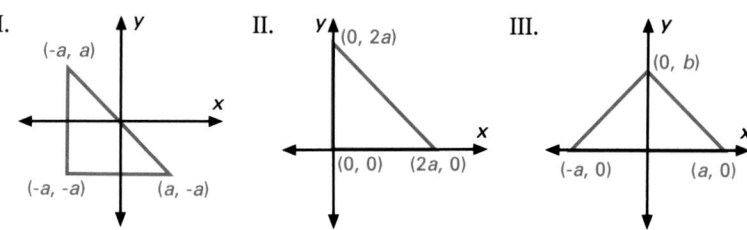

Solution   Placement II. The use of zero coordinates simplifies computation. The use of a coordinate such as $2a$ avoids fractions when midpoints must be found. Placement III is isosceles but not necessarily right.

EXAMPLE 2   What are the missing coordinates in parallelogram $ABCD$ if $AB = 2a$?

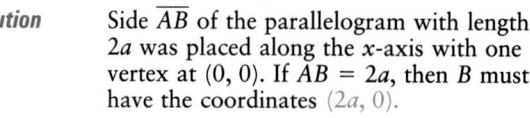

Solution   Side $\overline{AB}$ of the parallelogram with length $2a$ was placed along the x-axis with one vertex at $(0, 0)$. If $AB = 2a$, then B must have the coordinates $(2a, 0)$.

Since opposite sides of a parallelogram are congruent, $DC = 2a$. To show this, if the abscissa of D is $2b$, then the abscissa of C must be $2b + 2a$.

Since opposites sides of a parallelogram are parallel, the side opposite $\overline{AB}$ is horizontal. If the ordinate of D is $2c$, then the ordinate of C is $2c$.

The coordinates of C are $(2a + 2b, 2c)$.

404   Chapter 10 Coordinate Geometry

Additional Example 1

Which placement of the figure of Example 1 is better for proving the Distance Formula?   Right

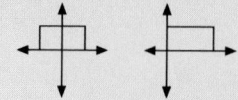

Additional Example 2

Explain why the given coordinates are those of a trapezoid.   Exactly 2 sides ∥

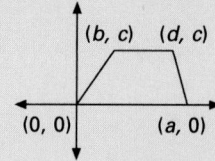

EXAMPLE 3 Using a coordinate proof, prove that the diagonals of a parallelogram bisect each other.

Given: ▱ABCD, with diagonals $\overline{AC}$ and $\overline{BD}$
Prove: $\overline{AC}$ and $\overline{BD}$ bisect each other.

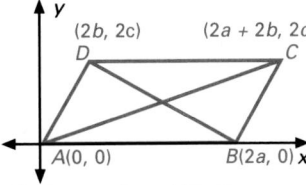

Proof

1. Place ▱ABCD in a coordinate plane with the vertices $A(0, 0)$, $B(2a, 0)$, $C(2a + 2b, 2c)$, and $D(2b, 2c)$.

2. $M_{\overline{AC}}$, the midpoint of $\overline{AC}$, has coordinates $(\frac{0 + 2a + 2b}{2}, \frac{0 + 2c}{2})$, or $(a + b, c)$.

3. $M_{\overline{BD}}$, the midpoint of $\overline{BD}$, has the coordinates $(\frac{2a + 2b}{2}, \frac{0 + 2c}{2})$, or $(a + b, c)$.

4. Since $M_{\overline{AC}} = M_{\overline{BD}}$, $\overline{AC}$ and $\overline{BD}$ have the same midpoint, $(a + b, c)$, and therefore bisect each other.

Classroom Exercises

What are the missing coordinates for the given figures?

1.
(0, 5), (5, 5)

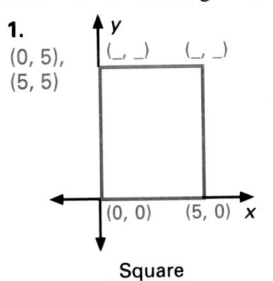

Square

2.
(0, 4), (6, 4)

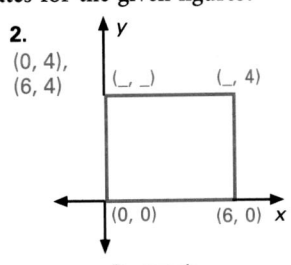

Rectangle

3.
(0, 2a)

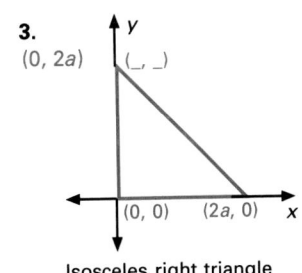

Isosceles right triangle

4.
(2d, 2c)

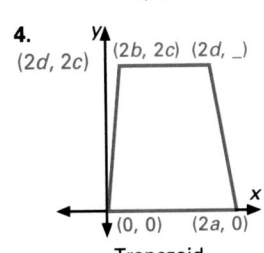

Trapezoid

5.
(−3, 2)

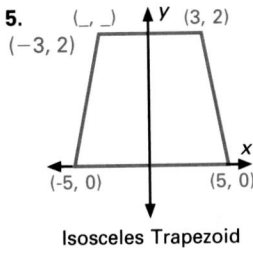

Isosceles Trapezoid

6.
(0, −b), (−b, 0)

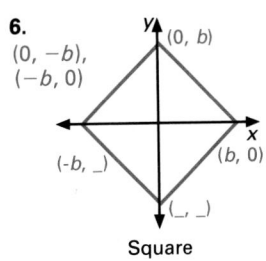

Square

7. Discuss a plan for a proof that the median of a trapezoid is parallel to the base.

10.6 Proofs with Coordinates **405**

Common Error Analysis

Error: Students have difficulty in writing coordinate proofs because of the placement of the required figure on the coordinate plane.

Before assigning Exercises 10–12 and Exercises 15–20 on pages 406–407, have student volunteers make drawings on the chalkboard that show the required figure for each exercise in the coordinate plane. Then have them discuss whether the placement can be changed to make computations easier or whether it is convenient as is.

Checkpoint

1. What are the missing coordinates if the given figure is an isosceles triangle?
(−a,0)

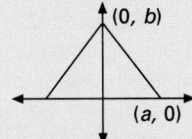

2. Discuss a plan to prove that the medians of an equilateral triangle are congruent. See answer to Written Exercise 21.

Closure

Ask the students how they determine the best set of coordinates to use for a figure in a coordinate proof. Steps 1–5 pages 403–404 Ask them what restriction there is on the choices they make for these coordinates. Do not assume more than is given in the hypothesis.

Additional Example 3

Using a coordinate proof, prove that the diagonals of a rectangle are congruent.
Place rectangle ABCD in the coordinate plane with vertices $A(0, 0)$, $B(a, 0)$, $C(a, b,)$, $D(0, b)$. Then the length of diagonal $\overline{AC}$ is $\sqrt{a^2 + b^2}$ and the length of diagonal $\overline{BD}$ is $\sqrt{a^2 + b^2}$. Therefore, $\overline{AC} \cong \overline{BD}$.

Guided Practice

Classroom Exercises 1–7

Independent Practice

A Ex. 1–12, **B** Ex. 13–20, **C** Ex. 21–25

Basic: WE 1–12, AR all
Average: WE 1–20, AR all
Above Average: WE 1–25 odd, AR all

Additional Answers

Classroom Exercises

7. Draw a trapezoid with its base on the x-axis. The y-coordinates of the endpoints of the base will both be zero. The parallel side opposite the base will have arbitrary x-coordinates for its endpoints, but the y-coordinates for these must be the same. Find the midpoints of the other two sides. If the theorem is true, then the midpoints must have the same y-coordinates.

Written Exercises

4.
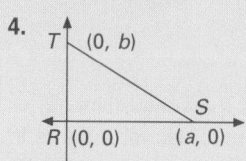

5.
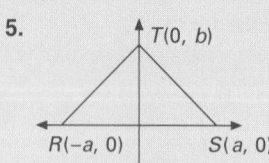

6.
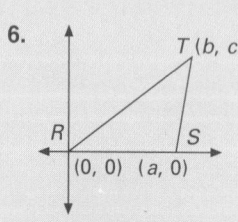

7.
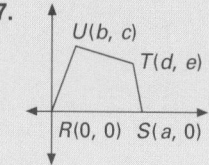

Written Exercises

What are the missing coordinates for the given figures?

1.
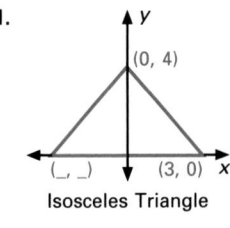
Isosceles Triangle
$(-3, 0)$

2.
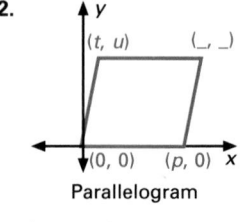
Parallelogram
$(t + p, u)$

3.
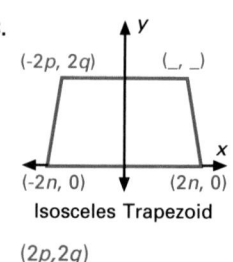
Isosceles Trapezoid
$(2p, 2q)$

Place each given figure in a coordinate plane. What are the coordinates of each vertex? (Exercises 4–7)

4. a right triangle
5. an isosceles triangle
6. a scalene triangle
7. a quadrilateral

8. Explain why the given coordinates are those of a rectangle.

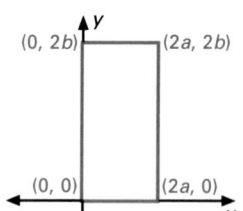

9. Explain why the given coordinates are those of a trapezoid.

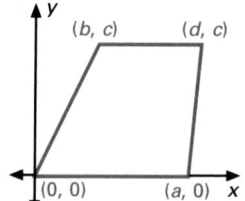

Prove the following statements using a coordinate proof.

10. The diagonals of a rectangle are congruent.
11. The diagonals of a square have the same midpoint.
12. The diagonals of a rectangle bisect each other.

Draw a set of coordinate axes. Place the given figure in the coordinate plane so that the coordinates of the vertices are as simple as possible.

13. a rhombus
14. a kite

Prove the following statements using a coordinate proof.

15. The diagonals of an isosceles trapezoid are congruent.
16. The midpoint of the hypotenuse of a right triangle is equidistant from the vertices.

Enrichment

As a summary activity on coordinate proofs, have students close their books and then state and prove everything they can about an isosceles trapezoid. If they have trouble getting started, suggest that they first draw the trapezoid on the coordinate system, label vertices appropriately using no more than three variables, and then experiment.

See how many of the students are able to list and prove without prompting:

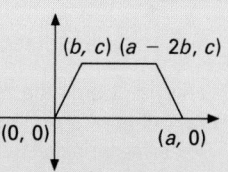

The median is parallel to the bases. The median is half the sum of the bases. The diagonals are congruent.

17. The median of a trapezoid is parallel to the bases.
18. The length of the median of a trapezoid is one-half the sum of the lengths of the two bases.
19. The segments joining the midpoints of the opposite sides of a quadrilateral bisect each other.
20. The segments determined by successive midpoints of any quadrilateral form a parallelogram.

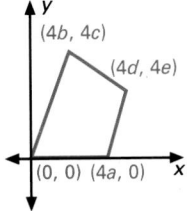

Ex. 19–20

21. The medians of an equilateral triangle are congruent. (See figure at right.)
22. The segments joining the midpoints of the sides of a triangle form another triangle which is similar to the given triangle.
23. The segments joining the midpoints of consecutive sides of a rectangle form a rhombus.
24. If the diagonals of a parallelogram are congruent, then the parallelogram is a rectangle.
25. If the diagonals of a quadrilateral bisect each other, then the quadrilateral is a parallelogram.

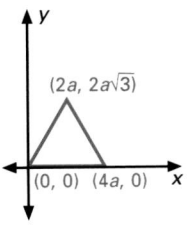

Ex. 21

Mixed Review

1. Find the midpoint of the segment joining the points $A(-1,-3)$ and $B(3, 5)$. **10.2** (1,1)
2. Find the length of the segment given in Exercise 1. **10.1** $4\sqrt{5}$
3. Find the slope and y-intercept of the line whose equation is $2y = 4x - 6$. **10.4** $m = 2, b = -3$
4. What is the slope of a line perpendicular to the line whose equation is given in Exercise 3? **10.5** $m = -\frac{1}{2}$

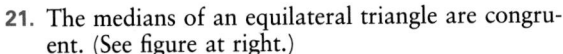 **Algebra Review**

To square a binomial, use $(a + b)^2 = a^2 + 2ab + b^2$.

Example: Simplify $(2x + 3)^2$.
$$(a + b)^2 = a^2 + 2ab + b^2$$
$$(2x + 3)^2 = (2x)^2 + 2(2x)(3) + 3^2$$
$$= 4x^2 + 12x + 9$$

Simplify.

1. $(x + 2)^2$
2. $(x - 3)^2$
3. $(y + 5)^2$
4. $(a - 8)^2$
5. $(2x + 1)^2$
6. $(3y - 2)^2$
7. $(2c - 5)^2$
8. $(4x + 1)^2$
9. $(3a + b)^2$
   $9a^2 + 6ab + b^2$
10. $(x + 3y)^2$
    $x^2 + 6xy + 9y^2$
11. $(2c - d)^2$
    $4c^2 - 4cd + d^2$
12. $(5x - 4y)^2$
    $25x^2 - 40xy + 16y^2$

8. Any 2 slopes are either equal (∥) or produce -1 (⊥).
9. Two of the 4 slopes are equal (2 ∥ sides).
10. $RT = SU = \sqrt{a^2 + b^2}$

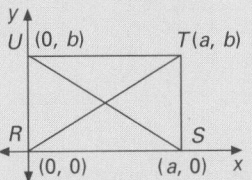

11. Midpt of $\overline{US}$ = midpt of $\overline{TR}$ = (a,a)

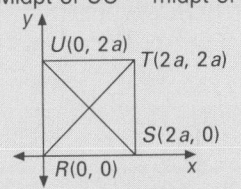

12. $UM = MS = \sqrt{a^2 + b^2}$;
    $RM = MT = \sqrt{a^2 + b^2}$

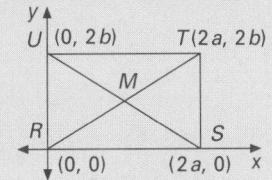

13.

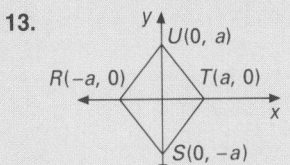

14.

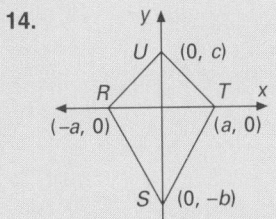

15. $US = TR = \sqrt{(b + a)^2 + c^2}$

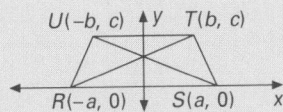

See page 409 for the answers to Ex. 16–21 and page 411 for the answers to Ex. 22–25 and Algebra Review Ex. 1–8.

1.

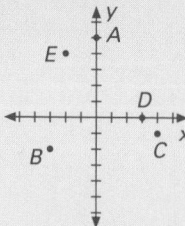

6. $m_{\overline{AB}} = \frac{1}{5}$, $m_{\overline{DC}} = \frac{1}{5}$, $m_{\overline{AD}} = 5$, $m_{\overline{BC}} = 5$;
$\overline{AB} \parallel \overline{DC}$, $\overline{AD} \parallel \overline{BC}$; $AB = BC = CD =$
$DA = \sqrt{26}$; $\therefore ABCD$ is rhom.

8. a. Zero **b.** Pos **c.** Undef **d.** Neg

9.

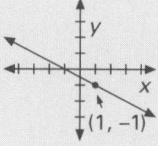

11.

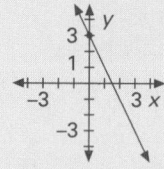

$m = -2$, $b = 3$

16.

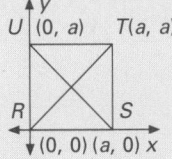

$m_{\overline{RT}} = m_{\overline{US}} = \frac{a}{a} \cdot -\frac{a}{a} = -1 \cdot$
$\therefore \overline{RT} \perp \overline{US}$

Key Terms

abscissa (p. 379)
coordinates (p. 379)
coordinate plane (p. 379)
Distance Formula (p. 380)
linear equation (p. 393)
Midpoint Formula (p. 384)
ordinate (p. 379)
origin (p. 379)

point-slope (p. 393)
quadrant (p. 379)
slope (p. 388)
slope-intercept (p. 394)
standard form (p. 393)
x-axis (p. 379)
y-axis (p. 379)
y-intercept (p. 394)

Key Ideas and Review Exercises

10.1 To graph ordered pairs and find the distance between two points
The distance between two points $P_1(x_1, y_1)$ and $P_2(x_2, y_2)$ is given by the formula $d = \sqrt{(x_2 - x_1)^2 + (y_2 - y_1)^2}$.

1. On the coordinate system, graph the points $A(0, 5)$, $B(-3, -2)$, $C(4, -1)$, $D(3, 0)$, and $E(-2, 4)$.
2. Find the distance between the points $P(6, 1)$ and $Q(-1, -5)$. $\sqrt{85}$
3. $\triangle ABC$ has coordinates $A(4, 1)$, $B(1, -6)$, $C(-2, 1)$. Is the triangle scalene, isosceles, or equilateral? Isosceles

10.2 To find the coordinates of the midpoint of a segment
The Midpoint Formula: Given $P(x_1, y_1)$ and $Q(x_2, y_2)$, the coordinates (x_m, y_m) of M, the midpoint, are $\left(\dfrac{x_1 + x_2}{2}, \dfrac{y_1 + y_2}{2}\right)$.

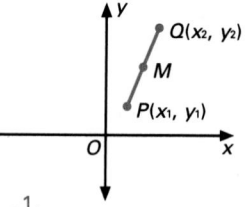

4. Given $P(-1, 2)$ and $Q(3, -5)$ find the midpoint of $\overline{PQ}$. $(1, -1\frac{1}{2})$
5. Given $M(2, 3)$ is the midpoint of $\overline{AB}$. The coordinates of A are $(-1, 0)$. Find the coordinates of B. (5, 6)
6. $ABCD$ is a quadrilateral with vertices $A(0, 0)$, $B(5, 1)$, $C(6, 6)$, $D(1, 5)$. Show that $ABCD$ is a rhombus.

10.3 To find the slope of a line and to draw a line with a given slope
The slope of a line with points $P_1(x_1, y_1)$ and $P_2(x_2, y_2)$

is the ratio $\dfrac{y_2 - y_1}{x_2 - x_1}$, $x_1 \neq x_2$.

7. Find the slope of a line with points $(4, -1)$ and $(-3, 6)$. $m = -1$

8. Tell whether the slope of each of the following is positive, negative, zero, or undefined.

a. 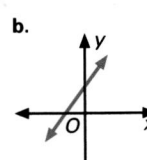   b. 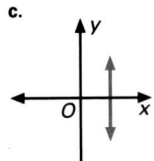   c. 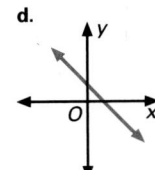   d.

9. Draw a line with slope $-\frac{1}{2}$ containing the point $(1,-1)$.

10.4 To draw the graph of an equation of a line and to write the equation of a line in three forms

A graph of an equation is the set of points whose coordinates satisfy the given equation.

10. Write an equation in standard form of the line containing the points $(-1, 1)$ and $(3, 5)$. $x - y = -2$

11. Find the slope and the y-intercept of the line whose equation is $y = -2x + 3$. Then draw the graph. $m = -2, b = 3$

12. Write an equation in slope-intercept form of the line pictured at the right. $y = 2x - 1$

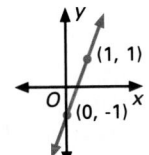

10.5 If two non-vertical lines have the same slope, then they are parallel. The product of the slope of two non-vertical perpendicular lines is -1.

13. Write an equation in slope-intercept form of a line containing $(2,-1)$ and perpendicular to the line whose equation is $y = \frac{3}{4}x + 1$. $y = -\frac{4}{3}x + \frac{5}{3}$

14. Determine whether the diagonals are perpendicular for the quadrilateral with vertices $A(3, 0)$, $B(8, 4)$, $C(4, 9)$, and $D(-1, 5)$.   Yes; $m_{\overline{AC}} = 9$, $m_{\overline{BD}} = -\frac{1}{9}$

10.6 To assign coordinates to the vertices of a polygon in a plane

To use coordinate geometry to prove statements

15. Find the missing coordinate for the parallelogram. $(r + t, s)$

16. Use a coordinate proof to prove that the diagonals of a square are perpendicular.

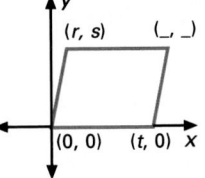

Chapter 10 Review   **409**

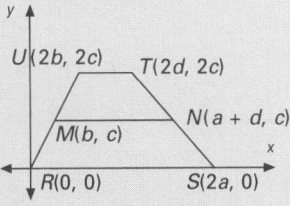
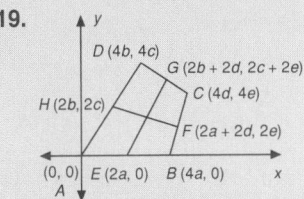
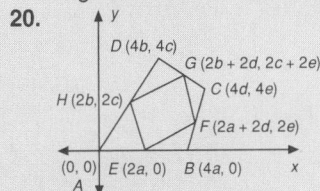
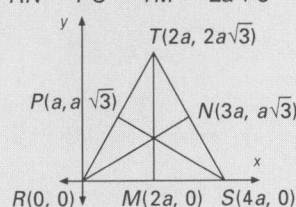

Chapter Test

11.

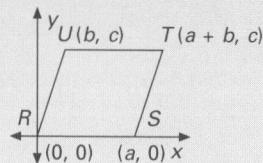

13.

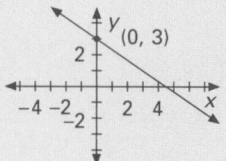

16.

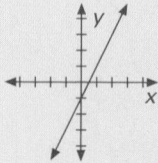

19. 1. Place rectangle ABCD with A(0,0),
   B(a,0), C(a,b), D(0,b)
   2. $AC = \sqrt{(a - 0)^2 + (b - 0)^2}$
   $= \sqrt{a^2 + b^2}$
   $BD = \sqrt{(0 - a)^2 + (b - 0)^2}$
   $= \sqrt{a^2 + b^2}$

20. 1. Place △ABC with A(0,0) B(2a,0)
   C(2b,2c)
   2. $m_{\overline{AC}} = \left(\dfrac{0 + 2b}{2}, \dfrac{0 + 2c}{2}\right) = (b, c)$

   $m_{\overline{BC}} = \left(\dfrac{2a + 2b}{2}, \dfrac{0 + 2c}{2}\right) = (a + b, c)$

   3. m (midsegment) $= \dfrac{c - c}{a + b - b} = 0$

   $m_{\overline{AB}} = \dfrac{0 - 0}{2a - 0} = 0$

   4. Slope of midsegment = slope of base.

21.

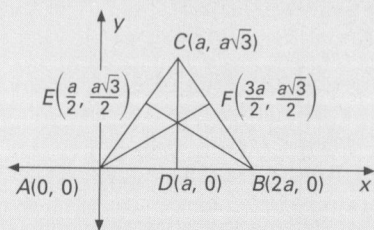

$AF = \sqrt{\left(\dfrac{3a}{2}\right)^2 + \left(\dfrac{a\sqrt{3}}{2}\right)^2} = a\sqrt{3}$

$BE = \sqrt{\left(\dfrac{a}{2} - 2a\right)^2 + \left(\dfrac{a\sqrt{3}}{2}\right)^2} = a\sqrt{3}$

$CD = \sqrt{0 + (a\sqrt{3})^2} = a\sqrt{3}$
So, $AF = BE = CD$.

410

Write the coordinates of each point.

1. A $(-2, 3)$

2. B $(2, -3)$

3. C $(-3, -1)$

4. D $(6, 1)$

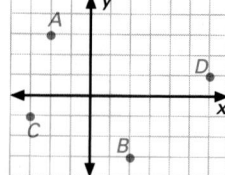

Find the length of $\overline{AB}$.

5. $A(-1, 0), B(2, -4)$ 5

6. $A(3, -2), B(7, 6)$ $4\sqrt{5}$

ABC is a triangle. Use the lengths AB, BC, and AC to tell whether or not △ABC is isosceles. (Exercises 7 and 8)

7. $A(1, -3)\ B(-3, 1)\ C(1, 1)$ yes

8. $A(2, 4)\ B(-1, 3)\ C(1, 0)$ no

9. Find the coordinates of the midpoint of $\overline{AB}$ for $A(7, -6)$ and $B(-1, 2)$ $(3, -2)$

10. If $M(-3, 4)$ is the midpoint of $\overline{XY}$ for $X(5, -2)$, find the coordinates of Y. $(-11, 10)$

11. Draw a set of coordinate axes. Place a parallelogram on these axes so that the simplest coordinates are assigned to the vertices.

12. Find the slope of a segment with endpoints $(-2, 1)$ and $(3, 5)$. $\frac{4}{5}$

13. Draw a set of coordinate axes. Draw a line through $(0, 3)$ with a slope of $-\frac{2}{3}$.

14. Write an equation in slope-intercept form of a line with a slope of 2 containing the point $(0, -3)$. $y = 2x - 3$

15. Write an equation in standard form of a line containing the points $(-1, 3)$ and $(3, 1)$. $x + 2y = 5$

16. Draw a set of coordinate axes. Draw a graph of $y = 2x - 1$.

17. Write an equation in slope-intercept form of a line parallel to $y = -5x + 3$ through the point $(1, 0)$. $y = -5x + 5$

18. Write an equation in slope-intercept form of a line perpendicular to $y = 2x + 1$ through the point $(-1, 3)$. $y = -\frac{1}{2}x + 2\frac{1}{2}$

19. Give a coordinate proof that the diagonals of a rectangle are congruent.

20. Give a coordinate proof that the slope of the midsegment of a triangle is equal to the slope of the base.

*21. Give a coordinate proof that the medians of an equilateral triangle are congruent.

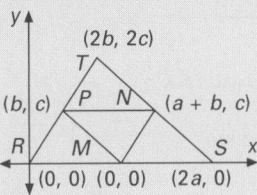

College Prep Test

Choose the one best answer to each question or problem.

1. A line segment has one endpoint at $(3,-2)$ and its midpoint at $(2,-5)$. What are the coordinates of the other endpoint?   A
   (A) $(1,-8)$ (B) $(5,-7)$
   (C) $(1,-3)$ (D) $(1, 8)$
   (E) $(-1, 8)$

2. What is the length of a line segment joining the points whose coordinates are $(-2,-7)$ and $(6, 8)$?   E
   (A) 4   (B) 5   (C) $7\frac{1}{2}$
   (D) $8\frac{1}{2}$ (E) 17

3. Find the slope of the line whose equation is $3x + 2y = 6$.   D
   (A) $\frac{2}{3}$ (B) $-\frac{2}{3}$ (C) $\frac{3}{2}$
   (D) $-\frac{3}{2}$ (E) 6

4. Which point lies at the greatest distance from the origin?   D
   (A) $(0,-9)$ (B) $(-2, 9)$
   (C) $(-7,-6)$ (D) $(8, 5)$ (E) $(0, 0)$

5. $\triangle AOB$ and $\triangle PCB$ are isosceles right triangles with equal areas. What are the coordinates of point P?   E

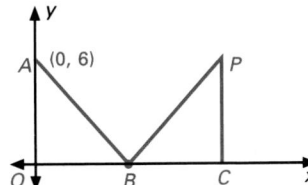

   (A) $(6, 0)$ (B) $(6, 12)$ (C) $(12, 0)$
   (D) $(0, 12)$ (E) $(12, 6)$

6. An equation of the line parallel to the x-axis and passing through point $(4, 2)$ is ___B___.
   (A) $x = -4$ (B) $y = 2$
   (C) $x = 2$ (D) $x = 4$
   (E) $y = 0$

7. An equation of the line passing through the origin and perpendicular to the line whose equation is $x - y = 2$ is ___A___.
   (A) $y = -x$ (B) $y = \frac{1}{2}x + 2$
   (C) $y = -2x$ (D) $y = x$
   (E) $y = -x + 1$

8. The coordinates of A are $(4, 0)$ and of C are $(15, 0)$. Find the area of $\triangle ABC$ if the equation of $\overline{AB}$ is $y = 2x - 8$.   B

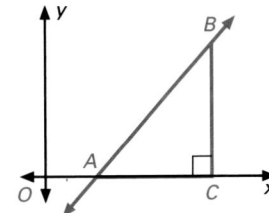

   (A) 100   (B) 121   (C) 144
   (D) 169   (E) 132

9. What is an equation of the line $\overleftrightarrow{AB}$?   A

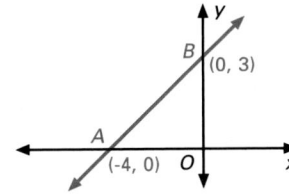

   (A) $y = \frac{3}{4}x + 3$ (B) $y = \frac{3}{4}x - 4$
   (C) $y = \frac{4}{3}x + 3$ (D) $y = -\frac{4}{3}x + 3$
   (E) $y = -\frac{3}{4}x + 3$

10. The vertices of rectangle $ABCD$ are the points $A(0, 0)$, $B(8, 0)$, $C(8, k)$, $D(0, 5)$. The value of k is ___B___.
   (A) 4   (B) 5   (C) 6   (D) 3   (E) 2

---

22. $\frac{MN}{TR} = \frac{PN}{SR} = \frac{PM}{TS} = \frac{1}{2}$

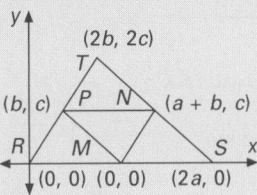

23.

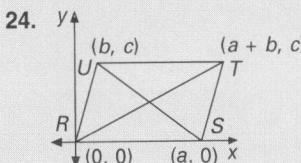

   $PQ = QM = MN = NP = \sqrt{a^2 + b^2}$;
   m = slope: $m_{\overline{PQ}} = m_{\overline{MN}} = \frac{b}{a}$, $m_{\overline{QM}} = m_{\overline{PN}} = -\frac{b}{a}$

24.

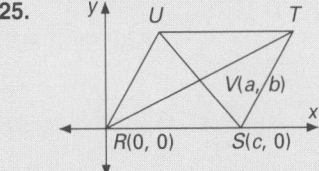

   If $\sqrt{(a + b)^2 + c^2} = \sqrt{(b - a)^2 + c^2}$, then $b = 0$, $\overline{UR} \perp \overline{RS}$, and $RSTU$ is a rectangle.

25.

   Since pt V is the midpt of $\overline{US}$ and midpt of $\overline{RT}$, $T = (2a, 2b)$ and $U = (2a - c, 2b)$. $m_{\overline{UR}} = m_{\overline{ST}} = \frac{2b}{2a - c}$; $m_{\overline{RS}} = m_{\overline{TU}} = 0$. Thus, $\overline{UR} \parallel \overline{ST}$ and $\overline{RS} \parallel \overline{TU}$; $\therefore$ Quad $RSTU$ is a $\square$.

Algebra Review

1. $x^2 + 4x + 4$
2. $x^2 - 6x + 9$
3. $y^2 + 10y + 25$
4. $a^2 - 16a + 64$
5. $4x^2 + 4x + 1$
6. $9y^2 - 12y + 4$
7. $4c^2 - 20c + 25$
8. $16x^2 + 8x + 1$

3. conjunction

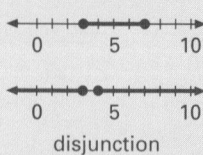

disjunction

(A) Line seg; (B) 2 rays

5. Corresponding angles are congruent if
    and only if the two lines are parallel;
    yes.

6. If two lines are not parallel, then
    corresponding angles are not
    congruent; yes.

7. If corresponding angles are not
    congruent, then the two lines are not
    parallel; yes.

9.
Statement	Reason
1. $\overline{DC}$ bis $\angle C$, $AC = CB$	1. Given
2. $\angle ACD \cong \angle BCD$	2. Def of $\angle$ bis
3. $\overline{DC} \cong \overline{DC}$	3. Reflex Prop of Congr
4. $\triangle ACD \cong \triangle BCD$	4. SAS
5. $\overline{AD} \cong \overline{DB}$	5. CPCTC

13. Hyp: $\overline{RS}$; legs: $\overline{TR}$, $\overline{TS}$; median: $\overline{TP}$;
    altitudes: $\overline{RT}$, $\overline{ST}$

16.
Statement	Reason
1. $\square ABCD$	1. Given
2. $\overline{BC} \cong \overline{AC}$, $\overline{BC} \parallel \overline{AD}$	2. Def of $\square$
3. $\angle BCA \cong \angle DAC$; $\angle BAC \cong \angle DCA$	3. Alt int $\angle$s of $\parallel$ lines are $\cong$.
4. $\overline{AC} \cong \overline{AC}$	4. Reflex Prop of Congr
5. $\triangle ABC \cong \triangle CDA$	5. ASA

Cumulative Review *(Chapters 1–10)*

1. Identify each of the following
    as line segment, ray, line, or
    the measure of a seg-
    ment. **A.** Ray **B.** Line seg
    (A) $\overrightarrow{AB}$ (B) $\overline{AB}$ (C) $\overleftrightarrow{AB}$
    (D) AB **C.** Line **D.** Meas of seg   *1.1*

2. Given: $\overline{RP}$ bisects $\angle SPQ$;
    m $\angle 1 = 4x - 5$, m $\angle 2 = 2x + 6$. Find m $\angle SPQ$.   34   *1.5*

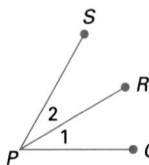

3. Graph the conjunction and the
    disjunction. Identify the result-
    ing figure.
    (A) $x \geq 3$ *and* $x \leq 7$
    (B) $x \geq 4$ *or* $x \leq 3$ *1.7*

4. Find m $\angle 1$.   45   *3.6*

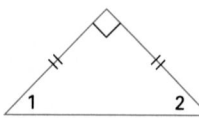

Use the following statement to answer Exercises 6–8: **"If two lines
are parallel, then corresponding
angles are congruent."**   *3.8*

5. Write a biconditional for the
    statement. Is it true?

6. Write the inverse of the state-
    ment. Is it true?

7. Write the contrapositive of the
    statement. Is it true?

8. Based on the markings, deter-
    mine whether you can prove
    the triangles are congruent. No   *4.6*

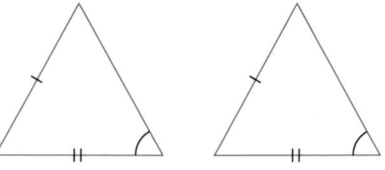

9. Given: $\overline{DC}$ bisects $\angle ACB$;
    $AC = CB$.
    Prove: $\overline{AD} \cong \overline{DB}$ *4.7*

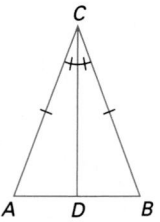

10. In $\triangle ABC$, $AB = 5$, $BC = 8$,
    $AC = 7$
    Name the largest angle. $\angle A$ *5.7*

**Complete the statement with =, <,
or >.**   *5.7*

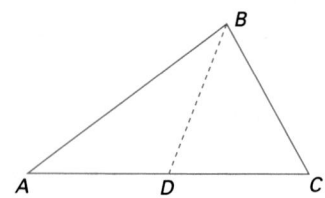

11. $\angle A < \angle C$, BC ____<____ AB

12. $\overline{BD}$ is a median; AD ____=____ DC.

13. Identify the hypotenuse, legs, median, and altitude(s) of right triangle *TRS*. *5.6*

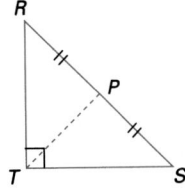

14. How many sides does a regular polygon have if each exterior angle has a measure of 72? 5 *6.2*

15. The measure of three exterior angles of a quadrilateral are 75, 85, and 95. What is the measure of the exterior angle at the fourth vertex? 105 *6.3*

16. Given: ▱*ABCD* *6.4*
Prove: △*ABC* ≅ △*CDA*

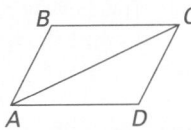

True or False? *7.2, 7.3*

17. The diagonals of a parallelogram bisect each other. T

18. The diagonals of a rectangle are perpendicular. F

19. The diagonals of a square are congruent. T

20. The diagonals of a rhombus are congruent. F

21. The diagonals of a square are perpendicular. T

22. The diagonals of a rhombus bisect the vertex angles. T

23. The diagonals of a parallelogram are congruent. F

24. Given: Isos trap *ABED*, *C* is the midpoint of $\overline{DE}$. *7.5*
Prove: △*ABC* is isosceles.

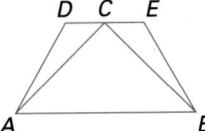

25. Find the geometric mean of 10 and 40. 20 *8.1*

26. Given: △*ARB* ~ △*DRC* *8.3*
Find *RD*. 12

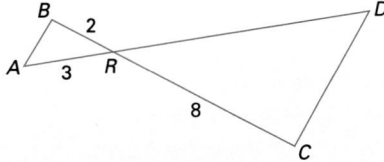

27. Prove: *8.1*
If $\frac{a}{b} = \frac{c}{d}$, then $\frac{a + b}{b} = \frac{c + d}{d}$

28. Find the length of the diagonal of a square with sides measuring 8 in. $8\sqrt{2}$ in *9.2*

29. A tree casts a 30-ft shadow. Find the height of the tree if the angle of elevation to the top of the tree has a degree measure of 51. 37 ft *9.7*

30. Find the distance *AB* for points *A*(−1,−2), and *B*(5,−3). $\sqrt{37}$ *10.1*

31. Find the slope and *y*-intercept of the line whose equation is $3y - 2x = 9$. $m = \frac{2}{3}$, (0, 3) *10.3, 10.4*

32. Write an equation, in standard form, of a line perpendicular to the line with equation $y = \frac{1}{2}x - 1$ and containing the point (0, 0). $2x + y = 0$ *10.5*

33. Use a coordinate proof to prove that the diagonals of a square bisect each other. *10.6*

Cumulative Review **413**

24.

Statement	Reason
1. Isos trap *ABED*, *C* is midpt of $\overline{DE}$	1. Given
2. $\overline{AD} \cong \overline{BE}$, $\overline{DE} \parallel \overline{AB}$	2. Def of isos trap
3. $\overline{DC} \cong \overline{CE}$	3. Def of midpt
4. ∠*DAB* ≅ ∠*EBA*	4. Base ∠s of isos trap are ≅.
5. ∠*D* and ∠*DAB* are supp, ∠*E* and ∠*EBA* are supp.	5. If lines are ∥, then int ∠s on same side of transv are supp.
6. ∠*D* ≅ ∠*E*	6. Supp ∠s of ≅ ∠s are ≅.
7. △*ADC* ≅ △*BEC*	7. SAS
8. $\overline{CA} \cong \overline{CB}$	8. CPCTC
9. △*ABC* is isos.	9. Def of isos △

27.

Statement	Reason
1. $\frac{a}{b} = \frac{c}{d}$	1. Given
2. $\frac{a}{b} + 1 = \frac{c}{d} + 1$	2. Add Prop of Eq
3. $\frac{b}{b} = 1, \frac{d}{d} = 1$	3. A number div by itself = 1.
4. $\frac{a}{b} + \frac{b}{b} = \frac{c}{d} + \frac{d}{d}$	4. Sub
5. $\frac{a + b}{b} = \frac{c + d}{d}$	5. Notation

33.

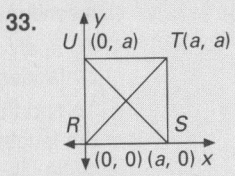

The midpt of $\overline{US}$ is $\left(\frac{a}{2}, \frac{a}{2}\right)$.

The midpt of $\overline{RT}$ is $\left(\frac{a}{2}, \frac{a}{2}\right)$.

These are the same point so the diagonals bisect each other.

11 CIRCLES

OVERVIEW

Students identify sets of points related to circles and spheres, such as radius, diameter, chord, secant, tangent, and arc. Students prove and apply theorems related to these sets of points as well as to central and inscribed angles. They prove theorems about angles formed by tangents and secants and two tangents. They also find lengths of segments related to chords, secants, and tangents. Students will construct circles as well as write and graph equations of circles.

CHAPTER OBJECTIVES

- To identify special segments and lines related to circles
- To prove and apply theorems about chords, tangents, and secants
- To find degree measures of arcs and inscribed angles
- To apply theorems about angles formed by tangents and secants
- To find the coordinates of the center and the length of the radius and to draw the circle given an equation of the circle

PROBLEM SOLVING

Lesson 11.10 emphasizes the use of constructions as a problem solving tool. Students are able to practice their constructions in Exercises 1–8, and then in Exercises 9–18 they are asked to solve problems or justify conclusions by employing the problem solving strategy, Using Constructions.

TECHNOLOGY

Computer: The Computer Investigation on page 431 will help students discover a pattern for finding the measures of angles formed by two chords. This activity can be extended to discover concepts discussed in Lessons 11.4–11.6.

Calculator: Students should be encouraged to use a calculator to facilitate computations in exercises throughout Chapter 11.

SPECIAL FEATURES

Focus on Reading pp. 417, 435, 463
Mixed Review pp. 419, 424, 430, 437, 441, 451, 456, 461, 465, 469
Computer Investigation: Arcs and Related Angles p. 431
Algebra Review p. 437
Midchapter Review p. 446
Brainteaser p. 451
Application: Geometric Designs p. 465
Key Terms p. 470
Key Ideas and Review Exercises pp. 470–471
Chapter 11 Test p. 472
College Prep Test p. 473

PLANNING GUIDE

Lesson	Basic	Average	Above Average	Resources
11.1 pp. 417–419	FR all, CE all WE 1–18 odd	FR all, CE all WE 1–28 odd	FR all, CE all WE 1–32 odd	Reteaching p. 149 Practice p. 150
11.2 pp. 423–424	CE all WE 1–12	CE all WE 1–16	CE all WE 1–21	Reteaching p. 151 Practice p. 152
11.3 pp. 428–431	CE all WE 1–11 odd CI all	CE all WE 1–17 odd CI all	CE all WE 1–21 odd CI all	Reteaching p. 153 Practice p. 154
11.4 pp. 435–437	FR all, CE all WE 1–14 odd AR all	FR all, CE all WE 1–25 odd AR all	FR all, CE all WE 1–28 odd AR all	Reteaching p. 155 Practice p. 156
11.5 pp. 440–441	CE all WE 1–7	CE all WE 1–10	CE all WE 1–12	Reteaching p. 157 Practice p. 158
11.6 pp. 444–446	CE all WE 1–26 odd Midchapter Review	CE all WE 1–34 odd Midchapter Review	CE all WE 1–38 odd Midchapter Review	Reteaching p. 159 Practice p. 160
11.7 pp. 449–451	CE all WE 1–14 odd Brainteaser	CE all WE 1–28 odd Brainteaser	CE all WE 1–37 odd Brainteaser	Reteaching p. 161 Practice p. 162
11.8 pp. 454–456	CE all WE 1–15 odd	CE all WE 1–27 odd	CE all WE 1–31 odd	Reteaching p. 163 Practice p. 164
11.9 pp. 460–461	CE all WE 1–10 odd	CE all WE 1–18 odd	CE all WE 1–23 odd	Reteaching p. 165 Practice p. 166
11.10 pp. 463–465	FR all, CE all WE 1–7 Application	FR all, CE all WE 1–14 odd Application	FR all, CE all WE 1–18 odd Application	Reteaching p. 167 Practice p. 168
11.11 pp. 468–469	CE all WE 1–18 odd	CE all WE 1–32 odd	CE all WE 1–40 odd	Reteaching p. 169 Practice p. 170
Chapter 11 Review pp. 470–471	1–28 odd	all	all	
Chapter 11 Test p. 472	1–7, 9–26 odd	1–20	all	
College Prep Test p. 473	all	all	all	

CE = Classroom Exercises   WE = Written Exercises   FR = Focus on Reading   AR = Algebra Review   CI = Computer Investigation

NOTE: For each level, all students should be assigned all Mixed Review exercises.

Investigation

Project: When a plane intersects a cone, the resulting curve is called a conic section. This investigation provides an informal preview of conic sections and allows students to explore the possibilities before they learn the more formal vocabulary.

Materials: scissors, glue or tape, and Project Worksheet 11

Procedure:

1. Have students cut out and assemble the shapes on Project Worksheet 11 according to the directions on the page.

2. Have students place the cone (Fig. 4 on Project Worksheet 11) on one side of their desks and the other three shapes (Figs. 1–3 on Project Worksheet 11) in a line on the other side of their desks.

3. Have students imagine that a plane is cutting or intersecting the cone in each of the 3 different ways listed in questions a–c. Read questions a–c aloud and have students decide which figure matches each intersection.

4. After each question is read, have students hold up 1, 2, or 3 fingers to indicate the correct figure number as an answer for you to quickly scan and check.

   a. Which conic section is formed when the cutting plane is parallel to the surface of the desk? Fig. 2 on Project Worksheet 11

   b. Which conic section is formed when the plane is perpendicular to the surface of the desk? Fig. 1 on Project Worksheet 11

   c. Which conic section is formed when the plane is neither parallel nor perpendicular to the desk? Fig. 3 on Project Worksheet 11

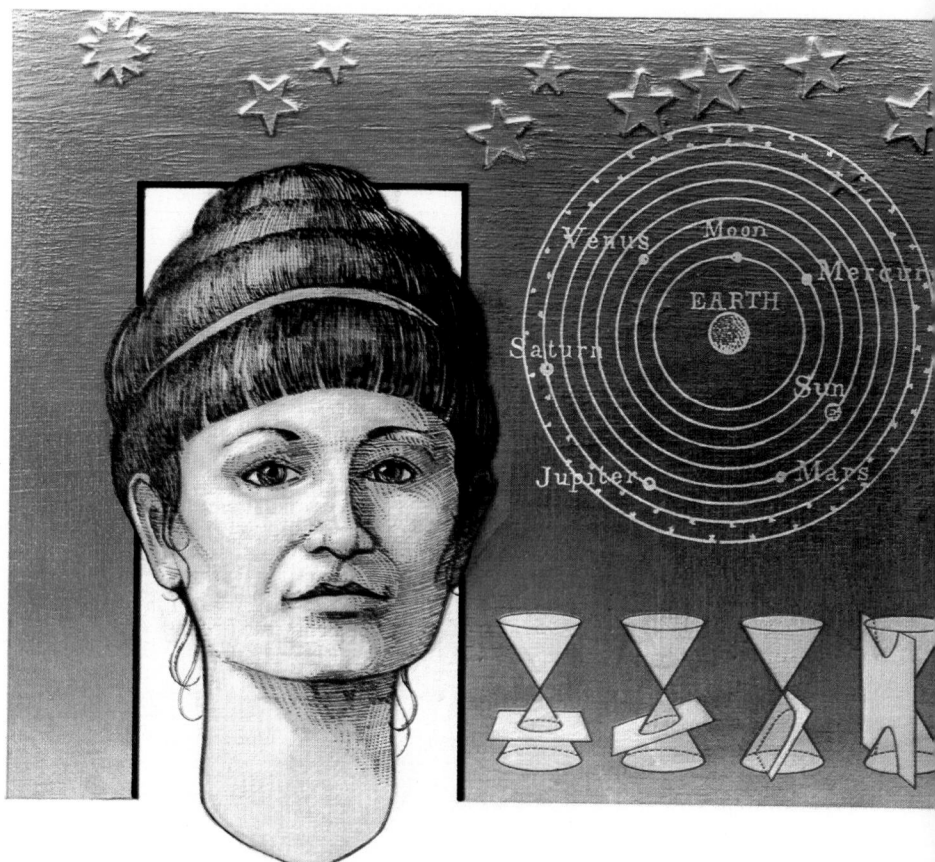

Hypatia (370?–415)—Hypatia lectured on mathematics, philosophy, and astronomy, as well as on simple mechanics. She designed scientific instruments such as the astrolabe, which measures the positions of planets and stars.

More About the Mathematician

Hypatia was one of the most popular lecturers at the Egyptian university at Alexandria, which was a center of Greek intellectual life. She wrote a book, *On the Conics of Apollonius*. The Greeks studied conic sections, formed by the intersection of a plane and a cone, as shapes and knew nothing of their equations. The conic sections were ignored after her death (which marked the end of the Greek period in mathematics) and interest in them did not revive again until the seventeenth century. Hypatia also invented instruments for distilling water, for measuring the level of liquids, and for determining the specific gravity of liquids.

11.1 Circles and Spheres: Basic Definitions

Objectives   To identify special segments, lines, and other sets of points related to circles and spheres
To find lengths of radii and diameters

The photograph at the right shows a bicycle wheel, which is a physical model of a *circle* in geometry. Every point of the rim of this wheel is the same distance, 13 in., from the hub or center of the wheel. *A*, *B*, *C*, and *D* are four such points.

Definitions

A **circle** is the set of all points in a plane that are a given distance from a fixed point in that plane. The fixed point is the **center** of the circle. A segment from the center to any point on the circle is a **radius** (plural: **radii**).

A circle is named by its center. The circle to the right is named Circle *O* or ⊙*O*, where the symbol for circle is ⊙.

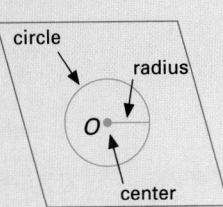

It follows from the definition of a circle that all radii of the same circle are congruent. It is now possible to define *congruent circles* in terms of their radii.

Definition

Congruent circles are circles whose radii are congruent.

A circle in a plane separates the plane into three sets of points: the circle itself, the interior of the circle, and the exterior of the circle.

Definitions

The **interior** of a circle is the set of all points in the plane of the circle whose distance from the center is *less* than the length of the radius. The **exterior** of a circle is the set of all points in the plane whose distance from the center is *greater* than the length of the radius.

11.1 Circles and Spheres: Basic Definitions   **415**

Teaching Resources

Application Worksheet 11
Project Worksheet 11
Quick Quizzes 75
Reteaching and Practice
 Worksheets, pp. 149, 150

GETTING STARTED

Prerequisite Quiz

Solve.

1. $4x + 2 = 2(3x - 1)$ $x = 2$
2. $3x - 6 = 18$ 8
3. $2x - 10 > 8$ $x > 9$
4. $4x - 3 < 9$ $x < 3$

Motivator

Ask students to label a point *O* somewhere in the middle of a sheet of paper. Then have them use a ruler to locate at least ten other points that are a distance of 4 cm from the point *O*. Ask them if this were to be continued with another ten points and then another ten, etc., what would the path of these points appear to be?   A circle   Ask them what the point *O* is called for the figure formed by the path of points.   The center

TEACHING SUGGESTIONS

Lesson Note

Use practical illustrations to explain definitions. Use the idea of a car tire touching the road surface as an illustration of a line tangent to a circle. Think of a porthole on a ship as a circular window. A window shade is halfway down. The bottom edge of the shade is a secant to the circle. As the shade is raised, the points of intersection of the secant and the circle get closer together. When the two points become one, then the edge of the shade is said to be tangent to the circle.

Math Connections

Balloons: A balloon can be a physical model for a sphere. Like a sphere, a balloon separates points in space into three sets—points on the balloon, points inside the balloon, and points outside the balloon.

Critical Thinking Questions

Application: Wanelle has four pieces of wood, each 6 inches wide and 2 feet long. Ask students to draw a diagram that illustrates how the wood can be arranged to form a round table top with a diameter of 24 inches.

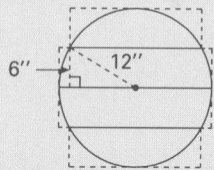

6"   12"

Ask students what the minimum length of the outer boards must be. $2\sqrt{12^2 - 6^2} = 2(6\sqrt{3}) \approx 20.8$ in

Common Error Analysis

Error: When finding the radius for a given diameter such as $6x + 8$, students tend to divide only the 6 by 2 rather than both 6 and 8.

Emphasize that $\frac{6x + 8}{2} = \frac{6x}{2} + \frac{8}{2} = 3x + 4$.

EXAMPLE 1   Given: $\odot O$, $OR = 2x - 10$, $OP = 8$
Find all values of x for which point R is in the exterior of the circle.

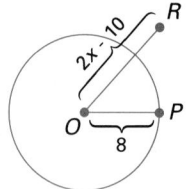

Plan   To meet the conditions, set $OR > OP$.

Solution
$$OR > 8$$
$$2x - 10 > 8$$
$$2x > 18, \text{ or } x > 9$$

Definitions   A **chord** of a circle is a segment whose endpoints are on the circle. A **diameter** is a *chord* that contains the center of the circle.

If the length of the radius is r and the length of the diameter is d, then $d = 2r$ and $r = \frac{d}{2}$.

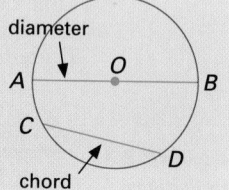

diameter

chord

EXAMPLE 2   For the given radius or diameter, find the indicated length.

a. Given: $d = 18$
Find r.

b. Given: $r = 2a + 4$
Find d.

c. Given: $d = 4x - 12$
Find r.

Solutions

$r = \dfrac{d}{2} = \dfrac{18}{2}$
Thus, $r = 9$.

$d = 2r = 2(2a + 4)$
Thus, $d = 4a + 8$.

$r = \dfrac{d}{2} = \dfrac{4x - 12}{2}$
Thus, $r = 2x - 6$.

Given a line and a circle that are coplanar, the possible number of points in which the line can intersect the circle are indicated by the following diagrams.

0 points   1 point   2 points

   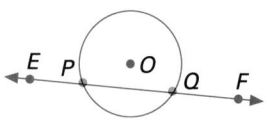

$\overleftrightarrow{AB}$ does not intersect the circle.

$\overleftrightarrow{CD}$ is a *tangent* to the circle; P is the point of *tangency*.

$\overleftrightarrow{EF}$ is a *secant* of the circle.

416   Chapter 11 Circles

Additional Example 1

Given: Circle O, $OP = 8$, $OR = 2x + 4$. Find all values of x for which point R is in the exterior of the circle. $x > 2$

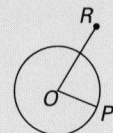

Additional Example 2

For the given radius (diameter), find the indicated length.

a. Given: $r = 12$
Find d.   24

b. Given: $d = 4x - 12$
Find r. $2x - 6$

c. Given: $r = 6x + 3$
Find d. $12x + 6$

Definitions

A **tangent** to a circle is a line that is coplanar with the circle and intersects the circle in exactly one point, called the **point of tangency**. A **secant** of a circle is a line that intersects the circle in two points.

EXAMPLE 3 Identify all radii, diameters, chords, secants, and tangents of $\odot O$ at the right.

Solution
Radii: $\overline{OQ}, \overline{ON}, \overline{OP}$
Diameter: $\overline{QN}$
Secant: $\overleftrightarrow{TU}$ (or $\overleftrightarrow{RS}$)
Chords: $\overline{TU}, \overline{QN}$ (also diameter), $\overline{QP}$
Tangent: $\overleftrightarrow{VW}$

The definition of circle can be extended to describe the three-dimensional counterpart of the circle, the *sphere*.

Definitions

A **sphere** is the set of all points in space that are a given distance from a fixed point. The fixed point is the **center** of the sphere. A segment from the center of the sphere to any point on the sphere is a **radius** of the sphere.

A **chord of a sphere** has endpoints on the sphere. A **secant of a sphere** is a line that intersects the sphere in two points.

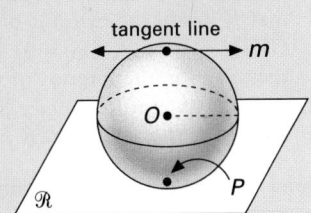

A line or plane that intersects a sphere in exactly one point is **tangent** to that sphere.

Focus on Reading

1. Rewrite the definition of a circle using the word *equidistant*.

Complete to make a true statement.

2. A tangent intersects a circle in <u>exactly one point</u>.
3. A secant intersects a circle in <u>two points</u>.
4. The length of the radius of a circle is <u>one-half</u> the length of the diameter.
5. Given: $\overline{OP}$ is the radius of a circle with center O; $OP = 5$; Q is such that $OQ = 9$. Point Q lies in the <u>exterior</u> of the circle.

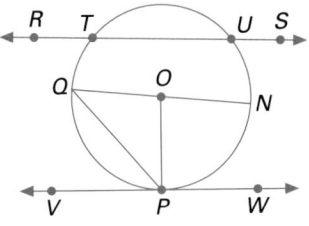

Identify each.

1. radius $\overline{OE}, \overline{OD}, \overline{OK}, \overline{OM}$
2. diameter $\overline{MK}$
3. chord $\overline{KB}, \overline{KM}, \overline{KJ}$
4. secant $\overleftrightarrow{HG}$
5. tangent $\overleftrightarrow{AC}$
6. Find the length of a diameter if $OD = 14$. 28
7. Given: $OD = 10$, $OF = 2x - 4$ Find all values of x for which point F is *in* the exterior of the circle. $x > 7$

Closure

Have the students describe each of the following terms.

Radius   Definition p. 415
Chord   Definition p. 416
Diameter   Definition p. 416
Secant   Definition p. 417
Tangent   Definition p. 417
Interior of a circle   Definition p. 415
Exterior of a circle   Definition p. 415
Ask students if a chord can also be a diameter.   Yes, if it contains the center of the circle.   Ask them what arithmetic relationship there is between the length of a radius and the length of a diameter of the same circle.   The length of the diameter is twice the length of the radius.   Have the students explain the difference between a circle and a sphere.   A circle is a set of points in a *plane*, and a sphere is a set of points in *space*.

Additional Example 3

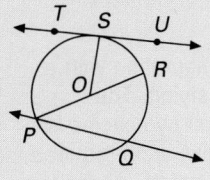

Identify all radii, diameters, chords, secants, and tangents of the circle.
Rad $\overline{OP}, \overline{OR}, \overline{OS}$, Diam $\overline{PR}$,
Chord $\overline{PQ}, \overline{PR}$, Tan $\overleftrightarrow{TU}$, Sec $\overleftrightarrow{PQ}$

FOLLOW UP

Guided Practice

Classroom Exercises 1–8

Independent Practice

A Ex. 1–18, **B** Ex. 19–28, **C** Ex. 29–32

Basic: FR all, WE 1–18 odd
Average: FR all, WE 1–28 odd
Above Average: FR all, WE 1–32 odd

Classroom Exercises

Indicate whether P is in the interior, the exterior, or on the circle.

1. P is 5 in. from the center of a circle with radius of length 4 in.   Exterior
2. P is 7 in. from the center of a circle with diameter of length 10 in.   Exterior
3. P is 4 ft from the center of a circle with radius of length 6 ft.   Interior
4. P is 7 in. from the center of a circle with diameter of length 14 in.   On the circle

For the given length of the radius or diameter, find the indicated length.

5. $r = 6,$
   $d = \underline{\quad 12 \quad}$

6. $d = 14,$
   $r = \underline{\quad 7 \quad}$

7. $r = 2.5,$
   $d = \underline{\quad 5 \quad}$

8. $d = 6.2,$
   $r = \underline{\quad 3.1 \quad}$

Written Exercises

Use the figure at the right for Exercises 1–3.

1. Given: $OP = 12$, $OR = 3x - 6$. Find all values of x such that R lies in the exterior of the circle. $x > 6$
2. Given: $OP = 13$, $OS = 2x - 7$. Find all values of x such that S lies in the interior of the circle. $x < 10$
3. Given: Length of diameter is 30; $OR = 5x - 10$. Find all values of x such that R is in the exterior of the circle.
   $x > 5$

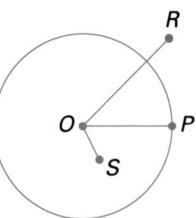

Name the given set of points, first for the circle with center O, then for the sphere with center O.

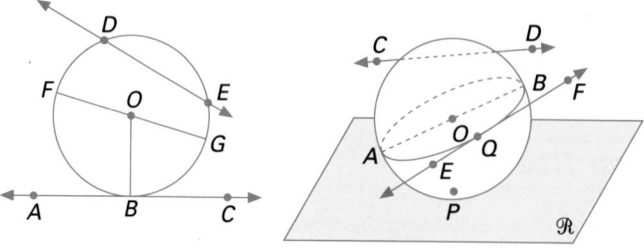

4. radius
   $\overline{OB}, \overline{OG}, \overline{OF}; \overline{OA}, \overline{OB}$

5. diameter
   $\overline{FG}; \overline{AB}$

6. chord
   $\overline{DE}, \overline{FG}; \overline{AB}$

7. secant
   $\overleftrightarrow{DE}; \overleftrightarrow{CD}$

8. tangent
   $\overleftrightarrow{AC}; \overleftrightarrow{EF}, \mathcal{R}$

Find the length of a diameter for the indicated length of a radius.

9. 3.7   7.4
10. $5\frac{1}{2}$ 11
11. $3x$ $6x$
12. $2a + 3$ $4a + 6$
13. $6x - 4$
    $12x - 8$

Find the length of a radius for the indicated length of a diameter.

14. 17.8   8.9
15. $10\sqrt{3}$ $5\sqrt{3}$
16. $6x$ $3x$
17. $4a - 12$ $2a - 6$
18. $8x + 10$
    $4x + 5$

418   Chapter 11 Circles

R and S are centers of the two intersecting circles. Identify the given set of points as a radius, diameter, chord, secant, or tangent, either for ⊙R or ⊙S or both.

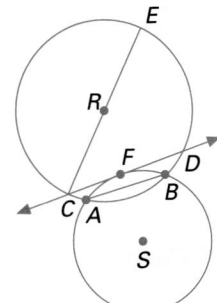

19. $\overline{AB}$ Chord for ⊙R and ⊙S
20. $\overline{CE}$ Diameter and chord for ⊙R
21. $\overleftrightarrow{CD}$ Secant for ⊙R, tangent to ⊙S
22. $\overline{RE}$ Radius for ⊙R
23. $\overline{CD}$ Chord for ⊙R, tangent to ⊙S

Use Circle O at the right, for Exercises 24–27.

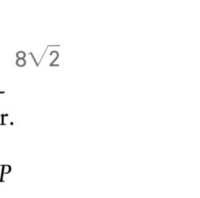

24. $QR = 4x + 6$, $OQ = x + 8$. Find QR.   26
25. $OP = 6x - 4$, $OR = -x + 10$. Find OQ.   8
26. $PQ = 10$. Would 9 be a reasonable length of the diameter? Why or why not?
27. Given: m $\angle POQ = 60$
Prove: $\triangle POQ$ is equilateral.

28. Write a definition of the interior of a sphere.
29. Draw a diagram of a circle with center O, radii $\overline{OA}$ and $\overline{OB}$, and chord $\overline{AB}$; $\overline{OA} \perp \overline{OB}$. Find the length of the diameter if $AB = 8$. $8\sqrt{2}$
30. The length of the diameter of a circle is 28. The length of the radius is the square of some number, increased by 5 times that number. Find the number.   2
31. The distance from a point P to a plane is 4. A sphere with center P intersects the plane in a circle with diameter of length 6. Find the length of a radius of the sphere.   5
32. Consider again the definitions of tangent and secant given in this lesson. Why was it not necessary to state that a secant is coplanar with the circle?   A secant intersects a circle at two points and is necessarily coplanar.

Mixed Review

1. Find the length of an altitude of an equilateral triangle with a side of length 4.   *9.4* $2\sqrt{3}$
2. The length of the hypotenuse of a 45-45-90 triangle is 20. Find the length of each leg of the right triangle.   *9.4* $10\sqrt{2}$
3. The measure of the vertex angle of an isosceles triangle is 70. Find the measure of a base angle.   *5.1* 55

11.1 Circles and Spheres: Basic Definitions   **419**

11.2 Properties of Chords

Objective To prove and apply theorems about chords

Recall that the distance from a point to a line is the length of a perpendicular segment from the point to the line.

Theorem 11.1 If a line or segment contains the center of a circle and is perpendicular to a chord, then it *bisects* the chord.

Given: ⊙O with $\overline{OD}$ ⊥ chord $\overline{AB}$
Prove: $\overline{OD}$ bisects $\overline{AB}$.

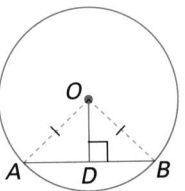

Proof

Statement	Reason
1. ⊙O with $\overline{OD}$ ⊥ chord $\overline{AB}$	1. Given
2. Draw radii $\overline{OA}$ and $\overline{OB}$.	2. Two points determine a line.
(H) 3. $\overline{OA} \cong \overline{OB}$	3. All radii of a ⊙ are ≅.
(L) 4. $\overline{OD} \cong \overline{OD}$	4. Reflex Prop
5. ∠ODA and ∠ODB are rt ∠s.	5. Def of ⊥ lines
6. △ODA and △ODB are rt △s.	6. Def of rt △s
7. △ODA ≅ △ODB	7. HL
8. $\overline{AD} \cong \overline{BD}$	8. CPCTC
9. ∴ $\overline{OD}$ bisects $\overline{AB}$.	9. Def of bis

EXAMPLE 1 A chord of length 8 is 2 units from the center of a circle. Find the length of a diameter.

Plan Draw a diagram. $\overline{OD}$ ⊥ $\overline{AB}$. Therefore, $\overline{OD}$ bisects $\overline{AB}$. $BD = \frac{1}{2} \cdot 8 = 4$; $OD = 2$. Let $OB = r$ and use the Pythagorean Theorem.

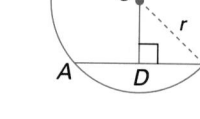

Solution
$$r^2 = 2^2 + 4^2$$
$$r^2 = 20$$
$$r = \sqrt{20}$$
$$= \sqrt{4} \cdot \sqrt{5} = 2\sqrt{5}$$

Thus, since $d = 2r$, the length of the diameter is $4\sqrt{5}$.

GETTING STARTED

Prerequisite Quiz

Assume $AC = BC$, $\overline{CD}$ ⊥ $\overline{AB}$.

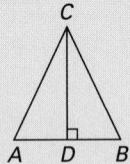

1. $AB = 8$, $CD = 4$, $AC =$ _____ $4\sqrt{2}$
2. $AC = 6$, m ∠$A = 30$, $CD =$ _____ 3
3. m ∠$ACB = 90$, $AC = 8$, $CD =$ _____ $4\sqrt{2}$
4. $AC = 10$, $AB = 6$, $CD =$ _____ $\sqrt{91}$
5. m ∠$ACB = 120$, $CD = 4$, $AB =$ _____ $8\sqrt{3}$
6. $AB = 12$, m ∠$A = 30$, $AC =$ _____ $4\sqrt{3}$

Motivator

Ask the students to construct a circle of any radius and label the center O. Have them construct any chord $\overline{AB}$, except a diameter. From the center, have them construct a perpendicular to the chord and label the point of the intersection C. Ask them what appears to be true about $\overline{AC}$ and $\overline{CB}$. They appear to be equal. Have the students generalize about the relationship of a perpendicular from the center of a circle to a chord. Theorem 11.1 page 420

TEACHING SUGGESTIONS

Lesson Note

As indicated in the Prerequisite Quiz above, it will probably be necessary to review the Pythagorean Theorem and the properties of 30–60–90 and 45–45–90 triangles before beginning the lesson.

Additional Example 1

Find the distance of an 8 in. chord from the center of a circle with a diameter of 10 in. 3 in

Applications of Theorem 11.1 may involve properties of special right triangles such as the 30-60-90 or 45-45-90 triangles.

EXAMPLE 2   Given: $\overline{OA}$ and $\overline{OB}$ are radii of circle O; $\overline{AB}$ is a chord; m $\angle AOB = 120$; $AB = 12$. Find the distance from O to $\overline{AB}$.

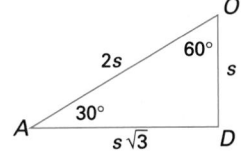

Plan   Draw a diagram. Since m $\angle AOB = 120$, m $\angle A$ + m $\angle B = 60$. Then, since $\triangle AOB$ is isosceles, m $\angle A$ = m $\angle B = 30$. With $\overline{OD} \perp \overline{AB}$, there are two 30-60-90 triangles.

Solution   Let $OD = s$.
Then $s\sqrt{3} = AD$
$s\sqrt{3} = 6$ [$\overline{OD}$ bisects $\overline{AB}$, $AD = 6$.]
$s = \dfrac{6}{\sqrt{3}} = \dfrac{6 \cdot \sqrt{3}}{\sqrt{3} \cdot \sqrt{3}} = \dfrac{6\sqrt{3}}{3} = 2\sqrt{3}$

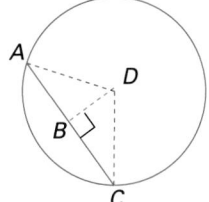

Thus, the chord is $2\sqrt{3}$ units from the center O.

Theorem 11.2   In the same circle or in congruent circles, congruent chords are equidistant from the center(s).

Given: $\odot D \cong \odot S$,
$\overline{AC} \cong \overline{PR}$
Prove: $\overline{BD} \cong \overline{QS}$

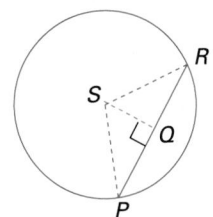

Plan   Prove $\triangle ADC \cong \triangle PSR$. Then the corresponding altitudes $\overline{BD}$ and $\overline{QS}$ are congruent.

Proof

Statement	Reason
1. $\odot D \cong \odot S$, $\overline{AC} \cong \overline{PR}$	(S) 1. Given
2. $\overline{AD} \cong \overline{PS}$, $\overline{CD} \cong \overline{RS}$	(S, S) 2. Def of $\cong \odot$s
3. $\triangle ADC \cong \triangle PSR$	3. SSS
4. $\therefore \overline{BD} \cong \overline{QS}$ ($BD = QS$)	4. Corr alt of $\cong \triangle$s are $\cong$.

Math Connections

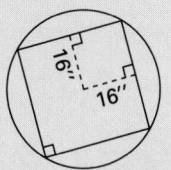

Carpentry: A square table is formed by folding down the four drop leaves of a circular table. As shown in the diagram, the sides of the square are congruent chords of the original circular table top.

Critical Thinking Question

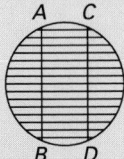

Application: Refer students to the circular barbecue grill shown above and ask them the following question.

If $\overline{AB}$ and $\overline{CD}$ are congruent chords, the diameter of the grill is 14 inches and $\overline{AB}$ is 3 inches from the center, how long is $\overline{CD}$?
$\overline{CD} \cong \overline{AB}$ and $AB = 2\sqrt{7^2 - 3^2} = 2\sqrt{40} \approx 12.65$, so $CD \approx 12.65$ in

Additional Example 2

$\overline{OA}$ and $\overline{OB}$ are radii of a circle O. $\overline{AB}$ is a chord and m $\angle A = 45$. $AB = 12$. Find the distance of the chord from the center of the circle.   6

Find the length of the diameter. $12\sqrt{2}$

Checkpoint

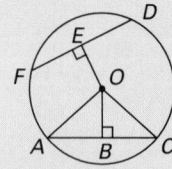

1. Given: $AC = 12$, $AO = 8$
   Find OB. $2\sqrt{7}$
2. Given: $AO = 4$, $OB = 2$
   Find AC. $4\sqrt{3}$
3. Given: $OB = 5$, $AO = 13$, $OE = 10$
   Show that $FD < AC$. $2\sqrt{69} < 24$

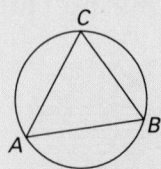

Name the chord closest to the center of the circle.

4. $m\angle A = 20$, $m\angle B = 100$ $\overline{AC}$
5. $m\angle B = m\angle C = 70$ $\overline{AB}, \overline{AC}$

Closure

Ask the students what property holds for a perpendicular from the center of a circle to a chord of the circle. It bisects the chord. Ask them to explain what is true about chords of the same circle that are equidistant from the center. The chords are congruent. Ask them which is closer to the center if two chords of a circle are of unequal length. The longer chord is nearer the center of the circle.

The converse of Theorem 11.2 also is true.

Theorem 11.3 In the same circle or in congruent circles, chords that are equidistant from the center(s) are congruent.

EXAMPLE 3 Given: $\odot O$, $\overline{OP} \cong \overline{OQ}$, $\overline{OP} \perp \overline{RT}$, $\overline{OQ} \perp \overline{ST}$,
   $m\angle T = 40$
   Find $m\angle R$.

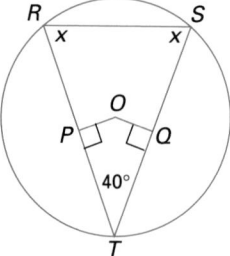

Plan Chords $\overline{TR}$ and $\overline{TS}$ are equidistant from the center O. So by Theorem 11.3, the chords are congruent. Then $\triangle RTS$ is isosceles and $m\angle R = m\angle S$.

Solution Let $m\angle R = m\angle S = x$.

Then, $x + x + 40 = 180$
$$2x + 40 = 180$$
$$2x = 140$$
$$x = 70 \quad \text{Thus, } m\angle R = 70$$

EXAMPLE 4 Given: $\odot O$ has radius of length 13;
   chords $\overline{AB}$ and $\overline{DF}$ are 5 and 12
   units from O, respectively.
   Show that $AB > DF$.

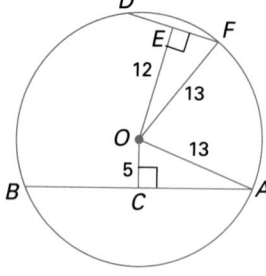

Solution Use the Pythagorean Theorem to compute the lengths.

$(AC)^2 + 5^2 = 13^2$ $(EF)^2 + 12^2 = 13^2$
$(AC)^2 + 25 = 169$ $(EF)^2 + 144 = 169$
$\quad (AC)^2 = 144$ $\quad (EF)^2 = 25$
$\qquad AC = \sqrt{144} = 12$ $\qquad EF = \sqrt{25} = 5$
$AB = 2 \cdot AC = 2 \cdot 12 = 24$ $DF = 2 \cdot EF = 2 \cdot 5 = 10$

Thus, $AB > DF$.

It is seen in Example 4 that the chord nearer the center of the circle is the longer chord. This suggests the following theorem and its converse.

Additional Example 3

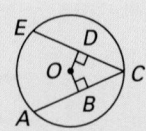

Given: $OD = OB$, $\overline{OD} \perp \overline{EC}$, $\overline{OB} \perp \overline{AC}$,
$m\angle E = 50$. Find $m\angle C$. 80

Additional Example 4

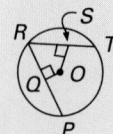

Given: Circle O with radius 10, $RT = 16$, $QR = 5\sqrt{3}$. Show that $\overline{PR}$ is closer to the center of the circle than chord $\overline{TR}$.

$OQ = 5$, $OS = 6$, $OQ < OS$

Theorem 11.4	In the same circle or congruent circles, if two chords are unequally distant from the center(s), then the chord nearer its corresponding center is the longer chord.

Theorem 11.5	In the same circle or congruent circles, if two chords are unequal in length, then the longer chord is nearer the center of its circle.

EXAMPLE 5 Determine which side of the triangle is farthest from the center.

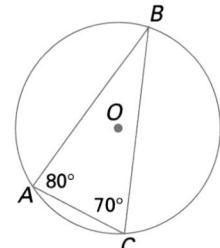

Solution
$$70 + 80 + m \angle B = 180$$
$$150 + m \angle B = 180$$
$$m \angle B = 30$$

$\overline{AC}$ is the shortest side of $\triangle ABC$, or the shortest chord of $\odot O$. It follows then from Theorem 11.5, that $\overline{AC}$ is the chord farthest from the center.

In the figure above, $\triangle ABC$ is said to be *inscribed* in the circle, because its sides are chords of the circle.

Definitions	A polygon whose sides are chords of a circle is an **inscribed polygon**. The polygon is **inscribed** in the circle; the circle is **circumscribed** about the polygon.

Classroom Exercises

1. What length represents the distance from chord $\overline{PR}$ to O?   *OQ*
2. Find m $\angle 1$ if m $\angle POR = 100$.   50
3. Given: $PQ = 2a$; $PR = \underline{4a}$
4. Given: $PR = 16$; $PQ = \underline{8}$

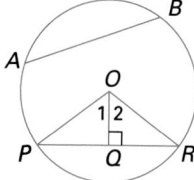

Written Exercises

1. Given: $PQ = 24$, $OR = 5$
   Find PO.   13
2. Given: $PQ = 8$, $OR = 4$
   Find PO. $4\sqrt{2}$

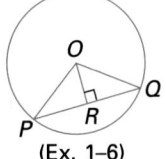

(Ex. 1–6)

11.2 Properties of Chords   **423**

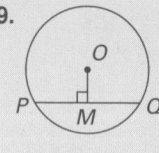

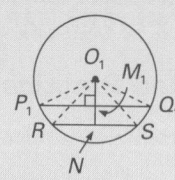

Statement	Reason
1. ⊙O with chord $\overline{PQ}$ ⊥ $\overline{OM}$, ⊙O_1 with chord $\overline{RS}$ ⊥ $\overline{O_1N}$, $OM \angle ON$, ⊙$O_1 \cong$ ⊙O	1. Given
2. Construct pt M_1 on $\overline{O_1N}$ such that $OM_1 = OM$.	2. Ruler Post
3. Construct chord $\overline{P_1Q_1}$ through pt M_1 ⊥ $\overline{O_1N}$.	3. Parallel Post
4. $PQ = P_1Q_1$	4. In $\cong$ ⊙s, equidistant chords are $\cong$.
5. m $\angle P_1O_1Q_1$ = m $\angle P_1O_1R$ + m $\angle RO_1S$ + m $\angle SO_1Q_1$	5. Angle Add Post
6. m $\angle P_1O_1Q_1$ > m $\angle RO_1S$	6. $a > b$ if there exists a positive number c such that $b + c = a$.
7. $O_1P_1 = O_1R = O_1S = O_1Q_1$	7. Radii of ⊙ are $\cong$.
8. $P_1Q_1 > RS$	8. SAS Ineq
9. $PQ > RS$	9. Sub

See page 431 for the answer to Ex. 20.

3. Given: $OR = 4$, $PO = 8$
Find PQ. $8\sqrt{3}$

4. Given: $PO = 6$, $OR = 4$
Find PQ. $4\sqrt{5}$

5. m $\angle POQ = 120$, $PQ = 10$
Find OP. $\frac{10}{3}\sqrt{3}$

6. Given: m $\angle POQ = 60$,
$OR = 6$
Find OQ. $4\sqrt{3}$

In circle O, $\overline{OU} \perp \overline{TS}$ and $\overline{OV} \perp \overline{RS}$.

7. Given: $OU = 4$, $OV = 4$,
$RV = 8$
Find TS. 16

8. Given: $TS = 12$, $RV = 6$,
and $OV = 4$
Find OU. 4

9. Given: $OU = OV$,
m $\angle S = 70$
Find m $\angle T$. 55

10. Given: $OU = OV$,
m $\angle T = 20$
Find m $\angle S$. 140

11. Given: m $\angle R = 25$,
m $\angle T = 55$
Name the shortest chord. $\overline{TS}$

12. Given: m $\angle T = 60$,
m $\angle S = 64$
Which chord is closest to the center? $\overline{TR}$

13. A chord of length 12 is 4 units from the center of a circle. Find the length of a diameter to the nearest tenth. 14.4

14. A chord of a circle is 12 units from the center. Draw a picture and estimate the length of the chord if the length of a radius is 14. Then find the length to the nearest hundredth. 14.42

15. A diameter of a circle is of length 16. The length of a chord is 12. Find the distance of the chord from the center of the circle. $2\sqrt{7}$

16. In circle O, $\overline{AB}$ is a chord. $\overline{OA}$ and $\overline{OB}$ are radii. m $\angle AOB = 120$, $OA = 4$. Find the distance of the chord from the center of the circle. 2

17. Write a biconditional for Theorem 11.2 and its converse.

18. Prove Theorem 11.3.   **19.** Prove Theorem 11.4.

20. Prove Theorem 11.5.

21. An archaeologist finds a piece of what appears to be a circular pipe. Holding the ends of a meter stick against the inside of the pipe, he finds that the midpoint of the 100-cm stick is 8 cm from the pipe wall. Find the diameter of the pipe. 320.5 cm

Mixed Review

1. Given: $\overline{DE} \parallel \overline{AB}$, m $\angle 2 = 30$, m $\angle 5 = 20$
Find m $\angle 4$. *3.2* 50

2. Given: m $\angle 1 = 40$, m $\angle 2 = 45$
Find m $\angle 3$. *3.6* 95

3. Given: m $\angle 1 = 60$, m $\angle 4 = 80$
Find m $\angle 2$. *3.7* 20

4. Given: $\overline{AD} \perp \overline{AB}$, m $\angle 5 = \frac{2}{3}$m $\angle 6$
Find m $\angle 6$. *5.4* 54

Enrichment

Have students draw two segments, one 2 in. long and the other 3 in. long, intersecting at an angle of measure 45. Ask them to construct a circle that contains these two segments as chords.

The perpendicular bisectors of the chords will intersect at the center of the desired circle.

11.3 Special Properties of Tangents to Circles

Objective	To prove and apply theorems about tangents to circles

This lesson develops basic relationships of tangents and circles.

Theorem 11.6

If a line is perpendicular to a radius at its endpoint on the circle, then the line is tangent to the circle.

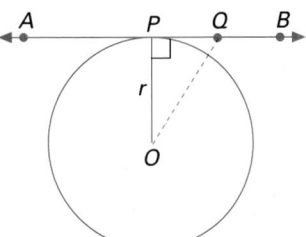

Given: $\overleftrightarrow{OP} \perp \overleftrightarrow{AB}$

Prove: $\overleftrightarrow{AB}$ is tangent to $\odot O$ at P.

Proof

Choose any point Q on $\overleftrightarrow{AB}$ other than P, and draw right $\triangle OPQ$. The hypotenuse of a right triangle is its longest side, so $OQ > r$. Therefore, Q is the exterior of the circle. This makes P the *only* point on $\overleftrightarrow{AB}$ lying on $\odot O$; thus, by definition $\overleftrightarrow{AB}$ is tangent to $\odot O$ at P.

Definition

A **tangent segment** is a segment that contains a point of tangency and another point of a tangent line to a circle.

Theorem 11.7

If a line is tangent to a circle, then the line is perpendicular to the radius drawn to the point of tangency.

EXAMPLE 1	Given: $\overline{AB}$ tangent to $\odot O$, $\overline{OA}$ is a radius, $OA = 6$, $OB = 8$.

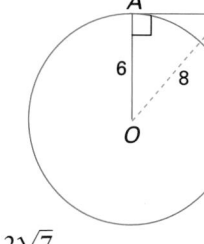

Find the length of segment $\overline{AB}$.

Plan	Angle OAB is a right angle since the radius is perpendicular to the tangent at A.
Solution	Let $AB = x$. $\quad x^2 + 6^2 = 8^2$

$$x^2 + 36 = 64$$
$$x^2 = 28$$
$$x = \sqrt{28} = \sqrt{4} \cdot \sqrt{7} = 2\sqrt{7}$$

Thus, the length of segment $\overline{AB}$ is $2\sqrt{7}$.

Additional Example 1

Use the figure for Example 1.
Given: $AB = 4$, $OB = 6$
Find: Length of the diameter. $4\sqrt{5}$

Teaching Resources

Manipulative Worksheet 20
Quick Quizzes 77
Reteaching and Practice Worksheets, pp. 153, 154

▰▰▰ GETTING STARTED

Prerequisite Quiz

1. Given: The legs of a right triangle have lengths 6 and 4. Find the length of the hypotenuse. $2\sqrt{13}$
2. Given: The length of the hypotenuse of a 60–30–90 triangle is 6. Find the length of the legs.   3 and $3\sqrt{3}$
3. $y = 2$, $x = 6$, $z =$ _____ $4\sqrt{2}$
4. $m \angle Z = 30$, $y = 4$, $z =$ _____ $\frac{4\sqrt{3}}{3}$
5. $m \angle Y = 45$, $x = 8$, $y =$ _____ $4\sqrt{2}$

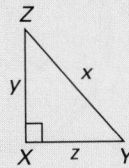

Motivator

Ask the students to draw a circle of any radius with center at some point O and label any point A on the circle. At point A, have them draw a line that appears to be tangent to the circle and draw the radius $\overline{OA}$. Ask them what relationship appears to exist between the radius and the tangent.
Theorem 11.7 page 425

▰▰▰ TEACHING SUGGESTIONS

Lesson Note

Try to get students to discover properties of tangents. Use physical illustrations. The straight segments of a fan belt are tangent to a circular pulley. See if students can guess the relationship between the length of tangent segments to a circle from an exterior point of the circle. Likewise, try to get students to discover the corollary to Theorem 11.8.

Math Connections

Design: Many common objects, such as the blades of a ceiling fan, the arms of an armchair and a triangular rack for pool balls, usually have rounded corners. The diagram below illustrates how congruent tangent segments $\overline{AB}$ and $\overline{BC}$ can be used in drawing the designs of such objects to round off corners.

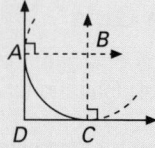

Critical Thinking Question

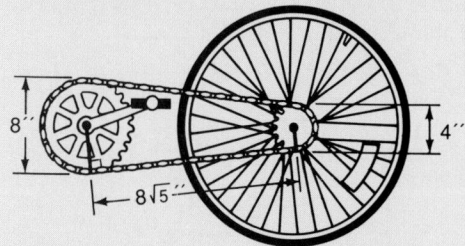

Application: A bike chain contains two external tangent segments that are $8\sqrt{5}$ inches long between the pedal gear and the hub of the rear wheel. If the diameter of the pedal gear is 8 inches and the diameter of the hub is 4 inches, how far apart are their centers? See Example 4 on page 427.

$\sqrt{(8\sqrt{5})^2 + 2^2} = \sqrt{324} = 18$ in

The figure at the right shows two segments tangent to a circle from an exterior point T. To prove that these segments are congruent segments, draw auxiliary lines $\overline{OA}$, $\overline{OB}$, and $\overline{OT}$. The right triangles formed can be proved congruent.

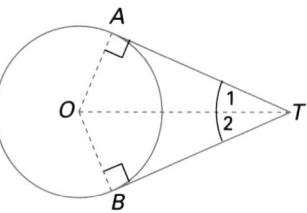

Theorem 11.8 Two segments drawn tangent to a circle from an exterior point are congruent.

Corollary The angle between two tangents to a circle from an exterior point is bisected by the segment joining its vertex to the center of the circle.

EXAMPLE 2 The measure of the angle between two tangents to a circle from a point P is 60. Find the length of a radius if the length of each tangent is 8.

Plan Make a sketch. By the corollary to Theorem 11.8, $\overline{OP}$ bisects $\angle APB$. So, m $\angle APO = 30$. $\triangle OAP$ is a 30-60-90 triangle.

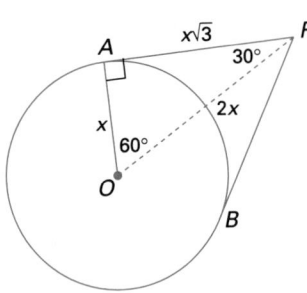

Solution
$$AP = x\sqrt{3} = 8$$
$$x = \frac{8}{\sqrt{3}} = \frac{8 \cdot \sqrt{3}}{\sqrt{3} \cdot \sqrt{3}}$$
$$x = \frac{8\sqrt{3}}{3}$$

Thus, the length of a radius is $\frac{8\sqrt{3}}{3}$.

EXAMPLE 3 Find x in the figure at the right if each side of $\triangle ABC$ is tangent to $\odot O$.

Plan Use the property that tangent segments from an exterior point are congruent. Label each segment in terms of x.

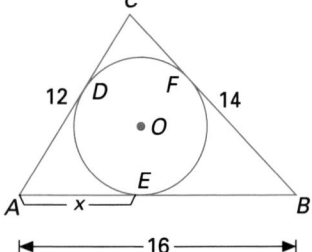

Additional Example 2

Two tangents to a circle from a point P are perpendicular to each other. The distance from P to the center of the circle is 8. Find the length of a radius. $4\sqrt{2}$

Additional Example 3

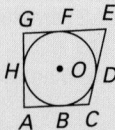

The sides of the quadrilateral are tangent to the circle. $AH = 5$, $CD = 4$, $GF = 3$, and $GE = 10$. Find $AC + CE + GE + GA$.   38

Solution

Let $AD = AE = x$.
Then, $EB = 16 - x$ and $DC = 12 - x$
Also, $FB = 16 - x$ and $CF = 12 - x$

By the Segment Addition Postulate,
$$CB = CF + FB$$
$$14 = (12 - x) + (16 - x)$$
$$14 = 28 - 2x$$
$$-14 = -2x$$

Thus, $x = 7$.

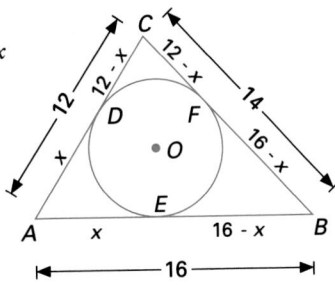

The figure for Example 3 above shows a circle inscribed in a polygon.

Definition

A polygon whose sides are tangent to a circle is a **circumscribed polygon**. The polygon is circumscribed about the circle, and the circle is *inscribed* in the polygon.

Definition

A line tangent to each of two coplanar circles is a **common tangent**.

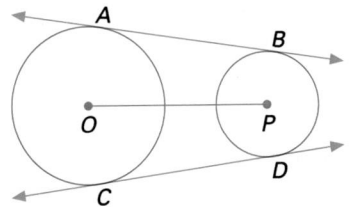

Common *external* tangents $\overleftrightarrow{AB}$ and $\overleftrightarrow{CD}$ do not intersect the segment $\overline{OP}$ joining the centers.

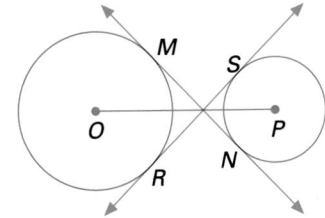

Common *internal* tangents $\overleftrightarrow{MN}$ and $\overleftrightarrow{RS}$ intersect the segment $\overline{OP}$ joining the centers.

EXAMPLE 4

The centers of two circles of radii of lengths 3 and 8 are 13 units apart. Find the length of a common external tangent segment.

Plan

Draw two circles with centers O and P. Draw $\overline{OP}$, common external tangent $\overline{AB}$, and radii $\overline{OA}$ and $\overline{PB}$. Draw $\overline{PQ} \perp \overline{OA}$ so that $ABPQ$ forms a rectangle. Since the opposite legs of a rectangle are congruent, $AQ = BP = 3$ and $AB = QP$. Note that $\overline{QP}$ is a leg of $\triangle OQP$. $OQ = 8 - 3 = 5$.

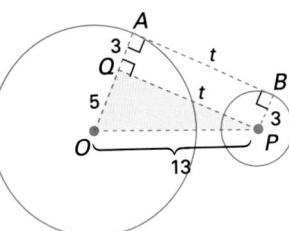

11.3 Special Properties of Tangents to Circles **427**

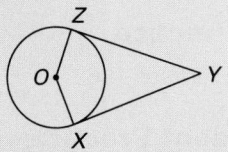

427

Guided Practice

Classroom Exercises 1–6

Independent Practice

A Ex. 1–11, **B** Ex. 12–17, **C** Ex. 18–21

Basic: WE 1–11 odd, Cl all
Average: WE 1–17 odd, Cl all
Above Average: WE 1–21 odd, Cl all

Additional Answers

Classroom Exercises

1. $\overline{OA} \perp \overline{FB}$, $\overline{OC} \perp \overline{DB}$; if a line is tan to ⊙, then it is ⊥ to radius drawn to pt of tan.
4. $AO = OC = 4$, since radii of ⊙ are ≅; △BAO is rt because $\overline{AO}$ is ⊥ to $\overline{FB}$ at pt of tan; by Pythagorean Thm $4^2 + 4^2 = (OB)^2$; $OB = 4\sqrt{2}$.
5. m ∠1 = 40, since ∠ between 2 tan to ⊙ from ext pt is bis by seg joining it and the ctr.

Solution Use the Pythagorean Theorem.
$$5^2 + t^2 = 13^2$$
$$25 + t^2 = 169$$
$$t^2 = 144$$
$$t = \sqrt{144} = 12$$

Thus, the length of tangent segment $\overline{AB}$ is 12.

Two circles can be tangent to the same line at the same point. At the right, circle O is in the interior of circle P, except for the point of tangency. The two circles are said to be *internally tangent* at point T.

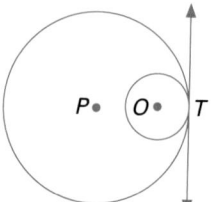

In the next diagram, circle S is in the exterior of circle R, except for the point of tangency V. The circles are said to be *externally tangent* at point V.

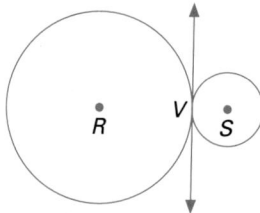

Definition Two coplanar circles are tangent to each other if they are tangent to the same line at the same point.

Classroom Exercises

Given: ⊙O is inscribed in △BFD.

1. Which segments are perpendicular? Why?
2. $\overline{CD} \cong$ _____ED_____. Why? Two segs tan to ⊙ are ≅.
3. $\overline{FE} \cong$ _____FA_____. Why? Two segs drawn to ⊙ are ≅.
4. $AB = 4$, $OC = 4$. Tell how to find OB.
5. m ∠$FBD = 80$; m ∠1 = _____. Why?
6. $CD = 5$, $FD = 12$, $FE =$ _____7_____

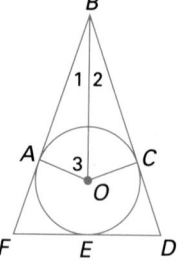

Enrichment

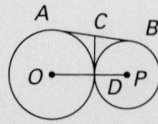

Given: $\overline{AB}$ is a common external tangent segment to the two circles; $\overline{CD}$ is a common internal tangent.

Prove: C is the midpoint of $\overline{AB}$.
$\overline{AB}$ is a common external tangent segment. (Given); $\overline{CD}$ is a common internal tangent segment. (Given); $AC = CD$, $CB = CD$ (Tan seg from ext pts are ≅.); $AC = BC$ (Sub); C is midpoint of $\overline{AB}$ (Def of midpt).

Written Exercises

Use the figures below for Exercises 1–6.

$\overline{PQ}$ is a tangent segment.

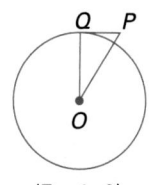

(Ex. 1–2)

$\overline{AB}$ and $\overline{AC}$ are tangent segments.

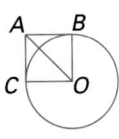

(Ex. 3–4)

Circle O is inscribed in $\triangle ABC$.

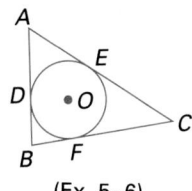

(Ex. 5–6)

1. $QP = 2$, $OP = 6$
   Find OQ. $4\sqrt{2}$

2. m $\angle P = 60$,
   $OQ = 2$
   Find PQ. $\frac{2}{3}\sqrt{3}$

3. m $\angle CAB = 120$,
   $OB = 6$
   Find AB and AC. $2\sqrt{3}$

4. m $\angle CAB = 90$,
   $OA = 10$
   Find OB and OC. $5\sqrt{2}$

5. $AC = 12$, $CB = 10$,
   $AB = 8$
   Find BF. 3

6. $AB = 6$, $AC = 8$,
   $BC = 9$
   Find EC. $5\frac{1}{2}$

In the figure at the right, $\overline{RQ}$ is an external tangent segment to circles O and P; $\overline{SP} \perp \overline{OR}$.

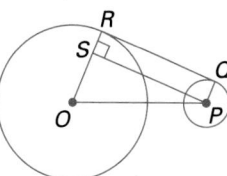

7. Given: $PQ = 6$, $OR = 10$, $OP = 20$
   Find RQ. $8\sqrt{6}$

8. Given: $RQ = 8$, $OP = 10$, $QP = 1$
   Find OS and OR. $OS = 6$, $OR = 7$

9. Given: Tangent segments $\overline{BA}$, $\overline{BD}$, and $\overline{BC}$;
   $BD = 17$
   Find AB. 17

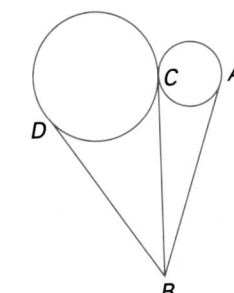

10. Given: Circumscribed $\triangle PQR$, $QU = 2$, $UR = 5$, and $PS = 4$
    Find $QR + RP + QP$. 22

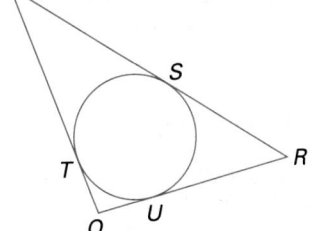

11. Given: Circumscribed polygon $ABCD$, $BE = 5$, $GD = 3$, $AH = 4$, and $CG = 6$
    Find $AB + BC + CD + DA$. 36

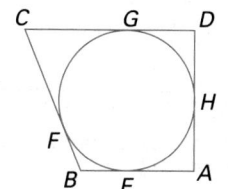

Written Exercises

12. Use the figure above Thm 11.8.

Statement	Reason
1. $\odot O$, with tan $\overline{TA}$ and $\overline{TB}$	1. Given
2. Draw seg $\overline{OA}$, $\overline{OB}$, $\overline{OT}$.	2. 2 pts determine a line.
3. $OA = OB$	3. Def of $\odot$
4. $\overline{OT} \cong \overline{OT}$	4. Reflex Prop of Congr
5. $\overline{OA} \perp \overline{TA}$ $\overline{OB} \perp \overline{BT}$	5. If line is tan, it is $\perp$ to radius at pt of tan.
6. $\angle OAT$ and $\angle OBT$ are rt $\angle$s.	6. Def of $\perp$
7. $\triangle OAT$ and $\triangle OBT$ are rt $\triangle$.	7. Def of rt $\triangle$
8. $\triangle OAT \cong \triangle OBT$	8. HL
9. $\overline{AT} \cong \overline{BT}$	9. CPCTC

13. Use the figure above Thm 11.8. Use Ex. 12 to justify $\triangle OAT \cong \triangle OBT$.

Statement	Reason
1. $\triangle OAT \cong \triangle OBT$	1. Ex. 12
2. $\angle 1 \cong \angle 2$.	2. CPCTC
3. $\overline{OT}$ bis $\angle ATB$	3. Def of $\angle$ bis

14.

Statement	Reason
1. $\odot O \cong \odot Q$, $\odot O$ and $\odot Q$ are tan to $\overleftrightarrow{SM}$ at M.	1. Given
2. $\overline{OM} \perp \overleftrightarrow{MS}$, $\overline{QM} \perp \overline{MS}$	2. Radii are $\perp$ to tan at pt of tan.
3. $\angle OMS$ and $\angle QMS$ are rt $\angle$s.	3. Def of $\perp$
4. $\angle OMS \cong \angle QMS$	4. All rt $\angle$s are $\cong$.
5. $\overline{OM} \cong \overline{QM}$	5. Radii of $\cong \odot$s are $\cong$.
6. $\overline{SM} \cong \overline{SM}$	6. Reflex Prop of Cong
7. $\triangle OMS \cong \triangle QMS$	7. SAS
8. $\overline{OS} \cong \overline{QS}$	8. CPCTC

15.

Statement	Reason
1. ⊙O and ⊙P are tan to $\overline{AC}$ and $\overline{AD}$.	1. Given
2. $AC = AD$, $AB = AE$	2. 2 segs tan to ⊙ from ext pt are ≅.
3. $AC = AB + CB$, $AD = AE + DE$	3. Seg Add Post
4. $AB + CB = AE + DE$	4. Sub
5. $CB = DE$	5. Equations may be subtracted.

19.

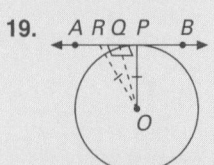

Given: $\overleftrightarrow{AB}$ is tan to ⊙O at pt P.
Assume: $\overleftrightarrow{AB}$ ∡ to $\overline{OP}$. Since there is exactly one ⊥ to a line from a pt not on that line (see proof in Ex. 9 on page 203), draw $\overline{OQ}$ ⊥ to $\overleftrightarrow{AB}$. By the Ruler Post, locate pt R on $\overleftrightarrow{AB}$ such that $RQ = PQ$. Then, by the Reflex Prop of Congr, $\overline{QO} \cong \overline{QO}$. Since $\overline{OQ}$ ⊥ $\overleftrightarrow{AB}$, ∠RQO and ∠PQO are rt ∠s. Since all rt ∠s are ≅, ∠RQO ≅ ∠PQO. Then, by SAS, △RQO ≅ △PQO. By CPCTC, $\overline{OR} \cong \overline{OP}$. Then, by Def of ⊙, pt R is on ⊙O. Thus, $\overleftrightarrow{AB}$ intersects ⊙O at 2 pts, R and P. But this is a contradiction of the Def of tan. So the assumption that $\overleftrightarrow{AB}$ ∡ to $\overline{OP}$ is false. Therefore, $\overleftrightarrow{AB}$ ⊥ $\overline{OP}$.

12. Prove Theorem 11.8.
14. Given: ⊙O ≅ ⊙Q, ⊙O and ⊙Q are tangent to $\overleftrightarrow{SM}$ at M.
Prove: $\overline{OS} \cong \overline{QS}$

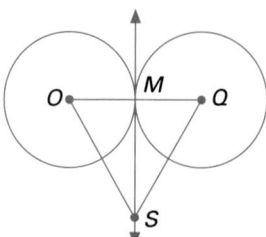

13. Prove the corollary to Theorem 11.8.
15. Given: Circles O and P are tangent to $\overline{AC}$ and $\overline{AD}$.
Prove: $CB = DE$

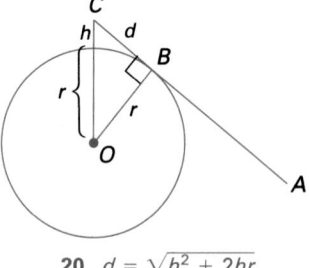

16. The measure of the angle between two tangent segments to a circle from a point R is 120. Find the length of a diameter if the length of each tangent segment is 10. $20\sqrt{3}$

17. The centers of two circles of radii lengths 10 and 2 are 17 units apart. Find the length of a common external tangent segment.   15

18. The lengths of the radii of two circles are 4 and 2. The centers are 10 units apart. Find the length of the common internal tangent segment.   8

19. Prove Theorem 11.7. (HINT: Use an indirect proof.)

The figure shows a point C at a height h above the earth. d is the distance from C to the visual horizon. r is the radius of the earth. $\overline{AC}$ is a tangent segment.

20. Write a formula for d in terms of h and r.

21. The formula of Exercise 20 can be used to calculate the range of VHF (very high frequency) ship-to-shore communications. A ship has a VHF transceiver on top of a mast 36 m tall. Find d, the maximum communication range of the ship to the nearest kilometer. (Use $r \approx 6{,}380$ km.)

20. $d = \sqrt{h^2 + 2hr}$
21. $d \approx 21$ km

Mixed Review

1. Find the number of sides of a regular polygon if the measure of each interior angle is 120. *6.2*   6

2. Find the measure of each of two complementary angles if their measures are in the ratio 2 to 3. *8.1*   36, 54

3. Find the length of an altitude of equilateral triangle ABC if AB = 8.   *9.4*
$4\sqrt{3}$

Computer Investigation

Arcs and Related Angles

Use a computer software program that draws circles of a given radius length, places random points on or in the exterior of the circle, labels points, constructs line segments between two given points, draws tangent segments to the circle at a given point, and measures angles.

The measure of a central angle equals the degree measure of its arc. Thus, in the figure, $\angle BAC$ is a central angle and m $\angle BAC = $ m$\overparen{BC}$. $\angle BDC$ is *not* a central angle. It is an angle formed by two chords, not radii. The computer activities below will help you discover a pattern for finding the measure of an angle such as $\angle BDC$, which intercepts $\overparen{BC}$, the arc of central angle BAC. This property will be formally stated and proved in the next lesson.

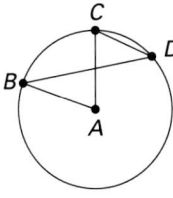

Activity 1

Draw a circle of radius 5, or some length that will fit on the screen. Label the center A. Draw three random points, B, C, and D on the circle as illustrated in the diagram above. Draw $\overline{AB}$, $\overline{AC}$, $\overline{DB}$, and $\overline{DC}$.

Find the following measures.

1. m $\angle BAC$ (measure of a central angle)
2. m $\angle BDC$

Repeat the steps above for the following three circles.

3. radius = 5, but with three new points B, C, and D on the circle
4. radius = 6
5. radius = 4
6. Draw $\odot A$. Place B and C randomly on the circle; draw $\overline{AB}$, $\overline{AC}$, and $\overline{BC}$, and a tangent segment $\overline{BD}$.
7. Find m $\angle BAC$ and m $\angle CBD$.
8. Repeat Exercises 6 and 7 for circles of radii 4 and 5.

Summary

What generalizations appear to hold true for the measures of:

1. an angle with its vertex on the circle and having its sides as chords of the circle? It is $\frac{1}{2}$ the measure of central $\angle$ of same arc.
2. an angle with its vertex on the circle and having its sides as a chord and tangent of the circle? It is $\frac{1}{2}$ the measure of central $\angle$ of same arc.

Additional Answer, page 424

20.

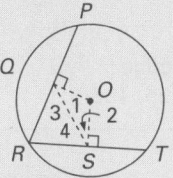

Statement	Reason
1. $\overline{RP} > \overline{RT}$, $\overline{OQ} \perp \overline{RP}$, $\overline{OS} \perp \overline{RT}$.	1. Given
2. $PQ + RQ = RP$, $TS + RS = RT$.	2. Seg Add Post
3. $PQ + RQ > TS + RS$.	3. Sub
4. $PQ = RQ$, $TS = RS$	4. If line contains ctr of $\odot$ and is $\perp$ to chord, it bis chord.
5. $2 \cdot RQ > 2 \cdot RS$.	5. Sub
6. $RQ > RS$	6. Div Prop of Ineq
7. m $\angle 4 >$ m $\angle 3$	7. If 1 side of $\triangle$ is longer than 2nd side, then $\angle$ meas opp longer side > $\angle$ meas opp shorter side.
8. m $\angle 1 +$ m $\angle 3 = 90$, m $\angle 2 +$ m $\angle 4 = 90$.	8. If outer rays of 2 adj acute $\angle$s $\perp$, sum of $\angle$ meas = 90.
9. m $\angle 3 = 90 - $ m $\angle 1$, m $\angle 4 = 90 - $ m $\angle 2$.	9. Subtr Prop of Eq
10. $90 - $ m $\angle 2 > 90 - $ m $\angle 1$	10. Sub
11. $-$m $\angle 2 > -$m $\angle 1$	11. Subtr Prop of Ineq
12. m $\angle 2 <$ m $\angle 1$	12. Mult Prop of Ineq
13. $OQ < OS$	13. If 1 $\angle$ of $\triangle$ has greater meas than 2nd $\angle$, then side opp greater $\angle$ is longer than side opp smaller $\angle$.

◼️GETTING STARTED

Prerequisite Quiz

Given: *AC* = *BC*

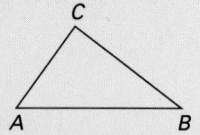

1. m ∠*A* = 40, m ∠*C* = _____ 100
2. m ∠*C* = 120, m ∠*B* = _____ 30
3. The measures of the angles of a triangle are in the ratio 3:4:5. Find the measure of the largest angle.   75
4. Given: ∠1 is a supplement of ∠2. m ∠1: m ∠2 = 4:5. Find m ∠1.   80

Motivator

Ask the students to draw a circle with center *O*. Ask them what they think is meant by a *central angle* of a circle.   An angle whose vertex is the center of the circle.   Have them draw a central acute angle *AOB*. Draw their attention to points *A* and *B* which divide the circle into two **arcs**. Ask them which arc they think will be called a minor arc. $\overarc{AB}$ A major arc.   The other arc. Ask them how they would define a minor arc and a major arc.   Definitions pages 432 and 433.

◼️TEACHING SUGGESTIONS

Lesson Note

Stress that the reason a central angle is measured by its *minor* arc is based upon the original agreement in Chapter 1 that in geometry an angle will not have a measure greater than 180.

11.4 Arcs and Central Angles

Objective   To find degree measures of arcs

You know that a segment may be considered part of a line. In the same way, an *arc* is an unbroken part of a circle. An arc is measured by referring to an angle whose vertex is the center of the circle.

Definitions

A **central angle** of a circle is an angle with measure less than 180 whose vertex is the center of the circle. The **minor arc** $\overarc{AB}$ of central angle *AOB* consists of points *A*, *B*, and all points on the circle that lie in the interior of the central angle. The **degree measure of a minor arc** is equal to the measure of its central angle: $m\overarc{AB}$ = m ∠*AOB*.

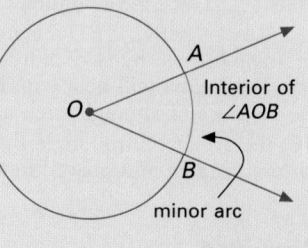

Since Chapter 1, you have assumed that measures of angles are given in degrees. Similarly, you may now assume that measures of arcs are given in degrees.

EXAMPLE 1   Given: m ∠*AOB* = 20, m ∠*BOC* = 40
Find $m\overarc{AC}$.

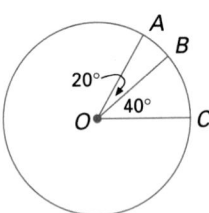

Solution   m ∠*AOC* = m ∠*AOB* + m ∠*BOC*
by the Angle Addition Postulate.
m ∠*AOC* = 20 + 40 = 60

Thus, by definition of the measure of a minor arc, $m\overarc{AC}$ = 60.

At the right, two adjacent central angles together form a straight angle. A straight angle separates the points of the circle into two special arcs, $\overarc{RTS}$ and $\overarc{RQS}$, called *semicircles*. The degree measure of a semicircle is 180.

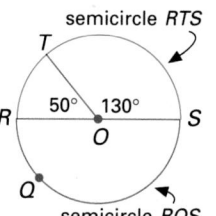

semicircle *RTS*

semicircle *RQS*

Additional Example 1

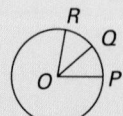

Given: $\overline{OQ}$ bisects ∠*ROP*, m ∠*POQ* = 35
Find $m\overarc{RP}$.   70

Definition	Semicircle $\overarc{RTS}$ consists of the endpoints R and S of diameter $\overline{RS}$ and all points of $\odot O$ that lie on the same side of $\overline{RS}$ as T.

Definition	The **major arc** $\overarc{ACB}$ consists of points A and B and all points of the circle in the *exterior* of central angle AOB. The degree measure of $\overarc{ACB}$ equals 360 minus the measure of central angle AOB.

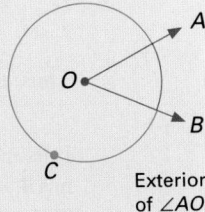

Exterior
of $\angle AOB$

Degree measures of arcs can be added in a manner very similar to that in which degree measures of angles are added. The following postulate permits this. It applies to minor arcs, major arcs, and semicircles.

Postulate 19	**Arc Addition Postulate** If P is a point on $\overarc{APB}$, then $m\overarc{AP} + m\overarc{PB} = m\overarc{APB}$.

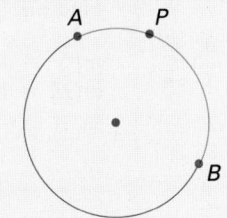

The sum of the degree measures of the two semicircles is $m\overarc{RTS} + m\overarc{RQS} = 180 + 180 = 360$. Thus, the degree measure of a circle is 360.

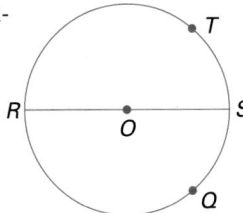

The following equations show addition of degree measures of arcs for the figure at the right.

$m\overarc{APB} = m\overarc{AP} + m\overarc{PB}$
$\qquad = 25 + 80 = 105$
$m\overarc{CAP} = m\overarc{CA} + m\overarc{AP}$
$\qquad = 100 + 25 = 125$

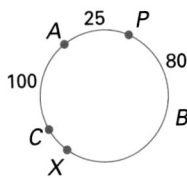

11.4 Arcs and Central Angles   **433**

Additional Example 2

Given: $m\overarc{ACB} = 290$
Find m $\angle A$.   55

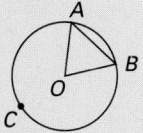

Math Connections

Probability: A game spinner has pins spaced around the circumference of a spinning wheel to divide the wheel into congruent sections. The probability of the pointer landing in any particular section of the spinner is 1 in 16. For example, the probability of landing on $1000 for the spinner shown is 1 ÷ 16 or .0625 (6.25%).

Critical Thinking Questions

MATH CLUB MEMBERSHIP

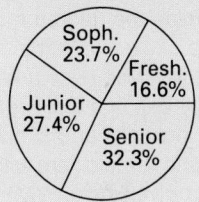

Application: Refer students to the pie chart, or circle graph, above and ask them to determine the measure of each arc using the percents shown.   16.6% = 59.76; 23.7% = 85.32; 27.4% = 98.64; 32.3% = 116.28

Common Error Analysis

Error: When solving exercises such as those illustrated in Examples 4 and 5 on pages 434–435, students tend to think they are finished when they have found x by solving the related equation.

Emphasize that students should read instructions carefully and use the value of x to find the unknown arc measures.

Checkpoint

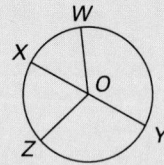

**Given: Circle O with diameter $\overline{XY}$.
Find the measure of the indicated arc.**

1. $m\overarc{WY} = 100$, $m\overarc{XW} = $ _____   80
2. $m\overarc{XZ} = 75$, $m\overarc{ZY} = $ _____   105
3. $m\overarc{XZ} = 80$, $m\overarc{YW} = 120$, $m\overarc{ZXW} = $ _____   140
4. $m\overarc{ZX}$: $\overarc{XW}$: $\overarc{WY}$: $\overarc{YZ} = 1:2:3:4$. $m\overarc{ZYW} = $ _____   252

Closure

Ask the students what postulate for circles is similar to the Angle Addition Postulate. Arc Addition Postulate, page 433   Ask them how to determine the degree measure of a minor arc of a circle.   It is the same as the degree measure of its central angle.   Ask them how they can find the degree measure of each arc if points A and B divide a circle into a minor and a major arc with degree measures in a given ratio.   Write an equation using the ratios, solve for the variable, and substitute into the ratio to determine the measure of each arc.

EXAMPLE 2   Given that m $\angle B = 20$, find $m\overarc{ACB}$.

Plan   Since radii are congruent, $\triangle AOB$ is isosceles and m $\angle A$ = m $\angle B$. Use the Triangle Sum Theorem.

Solution
$$m \angle A + m \angle B + m \angle AOB = 180$$
$$20 + 20 + m \angle AOB = 180$$
$$m \angle AOB = 140$$

Thus, $m\overarc{AB} = 140$, and $m\overarc{ACB} = 360 - 140 = 220$.

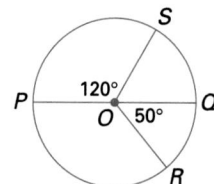

It is important to distinguish between two such symbols as $\overarc{AB}$ and $m\overarc{AB}$. $\overarc{AB}$ is *a set of points* that make up an arc; $m\overarc{AB}$ is the degree *measure* of the arc.

In this text, two letters under an arc sign, such as $\overarc{AB}$, refer *only* to minor arcs. Three letters, such as $\overarc{ACB}$, may refer either to a minor arc (for clarity), a major arc, or a semicircle.

EXAMPLE 3   Given: $\odot O$ with diameter $\overline{PQ}$, m $\angle POS = 120$, m $\angle QOR = 50$
Find $m\overarc{RQS}$.

Solution
$$m\overarc{PS} = 120 \text{ and } m\overarc{RQ} = 50 \text{ (by definition)}$$
$$m\overarc{QS} = 180 - 120 = 60 \text{ } (\overline{PQ} \text{ is a diameter.})$$
Thus, $m\overarc{RQS} = m\overarc{RQ} + m\overarc{QS}$
$$= 50 + 60 = 110.$$

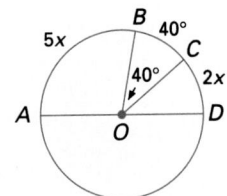

EXAMPLE 4   Given: $\overline{AD}$ is a diameter; m $\angle BOC = 40$, $m\overarc{AB}:m\overarc{CD} = 5:2$.
Find $m\overarc{AC}$.

Plan $\overarc{ABD}$ is a semicircle since $\overline{AD}$ is a diameter. Write an equation relating the three minor arcs. Let $m\overarc{AB} = 5x$, $m\overarc{CD} = 2x$ and $m\overarc{BC} = 40$, since m $\angle BOC = 40$.

Solution
$$m\overarc{ABD} = m\overarc{AB} + m\overarc{BC} + m\overarc{CD} = 180$$
$$5x + 40 + 2x = 180$$
$$7x = 140$$
$$x = 20$$
$$m\overarc{AB} = 5x = 5 \cdot 20, \text{ or } 100$$
Thus, $m\overarc{AC} = m\overarc{AB} + m\overarc{BC} = 100 + 40, \text{ or } 140.$

Additional Example 3

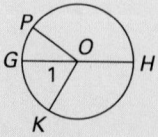

Given: $\overline{GH}$ is a diameter, $m\overarc{PH} = 150$, m $\angle 1 = 70$. Find $m\overarc{PGK}$.   100

Additional Example 4

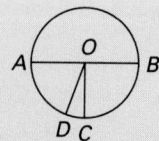

Given: $\overline{AB}$ is a diameter, $m\overarc{BC}:m\overarc{CD}:m\overarc{AD} = 5:2:3$.
Find $m\overarc{CD}$.   36

EXAMPLE 5 Find m$\widehat{ADC}$ for $\odot O$ at right.

Solution The degree measure of a circle is 360.

$$m\widehat{AB} + m\widehat{BC} + m\widehat{CD} + m\widehat{DA} = 360$$
$$40 + 2x + (x + 30) + (3x + 20) = 360$$
$$6x + 90 = 360$$
$$6x = 270$$
$$x = 45$$

Find m$\widehat{AD}$ and m$\widehat{DC}$.

$$m\widehat{AD} = 3x + 20 \qquad\qquad m\widehat{DC} = x + 30$$
$$= 3 \cdot 45 + 20, \text{ or } = 155 \qquad = 45 + 30 = 75$$

Thus, m$\widehat{ADC}$ = m$\widehat{AD}$ + m$\widehat{DC}$ = 155 + 75, or 230.

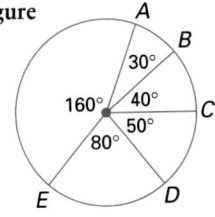

Focus on Reading

Indicate whether the statement is sometimes true, always true, or never true.

1. The degree measure of a minor arc equals the degree measure of its central angle. Always true
2. The degree measure of a semicircle is half the degree measure of the circle. Always true
3. If the degree measures of two minor arcs are added, the result is the degree measure of a major arc. Sometimes true
4. Given an arc of a circle that is not a minor arc, the arc is a major arc. Sometimes true

Classroom Exercises

The measures of some central angles of a circle are given in the figure at the right. Find the measure of the indicated arc.

1. m$\widehat{AC}$ 70 2. m$\widehat{BD}$ 90 3. m$\widehat{EC}$ 130 4. m$\widehat{AD}$ 120 5. m$\widehat{BE}$
6. m$\widehat{EAB}$ 190 7. m$\widehat{ECA}$ 200 8. m$\widehat{DAE}$ 280 9. m$\widehat{EDA}$ 200 10. m$\widehat{CBD}$ 310

Additional Example 5

Find m$\widehat{PRQ}$. 190

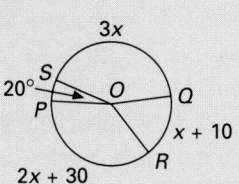

Guided Practice

Classroom Exercises 1–10

Independent Practice

A Ex. 1–14, **B** Ex. 15–25, **C** Ex. 26–28

Basic: FR all, WE 1–14 odd, AR all
Average: FR all, WE 1–25 odd, AR all
Above Average: FR all, WE 1–28 odd, AR all

24.

Statement	Reason
1. $\overline{RS}$ is tan to $\odot O$ at S.	1. Given
2. m $\angle OSR =$ 90.	2. If line is tan to $\odot$, it is $\perp$ to radius at pt of tan.
3. m $\overset{\frown}{ST}$ = m $\angle SOT$	3. Def of arc meas
4. m $\angle SOT$ + m $\angle R$ = 90.	4. In rt $\triangle$, $\angle$s other than rt $\angle$ are comp.
5. m $\overset{\frown}{ST}$ + m $\angle R$ = 90.	5. Sub
6. m $\overset{\frown}{ST}$ = 90 − m $\angle R$.	6. Subt Prop of Eq

25.

Statement	Reason
1. $\overline{AD}$ and $\overline{CD}$ are tan to $\odot O$, m $\overset{\frown}{ABC}$ = 270.	1. Given
2. m $\overset{\frown}{AC}$ = m $\angle AOC$	2. Def of arc meas
3. m $\overset{\frown}{AC}$ = 360 − m $\overset{\frown}{ABC}$.	3. Def of meas of major arc
4. m $\overset{\frown}{AC}$ = 360 − 270 = 90.	4. Sub, Subt Prop of Eq
5. m $\angle AOC$ = 90.	5. Sub
6. m $\angle OAD$ = m $\angle OCD$ = 90.	6. If line tan to $\odot$, line is $\perp$ to radius.
7. $\overline{OA} \parallel \overline{CD}$, $\overline{OC} \parallel \overline{AD}$	7. If int $\angle$s on same side of transv are supp, then lines are $\parallel$.
8. AOCD is $\square$.	8. Def of $\square$
9. AOCD is rect.	9. Def of rect
10. $\overline{OA} \cong \overline{OC}$	10. All radii are $\cong$.
11. AOCD is rhom.	11. $\square$ with 2 $\cong$ adj sides is rhom.
12. AOCD is square.	12. Square is rect that is a rhom.

Written Exercises

Given: $\odot O$ with diameter $\overline{AB}$. Find the measure of the indicated arc.

1. m$\overset{\frown}{AC}$ = 60, m$\overset{\frown}{CB}$ = __120__
2. m$\overset{\frown}{AD}$ = 50, m$\overset{\frown}{DB}$ = __130__
3. m$\overset{\frown}{CB}$ = 100, m$\overset{\frown}{AD}$ = 40, m$\overset{\frown}{ABC}$ = __280__
4. m$\overset{\frown}{DB}$ = 160, m$\overset{\frown}{AC}$ = 30, m$\overset{\frown}{BAC}$ = __210__
5. m$\overset{\frown}{AD}$ = 50, m$\overset{\frown}{CB}$ = 110, m$\overset{\frown}{DCB}$ = __230__
6. m $\angle B$ = 30
   m$\overset{\frown}{BAC}$ = __240__
7. m$\overset{\frown}{BC}$ = 100
   m $\angle B$ = __40__
8. m$\overset{\frown}{BAC}$ = 200
   m $\angle C$ = __10__

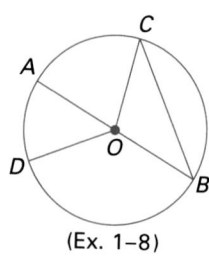

(Ex. 1–8)

9. m$\overset{\frown}{AD}$ = __120__
10. m$\overset{\frown}{DC}$ = __60__
11. m$\overset{\frown}{ADC}$ = __180__
12. m$\overset{\frown}{CB}$ = __110__
13. m$\overset{\frown}{ABC}$ = __180__
14. m$\overset{\frown}{ADB}$ = __290__

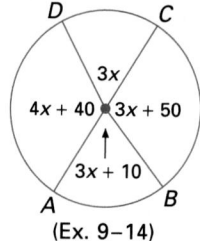

(Ex. 9–14)

For Exercises 15–23, $\odot O$ has diameter $\overline{AB}$.

Given: m$\overset{\frown}{AC}$:m$\overset{\frown}{CD}$ = 3:2, m$\overset{\frown}{DB}$ = 50

15. m$\overset{\frown}{CD}$ = __52__
16. m$\overset{\frown}{CB}$ = __102__
17. m$\overset{\frown}{CBA}$ = __282__

Given: m$\overset{\frown}{CD}$ = 90, m$\overset{\frown}{AC}$:m$\overset{\frown}{DB}$ = 5:4

18. m$\overset{\frown}{AC}$ = __50__
19. m$\overset{\frown}{DB}$ = __40__
20. m$\overset{\frown}{DAB}$ = __320__

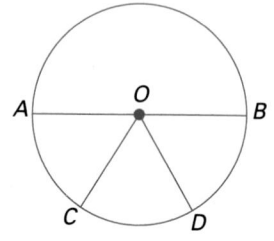

Given: m$\overset{\frown}{AC}$:m$\overset{\frown}{CD}$:m$\overset{\frown}{DB}$ = 3:2:4

21. m$\overset{\frown}{AC}$ = __60__
22. m$\overset{\frown}{DB}$ = __80__
23. m$\overset{\frown}{DBC}$ = __320__

Enrichment

Ask students for the times of the day at which the hands of the clock form a straight angle. Six o'clock is the first that comes to mind. The easiest way to explain the first straight angle after twelve is to consider that the minute hand must move through 30 minutes, plus the distance through which the hour hand has moved, to form the straight angle. Since the hour hand moves at one-twelfth the rate of the minute hand, the amount the minute hand moves will be: $x = 30 + \frac{x}{12}$. This gives a movement ($x =$ time in min.) of $32\frac{8}{11}$ min. or about 32 min. 43.6 sec. The time to the next straight angle will be twice this amount.

24. Given: $\odot O$ with tangent $\overline{RS}$
Prove: $m\widehat{ST} = 90 - m\angle R$

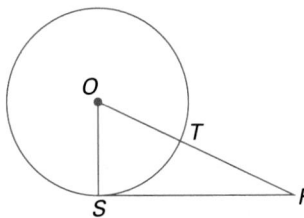

25. Given: $\overline{AD}$ and $\overline{CD}$ are tangent to $\odot O$; $m\widehat{ABC} = 270$.
Prove: $AOCD$ is a square.

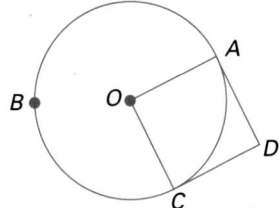

In $\odot O$ at the right, B and D are points of tangency, and $\overline{ED}$ is a diameter.

26. $m\widehat{EB}:m\widehat{BD} = 4:5$, $ED = 10$
Find AB to the nearest tenth. (HINT: Use a trig ratio.)   28.4

27. $AB = 4\sqrt{3}$, $OB = 4$
Find $m\widehat{BDE}$.   300

28. Prove: $m\angle C = m\widehat{EB}$

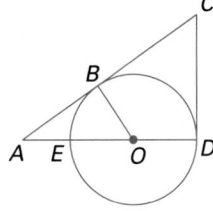

Mixed Review

1. The measure of a base angle of an isosceles triangle is 15 greater than the measure of the vertex angle. Find the measure of the vertex angle.   **5.1**   50

2. In quadrilateral $ABCD$, $\overline{AB} \cong \overline{DC}$ and $\overline{AD} \parallel \overline{BC}$. Is $ABCD$ a parallelogram? Justify your answer.   **6.4**   Not necessarily. $ABCD$ could be isos trap.

3. True or false? Skew lines are coplanar lines.   **3.1**   False

4. Draw an acute angle. Construct the bisector of the angle.   **1.5**   Check construction.

Algebra Review

To simplify real numbers and expressions involving square roots, the following properties are needed:

- $\sqrt{a^2} = |a|$,   for all real numbers a
- $\sqrt{a} \cdot \sqrt{a} = a$,   for all positive real numbers a
- $\sqrt{ab} = \sqrt{a} \cdot \sqrt{b}$,   for all positive real numbers a and b

Simplify the expression.

1. $(3\sqrt{5})^2$ 45   **2.** $\sqrt{27}$ $3\sqrt{3}$ **3.** $\sqrt{x^2y^2}$ $|xy|$ **4.** $3\sqrt{32}$ $12\sqrt{2}$

9 **5.** $(\sqrt{3})^2 + (\sqrt{6})^2$ **6.** $\sqrt{9 + 16}$ 5   **7.** $\sqrt{4x^2}$ $2|x|$ **8.** $\sqrt{24a^2}$ $2\sqrt{6}|a|$

9. $a^2\sqrt{a^2}$ $a^2|a|$ **10.** $(2\sqrt{2})^2$ 8   **11.** $\sqrt{5^2 + 12^2}$ 13   **12.** $\sqrt{k^4}$ k^2

11.4 Arcs and Central Angles   **437**

28. Statement	Reason
1. $\odot O$ with pts of tan at B and D, $\overline{ED}$ is diam	1. Given
2. $m\angle CBO + m\angle BOD + m\angle ODC + m\angle C = 360$.	2. Sum of meas of $\angle$s of polygon $= (n - 2)180$.
3. $\overline{BC} \perp \overline{BO}$, $\overline{CD} \perp \overline{DO}$	3. If line tan to $\odot$, then line is $\perp$ to radius.
4. $m\angle CBO = 90$, $m\angle ODC = 90$.	4. Def of $\perp$
5. $90 + 90 + m\angle BOD + m\angle C = 360$.	5. Sub
6. $m\angle BOD + m\angle C = 180$.	6. Subtr Prop of Eq
7. $m\angle BOE + m\angle BOD = 180$.	7. If outer rays of $\angle$s form st $\angle$, sum of meas $= 180$.
8. $m\angle BOE + m\angle BOD = m\angle BOD + m\angle C$	8. Sub
9. $m\angle BOE = m\angle C$	9. Subtr Prop of Eq
10. $m\widehat{EB} = m\angle BOE$	10. Def of arc meas
11. $m\widehat{EB} = m\angle C$	11. Sub

11.5 Arcs, Chords, and Central Angles

Objectives   To identify congruent arcs and congruent chords
To prove relationships among arcs, chords, and central angles

Definition   Concentric circles are coplanar circles with a common center.

Common center ———

Concentric circles

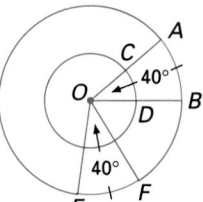

Definition   In the same circle or congruent circles, **congruent arcs** are arcs that have the same degree measure.

In the diagram of concentric circles at the right, some arcs with the same degree measure are congruent and some are not.

m $\angle AOB$ = m $\angle COD$ = m $\angle EOF$ = 40,

$\overset{\frown}{EF} \cong \overset{\frown}{AB}$ (same circle)

$\overset{\frown}{CD} \not\cong \overset{\frown}{AB}$ (circles are not congruent)

EXAMPLE 1   Given: m$\overset{\frown}{AE}$:m$\overset{\frown}{ED}$ = 4:5, m$\overset{\frown}{AB}$ = 40,
  m$\overset{\frown}{BC}$ = 80, m$\overset{\frown}{CD}$ = 150.
Determine which arcs are congruent.

Plan   Find m$\overset{\frown}{AE}$ and m$\overset{\frown}{ED}$, then compare all the arcs.
Let m$\overset{\frown}{AE}$ = 4x and m$\overset{\frown}{ED}$ = 5x.

Solution
$$4x + 5x + 150 + 80 + 40 = 360$$
$$9x + 270 = 360$$
$$9x = 90$$
$$x = 10$$

Therefore, m$\overset{\frown}{AE}$ = 4x = 40 and m$\overset{\frown}{ED}$ = 5x = 50.

Since m$\overset{\frown}{AB}$ = m$\overset{\frown}{AE}$ = 40 in the same circle, $\overset{\frown}{AB} \cong \overset{\frown}{AE}$.

To avoid confusion, whenever reference is made to the *arc of a chord,* it will be assumed to be the *minor* arc.

438   Chapter 11 Circles

GETTING STARTED

Prerequisite Quiz

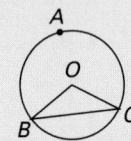

1. m$\overset{\frown}{BC}$ = 70, m $\angle BOC$ = _____   70
2. m$\overset{\frown}{BAC}$ = 300, m $\angle BOC$ = _____   60
3. m $\angle B$ = 40, m$\overset{\frown}{BAC}$ = _____   260
4. m$\overset{\frown}{BC}$:m$\overset{\frown}{BAC}$ = 2:3, m $\angle B$ = _____
   18
5. m $\angle B$ = 55, m $\angle BOC$ = _____   70

Motivator

Ask the students to draw a circle of any radius. Then have them use a protractor to draw two congruent central angles of measure 30, 45, or 60. Have them draw the chord of each resulting central angle. Ask them if the chords are congruent.   Yes
Ask the class what theorem this suggests.
Corresponding chords of congruent central angles are congruent.   Having students use different measures stresses that it is not one special case that is producing the desired result.

TEACHING SUGGESTIONS

Lesson Note

Proofs and applications in this lesson involve properties of isosceles triangles. Review the following: The bisector of the vertex angle of an isosceles triangle is the perpendicular bisector of the base. The base angles of an isosceles triangle are congruent.

Additional Example 1

Given: $\overline{OC} \perp \overline{OA}$, m$\overset{\frown}{AB}$: m$\overset{\frown}{BC}$ = 5:4,
m$\overset{\frown}{DA}$ = 40. Which arcs are congruent?
$\overset{\frown}{DA} \cong \overset{\frown}{BC}$

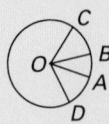

Theorem 11.9

In the same circle or in congruent circles:
1. if chords are congruent, then their corresponding arcs and central angles are congruent;
2. if arcs are congruent, then their corresponding chords and central angles are congruent;
3. if central angles are congruent, then their corresponding arcs and chords are congruent.

The proof of part 1 is given below.

Given: $\odot O \cong \odot M$, $\overline{PR} \cong \overline{QS}$
Prove: $\angle O \cong \angle M$ and $\overset{\frown}{PR} \cong \overset{\frown}{QS}$

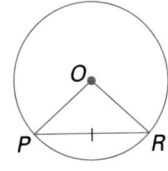
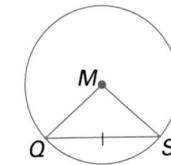

Proof

Statement	Reason
1. $\odot O \cong \odot M$, $\overline{PR} \cong \overline{QS}$	(S) 1. Given
2. $\overline{OP} \cong \overline{MQ}$, $\overline{OR} \cong \overline{MS}$	(S, S) 2. Radii of $\cong \odot$s are $\cong$.
3. $\triangle POR \cong \triangle QMS$	3. SSS
4. $\angle POR \cong \angle QMS$	4. CPCTC
$(m\angle POR = m\angle QMS)$	
5. $m\overset{\frown}{PR} = m\angle POR$,	5. Def of meas of minor arc
$m\overset{\frown}{QS} = m\angle QMS$	
6. $m\overset{\frown}{PR} = m\overset{\frown}{QS}$	6. Sub, Trans
7. $\overset{\frown}{PR} \cong \overset{\frown}{QS}$	7. Def of $\cong$ arcs

EXAMPLE 2 Which chords are congruent?

Plan Find the measure of the arc of each chord.
First, find $m\overset{\frown}{AB}$. (Use *minor* arcs, *not* major arcs.)

Solution
$m\overset{\frown}{AB} = 360 - (45 + 30 + 85 + 35 + 90)$
$= 75$
$m\overset{\frown}{FC} = 35 + 85 + 30 = 150$
$m\overset{\frown}{EB} = 85 + 30 + 45 = 160$
$m\overset{\frown}{AD} = m\overset{\frown}{AB} + 45 + 30 = 75 + 45 + 30 = 150$
Thus, $\overline{FC} \cong \overline{AD}$, since $m\overset{\frown}{FC} = m\overset{\frown}{AD} = 150$.

Definition

A point M is the **midpoint** of $\overset{\frown}{AMB}$ if $\overset{\frown}{AM} \cong \overset{\frown}{MB}$. A line, ray, or segment passing through point M *bisects* the arc.

11.5 Arcs, Chords, and Central Angles **439**

Math Connections

Seismology: A seismologist studies data about earthquakes and uses it to predict where and when earthquakes will occur. From the focus, or origin, of the earthquake, *seismic waves* travel in every direction. The seismologist charts the movement of these waves by drawing concentric circles from the focus to represent the area affected by the quake on a map.

Critical Thinking Questions

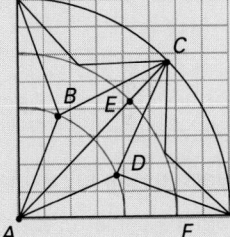

Application: Refer students to the quilt pattern above which consists of one-fourth of a circular design that is on a quilt. $\overline{AB}$, $\overline{AC}$ and $\overline{AD}$ bisect angles having A as a vertex. Ask students the following questions.

1. Find m $\angle BAC$. $90 \div 4 = 22.5$
2. Find the ratio of $m\overset{\frown}{EF}$ to $m\overset{\frown}{BD}$. 45:45, or 1
3. If segments are drawn from points B to D and from points E to F, show that triangles ABD and AEF are similar.
   m $\angle BAD$ = m $\angle AEF$ = 45; since $AB = AD$ and $AE = AF$, then $\frac{AB}{AE} = \frac{AD}{AF}$. Therefore, $\triangle ABD \sim \triangle AEF$ by SAS~.

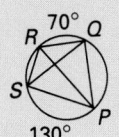

Checkpoint

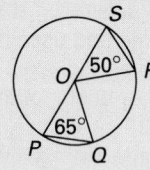

1. Are the two chords congruent? Why?
   Yes, the measure of each central
   ∠ is 50.

2.

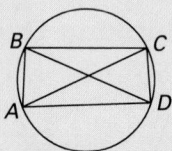

m⌢AB:m⌢BC:m⌢CD:m⌢DA = 1:3:2:4. Which
chords are congruent?   AC ≅ AD

Complete each of the following.

3. A line from the center of a circle
   perpendicular to a chord _____. is ⊥
   bis of the chord

4. In a circle, if two chords are congruent,
   then _____.   their arcs are ≅

Closure

Ask the students what congruency of a pair
of chords or arcs or central angles implies.
Theorem 11.9 page 439 Ask them what
conclusion they can draw if a radius is
perpendicular to a chord in a circle.   It
bisects the minor arc corresponding
to the chord.

▰▰▰FOLLOW UP

Guided Practice

Classroom Exercises 1–7

Independent Practice

A Ex. 1–7, **B** Ex. 8–10, **C** Ex. 11–12

Basic: WE 1–7
Average: WE 1–10
Above Average: WE 1–12

EXAMPLE 3   Prove that in a circle, a radius perpendicular
to a chord bisects the minor arc corresponding
to the chord.

Given: ⊙O with radius OM ⊥ chord AB at P
Prove: OM bisects ⌢AB, the arc of chord AB.

Proof

	Statement		Reason
1.	OM ⊥ AB at P	1.	Given
2.	OA ≅ OB	2.	Radii of the same ⊙ are ≅.
3.	△AOB is isos.	3.	Def of isos △
4.	OP is an altitude.	4.	Def of alt
5.	OP bis ∠AOB.	5.	Alt to base of isos △ bis vertex ∠.
6.	∠1 ≅ ∠2	6.	Def of ∠ bis
7.	⌢AM ≅ ⌢MB	7.	If central ∠s are ≅, then corr arcs are ≅.
8.	∴ OM bis ⌢AB.	8.	Def of arc bis

Classroom Exercises

For Exercises 1–4, complete the proof of part 2 of Theorem 11.9.

Given: ⊙O ≅ ⊙M, ⌢PR ≅ ⌢QS
Prove: ∠POR ≅ ∠QMS and PR ≅ QS

	Statement		Reason
1.	⊙O ≅ ⊙M, ⌢PR ≅ ⌢QS (m⌢PR = m⌢QS)	1.	___Given___
2.	m∠POR = m⌢PR, m ∠QMS = m⌢QS	2.	Def meas of minor arc
3.	m∠POR = m∠QMS (∠POR ≅ ∠QMS)	(A) 3.	Sub, Trans
4.	OP ≅ MQ, OR ≅ MS	(S, S) 4.	Radii of ≅ ⊙s are ≅.
5.	△POR ≅ △QMS	5.	SAS
6.	∴ PR ≅ QS	6.	CPCTC

7. Write the proof of part 3 of Theorem 11.9.

Written Exercises

Given: m⌢UT:m ⌢SR = 2:1, m ⌢TS = 70, m⌢RQ = 50,
   m⌢QP = 40, m⌢PW = 25, m⌢WV = 65, m⌢VU = 50

1. Which arc is congruent to ⌢QP?   ⌢UT
2. Which arc is congruent to ⌢QW?   ⌢WV
3. Which arc is congruent to ⌢WR?   ⌢UW
4. Which arc is congruent to ⌢PQS?   ⌢US
5. Which arcs are congruent to ⌢TR?   ⌢VT, ⌢VP, ⌢PR

Additional Example 3

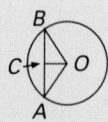

Given: OC bisects ∠BOA.
Prove: OC is the perpendicular bisector
of chord AB.

OB ≅ OA (Radii of ⊙ are ≅.); OC bis
∠AOB (Given); △AOB is isos. (Def
of isos △); OC is ⊥ bis of AB. (Bis of
vertex ∠ of isos △ is ⊥ bis of base.)

6. Given: $m\widehat{AC} = m\widehat{BA}$
Prove: $\triangle ABC$ is isosceles.

7. Given: $\angle B \cong \angle C$
Prove: $\widehat{AC} \cong \widehat{AB}$

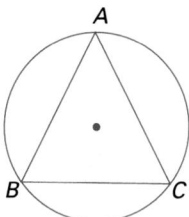

8. Given: $\odot O$ with $\widehat{AB} \cong \widehat{AD}$
Prove: $\angle 1 \cong \angle 2$
9. Given: $\odot O$, diameter $\overline{AC}$, $\overline{DC} \cong \overline{BC}$
Prove: $\widehat{AD} \cong \widehat{AB}$
10. Given: $\overline{AC}$ is a diameter of $\odot O$;
$\overline{AC}$ bisects $\angle DCB$.
Prove: $\widehat{AD} \cong \widehat{AB}$

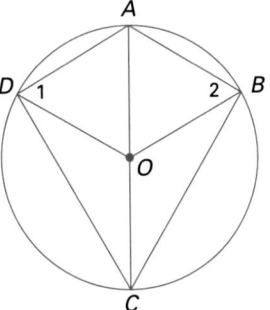

11. Given: $\overline{PQ}$ a diameter, trapezoid $PQRS$
Prove: The diagonals of the trapezoid are $\cong$.

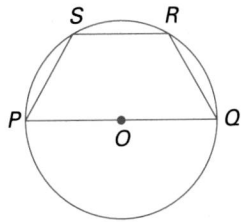

12. Given: $\odot O$, diameter $\overline{AB}$, $\overline{OD} \parallel \overline{BC}$
Prove: $\overline{OD}$ bisects $\widehat{AC}$.

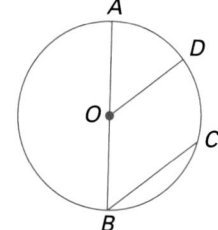

Mixed Review

Use the right triangle at the right to find each indicated length.

1. $AC = 6$, $AB = 4$, $BC = \underline{2\sqrt{13}}$ 9.2
2. $AC = 6$, $BC = 9$, $CD = \underline{4}$ 9.1
3. $CD = 4$, $BD = 9$, $AD = \underline{6}$
4. $m\angle C = 30$, $AC = 4$, $BC = \underline{\hspace{1cm}}$ 9.4 $\frac{8}{3}\sqrt{3}$
5. $m\angle B = 45$, $BC = 8$, $AB = \underline{4\sqrt{2}}$

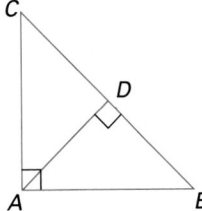

Enrichment

As a summary activity on central angles, show students the diagram at right with the information that it is a semicircle with radius 1 and $m\angle AOX = m\angle BOY = 45$. Ask them how much they can tell you about the triangle, arcs, angles, and segments.

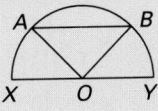

▰▰ GETTING STARTED

Prerequisite Quiz

1. The measures of the angles of a triangle are in the ratio 2:3:4. Find the measure of each angle.   40, 60, 80

2. Given: m$\overparen{CA}$ = 70   **3.** Find m$\overparen{RQ}$.
Find m $\angle BOC$.   115
110

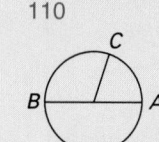

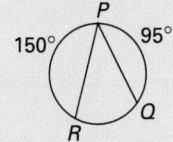

Motivator

Ask students to draw a circle of radius 10 cm. Have them draw a central $\angle AOB$ with measure 60. Ask them to label a point C anywhere on the circle, except on minor $\overparen{AB}$, and draw $\angle ACB$. Tell them this is called an **inscribed** angle. Have them measure $\angle ACB$ with a protractor. Then have them repeat the experiment using new central angle measures. Ask them what conclusion they can draw about the measure of the inscribed angle related to the measure of the central angle of the same arc.   It is one-half the measure of the central angle.

▰▰ TEACHING SUGGESTIONS

Lesson Note

Try to lead students to discover the properties of inscribed angles using the development below.

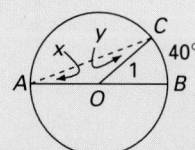

m $\angle A$ = m $\angle C$ = x since the triangle is isosceles. $\angle 1$ is an exterior angle. $x + x$ = 40. Therefore, x = 20. Repeat this for several other cases.

442

11.6 Inscribed Angles

Objective   To prove and apply theorems about inscribed angles

The angle illustrated at the right is not central. The sides of the angle contain chords of the circle, and the vertex is on the circle.

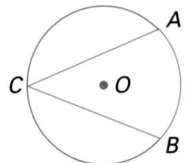

Definitions   An **inscribed angle** ACB is an angle whose vertex lies on a circle and whose sides contain chords of the circle.
The arc $\overparen{AB}$ is called the **intercepted arc** of inscribed angle ACB.

Theorem 11.10   **The Inscribed Angle Theorem:** The measure of an inscribed angle is one-half of the degree measure of its intercepted arc.

Given: ⊙O with inscribed $\angle ACB$

Prove: m $\angle ACB = \frac{1}{2}$m$\overparen{AB}$

Case 1
Point O lies *on* one of the sides of $\angle C$.

Case 2
Point O lies in the *interior* of $\angle C$.

Case 3
Point O lies in the *exterior* of $\angle C$.

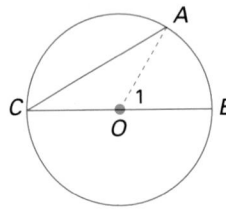

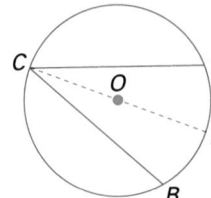

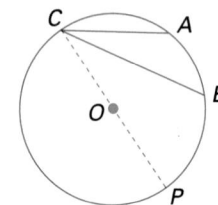

Proof   Case 1

Statement	Reason
1. Draw radius $\overline{OA}$.	1. Two pts determine a line.
2. $\overline{OA} \cong \overline{OC}$	2. Radii of the same ⊙ are ≅.
3. $\angle C \cong \angle A$ (m $\angle C$ = m $\angle A$)	3. $\angle$s opp ≅ sides of a △ are ≅.
4. m $\angle C$ + m $\angle A$ = m $\angle 1$	4. Ext Angle Thm

442   Chapter 11 Circles

5. m ∠C + m ∠C = 2(m ∠C) =   5. Sub
   m ∠1

6. m ∠C = $\frac{1}{2}$m ∠1   6. Div Prop of Equality

7. m ∠1 = m$\widehat{AB}$ 7. Def meas of minor arc

8. ∴ m ∠C = $\frac{1}{2}$m$\widehat{AB}$ 8. Sub

Corollary 1

If two inscribed angles intercept the same arc or congruent arcs, then the angles are congruent.

Given: ⊙O with inscribed angles A and E,
     $\widehat{BC} \cong \widehat{DF}$
Prove: ∠A ≅ ∠E

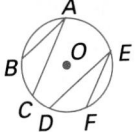

EXAMPLE 1

Given: m$\widehat{CA}$ = 50, m$\widehat{CB}$ = 130
Find m ∠C.

Solution

m$\widehat{BC}$ + m$\widehat{CA}$ + m$\widehat{ADB}$ = 360
130 + 50 + m$\widehat{ADB}$ = 360
180 + m$\widehat{ADB}$ = 360
m$\widehat{ADB}$ = 180

Therefore, m ∠C = $\frac{1}{2}$ · 180, or 90.

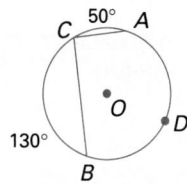

Corollary 2

An angle inscribed in a *semicircle* is a *right* angle.

EXAMPLE 2

Given: △ABC inscribed in ⊙O,
     m$\widehat{AB}$:m$\widehat{BC}$:m$\widehat{AC}$ = 3:4:2
     Find m ∠B.

Plan

Sketch △ABC inscribed in ⊙O. Use the ratio of arc measures to find m$\widehat{AC}$.
Apply the Inscribed Angle Theorem.

Solution

Let m$\widehat{AB}$ = 3x, m$\widehat{BC}$ = 4x, and m$\widehat{AC}$ = 2x.
Then, 3x + 4x + 2x = 360
        9x = 360
         x = 40

Then, m$\widehat{AC}$ = 2x = 80. Thus, m ∠B = $\frac{1}{2}$ · 80, or 40.

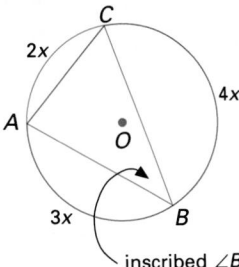

inscribed ∠B

11.6 Inscribed Angles **443**

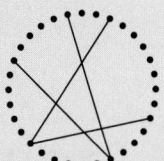
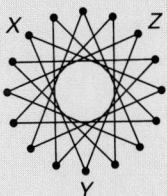

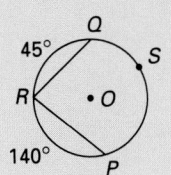
443

Checkpoint

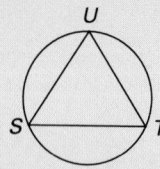

Find the indicated measures.

1. m$\overarc{UT}$ = 70, m ∠S = _____ 35

2. m$\overarc{ST}$ = 90, m$\overarc{UT}$ = 120, m ∠T = _____ 75

3. m ∠U = 40, m$\overarc{ST}$ = _____ 80

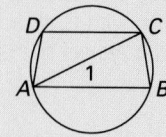

Given: $\overline{AB} \parallel \overline{CD}$

Find the indicated measures.

4. m ∠1 = 40, m$\overarc{AD}$ = _____ 80

5. m$\overarc{AB}$ = 160, m$\overarc{DC}$ = 80, m ∠1 = _____ 30

Closure

Ask the students to describe an inscribed angle. Definition page 442 Ask them to compare the measure of a central angle with the measure of an inscribed angle of the same arc. Theorem 11.10 page 442 Ask them to give the measure of an angle inscribed in a semicircle. 90 Ask them what relationship exists between the opposite angles of an inscribed quadrilateral. They are supplementary. Have them explain what conclusion they can make if two chords of a circle are parallel. The two arcs intercepted between the parallel chords are congruent.

▰▰▰ FOLLOW UP

Guided Practice

Classroom Exercises 1–8

Independent Practice

A Ex. 1–26, **B** Ex. 27–34, **C** Ex. 35–38

Basic: WE 1–26 odd, Midchapter Review

Average: WE 1–34 odd, Midchapter Review

Above Average: WE 1–38 odd, Midchapter Review

Corollary 3 If two arcs of a circle are included between parallel chords or secants, then the arcs are congruent.

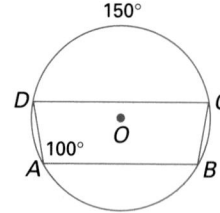

Proof

Statement	Reason
1. $\overline{SR} \parallel \overline{PQ}$	1. Given
2. Draw $\overline{PR}$.	2. Two pts determine a line.
3. ∠1 ≅ ∠2, (m ∠1 = m ∠2)	3. If lines are ∥, alt int ∠s are ≅.
4. m ∠1 = $\frac{1}{2}$m$\overarc{SP}$, m ∠2 = $\frac{1}{2}$m$\overarc{RQ}$	4. Inscr Angle Thm
5. $\frac{1}{2}$m$\overarc{SP}$ = $\frac{1}{2}$m$\overarc{RQ}$	5. Sub, Trans
6. ∴ m$\overarc{SP}$ = m$\overarc{RQ}$	6. Mult Prop of Eq

Corollary 4 The opposite angles of an inscribed quadrilateral are supplementary.

EXAMPLE 3 Given: Quadrilateral ABCD inscribed in ⊙O. $\overline{AB} \parallel \overline{CD}$, m ∠A = 100, m$\overarc{DC}$ = 150

Find (a) m ∠C, (b) m$\overarc{BC}$, and (c) m$\overarc{DA}$.

Solutions

a. m ∠C = 180 − 100 = 80 by Corollary 4.

b. Since m ∠A = 100, m$\overarc{BCD}$ = 200 by the Inscr Angle Thm. Then m$\overarc{BC}$ = m$\overarc{BCD}$ − m$\overarc{DC}$ = 200 − 150 = 50.

c. Finally, m$\overarc{DA}$ = 50 by Corollary 3.

Thus, m ∠C = 80, m$\overarc{BC}$ = 50 and m$\overarc{DA}$ = 50.

Classroom Exercises

Name the arc intercepted by the indicated inscribed angle.

1. ∠1 $\overarc{BCD}$ **2.** ∠2 $\overarc{AD}$

3. ∠3 $\overarc{AB}$ **4.** ∠4 $\overarc{BC}$

5. ∠5 $\overarc{DC}$ **6.** ∠6 $\overarc{BAD}$

7. ∠ABC $\overarc{ADC}$ **8.** ∠ADC $\overarc{ABC}$

Additional Example 3

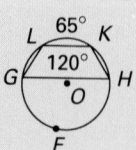

Given: Quadrilateral GHKL is inscribed in circle O with $\overline{GH} \parallel \overline{LK}$, m ∠L = 120, m$\overarc{LK}$ = 65. Find m ∠H and m$\overarc{GFH}$. 60, 185

Written Exercises

Use the figure at the right to find the indicated measure for Exercises 1–9.

1. m$\widehat{AC}$ = 36, m $\angle B$ = __18__   **2.** m$\widehat{AC}$ = 200, m $\angle B$ = __100__
3. m $\angle B$ = 65, m$\widehat{AC}$ = __130__   **4.** m $\angle B$ = 40, m$\widehat{AC}$ = __80__
5. m$\widehat{AC}$ = x, m $\angle B$ = __$\frac{x}{2}$__   **6.** m $\angle B$ = y, m$\widehat{AC}$ = __$2y$__
7. m$\widehat{BA}$ = 100, m$\widehat{BC}$ = 150, m $\angle B$ = __55__
8. m$\widehat{BA}$ = 110, m$\widehat{BC}$ = 200, m $\angle B$ = __25__
9. AB = CB, m$\widehat{AB}$ = 150, m $\angle B$ = __30__

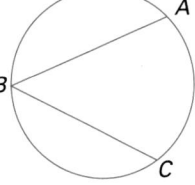

Use the figure at the right to find the indicated measure for Exercises 10–16.

10. m$\widehat{AB}$:m$\widehat{BC}$:m$\widehat{AC}$ = 2:5:3, m $\angle B$ = __54__
11. m$\widehat{AB}$:m$\widehat{AC}$:m$\widehat{BC}$ = 3:2:1, m $\angle C$ = __90__
12. $\triangle ABC$ is equilateral; m$\widehat{AC}$ = __120__ .
13. $\overline{AB}$ is a diameter; m $\angle C$ = __90__ .
14. $\overline{AB}$ is a diameter; m$\widehat{AC}$ = m$\widehat{BC}$, m $\angle A$ = __45__ .
15. AC = BC, m $\angle C$ = 70, m$\widehat{AC}$ = __110__
16. m$\widehat{AC}$ = 200, m $\angle C$ = 30, m$\widehat{CB}$ = __100__

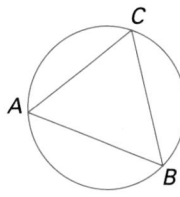

In the figure at the right, $\overline{PQ} \parallel \overline{RS}$. Find the indicated measure for Exercises 17–21.

17. m$\widehat{SP}$ = 90, m $\angle 2$ = __45__
18. m $\angle 1$ = 60, m$\widehat{RS}$ = 70, m$\widehat{PQ}$ = __50__
19. m$\widehat{SR}$ = 100, m$\widehat{PQ}$ = 140, m $\angle 1$ = __30__
20. m$\widehat{PS}$ = 140, m$\widehat{SR}$:m$\widehat{PQ}$ = 3:2, m$\widehat{PQ}$ = __32__
21. $\overline{PR}$ is a diameter; m$\widehat{PS}$:m$\widehat{SR}$ = 2:1, m $\angle 2$ = __60__ .

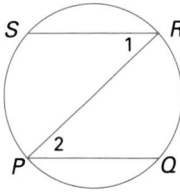

In the figure at the right, $ABCD$ is an inscribed quadrilateral. Find the indicated measure for Exercises 22–26.

22. m $\angle D$ = 100, m $\angle B$ = __80__
23. m $\angle C$ = 80, m $\angle A$ = __100__
24. m$\widehat{DCB}$ = 200, m $\angle C$ = __80__
25. m $\angle B$:m $\angle D$ = 4:5, m $\angle D$ = __100__
26. m $\angle A$ is twice m $\angle C$; m $\angle C$ = __60__ .

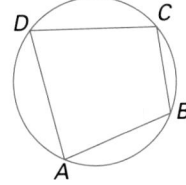

27. Prove Case 2 of Theorem 11.10. (HINT: Apply Case 1 to $\angle ACP$ and $\angle BCP$. Then add the measures.)

28. Prove Case 3 of Theorem 11.10. (HINT: Apply Case 1 to $\angle ACP$ and $\angle BCP$. Then subtract the measures.)

29. Prove Corollary 1.   **30.** Prove Corollary 2.   **31.** Prove Corollary 4.

Enrichment

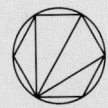

Have students prove that the four angles drawn from one vertex of the regular hexagon shown above are equal. Ask if this will be true in general for regular polygons. It is in fact true because the chords and arcs will always be equal. Have students use this fact to develop a new proof for the sum of the interior angles of a regular polygon. There will be n vertices, each having $(n-2)$ angles inscribed from it, each inscribed angle having measure $\frac{360}{2n}$. Thus, the sum is $180\,(n-1)$.

29.

Statement	Reason
1. ⊙ O with inscr ∠s BAC, DEF; $\overline{BC} \cong \overline{DF}$.	1. Given
2. m$\widehat{BC}$ = m$\widehat{DF}$	2. Def of ≅ arcs
3. $\frac{1}{2}$ m$\widehat{BC}$ = $\frac{1}{2}$ m$\widehat{DF}$	3. Mult Prop of Eq
4. m ∠A = $\frac{1}{2}$ m$\widehat{BC}$, m ∠E = $\frac{1}{2}$ m$\widehat{DF}$	4. Meas inscr ∠ = $\frac{1}{2}$ meas of intercepted arc.
5. m ∠A = m ∠E	5. Sub

30.

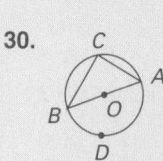

Statement	Reason
1. ⊙O with diam $\overline{AB}$ and inscr ∠ACB.	1. Given
2. m$\widehat{ADB}$ = 180.	2. Meas of semicircle = 180.
3. m ∠ACB = $\frac{1}{2}$ m$\widehat{ADB}$	3. Meas inscr ∠ = $\frac{1}{2}$ meas intercepted arc.
4. m ∠ACB = $\frac{1}{2}$ · 180 = 90.	4. Sub

31.

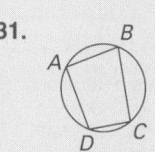

Statement	Reason
1. Inscr quad ABCD	1. Given
2. m$\widehat{ABC}$ + m$\widehat{ADC}$ = 360.	2. Arc meas ⊙ = 360.
3. $\frac{1}{2}$ m$\widehat{ABC}$ + $\frac{1}{2}$ m$\widehat{ADC}$ = 180.	3. Mult Prop of Eq
4. m ∠D = $\frac{1}{2}$ m$\widehat{ABC}$, m ∠B = $\frac{1}{2}$ m$\widehat{ADC}$.	4. Meas inscr ∠ = $\frac{1}{2}$ meas of intercepted arc.
5. m ∠D + m ∠B = 180.	5. Sub
6. ∠D and ∠B are supp.	6. Def of supp ∠s

See pages 662–663 for the answers to Ex. 32–38.

446

32. Given: ABCD is a trapezoid with $\overline{DC}$ and $\overline{AB}$ as bases.
Prove: Diagonals are congruent.

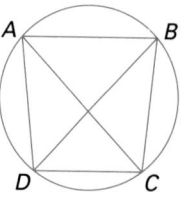

33. Given: m ∠1 = $\frac{1}{2}$m$\widehat{CD}$, $\overline{BD}$ is a diameter.
Prove: m$\widehat{BA}$ = m$\widehat{BC}$

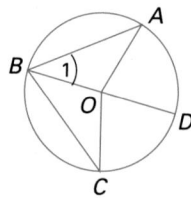

34. Given: ABCD is a parallelogram inscribed in a circle. Prove that the parallelogram is a rectangle.

35. Prove that a rhombus inscribed in a circle is a square.

36. Prove that if a pair of opposite sides of an inscribed quadrilateral are congruent, then the other pair of sides are parallel.

37. Given: △ABC is an equilateral triangle inscribed in a circle. R and S are midpoints of the arcs $\widehat{AB}$ and $\widehat{BC}$, respectively.
Prove: ARSC is a rectangle.

38. Prove that an equilateral hexagon inscribed in a circle is regular. Does this result extend to all equilateral *n*-gons (polygons of *n* sides)?

Midchapter Review

Identify the following for ⊙O. *11.1*

1. radius $\overline{OB}, \overline{OD}, \overline{OE}$
2. chord $\overline{HE}, \overline{DE}, \overline{DF}$
3. diameter $\overline{DE}$
4. secant $\overleftrightarrow{AE}$
5. tangent $\overleftrightarrow{AC}$

(Ex. 1–12)

6. Given: OB = 6x − 2, OD = 4x + 8. Find OE. *11.1* 28
7. Given: ED = 14, OA = 5x + 4. Find all values of x for which A is in the exterior of the circle. *11.1* x > $\frac{3}{5}$
8. Given: $\widehat{HE} \parallel \overline{DF}$, m$\widehat{HE}$ = 70, m$\widehat{DF}$ = 100. Find m$\widehat{FE}$. *11.6* 95
9. Given: m ∠1 = 30, OG = 4. Find the length of the radius. *11.2* $\frac{8}{3}\sqrt{3}$
10. Given: m ∠2 = 25. Find m$\widehat{FE}$. *11.6* 50
11. Given: m$\widehat{DF}$:m$\widehat{FE}$ = 2:3. Find m ∠1. *11.6* 36
12. Given: DF = 24, OD = 13. Find OG. *11.2* 5

13. △ABC is inscribed in a circle. m ∠A = 50, m ∠C = 70. Name the longest chord of the circle. *11.5* $\overline{AB}$

14. The measure of the angle between two tangents to a circle from P is 60. Find the length of each radius if the length of a tangent segment is 8. *11.3* $\frac{8}{3}\sqrt{3}$

446 Chapter 11 Circles

11.7 Angles Formed by Secants and Chords

Objective   To prove and apply theorems about secants and chords

In this lesson the Inscribed Angle Theorem is used to explore various types of angles related to circles. In the figure at the right, notice that the vertical angles 1 and 2 are neither central nor inscribed. Yet their measures can be found by working from the given measures of their intercepted arcs. In the next figure, an auxiliary segment $\overline{BC}$ has been drawn.

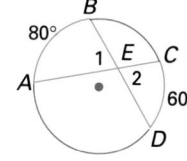

$$m \angle 1 = m \angle B + m \angle C$$
$$= \tfrac{1}{2}m\widehat{CD} + \tfrac{1}{2}m\widehat{AB}$$
$$= \tfrac{1}{2}(m\widehat{CD} + m\widehat{AB})$$

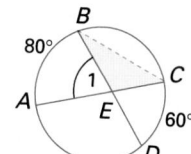

Thus, $m \angle 1 = \tfrac{1}{2}(60 + 80) = 70$.

Theorem 11.11   The measure of an angle formed by two secants or chords intersecting in the interior of a circle is one-half the sum of the measures of the arcs intercepted by the angle and its vertical angle.

EXAMPLE 1   Given: $m \angle PTQ = 120$, $m\widehat{PQ} = 100$
Find $m\widehat{RS}$.

Solution   Write an equation based on Theorem 11.11.
$$m \angle PTQ = \tfrac{1}{2}(m\widehat{PQ} + m\widehat{RS})$$
$$120 = \tfrac{1}{2}(100 + m\widehat{RS})$$
$$240 = 100 + m\widehat{RS}$$
$$m\widehat{RS} = 140$$

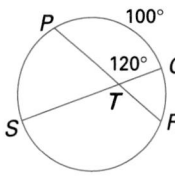

EXAMPLE 2   Given: $m\widehat{AB} = 3x + 40$, $m\widehat{BC} = 2x + 10$,
$m\widehat{CD} = 2x + 30$, $m\widehat{DA} = 3x - 20$
Find $m \angle DEC$.

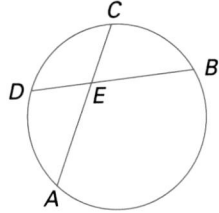

Additional Example 1

Find $m\widehat{AB}$.   68

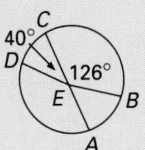

▰ GETTING STARTED

Prerequisite Quiz

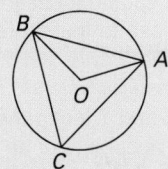

Find the indicated measures.

1. $m\widehat{AB} = 50$, $m \angle C = $ _____   25
2. $m\widehat{BC} = 100$, $m\widehat{CA}$ 140, $m \angle BOA = $ _____   120
3. $m\widehat{BC}{:}m\widehat{BA}{:}m\widehat{AC} = 1{:}2{:}3$, $m \angle CBA = $ _____   90
4. $m \angle BOA = 120$, $m \angle C = $ _____   60
5. $m \angle OBA = 20$, $m \angle C = $ _____   70
6. $m\widehat{BC} = m\widehat{CA} = 150$, $m \angle BOA = $ _____   60

Motivator

Have the students draw circle O with two intersecting chords that do not pass through the center of the circle. Ask them how this example differs from other examples already described.   The angles formed by the intersecting lines are neither central nor inscribed.

▰ TEACHING SUGGESTIONS

Lesson Note

It may be helpful to the students if the proofs of Theorems 11.11 and 11.12 (see Written Exercises 15 and 16, on page 450) are done in class by students working in small groups. This will enable students to see how these Theorems relate to the Inscribed Angle Theorem, taught in Lesson 11.6, and how the Inscribed Angle Theorem is used to prove Theorems 11.11 and 11.12.

Math Connections

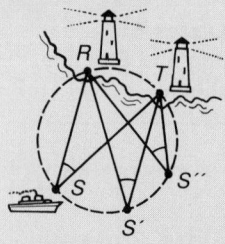

Navigation: Seamen identify areas that are dangerously close to the shoreline by locating a *danger circle*. Inside the danger circle is too close for ships to travel safely. Lighthouses *R* and *T* along the shore determine the endpoints of an arc. The danger circle is defined by a published angle measure of the inscribed angles that intercept the arc. Navigators can compare the location, *S*, of their ship and the danger circle by finding the measure of the angle that the location of their ship forms with the two lighthouses on shore.

Critical Thinking Questions

Analysis: Refer students to the diagram of a danger circle from Math Connections above and ask the following question.
If the navigator finds the measure of the angle formed by the ship and the lighthouses to be greater than the published danger angle, where is the ship located in comparison to the danger circle? Inside the danger circle If the measure of the angle is less than the danger angle, is the ship inside or outside the danger circle?
Outside

Common Error Analysis

Error: Students may forget whether to add or subtract the measures of the arcs in applying the theorems. Summarize the results of the definitions and theorems related to the measures of angles formed by two lines which intersect on, in, or outside a circle in this way.

Point of Intersection	Measure of Angle Formed
On the ⊙	$\frac{1}{2}$ the meas. of intercepted arc
Inside the ⊙	$\frac{1}{2}$ the sum of intercepted arcs
Outside the ⊙	$\frac{1}{2}$ the diff. of intercepted arcs

448

Plan Use the fact that a circle has measure of 360 to write and solve an equation for *x*. Find m$\widehat{CD}$ and m$\widehat{AB}$, then apply Theorem 11.11.

Solution
$$m\widehat{AB} + m\widehat{BC} + m\widehat{CD} + m\widehat{DA} = 360$$
$$(3x + 40) + (2x + 10) + (2x + 30) + (3x - 20) = 360$$
$$10x + 60 = 360$$
$$10x = 300$$
$$x = 30$$

$$m\widehat{CD} = 2x + 30 \qquad\qquad m\widehat{AB} = 3x + 40$$
$$= (2 \cdot 30) + 30 = 90 \qquad\qquad = (3 \cdot 30) + 40 = 130$$

Thus, m $\angle DEC = \frac{1}{2}(m\widehat{CD} + m\widehat{AB})$
$$= \frac{1}{2}(90 + 130) = 110$$

In the figure at the right, two secant segments are drawn to ⊙O from a point *A* in the exterior of the circle. Angle *A* intercepts two arcs, $\widehat{EB}$ and $\widehat{CD}$. The relationship between $\angle A$ and these arcs can be seen by drawing chord $\overline{DB}$. Again, the property of exterior angles of a triangle is applicable.

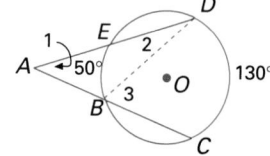

$$m \angle 1 + m \angle 2 = m \angle 3$$
$$m \angle 1 = m \angle 3 - m \angle 2$$
$$m \angle 1 = \frac{1}{2} m\widehat{DC} - \frac{1}{2} m\widehat{EB}$$
$$= \frac{1}{2}(m\widehat{DC} - m\widehat{EB})$$
Thus, m $\angle 1 = \frac{1}{2}(130 - 50) = 40$,

or m $\angle 1 = \frac{1}{2}(\textit{difference}$ of the measures of the intercepted arcs).

The following theorem is suggested.

Theorem 11.12   The measure of an angle formed by two secants intersecting in the exterior of the circle is one-half the difference of the measures of the intercepted arcs.

EXAMPLE 3   Given: m$\widehat{BC} = 85$, m$\widehat{ED} = 95$,
m$\widehat{BE}$:m$\widehat{CD}$ = 4:5
Find m $\angle A$.

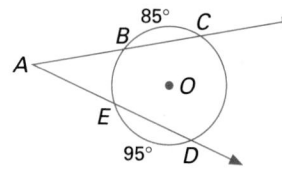

Additional Example 2

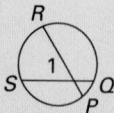

m$\widehat{SR} = 3x + 20$, m$\widehat{RQ} = x + 40$,
m$\widehat{QP} = x - 30$, m$\widehat{PS} = 5x$,
m $\angle 1 = $ _____? 61

Solution

Let $m\overset{\frown}{BE} = 4x$, $m\overset{\frown}{CD} = 5x$.

Then, $m\overset{\frown}{BC} + m\overset{\frown}{CD} + m\overset{\frown}{ED} + m\overset{\frown}{BE} = 360$

$$85 + 5x + 95 + 4x = 360$$
$$9x + 180 = 360$$
$$9x = 180$$
$$x = 20$$

$m\overset{\frown}{BE} = 4x = 4 \cdot 20 = 80$ and
$m\overset{\frown}{CD} = 5x = 5 \cdot 20 = 100$

$m \angle A = \frac{1}{2}(m\overset{\frown}{CD} - m\overset{\frown}{BE})$
$= \frac{1}{2}(100 - 80) = 10$

Classroom Exercises

Find each indicated measure.

1. $m\overset{\frown}{QR} = 100$, $m\overset{\frown}{TS} = 60$, m $\angle 2 =$ ___ 80
2. $m\overset{\frown}{QT} = 50$, $m\overset{\frown}{RS} = 40$, m $\angle 1 =$ ___ 45
3. $m\overset{\frown}{QR} = 60$, $m\overset{\frown}{TS} = 40$, m $\angle P =$ ___ 10
4. $m\overset{\frown}{QR} = 90$, $m\overset{\frown}{TS} = 40$, m $\angle P =$ ___ 25
5. $m\overset{\frown}{RQ} = m\overset{\frown}{QT} = m\overset{\frown}{RS} = 100$, m $\angle P =$ ___ 20

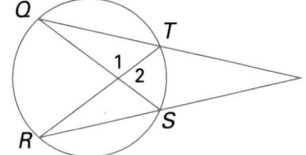

Written Exercises

Find each indicated measure.

1. $m\overset{\frown}{PS} = 138$, $m\overset{\frown}{QR} = 76$, m $\angle 1 =$ ___ 107
2. $m\overset{\frown}{PQ} = 119$, $m\overset{\frown}{SR} = 65$, m $\angle 2 =$ ___ 92
3. $m\overset{\frown}{PS} = 100$, m $\angle 1 = 80$, $m\overset{\frown}{QR} =$ ___ 60
4. $m\overset{\frown}{SR} = 240$, m $\angle 2 = 175$, $m\overset{\frown}{PQ} =$ ___ 110
5. $m\overset{\frown}{SP} = x + 30$, $m\overset{\frown}{PQ} = x + 50$, $m\overset{\frown}{QR} = 2x + 60$,
   $m\overset{\frown}{RS} = 5x - 50$, m $\angle 1 =$ ___ 90
6. $m\overset{\frown}{SP}:m\overset{\frown}{PQ}:m\overset{\frown}{QR}:m\overset{\frown}{RS} = 5:4:3:6$, m $\angle 1 =$ ___ 80

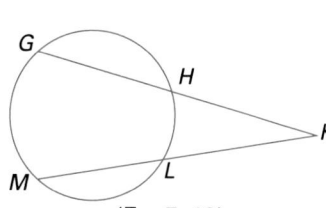
(Ex. 1–6)

7. $m\overset{\frown}{GM} = 163$, $m\overset{\frown}{HL} = 39$, m $\angle K =$ ___ 62
8. $m\overset{\frown}{GM} = 103$, $m\overset{\frown}{HL} = 47$, m $\angle K =$ ___ 28
9. $m\overset{\frown}{GH} = 90$, $m\overset{\frown}{HL} = 80$, $m\overset{\frown}{ML} = 100$,
   m $\angle K =$ ___ 5
10. $m\overset{\frown}{GM} = 100$, m $\angle K = 30$, $m\overset{\frown}{HL} =$ ___ 40
11. $m\overset{\frown}{GM}:m\overset{\frown}{HL} = 5:3$, $m\overset{\frown}{GH} = 150$, $m\overset{\frown}{ML} = 50$,
    m $\angle K =$ ___ 20
12. $m\overset{\frown}{GM} = 3x + 40$, $m\overset{\frown}{GH} = 2x - 20$,
    $m\overset{\frown}{HL} = x - 30$, $m\overset{\frown}{LM} = 2x + 10$, m $\angle K =$ ___ 80

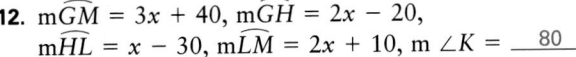
(Ex. 7–12)

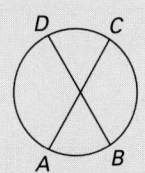 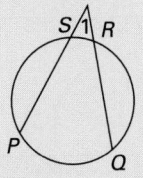

Find the indicated measures.

1. $m\overset{\frown}{DC} = 50$, $m\overset{\frown}{AB} = 90$, m $\angle 2 =$ ___
   70
2. $m\overset{\frown}{PQ} = 100$, $m\overset{\frown}{SR} = 80$, m $\angle 1 =$
   ___ 10
3. $m\overset{\frown}{CD} = 40$, m $\angle 2 = 30$, $m\overset{\frown}{AB} =$ ___
   20
4. $m\overset{\frown}{PQ} = 160$, m $\angle 1 = 50$, $m\overset{\frown}{SR} =$
   ___ 60
5. $m\overset{\frown}{PQ}:m\overset{\frown}{SP}:m\overset{\frown}{SR}:m\overset{\frown}{QR} = 3:2:1:2$, m $\angle 1$
   = ___ 45

Closure

Ask the students how to determine the measure of an angle formed by two chords intersecting in the interior of a circle. Theorem 11.11 page 447   Ask them how to determine the measure of an angle formed by two chords intersecting in the exterior of a circle.   Theorem 11.12 page 448.

◼️◼️◼️ FOLLOW UP

Guided Practice

Classroom Exercises 1–5

Independent Practice

🅐 Ex. 1–14, 🅑 Ex. 15–28, 🅒 Ex. 29–37

Basic: WE 1–14 odd, Brainteaser
Average: WE 1–28 odd, Brainteaser
Above Average: WE 1–37 odd, Brainteaser

Additional Example 3

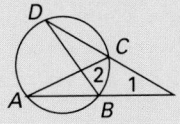

$m\overset{\frown}{AD}:m\overset{\frown}{DC}:m\overset{\frown}{CB}:m\overset{\frown}{BA} = 4:1:3:1$.
Find m $\angle 1$ and m $\angle 2$.   20, 140

Additional Answers

Written Exercises

15.

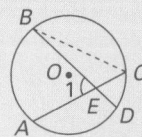

Statement	Reason
1. $\odot O$ with sec $\overleftrightarrow{AC}$ and $\overleftrightarrow{BD}$ intersecting at int pt E	1. Given
2. m $\angle 1$ = m $\angle CBE$ + m $\angle BCE$	2. Meas ext $\angle$ = sum of meas of 2 remote int $\angle$s.
3. m $\angle CBE$ = $\frac{1}{2}$ m$\widehat{CD}$, m $\angle BCE$ = $\frac{1}{2}$ m$\widehat{AB}$	3. Meas inscr $\angle$ = $\frac{1}{2}$ meas intercepted arc.
4. m $\angle 1$ = $\frac{1}{2}$ m$\widehat{CD}$ + $\frac{1}{2}$ m$\widehat{AB}$	4. Sub
5. m $\angle 1$ = $\frac{1}{2}$ (m$\widehat{CD}$ + m$\widehat{AB}$)	5. Distr Prop

16.

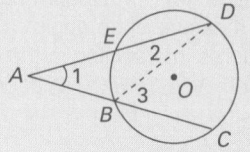

Statement	Reason
1. $\odot O$ with sec $\overleftrightarrow{AB}$ and $\overleftrightarrow{AE}$ intercepting $\widehat{EB}$ and $\widehat{DC}$	1. Given
2. Draw $\overline{BD}$	2. 2 pts determine a line.
3. m $\angle 3$ = m $\angle 1$ + m $\angle 2$	3. Meas of ext $\angle$ = sum of meas int $\angle$s.
4. m $\angle 1$ = m $\angle 3$ − m $\angle 2$	4. Subt Prop of Eq
5. m $\angle 3$ = $\frac{1}{2}$(m$\widehat{DC}$), m $\angle 2$ = $\frac{1}{2}$(m$\widehat{EB}$)	5. Meas inscr $\angle$ = $\frac{1}{2}$ meas intercepted arc.
6. m$\angle 1$ = $\frac{1}{2}$(m$\widehat{DC}$) − $\frac{1}{2}$(m$\widehat{EB}$)	6. Sub
7. m $\angle 1$ = $\frac{1}{2}$(m$\widehat{DC}$ − m$\widehat{EB}$)	7. Distr Prop

29. No. If m $\widehat{AF}$ = 60 and m $\angle 4$ = 30, then 30 = $\frac{1}{2}$(60 − m$\widehat{EB}$), or m$\widehat{EB}$ = 0. This is a contradiction.

---

13. Given: $\overline{AB} \parallel \overline{CD}$,
m$\widehat{AD}$ = 60
Find m $\angle 1$.   60

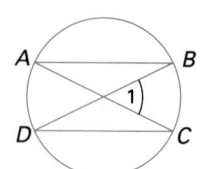

14. Given: diameter $\overline{SQ}$,
m$\widehat{SP}$ = 140,
m$\widehat{RS}$ = 108
Find m $\angle 1$.   74

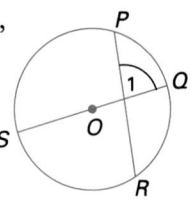

15. Prove Theorem 11.11.

16. Prove Theorem 11.12.

17. Given: m$\widehat{AE}$ = 140, m $\angle C$ = 30
Find m $\angle 1$.   110

18. Given: m$\widehat{AE}$ = 100, m $\angle 1$ = 70
Find m $\angle C$.   30

19. Given: m $\angle BAD$ = 30, m $\angle ABE$ = 40
Find m $\angle C$.   10

20. Given: m$\widehat{AE}$ is 3 times m$\widehat{BD}$; m$\widehat{BA}$ = 130,
m$\widehat{ED}$ = 130. Find m $\angle C$.   25

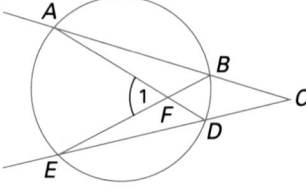

(Ex. 17–20)

Given: $\overline{AB}$ is a diameter; m$\widehat{AF}$:m$\widehat{FE}$:m$\widehat{EB}$ = 3:4:2, m$\widehat{BC}$ = 140. Find the following.

21. m $\angle 6$ 60

22. m $\angle 2$ 90

23. m $\angle 5$ 60

24. m $\angle 3$ 100

Given: $\overline{AB}$ is a diameter; m$\widehat{BC}$ = $\frac{2}{3}$m$\widehat{CA}$,
m$\widehat{AF}$ = 60, m$\widehat{EB}$ = 30. Find the following.

25. m $\angle 7$ 54

26. m $\angle 3$ 66

27. m $\angle 1$ 54

28. m $\angle 4$ 15

29. Given: m$\widehat{AF}$ = 60. Would it be reasonable to expect $\angle 4$ to have a measure of 30? Why?

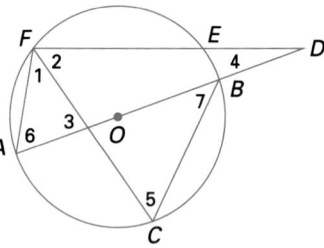

(Ex. 21–29)

30. Two chords $\overline{PQ}$ and $\overline{RS}$ meet at a point Y in the interior of a circle. m $\angle PYR$ = 54 and m$\widehat{PR}$ = 58. Find m$\widehat{SQ}$.   50

31. Given: chords $\overline{AB}$ and $\overline{CD}$ of $\odot O$ are perpendicular at point E.
Prove: m$\widehat{BD}$ + m$\widehat{AC}$ = 180

32. The photo at the right shows a carpenter's square and a metal disk. How can the carpenter's square be used to find the center of the circular disk? Explain why the method works.

450   Chapter 11 Circles

Enrichment

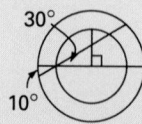

In this activity, the same angle is in different positions for two different concentric circles. Using the arc and angle measures shown, students should be able to find the measures of all the arcs in the figure. Answers, reading clockwise from 9 o'clock, are as follows: inner circle = 90, 30, 60, 180; outer circle = 130, 50, 170, 10.

33. Prove: If P is any point in the exterior of $\odot O$, then $m \angle P < m \angle C$.

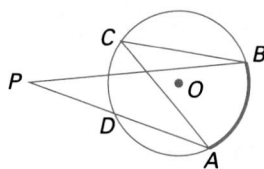

34. The figure at the right shows a ship at point F. The "horizontal danger angle" is used by navigators to chart a course that avoids rocks and shoals close to the shore. Points A and B represent two lighthouses. It is known that within the circle passing through A, B, and C, there are dangerous rocks. If no rocks are located outside the circle, why is the ship safe when $m \angle F < m \angle C$? ($\angle C$ is the "Horizontal danger angle.")

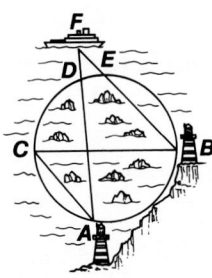

35. Given: $m \angle 1 = 20$, $m \angle 2 = 70$ $m\widehat{DE} = 50$, $m\widehat{CB} = 90$
Find $m\widehat{DE}$ and $m\widehat{CB}$. (HINT: Use simultaneous equations.)

36. Given: $\overline{EC}$ is the perpendicular bisector of chord $\overline{DB}$; $m \angle 1 = 40$.
Find $m\widehat{DE}$ and $m\widehat{CB}$. $m\widehat{DE} = 50$, $m\widehat{CB} = 130$

37. Given: $\overline{CD}$ and $\overline{CE}$ are congruent chords in a circle. W is a point on $\overline{CE}$; $\overleftrightarrow{CW}$ and $\overline{DE}$ meet at some point Y.
Prove: $\angle CYD \cong \angle CDW$

Mixed Review

In $\odot O$ at the right, $\overline{AB}$ is a diameter.

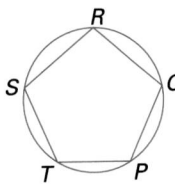

1. What kind of arc is $\widehat{ACB}$? **11.4** Semicircle
2. Find $m \angle C$. **11.6** 90
3. $m\widehat{CB} = 70$, $m \angle A =$ ___35___   **11.6**
4. Given: $\overline{OD} \perp \overline{BC}$, $AB = 10$, $BC = 8$. Find OD.   **11.2** 3

Brainteaser

In the figure at the right, $PQRST$ is an inscribed pentagon and $m\widehat{RQ} = 80$.
Find $m \angle S + m \angle P$.   220

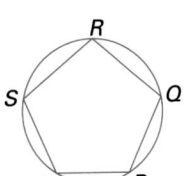

31.

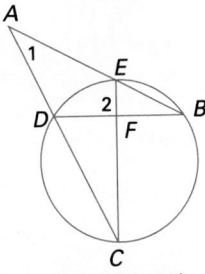

Statement	Reason
1. Chords $\overline{AB}$ and $\overline{CD}$ of $\odot O$ are $\perp$ at pt E.	1. Given
2. $m \angle AEC = 90$.	2. Def of $\perp$ lines
3. $m \angle AEC = \frac{1}{2}(m\widehat{AC} + m\widehat{BD})$	3. Meas of $\angle$ formed by 2 chords is $\frac{1}{2}$ sum of meas of arcs intercepted.
4. $90 = \frac{1}{2}(m\widehat{AC} + m\widehat{BD})$	4. Sub
5. $180 = m\widehat{AC} + m\widehat{BD}$	5. Mult Prop of Eq

32. Since any angle inscribed in a semicircle is right, the carpenter's square can be used to find the endpts of a diameter.

33.

Statement	Reason
1. P is in ext of $\odot O$, $\overline{PB}$ and $\overline{PA}$ are sec, C is on $\odot$.	1. Given
2. $m \angle C = \frac{1}{2}(m\widehat{AB})$	2. Meas of inscr $\angle = \frac{1}{2}$ meas intercepted arc.
3. $m \angle P = \frac{1}{2}(m\widehat{AB} - m\widehat{CD})$	3. Meas of $\angle$ formed by 2 secs is $\frac{1}{2}$ diff of meas of intercepted arcs. $\frac{1}{2}(m\widehat{AB} - m\widehat{CD}) < \frac{1}{2}(m\widehat{AB}) \cdot (m\widehat{CD} > 0)$

34. If $m \angle F < m \angle C$, then F lies outside the $\odot$ of danger.

35. $m\widehat{DE} = 50$, $m\widehat{CB} = 90$

36. $m\widehat{DE} = 50$, $m\widehat{CB} = 130$

See page 471 for the answer to Ex. 37.

11.8 Angles Formed by Tangents and Secants

Objective To prove and apply theorems about angles formed by tangents and secants or by two tangents

---

▰▰▰ **GETTING STARTED**

Prerequisite Quiz

1. What is the relationship between a tangent to a circle and a radius drawn to the point of contact of the tangent and the circle?   Perpendicular

Find the indicated measures.

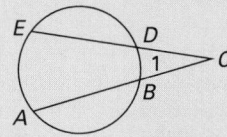

2. $m\overarc{AE} = 130$, $m\overarc{DB} = 60$, $m\angle 1 = $ _____ 35

3. $m\overarc{DB} = 50$, $m\angle 1 = 25$, $m\overarc{AE} = $ _____ 100

4. $m\overarc{ED} = 130$, $m\overarc{AB} = 90$, $m\overarc{AE}:m\overarc{BD} = $ 3:1, $m\angle 1 = $ _____ 35

Motivator

Ask students to list the angles related to circles previously studied in this chapter and the location of the vertex of each angle. central angle: vertex at the center of the circle; inscribed angle: vertex on the circle; tangent-radius: vertex on the circle; chord-chord: vertex inside or on the circle; secant-secant: vertex inside or outside the circle

▰▰▰ **TEACHING SUGGESTIONS**

Lesson Note

Stress the similarity of the theorem about the measure of an angle formed by a chord and tangent to the theorem of an inscribed angle. Also point out the similarity of the new theorem about the measure of an angle between a tangent and a secant or two tangents, to the theorem for the measure of the angle formed by two secants.

Theorem 11.13 If a tangent and a secant (or a chord) intersect at the point of tangency on a circle, then the measure of the angle formed is one-half the measure of its intercepted arc.

Given: Secant $\overleftrightarrow{AC}$, tangent $\overleftrightarrow{AB}$ intersecting at point of tangency A on $\odot O$.
Prove: $m\angle 1 = \frac{1}{2}m\overarc{AC}$

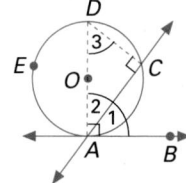

Proof

	Statement		Reason
1.	Sec $\overleftrightarrow{AC}$ and tan $\overleftrightarrow{AB}$ intersecting at A	1.	Given
2.	Draw diam $\overline{AD}$. Draw $\overline{DC}$.	2.	Two pts determine a line.
3.	$\overline{DA} \perp \overleftrightarrow{AB}$	3.	Radius is $\perp$ to tan at pt of tangency.
4.	$\overarc{DCA}$ is a semicircle.	4.	Def of semicircle
5.	$\angle DCA$ is a rt $\angle$.	5.	$\angle$ inscr in semicircle is rt.
6.	$\triangle DCA$ is a rt $\triangle$.	6.	Def of rt $\triangle$
7.	$\angle 3$ and $\angle 2$ are comp.	7.	Acute $\angle$s of rt $\triangle$s are comp.
8.	$\angle 1$ and $\angle 2$ are comp.	8.	Acute adj $\angle$s are comp if outer rays are $\perp$.
9.	$\angle 3 \cong \angle 1$ ($m\angle 3 = m\angle 1$)	9.	Comps of the same $\angle$ are $\cong$.
10.	$m\angle 3 = \frac{1}{2}m\overarc{AC}$	10.	Inscr Angle Theorem
11.	$\therefore m\angle 1 = \frac{1}{2}m\overarc{AC}$	11.	Sub

EXAMPLE 1 Given: Tangent $\overleftrightarrow{QR}$ and chord $\overline{PQ}$, $m\overarc{PSQ} = 300$
Find $m\angle PQR$.

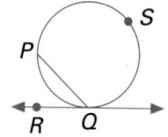

Solution $m\overarc{PQ} = 360 - 300 = 60$

Then, $m\angle PQR = \frac{1}{2}m\overarc{PQ} = \frac{1}{2} \cdot 60 = 30$.

Additional Example 1

$m\overarc{PSQ}:m\overarc{PQ} = 4:1$. Find $m\angle PQR$.   36

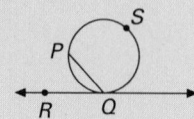

The theorem for finding the measure of an angle formed by two secants intersecting at an exterior point to a circle can be extended to two similar cases.

Theorem 11.14

The measure of an angle formed either by
1. a tangent and a secant intersecting at a point exterior to a circle, or
2. two tangents intersecting at a point exterior to a circle equals one-half the *difference* of the measures of the intercepted arcs.

Proof

Case 1

Given: Tangent $\overleftrightarrow{QR}$ and secant $\overleftrightarrow{QS}$ intersecting at point Q

Prove: m $\angle 1 = \frac{1}{2}$ (m$\widehat{SUR}$ − m$\widehat{TR}$)

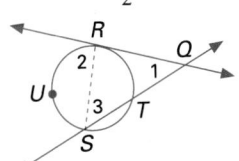

Case 2

Given: Tangents $\overleftrightarrow{QR}$ and $\overleftrightarrow{QS}$ intersecting at point Q

Prove: m $\angle 1 = \frac{1}{2}$(m$\widehat{RTS}$ − m$\widehat{RS}$)

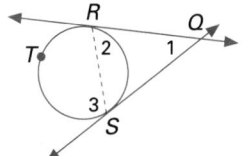

EXAMPLE 2

Given: Tangent $\overleftrightarrow{EA}$, secant $\overleftrightarrow{ED}$, m$\widehat{AB}$ = 60, m$\widehat{BCD}$ = 200 Find m $\angle E$.

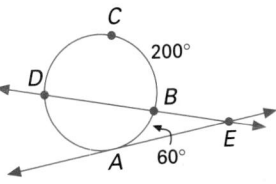

Solution

m $\angle E = \frac{1}{2}$(m$\widehat{AD}$ − m$\widehat{AB}$) = $\frac{1}{2}$(m$\widehat{AD}$ − 60)

m$\widehat{AD}$ = 360 − (200 + 60) = 100

Therefore, m $\angle E = \frac{1}{2}$(100 − 60) = 20.

EXAMPLE 3

Two tangents intersect at point Q so that m $\angle Q$ = 40. Find the measures of the intercepted arcs of $\angle O$.

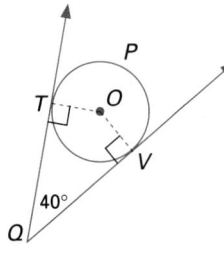

Solution

Let m$\widehat{TV}$ = x. Then m$\widehat{TPV}$ = 360 − x, since the measure of a circle is 360.

$$m \angle Q = \frac{1}{2}[(360 - x) - x]$$

$$40 = \frac{1}{2}(360 - 2x)$$

$$40 = 180 - x$$

$$-140 = -x, \text{ or } x = 140$$

Thus, m$\widehat{TV}$ = 140, and m$\widehat{TPV}$ = 360 − 140 = 220.

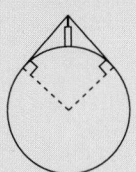

Radio Signals: Radio towers are tall structures which send signals out in all directions. Since the surface of the earth is curved and the radio signals travel in straight lines, the signals can not reach beyond the horizon (see diagram). The radio signals are tangent to the surface of the earth.

Critical Thinking Question

Synthesis: Ask students the following question. It may be helpful to remind them of a 30–60–90 triangle.

If the measure of the angle formed by two tangents intersecting at a point exterior to a circle is 60 and the distance from the exterior point to either point of tangency is $4\sqrt{3}$ units, what is the diameter of the circle? Radius, tangent and angle bis from center to ext pt form a 30–60–90 △ whose side lengths are in the ratio $1:\sqrt{3}:2$, with respect to the sides opp to those ∠s. Therefore, by Pythagorean Thm, the radius = 4 and the diameter = 8.

Additional Example 2

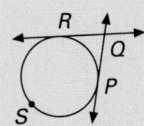

Given: tangents $\overleftrightarrow{QR}$ and $\overleftrightarrow{QP}$, m$\widehat{PSR}$:m$\widehat{PR}$ = 2:1. Find m $\angle Q$. 60

Additional Example 3

The measure of the minor arc intercepted by two tangents to a circle from a point P is 70. Find the measure of the angle between the two tangents. 110

453

Checkpoint

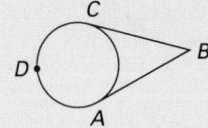

Find the indicated measures.

1. $m\widehat{AC} = 135$, m $\angle B =$ _____ 45
2. $m\widehat{ADC}:m\widehat{CA} = 7:3$, m $\angle B =$ _____
   72
3. m $\angle B = 50$, $m\widehat{AC} =$ _____ 130

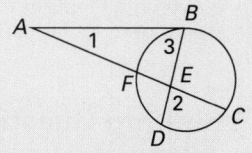

Given: $\overline{AB}$ is a tangent segment, $m\widehat{FB} = 80$, $m\widehat{FD} = 40$, $m\widehat{DC} = 50$.

Find the indicated measure.

4. m $\angle 1$ 55
5. m $\angle 2$ 65
6. m $\angle 3$ 60

Closure

Ask the students how to find the measure of the angle formed when a tangent and a secant intersect at the point of tangency. It is one-half the measure of the intercepted arc. Ask them what conclusions they can draw if the measure of an angle formed is one-half the difference of the measures of the intercepted arcs. The angle is formed by the intersection of a tangent and a secant or two tangents intersecting in the exterior of the circle.

EXAMPLE 4 Given: Tangent $\overleftrightarrow{AB}$, secant $\overleftrightarrow{AC}$, diameter $\overline{DC}$, $m\widehat{CF} = 65$, $m\widehat{GF} = 40$, $m\widehat{BC} = 135$
Find m $\angle 1$, m $\angle 2$, and m $\angle 3$.

Solution First find the measures of $\widehat{DG}$ and $\widehat{BD}$.
$\overline{CD}$ is a diameter. So, $m\widehat{CGD} = m\widehat{CBD} = 180$.

$m\widehat{DG} = 180 - (m\widehat{CF} + m\widehat{GF})$
$m\widehat{DG} = 180 - (65 + 40)$
$m\widehat{DG} = 75$
$m\widehat{BD} = 180 - m\widehat{BC}$
$m\widehat{BD} = 180 - 135$
$m\widehat{BD} = 45$

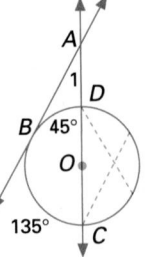

$\angle 1$ is formed by a tangent and a secant with vertex in the exterior.	$\angle 2$ is an inscribed angle.	$\angle 3$ is formed by two intersecting chords.
m $\angle 1 = \frac{1}{2}(m\widehat{BC} - m\widehat{BD})$	m $\angle 2 = \frac{1}{2}m\widehat{DG}$	m $\angle 3 = \frac{1}{2}(m\widehat{DG} + m\widehat{CF})$
m $\angle 1 = \frac{1}{2}(135 - 45)$	m $\angle 2 = \frac{1}{2} \cdot 75$	m $\angle 3 = \frac{1}{2}(75 + 65)$
m $\angle 1 = 45$	m $\angle 2 = 37\frac{1}{2}$	m $\angle 3 = 70$

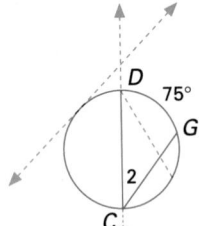
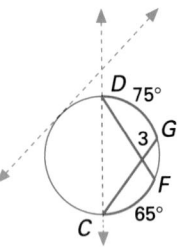

Classroom Exercises

Find the indicated measure.

1. $m\widehat{AC} = 150$, m $\angle D =$ ___75___
2. $m\widehat{AD} = 200$, $m\widehat{AC} = 100$, m $\angle B =$ ___50___
3. $m\widehat{AD} = 40$, m $\angle EAD =$ ___20___
4. $m\widehat{AD} = x$, $m\widehat{AC} = y$, m $\angle B =$ ___$\frac{x-y}{2}$___

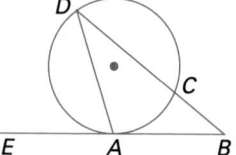

Additional Example 4

Given: Tangent segment $\overline{AB}$, $m\widehat{ED} = 40$, $m\widehat{DC} = 100$, $m\widehat{AE}:m\widehat{AC} = 6:5$.
Find m $\angle 1$, m $\angle 2$, and m $\angle 3$.   10, 70, 50

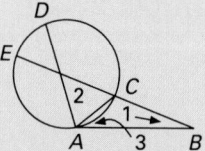

5. m$\overarc{PT}$ = 80, m$\overarc{SQ}$ = 60, m ∠R = ___10___
6. m$\overarc{SQ}$ = 90, m ∠1 = ___45___
7. m$\overarc{PT}$ = 70, m$\overarc{SQ}$ = 30, m ∠3 = ___50___

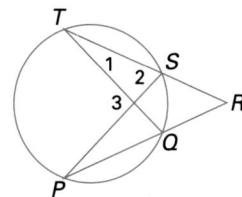

Written Exercises

1. Given: Tangent $\overleftrightarrow{AB}$, m$\overarc{BDC}$ = 240
Find m ∠ABC.   60

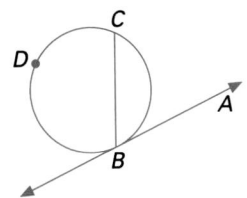

2. Given: Tangent $\overleftrightarrow{PS}$, m$\overarc{RQ}$ = 160, m$\overarc{RS}$ = 60
Find m ∠P.   40

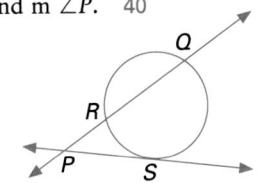

3. Given: Tangents $\overleftrightarrow{AB}$ and $\overleftrightarrow{AC}$,
m$\overarc{BC}$:m$\overarc{BDC}$ = 1:2
Find m ∠1.   60

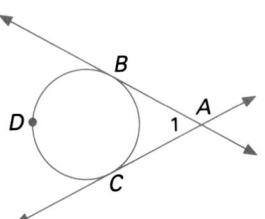

4. Given: Tangent $\overleftrightarrow{RS}$ and $\overleftrightarrow{RT}$,
m$\overarc{SUT}$ = 235
Find m ∠SRT.   55

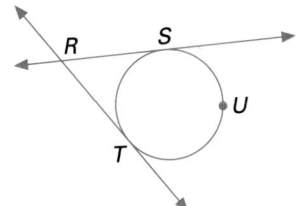

In the figure at the right, $\overleftrightarrow{AB}$ and $\overleftrightarrow{AC}$ are tangents.

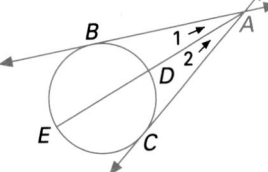

5. m$\overarc{EB}$ = 60, m$\overarc{BD}$ = 40, m ∠1 = ___10___
6. m$\overarc{EC}$ = 100, m ∠2 = 35, m$\overarc{DC}$ = ___30___
7. m$\overarc{EC}$ = 80, m ∠2 = 5, m$\overarc{DC}$ = ___70___
8. m$\overarc{BEC}$ = 250, m ∠BAC = ___70___
9. m$\overarc{EB}$:m$\overarc{BD}$:m$\overarc{DC}$:m$\overarc{CE}$ = 5:4:4:5, m ∠1 = ___10___
10. The measure of an angle formed by two tangents to a circle is 70. Find the measure of the central angle formed by radii to the points of tangency.   110
11. A central angle measures 120. Find the measure of the angle formed by two tangents to the circle at the intersections of the central angle and the circle.   60

Enrichment

Ask students to prove or disprove: If parallel secants are drawn through a circle each at a distance of half the radius from the center, then they divide the circle into four equal arcs.

The first step is to draw a diagram correctly. Students will soon be able to see that the statement is not true.

Without using trigonometry, the methods of proof rely on the contrapositive. In a circle of radius r, parallel secants dividing the circle into four equal arcs will be at a distance of $\frac{r}{\sqrt{2}}$ from the center, not $\frac{r}{2}$.

Guided Practice

Classroom Exercises 1–7

Independent Practice

A Ex. 1–15, **B** Ex. 16–27, **C** Ex. 28–31

Basic: WE 11–15 odd
Average: WE 1–27 odd
Above Average: WE 1–31 odd

Additional Answers

Written Exercises

20.

Statement	Reason
1. Tan $\overleftrightarrow{QR}$ and sec $\overleftrightarrow{QS}$ intersecting at Q	1. Given
2. m ∠1 + m ∠3 = m ∠2	2. Meas of ext ∠ = sum of meas of 2 remote int ∠s.
3. m ∠1 = m ∠2 − m ∠3	3. Subtr Prop of Eq
4. m ∠2 = $\frac{1}{2}$ m$\overarc{SUR}$	4. If tan and sec intersect at pt of tan, then ∠ formed is $\frac{1}{2}$ meas of intercepted arc.
5. m ∠3 = $\frac{1}{2}$ m$\overarc{TR}$	5. Meas of inscr ∠ = $\frac{1}{2}$ meas of intercepted arc.
6. m ∠1 = $\frac{1}{2}$ m$\overarc{SUR}$ − $\frac{1}{2}$ m$\overarc{TR}$	6. Sub
7. m ∠1 = $\frac{1}{2}$[m$\overarc{SUR}$ − m$\overarc{TR}$]	7. Distr Prop

21.

Statement	Reason
1. Tan $\overleftrightarrow{QR}$ and $\overleftrightarrow{QS}$ intersecting at Q	1. Given
2. m ∠1 + m ∠2 = m ∠3	2. Meas of ext ∠ = sum of meas of 2 remote int ∠s.
3. m ∠1 = m ∠3 − m ∠2	3. Subtr Prop of Eq
4. m ∠3 = $\frac{1}{2}$ m$\overarc{RTS}$, m ∠2 = $\frac{1}{2}$ m$\overarc{RS}$	4. If tan and sec intersect at pt of tan, then ∠ formed is $\frac{1}{2}$ meas of intercepted arc.
5. m ∠1 = $\frac{1}{2}$ m$\overarc{RTS}$ − $\frac{1}{2}$ m$\overarc{RS}$	5. Sub
6. m ∠1 = $\frac{1}{2}$(m$\overarc{RTS}$ − m$\overarc{RS}$)	6. Distr Prop

28.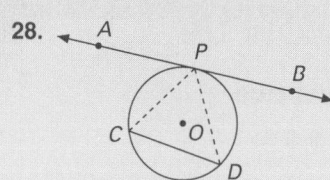

Statement	Reason
1. $\overleftrightarrow{AB}$ tan to $\odot$ O at P, chord $\overline{CD} \parallel \overline{AB}$.	1. Given
2. m $\angle APC$ = m $\angle PCD$	2. If $\parallel$ lines are cut by transv, alt int $\angle$s are $\cong$.
3. m $\angle APC$ = $\frac{1}{2}$ m$\overparen{CP}$	3. If tan and sec intersect at pt of tan, $\angle$ formed = $\frac{1}{2}$ meas of intercepted arc.
4. m $\angle PCD$ = $\frac{1}{2}$ m$\overparen{DP}$	4. Meas inscr $\angle$ = $\frac{1}{2}$ meas intercepted arc.
5. $\frac{1}{2}$ m$\overparen{CP}$ = $\frac{1}{2}$ m$\overparen{DP}$	5. Sub
6. m$\overparen{CP}$ = m$\overparen{DP}$	6. Mult Prop of Eq

29.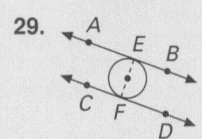

Statement	Reason
1. $\overleftrightarrow{AB}$ and $\overleftrightarrow{CD}$ tan to $\odot$ at E and F, resp. $\overleftrightarrow{AB} \parallel$ $\overleftrightarrow{CD}$.	1. Given
2. Draw $\overline{EF}$.	2. 2 pts determine line.
3. m $\angle BEF$ = m $\angle CFE$	3. If 2 $\parallel$ lines are cut by transv, alt int $\angle$s are $\cong$.
4. m $\angle BEF$ = $\frac{1}{2}$ m$\overparen{EF}$, m $\angle CFE$ = $\frac{1}{2}$ m$\overparen{FE}$	4. If tan and sec intersect at pt of tan, $\angle$ formed = $\frac{1}{2}$ meas of intercepted arc.
5. $\frac{1}{2}$ m$\overparen{EF}$ = $\frac{1}{2}$ m$\overparen{FE}$	5. Sub
6. m$\overparen{EF}$ = m$\overparen{FE}$	6. Mult Prop of Eq

See page 472 for the answer to Ex. 31.

456

In the figure at the right, $\overleftrightarrow{AF}$ is a tangent.

12. m $\angle 2$ = 35, m$\overparen{FB}$ = 45, m$\overparen{FD}$ = _115_

13. m$\overparen{FD}$ = 140, m $\angle 1$ = 125, m$\overparen{BE}$ = _110_

14. m$\overparen{FB}$ = 115, m $\angle 3$ = 95, m$\overparen{DE}$ = _75_

15. $\overline{BD}$ is a diameter; $\dfrac{m\overparen{FB}}{m\overparen{FD}} = \dfrac{4}{5}$, m $\angle 2$ = _10_.

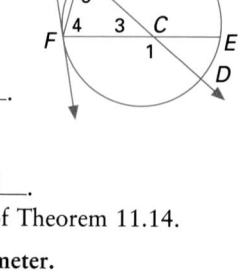

16. $\overline{FE}$ is a diameter; m$\overparen{BE}$ = 120, m$\overparen{FD}$ = 80, m $\angle 2$ = _10_.

17. m $\angle 2$ = 30, m$\overparen{FD}$ = 140, m$\overparen{DE}$ = 50, m $\angle 3$ = _65_

18. m $\angle 5$ = 65, m$\overparen{FB}$ = 60, m $\angle 2$ = _35_

19. $\overline{FE}$ is a diameter; m$\overparen{BE}$ = 160, m $\angle 1$ = 155, m $\angle 2$ = _65_.

20. Prove Case 1 of Theorem 11.14. **21.** Prove Case 2 of Theorem 11.14.

In the figure at the right $\overleftrightarrow{AC}$ is a tangent, $\overline{BD} \parallel \overline{FG}$, $\overline{BD}$ is a diameter. Find the indicated measure.

22. m$\overparen{DG}$ = 50, m $\angle 2$ = _25_

23. m $\angle 5$ = 70, m $\angle 1$ = _20_

24. m $\angle 1$ = 30, m $\angle 3$ = 105, m$\overparen{BF}$ = _90_

25. m $\angle 1$ = 50, m $\angle 6$ = 20, m $\angle 7$ = _120_

26. m $\angle 5$ = 85, m $\angle 1$ = _5_

27. m $\angle 4$ = 30, m $\angle 1$ = 60, m$\overparen{FDE}$ = _180_

28. Given: $\overleftrightarrow{AB}$ tangent to $\odot O$ at P, chord $\overline{CD} \parallel \overleftrightarrow{AB}$
Prove: $\overparen{CP} \cong \overparen{DP}$

29. Prove: If two tangents to a circle are parallel, then the points of tangency divide the circle into two congruent arcs.

30. Triangle PQR is circumscribed about a circle. m $\angle P$:m $\angle Q$: m $\angle R$ = 3:2:1. The three points of tangency are joined to form an inscribed triangle. Find the measure of each angle of the inscribed triangle.   75, 60, 45

31. From a point P exterior to $\odot O$, two tangents are drawn intersecting the circle at points A and B. Prove that m $\angle APB$ + m$\overparen{AB}$ = 180.

Mixed Review

1. Find the sum of the measures of the interior angles of a decagon.   *6.2*   1,440

2. Given: Parallelogram ABCD, m $\angle A$ = 120. Find m $\angle B$.   *6.4*   60

3. Find the length of a leg of an isosceles right triangle if the length of the hypotenuse is 6.   *9.2* $3\sqrt{2}$

4. Find the measure of each base angle of an isosceles triangle if the measure of the vertex angle is 50.   *5.1*   65

456   Chapter 11 Circles

11.9 Lengths of Segments Formed by Secants, Chords, and Tangents

Objective   To find lengths of segments related to chords, secants, and tangents

Besides the measures of angles and arcs, the lengths of various segments relating to circles can also be determined. In the figure at the right, it can be shown by AA Similarity that $\triangle ABE \sim \triangle DCE$. First, $\angle 1 \cong \angle 2$ by the vertical angles property; then, $\angle A \cong \angle D$ (or $\angle B \cong \angle C$), because inscribed angles intercepting the same arc are congruent. As a result:

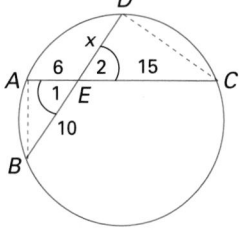

$$\frac{BE}{EC} = \frac{AE}{ED} \longrightarrow \frac{10}{15} = \frac{6}{x}$$
$$BE \cdot ED = EC \cdot AE \longrightarrow 10x = 90$$
$$x = 9$$

Thus, $ED = 9$. It is also seen that the *product* of the lengths of the segments of one chord equals the *product* of the lengths of the segments of the other chord. This suggests the following theorem.

Theorem 11.15

> If two chords of a circle intersect, then the product of the lengths of the segments of one chord equals the product of the lengths of the segments of the other chord.

Given: Chords $\overline{AC}$ and $\overline{BD}$ intersect at E.
Prove: $BE \cdot ED = AE \cdot EC$

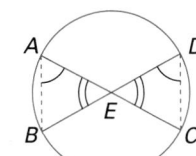

EXAMPLE 1   Given: $BE = 3(EA)$, $DC = 7$, $EC = 3$
Find EA.

Solution   Let $EA = x$; then $BE = 3x$.
$EA \cdot BE = ED \cdot EC$ ← (Theorem 11.15)
$$x \cdot 3x = (7 - 3) \cdot 3$$
$$3x^2 = 12$$
$$x^2 = 4$$
$$x = \sqrt{4} = 2$$
(Use the principal square root.)

Thus, $EA = 2$.

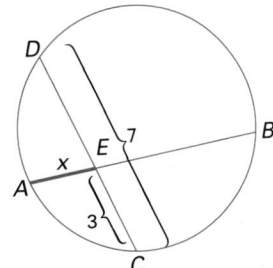

Teaching Resources

Quick Quizzes 83
Reteaching and Practice
    Worksheets, pp. 165, 166
Transparency 31

▰▰▰ GETTING STARTED

Prerequisite Quiz

Solve each equation.

1. $\frac{10}{15} = \frac{6}{x}$ 9
2. $3x^2 = 12$ ± 2
3. $36 = x^2 + 5x$ $-9, 4$
4. $x^2 + 7x = 18$ $+2, -9$

Motivator

Have the students evaluate the following statement and determine if it is always true, sometimes true, or never true. Have them explain their answer. If two chords intersect, the sum of the segments of one chord are equal to the sum of the segments of the other.   It is sometimes true. When two chords are the same length then the statement is true. For all other instances, the statement is false.

▰▰▰ TEACHING SUGGESTIONS

Lesson Note

First review the AA Similarity Postulate for triangles. Try to get the students to inductively discover the two theorems of this lesson by working through three or four concrete examples using similar triangles.

Additional Example 1

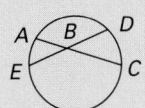

Given: $AC = 16$, $EB = 4$, $BD = 7$
Find AB and BC 2, 14

Math Connections

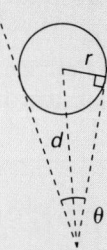

Outer Space: The diameter of a planet at a known distance *d* from the Earth can be determined by measuring the angle between the lines of sight tangent to the outer edges of the planetary disc. The radius *r* will be $d[\sin(\frac{1}{2}\theta)]$.

Critical Thinking Questions

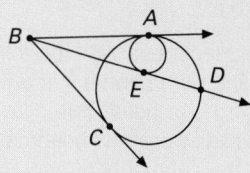

Analysis: If $\overrightarrow{BA}$ and $\overrightarrow{BD}$ are tangent to the smaller circle at points *A* and *E*, respectively, and $\overrightarrow{BA}$ and $\overrightarrow{BC}$ are tangent to the larger circle at points *A* and *C*, respectively, show that *BE* = *BC*. *BE* = *BA* and *BA* = *BC* by Thm 11.8, so by the Trans Prop of Eq, *BE* = *BC*.

Checkpoint

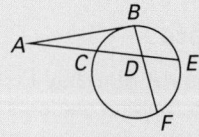

Given: $\overline{AB}$ is a tangent segment. Find the indicated lengths.

1. *CD* = 8, *DE* = 9, *BD* = 12, *DF* = _____ 6
2. *AC* = 4, *CE* = 21, *AB* = _____ 10
3. *AB* = 6, *AE* = 12, *AC* = _____ 3
4. *DF* = 8, *DE* = 22, *DC* = 4, *DB* = _____ 11
5. *AB* = 10, *AE* = 50, *CE* = _____ 48

In the figure below, $\overline{PQ}$ is a *tangent* segment, $\overline{PS}$ is a *secant* segment, and $\overline{PR}$ is an *external secant* segment.

Theorem 11.16

If a tangent and a secant intersect in the exterior of a circle, then the square of the length of the tangent segment equals the product of the lengths of the secant segment and the external secant segment.

Given: $\odot O$, tangent $\overleftrightarrow{PQ}$ and secant $\overleftrightarrow{PS}$ intersecting at point *P*
Prove: $(PQ)^2 = PS \cdot PR$

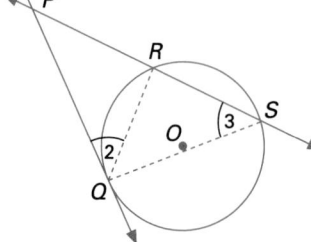

Plan

m ∠2 and m ∠3 each equal $\frac{1}{2}m\overarc{QR}$, so m ∠2 = m ∠3. Also, $\angle P \cong \angle P$. Therefore $\triangle PQR \sim \triangle PSQ$ by AA Similarity. Then, by similar triangles, $\frac{PQ}{PS} = \frac{PR}{PQ}$, and by a proportion property $PQ \cdot PQ = PS \cdot PR$, or $(PQ)^2 = PS \cdot PR$.

In the figure at the right, $\overleftrightarrow{AC}$ and $\overleftrightarrow{AE}$ are secants and $\overleftrightarrow{AF}$ is a *tangent*. By Theorem 11.16:

$$(AF)^2 = AC \cdot AB,$$
$$\text{and } (AF)^2 = AE \cdot AD.$$

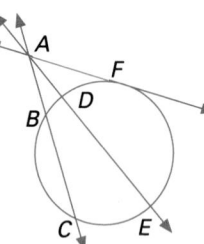

So, by substitution, $AC \cdot AB = AE \cdot AD$. This result is stated as a corollary to Theorem 11.16.

Corollary

If two secants intersect in the exterior of a circle, then the product of the lengths of one secant segment and its external segment equals the product of the lengths of the other secant segment and its external segment.

Given: $\odot O$ with secants $\overleftrightarrow{BC}$ and $\overleftrightarrow{DE}$ intersecting at point *A*
Prove: $AC \cdot AB = AE \cdot AD$

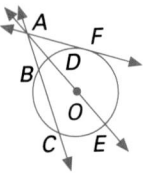

458 Chapter 11 Circles

Additional Example 2

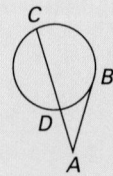

Given: $\overline{AB}$ is a tangent segment, *AC* = 9, *CD* = 5. Find *AB*. 6

Given: *RT* = 10, *RV* = 5, *VU* = 3
Find *RS* and *ST*. 4, 6

EXAMPLE 2 Given: Tangent $\overleftrightarrow{AB}$, secant $\overleftrightarrow{AD}$,
  $AC = 4$, $CD = 8$
Find AB.

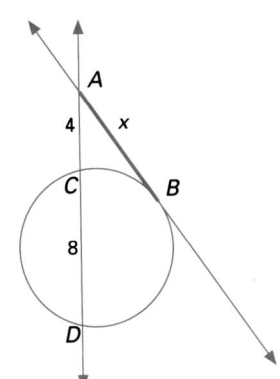

Given: Secants $\overleftrightarrow{PT}$ and $\overleftrightarrow{PR}$,
  $PQ = 6$, $PS = 4$, $TS = 10$
Find QR.

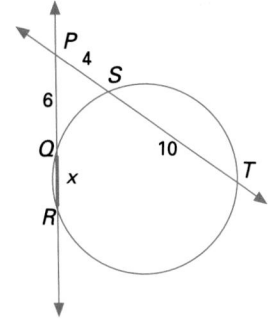

Solution

Let $x = AB$.

Then by Theorem 11.16:

$(AB)^2 = AD \cdot AC$
$x^2 = (4 + 8)4$
$x^2 = 48$
$x = \sqrt{48} = \sqrt{16} \cdot \sqrt{3} = 4\sqrt{3}$

Thus, $AB = 4\sqrt{3}$.

Let $x = QR$.

Then by Corollary 1:

$PR \cdot PQ = PT \cdot PS$
$(x + 6)6 = (4 + 10)4$
$6x + 36 = 56$
$6x = 20$
$x = 3\frac{1}{3}$ Thus, $QR = 3\frac{1}{3}$.

EXAMPLE 3 From exterior point A, a tangent segment $\overline{AB}$ is drawn to $\odot O$ meeting the circle at B. A secant $\overline{AC}$ is drawn, meeting the circle at D and C. If $AB = 6$ and $DC = 5$, find AD.

Plan First draw the figure.
Let $AD = x$.

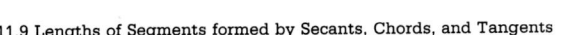

Solution

$(AB)^2 = AD \cdot AC$ by Theorem 11.16
$6^2 = x(x + 5)$
$36 = x^2 + 5x$

$x^2 + 5x - 36 = 0$
$(x + 9)(x - 4) = 0$
$x + 9 = 0 \quad or \quad x - 4 = 0$
$x = -9 \quad or \quad x = 4$

Since $x = AD$, a length, $x = 4$ is the only possible solution.
Therefore, $AD = 4$.

Closure

Have the students explain the three theorems that relate lengths of segments of two intersecting chords, intersecting tangents and secants, and two intersecting secants. Theorem 11.15 page 457, Theorem 11.16 page 458 and Corollary page 458

■■■FOLLOW UP

Guided Practice

Classroom Exercises 1–3

Independent Practice

A Ex. 1–10, **B** Ex. 10–18, **C** Ex. 19–23

Basic: WE 1–10 odd
Average: WE 1–18 odd
Above Average: WE 1–23 odd

Additional Example 3

From exterior point P, tangent $\overline{PQ}$ is drawn to circle O, meeting the circle at Q. A secant $\overline{PS}$ is drawn, meeting the circle at R and S. If $PQ = 8$ and $RS = 30$, find PR. 2

Additional Answers

Classroom Exercises

1. $4 \cdot 5 = 10 \cdot x$

2. $x^2 = 4(21 + 4)$

3. $x(13 + x) = 4(8 + 4)$

Written Exercises

15.

Statement	Reason
1. Chords $\overline{AC}$ and $\overline{BD}$ intersecting at E	1. Given
2. m $\angle AEB$ = m $\angle DEC$	2. Vertical $\angle$s are $\cong$.
3. m $\angle A$ = m $\angle D$	3. If 2 inscr $\angle$s intercept same arc, $\angle$s are = in meas.
4. $\triangle AEB \sim \triangle DEC$	4. AA for $\sim \triangle$s
5. $\frac{BE}{AE} = \frac{EC}{ED}$	5. Def of $\sim \triangle$s
6. $BE \cdot ED = AE \cdot EC$	6. In true proportion, prod of extremes = prod of means.

16.

Statement	Reason
1. $\odot O$ with tan $\overleftrightarrow{PQ}$ and sec $\overleftrightarrow{PS}$ intersecting at P	1. Given
2. m $\angle 3$ = $\frac{1}{2}$ m$\widehat{QR}$	2. Meas of inscr $\angle$ = $\frac{1}{2}$ meas of intercepted arc.
3. $\frac{1}{2}$ m$\widehat{QR}$ = m $\angle 2$	3. Angle formed by tan and sec at pt of tan = $\frac{1}{2}$ meas of intercepted arc.
4. m $\angle 3$ = m $\angle 2$	4. Sub
5. m $\angle P$ = m $\angle P$	5. Reflex Prop of Eq
6. $\triangle PRQ \sim \triangle PQS$	6. AA for $\sim \triangle$s
7. $\frac{PQ}{PS} = \frac{PR}{PQ}$	7. Def of $\sim \triangle$s
8. $(PQ)^2 = PS \cdot PR$	8. In true proportion, prod of extremes = prod of means.

Classroom Exercises

Write an equation you could use to find the value of x.

1.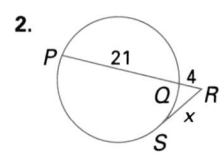

2.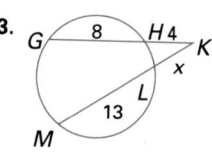

3.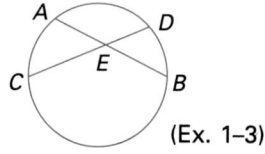

Written Exercises

For Exercises 1–10, assume that lines that appear to be tangent are tangent.

1. Given: $AE = 8$, $EB = 10$, $DE = 5$
Find EC.   16

2. Given: $DE = 4(EC)$, $AE = 4$, $EB = 25$
Find DE.   20

3. Given: $DE = 2(EC)$, $AE = 2$, $BE = 10$
Find DE. $2\sqrt{10}$

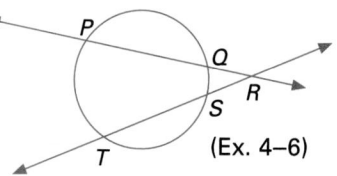

(Ex. 1–3)

4. Given: $PQ = 7$, $QR = 8$, $RS = 2$
Find TS.   58

5. Given: $PR = 9$, $PQ = 1$, $TR = 24$
Find TS.   21

6. Given: $PQ = 3$, $QR = 5$, $SR = 4$
Find TS.   6

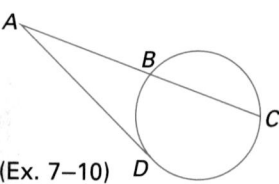

(Ex. 4–6)

7. Given: $AB = 4$, $AC = 10$
Find AD. $2\sqrt{10}$

8. Given: $AB = 4$, $BC = 16$
Find AD. $4\sqrt{5}$

9. Given: $AD = 4$, $AC = 2(AB)$
Find AB. $2\sqrt{2}$

10. $AD = 8$, $BC = 12$. Find AB.   4

(Ex. 7–10)

11. Two chords $\overline{AB}$ and $\overline{CD}$ intersect at a point E in the interior of a circle. $BE = 15.4$, $DE = 8.5$, $EC = 3.1$. Find AE to the nearest tenth.   1.7

12. From exterior point P, tangent $\overline{PT}$ is drawn to $\odot O$ meeting the circle at T. A secant $\overline{PB}$ is drawn meeting the circle at A and B. $PT = 6$ and $AB = 9$. Find PB.   12

13. Two chords $\overline{PO}$ and $\overline{RS}$ of a circle meet, when extended through O and S, at point T. $PO = 8$, $RS = 22$, $ST = 2$. Find OT.   4

14. Two chords $\overline{AB}$ and $\overline{CD}$ intersect at a point E in the interior of a circle. $AB = 16$, $DE = 12$, $EC = 4$. Find AE.   12 or 4

15. Prove Theorem 11.15.

16. Prove Theorem 11.16.

17. Prove the corollary to Theorem 11.16.

Enrichment

When the product of two quantities is a constant, they are said to *vary inversely*. For example, if $xy = k$, where k is a constant, then x varies inversely as y and y varies inversely as x. Discuss the quantities that vary inversely in the figure at the right.

In the figure, AE varies inversely as ED (and vice versa), where the constant is $CE \cdot EB$. Also, CE varies inversely as EB (and vice versa), where the constant is $AE \cdot ED$.

18. From exterior point P, tangent $\overline{PA}$ is drawn to $\odot O$ meeting the circle at A. From P, secant $\overrightarrow{PR}$ is drawn meeting the circle at M and R and passing through O. $\odot O$ has a radius of length 4. $PA = 3$. Find PR.   9

19. A piece of a broken gear is brought to a machinist. In order to build a new gear, he must first compute the diameter of the original gear. He finds the distance AB to be 8 in. He then measures ED, which is the perpendicular distance from E (the midpoint of $\overline{AB}$) to $\overset{\frown}{AB}$. $DE = 2$ in. Find the diameter.   10 in.

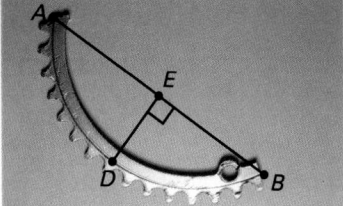

20. From exterior point T, tangent $\overline{TQ}$ and secant $\overline{TS}$ are drawn to a circle meeting the circle at Q, R, and S. The two segments are at right angles to each other. $TQ = 4$ and $TR = 2$. Find the length of a radius of the circle.   5

21. Given: $\overrightarrow{PR}$ is tangent to $\odot O$ at Q; $\overline{VU} \parallel \overline{PR}$.
Prove: $VQ \cdot QS = QU \cdot TQ$

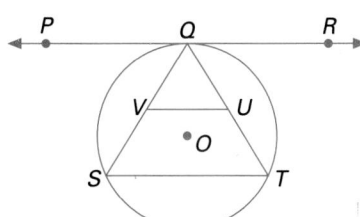

22. Given: $AB = 16$, $BC = 6$, $CD = 8$
Find BE.   4

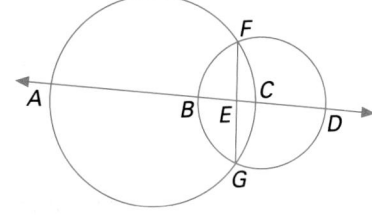

23. Given: Two noncongruent circles externally tangent at A with points C on the larger circle, D on the smaller circle such that $\overleftrightarrow{CD}$ passes through A, E on the larger circle, and F on the smaller circle such that $\overleftrightarrow{EF}$ passes through A.
Prove: $CA \cdot AF = EA \cdot AD$

Mixed Review

1. Given: A segment $\overline{AB}$. Construct the midpoint.   *1.3*   Check student construction.

2. A 14-ft ladder leans against a wall. The foot of the ladder is 5 ft from the bottom of the wall. At what angle (measured to the nearest degree) is the ladder to the ground?   *9.7*   69

3. How many diagonals does a convex pentagon have?   *6.1*   5

4. Given: $\triangle ABC$ with D on $\overline{AC}$ and E on $\overline{CB}$ such that $\overline{DE} \parallel \overline{AB}$, $DC = 4$, $AD = 5$, $CE = 12$. Find EB.   *8.2*   15

17.

	Statement	Reason
1.	$\odot O$ with sec $\overleftrightarrow{BC}$ and $\overleftrightarrow{DE}$ intersecting at A	1. Given
2.	Draw $\overleftrightarrow{AF}$ tan at F.	2. 2 pts determine line.
3.	$AC \cdot AB = (AF)^2$, $(AF)^2 = AE \cdot AD$	3. If tan and sec intersect, then square of tan seg = prod of sec seg and external sec seg.
4.	$AC \cdot AB = AE \cdot AD$	4. Sub

21.

	Statement	Reason
1.	$\overleftrightarrow{PR}$ tan to $\odot O$ at Q, $\overline{VU} \parallel \overline{PR}$.	1. Given
2.	$m \angle S = \frac{1}{2} \, m\overset{\frown}{QT}$	2. Meas of inscr $\angle = \frac{1}{2}$ meas of intercepted arc.
3.	$m \angle RQT = \frac{1}{2} \, m\overset{\frown}{QT}$	3. Angle formed by tan and sec at pt of tan $= \frac{1}{2}$ meas intercepted arc.
4.	$m \angle S = m \angle RQT$	4. Sub
5.	$m \angle RQT = m \angle VUQ$	5. If lines $\parallel$, alt int $\angle$s are $\cong$.
6.	$m \angle S = m \angle VUQ$	6. Sub
7.	$m \angle Q = m \angle Q$	7. Reflex
8.	$\triangle QVU \sim \triangle QTS$	8. AA $\triangle \sim$
9.	$\frac{VQ}{QU} = \frac{TQ}{QS}$	9. Def of $\sim$
10.	$VQ \cdot QS = QU \cdot TQ$	10. In true proportion, prod of extremes = prod of means.

See page 473 for the answer to Ex. 23.

◢◣GETTING STARTED

Prerequisite Quiz

1. Draw $\overline{AB}$ and construct its midpoint M.
2. Draw an angle. Construct its bisector.
3. Draw a line. At some point P on this line, construct a line perpendicular to the line.
4. How is a tangent related to a radius drawn to the point of contact of the tangent and the circle?   Perpendicular
5. Find the measure of an angle inscribed in a semicircle.   90

Motivator

Lesson 11.10 introduces the student to various constructions involving circles. Have the students describe a way to construct two circles of the same size, such that the circles are tangent to each other.

1. Draw a circle O of arbitrary radius r with its center at point O.
2. Select a point P at some distance greater than $2r$ outside the circle.
3. Draw $\overline{OP}$.
4. With compass still set at radius r place it at the point of intersection of circle O and $\overline{OP}$ and draw an arc intersecting $\overline{OP}$ near P. Label this pt Q.
5. Place the compass at Q and draw a circle of radius r.

◢◣TEACHING
SUGGESTIONS

Lesson Note

It will be helpful to review basic constructions such as the following:
1. angle bisector
2. midpoint of a segment
3. perpendicular to a line at a point on the line

11.10 Circles and Constructions

Objective   To perform and explain constructions involving circles

Two properties of circles form the basis for constructing two tangents to a circle from an exterior point:
- The sides of an angle inscribed in a semicircle are perpendicular.
- A line perpendicular to a radius at its endpoint on the circle is tangent to the circle.

Construction   Construct two tangents to a circle from a point in the exterior.

Procedure

Draw $\overline{OT}$. Construct M, the midpoint of $\overline{OT}$.	Use MT as the length of a radius to draw a circle intersecting $\odot O$ at points A and B.	Draw $\overline{TA}$ and $\overline{TB}$. Result: $\overrightarrow{TA}$ and $\overrightarrow{TB}$ are tangents.

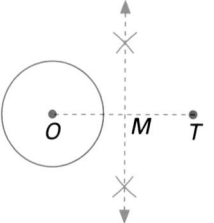 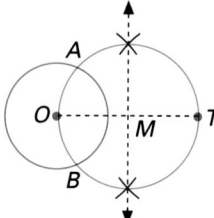 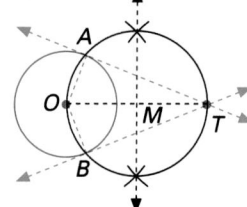

In the figure at the right, $\overline{AB}$ is constructed congruent to radius $\overline{OA}$. A radius $\overline{OB}$ is drawn; $OA = AB = OB$. Therefore, $\triangle OAB$ is equilateral. m $\angle O = 60$, and it follows that m$\overarc{AB} = 60$. Since the measure of a circle is 360, *six* of such arcs can be marked off around the circle. This suggests the following construction.

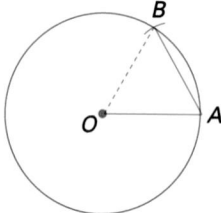

Construction   Inscribe a regular hexagon in a circle.

Procedure   Open compass to the length of a radius. Beginning at any point A on the circle, mark successive arcs on the circle. Draw chords to form a regular hexagon. Thus, $ABCDEF$ is a regular hexagon inscribed in $\odot O$.

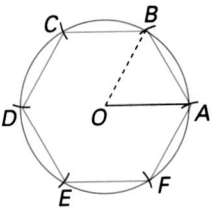

462   Chapter 11 Circles

462

Once it is determined how to construct a central angle and arc of given measure, other arcs of equal measure may be marked off around the circle. Chords between them can then be drawn. After marking off the six points, a chord can be drawn between every other point: $\overline{AC}$, $\overline{CE}$, and $\overline{EA}$. This results in an equilateral triangle.

EXAMPLE   Construct: A regular inscribed octagon.

Plan   An octagon has eight sides, so the measure of each central angle of the inscribed regular octagon is $\frac{360}{8}$, or 45.

Solution

Draw radius $\overline{OA}$. At O, construct $\overline{OP} \perp \overline{OA}$.

Construct $\overline{OB}$, the bisector of right $\angle POA$.

Starting at B, mark off successive arcs congruent to $\overgroup{AB}$. Draw chords.

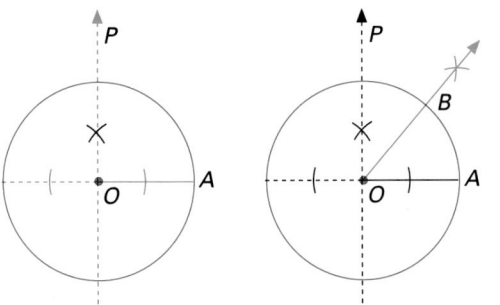
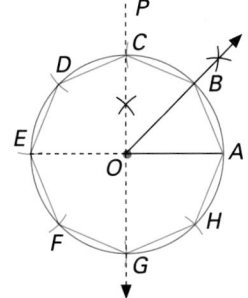

Result: $ABCDEFGH$ is a regular inscribed octagon.

Focus on Reading

List the following four steps in proper sequence for constructing two tangents to a circle O from an exterior point T.   d, b, a, c

a. Using $\overline{TQ}$ as a radius, draw a circle intersecting $\odot O$ at points A and B.
b. Construct Q, the midpoint of $\overline{OT}$.
c. Draw $\overline{TA}$ and $\overline{TB}$.
d. Draw $\overline{OT}$.

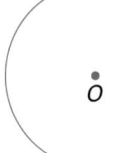

Additional Example

Inscribe an equilateral triangle in a circle.

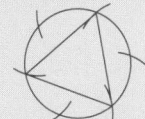

Mark off a radius 6 times around the circle. Join every other point.

Geometric Designs: Geometric designs are often results of constructions that begin with only one circle. The patterns that emerge are often pleasing to the eye because of the connections between properties of circles, regular polygons, and perpendicular and parallel line segments. Ask students to create their own geometric designs using any combination of the constructions introduced in this lesson with those presented in other lessons.

Critical Thinking Questions

Application: How can Thio find the diameter of his hockey puck using only a postcard?   See diagram.

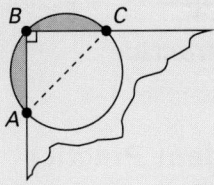

He can place the corner of the card, B, on the puck and mark the pts, A and C, on the puck where the edges of the postcard intersect the circle. These are the endpts of the diameter because $\angle$s inscr in a semicircle are rt $\angle$s.

Checkpoint

1. Draw a circle of radius 4 in. From a point P exterior to the circle, construct two tangents to the circle.
2. Inscribe a regular hexagon in a circle of radius 10 cm.
3. Inscribe a regular octagon in a circle of diameter 10 in.
4. Draw a 10 in. segment, $\overline{AB}$. Construct a circle with $\overline{AB}$ as the diameter. At B, construct a tangent to the circle.
5. Draw a circle of radius 8 in. From a point P, 17 in. from the center of the circle, construct a tangent segment to the circle. Predict its length. Then measure it to confirm your answer.   15 in

Closure

Have the students explain how they would construct the following figure. Within circle O of radius r draw circle H of radius $\frac{1}{2}r$, such that circle H is internally tangent to circle O.

1. Draw a circle of radius r and with center O.
2. Select some exterior point P.
3. Draw $\overline{OP}$ to form point L, the intersection of $\overline{OP}$ and circle O.
4. Draw a perpendicular bisector to $\overline{OL}$ and let point H, be the point of intersection.
5. Place compass at point H and reset compass to equal $\overline{HL}$.
6. Draw circle H.

◼◼◼FOLLOW UP

Guided Practice

Classroom Exercises 1–6

Independent Practice

🅐 Ex. 1–7, 🅑 Ex. 8–14, 🅒 Ex. 15–18

Basic: FR all, WE 1–7, Application

Average: FR all, WE 1–14 odd, Application

Above Average: FR all, WE 1–18 odd, Application

Additional Answers

Classroom Exercises

1. A tan is ⊥ to a radius drawn to the pt of tan.
2. Draw a radius to pt T. Then construct a line through T ⊥ to the radius.
3. Follow procedure for construction of 2 tan to a ⊙ from an ext pt on page 462.
4. Follow procedure for inscribing a regular hexagon in a ⊙ on page 462.
5. Follow procedure for inscribing a regular hexagon until six arcs are marked off. Then draw chords between every other pt.
6. Follow procedure for inscribing a regular octagon on page 463.

Classroom Exercises

1. How is a tangent related to the radius of a circle drawn to the point of tangency?　⊥
2. Tell how to construct a tangent to a circle at a point on the circle.
3. Tell how to construct a tangent to a circle from an exterior point.
4. Tell how to inscribe a regular hexagon in a circle.
5. Tell how to inscribe an equilateral triangle in a circle.
6. Tell how to inscribe a regular octagon in a circle.

Written Exercises

1. Draw a segment $\overline{AB}$. Construct a circle with $\overline{AB}$ as diameter.
2. Draw a circle and a point P in the exterior. From P, construct two tangents to the circle.
3. Draw a segment $\overline{PQ}$ of length 4 units. Construct a circle with $\overline{PQ}$ as diameter. Construct a tangent at Q.
4. Draw a circle with a 6-cm diameter. From a point in the exterior, construct 2 tangents to the circle.
5. Inscribe a regular hexagon in a circle.
6. Inscribe an equilateral triangle in a circle. Justify the construction.
7. Inscribe a regular octagon in a circle.
8. Inscribe a regular 12-sided polygon in a circle.
9. Construct a circle with radius of 3 in. From a point 5 in. from the center of the circle, construct two tangents to the circle. Predict the length of each tangent segment using the properties of tangents and the Pythagorean Theorem. Then verify the length of each tangent with a ruler.
10. Draw a circle and a point A in the exterior of the circle. From A, construct two tangents to the circle. Construct the bisector of $\angle A$. Does this line pass through the center of the circle? Explain.
11. Draw a segment $\overline{PQ}$. Construct a circle with $\overline{PQ}$ as diameter. Construct a diameter $\overline{ST} \perp \overline{PQ}$. Draw the four chords $\overline{SP}$, $\overline{PT}$, $\overline{TQ}$, and $\overline{QS}$. Prove that the quadrilateral $PTQS$ is a square.
12. Use the results of Exercise 11 to construct a square inscribed in a circle.
13. At any point T on a circle, construct a tangent to the circle. At this point T, construct another circle externally tangent to the first circle. From any point on the common tangent to the two circles, construct two other tangents to the circles.
14. Construct a right triangle. Construct a circle that circumscribes it. Justify why the hypotenuse will be the diameter of the circle.

Enrichment

If three circles with radii of different lengths are drawn and then external tangents are drawn for the circles, taken two at a time, these three points where the tangents meet will lie on a straight line. Have the students do this construction. Best results will be obtained if the circles are significantly different in size and if they are placed in a triangular arrangement. The result of this construction, even without proof, is pleasing and rather mysterious.

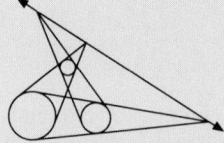

15. Consider the construction of tangents from an external point, as given in the lesson (see p. 462).
    a. Explain why $\overline{OT}$ is a diameter of $\odot M$.
    b. Explain why $\angle OAT$ is a right angle.
    c. Explain why $\overline{TA}$ is perpendicular to $\odot O$ at point A.

16. Consider the construction of a regular hexagon, as given in the lesson. Explain why the hexagon is
    a. equilateral; and b. equiangular.

17. Construct two circles with radii of lengths 3 units and 8 units, and centers 13 units apart. Then construct two external tangents to the circles. Predict the lengths of these two tangent segments. After performing the construction, confirm by measurement.

18. Construct $\overline{AB}$ with length of 4 units and point C between A and B such that $AC = 1$ and $CB = 3$. Construct a circle with $\overline{AB}$ as diameter. At C, construct a perpendicular to the diameter meeting the circle at points D and E. Prove that $CD = \sqrt{3}$. This provides a procedure for constructing the square root of a number, for example, $\sqrt{3}$.

Mixed Review

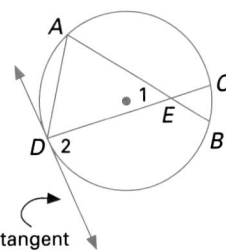

tangent

1. $m\widehat{AD} = 80$, $m\widehat{BC} = 40$, m $\angle 1 =$ ___60___ **11.7**
2. $\overline{DC}$ is a diameter; $m\widehat{CB} = 30$, m $\angle A =$ ___75___. **11.6**
3. $AE = 20$, $EB = 4$, $DE = 16$, $EC =$ ___5___ **11.9**
4. $m\widehat{DBC} = 150$, m $\angle 2 =$ ___75___ **11.8**

Application: *Geometric Designs*

Using the method of inscribing a regular hexagon in a circle, construct the following mathematical designs.

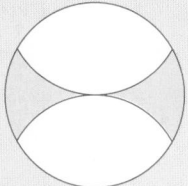 

1. Explain your step-by-step procedure for constructing these designs.
2. Design another pattern using circles, central angles, and arcs. Write out a procedure for constructing it.   Answers will vary.

13.

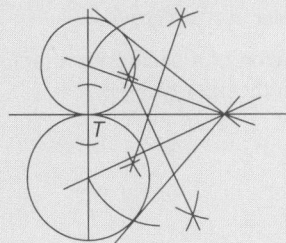

Construct a line containing ctr O of original $\odot$ and pt T. Construct a $\perp$ to $\overline{OT}$ at T, including a pt A. Mark off any pt Q on $\overrightarrow{OT}$ on the other side of T from O. Draw $\odot O$ with radius QT. From pt A on $\overleftrightarrow{TA}$ construct a tan to $\odot O$ and a tan to $\odot Q$, using the procedure on page 462.

---

1. Construct midpt M of $\overline{AB}$. Using $\overline{MB}$ as radius, draw a $\odot$ with M as ctr.
2. See procedure on page 462.
3. Construct midpt M of $\overline{PQ}$. Using $\overline{MQ}$ as radius, draw a $\odot$ with M as ctr. Then construct a line through $Q \perp \overline{MQ}$.
4. Follow procedure for construction of 2 tan to a $\odot$ from an ext pt on page 462.
5. Follow procedure for inscribing a regular hexagon in a $\odot$ on page 462.
6. Mark off 6 arcs of equal measure as per procedure for inscribing regular hexagon. Then draw chords between every other pt. An equil $\triangle$ results because the 3 central $\angle$s corr to the chords are equal in meas and form isos $\triangle$s intersecting the $\odot$.
7. See procedure on page 463.
8. Inscribe a regular hexagon in a $\odot$. Construct diam bisecting the three vertical pairs of central angles. Between the 6 endpts of these diameters and the vertices of the hexagon draw 12 chords.
9. Tan seg should measure 4 in.
10. Yes. A seg containing pt A and the ctr forms the hyp of 2 rt triangles. Angle A is necessarily divided into 2 $\cong \angle$s by this seg.
11.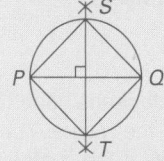

    The diameters $\overline{PQ}$ and $\overline{ST}$, together with $\overline{SP}$, $\overline{PT}$, $\overline{TQ}$, and $\overline{QS}$, form four rt $\triangle$s, since $\overline{ST} \perp \overline{PQ}$. The legs are all radii of the $\odot$, and thus congr. So the 4 $\triangle$s are $\cong$ by SAS. By CPCTC the hyp (sides of the quad $PTQS$) are $\cong$. The 2 acute $\angle$s of each $\triangle$ meas 45, since the legs are equal in length. Thus, each angle of $PTQS$ is $45 + 45 = 90$. As a result $PTQS$ is a rect and a square.
12. Check students' constructions.

See page 664 for the answers to Written Ex. 14–18 and Application.

Teaching Resources

Quick Quizzes 85
Reteaching and Practice
  Worksheets, pp. 169, 170
Technology Worksheet 11
Transparencies 32A, 32B

**GETTING STARTED**

Prerequisite Quiz

1. Find the distance *PQ* between the points
   P(3, −2) and *Q*(−1, 1).   5
2. Write the formula for the distance
   between (x_1, y_1) and (x_2, y_2).
   $d = \sqrt{(x_2 - x_1)^2 + (y_2 - y_1)^2}$

Motivator

Lesson 11.11 introduces the students to
equations of circles. Ask the students how a
circle is defined.   The set of all points in a
plane a given distance from a point.   Ask
them what formula is implied in the
definition of a circle.   Distance Formula

**TEACHING
  SUGGESTIONS**

Lesson Note

The equation for a circle depends upon the
location of the center and the radius. The
students should realize that if the center and
the length of the radius are given, then
there is exactly one circle determined.
You might begin the lesson by using a chart
or screen to show several circles, some of
which are congruent and some of which are
concentric.

Ask the students how any one of the circles
is different from any other one. The
differences will be in the location of the
center and the length of a radius. A review
of the Distance Formula may be needed.

11.11 Equations of Circles

Objectives   To write equations of circles
To graph equations of circles

A circle is the set of all points in a plane a given distance from a fixed
point. Using this definition, an equation of a circle can be derived.

Suppose that a circle has its center at some point *P*(*h*,*k*) and has a radius of length *r*. Let *P*(*x*,*y*) be any point on the circle. The Distance
Formula leads to this equation.

$$\sqrt{(x - h)^2 + (y - k)^2} = r, \text{ or}$$
$$(x - h)^2 + (y - k)^2 = r^2$$

This proves the following theorem.

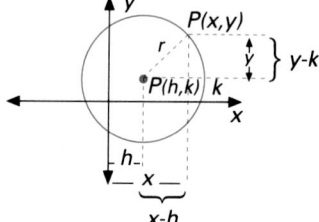

Theorem 11.17   The equation of a circle with the coordinates of the center (*h*,*k*) and
a radius of length *r* is $(x - h)^2 + (y - k)^2 = r^2$.

When the equation of a circle is written in the form
$(x - h)^2 + (y - k)^2 = r^2$, it is easy to identify the coordinates of the
center and the length of the radius. The graph of the circle can then be
drawn.

EXAMPLE 1   Draw the graph of $(x - 1)^2 + (y + 3)^2 = 16$.

Plan   Rewrite the equation to show the center
and radius.

Solution $(x - 1)^2 + [y - (-3)]^2 = 4^2$
$(h,k) = (1, -3) \ r = 4$

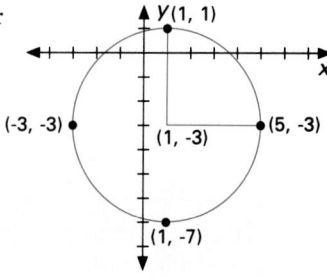

466   Chapter 11 Circles

Additional Example 1

Draw the graph of
$(x + 2)^2 + (y - 2)^2 = 9$.

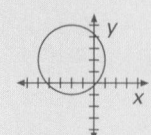

Additional Example 2

Find an equation of a circle with a radius
of length 2 and center (3, −4).
$(x - 3)^2 + [y - (-4)]^2 = 2^2$
$(x - 3)^2 + (y + 4)^2 = 4$

EXAMPLE 2 Find an equation of a circle with a radius of length 3 and center $(-2, 3)$.

Solution

$(x - h)^2 + (y - k)^2 = r^2$
$[x - (-2)]^2 + (y - 3)^2 = 3^2$
$(x + 2)^2 + (y - 3)^2 = 9$

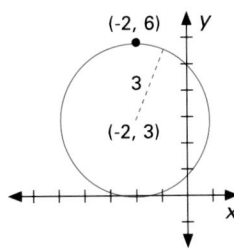

If the center of a circle is the origin, the equation $(x - 0)^2 + (y - 0)^2 = r^2$ simplifies to $x^2 + y^2 = r^2$. For example, the graph of $x^2 + y^2 = 16$ is the circle with the origin as its center and with a radius of length 4.

To simplify an equation of a circle means to write it in the form $x^2 + y^2 + dx + ey + f = 0$. Such an equation is a *quadratic equation*.

EXAMPLE 3 Simplify the equation in Example 2.

Solution

$(x + 2)^2 + (y - 3)^2 = 9$
$x^2 + 4x + 4 + y^2 - 6y + 9 = 9$
$x^2 + y^2 + 4x - 6y + 4 = 0$

EXAMPLE 4 Find the coordinates of the center and the length of a radius of a circle whose equation is $x^2 + y^2 + 2x - 4y + 1 = 0$.

Plan Change the given equation to form $(x - h)^2 + (y - k)^2 = r^2$.

Solution

$x^2 + y^2 + 2x - 4y + 1 = 0$
$(x^2 + 2x + \underline{}) + (y^2 - 4y + \underline{}) = -1$
$(x^2 + 2x + 1) + (y^2 - 4y + 4) = -1 + 1 + 4$
$(x + 1)^2 + (y - 2)^2 = 4$
$[x - (-1)]^2 + (y - 2)^2 = 2^2$

Therefore, the coordinates of the center are $(-1, 2)$, and the length of a radius is 2.

11.11 Equations of Circles **467**

Math Connections

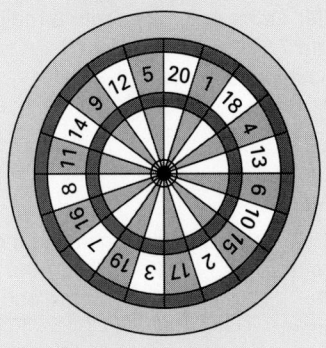

Darts: A dartboard consists of seven concentric circles that increase in size. Each circle, except the outermost and the innermost, are divided by congruent angles into 20 congruent sections. The circles and the angles are regulated by standard measures for the sport. When the standard measures are entered into a computer, official-size dartboards can be mass produced.

Critical Thinking Question

Application: Refer students to the dartboard in Math Connections above and ask the following questions.

The diameters of the circles on a dartboard, from the outermost circle to the innermost circle, are 17″, 13.25″, 12.5″, 8.25″, 7.5″, 1.25″, and 0.5″. If the center of the dartboard is placed at the origin of a coordinate plane, what are equations for these circles? $x^2 + y^2 = r^2$, where $r^2 = 72.25, 43.89, 39.06, 17.02, 14.06, 0.39$ and 0.25

Common Error Analysis

Error: Students state that the equation of a circle of the form $(x + h)^2 + (y + k)^2 = r^2$ has its center at (h, k) rather than at $(-h, -k)$.

Remind students that the equation of a circle with center (h, k), and radius, r, is an application of the Distance Formula. That is, $(x + h)^2 + (y + h)^2 = [x - (-h)]^2 + [y - (-k)]^2$.

Therefore, the center of a circle with equation $(x + h)^2 + (y + k)^2 = r^2$ is at $(-h, -k)$.

Checkpoint

1. Write an equation of a circle with a center at the origin and with a radius of length 7. $x^2 + y^2 = 49$
2. Simplify the equation $(x - 3)^2 + (y + 1)^2 = 9$.
   $x^2 - 6x + y^2 + 2y + 1 = 0$

Closure

Ask the students how they can find the radius of a circle.   Use the Distance Formula.   Have them tell you the equation for a circle.   Theorem 11.17 page 466.
Ask them what the coordinates of the center are. (h, k) Have them give the equation for a circle. $x^2 + y^2 + dx + ey + f = 0$

**FOLLOW UP**

Guided Practice

Classroom Exercises 1–6

Independent Practice

A Ex. 1–18, **B** Ex. 19–32, **C** Ex. 33–40

Basic: WE 1–18 odd

Average: WE 1–32 odd

Above Average: WE 1–40 odd

EXAMPLE 5   Write an equation of the form $x^2 + y^2 + dx + ey = 0$ for a circle with center $A(4,-3)$ if the circle passes through the origin.

Plan   Find the length of a radius by finding the distance from the center to the origin.

Solution
$$r = \sqrt{(4 - 0)^2 + (-3 - 0)^2}$$
$$= \sqrt{16 + 9}$$
$$= \sqrt{25}, \text{ or } 5$$
$$(x - 4)^2 + [y - (-3)]^2 = 5^2$$
$$x^2 - 8x + 16 + y^2 + 6y + 9 = 25$$
$$x^2 + y^2 - 8x + 6y = 0$$

Classroom Exercises

Give the coordinates of the center and the length of a radius.

1. $(x - 2)^2 + (y - 5)^2 = 9$ $(2,5), r = 3$ 2. $(x - 4)^2 + (y - 1)^2 = 4$ $(4,1), r = 2$
3. $(x + 3)^2 + (y - 9)^2 = 1$ $(-3,9), r = 1$ 4. $(x - 7)^2 + (y + 1)^2 = 16$ $(7,-1), r = 4$
5. $(x + 7)^2 + (y + 2)^2 = 4$ $(-7,-2), r = 2$ 6. $(x + 1)^2 + (y + 8)^2 = 25$
                                                                                          $(-1,-8), r = 5$

Written Exercises

4. $(x + 2)^2 + (y + 2)^2 = 1$
6. $(x + 5)^2 + (y - 3)^2 = 81$

Write an equation of the form $(x - h)^2 + (y - k)^2 = r^2$ for the circle with center O and radius of length r.

1. $O(0,0), r = 5$ $x^2 + y^2 = 25$ 2. $O(0,0), r = 2$ $x^2 + y^2 = 4$
3. $O(1,1), r = 2$ $(x - 1)^2 + (y - 1)^2 = 4$ 4. $O(-2,-2), r = 1$
5. $O(1,-3), r = 4$ $(x - 1)^2 + (y + 3)^2 = 16$ 6. $O(-5,3), r = 9$
7. $O(-2,-4), r = 1$ 8. $O(-4,-2), r = 4$
   $(x + 2)^2 + (y + 4)^2 = 1$ $(x + 4)^2 + (y + 2)^2 = 16$

Write an equation of the form $x^2 + y^2 + dx + ey + f = 0$ for the circle with center Q and radius of length r.

   $x^2 + y^2 - 2x - 6y - 6 = 0$

9. $Q(0,0), r = 2$ $x^2 + y^2 - 4 = 0$ 10. $Q(1,3), r = 4$
11. $Q(-3,4), r = 5$ $x^2 + y^2 + 6x - 8y = 0$ 12. $Q(4,-2), r = 4$
                                                                                     $x^2 + y^2 - 8x + 4y + 4 = 0$

Draw the graph of each circle.

                                                                     Ctr$(-2,1), r = 2$ Ctr$(4,-2), r = 4$
13. $(x + 2)^2 + (y - 1)^2 = 4$ 14. $(x - 4)^2 + (y + 2)^2 = 16$
15. $(x - 2)^2 + y^2 = 1$ Ctr$(2,0), r = 1$ 16. $x^2 + (y - 2)^2 = 9$ Ctr$(0,2), r = 3$
17. $x^2 + y^2 = 4$ Ctr$(0,0), r = 2$ 18. $x^2 + y^2 = 16$ Ctr$(0,0), r = 4$

468   Chapter 11 Circles

Additional Example 5

Write an equation of a circle with center $(-2, 3)$ passing through the origin.
$$r = \sqrt{(-2 - 0)^2 + (3 - 0)^2}$$
$$r = \sqrt{4 + 9} = \sqrt{13}$$
$$[x - (-2)]^2 + (y - 3)^2 = (\sqrt{13})^2$$
$$x^2 + 4x + 4 + y^2 - 6y + 9 = 13$$
$$x^2 + 4x + y^2 - 6y = 0$$

Find the coordinates of the center and the length of a radius of each circle. Ctr(0,2), r = 2√2

19. $x^2 + y^2 - 6x + 5 = 0$ Ctr(3,0), r = 2 **20.** $x^2 + y^2 - 4y - 4 = 0$

21. $x^2 + y^2 - 2x - 8y + 13 = 0$ **22.** $x^2 + y^2 - 16x - 6y + 72 = 0$

23. $x^2 + y^2 + 18x - 20y + 177 = 0$ **24.** $x^2 + y^2 + 4x + 2y - 145 = 0$

Ctr(−9,10), r = 2 **21.** Ctr(1,4), r = 2 Ctr(−2,−1), r = √19

Write an equation of the circle with center A if the circle passes through the origin. **22.** Ctr(8,3), r = 1

25. $A(3,4)$ $(x - 3)^2 + (y - 4)^2 = 25$ **26.** $A(5,12)$ $(x - 5)^2 + (y - 12)^2 = 169$

27. $A(-4,-3)$ $(x + 4)^2 + (y + 3)^2 = 25$ **28.** $A(-12,-5)$ $(x + 12)^2 + (y + 5)^2 = 169$

Tell whether point A is on the circle with the given equation.

Yes

29. $A(0,4)$, $x^2 + y^2 = 16$ Yes **30.** $A(-1,-1)$, $(x - 2)^2 + (y + 1)^2 = 9$

31. $A(-2,3)$, $x^2 + y^2 + 4x = 0$ No **32.** $A(2,5)$, $x^2 + y^2 + 2x - 8y + 13 = 0$

No

Write an equation of the circle containing the two given points as end-points of a diameter.

33. $(1,2)$ and $(-3,-4)$ **34.** $(4,5)$ and $(-4,-1)$

$(x + 1)^2 + (y + 1)^2 = 13$ $x^2 + (y - 2)^2 = 25$

Tell whether the two circles are congruent.

35. $x^2 + y^2 - 4x + 2y - 4 = 0$ and $x^2 + y^2 + 2x - 6y - 6 = 0$ No; $r_1 = 3$, $r_2 = 4$

36. $x^2 + y^2 + 10x + 8 = 0$ and $x^2 + y^2 - 8y - 1 = 0$ Yes

Write an equation of the line that is tangent to the given circle at the given point P.

37. $x^2 + y^2 + 4x - 6y - 12 = 0$, $P(-6,6)$ $4x - 3y + 42 = 0$

38. $x^2 + y^2 + 10x + 8y + 21 = 0$, $P(-7,-8)$ $x + 2y + 23 = 0$

Which of the following sets of points is a circle?

39. the set of all points in a plane equidistant from two points No

40. the set of all points in a plane twice as far from one point as from another Yes

Mixed Review

1. Write an equation of a line containing the points $P(0,1)$ and $Q(-3,-5)$. **10.4** $2x - y + 1 = 0$

2. Find the distance between the points $R(2,-5)$ and $S(-3,0)$. **10.1** $5\sqrt{2}$

3. Find the midpoint of $\overline{MN}$ for $M(-1,-3)$ and $N(5,9)$. **10.2** $(2,3)$

4. Find the slope of the line $2x + y = 5$. **10.4** -2

11.

Statement	Reason
1. Trap PQRS inscribed in ⊙O, $\overline{PQ}$ is diam.	1. Given
2. $\overline{PO} \cong \overline{SO} \cong \overline{RO} \cong \overline{QO}$	2. Radii of ⊙ are ≅.
3. ∠RSO ≅ ∠SRO	3. If 2 sides of △ ≅, ∠s opp sides are ≅.
4. $\overline{SR} \parallel \overline{PQ}$	4. Def of trap
5. ∠SOP ≅ ∠RSO, ∠ROQ ≅ ∠SRO	5. If transv intersects ∥ lines, alt int ∠s are ≅.
6. ∠SOP ≅ ∠ROQ	6. Sub
7. △SOP ≅ △ROQ	7. SAS
8. m∠SOR = m∠SOR	8. Reflex Prop of Eq
9. m∠SOP + m∠SOR = m∠ROQ + m∠SOR	9. Equations may be added.
10. m∠SOP + m∠SOR = m∠POR, m∠ROQ + m∠SOR = m∠QOS	10. Angle Add Post
11. m∠POR = m∠QOS	11. Sub
12. $\overline{PR} \cong \overline{SQ}$	12. If central ∠s ≅, corr chords are ≅.

12.

Statement	Reason
1. $\overline{AB}$ is diam of ⊙O, $\overline{OD} \parallel \overline{BC}$	1. Given
2. OA = OB = OC	2. Radii of ⊙ are ≅.
3. O is midpt of $\overline{AB}$.	3. Def of midpt
4. Draw $\overline{AC}$ and $\overline{CO}$.	4. 2 pts determine line.
5. $\overline{OD}$ bisects $\overline{AC}$.	5. If seg is ∥ to one side of △ and contains midpt of 2nd side, it bis 3rd side.
6. ∠OAC ≅ ∠OCA	6. If 2 sides of △ are ≅, ∠s opp sides are ≅.
7. △AOD ≅ △COD	7. SAS
8. ∠AOD ≅ ∠COD	8. CPCTC
9. $\overparen{AD} \cong \overparen{DC}$	9. If central ∠s are ≅, corr arcs are ≅.
10. $\overline{OD}$ bis $\overparen{AC}$.	10. Def of bis

Chapter 11 Review

Key Terms

arc (p. 432)	inscribed polygon (p. 423)
central angle (p. 432)	intercepted arc (p. 442)
chord (p. 416)	major arc (p. 433)
circle (p. 415)	minor arc (p. 432)
circumscribed polygon (p. 427)	radius (p. 415)
common tangent (p. 427)	secant (p. 417)
concentric circles (p. 438)	semicircle (p. 432)
diameter (p. 416)	tangent (p. 417)
external secant segment (p. 458)	tangent circles (p. 428)
inscribed angle (p. 442)	tangent segment (p. 425)

Key Ideas and Review Exercises

11.1   To identify lines and segments related to circles

Identify the following for the circle.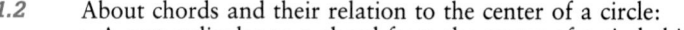
1. $\overline{OU}, \overline{OT}, \overline{OS}$ 2. $\overline{US}$ 3. $\overline{QS}, \overline{US}, \overline{TS}$
1. radius   2. diameter   3. chord
4. secant $\overleftrightarrow{QS}$ 5. tangent $\overleftrightarrow{PR}$
6. Given: OU = 7, OW = 3x − 5.
   Find all the values of x such that W lies in the exterior of the circle.   x > 4
7. OT = 2x − 8, OU = 4x − 20, OS = ___4___

11.2   About chords and their relation to the center of a circle:
   • A perpendicular to a chord from the center of a circle bisects the chord.
   • Congruent chords are equidistant from the center, and conversely.
   • Chords of greater length are closer to the center of a circle.

8. AB = 12, AO = 10, OF = ___8___ $\frac{8}{3}\sqrt{3}$
9. m∠AOB = 120, AB = 8, OA = _____
10. OD = 5, OF = 4. Name the longer of the two chords $\overline{AB}$ and $\overline{EC}$. $\overline{AB}$

11. A chord of length 16 is 4 units from the center of a circle. Find the length of a diameter. $8\sqrt{5}$
12. △ABC is inscribed in a circle. m∠A = 30, m∠B = 70. Name the longest chord. $\overline{AB}$

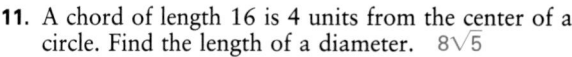

(Ex. 8–10)

11.3   Properties of tangents and radii: If $\overline{TA}$ and $\overline{TB}$ are tangent segments to ⊙O with radius $\overline{OA}$, then $\overline{OA} \perp \overline{AT}$, AT = BT, and $\overline{OT}$ bisects ∠T (see p. 426).

13. TB = 6, OT = 8, OA = ___$2\sqrt{7}$___
14. m∠ATB = 120, AT = 10, OA = ___$10\sqrt{3}$___

Arcs and Angles of Circles

In the diagrams below, $\overline{RV} \parallel \overline{PQ}$, $\overline{PQ}$ is a diameter, $\overline{UT}$ is tangent to $\odot O$, and $\overline{FE}$ is tangent to $\odot P$.

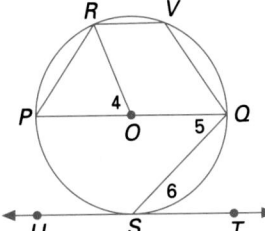
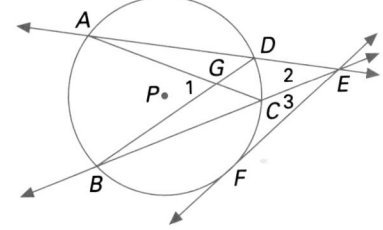

$m \angle 4 = m\overset{\frown}{PR}$ *11.4*

$m \angle 5 = \frac{1}{2}m\overset{\frown}{PS}$ *11.6*

$m \angle 6 = \frac{1}{2}m\overset{\frown}{SQ}$ *11.8*

$m\overset{\frown}{PR} = m\overset{\frown}{VQ}$ (since $\overline{RV} \parallel \overline{PQ}$)   *11.6*

15. $m\overset{\frown}{RQ} = 160$, m $\angle 4 = $ ___20___
16. $m\overset{\frown}{PR}{:}m\overset{\frown}{RQ} = 5{:}4$, m $\angle 4 = $ ___100___
17. $m\overset{\frown}{PS} = 100$, m $\angle 6 = $ ___40___
18. $m\overset{\frown}{RV} = 60$, m $\angle 4 = $ ___60___
19. m $\angle V = 150$, m $\angle P = $ ___30___

$m \angle 1 = \frac{1}{2}(m\overset{\frown}{AB} + m\overset{\frown}{CD})$ *11.7*

$m \angle 2 = \frac{1}{2}(m\overset{\frown}{AB} - m\overset{\frown}{CD})$

$m \angle 3 = \frac{1}{2}(m\overset{\frown}{BF} - m\overset{\frown}{FC})$ *11.8*

20. $m\overset{\frown}{AB} = 100$, m $\angle 1 = 75$, $m\overset{\frown}{CD} = $ ___50___
21. $m\overset{\frown}{AB}{:}m\overset{\frown}{CD} = 4{:}1$, $m\overset{\frown}{AD} = 130 = m\overset{\frown}{BC}$, m $\angle 2 = $ ___30___
22. $m\overset{\frown}{AB} = 100$, m $\angle 1 = 60$, m $\angle 2 = $ ___40___

11.9   **Products of Segments**

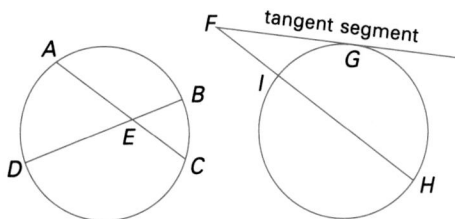
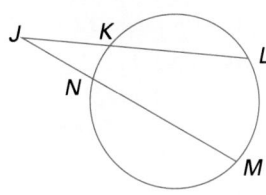

$AE \cdot EC = DE \cdot EB$ $(FG)^2 = FH \cdot FI$ $JM \cdot JN = JL \cdot JK$

23. $AE = 8$, $EC = 10$, $ED = 16$, $EB = $ ___5___
24. $FG = 4$, $IH = 6$, $FH = $ ___8___
25. $JK = 4$, $JL = 10$, $MN = 3$, $JN = $ ___5___
26. $FI = 4$, $IH = 6$, $FG = $ ___$2\sqrt{10}$___

27. Construct two tangents to a circle from a point exterior to the circle.   *11.10*

11.11   The equation of a circle with center at point $P(h,k)$ and radius r is $(x - h)^2 + (y - k)^2 = r^2$.

28. Graph the equation $(x - 3)^2 + (y + 2)^2 = 9$.   Circle with Ctr(3,−2), r = 3

---

27. Follow procedure on page 462.

Additional Answer, page 451

37.

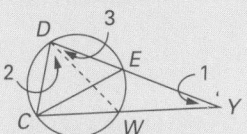

Statement	Reason
1. $\overline{CD}$ and $\overline{CE}$ are $\cong$ chords in $\odot$; W is on $\overline{CE}$; $\overleftrightarrow{CW}$ and $\overleftrightarrow{DE}$ meet at Y.	1. Given
2. $m\overset{\frown}{CD} = m\overset{\frown}{CE}$	2. If chords $\cong$, then corr arcs are $\cong$.
3. m $\angle 2 +$ m $\angle 3 = \frac{1}{2} m\overset{\frown}{CE}$	3. Meas of inscr $\angle = \frac{1}{2}$ meas of intercepted arc.
4. m $\angle 2 +$ m $\angle 3 = \frac{1}{2}$ m $\overset{\frown}{CD}$	4. Sub
5. m $\angle 3 = \frac{1}{2}$ m$\overset{\frown}{WE}$	5. Meas of inscr $\angle = \frac{1}{2}$ meas of intercepted arc.
6. m $\angle 1 = \frac{1}{2}$ ($m\overset{\frown}{CD} - m\overset{\frown}{WE}$)	6. Meas of $\angle$ formed by 2 sec in ext $= \frac{1}{2}$ diff of meas of intercepted arcs.
7. m $\angle 1 = \frac{1}{2}$ m$\overset{\frown}{CD} - \frac{1}{2}$ m$\overset{\frown}{WE}$	7. Distr Prop
8. m $\angle 1 =$ m $\angle 2 +$ m $\angle 3 -$ m $\angle 3 =$ m $\angle 2$, or m $\angle CYD =$ m $\angle CDW$.	8. Sub

8.

Statement	Reason
1. Diam $\overline{GH} \perp$ to chords $\overline{AC}$ and $\overline{DF}$.	1. Given
2. $\overline{AC} \parallel \overline{DF}$	2. (If 2 lines are $\perp$ to same line, they are $\parallel$.
3. m $\overarc{AD}$ = m $\overarc{CF}$	3. If arcs of $\odot$ are included between $\parallel$ chords, then arcs have = meas.

Additional Answer, page 456

31.

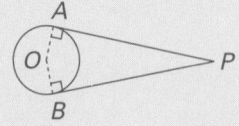

Statement	Reason
1. From pt P in ext $\odot$, two tan intersect $\odot O$ at A and B.	1. Given
2. Draw $\overline{OA}$, $\overline{OB}$.	2. 2 pts determine line.
3. $\overline{AO} \perp \overline{AP}$, $\overline{OB} \perp \overline{BP}$	3. Line tan to $\odot$ is $\perp$ to radius at pt of tan.
4. m $\angle OAP$ = m $\angle OBP$ = 90.	4. Def of $\perp$
5. m $\angle OAP$ + m $\angle OBP$ + m $\angle AOB$ + m $\angle P$ = 360.	5. Sum of $\angle$ meas of polygon = $(N-2)$ 180.
6. 90 + 90 + m $\angle AOB$ + m $\angle P$ = 360.	6. Sub
7. m $\angle AOB$ + m $\angle P$ = 180.	7. Subt prop of Eq
8. m $\overarc{AB}$ = m $\angle AOB$	8. Arc meas = meas of central $\angle$.
9. m $\overarc{AB}$ + m $\angle P$ = 180.	9. Sub

Chapter 11 Test

A-Level Ex.: 1–7, 9, 11, 12, 14, 16, 17, 21
B-Level Ex.: 8, 10, 13, 15, 18–20, 22

Identify the following for $\odot O$ at the right.

1. radius   **2.** chord   **3.** diameter
4. tangent $\overleftrightarrow{PQ}$ **5.** secant $\overleftrightarrow{TS}$
6. Given: $OQ = 4x - 2$, $OS = 2x + 8$. Find OU.   18
7. Given: $SU = 12$, $OV = 3x - 9$. Find all values of x for which V is in the exterior of the circle. $x > 5$
1. $\overline{OU}, \overline{OQ}, \overline{OR}, \overline{OS}$ **2.** $\overline{QR}, \overline{US}, \overline{TS}$ **3.** $\overline{US}$

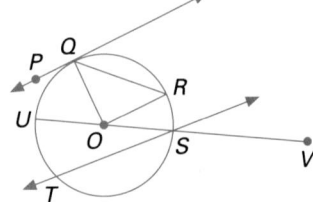

Use $\odot O$ with radius $\overline{OD}$ for Exercises 8–10.

8. Given: Diameter $\overline{GH} \perp$ chords $\overline{AC}$ and $\overline{DF}$
   Prove: m$\overarc{AD}$ = m$\overarc{CF}$
9. Given: m$\overarc{AHC}$ = 115, m$\overarc{DGF}$ = 185
   Find m$\overarc{AD}$.   30
10. Given: m $\angle DOE$ = 30, $OE = 4$
   Find the length of the radius. $\frac{8}{3}\sqrt{3}$

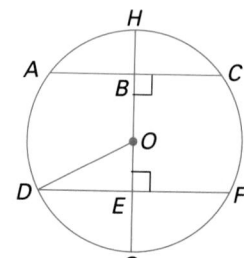

Use $\odot O$ with tangent $\overleftrightarrow{EF}$ for Exercises 11–14.

11. m$\overarc{AE}$ = 100, m$\overarc{DC}$ = 70, m $\angle 1$ = __85__
12. $AB = 24$, $BC = 4$, $EB = 6$, $BD =$ __16__
13. m$\overarc{AE}$:m$\overarc{AD}$:m$\overarc{DC}$:m$\overarc{EC}$ = 4:3:2:1,
   m $\angle 2$ = __72__
14. m$\overarc{AE}$ = 100, m $\angle 1$ = 70, m$\overarc{DC}$ = __40__

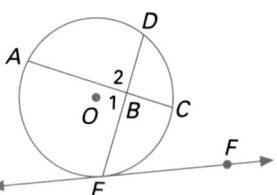

Use $\odot O$ with tangent segment $\overline{AB}$ for Exercises 15–18.

15. m$\overarc{DC}$ = 50, m $\angle 2$ = 5, m$\overarc{FE}$ = __40__
16. m$\overarc{FBC}$ = 160, m$\overarc{BC}$:m$\overarc{FB}$ = 3:2, m $\angle 1$ = __16__
17. $AF = 2$, $FC = 4$, $AB =$ __$2\sqrt{3}$__
18. $ED = 1$, $AF = 2$, $FC = 8$, $AE =$ __4__
19. Find the length of a chord of a circle if the chord is 6 units from the center and the length of the radius is 10 units.   16
20. The length of the common external tangent to two circles is 12. Find the distance between the centers if the lengths of the radii are 7 and 2.   13
21. $\triangle ABC$ is inscribed in a circle. m $\angle A$ = 40, m $\angle C$ = 80. Name the shortest chord. $\overline{BC}$
22. Find the coordinates of the center and the length of the radius of the circle. Draw the circle $x^2 + y^2 - 4x - 5 = 0$.   Ctr(2,0), $r = 3$

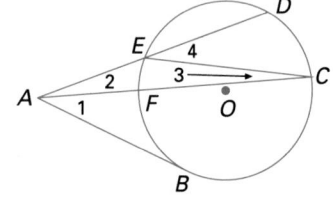

College Prep Test

Strategy for Achievement in Testing

Each item consists of two quantities, one in Column A and one in Column B. You are to compare the two quantities and choose an answer from the following list.

A—The quantity in Column A is greater.
B—The quantity in Column B is greater.
C—The two quantities are equal.
D—The relationship cannot be determined from the information given.

Column 1	Column 2
1. the length of the hypotenuse of a right triangle with legs of lengths 16 and 12	the length of the hypotenuse of a right triangle with legs of lengths 5 and 12

A

2. Lines *a* and *b* are parallel. Lines *m* and *n* are parallel.

C

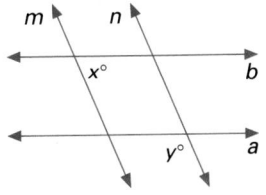

$180 - y$	x

3.

C

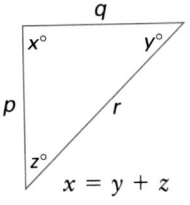

$x = y + z$

NOTE: The figure is not drawn to scale.

$p^2 + q^2$	r^2

Column 1	Column 2
4.	

B

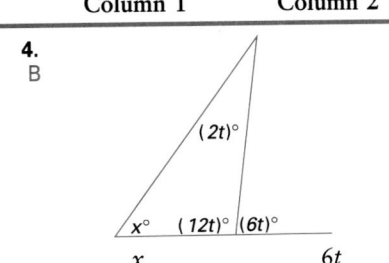

x	$6t$

5. Circle with center O.

C

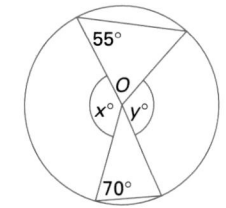

$x + y$	250

6. $a \parallel b \parallel c$

C

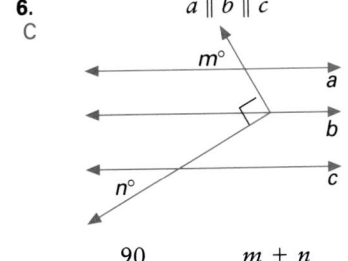

90	$m + n$

Additional Answer, page 461

23.

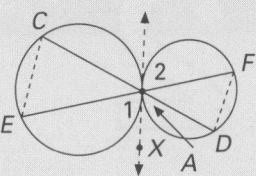

Statement	Reason
1. Two non ≅ ⊙s externally tan at A with $\overleftrightarrow{CAD}$ and $\overleftrightarrow{EAF}$	1. Given
2. Draw $\overleftrightarrow{AX}$ tan to both ⊙s.	2. There is 1 line tan at given pt.
3. m ∠1 = $\frac{1}{2}$ m $\widehat{AE}$, m ∠2 = $\frac{1}{2}$ m$\widehat{AF}$	3. Angle formed by sec and tan at pt of tan = $\frac{1}{2}$ meas intercepted arc.
4. m ∠C = $\frac{1}{2}$ m $\widehat{AE}$, m∠D = $\frac{1}{2}$ m$\widehat{AF}$	4. Meas inscr ∠ = $\frac{1}{2}$ meas intercepted arc.
5. m ∠C = m ∠1, m ∠D = m ∠2	5. Sub
6. m ∠CAE = m ∠FAD, m ∠1 = m ∠2	6. Vert ∠s are ≅.
7. m ∠C = m ∠D	7. Trans Prop of Congr
8. △ACE ~ △ADF	8. AA △ ~
9. $\frac{CA}{EA} = \frac{AD}{AF}$	9. Def of ~ △s
10. CA · AF = EA · AD	10. In true proportion, prod of extremes = prod of means.

473

12 AREA

OVERVIEW

In this chapter, students use formulas to find the areas of basic polygons, circles, and parts of circles. Students find perimeters, areas, and apothems of regular polygons, as well as ratios of perimeters and areas of similar polygons. They study techniques for finding circumferences, areas, and arc lengths, and find the areas of annuli.

OBJECTIVES

- To find the areas of the following: a rectangle, parallelogram, triangle, rhombus, kite, trapezoid, square, circle, regular octagon, sector and segment of a circle, and an annulus bounded by circles
- To find the length of an arc given its degree measure and the length of the radius of its circle
- To find the perimeter of a square inscribed in a circle with a given circumference
- To find the area of a complex region formed by rectangles

PROBLEM SOLVING

The problem solving strategy, Using a Formula, is used throughout Chapter 12 to find the area of polygons. The problem solving strategy, More Than One Way, is taught on page 487 and will be useful to students as they do the exercises in Chapter 12.

TECHNOLOGY

Computer: The Computer Investigation on page 509 enables students to use the computer to approximate the value of π. This activity can be extended so that students can discover the formulas for finding the circumference and area of a circle.

Calculator: A calculator may be used by students throughout Chapter 12 to facilitate computations.

SPECIAL FEATURES

Mixed Review pp. 477, 482, 486, 492, 502, 508, 514, 519,
Focus on Reading p. 476
Brainteaser p. 482
Problem Solving Strategies: More than One Way p. 487
Midchapter Review p. 497
Application: Areas of Irregularly-Shaped Regions p. 497
Application: Polygons in Structures p. 503
Algebra Review p. 503
Application: Enlarging Drawings p. 508
Computer Investigation: Estimating π p. 509
Application: Estimating Area pp 520–521
Key Terms p. 522
Key Ideas and Review Exercises pp. 522–523
Chapter 12 Test p. 524
College Prep Test p. 525
Cumulative Review (Chapters 1–12) pp. 526–527

PLANNING GUIDE

Lesson	Basic	Average	Above Average	Resources
12.1 pp. 476–477	FR all, CE all WE 1–3	FR all, CE all WE 1–15	FR all, CE all WE 1–19	Reteaching p. 171 Practice p. 172
12.2 pp. 479–482	CE all WE 1–14 Brainteaser	CE all WE 1–35 odd Brainteaser	CE all WE 1–42 odd Brainteaser	Reteaching p. 173 Practice p. 174
12.3 pp. 484–487	CE all WE 1–12 odd Problem Solving	CE all WE 1–26 odd Problem Solving	CE all WE 1–33 odd Problem Solving	Reteaching p. 175 Practice p. 176
12.4 pp. 490–492	CE all WE 1–13	CE all WE 1–29 odd	CE all WE 1–34 odd	Reteaching p. 177 Practice p. 178
12.5 pp. 495–497	CE all WE 1–9 Midchapter Review Application	CE all WE 1–30 odd Midchapter Review Application	CE all WE 1–33 odd Midchapter Review Application	Reteaching p. 179 Practice p. 180
12.6 pp. 501–503	CE all WE 1–14 odd AR all, Application	CE all WE 1–27 odd AR all, Application	CE all WE 1–33 odd AR all, Application	Reteaching p. 181 Practice p. 182
12.7 pp. 506–509	CE 1–13 odd WE 1–16 odd Application, CI 1	CE all WE 1–21 odd Application, CI all	CE all WE 1–29 odd Application, CI all	Reteaching p. 183 Practice p. 184
12.8 pp. 512–514	CE 1–10 odd WE 1–18 odd	CE all WE 1–30 odd	CE all WE 1–37 odd	Reteaching p. 185 Practice p. 186
12.9 pp. 518–521	CE all WE 1–16 odd Application	CE all WE 1–29 odd Application	CE all WE 1–33 odd Application	Reteaching p. 187 Practice p. 188
Chapter 12 Review pp. 522–523	1–17 odd	all	all	
Chapter 12 Test p. 524	1–8, 10–18 even	all	all	
College Prep Test p. 525	all	all	all	
Cumulative Review pp. 526–527	1–22 odd	all	all	

CE = Classroom Exercises   WE = Written Exercises   FR = Focus on Reading   AR = Algebra Review   CI = Computer Investigation

NOTE: For each level, all students should be assigned all Mixed Review exercises.

▬▬ INVESTIGATION

Manipulative: This investigation of tessellating figures helps students discover some of the principles discussed in Lesson 12.1 and to use facts about polygons from Chapter 6.

Each pair of students needs a copy of Manipulative Worksheet 22, some heavy paper or light cardboard, scissors, paper, and pencil. This investigation may be extended over more than one class period.

Tell students that tile patterns, such as those shown on the worksheet and those they will create, are called tessellations. A **tessellation** is a pattern of closed shapes that completely covers a plane and leaves no gaps and has no overlaps. Many streets and floors in Spain and Italy are covered in tessellations. Students may see tessellations at home in bathroom and kitchen tiles, or in the tiles that are used at a swimming pool. Nature also provides examples of tessellations in the molecular structure of crystals and in the wax honeycomb that bees use to store honey. The honeycomb is a tessellation of regular hexagons. The honeycomb is called a pure tessellation because it uses only one shape. After the students complete the investigation with the worksheet, you might challenge them to create some tessellations that involve more than one shape. Some possible patterns are shown below.

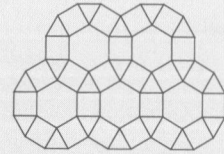

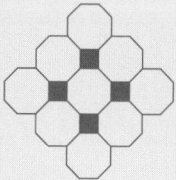

"This statement is false."

Kurt Gödel (1906–1978)—Gödel, at age 25, showed that some mathematical statements cannot be proved either true or false. This shook the foundations of mathematics and logic which had existed for thousands of years.

More About the Mathematician

Gödel's ideas were the result of thought experiments, and dealt with abstractions. He showed that for a mathematical statement, "true" and "capable of being proved true" are not the same. He also proved that no consistent system can prove its own consistency. These two ideas are expressed in his famous theorems on undecidability and incompleteness. Gödel also worked on problems that occur when statements refer to themselves. Both the sentence and the drawing in the picture are self-referent; that is, the hand is drawing a hand that is drawing a hand and the sentence is saying something about itself.

12.1 Standard Units

Objective   To determine appropriate standard units of area measurement

All measurement systems use standard units. To measure distance or length, a unit such as the *centimeter* is used. If a segment is 5 cm long, the segment can be divided into five adjacent, congruent, nonoverlapping segments, each of which is one centimeter long.

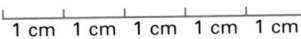
1 cm   1 cm   1 cm   1 cm   1 cm

To measure area, a standard unit is also used. People have used many different units over the years. The square foot, the acre, and the hectare are three such units. Each has its own history. For example, an acre was originally the amount of land that could be plowed in a single day by a man and a team of oxen. Since different men and different oxen moved at different speeds, this was not a very accurate unit of measure.

Standard units of area today are based on congruent figures that will cover a surface without overlapping. The square is usually used. However, many other figures have these properties.

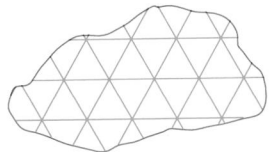   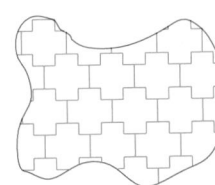

EXAMPLE   Copy the figure. Show how it can be used as a standard unit of area that covers the surface with no overlapping.

Solution

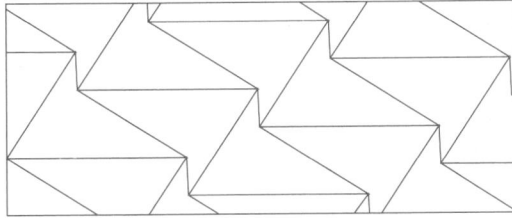

12.1 Standard Units   **475**

Additional Example

Copy the figure. Show how it can be used as a standard unit which covers the surface with no overlapping.

   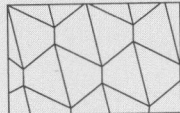

Teaching Resources

Manipulative Worksheet 22
Project Worksheet 12
Quick Quizzes 86
Reteaching and Practice Worksheets, pp. 171, 172

▰▰▰GETTING STARTED

Prerequisite Quiz

Given: Rhombus *ABCD*; m ∠*A* = 60, *E*, *F*, *G* and *H* are midpoints of $\overline{AB}$, $\overline{BC}$, $\overline{CD}$, and $\overline{AD}$, respectively.

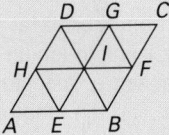

Find the length of each segment if *AE* = 1 cm. (Exercises 1-3)

1. $\overline{HE}$ 1 cm **2.** $\overline{DB}$ 2 cm **3.** $\overline{BC}$ 2 cm

Paper triangles are cut congruent to △*AEH*. How many are needed to fit over each figure?

4. △*ABD*   4
5. rhombus *DIFG*   2
6. trapezoid *EBFH*   3

Motivator

Ask the students how they determine which of two things is larger.   Comparison or measurement   Ask them what the measurements are based on.   Standard units   Ask them to give examples of a standard unit used to measure length. Meter, foot

▰▰▰TEACHING SUGGESTIONS

Lesson Note

In this lesson, students learn that some geometric figures other than the square fit together to cover a surface without overlapping, and so could be used as standard units of area measurement. However, emphasize that in everyday life, square units are used as units of area measurement.

Math Connections

History: Units of measurement used in ancient times were inconsistent. Some systems of measurement were based on parts of the body. The *cubit*, for example, was the average length of a man's forearm from the elbow to the tip of the middle finger. The Romans used the *uncia*, which was the average width of a man's thumb. These units were inconsistent because of variation in body size. Modern systems of measurement are based on standard units that have been defined and remain constant.

Critical Thinking Questions

Analysis: Let an equilateral triangle with sides that are 1 unit in length be a standard unit of area measure. Ask students to draw a diagram and find the lengths of all sides of an isosceles trapezoid whose area is 3 triangular units. Ask students to draw a diagram and find the lengths of the sides of a rectangle whose area is 3 triangular units.

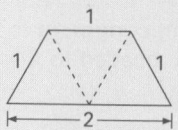 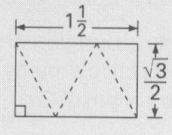

Other answers are possible.

Checkpoint

Each angle of each given regular polygon has the given measure. Can the polygon be used as a standard unit of area?

1. hexagon, 120   Yes
2. nonagon, 140   No

Copy each figure. Show how each figure can be used as a standard unit that covers a surface with no overlapping.

3.

4.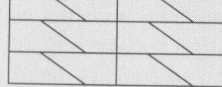

Other patterns are possible.

476

Not all shapes have these properties. For example, a circle cannot be used as a standard unit for area. The rule for covering a piece of surface with a given shape is that the shapes cannot overlap and yet must cover all of the surface. This is not possible with circles.

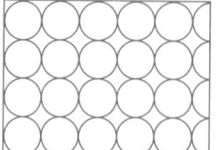 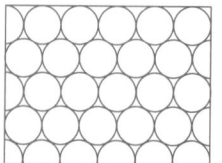 

For a regular polygon to serve as a standard unit of area, the measure of each of its angles must be a factor of 360. Each angle of the square below measures 90. Since 4 · 90 = 360, four such angles fit around a common vertex without overlapping. Each angle of the regular pentagon measures 108. Since 108 is not a factor of 360, three pentagons leave space uncovered and four pentagons overlap.

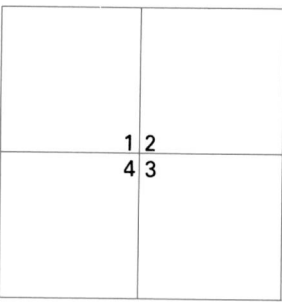 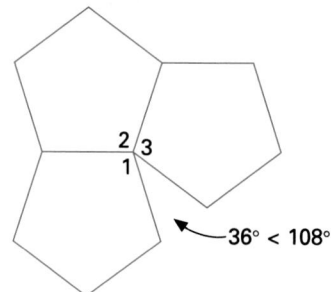

36° < 108°

The concept of a standard unit is based on the assumption that standard units have constant areas.

Postulate 20	Congruent polygons have equal areas.

Focus on Reading

Choose the best answer to complete each statement.

1. The square mile is a unit of measure that might be used to indicate
   a. the distance from the earth to the moon.
   b. the area of Indiana.
   c. the area of your classroom.

Enrichment

Many different tile patterns are used for walls and floors. Ask students to notice the brickwork and tile patterns used in streets and public buildings. Ask them to design different patterns using squares and/or rectangles. Two are shown at the right.

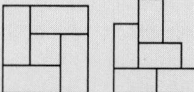

2. The square centimeter might be used to give
  a. the area of this page.
  b. the area of a ball field.
  c. the area of the United States. a

3. The area of this page is approximately
  a. 4 in^2.
  b. 4 ft^2.
  c. 63 in^2. c

Classroom Exercises

Each angle of the given regular polygon has the given measure. Can the polygon be used as a standard unit of area?

1. triangle, 60 Yes
2. quadrilateral, 90 Yes
3. decagon, 144 No
4. pentagon, 108 No
5. hexagon, 120 Yes
6. octagon, 135 No

Written Exercises

Copy the figure. Show how it can be used as a standard unit that covers a surface with no overlapping.

1.
2.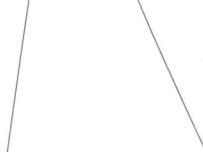
3.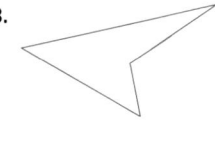

Can the shape described be used as a standard unit of area? If so, draw a picture to show how.

4. a right triangle Yes
5. an acute triangle Yes
6. an obtuse triangle Yes
7. any rectangle Yes
8. any parallelogram Yes
9. a trapezoid Yes
10. an isosceles trapezoid Yes
11. any convex quadrilateral Yes
12. a regular hexagon Yes
13. any convex hexagon No
14. any convex pentagon No
15. a regular octagon No
16. any concave quadrilateral No
17. any equilateral polygon No
18. any equiangular polygon No
19. any regular polygon No

Mixed Review

What is the measure of each interior angle and each exterior angle of the polygon? *6.2, 6.3*

176.4, 3.6

1. a regular pentagon 108, 72
2. a regular octagon 135, 45
3. a regular 100-gon

12.1 Standard Units **477**

 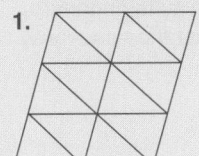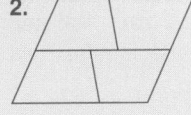

**GETTING STARTED**

Prerequisite Quiz

Multiply.

1. $(30)(1.5)$ 45
2. $(3.5)(3.5)$ 12.25
3. $5\frac{1}{2} \cdot 8$ 44
4. $\sqrt{2} \cdot 6\sqrt{2}$ 12
5. $(x - 6)(x - 12)$ $x^2 - 18x + 72$

Simplify.

6. $6 \cdot 2 + 3 \cdot 4 + 3 \cdot 16$ 74
7. $8 \cdot 12 + 4 \cdot 12 + 6 \cdot 6$ 180

Motivator

If a surface is divided into squares, ask the students if they can count the squares to find the area of the surface.   Yes   Ask them if there is an easier way to find the area.   Yes   Ask them what the easier way is.   Use area formulas.

**TEACHING
SUGGESTIONS**

Lesson Note

Emphasize the pattern of logical development in this lesson. The definition of the area of a polygon leads to Postulate 21, the formula for the area of a rectangle. This leads to Theorem 12.1, the formula for the area of the square. This in turn is followed by Postulate 22, the Area Addition Postulate. These theorems and postulates will be used in developing subsequent theorems involving areas.

12.2 Areas of Rectangles and Squares

Objective   To apply the formulas for the areas of a rectangle and a square

A **polygonal region** is the union of a polygon and its interior. For example, the *triangular region* below is the union of a triangle and its interior.

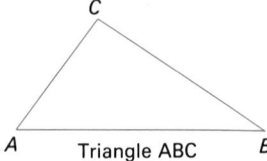
Triangle ABC

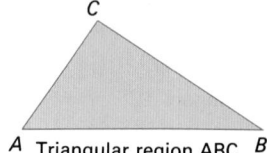
Triangular region ABC

The relative sizes of some regions can be compared directly. If one region can fit in the interior of a second region, then the second is larger than the first.

It is not practical to compare most regions in this manner. However, their areas can be used to compare their sizes. Using the small square as a standard unit, the area of the 4-by-8 rectangle at right is 32 square units. When the unit of length used is the inch, the unit of area is the square inch (in²). When the unit of length is the centimeter, the unit of area is the square centimeter (cm²).

Definition   The **area of a polygon** is the number of square units in the region bounded by the polygon.

EXAMPLE 1   Using the small square region as the unit of area, find the area of the given rectangle.

Solution   The rectangle is $2\frac{1}{2}$ units wide and $2\frac{1}{2}$ units long.

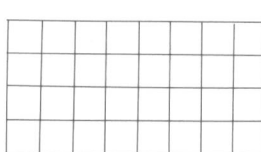

$2\frac{1}{2} \cdot 2\frac{1}{2} = \frac{5}{2} \cdot \frac{5}{2} = \frac{25}{4} = 6\frac{1}{4}$

It contains $6\frac{1}{4}$ square units.

Postulate 21   The area of a rectangle is the product of the lengths of a base and a corresponding altitude (Area of rectangle = bh).

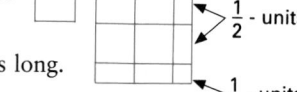

478   Chapter 12 Area

Additional Example 1

Using the small square as the unit of measure, find the area of the given rectangle.

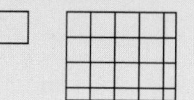

The rectangle is $4\frac{1}{2}$ units long and $3\frac{1}{2}$ units wide. It contains $15\frac{3}{4}$ square units.

Additional Example 2

If the area of a rectangle is 144 ft² and the length of an altitude is 8 ft, what is the length of the base?   18ft

EXAMPLE 2 If the area of a rectangle is 42 cm², and the length of an altitude is 6 cm, what is the length of the base?

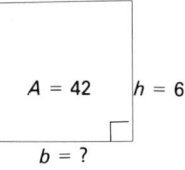

Solution
$$A = bh$$
$$42 = b \cdot 6, \text{ or } 6b$$
$$b = 7 \text{ cm}$$

Since a square is a rectangle, you can use the formula $A = bh$ to find its area. However, a special formula is often used.

Theorem 12.1 The area of a square is the square of the length s of a side ($A = s^2$).

EXAMPLE 3 What is the area of a square with sides 34.5 cm long?

Solution
$$A = s^2$$
$$= (34.5)^2 = 1,190.25 \text{ cm}^2$$

Postulate 22 **Area Addition Postulate:** If a region is the union of two or more nonoverlapping regions, then its area is the sum of the areas of these nonoverlapping regions.

EXAMPLE 4 Find the area of the polygon. Each angle is a right angle.

Plan Divide the region into three rectangular regions. Apply Postulate 22.

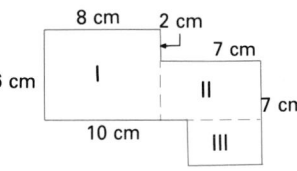

Solution
Area of rectangle I = 8 · 6, or 48 cm²
Area of rectangle II = 7 · 4, or 28 cm²
Area of rectangle III = 5 · 3, or 15 cm²
The total area is 91 cm².

Classroom Exercises

What is the area of the rectangle or square?

1. $b = 5$ cm, $h = 2$ cm   10 cm²
2. $b = 3$ in., $h = 7$ in.   21 in²
3. $b = 9$ yd, $h = 7$ yd   63 yd²
4. $b = 8$ m, $h = 6$ m   48 m²
5. $s = 8$ cm   64 cm²
6. $s = 6$ ft   36 ft²

Math Connections

Farming: Since many plots of land are rectangular, the formula for the area of a rectangle can often be used when estimating the yield of a plot of land. Sometimes the entire plot is not arable land, or for some reason cannot be used for agricultural purposes. In this case, an estimate of the amount of profit after harvest may involve adding or subtracting areas.

Critical Thinking Questions

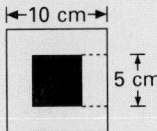

Application: The tile shown above consists of a black square centered in a white square. Refer students to the tile and ask them how many square meters of a floor that is 2.5 m long and 1.75 m wide will be covered with black if it is covered with this tile.   Approximately 1.09 m²

Checkpoint

What is the area of each square?

1. $s = 15$ in.   225 in²
2. $s = 14$ m   196 m²

What is the area of each rectangle?

3. $b = 16$ cm, $h = 5$ cm   80 cm²
4. $b = 4.5$ ft, $h = 8$ ft   36 ft²
5. $b = x + 8$, $h = x - 2$ $x^2 + 6x - 16$
6. $b = 2x + 3$, $h = x - 1$ $2x^2 + x - 3$
7. Find the area of the polygon. Each angle is a right angle.   320 cm²

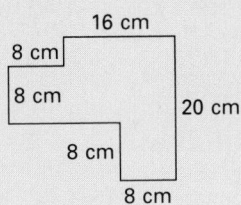

Additional Example 3

What is the area of a square with sides 8.8 cm long?   77.44 cm²

Additional Example 4

Find the area of the polygon. Each angle is a right angle.   162 cm²

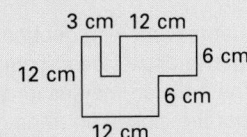

Ask the students what measurements must be known to find the area of a rectangle. The lengths of a base and a corresponding altitude A square. The length of a side Ask them if these measurements provide the area directly.   No If not, ask them how they are used. They are factors, and the area is the product of these factors. If a region is the union of two or more nonoverlapping regions, ask the class how they would find the total area of the region. It is the sum of the areas of the nonoverlapping regions.

◤◤◤◤ FOLLOW UP

Guided Practice

Classroom Exercises 1–6

Independent Practice

A Ex. 1–14, **B** Ex. 15–35, **C** Ex. 36–42

Basic: WE 1–14, Brainteaser

Average: WE 1–35 odd, Brainteaser

Above Average: WE 1–42 odd, Brainteaser

Additional Answers

Written Exercises

14.

Statement	Reason
1. Square with side length s.	1. Given
2. Sq is a rect.	2. Def of sq
3. Area of sq $= bh$	3. Area of rect $= bh$
4. $b = s, h = s$	4. Def of base
5. Area of sq $= s \cdot s$	5. Sub
6. Area of sq $= s^2$	6. Notation

Written Exercises

Copy the rectangle. Using the small square region as the unit of area, find the area.

1.

17.5 sq units

2.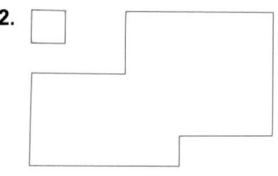

31 sq units

What is the area of the rectangle or square?

3. $b = 4.5$ m, $h = 3.2$ m   14.4 m²
4. $b = 2.6$ cm, $h = 6.5$ cm   16.9 cm²
5. $b = 2.7$ in., $h = 4.6$ in.   12.42 in²
6. $b = 8.3$ ft, $h = 1.9$ ft   15.77 ft²
7. $s = 2.5$ mm   6.25 mm²
8. $s = 4.5$ cm   20.25 cm²
9. $s = 8.7$ yd   75.69 yd²
10. $s = 9.4$ ft   88.36 ft²

Find the unknown length of a side of the rectangle.

11. $b = 9$ in., $h = \underline{\ 8\ in\ }$, $A = 72$ in²
12. $b = \underline{\ 5\ cm\ }$, $h = 13$ cm, $A = 65$ cm²
13. $b = 1.5$ yd, $h = \underline{\ 4\ yd\ }$, $A = 6$ yd²
14. Prove Theorem 12.1.

What is the area of the rectangle?

15. $b = x + 3$, $h = x - 3$ $x^2 - 9$
16. $b = y + 4$, $h = y + 4$ $y^2 + 8y + 16$
17. $b = y + z$, $h = y - z$ $y^2 - z^2$
18. $b = 2x + y$, $h = 3x - 2y$ $6x^2 - xy - 2y^2$

Find the unknown length of a side of the rectangle.

19. $h = 4.5$ cm, $A = 13.5$ cm² $b = 3$ cm
20. $b = 3.6$ m, $A = 18$ m² $h = 5$ m
21. $b = \sqrt{6}$ m, $A = 6$ m² $h = \sqrt{6}$ m
22. $h = \sqrt{3}$ cm, $A = \sqrt{18}$ cm² $b = \sqrt{6}$ cm
23. $b = 2x + y$, $A = 4x^2 - y^2$ $h = 2x - y$
24. $h = x - 2y$, $A = x^2 - 4yx + 4y^2$ $b = x - 2y$

Find the area of the polygon. Each angle is a right angle.

25.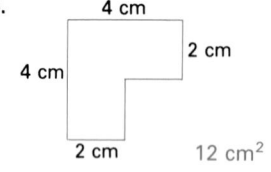

4 cm
2 cm
4 cm
2 cm
12 cm²

26.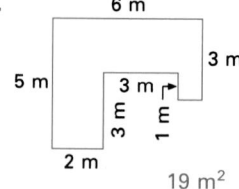

6 m
3 m
5 m
3 m
3 m
1 m
2 m
19 m²

Enrichment

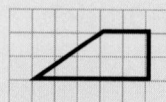

Have students work with geoboards or graph paper to draw closed figures that connect the nails or intersections. The area of the closed figure is given by the formula:

$$A = \frac{x}{2} + y - 1$$

where x is the number of nails or intersections that the border touches and y is the number of nails or intersections in the interior. For the figure at the left, $x = 10$, and $y = 3$. The area is $5 + 3 - 1 = 7$ sq units. Have the students construct several figures and compare the areas of the figures with the results of the formula.

27. 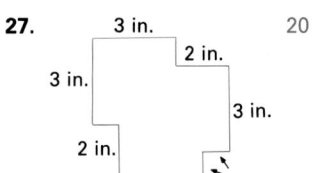 20 in²

28. 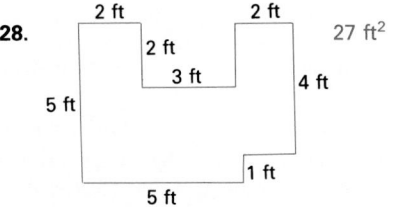 27 ft²

For Exercises 29–31, find the area of the rectangle. (HINT: Use the Pythagorean Theorem.)

29.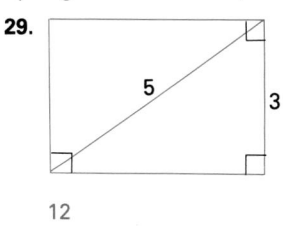

30.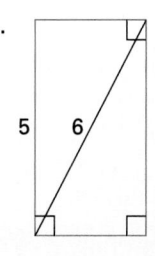

31.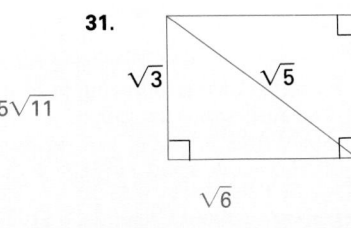

32. The length of a diagonal of a square is 20 in. What is the area of the square?   200 in²

33. The length of a diagonal of a square is 16 ft. What is the area of the square?   128 ft²

34. Two sides of a rectangle have lengths of 5 cm and 6 cm. What is the length of a side of a square with the same area? $\sqrt{30}$ cm

35. Two sides of a rectangle have lengths of 3.4 m and 2.6 m. What is the length, to the nearest hundredth of a meter, of a side of a square with the same area?   2.97 m

36. The lengths of two consecutive sides of a rectangle have the ratio of 3 to 4. The area is 108 cm². What are the lengths?   9 cm, 12 cm

37. The lengths of two consecutive sides of a rectangle have the ratio of 3 to 2. The area is 216 mm². What are the lengths?   18 mm, 12 mm

Prove or disprove.

38. The ratio of the areas of two rectangles with congruent bases is equal to the ratio of the lengths of their altitudes.

39. If the length of a base of a rectangle is doubled and the length of a corresponding altitude remains constant, then the area is doubled.

40. If the lengths of the sides of one square are twice the lengths of the sides of a second square, then the area of the first square is twice that of the second.

41. If the lengths of the sides of a rectangle are multiplied by any constant, then the area of the original rectangle is multiplied by the same constant.

38. True. Area 1 = bh_1, Area 2 = bh_2 (Area of rect = bh); $\frac{\text{Area 1}}{\text{Area 2}} = \frac{bh_1}{bh_2} = \frac{h_1}{h_2}$

39. True. Area 1 = bh, Area 2 = $(2b)h$ (Area of rect = bh); $\frac{\text{Area 1}}{\text{Area 2}} = \frac{bh}{2bh} = \frac{1}{2}$; Area 2 = 2 · Area 1

40. False. $A_1 = (2s)^2 = 4s^2$, $A_2 = s^2$; $A_1 = 4 \cdot A_2$.

41. False. $A_1 = (cb)(ch) = c^2bh$, $A_2 = c(bh) = cbh$; $c^2bh \neq cbh$, if $c \neq 0$.

42. A painter is buying paint for the walls and ceiling of a room of a house. The room is 14 ft wide by 16 ft long, with walls 8 ft tall. There are two doors in the room, each $2\frac{1}{2}$ ft wide and 6 ft tall. There are two windows, each 4 ft wide and 3 ft tall. If one gallon of paint covers approximately 500 ft² of space, how many gallons must be bought? How much paint will be left over?   2 gal; $\frac{7}{10}$ gal

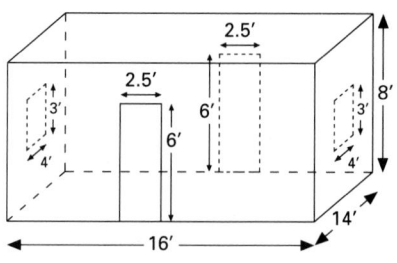

Mixed Review

1. Which of the following patterns can be used to prove that two quadrilaterals are congruent: SSSS, AAAA, SASAS, ASASA, AASAA? *7.1*   SASAS, ASASA

True or False?

2. A necessary condition for a quadrilateral to be a parallelogram is that opposite angles are congruent.   *6.4* T
3. A sufficient condition for a parallelogram to be a rectangle is for diagonals to be congruent.   *7.3* T

Brainteaser

The rectangular region shown has been divided into nine square regions, all different in size. The area of Square III is 64 cm². The area of Square IV is 81 cm². What is the area of the rectangle?
1,056 cm²

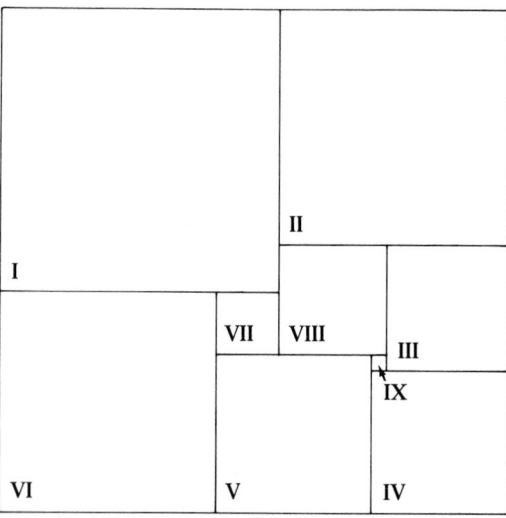

12.3 Areas of Parallelograms

Objective To apply the formula for the area of a parallelogram

If you formed a parallelogram out of paper, you could cut it and rearrange the pieces to form a rectangle. The base and the altitude of this rectangle are congruent to those of the original parallelogram. This fact can be used to prove the following theorem.

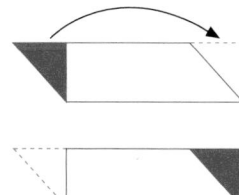

Theorem 12.2 The area of a parallelogram is the product of the length of a *base* and the length of a corresponding *altitude* (Area of parallelogram = bh).

Given: $\square ABCD$ with base $\overline{AB}$
and altitude $\overline{CE}$, $AB = b$,
$CE = h$
Prove: Area($\square ABCD$) = bh

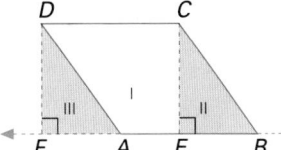

Plan for proof:
The region $ABCD$ can be separated into two parts, which can be rearranged to form a rectangular region $FECD$. By constructing $\overline{DF} \perp \overleftrightarrow{AB}$ and $\overline{CE} \perp \overleftrightarrow{AB}$, right triangles DFA and CEB are formed. $\triangle DFA \cong \triangle CEB$ by HL, and Area II = Area III. Therefore, Area I + Area II = Area I + Area III, or Area ($\square ABCD$) = Area(rect $FECD$) = bh.

EXAMPLE 1 Find the area of $\square ABCD$.

Solution
$b = 3.4$ cm, $h = 2.5$ cm
Area($\square ABCD$) = bh = $3.4 \cdot 2.5 = 8.5$ cm^2

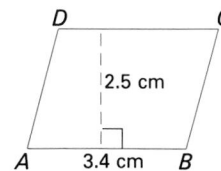

If the area of a parallelogram and the length of either a base or an altitude are known, then the other length can be found.

Teaching Resources

Quick Quizzes 88
Reteaching and Practice
  Worksheets, pp. 175, 176

▰ GETTING STARTED

Prerequisite Quiz

Find the length x in each right triangle.

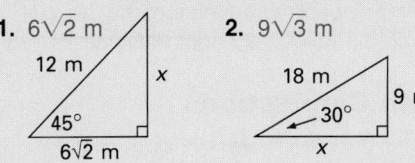

1. $6\sqrt{2}$ m

2. $9\sqrt{3}$ m

3. A rectangle with a height of 8 in. has an area of 120 in^2. How long is the base?
    15 in
4. Rectangles I and II have equal areas. In rectangle I, $b = 10$ cm and $h = 9$ cm. In rectangle II, $b = 12$ cm. What is the height? 7.5 cm

Motivator

Use a deck of cards to perform the following demonstration. First show the profile of the cards in a rectangular pile like they are when they are in their carton. Then gently slide the top portion of the deck in one direction and the bottom portion in the other direction, forming a parallelogram. Ask the students which dimensions of the deck stay the same. Altitude and base Then ask them which ones change. The length of the sides Ask them which of these dimensions are used to find the area of a rectangle. The altitude and base Ask them if there is any possible relationship between these two figures. Yes, the area formula for both is the same.

Additional Example 1

Find the area of $\square ABCD$.
26.1 cm^2

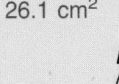
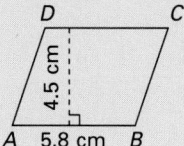

Lesson Note

In this lesson, the theorem for finding the area of a parallelogram is deduced from the postulate for the area of a rectangle and the Area Addition Postulate. Be certain to emphasize the building–block structure involved in the sequence of theorems. Some exercises require that the students recall the special properties of the 30–60–90 and the 45–45–90 right triangles.

Math Connections

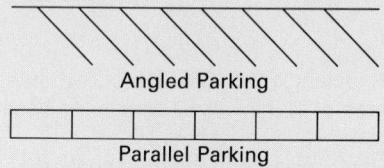

Angled Parking

Parallel Parking

City Planning: There are usually many parking places along a street in the downtown area of a large city. Two main types of parking along a street—parallel parking and angled parking—are shown above. Since the efficient use of space is important in heavily populated areas, deciding on which parking arrangement to adopt for new roads involves comparing the area used for each parking plan with the number of cars allowed to park there.

Critical Thinking Questions

Analysis: Ask students to draw a diagram and find the dimensions of a parallelogram that consists of two adjacent isosceles right triangles if the parallelogram has an area of 36 cm².

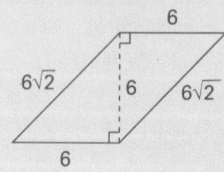

EXAMPLE 2   Given: Area($\square PQRS$) = 36 cm², length of altitude $\overline{HS}$ = 4 cm
Find the length of base $\overline{PQ}$.

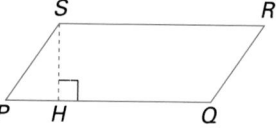

Solution $A = bh$
$36 = 4 \cdot PQ$
$PQ = 9$ cm

EXAMPLE 3   Given: $\square ABCD$, $\overline{DE} \perp \overline{AB}$, $\overline{BF} \perp \overline{AD}$,
$AB = 15$, $DE = 10$, $AD = 12$
Find BF.

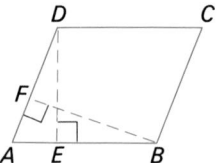

Plan   Use both $\overline{AB}$ and $\overline{DA}$ as bases, with corresponding altitudes $\overline{DE}$ and $\overline{BF}$.

Solution   Area($\square ABCD$) = $AB \cdot DE$ Area($\square ABCD$) = $AD \cdot BF$
= $15 \cdot 10 = 150$ = $12 \cdot BF$
Therefore, $150 = 12 \cdot BF$ and $BF = 12\frac{1}{2}$.

EXAMPLE 4   The lengths of two sides of a parallelogram are 12 cm and 10 cm. The measure of the angle between the two sides is 30. Find the area of the parallelogram.

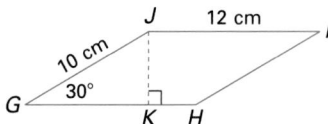

Plan   Draw altitude $\overline{JK}$ to make a 30-60-90 triangle. The length of the side opposite the angle of measure 30 is half that of the hypotenuse.

Solution $JK = \frac{1}{2} \cdot GJ = 5$
Therefore, Area($\square GHIJ$) = $5 \cdot 12$, or 60 cm².

Classroom Exercises

Tell whether enough information is given to find the area.

1.
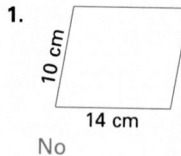
No

2.   20 cm
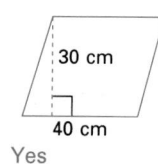
No   20 cm

3.
30 cm
40 cm
Yes

Additional Example 2

Given: Area $\square WXYZ$ = 42 cm², length of altitude $\overline{ZR}$ = 6 cm. Find: length of base $\overline{WX}$.   7 cm

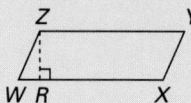

Additional Example 3

Given: $\square ABCD$, $\overline{DE} \perp \overline{AB}$, $\overline{BF} \perp \overrightarrow{AD}$,
$AB = 10$, $AD = 4$, $BF = 7$
Find DE. $2\frac{4}{5}$, or 2.8

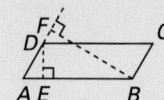

Find the area of the parallelogram.

4.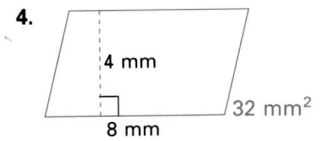
32 mm²

5.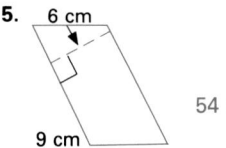
54 cm²

6.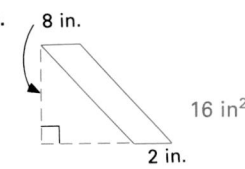
16 in²

Written Exercises

Find the missing measurement for the indicated parallelogram.

	Length of base *b*	Length of altitude *h*	Area
1.	8 cm	7 cm	56 cm²
2.	9 m	3 m	27 m²
3.	5.2 in.	3.4 in.	17.68 in²
4.	2.7 ft	1.2 ft	3.24 ft²
5.	6 mm	5 mm	30 mm²
6.	8 cm	9 cm	72 cm²
7.	2.7 ft	3.4 ft	9.18 ft²
8.	6.9 in.	7.8 in.	53.82 in²
9.	$\sqrt{5}$ m	$\sqrt{5}$ m²	5 m²
10.	$\sqrt{2}$ mm	$\sqrt{7}$ mm	$\sqrt{14}$ mm²
11.	2.6 cm	3.4 cm	8.84 cm²
12.	8.3 in	9.2 in.	76.36 in²

In $\square ABCD$, $\overline{FD} \perp \overline{AB}$ and $\overline{EG} \perp \overline{DA}$. **Find the missing measurement.**

13. $AB = 12$, $FD = 6$, $DA = 8$, $EG = \underline{9}$
14. $AB = 30$, $FD = \underline{2}$, $DA = 12$, $EG = 5$
15. $AB = \underline{20}$, $FD = 2.5$, $DA = 10$, $EG = 5$
16. $AB = 24$, $FD = 1.2$, $DA = \underline{8}$, $EG = 3.6$
17. $AB = \sqrt{6}$, $FD = 2$, $DA = \underline{2\sqrt{2}}$, $EG = \sqrt{3}$
18. $AB = \underline{6}$, $FD = AB$, $DA = 4$, $EG = 9$

19. 36 sq units
20. 40 sq units

Find the area of $\square PQRS$.

19. $p = 6$, $q = 12$, m $\angle P = 30$
20. $p = 10$, $q = 8$, m $\angle P = 30$
21. $p = 10$, $q = 8$, m $\angle P = 45$
  40$\sqrt{2}$ sq units
22. $p = 10$, $q = 8$, m $\angle P = 60$
  40$\sqrt{3}$ sq units

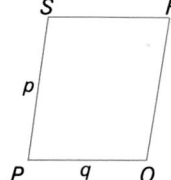

23. Solve Example 3 of this lesson using similar triangles instead of area formulas. 12.5

Additional Example 4

The lengths of two sides of a parallelogram are 15 cm and 8 cm. The measure of an angle between two sides is 45. Find the area of the parallelogram.
60$\sqrt{2}$ cm² or approximately 84.85 cm²

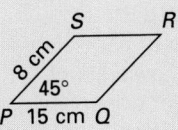

Common Error Analysis

Error: Some students confuse the altitude of a parallelogram with one of its sides. Using a drawing, show students that an altitude of a parallelogram may lie both inside or outside the parallelogram. Show them also that when the sides of a parallelogram are altitudes, the parallelogram is a rectangle.

Checkpoint

Find the missing information for each parallelogram. (Exercises 1–5)

	b	*h*	Area	
1.	4 in.	6 in.	_?_	24 in²
2.	9 cm	5 cm	_?_	45 cm²
3.	_?_	12 mm	180 mm²	15 mm
4.	_?_	8 ft	100 ft²	12.5 ft
5.	20 ft	_?_	320 ft²	16 ft

6. In $\square ABCD$, m $\angle A = 30$, $AB = 18$ cm, and $AD = 4$ cm. Find the area. 36 cm²

Closure

Ask the students how they determine the area of a parallelogram. Theorem 12.2 page 483 Ask them what other measurement is needed to find the length of the base of a parallelogram if they know the area. The altitude Ask them what measurement is needed to find the altitude if they know the area. The length of the base

■■■FOLLOW UP

Guided Practice

Classroom Exercises 1–6

Independent Practice

A Ex. 1–12, **B** Ex. 13–26, **C** Ex. 27–33

Basic: WE 1–12 odd, Problem Solving
Average: WE 1–26 odd, Problem Solving
Above Average: WE 1–33 odd, Problem Solving

Written Exercises

31.

Statement	Reason
1. $\square ABCD$ with base $\overline{AB}$ and alt $\overline{CE}$, $AB = b$, $CE = h$	1. Given
2. Construct $\overline{DF} \perp$ to $\overleftrightarrow{AB}$ with F on $\overleftrightarrow{AB}$ and $\overline{CE} \perp \overline{AB}$ with E on $\overline{AB}$.	2. From pt to line there is unique $\perp$.
3. $\overline{AB} \parallel \overline{CD}$	3. Def of $\square$
4. $DF = CE$	4. Distance between $\parallel$ lines is constant.
5. $DA = CB$	5. Opp sides of $\square$ have eq length.
6. $\angle DFA$ and $\angle CEB$ are rt $\angle$s.	6. Def of $\perp$
7. $\triangle DFA \cong \triangle CEB$	7. HL
8. $FA = EB$	8. CPCTC
9. Area $\triangle DFA =$ Area $\triangle CEB$	9. $\cong$ polygons have eq area.
10. Area $\square ABCD =$ Area $AECD$ + Area $\triangle CEB$	10. Area Add Post
11. Area $\square ABCD =$ Area $AECD$ + Area $\triangle DFA$	11. Sub
12. Area $AECD$ + Area $\triangle DFA =$ Area rect $FECD$	12. Area Add Post
13. Area $\square ABCD =$ Area rect $FECD$	13. Trans Prop
14. Area rect $FECD =$ $FE \cdot CE$	14. Area rect $= bh$
15. $FE = FA + AE$, $AB = AE + EB$	15. Seg Add Post
16. $FE = EB + AE$	16. Sub
17. $FE = AB$	17. Sub
18. Area rect $FECD =$ $AB \cdot CE = b \cdot h$	18. Sub
19. Area $\square ABCD =$ bh	19. Sub

See page 487 for the answer to Ex. 32.

486

Find the area of the parallelogram.

24.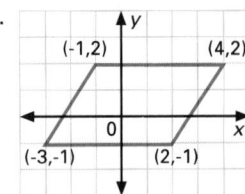

15 sq units

25.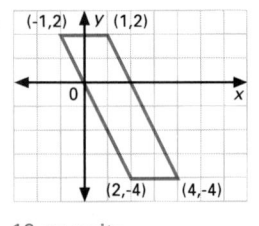

12 sq units

26.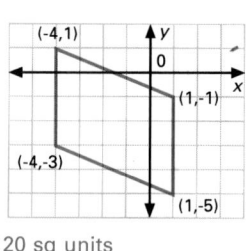

20 sq units

Use the appropriate trigonometric ratio to find the area, to the nearest tenth, of a parallelogram with the indicated measures.

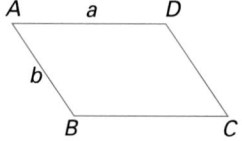

27. $a = 3.8$, $b = 2.7$, m $\angle A = 49$ 7.7 sq units

28. $a = 5.9$, $b = 4.2$, m $\angle A = 67$ 22.8 sq units

29. $a = 3.6$, $b = 4.7$, m $\angle A = 68$ 15.7 sq units

30. $a = 5.8$, $b = 3.1$, m $\angle A = 31$ 9.3 sq units

31. Prove Theorem 12.2.

32. Prove: Area($\square ABCD$) $= ab \cdot \sin A$, where a and b are the lengths of the sides that include $\angle A$, and A is an acute angle.

33. Several parallelograms are shown on the dot grid below. The horizontal or vertical distance between adjacent dots is 1 cm. Find the area of each parallelogram. Determine if there is any relationship between the area, the number of dots on the sides of the parallelogram, and the number of dots in the interior of the parallelogram. 3, 16, 6; $A = \frac{1}{2}D_s + D_i - 1$

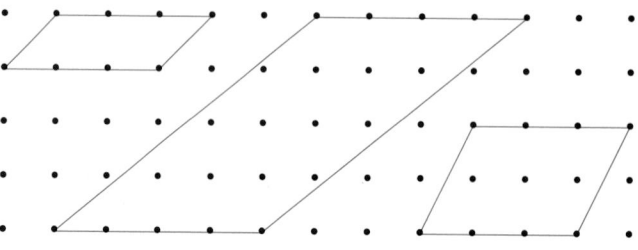

Mixed Review

Do the three numbers form a Pythagorean triple?   *9.2*

1. 4, 5, 6   No   **2.** 8, 9, 10   No   **3.** 6, 8, 10   Yes

4. If the length of each leg of an isosceles right triangle is 5 in., find the measure of the hypotenuse.   *9.4* $5\sqrt{2}$ in

Enrichment

Many states in the United States approximate rectangles or parallelograms in shape. Using their knowledge of distances between major cities, have students estimate the dimensions of your state or a neighboring state. Using these dimensions, have them estimate areas.

Finally, have them check their answers using an atlas or almanac.

Students will gain an appreciation of how large numbers work and how errors are magnified when estimated numbers are multiplied.

Problem Solving Strategies

More than One Way

The Pythagorean Theorem can be proved in a number of ways. Three different ways are suggested here. Each one requires that you compare different areas and use algebra. You will need to use the formulas for the areas of a square, a triangle, and a trapezoid. You will have to add or subtract areas in order to find areas in different ways and then compare them.

1. Four identical right triangles are cut out and arranged as shown. Prove the Pythagorean Theorem by finding the area of the large square in two ways—side squared and sum of five pieces. Set the two areas equal to each other.

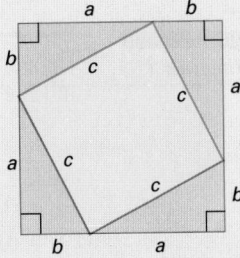

2. The figure to the right shows four right triangles arranged differently. This time there is a small square in the center. Again, use the area of the large square in two ways to prove the Pythagorean Theorem.

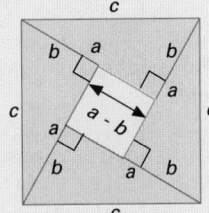

3. This method was discovered by President Garfield. Two identical right triangles are arranged as shown. Follow the steps below to prove the Pythagorean Theorem.
   a. Verify that the figure in the center is a right triangle.
   b. Verify that the large figure is a trapezoid with a and b the lengths of the bases and $(a + b)$ the length of an altitude.
   c. Find the area of the trapezoid in two different ways.

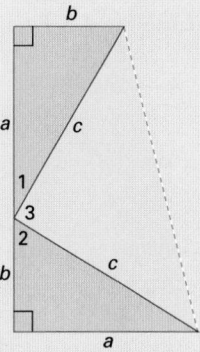

Lesson Note

You may want to use Project Worksheet 9 with this Lesson.

Additional Answers

Problem Solving Strategies

1. $(a + b)^2 = c^2 + 4 \cdot \frac{1}{2}ab$; $a^2 + 2ab + b^2 = c^2 + 2ab$; $a^2 + b^2 = c^2$
2. $c^2 = 4 \cdot \frac{1}{2}ab + (a - b)^2$; $c^2 = 2ab + a^2 -2ab + b^2$; $c^2 = a^2 + b^2$
3. (a) m $\angle 1$ + m $\angle 2$ = 90 ($\angle 1$ and $\angle 2$ are comp) m $\angle 3$ = 90 (The 3 $\angle$s = a st $\angle$)
   (b) The bases are ∥ so the figure is a trap.
   (c) $\frac{a + b}{2}(a + b) = \frac{1}{2}ab + \frac{1}{2}ab + \frac{1}{2}c \cdot c$; $\frac{1}{2}(a^2 + 2ab + b^2) = ab + \frac{1}{2}c^2$; $a^2 + 2ab + b^2 = 2ab + c^2$; $a^2 + b^2 = c^2$.

Additional Answer, page 486

32.

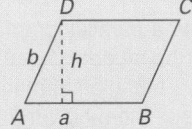

Statement	Reason
1. ▱ABCD, with a and b the lengths of sides including $\angle A < 90$	1. Given
2. Construct seg from D $\perp$ to $\overline{AB}$ of length h.	2. From pt to line there is unique $\perp$.
3. Area ABCD $= a \cdot h$	3. Area ▱ = bh
4. $\sin A = \frac{h}{b}$	4. Def of sine
5. $h = b \cdot \sin A$	5. Mult Prop of Eq
6. Area ABCD $= ab \cdot \sin A$	6. Sub

▰▰▰ GETTING STARTED

Prerequisite Quiz

1. Find the height of a rectangle with a base of 10 m and an area of 35 m².
   3.5 m
2. Find the area of a parallelogram with an angle of measure 30 and sides of lengths 16 ft and 20 ft. 160 ft²
3. In ▱*PQRS*, *PQ* = 6 cm, *QR* = 8 cm, and the altitude to $\overline{PQ}$ is 4 cm long. Find the length of the altitude to $\overline{QR}$. 3 cm

Motivator

On the chalkboard, draw a parallelogram with one diagonal. Ask the students what they know about the two triangles formed. They are congruent. Ask them how the area of one of these triangles would compare with the area of the entire parallelogram. The area of the triangle is one-half the area of the parallelogram.

▰▰▰ TEACHING SUGGESTIONS

Lesson Note

The formula for the area of a triangle, $A = \frac{1}{2}bh$, already familiar to students, is readily derived from the formula for the area of a parallelogram. More challenging are the derivations of the area formulas for the kite and rhombus with known diagonals, and for the equilateral triangle. The formula for finding the area of a triangle (with three sides known) is presented without proof.

12.4 Areas of Triangles

Objectives To apply the formula for the area of a triangle
To apply the formulas for the areas of a rhombus and a kite

If a parallelogram region is divided by a diagonal, two congruent triangles are formed. This fact suggests a way of finding the formula for the area of any triangle.

Theorem 12.3 The area of a triangle is one-half the product of the length of a base and the length of a corresponding altitude (Area of triangle = $\frac{1}{2}bh$).

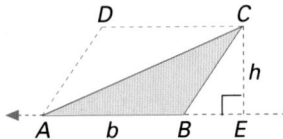

Given: △*ABC*, altitude $\overleftrightarrow{CE} \perp \overleftrightarrow{AB}$,
  AB = *b*, *CE* = *h*
Prove: Area(△*ABC*) = $\frac{1}{2}bh$

Proof:
(Paragraph Form)

Construct a parallelogram, the area of which is *bh*. This is done by using the Parallel Postulate: Through points *A* and *C* construct $\overline{AD} \parallel \overline{BC}$ and $\overline{CD} \parallel \overline{AB}$. By the Area Addition Postulate, Area(▱*ABCD*) = Area(△*ABC*) + Area(△*CDA*). Since the diagonal of a parallelogram forms two congruent triangles, △*ABC* ≅ △*CDA* so that Area(△*ABC*) = Area(△*CDA*). Then Area(▱*ABCD*) = 2 · Area(△*ABC*) = *bh*, or Area(△*ABC*) = $\frac{1}{2}bh$.

EXAMPLE 1 Find the area of the given triangles.

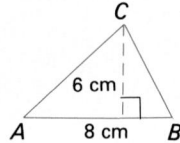

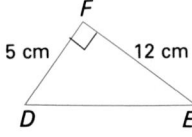

Solutions
Area = $\frac{1}{2}bh$

  = $\frac{1}{2} \cdot 8 \cdot 6$ = 24 cm²

Area = $\frac{1}{2}bh$

  = $\frac{1}{2} \cdot 5 \cdot 12$ = 30 cm²

Additional Example 1

Find the area of each given triangle.

a.
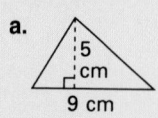

b.
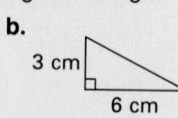

22.5 cm²   9 cm²

EXAMPLE 2   Given: Area($\triangle GHI$) = 36 mm², IM = 6 mm
Find GH.

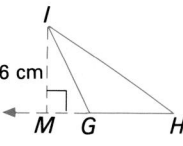

Solution
$$\text{Area}(\triangle GHI) = \tfrac{1}{2}bh$$
$$36 = \tfrac{1}{2} \cdot GH \cdot 6$$
$$36 = 3(GH)$$
$$GH = 12 \text{ mm}$$

A **kite** is a quadrilateral in which one diagonal is the perpendicular bisector of the other. The region bounded by a kite can be thought of as two triangular regions. This leads to a formula for the area of a kite.

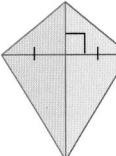

Theorem 12.4   The area of a kite is one-half the product of the lengths of the diagonals. (Area of Kite = $\tfrac{1}{2}d_1 d_2$)

Given: Kite $ABCD$; diagonal $\overline{AC}$ is the perpendicular bisector of diagonal $\overline{BD}$, $AC = d_1$, $BD = d_2$
Prove: Area(Kite $ABCD$) = $\tfrac{1}{2}d_1 d_2$

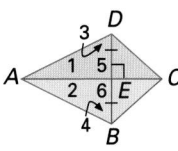

Plan
Flow Diagram

Kite $ABCD \longrightarrow \overline{AC} \perp \overline{DB} \longrightarrow \begin{cases} \text{Area}(\triangle ADC) = \tfrac{1}{2}\,d_1 \cdot DE \\ \text{Area}(\triangle ABC) = \tfrac{1}{2}\,d_1 \cdot EB \end{cases}$

$$\text{Area(Kite } ABCD) = \text{Area}(\triangle ADC) + \text{Area}(\triangle ABC)$$
$$= \tfrac{1}{2}\,d_1(DE + EB) = \tfrac{1}{2}\,d_1 d_2$$

Corollary   The area of a rhombus is one-half the product of the lengths of the two diagonals.

EXAMPLE 3   Find the area of a rhombus with diagonals 16 cm and 12 cm long.

Solution
$$\text{Area} = \tfrac{1}{2}d_1 d_2$$
$$= \tfrac{1}{2} \cdot 16 \cdot 12 = 96 \text{ cm}^2$$

12.4 Areas of Triangles   **489**

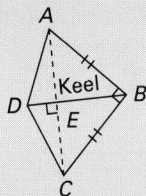

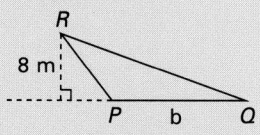

There is a special formula for the area of an equilateral triangle.

Theorem 12.5

If s is the length of a side of an equilateral triangle, then the area is $\frac{s^2}{4}\sqrt{3}$.

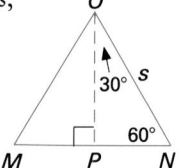

Given: Equilateral $\triangle MNO$, $MN = NO = OM = s$, altitude $\overline{OP}$ of length h

Prove: Area($\triangle MNO$) $= \frac{s^2}{4}\sqrt{3}$

Plan   The altitude of an equilateral triangle forms two congruent triangles. Use the properties of 30-60-90 triangles to find OP in terms of s.

Another famous formula is ascribed to the ancient Greek mathematician, Heron. *Heron's Formula* uses the lengths of the sides of a triangle to find the area. The formula is stated here without proof. Note that the formula uses half the perimeter, called the *semiperimeter*.

Theorem 12.6

Heron's Formula: If a, b, and c are the lengths of the sides of a triangle and s is the semiperimeter, such that $s = \frac{1}{2}(a + b + c)$, then Area(triangle) $= \sqrt{s(s - a)(s - b)(s - c)}$.

EXAMPLE 4 Find the area of a triangle with the sides 5 cm, 12 cm, and 13 cm long.

Solution   Let $a = 5$, $b = 12$, $c = 13$.
$s = \frac{1}{2}(5 + 12 + 13) = 15$
$A = \sqrt{15(15 - 5)(15 - 12)(15 - 13)}$
$A = \sqrt{15 \cdot 10 \cdot 3 \cdot 2} = \sqrt{900} = 30$

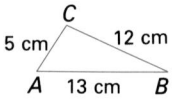

Therefore, the area is 30 cm².

Classroom Exercises

Find the area of the triangle, rhombus, or kite.

1.

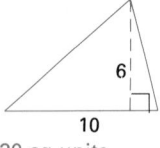

30 sq units

2.

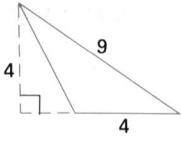

8 sq units

3.

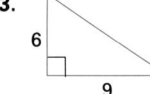

27 sq units

4.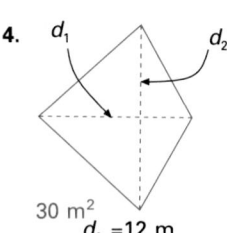

d_1 d_2

30 m²
d_1 =12 m
d_2 = 5 m

5.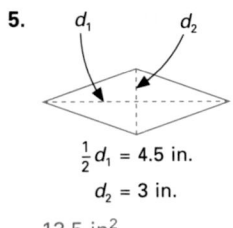

d_1 d_2

$\frac{1}{2}d_1$ = 4.5 in.
d_2 = 3 in.

13.5 in²

Written Exercises

Find the area of the triangle.

1.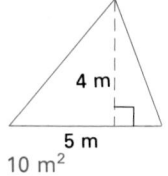

4 m

5 m
10 m²

2.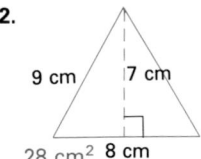

9 cm 7 cm

28 cm² 8 cm

3.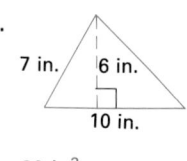

7 in. 6 in.

10 in.

30 in²

4.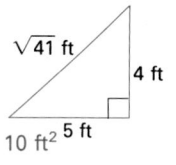

$\sqrt{41}$ ft

4 ft

10 ft² 5 ft

5.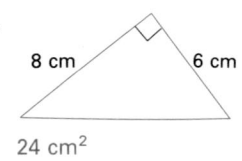

8 cm 6 cm

24 cm²

6.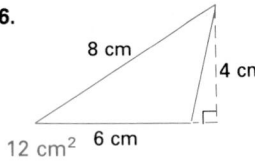

8 cm

4 cm

12 cm² 6 cm

Find the missing measurement for the indicated triangle.

	b	h	Area
7.	8 cm	20 cm	80 cm²
8.	10 m	12 m	60 m²

	b	h	Area
9.	3.4 mm	10 mm	17 mm²
10.	5 in	1.6 in.	4 in²

11. Find the area of a rhombus with diagonals 9 mm and 14 mm long.   63 mm²

Find the area of each of the following polygons.

12. Find the area of a kite with diagonals 12 in. and 23 in. long.   138 in²

13. Find the area of an equilateral triangle with sides 10 in. long. $25\sqrt{3}$ in²

In $\triangle ABC$, $\overline{CF} \perp \overline{AB}$, $\overline{AD} \perp \overline{BC}$, and $\overline{BE} \perp \overline{CA}$. **Find the missing length.**

14. $AB = 15$, $CF = 8$, $BC = 12$, $AD = $ __10__
15. $AB = 24$, $CF = 18$, $CA = $ __27__, $BE = 16$
16. $CA = 15$, $BE = 7$, $BC = 10$, $AD = $ __10.5__
17. $CA = 6$, $BE = 9$, $BC = $ __6.75__, $AD = 8$

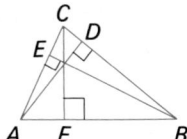

C D
E
A F B

12.4 Areas of Triangles   **491**

30.

Statement	Reason
1. Kite *ABCD*, diag $\overline{AC}$ is $\perp$ bis of diag $\overline{BD}$ at pt *E*, $AC = d_1$, $BD = d_2$.	1. Given
2. $DE = EB$	2. Def of bis
3. $\overline{DE} \perp \overline{AC}$, $\overline{EB} \perp \overline{AC}$	3. Def of $\perp$
4. $DE = $ alt of $\triangle ACD$, $EB = $ alt of $\triangle ACB$	4. Def of alt
5. Area $(\triangle ADC) = \frac{1}{2} \cdot AC \cdot DE$, Area $(\triangle ABC) = \frac{1}{2} \cdot AC \cdot EB$	5. Area of $\triangle$ $= \frac{1}{2} bh$
6. Area $(ABCD) = $ Area $(\triangle ADC) + $ Area $(\triangle ABC)$	6. Area Add Post
7. Area $(ABCD) = \frac{1}{2} \cdot AC \cdot EB + \frac{1}{2}AC \cdot DE$	7. Sub
8. Area $(ABCD) = \frac{1}{2} AC(EB + DE)$	8. Distr Prop
9. $EB + DE = BD$	9. Seg Add Post
10. Area $(ABCD) = \frac{1}{2} \cdot AC \cdot BD$	10. Sub
11. Area $(ABCD) = \frac{1}{2} \cdot d_1 d_2$	11. Sub

31.

Statement	Reason
1. Square *ABCD* with side of length *s* and diag *d*	1. Given
2. Area $(ABCD) = s^2$	2. $A = bh$
3. $d = s\sqrt{2}$	3. In 45-45-90 $\triangle$, hyp $= s\sqrt{2}$
4. $s = \frac{d}{\sqrt{2}} = \frac{d\sqrt{2}}{2}$	4. Rationalization of denominator
5. Area $(ABCD) = \left(\frac{d\sqrt{2}}{2}\right)^2$	5. Sub
6. Area $(ABCD) = \frac{1}{4} \cdot 2d^2 = \frac{1}{2}d^2$	6. Diagonals of a square are equal.
7. $A = \frac{1}{2}d^2$	7. Sub

Alternate proof using rhombus: Since a square is a rhombus, $A = \frac{1}{2}d_1 \cdot \frac{1}{2}d_2$; $d_1 = d_2 = d_1$

Enrichment

Ask students to find the area of a regular hexagon with a side of length 2. This will be taught in future lessons, but with their present knowledge it can be an interesting challenge. There are a number of different ways to divide the hexagon into regions whose areas can be found.

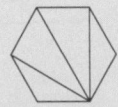 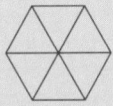 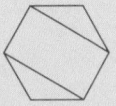

Using the 30–60–90 triangle relationships, the area of the hexagon is found to be $6\sqrt{3}$. Dividing the hexagon into six equilateral triangles and using an area formula, the same area is found.

32. $\triangle OPN \cong \triangle OPM$ Area $(\triangle MNO) = \frac{1}{2} \cdot MN \cdot OP$, m $\angle NOP = 30$, m $\angle ONP = 60$ $OP = \frac{1}{2}s\sqrt{3}$ Area $(\triangle MNO) = \frac{1}{2}s$ $(\frac{1}{2}s\sqrt{3}) = \frac{s^2}{4}\sqrt{3}$

33.

Statement	Reason
1. $\triangle MNO$, $MN = NO = OM = s$	1. Given
2. Draw $\overline{OP} \perp \overline{MN}$, with P on $\overline{MN}$ and $OP = h$.	2. From pt to line there is unique $\perp$.
3. $\overline{OP} \cong \overline{OP}$	3. Reflex
4. $\angle OPM \cong \angle OPN$	4. Rt $\angle$s are $\cong$.
5. $\triangle OPM \cong \triangle OPN$	5. HL
6. m $\angle NOM = 60$.	6. Meas of $\angle$ of equil $\triangle$ is 60.
7. m $\angle NOM = $ m $\angle NOP + $ m $\angle MOP$	7. Angle Add Post
8. m $\angle NOP = $ m $\angle MOP$	8. CPCTC
9. m $\angle NOM = 2 \cdot$ m $\angle NOP$	9. Sub
10. $2 \cdot$ m $\angle NOP = 60$.	10. Sub
11. m $\angle NOP = 30$.	11. Div Prop of Eq
12. $OP = \frac{1}{2}s\sqrt{3}$	12. Prop of 30–60–90 $\triangle$
13. Area $(\triangle MNO) = \frac{1}{2}$ $MN \cdot OP$	13. Area $\triangle = \frac{1}{2}bh$.
14. Area $(\triangle MNO) = \frac{1}{2}s$ $\frac{1}{2}s\sqrt{3}$	14. Sub
15. Area $(\triangle MNO) = \frac{s^2}{4}\sqrt{3}$	15. Comm Assoc Prop

34. False: Area $(\triangle 1) = $ Area $(\triangle 2)$ (Given); Area $(\triangle 1) = \frac{1}{2}b_1h_1$, Area $(\triangle 2) = \frac{1}{2}b_2h_2$ (Area of $\triangle = \frac{1}{2}bh$); $\frac{1}{2}b_1h_1 = \frac{1}{2}b_2h_2$ (Sub); $b_1h_1 = b_2h_2$ (Mult Prop of Eq); $\frac{b_1}{b_2} = \frac{h_2}{h_1}$ (Div Prop); So $\frac{b_1}{b_2} \neq \frac{h_1}{h_2}$

Mixed Review

1. Sufficient for $\square$; not sufficient for any trap
2. Not sufficient for $\square$ or any trap
3. Not sufficient for $\square$; sufficient for trap, but not for isosceles trap
4. Not sufficient for $\square$ or any trap
5. Not sufficient for $\square$; not sufficient for any trap
6. Not sufficient for $\square$; not sufficient for any trap

Use Heron's Formula to find the area of a triangle with sides of the given lengths.

18. $a = 6$, $b = 8$, $c = 10$ 24 sq units
19. $a = 5$, $b = 12$, $c = 13$ 30 sq units
20. $a = 4$, $b = 5$, $c = 6$ $\frac{15}{4}\sqrt{7}$ sq units
21. $a = 6$, $b = 9$, $c = 12$ $\frac{27}{4}\sqrt{15}$ sq units

Find the length of an altitude and the area of an equilateral triangle with sides of each length. (Exercises 22–24)

22. $s = 10$ **23.** $s = 8$ **24.** $s = t$
25. The area of a rhombus is 25 cm^2. The length of one diagonal is 8 cm. What is the length of the second diagonal?   6.25 cm
26. The area of a kite is 32 ft^2. The length of one diagonal is 12 ft. What is the length of the second diagonal? $5\frac{1}{3}$ ft

Find the area of the triangle or rhombus.

27.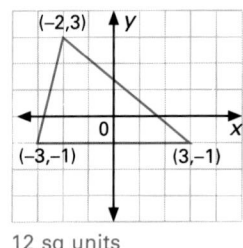
12 sq units

28.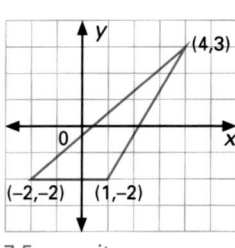
7.5 sq units

29.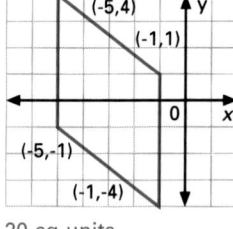
20 sq units

30. Write the proof of Theorem 12.4.
31. Prove: The area of a square is one-half the square of the length of the diagonal.
32. Write a plan for the proof of Theorem 12.5.
33. Prove Theorem 12.5.

Prove or disprove the statement.

34. If two triangles have equal areas, then the ratio of the lengths of their bases is equal to the ratio of the lengths of their altitudes.

Mixed Review

Which of the properties listed is a sufficient reason to prove that a quadrilateral is a parallelogram? To prove that it is a trapezoid? To prove that it is an isosceles trapezoid?   6.5, 7.4, 7.5

1. Diagonals bisect each other.
2. A pair of opposite sides are congruent.
3. A pair of opposite sides are parallel.
4. Diagonals are congruent.
5. Opposite angles are congruent.
6. Opposite angles are supplementary.

12.5 Areas of Trapezoids

Objective

To apply the formula for the area of a trapezoid

The roof of the New York Life Insurance Building is composed of trapezoidal sections. To find how much roofing material is needed for the building, the area of each section must be found.

Theorem 12.7

The area of a trapezoid is one-half the product of the length of an altitude and the sum of the lengths of the upper and lower bases [Area = $\frac{1}{2}h(b_1 + b_2)$].

Given: Trapezoid $ABCD$, $b_2 = DC$, $b_1 = AB$, length of altitude is h.

Prove: Area(trap $ABCD$) = $\frac{1}{2}h(b_1 + b_2)$

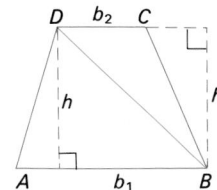

Statement	Reason
1. Trap $ABCD$, $AB = b_1$, $DC = b_2$, length of alt is h.	1. Given
2. Draw $\overline{DB}$.	2. Two points determine a line.
3. Area($\triangle ADB$) = $\frac{1}{2}b_1 h$, Area($\triangle CBD$) = $\frac{1}{2}b_2 h$	3. Area($\triangle$) = $\frac{1}{2}bh$
4. Area(trap $ABCD$) = Area($\triangle ADB$) + Area($\triangle CBD$)	4. Area Add Post
5. Area($\triangle ADB$) + Area($\triangle CBD$) = $\frac{1}{2}b_1 h + \frac{1}{2}b_2 h$	5. Add Prop of Eq
6. Area($\triangle ADB$) + Area($\triangle CBD$) = $\frac{1}{2}h(b_1 + b_2)$	6. Distr Prop
7. $\therefore$ Area(trap $ABCD$) = $\frac{1}{2}h(b_1 + b_2)$	7. Trans Prop of Eq

12.5 Areas of Trapezoids **493**

GETTING STARTED

Prerequisite Quiz

Solve.

1. $\frac{1}{2}x = 24$ 48
2. $16(x + 2) = 80$ 3
3. $24 = \frac{1}{2}x(7 + 9)$ 3

Simplify.

4. $\frac{1}{2} \cdot \frac{\sqrt{3}}{2} \cdot 5(4 + 20)$ $30\sqrt{3}$
5. $\frac{1}{2} \cdot \frac{\sqrt{2}}{2} \cdot 3(1 + 15)$ $12\sqrt{2}$

Motivator

Draw a trapezoid and one of its diagonals on the chalkboard. Ask the students if they can find the areas of the two triangles formed. Yes Ask them what measurements would be needed to find these two areas. The lengths of the upper and lower bases and the altitude Ask them if these are also measurements of the trapezoid. Yes Have them explain how these two areas are related to the area of the trapezoid. The sum of the two areas of the triangles equals the area of the trapezoid.

Lesson Note

Before presenting the formal proof for the area of a trapezoid, Theorem 12.7, use several chalkboard drawings of trapezoids with specific dimensions to demonstrate that the area of a trapezoid can be found by dividing the trapezoid into two triangles and using the area addition postulate.

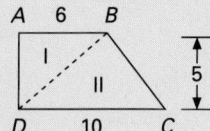

Area $\triangle I = \frac{1}{2}(6)(5)$

Area $\triangle II = \frac{1}{2}(10)(5)$

Area trap $ABCD = 15 + 25 = 40$ sq units
Be certain that students *see* that the bases of the trapezoid are the bases of the triangles, and that both triangles and the trapezoid have the same altitude.

Math Connections

Art: Picture frames can be purchased in pieces at a hobby shop. The pieces that form a rectangular frame are isosceles trapezoids. When the frame is assembled, opposite sides of the rectangular frame are congruent.

Critical Thinking Questions

Application: Suppose that you have a secret safe in the wall that you want to cover with a painting. The safe has a square front with 11-inch sides, but the dimensions of the only painting you have to cover the safe are 8 in. × 10 in. If all pieces of the frame are to be the same width and frame pieces are sold in only 1-, 2-, and 3-inch widths, what is the area of each piece of the smallest frame available that will allow you to completely cover the safe? For a 2-in wide frame, the area of each of the two shorter congruent pieces is 20 in² and the area of each of the two longer pieces is 24 in².

EXAMPLE 1   Find the area of trapezoid $ABCD$.

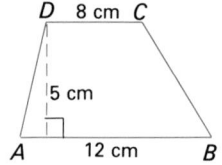

Solution $b_1 = 12$ cm, $b_2 = 8$ cm, $h = 5$ cm
Area $= \frac{1}{2} \cdot 5(12 + 8) = 50$ cm²

EXAMPLE 2   The area of a trapezoid is 80 mm². The length of one base is 6 mm. The length of an altitude is 8 mm. What is the length of the other base?

Solution   Area $= 80$ mm², $b_2 = 6$ mm, $h = 8$ mm
$$80 = \frac{1}{2} \cdot 8(b_1 + 6)$$
$$80 = 4b_1 + 24$$
$$56 = 4b_1$$
$$b_1 = 14$$

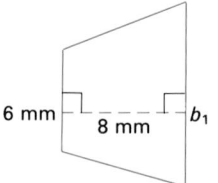

The length of the other base is 14 mm.

The lengths of the two nonparallel sides, or legs, are not used in the formula for the area of a trapezoid. However, for certain angle measures, the area can be found if one of the legs and the measure of an angle are known.

EXAMPLE 3   Find the area of trapezoid $IJKL$.

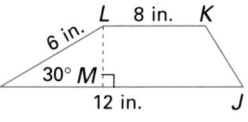

Solution $LM = 3$ in., since $\triangle ILM$ is 30-60-90.
Then, Area(trap $IJKL$) $= \frac{1}{2} \cdot 3(8 + 12)$
$$= 30 \text{ in}^2$$

The legs of an isosceles trapezoid are congruent. If the lengths of the bases are known, then the area of the isosceles trapezoid can be found.

EXAMPLE 4   Find the area of isosceles trapezoid $ABCD$.

Solution   When $\overline{DX}$ and $\overline{CY}$ are drawn perpendicular to $\overline{AB}$, rectangle $CDXY$ is formed. Therefore, $XY = DC = 6$. $\triangle AXD \cong \triangle BYC$ by HL, so that $AX = BY = 3$.

Then $h = 4$ by the Pythagorean Theorem.
Area($ABCD$) $= \frac{1}{2}h(b_1 + b_2)$
$$= \frac{1}{2}(4)(18) = 36$$

494   Chapter 12 Area

Additional Example 1

Find the area of trapezoid $PQRS$.
18 cm²

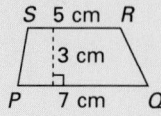

Additional Example 2

The area of a trapezoid is 45 cm². The length of one base is 9 cm and the length of the altitude is 3 cm. What is the length of the other base?
21 cm

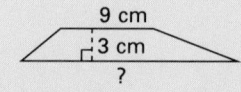

Classroom Exercises

State whether enough information is given to find the area.

1. 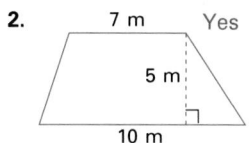 No

2. 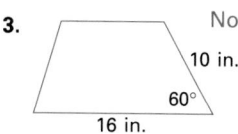 Yes

3. 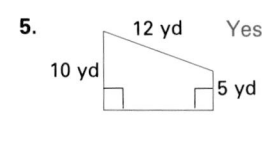 No

4. 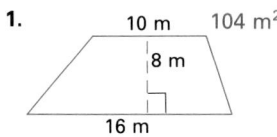 Yes

5. 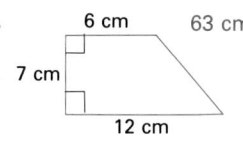 Yes

6. 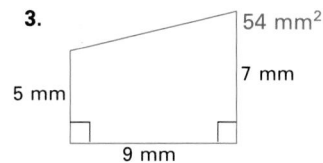 No

Written Exercises

Find the area of the trapezoid.

1. 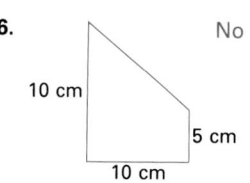 104 m²

2. 63 cm²

3. 54 mm²

Find the missing measurement for the indicated trapezoid.

	b_1	b_2	h	Area
4.	8 cm	12 cm	5 cm	50 cm²
5.	10 mm	15 mm	9 mm	112.5 mm²
6.	8 in.	12 in	7 in.	70 in²
7.	18 ft	12 ft	5 ft	75 ft²
8.	7 m	10 m	3 m	25.5 m²
9.	16 cm	9 cm	7 cm	87.5 cm²

$\overline{AB}$ and $\overline{CD}$ are the bases of trapezoid $ABCD$. Find the area.

10. $AB = 16$, $CD = 12$, $AD = 8$, m $\angle A = 30$ 56 sq units

11. $AB = 19$, $CD = 12$, $AD = 7$, m $\angle A = 30$ 54.25 sq units

12. $AB = 12$, $CD = 8$, $AD = 5$, m $\angle A = 60$ 25√3 sq units

13. $AB = 17$, $CD = 8$, $AD = 7$, m $\angle A = 60$

14. $AB = 20$, $CD = 14$, $AD = 10$, m $\angle A = 45$

15. $AB = 13$, $CD = 6$, $AD = 3$, m $\angle A = 45$

13. 43.75√3 sq units 14. 85√2 sq units 15. 14.25√2 sq units

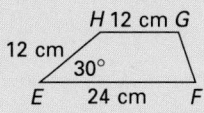

Additional Example 3

Find the area of trapezoid $EFGH$.
108 cm²

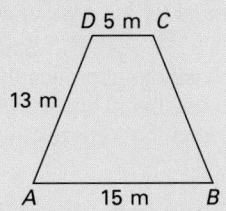

Additional Example 4

Find the area of isosceles trapezoid $ABCD$. 120 m²

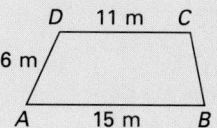

Common Error Analysis

Error: In most representations of trapezoids, the bases are horizontal segments. When this orientation is reversed, students sometimes become confused.

Remind students that, by definition, the parallel sides of a trapezoid are its bases, and an altitude is the distance between these bases (see Example 2 on page 494).

Checkpoint

Find the missing information for each trapezoid.

	b_1	b_2	h	Area	
1.	4 cm	5 cm	6 cm	?	27 cm²
2.	7 ft	11 ft	5 ft	?	45 ft²
3.	15 m	20 m	?	140 m²	8 m
4.	?	11 cm	7 cm	56 cm²	5 cm
5.	14 m	?	5 m	52.5 m²	7 m

6. In trapezoid $ABCD$, m $\angle A = 60$, $AD = 6$ m, $AB = 15$ m and $DC = 11$ m. Find the area. 39√3 m² or about 67.5 m²

Closure

Ask the students how to determine the area of a trapezoid. Theorem 12.7 page 493
Ask the students how they can determine the area of an isosceles trapezoid if they know the lengths of the bases and a leg.
Use the Pythagorean Theorem to find the length of the altitude. Then use the formula for the area of a trapezoid.

◼️◼️FOLLOW UP

Guided Practice

Classroom Exercises 1–6

Independent Practice

A Ex. 1–9, **B** Ex. 10–25, **C** Ex. 26–33

Basic: WE 1–9, Application, Midchapter Review

Average: WE 1–30 odd, Application, Midchapter Review

Above Average: WE 1–33 odd, Application, Midchapter Review

Written Exercises

24. The trap formula is sufficient for finding all areas, since $\frac{1}{2}(b_1 + b_2)$ is equivalent to the base (b) used in the formulas for rectangle and parallelogram, and equivalent to $\frac{1}{2}b$ in triangle formula. The altitude (h) is equivalent in all the figures.

33.

Statement	Reason
1. Trap with bases b_1 and b_2, s the length of a side adj to $\angle A$	1. Given
2. Let h be the length of alt of trap.	2. Distance between $\parallel$ lines is constant
3. $\sin A = \frac{h}{s}$	3. Def of sine
4. $h = s(\sin A)$	4. Mult Prop of Eq
5. Area (trap) $= \frac{1}{2}(b_1 + b_2)h$	5. Formula for area of trap
6. Area (trap) $= \frac{1}{2}h(b_1 + b_2)$	6. Comm Prop
7. Area (Trap) $= \frac{1}{2}s(\sin A) \cdot (b_1 + b_2)$	7. Sub

---

Find the area of the indicated isosceles trapezoid $ABCD$.

16. $AB = 12$, $CD = 6$, $AD = 5$ 36 sq units
17. $AB = 40$, $CD = 30$, $AD = 13$ 420 sq units
18. $AB = 16$, $CD = 8$, m $\angle A = 30$ $16\sqrt{3}$ sq units
19. $AB = 12$, $CD = 6$, m $\angle A = 60$ $27\sqrt{3}$ sq units
20. $AB = 20$, $CD = 12$, m $\angle A = 45$ 64 sq units

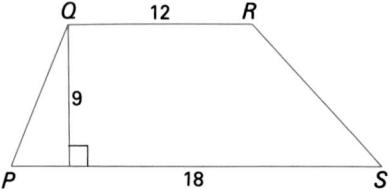

Find the dimensions of a figure with the same altitude and the same area as trapezoid $PQRS$. Then draw the figure. (Exercises 21–23)

21. a triangle $b = 30$, $h = 9$
22. a rectangle $b = 15$, $h = 9$
23. a parallelogram $b = 15$, $h = 9$

24. It has been suggested that the trapezoid formula is the only formula needed for finding the areas of all of the following: rectangle, square, parallelogram, triangle, and trapezoid. Give an argument for or against this.

Find the area of the trapezoid.

25.
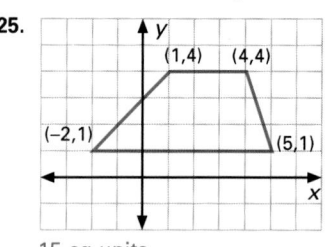
15 sq units

26.
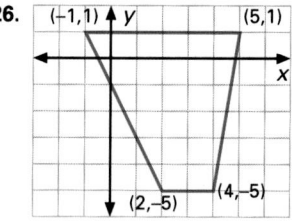
24 sq units

Find the area, to the nearest hundredth, of the indicated trapezoid using the appropriate trigonometric ratio.

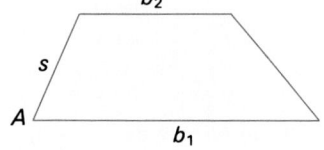

	b_1	b_2	s	m $\angle A$
27.	10 cm	8 cm	5 cm	56
28.	16 mm	12 mm	6 mm	87
29.	17.6 m	12.3 m	11.7 m	77
30.	16.2 cm	13.7 cm	9.6 cm	58

27. 37.31 cm^2 **28.** 83.88 mm^2 **29.** 170.43 m^2 **30.** 121.71cm^2

31. An altitude of a trapezoid is 8 cm long. The area is 168 cm^2. If one base is 12 cm longer than twice the other base, what is the length of each base?   10 cm, 32 cm

32. An altitude of a trapezoid is 11 m long. The area is 143 m^2. If the lower base is 3 m shorter than three times the upper base, what is the length of each base?   7.25 m, 18.75 m

Enrichment

Challenge the students to use a coordinate system to prove that an alternate formula for the area of a trapezoid is $A = \frac{(a + d - c)x}{2}$. Explain to them that one way to approach this problem is to use the midpoints P_1 and P_2 of the sides of the trapezoid.

$P_1 = (\frac{c}{2}, \frac{x}{2})$,
$P_2 = (\frac{a + d}{2}, \frac{x}{2})$.   They can construct a rectangle of equal area by drawing vertical sides through P_1 and P_2.

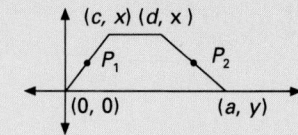

Use the trapezoid from Exercises 27–30 to prove the following.

33. If b_1 and b_2 are the lengths of the bases of a trapezoid, and s is the length of a side adjacent to $\angle A$, then the area of the trapezoid is $\frac{1}{2} \cdot s(\sin A)(b_1 + b_2)$.

Midchapter Review

Find the area of each figure.

1.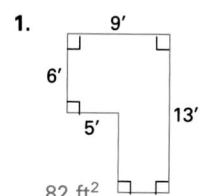
   82 ft²

2. parallelogram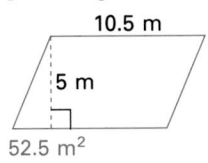
   52.5 m²

3.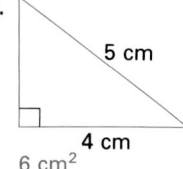
   6 cm²

4. equilateral triangle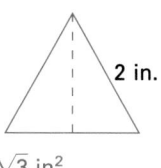

   √3 in²

5.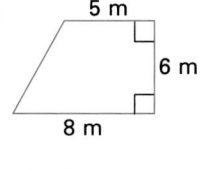
   39 m²

6.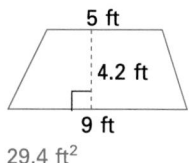
   29.4 ft²

Application: *Areas of Irregularly-Shaped Regions*

The area of an irregularly-shaped region can be estimated by dividing the region into narrow strips, all the same width. Each strip is approximately trapezoidal in shape. A surveyor's map shows the region below. Each strip is 10 m wide. What is the approximate area of the region in square meters? If a hectare is 10,000 m², what is the approximate area in hectares?   1,600 m², or 0.16 hectares

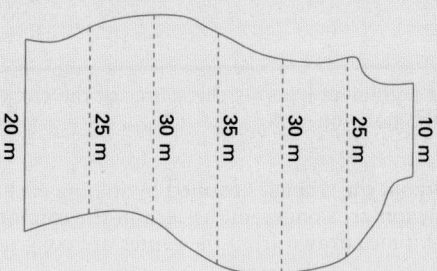

▰▰▰GETTING STARTED

Prerequisite Quiz

Solve.

1. $\frac{1}{2}(18x) = 288$ 32
2. $\frac{1}{2}(7x) = 28$ 8
3. $75 = \frac{1}{2}x(15)$ 10

**In each right triangle, find the value of *x*
to the nearest hundredth.**

4.   1.62   **5.**   4.13

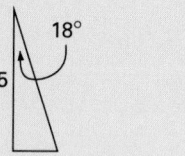
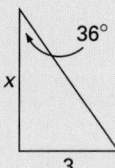

Motivator

Ask the students to define an equilateral
triangle.   A triangle with three congruent
sides   An equiangular triangle.   A triangle
with three congruent angles   Ask them to
define an equilateral polygon.   A polygon
with congruent sides   An equiangular
polygon.   A polygon with congruent angles
Ask the class if there are any polygons that
are both equilateral and equiangular.   Yes
Ask them what they are called.   Regular
polygons   Ask them if they can separate a
regular polygon into nonoverlapping
congruent triangles.   Yes   Have them tell
you how many of these triangles there
would be.   The same number as the
number of sides of the polygon.   Ask them
if they could find the area of the polygon if
they knew the area of one of the triangles.
Yes

12.6 Measuring the Regular Polygons

Objectives
To find the perimeter and area of a regular polygon
To find the apothem of a regular polygon

A regular polygon is both equilateral and equiangular. These properties
allow you to find the circle that circumscribes the regular polygon.

Suppose *ABCDEF* is a regular hexagon with $\overline{AO}$
and $\overline{BO}$ the angle bisectors of $\angle FAB$ and $\angle ABC$,
respectively. Since the hexagon is equiangular,
$\angle 2 \cong \angle 3$. Therefore, $OA = OB$.

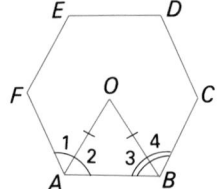

Draw $\overline{OC}$. $\triangle OBA \cong \triangle OBC$ by *SAS*. Therefore,
$OA = OC$ by *CPCTC*. So $OA = OB = OC$.
This process can be continued for each vertex of
the hexagon so that $OA = OB = OC = OD =
OE = OF$. As a result, O is the center of the cir-
cle containing points A, B, C, D, E, and F.

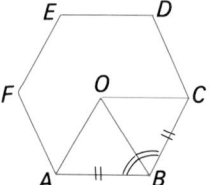

The above method of constructing the center of
the circumscribed circle can be extended to any
regular polygon. This essentially proves the following theorem.

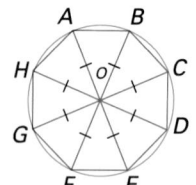

Theorem 12.8   A circle can be circumscribed about any regular polygon.

Definition   The **center of a regular polygon** is the center of the circle circumscribed about the polygon.

In a regular polygon, the triangles formed by joining each vertex to the
center of the polygon are congruent, as seen in the case of the hexagon.
Therefore, the altitudes drawn from the center are congruent for each
triangle.

498   Chapter 12 Area

The **apothem of a regular polygon** is the length of a perpendicular segment from the center of the polygon to a side. A **radius of the regular polygon** is a segment from the center to a vertex of the polygon.

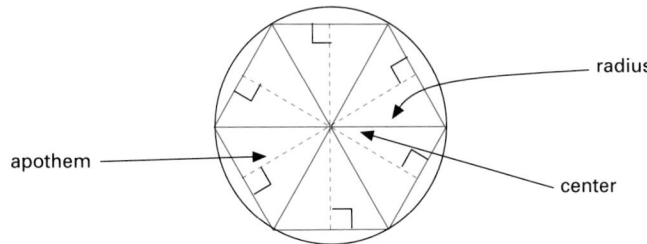

apothem — radius — center

EXAMPLE 1 Find the apothem of a regular hexagon with a side 10 cm long.

Plan In the regular hexagon, $\triangle ABO$ is equiangular (equilateral) since $m \angle OAB = m \angle OBA = \frac{1}{2} \cdot 120 = 60$.
Find the length of the altitude of $\triangle ABO$.

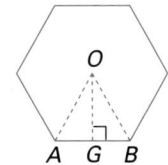

Solution The length h of an altitude of an equilateral triangle is $\frac{s\sqrt{3}}{2}$. In $\triangle ABO$, $s = AB = 10$.
Therefore, the apothem is $\frac{10\sqrt{3}}{2} = 5\sqrt{3}$ cm.

Definition The **perimeter** of a polygon is the sum of the lengths of its sides.

The radii of a regular *n*-gon, or polygon with *n* sides, form *n* congruent triangles, where *a* is the apothem and *s* is the length of a side of the polygon.

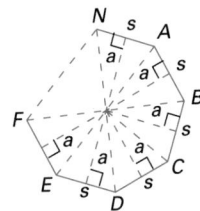

The area of the polygon is equal to the sum of the areas of all the triangles. Thus, Area(*n*-gon) $= n \cdot \frac{1}{2} a \cdot s$, where *n* is the number of sides. Since *ns* is equal to the perimeter *p*, Area(*n*-gon) $= \frac{1}{2} ap$.

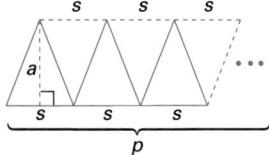

Additional Example 1

Find the apothem of a regular hexagon with a side 8 cm long. $4\sqrt{3}$ cm

Lesson Note

Carefully establish the definitions of regular polygon, perimeter, center, and apothem. The last two terms will be unfamiliar, and several chalkboard drawings should be used to illustrate them. Theorem 12.9, the formula for the area of a regular polygon, is important; it is used in Lesson 12.8 in developing the formula for the area of a circle.

Math Connections

Architecture: Garden plots, water fountains, sidewalks, and buildings often have a regular polygonal shape because the congruent sides and angles of these shapes offer a symmetry that is pleasing to the eye.

Critical Thinking Questions

Analysis: Ask students what happens to the length of the apothem as the number of sides of an inscribed regular polygon increases.   The length increases.   Ask students to generalize what happens to the ratio of the length of the apothem to the length of the radius as the number of sides increases.   The ratio approaches 1.

Checkpoint

1. Find the area of a regular polygon with an apothem of 6 m and a perimeter of 48 m.   144 m²

Find the missing information for each regular polygon, where *a* is the apothem, *p* is the perimeter, and *s* is the length of a side.

2. pentagon: $s = 1.25$ cm, $p =$ _____
   6.25 cm

3. hexagon: $a = 9\sqrt{3}$ ft, area = _____
   4.86$\sqrt{3}$ ft²

4. hexagon: $p = 72$ mm, area = _____
   216$\sqrt{3}$ mm²

5. octagon: Area = 55.76 cm², $p = 27.2$ cm, $a =$ _____   4.1 cm

Closure

Ask the students how they find the perimeter of a regular polygon.   It is the sum of the lengths of the sides.   Ask them to define the apothem of a regular polygon.   It is the length of a perpendicular segment from the center of a polygon to a side.   Ask them how they determine the area of a regular polygon.   Theorem 12.9 page 500

████FOLLOW UP

Guided Practice

Classroom Exercises 1–4

Independent Practice

A Ex. 1–14, **B** Ex. 15–27, **C** Ex. 28–33

Basic: WE 1–14 odd, AR all, Application

Average: WE 1–27 odd, AR all, Application

Above Average: WE 1–33 odd, AR all, Application

Theorem 12.9 — The area of a regular polygon is one-half the product of the apothem and the perimeter [Area (n-gon) = $\frac{1}{2}ap$].

EXAMPLE 2 The perimeter of a regular pentagon is 21.8 mm. Its apothem is 3 mm. Find its area.

Solution Area = $\frac{1}{2}ap$

Area = $\frac{1}{2}(3 \cdot 21.8)$ = 32.7 mm^2

EXAMPLE 3 The area of a regular octagon is 482.84 cm^2. The length of a side is 10 cm. Find the apothem.

Solution
$$\text{Area} = \frac{1}{2}ap = \frac{1}{2}a \cdot ns$$
$$482.84 = \frac{1}{2}a(8 \cdot 10)$$
$$482.84 = 40a$$
$$a = \frac{482.84}{40} \approx 12.071 \text{ cm}$$

10 cm
a

Other formulas relate the area of a regular polygon directly to the length of a side or to the length of a radius.

Theorem 12.10 — The area of a regular polygon is $n\left[\sin\left(\frac{180}{n}\right)\right]\left[\cos\left(\frac{180}{n}\right)\right]r^2$, or $\dfrac{ns^2}{4\tan\left(\frac{180}{n}\right)}$, where n is the number of sides, s is the length of a side, and r is the length of a radius.

EXAMPLE 4 Find the area, to the nearest hundredth, of a regular hexagon with sides 5 cm long.

Solution
$$n = 6, s = 5$$
$$A = \frac{6 \cdot 5^2}{4\tan\left(\frac{180}{6}\right)}$$
$$A = \frac{150}{4 \cdot 0.5774} \approx 64.95 \text{ cm}^2$$

5 cm

500   Chapter 12 Area

Additional Example 2

The perimeter of a regular pentagon is 36.30 mm. Its apothem is 5 mm. Find its area.   90.75 mm^2

Additional Example 3

The area of a regular decagon is 492.40 cm^2. The length of a side is 8 cm. Find the apothem.   12.31 cm

Classroom Exercises

Find the area of the regular polygon.

1. perimeter = 40 in., apothem = 5 in.
100 in^2

2. perimeter = 8 yd, apothem = 4 yd
16 yd^2

Find the area of the regular polygon if the length of a side is 10 cm.
261 cm^2

3. a square, apothem = 5 cm 100 cm^2

4. a hexagon, apothem = 8.7 cm

Written Exercises

Find the apothem of the regular polygon.

5√3 cm

1. a hexagon with a side 2 cm long √3 cm

2. a hexagon with a side 10 cm long

3. a square with a side 6 mm long 3 mm

4. a square with a side 10 mm long 5 mm

Find, to the nearest tenth, the perimeter of the regular polygon.

204 mm

65 cm

5. a pentagon with a side 13 cm long

6. a dodecagon with a side 17 mm long

7. a 130-gon with a side 14 in. long

8. a 150-gon with a side 29 ft long 4,350 ft

1,820 in.

Find the area of the regular polygon.

	Polygon	Perimeter	Apothem	Area
9.	square	16 cm	2 cm	16.0 cm^2
10.	pentagon	10 m	1.38 m	6.9 m^2
11.	pentagon	7.28 in.	1 in.	3.6 in^2
12.	hexagon	60 mm	8.66 mm	259.8 mm^2

13. The area of a regular hexagon is 64.95 cm^2. The perimeter is 30 cm. Find the apothem. 4.33 cm

14. The area of a regular octagon is 43.46 in^2. The apothem is 3.62 in. Find the perimeter. 24.01 in

Find the missing information for the regular polygon.

	Polygon	Length of side	Perimeter	Apothem	Area
15.	square	8 m	32 m	4 m	64 m^2
16.	hexagon	10 cm	60 cm	5√3 cm	150√3 cm^2
17.	hexagon	$\frac{2}{3}$√30 ft	4√30 ft	√10 ft	20√3 ft^2
18.	octagon	1 in.	8 in	1.207 in	4.828 in^2

Find, to the nearest tenth, the area of the regular polygon.

19. 172.0 cm^2
20. 166.3 m^2

19. a pentagon with sides 10 cm long

20. a hexagon with sides 8 m long

21. an octagon with sides 12 yd long

22. a dodecagon with sides 20 ft long

23. a 30-gon with sides 10 mm long

24. a 100-gon with sides 50 cm long

7,135.8 mm^2 21.695.3 yd^2

1,988,782.0 cm^2 **22.** 4,478.5 ft^2

Additional Example 4

Find the area of a regular dodecagon with sides 4 cm long. 179.17 cm^2

28.

Statement	Reason
1. Reg polygon with ctr O, n sides, radius of length r, side $\overline{AB}$.	1. Given
2. O is ctr of circumscribing $\odot$.	2. Def of ctr of reg polygon.
3. $OA = OB$	3. Radii of $\odot$ are $\cong$.
4. $\triangle OAB$ is isos.	4. Def of isos $\triangle$.
5. Draw bis $\overrightarrow{OM}$ of $\angle AOB$ inters $\overline{AB}$ at M.	5. Protractor, Ruler Post
6. $\overline{OM} \perp \overline{AB}$, $AM = BM$	6. Bis of vertex $\angle$ of isos $\triangle$ is $\perp$ bis of base.
7. $MB = \frac{1}{2}AB$	7. Def of bis
8. Area $(\triangle OAB) = \frac{1}{2}AB \cdot OM$	8. Area $\triangle = \frac{1}{2}(bh)$
9. Area $(\triangle OAB) = MB \cdot OM$	9. Sub
10. Area $(\triangle OAB) = \frac{MB}{r} \cdot \frac{OM}{r} \cdot r^2$	10. Real number div by self = 1, or $r^2 = 1$
11. $\angle OMB$ is rt $\angle$.	11. Def of $\perp$
12. $\triangle OMB$ is rt	12. m $\angle OMB = 90$.
13. $\sin \angle BOM = \frac{MB}{r}$, $\cos \angle BOM = \frac{OM}{r}$	13. Def of sin, cos
14. Area $(\triangle OAB) = \sin \angle BOM \cdot \cos \angle BOM \cdot r^2$	14. Sub
15. m $\angle BOM = \frac{1}{2} \cdot$ m $\angle BOA$	15. $\overrightarrow{OM}$ is bis of $\angle BOA$.
16. m $\angle BOA = \frac{360}{n}$.	16. Meas of $\odot$ = 360.
17. m $\angle BOM = \frac{1}{2} \cdot \frac{360}{n} = \frac{180}{n}$.	17. Sub
18. Area $(\triangle OAB) = \sin \left(\frac{180}{n}\right) \cdot \cos \left(\frac{180}{n}\right) \cdot r^2$	18. Sub
19. Total area of reg polygon = $n \cdot$ Area $(\triangle OAB)$	19. All $\triangle$s formed by sides of reg polygon are $\cong$.
20. Total area = $n \left[\sin \left(\frac{180}{n}\right)\right] \left[\cos \left(\frac{180}{n}\right)\right] r^2$	20. Sub

See page 503 for the answer to Ex. 29.

Prove the following.

25. An apothem of a regular polygon bisects a side of the polygon.

26. A radius of a regular polygon bisects an angle of the polygon.

27. All triangles formed by the sides of a regular polygon and the radii of the polygon are congruent.

28. Prove that the area of a regular polygon is $n\left[\sin\left(\frac{180}{n}\right)\right]\left[\cos\left(\frac{180}{n}\right)\right]r^2$.

29. Prove that the area of a regular polygon is $\dfrac{ns^2}{4 \tan \left(\frac{180}{n}\right)}$.

$\left(\text{HINT: } \tan A = \dfrac{\sin A}{\cos A}.\right)$

30. Find the ratio of the area of an equilateral triangle inscribed in a circle to an equilateral triangle circumscribed about the same circle.   1:4

31. Find the ratio of the area of a square inscribed in a circle to a second square circumscribed about the same circle.   1:2

32. A surveyor was instructed to start at a given point, walk 10 ft due north, turn two degrees to the left, walk 10 ft, turn two more degrees to the left, and repeat the process until he was back at the starting point. How many acres of land did he walk around? (1 acre = 43,560 ft^2)   5.92 acres

33. A 20-ft by 30-ft courtyard is to be surfaced with hexagonal concrete paving tiles. If each edge of each tile is 3 in. long, how many tiles will it take to cover the floor of the courtyard?   3,696 tiles

Mixed Review

Find the area.   *12.2–12.5*

1.

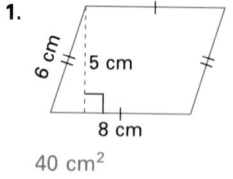

6 cm · 5 cm · 8 cm

40 cm^2

2.

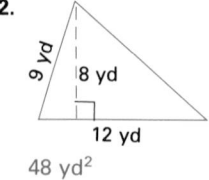

9 yd · 8 yd · 12 yd

48 yd^2

3.

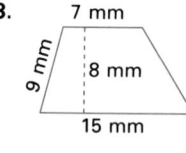

7 mm · 9 mm · 8 mm · 15 mm

88 mm^2

502   Chapter 12 Area

Enrichment

Square numbers are, of course, the perfect squares given by the formula n^2. Triangular numbers are found by counting the dots in successive triangles, as suggested by the illustration.

The first few triangular numbers are 1, 3, 6, 10, . . . Challenge the students to state a rule for this series in relation to the illustration.   Answers may vary. Triangular numbers can also be found by the formula $\frac{1}{2}n(n + 1)$. Have the students find the first few terms of the series by substituting values for n.

To simplify radical expressions:
(1) Rationalize the denominator.
(2) Write in simplest form.

Example: Simplify $\dfrac{3}{2 + \sqrt{3}}$.

$$\frac{3}{2 + \sqrt{3}} = \frac{3}{2 + \sqrt{3}} \cdot \frac{2 - \sqrt{3}}{2 - \sqrt{3}} = \frac{3(2 - \sqrt{3})}{4 - 3} = 6 - 3\sqrt{3}$$

Simplify.

1. $\dfrac{4}{2 - \sqrt{5}}$ $-8 - 4\sqrt{5}$

2. $\dfrac{3}{3 + \sqrt{2}}$ $\dfrac{9}{7} - \dfrac{3}{7}\sqrt{2}$

3. $\dfrac{5}{\sqrt{7} - 2}$ $\dfrac{5}{3}\sqrt{7} + \dfrac{10}{3}$

4. $\dfrac{-2}{\sqrt{5} - 4}$ $\dfrac{2}{11}\sqrt{5} + \dfrac{8}{11}$

5. $\dfrac{3\sqrt{2} - 1}{3\sqrt{2} + 1}$ $\dfrac{19}{17} - \dfrac{6\sqrt{2}}{17}$

6. $\dfrac{2\sqrt{8} - 3}{3\sqrt{12} + 1}$ $\dfrac{24\sqrt{6} - 18\sqrt{3} - 4\sqrt{2} + 3}{107}$

Application: *Polygons in Structures*

A certain zoo constructed buildings and outdoor areas in polygonal shapes.

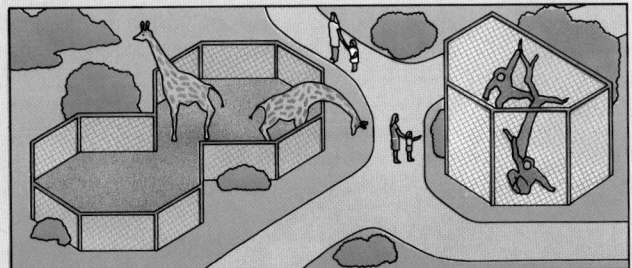

1. The area in the giraffe cage is made up of parts of hexagons. Each side is 8 ft long. How much fencing was needed for the enclosure? What is the area of the enclosure?   80 ft of fencing; $A = 333$ ft^2

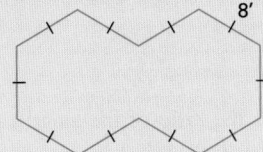

8'

2. The gorilla cage is a partial regular octagon. Find the total area of the floor space if each of the shorter sides has a length of 10 ft.   362.1 ft^2

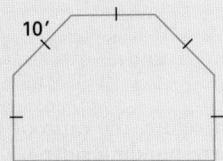

10'

Additional Answer, page 502

29.

Statement	Reason
1. Reg polygon with ctr O, n sides, radius r, side of length s, $\overline{OM}$ a $\perp$ bis of side $\overline{AB}$ at pt M	1. Given
2. Area (reg polygon) $= n\,[\sin(\frac{180}{n})]\,[\cos(\frac{180}{n})]\,r^2$	2. Proved in Ex 28
3. Area (reg polygon) $= n\,[r \cdot \sin(\frac{180}{n})]\,[r \cdot \cos(\frac{180}{n})]$	3. Comm Prop
4. $r \cdot \sin\frac{180}{n} = r \cdot \frac{MB}{r} = MB$	4. Def of sine
5. $MB = \frac{1}{2}s$	5. Def of seg bis
6. Area (reg polygon) $= n\,(\frac{1}{2}s)[r \cdot \cos(\frac{180}{n})]$	6. Sub
7. $\tan\frac{180}{n} = \sin\frac{180}{n} \div \cos\frac{180}{n}$	7. Trig Iden
8. $\cos\frac{180}{n} = \sin\frac{180}{n} \div \tan\frac{180}{n}$	8. Mult Prop of Eq
9. Area (reg polygon) $= n\,(\frac{1}{2}s)[r \cdot \sin(\frac{180}{n}) \div \tan\frac{180}{n}]$	9. Sub
10. Area (reg polygon) $= n\,(\frac{1}{2}s)[\frac{1}{2}s \div \tan(\frac{180}{n})]$	10. Sub

Area (reg polygon) $= \dfrac{ns^2}{4\tan\left(\frac{180}{n}\right)}$

Prerequisite Quiz

Simplify.

1. $\frac{6}{16}(56)$ 21

2. $\frac{14}{63}(18)$ 4

3. $\frac{4^2}{7^2}$ $\frac{16}{49}$

4. $\frac{8^2}{10^2}$ $\frac{16}{25}$

5. $\sqrt{\frac{49}{169}}$ $\frac{7}{13}$

6. $\sqrt{\frac{18}{50}}$ $\frac{3}{5}$

7. $\frac{9 + 12 + 15}{12 + 16 + 20}$ $\frac{3}{4}$

Motivator

Ask the students what they think would happen to the perimeter of a triangle if the lengths of its sides were doubled. The perimeter would double. Ask them what would happen to the perimeter if they were to triple the lengths of the sides of a rectangle. It would triple.

▰▰▰TEACHING SUGGESTIONS

Lesson Note

This lesson establishes two important properties of similar polygons. First, the ratio of their perimeters is the same as the ratio of the lengths of any two corresponding sides. Secondly, the ratio of their areas is the square of the ratio of the lengths of any two corresponding sides. The proofs of these theorems are left for the exercises. Accept informal proofs, since these proofs do not lend themselves to a two-column format.

504

12.7 Areas and Perimeters of Similar Polygons

Objectives To find the ratio of the perimeters of similar polygons
To find the ratio of the areas of similar polygons

A window in an old house was originally a square with sides 1 yd long. The owner nailed boards covering half the window. The resulting window was still a square, 1 yd tall and 1 yd wide. The new window and the original window were similar shapes, although turned in different directions. What is the ratio of their perimeters?

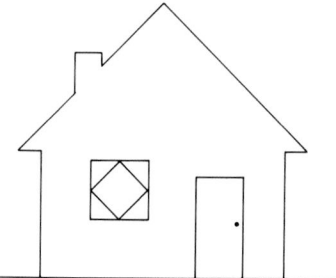

If two triangles are similar, then the ratios of the lengths of corresponding sides are equal. A triangle with sides of lengths 4 cm, 5 cm, and 6 cm is similar to a triangle with sides of lengths 8 cm, 10 cm, and 12 cm. The perimeters are 15 cm and 30 cm. Notice that the ratio of the perimeters of the triangles, 15:30, equals the ratio of the lengths of corresponding sides, 1:2.

Theorem 12.11 The ratio of the perimeters of two similar polygons is the same as the ratio of the lengths of any two corresponding sides.

Proof Suppose two similar polygons have sides of length $a_1, a_2, \cdots, a_n$ and $b_1, b_2, \cdots, b_n$ respectively. By the properties of similar triangles, $\frac{a_1}{b_1} = \frac{a_2}{b_2} = \cdots = \frac{a_n}{b_n} = s$, where s is the ratio of corresponding sides.

By multiplication, $a_1 = sb_1, a_2 = sb_2, \cdots, an = sb_n$.

The perimeter $a_1 + a_2 + \cdots + a_n = sb_1 + sb_2 + \cdots + sb_n$
$= s(b_1 + b_2 + \cdots + b_n)$

and $\dfrac{a_1 + a_2 + \cdots + a_n}{b_1 + b_2 + \cdots + b_n} = s$

Therefore, the ratio of the perimeters equals the ratio of the lengths of corresponding sides.

Additional Example 1

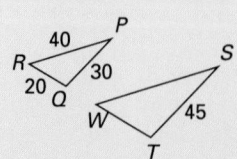

Given: $\triangle PQR \sim \triangle STW$, $PQ = 30$, $QR = 20$, $PR = 40$, $ST = 45$

Find the perimeters of $\triangle PQR$ and $\triangle STW$. What is the ratio of the perimeter of $\triangle STW$ to $\triangle PQR$? $\triangle PQR{:}P = 90$; $\triangle STW{:}P = 135$; $\frac{3}{2}$

EXAMPLE 1   Given: $\triangle DEF \sim \triangle ABC$, $AB = 7$, $BC = 9$,
$CA = 11$, $DE = 21$
Find the perimeters of $\triangle ABC$ and $\triangle DEF$.

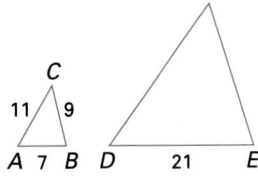

Plan $\dfrac{DE}{AB} = \dfrac{21}{7} = 3$. The ratio of the perimeters is 3:1.

Solution   Perimeter of $\triangle ABC = 7 + 9 + 11 = 27$
Perimeter of $\triangle DEF = 3(\text{perimeter of } \triangle ABC)$
$= 3 \cdot 27 = 81$

EXAMPLE 2   The perimeter of hexagon $ABCDEF$ is 64 m. The length of a side of similar hexagon $GHIJKL$ is $\frac{3}{4}$ that of the corresponding side of hexagon $ABCDEF$. Find the perimeter of hexagon $GHIJKL$.

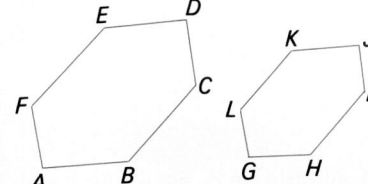

Solution   Perimeter of hex $GHIJKL = \frac{3}{4}(\text{perimeter of hex } ABCDEF)$
$= \frac{3}{4} \cdot 64 = 48$ m

Theorem 12.12   The **ratio** of the areas of two similar triangles is the square of the ratio of the lengths of any two corresponding sides.

EXAMPLE 3   Find the ratios of the perimeters and of the areas of the similar triangles.

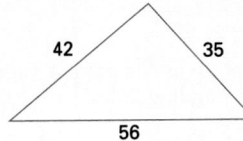
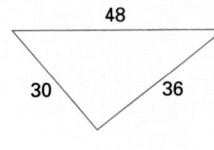

Solution   The ratio of the lengths of two corresponding sides is $\frac{56}{48}$, or $\frac{7}{6}$.
Therefore, the ratio of the perimeters is $\frac{7}{6}$,
and the ratio of the areas is $\frac{7^2}{6^2} = \frac{49}{36}$.

Math Connections

Engineering: When designing a cooling system, an engineer can compare areas and perimeters of an air duct to determine whether one duct or two smaller ducts will be more cost efficient. For example, if a cross section of the smaller air duct is 4 in. × 9 in. and a cross section of the larger duct is 6 in. × 13.5 in., then by comparing the ratio of areas with the ratio of perimeters, the engineer can determine which plan allows more air to be transported (through the area) per material used (for the perimeter). In this case, the ratio of areas, $(\frac{2}{3})^2 = \frac{4}{9}$, is more than 2 times the ratio of perimeters, $\frac{2}{3}$. Therefore, since over twice as much air can be transported for less than twice the cost, the engineer chooses to use the two smaller ducts instead of the larger one.

Critical Thinking Questions

Analysis: Refer students to Theorem 12.11 on page 504 and, given any two similar polygons, ask them which of the following ratios is equal to the ratio of their perimeters. the lengths of their corresponding altitudes? the lengths of their corresponding medians? the lengths of their corresponding midsegments? the lengths of corresponding apothems of similar regular polygons?   All are equal to the ratio of perimeters of similar polygons.

Additional Example 2

The perimeter of quad $ABCD$ is 36 m. The length of a side of a similar quad $EFGH$ is $\frac{4}{9}$ that of the corresponding side of quad $ABCD$. Find the perimeter of quad $EFGH$.   16 m

Additional Example 3

Find the ratio of the perimeters and of the areas of the two triangles.
Ratio of perimeters $= \frac{4}{5}$,
ratio of areas $= \frac{16}{25}$

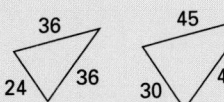

Common Error Analysis

Error: Some students confuse the concepts of similarity and congruence. Remind students that congruent triangles are similar triangles in which the ratio of similarity is 1:1. Thus, in two congruent figures, the ratio of their areas is $1^2 : 1^2$, or 1:1. (The areas of congruent polygons are equal.)

Checkpoint

Find the ratios of the perimeters and of the areas of each pair of similar polygons, given the lengths of a pair of corresponding sides.

1. 8 in. and 12 in. $\frac{2}{3}; \frac{4}{9}$

2. 15 m and 18 m $\frac{5}{6}; \frac{25}{36}$

3. 14 cm and 63 cm $\frac{2}{9}; \frac{4}{81}$

4. 33 cm and 24 cm $\frac{11}{8}; \frac{121}{64}$

5. Find the ratios of the perimeters and of the areas of the two triangles, larger to smaller. $P: \frac{8}{5}, A: \frac{64}{25}$

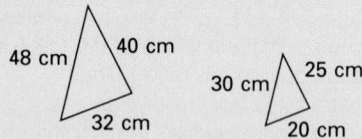

6. The sides of △ABC are 72 cm, 72 cm, and 18 cm long. The shortest side of similar △DEF is 22 cm long. What is the perimeter of △DEF? 198 cm

Closure

Ask the students how they find the ratio of the perimeters of two similar polygons if they know the lengths of corresponding sides. Theorem 12.11 page 504 Ask them how they determine the ratio of the areas of two similar polygons. Theorem 12.13 page 506

Any polygonal region can be broken up into triangular regions. This fact suggests a generalization of Theorem 12.12.

Theorem 12.13 | The ratio of the areas of two similar polygons is the square of the ratio of the lengths of any two corresponding sides.

EXAMPLE 4 The length of a side of square *ABCD* is 5mm. The length of a side of square *EFGH* is 8 mm. What is the ratio of the areas of the two squares?

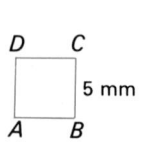
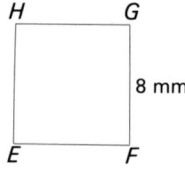

Solution $\frac{AB}{EF} = \frac{5}{8}$

Therefore, $\frac{\text{Area}(ABCD)}{\text{Area}(EFGH)} = \left(\frac{5}{8}\right)^2 = \frac{25}{64}.$

Classroom Exercises

Find the ratio of the perimeters of two similar polygons, given the ratio of the lengths of two corresponding sides.

1. $\frac{2}{3}$ $\frac{2}{3}$

2. $\frac{5}{9}$ $\frac{5}{9}$

3. $\frac{7}{4}$ $\frac{7}{4}$

4. $\frac{3}{11}$ $\frac{3}{11}$

Find the ratio of the areas of two similar polygons, given the ratio of the lengths of two corresponding sides.

5. $\frac{4}{9}$ $\frac{16}{81}$

6. $\frac{2}{9}$ $\frac{4}{81}$

7. $\frac{8}{5}$ $\frac{64}{25}$

8. $\frac{9}{11}$ $\frac{81}{121}$

Find the ratio of the lengths of corresponding sides of the two similar polygons, given the ratio of areas.

9. $\frac{16}{9}$ $\frac{4}{3}$

10. 4:4 1:1

11. $\frac{4}{25}$ $\frac{2}{5}$

12. 9:1 3:1

13. 49 to 100 7 to 10

Written Exercises

Find the ratio of perimeters and the ratio of areas for pairs of similar triangles, given the measures of their corresponding sides.

1.
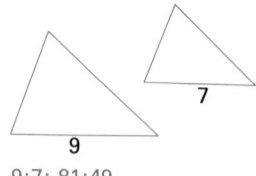
9:7; 81:49

2.
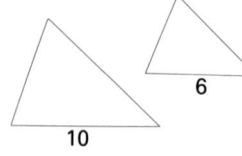
5:3; 25:9

3.
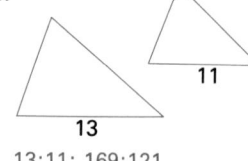
13:11; 169:121

Additional Example 4

The length of a side of square *ABCD* is 9 mm. The length of a side of square *EFGH* is 11 mm. What is the ratio of the areas of the two squares? $\frac{81}{121}$

4.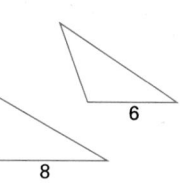

$\frac{4}{3}, \frac{16}{9}$

5.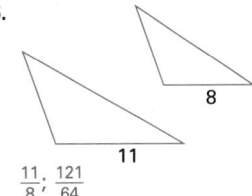

$\frac{5}{3}, \frac{25}{9}$

6.

$\frac{11}{8}, \frac{121}{64}$

7. The perimeter of pentagon *ABCDE* is 35 cm. The length of a side of similar pentagon *FGHIJ* is $\frac{2}{5}$ that of the corresponding side of *ABCDE*. Find the perimeter of *FGHIJ*.   14 cm

8. The perimeter of heptagon *KLMNOPQ* is 33 in. The length of a side of similar heptagon *RSTUVWX* is $\frac{9}{11}$ that of the corresponding side of *KLMNOPQ*. Find the perimeter of *RSTUVWX*.   27 in

9. The perimeter of octagon *ABCDEFGH* is 54 ft. The length of a side of similar octagon *IJKLMNOP* is $\frac{13}{9}$ that of the corresponding side of *ABCDEFGH*. Find the perimeter of *IJKLMNOP*.   78 ft

For the given ratio of areas for two similar polygons, find the ratio of lengths of corresponding sides.

10. $\frac{25}{81}$ $\frac{5}{9}$ **11.** $\frac{49}{64}$ $\frac{7}{8}$ **12.** $\frac{121}{225}$ $\frac{11}{15}$ **13.** $\frac{289}{361}$ $\frac{17}{19}$

Find the ratios of the perimeters and the ratios of the areas for the similar triangles.

14.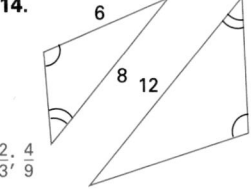

$\frac{2}{3}, \frac{4}{9}$

15.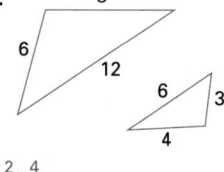

$\frac{2}{1}, \frac{4}{1}$

16.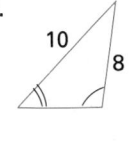

$\frac{8}{5}, \frac{64}{25}$

$\triangle ABC \sim \triangle DEF$. Find the missing length or ratio.

	$\dfrac{\text{Area}(\triangle ABC)}{\text{Area}(\triangle DEF)}$	AB	DE	Perimeter of $\triangle ABC$	Perimeter of $\triangle DEF$
17.	$\frac{16}{9}$	12 cm	9 cm	36 cm	27 cm
18.	$\frac{49}{36}$	21 mm	18 mm	56 mm	48 mm
19.	$\frac{2}{3}$	5 in.	$\frac{5}{2}\sqrt{6}$ in.	12 in.	$6\sqrt{6}$ in.
20.	$\frac{7}{3}$	$\frac{7}{3}\sqrt{21}$ ft	7 ft	$\frac{17}{3}\sqrt{21}$ ft	17 ft

Guided Practice

Classroom Exercises 1–13

Independent Practice

Ⓐ Ex. 1–16, **Ⓑ** Ex. 17–26, **Ⓒ** Ex. 27–29

Basic: WE 1–16 odd, Application, CI 1

Average: WE 1–26 odd, Application, CI all

Above Average: WE 1–29 odd, Application, CI all

Additional Answers

Written Exercises

22.

Statement	Reason
1. $\triangle ABC \sim \triangle DEF$	1. Given
2. Let $\frac{AB}{DE} = \frac{a}{b}$; draw alt $\overline{CD}$ of $\triangle ABC$, alt $\overline{FG}$ of $\triangle DEF$.	2. ⊥ from pt to line is unique.
3. $\frac{CD}{FG} = \frac{a}{b}$	3. Corr alt of △s are proport to corr sides.
4. $A(\triangle ABC) = \frac{1}{2} AB \cdot CD$, $A(\triangle DEF) = \frac{1}{2} DE \cdot FG$	4. Area △ = $\frac{1}{2}(bh)$.
5. $\frac{A(\triangle ABC)}{A(\triangle DEF)} = \left(\frac{a}{b}\right)^2$	5. Sub

23.

Statement	Reason
1. Similar rects A and B, with corr adj sides of lengths a and b, a_1 and b_1, respectively	1. Given
2. $a = b$	2. Def of ~ poly
3. $A(\text{rect } A) = ab$, $A(\text{rect } B) = a_1 b_1$	3. $A(\text{rect}) = bh$
4. $\frac{A(\text{rect } A)}{A(\text{rect } B)} = \frac{ab}{a_1 b_1} = \frac{a}{a_1} \cdot \frac{b}{b_1} = \frac{a}{a_1} \cdot \frac{a}{a_1}$	4. Sub
5. $\frac{A(\text{rect } A)}{A(\text{rect } B)} = \left(\frac{a}{a_1}\right)^2$	5. Notation

24. False. Counterexample: 2 squares with $a = 1, s = 2$, and $a = 2, s = 4$, respectively. Ratio of apothems $= \frac{1}{2}$, ratios of areas $= \frac{4}{16} = \frac{1}{4}$.

Enrichment

If two regular hexagons are successively inscribed in a regular hexagon, what is the ratio of the area of the inner hexagon to that of the outer hexagon?

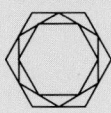

Starting with a regular hexagon of side 2, use repeated applications of the 30–60–90 relationships. The outer hexagon has an apothem of $\sqrt{3}$, which is the length of a side of the first inscribed hexagon. Similarly, the apothem of the first inscribed hexagon is $\frac{3}{2}$, which is the length of a side of a second inscribed (inner) hexagon. The ratio of the areas is the ratio of the squares of the sides, $\left(\frac{3}{2}\right)^2 : 2^2$ or 9:16.

507

29.

Statement	Reason
1. $\triangle ABC \sim \triangle DEF$, with corr sides of lengths a, b, and c, and d, e, and f, respectively, and semiperimeters s_1 and s_2.	1. Given
2. $\dfrac{A(\triangle ABC)}{A(\triangle DEF)} = \dfrac{\sqrt{s_1(s_1-a)(s_1-b)(s_1-c)}}{\sqrt{s_2(s_2-d)(s_2-e)(s_2-f)}}$	2. Heron's formula; Div Prop of Eq
3. $\dfrac{p(\triangle ABC)}{p(\triangle DEF)} = \dfrac{a}{d}$	3. Ratio of perim of $\sim$ poly = ratio of corr sides.
4. $\dfrac{\frac{1}{2}p(\triangle ABC)}{\frac{1}{2}p(\triangle DEF)} = \dfrac{a}{d}$	4. Any nonzero real div by self = 1.
5. $\dfrac{s_1}{s_2} = \dfrac{a}{d}$	5. Def of semiperimeter, sub
6. $\dfrac{s_1}{a} = \dfrac{s_2}{d}$	6. If $\frac{a}{b} = \frac{c}{d}$, then $\frac{a}{c} = \frac{b}{d}$.
7. $\dfrac{s_1-a}{a} = \dfrac{s_2-d}{d}$	7. If $\frac{a}{b} = \frac{c}{d}$, then $\frac{a-b}{b} = \frac{c-d}{d}$
8. $\dfrac{s_1-a}{s_2-d} = \dfrac{a}{d}$	8. If $\frac{a}{b} = \frac{c}{d}$, then $\frac{a}{c} = \frac{b}{d}$
9. $\dfrac{s_1-b}{s_2-e} = \dfrac{b}{e}$ and $\dfrac{s_1-c}{s_2-f} = \dfrac{c}{f}$	9. Steps 5–8
10. $\dfrac{s_1}{s_2} \cdot \dfrac{s_1-a}{s_2-d} \cdot \dfrac{s_1-b}{s_2-e} \cdot \dfrac{s_1-c}{s_2-f} = \dfrac{a}{d} \cdot \dfrac{a}{d} \cdot \dfrac{a}{c} \cdot \dfrac{b}{e} \cdot \dfrac{c}{f}$	10. Mult Prop of Eq; Sub
11. $\dfrac{a}{d} = \dfrac{b}{e} = \dfrac{c}{f}$	11. Def of $\sim\triangle$
12. $\dfrac{s_1(s_1-a)(s_1-b)(s_1-c)}{s_2(s_2-d)(s_2-e)(s_2-f)} = \dfrac{a}{d} \cdot \dfrac{a}{d} \cdot \dfrac{a}{d} \cdot \dfrac{a}{d}$	12. Sub
13. $\dfrac{\sqrt{s_1(s_1-a)(s_1-b)(s_1-c)}}{\sqrt{s_2(s_2-d)(s_2-e)(s_2-f)}} = \sqrt{\left(\dfrac{a}{d}\right)^4} = \dfrac{a^2}{d^2}$	13. Prop of roots
14. $\dfrac{A(\triangle ABC)}{A(\triangle DEF)} = \dfrac{a^2}{d^2}$	14. Trans Prop of Eq

21. An equilateral triangle has an area of 1 cm². What is the area of an equilateral triangle with sides twice as long? **4 cm²**

22. Prove Theorem 12.12 using the standard formula for the area of a triangle.

23. Prove Theorem 12.13 for rectangles.

24. Prove or disprove: The ratio of the areas of two similar regular polygons is equal to the ratio of their apothems.

25. Area($\triangle ABC$) = 16 cm²; Area($\triangle DEF$) = 25 cm²; the perimeter of $\triangle ABC$ is $2x + 8$; the perimeter of $\triangle DEF$ is $9x - 3$. Find each perimeter. **12 cm, 15 cm**

26. Area($\square GHIJ$) = 18 mm²; Area($\square KLMN$) = 50 mm²; the perimeter of $\square GHIJ$ is $5x - 3$; the perimeter of $\square KLMN$ is $7x + 11$. Find each perimeter. **57 mm, 95 mm**

27. A side of an equilateral triangle is 1 cm long. What is the length of the side of an equilateral triangle with twice the area? **$\sqrt{2}$ cm**

28. A home had a window in the shape of a regular hexagon 1 yd tall. The owner nailed boards over it in such a way that the new window was still a regular hexagon 1 yd tall. What is the ratio of the area of this new window to the original window? **3:4**

29. Prove Theorem 12.12 using Heron's formula.

Mixed Review

Define the following.

1. coplanar lines *1.1*
2. parallel lines *3.1*
3. supplementary angles *1.6*
4. Transitive Property of Congruence *2.3*
5. alternate interior angles *3.2*

█ **Application:** *Enlarging Drawings*

A pantograph is a device used for enlarging (or reducing) drawings. The entire device is free to move, except for a fixed pivot point at P. The adjustment of the pivots determines the ratio of the lengths of corresponding parts in the two drawings. Explain how to set the pivots so that the area of the larger figure is twice that of the smaller.

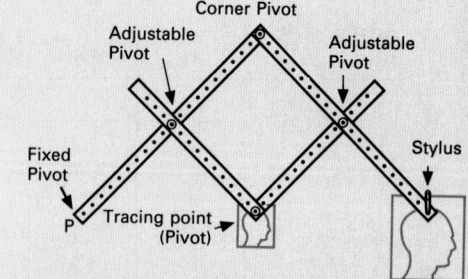

Mixed Review

1. Coplanar lines lie within the same plane.
2. Parallel lines are copl lines that do not intersect.
3. Supplementary angles have meas whose sum is 180.
4. If $\overline{AB} \cong \overline{CD}$, and $\overline{DC} \cong \overline{EF}$, then $\overline{AB} \cong \overline{EF}$; and if $\angle A \cong \angle B$ and $\angle B \cong \angle C$, then $\angle A \cong \angle C$.
5. Given two lines and a transv, alternate int angles are formed in the int of the two lines on opposite, non-adjacent sides of the transv.

Computer Investigation

Estimating π

Use a computer software program that draws regular polygons of a given size, defined by the distance from the center to a vertex, and that measures its area.

Notice in the figure at the right, the area of the circle appears to be a little larger than 3 times the area of a square with side lengths equal to the length of the radius of the circle. Greek mathematicians used this intuitive idea to guess that the area of a circle, of radius length R, is R^2 times a number a little larger than 3. They decided to call this number pi, or π. The computer activity below illustrates a method for approximating the value of π.

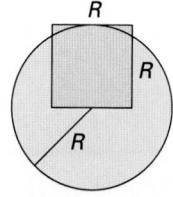

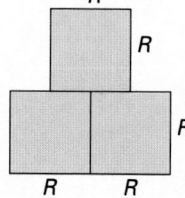

Activity 1

Draw several regular polygons, each with a distance of length 3 from the center to a vertex with the following number of sides.

1. 3 sides	**2.** 4 sides	**3.** 5 sides
4. 6 sides	**5.** 7 sides	**6.** 9 sides
7. 11 sides	**8.** 13 sides	**9.** 15 sides
10. 16 sides	**11.** 17 sides	**12.** 18 sides

Activity 2

The figure at the right shows an 8-sided regular polygon inscribed in a circle. The length of a radius is 3.

13. Construct the polygon shown at the right.

14. Find the area of this polygon.

15. Divide the area by R^2, in this case 3^2, or 9.

16. Repeat the steps of Exercises 13-15 for each of the polygons whose number of sides is given in Exercises 1-12. Keep a record of the results.

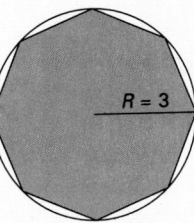

Summary

As the number of sides of a regular polygon of a given radius increases, the shapes of successive polygons approach a _____. The corresponding quotients, $\dfrac{\text{Area of polygon}}{R^2}$, approach the number _____.   circle, π

Teaching Resources

Application Worksheet 12
Manipulative Worksheet 23
Quick Quizzes 93
**Reteaching and Practice
Worksheets,** pp. 185, 186

◢◢◢ GETTING STARTED

Prerequisite Quiz

**Find the perimeter and area of each
square, where s is the length of a side
and a is the apothem.**

1. $s = 3$ cm $\quad P = 12$ cm, $A = 9$ cm^2
2. $s = 9$ cm $\quad P = 36$ cm, $A = 81$ cm^2
3. $a = 5$ cm $\quad P = 40$ cm, $A = 100$ cm^2

**Find the perimeter and area of each
equilateral triangle, where s is the length
of a side and a is the apothem.**

4. $s = 6$ in. $\quad P = 18$ in, $A = 9\sqrt{3}$ in^2
5. $s = 3$ m $\quad P = 9$ m, $A = \frac{9\sqrt{3}}{4}$
6. $a = 4$ cm $\quad P = 24\sqrt{3}$ cm, $A =$
$48\sqrt{3}$ cm^2

Motivator

Ask the students if a square looks much like
a circle.   No   Ask them if a regular
octagon looks more like a circle than the
square does.   Yes   Ask them if a regular
16–gon would look even more like a circle.
Yes   A regular 32–gon.   Yes   Ask the
class if doubling the number of sides of a
polygon makes it look more and more like a
circle.   Yes   Ask them if they think they
could approximate the circumference and
area of a circle by finding the area of a
many-sided regular polygon since they know
how to find the perimeter and area of a
regular polygon.   Yes

12.8 Circumferences and Areas of Circles

Objectives   To find the circumference of a circle
To find the area of a circle

Consider a circle with a radius 1 unit long. Regular polygons can be inscribed in it. As the number of sides of the polygon increases, the perimeter also increases, but not indefinitely.

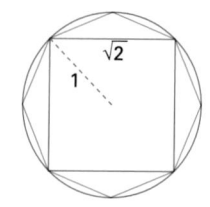

Number of sides	Approximate perimeter
4	5.6569
8	6.1229
16	6.2429
32	6.2731
64	6.2807
128	6.2826
256	6.2830

As the number of sides of the inscribed polygons increases, the polygons more closely approximate a circle. The perimeters approach a
value of approximately 6.2832. A good measure of the distance around
the circle seems to be the perimeter of an inscribed regular polygon
with a very large number of sides.

These perimeters are approaching a *limit.* A sequence of numbers approaches a limit if there is a number *b* to which the numbers of the
sequence get closer and closer. The number *b* is called the *limit of the
sequence.*

Definition   The **circumference of a circle** is the limit of the perimeters of inscribed regular polygons of the circle as the number of sides of the
polygon increases indefinitely.

If you wrapped a piece of string around a can, you would find that the
string is slightly more than three times the length of a diameter of the
can. This would be true for any can that is a circular cylinder, regardless of its size. It appears that the ratio of the distance around a circle
to the length of the diameter is a constant.

Theorem 12.14   The ratio of the circumference to the length of a diameter is the same
for all circles.

This constant ratio of the circumference to the length of a diameter is called **pi**, denoted by the symbol π. π is an irrational number that can be approximated to any degree of accuracy. To fifteen decimal places $\pi = 3.141592653589793$. Commonly, 3.14, 3.1416, or $\frac{22}{7}$ is used as an approximation. Unless otherwise specified, you may leave answers in terms of π rather than approximating.

Let C be the circumference of a circle, and d the length of its diameter. Since $\frac{C}{d} = \pi$, then $C = \pi d$. Since a diameter has twice the length of a radius, $C = 2\pi r$.

Corollary

The circumference of a circle with radius of length r is $2\pi r$.

EXAMPLE 1 The diameter of a circle is 5 cm long. What is the circumference?

Solution
$C = \pi d$
$C = \pi \cdot 5$, or 5π

EXAMPLE 2 If the circumference of a circle is 30π, what is the length of a radius and the length of a diameter?

Solution
$C = 2\pi r$
$30\pi = 2\pi r$

Therefore, $r = \frac{30\pi}{2\pi} = 15$, and $d = 2r = 30$.

The area of a circle can be found through a process similar to that used for the circumference. As the number of sides of a regular polygon inscribed in a circle increases, the area also increases, but not indefinitely.

Sides (n)	Approximate area
4	2
8	2.8284
16	3.0615
32	3.1214
64	3.1365
128	3.1403
256	3.1413

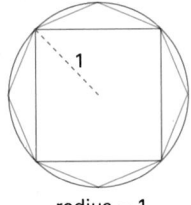

radius = 1

These areas, like the perimeters earlier, approach a limit.

Lesson Note

This lesson introduces the mathematical concept of *limits* in an informal way. First, the circumference of a circle is seen as the limit of the perimeters of inscribed regular polygons as the number of sides increases indefinitely. Secondly, the area formula for a circle is deduced by visualizing the circle as a regular polygon with an infinite number of sides. Use ample chalkboard drawings to illustrate that three things happen as the number of sides of a regular polygon increases.

1. The polygon approaches the shape of a circle.
2. The perimeter of the polygon approaches the circumference of the circle.
3. The apothem of the polygon approaches the length of the radius of the circle.

By substitution of terms, then, the formula for the area of a circle is readily deduced from the formula for the area of a regular polygon.

Math Connections

History: The ancient Chinese used 3 as the value of π. The Egyptians improved this approximation around 1650 B.C.. Ptolemy of Alexandria was a Greek astronomer who calculated 3.1416 as the value of π. After the 1600s, when using decimals became more common, a more exact value for π with repeating decimals was used. Today, mathematicians know that π is irrational; it is a nonterminating decimal without a repeating pattern.

Additional Example 1

If the length of a diameter of a circle is 7 cm, what is the circumference? 7π, or approximately 21.98 cm (using 3.14 for π)

Additional Example 2

If the circumference of a circle is 18π, what is the length of a radius and the length of a diameter? $r = 9$, $d = 18$

Critical Thinking Questions

Application: Ask students the following question. Suppose you want to buy a garden hose. At the store you find that all except one of the garden hoses on the shelf have their lengths marked on the labels. If the unmarked hose is coiled uniformly, how can you estimate its length? First estimate the diameter of the top circle of the coiled hose and calculate the circumference of that circle. Then multiply by the number of rings stacked in the coil.

Common Error Analysis

Error: Students tend not to associate answers expressed in terms of π with real numbers. Ask students to estimate the value of expressions such as $(4\pi - 8)$ m^2. Accept estimates such as $(4 \times 3 - 8)$ m^2 = 4 m^2.

Checkpoint

Find the unknown measurements.

	r	d	c	Area of $\odot$
1.	8 cm	?	?	?
2.	?	50 m	?	?
3.	?	?	6π m	?
4.	?	?	?	81π m^2
5.	?	?	15π m	?

1. 16 cm, 16π cm, 64π cm^2
2. 25 m, 50π m, 625π m^2
3. 3 m, 6 m, 9π m^2
4. 9 m, 18 m, 18π m
5. 7.5 m, 15 m, 56.25π m^2

6. Find the area of the shaded region between the circle and the inscribed square. The circle has a radius of 8 cm.
   $64\pi - 128$, or approximately 72.96 cm^2

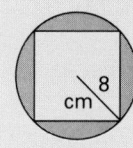

Definition
The **area of a circle** is the limit approached by the areas of inscribed regular polygons as the number of sides of the polygons increases indefinitely. The length of the apothem of a polygon with many sides approaches the length of a radius of the circumscribing circle.

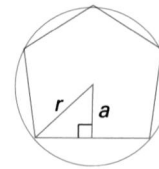 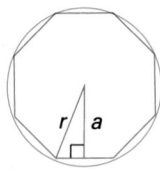 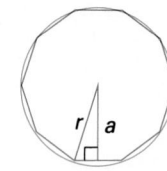

Area(regular polygon) = $\frac{1}{2}ap$. This suggests that as a approaches r,
Area($\odot$) = $\frac{1}{2}rC = \frac{1}{2}r(2\pi r)$, or πr^2.

Theorem 12.15 The area of a circle with radius of length r is πr^2.

EXAMPLE 3 Find the area of a circle with a radius of 6 mm. Use 3.14 for π.

Solution Area = πr^2 = $3.14 \cdot 6^2$ = 113.04 mm^2

EXAMPLE 4 Find the area of a circle with a diameter 20 cm long.

Solution $A(\odot) = \pi r^2$
$A(\odot) = \pi(\frac{1}{2} \cdot 20)^2$ = 100π cm^2

EXAMPLE 5 Find the area of the shaded region between $\odot O$ and the inscribed square $ABCD$. The radius of the circle is 2 cm long.

Solution Area($\odot O$) = $\pi \cdot 2^2$ = 4π cm^2
$\triangle AOB$ is a 45-45-90 triangle,
so $AB = \sqrt{2} \cdot AO = 2\sqrt{2}$ cm.

The area of the square is $(2\sqrt{2})^2$ = 8 cm^2.
Therefore, the area of the shaded region is $(4\pi - 8)$ cm^2.

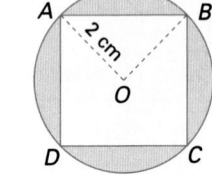

Classroom Exercises

Find the circumference of the circle in terms of π.

1. d = 10 m
10π m

2. d = 2 cm
2π cm

3. $d = \frac{1}{4}$ ft
$\frac{1}{4}\pi$ ft

4. d = 2 mi
2π mi

5. d = 4 yd
4π yd

Additional Example 3

Find the area of a circle with a radius of 11 cm. Use 3.14 as an approximation of π. 379.94 cm^2

Additional Example 4

Find the area of a circle with a diameter of 6 cm. $A\odot = 9\pi$ cm^2

Find the area of the circle in terms of π.

6. $r = 10.1$ mm
102.01π mm^2

7. $r = 3$ cm
9π cm^2

8. $r = 12$ in.
144π in^2

9. $r = 1\frac{1}{2}$ in.
$2\frac{1}{4}\pi$ in^2

10. $r = 7$ ft
49π ft^2

Written Exercises

Find, to the nearest tenth, the circumference of the circle.

1. $d = 20$ cm
62.8 cm

2. $d = 6.7$ m
21.0 m

3. $d = 100$ yd
314.0 yd

4. $d = 30$ ft
94.2 ft

Find, to the nearest tenth, the area of the circle.

5. $r = 2$ mm
12.6 mm^2

6. $r = 5$ cm
78.5 cm^2

7. $r = 10$ in.
314.0 in^2

8. $r = 20\frac{1}{2}$ ft
1,319.6 ft^2

Find the missing measurements.

	r	d	C	Area of circle
9.	5 cm	10 cm	10π cm	25π cm^2
10.	10 m	20 m	20π m	100π m^2
11.	4 in.	8 in	8π in	16π in^2
12.	12 ft	24 ft	24π ft	144π ft^2
13.	5 m	10 m	10π m	25π m^2
14.	8 cm	16 cm	16π cm	64π cm^2
15.	5 in	10 in	10π in.	25π in^2
16.	8 yd	16 yd	16π yd	64π yd^2
17.	10 cm	20 cm	20π cm	100π cm^2
18.	8 m	16 m	16π m	64π m^2
19.	2.5 m	5 m	5π m	6.25π m^2
20.	3.4 cm	6.8 cm	6.8π cm	11.56π cm^2
21.	$\frac{50}{\pi}$ in,		100 in.	$\frac{100}{\pi}$ in, $\frac{2,500}{\pi}$ in^2
22.	$\frac{10\sqrt{\pi}}{\pi}$ in,	$\frac{20\sqrt{\pi}}{\pi}$ in, $20\sqrt{\pi}$ in		100 in^2

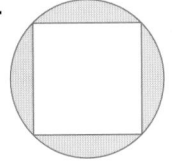 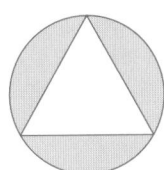 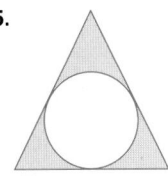

Find, to the nearest tenth, the area of the shaded region. The circle has a radius 10 cm long, and the polygon is a regular polygon.

23.

24.

25.

$100(\pi - 2) \approx 114.2$ cm^2 $(100\pi - 75\sqrt{3}) \approx 184.1$ cm^2 $(300\sqrt{3} - 100\pi) \approx 205.6$ cm^2

26. Two approximations, 3.14 and $\frac{22}{7}$, are often used for π. Which is closer to π? $\frac{22}{7}$

27. π is known to be between $3\frac{1}{7}$ and $3\frac{10}{71}$. Which is closer to π? $3\frac{10}{71}$

12.8 Circumferences and Areas of Circles   **513**

Closure

Ask the students what the number π represents. It is the constant ratio of the circumference to the length of a diameter. Have them tell you the approximate value of π. 3.14 or $\frac{22}{7}$ Ask them how π is used in finding the area and the circumference of a circle. Theorem 12.15 page 512 and Corollary page 511

◼◼◼ FOLLOW UP

Guided Practice

Classroom Exercises 1–10

Independent Practice

A Ex. 1–18, **B** Ex. 19–30, **C** Ex. 31–37

Basic: WE 1–18 odd
Average: WE 1–30 odd
Above Average: WE 1–37 odd

Additional Answers

Written Exercises

34.

Statement	Reason
1. $\odot M$ with radius of length a, $\odot N$ with rad of length b	1. Given
2. $A(\odot M) = \pi a^2$, $A(\odot N) = \pi b^2$	2. $A \odot = \pi r^2$.
3. $\frac{A(\odot M)}{A(\odot N)} = \frac{\pi a^2}{\pi b^2}$	3. Equals may be div by equals.
4. $\frac{A(\odot M)}{A(\odot N)} = \frac{a^2}{b^2}$	4. Any non-zero real number div by itself = 1.
5. $\frac{A(\odot M)}{A(\odot N)} = \left(\frac{a}{b}\right)^2$	5. Notation

35. False. Counterexample: $\odot O$ with $r = 1$, $\odot P$ with $r = 2$.

$\frac{C(\odot O)}{C(\odot P)} = \frac{2\pi}{4\pi} = \frac{1}{2}$, $\frac{A(\odot O)}{A(\odot P)} = \frac{\pi}{\pi 4} = \frac{1}{4}$; $\frac{1}{2} \neq \frac{1}{4}$.

Additional Example 5

Find the area of the shaded region between the equilateral triangle and the inscribed circle. The radius of the circle is 3 cm. $27\sqrt{3} - 9\pi$, or approximately 18.50 cm^2

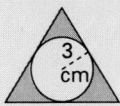

36.

Statement	Reason
1. $\odot O$ with r_1 and inscrib sq O, $\odot P$ with r_2 and inscrib sq P	1. Given
2. $A(\odot O) = \pi r_1{}^2$, $A(\odot P) = \pi r_2{}^2$	2. $A(\odot) = \pi r^2$
3. Diag of sq O and sq P form 45-45-90 $\triangle$s.	3. Diag of a sq are $\perp$ bis.
4. Side of sq $O = \sqrt{2}$ r_1, side of sq $P = \sqrt{2}$ r_2.	4. Hyp of 45-45-90 $\triangle$ = $\sqrt{2}$s.
5. $A(\text{sq } O) = (\sqrt{2}r_1)^2$, $A(\text{sq } P) = (\sqrt{2}r_2)^2$	5. $A(\text{sq}) = s^2$
6. $\frac{A(\odot O)}{A(\odot P)} \frac{\pi r_1{}^2}{\pi r_2{}^2}, \frac{A(\text{sq } O)}{A(\text{sq } P)} = \frac{2r_1{}^2}{2r_2{}^2}$	6. Div Prop for Eq.
7. $\frac{\pi r_1{}^2}{\pi r_2{}^2} = \frac{r_1{}^2}{r_2{}^2}, \frac{2r_1{}^2}{2r_2{}^2} = \frac{r_1{}^2}{r_2{}^2}$	7. Any non-zero real div by itself = 1.
8. $\frac{A(\odot O)}{A(\odot P)} = \frac{r_1{}^2}{r_2{}^2}, \frac{A(\text{sq } O)}{A(\text{sq } O)} = \frac{r_1{}^2}{r_2{}^2}$	8. Trans
9. $\frac{A(\odot O)}{A(\odot P)} = \frac{A(\text{sq } O)}{A(\text{sq } P)}$	9. Sub

37. True.

Statement	Reason
1. $\odot Q$ with r_1 and circumscr sq Q, $\odot R$ with r_2 and circumscr sq R	1. Given
2. $A(\odot Q) = \pi r_1{}^2$, $A(\odot R) = \pi r_2{}^2$	2. $A(\odot) = \pi r^2$
3. Sides of sq R are tan to $\odot R$	3. Def of inscr $\odot$
4. Diam of $\odot R$ including pts of tan to sq R is $\perp$ to sides of sq R.	4. Radii are $\perp$ to tan at pt of tan.
5. Such diam of $\odot R$ is $\parallel$ to side of sq R.	5. Lines $\perp$ to same line are $\parallel$.
6. Length of diam of $\odot R$ = length of side of sq R.	6. Opp sides of rect are $\cong$.
7. Length of diam of $\odot R$ = 2 · r_2.	7. Def of diam
8. $2r_2$ = length of s of sq R.	8. Sub
9. $A(\text{sq } R) = 4r_2{}^2$	9. $A (\text{sq}) = s^2$
10. $A(\text{sq } Q) = 4r_1{}^2$ $\frac{A(\odot Q)}{A(\odot R)} = \frac{\pi r_1{}^2}{\pi r_2{}^2} = \frac{r_1{}^2}{r_2{}^2}$ $\frac{A(\text{sq } Q)}{A(\text{sq } R)} = \frac{4r_1{}^2}{4r_2{}^2} = \frac{r_1{}^2}{r_2{}^2};$ $\frac{A(\odot Q)}{A(\odot R)} = \frac{A(\text{sq}Q)}{A(\text{sq}R)}$	10. Proceed similarly

For Exercises 28–30, $ABCD$ is a square with a side of length 10. All circles in the square are congruent and tangent to adjacent circles and/or the square. Find the area of the shaded region in terms of π.

28.
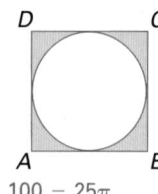
$100 - 25\pi$

29.
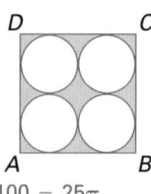
$100 - 25\pi$

30.
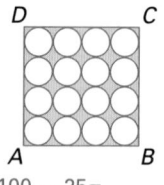
$100 - 25\pi$

31. Find the ratio of the area of a circle inscribed in an equilateral triangle to the area of the circle circumscribed about the same triangle. $\frac{1}{4}$

32. Find the ratio of the area of a circle inscribed in a square to the area of the circle circumscribed about the same square. $\frac{1}{2}$

33. The radius of the earth is approximately 6,400 km. If a rope were stretched around the earth at the equator, how long would the rope be? Suppose the rope were lifted 3 km off the surface of the earth all the way around. Approximately how much longer would the rope have to be? The radius of the moon is approximately 1,700 km. If the same experiment were repeated on the moon, how much longer would the rope be? 40,192 km; 19 km; 19 km

Prove or disprove the statement.

34. The ratio of the areas of two circles equals the square of the ratio of the lengths of the radii.

35. The ratio of the areas of two circles equals the ratio of their circumferences.

36. The ratio of the areas of two circles equals the ratio of the areas of their inscribed squares.

37. The ratio of the areas of two circles equals the ratio of the areas of their circumscribed squares.

Mixed Review

Find the degree measure of the arc. 11.4–11.7

1. m $\angle O = 42$, m$\overarc{AC}$ = ___42___
2. m $\angle ABC = 31$, m$\overarc{AC}$ = ___62___
3. m$\overarc{AC} = 64$, m$\overarc{ABC}$ = ___296___
4. m$\overarc{AB} = 108$, m$\overarc{BC} = 113$, m$\overarc{AC}$ = ___139___
5. Find the area of an equilateral triangle with a side 4 cm long. 12.4 $4\sqrt{3}$ cm^2

Enrichment

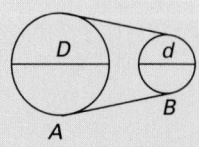

Challenge students to find a formula that will relate the number of rotations per minute (rpms) of pulleys D and d. First, all points on the belt move at the same rate. Second, the number of rotations the two pulleys make in a minute is not the same. When a point on the smaller pulley moves through a complete rotation (πd), the amount the larger pulley moves is a fraction of a complete rotation equal to $\frac{\pi d}{\pi D}$ or just $\frac{d}{D}$. So if the small pulley moves through x rotations in a minute, the number of rotations X the larger one makes will be $x(\frac{d}{D})$.

12.9 Measuring Arcs and Sectors of Circles

Objectives

To find the lengths of arcs
To find the areas of sectors and segments
To find the areas of annuli

The measure of an arc was defined as the measure of its corresponding central angle. The concept of arc measure may also serve to develop the meaning of the *length* of an arc of a circle. The diagram at right shows twelve arcs, each of which, like $\overarc{AB}$, has a degree measure of 30. One may think of $\overarc{AB}$ as having a length equal to $\frac{1}{12}$ the circumference of the circle.

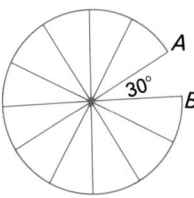

Definition

In a circle of radius r, an arc of degree measure m has **arc length** equal to $\frac{m}{360} \cdot 2\pi r$.

In other words, the length of the arc is that fraction $\left(\frac{m}{360}\right)$ of the circumference $(2\pi r)$ that corresponds to the arc.

EXAMPLE 1   A circle has a circumference of 24 cm. The degree measure of an arc of the circle is 36. What is the length of the arc?

Solution   Let l = length of the arc.
Then $l = \frac{m}{360} \cdot C$

$= \frac{36}{360} \cdot 24 = \frac{1}{10} \cdot 24 = 2.4$ cm

EXAMPLE 2   Given: Diameter $\overline{AB}$, $AB = 12$ cm, $m\overarc{BC} = 60$
Find the length of $\overarc{BC}$.

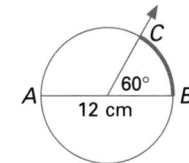

Solution   Let l = length of $\overarc{BC}$.
If $AB = 12$, then $r = 6$

$l = \frac{60}{360} \cdot 2\pi \cdot 6 = 2\pi$ cm

Additional Example 1

A circle has a circumference of 30 cm. The degree measure of an arc of the circle is 150. What is the length of the arc?   12.5 cm

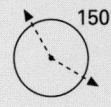

Additional Example 2

Given: Diameter $RS = 40$ cm,
  m $\overarc{TS} = 45$
Find the length of $\overarc{TS}$. 5π cm

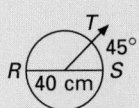

GETTING STARTED

Prerequisite Quiz

Find the circumference and the area of each circle. Leave answers in terms of π.

1. $d = 12$ cm $C = 12\pi$ cm, $A = 36\pi$ cm^2
2. $r = 70$ mm $C = 140\pi$ mm,
  $A = 4{,}900\pi$ mm^2
3. $r = 2\sqrt{3}$ m $C = 4\sqrt{3}\pi$ m, $A = 12\pi$ m^2

Solve.

4. $\frac{x}{8} = \frac{2}{5}$ $3\frac{1}{5}$
5. $56 = \frac{1}{2}(7x)$ 16

Motivator

Explain to the students that a portion of a circular region shaped like a piece of pie is called a *sector*. Ask them what part of the total area of the circle would each sector contain if they were to divide a circular region into four congruent sectors.
One-fourth   Ask them what the measure of the angle formed by the sides of the sector would be.   90   Ask them what the ratio of this measure is to 360.   One to four   Ask the class what part of the total area would each sector contain if they divided the circular region into 10 congruent pieces. $\frac{1}{10}$
Into n pieces. $\frac{1}{n}$

515

Lesson Note

The formulas in this lesson are based on earlier formulas for the circumference and area of a circle. Arc length is the product of circumference, $2\pi r$, and $\frac{m}{360}$, in which m is the degree measure of the arc.
The area of a sector is the product of the area of the circle, πr^2, and $\frac{m}{360}$.

The area of a segment of a circle is the difference between the area of a sector and a triangle formed by two radii and a chord. The area of an annulus is the difference between the areas of two concentric circles. Cardboard cutouts can be used effectively to illustrate these relationships.

Encourage students to see the logic of the formulas so that, instead of memorizing them, the students can derive the formulas whenever they are needed.

Math Connections

Home Repair: A washer or the cross section of a pipe can be compared to the concept of the annulus. Both are formed by two concentric circles.

Critical Thinking Questions:

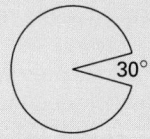 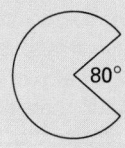

Application: Pacman's® mouth is formed by a central angle. On a video screen, his diameter is $\frac{3}{4}$ in. Find the area of Pacman when his mouth is open at an angle of 30 and when his mouth is open at an angle of 80. $\frac{33}{256}\pi \approx 0.40 \text{ in}^2$; $\frac{7}{64}\pi \approx 0.34 \text{ in}^2$

Two arcs may have the same lengths, but not the same degree measure. For example, in a circle with diameter 8, an arc of degree measure 90 has a length of 2π. In a circle with diameter 16, an arc of degree measure 45 has the same length, 2π.

Definition

> A **sector of a circle** is the region bounded by an arc of the circle and two radii whose endpoints are endpoints of that arc.

When you find the area of a sector, you actually are finding part of the area of the circle. For example, if two radii form an angle of 60, the area of the corresponding sector is $\frac{60}{360} = \frac{1}{6}$ of the area of the circle.

Definition

> In a circle of radius r, where a sector has an arc of degree measure m, the **area of the sector** is $\frac{m}{360} \cdot \pi r^2$.

In other words, the area of a sector is that fraction $\left(\frac{m}{360}\right)$ of the total area (πr^2) that corresponds to the sector.

EXAMPLE 3 A circle has circumference of 24π. Find the area of a sector with arc of degree measure 50.

Plan Use the circumference to compute the radius. Then use the formula for the area of a sector.

Solution

$$2\pi r = C$$
$$2\pi r = 24\pi$$
$$r = \frac{24\pi}{2\pi}$$
$$= 12$$

$$\text{area of sector} = \frac{m}{360} \cdot \pi r^2$$
$$= \frac{50}{360} \cdot \pi (12)^2$$
$$= \frac{7{,}200}{360}\pi = 20\pi$$

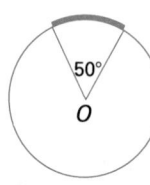

Theorem 12.16

> The area of a sector of a circle is one-half the product of the length s of the arc and the length r of its radius (Area $= \frac{1}{2}rs$).

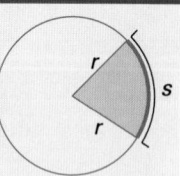

Additional Example 3

The length of a radius of a circle is 15 in. The degree measure of the arc of a sector is 120. Find the area of the sector.
$75\pi \text{ in}^2$

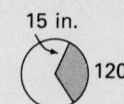

Additional Example 4

The radius of a circle is 9 in. If the area of a sector of this circle is 54 in², what is the length of the arc of the sector?
12 in

EXAMPLE 4 The length of a radius of a circle is 12 in. If the area of a sector of this circle is 36 in², what is the length of the arc of the sector?

Solution
$$Area = \frac{1}{2}rs$$
$$36 = \frac{1}{2} \cdot 12s$$
$$s = \frac{36}{6} = 6 \text{ in.}$$

Areas of several other regions can be found as parts of circles.

Definition A **segment of a circle** is a region bounded by a minor arc of the circle and the chord determined by the arc.

EXAMPLE 5 The length of a radius of a circle is 6 cm. Find the area of a segment bounded by an arc of degree measure 60 and its corresponding chord.

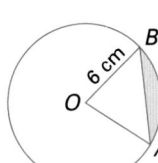

Plan From the area of the sector subtract the area of equilateral $\triangle AOB$. The sides of the triangle are the same length as the radius.

Solution
$$\text{Area(segment)} = \text{Area(sector)} - \text{Area}(\triangle AOB)$$
$$= \frac{m}{360}\pi r^2 - \frac{1}{2} \cdot 6 \cdot 3\sqrt{3}$$
$$= \frac{60}{360}\pi 6^2 - 9\sqrt{3}$$
$$= (6\pi - 9\sqrt{3}) \text{ cm}^2$$

Definition An **annulus** is a region bounded by two concentric circles.

EXAMPLE 6 Find the area of the annulus bounded by concentric circles with radii 12 cm and 8 cm long.

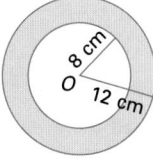

Plan Subtract the area of the smaller circle from the area of the larger circle.

Solution
Area(annulus) = Area(outer circle) − Area(inner circle)
Area(annulus) = $\pi \cdot 12^2 - \pi \cdot 8^2$
Area(annulus) = $144\pi - 64\pi$
$$= 80\pi \text{ cm}^2$$

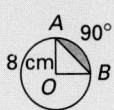
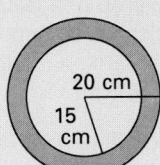
517

Guided Practice

Classroom Exercises 1–8

Independent Practice

A Ex. 1–16, **B** Ex. 17–29, **C** Ex. 30–36

Basic: WE 1–16 odd, Application
Average: WE 1–29 odd, Application
Above Average: WE 1–36 odd, Application

Additional Answers

Written Exercises

1. Yes; arcs of eq meas in same ⊙ have eq length.
2. No; arcs of uneq meas in same ⊙ have uneq length.
3. Yes; arcs of eq meas in ≅ ⊙ have eq length.
4. No; arcs of eq meas in non-≅ ⊙ have uneq length.
5. No; arcs of eq meas in non-≅ ⊙ have uneq length.
6. No; arcs of uneq meas in ≅ ⊙ have uneq length.

Classroom Exercises

What part of the area of a circle is a sector with the given degree measure of the arc?

1. 60 $\frac{1}{6}$ **2.** 12 $\frac{1}{30}$ **3.** 45 $\frac{1}{8}$ **4.** 120 $\frac{1}{3}$

What fractional part of the circumference is an arc with the given degree measure?

5. 30 $\frac{1}{12}$ **6.** 90 $\frac{1}{4}$ **7.** 120 $\frac{1}{3}$ **8.** 180 $\frac{1}{2}$

Written Exercises

$\odot O \cong \odot P$; $\odot O \not\cong \odot Q$. Do the arcs have equal length? Why or why not?

1. $m\overarc{AB} = 50$, $m\overarc{BC} = 50$
2. $m\overarc{DE} = 58$, $m\overarc{EF} = 59$
3. $m\overarc{AC} = 112$, $m\overarc{DF} = 112$
4. $m\overarc{AB} = 50$, $m\overarc{IH} = 50$
5. $m\overarc{GI} = 115$, $m\overarc{DF} = 115$
6. $m\overarc{BC} = 53$, $m\overarc{ED} = 54$

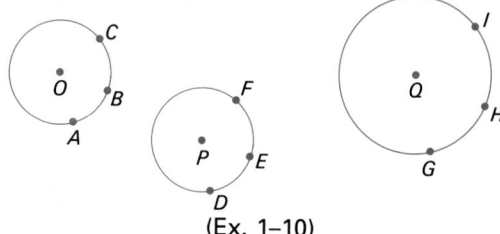

(Ex. 1–10)

Find the arc length.

7. Circumference = 24, $m\overarc{AB} = 60$ 4   **8.** Circumference = 36, $m\overarc{CD} = 45$ 4.5
9. Circumference = 18π, $m\overarc{EF} = 120$ 6π **10.** Circumference = 25π, $m\overarc{GH} = 36$ 2.5π

Find the arc length and the area of the indicated sector.

11. $m\overarc{AB} = 60$, $r = 10$ m   **12.** $m\overarc{AB} = 90$, $r = 4$ cm
13. $m\overarc{AB} = 120$, $r = 8$ ft   **14.** $m\overarc{AB} = 180$, $r = 9$ in.
15. $m\overarc{AB} = 210$, $r = 3$ mm   **16.** $m\overarc{AB} = 270$, $r = 12$ cm

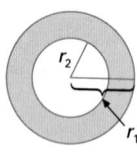

11. $\frac{10}{3}\pi$ m, $\frac{50}{3}\pi$ m^2 **12.** 2π cm, 4π cm^2 **13.** $\frac{16}{3}\pi$ ft, $\frac{64}{3}\pi$ ft^2
14. 9π in, 40.5π in^2 **15.** $\frac{7}{2}\pi$ mm, $\frac{21}{4}\pi$ mm^2 **16.** 18π cm, 108π cm^2

Find the area of the annulus bounded by concentric circles with the given radii.

                  75π cm^2 108π cm^2
17. $r_1 = 10$ cm, $r_2 = 5$ cm   **18.** $r_1 = 12$ cm, $r_2 = 6$ cm
19. $r_1 = 9$ mm, $r_2 = 3$ mm   **20.** $r_1 = 5$ mm, $r_2 = 2$ mm
    72π mm^2 21π mm^2

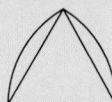

Enrichment

The Gothic arch illustrated in the diagram is formed from two arcs, each of which is one-sixth of a circle. The centers of the circles are at opposite ends of the base of the arch. Find the area of an arch with radius 6 m^2.

The area of one sector is one-sixth the area of a circle with radius 6 m, that is,

6π m^2. The area of the remaining segment is the area of the sector less the area of the triangle. The total area of the arch is thus: $6\pi + (6\pi - 9\sqrt{3})$ m^2.

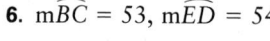

Find the missing length of arc s, length of radius r, or area of sector A for the indicated circle.

21. $s = 4$ ft, $r = 3$ ft, $A =$ ___6 ft²___   22. $s = 8$ m, $A = 48$ m², $r =$ ___12 m___
23. $A = 1$ in², $r = 8$ in., $s =$ _____ ¼ in   24. $A = 2$ ft², $r = 24$ in., $s =$ ___2 ft___

$\overline{OA}$ and $\overline{OB}$ are radii of $\odot O$. Find the missing measures.

	m $\angle O$	r	m$\widehat{AB}$	Area (sector AB)	Area (segment AB)
25.	___	6	60	___	___
26.	120	12	___	___	___
27.	___	10	___	25π sq units	___
28.	___	___	120	24π sq units	___

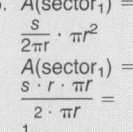

29. Prove Theorem 12.16.

Find the area of the shaded region if the length of the side of the square is 10. (Curved lines are arcs of circles.)

30.
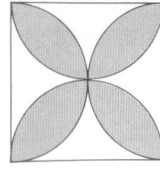
$(50\pi - 100)$ sq units

31.
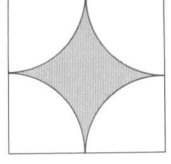
$(100 - 25\pi)$ sq units

32.
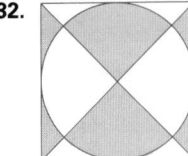
50 sq units

33. Write a formula for the area of an annulus in terms of radii r_1 and r_2, where $r_1 \geq r_2$. $\pi(r_1{}^2 - r_2{}^2)$

C 34. Find, to the nearest hundredth, the area of a segment of a circle with 10-cm radius and an arc of degree measure 54. (HINT: Use appropriate trigonometric ratios.)   6.65 cm²

C 35. Find, to the nearest tenth, the area of a segment of a circle with a 16-cm radius and an arc of degree measure 137.   218.6 cm²

36. Write a formula for the area of a segment of a circle.
$r^2 \left[\frac{m}{360}\pi - \left(\sin \frac{m}{2}\right)\left(\cos \frac{m}{2}\right) \right]$

Mixed Review

1. Find the area of a regular hexagon with sides 12 in. long.   *12.6* $216\sqrt{3}$ in²
2. Find the area of a trapezoid with a 12-cm upper base, a 16-cm lower base, and a 5-cm altitude.   *12.5*   70 cm²
3. What is the distance between two points with coordinates -9 and 17?   *1.2*   26
4. The measure of an angle is 50 more than its supplement. Find the measure of the angle.   *1.6*   115

---

25. 60, 6π sq units, $6\pi - 9\sqrt{3}$ sq units
26. 120, 48π sq units, $48\pi - 36\sqrt{3}$ sq units
27. 90, 90, $25\pi - 50$ sq units
28. 120, $6\sqrt{2}$, $24\pi - 18\sqrt{3}$ sq units

29.

Statement	Reason
1. $\odot P$ with radius r and arc of meas m with arc length s	1. Given
2. $s = \frac{m}{360} \cdot 2\pi r$	2. Def of arc length
3. $\frac{s}{2\pi r} = \frac{m}{360}$	3. Div Prop of Eq
4. Let sector$_1$ of $\odot P$ be bounded by arc s and radii with endpts on s.	4. Def of sector of a $\odot$
5. A(sector$_1$) = $\frac{m}{360} \cdot \pi r^2$	5. Def of area of sector
6. A(sector$_1$) = $\frac{s}{2\pi r} \cdot \pi r^2$ A(sector$_1$) = $\frac{s \cdot r \cdot \pi r}{2 \cdot \pi r}$ = $\frac{1}{2}rs$	6. Sub

Example 1

The aerial photograph below, which was taken from an altitude of 20,000 ft, shows a piece of lakefront property which is being considered for development. How can the photograph be used to estimate the area of the land in acres?

First, it is necessary to know the scale ratio. In the enlargement below, 1 in. = 1,650 ft or .3125 mi. Therefore, one square inch represents $(.3125)^2$ or .098 mi^2.

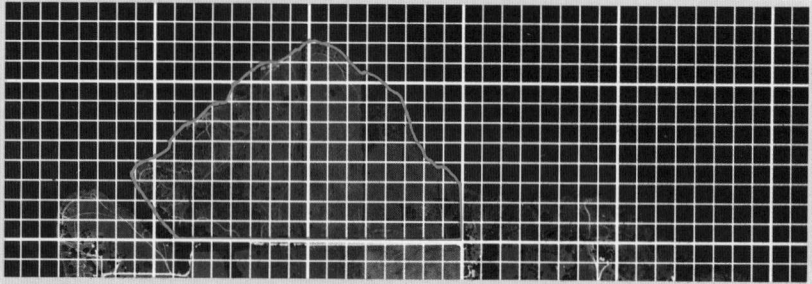

Place a grid over the photograph as shown and count the number of squares that fall inside the boundary of the property. Then count the number of squares that straddle the boundary and divide that number by 2:

$$105 + (24 \div 2) = 117 \text{ squares}$$

The grid shown has 64 squares to the square inch, so the area of each square is $\frac{1}{64}$ in^2. Thus, an estimate of the area of the property as shown in the photograph is $117 \times \frac{1}{64} \approx 1.83$ in^2. Since 1 in^2 represents .098 mi^2, the actual area of the property is about $1.83 \times .098 \approx .179$ mi^2. There are 640 acres in a square mile, so the estimated area in acres of the property is $.179 \times 640 \approx 115$ acres.

Other methods are used to approximate irregular areas. In one laboratory, the weight of a plywood cutout of a shape is compared with the weight of a piece of plywood of known area.

Example 2

Another method used to approximate area is the *trapezoid method*. To find the approximate area of $ABCD$ below, divide $\overline{AB}$ into n congruent segments each d units long. Draw parallel segments across the region at these points. Each region is approximately a trapezoid.

$$\text{Area} = \tfrac{1}{2}(b_1 + b_2)d + \tfrac{1}{2}(b_2 + b_3)d + \cdots + \tfrac{1}{2}(b_{n-1} + b_n)$$
$$= d(\tfrac{1}{2}b_1 + b_2 + b_3 + \cdots + b_{n-1} + \tfrac{1}{2}b_n)$$

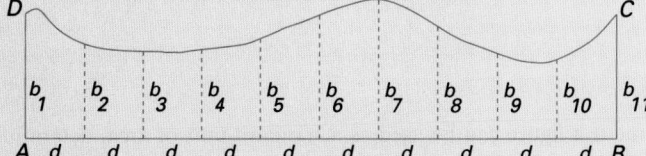

Use the trapezoid method to estimate the area of the region shown at right.

$$\text{Area} = \tfrac{12}{5}[\tfrac{1}{2}(0) + 6 + 10 + 11 + 12 + \tfrac{1}{2}(12)]$$
$$= \tfrac{12}{5}(45)$$
$$= 108 \text{ sq units}$$

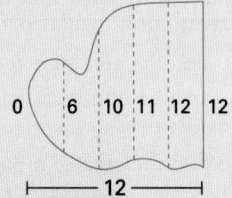

A plowed field is outlined in the photograph below. If 1 ton of limestone per acre is required to treat the soil for planting hay, what is the weight of limestone required for the field shown? Use two different methods of area approximation and compare the results. (The scale ratio is the same as in Example 1.) Accept estimates close to 89 tons.

Chapter Review

1.

90°	90°	
90°	90°	

135° 135°
90° 135°
90° 135°
90° 135°
135° 135°

Chapter 12 Review

Key Terms

altitude (p. 483)	length of an arc (p. 515)
annulus (p. 517)	limit (p. 510)
apothem of a regular polygon (p. 499)	perimeter (p. 499)
area of a circle (p. 512)	pi (π) (p. 511)
area of a polygon (p. 478)	polygonal region (p. 478)
area of a sector (p. 516)	radius of a regular polygon (p. 499)
base (p. 483)	sector of a circle (p. 516)
center of a regular polygon (p. 498)	segment of a circle (p. 517)
circumference of a circle (p. 510)	standard unit (p. 475)
Heron's Formula (p. 490)	triangular region (p. 478)
kite (p. 489)	

Key Ideas and Review Exercises

12.1   To determine if a figure can be used as a standard unit of area, determine if it can completely cover a plane surface without overlapping.

1. Make a sketch to show that a rectangle can be used as a standard unit of area, but a regular octagon cannot be used.

12.2   To find the area of a rectangle, use the formula: Area = bh.

2. What is the area of a rectangle with a 3.5-cm base and a 4.7-cm altitude?   16.45 cm²

12.3   To find the area of a parallelogram, use the formula: Area = bh.

3. What is the area of a parallelogram with a 7.9-mm base and a 2.3-mm altitude?   18.17 mm²

12.4   To find the area of a triangle, use the formula: Area = $\frac{1}{2}bh$.

4. Find the area of a triangle with a 4.6-in. base and 2.3-in. altitude.   5.29 in²
5. Find the area of a right triangle with sides of length 5, 12, and 13.   30 sq units

12.5   To find the area of a trapezoid, use the formula: Area = $\frac{1}{2}h(b_1 + b_2)$.

6. Find the area of a trapezoid with a 9-cm lower base, a 17-cm upper base, and a 6-cm altitude.   78 cm²
7. Find the area of an isosceles trapezoid with bases of length 10 and 20 and a base angle measuring 60. $75\sqrt{3}$ sq units

12.6   To find the perimeter of a regular polygon, add the lengths of the sides.
To find the area of a regular polygon, use the formula: Area = $\frac{1}{2}ap$.

8. Find the perimeter of a regular nonagon with sides 4 m long.   36 m
9. Find the area of a regular octagon with a 160-in. perimeter and an apothem of 24.14 in.   1,931.2 in²

12.7   To find the ratio of the perimeters of similar polygons, find the ratio of the lengths of corresponding sides.

To find the ratio of the areas of two similar polygons, find the square of the ratio of the lengths of corresponding sides.

10. The lengths of two corresponding sides of two similar octagons are 8 cm and 12 cm. If the perimeter of the first is 70 cm, what is the perimeter of the second?   105 cm

11. The lengths of two corresponding sides of two similar hexagons are 12 mm and 15 mm. If the area of the first hexagon is 40 mm², what is the area of the second?   62.5 mm²

12.8   To find the circumference of a circle, use the formula $C = \pi d$. To find the area of a circle, use the formula: $A = \pi r^2$.

12. What is the circumference of a circle with a 5-in. radius? Use 3.14 as an approximation of π.   31.4 in

13. What is the area of a circle with a 3-in. radius? Use 3.14 as an approximation of π.   28.26 in²

12.9   To find the length of an arc of a circle with measure m, use $l = \dfrac{m}{360} \cdot 2\pi r$.

To find the area of a sector of a circle with an arc of degree measure m, use

$A = \dfrac{m}{360} \cdot \pi r^2$.

To find the areas of the segments of circles and annuli, subtract appropriate portions of the circle.

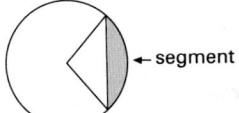
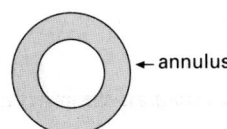

$A(\text{segment}) = A(\text{sector}) - A(\text{triangle})$ $A(\text{annulus}) = A(\text{outer } \odot) - A(\text{inner } \odot)$

14. What is the length of an arc, with measure 30, of a circle with a 12-cm radius? 2π cm

15. Find the area of a sector of a circle with a 10-in. radius determined by an arc of degree measure 30. $\dfrac{25}{3} \pi$ in²

16. Find the area of a segment of a circle with a 6-cm radius determined by an arc of degree measure 60. $6\pi - 9\sqrt{3}$ cm²

17. Find the area of an annulus bounded by concentric circles with the radii of 10 cm and 8 cm. 36π cm²

Find the area of the figure. (Items 1–12)

1. rectangle *ABCD*

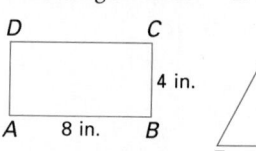

32 in²

2. parallelogram *EFGH*

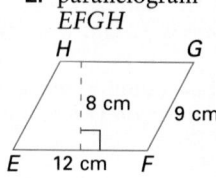

96 cm²

3. triangle *IJK*

12 m²

4. triangle *LMN*

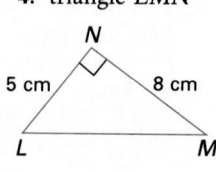

20 cm²

5. rhombus *OPQR*

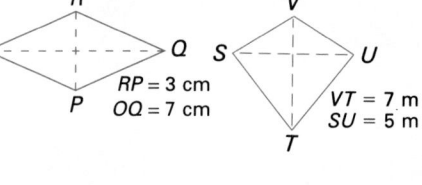

RP = 3 cm
OQ = 7 cm

10.5 cm²

6. kite *STUV*

VT = 7 m
SU = 5 m

17.5 m²

7. equilateral △*WXY*

9 mm

$\frac{81}{4}\sqrt{3}$ mm²

8. trapezoid *ZABC*

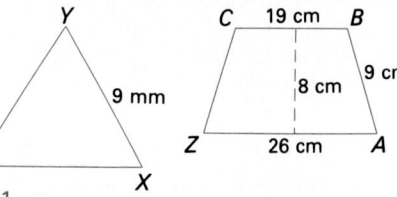

180 cm²

9. trapezoid *DEFG*

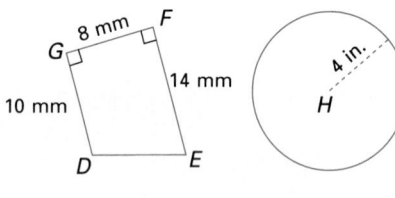

96 mm²

10. circle *H*

4 in.

16π in²

11. regular octagon *IJKLMNOP*

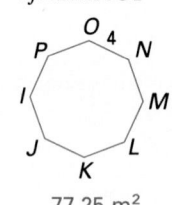

77.25 m²

12. sector *QRS*

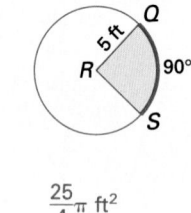

90°
5 ft

$\frac{25}{4}\pi$ ft²

13. Find the area of a square with 9-in. sides.   81 in²

14. Find the area of a circle with a 12-cm diameter.   36π cm²

15. Find the area of an annulus bounded by circles with 10-cm and 12-cm diameters.   44π cm²

16. Find the area of a regular octagon with an apothem of 10 m and a side length of 8.28 m.   331.2 m²

17. Find the area of a segment of a circle with 10-mm radii determined by an arc of degree measure 60. $(\frac{50}{3}\pi - 25\sqrt{3})$ mm²

18. Find the length of an arc measuring 60 in a circle with 12-cm radii.   4π cm

19. The circumference of a circle is 8π. Find the perimeter of an inscribed square.   16√2 units

1. The area of square *DEFH* is 64, and
D the area of square AHGK is 25.
Find the area of square *ABCD*.
(A) 9
(B) 89
(C) 144
(D) 169
(E) 225

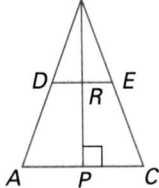

2. In $\triangle ABC$, $BR = RP$, $AC = 12$,
A $RP = 4$. The area of $ACED = 40$.
Find the area of $\triangle DEB$.
(A) 8
(B) 10
(C) 12
(D) 14
(E) 16

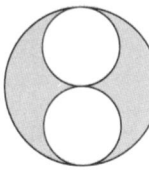

3. In $\odot O$, $OA = 6$ and $\overline{AO} \perp \overline{OB}$
D Find the area of the shaded portion.
(A) 2π
(B) $\pi - 2$
(C) $6\pi - 9\sqrt{3}$
(D) $9\pi - 18$
(E) $36\pi - 9\sqrt{3}$

4. The larger circle has a
B diameter of 4 ft. The
smaller circles are con-
gruent and touch at
the center of the larger
circle. Find the area of
the shaded portion.
(A) π (B) 2π (C) 4π
(D) 6π (E) 8π

5. A picture is 6 ft wide by 8 ft long. If
A the frame has a width of 6 in., what
is the ratio of the area of the frame
to the area of the picture?
(A) $\frac{5}{16}$ (B) $\frac{5}{4}$ (C) $\frac{4}{5}$
(D) $\frac{5}{12}$ (E) $\frac{16}{5}$

6. On each side of a square, an isosce-
D les right triangle is drawn. If the pe-
rimeter of the square is 16, what is
the area of the shaded portion?
(A) 4
(B) 8
(C) 12
(D) 16
(E) 32

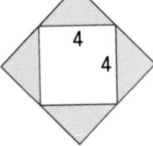

7. Given: $\overline{AB} \parallel \overline{CD}$, $DC = 4$, $AB = 12$,
B m $\angle A$ = m $\angle B$ = 45. Find the area
of trapezoid *ABCD*.
(A) 16
(B) 32
(C) $16\sqrt{2}$
(D) $32\sqrt{2}$
(E) 64

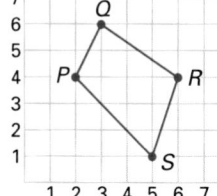

8. A rectangular field is half as wide as
C it is long. It is enclosed by x yards
of fencing. What is the area of the
field?
(A) $\frac{x^2}{2}$ (B) $2x^2$ (C) $\frac{x^2}{18}$
(D) $3\frac{1}{2}$ (E) $\frac{x^2}{72}$

9. What is the area of *PQRS*?
E (A) 5
(B) 8
(C) 20
(D) 16
(E) 10

5.

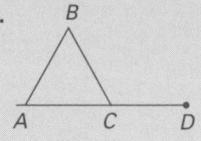

Statement	Reason
1. m ∠ACB + m ∠DCB = 180.	1. Def of st ∠
2. m ∠A + m ∠B + m ∠ACB = 180.	2. Sum of meas of ∠s of △ = 180.
3. m ∠ACB + m ∠DCB = m ∠A + m ∠B + m ∠ACB	3. Trans Prop
4. m ∠DCB = m ∠A + m ∠B	4. Subtr Prop of Eq

6.

Statement	Reason
1. $\overline{DB}$ and $\overline{AE}$ bis each other.	1. Given
2. AC = CE, BC = CD	2. Def of bis
3. m ∠BCA = m ∠DCE	3. Vert ∠s are ≅.
4. △ACB ≅ △ECD	4. SAS

7.

Statement	Reason
1. AD = CE, AB = CB, DB = EB	1. Given
2. AB + EB = CB + DB	2. Eq may be added
3. AC = AC	3. Reflex Prop
4. AB + BE = AE, CB + DB = CD	4. Seg Add Post
5. △ACD ≅ △CAE	5. SSS

Cumulative Review *(Chapters 1–12)*

1. State whether true or false.   *1.2*
  (A) AB = BC   F
  (B) AB + BC = CD   T
  (C) BD = −7   F

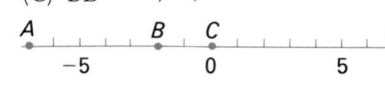

2. Based on the diagram,   *1.4, 1.6*
  (A) m ∠AOC = __110__.
  (B) m ∠DOC = __70__.
  (C) supp of ∠DOX is ∠ _AOX_.

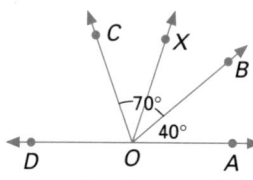

3. Based on the diagram, find: *1.8*
  (A) m ∠COD.   45
  (B) m ∠COB.   135
  (C) m ∠XOD.   150

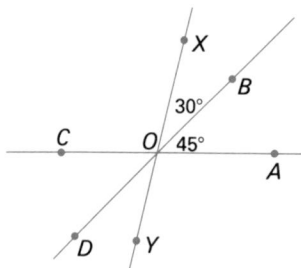

4. *m* and *n* are parallel, find: *3.2*
  (A) m ∠5.   70   (B) m ∠2.   70
  (C) m ∠4.   110   (D) m ∠3.   110

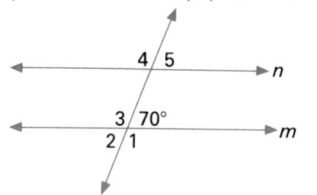

5. Prove that in any triangle the   *3.7*
measure of an exterior angle is
equal to the sum of the measures of the two remote interi-
or angles.

6. Given: $\overline{DB}$ and $\overline{AE}$ bisect each   *4.3*
other.
  Prove: △ACB ≅ △ECD

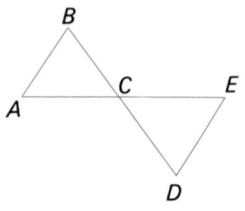

7. Given: AD = CE, AB = CB,   *5.3*
DB = EB
  Prove: △ACD ≅ △CAE

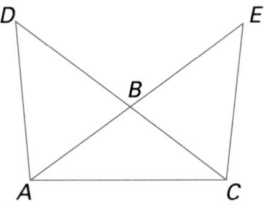

8. If △ABC is equilateral and the   *5.5*
length of each side is 10, find
the length of the altitude.   5√3

9. For each of the following, state   *5.8*
whether a triangle can be
formed. If so, is it right, ob-
tuse, or acute? Sides are:
  (A) 2, 3, 2.   Yes, obtuse
  (B) 7, 8, 17.   No
  (C) 6, 8, 10.   Yes, right
  (D) 4, 4, 4.   Yes, acute

10. What is the sum of the mea-   *6.2*
sures of the interior angles of a
pentagon?   540

11. State whether true or false: *7.3*
    (A) Every rhombus is a square.   F
    (B) Every square is a rhombus.   T
    (C) Every trapezoid can be divided into two con-gruent triangles.   F

12. Given: $\overline{AB} \parallel \overline{CD}$ *8.3*
    Prove: $\triangle AXB \sim \triangle CXD$

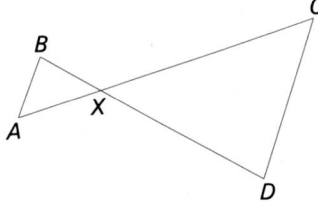

13. Given the triangles in Exercise 12, complete the proportion: *8.4*
$\dfrac{AB}{CD} = \dfrac{BX}{?}$.   *XD*

14. For the triangle below, find: *9.5, 9.6*
    (A) $\sin \angle A$. $\frac{4}{5}$ (B) $\cos \angle A$. $\frac{3}{5}$
    (C) $\tan \angle B$. $\frac{3}{4}$

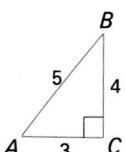

15. Find the midpoint of the segment joining the points $A(2,3)$ and $B(-4,-7)$. $(-1,-2)$ *10.2*

16. What is the slope of the line whose equation is $2y + 2x = 5$? -1 *10.3*

17. Find m$\overset{\frown}{AB}$, for $\odot O$.   90   *11.4*

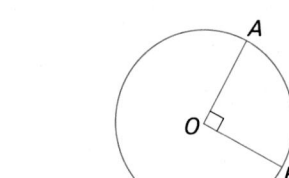

18. Find m$\overset{\frown}{AB}$.   100   *11.6*

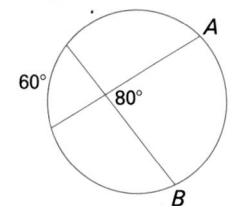

19. $\overline{OA}$ is a tangent and $\overline{OB}$ is a secant. Find m $\angle AOB$.   76   *11.8*

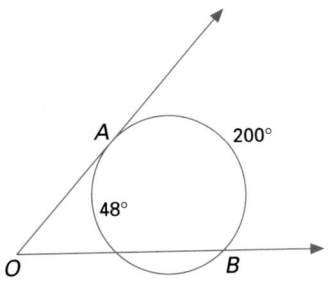

20. Find the area of a triangle with base 12 ft and altitude 6 ft.   36 ft^2 *12.4*

21. Find the area of the trapezoid.   80.5 m^2 *12.5*

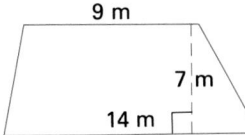

22. Find the area of the shaded portion if the measure of the central angle is 120.   12π in^2 *12.9*

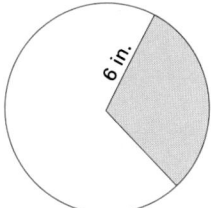

12.	Statement	Reason
	1. $AB \parallel CD$	1. Given
	2. m $\angle BAX =$ m $\angle DCX$	2. Alt int $\angle$s
	3. m $\angle BXA =$ m $\angle DXC$	3. Vert $\angle$s
	4. $\triangle AXB \sim \triangle CXD$	4. AA Post

13 LOCI

OVERVIEW

In this chapter, students will sketch and describe loci in space. They locate and use incenters and circumcenters of triangles, and construct orthocenters and circumcenters of triangles. They will consider concurrency of bisectors of angles, and of perpendicular bisectors of sides, altitudes, and medians.

OBJECTIVES

- To sketch and describe loci in a plane satisfying one condition
- To sketch and describe loci in a plane satisfying two or more conditions
- To sketch and describe loci in space
- To construct the incenter and centroid of a triangle
- To apply altitude and median concurrency theorems

PROBLEM SOLVING

To describe a locus of points that satisfy given conditions, the problem solving strategy, Drawing a Diagram, can be used. The Written Exercises in Lessons 13.1, 13.2, and 13.3 ask students to sketch the locus of points and then describe its location. The problem solving strategy, Guess and Check, can be used to solve the Application on page 540.

TECHNOLOGY

Computer: The Computer Investigation on page 546 enables students to discover Theorem 13.7 on the properties of concurrency for medians of a triangle. This lesson can be extended to explore altitude concurrency as well.

Calculator: Encourage students to use a calculator to facilitate computations for the Written Exercises in Lesson 13.5.

SPECIAL FEATURES

Mixed Review pp. 532, 537, 545, 550
Midchapter Review p. 540
Application: Site for a House p. 540
Focus on Reading p. 544
Computer Investigation: Concurrence in Triangles p. 546
Brainteaser p. 550
Key Terms p. 551
Key Ideas and Review Exercises p. 551
Chapter 13 Test p. 552
College Prep Test p. 553

PLANNING GUIDE

Lesson		Basic	Average	Above Average	Resources
13.1	pp. 531–532	CE all WE 1–7	CE all WE 1–18	CE all WE 1–23	Reteaching p. 189 Practice p. 190
13.2	pp. 536–537	CE all WE 1–10	CE all WE 1–24 odd	CE all WE 1–30 odd	Reteaching p. 191 Practice p. 192
13.3	pp. 539–540	CE all WE 1–7 Midchapter Review Application	CE all WE 1–11 Midchapter Review Application	CE all WE 1–16 Midchapter Review Application	Reteaching p. 193 Practice p. 194
13.4	pp. 544–546	FR all CE all WE 1–8 CI 1	FR all CE all WE 1–15 CI all	FR all CE all WE 1–20 CI all	Reteaching p. 195 Practice p. 196
13.5	pp. 549–550	CE all WE 1–12 Brainteaser	CE all WE 1–19 odd Brainteaser	CE all WE 1–23 odd Brainteaser	Reteaching p. 197 Practice p. 198
Chapter 13 Review p. 551		1–9 odd	all	all	
Chapter 13 Test p. 552		1–6, 8, 10–14	1–14	all	
College Prep Test p. 553		all	all	all	

CE = Classroom Exercises   WE = Written Exercises   FR = Focus on Reading   AR = Algebra Review   CI = Computer Investigation

NOTE: For each level, all students should be assigned all Mixed Review exercises.

INVESTIGATION

Manipulative: This informal hands-on investigation of the medians and angle bisectors of a triangle helps the students to discover how these lines relate.

Use this investigation to introduce Chapter 13 or before Lesson 13.4.

Students can work individually or in groups on this investigation. Cooperative learning in groups will help students teach each other and formulate their discoveries in words.

Students need a copy of Manipulative Worksheet 25, scissors, a straightedge, a compass, paper, and pencils or pens of two different colors.

(Note that Manipulative Worksheet 26 provides a related investigation of the altitudes of a triangle.)

After the students complete the worksheet, encourage them to compare their results with those obtained by other groups.

Lead them to the conjecture that these statements are probably true for all triangles, although this experimental approach does not *prove* the conclusions written in the summary statements.

The discussion that follows completion of the worksheet gives the students practice in using some of the vocabulary of this chapter: median, center of gravity, centroid, concurrent, angle bisector, tangent, incenter, inscribed circle.

$$F(n) = 2^{2^n} + 1$$

Karl Friedrich Gauss (1777–1855)—Gauss, the son of uneducated parents, is widely regarded as one of the three or four greatest mathematicians of all time. At age three, he corrected an arithmetic mistake in his father's payroll.

More About the Mathematician

When Gauss was nineteen, he solved this 2000-year old problem: Which regular polygons with an odd number of sides can be constructed with only a straightedge and compass? The Greeks could construct such a polygon only for 3 and 5 sides. Gauss proved that the next one that can be constructed has 17 sides. The formula for finding such polygons is shown here. Gauss also developed the $a + bi$ form for complex numbers (called Gaussian integers) and the normal curve. Publication of his work, *Theory of the Motion of the Celestial Bodies*, led to his appointment as the director of the new Göttingen observatory. The telescope he used there is shown in the picture.

13.1 Finding Locations

Teaching Resources

Manipulative Worksheet 25
Quick Quizzes 95
Reteaching and Practice
  Worksheets, pp. 189, 190

Objective   To sketch and describe loci

In baseball, a pitch is considered a strike if it passes over any part of home plate at a height between the knee and the shoulder of the batter. Since home plate is 17 in. wide, the strike zone is a rectangle 17 in. wide with a height depending on the batter's height.

Definition   A **locus** is the set of all points that satisfy a given set of conditions. (The plural of *locus* is *loci*.)

EXAMPLE 1   Find the locus of points in a plane exactly 2 cm from a given point *P*. Give a reason for your answer.

Solution   (1) Draw point *P*.
(2) Locate several points 2 cm from *P*.
(3) Draw a smooth curve through these points.

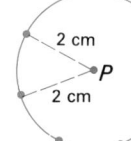

The locus is a circle with *P* as its center and a 2-cm radius.

EXAMPLE 2   Find the locus of points in a plane less than 3 cm from a given point.

Solution   The locus is the interior of a circle with point *Q* as the center and with a 3-cm radius. (A dashed circle is drawn to show that the points of the circle are not included in the locus.)

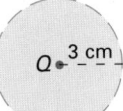

To determine a locus: (1) Start with the given figure.
(2) Locate several points that satisfy the specified conditions.
(3) Draw a smooth curve through the points.
(4) Describe the figure and its location.

13.1 Finding Locations   **529**

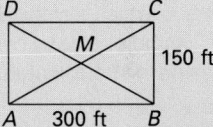

Additional Example 1

In the coordinate plane, find the locus of points that are 5 units from the origin. The locus is a circle with the origin as the center and with a radius of 5 units.

Additional Example 2

Find the locus of points in a plane that are more than 1 cm from a given point *P*. The locus is the exterior of a circle with point *P* as the center and with a 1 cm radius.

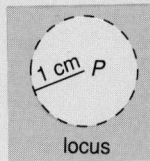

locus

■ TEACHING SUGGESTIONS

Lesson Note

Students often find it difficult to understand why the proof of a Locus Theorem has two parts. Point out that it is necessary to prove that any point which satisfies the given condition must be on the locus, and conversely, any point on the locus must satisfy the given condition.

Math Connections

Traffic Laws: Many traffic laws describe a locus of points in which a vehicle may park or drive. For example, unless changing lanes or crossing a street, automobiles on a public road are required by law to remain in their lanes. Parking laws prohibit vehicles from parking less than a specific distance from a fire hydrant or an intersection.

Critical Thinking Questions

Analysis: Ask students to draw a 4–inch long line segment, $\overline{AB}$, and find the locus of all points P such that $AP + PB = 5$ in. Sample values for AP and PB are shown below.

$$5 = AP + PB$$
$$5 = 0.5 + 4.5$$
$$5 = 1 + 4$$
$$5 = 2 + 3$$
$$5 = 2.5 + 2.5$$
$$5 = 3 + 2$$
$$5 = 4 + 1$$
$$5 = 4.5 + 0.5$$

The locus is an ellipse.

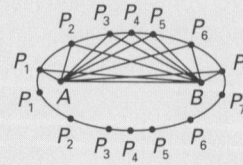

EXAMPLE 3   Sketch and describe the locus of points in a plane that are equidistant from two given parallel lines m and n.

Solution   (1) Draw two parallel lines m and n.

(2) Locate several points the same distance from m as from n.
(Distance is measured along the perpendicular from a point to the line.)

(3) Draw a line through these points.

The locus is a line parallel to m and n and halfway between them.

Drawing a picture does not prove that a locus is correct. To establish a locus, it is necessary to prove that the locus contains all points, and only those points, that satisfy the given conditions. The statement of a locus theorem is a biconditional. Its proof involves two parts.

Theorem 13.1   The **locus of points** in a plane equidistant from two given points is the perpendicular bisector of the segment having the two points as endpoints.

Part I: If a point lies on the perpendicular bisector of a segment, then it is equidistant from the endpoints of the segment.
Given: P is on the perpendicular bisector of $\overline{AB}$.
Prove: $PA = PB$

Part II: If a point is equidistant from the endpoints of a segment, then it lies on the perpendicular bisector of the segment.
Given: $PA = PB$
Prove: P is on the perpendicular bisector of $\overline{AB}$.

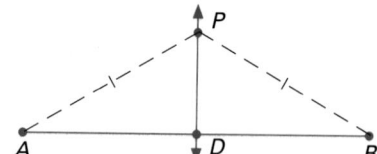

Additional Example 3

Sketch and describe the locus of points in a plane that are 5 cm from a given line k. The locus is a pair of lines, each parallel to line k and at a distance of 5 cm from k.

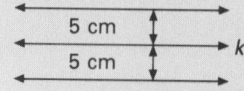

Theorem 13.2 In a plane, the locus of points equidistant from the sides of an angle is the bisector of the angle.

Part I: If a point is on the bisector of an angle, then it is equidistant from the sides of the angle.

Given: $\overrightarrow{OC}$ bisects $\angle AOB$;
P is on $\overrightarrow{OC}$; $\overline{PD} \perp \overrightarrow{OA}$,
$\overline{PE} \perp \overrightarrow{OB}$.
Prove: $PD = PE$

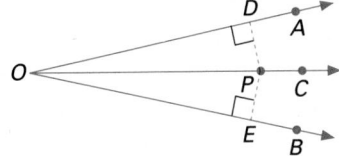

Plan

The distance from a point to a line is measured along the perpendicular to the line. Show that $\overline{PD}$ and $\overline{PE}$ are congruent corresponding sides of congruent triangles ODP and OEP.

Part II: If a point is equidistant from the sides of an angle, then it is on the bisector of the angle.

Given: P is in the interior of $\angle AOB$; $\overline{PD} \perp \overrightarrow{OA}$,
$\overline{PE} \perp \overrightarrow{OB}$, and $PD = PE$.
Prove: $\overrightarrow{OP}$ bisects $\angle AOB$.

Plan

To show that $\overrightarrow{OP}$ is the bisector, show that $\angle DOP$ and $\angle EOP$ are congruent corresponding angles of congruent triangles ODP and OEP.

Classroom Exercises

Describe the locus of points in a plane. (Exercises 1–4)

1. equidistant from two parallel lines that are 10 cm apart
2. 7 cm from a given point
3. equidistant from two points 8 cm apart
4. less than 5 mm from a given point
5. Explain the meaning of the word *locus*.

Written Exercises

Sketch and describe the given locus of points in a plane. (Exercises 1–14)

1. 5 cm from a given point
2. less than 4 cm from a given point
3. greater than 3 cm from a given point
4. equidistant from two points that are 4 cm apart

Enrichment

Ask the students to describe the following loci:
1. The locus of points equidistant from the two points $A(4,0)$ and $B(10,0)$ in the Cartesian coordinate system.
   The perpendicular bisector of the line segment connecting them. It passes through $(7,0)$; its equation is $x = 7$.

2. What is the locus of points that are equidistant from the two longer sides of a rectangle? The locus is the segment whose endpoints are the midpoints of the two shorter sides of the rectangle.

Common Error Analysis

Error: Sketching a locus may be difficult for some students.

Emphasize the need to locate several points close enough together, in order to see how the entire locus should be completed. It is also helpful to use two colors to distinguish between the given figure and the locus.

Checkpoint

Sketch and describe each of the following loci in a plane.

1. The locus of points less than 6 cm from a given point. The locus is the interior of a circle with the given point as center and with a 6 cm radius.

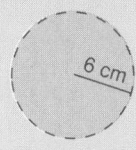

2. The locus of points in the interior of a square that are equidistant from a pair of adjacent sides of a square. The locus is the pair of diagonals of the square, excluding their endpoints.

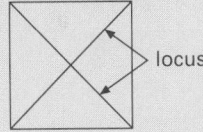

Closure

Ask the students to describe a locus that is a line. Example 3 page 530 Ask them to describe a locus that is a circle. Example 1 page 529

◤◤◤◤ FOLLOW UP

Guided Practice

Classroom Exercises 1–5

Independent Practice

A Ex. 1–7, **B** Ex. 8–18, **C** Ex. 19–23

Basic: WE 1–7
Average: WE 1–18
Above Average: WE 1–23

Additional Answers

Classroom Exercises

1. Line ∥ to the given lines, 5 cm from each, halfway between them.
2. ⊙ with a given point as ctr and radius of 7 cm.
3. Line ⊥ to the seg connecting the pts 4 cm from each pt.
4. Int of a ⊙ with ctr at the given point and radius of 5 mm.
5. A locus is a set of all pts that satisfy a given condition or conditions.

Written Exercises

1. ⊙ with radius of 5 cm.
2. All pts in the int of a ⊙ with radius of 4 cm.
3. All pts in the ext of a ⊙ with radius of 3 cm.
4. ⊥ bis of seg joining the two pts.
5. The bis of the ∠ through the vertex of the rt ∠.
6. The ∠ bis.
7. Two ∥ lines 3 cm from the given line.
8. The region between two ∥ lines that are 8 cm apart, excluding the lines.
9. The region outside two ∥ lines that are 14 cm apart, excluding the lines.
10. The region that is closer to the pt A on one side of the ⊥ bis of $\overline{AB}$, excluding the line which makes up the bis.
11. The region that is closer to side $\overrightarrow{OB}$ on one side of the ∠ bis of ∠AOB, excluding the line which makes up the bis.
12. A line at a distance from the given line equal to the radius of the ⊙ and ∥ to the line.
13. A ⊙ concentric to the int ⊙ and having a radius that is the sum of the radii of the two given circles.
14. The pt in the int of the square where its diagonals intersect.
15. The boundary is a pair of ∥ lines that are 20 ft apart, each 10 ft from the power line.
16. The wire is laid in a line 12 ft from each edge of the lane.

17.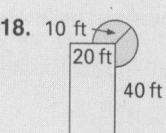
    10 ft

18. 10 ft
    20 ft
    40 ft

See pages 664 and 665 for the answers to Written Ex. 19–23 and Mixed Review Ex. 1–3.

532

---

5. equidistant from the sides of a right angle
6. equidistant from the sides of an obtuse angle
7. 3 cm from a given line
8. less than 4 cm from a given line
9. more than 7 cm from a given line
10. closer to point A than to point B, if A and B are 8 cm apart
11. closer to side $\overrightarrow{OB}$ of ∠AOB than to side $\overrightarrow{OA}$
12. of the center of a circle rolling along a line
13. of the center of a circle rolling around another circle
14. equidistant from the four vertices of a square

15. A power line runs in a straight line for several miles. According to the description of the right-of-way, the power company has a right-of-way that extends 10 ft on either side of the line. Describe in mathematical terms the boundary of this right-of-way.

16. An experimental steering system for automobiles uses a magnetic wire buried in the center of a lane of the highway. Such a wire is placed in a perfectly straight stretch of highway with a lane 24 ft wide. Describe in mathematical terms the location of this wire.

17. A goat is tied to a stake with a 10-ft rope. While tied, the goat eats all the grass that it can reach. If there are no obstacles in the way, describe the patch of ground from which the goat could eat.

18. Suppose that the same goat in Exercise 17 was tied to a corner of a barn that was 20 ft by 40 ft. Describe the patch of ground from which the goat could eat the grass.

19. Prove Theorem 13.1.
20. Prove Theorem 13.2, (Part I).
21. Prove Theorem 13.2, (Part II).
22. Prove the locus of the midpoints of all congruent chords of a given circle is a circle concentric to the original circle.
23. Prove the locus of vertices of all right triangles sharing a given segment as hypotenuse is a circle with the hypotenuse as a diameter. The endpoints of the diameter are not part of the locus.

Mixed Review

Construct the following.

1. the perpendicular bisector of a given segment   *5.6*
2. the bisector of a given obtuse angle   *1.5*
3. a perpendicular to a line from a point not on the line   *5.6*

13.2 Multiple Conditions for Loci

Objective   To sketch and describe loci satisfying two or more conditions

Descriptions of most loci involve distances. The distance between two points is the length of the segment that has the two points as endpoints. The distance from a point to a line is measured along the perpendicular from the point to the line. The distance between two parallel lines is the distance from any point on one of the lines to the other line along the perpendicular. In general, the distance between two geometric figures is the length of the shortest segment between the figures.

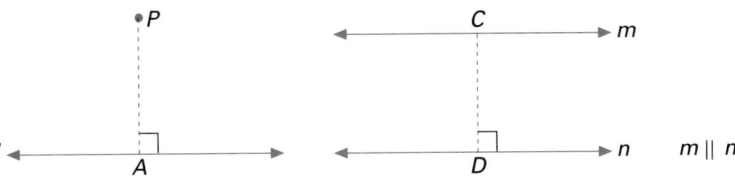

Definition

The distance from a point P to a circle O having radii of length r is $|OP - r|$.

Case 1: P is in the
exterior of $\odot O$.
$PA = |OP - r|$

Case 2: P is the
interior of $\odot O$.
$PA = |r - OP| = |OP - r|$

Case 3: If P is on $\odot O$, then $PA = |OP - r| = 0$.

EXAMPLE 1   Find the locus of points in a plane that are 6 cm from a circle having a 9-cm radius.

Solution
$$6 = |OP - r|$$
$$6 = |OP - 9|$$
$$OP + 9 = 6 \text{ or } OP - 9 = -6$$
$$OP = 15 \qquad OP = 3$$

The locus is two circles concentric to the given circle, one with a 3-cm radius and one with a 15-cm radius.

Additional Example 1

Find the locus of points in a plane that are 1 cm from a circle having a 3 cm radius.

The locus is two circles concentric to the given circle, one with a 2 cm radius and one with a 4 cm radius.

Teaching Resources

Quick Quizzes 96
**Reteaching and Practice
Worksheets** pp. 191, 192

▮▮▮GETTING STARTED

Prerequisite Quiz

Identify each locus in a plane as a point, a line, two lines, a circle, or a circular region.

1. The locus of points less than 3 cm from a point P.   Circular region
2. The locus of points 8 units from a given line, k.   Two lines
3. The locus of points equidistant from two given points, R and S.   A line
4. The locus of points equidistant from the four corners of a rectangle.   A point

Motivator

Explain to the students that most descriptions need more than one condition. For example, to tell someone where to go, they must tell them how far, as well as which direction. Ask students to name one geometric figure that is defined by using distance as part of the definition.   A circle

▮▮▮TEACHING SUGGESTIONS

Lesson Note

After reviewing Example 1, have the students consider how the locus changes if the distance from the given circle were 9 cm or 12 cm, while r remains 9. For loci in which measures are not specified, as in Example 2, prepare the students for various cases by considering specific replacements for the variables.

For loci that must satisfy more than one condition, have the students sketch each individual locus completely and find their points of intersection, to find the required locus.

Math Connections

Sports: Some sports, such as tennis, racquetball, and volleyball, are played on a court involving boundary lines in which the ball must be served or remain in play.

Critical Thinking Questions

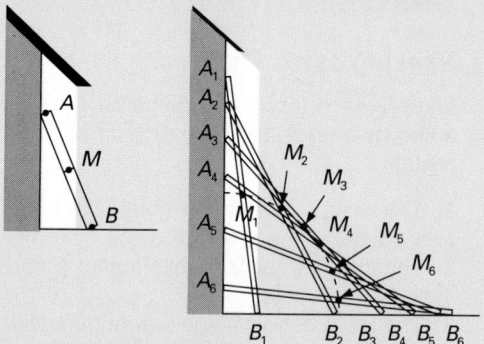

Application: Refer students to the diagram at the left and ask the following question.

As the top of the ladder, *A*, moves up and down the wall, the bottom of the ladder, *B*, moves closer to or further away from the wall along the ground. Find the locus of the middle step *M* on the ladder.   The locus is an arc with a measure of 90 and a radius of length *AM*. (See diagram at right above.)

Checkpoint

Describe each locus in a plane.

1. The locus of points that are 7 cm from a circle having a 5 cm radius.   A circle concentric to the given circle and with a 12 cm radius

2. The locus of points 2 cm from a given line and 5 cm from a point on the line. Four points; the intersection of two lines parallel to the given line (2 cm away) and a circle (of radius 5) with the given point as center

3. The locus of points that are equidistant from two concentric circles with radii 4 cm and 14 cm and at a distance of 9 cm from a line through the center of the circles.   Two points; the points of tangency of a circle concentric to the two lines parallel to the given line and 9 cm from it.

EXAMPLE 2 Sketch and describe the locus of points in a plane that are a distance *d* from a circle having a radius of length *r*.

Plan Because the lengths *r* and *d* are variables, three cases must be investigated: $d > r$, $d = r$, $d < r$.

Solution

Case 1: $d > r$
The locus is a circle concentric to the given circle, having radius of length $r + d$.

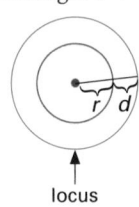

locus

Case 2: $d = r$
The locus is the center of the given circle and a circle concentric to the given circle, having radius of length $r + d$, or $2d$.

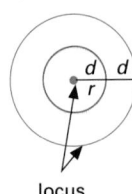

locus

Case 3: $d < r$
The locus is two circles concentric to the given circle. The two circles have radii of lengths $r + d$ and $r - d$.

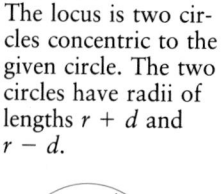
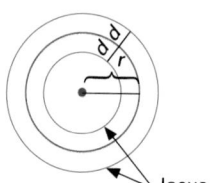

locus

A locus may involve more than one condition.

EXAMPLE 3 Find the locus of points in a plane that are 3 cm from line *m* and 4 cm from point *P* on line *m*.

Plan (1) The locus of points 3 cm from line *m* is two lines, each parallel to *m* and 3 cm from *m*.
(2) The locus of points 4 cm from point *P* is a circle with *P* as center and radius of length 4 cm.
(3) The intersection of these two loci is the required locus.

Solution

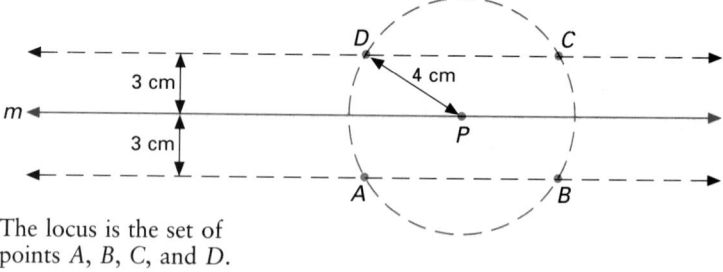

The locus is the set of points *A*, *B*, *C*, and *D*.

Additional Example 2

Sketch and describe the locus of points at a distance *d* from a circle having a radius of length *r* for each of the three cases. See Example 2 for sketches.

Case I: $d = 5$ and $r = 2$.
The locus is a circle concentric to the given circle and with a radius of 7 units.

Case II: $d = 4$ and $r = 4$
The locus is the center of the given circle and a circle concentric to the given circle with a radius of 8 units.

Case III: $d = 3$ and $r = 7$
The locus is two circles concentric to the given circle. The circles have radii of lengths 4 units and 10 units.

EXAMPLE 4

Find the locus of points in a plane that are 3 cm from point P *and* 4 cm from point Q, if $PQ = 6$ cm.

Solution

$\odot P$ is the locus of points 3 cm from P.
$\odot Q$ is the locus of points 4 cm from Q.

The locus is the set of points A and B, the intersections of $\odot P$ and $\odot Q$.

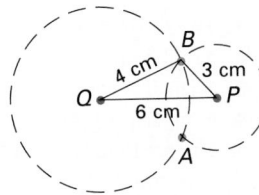

There may be several possibilities for the intersection of two or more loci, depending on the distance involved. If they do not intersect at all, then there are no points that satisfy all of the given conditions.

EXAMPLE 5

Find the locus of points in a plane that are a distance r from a point P and a distance s from point Q, if P and Q are a distance d apart $(d \neq 0)$.

Solution

Case 1: $r + s < d$
There are no points in the locus.

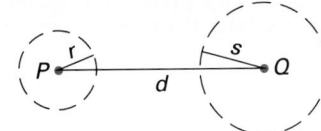

Case 2: $r + s = d$
The locus is one point, the point of tangency of the two circles with centers P and Q having radii of lengths r and s, respectively.

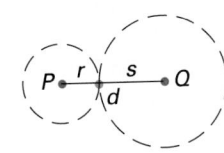

Case 3: $r + s > d$
The locus is the set of two points, which are the intersections of the circles with centers P and Q having radii of lengths r and s, respectively.

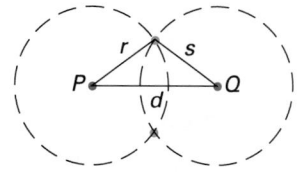

There are two other possibilities for the locus. If $s > r + d$ or $r > s + d$, then one circle lies in the interior of the other, with no intersection. If $s = r + d$ or $r = s + d$, then one circle is internally tangent to the other with one point of intersection.

13.2 Multiple Conditions for Loci **535**

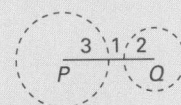

11. Inters of ext of ⊙s with P as ctr, radius of 8 cm and with Q as ctr, radius of 7 cm.

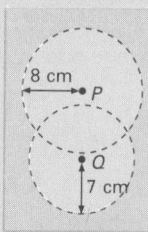

12. Inters of int of ⊙s with P as ctr, radius of 7 cm and with Q as ctr, radius of 6 cm.

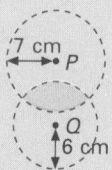

13. 2 pts; inters of l and the ⊙ with P as ctr and radius of 5 cm.

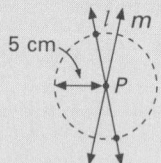

14. 2 pts; inters of m and the ⊙ with P as ctr and radius of 8 cm.

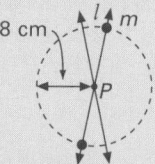

15. 4 pts; inters of 2 pairs of ∥ lines, 1 on either side of l, ∥ to and 5 cm from l, the other pair the same with respect to m.

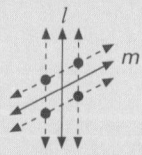

16. 2 pts; inters of ⊙ with P as ctr, radius of 4 cm and 2 lines ∥ to l, 1 on each side, 4 cm from l.

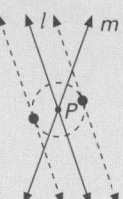

536

Classroom Exercises

Describe the locus of points in a plane.

1. 10 cm from a given point
2. equidistant from two parallel lines
3. 5 cm from a circle with a 10-cm radius
4. 4 cm from P and 6 cm from Q, where P and Q are 12 cm apart
5. greater than 6 cm from a circle having a 3-cm radius

Written Exercises

Sketch and describe the locus of points in a plane.

1. 2 cm from ⊙O
2. 5 cm from ⊙O
3. 8 cm from ⊙O
4. less than 3 cm from ⊙O
5. in the exterior of ⊙O *and* greater than 2 cm from the circle

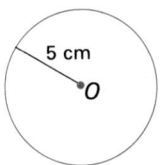

6. 2 cm from P *and* 5 cm from Q
7. 6 cm from P *and* 6 cm from Q
8. 5 cm from P *and* 5 cm from Q
9. 9 cm from P *and* 6 cm from Q
10. equidistant from P and Q
11. more than 8 cm from P *and* more than 7 cm from Q
12. less than 7 cm from P *and* less than 6 cm from Q

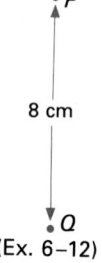

(Ex. 6–12)

13. on l *and* 5 cm from P
14. on m *and* 8 cm from P
15. 5 cm from l *and* 5 cm from m
16. 4 cm from P *and* 4 cm from l
17. 2 cm from m *and* 9 cm from P
18. less than 8 cm from P *and* less than 3 cm from m

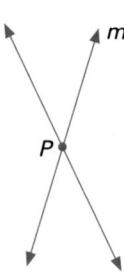

Additional Example 5

Points P and Q are d units apart. Describe the locus of points r units from P and s units from Q for each of the three cases.

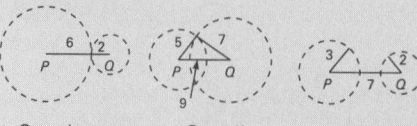

Case I   Case II   Case III

Case I: r = 6, s = 2, d = 8
The locus is one point.

Case II: r = 5, s = 7, d = 9
The locus is the two points of intersection of the circles.

Case III: r = 3, s = 2, d = 7
There are no points in the locus.

Sketch and describe the locus of points in a plane. Give all possible cases.

19. 3 cm from point *P and* 5 cm from point *Q*, given that *P* and *Q* are a distance *d* apart
20. 8 cm from *P and* *t* cm from point *Q*, given that *P and Q* are 8 cm apart
21. in the interior of an angle *and* equidistant from the sides and a distance *d* from the vertex of the angle
22. equidistant from two intersecting lines
23. equidistant from all points of ⊙*O*
24. equidistant from two concentric circles
25. a distance *d* from a given line *and* a distance *r* from a point on the line
26. a distance *d* from each of two points that are a distance *t* from each other
27. less than a distance *d* from each of two points that are a distance *t* from each other

Sketch and describe the locus in a plane. (Exercises 28–29)

28. determined by a point on a circle as it rolls along a line
29. determined by a point on a circle as it rolls around outside another circle

30. A Spirograph® is a device for drawing designs using a pen mounted in a moving wheel. The designs are actually loci. The locus of points illustrated below is determined by an interior point that is a distance *d* from a circle as the circle rolls along a line. How does the design differ as the distance *d* changes?

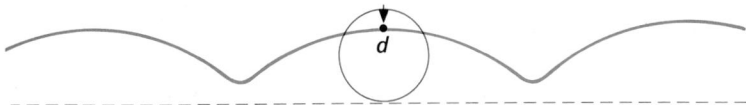

Mixed Review

Is the statement true or false?

1. Two triangles are similar if two angles of one are congruent to two angles of the other. *8.3* T
2. Two triangles are similar if two sides of one are congruent to two sides of the other. *8.4* F
3. Two triangles are congruent if two angles of one are congruent to two angles of the other. *4.6* F

Enrichment

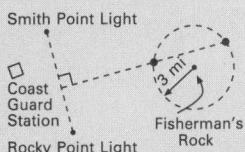

A Coast Guard station receives a distress call from a ship on fire. The Captain describes his position as being equidistant from the Smith Point and Rocky Point Lights, but he isn't able to tell how far he is from them. A separate report from a helicopter places the ship's position at 3 miles from Fisherman's Rock but fails to specify the direction.

Sketch the possible positions of the ship. In the above drawing, there are two possibilities. Challenge the students to find other arrangements.

17. 4 pts; inters of 2 lines ∥ *m*, 1 on each side 2 cm from *m* and ⊙ with *P* as ctr, radius of 9 cm.

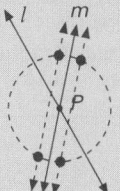

18. Inters of region between 2 lines, each ∥ to and 3 cm from *m* and int of ⊙ with *P* as ctr, radius of 8 cm.

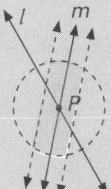

19. Inters of ⊙ with *P* as ctr, radius of 3 cm and ⊙ with *Q* as ctr, radius of 5 cm.
    If *d* > 8 cm, locus contains no pts.
    If *d* = 8 cm, locus is pt of tangency.
    If 2 cm < *d* < 8 cm, locus is 2 pts.
    If *d* = 2 cm, locus is pt of tangency.
    If *d* < 2 cm, locus contains no pts.

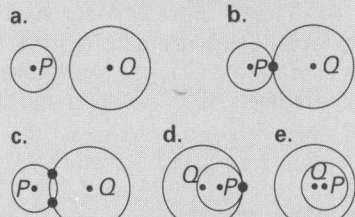

20. If *t* = 0, locus is pt *Q*.
    If *t* < 16, locus is 2 pts.
    If *t* = 16, locus is 1 pt.
    If *t* > 16, locus contains no pts.
21. If *d* = 0, then locus has no pts.
    If *d* > 0, then locus is pt; inters of ∠ bis and ⊙ with vert as ctr, *r* of *d*.

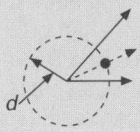

See page 546 for the answers to Ex. 22–30.

▰▰GETTING STARTED

Prerequisite Quiz

**Describe the intersection of the
given figures.**

1. a circle and a line through its interior
   Two points
2. a sphere and a plane through it its
   interior   A circle
3. two nonparallel planes   A line
4. a pair of parallel planes and a plane
   perpendicular to either of the first two
   Two parallel lines

Motivator

Explain to the students that most of the
directions they give are based on the
assumption that they will be moving in a
plane. Ask them how many directions it
usually takes to describe such a move.   It
usually takes two directions.   **Have them
give an example.**   100 feet north and then
50 feet east

▰▰TEACHING
SUGGESTIONS

Lesson Note

Review examples and theorems about loci
in a plane and consider the corresponding
loci in space. Discuss "distance" (shortest
distance) in space, between two points,
from a point to a sphere. The walls and
edges of the room can be used as physical
models of planes and lines; paper taped
end-to-end can represent a cylindrical
surface; a ball can represent a sphere.
Have students distinguish between a
sphere, its interior, and its exterior by asking
them for the loci in space of points that are
equal to, less than, or greater than a given
distance d from a fixed point P.

13.3 Loci in Space

Objective   To sketch and describe loci in space

Loci (plural of locus) can be
restricted to one plane, or they
can be generalized to space.
The locus of points a given
distance r from a fixed point
was described as a circle. This
is true only if the locus is
restricted to a plane. If this restric-
tion is removed, the locus is a sphere with the fixed point as its center.

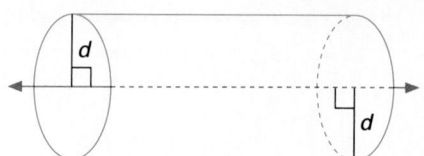

EXAMPLE 1   Sketch and describe the locus of points in space a given distance *d* from
a fixed line *l*. How is this different from the locus when restricted to
one plane?

Solution

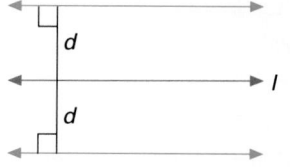

The locus in one plane is a pair of parallel lines with *l* halfway between
them. In space the locus is a cylindrical surface with *l* as its axis and *d*
as the length of a radius.

Sometimes an inequality is used to describe a locus. The locus of points
in space a given distance from a fixed point is a sphere. A sphere is a
surface. The locus of points in space less than a given distance from a
fixed point is the *interior* of a sphere, a solid. The locus of points in
a space greater than a given distance from a fixed point is the *exterior*
of a sphere.

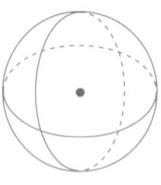 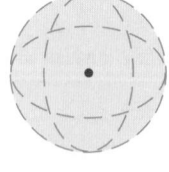 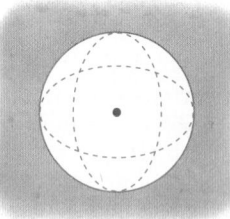

Sphere   Interior of sphere   Exterior of sphere

Additional Example 1

Sketch and describe the locus of points in
space that are equidistant from the sides
of an angle. How is this different from
the locus when restricted to one plane?

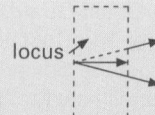

locus

The locus is a plane perpendicular to the
plane of the angle and containing the
bisector of the angle. The locus in a
plane is the bisector of the angle.

EXAMPLE 2 Points A and B are 10 cm apart. Point P is the midpoint of $\overline{AB}$. Sketch and describe the locus of points in space equidistant from A and B *and* more than 4 cm from point P.

Solution
(1) In a plane, the locus of points equidistant from A and B is the perpendicular bisector of $\overline{AB}$. In space, the locus of points equidistant from A and B is a plane $\mathfrak{Q}$ that is the perpendicular bisector of $\overline{AB}$. Note that P is in $\mathfrak{Q}$, since P is the midpoint of $\overline{AB}$.

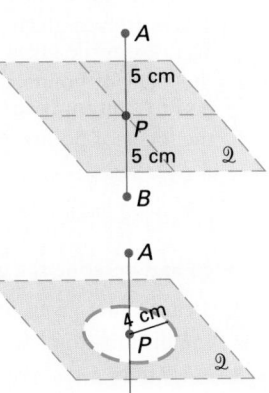

(2) The locus of points in $\mathfrak{Q}$ that are more than 4 cm from point P is the exterior of the circle in $\mathfrak{Q}$ with a radius of length 4 cm and P as its center.

Thus, the required locus is the set of points in the perpendicular bisector $\mathfrak{Q}$ of $\overline{AB}$ and in the exterior of $\odot P$, where $\odot P$ has a 4-cm radius.

Classroom Exercises

Describe the locus of points in space.

1. 5 cm from a given point
2. 5 cm from a given line
3. 5 cm from a given plane
4. equidistant from two given points

Written Exercises

Sketch and describe the locus of points in space.

1. 6 m from a given point
2. 10 m from a given line
3. less than 5 cm from a given point
4. greater than 5 cm from a given point
5. less than 5 cm from a given line
6. greater than 5 cm from a given line
7. equidistant from two points that are 10 cm apart
8. equidistant from two parallel lines
9. equidistant from two parallel planes
10. less than 5 cm from a given plane
11. equidistant from the sides of given angle

13.3 Loci in Space **539**

Additional Example 2

Sketch and describe the locus of points in space that are 2 cm from plane $\mathcal{W}$ and equidistant from two points R and S in plane $\mathcal{W}$.

The locus is a pair of parallel lines, formed by the intersections of two parallel planes 2 cm distant from plane $\mathcal{W}$, and a plane perpendicular to $\overline{RS}$ and bisecting it.

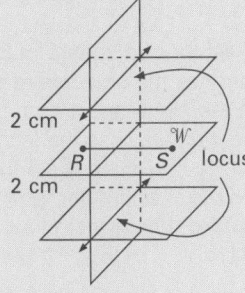

Math Connections

Navigation: A pilot must guide an airplane along a course according to instructions given by a navigator. Based on conditions such as the weather, the location, and the destination, the navigator describes a locus in space in which it is safe to fly.

Critical Thinking Questions

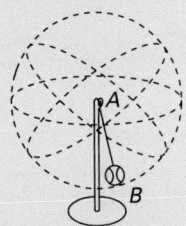

Tetherball is played between two opponents who try to wrap a suspended ball in opposing directions until it is wrapped competely around the pole. Refer students to the diagram and ask them to find the locus of the ball in play. The locus is a sphere whose radius is the length of the rope and ball, AB, incl the int pts of the sphere, but not incl the pts of or inside the pole.

Checkpoint

Describe the locus in space.

1. The locus of points that are 5 meters from a sphere with a radius of 2 meters. A sphere; concentric to the given sphere, with a 7 meter radius.
2. The locus of points more than 4 cm from line t. The locus is the exterior of a cylindrical surface with radius 4 cm and with line t as the axis.
3. The locus of points 2 cm from a plane and 5 cm from a line in that plane. The locus is four lines parallel to the given line, two of them in each of two planes that are 2 cm from the given plane.

Closure

Ask students to describe a sphere. It is the locus of points in space at a given distance from a fixed point Ask them to describe the exterior of a sphere. The locus of points in space greater than a given distance from a fixed point. The interior of a sphere. The locus of points in space less than a given distance from a fixed point.

■ FOLLOW UP

Guided Practice

Classroom Exercises 1–4

Independent Practice

A Ex. 1–7, **B** Ex. 8–11, **C** Ex. 12–16

Basic: WE 1–7, Midchapter Review, Application

Average: WE 1–11, Midchapter Review, Application

Above Average: WE 1–16, Midchapter Review, Application

Additional Answers

Classroom Exercises

1. Sphere with pt as ctr, radius of 5 cm.
2. Cyl surface with line as axis, radius of 5 cm.
3. 2 planes each ∥ to and 5 cm from given plane.
4. Plane that is the ⊥ bis of the seg between the pts.

Written Exercises

1. Sphere with radius of 6 m.
2. Cyl surface with line as axis, radius of 10 m.
3. Int of sphere with pt as ctr, radius of 5 cm.
4. Ext of sphere with pt as ctr, radius of 5 cm.
5. Int of cyl surface with given line axis, radius of 5 cm.
6. Ext of cyl surface with given lines as axis, radius of 5 cm.
7. Plane that is the ⊥ bis of $\overline{AB}$.
8. Plane that is ⊥ bis of seg ⊥ both lines.
9. A third plane that is ∥ to the other two and halfway between them.
10. All pts between 2 ∥ planes, each 5 cm from the given plane.
11. Plane that bis the angle.
12. Circle of radius 4 cm.

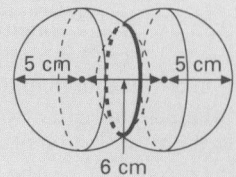

5 cm   5 cm

6 cm

See page 553 for the answers to Written Ex. 13–16, Midchapter Review Ex. 1–5, and Application Ex. 1–2.

Sketch and describe the locus of points in space.

12. 5 cm from each of two points that are 6 cm apart
13. 5 cm from each of two parallel lines that are 6 cm apart
14. 10 cm from a given point *and* 10 cm from a given plane that is 5 cm from the given point
15. the locus of points a distance *d* from point *P and* a distance *t* from point *Q*, given that the distance between *P* and *Q* is *s*
16. the locus of points 5 cm from a given segment that is 4 cm long

Midchapter Review

Sketch and describe the given locus of points in a plane.

1. 3 cm from a given point   *13.1*
2. equidistant from two rays that form a 60-degree angle   *13.1*
3. more than 5 cm from the circumference of a circle with radius 4 cm   *13.2*

Sketch and describe the given locus of points in space.   *13.3*

4. equidistant from two points that are 8 cm apart
5. 5 in. from line *m and* 10 in. from point *P* on *m*

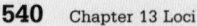

■/ Application: *Site for a House*

A landscape architect is developing a site plan for a new house. The local building code specifies that no part of the house can be closer than 30 ft to the street right-of-way line, nor 10 ft to any of the other property lines. A tree is to remain in the front yard. The house must be at least 10 ft away from the trunk of the tree.

1. Suppose that the house is rectangular in shape, with a front 50 ft long and parallel to the street and a depth of 40 ft. Copy the map and show where this house could be built.

2. Suppose that the house could be turned in any direction. Show on the map where it now could be located.

Property

15 ft

Tree   35 ft

80 ft   Street

Property

110 ft

Property

540   Chapter 13 Loci

Enrichment

The center of a sphere of 5 cm radius is 16 cm from a plane. Ask the students to describe the locus of points 8 cm from the plane and 10 cm from the center of the sphere. Have them discuss the length of the locus formed.
The locus will be a circle formed on a plane parallel to the given plane and 8 cm from it. The radius of the circle will be 6 cm and the "length of the locus" will be its circumference, 12π cm.

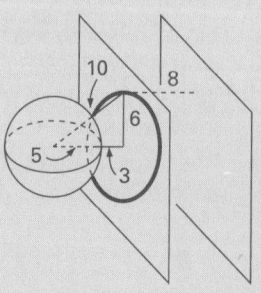

10   8

5   6

3

13.4 Concurrent Bisectors in Triangles

Teaching Resources

Manipulative Worksheet 26
Project Worksheet 13
Quick Quizzes 98
**Reteaching and Practice
   Worksheets,** pp. 195, 196
Technology Worksheet 13

Objectives   To use constructions to find loci
To apply perpendicular bisector and angle bisector concurrency
theorems
To locate and use incenters and circumcenters

A piece of machinery has a broken gear. To replace the
gear, it is necessary to know the diameter of the gear
and the total number of teeth on the gear. A piece of the
broken gear was salvaged. How can the diameter and
the number of teeth be determined from this piece?

In Chapter 11, one approach to this problem was suggested. In this lesson you will investigate geometric constructions and loci that will reveal an alternate method.

EXAMPLE 1   Use constructions to find the locus of points equidistant from the three
vertices of △ABC.

Plan   In a plane, the locus of points equidistant from two points is the perpendicular bisector of the segment having the points as endpoints. Construct the perpendicular bisectors of $\overline{AB}$ and $\overline{AC}$.

Solution   Notice that $DB = AD$ and $AD = DC$. So,
by the Transitive Property, $DB = DC$. This
means that D will also be on the perpendicular bisector of $\overline{BC}$. Therefore, the locus of
points equidistant from the three vertices of
△ABC is a single point D.

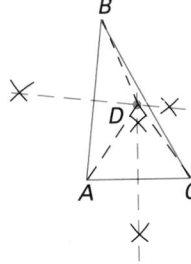

Three or more lines that intersect or converge at a common point are
concurrent lines.

Theorem 13.3   The perpendicular bisectors of the sides of a triangle are concurrent
at a point equidistant from the vertices of the triangle.

▰ GETTING STARTED

Prerequisite Quiz

Draw an acute triangle *ABC* (not an
equilateral triangle). Then construct the
perpendicular bisectors of the three sides.
Check student constructions.

Motivator

Ask the students to describe the intersection
of two lines.   A point   Ask them to find an
example in the classroom where three lines
in space intersect in a single point.   A
corner of the classroom

▰ TEACHING
   SUGGESTIONS

Lesson Note

The students will notice that the
perpendicular bisectors constructed for the
Prerequisite Quiz meet at a point. Call it *D*.
Using the compass points, the students can
see that $DA = DB = DC$. Have them
construct the circumscribed circle, with *D* as
center and $\overline{DA}$ as radius.

Additional Example 1

Use constructions to find the locus of
points equidistant from any three points
of a circle.

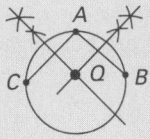

The locus is point *Q*, the point of
intersection of the perpendicular
bisectors of $\overline{AB}$ and $\overline{AC}$.

Math Connections

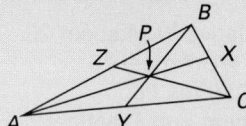

History: Around the middle of the 17th century, an Italian mathematician named Giovanni Ceva discovered a property for any three concurrent lines from the vertices of a triangle. Ceva's Theorem states that if three lines from the vertices *A*, *B* and *C* of a triangle *ABC* are concurrent at a point *P* and meet the opposite sides at *X*, *Y* and *Z*, as shown above, then $\frac{AZ}{ZB} \cdot \frac{BX}{XC} \cdot \frac{CY}{YA} = 1$. Have students construct a triangle and measure the distances mentioned in Ceva's Theorem, to confirm that the theorem is true.

Critical Thinking Questions

Analysis: Ask students to make a generalization about the location of the following points.

1. the incenter of any acute triangle, of any obtuse triangle, of any right triangle, and of any equilateral triangle.   For each type of △, the incenter is located in the int of the △.
2. the circumcenter of any acute triangle, of any obtuse triangle, of any right triangle, and of any equilateral triangle.   Int of △, ext of △, on the △, int of △, respectively
3. the incenter and the circumcenter of any equilateral triangle   They coincide.

Given: △*ABC*; *l*, *m*, and *n* are the perpendicular bisectors of $\overline{AB}$, $\overline{BC}$, and $\overline{CA}$, respectively.

Prove: *l*, *m*, and *n* are concurrent at a point equidistant from *A*, *B*, and *C*.

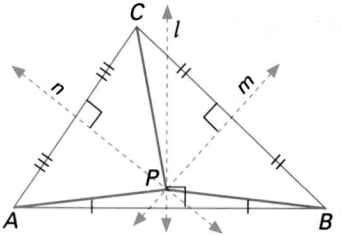

Proof

Statement	Reason
1. *l*, *m*, and *n* are the ⊥ bis of $\overline{AB}$, $\overline{BC}$, and $\overline{CA}$, respectively.	1. Given
2. $\overline{AB} \nparallel \overline{CA}$	2. Sides of △ intersect.
3. $l \nparallel n$	3. Coplanar lines ⊥ to nonparallel lines are ∦.
4. *l* and *n* intersect at point *P*.	4. ∦ lines intersect.
5. $PB = PA$, $PA = PC$	5. A point on ⊥ bis of seg is equidist from its endpts.
6. $PB = PC$	6. Trans Prop
7. *m* contains *P*.	7. A point equidist from endpts of seg lies on its ⊥ bis.
8. *l*, *m*, and *n* are concurrent at a point equidistant from *A*, *B*, and *C*.	8. Def of concurrent lines

---

Definition   The **circumcenter** of a triangle is the intersection of the perpendicular bisectors of the sides of the triangle.

EXAMPLE 2   Construct the locus of points equidistant from the sides of △*EFG*.

Plan   The locus of points equidistant from the sides of an angle is the bisector of the angle. Construct the bisector of ∠*E* and ∠*F*.

Solution   Point *H* is equidistant from sides $\overline{EG}$ and $\overline{EF}$. It is equidistant from sides $\overline{EF}$ and $\overline{FG}$. Therefore, it is equidistant from sides $\overline{EG}$ and $\overline{FG}$.

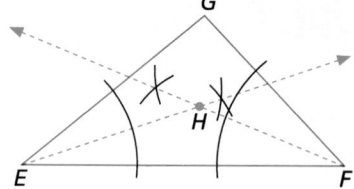

The locus is point *H*.

Additional Example 2

Use constructions to find the locus of points outside △*ABC* that are equidistant from $\overleftrightarrow{AC}$, $\overrightarrow{BA}$, and $\overrightarrow{BC}$.

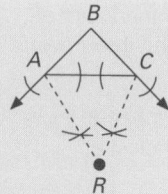

The locus is point *R*, the intersection of the bisectors of exterior angles at *A* and *C*.

The locus of points equidistant from the sides of the triangle is the intersection of the three angle bisectors of the triangle.

Theorem 13.4

The bisectors of the angles of a triangle are concurrent at a point equidistant from the sides of the triangle.

Given: △ABC, $\overrightarrow{AD}$, $\overrightarrow{BE}$, and $\overrightarrow{CF}$ are the bisectors of ∠A, ∠B, and ∠C, respectively.

Prove: $\overrightarrow{AD}$, $\overrightarrow{BE}$, and $\overrightarrow{CF}$ are concurrent at a point equidistant from $\overline{AB}$, $\overline{BC}$, and $\overline{CA}$.

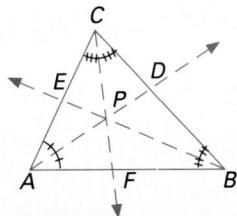

Definition

The **incenter** of a triangle is the intersection of the bisectors of the angles of the triangle.

The two previous theorems prove that an incenter and a circumcenter exist in every triangle. This fact suggests some useful constructions.

EXAMPLE 3

Draw a triangle. Construct a circle that contains the three vertices of the triangle.

Construction

(1) Draw △ABC. Construct perpendicular bisectors of $\overline{AB}$ and $\overline{BC}$, intersecting at point P.

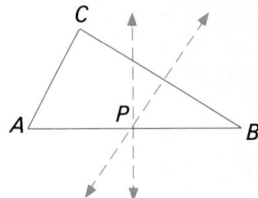

(2) Using PA as the length of a radius, draw ⊙P.

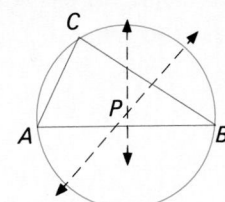

Result: ⊙P is the desired circle.

The circle constructed is the **circumcircle**, which **circumscribes** the triangle. P is the circumcenter of the triangle.

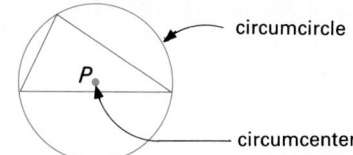
circumcircle
circumcenter

Error: Students sometimes confuse the circumcenter and the incenter of a triangle.

Remind students that the circumcenter of a triangle must be equidistant from the vertices of the triangle while the incenter must be equidistant from the sides. Ask the students to keep in mind which locus they are constructing with the perpendicular bisector of a side of a triangle or the bisector of an angle.

Checkpoint

True or false?

1. In an isosceles triangle, the altitude to the base contains both the circumcenter and the incenter.   T
2. A circle can be constructed through any three points.   F (not if the points are collinear)
3. Just one point inside a triangle ABC is equidistant from points A and B.   F
4. The circumcircle and incircle of a triangle are concentric only if the triangle is equilateral.   T

Additional Example 3

Construct a circle which contains three noncollinear points, R, S, and T. When $\overline{RS}$, $\overline{ST}$, and $\overline{RT}$ are drawn, the construction is the same as that shown in Example 3.

544

Closure

Ask the students if the three angle bisectors of a triangle will always be concurrent. **Yes** Ask them what this means. **They will intersect at one point.** Then ask them if this point will be inside the triangle, outside the triangle, or on the triangle itself and whether it has a specific name. **Inside—equidistant from the sides of the triangle; incenter** Ask them if the three perpendicular bisectors of the sides of a triangle will always be concurrent. **Yes** Then ask them where this point will be and what it is called. **Inside—equidistant from the vertices of the triangle; circumcenter**

◼◼◼FOLLOW UP

Guided Practice

Classroom Exercises, 1–4

Independent Practice

A Ex. 1–8, **B** Ex. 9–15, **C** Ex. 16–20

Basic: FR all, WE 1–8, CI 1
Average: FR all, WE 1–15, CI all
Above Average: FR all, WE 1–20 CI all

Additional Answers

Focus on Reading

1. Circum: around, scribere: to write or draw.
2. Circum: around, ⊙: ring.
3. Circum: around, ctr: sharp pt.
4. In: within, scribere: to write or draw.
5. In: within, ⊙: ring.
6. In: within, ctr: sharp pt.

EXAMPLE 4 Draw a triangle. Inscribe a circle in the triangle.

Construction (1) Draw △ABC. Construct the bisectors of ∠A and ∠B, intersecting at point P.

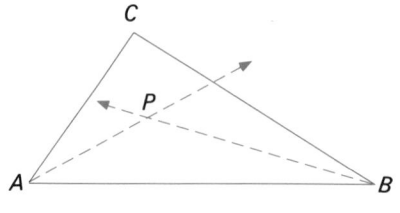

(2) Construct $\overline{PD} \perp \overline{AB}$. Using PD as the length of a radius, draw ⊙P.

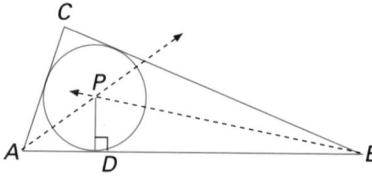

Result: ⊙P is inscribed in △ABC.

The circle constructed is called the **incircle** of the triangle. It is *inscribed* in the triangle. P is the **incenter** of the triangle.

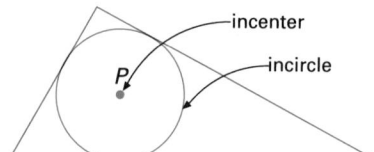

◼◼ *Focus on Reading*

The given word has both a prefix and a root word. In a dictionary that has derivations, find the meanings of the prefix and root word.

1. circumscribed
2. circumcircle
3. circumcenter
4. inscribed
5. incircle
6. incenter

Classroom Exercises

Is the statement true or false?

1. The point of intersection of the bisectors of the sides of a triangle is a circumcenter.   T
2. The bisectors of the angles of a triangle are concurrent at a point equidistant from the vertices.   F
3. The locus of points equidistant from the three vertices of a triangle is the intersection of the perpendicular bisectors of the sides.   T
4. The perpendicular bisectors of the sides of an obtuse triangle are not concurrent.   F

544 Chapter 13 Loci

Additional Example 4

Draw a triangle *ABC*. Outside the triangle, construct a circle that is tangent to $\overleftrightarrow{AC}$, $\overrightarrow{BA}$, and $\overleftrightarrow{BC}$.

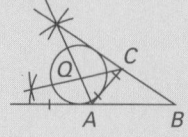

The center *Q* of the required circle (*excenter* of the triangle) is the intersection of the bisectors of two exterior angles. The radius is the length of the perpendicular from *Q* to $\overleftrightarrow{AB}$. Note: A triangle has three such excircles.

Written Exercises

1. Draw an acute triangle. Construct the incenter of the triangle.
2. Draw an obtuse triangle. Construct the circumcenter of the triangle.
3. Draw a right triangle. Construct the incenter of the triangle.
4. Draw a right triangle. Construct the circumcenter of the triangle.
5. Draw an obtuse triangle. Construct the inscribed circle.
6. Draw a right triangle. Construct the circumscribed circle.
7. Construct an isosceles triangle. Construct both the incenter and the circumcenter of the triangle. Where do the two centers lie?
8. Construct an equilateral triangle. Construct both the incenter and the circumcenter of the triangle. Where do the two centers lie?
9. Trace a circle. Use a construction to locate the center of the circle.
10. Using the diagram in the lesson introduction, draw the circle containing the gear. Find the number of gear teeth.

True or false? Draw a diagram to justify your answer.

11. No triangle has its incenter on one side of the triangle. T
12. No triangle has its circumcenter on one side of the triangle. F
13. The incenter of a triangle can never be the circumcenter. F
14. The circumcenter of a triangle can never be in the exterior of the triangle. F
15. Prove Theorem 13.4.

The equilateral triangle shown has sides of length 10. (Exercises 16–20)

16. Find the radius of the circumcircle of the triangle. $\frac{10}{3}\sqrt{3}$
17. Find the radius of the incircle of the triangle. $\frac{5}{3}\sqrt{3}$
18. Find the difference between the areas of the triangle and its incircle. $\frac{25}{3}(3\sqrt{3} - \pi)$ sq units
19. Find the difference between the areas of the triangle and its circumcircle. $\frac{25}{3}(4\pi - 3\sqrt{3})$ sq units
20. Find the difference between the areas of the two circles. 25π sq units

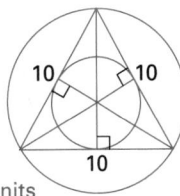

Mixed Review

1. Draw a line and a point approximately 5 cm away from the line. Construct the line through the point parallel to the given line. *3.3*

Find the value of *x* that makes a true proportion. *8.1*

2. $\frac{2}{3} = \frac{x}{12}$ 8

3. $\frac{8}{12} = \frac{12}{x}$ 18

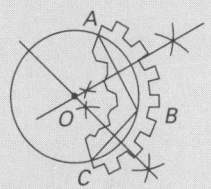
Enrichment

Present the students with the following challenges.

1. For a given perimeter, describe the triangle that will have the largest incircle. Equilateral
2. How would you draw a triangle of a given perimeter so that it would have the smallest incircle? Make the measure of at least one angle very close to zero. There is no limit to how small the incircle can be, but it can be made as small as desired by making an angle of the triangle sufficiently close to zero in measure.
3. For a given perimeter, describe the triangle that has the smallest circumcircle. Equilateral.

22. Bis of the 4 ∠s formed by the inters lines.

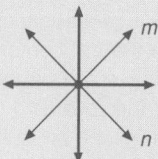

23. Center of ⊙ O.

24. A third concentric ⊙ that is positioned between the given ⊙s and equidistant from them.

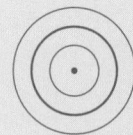

25. If $d = 0$ and $r = 0$ locus is a pt.
If $d = 0$ and $r > 0$ then locus is 2 pts.
If $d > r$ locus contains no pts.
If $d = r$ locus is 2 pts.
If $d < r$ locus is 4 pts.

26. If $2d < t$, locus contains no pts.
If $2d = t$, locus is pt of tangency.
If $2d > t$, locus is 2 pts.

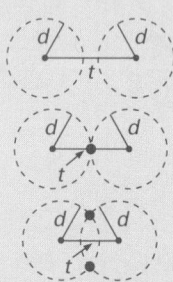

27. If $2d < t$ or $2d = t$, locus contains no pts.

a.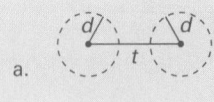

b.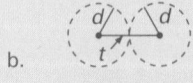

c.

If $2d > t$, locus is region in int of both ⊙s.

Computer Investigation

Concurrence in Triangles

Use a computer software program that draws random triangles by classification (acute, obtuse, or right), constructs medians, labels points, and measures line segments.

In the triangle shown at the right, three *medians* meet at one point, G. In the computer investigation of Chapter 5, you discovered that the medians of any triangle always meet, or *concur*. This will be formally stated in the next lesson. However, there is an even more interesting relationship between the three medians of any triangle and their point of intersection. The activities below will lead you to discover this important numerical relationship.

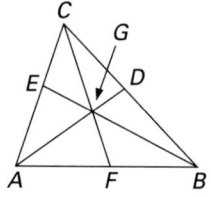

Activity 1

Draw an acute triangle and the indicated medians.

1. median $\overline{AD}$ from A to $\overline{BC}$
2. median $\overline{BE}$ from B to $\overline{AC}$
3. median $\overline{CF}$ from C to $\overline{AB}$
4. Label the point of intersection of the three medians G.

Activity 2

Find the indicated measures for the segments of the medians constructed in Activity 1. Then compute the indicated ratio to the nearest tenth. Keep a record of the ratios.

5. DG, GA; $\dfrac{DG}{GA}$ **6.** FG, GC; $\dfrac{FG}{GC}$ **7.** EG, GB; $\dfrac{EG}{GB}$

Activity 3

Repeat the steps of Exercises 1–7 for the following new triangles.

8. obtuse triangle   **9.** right triangle   **10.** isosceles triangle

Summary

The three medians of any triangle *concur* at one point. The point of *concurrence* of the three medians of any triangle divides each median into two segments with lengths in the ratio 2:1 .

---

28. Series of arcs with height = diam, length = circum of ⊙.

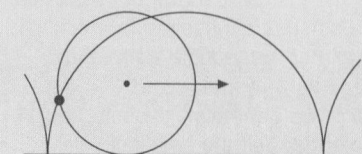

29. Locus is a series of arcs with endpts on ⊙.

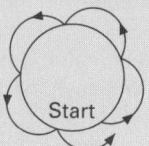

30. As pt moves closer to ctr. design flattens out, becomes line if pt is ctr.

13.5 Concurrent Altitudes and Medians

Objectives
To apply altitude and median concurrency theorems
To construct orthocenters and centroids of triangles

Three lines usually do not intersect in a point. However, in the last lesson, two cases of concurrency were developed. The altitudes and medians of a triangle are also concurrent.

Theorem 13.5
The lines containing the altitudes of a triangle are concurrent.

Given: △ABC,
$\overline{AD}$, $\overline{BE}$,
and $\overline{CF}$ are
altitudes.
Prove: $\overline{AD}$, $\overline{BE}$,
and $\overline{CF}$ are
concurrent.

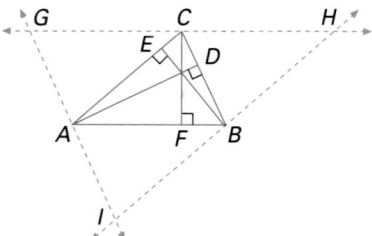

Plan
Draw $\overleftrightarrow{GH} \parallel \overline{AB}$ through C, $\overleftrightarrow{HI} \parallel \overline{AC}$ through B, and $\overleftrightarrow{IG} \parallel \overline{BC}$ through A, to form △GHI. By definition, $ABHC$ and $ABCG$ are parallelograms, and $CH = AB = GC$. Since a line perpendicular to one of two parallel lines is perpendicular to the other, $\overline{CF} \perp \overline{GH}$. Therefore, $\overline{CF}$ is the perpendicular bisector of $\overline{GH}$.

Similarly, it can be shown that $\overline{AD}$ and $\overline{BE}$ are perpendicular bisectors of $\overline{GI}$ and $\overline{HI}$, respectively. So, $\overline{AD}$, $\overline{BE}$, and $\overline{CF}$ are concurrent, being the perpendicular bisectors of the sides of △GHI.

Definition
The **orthocenter** of a triangle is the intersection of the lines containing the altitudes of the triangle.

The orthocenter of a triangle can be either in the interior of the triangle, on the triangle itself, or in the exterior of the triangle, depending on whether the triangle is acute, right, or obtuse, respectively.

13.5 Concurrent Altitudes and Medians **547**

Teaching Resources

Quick Quizzes 99
Reteaching and Practice
 Worksheets, pp. 197, 198
Transparencies 36, 37

GETTING STARTED

Prerequisite Quiz

Draw acute triangles *ABC* and *DEF* (not equilateral).

1. In △*ABC*, construct the altitude to side $\overline{BC}$.

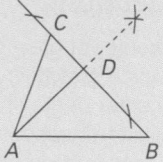

2. In △*DEF*, construct the median to side $\overline{EF}$.

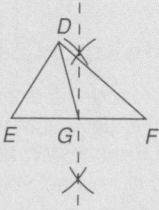

Motivator

Ask the students what they know about the perpendicular bisectors of the sides of a triangle and the angle bisectors. They intersect in a single point. Ask them to name two other lines or segments commonly associated with triangles. Altitudes and medians Ask them if the altitudes meet. Yes Ask them if the medians meet. Yes Ask them what their intersection is. A point

TEACHING SUGGESTIONS

Lesson Note

In this lesson, two more points of concurrency are examined, the orthocenter and the centroid. The orthocenter is the point of concurrency of the altitudes of a triangle, and the centroid is the point of concurrency of the medians. The centroid is also called the center of gravity of the triangle. This is because a triangular board of uniform thickness will balance if it is supported by a pencil tip, or some such object, at that one point.

Math Connections

Physics: The center of gravity of an object is the point at which gravity is focused. The centroid is the center of gravity for a triangle. Since lifting or supporting an object is easiest at its center of gravity, architects, sculptors, and other builders apply this principle to build stable structures.

Critical Thinking Questions

Analysis: Have students draw a large acute triangle; find its circumcenter, its centroid and its orthocenter, and label these points *A*, *B* and *C*, respectively. Ask students if these three points *A*, *B* and *C* are collinear.   Yes. See diagram.

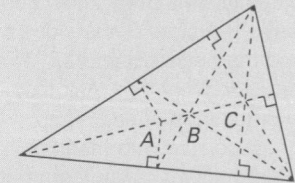

Common Error Analysis

Error: Some students have trouble constructing the orthocenter of an obtuse triangle. They may try to find a way to construct all three altitudes inside the triangle.

Construct the orthocenter of an obtuse triangle on the chalkboard or by using an overhead projector. Have students perform the same construction at their desks as you work on the chalkboard or overhead.

All triangles have three perpendicular bisectors of sides, three angle bisectors, and three altitudes. For each of these, the three intersect in a common point. All triangles also have three medians. These also intersect in a common point.

Theorem 13.6

Two medians of a triangle intersect at a point two-thirds of the distance from each vertex to the midpoint of the opposite side.

Given: Medians $\overline{AD}$ and $\overline{BE}$ of $\triangle ABC$ intersect at *O*.
Prove: $AO = \frac{2}{3}(AD)$, $BO = \frac{2}{3}(BE)$

Plan

Diagonals of a parallelogram bisect each other. Construct a parallelogram with two of its vertices at the points where the medians intersect the sides.

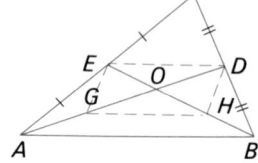

Flow diagram

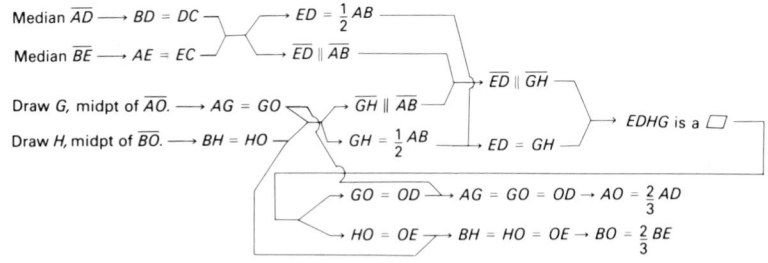

Theorem 13.7

The medians of a triangle are concurrent at a point that is two-thirds the distance from each vertex to the midpoint of the opposite side.

Definition

The **centroid** of a triangle is the point of intersection of the medians of the triangle.

EXAMPLE 1 Construct the centroid of $\triangle XYZ$.

Solution A median is a segment from a vertex of the triangle to the midpoint of the opposite side. Bisect $\overline{XY}$ and $\overline{YZ}$. Draw the two medians.

The centroid is the intersection, point *C*.

548 Chapter 13 Loci

Additional Example 1

Construct the orthocenter of obtuse triangle *XYZ*.

The orthocenter is point *P*, the intersection of two of the lines containing altitudes.

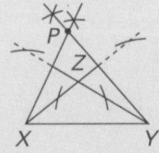

548

EXAMPLE 2 Medians $\overline{IM}$ and $\overline{JL}$ of $\triangle IJK$ intersect at N; $JN = 10x + 4$, $NL = 7x + 1$. Find JN.

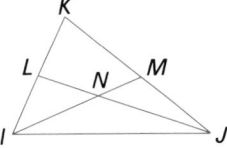

Plan By Theorem 13.6, $JN = \frac{2}{3}(JL)$ and $NL = \frac{1}{3}(JL)$. So, $JN = 2(NL)$.

Solution
$$JN = 2(NL)$$
$$10x + 4 = 2(7x + 1)$$
$$10x + 4 = 14x + 2$$
$$x = \frac{1}{2}$$

Therefore, $JN = 10x + 4 = 10 \cdot \frac{1}{2} + 4$, or 9.

Classroom Exercises

$\overline{AD}$, $\overline{BE}$, and $\overline{CF}$ are medians of $\triangle ABC$.

1. If $AP = 16$, find PD. 8
2. If $BP = 9$, find PE. 4.5
3. If $PD = 7$, find AP. 14
4. If $PF = 6$, find CP. 12
5. If $PE = 3$, find BE. 9
6. If $PF = 8$, find CF. 24
7. If $BE = 9$, find PE. 3
8. If $AD = 24$, find PD. 8

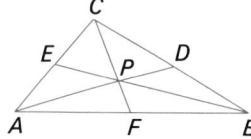

Written Exercises

1. Draw an obtuse triangle. Construct the centroid of the triangle.
2. Draw a right triangle. Construct the centroid of the triangle.
3. Draw an acute triangle. Construct the orthocenter of the triangle.
4. Draw a right triangle. Construct the orthocenter of the triangle.
5. Construct an equilateral triangle. Construct both the orthocenter and the centroid of the triangle. Where do these two points lie?

$\triangle ABC$ has medians $\overline{AD}$, $\overline{BE}$, and $\overline{CF}$ intersecting at P.

6. If $CP = 4x - 6$ and $PF = x + 1$, find PF. 5
7. If $DP = 2x + 3$ and $PA = 6x - 14$, find PA. 46
8. If $PF = x - 3$ and $CF = 2x + 1$, find PC. 14
9. If $BE = 2x + 9$ and $EP = x - 2$, find BP. 26
10. If $BP = 4x$ and $BE = 5x + 5$, find EP. 10
11. If $AP = 2x - 6$ and $AD = 2x + 9$, find DP. 15

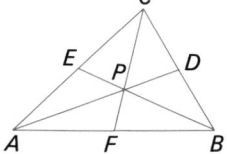

12. Draw an acute, a right, and an obtuse triangle. Show that the orthocenter may lie within, on, or outside such triangles, respectively. Then show that the centroid always lies in the interior of the triangle.

Additional Example 2

Medians $\overline{CE}$ and $\overline{AD}$ of $\triangle ABC$ intersect at F; $EF = x + 3$, $CF = 7x - 9$.
Find CF. 12

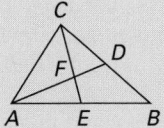

In $\triangle ABC$, medians $\overline{BD}$ and $\overline{AE}$ intersect at P. (Ex 1–3)

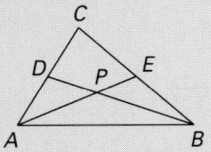

1. If $PB = 12$, find DB 18
2. If $DP = x + 5$ and $PB = 3x + 8$, find PB. 14
3. If $AP = 5x - 13$ and $PE = x + 1$, find AE. 18
4. In $\triangle RST$, m $\angle S = 90$, $RS = 12$, $SK = 9$, and $KT = 9$. How far is the centroid, C, from K? 5

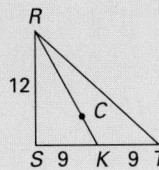

Closure

Ask the students if the three altitudes of a triangle will always be concurrent. Yes
Ask them what their intersection is called. The orthocenter Ask them if this point will be inside the triangle, outside the triangle, or on the triangle itself. It can be in any of the three regions depending on whether the triangle is right, acute, or obtuse. Ask them if the three medians of a triangle will always be concurrent. Yes If so, ask them where this point will be. It is two thirds the distance from each vertex to the midpoint of the opposite side. Ask them what this point is called. The centroid

◼◼◼FOLLOW UP

Guided Practice

Classroom Exercises 1–8

Independent Practice

🅰 Ex. 1–12, 🅱 Ex. 13–19, 🅲 Ex. 20–23

Basic: WE 1–12 Brainteaser

Average: WE 1–19 odd, Brainteaser

Above Average: WE 1–23 odd, Brainteaser

Written Exercises

1. Construct ⊥ bis to 2 sides. Draw median to each of the 2 sides. Pt of inters will be centroid.

2. Same as Ex 1.

3. Constr alt to 2 sides; pt of inters is orthocenter.

4. Constr alt to 2 sides; pt of inters will be vert of rt ∠, the orthoctr.

5. In equil △, alt is also median; centroid, orthocenter same pt.

12.

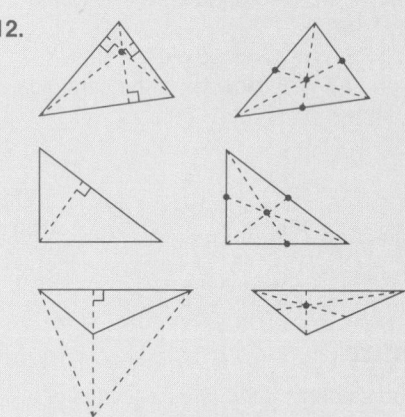

19. Except for its endpts, a median contains only pts in int of △. Intersection of medians must be int pt.

See pages 665 and 666 for the answers to Written Ex. 20–23, Mixed Review Ex. 1–2, and Brainteaser.

△*GHI* has altitude $\overline{IJ}$, median $\overline{IK}$, and centroid *C*.

13. If *JK* = 9 and *IJ* = 12, find *IC*.   10
14. If *JK* = 15 and *IJ* = 36, find *CK*.   13
15. If *CK* = 2 and *JK* = 3, find *IJ*.   3√3
16. If *IC* = 6 and *IJ* = 8, find *JK*.   √17

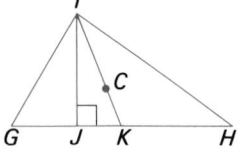

17. An isosceles triangle has sides measuring 10 cm, 10 cm, and 12 cm. How far above the base is the centroid? $2\frac{2}{3}$ cm

18. An equilateral triangle has sides measuring 12 cm. How far above the base is the centroid?   2√3 cm

19. Based on Theorem 13.7, present an argument explaining why a centroid could never lie on or outside a triangle.

20. Prove Theorem 13.5.   21. Prove Theorem 13.7.

22. Prove that the three medians of an equilateral triangle form, with the sides, six congruent triangles.

23. Prove that in an equilateral triangle the incenter, the circumcenter, the orthocenter, and the centroid are the same point.

Mixed Review

1. Draw two parallel lines 2 in. apart. Sketch and describe the locus of points in a plane equidistant from these two lines.   *13.1*

2. Draw two points 2 in. apart. Sketch and describe the locus of points in a plane equidistant from these two points.   *13.1*

3. If the lengths of two legs of a right triangle are 5 cm and 12 cm, what is the length of the hypotenuse?   *9.2*   13 cm

4. If the lengths of the hypotenuse of a right angle is 12 cm and the length of one leg is 4 cm, what is the length of the other leg?   *9.2*   8√2 cm

◢◢ *Brainteaser*

Shadows are a special type of locus. Consider shadows formed by an ordinary light such as a street light. In the illustration, the shadow shown is the intersection of the pavement with a cone, where the light source is considered to be a point. If the shadow had been formed by sunlight, it would be (for all practical purposes) the intersection of the pavement with a cylinder. Why would there be this difference for the two light sources? What shapes of shadows would be formed on a flat surface by holding a circular disk in different positions? A square piece of cardboard? A right triangular piece of cardboard?

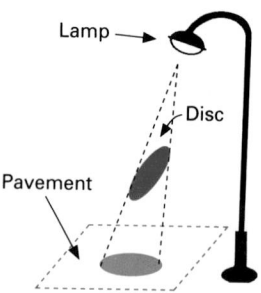

Enrichment

Challenge the students to draw a triangle *ABC* such that *CA* = 5 cm, *CB* = 3 cm and median *CE* to side $\overline{AB}$ = $3\frac{1}{2}$ cm. First draw any triangle *A'B'C'* with *C'A'* = 5 and *C'B'* = 3. Then draw median

$\overline{C'D'}$. Extend $\overline{C'D'}$ to *E'*, where *C'D'* = *D'E'*. Then *A'E'B'C'* is a parallelogram, since diagonals $\overline{C'E'}$ and $\overline{A'B'}$ bisect each other, and *B'E'* = 5. So if *C'E'* were 7, then *C'D'* = $3\frac{1}{2}$, as desired. Therefore, the solution lies in first constructing a triangle *CBE* whose sides are 3, 5, and 7. Bisect $\overline{CE}$ to locate *D*. Extend $\overline{BD}$ to *A*, where *BD* = *DA*. Then △ *ABC* is the required triangle.

Chapter 13 Review

Key Terms

centroid   (p. 548)
circumcenter   (p. 542)
circumcircle   (p. 543)
circumscribed   (p. 543)
concurrent lines   (p. 541)

incenter   (p. 543)
incircle   (p. 544)
inscribed   (p. 544)
locus, loci   (p. 529)
orthocenter   (p. 547)

Key Ideas and Review Exercises

13.1   To determine a locus, locate those points and only those points that satisfy the condition of the locus.

1. Sketch and describe the locus of points in a plane that is 9 cm from a given point P.

2. Sketch and describe the locus of points in a plane that is less than 2 cm from a given line.

13.2   To determine a locus that satisfies two or more given conditions, locate the set of points that satisfies each condition. The intersection of these sets of points is the desired locus.

3. Sketch and describe the locus of points in a plane that are 5 cm from point P *and* 12 cm from point Q if P and Q are 13 cm apart.

13.3   To determine a locus in space, locate those points in all planes that satisfy the given conditions.

4. Sketch and describe the locus of points in space that are 2 cm from a given line.

5. Sketch and describe the locus of points in space that are less than 2 cm from a given point.

13.4   To locate the *circumcenter* of a triangle, construct the perpendicular bisectors of two sides. To locate the *incenter* of a triangle, construct the bisectors of two angles.

6. Draw an obtuse triangle. Construct the circumcenter of the triangle.

7. Draw a right triangle. Construct the incenter of the triangle.

13.5   To construct the *orthocenter* of a triangle, construct two altitudes. The intersection of the lines containing the altitudes is the orthocenter.

To construct the *centroid* of a triangle, construct two medians. The point of intersection is the centroid.

8. Draw an obtuse triangle. Construct the orthocenter of the triangle.

9. Draw an acute triangle. Construct the centroid of the triangle.

Chapter Review

1. Circle with pt P as ctr, radius of 9 cm.
2. Region between two ∥ lines 4 cm apart, each ∥ to and 2 cm from the given line.
3. 2 pts; inters of ⊙ with P as ctr, radius of 5 cm and ⊙ with Q as ctr, r of 12.
4. Cylindrical surface with given line as axis, radius of 2 cm.
5. Int of sphere with given pt as ctr, radius of 2 cm.
6–9. Check student constructions.

Chapter Test

1. ⊥ bis of the seg det by the 2 pts.
2. Line ∥ to given lines and 2 cm from each.
3. Int of ⊙ with pt as ctr, radius of 4 cm.
4. 2 ⊙s concentric with given ⊙, with radii of 7 cm and 3 cm, respectively.
5. 2 pts determined by inters of ⊙ with radius of 3 cm about *P* and line ∥ to *l*, midway between *P* and *l*.

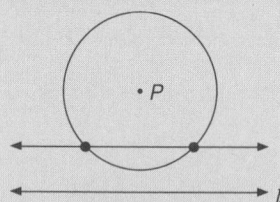

6. Diam $\overline{AB}$, a seg of length 4 cm; all points in int of circle with radius of 2 cm about *P* on angle bis of ∠*O*.

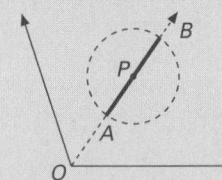

7. Line seg with the vertex as one endpt and a point on the ∠ bis that is a distance *r* from the vertex as the other endpt. The endpt is not included in the locus.
8. Sphere with *P* as ctr, r of 3 cm.
9. Locus is all poss inters of circle with 2 ∥ lines. Depending on the distance *d*, the radii of the conc ⊙s, and the dist of the center of the ⊙s from *l*, there will be 0, 1, 2, 3 or 4 pts of intersection.
10. Check student constructions.
11. Check student constructions.
*15.

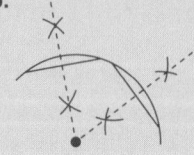

Sketch and describe the locus of points in a plane (Ex. 1–6).

1. equidistant from two points 6 cm apart
2. equidistant from two parallel lines 4 cm apart
3. less than 5 cm from a given point
4. 2 cm from a circle with a 5-cm radius
5. 3 cm from point *P* *and* 2 cm from line *l*

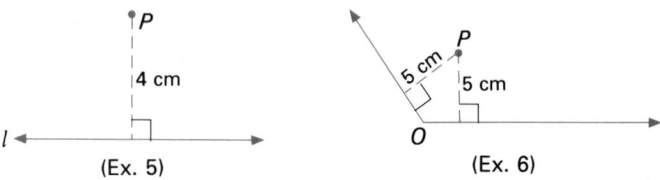

(Ex. 5)   (Ex. 6)

6. less than 2 cm from point *P* *and* equidistant from the sides of ∠*O*

7. Describe the locus of points in a plane equidistant from the sides *and* in the interior of an obtuse angle *and* less than a given distance *r* from the vertex of the angle.

8. Describe the locus of points in space 3 cm from a given point *P*.

9. Describe the locus of points in a plane equidistant from two given concentric circles *and* at a given distance *d* from a given line *l*.

10. Copy △*ABC*. Construct the incenter of the triangle.

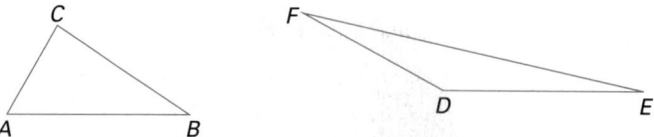

11. Copy △*DEF*. Construct the centroid of the triangle.

$\overline{GJ}$. $\overline{HK}$, and $\overline{IL}$ are medians of △*GHI*. (Ex. 12–14)

12. If *HK* = 24, find *HM*.   16
13. If *IM* = 12, find *IL*.   18
14. If *GM* = 8*x* −12 and *MJ* = *x* + 3, find *GM*.   12

*15. Copy the arc of the circle shown. Using appropriate constructions, find its center.

College Prep Test

Choose the one best answer to each question or problem.

1. The locus of points 3 in. from a
D given line and 5 in. from a point on
that line is exactly
(A) 1 point. (B) 2 points.
(C) 3 points. (D) 4 points.
(E) 5 points.

2. The locus of points in a plane at a
C given distance d from a given point
in the plane represents which of the
following?
(A) One line (B) Two lines
(C) A circle (D) Two circles
(E) A point

3. Which of the following is an equa-
D tion of the locus of points a distance
of 6 from the origin?
(A) $x = 6$ (B) $y = 6$
(C) $x^2 + y^2 = 6$ (D) $x^2 + y^2 = 36$
(E) $x^2 + y^2 = 0$

4. The locus of points in a plane of a
A given triangle equidistant from the 3
vertices of the triangle depicts which
of the following?
(A) One point (B) One circle
(C) One line (D) Two lines
(E) Two circles

5. How many points in a plane are
C equidistant from 2 given points A
and B *and* 3 in. from the line $\overleftrightarrow{AB}$?
(A) 0 (B) 1 (C) 2
(D) 3 (E) 4

6. The midpoint of the hypotenuse of a
A right triangle is
(A) equidistant from all 3 vertices.
(B) the intersection of the 3 angle
bisectors.
(C) the intersection of the 3
medians.
(D) the center of the incircle.
(E) the incenter.

7. Given: $\triangle ABC$, with medians $\overline{AY}$,
C $\overline{BX}$, and $\overline{CZ}$ concurrent at P,
$XP = 4$. Find BX.
(A) 4
(B) 8
(C) 12
(D) 6
(E) 10

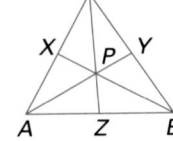

8. What is the equation of the locus
E of points equidistant from A and B?
(A) $y = x + 1$
(B) $y = x - 1$
(C) $y = 1$
(D) $y = -x$
(E) $y = x$

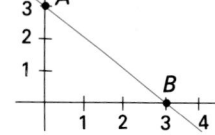

9. Given: Point P on line l
A How many points are 2 in.
from P *and* 4 in. from l?
(A) 0
(B) 1
(C) 2
(D) 3
(E) 4

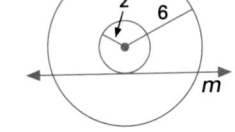

10. Two concentric circles have radii of
D 2 cm and 6 cm. Line m is tangent to
the smaller circle. How many points
are equidistant from the circles and
2 in. from m?
(A) 0
(B) 1
(C) 2
(D) 3
(E) 4

Additional Answers, page 540

13. Two lines formed by the inters of cyl
surfaces around the given lines.

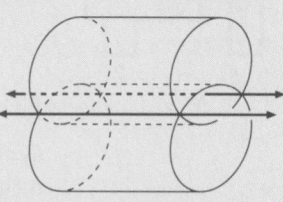

14. Circle where sphere about the pt
intersects plane nearer to given plane.

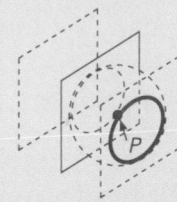

15. Locus is inters of sphere with P as ctr
and sphere with Q as ctr. Depending on
the radii of the spheres and the
distance apart of P and Q, there will be
no inters, a point of inters, a circle of
inters, or a sphere of inters. One
sphere may be inside the other.

16. A rt cylinder with alt = 4 and
diam = 10, with half-sphere of radius 5
on each end.

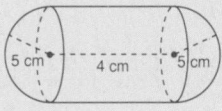

Midchapter Review

1. Circle with $r = 3$ cm.
2. Ray bisecting the angle.
3. Pts ext to a second $\odot$, concentric with
the first $\odot$, $r = 9$.
4. A plane halfway between the two pts and
$\perp$ to the seg joining them.
5. Two $\odot$s on the cyl with radius 5 in.
around line m. The $\odot$s are 10 in. from p.

Application: Site for a House

1. House may be
built anywhere
in shaded area.
2. House may be
built anywhere
in shaded area.

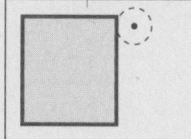 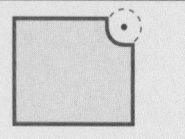

14 FIGURES IN SPACE

OVERVIEW

In this chapter, students determine whether a figure is rigid. They determine the number of faces, edges, and vertices of a polyhedron and find the volumes and areas of right prisms, cylinders, pyramids, cones, and spheres. Students consider ratios of areas and volumes of these solids, locate points in space, and compute distances between two points.

OBJECTIVES

- To determine whether given figures are rigid
- To sketch solids and to determine the number of faces, edges, and vertices in a solid
- To find the lateral areas, total areas, and volumes of solids
- To explain and use Cavalieri's Principle
- To find the volume of a cube, given its area
- To find the ratio of volumes of cylinders, given the changes in dimensions
- To find the distance between two points in space, given their coordinates

PROBLEM SOLVING

The problem solving strategy, Using a Formula, is used throughout Chapter 14 so that students can find the area and volume of three-dimensional figures. In addition, the problem solving strategy, Write an Equation, is applied in solving Exercise 16 on page 559, Exercises 21 and 22 on page 563, Exercise 27 on page 568 and Exercise 20 on page 578.

TECHNOLOGY

Calculator: Encourage students to use calculators throughout Chapter 14. Exercises marked with the calculator logo, such as Exercises 16–20 on page 567, Exercises 8–11 on page 581, and Exercise 24 on page 582, may be solved with a calculator.

SPECIAL FEATURES

Focus on Reading p. 558
Mixed Review pp. 559, 563, 568, 578, 582, 588, 591
Brainteaser p. 559
Midchapter Review p. 573
Application: Manufacturing Paper Cups p. 583
Algebra Review p. 588
Key Terms p. 592
Key Ideas and Review Exercises pp. 592–593
Chapter 14 Test p. 594
College Prep Test p. 595
Cumulative Review (Chapters 1–14) pp. 596–597

PLANNING GUIDE

Lesson	Basic	Average	Above Average	Resources
14.1 pp. 558–559	FR all CE all WE 1–14 odd Brainteaser	FR all CE all WE 1–17 odd Brainteaser	FR all CE all WE 1–20 odd Brainteaser	Reteaching p. 199 Practice p. 200
14.2 pp. 562–563	CE all WE 1–13 odd	CE all WE 1–20 odd	CE all WE 1–24 odd	Reteaching p. 201 Practice p. 202
14.3 pp. 566–568	CE all WE 1–15 odd	CE all WE 1–24 odd	CE all WE 1–27 odd	Reteaching p. 203 Practice p. 204
14.4 pp. 572–573	CE all WE 1–13 odd Midchapter Review	CE all WE 1–20 odd Midchapter Review	CE all WE 1–23 odd Midchapter Review	Reteaching p. 205 Practice p. 206
14.5 pp. 577–578	CE all WE 1–10 odd	CE all WE 1–18 odd	CE all WE 1–21 odd	Reteaching p. 207 Practice p. 208
14.6 pp. 580–583	CE all WE 1–17 odd Application	CE all WE 1–24 odd Application	CE all WE 1–28 odd Application	Reteaching p. 209 Practice p. 210
14.7 pp. 586–588	CE all WE 1–12 odd AR all	CE all WE 1–28 odd AR all	CE all WE 1–32 odd AR all	Reteaching p. 211 Practice p. 212
14.8 p. 591	CE all WE 1–12	CE all WE 1–16	CE all WE 1–18	Reteaching p. 213 Practice p. 214
Chapter 14 Review pp. 592–593	1–21 odd	all	all	
Chapter 14 Test p. 594	1–12, 16	1–12, 15–16	all	
College Prep Test p. 595	all	all	all	
Cumulative Review pp. 596–597	1–31 odd	all	all	

CE = Classroom Exercises   WE = Written Exercises   FR = Focus on Reading   AR = Algebra Review

NOTE: For each level, all students should be assigned all Mixed Review exercises.

INVESTIGATION

Manipulative: Use this investigation to introduce Chapter 14. This investigation leads the students to explore polyhedra and the property of rigidity through building models for three of the five regular polyhedra—a tetrahedron (4 faces), a dodecahedron (12 faces) and an icosahedron (20 faces).

Project Worksheet 14 helps students build paper models of the other two regular polyhedra—a cube (6 faces) and an octahedron (8 faces). This investigation can be done individually or by students working in pairs.

Students need a copy of Manipulative Worksheet 27, and the materials listed there. These same models may be made in a more permanent form with plastic or cellophane soda straws using thread and a long blunt needle, such as a yarn needle, or a hairpin to serve as a needle.

After the model is completed, use thread or string to hang it from the ceiling. With the help of the completed models, have the students practice naming the various polyhedra, the number of faces each has, and the shape that makes up each face. Also ask students to identify the edges and vertices, and say how many there are of each. Ask students to explain why polyhedra of 20 and 12 faces both have 30 edges and why a polyhedron of 20 faces has fewer vertices (12) than a polyhedron of 12 faces (20 vertices). The face of a polyhedron with 12 faces has more sides (5) than the face of a polyhedron with 20 faces (3 sides each).

Grace Chisholm Young (1868–1944)—Grace Young wrote a geometry book that, like all the texts of that time, dealt only with two-dimensional figures in a plane. However, she included patterns for folding paper to make three-dimensional models.

More About the Mathematician

After qualifying for a first-class degree at Cambridge, Grace Young left England, where women were not admitted to graduate schools, and studied at Göttingen University where she was the first woman in Germany to pass the examination for a doctorate in any field. (Kovalevskaya had received an informal degree, without examination.) The picture shows her at a meeting of the Mathematics Club. Among her many publications were a prize-winning paper on the foundations of the differential calculus and a paper on the history of the Pythagorean Theorem. She published, jointly with W.H. Young, *The First Book of Geometry*, reprinted in 1969.

14.1 Polyhedra and Rigidity

Objectives

To determine whether a given figure is rigid

To determine the number of faces, edges, and vertices of a polyhedron

Teaching Resources

Manipulative Worksheet 27
Problem Solving Worksheet 14
Project Worksheet 14
Quick Quizzes 100
**Reteaching and Practice
  Worksheets,** pp. 199, 200
Transparency 38

A newly built garage looked like the picture on the left below when it was first built. Several months later, after a heavy snow, it looked like the one on the right.

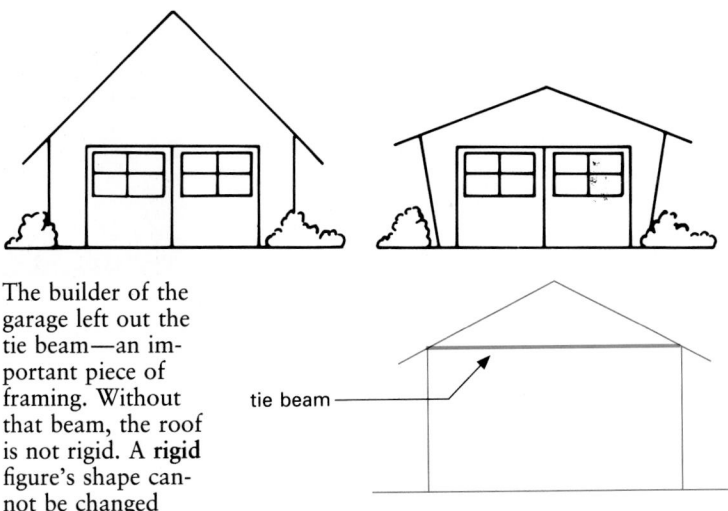

The builder of the garage left out the tie beam—an important piece of framing. Without that beam, the roof is not rigid. A **rigid** figure's shape cannot be changed without changing the lengths of the sides.

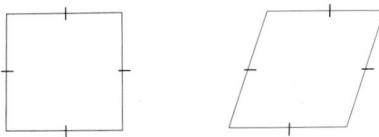

tie beam

A triangle is a rigid figure. To change the shape of a triangle, the lengths of the sides must be changed. A quadrilateral is not a rigid figure. The shape of a quadrilateral can change without changing the lengths of the sides. This is to be expected since triangles with congruent corresponding sides are congruent, while quadrilaterals with congruent corresponding sides are not necessarily congruent.

You can demonstrate that a plane figure is not rigid if there is a non-congruent figure with congruent corresponding sides.

14.1 Polyhedra and Rigidity **555**

▰▰▰GETTING STARTED

Prerequisite Quiz

1. For which of the following correspondence patterns are two triangles congruent?
AAA, SSS, ASA, SAS, AAS, SSA   SSS, ASA, SAS, AAS

2. For which of the following correspondence patterns are two quadrilaterals congruent?
AAAA, SSSS, ASASA, SASAS, AASA, SSAS   ASASA, SASAS

Motivator

Ask the students to list things they would do to ensure that a structure they are building will not fall down.   Brace the structure, keep the shape rigid, build a sturdy frame, etc.   Ask students if braces in the form of a rectangle will help and have them explain their answer.   No. The shape of a rectangle can change without changing the lengths of the sides.   Ask the class what shape of braces will make the building rigid and have them explain their answers.   A triangular shape because its shape will not change unless the lengths of the sides are changed.

Lesson Note

Review triangle congruence properties. Stress the side-side-side pattern for triangles. Demonstrate the relationship of the SSS pattern to rigidity using two matching sets of three sticks of different lengths. Form a triangle from one set of the sticks. Randomly placing the three sticks of the second set to form a triangle will result in a triangle that is always congruent to the first one. Congruent figures have the same size and shape. A rigid figure always maintains the same size and shape. Explore to find the minimum necessary conditions for a figure to be rigid. All triangular shapes are rigid. No quadrilateral shapes are rigid. In space, a polyhedron with triangular faces will be rigid.

Encourage students to learn the standard names of basic polyhedra.

Math Connections

History: Regular polyhedra are sometimes referred to as the *Platonic Solids*, named after Plato (429–328 B.C.), who believed that the atoms of the elements had these shapes. Today chemists know that this is not true. However, the arrangements of atoms into the unit cells of many crystals *do* have the shape of polyhedra.

Critical Thinking Questions

Refer students to Example 4 on page 557 and ask them to state general formulas for the number of edges and for the number of vertices that a convex regular polyhedron has. Let e = the number of edges, v = the number of vertices, f = the number of faces, s = the number of sides of each face, and t = the number of faces that share an edge (for edges formula) or a vertex (for vertex formula). Then $e = \frac{sf}{t}$ and $v = \frac{sf}{t}$.

EXAMPLE 1 Use sketches to show that a trapezoid is not a rigid figure.

Solution

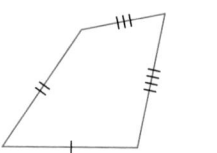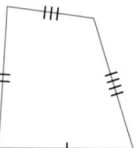

The counterexample at the right has congruent corresponding sides but not the same shape as the trapezoid at the left.

Three-dimensional or *space* figures can also be rigid. A geodesic dome is an example of a rigid structure. Its frame consists entirely of triangular shapes. If all the faces of a figure are triangular, then the figure is rigid. If any of the faces are not triangles, the figure may not be rigid.

EXAMPLE 2 Is the solid rigid? Explain why or why not.

Solution

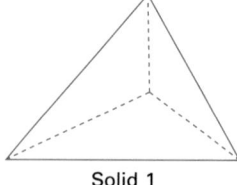 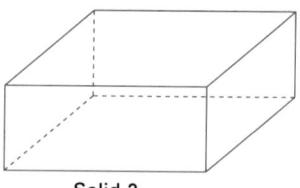

| Solid 1 | Solid 2 |

All faces of Solid 1 are triangles. Triangles are rigid. The solid is rigid.

The faces of Solid 2 are not rigid. There exists a solid with a different shape and congruent corresponding sides of faces. Solid 2 is not rigid.

Definition

A **polyhedron** is a three-dimensional figure with polygonal regions as its faces. The plural of polyhedron is *polyhedra*.

To simplify discussion, the faces of polyhedra will be referred to as polygons, rather than polygonal regions.

556 Chapter 14 Figures in Space

Additional Example 1

Draw pictures to show that a quadrilateral with one fixed right angle is not a rigid figure.

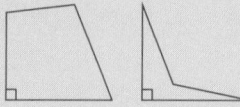

Additional Example 2

Tell whether the three dimensional figure shown is rigid. Why?   Not rigid. Quadrilateral faces are not rigid.

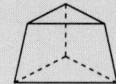

EXAMPLE 3 State whether the space figure is a polyhedron.

Solution

A polyhedron

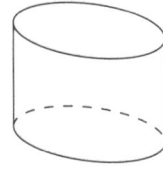

Not a polyhedron

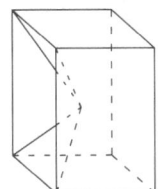

A polyhedron

Polyhedra are named by the number of faces they contain. If all faces of a polyhedron are congruent regular polygons, the polyhedron is called a **regular polyhedron**. There are exactly five regular polyhedra, as illustrated below.

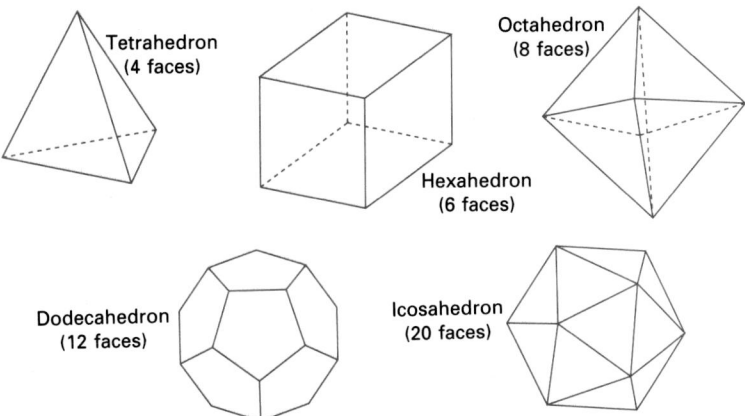

Tetrahedron
(4 faces)

Octahedron
(8 faces)

Hexahedron
(6 faces)

Dodecahedron
(12 faces)

Icosahedron
(20 faces)

The faces of a polyhedron intersect in segments called **edges**. The intersections of the edges are called the **vertices**.

EXAMPLE 4 What shape are the faces of a regular dodecahedron? How many edges and how many vertices does it have?

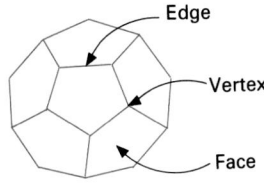

Edge

Vertex

Face

Solution The faces of a regular dodecahedron are regular pentagons. A dodecahedron has 12 faces, each of which has 5 sides. Since each edge is shared by two faces, there are $(12 \cdot 5) \div 2 = 30$ edges. Each pentagon has 5 vertices. Since each vertex is shared by three faces, there are $(12 \cdot 5) \div 3 = 20$ vertices.

14.1 Polyhedra and Rigidity **557**

Checkpoint

1. How many faces does a regular octahedron have? 8
2. What shape are the faces of a regular hexahedron? Square

Tell whether each figure is rigid. Why or why not?

3.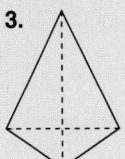

Yes, faces are △s.

4.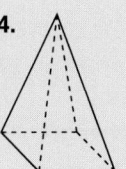

No, faces are not all △s.

Closure

Ask the students to describe the shapes that they found to be rigid. Triangles and figures in which all the faces are triangular. Ask them to explain what a rigid figure is. It is a figure whose shape cannot be changed without changing the lengths of the sides. Ask them to define a polyhedron. Definition page 556. Have them describe the following parts of a polyhedron:

faces — the sides
edges — segments which are the intersections of the faces
vertices — the intersections of the edges

Additional Example 3

Which of the following is a polyhedron?

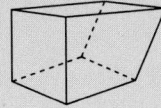

A polyhedron

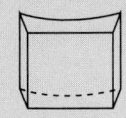

Not a polyhedron

Additional Example 4

What shape are the faces of a regular tetrahedron? How many edges and how many vertices does it have? Triangular, 6 edges, 4 vertices

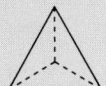

557

Guided Practice

Classroom Exercises 1–4

Independent Practice

A Ex. 1–14, **B** Ex. 15–17, **C** Ex. 18–20

Basic: FR all, WE 1–14 odd, Brainteaser

Average: FR all, WE 1–17 odd, Brainteaser

Above Average: FR all, WE 1–20 odd, Brainteaser

Additional Answers

Classroom Exercises

4. Tetrahedron, 4 ≅ equilateral triangles; Hexahedron, 6 ≅ squares; Octahedron, 8 ≅ equilateral triangles; Dodecahedron, 12 ≅ regular pentagons; Icosahedron, 20 ≅ equilateral triangles.

Written Exercises

1.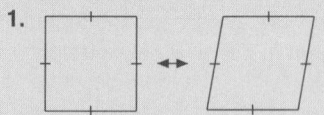

2.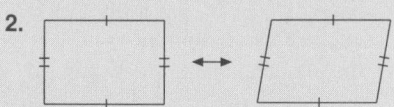

3.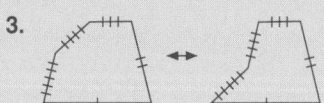

4.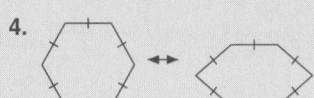

Focus on Reading

Look up the word *polygon* in a dictionary. Underline the prefix. Using this prefix, write the name for the polyhedron with the same number of faces. From your reading of this lesson, tell whether there is a regular polyhedron of that name. (Exercises 1–6)

1. pentagon Pentahedron, no
2. heptagon Heptahedron, no
3. nonagon Nonahedron, no
4. decagon Decahedron, no
5. undecagon Undecahedron, no
6. dodecagon Dodecahedron, yes
7. Find several other words, from outside the area of geometry, that use the same prefixes as in Exercises 1–6. Answers will vary.

Classroom Exercises

Is the figure a polyhedron? Explain why or why not.

1.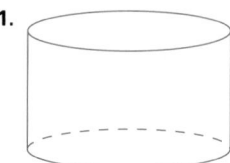

2.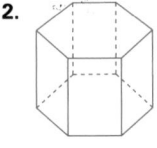

3.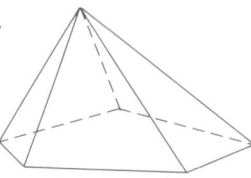

No, curved surface   Yes, all faces are polygons.   Yes, all faces are polygons.

4. Name the five regular polyhedrons. Describe each polyhedron.

Written Exercises

Make sketches to show that the polygon is not rigid.

1. a square
2. a rectangle
3. a pentagon
4. a hexagon

Tell whether the figure is rigid. Explain why or why not.

5.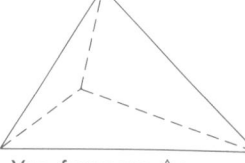

6.

7.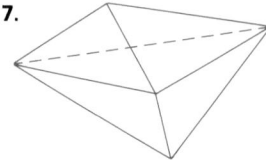

Yes, faces are △s.   No, three faces are quad.   Yes, faces are △s.

Do the edges form a rigid figure for the given regular polyhedron? Why or why not?

8. a regular tetrahedron Yes, faces are △s.
9. a regular hexahedron No, faces are quad.
10. a regular octahedron Yes, faces are △s.
11. a regular icosahedron Yes, faces are △s.

Enrichment

If a cube is painted and then cut twice, both vertically and horizontally, so that 27 congruent cubes result, how many of the resulting cubes are painted on 3 faces? on 2 faces? on 1 face? are not painted at all? 3 faces—8; 2 faces—12; 1 face—6; no faces—1.

A further challenge is to try this with three or more cuts, and finally to look for the general rule where n is the number of cuts. 3 faces—always 8; 2 faces—$12(n - 1)$, 1 face—$6(n - 1)^2$, no faces—$(n - 1)^3$

For the regular polyhedron, find the number of faces (F), the number of edges (E), and the number of vertices (V).

12. a regular tetrahedron *F:4, E:6, V:4*   **13.** a regular octahedron *F:8, E:12, V:6*

14. a regular dodecahedron *F:12, E:30, V:20* **15.** a regular icosahedron *F:20, E:30, V:12*

16. Using your results from the Exercises 12–15, find a formula relating F, E, and V for regular polyhedra. Does this formula work for polyhedra that are not regular? *V + F = E + 2; yes*

17. Explain how SSS congruency guarantees that any given triangle is rigid.

18. A regular icosahedron can be assembled using the pattern shown. Copy the pattern on a larger scale, cut it out, and build your own icosahedron.

19. Design a pattern for building a regular hexahedron. Verify the pattern by assembling it.

20. The sum of the angle measures around a common vertex in a plane is 360. What is the sum of the measures of the angles of each regular polyhedron around a common vertex? Does this sum appear to be proportional to the number of faces of the polyhedron?

Mixed Review

Find the area of the polygon. *12.3, 12.5, 12.6*

1. parallelogram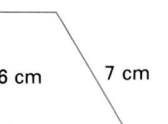

6 cm 7 cm

8 cm

48 cm²

2. trapezoid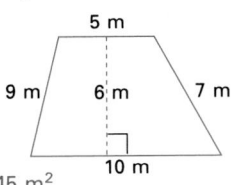

5 m

9 m 6 m 7 m

10 m

45 m²

3. regular hexagon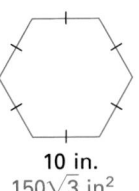

10 in.

150√3 in²

4. Write the truth table for $(p \lor q) \land \sim p$. *1.7, 3.8*

◤◤◤ *Brainteaser*

In the figure below, can you reposition two matches to make three triangles?

In the figure at the right, can you reposition just three matches to make three squares?

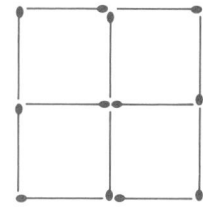

---

17. No noncongruent counterexample of a triangle with congruent sides can be provided.

19. Check students' patterns and models.

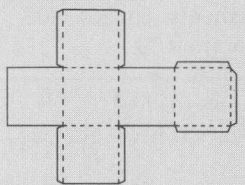

20. Tetrahedron 180; hexahedron 270; octahedron 240; dodecahedron 324; icosahedron 300. No.

Mixed Review

4.

p	q	$p \lor q$	$\sim p$	$(p \lor q) \land \sim p$
T	T	T	F	F
T	F	T	F	F
F	T	T	T	T
F	F	F	T	F

Brainteaser

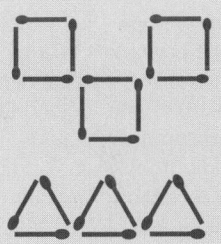

◤■ GETTING STARTED

Prerequisite Quiz

1. How many sides does a pentagon have?
   5
2. How many sides does an octagon have?
   8
3. How many vertices does a hexagon have? 6
4. How many faces does a hexahedron have? 6
5. Draw a tetrahedron. See page 557.

Motivator

Ask the students how they would describe the geometric properties of a typical cardboard box. It is a three dimensional figure with square and rectangular sides. Of a typical soup can. It is a three dimensional figure with curved surfaces and circles for the bases. Ask them if there are any congruent parts. Yes Any parallel parts. Yes Ask the students what the box and the can have in common. They both have congruent parts and parallel parts.

◤■ TEACHING SUGGESTIONS

Lesson Note

Certain three-dimensional figures have common names. Prisms and cylinders are of special importance, since many practical applications are based on them. Cans and boxes are examples of right circular cylinders and rectangular prisms.

14.2 Prisms and Cylinders

Objectives   To classify and draw prisms and cylinders
To identify parts of prisms and cylinders

Geometric figures contained in a plane, such as polygons or circles, are *two-dimensional* figures. *Three-dimensional* figures, or solids, cannot be contained in a plane. Prisms, which have polygonal faces, and cylinders, which have curved surfaces, are special three-dimensional figures.

◤ Definitions

> A **prism** is a polyhedron with two congruent polygonal faces, called the **bases**, in parallel planes. The remaining faces, called the **lateral faces**, are parallelograms. Each lateral face shares an edge with each of the bases.

Prisms are often classified by their bases. For example, a triangular prism has triangular bases.

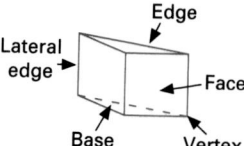

EXAMPLE 1   How many faces, edges, and vertices does a quadrilateral prism have?

Solution   Sketch such a prism and count.
The prism has 6 faces, 12 edges, and 8 vertices.

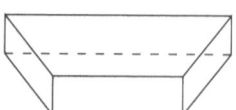

In a **right prism,** the lateral edges are *perpendicular* to the bases. The lateral faces are rectangular. A prism that is not a right prism is called **oblique prism.** A **rectangular solid** is a right prism in which the bases and lateral faces are rectangles. A **parallelepiped** is a prism in which the bases and the lateral faces are parallelograms.

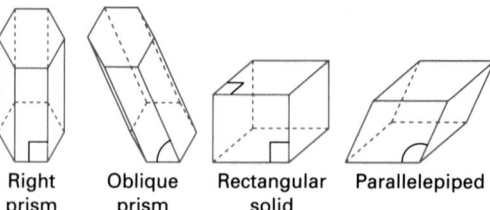

Right prism   Oblique prism   Rectangular solid   Parallelepiped

560   Chapter 14 Figures in Space

Additional Example 1

How many faces, edges, and vertices does a pentagonal prism have?
7 faces, 15 edges, 10 vertices

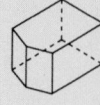

EXAMPLE 2 Draw a right hexagonal prism.

Solution

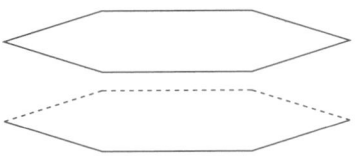 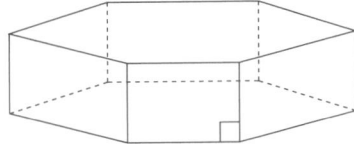

Draw two congruent hexagons. The corresponding sides should be parallel and vertically aligned.

Connect the corresponding vertices. Use dashed lines to indicate edges hidden from view.

Like a prism, a **cylinder** consists of two congruent, parallel bases. However, the bases of a cylinder are circles instead of polygons. The *lateral surface* of the cylinder is the curved surface between the bases. It can be thought of as consisting of an infinite number of parallel segments with their endpoints on the bases.

If a segment connecting a pair of corresponding points of the bases is perpendicular to the bases, then the cylinder is a *right cylinder*. A cylinder that is not a right cylinder is an *oblique cylinder*.

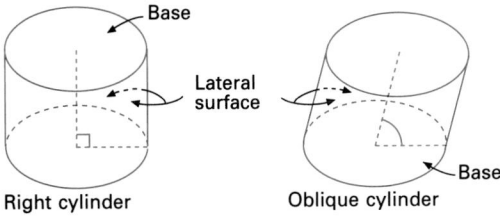

Right cylinder   Oblique cylinder

EXAMPLE 3 Draw a right cylinder.

Solution

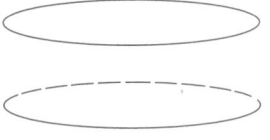 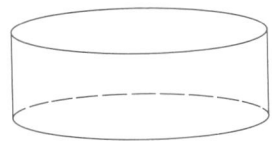

Draw two congruent ellipses, with corresponding points aligned vertically.

Connect two pairs of corresponding points.

Math Connections

Physics: In 1666, Isaac Newton discovered that white light contains all of the colors in the spectrum. When passing through a prism, the wavelengths of the light slow down in varying degrees—violet the most, red the least—causing the colored rays to bend in different amounts and spread out in a rainbow pattern. He also found that with a second pass through another prism that is set up to invert the first refraction, the resulting beam appears pure white again. Since diamonds are like many separate prisms, they can bend light farther than any other transparent substance, which allows them to produce more vivid rainbows.

Critical Thinking Questions

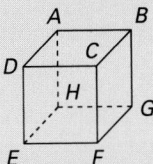

Analysis: Refer students to the diagram above and ask them to name the lateral faces if *CBGF* is a base. *ABCD, ABGH, EFGH,* and *CDEF* if *ABGH* is a base. *FGBC, ABCD, ADEH,* and *EFGH* Ask students to write a formula that relates the number of sides, *S*, of each base of a prism to the number of faces, *F*, of the prism. *S = F − 2, or F = S + 2*

Common Error Analysis

Error: Students have difficulty drawing two-dimensional diagrams showing three-dimensional figures.

To draw prisms and cylinders, have students follow the procedures shown in Examples 2 and 3 on page 561.

Additional Example 2

Draw a right octagonal prism.

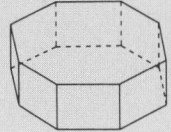

Additional Example 3

Draw an oblique circular cylinder.

561

Checkpoint

Which of the following appears to be a prism? A cylinder? Neither?

1. prism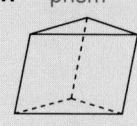

2. prism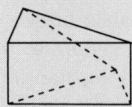

3. neither

4. neither

Draw each of the following.

5. a right equilateral triangular prism. See figure in Written Ex. 1.

6. an oblique circular cylinder. See page 561.

Closure

Ask the students how a prism and a cylinder are alike. They both have two congruent, parallel bases. Ask them how they are different. A prism has polygonal bases and the lateral faces are parallelograms. A cylinder has circular bases and the lateral surface is a curved surface between the bases. Ask them how prisms are classified. By their bases Ask the students to describe a right cylinder. It is a cylinder that contains a segment connecting a pair of corresponding points of the bases which is perpendicular to the bases. A right prism. The lateral edges are perpendicular to the bases.

◤◥ FOLLOW UP

Guided Practice

Classroom Exercises 1–9

Independent Practice

A Ex. 1–13, **B** Ex. 14–20, **C** Ex. 21–24

Basic: WE 1–13 odd
Average: WE 1–20 odd
Above Average: WE 1–24 odd

Classroom Exercises

Does the figure appear to be a prism, a cylinder, or neither?

1.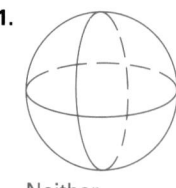
Neither

2.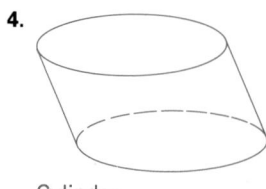
Cylinder

3.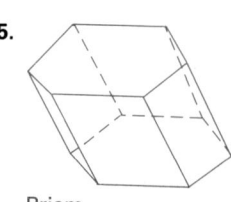
Prism

4.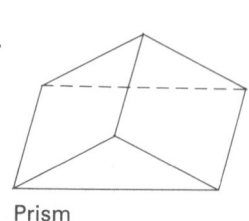
Cylinder

5.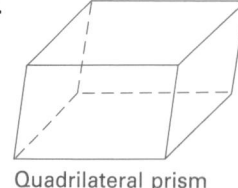
Prism

6.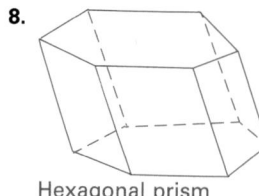
Prism

Classify the prism by its base.

7.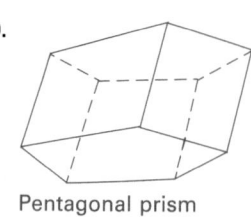
Quadrilateral prism

8.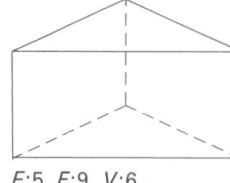
Hexagonal prism

9.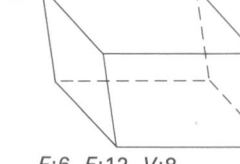
Pentagonal prism

Written Exercises

How many faces, edges, and vertices does the prism have?

1.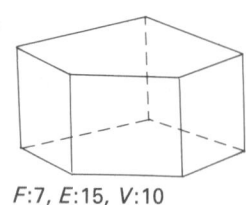
F:5, *E*:9, *V*:6

2.
F:6, *E*:12, *V*:8

3.
F:7, *E*:15, *V*:10

How many lateral faces and lateral edges does the prism have?

4. a triangular prism *F*:3, *E*:3
5. a pentagonal prism *F*:5, *E*:5
6. an octagonal prism *F*:8, *E*:8
7. a dodecagonal prism *F*:12, *E*:12

Enrichment

Triangular prisms made of glass or other transparent materials should be available from the science department. Demonstrate to students how a prism separates light into its various colors.

Have selected students prepare a report on the laws of refraction and the relationship between the color of light and its wave length.

Draw an example of the described figure.

8. a right triangular prism
9. an oblique rectangular prism
10. an oblique pentagonal prism
11. a right heptagonal prism
12. a right cylinder
13. an oblique cylinder

True or false? (Exercises 14–17)

14. A prism has the same number of lateral faces as lateral edges.   T
15. A prism has the same number of faces as vertices.   F
16. An octagonal prism has twice the number of faces as a quadrilateral prism.   F
17. An octagonal prism has twice the number of vertices as a quadrilateral prism.   T
18. Prove that any two lateral edges of a prism are parallel and congruent.
19. Prove that any two lateral faces of a right prism with square bases are congruent.
20. Prove that two opposite lateral faces of a rectangular solid are congruent.
21. Find a formula that relates the number of vertices in a prism to the number of sides in each base. $V = 2 \cdot S$
22. Find a formula that relates the number of edges in a prism to the number of sides in each base. $E = 3 \cdot S$
23. A diagonal of a prism is a segment with two vertices in different faces as endpoints. Prove that any two diagonals of a rectangular solid are congruent.
24. A right triangular prism can be assembled using the pattern shown. Design a similar pattern for a rectangular solid, a parallelepiped, a right hexagonal prism, an oblique hexagonal prism, and a right cylinder.

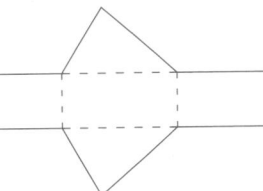

Mixed Review

1. The measure of an angle is 3 less than twice the measure of its complement. What is the measure of the angle?   *1.6*   59
2. The measure of an angle is 12 more than twice the measure of its supplement. What is the measure of the angle?   *1.6*   124
3. Write the conjunction of the following statements: Lines are long. Segments are short.   *1.7*   Lines are long and segments are short.
4. Write an indirect proof for this statement: A triangle cannot contain two obtuse angles.   *3.6*
5. Prove that the measure of an exterior angle of a triangle is equal to the sum of the measures of the two remote interior angles.   *3.7*

14.2 Prisms and Cylinders   **563**

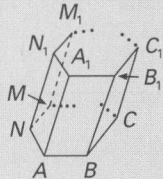

Teaching Resources

Application Worksheet 14
Quick Quizzes 102
Reteaching and Practice
   Worksheets, pp. 203, 204
Transparency 39

▰▰▰ GETTING STARTED

Prerequisite Quiz

Find the area of the given figure.

1. 24 cm²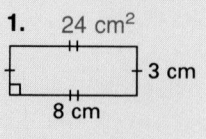

2. 35 m²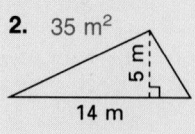

3. 63 in²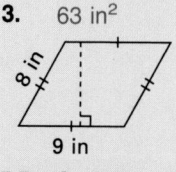

4. 126 ft²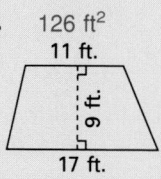

Motivator

Ask the students to suppose that they are in charge of manufacturing cans and boxes for a grocery company. As part of their job, they order the sheet metal from which the cans are made and the cardboard from which the boxes are made. Ask them what measurements they must know in order to know how much material is needed. The length, width, altitudes and radius.

▰▰▰ TEACHING SUGGESTIONS

Lesson Note

Prepare paper cutouts for several solids that can be unfolded to show the surface clearly. Use these to illustrate what is meant by the surface of a polyhedron or cylinder. Have students help find the areas of the parts of these unfolded surfaces using standard area formulas.

14.3 Areas of Prisms and Cylinders

Objectives To find the areas of right prisms and cylinders
To compare areas of similar solids

The *total area of a prism* is the sum of the areas of its faces. This is called the *surface area*. The sum of the areas of the lateral faces only is called the *lateral area* of the prism.

EXAMPLE 1 Find the total area of the rectangular solid.

Plan All faces are rectangles. Find the sum of the areas of the six faces.

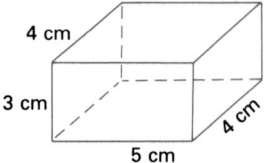

Solution $A = 2 \cdot (5 \cdot 4) + 2 \cdot (4 \cdot 3) + 2 \cdot (5 \cdot 3)$
$A = 40 + 24 + 30 = 94 \text{ cm}^2$

In finding areas of prisms it is helpful to visualize the surface as being flattened out. The figure below shows how a right hexagonal prism would appear when unfolded. The resulting surface consists of one large rectangle connected to two hexagons.

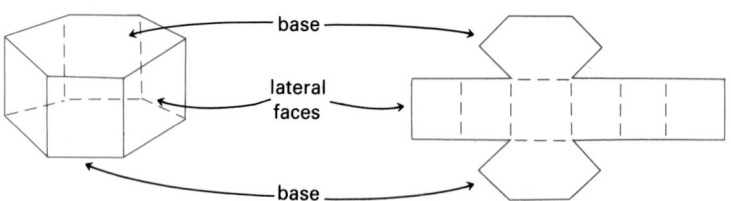

Definition An **altitude** of a prism or cylinder is any segment perpendicular to the planes containing the two bases with endpoints in these planes.

Theorem 14.1 The lateral area of a right prism is the product of the perimeter of a base and the length of an altitude ($L = ph$).

EXAMPLE 2 Find the lateral area of a right square prism with 6-cm base edges and a 10-cm altitude.

Solution The perimeter of the base is $4 \cdot 6 = 24$ cm.
$L = ph$
$L = (4 \cdot 6) \cdot 10 = 240 \text{ cm}^2$

564 Chapter 14 Figures in Space

Additional Example 1

Find the area of the right triangular prism. $A = (10 \cdot 5) + (6 \cdot 5) + (8 \cdot 5) + 2[\frac{1}{2}(6 \cdot 8)]$
$A = 50 + 30 + 40 + 48$, or 168 cm²

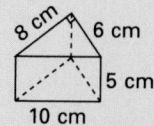

Additional Example 2

Find the lateral area of a right equilateral triangular prism with a 7 cm altitude and 5 cm base edges.
$L = ph$
$L = (3 \cdot 5) \cdot 7$, or 105 cm²

If a right cylinder is cut open and unwrapped, the resulting surface consists of a rectangle and two circles. The length of one side of the rectangle equals the circumference of the circle.

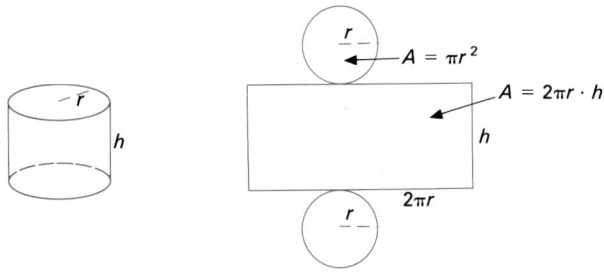

$$A = \pi r^2$$
$$A = 2\pi r \cdot h$$

Theorem 14.2

The lateral area of a right cylinder is the product of the circumference of a base and the length of an altitude. The lateral area of a right cylinder with radius r and altitude of length h is $2\pi rh$.

Theorem 14.3

The total area of a right cylinder with radius of length r and altitude of length h is $2\pi r^2 + 2\pi rh$, or $2\pi r(r + h)$.

EXAMPLE 3   A diameter of a base of a right cylinder is 16 cm long. An altitude is 5 cm long. Find the lateral area and the total area.

Solution

Since $d = 16$ cm, $r = 8$ cm
$$
\begin{aligned}
\text{lateral area} &= 2\pi rh \\
&= 2\pi \cdot 8 \cdot 5 \\
&= 80\pi \text{ cm}^2 \\
\text{total area} &= 2\pi r(r + h) \\
&= 2\pi \cdot 8 \cdot (8 + 5) \\
&= 208\pi \text{ cm}^2
\end{aligned}
$$

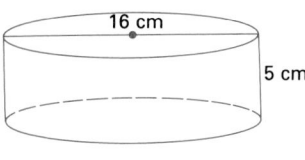

16 cm

5 cm

Similar solids have the same shape but may have different sizes.

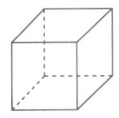
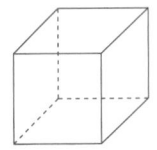

Definition

Two polyhedra are **similar** if all corresponding angles between faces and between edges are congruent, and the lengths of corresponding edges are proportional.

14.3 Areas of Prisms and Cylinders   **565**

Additional Example 3

A base of a right circular cylinder has a radius of 5 cm. The altitude is 8 cm. Find the lateral area and total area.
$L = 2\pi rh$
$\quad = 2\pi \cdot 5 \cdot 8$, or 80π cm^2
$T = 2\pi r(r + h)$
$\quad = 2\pi \cdot 5 \cdot (5 + 8)$, or 130π cm^2

Checkpoint

1. Find the lateral area and total area of the rectangular solid. The bases are shaded.

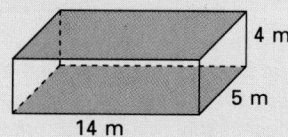

Lat. $A = 40 \text{ m}^2 + 112 \text{ m}^2 = 152 \text{ m}^2$
Tot. $A = 152 \text{ m}^2 + 140 \text{ m}^2 = 292 \text{ m}^2$

2. Draw a picture to illustrate how the surface of the given right prism would appear if unfolded. See page 564.

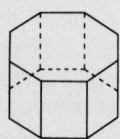

3. Find the lateral area and the total area of the right circular cylinder.

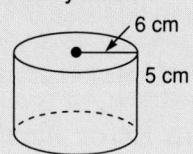

Lat. $A = 60\pi \text{ cm}^2$
Tot. $A = (60\pi + 72\pi) = 132 \text{ cm}^2$

Closure

Ask the students to give the formulas for the following areas:

lateral area of a right prism   Theorem 14.1 page 564
lateral area of a right cylinder   Theorem 14.2 page 565
total area of a prism   the sum of the areas of its faces
total area of a right cylinder   Theorem 14.3 page 565

▰FOLLOW UP

Guided Practice

Classroom Exercises 1–6

Independent Practice

A Ex. 1–15, **B** Ex. 16–24, **C** Ex. 25–27

Basic: WE 1–15 odd
Average: WE 1–24 odd
Above Average: WE 1–27 odd

EXAMPLE 4   The lengths of edges of two rectangular solids have the ratio 3 to 2. What is the ratio of their areas?

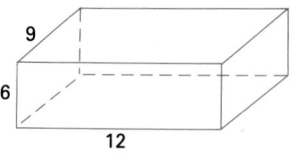
Prism 1

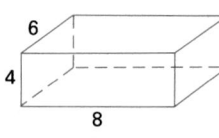
Prism 2

Solution $\text{Area}_1 = 2 \cdot (6 \cdot 12) + 2 \cdot (9 \cdot 12) + 2 \cdot (6 \cdot 9) = 468$
$\text{Area}_2 = 2 \cdot (4 \cdot 8) + 2 \cdot (6 \cdot 8) + 2 \cdot (4 \cdot 6) = 208$

$$\frac{\text{Area}_1}{\text{Area}_2} = \frac{468}{208} = \frac{9}{4}$$

The ratio of the areas in Example 4 is the square of the ratio of the lengths of the edges of the two similar polyhedra: $(\frac{3}{2})^2 = \frac{9}{4}$. This example suggests a more general relationship of linear and area measures.

Theorem 14.4   If the ratio of the lengths of corresponding edges of two similar polyhedra is $\frac{a}{b}$, then the ratio of the lateral areas and total areas is $(\frac{a}{b})^2$.

Classroom Exercises

Find the lateral area of the right prism. (The bases are shaded.)

1.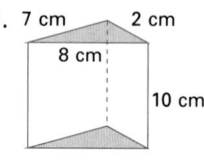
170 cm²

2.
156 in²

3.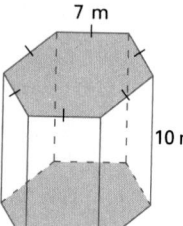
420 m²

For Exercises 4–6, use the right cylinder.

4. Find the lateral area. 48π m²
5. Find the area of the bases. 32π m²
6. Find the total area. 80π m²

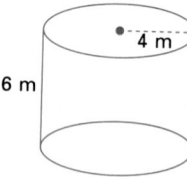

Additional Example 4

The lengths of the edges of two right triangular prisms are in the the ratio 4 to 7. What is the ratio of their total areas?

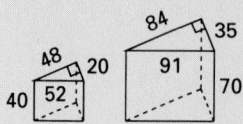

First area = $40 \cdot 52 + 40 \cdot 20 + 40 \cdot 48 + 2 \cdot \frac{1}{2}(20 \cdot 48)$ or 5760
Second area = $70 \cdot 91 + 70 \cdot 35 + 70 \cdot 84 + 2 \cdot \frac{1}{2}(35 \cdot 84)$ or 17640
$\frac{\text{First area}}{\text{Second area}} = \frac{5760}{17640}$, or $\frac{16}{49}$

Written Exercises

Find the lateral area and the total area of the rectangular solid, the bases of which are shaded.

1.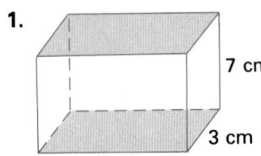
7 cm
3 cm
8 cm
154 cm², 202 cm²

2.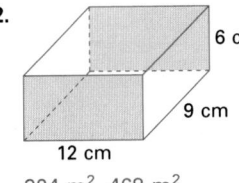
6 cm
9 cm
12 cm
324 m², 468 m²

3.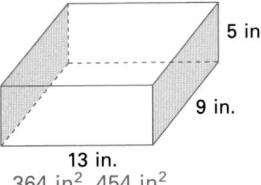
5 in.
9 in.
13 in.
364 in², 454 in²

Draw a picture to illustrate how the surface of the right prism would appear when unfolded.

4.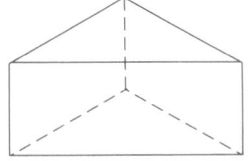

5.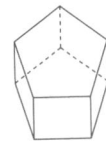

6.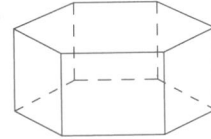

Find the lateral area and total area of the right cylinder given the radius and height.

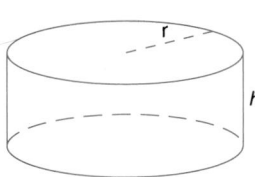
r
h

7. $r = 10$ cm, $h = 6$ cm 120π cm², 320π cm²
8. $r = 5$ in., $h = 8$ in. 80π in², 130π in²
9. $r = 7$ m, $h = 9$ m 126π m², 224π m²
10. $r = 9$ cm, $h = 7$ cm 126π cm², 288π cm²

Find the ratio of the areas of two similar prisms for the given ratio of lengths of corresponding edges.

11. 3 to 5   9 to 25
12. 9 to 5   81 to 25
13. 1 to 4   1 to 16
14. 15 to 21   25 to 49

15. The walls and ceiling of a warehouse are to be painted. How many square meters must be covered if the warehouse is 120 m by 96 m with a ceiling 3 m high?   12,816 m²

Make a sketch of the right prism described. Find the lateral area and the total area. (Exercises 16–19)

16. The length of an altitude is 12 cm. Each base is a right triangle with legs measuring 6 cm and 8 cm.   288 cm², 336 cm²

17. The length of an altitude is 15 cm. Each base is a right triangle with legs measuring 8 cm and 9 cm. (Round to the nearest tenth.)   435.6 cm², 507.6 cm²

18. The length of an altitude is 10 cm. Each base is a regular hexagon with an edge measuring 4 cm.   240 cm², $(240 + 48\sqrt{3})$ cm², or 323 cm²

Enrichment

The following can provide additional practice in visualization and an intuitive introduction to volume. Equal amounts of water are poured into the two containers shown. In which container will the water level be higher?   Container 2

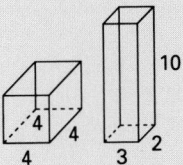
10
4
4
4
3
2

Additional Answers

Written Exercises

4–6. Check student drawings.

21.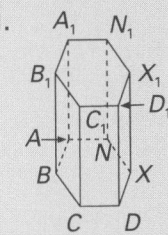
A_1 N_1
B_1 X_1
C_1 D_1
A N
B X
C D

Statement	Reason
1. Rt prism with alt h and bases $ABCDXN$ and $A_1B_1C_1D_1X_1N_1$	1. Given
2. Lateral area $L =$ Area (ABB_1A_1) + Area (BCC_1B_1) + Area (CDD_1C_1) + Area (DXX_1D_1) + Area (XNN_1X_1) + Area (NAA_1N_1)	2. Def of lateral area of prism
3. $\overline{AA_1}, \overline{BB_1}, \overline{CC_1}, \overline{DD_1}, \overline{XX_1}, \overline{NN_1}$ are $\perp$ to bases.	3. Def of rt prism
4. $h = AA_1 = BB_1 = CC_1 = DD_1, = XX_1 = NN_1$	4. Def of alt of prism
5. $\overline{AB}, \overline{BC}, \overline{CD}, \overline{DX}, \overline{XN},$ are $\overline{NA}$ and bases of $\square$s	5. Def of prism
6. Area $(ABB_1A_1) = AB \cdot h$, Area $(BCC_1B_1) = BC \cdot h$, Area $(CDD_1 C_1) = CD \cdot h$, Area $(DXX_1 D_1) = DX \cdot h$, Area $(XNN_1X_1) = XN \cdot h$, Area $(NAA_1 N_1) = NA \cdot h$	6. Area $\square = bh$
7. $L = AB \cdot h + BC \cdot h + CD \cdot h + DX \cdot h + XN \cdot h + AN \cdot h$	7. Sub
8. $L = (AB + BC + CD + DX + XN + AN)h$	8. Distr Prop
9. $AB + BC + CD + DX + XN + AN =$ perim of base of prism.	9. Def of perim
10. $L = ph$	10. Sub

22.

Statement	Reason
1. Rt cyl with radius r and alt h	1. Given
2. Lateral area $= 2\pi rh$.	2. Lateral area of rt cyl $= 2\pi rh$.
3. Area of base cyl $= \pi r^2$.	3. Base of cyl is $\odot$.
4. Total area of cyl $= 2\pi rh + 2\pi r^2$.	4. Area Add Post
5. Total area $= 2\pi r(r + h)$.	5. Distr Prop

23.

Statement	Reason
1. Cubes 1 and 2 with edges of lengths a and b, respectively	1. Given
2. L (Cube 1) $= 4a \cdot a = 4a^2$, L (Cube 2) $= 4b \cdot b = 4b^2$	2. $L = ph$
3. $\frac{L \,(\text{Cube 1})}{L \,(\text{Cube 2})} = \frac{4a^2}{4b^2} = \frac{a^2}{b^2}$	3. Equations may be divided.
4. $\frac{L \,(\text{Cube 1})}{L \,(\text{Cube 2})} = \left(\frac{a}{b}\right)^2$	4. Algebra
5. A (Cube 1) $= 4a^2 + 2a^2 = 6a^2$, A (Cube 2) $= 4b^2 + 2b^2 = 6b^2$	5. Def of total area of prism
6. $\frac{A \,(\text{Cube 1})}{A \,(\text{Cube 2})} = \frac{6a^2}{6b^2} = \frac{a^2}{b^2}$	6. Equations may be divided.
7. $\frac{A \,(\text{Cube 1})}{A \,(\text{Cube 2})} = \left(\frac{a}{b}\right)^2$	7. Algebra

19. The length of an altitude is 5 m. Each base is a regular pentagon with an edge measuring 2 m and an apothem measuring approximately 1.4 m.   50 m², 64 m²

20. A concrete roller is 3 m long. Its diameter is 2 m long. How much area will it cover in 400 revolutions?   2,400π m², or 7,539.8 m²

21. Prove Theorem 14.1.   **22.** Prove Theorem 14.3.

23. Prove that Theorem 14.4 holds for the case of two cubes with edges of lengths a and b.

24. If one gallon of paint will cover 400 ft² of surface, how many gallons will be needed to paint the swimming pool shown below?   28.7 gal

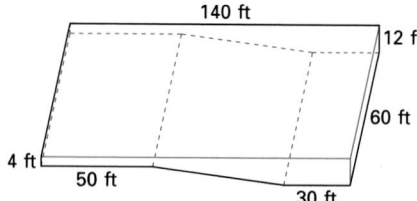

25. A right square prism is inscribed in a right circular cylinder of the same height. A radius and an altitude of the cylinder are each 10 cm long. What is the ratio of their total areas? $(1 + \sqrt{2})$ to π

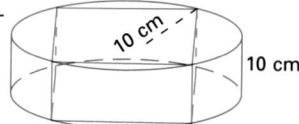

26. A right cylinder is inscribed in a right square prism of the same height. A radius and an altitude of the cylinder are each 10 cm long. What is the ratio of their total areas?   4 to π

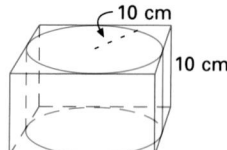

27. If the lateral surface of an oblique cylinder were "unrolled," what would its shape be? Is there a simple formula for its lateral area?   Parallelogram, $L = 2\pi rh$

Mixed Review

If a and b are the lengths of the two legs of a right triangle and c is the length of the hypotenuse, find the missing length.   9.2

1. $a = 3$, $b = 4$, $c = \underline{ 5 }$
2. $a = \underline{ 5 }$, $b = 12$, $c = 13$
3. $a = 9$, $b = \underline{ 40 }$, $c = 41$
4. $a = 2$, $b = 3$, $c = \underline{ \sqrt{13} }$

14.4 Volumes of Prisms and Cylinders

Objectives   To find the volume of a prism
To find the volume of a cylinder

Direct comparison can be used to compare
the sizes of some solids. If one solid can fit
inside a second solid, then the second is
larger than the first. Many solids cannot be
compared in this way, however. Their *volumes* can be used to compare their sizes.

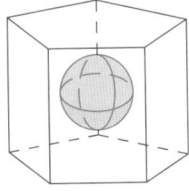

The figure below consists of a number of cubes of the same size. If one
of these cubes is considered to be a *unit* of measure, then the total
number of these cubes can be considered to be a *measure* of the solid.
This measure is the **volume** of the solid.

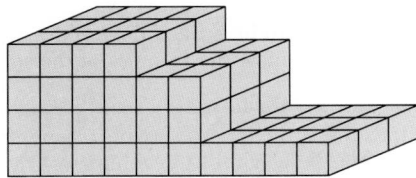

To find any measure, an appropriate unit must be used. *Square units*
are used to measure *area*. *Cubic units* are used to measure volume. One
cubic unit is the volume of a cube in which each edge measures one
unit. A cube with each edge one centimeter (1 cm) long has a volume of
one *cubic centimeter* (cm^3).

The rectangular solid below contains four layers of unit cubes. Each
layer contains 5 rows of 6, or 30 unit cubes. (Thirty also is the area of
a five-by-six unit rectangle.) Four layers of 30 cubic units makes a total
of 120 cubic units. This suggests a formula for the volume V of a rectangular solid.

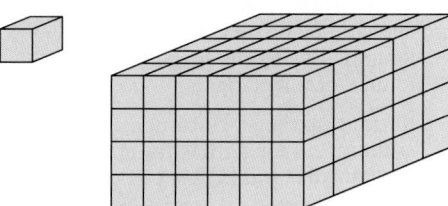

Additional Example 1

Find the volume of a rectangular solid 4
in. by 7 in. by 9 in. $V = 4 \cdot 7 \cdot 9$, or 252 in^3

Teaching Resources

Quick Quizzes, p. 103
**Reteaching and Practice
Worksheets,** pp. 205, 206

■■■ GETTING STARTED

Prerequisite Quiz

1. If length is measured in centimeters,
   what is the unit of area measurement?
   sq cm
2. If length is measured in inches, what is
   the unit of area measurement?   sq in
3. Interpret the formula:
   $A = \frac{1}{2}bh$ Area of $\triangle = \frac{1}{2}$base × height
4. Does the area of a regular polygon
   depend upon its perimeter? If so, how?
   Yes. $A = \frac{1}{2}ap$

Motivator

Have the students imagine that they are in
charge of designing cans for a grocery
company. They want to use as little metal
as possible, yet have the can hold the
greatest possible amount. Ask them what
information will be needed to design such a
can.   Total area; Volume   Ask them what
measures must be determined.   The
measures of the altitude and the radius

■■■ TEACHING SUGGESTIONS

Lesson Note

If students have difficulty visualizing
three-dimensional figures, use wooden or
plastic cubes to build models of several of
these solids. Show how the volume can be
determined by counting these cubes. If the
cubes are one centimeter on edge, then
each cube is 1 cm^3 in volume. A memory
trick useful for some students is that finding
the area of two-dimensional objects involves
multiplying two factors, whereas the volume
of three-dimensional objects involves
multiplying three factors.

Math Connections

Homemaking: Recipes typically suggest that a specific size pot or pan be used to prepare the dish. The sizes of pots and pans are classified by the maximum volume they contain. Discuss with students that cubic units describe measurements of volume just as gallons, quarts, liters, cups, and tablespoons are measurements of volume.

Critical Thinking Questions

Analysis: Ask students the following questions.

Which can hold more liquid—a 4-inch tall prism with a 3-inch square base or a 5-inch tall regular hexagonal container with a 1.5-inch radius? The container with a square base holds about 6.8 in^3 more. Suppose that the hexagonal container is half full. If you pour the contents into the square container, to what height will it be filled? To a height of about 1.63 in

Common Error Analysis

Error: Some students confuse volume with surface area.

Use models to emphasize the difference between volume and surface area. Emphasize also that area is given in square units and volume is given in cubic units.

Postulate 23 For any rectangular solid, the volume $V = lwh$, where l, w, and h are the lengths of three edges with a common vertex.

EXAMPLE 1 Find the volume of a rectangular solid measuring 6 cm by 8 cm by 10 cm.

Solution Volume = $6 \cdot 8 \cdot 10 = 480$ cm^3

A cube is a special rectangular solid in which all edges are congruent.

Theorem 14.5 The volume of a cube with edges of length s is s^3.

EXAMPLE 2 If the volume of a cube is 8 m^3, what is the length of an edge?

Solution
$$\text{Volume} = s^3 = 8$$
$$s = \sqrt[3]{8}$$
$$s = 2 \text{ m}$$

Some solids can be broken down into smaller solids. The volume of the solid is the sum of the volumes of the smaller solids.

Postulate 24 If a solid is the union of two or more nonoverlapping solids, then its volume is the sum of the volumes of these nonoverlapping parts.

EXAMPLE 3 Find the volume of the solid. Each angle is a right angle.

Solution
Volume (Solid I) = $8 \cdot 10 \cdot 12 = 960$ cm^3
Volume (Solid II) = $6 \cdot 4 \cdot 12 = 288$ cm^3
By Postulate 24, total volume = $1,248$ cm^3.

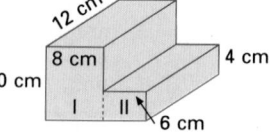

A stack of index cards closely approximates a rectangular solid. The volume of such a stack can be computed using the formula for the volume of a rectangular solid.

Notice that the stack can be slanted so that it no longer has the shape of a rectangular solid. This can be done in many ways.

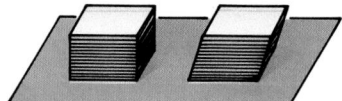

570 Chapter 14 Figures in Space

Additional Example 2

If the volume of a cube is 12 ft^3, what is the length of a side?

$s = \sqrt[3]{12}$ ft ≈ 2.29 ft

Additional Example 3

Find the volume of the solid. Each angle is a right angle.

Volume I = $5 \cdot 9 \cdot 12$, or 540 cm^3
Volume II = $14 \cdot 9 \cdot 8$, or 1008 cm^3
Total Volume = 1548 cm^3

Even though the shape of the stack of cards has changed, it should be clear that the volume has not. The area of every layer (the *cross-sectional area* at a particular level) has stayed the same. Based on the example of the cards, a generalization can be made.

Postulate 25

Cavalieri's Principle: If two solids have equal heights, and if the cross sections formed by every plane parallel to the bases of both solids have equal areas, then the volumes of the solids are equal.

The solids in the illustration will have equal volumes if all their cross-sectional areas are equal.

Using Cavalieri's Principle, a formula for the volume of any prism or cylinder can be obtained. It is stated here without proof.

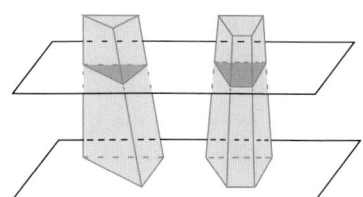

Theorem 14.6

For any prism or cylinder, the volume is the product of the area of a base and the length of an altitude ($V = Bh$, where B = area of base and h = altitude).

EXAMPLE 4 Find the volume of the prism if the area of its base is 30 cm² and its altitude is 4 cm.

Solution
$V = Bh$
$= 30 \cdot 4$
$= 120 \text{ cm}^3$

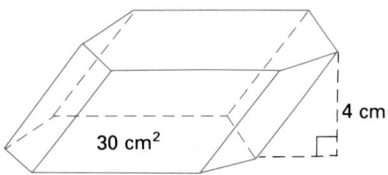

Since the area of a circle is πr^2, the following is easily proved.

Corollary

The volume of a cylinder is Bh, or $\pi r^2 h$.

EXAMPLE 5 A radius of a base of a right cylinder is 5 cm long. An altitude is 6 cm long. Find the volume.

Solution
$V = Bh = \pi r^2 \cdot h$
$V = \pi \cdot 5^2 \cdot 6$
$V = 150\pi \text{ cm}^3$

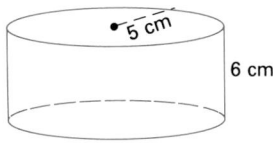

Checkpoint

1. Find the volume of a cube with sides 5 mm long. 125 mm³
2. Find the volume of the rectangular solid. 180 in³

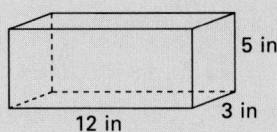

3. Find the volume of a right circular cylinder with a radius of length 10 cm and an altitude of length 3 cm. 300π cm³
4. Find the volume of the prism if the area of the base is 30 cm² and the length of an altitude is 9 cm. 270 cm³

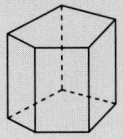

Closure

Ask the students how they would determine the volume for the following figures:
a rectangular solid Postulate 23 page 570
a cube Theorem 14.5 page 570
a prism Theorem 14.6 page 571
a cylinder Corollary page 571
How are the formulas for the volumes of prisms and cylinders alike? for each, $V = Bh$ How are they different? derivation of B

Additional Example 4

Find the volume of the prism if the area of its base is 50 m² and the altitude is 7 m. $V = 50 \cdot 7$, or 350 m³

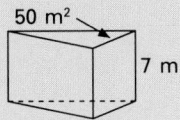

Additional Example 5

The radius of a base of a right circular cylinder is 3 cm. Find the volume.
$V = \pi r^2 h$
$V = \pi \cdot 3^2 \cdot 8$, or 72π cm³

Guided Practice

Classroom Exercises 1–7

Independent Practice

A Ex. 1–13, **B** Ex. 14–20, **C** Ex. 21–23

Basic: WE 1–13 odd, Midchapter Review

Average: WE 1–20 odd, Midchapter Review

Above Average: WE 1–23 odd, Midchapter Review

Additional Answers

Written Exercises

17.

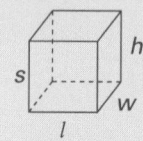

Statement	Reason
1. Cube with edges of length s	1. Given
2. Each face of cube is rect.	2. A sq is a rect.
3. Cube is rect solid.	3. Def of rect solid
4. Vol of cube $= lwh$	4. For rect solid, $V = lwh$
5. $s = l = w = h$	5. All edges of cube are $\cong$.
6. Vol of cube $= s \cdot s \cdot s = s^3$	6. Sub, alg

Classroom Exercises

Find the volume of the rectangular solid.

1.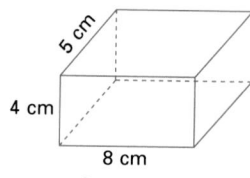

160 cm³

2.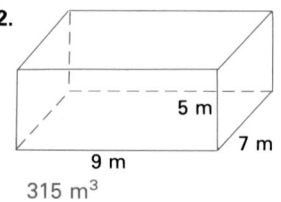

315 m³

3.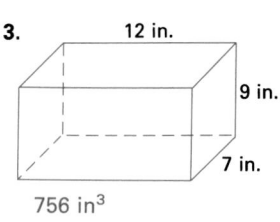

756 in³

Find the volume of the prism.

4. $B = 7$ cm², $h = 5$ cm   35 cm³

5. $B = 9$ m², $h = 6$ m   54 m³

Find the volume of the cylinder.

6. $r = 10$ mm, $h = 5$ mm 500π mm³

7. $r = 2$ cm, $h = 3$ cm 12π cm³

Written Exercises

Find the volume of the prism.

1. $B = 17$ cm², $h = 23$ cm   391 cm³

2. $B = 32$ ft², $h = 17$ ft   544 ft³

Find the volume of the cylinder.

3. $r = 12$ in., $h = 7$ in. $1{,}008\pi$ in³

4. $r = 15$ ft, $h = 8$ ft $1{,}800\pi$ ft³

5. $d = 8$ mm, $h = 3$ mm 48π mm³

6. $d = 16$ cm, $h = 9$ cm 576π cm³

Find the volume of the rectangular solid.

7. 5 cm by 8 cm by 7 cm   280 cm³

8. 9 mm by 4 mm by 6 mm   216 mm³

9. 13 in. by 12 in. by 17 in.   2,652 in³

10. 27 yd by 13 yd by 7 yd   2,457 yd³

Find the volume of the solid. Each angle is a right angle.

11.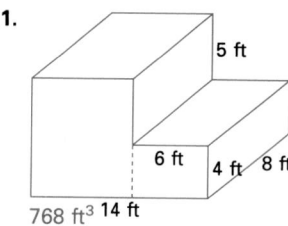

768 ft³

Find the volume.

12. a cube with 6-cm edges   216 cm³

13. a cube with 12-mm edges   1,728 mm³

Enrichment

Manufacturers often want to maximize volume while minimizing surface area. Ask students to experiment to find the best shape of a rectangular prism that will hold the greatest volume while using the least material for the surface.   Cube

Try to solve the same problem for a right cylinder. $h = d$ These would be good computer problems.

14. a prism with an isosceles right triangular base with 10-cm legs and a 17-cm altitude   850 cm³

15. a prism with an equilateral triangle as its base with 10-cm sides and a 17-cm altitude   425√3 cm³

16. a prism with a regular hexagonal base with 10-cm sides and a 17-cm altitude   2,550√3 cm³

17. Prove Theorem 14.5.   **18.** Prove the corollary to Theorem 14.6.

A volume of 231 in³ has a capacity of 1 gallon. (Exercises 19–20)

19. A large bucket in the shape of a right cylinder has a diameter of 16 in. and a height of 20 in. How many gallons of liquid can the bucket hold?   17.4 gal

20. A can of juice has a diameter of 4 in. and a height of 7 in. How many pints of juice can the can hold?   3 pt

21. Find the ratio of the area of a cube to its volume.   6 to s

22. A right square prism is inscribed in a right circular cylinder of the same height. A radius and an altitude of the cylinder are each 10 cm long. What is the ratio of their volumes? $\dfrac{V_p}{V_c} = \dfrac{2}{\pi}$

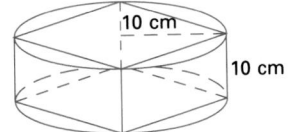

23. A cylinder of revolution is formed when a rectangle with sides of length m and n is revolved about one of its sides. Find the ratio of the volumes of the cylinders formed by the revolution in the diagram at the right. n to m

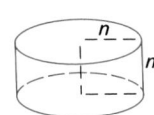 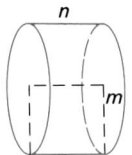

Midchapter Review

1–2. Find the number of faces (F), the number of edges (E), and the number of vertices (V) for each of the polyhedra shown.   **14.1**

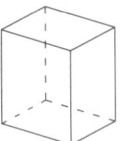 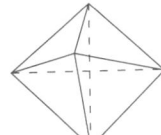

3. Draw a triangular prism. How many lateral faces and lateral edges does it have?   **14.2**

4. Find the lateral area of the prism. The bases are shaded.   **14.3** 168 in²

5. Find the lateral area of the cylinder.   **14.3** 120π in³

6. Find the volume of the prism.   **14.4** 270 in³

7. Find the volume of the cylinder.   **14.4** 300π in³

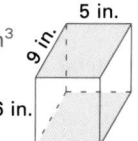 

14.4 Volumes of Prisms and Cylinders   **573**

18. Cylinder with radius of length r (Given); base of cyl is circle (Def of cyl); base has area πr^2 (A of ⊙ = πr^2); V (cyl) = Bh (For prism or cyl, $V = Bh$); V (cyl) = $\pi r^2 h$ (Sub).

Midchapter Review

1. $F = 6$, $E = 12$, $V = 8$
2. $F = 8$, $E = 12$, $V = 6$
3. L faces = 3, L edges = 3.

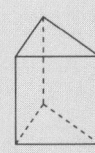

GETTING STARTED

Prerequisite Quiz

Find the area of the given figure.

1. $4\sqrt{3}$ cm²

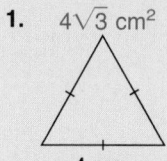

2. 16π cm²

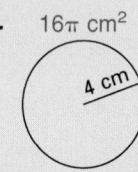

4 cm

4 cm

3. If length is measured in centimeters, what is the unit of area measurement? cm²

4. If volume is measured in cubic inches, what is the unit of area measurement? in²

Motivator

Have the students look at the picture of the pyramids on page 574. Ask them what geometric figures form the pyramids. Triangles and a polygonal base Ask them what area they must find in order to buy cardboard to build a pyramid model. total area

TEACHING SUGGESTIONS

Lesson Note

Review standard area formulas, especially those for triangles, circles, and sectors of circles. Stress the position of the altitude and radius of these figures.

Prepare paper cutouts for several pyramids or cones that can be unfolded to show the surface clearly. Use these to illustrate what is meant by the surface of the pyramid or cone. Have students help find the areas of the parts of these unfolded surfaces using standard area formulas. You may find it helpful to make these cutouts out of graph paper to indicate square units more clearly.

574

14.5 Areas of Pyramids and Cones

Objectives
To find the lateral and total areas of a regular pyramid
To find the lateral and total areas of a right cone

The pyramids of Egypt have fascinated people for centuries. Because of the rigidity of the triangular faces of pyramids, such shapes are sometimes used in modern building designs.

Definitions

A **pyramid** is a polyhedron composed of a polygonal region, called the *base*, and triangular regions, called the *lateral faces*. The lateral faces intersect in a common point called the *vertex*. The intersections of each pair of lateral faces are the **lateral edges**. The **altitude** of the pyramid is the segment from the vertex to the plane containing and perpendicular to the base.

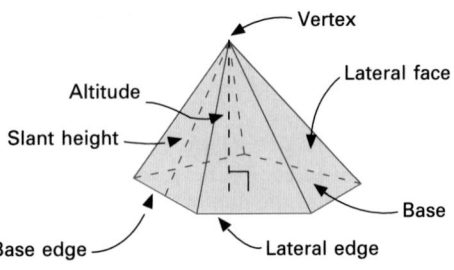

In a *right pyramid*, the altitude contains and is perpendicular to the center of the base. A **regular pyramid** is a right pyramid where the base is a regular polygon and the lateral faces are congruent isosceles triangles.

The **lateral area** of a pyramid is the sum of the areas of its lateral faces. The length of the altitude of each lateral face is called the **slant height** of the regular pyramid.

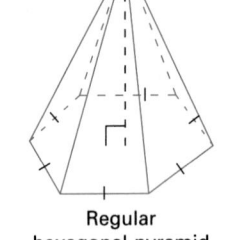

Regular
hexagonal pyramid

574 Chapter 14 Figures in Space

In finding areas of pyramids, it may be helpful to think of the lateral faces of the solid as flattened out into a plane with the base.

EXAMPLE 1 Find the lateral area of a regular square pyramid with 4-cm base edges and a 5-cm slant height.

Plan Multiply the area of a triangular face by 4.

Solution Lateral area = $4 \cdot (\frac{1}{2} \cdot 4 \cdot 5)$
$= 4 \cdot 10 = 40$ cm^2

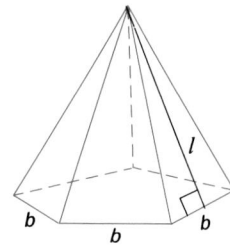

As suggested by Example 1, the lateral area of a regular pyramid is the product of the number of lateral faces and the area of a lateral face. The lateral area $L = n \cdot (\frac{1}{2}bl)$, or $\frac{1}{2}l(bn)$, where b is the length of a side of the base, l is the slant height, and n is the number of sides of the base. But bn is the perimeter p of the base. Therefore, $L = \frac{1}{2}pl$. This reasoning constitutes a proof of the theorem that follows.

Theorem 14.7 The lateral area of a regular pyramid is one-half the product of the perimeter of the base and the slant height ($L = \frac{1}{2}pl$).

The *total area* of any pyramid is the sum of the lateral area and the area of the base. The base of a regular pyramid is a regular polygon, the area of which is one-half the product of its perimeter and its apothem.

EXAMPLE 2 Find the total area of a regular pentagonal pyramid if the length of a side of the base is 2 m, the length of an apothem a of the base is 1.4 m, and the slant height is 4 m.

Solution Total area = Lateral area + Base area
$= \frac{1}{2}pl + \frac{1}{2}pa$
$= \frac{1}{2}(5 \cdot 2) \cdot 4 + \frac{1}{2}(5 \cdot 2) \cdot 1.4$
$= 20 + 7 = 27$ m^2

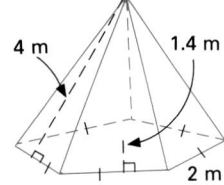

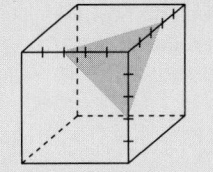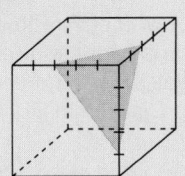

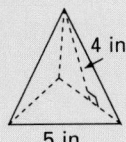

Checkpoint

1. Draw a picture to illustrate how the surface of the given regular pyramid would appear if unfolded.   See answer to Classroom Ex. 4.

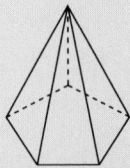

2. Find the lateral area of a regular quadrilateral pyramid if the edge length of the base is 8 cm, the length of an apothem is 4 cm, and the slant height is 5 cm.   80 cm²

3. Find the total area of a right circular cone with a 10 cm radius and a 6 cm slant height.   160 cm²

Closure

Ask the students how they determine the lateral area of a regular pyramid.   Theorem 14.7 page 575   The total area.   It is the sum of the lateral area and the base area which is one-half the product of its perimeter and its apothem   The lateral area and the total area of a right cone.   Theorem 14.8 page 577

EXAMPLE 3   Find the total area of a regular square pyramid whose lateral sides are equilateral triangles with 4-cm edges.

Plan   To find the area of one lateral face, use the formula for the area of an equilateral triangle.

Solution

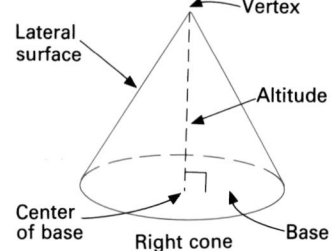

Total area = Lateral area + Base area

$$= 4 \cdot \left(\frac{s^2}{4}\sqrt{3}\right) + s^2$$

$$= 4 \cdot \left(\frac{4^2}{4}\sqrt{3}\right) + 4^2$$

$$= 4 \cdot (4\sqrt{3}) + 16 = (16\sqrt{3} + 16) \text{ cm}^2$$

As a cylinder resembles a prism, so a *cone* resembles a pyramid. The *base* of a cone is a circle. The *lateral surface* of a cone is made up of an infinite number of segments that connect a single point, called the *vertex*, and all points on the edge of the base. The *altitude* of the cone is the segment from the vertex to the base and perpendicular to the base. If an endpoint of the altitude is the center of the circle, then it is a *right cone*. If it is not a right cone, then it is an *oblique cone*.

The approach used to find the lateral area of a regular pyramid can be generalized to find the lateral area of a right cone. This formula will be presented without proof.

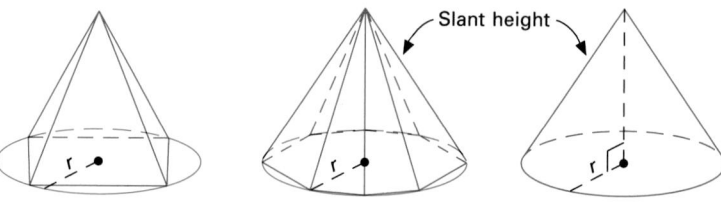

A regular pyramid has lateral area of $\frac{1}{2}pl$, where p is the perimeter of the base and l is the slant height. In a right cone, $p = 2\pi r$, where r is the radius of the base because the base is a circle. Therefore, a cone has a lateral area of $\frac{1}{2}(2\pi r)l$, or πrl.

Additional Example 3

Find the lateral area and total area of a regular triangular pyramid whose lateral sides are equilateral triangles with 3 mm edges.

Total area = Lateral area + base area

$$= 3 \cdot \frac{3^2}{4}\sqrt{3} + \frac{3^2}{4}\sqrt{3}$$

$$= 9\sqrt{3} \text{ mm}^2$$

Theorem 14.8	The lateral area (L) of a right cone is $\pi r l$. The total area (A) is $\pi r l + \pi r^2 = \pi r(l + r)$.

EXAMPLE 4 Find the total area of a right cone with a 7-cm radius and a 12-cm slant height.

Solution
$$\text{Total area} = \pi r(l + r)$$
$$= \pi \cdot 7(12 + 7) = 133\pi \text{ cm}^2$$

Classroom Exercises

Use sketches to illustrate how the surface of the right pyramid would appear when unfolded and flattened.

1. a square pyramid
2. an equilateral triangular pyramid
3. a regular hexagonal pyramid
4. a regular pentagonal pyramid
5. Find the lateral area of a hexagonal pyramid with 10-cm base edges and an 8-cm slant height. 240 cm²

Written Exercises

Find the lateral area of the regular pyramid.

1.
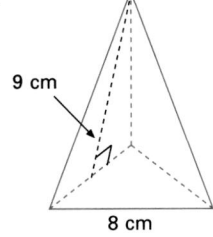
9 cm
8 cm
108 cm²

2.
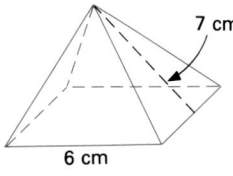
7 cm
6 cm
84 cm²

3.
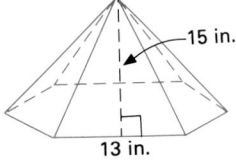
15 in.
13 in.
585 in²

Find the total area of the regular pyramid.

4.
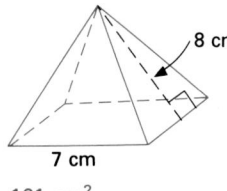
8 cm
7 cm
161 cm²

5.
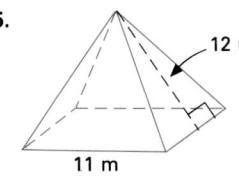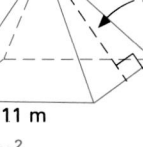
12 m
11 m
385 m²

6.
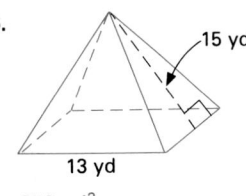
15 yd
13 yd
559 yd²

14.5 Areas of Pyramids and Cones   **577**

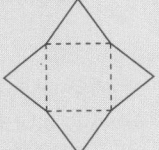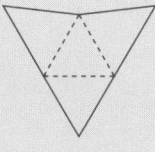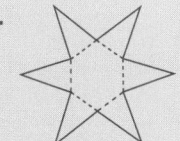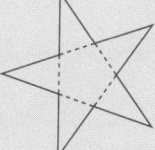
Additional Example 4

Find the total area of a right circular cone with a 5 in. radius and a 6 in. slant height.

$$\text{Total area} = \pi r l + \pi r^2$$
$$= \pi 5 \cdot 6 + \pi 5^2$$
$$= 55\pi \text{ in}^2$$

Written Exercises

19.

Statement	Reason
1. Pyramid with reg poly base $ABC...N$, alt $\overline{VX}$, X at ctr of $ABC...N$	1. Given
2. $\overline{VX}$ is $\perp$ to $\overline{AX}$, $\overline{BX}$, $\overline{CX}$,...$\overline{NX}$	2. Alt of pyramid $\perp$ to base.
3. $\angle VXA$, $\angle VXB$, $\angle VXC$,... $\angle VXN$ are rt $\angle$s.	3. Def of $\perp$.
4. $\angle VXA$, $\angle VXB$, $\angle VXC$... $\angle VXN$ are $\cong$.	4. Rt $\angle$s are $\cong$.
5. $\overline{VX} \cong \overline{VX}$	5. Reflex Prop of Congr
6. $\overline{AX}$, $\overline{BX}$, $\overline{CX}$ ...$\overline{NX}$ are $\cong$.	6. Radii reg poly are $\cong$.
7. $\triangle VXA$, $\triangle VXB$, $\triangle VXC$ ...$\triangle VXN$ are $\cong$.	7. SAS
8. $\overline{VA}$, $\overline{VB}$, $\overline{VC}$...$\overline{VN}$ are $\cong$.	8. CPCTC
9. $AB = BC$ $= CD = ...$	9. Def of reg poly
10. $\triangle AVB \cong$ $\triangle BVC \cong$ $\triangle CVD \cong$ ...	10. SSS
11. $\triangle AVB$, $\triangle BVC$,... are isos $\triangle$s.	11. Def of isos $\triangle$

Find the lateral area and the total area of the right cone.

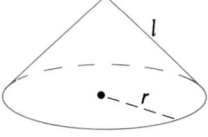

7. $r = 10$ cm, $l = 6$ cm   **8.** $r = 8$ ft, $l = 4$ ft

9. $r = 12$ m, $l = 11$ m 132π **10.** $r = 17$ in., $l = 13$ in.
m², 276π m² 221π in², 510π in²

7. 60π cm², 160π cm²   **8.** 32π ft², 96π ft²

Find the lateral area and the total area of the pyramid. (Exercises 11–14)

11. a regular triangular pyramid with 5-cm base edges and 8-cm slant height 60 cm², $(60 + \frac{25}{4}\sqrt{3})$ cm²

12. a regular triangular pyramid with 5-cm base edges and 8-cm lateral edges 57 cm², $(57 + \frac{25}{4}\sqrt{3})$ cm²

13. a regular hexagonal pyramid with 5-cm base edges and 8-cm slant heights 120 cm², $(120 + \frac{75}{2}\sqrt{3})$ cm²

14. a regular hexagonal pyramid with 5-cm base edges and 8-cm lateral edges 114 cm², $(114 + \frac{75}{2}\sqrt{3})$ cm²

15. The altitude of a regular square pyramid is 8 cm long, and the length of a side if the base is 20 cm. What is the lateral area of the pyramid? $80\sqrt{41}$ cm²

16. The altitude of a right cone is 12 cm long, and the length of a radius of the base is 16 cm. What is the lateral area of the cone? 320π cm²

17. The lateral area of a regular pyramid is 144 cm², and the slant height is 16 cm. What is the perimeter of the base?   18 cm

18. The lateral area of a regular pyramid is 196 mm², and the perimeter of the base is 14 mm. What is the slant height of the pyramid?   28 mm

19. Prove: If the base of a pyramid is a regular polygon and the foot of the altitude is at the center of the base, then all the lateral faces are congruent isosceles triangles.   **20.** $A = \pi r(r + \sqrt{r^2 + h^2})$

20. Derive a formula for the total area of a right cone in terms of the length of its altitude h and the length of the radius r of the base.

21. A right cone with radius of length r and altitude of length h is inscribed in a right cylinder of the same radius and altitude lengths. What is the ratio of their lateral areas? What is the ratio of their total areas?

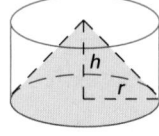

$$\frac{LA_{cyl}}{LA_{cone}} = \frac{2h}{\sqrt{r^2 + h^2}}; \frac{TA_{cyl}}{TA_{cone}} = \frac{2(h + r)}{\sqrt{r^2 + h^2} + r}$$

Mixed Review

1. What is the lateral area of a right cylinder with a 25-cm base radius and a 12-cm altitude?   *14.3* 600π cm²

2. What is the total area of a right regular triangular prism with 5-cm base edges and a 10-cm altitude?   *14.3* $(150 + \frac{25}{2}\sqrt{3})$ cm²

Enrichment

The following problem can be solved by
(1) visualizing the entire object;
(2) drawing in three dimensions;
(3) making a model.

Sketch or make a model of a five-sided figure such that two faces are equilateral triangles with 10 cm sides, two faces are isosceles trapezoids with sides 10 cm, 10 cm, 10 cm, 20 cm, and one face is a square with 10 cm sides.

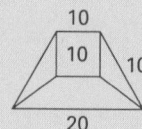

14.6 Volumes of Pyramids and Cones

Teaching Resources

Quick Quizzes 105
**Reteaching and Practice
    Worksheets,** pp. 209, 210

Objectives   To find the volume of a pyramid and a cone
To find the ratios of volumes of solids

Taking models of a cone and a cylinder with the same base and height,
fill up the cone with sand or water
and pour the contents of that into
the cylinder. Repeat this process until the cylinder is full. How do the
volumes compare? Try this for other
such pairs of cones and cylinders.
Does the same relation hold? Try
this experiment for a pyramid and a
prism with the same base and height.

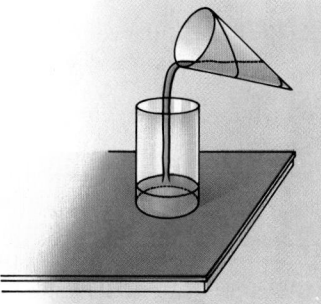

Theorem 14.9   The volume of a pyramid is one-third the volume of a prism with the
same base and altitude as the pyramid. The volume of a cone is one-
third the volume of a cylinder with the same base and altitude as the
cone ($V = \frac{1}{3}Bh$).

Corollary 1   The volume of a pyramid or cone is one-third the product of the area
of its base and the length of its altitude ($V = \frac{1}{3}Bh$).

EXAMPLE 1   Find the volume of the triangular pyramid.

Solution $V = \frac{1}{3}Bh$

$V = \frac{1}{3}(\frac{1}{2} \cdot 5 \cdot 6) \cdot 8$

$V = \frac{1}{3} \cdot 15 \cdot 8$

$= 40 \text{ cm}^3$

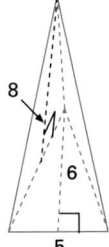

Corollary 2   The volume of a cone with a base radius of length r and an altitude
of length h is $\frac{1}{3}\pi r^2 h$.

14.6 Volumes of Pyramids and Cones   **579**

Additional Example 1

Find the volume of the rectangular
pyramid.

$V = \frac{1}{3}Bh$

$= \frac{1}{3} \cdot (5 \cdot 2) \cdot 3$

$= 10 \text{ cm}^3$

3 cm

2 cm

5 cm

GETTING STARTED

Prerequisite Quiz

1. If length is measured in centimeters,
   what is the unit of volume measurement?
   cm^3
2. If area is measured in ft^2, what is the unit
   of volume measurement? ft^3
3. What is the volume of a right circular
   cylinder with a base radius of 4 cm and
   an altitude of 5 cm? $80\pi \text{ cm}^3$

Motivator

Ask the students to imagine that they work
for a company that makes paper cups. One
type of cup made is shaped like a cone.
The company wants to use as little paper as
possible, yet have the cup hold the greatest
quantity of liquid possible. Ask them what
information they would need to design such
a cup.   Volume.   What measurements
must be determined?   Altitude and radius.

TEACHING SUGGESTIONS

Lesson Note

Review basic area formulas. Go over the
formulas developed for volumes of prisms
and cylinders.

Hollow models are especially helpful in
demonstrating the relationship of the
volumes of pyramids and cones to those of
prisms and cylinders. Pouring water or sand
from a hollow pyramid into a prism of
congruent base and altitude reveals that the
prism holds three times as much as the
pyramid. A similar demonstration with
cylinders and cones reveals the same ratio
of volumes. In other words, the volume of a
pyramid or cone is one-third that of the
corresponding prism or cylinder.

579

Math Connections

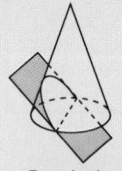 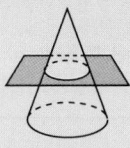 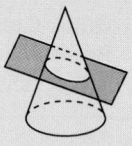

Parabola   Circle   Ellipse

Conic Sections: When a cone is intersected by a plane, the resulting curve is called a conic section. Three different cross sections are shown above. The properties and equations of conic sections are studied in greater depth in algebra courses.

Critical Thinking Questions

Application: Ask students to use their calculators to answer the following question. A conical paper cup has a height of 3.5 inches and a 2.5-in diameter. Find the number of cubic inches of liquid it contains when it is filled to a point half way to the top of the cup along the side.   About 0.23π in^3, or 0.72 in^3

Checkpoint

Find the volume of the cone.

1. $B = 24\pi$ ft^2, $h = 7$ ft.
   56π ft^3
2. $r = 3$ m, $h = 10$ m 30π m^3

Find the radius of the base of the cone.

3. $V = 60\pi$ in.^{3}, $h = 20$ in.
   3 in.
4. $V = 6$ cm^3, $h = 2$ cm
   $\frac{3\sqrt{\pi}}{\pi}$

EXAMPLE 2   Find the volume of a cone with a 5-cm radius and a 6-cm altitude.

Solution $V = \frac{1}{3}\pi r^2 h$

$\quad\quad\quad = \frac{1}{3}\pi(5^2) \cdot 6$

$\quad\quad\quad = \frac{1}{3}\pi 25 \cdot 6 = 50\pi$ cm^3

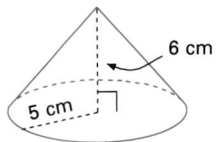

EXAMPLE 3   If the volume of a pyramid is 20 cm^3, and the area of the base is 5 cm^2, what is the length of the altitude?

Solution $V = \frac{1}{3}Bh$

$\quad\quad 20 = \frac{1}{3}(5)h$

$\quad\quad 60 = 5h$

$\quad\quad\; h = \frac{60}{5} = 12$ cm

EXAMPLE 4   A right triangle has legs 3 cm and 4 cm long. Find the volumes of the cones formed by revolving the triangle around each leg.

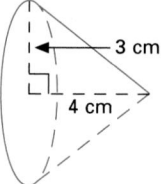 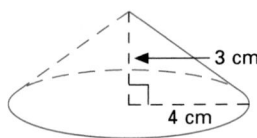

Solution   Revolution around 4-cm leg   Revolution around 3-cm leg

$V = \frac{1}{3}\pi r^2 h$ $V = \frac{1}{3}\pi r^2 h$

$V = \frac{1}{3}\pi(3^2)4$ $V = \frac{1}{3}\pi(4^2)3$

$V = 12\pi$ cm^3 $V = 16\pi$ cm^3

Classroom Exercises

Find the volume of the pyramid.

1. $B = 20$ cm^2, $h = 6$ cm   40 cm^3 2. $B = 15$ mm^2, $h = 8$ mm   40 mm^3

Find the volume of the cone.

3. $B = 15\pi$ in^2, $h = 2$ in. 10π in^3 4. $r = 2$ mm, $h = 6$ mm 8π mm^3

Find the altitude of the pyramid.

5. $V = 100$ cm^3, $B = 60$ cm^2 5 cm   6. $V = 144$ in^3, $B = 3$ in^2 144 in

580   Chapter 14 Figures in Space

Additional Example 2

Find the volume of a circular cone with a 3 ft base radius and a 7 ft altitude.

$V = \frac{1}{3}\pi r^2 h$

$\quad = \frac{1}{3} \cdot \pi \cdot (3^2) \cdot 7$

$\quad = 21\pi$ ft^3

Additional Example 3

If the volume of a pyramid is 50 cm^3 and the altitude is 5 cm, what is the area of the base?

$\quad 50 = \frac{1}{3}B(5)$

$\quad 150 = 5B$

$\quad\quad B = 30$ cm^2

Written Exercises

Find the volume of the rectangular pyramid.

1.
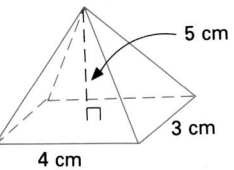
5 cm
3 cm
4 cm
20 cm³

2.
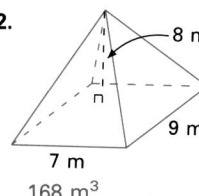
8 m
9 m
7 m
168 m³

3.
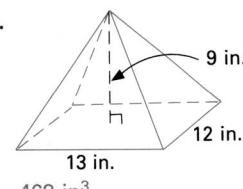
9 in.
12 in.
13 in.
468 in³

Find the volume of the pyramid.

4. $B = 9$ cm², $h = 7$ cm, 21 cm³   **5.** $B = 7$ mm², $h = 12$ mm   28 mm³

6. $B = 5$ in², $h = 8$ in. $\frac{40}{3}$ in³   **7.** $B = 10$ ft², $h = 7$ ft $\frac{70}{3}$ ft³

Using a calculator, find the volume of the cone, correct to the nearest tenth. For π, use 3.14. (Exercises 8–11)

8. $r = 9$ mm, $h = 10$ mm   847.8 mm³   **9.** $r = 5$ cm, $h = 12$ cm   314.0 cm³

10. $r = 8.5$ m, $h = 5.2$ m   393.2 m³   **11.** $r = 7.7$ cm , $h = 4.6$ cm   285.5 cm³

12. If the volume of a pyramid is 12 cm³, and the area of the base is 4 cm², what is the length of the altitude?   9 cm

13. If the volume of a square pyramid is 18 mm³, and the length of the altitude is 6 mm, what is the length of an edge of the square base?   3 mm

14. If the volume of a right cone is 32π m³, and the length of an altitude is 6 m, what is the length of the radius of the base?   4 m

15. If the volume of a right cone is 75π ft³, and the area of the base is 25π ft², what is the length of the altitude?   9 ft

Find the volume.

16. In a pyramid with a rectangular base, adjacent sides of the base are 7 cm and 9 cm long. The length of the altitude of the pyramid is 8 cm.   168 cm³

17. In a pyramid with a square base, a side of the square is 5 mm long. The length of the altitude of the pyramid is 15 mm.   125 mm³

18. A pyramid has an isosceles right-triangular base. The legs of the triangle are 12 cm long. The length of the altitude of the pyramid is 7 cm.   168 cm³

19. A pyramid has a right-triangular base. The two legs are 6 cm and 8 cm long. The length of the altitude of the pyramid is 10 cm.   80 cm³

20. A pyramid has an equilateral triangular base. The sides of the triangle are 10 cm long. The length of the altitude of the pyramid is 18 cm. $150\sqrt{3}$ cm³

21. A pyramid has a regular hexagonal base. The sides of the hexagon are 6 mm long. The length of the altitude is 10 mm. $180\sqrt{3}$ mm³

14.6 Volumes of Pyramids and Cones   **581**

Closure

Ask the students how they determine the volume of a pyramid.   Corollary 1 page 579 The volume of a cone.   Corollary 2 page 579   Ask them what relationship exists between the volume of a pyramid and the volume of a prism. Between the volume of a cone and the volume of a cylinder. Theorem 14.9 page 579

▰▰▰ FOLLOW UP

Guided Practice

Classroom Exercises 1–6

Independent Practice

A Ex. 1–17, **B** Ex. 18–24, **C** Ex. 25–28

Basic: WE 1–17 odd, Application

Average: WE 1–24 odd, Application

Above Average: WE 1–28 odd, Application

Additional Example 4

Find the volume of the cone of revolution formed if a right triangle with 2 mm and 7 mm legs is revolved around the 7 mm leg.

$V = \frac{1}{3} \cdot \pi \cdot (2^2) \cdot 7$

$\quad = \frac{28}{3}\pi$ mm³

Find the ratio of the volumes of the two solids.

22. a square pyramid with 2-cm base edges and a 2-cm altitude and a square pyramid with 3-cm base edges and a 3-cm altitude   8 to 27

23. a right cone with a 2-mm base radius and a 2-mm altitude and a right cone with a 3-mm base radius and a 3-mm altitude   8 to 27

24. The owner of an ice cream parlor is choosing cups for her new store. Of the two cups shown, which has the greater volume? Express the difference in the volumes in terms of π. How do the amounts of ice cream heaped above the tops of the cups compare?

25. Find the ratio of the total area of a regular tetrahedron with edge length s to its volume. $6\sqrt{6}$ to s

26. A right square pyramid is inscribed in a right cone of the same height. A radius of the base and the altitude of the cone are each 10 cm long. What is the ratio of the volume of the pyramid to that of the cone?   2 to π

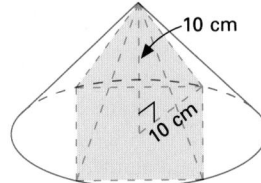

27. A right cone is inscribed in a right square pyramid of the same height. A radius and the altitude of the cone are each 10 cm long. What is the ratio of the volumes of the two solids? π to 4

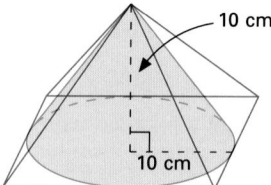

28. A right cone is inscribed in a cube with edge length s. What is the ratio of the volumes? π to 12

Mixed Review

1. Draw a segment 2 in. long. Construct the bisector of the segment.   *1.3*

2. Draw an obtuse angle. Construct the angle bisector.   *1.5*

3. Draw a line and a point not on the line. Construct a line through the given point parallel to the given line.   *3.3*

Enrichment

The following outline of the classic proof of the formula for the volume of a pyramid may interest some students in investigating the detailed version. Given any triangular pyramid with base ABC, construct a triangular prism with $\triangle ABC$ as a base (See first figure.) There are three pyramids in the prism, each having the same volume. (Pyramids with congruent bases and equal altitudes have the same volume.) This leads to the formula. The general case can be proved by dividing any given pyramid into triangular pyramids (second figure).

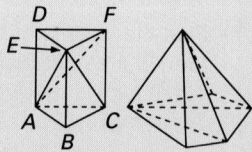

Application: *Manufacturing Paper Cups*

Cone-shaped paper cups can be manufactured from patterns in the shape of sectors of circles. The figures below show two patterns. The first is cut with a straight angle. The second is cut with an angle measuring 120.

1. If each sector has a radius of 6 cm, what will be the area of each sector? (HINT: What part of the circle is used by the sector?)

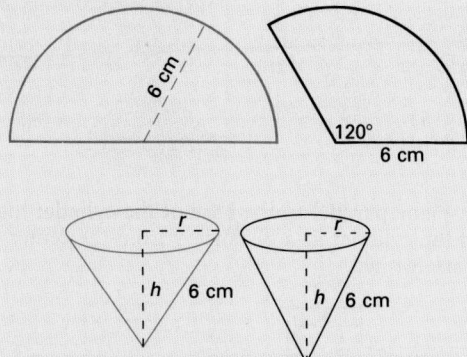

To find the volume of a cone-shaped cup formed from the first sector, we must first consider that the length of the arc of the sector becomes the circumference of the base of the cone.

$$\text{Arc length} = \tfrac{1}{2} \cdot 2\pi \cdot 6 = 6\pi = \text{Circumference of cone base}$$

Using the circumference formula again, we find the radius of the base.

$$\text{If } C = 2\pi r = 6\pi, \text{ then } r = 3$$

Using the Pythagorean Theorem, we can find the height of the cone.

$$h = \sqrt{36 - 9} = \sqrt{27} = 3\sqrt{3}$$

We can now find the volume of the cone.

$$V = \tfrac{1}{3}\pi r^2 h = \tfrac{1}{3}\pi \cdot 3^2 \cdot 3\sqrt{3}$$
$$= \pi \cdot 9 \cdot \sqrt{3}$$
$$= 48.97 \text{ cm}^3$$

2. Find the volume of the cup formed by the sector measuring 120.   23.70 cm³
3. As the area increases by 50% from the smaller to the larger sector, by how much does the volume of the cup increase?   Volume increases 106.6%

▰▰▰ GETTING STARTED

Prerequisite Quiz

1. What is the area of a circle with a 6 m radius? 36π m²
2. If the area of a circle is 64π in², what is the radius?   8 in
3. What is the approximate value of π?
   3.14

Motivator

Ask the students how they can find the amount of leather it takes to cover a baseball or how much rubber it takes to make a basketball.   Find the area   Ask them what measurements they need to find this out.   The length of the radius

▰▰▰ TEACHING SUGGESTIONS

Lesson Note

Many of the concepts of this lesson depend upon limits. Do not stress details at this time. Hollow models of spheres can be used effectively to show these same concepts.

14.7 Areas and Volumes of Spheres

Objectives   To find areas and volumes of spheres
             To find the ratios of areas and volumes of spheres

A **sphere** is the set of all points in space at a given distance from a given point called the *center*. A formula for the volume of a sphere can be found by comparing a sphere to a cylinder with a double cone removed. Let the sphere and the cylinder have equal radii r and equal heights $2r$.

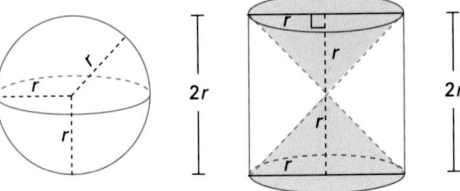

Suppose that a plane parallel to the base of the cylinder intersects both the sphere and the cylinder at a distance y from the center of each. Two cross sections are formed.

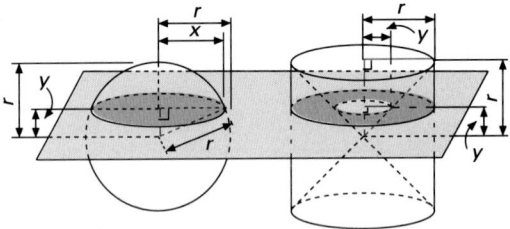

The plane cuts a circle in the sphere, and a ring-shaped figure called an *annulus* in the cylinder with the double cone removed. Let the radius of the circle be x. To find the dimensions of the annulus, notice that the altitude r of one of the cones forms an isosceles right triangle with a radius and the side of the cone. A similar triangle is formed by a radius of the inner circle of the annulus with the side and altitude of the cone. By similarity the triangle is isosceles, and so the radius of the inner circle is equal to the distance y.

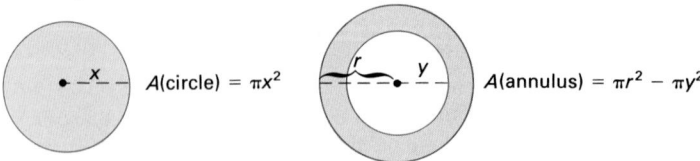

$A(\text{circle}) = \pi x^2$ $A(\text{annulus}) = \pi r^2 - \pi y^2$

584   Chapter 14 Figures in Space

In the sphere, a right triangle relationship exists between the distances *x*, *y*, and the radius *r*. By the Pythagorean Theorem, $r^2 = x^2 + y^2$, or $x^2 = r^2 - y^2$. Thus the area πx^2 of the circle may be expressed as $\pi(r^2 - y^2)$ or $\pi r^2 - \pi y^2$. But this is the same as the area of the annulus. Therefore, *all* corresponding cross sections of the sphere and the solid formed by the cylinder minus the double cone have equal areas. By Cavalieri's Principle, the sphere and this solid have equal volumes.

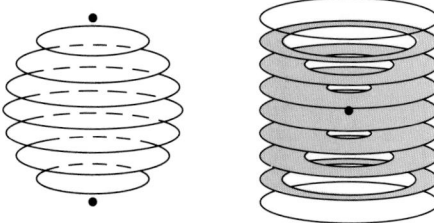

Corresponding cross sections.

The volume of this solid is the difference between the volumes of the cylinder and the double cone cut out of it.

$$\text{Volume(sphere)} = \text{Volume(cylinder)} - \text{Volume(double cone)}$$
$$= \pi r^2 \cdot (2r) - 2 \cdot (\tfrac{1}{3}\pi r^2 \cdot r)$$
$$= 2\pi r^3 - \tfrac{2}{3}\pi r^3$$
$$= \tfrac{4}{3}\pi r^3$$

This essentially proves the theorem that follows.

Theorem 14.10 The volume of a sphere with radius of length *r* is $\frac{4}{3}\pi r^3$.

EXAMPLE 1 Find the volume of a sphere with a 3-cm radius.
$$V = \tfrac{4}{3}\pi r^3$$
$$= \tfrac{4}{3}\pi \cdot 3^3$$
$$= 36\pi \text{ cm}^3$$

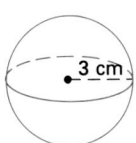

3 cm

A *great circle of a sphere* is the intersection of the sphere and any plane that contains the center of the sphere. Thus, a radius of a great circle is also a radius of the sphere.

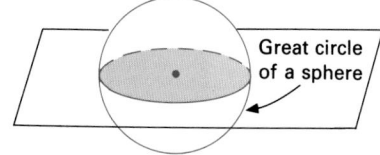

Great circle of a sphere

Plane intersecting sphere and passing through its center

Globe: Although the earth is not a perfect sphere, a globe is a good model of the earth. Two of the great circles on the globe are used for reference, the equator and the Greenwich Meridian.

Critical Thinking Questions

Analysis: Ask students to find the surface area and volume of a sphere with a diameter that is 6 units long. 36π sq units and 36π cubic units Ask students to generalize when the number of cubic units of volume is less than the number of square units of surface area. When the diameter is less than 6 units long

Checkpoint

1. Find the volume of a sphere with a 9 cm radius. 972π cm^3
2. Find the area of a sphere with a 12 in. radius. 576π in^2
3. If the volume of a sphere is 36π mm^3, what is the length of a radius? 3 mm

Closure

Ask the students to give the formulas for the area and the volume of a sphere. Theorem 14.10 page 585 and Theorem 14.11 page 586 Ask them what measurement must be known in order to find the area and volume of a given sphere. The length of the radius

Additional Example 1

Find the volume of a sphere with a 10 in. radius.

$$V = \tfrac{4}{3} \cdot \pi \cdot (10^3)$$
$$= \tfrac{4000}{3}\pi \text{ in}^3$$

Guided Practice

Classroom Exercises 1–6

Independent Practice

A Ex. 1–12, **B** Ex. 13–28, **C** Ex. 29–32

Basic: WE 1–12 odd, AR all

Average: WE 1–28 odd, AR all

Above Average: WE 1–32 odd, AR all

Two intersecting great circles formed by perpendicular planes divide a sphere into four *quadrants*. It is proven in advanced geometry that the surface area of a quadrant is equal to the area of a great circle. Thus, the area of a quadrant is πr^2. This suggests the next theorem.

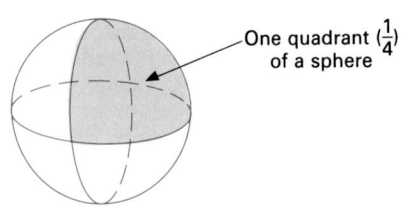

One quadrant $(\frac{1}{4})$ of a sphere

Two intersecting great circles formed by perpendicular planes

Theorem 14.11 | The area of a sphere is $4\pi r^2$.

EXAMPLE 2 Find the area of a sphere with a 6-cm radius.

Solution
$$\text{Area} = 4\pi r^2$$
$$= 4\pi(6^2) = 4\pi(36)$$
$$= 144\pi \text{ cm}^2$$

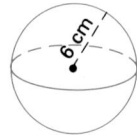

EXAMPLE 3 Find the length of a radius of a sphere with an area of 64π cm^2.

$$\text{Area} = 4\pi r^2$$
$$64\pi \text{ cm}^2 = 4\pi r^2$$
$$r^2 = \frac{64\pi}{4\pi} = 16$$
$$r = \sqrt{16} = 4 \text{ cm}$$

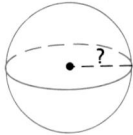

Classroom Exercises

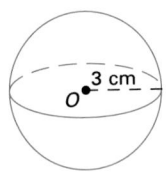

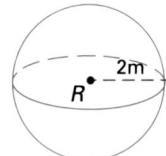

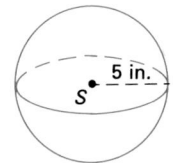

1. Find the area of Sphere O. 36π cm^2
2. Find the volume of Sphere O. 36π cm^3

3. Find the area of Sphere R. 16π m^2
4. Find the volume of Sphere R. $\frac{32}{3}\pi$ m^3

5. Find the area of Sphere S. 100π in^2
6. Find the volume of Sphere S. $\frac{500}{3}\pi$ in^3

Additional Example 2

Find the area of a sphere with a 5 m radius.

$A = 4 \cdot \pi \cdot (5^2)$
$= 100\pi$ m^2

Additional Example 3

Find the length of a radius of a sphere with an area of 20π in^2.

20π in$^2 = 4\pi r^2$

$r^2 = 5$ in
$r = \sqrt{5}$ in

Written Exercises

Find the area and volume of the sphere.

1.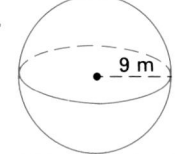

$A = 324\pi$ m², $V = 972\pi$ m³

2.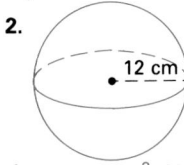

$A = 576\pi$ cm², $V = 2,304\pi$ cm³

3.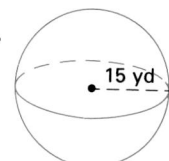

$A = 900\pi$ yd²,
$V = 4,500\pi$ yd³

Find the area and volume of a sphere with the given radius.

4. $r = 7$ cm   **5.** $r = 4$ mm   **6.** $r = 13$ in.
7. $r = 2.3$ ft   **8.** $r = 1.5$ m   **9.** $r = 4.3$ yd

Find the area and volume of a sphere with the given diameter.

10. $d = 4$ m $A = 16\pi$ m²,   **11.** $d = 12$ mm $A = 144\pi$ **12.** $d = 16$ ft $A = 256\pi$ ft²,
$V = 10\frac{2}{3}\pi$ m³   mm², $V = 288\pi$ mm³ $V = 682\frac{2}{3}\pi$ ft³

Find the length of a radius of the sphere.

13. volume $= 36\pi$ cm³ 3 cm   **14.** volume $= 324\pi$ m³ $3\sqrt[3]{9}$ m   **15.** area $= 100\pi$ cm²   5 cm
16. area $= 10$ mm²   **17.** volume $= 64$ ft³   **18.** area $= 100$ in²

19. The radius of a sphere is 10 cm. The radius of a second sphere is 20 cm. What is the ratio of the volumes of the two spheres?   1 to 8

20. The radius of a sphere is 5 cm. The radius of a second sphere is 15 cm. What is the ratio of their volumes?   1 to 27

21. The radius of a sphere is 4 cm. The radius of a second sphere is 8 cm. What is the ratio of their areas?   1 to 4

22. The radius of a sphere is 2 cm. The radius of a second sphere is 6 cm. What is the ratio of their areas?   1 to 9

23. What is the volume of a hollow spherical shell with a 5-cm inner radius and a 6-cm outer radius? $121\frac{1}{3}\pi$ cm³

24. What is the volume of a hollow spherical shell with a 6-cm inner radius and a 7-cm outer radius? $169\frac{1}{3}\pi$ cm³

25. If the area of the great circle is 64π cm², what is the area of the sphere? What is the volume of the sphere? $A = 256\pi$ cm², $V = 682\frac{2}{3}\pi$ cm³

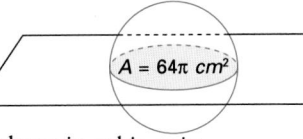

26. Find the length of a radius of a sphere whose volume in cubic units is equal to its area in square units.   3

27. The earth is almost spherical in shape. If the circumference of a great circle of the earth is about 40,000 km, what is the surface area of the earth? $\frac{1.6}{\pi} \times 10^9$ km²

28. The atmosphere of the earth has an altitude of about 550 km. Use Exercise 27 to find the volume of the earth and its atmosphere. 1.387×10^{12} km³

14.7 Areas and Volumes of Spheres   **587**

Enrichment

The formula for the area and volume of a *torus* (doughnut shape) are not difficult to grasp informally. A torus can be thought of as a curved cylinder. In the diagram, the radius R from the pt O to the center of the torus has the length OX, and the radius r of the torus itself has the length XP. Intuitively, the cylinder equal to the torus in area and volume would seem to have an altitude of $2\pi R$ and a radius of r. Thus, the area and volume of a torus are given by

$A = (2\pi R)(2\pi r)$
$V = (2\pi R)(\pi r^2)$.

587

29. A steel gas tank is in the shape of a sphere. A radius to the inner surface of the tank is 2 ft long. The tank itself is made of $\frac{1}{4}$-in. thick steel. Find the difference between the outside area of the tank and the inside area.   151 in²

30. If a gallon of paint will cover 400 ft² of surface, how many gallons of paint will be needed to paint both the inside and the outside of the tank described in Exercise 29?   Approximately $\frac{1}{4}$ gal

31. A sphere with a radius of length r is inscribed in a cube. What is the ratio of the volumes? $\frac{V_s}{V_c} = \frac{\pi}{6}$

32. A sphere with 13-cm radius is intersected by a plane. How far from the plane must the center of the sphere be so that the intersection is a circle with an area of 25π cm²?   12 cm

Mixed Review

Using the number line below, find the distance between the pair of points.   *1.2*

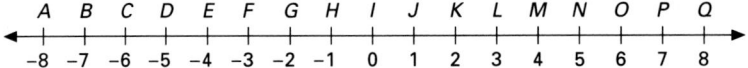

1. J and M 3   2. I and Q 8   3. A and F 5   4. C and H 5
5. State the symmetric property of congruence.   *2.3*   If $\overline{AB} \cong \overline{CD}$, then $\overline{CD} \cong \overline{AB}$

Algebra Review

To solve inequalities of the form $ax + b < cx + d$:
(1) Combine like terms.
(2) Solve for x.

Example: $3x - 5 < 5x + 7$
                  $-2x < 12$
                  $x > -6$ ⟵ Direction of inequality changes when multiplying or dividing by a negative number.

Solve the inequality.

3. $x > -3$ 6. $x > -3$

1. $3x + 5 > 2x + 1$ $x > -4$ 2. $7x - 1 < 6x + 3$ $x < 4$ 3. $-2x - 3 > -3x - 6$
4. $5x + 4 < 2x + 1$ $x < -1$ 5. $9x + 2 < 4x - 8$ $x < -2$ 6. $-4x + 1 > -7x - 8$
7. $3x + 5 > 5x - 1$ $x < 3$ 8. $4x - 4 > 7x - 10$ $x < 2$ 9. $x - 9 < 5x - 1$
10. $3 + 2x < 4 - 3x$ $x < \frac{1}{5}$ 11. $4 - x > x - 4$ $x < 4$ 12. $3 - 2x < 4x + 7$

9. $x > -2$ 12. $x > -\frac{2}{3}$

14.8 Coordinates in Space

Objectives To locate points in space
To find the distance between two points in space

Two coordinates are needed to locate a point in a plane. The *x*-coordinate, or *abscissa*, shows the distance and direction from the *y*-axis to the point. The *y*-coordinate, or *ordinate*, shows the distance and direction from the *x*-axis to the point.

Coordinates can also be used to describe figures in space. To locate points in space, a third coordinate is necessary. In space, three lines can be perpendicular to each other at a given point. Three such perpendicular lines, an *x*-axis, a *y*-axis, and a *z*-axis, are used to set up the coordinate system.

An **ordered triple** of numbers (*x*,*y*,*z*) locates a point with reference to the *x*-, *y*-, and *z*-axes. The *x*-coordinate gives the distance of the point from the plane containing the *y*- and *z*-axes (usually called the *yz*-plane). The *y*-coordinate gives the distance from the *xz*-plane, and the *z*-coordinate gives the distance from the *xy*-plane. Positive and negative directions are indicated on the axes.

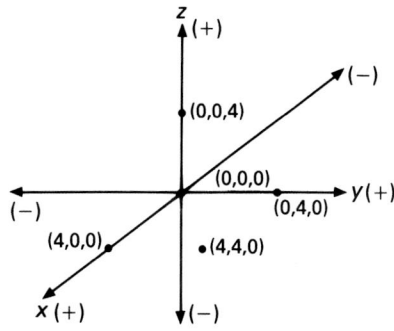

EXAMPLE 1 Graph $P(3,-2,4)$.

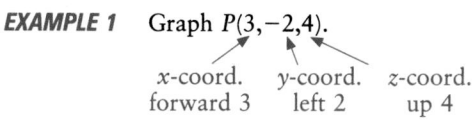

x-coord. y-coord. z-coord.
forward 3   left 2   up 4

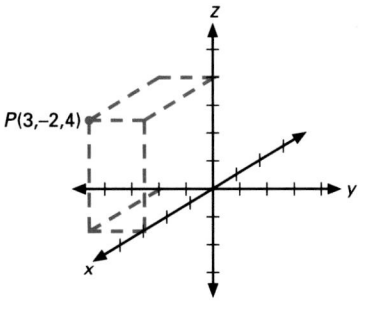

$P(3,-2,4)$

14.8 Coordinates in Space **589**

Additional Example 1

Graph $P(1,2,-3)$

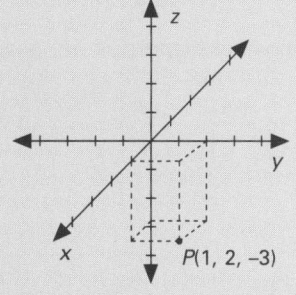

$P(1, 2, -3)$

Teaching Resources

Quick Quizzes 107
Reteaching and Practice Worksheets, pp. 213, 214
Transparencies 43A, 43B, 44

▬▬ GETTING STARTED

Prerequisite Quiz

1. Find the distance between the points $A(3,4)$ and $B(-2,5)$. $\sqrt{26}$
2. Find the slope of the line passing through the points $M(-2,2)$ and $N(3,-5)$. $-\frac{7}{5}$
3. Graph the line with equation $y = 2x - 2$.

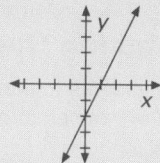

4. Write an equation for the line shown in the figure. $y = -x + 2$

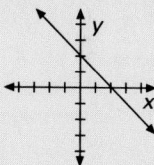

Motivator

Ask the students how many coordinates it takes to locate a point in a plane. It takes two coordinates, usually labeled the x- and y-coordinates. Ask them how many coordinates they think it would take to locate a point in space. Three Ask them to explain their answer. Figures in space are three dimensional whereas figures in a plane are two dimensional. Since it takes two coordinates to locate a point in a plane, then it will take three coordinates to locate a point in space.

TEACHING SUGGESTIONS

Lesson Note

It may be helpful to begin this lesson by reviewing concepts and terminology from Chapter 10 on Coordinate Geometry. A review of the Pythagorean Theorem will also be helpful for finding distance between points in space.

Math Connections

Algebra: All geometric figures and their properties can be described with algebraic equations and inequalities. The coordinate system is the base of the connection between algebra and geometry.

Critical Thinking Questions

Analysis: Ask students to name the coordinates of the vertices of each cube described below.

1. a cube whose sides are 6 units long and whose center is at the origin $(3,3,3)$, $(3,-3,3)$, $(3,3,-3)$, $(3,-3,-3)$, $(-3,3,3)$, $(-3,-3,3)$, $(-3,-3,3)$, and $(-3,-3,-3)$
2. a cube whose sides are 4 units long, whose coordinates are all positive, and with one vertex at the origin $(0,0,0)$, $(0,0,4)$, $(0,4,4)$, $(0,4,0)$, $(4,4,0)$, $(4,4,4)$, $(4,0,4)$, and $(4,0,0)$

Checkpoint

Graph the ordered triples.

1. $(1,3,1)$

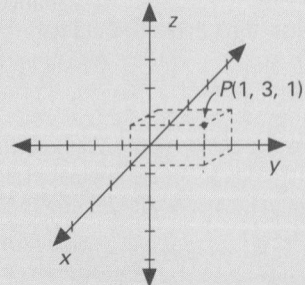

2. Find the distance between the points $P(2,3,-1)$ and $Q(-2,4,0)$ $3\sqrt{2}$

Find the coordinates of the midpoints.

3. $M(1,1,1)$ $N(3,-3,0)$ $(2,-1\frac{1}{2})$
4. $A(2,-1,4)$ $B(-4,0,6)$ $(-1,-\frac{1}{2},5)$

The Pythagorean Theorem gives a formula for the length of the hypotenuse of a right triangle in terms of the lengths of the two legs. The length of a diagonal of a rectangular solid can be found by applying the Pythagorean Theorem twice. Consider the diagonal of a rectangular solid with edges of lengths a, b, and c.

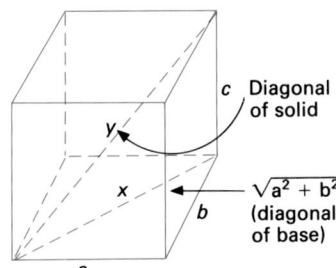

$$x^2 = a^2 + b^2$$
$$y^2 = x^2 + c^2$$
$$y^2 = a^2 + b^2 + c^2$$
$$y = \sqrt{a^2 + b^2 + c^2}$$

This extension of the Pythagorean Theorem can be used to prove a formula for the distance between points in space.

Theorem 14.12   The distance between points $P(x_1,y_1,z_1)$ and $Q(x_2,y_2,z_2)$ is given by the formula $d = \sqrt{(x_2 - x_1)^2 + (y_2 - y_1)^2 + (z_2 - z_1)^2}$.

EXAMPLE 2   Find the distance between $P(3,-2,4)$ and $Q(-3,3,1)$.

$P(3,-2,4)$ $Q(-3,3,1)$
  x_1,y_1,z_1 x_2,y_2,z_2

$$d = \sqrt{(x_2 - x_1)^2 + (y_2 - y_1)^2 + (z_2 - z_1)^2}$$
$$PQ = \sqrt{(-3 - 3)^2 + (3 - (-2))^2 + (1 - 4)^2}$$
$$= \sqrt{(-6)^2 + (5)^2 + (-3)^2}$$
$$= \sqrt{36 + 25 + 9} = \sqrt{70}$$

So, $PQ = \sqrt{70}$.

Additional Example 2

Find the distance between $P(2,1,-2)$ and $Q(-1,3,2)$ $\sqrt{29}$

Classroom Exercises

On which axis does the point lie?

1. $A(0,0,2)$ *z-axis*

2. $B(-5,0,0)$ *x-axis*

3. $C(0,-1,0)$
y-axis

Which coordinate plane contains the point?

4. $(0,7,-3)$ *yz-plane*

5. $(1,0,-4)$ *xz-plane*

6. $(2,-5,0)$
xy-plane

Written Exercises

Graph the ordered triple.

1. $(0,0,0)$ **2.** $(3,1,4)$ **3.** $(0,3,4)$ **4.** $(-5,2,3)$ **5.** $(-2,-3,-4)$ **6.** $(0,-5,3)$

Find the distance between the pair of points.

7. $J(3,-1,2)$ and $K(0,2,-1)$ $3\sqrt{3}$

8. $L(0,3,-2)$ and $M(2,0,-1)$ $\sqrt{14}$

9. $P(8,-2,6)$ and $Q(2,5,3)$ $\sqrt{94}$

10. $R(2,8,6)$ and $S(4,-2,-3)$ $\sqrt{185}$

11. $T(-2,1,-1)$ and $U(-9,6,2)$ $\sqrt{83}$

12. $V(7,3,0)$ and $W(-4,1,6)$ $\sqrt{161}$

Find the coordinates of the midpoint of the segment $\overline{PQ}$.

13. $P(0,0,0)$, $Q(-2,6,4)$ $(-1,3,2)$

14. $P(-2,4,11)$, $Q(8,0,3)$ $(3,2,7)$

15. $P(-2,1,3)$, $Q(1,-5,4)$ $(-\frac{1}{2},-2,\frac{7}{2})$

16. $P(0,5,-9)$, $Q(-7,2,1)$ $(-\frac{7}{2},\frac{7}{2},-4)$

17. A certain sphere is the set of all points in space a given distance r from a point (h,j,k). Use the distance formula for space to write an equation of a sphere. $r = \sqrt{(x-h)^2 + (y-j)^2 + (z-k)^2}$

18. Write an equation of a sphere with center $(3,-1,2)$ and a radius of length 2. $2 = \sqrt{(x-3)^2 + (y+1)^2 + (z-2)^2}$

Mixed Review

1. What is the area of a rectangle with sides 3 cm and 7 cm long? *12.2* 21 cm^2

2. What is the area of a triangle with base 7 in. long and an altitude of 5 in.? *12.4* 17.5 in^2

3. What is the area of a trapezoid with bases 7 m and 13 m long and an altitude of 3 m? *12.5* 30 m^2

4. What is the distance between the points $P(2,-7)$ and $Q(-4,7)$? *10.1* $2\sqrt{58}$

5. What is the midpoint of the segment with endpoints $A(2,-5)$ and $B(-8,7)$? *10.2* $(-3,1)$

Closure

Ask the students how to locate a point in space.   Use an ordered triple of numbers which describes the location of the point with reference to the *x*-, *y*-, and *z*-axes. Ask them how the formula for the distance between two points in space is different from the formula for the distance between two points in a plane.   The distance formula for two points in space includes the square of the difference of the z-coordinates.

▰▰▰▰ FOLLOW UP

Guided Practice

Classroom Exercises 1–6

Independent Practice

🅐 Ex. 1–12, 🅑 Ex. 13–16, 🅒 Ex. 17–18

Basic: WE 1–12
Average: WE 1–16
Above Average: WE 1–18

Additional Answers

Written Exercises

1–6. Check student graphs.

Enrichment

The following would be a good calculator activity. Find the area of the triangle formed by connecting the points $P(1,3,2)$, $Q(-1,4,-2)$, $R(3,-1,2)$. Using the Distance Formula, and rounding to the nearest tenth, we have $PQ = 4.6$, $PR = 4.5$, $RQ = 7.5$. Using Heron's Formula, the area is found to be 9.7.

1. 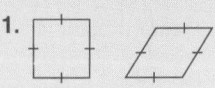   2.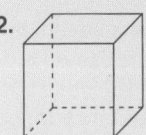

3. 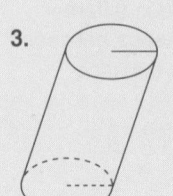   20.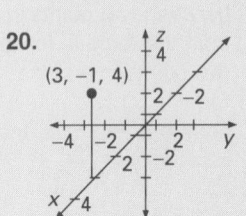

Chapter 14 Review

Key Terms

altitude (p. 564)	polyhedron (p. 556)
area of a prism (p. 564)	prism (p. 560)
base (p. 560)	pyramid (p. 574)
cone (p. 579)	rectangular solid (p. 560)
cone of revolution (p. 580)	regular polyhedron (p. 557)
cylinder (p. 561)	right prism (p. 560)
distance in space (p. 589)	rigid (p. 555)
edge (p. 557)	similar polyhedra (p. 565)
face (p. 560)	slant height (p. 574)
lateral face (p. 560)	sphere (p. 584)
lateral surface (p. 561)	three-dimensional (p. 560)
oblique (p. 560)	two-dimensional (p. 560)
ordered triple (p. 589)	vertices (p. 557)
parallelepiped (p. 560)	volume (p. 569)

Key Ideas and Review Exercises

14.1   To show that a polyhedron is rigid, show that all faces are triangular.

To show that a polyhedron is not rigid, show a counterexample.

1. Use sketches to show that a square is not rigid.

14.2   A prism has lateral faces that are parallelograms.

To draw a figure that is a prism or a cylinder, draw congruent parallel bases. For a rigid prism or cylinder, the corresponding sides should be parallel and vertically aligned.

2. Draw a right square prism.   **3.** Draw an oblique cylinder.

How many faces, edges, and vertices does the prism have?

4. a triangular prism $F = 5$, $E = 9$, $V = 6$ **5.** a pentagonal prism $F = 7$, $E = 15$, $V = 10$

14.3   In the following formulas, p is the perimeter of the base h is the length of the altitude, and r is the radius of the base.

To find the lateral area of a right prism, use the formula $L = ph$.

To find the lateral area of a right cylinder, use the formula $L = 2\pi rh$.

To find the total area of a right prism or cylinder, add the lateral area to the sum of the areas of the bases.

6. Find the lateral area of a right-square prism with a 12-cm altitude and 5-cm base edges.   240 cm²

7. Find the total area of a right cylinder with a 4-cm altitude and a 5-cm base radius.   90π cm²

14.4   To find the volume of a prism or a cylinder, use the formula $V = Bh$.

8. Find the volume of a prism with a base area of 8 cm² and a 5-cm altitude.   40 cm³

9. Find the volume of a right cylinder with a 3-m base radius and a 8-m altitude. 72π m³

10. Find the height of a right cylinder with a 4-m radius and a volume of 48π m³.   3 m

14.5   In the following formulas, p is the perimeter of the base, l is the slant height, and r is the radius of the base.

To find the lateral area of a regular pyramid, use the formula $L = \frac{1}{2}pl$.

To find the lateral area of a right cone, use the formula $L = \pi rl$.

To find the total area of a pyramid or cone, add the lateral area and the base area.

11. Find the lateral area of a regular pyramid with a 20-cm base perimeter and a 3-cm slant height.   30 cm²

12. Find the lateral area of a right cone with a 6-cm base radius and a 9-cm slant height. 54π cm²

13. Find the total area of a regular pyramid with a 3-mm base edge and an 8-mm slant height.   57 mm²

14. Find the total area of a right cone with a 5-mm base radius and a 7-mm slant height. 60π mm²

14.6   To find the volume of a pyramid or cone, use the formula $V = \frac{1}{3}Bh$.

15. Find the volume of a pyramid with a 25-cm² base area and a 9-cm altitude.   75 cm³

16. Find the volume of a cone with a 3-cm base radius and a 5-cm altitude. 15π cm³

17. Find the radius of a cone with height of 10 in. and a volume of 30π in³.   3 in

14.7   To find the volume of a sphere, use the formula $V = \frac{4}{3}\pi r^3$.

To find the area of a sphere, use the formula $A = 4\pi r^2$.

18. Find the volume and area of a sphere with a 6-mm radius. $V = 288\pi$ mm³ $A = 144\pi$ mm²

19. Find the radius of a sphere with a volume of 972π cm³.   9 cm

14.8   To locate a point in space, three coordinates (x,y,z) are needed.

To find the distance between two points in space, use the formula $d = \sqrt{(x_2 - x_1)^2 + (y_2 - y_1)^2 + (z_2 - z_1)^2}$.

20. Draw a set of x-, y-, and z-axes. Draw the point $(3,-1,4)$.   Check student drawings.

21. Find the distance between $A(3,-5,2)$ and $B(-5,3,4)$. $2\sqrt{33}$

Chapter 14 Review   **593**

19.

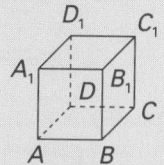

Statement	Reason
1. Rt prism with sq bases $ABCD \cong A_1B_1C_1D_1$	1. Given
2. ABB_1A_1, BCC_1B_1, CDD_1C_1, DAA_1D_1 are $\square$s.	2. Def of prism
3. $\overline{AA_1} \perp \overline{AB}$, $\overline{BB_1} \perp \overline{AB}$, $\overline{BB_1} \perp \overline{BC}$, $\overline{CC_1} \perp \overline{BC}$, $\overline{CC_1} \perp \overline{CD}$, $\overline{DD_1} \perp \overline{CD}$, $\overline{DD_1} \perp \overline{DA}$, $\overline{AA_1} \perp \overline{DA}$	3. In rt prism, lateral edges are $\perp$ to bases.
4. $\angle A_1AB$, $\angle ABB_1$, $\angle B_1BC$, $\angle BCC_1$, $\angle C_1CD$, $\angle CDD_1$, $\angle D_1DA$, $\angle DAA_1$ are rt $\angle$s.	4. Def of $\perp$
5. $AA_1 \cong BB_1 \cong CC, \cong DD_1$	5. Opp sides of $\square$ are $\cong$.
6. $\overline{AB} \cong \overline{BC} \cong \overline{CD} \cong \overline{DA}$	6. Def of square
7. $ABB_1A_1 \cong BCC_1B_1 \cong CDD_1C_1 \cong DAA_1D_1$	7. SASAS

20.

Statement	Reason
1. Rect solid with bases $ABCD \cong EFGH$	1. Given
2. Lateral faces $BCGF$ and $ADHE$ are rect.	2. Def of rect solid
3. $\angle CBF$, $\angle BFG$, $\angle DAE$, and $\angle AEH$ are rt $\angle$s.	3. Def of rect
4. $BCGF$, $ADHF$, and $ABCD$ are $\square$s.	4. Rects are $\square$s.
5. $\overline{BC} \cong \overline{AD}$, $\overline{BF} \cong \overline{AE}$, $\overline{GF} \cong \overline{HE}$	5. Opp sides of $\square$ are $\cong$.
6. $BCGF \cong ADHF$	6. SASAS

22. $E = 3 \cdot S$

593

13. Answers will vary.

Additional Answers, page 563

23.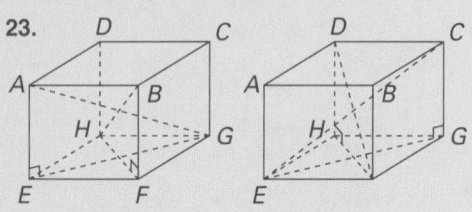

Statement	Reason
1. Rect solid *ABCDEFGH* with bases *ABCD* and *EFGH*	1. Given
2. ∠*AEG*, ∠*BFH*, ∠*CGE*, ∠*DHF* are rt ∠s.	2. In rt prism, lateral edges are ⊥ to bases.
3. *ABCD* and *EFGH* lie in ∥ planes.	3. Def of prism
4. *AE* = *BF* = *CG* = *DH*	4. Distance between ∥ planes is constant.
5. *EG* = *FH*	5. Diags of rect are ≅.
6. △*AEG* ≅ △*BFH* ≅ △*CGE* ≅ △*DHF*	6. SAS
7. $\overline{AG} ≅ \overline{BH} ≅ \overline{CE} ≅ \overline{DF}$	7. CPCTC

24.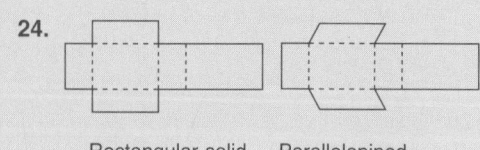

Rectangular solid   Parallelepiped

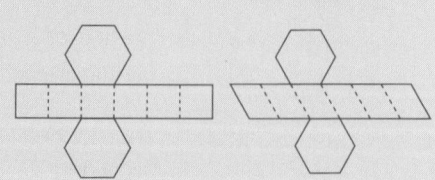

Right hexagonal prism   Oblique hexagonal prism

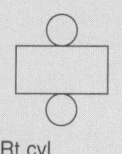

Rt cyl

Chapter 14 Test   A–Level Ex.: 1–12, 16;   C–Level Ex.: starred
B–Level Ex.: 15

Sketch the solid. Tell how many faces, edges, and vertices it has.

1. a cube
   $F = 6$, $E = 12$, $V = 8$

2. a regular tetrahedron
   $F = 4$, $E = 6$, $V = 4$

3. a regular quadrilateral pyramid $F = 5$, $E = 8$, $V = 5$

Find the total area of the solid. State the units of measure.

4. rectangular solid
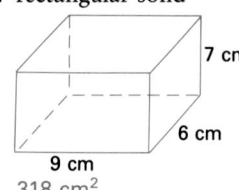
7 cm
6 cm
9 cm
318 cm²

5. sphere
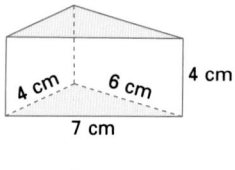
5 ft
100π ft²

6. regular square pyramid
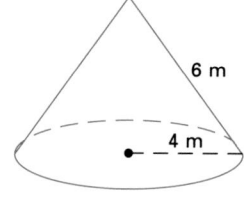
13 in.
10 in.
360 in²

Find the lateral area of the solid. State the units of measure.

7. right triangular prism
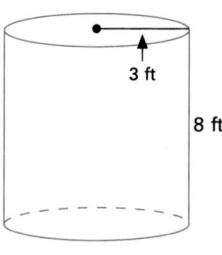
4 cm   6 cm   4 cm
7 cm
68 cm²

8. right cone
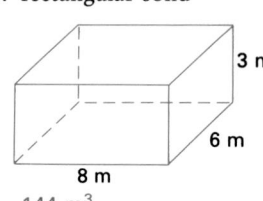
6 m
4 m
24π m²

9. right cylinder
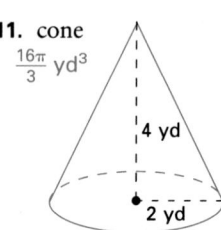
3 ft
8 ft
48π ft²

Find the volume of the solid.

10. rectangular solid
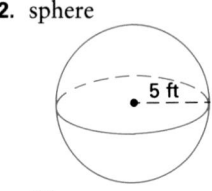
3 m
6 m
8 m
144 m³

11. cone
$\frac{16\pi}{3}$ yd³
4 yd
2 yd

12. sphere
5 ft
$\frac{500\pi}{3}$ ft³

*13. Use a sketch to illustrate Cavalieri's Principle. In your own words, write a brief, clear explanation of the principle.

*14. The volume of a cube is 64π cm³. Find its area. $96\sqrt[3]{\pi^2}$ cm²

15. If the lengths of the altitude and the radius of a right cylinder are multiplied by 5, what is the ratio of the new volume to the original? 125 to 1

16. Find the distance between $P(-1,3,5)$ and $Q(4,2,-1)$. $\sqrt{62}$

College Prep Test

In each item compare a quantity in Column 1 with a quantity in Column 2. Write the letter of the correct answer from the following choices.

A—The quantity in Column 1 is greater than the quantity in Column 2.
B—The quantity in Column 2 is greater than the quantity in Column 1.
C—The quantity in Column 1 is equal to the quantity in Column 2.
D—The relationship cannot be determined from the given information.

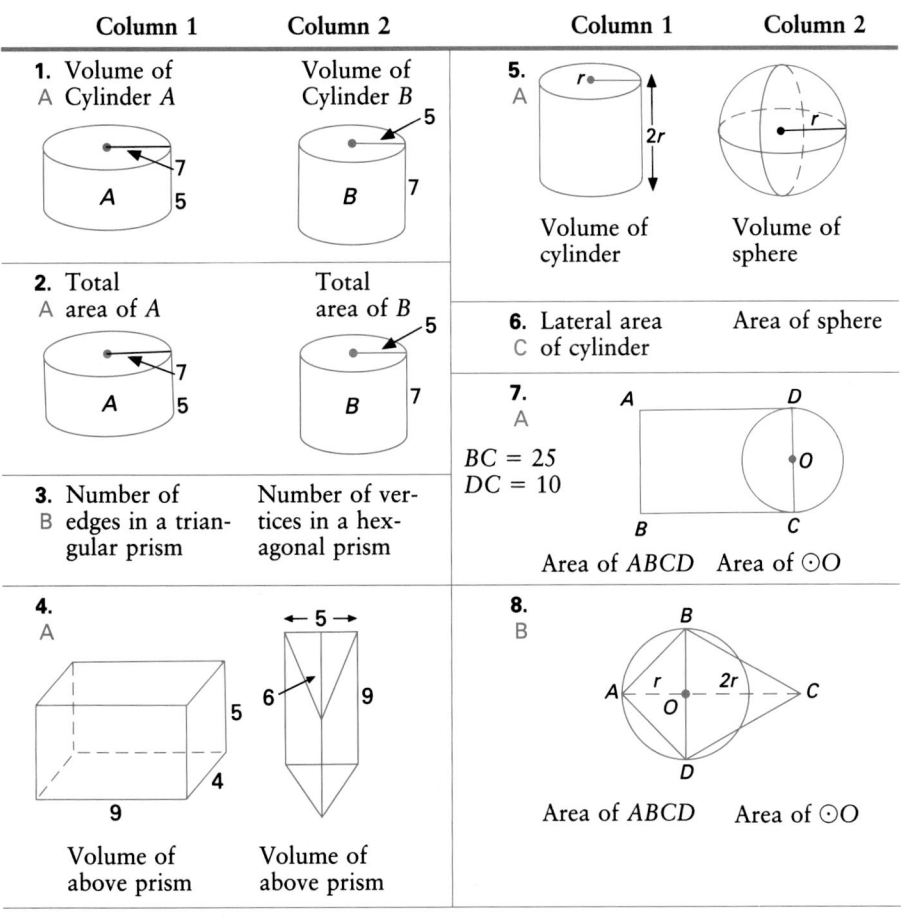

Column 1	Column 2
1. Volume of Cylinder A	Volume of Cylinder B
2. Total area of A	Total area of B
3. Number of edges in a triangular prism	Number of vertices in a hexagonal prism
4. Volume of above prism	Volume of above prism

Column 1	Column 2
5. Volume of cylinder	Volume of sphere
6. Lateral area of cylinder	Area of sphere
7. $BC = 25$ $DC = 10$ Area of $ABCD$	Area of $\odot O$
8. Area of $ABCD$	Area of $\odot O$

1. A
2. A
3. B
4. A
5. A
6. C
7. A
8. B

Mixed Review

4. Proof: Assume $\triangle ABC$ contains 2 obtuse $\angle$s, $\angle A$ and $\angle B$; m $\angle A > 90$, m $\angle B > 90$ (def of obt $\angle$); m $\angle A$ + m $\angle B > 180$ (If $a > b$ and $c > d$, $a + c > b + d$); m $\angle C > 0$ (Protractor Post); m $\angle A$ + m $\angle B$ + m $\angle C > 180$ (If $a > b$ and $c > d$, $a + c > b + d$); m $\angle A$ + m $\angle B$ + m $\angle C = 180$ (Sum of meas of $\angle$s in a $\triangle$ is 180.); But m $\angle A$ + m $\angle B$ + m $\angle C > 180$ and m $\angle A$ + m $\angle B$ + m C = 180 is a contradiction. So, the assumption that $\triangle ABC$ contains 2 obtuse $\angle$s is false. Therefore, $\triangle ABC$ cannot contain 2 obtuse $\angle$s.

5.

Statement	Reason
1. $\triangle ABC$	1. Given
2. Extend $\overline{BC}$ through C to D.	2. 2 pts determine line.
3. Draw $\overleftrightarrow{CE} \parallel$ to $\overline{BA}$ through C.	3. There is exactly 1 line $\parallel$ to given line through pt not on line.
4. m $\angle B$ = m $\angle ECD$	4. Corr $\angle$s are $\cong$.
5. m $\angle A$ = m $\angle ECA$	5. Alt int $\angle$s are $\cong$.
6. m $\angle A$ + m $\angle B$ = m $\angle ECD$ + m $\angle ECA$	6. Equations may be added.
7. m $\angle A$ + m $\angle B$ = m $\angle ACD$	7. Angle Add Post

3. If two ∠s are not ≅, then their supp are not ≅.

4.

Statement	Reason
1. △ABC ≅ △DEF, G is midpt $\overline{AB}$, H is midpt $\overline{DE}$.	1. Given
2. CB = FE, m∠B = m∠E, AB = DE	2. CPCTC
3. $\frac{1}{2}AB = \frac{1}{2}DE$	3. Mult Prop of Eq
4. GB = $\frac{1}{2}AB$, HE = $\frac{1}{2}DE$	4. Def of midpt
5. GB = HE	5. Sub
6. △CBG ≅ △FEH	6. SAS
7. $\overline{CG} \cong \overline{FH}$	7. CPCTC

5. From endpts of one seg, swing arcs of lengths of other 2 seg; intersection of arcs forms 3rd vertex of desired △.

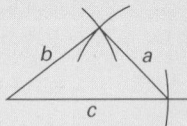

9.

Statement	Reason
1. Trap ABCD, $\overline{AB} \parallel \overline{DC}$, $\overline{AD} \cong \overline{BC}$, alts $\overline{DE}$ and $\overline{CF}$	1. Given
2. $\overline{DE} \perp \overline{AB}$, $\overline{CF} \perp \overline{AB}$	2. Def of alt
3. $\overline{DE} \parallel \overline{CF}$	3. If 2 lines ⊥ to same line, they are ∥.
4. Quad CDEF is ▱.	4. Def of ▱
5. $\overline{DE} \cong \overline{CF}$	5. Opp sides ▱ are ≅.

Cumulative Review (Chapters 1–14)

1. Find the measure of a supplement of an angle with a measure of 116. 64 *1.6*

2. The measure of an angle is $3x + 40$. The measure of its vertical angle is $2x + 50$. What is the measure of the angle? 70 *2.7*

3. Write the inverse of the statement: If two angles are congruent, then their supplements are congruent. *3.8*

4. Given: △ABC ≅ △DEF, G is the midpoint of $\overline{AB}$; H is the midpoint of $\overline{DE}$. *5.6*
Prove: $\overline{CG} \cong \overline{FH}$

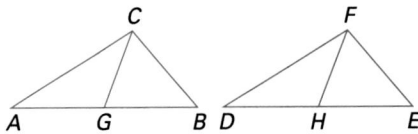

5. Construct a triangle with sides congruent to the three segments below. *5.8*

_____ _____

_____

6. True or false? If ∠A of △ABC is an obtuse angle, then BC > AC. True *5.8*

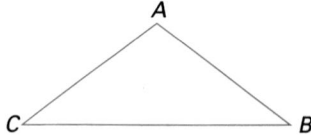

7. True or false? The property that *consecutive angles of a quadrilateral are supplementary* is a sufficient condition for the quadrilateral to be a parallelogram. True *6.5*

8. Given: Trap ABCD with median $\overline{EF}$, AB = 26, DC = 13 *7.4*
Find EF. $19\frac{1}{2}$

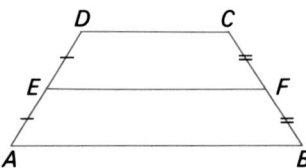

9. Given: Trap ABCD, $\overline{AB} \parallel \overline{DC}$, $\overline{AD} \cong \overline{BC}$, altitudes $\overline{DE}$ and $\overline{CF}$ *7.5*
Prove: $\overline{DE} \cong \overline{CF}$

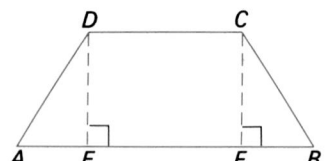

10. Copy the segment $\overline{AB}$ below. Construct the point P such that $\frac{AP}{PB} = \frac{3}{5}$. *8.5*

A _____ B

11. If the lengths of the sides of a triangle are 9, 40, and 41, is the triangle a right triangle? Yes *9.2*

12. Find the length of an altitude of an equilateral triangle if the length of a side is 4. $2\sqrt{3}$ *9.4*

13. Given: Right triangle ABC,   *9.5*
m $\angle A = 38$ and $AB = 10$.
Find AC.   7.9

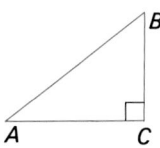

14. Prove using coordinate geometry   *10.6*
that the diagonals of a rectangle
bisect each other.

15. A chord of a circle is 2 cm long   *11.2*
and is 2 cm from the center of the
circle. What is the length of a radius
of the circle? $\sqrt{5}$ cm

16. Find the measure of an inscribed   *11.6*
angle of a circle if the angle intercepts an arc of measure 42.   21

17. What is the area of triangle ABC *12.4*
below?   42 cm²

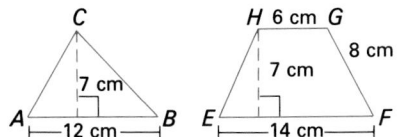

18. Given: Trapezoid $EFGH$ with   *12.5*
$\overline{EF} \parallel \overline{HG}$
Find the area of $EFGH$.   70 cm²

19. Sketch and describe the locus of   *13.2*
points in a plane 3 cm from a
circle with a 7-cm radius.

20. Find the total area of the cylin-   *14.3*
der.   108π m²

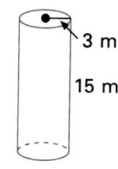

21. Find an equation in standard   *10.5*
form of a line that passes through
$(4, -2)$ and is perpendicular to the
line $3x + 2y = 6$. $2x - 3y = 14$

22. Find the area of $\square ABCD$ if   *12.3*
m $\angle A = 45$, $AB = 12$, and
$AD = 8$. $48\sqrt{2}$ sq units

23. Tangents $\overline{PA}$ and $\overline{PB}$ to a circle   *11.7*
from external point P form an angle
measuring 70. What is the measure
of the minor arc $\overset{\frown}{AB}$?   110

24. If secant $PC = 9$ cm and chord   *11.9*
$CB = 5$ cm, how long is tangent
$\overline{PA}$?   6 cm

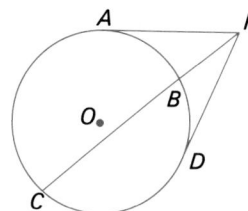

25. The apothem of a regular hexa-   *12.6*
gon is $6\sqrt{3}$ cm. What is its
area? $216\sqrt{3}$ cm²

26. A regular square pyramid has a   *14.5*
base 6 cm on a side and an altitude
of 4 cm. What is its total area?   96 cm²

27. Find the diagonal of a room that   *14.7*
is 16 ft × 12 ft × 8 ft. $4\sqrt{29}$ ft

28. Two circles with radii 8 cm and   *11.3*
3 cm have their centers 13 cm apart.
How long is their common external tangent?   12 cm

29. Find the area of a segment that   *12.9*
intercepts an arc of measure
60 in a circle with a radius of
12 in. $4(6\pi - 9\sqrt{3})$ in²

30. A balloon is inflated to a radius   *14.7*
of 12 in. If more air is added to increase the radius by 3 in., what will
be the change in volume?   2,196π in³

31. Describe the triangle formed by   *10.5*
$A(-6,5)$, $B(-2,-3)$, $C(6,1)$.   Isos rt $\triangle$

10. From A construct seg $\overline{AC}$ of length $8x$;
draw $\overline{CB}$, and $\parallel$ to $\overline{CB}$ draw 7 seg inters
$\overline{AC}$ at intervals of x; these 7 seg divide
$\overline{AB}$ into 8 seg of equal length; locate P
at third pt of intersection from A on $\overline{AB}$;
$\frac{AP}{PB} = \frac{3}{5}$.

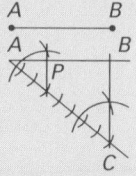

14. Given rect $ABCD$ with coordinates as
shown:
Midpt of $\overline{BD}$ is $\left(\frac{0 + 2a}{2}, \frac{2b + 0}{2}\right)$ or (a, b).
Midpt of $\overline{AC}$ is $\left(\frac{0 + 2a}{2}, \frac{0 + 2b}{2}\right)$ or (a, b).
Since $\overline{BD}$ and $\overline{AC}$ have same midpt,
$\overline{BD}$ and $\overline{AC}$ bis each other.

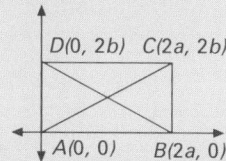

19. The locus is 2 circles concentric to the
given circle and 3 cm from it.

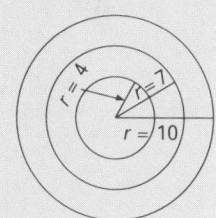

15 TRANSFORMATIONS

OVERVIEW

In this chapter, students construct reflections, translations, rotations, combined transformations of figures, and lines of symmetry. Students determine centers of rotation, and transformations that map a figure into another figure. They also draw dilations of figures with specified coordinate changes and shear images of figures.

OBJECTIVES

- To tell whether a figure is a reflection, translation, or rotation image, of a given figure
- To find the line of symmetry of a figure
- To slide a figure a given distance
- To find the center of rotation of a figure and its image
- To determine the number of lines of symmetry for a figure
- To construct a reflection of a figure
- To determine the transformations used in obtaining an image of a triangle
- To prove that a rotation is an isometry

PROBLEM SOLVING

The problem solving strategy, Making a Model, can be used by students to visualize and understand the properties of reflections, translations, and rotations. Encourage students to make cut-outs of geometric figures to help them solve problems.

TECHNOLOGY

Computer: The Computer Investigation on page 605 enables students to discover that the slide image of a given figure is congruent to the original figure. If the computer software can rotate a geometric figure about a given point, this investigation can be extended to Lesson 15.3 to allow students to discover properties of rotation images.

SPECIAL FEATURES

Mixed Review pp. 604, 611, 621, 625
Computer Investigation: Reflections p. 605
Application: Designs p. 611
Midchapter Review p. 617
Application: Stadium Lights p. 621
Focus on Reading p. 623
Key Terms p. 626
Key Ideas and Review Exercises pp. 626–627
Chapter 15 Test p. 628
College Prep Test p.629

PLANNING GUIDE

Lesson	Basic	Average	Above Average	Resources
15.1 pp. 602–605	CE all WE 1–7 CI 1	CE all WE 1–16 CI all	CE all WE 1–20 odd CI all	Reteaching p. 215 Practice p. 216
15.2 pp. 609–611	CE all WE 1–12 odd Application	CE all WE 1–15 odd Application	CE all WE 1–18 odd Application	Reteaching p. 217 Practice p. 218
15.3 pp. 615–617	CE all WE 1–9 Midchapter Review	CE all WE 1–18 Midchapter Review	CE all WE 1–24 odd Midchapter Review	Reteaching p. 219 Practice p. 220
15.4 pp. 620–621	CE all WE 1–16 odd Application	CE all WE 1–22 odd Application	CE all WE 1–28 odd Application	Reteaching p. 221 Practice p. 222
15.5 pp. 623–625	FR all CE all WE 1–8 odd	FR all CE all WE 1–21 odd	FR all CE all WE 1–23 odd	Reteaching p. 223 Practice p. 224
Chapter 15 Review pp. 626–627	1–8 odd	all	all	
Chapter 15 Test p. 628	1–5, 7–9	1–9	all	
College Prep Test p. 629	all	all	all	

CE = Classroom Exercises   WE = Written Exercises   FR = Focus on Reading   AR = Algebra Review   CI = Computer Investigation

NOTE: For each level, all students should be assigned all Mixed Review exercises.

Manipulative: Use this investigation to introduce Chapter 15. This investigation leads the students to an understanding of reflections and a method for finding a line of reflection.

This investigation can be done individually or with students working in pairs. Students need a copy of Manipulative Worksheet 29, a small pocket mirror (glass or metal), pencil, and straightedge. An unframed mirror works best, but a mirror with a narrow frame can be used.

After the students have completed the worksheet, ask them what is the same about a figure and its reflection, and what is different about a figure and its reflection. Ask them if distances are preserved in the reflections. Distances are the same; figures are congruent; position on the plane has changed.

Discuss with them a more formal definition for reflection and line of reflection.

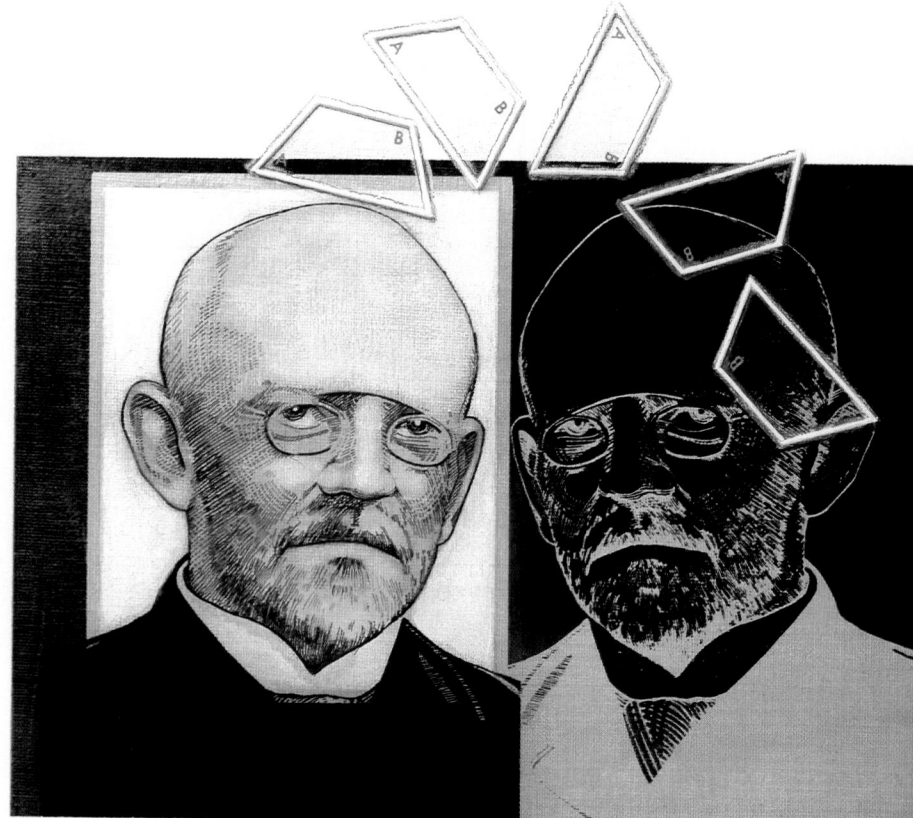

David Hilbert (1862–1943)—Hilbert studied mathematical properties called invariants that stay the same when a geometric figure is rotated, stretched, and reflected. When a segment is stretched, its length varies, but the "betweenness" of its points stays the same.

More About the Mathematician

David Hilbert (1862–1943) believed that every problem in mathematics can be resolved either by an answer or a proof that no answer is possible. Even though Gödel showed that this belief was wrong, Hilbert's work on the question influenced the whole world of modern mathematics and greatly extended the role of mathematics in the sciences. He closed loopholes in Euclid's geometry and established that if there are three points on a line, one of them is between the other two. The picture shows a negative reflection of his image and a trapezoid that is being rotated and translated.

15.1 Reflections

Teaching Resources

Manipulative Worksheet 29
Project Worksheet 15
Quick Quizzes 108
**Reteaching and Practice
    Worksheets** pp. 215, 216
Transparencies 45A, 45B

Objectives

To find reflections of figures
To locate and construct lines of symmetry
To determine whether or not figures are reflections of each other

If you were to draw a triangle on a balloon, blowing up the balloon would stretch the figure into a new shape. However, each point of the new figure, or *image*, would still correspond to exactly one point of the original triangle. Such a matching of points is a **geometric transformation.**

One kind of transformation is a *reflection*.

Definition

In a plane, a **reflection** about line *l* is a transformation which maps each point P into a point P' such that (1) if P is on l, then $P' = P$; and (2) if P is not on l, then l is the perpendicular bisector of $\overline{PP'}$.

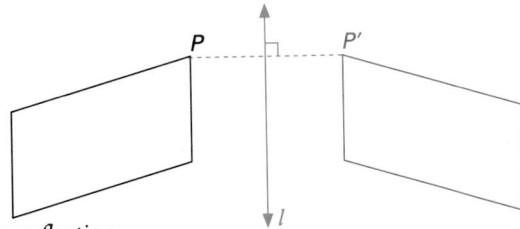

In this book, the term *reflection* will also be used to refer to the reflection image itself.

When you see a reflection of yourself in a mirror, your image seems to be the same size as your body. One way to show that the size of the image is the same size as the original figure is to use a transparent mirror. Using this, you can see both the original figure and its mirror image at the same time. It is possible to reach behind the mirror and draw the image that you see. The image drawn will be the same size as the original figure.

15.1 Reflections **599**

GETTING STARTED

Prerequisite Quiz

1. Draw a segment. Construct the perpendicular bisector of the segment.
2. Draw a triangle. Construct a second triangle congruent to the first.
3. Given a line and a point not on the line, construct a perpendicular from the point to the line.

Check student constructions.

Motivator

Ask the students what they see when they look in a mirror.   They see a reflection of themselves.   Ask them what the difference is between themselves and their reflections.   The position of things is opposite. For instance, the left part of the body is reflected on the right side.   Ask them how they are alike.   The size is the same.

TEACHING SUGGESTIONS

Lesson Note

Students may need to review simple transformations at an intuitive level before studying their properties. Have them think about what they see when they look at themselves in a mirror. How is what they see different from the "real" person? How tall is the reflection seen? Help them realize that what they are seeing appears to be as far behind the mirror as they are in front of it. Remind them that when they see something at a distance, it appears smaller than it actually is. From this discussion, lead into the formal definition of a reflection.

Math Connections

Palindromes: Words, phrases, or sentences which read the same backward or forward are called *palindromes*. Some examples are the word "madam" and the sentence "Name no one man." Both of these examples have a reflection line about which, except for the spacing between words, each side is reflected.

Critical Thinking Questions

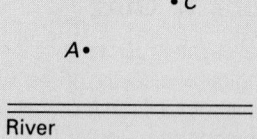

River

Application: Refer students to the diagram above and ask the following question.

Two neighbors who live in Town *A* are going fishing, and then will go to Town *C* to sell what they catch. Find the point *B*, where they will go fishing along the river, such that the total distance they travel, *AB* + *BC*, is the shortest distance possible. Find the reflection of *C*, *C'*, about the river. The pt at which $\overline{AC'}$ meets the river is *B* such that *AB* + *BC* is the shortest distance possible.

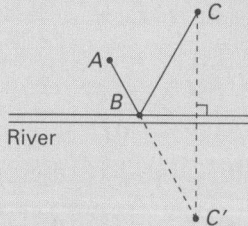

River

Common Error Analysis

Error: Some students have difficulty visualizing the number of lines of symmetry of a given polygon.

Have students draw the polygon on a sheet of paper. Then have them cut out the polygon and explore ways of folding it in half so that the two halves fit over each other exactly. Relate the lines of folding to lines of symmetry.

A transformation is distance-preserving if and only if the distance between two points is the same as the distance between the corresponding images of these points. Such a transformation is called an **isometry**.

Theorem 15.1   A reflection is an isometry.

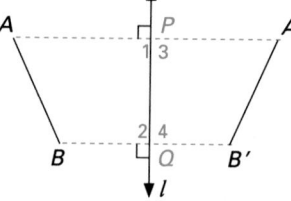

Given: Points *A* and *B* with reflections *A'* and *B'*, respectively, about *l*
Prove: *AB* = *A'B'*

Plan   Show that $\overline{AB}$ and $\overline{A'B'}$ are corresponding sides of congruent quadrilaterals.

Proof   Since *A'* is a reflection of *A* about *l* and *B'* is a reflection of *B* about *l*, then *l* is the perpendicular bisector of $\overline{AA'}$ and $\overline{BB'}$. This means that *AP* = *A'P*, *BQ* = *B'Q*, and that ∠1, ∠2, ∠3, and ∠4 are congruent right angles. *PQ* = *PQ*. Therefore, quad *APQB* ≅ quad *A'PQB'* by SASAS, and *AB* = *A'B'* for any two points *A* and *B* of the figure.

If a property is preserved, it is called an **invariant condition**. Therefore, congruence of segments, or distance, is an invariant condition under a reflection.

Using the definition of a reflection, you can construct a figure which is the reflection of a given figure about a given line.

EXAMPLE 1   Construct a figure which is a reflection about *l* of △*ABC*.

Plan   Construct a perpendicular segment from each vertex to *l* such that *l* bisects the segment.

Solution
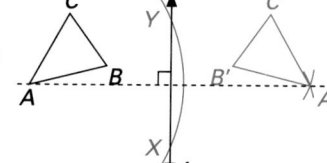

(1) Swing an arc with center *A*, intersecting *l* at *X* and *Y*.
(2) Swing arcs, with the same radius as the first arc, with centers *X* and *Y*, intersecting at *A'* on the opposite side of *l*.
(3) Repeat this for points *B* and *C*.
(4) Draw $\overline{A'B'}$, $\overline{B'C'}$, and $\overline{C'A'}$.
(5) △*A'B'C'* is the reflection of △*ABC* about the line *l*.

The reflection or mirror line is called the **line of symmetry**.

Additional Example 1

Draw a line and a quadrilateral about two inches from the line. Construct a figure which is the reflection of the quadrilateral about the line.   Check student constructions.

EXAMPLE 2 Find the line of symmetry of the figure *ABCD* and its reflection *A'B'C'D'*.

Plan Construct the perpendicular bisector to the segment joining a point and its image.

Solution

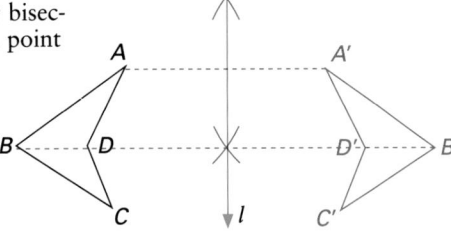

(1) Draw $\overline{AA'}$. Using compass and straightedge, construct its perpendicular bisector *l*. *l* is the line of symmetry.

(2) Verify by drawing $\overline{BB'}$ and constructing its perpendicular bisector. The two perpendicular bisectors should be the same line.

EXAMPLE 3 Determine whether or not the given pair of figures are reflections. If they are, then find the line of symmetry.

a.

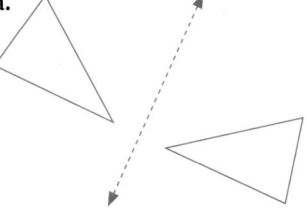

b.

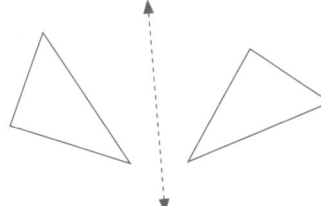

Plan
(1) Trace the pair of figures.
(2) Fold the paper so that the two figures come together.
(3) If the figures coincide, then the crease in the paper is the line of symmetry of the two figures.

Solutions
a. The figures coincide. They are reflections of each other, and the crease is the line of symmetry.

b. The figures do *not* coincide. They are not reflections of each other.

Some figures are their own mirror images about some line. The pentagon is its own image about *l*. A hexagon is its own mirror image about any diagonal and about each line connecting the midpoints of opposite sides. Such figures are called **symmetric figures.** A symmetric polygon has *line symmetry*. A symmetric polyhedron has *plane symmetry*.

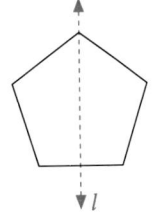

15.1 Reflections   **601**

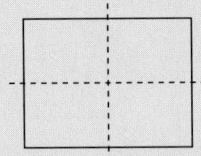
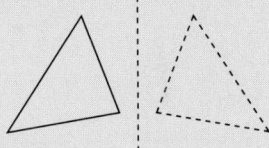
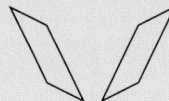
601

Additional Answers

1.

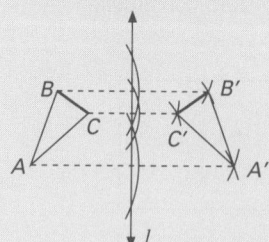

2.

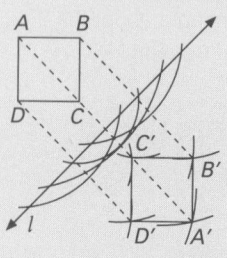

3.

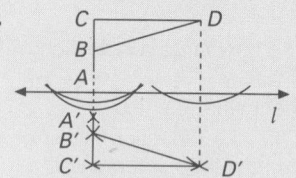

4.

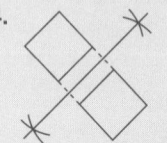

5.

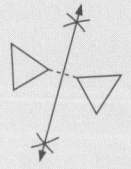

8.

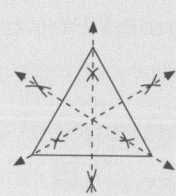

9.

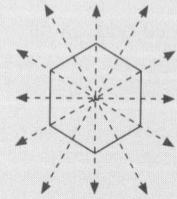

EXAMPLE 4 How many lines or planes of symmetry does each figure have?

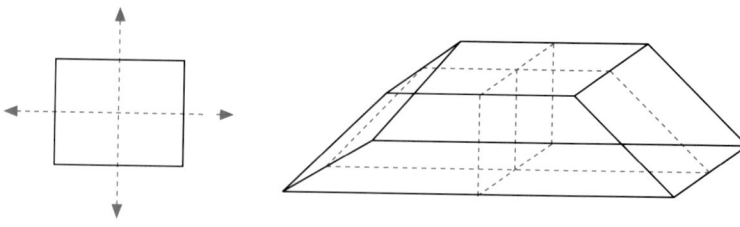

The rectangle has two lines of symmetry.

This polyhedron has two planes of symmetry.

Classroom Exercises

Do the pairs of figures below appear to be reflections about the given line?

1. No

2. No

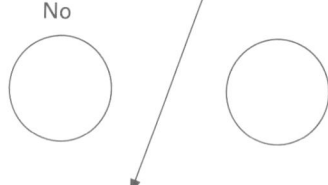

Which of the figures below have at least one line of symmetry?

3. No   **4.** Yes   **5.** Yes

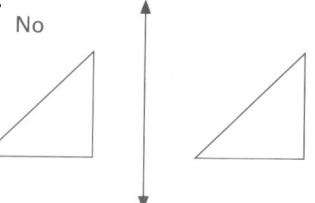

Which of the polyhedra below appear to have planes of symmetry?

6. tetrahedron Yes   **7.** rectangular solid Yes   **8.** volleyball Yes

Tell whether the indicated property is an invariant condition under a reflection.

9. length of a side Yes **10.** angle measure Yes **11.** parallel sides Yes **12.** position No

602   Chapter 15 Transformations

Additional Example 4

How many lines of symmetry does the ellipse below have? Two

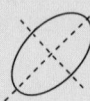

Written Exercises

Copy each given figure and line *l*. Construct its reflection using the method shown in Example 1.

1.
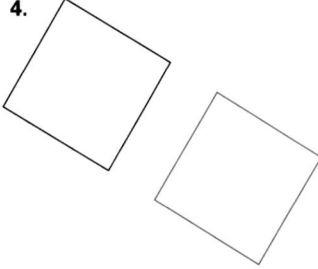

2.
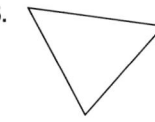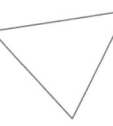

3.
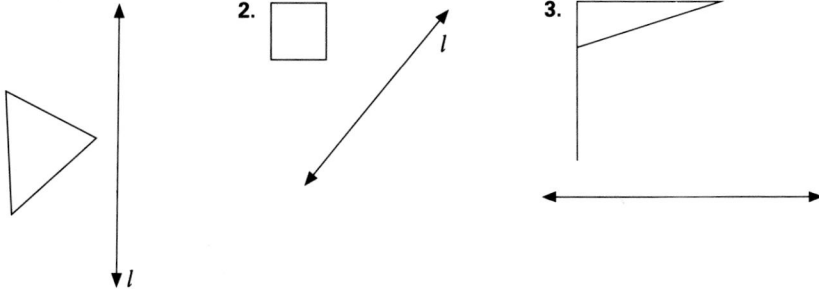

Copy each figure and its reflection. Find the line of symmetry using geometric constructions.

4.
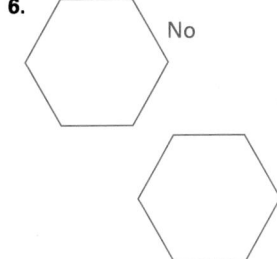

5.
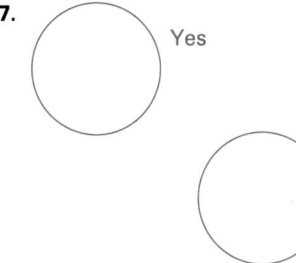

Copy each pair of figures. Determine which figures are reflections of each other.

6. No

7. Yes

15.1 Reflections **603**

10.
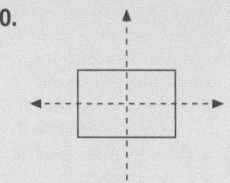

11.
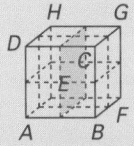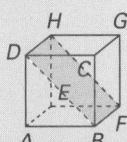

3 planes   5 more diagonal planes possible

12.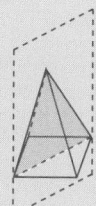

one other possible   one other possible

13.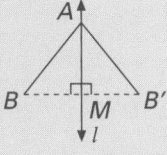

Statement	Reason
1. Pts *A* & *B* with reflec $\overline{A'B'}$ about *l* with *A* on *l* and *B* not on *l*	1. Given
2. Since *A* is on *l*, *A* = *A'*, *l* is ⊥ bis at *M* of $\overline{BB'}$	2. Def of reflec
3. *BM* = *B'M*, *BB'* ⊥ *l*	3. Def of ⊥ bis
4. ∠*AMB'* and ∠*AMB* are rt ∠s.	4. Def of ⊥
5. ∠*AMB'* ≅ ∠*AMB*	5. Rt ∠s are ≅.
6. $\overline{AM} ≅ \overline{AM}$	6. Reflex Prop of Congr
7. △*AMB* ≅ △*AMB'*	7. SAS
8. *AB* = *AB'*	8. CPCTC

14. Statement	Reason
1. Pts *A* and *B* with reflec *A'B'* about *l* with *A* and *B* on line *l*	1. Given
2. *A* = *A'*, *B* = *B'*	2. Def of Reflec
3. *AB* = *AB*	3. Reflex
4. *A'B'* = *AB*	4. Sub

15.

Statement	Reason
1. △ABC with its reflec △A'B'C'	1. Given
2. A' is reflec of A, B' is reflec of B, C' is reflec of C.	2. Def of reflec
3. AB = A'B', BC = B'C', AC = A'C'	3. Reflec is an isometry.
4. △ABC ≅ △A'B'C'	4. SSS

16. Under reflec congr is preserved, so figures with ≅ sides and ∠s will have the same area.

17.

Statement	Reason
1. $\overline{AY} \cong \overline{A'Y}$, $\overline{AX} \cong \overline{A'X}$	1. Constr with ≅ arcs
2. l is ⊥ bis $\overline{AA'}$	2. 2 pts each equidist from endpts determ ⊥ bis of seg.
3. Repeat for pts B, B' and C, C'; $\overline{A'B'}$ reflec of $\overline{AB}$, $\overline{B'C'}$ reflec of $\overline{BC}$, $\overline{A'C'}$ reflec of $\overline{AC}$	3. Def of reflec
4. △A'B'C' is reflec of △ABC.	4. Def of reflec

18. Yes, the line of reflec for the ctrs will be the line of reflec for the ⊙s because all pts of the ⊙ are equidist from ctr.

19. No, see counterexample.

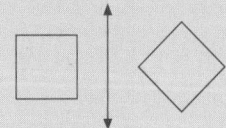

20. Answers will vary.

Copy each figure. Construct all lines of symmetry for each figure. How many lines of symmetry are there?

8. 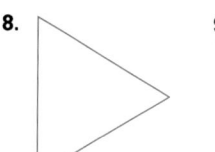   **9.** 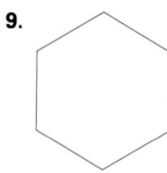   **10.**

Copy each figure. Construct all planes of symmetry.

11. 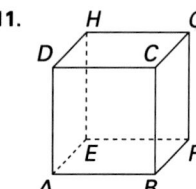   **12.**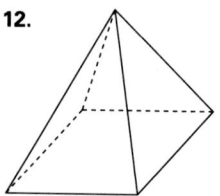

13. Prove Theorem 15.1 for the case where A lies on line l and B does not lie on line l.

14. Prove Theorem 15.1 for the case where A and B both lie on line l.

15. Prove that a triangle and its reflection are congruent.

16. Explain why area is an invariant condition under reflections.

17. Justify the construction used in Example 1 by proving that l is the perpendicular bisector of $\overline{AA'}$.

18. Will two congruent circles always be reflections of each other? Why or why not?

19. Will two congruent squares always be reflections of each other? Why or why not?

20. Name some man-made products that do not have line or plane symmetry.

Mixed Review

Given: $\overline{AB} \cong \overline{DC}$, $\overline{HG} \cong \overline{EF}$, $\overline{HG} \perp \overline{GB}$, $\overline{EF} \perp \overline{FC}$, $\overline{AB} \perp \overline{GB}$, $\overline{DC} \perp \overline{FC}$, $\overline{AG} \cong \overline{DF}$
Prove each of the following.   *4.3, 5.4*

1. △ABG ≅ △DCF   HL
2. $\overline{BG} \cong \overline{CF}$ CPCTC
3. ∠AGH ≅ ∠DFC   CPCTC
4. △AGH ≅ △DFE   SAS
5. $\overline{AH} \cong \overline{DE}$ CPCTC

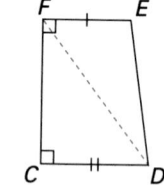

Computer Investigation

Reflections

Use a computer software program that draws random triangles and quadrilaterals by classification (acute, obtuse, etc.); moves and labels points; constructs line segments between two given points; draws reflections about a given line segment; and measures line segments and angles.

The following activities make use of a computer software program to draw reflected polygons. Then the program's measuring tool can be used to verify congruence to the nearest tenth of a unit.

Activity 1

Draw and label an obtuse triangle *ABC*. Draw and label any segment $\overline{ED}$ that is in the exterior of the triangle. $\overline{ED}$ will be the axis of symmetry for drawing the reflection of $\triangle ABC$.

1. Draw the reflection of $\overline{AC}$ about $\overline{ED}$.
2. Draw the reflection of $\overline{AB}$ about $\overline{ED}$.
3. Draw the reflection of $\overline{BC}$ about $\overline{ED}$.

The new triangle is the reflection of $\triangle ABC$ about $\overline{ED}$.

4. Verify that all corresponding points are symmetric with respect to $\overline{ED}$. (Use the measuring tool to show that $\overline{ED}$ is the perpendicular bisector of a segment formed by two corresponding points of the triangles.)
5. Use the measuring tool of the software program to verify that all corresponding parts of the two triangles are indeed congruent.

Activity 2

Repeat the steps of Exercises 1–5 of Activity 1 above for each of the following new triangles.

6. acute triangle   7. right triangle   8. isosceles triangle

Activity 3

Draw and label a random quadrilateral *ABCD*. Draw and label any segment $\overline{EF}$ that is in the exterior of *ABCD*. Draw the reflection of this quadrilateral about $\overline{EF}$.

9. Verify that a pair of corresponding points are symmetric with respect to $\overline{EF}$.
10. Verify that the two quadrilaterals are congruent.

◼ GETTING STARTED

Prerequisite Quiz

1. Draw a segment. Construct a second segment through a given point parallel to the first segment.
2. Draw a segment. On a given line, construct a segment congruent to the first segment.
3. Draw a triangle and a line. Construct a triangle which is the reflection of the first triangle about the line.
   For Ex. 1–3, check student constructions.

Motivator

Ask the students to describe what does not change about a picture if they slide the picture along a straight line path. Nothing about the picture changes. Ask them what does change. Its position Ask the class if all parts of the picture have moved the same distance in the same direction. Yes

◼ TEACHING SUGGESTIONS

Lesson Note

Use the activity described in the Motivator above to lead to a formal definition of a translation, or slide. This can also be used to introduce Theorem 15.2 and its proof on page 607.

15.2 Translations

Objectives

To draw translation images of figures
To find the resultant image of two reflections

If you were to push a box along a straight path, the box in its final position would be the *slide image* of the box in its original position. Any such motion of an object along a line that does not result in the object's turning in any way is a slide, or *translation*. A translation is another kind of transformation.

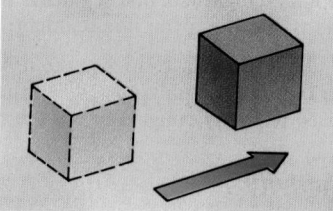

To describe a translation, you need to know which way and how far to *slide* the given figure. An informal way to show this is to use a **slide arrow** to indicate direction and distance.

EXAMPLE 1 Draw the slide image of the given figure using the given slide arrow. Do lengths of sides appear to be invariant?

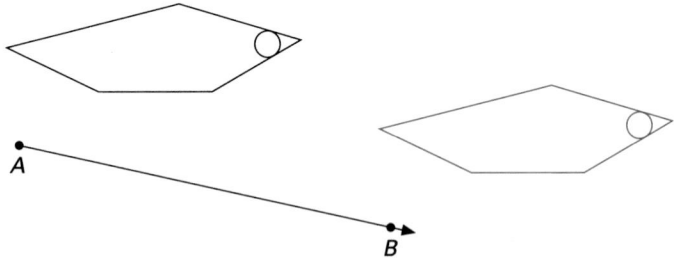

Solution

(1) Trace the given figure (black) on a sheet of paper. Draw a dot on your tracing at point *A* of the slide arrow.
(2) Without turning your paper, slide your drawing so that the dot moves along the arrow to point *B*.
(3) Notice that your traced image is now positioned directly over the slide image (red).
(4) Make a drawing of your own, with a given image and slide arrow of your choice. Repeat steps (1) and (2).
(5) Using carbon paper, or other blackened material, transfer your traced image in its new position onto your original drawing. The resulting image is the slide image of the original image with respect to the slide arrow.

606 Chapter 15 Transformations

Additional Example 1

Draw a triangle. Draw an arrow approximately two inches long. Draw the slide image of the triangle using the given arrow as the slide arrow.

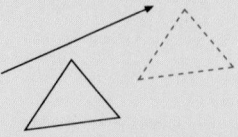

To construct the image of *ABCD* using the slide arrow $\overrightarrow{OP}$, construct a line through each vertex of *ABCD* parallel to $\overrightarrow{OP}$. Construct $\overline{AA'} \cong \overline{OP}$, $\overline{BB'} \cong \overline{OP}$, $\overline{CC'} \cong \overline{OP}$, and $\overline{DD'} \cong \overline{OP}$. Draw $\overline{A'B'}$, $\overline{B'C'}$, $\overline{C'D'}$, and $\overline{D'A'}$. *A'B'C'D'* is the image of *ABCD* with respect to the given slide arrow.

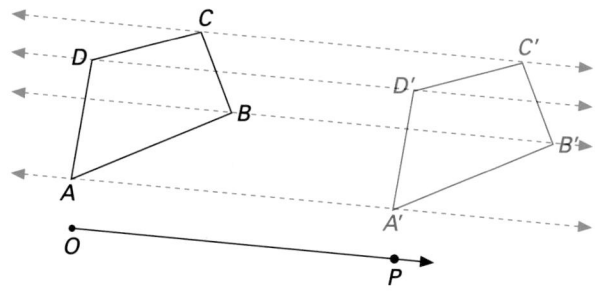

Definition

A **translation** in a plane from *A* to *A'* is a transformation which maps any point *P* into a point *P'* such that $\overline{PP'} \cong \overline{AA'}$ and $\overline{PP'} \parallel \overline{AA'}$.

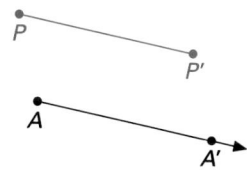

Theorem 15.2

A translation is an isometry.

Given: Points *A* and *B*, with translation images *A'* and *B'* from *O* to *P*
Prove: *AB* = *A'B'*

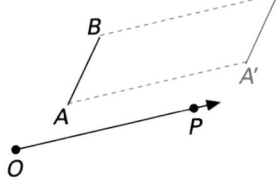

Plan

Show that $\overline{AB}$ and $\overline{A'B'}$ are opposite sides of a parallelogram.

Proof (Paragraph Form)

Since *A'* is the image of *A*, and *B'* is the image of *B* under the translation from *O* to *P*, then $\overline{AA'} \cong \overline{OP}$, $\overline{BB'} \cong \overline{OP}$, $\overline{AA'} \parallel \overline{OP}$, and $\overline{BB'} \parallel \overline{OP}$. Then by transitivity, $\overline{AA'} \cong \overline{BB'}$ and $\overline{AA'} \parallel \overline{BB'}$. These are congruent, parallel sides of *AA'B'B*. Therefore, *AA'B'B* is a parallelogram, and opposite sides $\overline{AB}$ and $\overline{A'B'}$ are congruent. Thus, *AB* = *A'B'*.

15.2 Translations **607**

Math Connections

Mass Production: Products that are mass produced and transported on a conveyor belt provide a good example of translation, since all parts of the product are moved the same distance in the same direction.

Critical Thinking Questions

Analysis: If the translation of △*ABC* is △*XYZ*, then what is m ∠*CAB* + m ∠*ACB* + m ∠*XYZ*? Explain. ∠*CAB* and ∠*ACB* are 2 ∠s of the original △*ABC*, and ∠*XYZ*, of the translated △, is the ≅ corr ∠ of ∠*ABC*, the third ∠ of △*ABC*. Since the sum of meas of all ∠s of a △ is 180, by substitution, m ∠*CAB* + m ∠*ACB* + m ∠*XYZ* = 180.

Checkpoint

1. Do the figures shown appear to be translations of each other?   No

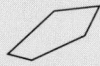

2. Draw the slide image using the given slide arrow.

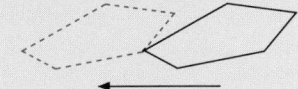

Closure

Ask the students how large a slide image of a geometric figure is. *It is the same size as the original geometric figure.* Ask them how they show the direction and the distance of the slide. *Use a slide arrow.* Ask them to explain the relationship between translations and reflections. Theorem 15.3 page 608.

▰▰▰ FOLLOW UP

Guided Practice

Classroom Exercises 1–8

Independent Practice

A Ex. 1–12, **B** Ex. 13–15, **C** Ex. 16–18

Basic: WE 1–12 odd, Application

Average: WE 1–15 odd, Application

Above Average: WE 1–18 odd, Application

Since a translation is an isometry, lengths of sides and angle measures of polygons are invariant. Different slide arrows may lead to the same slide image. Slide arrows m, n, and t below all result in the same image. All three slide arrows are congruent, parallel, and point in the same direction.

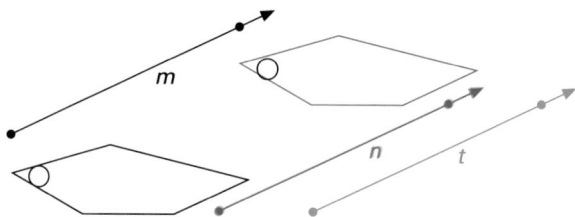

Translations can be determined by a pair of reflections about parallel lines.

EXAMPLE 2 Find the translation image determined by two reflections, first about line l and then about line m, where l and m are parallel.

Plan Construct the reflection $\triangle A'B'C'$ of $\triangle ABC$ about line l.

Construct the reflection $\triangle A''B''C''$ of $\triangle A'B'C'$ about line m.

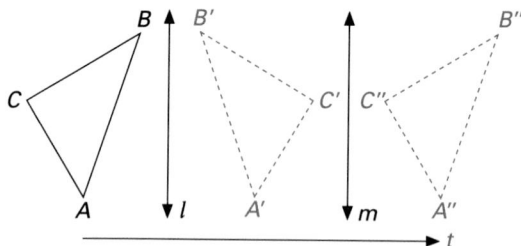

Solution $\triangle A''B''C''$ is the translation image of $\triangle ABC$ using slide arrow t, which is parallel and congruent to $\overline{AA''}$.

Example 2 suggests the following theorem.

Theorem 15.3 The resultant image determined by two successive reflections about parallel lines is a translation.

Successive reflections are sometimes referred to as the *product* of reflections.

608 Chapter 15 Transformations

Additional Example 2

Draw two parallel lines approximately one inch apart. Draw a quadrilateral to the left of the two lines. Find the translation image determined by two reflections with that image then reflected about the right hand line.

Classroom Exercises

Which pairs of figures below appear to be translations?

1.
Yes

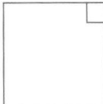
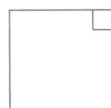

2.
No

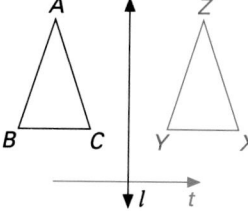

Using slide arrow *t*, find the translation image of each point.

3. *A*   *Z*   **4.** *B*   *Y*   **5.** *C*   *X*

Using line of symmetry *l*, find the reflection of each point.

6. *A*   *Z*   **7.** *B*   *X*   **8.** *C*   *Y*

Written Exercises

Copy each given figure and the slide arrow. Draw a slide image for each.

1.
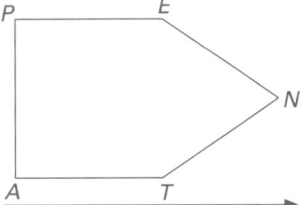

2.
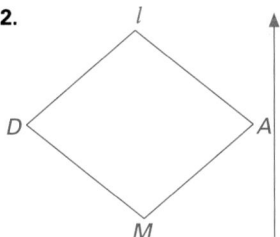

3.
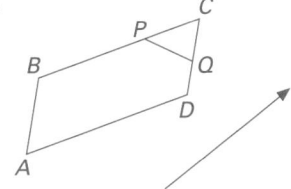

4.
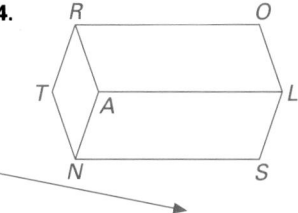

Enrichment

Bring to class pairs of congruent polygons and circles made from plastic or cardboard. Distribute these by pairs to the students. Have the students toss these randomly on a sheet of paper and trace their positions. Determine whether the tracings are translation images of each other.   The tracings of the pairs of congruent circles will always be translations of each other.

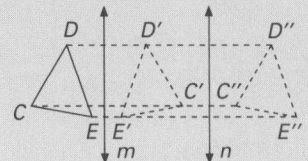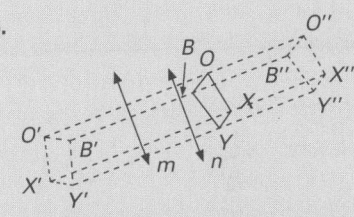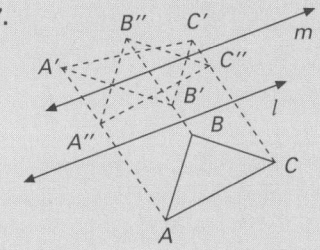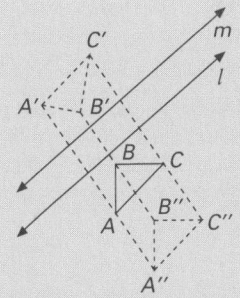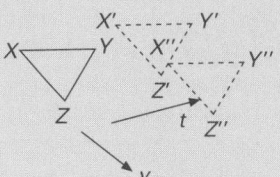

13.

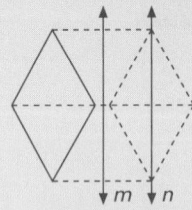

14.

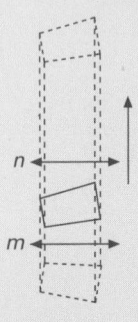

16. Length of sides, measures of ∠s, congruence, area.

17.

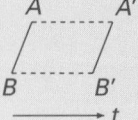

Statement	Reason
1. $\overline{AA'} \parallel \overline{BB'}$ and $\overline{AA'} \cong \overline{BB'}$	1. Def transl
2. $ABB'A'$ is ▱.	2. Quad with 1 pair opp sides ≅ and ∥ is ▱.
3. $\overline{AB} \parallel \overline{A'B'}$	3. Opp sides ▱ are ∥.

Find the slide image determined by two reflections, first about line _m_, and then about line _n_.

5.

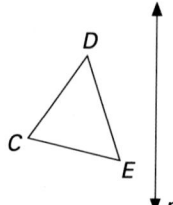

6.

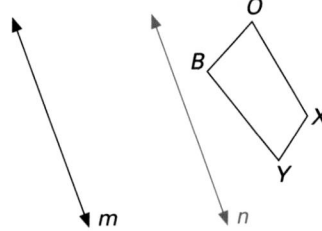

7. Find the image of △ABC determined by two reflections, first about line _l_, and then about line _m_.

8. Find the image of △ABC determined by two reflections, first about line _m_, and then about line _l_.

9. Based on the results of Exercises 7 and 8, does it appear that order is important when performing successive reflections? Explain.

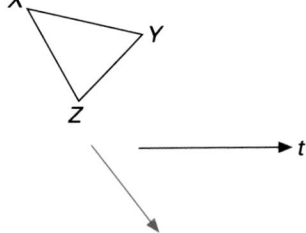

10. Find the translation image of △XYZ using _t_. Then find the translation image of the result using _v_.

11. Find the translation image of △XYZ using _v_. Then find the translation image of the result using _t_.

12. Based on the results of Exercises 10 and 11, does it appear that order is important when performing successive translations? Explain.   No. Translation is commutative

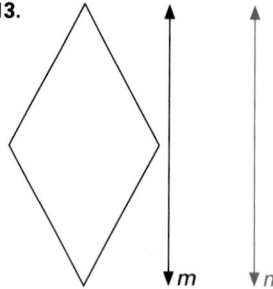

Find the reflection of the figure about _m_. Find the reflection of the image about _n_. Draw a slide arrow that results in the same final image.

13.

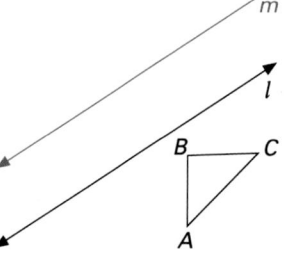

14.

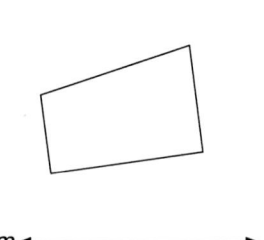

15. Copy the given figures and lines. Find the image of *ABCD* determined by two reflections, first about line *l* and then about line *m*. Is the resultant image a reflection? Is the resultant image a translation?   Neither

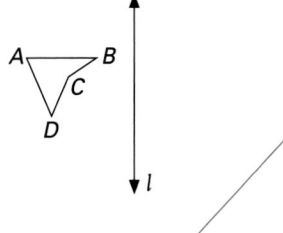

16. What properties of figures can be proved invariant under translations?

17. Prove that the translation image of a segment is parallel to that segment.

18. Prove Theorem 15.3 for $\overline{AB}$ about parallel lines *m* and *n*.

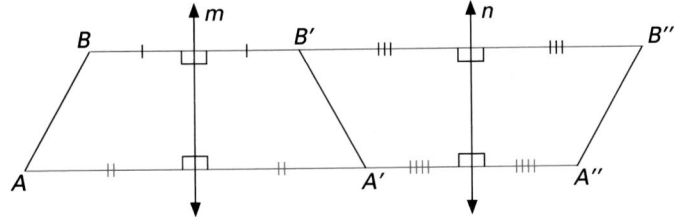

Mixed Review

1. Draw an angle and then construct an angle congruent to it.   *1.4*

2. Construct an angle with degree measure 90.   *5.4*

3. Draw a line segment $\overline{AB}$ and construct the perpendicular bisector of the segment.   *5.6*

Application: *Designs*

Some designs are the result of translating a simple figure several times. For each of the following basic designs, draw the design that is found by sliding each given basic design 1 in. to the right a number of times.

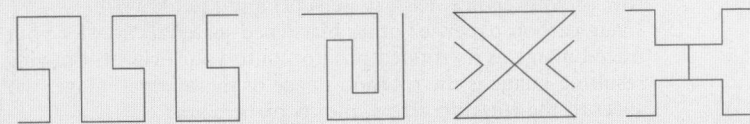

18.

Statement	Reason
1. $\overline{A'B'}$ is a reflec of $\overline{AB}$ about line *m*. $\overline{A''B''}$ is a reflec of $\overline{A'B'}$ about line *n*. *m* ∥ *n*	1. Given
2. $\overline{AB} \cong \overline{A'B'} \cong \overline{A''B''}$	2. A reflec is an isometry.
3. $\overline{AB} \cong \overline{A''B''}$	3. Trans Prop of ≅
4. $\overline{AA'}$ and $\overline{BB'} \perp m$; $\overline{A'A}$ and $\overline{B'B''} \perp n$	4. Def of reflec
5. $\overline{AA^1} \parallel BB^1$; $\overline{A'A''} \parallel \overline{B'B''}$	5. Two lines ⊥ same line or ∥ lines are ∥.
6. AA'B'B and A'B'B''A'' are isos trap.	6. Def of isos trap
7. ∠A ≅ ∠AA'B'; ∠A'B'B'' ≅ ∠B'B''A''; ∠B ≅ ∠BB'A'; ∠B'A'A'' ≅ ∠A'A''B''	7. Base ∠s of isos trap are ≅.
8. ∠AA'B' ≅ ∠A'B'B''; ∠BB'A' ≅ ∠B'A'A''	8. Alt int ∠s of ∥ lines are ≅.
9. ∠B ≅ A'A''B''; ∠A ≅ ∠A''B''B'	9. Trans Prop of ≅
10. AA''B''B is a ▱.	10. A quad with opp ∠s ≅ is a ▱.
11. $\overline{AB} \parallel \overline{A''B''}$	11. Opp sides of a ▱ are ∥.
12. $\overline{A''B''}$ is a transl of $\overline{AB}$.	12. Def of Transl

Mixed Review

1. See construction in Lesson 1.4.

2. See construction in Lesson 5.6.

3. See construction in Lesson 5.6.

▰▰▰GETTING STARTED

Prerequisite Quiz

1. Define *circle*.   The set of all pts in a plane equidist from a fixed pt.
2. What is the total degree measure of all angles around a common vertex?   360
3. Each ratio represents the length of an arc of a circle compared to its circumference. What is the degree measure of the arc?
   a. $\frac{1}{6}$ 30° b. $\frac{1}{4}$ 90° c. $\frac{5}{9}$ 200°

Motivator

Ask the students to describe the characteristics of a figure that do not change if it is turned or rotated around a given point.   The size and shape   Ask them to describe what does change.   Its position and its orientation.   Ask them if all parts of the figure move the same distance. Yes   Ask the class what instructions they would give to someone who was rotating an object so that the person would turn the object exactly the way they wanted it. They would have to give the person a reference point and tell him/her how far to rotate the object and in which direction to go.

15.3 Rotations

Objectives To draw rotation images of figures
To find the centers of rotations
To determine measures of angles of rotations

All wheels share a common property. Points on a wheel rotate in a circular path as the wheel turns on its axis. A pendulum has a similar property. The weight on the end of the pendulum moves in a path which is an arc of a circle. This motion is a *rotation*. The weight *rotates* around the pivot of the pendulum, which is the **center of rotation**. The center of rotation will be a point for coplanar figures or a line for figures in space.

To describe a rotation, you need to know the center of rotation and the magnitude of the rotation. To show this, use a turn arrow.

EXAMPLE 1 Draw a figure that is the rotation image of the given figure about P using the given turn arrow, which is an arc.

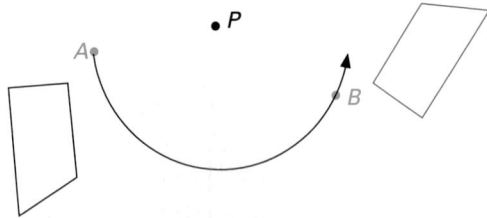

Solution (1) Trace the given figure (black) on a sheet of paper. Draw a dot on your tracing at point A of the rotation arrow.
(2) Place a pen or pencil on your tracing at point P and hold it firmly in place. Turn your copy around point P until the dot is over the point B.
(3) Notice that your traced image is now positioned over the rotation image (red).
(4) Make a drawing of your own, with a given image and rotation arrow of your choice. Repeat steps (1) and (2).
(5) Using carbon paper, or other blackened material, transfer your traced image in its rotated position onto your original drawing. The resulting image is the rotation image of the original image, with respect to the rotation arrow and its pivot point.

Additional Example 1

Draw a figure which is the rotation image of the given figure about P using the given turn arrow.

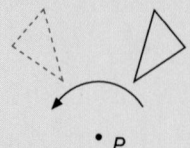

The measure of an angle (normally considered to be counterclockwise) can be used to specify the magnitude of the rotation. For example, the figure at the left below is rotated 90 degrees about P. This can be written as $R_{P,90}$. The rotation at the right is $R_{Q,120}$. The figure is rotated 120 degrees about Q.

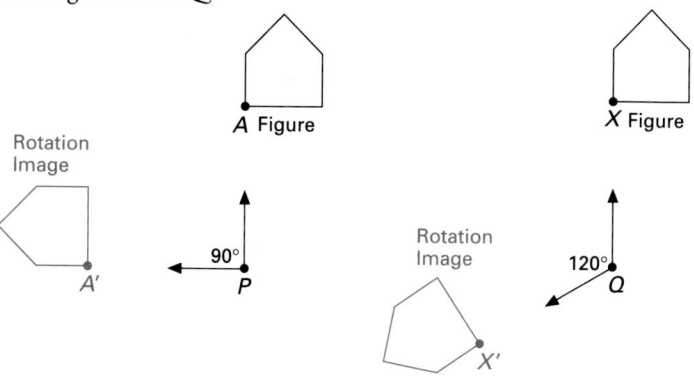

Definition

A **rotation** $R_{P,m}$ of point A about point P through an angle of measure m is a transformation which *maps* A into its *rotation image* A' such that $PA = PA'$ and m $\angle APA' = m$.

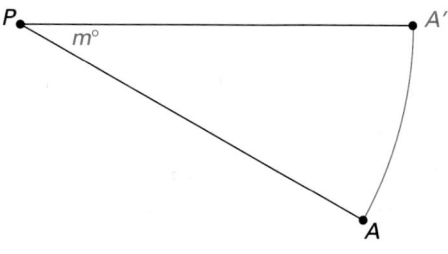

Theorem 15.4

A rotation is an isometry.

Given: Points A and B, with rotation images A' and B' about P; m $\angle APA' = $ m $\angle BPB'$

Prove: $AB = A'B'$

Plan

Show that $\overline{A'B'}$ and $\overline{AB}$ are corresponding congruent parts of $\cong \triangle$s $PA'B'$ and PAB.

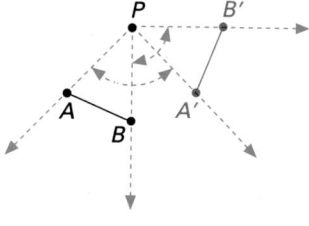

15.3 Rotations **613**

Construction: Doors, gates, lids, etc., which swing on a hinge, are good models for rotations in space. They represent a plane whose rotation depends on a center of rotation which is a line that is determined by the hinges.

Critical Thinking Questions

Analysis: Refer students to the diagram above and ask them to give a convincing argument that the lower figure to the right is not the rotational image of the upper figure to the left.   By def of rotation, each seg from P to corr pts of the figure should be $=$ and the meas of each $\angle$ formed by P to corr pts should be $=$, but neither of these is true in the diagram.

Common Error Analysis

Error: Rotations of 90 degrees are often difficult for students to visualize.

Have students draw a pentagon such as the one at the top of page 613, label or number its vertices, and cut it out. Then have them rotate the pentagon 90 degrees about one or more of its vertices and note the results.

Checkpoint

1. Do the figures shown appear to be rotations of each other?   No

2. Draw the rotation image using the given turn arrow.

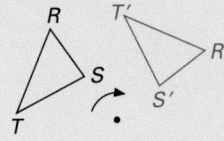

You learned earlier that a translation can be determined by two successive reflections about parallel lines. If the reflection lines are not parallel, then the result of the reflections is a rotation. The intersection of the two reflection lines is the center of rotation.

EXAMPLE 2   Find the rotation image of $ABCD$ about P by finding successive reflections about m and n. Using a protractor, measure the angle of rotation (in blue). How does this measure compare with the measure of the angle formed by the two lines of symmetry (in black)?

Solution   m $\angle APA'' = 60$, m $\angle XPZ = 30$

Therefore, the measure of the angle of rotation has a measure twice that of the angle formed by the two lines of symmetry.

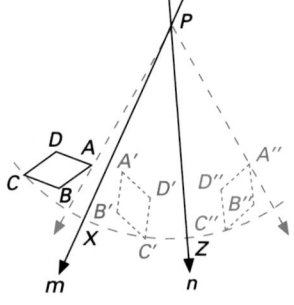

Theorem 15.5   The measure of the angle of rotation formed by two successive reflections (or by the product of two reflections) is twice the measure of the non-obtuse angle between the two lines of symmetry.

EXAMPLE 3   Find the center of rotation of $\triangle DEF$ and its rotation image $\triangle D'E'F'$ by paper folding.

Solution   (1) Trace the figures onto a sheet of paper. Fold the paper so that D and D' coincide. Crease the paper.
(2) Refold the paper so that E and E' coincide. Crease the paper again.
(3) The point where these two creases intersect is the center of rotation.

Additional Example 2

Find the rotation image of $\triangle ABC$ about P by finding successive reflections about m and n.

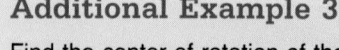

Additional Example 3

Find the center of rotation of the figure and its rotation image by paper folding.

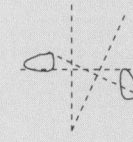

Some figures are their own rotation images. If the square below is rotated 90 degrees or 180 degrees about P, the point of intersection of the two diagonals, the rotation image coincides with the original square. If the figure at the right below is rotated 120 degrees about O, the rotation image coincides with the original figure. Each figure has **rotational symmetry.**

EXAMPLE 4 A regular pentagonal prism has rotational symmetry about a line l. What is the measure of the angle of rotation necessary for the figure to coincide with itself?

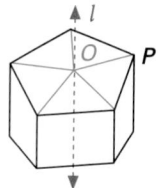

Solution Since each of the angles shown at O has a measure of 72, then the necessary angle of rotation measures 72, or any multiple of 72.

The figure at the right has *point symmetry.* The bisector of the segment determined by a point and its image is P for all points of the figure. Each point is rotated about P into its image. Point symmetry is a special case of rotational symmetry in which the angle of rotation is 180.

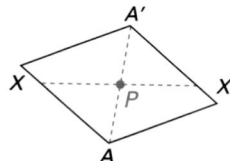

Classroom Exercises

In the diagram, *ABC* and *JKLM* have been reflected about lines l and m, successively.

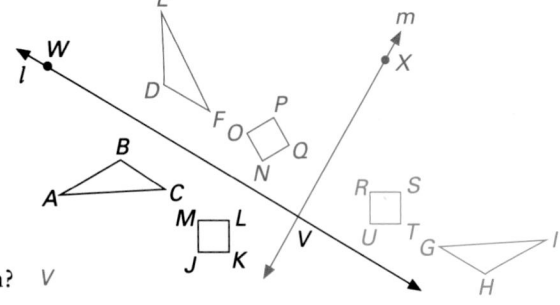

1. What is the center of rotation?   *V*

Find the rotational image of each point.

2. *A* *I*   **3.** *B* *H*   **4.** *C* *G*   **5.** *J* *S*   **6.** *K* *R*   **7.** *L* *U*   **8.** *M* *T*

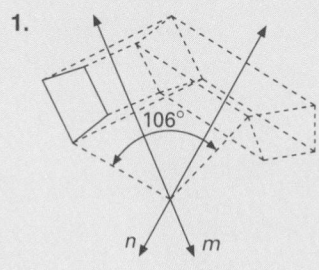
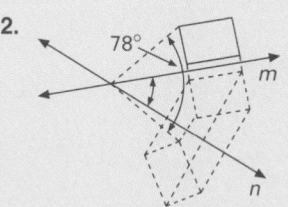
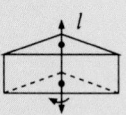

3.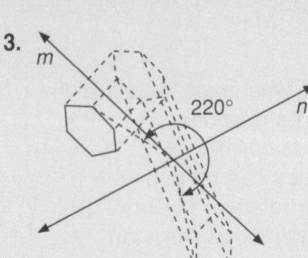

4. 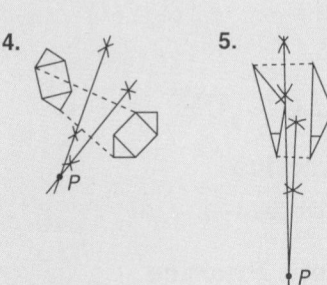   **5.**

6.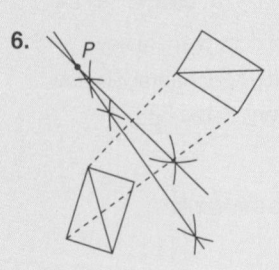

7.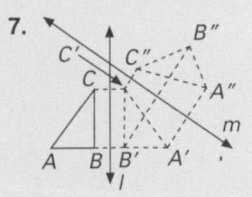

8.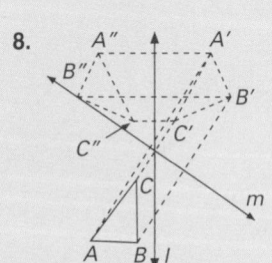

616

Written Exercises

Find the rotation image of each given figure about *P* by finding successive reflections about *m* and *n*. Measure the angle of rotation.

1. 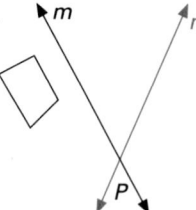   **2.**    **3.**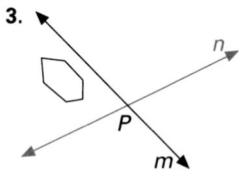

Copy each pair of figures. Find the center of rotation.

4.    **5.**    **6.**

7. Draw the rotational image obtained by reflecting △*ABC* successively about *l* and *m*.

8. Draw the rotational image obtained by reflecting △*ABC* successively about *m* and *l*.

9. Based on the results of Exercises 7 and 8, is order of reflection important when performing a rotation?   Yes

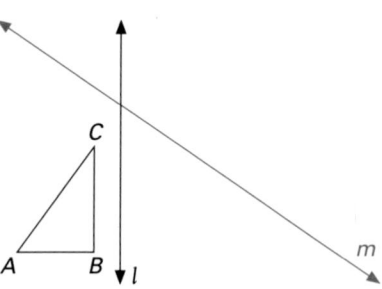

Each figure has rotational symmetry. What is the measure of the angle of rotation necessary for the figure to coincide with itself? (Exercises 10–15)

10.    60   **11.**    180   **12.** 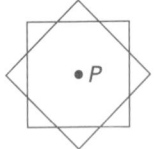   45

616   Chapter 15 Transformations

13.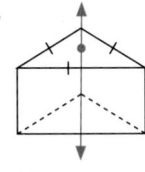

120

14.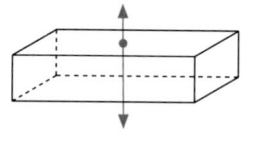

180

15.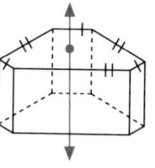

120

What product of two transformations will result in the image for each figure?

16.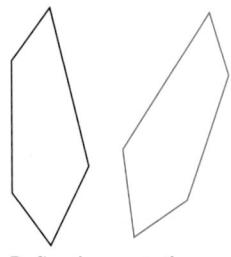

Reflection, rotation

17.

Rotation, reflection

18.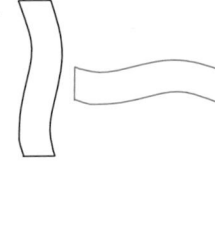

Reflection, rotation

19. A segment is rotated 180 degrees about a point not on the line containing the segment. Prove that the segment and its rotation image are parallel.

20. Prove that an equilateral triangle has rotational symmetry.

21. Prove that a parallelogram has point symmetry.

22. Prove Theorem 15.4.   **23.** Prove Theorem 15.5.

24. Prove or disprove: If a polygon has both line symmetry and rotational symmetry, then the polygon is regular.

Midchapter Review

Complete each statement. *15.1, 15.2, 15.3*

1. A transformation that preserves distances is called an ___isometry___.

2. For a reflection the perpendicular bisectors of segments joining each point and its image form the line of ___symmetry___.

3. The distance between any two points and their image will be equal for a ___translation___.

4. An equilateral triangle has ___rotational___ symmetry.

5. Two successive reflections about lines that are not parallel result in a ___rotation___.

15.3 Rotations   **617**

19.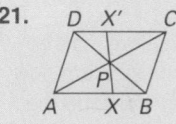

Statement	Reason
1. $\overline{AB} \cong \overline{A'B'}$	1. Rot is an isometry.
2. $\overline{AP} \cong \overline{A'P'}$, $\overline{BP} \cong \overline{B'P}$	2. Def of rot
3. $\triangle ABP \cong \triangle A'B'P$	3. SSS
4. $\angle PBA \cong \angle PB'A'$	4. CPCTC
5. $\overline{AB} \parallel \overline{A'B'}$	5. If 2 alt int $\angle$s are $\cong$, lines are $\parallel$.

20.

Statement	Reason
1. Rotate $\triangle ABC$ 60 deg clockwise, $\triangle ABC \cong \triangle BAC$	1. SSS
2. Rotate another 60, $\angle ABC \cong \angle CBA$, figures coincide and then have rotational symmetry.	2. SSS

21.

Statement	Reason
1. $\square ABCD$ with diags $\overline{AC}$ and $\overline{BD}$ which intersect at P.	1. Def of parallelogram
2. From any pt X of $\square ABCD$ draw $\overleftrightarrow{XP}$. Label the other pt of intersection of $\square ABCD$ and $\overleftrightarrow{XP}$ as X'.	2. Ruler Post
3. $\overline{AB} \parallel \overline{DC}$	3. Def of $\square$
4. $\angle PAX \cong \angle PCX'$	4. If lines $\parallel$, alt int $\angle$s are $\cong$.
5. $\angle APX \cong \angle CPX'$	5. Vert $\angle$s are $\cong$.
6. $AP = CP$	6. Diags of $\square$ bis each other.
7. $\triangle APX \cong \triangle CPX'$	7. ASA
8. $\overline{PX} \cong \overline{PX'}$	8. CPCTC
9. $\square ABCD$ has pt symmetry.	9. Def of pt symmetry

See page 627 for the answers to Ex. 22–24.

▰▰▰ GETTING STARTED

Prerequisite Quiz

1. Plot the following points: $P_1(5,1)$, $P_2(-4,0)$.   Check student graphs.
2. Find the distance AB between the points $A(-3,2)$ and $B(1,-1)$.

$$AB = \sqrt{(1 - (-3))^2 + (-1 - 2)^2}$$
$$= \sqrt{4^2 + 3^2}$$
$$= 5$$

Motivator

Ask the students if transformations can be thought of as moves.   Yes   Ask them how they describe a move.   They describe the direction and the distance.   Ask them if a coordinate plane could be used to show these moves.   Yes   Ask them how they would describe a move on a coordinate plane.   They would use the coordinates of the points.

▰▰▰ TEACHING SUGGESTIONS

Lesson Note

More advanced studies of mathematics use coordinates to describe transformations. Studies involving vectors deal extensively with such an approach.

Although pictures are helpful in setting up coordinate problems, the use of coordinates allows algebraic rather than geometric solutions of transformation problems. At first, encourage students to plot the points involved and to verify that the result looks appropriate. After several examples, encourage them to attempt the algebraic solution without drawing the picture.

15.4 Transformations and Coordinates

Objectives
To describe transformations using coordinates
To draw transformation images using coordinates

Transformations can be described using coordinate geometry. When describing a transformation, or *mapping*, the notation $(x, y) \rightarrow (x + 6, y + 3)$ means that a point in the original figure with coordinates x and y maps onto the point with coordinates $x + 6$ and $y + 3$, respectively.

EXAMPLE 1   Draw the image of $ABCD$ under the mapping $(x, y) \rightarrow (x + 6, y + 3)$.

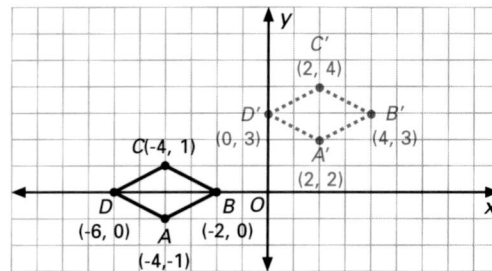

(1) Draw $ABCD$ on graph paper.
(2) For each vertex, add 6 to the abscissa and 3 to the ordinate.
(3) Mark the points corresponding to the new coordinates and label them A', B', C', and D'.
(4) Draw $A'B'C'D'$, the image of $ABCD$.

EXAMPLE 2   Using Example 1, show that the distance between any point of the figure and its image is constant. Show that the slope of the segment, determined by any point of this figure and its image, is also constant.

Plan   Choose two points $P_1(x_1, y_1)$ and $P_2(x_2, y_2)$ and their images $P_1'(x_1 + 6, y_1 + 3)$ and $P_2'(x_2 + 6, y_2 + 3)$. Use the distance formula and the slope formula.

Proof $P_1P_1' = \sqrt{(x_1 + 6 - x_1)^2 + (y_1 + 3 - y_1)^2}$, or $\sqrt{6^2 + 3^2}$
$P_2P_2' = \sqrt{(x_2 + 6 - x_2)^2 + (y_2 + 3 - y_2)^2}$, or $\sqrt{6^2 + 3^2}$

Therefore, $P_1P' = P_2P_2'$.

$$m_1 = \frac{y_1 + 3 - y_1}{x_1 + 6 - x_1}, \text{ or } \frac{3}{6} \qquad m_2 = \frac{y_2 + 3 - y_2}{x_2 + 6 - x_2}, \text{ or } \frac{3}{6}$$

Therefore, $m_1 = m_2$.

Additional Example 1

1. Draw a rectangle with vertices $A(2,1)$, $B(8,1)$, $C(8,5)$, and $D(2, 5)$. Draw the image of $ABCD$ under the mapping $(x, y) \rightarrow (x-2, y-3)$.

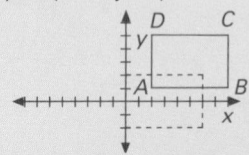

Additional Example 2

Using the data of Additional Example 1, show that the slope of the segment determined by any point of the rectangle and its image is a constant.

$$\frac{(y - 3) - y}{(x - 2) - x} = \frac{3}{2}$$

Theorem 15.6	The transformation defined by adding a constant to the coordinates of each point is a translation.

Plan   Show that the distance between each pair of points is equal to the distance between their images. Show that the segments connecting all points and their images are parallel.

EXAMPLE 3   Draw the reflection of $\triangle ABC$ about the y-axis. Find the coordinates of the vertices and describe the mapping using coordinate notation.

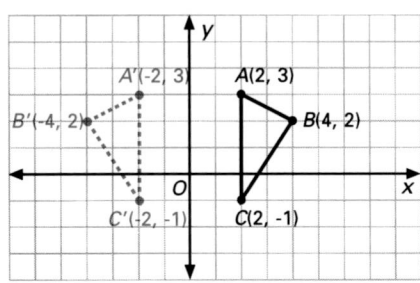

Solution   The coordinates of $\triangle A'B'C'$ are $(-2, 3)$, $(-4, 2)$, and $(-2, -1)$. This reflection is described by $(x, y) \rightarrow (-x, y)$.

EXAMPLE 4   Classify the transformation shown on the graph below. Describe the mapping using coordinate notation.

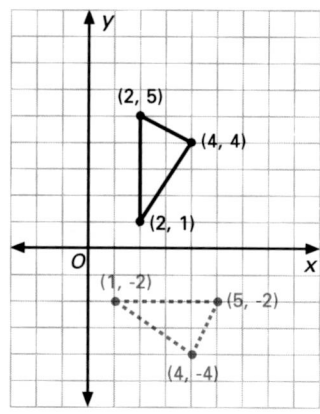

Solution   The transformation is a 90-degree clockwise rotation about the origin. This rotation is described by $(x, y) \rightarrow (y, -x)$.

Math Connections

Algebra: Since a function is a one-to-one correspondence that maps each element of the domain onto exactly one element of the range, a transformation can be described in functional notation. Although every transformation is a function, not every function is a transformation.

Critical Thinking Questions

Analysis: Ask students the following questions.

The line described by $y = a_2 x + b_2$ is the result of two reflections of the line with equation $y = a_1 x + b_1$. Write an equation that relates a_1 and a_2.   Reflected lines are parallel, so $a_1 = a_2$.   The vertices of $\triangle ABC$ in a coordinate plane are $A(-4, 0)$, $B(0, 4)$ and $C(4, 0)$. What type of triangle is $\triangle ABC$?   Isos   What are the coordinates of the vertices of the reflection image of $\triangle ABC$ about the y-axis? $A'(4, 0)$, $B'(0, 4)$, $C'(-4, 0)$ What are the coordinates of the vertices of a 180 degree rotation of the new image about its vertex B'? $A''(-4, 8)$, $B''(0, 4)$, $C''(4, 8)$

Common Error Analysis

Error: Some students may have difficulty interpreting mappings such as $(x, y) \rightarrow (x + 6, y + 3)$.

Have students interpret each mapping orally. For example, $(x, y) \rightarrow (x + 6, y + 3)$ means that for each original point (x, y), add 6 to the x-coordinate and add 3 to the y-coordinate to determine its image.

Additional Example 3

Draw a triangle with vertices $A(-3, -2)$, $B(7, -3)$, and $C(5, 4)$. Draw the reflection of the triangle about the x-axis.

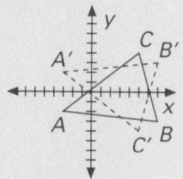

Additional Example 4

Classify the transformation shown below. Describe the mapping using coordinate notation.   reflection about y-axis: $(x, y) \rightarrow (-x, y)$

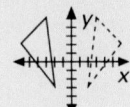

Checkpoint

Tell whether each transformation is a reflection, a translation, a rotation, or none of these.

1. $(x,y) \rightarrow (x - 2, y + 4)$ translation
2. $(x,y) \rightarrow (2x, -y)$ neither

Given $\triangle ABC$, with $A(0,0)$, $B(5,-3)$, and $C(2,3)$. Draw the image of $\triangle ABC$ according to each given coordinate change.

3. $(x,y) \rightarrow (x + 3, y - 1)$
4. $(x,y) \rightarrow (-x, y)$

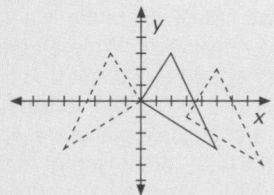

Closure

Ask the students what changes in coordinates are involved in a translation.   A constant is added to the coordinates of each point.   Ask them if reflections and rotations can be described using coordinates.   Yes Ask them what changes in coordinates were used in the examples for these transformations.   In Example 3 on page 619, the reflection is described by $(x, y) \rightarrow (-x, y)$. In Example 4 on page 619, the rotation is described by $(x, y) \rightarrow (y, -x)$.

◼◼◼FOLLOW UP

Guided Practice

Classroom Exercises 1–3

Independent Practice

A Ex. 1–16, **B** Ex. 17–22, **C** Ex. 23–28

Basic: WE 1–16 odd, Application

Average: WE 1–22 odd, Application

Above Average: WE 1–28 odd, Application

Classroom Exercises

Classify the transformation shown on the graph. Describe the mapping using coordinate notation.

1.
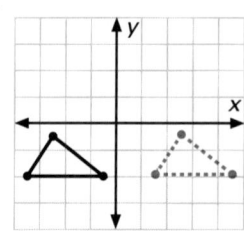

2.
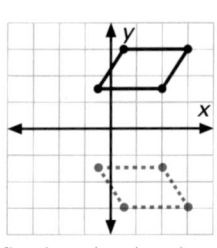

3.
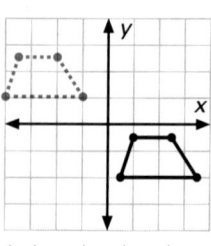

Translation; $(x, y) \rightarrow (x + 5, y)$ Reflection; $(x, y) \rightarrow (x, -y)$ Translation; $(x, y) \rightarrow (x - 4.5, y + 3)$

Written Exercises

Copy $\triangle ABC$ and the coordinate axes. Draw the image of $\triangle ABC$ according to the given coordinate change. Classify the transformation.

1. $(x, y) \rightarrow (x + 5, y)$
2. $(x, y) \rightarrow (x, y - 3)$
3. $(x, y) \rightarrow (x + 2, y + 1)$
4. $(x, y) \rightarrow (x, -y)$
5. $(x, y) \rightarrow (y, x)$
6. $(x, y) \rightarrow (y, -x)$
7. $(x, y) \rightarrow (-y, -x)$
8. $(x, y) \rightarrow (x, -y + 4)$
9. $(x, y) \rightarrow (-x - 3, -y - 1)$
10. $(x, y) \rightarrow (-y, -x + 2)$

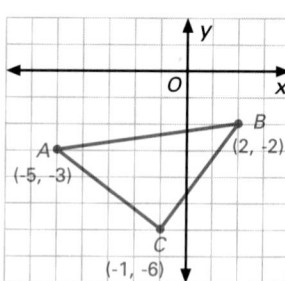

Show that the distances between two points and the distances between the corresponding images are equal for each transformation.

11. $(x, y) \rightarrow (x + 1, y + 2)$ 12. $(x, y) \rightarrow (x - 4, y + 5)$ 13. $(x, y) \rightarrow (x + 6, y - 7)$
14. $(x, y) \rightarrow (x - 9, y - 7)$ 15. $(x, y) \rightarrow (y, x)$ 16. $(x, y) \rightarrow (-y, x)$

Classify the resulting transformation.

17. adding a constant to each x-coordinate   Translation
18. replacing each x-coordinate by its opposite   Reflection
19. replacing both the x- and the y-coordinates by their opposites   Rotation
20. interchanging the x- and y-coordinates of each point   Reflection

620   Chapter 15 Transformations

Enrichment

A standard method of dealing with rotations is to use polar coordinates instead of rectangular coordinates. Introduce the basic concept of polar coordinates and the notation (r, θ). The transformation $(r, \theta) \rightarrow (r, \theta + c)$ describes a rotation about the origin.

Describe the mapping using coordinate notation.

21. a rotation of 180 degrees about the origin $(x, y) \rightarrow (-x, -y)$

22. a reflection about the x-axis $(x, y) \rightarrow (x, -y)$

23. a reflection about the line $y = x$

24. a reflection about the line $y = -x$

25. a reflection about the line $x = 5$

26. a reflection about the line $y = -3$

27. a reflection about the line $x = c$
$(x, y) \rightarrow (2c - x, y)$ **23.** $(x, y) \rightarrow (y, x)$

28. a rotation of 180 degrees about the point $(3, 4)$ $(x, y) \rightarrow (6 - x, 8 - y)$

25. $(x, y) \rightarrow (10 - x, y)$ **24.** $(x, y) \rightarrow (-y, -x)$ **26.** $(x, y) \rightarrow (x, 6 + y)$

Mixed Review

Write the truth table for each of the following.   *1.7*

1. $(p \vee q) \rightarrow p$

2. $(p \wedge q) \rightarrow q$

3. $[(p \rightarrow q) \wedge p] \rightarrow q$

4. Write an equation of a circle with a radius of 3 having its center at the origin.   *11.11* $x^2 + y^2 = 9$

5. Write an equation in standard form of a circle with a radius of 5 having its center at the point $(-3, 2)$.   *11.11* $x^2 + y^2 + 6x - 4y = 12$

 Application: *Stadium Lights*

A new lighting system is being installed at the Superior High School football stadium. Four large banks of lights mounted on 60-ft-tall poles are being placed around the field, two on each side. The poles, with the lights already installed on them, have been delivered to the stadium and are lying on the ground as shown. The holes for the foundations have been dug and the concrete footings poured. A large crane has arrived to position the poles in place. What combination of rotations, translations, and/or reflections will it take to position the poles properly? What instructions would you give the crane operator?

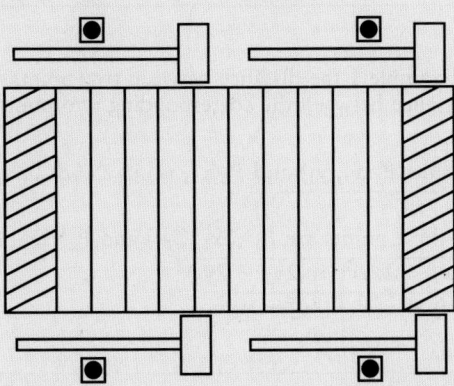

15.4 Transformations and Coordinates   **621**

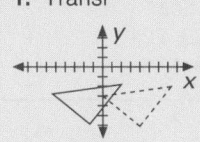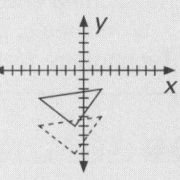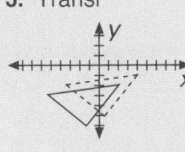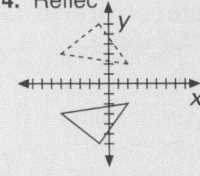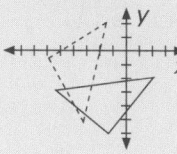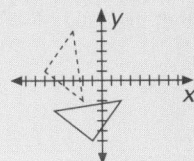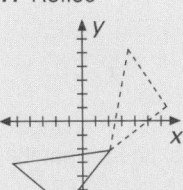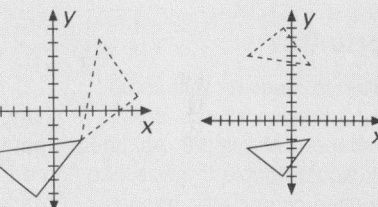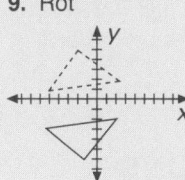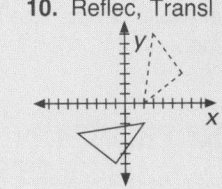

15.5 Dilations and Other Transformations

Objectives   To find a dilation of a figure
                 To find a shear image of a figure

The map shown to the right represents the state of Iowa. If we disre-
gard the curvature of the earth, this
map is similar to the actual state.
That is, distances are proportional
and the measures of angles are
equal. Such a map is a dilation.

Definition   A **dilation** is a transformation which produces an image similar to
                 the original figure.

EXAMPLE 1   Draw the image of $\triangle ABC$ under the mapping $(x, y) \rightarrow (2x, 2y)$.

(1) Draw the figure on graph paper.
(2) Replace each x-coordinate with $2x$ and each y-coordinate with $2y$.
(3) Draw the points corresponding to the new coordinates.
(4) Connect these points.

EXAMPLE 2   Show that in Example 1 the distance between two points of the image
                is twice the distance between the corresponding two points of the
                original figure.

Solution   Choose two points $P_1(x_1, y_1)$ and $P_2(x_2, y_2)$ in $\triangle ABC$.
                $$P_1P_2 = \sqrt{(x_2 - x_1)^2 + (y_2 - y_1)^2}$$
                The images of these points are $P_1'(2x_1, 2y_1)$ and $P_2'(2x_2, 2y_2)$.
                $$P_1'P_2' = \sqrt{(2x_2 - 2x_1)^2 + (2y_2 - 2y_1)^2}$$
                $$= 2\sqrt{(x_2 - x_1)^2 - (y_2 - y_1)^2}$$
                Therefore, $P_1'P_2' = 2P_1P_2$.

622   Chapter 15 Transformations

GETTING STARTED

Prerequisite Quiz

1. For the points (3,5), (−2,1), and (−5,3), find the image to the transformation $(x,y) \rightarrow (x + 2, y − 1)$.
   △ moved 2 units right and 1 down.
2. Draw a triangle *ABC*. Construct a triangle *A'B'C'* similar to △*ABC* such that the ratio of *AB* to *A'B'* is 1 to 2.

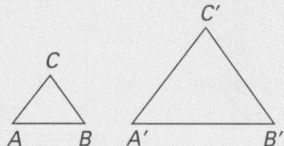

Motivator

Explain to students that some transformations stretch, shrink, or change the shape of the figure. Ask them what they think would double the size of a figure if they used coordinate changes.   Doubling all the *x*- and *y*- coordinates   Ask them what would cut the size in half.   Multiplying all the *x*- and *y*- coordinates by one-half.

TEACHING SUGGESTIONS

Lesson Note

If two triangles are similar, there is some transformation that maps the first triangle into the second. Such a transformation is an example of a dilation. If the corresponding sides of the two similar triangles are parallel, then, $(x,y) \rightarrow (ax, by)$, where *a* and *b* are constants, is such a transformation. A good example of a shear transformation involves a deck of playing cards. The cards can be positioned in either of the following patterns (along with many others).

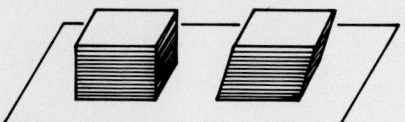

Additional Example 1

Draw a triangle with vertices $A(-1,-3)$, $B(3,-2)$, and $C(1,3)$. Draw the image under the mapping $(x,y) \rightarrow (3x,3x)$.

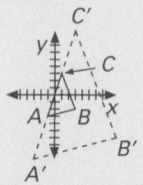

Additional Example 2

Show that in Additional Example 1 the distance between two points of the image is three times the distance between the corresponding two points of the original.
$$d_1 = \sqrt{(x_2 - x_1)^2 + (y_2 - y_1)^2}$$
$$d_2 = \sqrt{(3x_2 - 3x_1)^2 + (3y^2 - 3y_1)^2}$$
$$= \sqrt{3^2(x_2 - x_1)^2 + 3^2(y_2 - y_1)^2}$$
$$= 3\sqrt{(x_2 - x_1)^2 + (y_2 - y_1)^2}$$

Theorem 15.7

The transformation defined by multiplying each coordinate of each point by a constant is a dilation.

EXAMPLE 3 What changes in coordinates will determine a dilation image with lengths of sides that are half the original lengths?

(1) Draw △ABC on graph paper.
(2) Take half of each coordinate for each vertex.
(3) Draw the points corresponding to the new coordinates.
(4) Connect these points.
(5) The new triangle is similar to the original but with sides half as long.

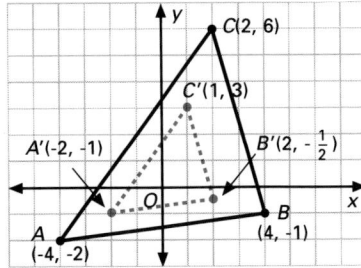

Some transformations do not preserve either size or shape. The **shear** transformation preserves horizontal distances or vertical distances, but not both. Also, the images of angles formed by perpendiculars to horizontal segments will be congruent to one another.

EXAMPLE 4 Draw the image of quad ABCD under the mapping $(x, y) \rightarrow (x + y, y)$.

(1) Draw the figure on graph paper. Mark the axes.
(2) Replace each x-coordinate with $x + y$.
(3) Draw the points corresponding to the new coordinates.
(4) Connect these points.

Quadrilateral $A'B'C'D'$ is the shear image of quadrilateral $ABCD$.

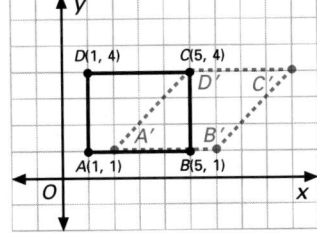

Focus on Reading

For each of the following, indicate whether the statement is **always**, **sometimes**, or **never** true.

1. The product of two reflections is a translation. Sometimes true
2. A square will remain a square after a rotation. Always true
3. A trapezoid has point symmetry. Sometimes true
4. The product of two reflections is a rotation. Sometimes true

---

Additional Example 3

What changes in coordinates will determine a dilation image which has lengths of sides one third the original lengths?

$(x, y) \rightarrow (\frac{1}{3}x, \frac{1}{3}y)$

Additional Example 4

Draw a rectangle with vertices $P(-1,-1)$, $Q(4,-1)$, $R(4,3)$, and $S(-1,3)$. Draw the image of quad $PQRS$ under the mapping $(x,y) \rightarrow (x,x+y)$.

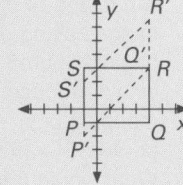

Math Connections

Party Balloons: A balloon with words or an image printed on it can be used to demonstrate a dilation. When it is inflated, the resulting image is a dilation of the image when it is deflated, and vice versa.

Critical Thinking Questions

Analysis: Ask students the following questions.

Compare the properties of a dilation to those of reflections, rotations, and translations. All of the above transformations have a one-to-one correspondence, but all except dilations preserve distance. Are parallel and perpendicular lines preserved in dilations? Yes

Checkpoint

Draw a triangle with vertices A(1,1), B(−3,2), and C(3,3). Draw the image of △ABC under the mappings:

1. $(x,y) \rightarrow (2x,2y)$

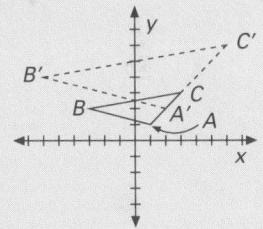

2. $(x,y) \rightarrow (x+y,y)$
Vertices of new △ are (2,1), (−1,2), (6,3).

Closure

Ask the students to explain what a dilation is. Definition page 622 Ask them what coordinate changes result in a dilation. Each coordinate of each point is multiplied by a constant. Ask them what coordinate changes produced a shear tranformation. Changes in the x- coordinates or y-coordinates but not both. Ask the class what properties of the original figure are retained in a dilation. The shape Ask them what properties change. The size. Ask them what properties of the original figure change in a shear. The size and shape.

Guided Practice

Classroom Exercises 1–4

Independent Practice

A Ex. 1–8, **B** Ex. 9–21, **C** Ex. 22–23

Basic: FR all, WE 1–8 odd

Average: FR all, WE 1–21 odd

Above Average: FR all, WE 1–23 odd

Additional Answers

Written Exercises

1. 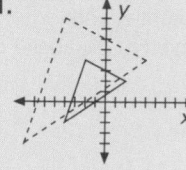   2.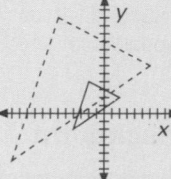

3. 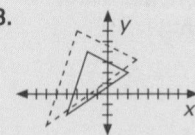   4.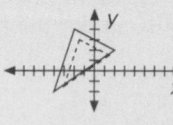

5. 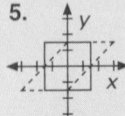   6. 7.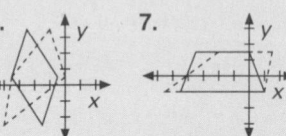

8. $PQ = \sqrt{(x_1 - x_2)^2 + (y_1 - y_2)^2}$; $P'Q' = \sqrt{(3x_1 - 3x_2)^2 + (3y_1 - 3y_2)^2} = 3\sqrt{(x_1 - x_2)^2 + (y_1 - y_2)^2} = 3PQ$

9. $PQ = \sqrt{(x_1 - x_2)^2 + (y_1 - y_2)^2}$; $P'Q' = \sqrt{(kx_1 - kx_2)^2 + (ky_1 - ky_2)^2} = k\sqrt{(x_1 - x_2)^2 + (y_1 - y_2)^2} = kPQ$ (Distr form); In the same way, prove $Q'R' = kQR$, $P'R' = kPR$; $\frac{PQ}{P'Q'} = \frac{QR}{Q'R'} = \frac{PR}{P'R'}$ (Div Prop of Eq); $\triangle PQR \sim \triangle P'Q'R'$ (SSS~); Transl is a dilation (Def of dilation)

Classroom Exercises

1. What changes in coordinates will determine the given dilation image of $\triangle ABC$ if sides of the image are three times as long as the sides of the original figure. $(x, y) \rightarrow (3x, 3y)$

What changes in coordinates determine the given dilation from the solid line to the dashed line figure?

2.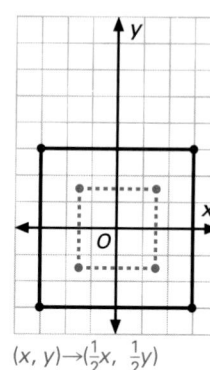

$(x, y) \rightarrow (\frac{1}{2}x, \frac{1}{2}y)$

3.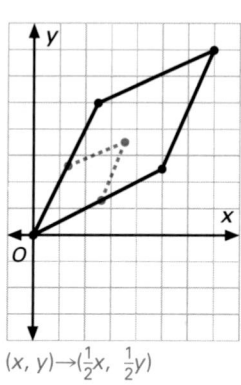

$(x, y) \rightarrow (\frac{1}{2}x, \frac{1}{2}y)$

4.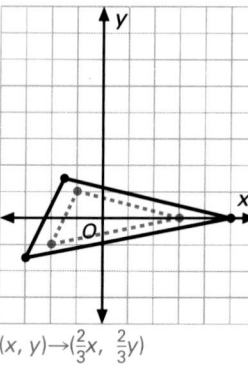

$(x, y) \rightarrow (\frac{2}{3}x, \frac{2}{3}y)$

Written Exercises

Draw the image of $\triangle ABC$ under the given mapping.

1. $(x, y) \rightarrow (2x, 2y)$
2. $(x, y) \rightarrow (3x, 3y)$
3. $(x, y) \rightarrow (1\frac{1}{2}x, 1\frac{1}{2}y)$
4. $(x, y) \rightarrow (\frac{3}{4}x, \frac{3}{4}y)$

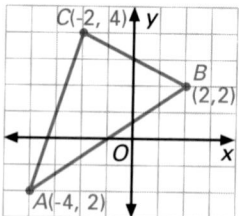

Draw the image of quad $ABCD$, replacing x with $x + y$.

5. 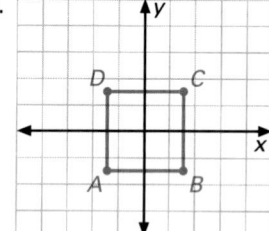   6. 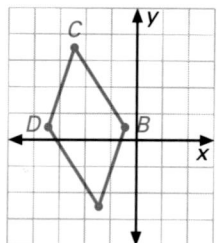   7.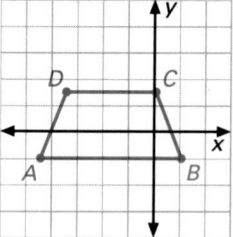

Enrichment

What type of mapping would produce an image having point symmetry with the original figure? What single mapping would produce the same image as mapping $(x,y) \rightarrow (2x,2y)$, followed by $(x,y) \rightarrow (-x,y)$? $(x,y) \rightarrow (-2x,2y)$
What single mapping would produce the same image as the mapping $(x,y) \rightarrow$

$(x,-y)$, followed by $(x,y) \rightarrow (x,x+y)$?
$(x,y) \rightarrow (x,x-y)$

8. Show that if $(x, y) \rightarrow (3x, 3y)$, the distance between two points of the image is three times the distance between the corresponding two points of the original figure.

9. Prove Theorem 15.7.

Draw the image $\triangle A'B'C'$ of $\triangle ABC$ under the given transformation. Is the transformation a dilation, a shear, or neither of these?

10. $(x, y) \rightarrow (3x, y)$
11. $(x, y) \rightarrow (x, 3y)$
12. $(x, y) \rightarrow (2x, 2y)$
13. $(x, y) \rightarrow (3x, 2y)$
14. $(x, y) \rightarrow (\frac{1}{2}x, \frac{1}{2}y)$
15. $(x, y) \rightarrow (0, 2y)$

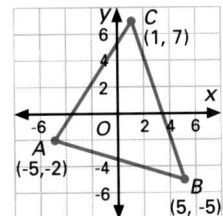

Draw the image of $A'B'C'D'$ of $ABCD$ under the given transformation. Is the transformation a dilation, a shear, or neither of these?

16. $(x, y) \rightarrow (x + y, y)$
17. $(x, y) \rightarrow (x, x + y)$
18. $(x, y) \rightarrow (x + y, x + y)$
19. $(x, y) \rightarrow (x - y, y)$
20. $(x, y) \rightarrow (x, y - x)$
21. $(x, y) \rightarrow (x + y, y - x)$
22. $(x, y) \rightarrow (x^2, y)$
23. $(x, y) \rightarrow (x^2, 2y)$

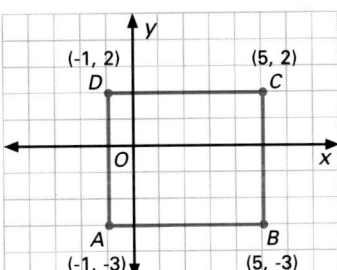

Mixed Review

1. Draw an obtuse angle. Construct the bisector of the angle. *1.5*
2. Draw a segment approximately 4 in. long. Construct a segment two-thirds as long. *8.6* 1. and 2. Check student constructions.

In $\odot O$, $m\widehat{AD} = 74$ and $m\widehat{BC} = 28$.

3. Find m $\angle DBA$. 37
4. Find m $\angle DFA$. 23
5. Find m $\angle DEA$. 51

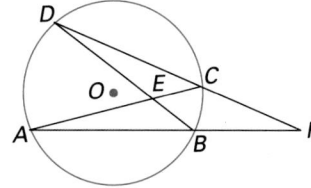

10. Shear
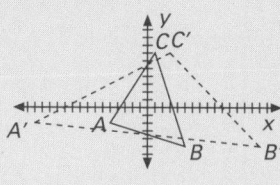

11. Shear
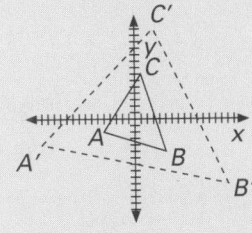

12. Dilation
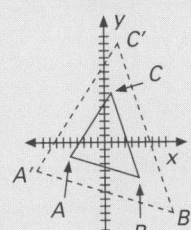

13. Neither
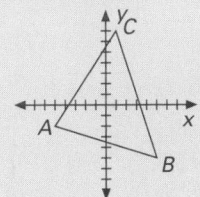

14. Dilation
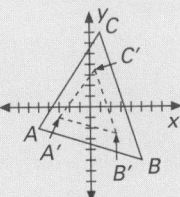

15. Neither
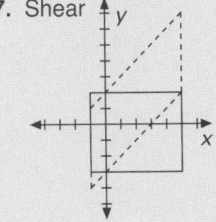

16. Shear
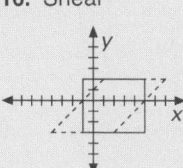

17. Shear
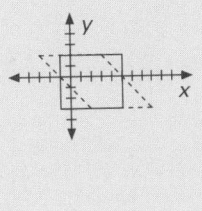

18. Neither
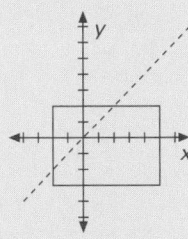

19. Shear

See page 629 for the answers to Ex. 20–23.

Chapter Review

1.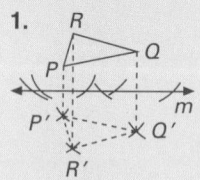

2.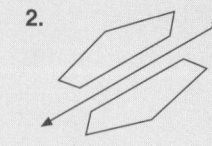

3.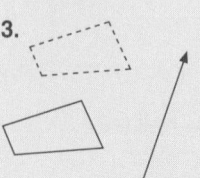

4.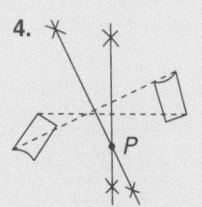

5.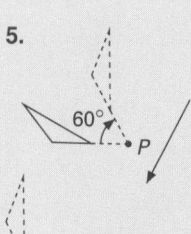

6.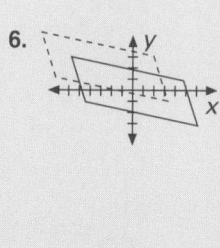

Chapter 15 Review

Key Terms

center of rotation (p. 612)
dilation (p. 622)
geometric transformation (p. 599)
image (p. 599)
invariant condition (p. 600)
isometry (p. 600)
line of symmetry (p. 600)
mapping (p. 618)

reflection (p. 599)
rotation (p. 613)
rotational symmetry (p. 615)
shear (p. 623)
slide arrow (p. 606)
slide image (p. 606)
symmetric figure (p. 601)
translation (p. 607)

Key Ideas and Review Exercises

15.1   To find the reflection of a figure, construct the image so that the reflection line is the perpendicular bisector of the segment joining corresponding points.
To locate the line of symmetry of a figure and its image, construct the perpendicular bisector of the segment joining corresponding points.

1. Construct a figure that is the reflection image of △PQR about line m.

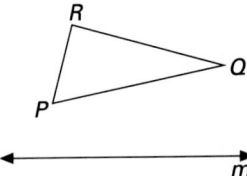

2. Find the line of symmetry of the given figure and its image.

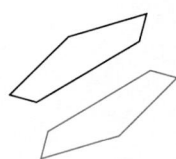

15.2   To find the translation image of a figure, construct the image so that segments joining corresponding points are congruent and parallel.
To find the product of two reflections, find the image of the first reflection; then find the image of that image by the second reflection.

3. Find the translation image of the given figure using the given slide arrow.

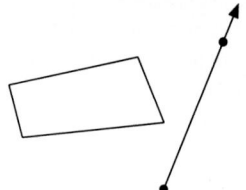

15.3   To find the rotation image of a figure, construct the image so that the distances from the center of rotation to corresponding points are equal.
To find the center of rotation, trace the figures on paper. Fold the paper so that a point and its image coincide. Fold again.

626   Chapter 15 Review

626

4. Find the center of rotation of the given figure and its image.

15.4 To find the product of two transformations, find the image of the original figure by the first transformation. Then find the image of that image by the second transformation.

5. Find the image of the given figure through a 60 degree clockwise rotation about *P*. Find the translation of the rotation image using the given slide arrow.

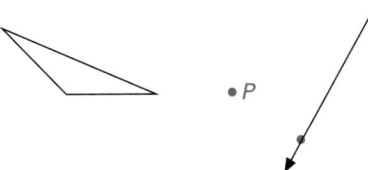

15.5 To map a transformation using coordinates, determine the necessary change in coordinates needed to describe the image.

To draw a transformation image using coordinates, draw the points corresponding to the new coordinates.

To describe a dilation, multiply each coordinate of the figure by a constant.

To describe a horizontal shear image, replace each *x*-coordinate with a new coordinate that is the sum of multiples of the *x*- and *y*-coordinates.

6. Draw the image of quad *ABCD* so that each *x*-coordinate is increased by 2 and each *y*-coordinate is decreased by 3.
7. What changes in coordinates will give a dilation image with sides 4 times as long as those of the original polygon?
8. Draw the image of quad *ABCD* replacing *x* with *x* + *y*. $(x, y) \rightarrow (x + y, y)$
    7. $(x, y) \rightarrow (4x, 4y)$

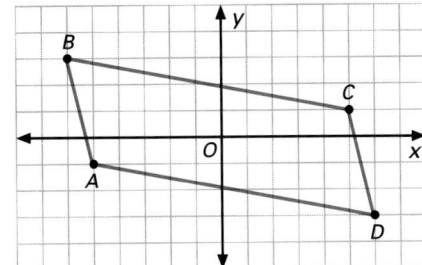

22.

Statement	Reason
1. $\overline{AP} \cong \overline{A'P}$, $\overline{BP} \cong \overline{B'P}$, m $\angle APA'$ = *n*, m $\angle BPB'$ = *n*	1. Def of rotation
2. m $\angle APB$ = m $\angle A'PB'$	2. Subt Prop Eq
3. $\triangle APB \cong$ $\triangle A'PB'$	3. SAS
4. $\overline{A'B'} \cong \overline{AB}$	4. CPCTC

23.

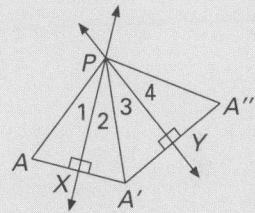

Statement	Reason
1. $\overline{AX} \cong \overline{A'X}$, $\overline{A'Y} \cong \overline{A''Y}$	1. Def of reflec
2. $\overline{PX} \cong \overline{PX}$, $\overline{PY} \cong \overline{PY}$	2. Reflex Prop of Congr
3. $\angle AXP \cong$ $\angle A'XP$, $\angle A'YP \cong$ $\angle A''YP$	3. Def of reflec, rt $\angle$s are $\cong$.
4. $\triangle AXP \cong$ $\triangle A'XP$, $\triangle P'AY \cong$ $\triangle A''YP$	4. SAS
5. m $\angle 1$ = m $\angle 2$, m $\angle 3$ = m $\angle 4$	5. CPCTC
6. m $\angle APA''$ = $2 \cdot$ m $\angle 2$ + $2 \cdot$ m $\angle 3$	6. Angle Add Post, Sub
7. m $\angle APA''$ = $2 \cdot$ m $\angle XPY$	7. Distr Prop, Angle Add Post

24.

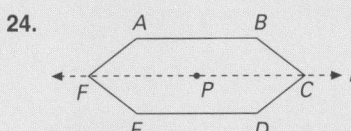

False. Hexagon *ABCDEF* has line symmetry about *l* and rotational symmetry about *P* (Angle of rotation = 180), but $\overline{AB} \not\cong \overline{BC}$.

627

4. None of the figures have a line of symmetry.

5. 2, 2,

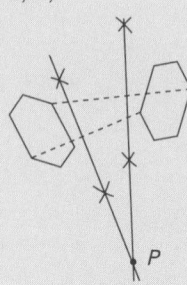

7. Transl

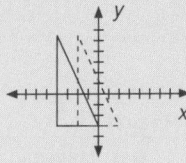

8. Dilation   9. Neither

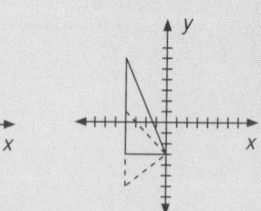

10. See proof of Ex. 22, page 617.

628

Chapter 15 Test   A—Level Ex.: 1–5, 7–9   C—Level Ex.: starred
B—Level Ex.: 6

For each figure below, tell whether the figure on the right appears to be a reflection, a translation image, a rotation image, or neither of the figure on the left.

1.
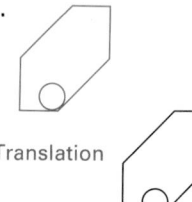
Translation

2.
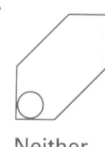 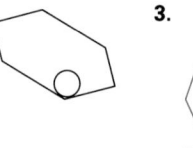
Neither

3.
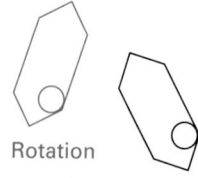
Rotation

4. Find the line of symmetry or center of rotation of the figures in Exercise 1–3.

5. Find the center of rotation of the figure and its image.

6. Each triangle is the image of the other by a combination of two transformations. Tell what these most likely are.   Reflection, rotation

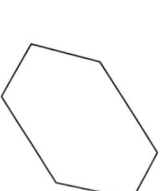 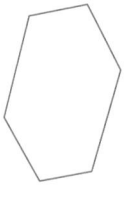

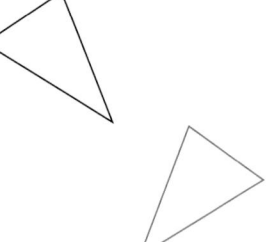

Replace the vertices of the given triangle with each set of vertices. Is the new figure a reflection, a translation, a rotation, a shear image, a dilation, or neither of the original triangle?

7. $(x, y) \rightarrow (x + 2, y)$
8. $(x, y) \rightarrow (2x, 2y)$
9. $(x, y) \rightarrow (x, x + y)$

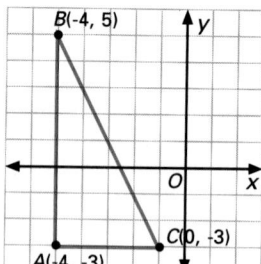

*10. Prove that a rotation is an isometry.

1. Which of the following is not true
C for a translation?
  (A) Figure and image are congruent.
  (B) Figure slides to become an image.
  (C) Figure flips over to become an image.
  (D) It is an isometry.

2. To turn a figure upside down and
move it 60 degrees down and left requires ___B___.
  (A) a translation and a rotation
  (B) a reflection and a translation
  (C) a reflection and a rotation
  (D) a rotation and a dilation

3.

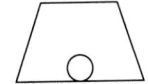 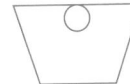

The image is obtained through
___B___.
  (A) a translation   (B) a rotation
  (C) a dilation   (D) a reflection

4. $(x, y) \rightarrow (3x, 3y)$ will give ___A___.
  (A) a dilation   (B) a reflection
  (C) a rotation   (D) a translation

5. Which of the figures below have
D point symmetry?

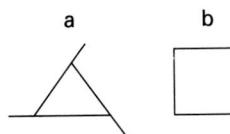

a   b

  (A) Neither a nor b
  (B) Both a and b
  (C) a only
  (D) b only

6.

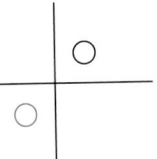

The mapping for the figure and
its image is ___C___.
  (A) $(x, y) \rightarrow (-x, y)$
  (B) $(x, y) \rightarrow (x, -y)$
  (C) $(x, y) \rightarrow (-x, -y)$
  (D) $(x, y) \rightarrow (2x, y)$

7. The figure below has ___B___.

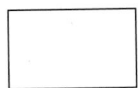

  (A) one line of symmetry
  (B) two lines of symmetry
  (C) three lines of symmetry
  (D) four lines of symmetry

8. The following is not an isometry.
D (A) a reflection   (B) a translation
   (C) a rotation   (D) a dilation

9.

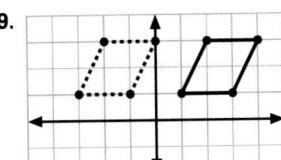

The mapping above is ___A___.
  (A) $(x - 4, y)$ (B) $(-x, y)$
  (C) $(x - 2, y)$ (D) $(x, -y)$

College Prep Test   **629**

Additional Answers, page 625

20. Shear   **21.** Neither

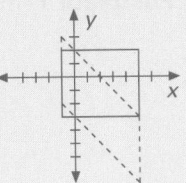   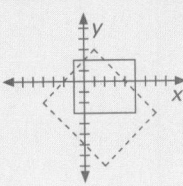

22. Shear

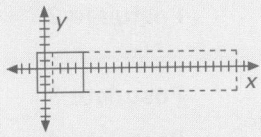

23. Neither

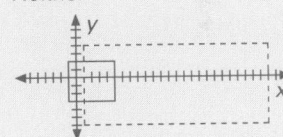

Postulates, Theorems, and Corollaries

Postulate 1   *Ruler Postulate:*
1. Any two distinct points on a line can be assigned coordinates 0 and 1.
2. There is a one-to-one correspondence between the real numbers and all points on the line.
3. To every pair of points, there corresponds exactly one positive number called the distance between the two points.

Postulate 2   *Segment Addition Postulate:* If C is between A and B, then $AC + CB = AB$.

Postulate 3   Any segment has exactly one midpoint.

Postulate 4   *Protractor Postulate:* In a given plane, select any line $\overleftrightarrow{AB}$ and any point C between A and B. Also select any two points R and S on the same side of $\overleftrightarrow{AB}$ such that S is not on $\overrightarrow{CR}$. Then there is a pairing of rays to real numbers from 0 to 180 as follows.
1. $\overrightarrow{CA}$ is paired with 0 and $\overrightarrow{CB}$ is paired with 180.
2. If $\overrightarrow{CR}$ is paired with x, then $0 < x < 180$.
3. If $\overrightarrow{CR}$ is paired with x and $\overrightarrow{CS}$ is paired with y, then m $\angle RCS = |x - y|$.

Postulate 5   *Angle Addition Postulate:* If D is in the interior of $\angle ABC$, then m $\angle ABC =$ m $\angle ABD +$ m $\angle DBC$.

Postulate 6   Every angle, except a straight angle, has exactly one bisector.

Postulate 7   If the outer rays of two adjacent angles form a straight angle, then the sum of the measures of the angles is 180.

Theorem 1.1   If the outer rays of two acute adjacent angles are perpendicular, then the sum of the measures of the angles is 90.

Theorem 2.1   If two angles are supplements of congruent angles, then they are congruent. (Supplements of congruent angles are congruent.)

Corollary   If two angles are supplements of the same angle, then they are congruent. (Supplements of the same angle are congruent.)

Theorem 2.2   If two angles are complements of congruent angles, then they are congruent. (Complements of congruent angles are congruent.)

Corollary   If two angles are complements of the same angle, then they are congruent. (Complements of congruent angles are congruent.)

Theorem 2.3   If two angles are right angles, then they are congruent.

Theorem 2.4   *Vertical Angles Theorem:* Vertical angles are congruent.

Corollary   If two lines are perpendicular, then four right angles are formed.

Postulate 8   A line contains at least two points. A plane contains at least three noncollinear points. Space contains at least four noncoplaner points.

Postulate 9	For any two points, there is exactly one line containing them.
Theorem 2.5	Two lines intersect at exactly one point.
Postulate 10	If two points of a line are in a given plane, then the line itself is in the plane.
Theorem 2.6	If a line intersects a plane, but is not contained in the plane, then the intersection is exactly one point.
Postulate 11	If two planes intersect, then they intersect in exactly one line.
Postulate 12	Three noncollinear points are contained in exactly one plane.
Theorem 2.7	A line and a point not on the line are contained in exactly one plane.
Theorem 2.8	Two intersecting lines are contained in exactly one plane.
Postulate 13	*Alternate Interior Angles Postulate:* If a transversal intersects two lines such that alternate interior angles are congruent (equal in measure), then the lines are parallel.
Theorem 3.1	If a transversal intersects two lines such that corresponding angles are congruent, then the lines are parallel.
Theorem 3.2	If two lines are intersected by a transversal such that interior angles on the same side of the transversal are supplementary, then the lines are parallel.
Theorem 3.3	In a plane, if two lines are perpendicular to the same line, then they are parallel.
Postulate 14	*Parallel Postulate:* Through a point not on a line, there is exactly one line parallel to the given line.
Theorem 3.4	If two parallel lines are intersected by a transversal, then alternate interior angles are congruent.
Theorem 3.5	If two parallel lines are intersected by a transversal, then corresponding angles are congruent.
Theorem 3.6	If two parallel lines are intersected by a transversal, then interior angles on the same side of the transversal are supplementary.
Theorem 3.7	If a transversal is perpendicular to one of two parallel lines, then it is perpendicular to the other.
Theorem 3.8	In a plane, if two lines are parallel to the line, then they are parallel to each other.
Theorem 3.9	The sum of the measures of the angles of a triangle is 180.
Theorem 3.10	*Exterior Angle Theorem:* The measure of an exterior angle of a triangle is equal to the sum of the measures of its two remote interior angles.
Theorem 3.11	If two parallel planes are intersected by a third plane, then the lines of intersection are parallel.

Theorem 4.1	In a right triangle, the two angles other than the right angle are complementary and acute.
Postulate 15	*SAS Postulate for Congruence of Triangles:* If two sides and the included angle of one triangle are congruent to the corresponding two sides and included angle of a second triangle, then the triangles are congruent.
Postulate 16	*SSS Postulate for Congruence of Triangles:* If the three sides of one triangle are congruent to the corresponding three sides of a second triangle, then the triangles are congruent.
Postulate 17	*ASA Postulate for Congruence of Triangles:* If two angles and the included side of one triangle are congruent to the corresponding two angles and included side of a second triangle, then the triangles are congruent.
Theorem 4.2	*Third Angle Theorem:* If two angles of one triangle are congruent to two angles of a second triangle, then the third angles of the triangles are congruent.
Theorem 4.3	*AAS Theorem:* If two angles and a nonincluded side of one triangle are congruent to the corresponding two angles and side of a second triangle, then the triangles are congruent.
Theorem 5.1	If two sides of a triangle are congruent, then angles opposite these sides are congruent. (The base angles of an isosceles triangle are congruent.)
Corollary	If a triangle is equilateral, then it is also equiangular, and the measure of each angle is 60.
Theorem 5.2	If two angles of a triangle are congruent, then the sides opposite these angles are congruent.
Corollary	If a triangle is equiangular, then it is also equilateral.
Theorem 5.3	*Hypotenuse-Leg (HL) Theorem:* Two right triangles are congruent if the hypotenuse and a leg of one are congruent, respectively, to the hypotenuse and corresponding leg of the other.
Theorem 5.4	The altitude from the vertex angle to the base of an isosceles triangle is a median. (The altitude bisects the base.)
Theorem 5.5	Corresponding medians of congruent triangles are congruent.
Theorem 5.6	Corresponding altitudes of congruent triangles are congruent.
Theorem 5.7	The bisector of the vertex angle of an isosceles triangle is the perpendicular bisector of the base.
Corollary	The bisector of the vertex angle of an isosceles triangle is also a median and an altitude of the triangle.
Theorem 5.8	A line containing two points, each equidistant from the endpoints of a given segment, is the perpendicular bisector of the segment.

Theorem 5.9	Any point on the perpendicular bisector of a segment is equidistant from the endpoints of the segment.
Theorem 5.10	*Exterior Angle Inequality Theorem:* The measure of an exterior angle of a triangle is greater than the measure of either of its remote interior angles.
Theorem 5.11	If one side of a triangle is longer than another side, then the measure of the angle opposite the longer side is greater than the measure of the angle opposite the shorter side.
Theorem 5.12	If one angle of a triangle has a greater measure than a second angle, then the side opposite the greater angle is longer than the side opposite the smaller angle.
Theorem 5.13	In a scalene triangle, the longest side is opposite the largest angle and the largest angle is opposite the longest side.
Theorem 5.14	The perpendicular segment from a point to a line is the shortest segment from the point to the line.
Corollary	The longest side of a right triangle is the hypotenuse.
Theorem 5.15	*Triangle Inequality Theorem:* The sum of the lengths of any two sides of a triangle is greater than the length of the third side.
Theorem 5.16	*SAS Inequality Theorem:* If two sides of one triangle are congruent, respectively, to two sides of a second triangle, and the included angle of the first triangle has a greater measure than the included angle of the second triangle, then the third side of the first triangle is longer than the third side of the second triangle.
Theorem 5.17	*SSS Inequality Theorem:* If two sides of one triangle are congruent, respectively, to two sides of a second triangle, and the length of the third side of the first triangle is greater than the length of the third side of the second triangle, then the angle opposite the third side of the first triangle has a greater measure than the angle opposite the third side of the second triangle.
Theorem 6.1	The sum of the measures of the interior angles of a convex polygon with n sides is $(n - 2)180$.
Corollary 1	The sum of the measures of the interior angles of a convex quadrilateral is 360.
Corollary 2	The measure of an angle of a regular polygon with n sides is $\dfrac{(n - 2)180}{n}$.
Theorem 6.2	The sum of the measures of the exterior angles, one at each vertex, of any convex polygon is 360.
Corollary	The measure of an exterior angle of a regular polygon with n sides is $\dfrac{360}{n}$.

Theorem 6.3	A diagonal of a parallelogram forms two congruent triangles.
Corollary 1	Opposite sides of a parallelogram are congruent.
Corollary 2	Opposite angles of a parallelogram are congruent.
Theorem 6.4	Consecutive angles of a parallelogram are supplementary.
Theorem 6.5	The diagonals of a parallelogram bisect each other.
Theorem 6.6	If both pairs of opposite sides of a quadrilateral are congruent, then the quadrilateral is a parallelogram.
Theorem 6.7	If the diagonals of a quadrilateral bisect each other, then the quadrilateral is a parallelogram.
Theorem 6.8	If two sides of a quadrilateral are parallel and congruent, then the quadrilateral is a parallelogram.
Theorem 6.9	If both pairs of opposite angles of a quadrilateral are congruent, then the quadrilateral is a parallelogram.
Theorem 6.10	*Midsegment Theorem:* The segment joining the midpoints of two sides of a triangle is parallel to the third side, and its length is half the length of the third side.
Theorem 6.11	If two lines are parallel, then all points of each line are equidistant from the other line.
Theorem 6.12	If three parallel lines cut off congruent segments on one transversal, then they cut off congruent segments on every transversal.
Corollary	If any number of parallel lines cut off congruent segments on one transversal, then they cut off congruent segments on every transversal.
Theorem 6.13	If a segment is parallel to one side of a triangle and contains the midpoint of a second side, then this segment bisects the third side.
Theorem 7.1	*SASAS for Congruent Quadrilaterals:* Two quadrilaterals are congruent if any three sides and the included angles of one are congruent, respectively, to three sides and the included angles of the other.
Theorem 7.2	*ASASA for Congruent Quadrilaterals:* Two quadrilaterals are congruent if any three angles and the included sides of one are congruent, respectively, to three angles and the included sides of the other.
Theorem 7.3	The diagonals of a rhombus are perpendicular.
Theorem 7.4	The diagonals of a rectangle are congruent.
Theorem 7.5	Each diagonal of a rhombus bisects two angles of the rhombus.
Theorem 7.6	A parallelogram with one right angle is a rectangle.
Theorem 7.7	A parallelogram with two adjacent, congruent sides is a rhombus.
Theorem 7.8	A parallelogram with perpendicular diagonals is a rhombus.
Theorem 7.9	A parallelogram with congruent diagonals is a rectangle.

Theorem 7.10	A parallelogram with a diagonal that bisects opposite angles is a rhombus.
Theorem 7.11	A quadrilateral with four congruent sides is a rhombus.
Theorem 7.12	All altitudes of a trapezoid are congruent.
Theorem 7.13	The median of a trapezoid is parallel to its bases. Its length is one-half the sum of the lengths of the two bases.
Theorem 7.14	The base angles of an isosceles trapezoid are congruent.
Theorem 7.15	If the base angles of a trapezoid are congruent, then the trapezoid is isosceles.
Theorem 7.16	The diagonals of an isosceles trapezoid are congruent.
Theorem 7.17	If the diagonals of a trapezoid are congruent, then the trapezoid is isosceles.
Theorem 8.1	In a proportion, the product of the extremes equals the product of the means.
Corollary	If the product of the extremes equals the product of the means, then a proportion exists.

Theorem 8.2 If $\frac{a}{b} = \frac{c}{d}$ (assume no denominator equals zero), then:

1. $\frac{b}{d} = \frac{d}{c}$ 4. $\frac{a - b}{b} = \frac{c - d}{d}$

2. $\frac{a}{c} = \frac{b}{d}$ 5. $\frac{a}{b} = \frac{a + c}{b + d}$

3. $\frac{a + b}{b} = \frac{c + d}{d}$

Theorem 8.3	Congruent triangles are similar.
Theorem 8.4	*Transitive Property of Triangle Similarity:* If $\triangle ABC \sim \triangle DEF$ and $\triangle DEF \sim \triangle GHI$, then $\triangle ABC \sim \triangle GHI$.
Postulate 18	*AA Similarity Postulate:* If two angles of a triangle are congruent to two angles of another triangle, then the two triangles are similar.
Theorem 8.5	*Triangle Proportionality Theorem:* If a line is parallel to one side of a triangle and intersects the other two sides, then it divides the two sides proportionally.
Theorem 8.6	If a line divides two sides of a triangle proportionally, then the line is parallel to the third side of the triangle.
Theorem 8.7	*SAS Similarity Theorem:* If an angle of one triangle is congruent to an angle of another triangle and the corresponding sides that include these angles are proportional, then the triangles are similar.
Theorem 8.8	*SSS Similarity Theorem:* If all three pairs of corresponding sides of two triangles are proportional, then the two triangles are similar.

Theorem 8.9 Corresponding altitudes of similar triangles are proportional to corresponding sides.

Theorem 8.10 Corresponding medians of similar triangles are proportional to corresponding sides.

Theorem 8.11 The bisector of an angle of a triangle divides the opposite side of the triangle into segments proportional to the other two sides.

Theorem 9.1 In a right triangle, the altitude to the hypotenuse forms two similar right triangles, each of which is also similar to the original triangle.

Corollary 1 In a right triangle, the square of the length of the altitude to the hypotenuse equals the product of the lengths of the segments formed on the hypotenuse.

Corollary 2 If the altitude is drawn to the hypotenuse of a right triangle, then the square of the length of either leg equals the product of the lengths of the hypotenuse and the segment of the hypotenuse adjacent to that leg.

Theorem 9.2 *Pythagorean Theorem:* In any right triangle, the sum of the squares of the lengths of the legs is equal to the square of the length of the hypotenuse.

Theorem 9.3 *Converse of the Pythagorean Theorem:* If the sum of the squares of the lengths of two sides of a triangle is equal to the square of the length of the third side, then the triangle is a right triangle.

Theorem 9.4 If the square of the longest side of a triangle is greater (less) than the sum of the squares of the lengths of the other two sides, then the triangle is obtuse (acute).

Theorem 9.5 In a 45-45-90 triangle, the hypotenuse is $\sqrt{2}$ times as long as a leg.

Theorem 9.6 In a 30-60-90 triangle, the hypotenuse is twice as long as the leg opposite the 30 angle. The leg opposite the 60 angle is $\sqrt{3}$ times as long as the leg opposite the 30 angle.

Theorem 10.1 *Distance Formula:* The distance d between $P_1(x_1, y_1)$ and $P_2(x_2, y_2)$ is given by the formula $d = \sqrt{(x_2 - x_1)^2 + (y_2 - y_1)^2}$.

Theorem 10.2 *Midpoint Formula:* Given $P(x_1, y_1)$ and $Q(x_2, y_2)$, the coordinates (x_m, y_m) of M, the midpoint of $\overline{PQ}$, are $\left(\dfrac{x_1 + x_2}{2}, \dfrac{y_1 + y_2}{2} \right)$.

Theorem 10.3 All segments of a non-vertical line have equal slopes.

Theorem 10.4 An equation of a line with slope m containing the point $P_1(x_1, y_1)$ is $y - y_1 = m(x - x_1)$.

Theorem 10.5 If a line has slope m and y-intercept b, then an equation of the line is $y = mx + b$.

Theorem 10.6 If two non-vertical lines have the same slope, then they are parallel.

Theorem 10.7	If two non-vertical lines are parallel, then they have equal slopes.
Theorem 10.8	If the product of the slopes of two non-vertical lines is -1, then the lines are perpendicular.
Theorem 10.9	The product of the slopes of two non-vertical perpendicular lines is -1.
Theorem 11.1	If a line or segment contains the center of a circle and is perpendicular to a chord, then it *bisects* the chord.
Theorem 11.2	In the same circle or in congruent circles, congruent chords are equidistant from the center(s).
Theorem 11.3	In the same circle or in congruent circles, chords that are equidistant from the center(s) are congruent.
Theorem 11.4	In the same circle or congruent circles, if two chords are unequally distant from the center(s), then the chord nearer its corresponding center is the longer chord.
Theorem 11.5	In the same circle or congruent circles, if two chords are unequal in length, then the longer chord is nearer the center of its circle.
Theorem 11.6	If a line is perpendicular to a radius at its endpoint on the circle, then the line is tangent to the circle.
Theorem 11.7	If a line is tangent to a circle, then the line is perpendicular to the radius drawn to the point of tangency.
Theorem 11.8	Two segments drawn tangent to a circle from an exterior point are congruent.
Corollary	The angle between two tangents to a circle from an exterior point is bisected by the segment joining its vertex and the center of the circle.
Postulate 19	If P is a point on $\overset{\frown}{APB}$, then $m\overset{\frown}{AP} + m\overset{\frown}{PB} = m\overset{\frown}{APB}$.
Theorem 11.9	In the same circle or in congruent circles: 1. If chords are congruent, then their corresponding arcs and central angles are congruent; 2. If arcs are congruent, then their corresponding chords and central angles are congruent; 3. If central angles are congruent, then their corresponding arcs and chords are congruent.
Theorem 11.10	*Inscribed Angle Theorem:* The measure of an inscribed angle is one-half of the degree measure of its intercepted arc.
Corollary 1	If two inscribed angles intercept the same arc or congruent arcs, then the angles are congruent.
Corollary 2	An angle inscribed in a *semicircle* is a *right* angle.
Corollary 3	If two arcs of a circle are included between parallel chords or secants, then the arcs are congruent.

Corollary 4	The opposite angles of an inscribed quadrilateral are supplementary.
Theorem 11.11	The measure of an angle formed by two secants or chords intersecting in the interior of a circle is one-half the sum of the measures of the arcs intercepted by the angle and its vertical angle.
Theorem 11.12	The measure of an angle formed by two secants intersecting in the exterior of the circle is one-half the difference of the measures of the intercepted arcs.
Theorem 11.13	If a tangent and a secant (or a chord) intersect at the point of tangency on a circle, then the measure of the angle formed is one-half the measure of its intercepted arc.
Theorem 11.14	The measure of an angle formed either by (1) a tangent and secant intersecting at a point exterior to a circle, or (2) two tangents intersecting at a point exterior to a circle equals one-half the difference of the measures of the intercepted arcs.
Theorem 11.15	If two chords of a circle intersect, then the product of the lengths of the segments of one chord equals the product of the lengths of the segments of the other chord.
Theorem 11.16	If a tangent and a secant intersect in the exterior of a circle, then the square of the length of the tangent segment equals the product of the lengths of the secant segment and the external secant segment.
Corollary	If two secants intersect in the exterior of a circle, then the product of the lengths of one secant segment and its external segment equals the product of the lengths of the other secant segment and its external segment.
Theorem 11.17	The equation of a circle with the coordinates of the center (h,k) and a radius of length r is $(x - h)^2 + (y - k)^2 = r^2$.
Postulate 20	Congruent polygons have equal areas.
Postulate 21	The area of a rectangle is the product of the lengths of a base and a corresponding altitude (Area of rectangle = bh).
Theorem 12.1	The area of a square is the square of the length s of a side ($A = s^2$).
Postulate 22	*Area Addition Postulate:* If a region is the union of two or more nonoverlapping regions, then its area is the sum of the areas of these nonoverlapping regions.
Theorem 12.2	The area of a parallelogram is the product of the lengths of a base and a corresponding altitude (Area of parallelogram = bh).
Theorem 12.3	The area of a triangle is one-half the product of the lengths of a base and a corresponding altitude (Area of triangle = $\frac{1}{2}bh$).

Theorem 12.4	The area of a kite is one-half the product of the lengths of the diagonals (Area of kite $= \frac{1}{2}d_1d_2$).
Corollary	The area of a rhombus is one-half the product of the lengths of the two diagonals.
Theorem 12.5	If s is the length of a side of an equilateral triangle, then the area is $\frac{s^2}{4}\sqrt{3}$.
Theorem 12.6	*Heron's Formula:* If a, b, and c are the lengths of the sides of a triangle and s is the semiperimeter, such that $s = \frac{1}{2}(a + b + c)$, then Area(triangle) $= \sqrt{s(s - a)(s - b)(s - c)}$.
Theorem 12.7	The area of a trapezoid is one-half the product of the sum of the lengths of the upper and lower bases and the length of an altitude.
Theorem 12.8	A circle can be circumscribed about any regular polygon.
Theorem 12.9	The area of a regular polygon is one-half the product of the apothem and the perimeter [Area (n-gon) $= \frac{1}{2}ap)$].
Theorem 12.10	The area of a regular polygon is $n[\sin(\frac{180}{n})]\,[\cos(\frac{180}{n})]r^2$, or $\frac{ns^2}{4\tan\left(\frac{180}{n}\right)}$, where n is the number of sides, s is the length of a side, and r is the length of a radius.
Theorem 12.11	The ratio of the perimeters of two similar polygons is the same as the ratio of the lengths of any two corresponding sides.
Theorem 12.12	The ratio of the areas of two similar triangles is the square of the ratio of the lengths of any two corresponding sides.
Theorem 12.13	The ratio of the areas of two similar polygons is the square of the ratio of the lengths of any two corresponding sides.
Theorem 12.14	The ratio of the circumference to the length of a diameter is the same for all circles.
Corollary	The circumference of a circle with radius of length r is $2\pi r$.
Theorem 12.15	The area of a circle with radius of length r is πr^2.
Theorem 12.16	The area of a sector of a circle is one-half the product of the length s of the arc and the length r of its radius ($A = \frac{1}{2}rs$).
Theorem 13.1	The locus of points in a plane equidistant from two given points is the perpendicular bisector of the segment having the two points as endpoints.
Theorem 13.2	In a plane, the locus of points equidistant from the sides of an angle is the bisector of the angle.
Theorem 13.3	The perpendicular bisectors of the sides of a triangle are concurrent at a point equidistant from the vertices of the triangle.
Theorem 13.4	The bisectors of the angles of a triangle are concurrent at a point equidistant from the sides of the triangle.

Theorem 13.5 The lines containing the altitudes of a triangle are concurrent.

Theorem 13.6 Two medians of a triangle intersect at a point two-thirds of the distance from each vertex to the midpoint of the opposite side.

Theorem 13.7 The medians of a triangle are concurrent at a point that is two-thirds the distance from each vertex to the midpoint of the opposite side.

Theorem 14.1 The lateral area of a right prism is the product of the perimeter of a base and the length of an altitude ($L = ph$).

Theorem 14.2 The lateral area of a right cylinder is the product of the circumference of a base and the length of an altitude. The lateral area of a right cylinder with radius r and altitude of length h is $2\pi rh$.

Theorem 14.3 The total area of a right cylinder with radius of length r and altitude of length h is $2\pi r^2 + 2\pi rh$, or $2\pi r(r + h)$.

Theorem 14.4 If the ratio of the lengths of corresponding edges of two similar polyhedra is $\frac{a}{b}$, then the ratio of the lateral areas and of the total areas is $(\frac{a}{b})^2$.

Postulate 23 For any rectangular solid, the volume $V = lwh$, where l, w, and h are the lengths of three edges with a common vertex.

Theorem 14.5 The volume of a cube with edges of length s is s^3.

Postulate 24 If a solid is the union of two or more nonoverlapping solids, then its volume is the sum of the volumes of these nonoverlapping parts.

Postulate 25 *Cavalieri's Principle:* If two solids have equal heights, and if the cross sections formed by any plane parallel to the bases of both solids have equal areas, then the volumes of the solids are equal.

Theorem 14.6 For any prism or cylinder, the volume is the product of the area of a base and the length of an altitude ($V = Bh$, where B = area of base and h = altitude).

Corollary The volume of a cylinder is Bh, or $\pi r^2 h$.

Theorem 14.7 The lateral area of a regular pyramid is one-half the product of the perimeter of the base and the slant height ($L = \frac{1}{2}pl$).

Theorem 14.8 The lateral area L of a right cone is πrl. The total area (A) is $\pi rl + \pi r^2 = \pi r(l + r)$.

Theorem 14.9 The volume of a pyramid is one-third the volume of a prism with the same base and altitude as the pyramid. The volume of a cone is one-third the volume of a cylinder with the same base and altitude as the cone ($V = \frac{1}{3}Bh$).

Corollary 1 The volume of a pyramid or cone is one-third the product of the area of its base and the length of its altitude ($V = \frac{1}{3}Bh$).

Corollary 2	The volume of a cone with a base radius of length r and an altitude of length h is $\frac{1}{3}\pi r^2 h$.
Theorem 14.10	The volume of a sphere with radius of length r is $\frac{4}{3}\pi r^3$.
Theorem 14.11	The area of a sphere is $4\pi r^2$.
Theorem 14.12	The distance between points $P(x_1, y_1, z_1)$ and $Q(x_2, y_2, z_2)$ is given by the formula $d = \sqrt{(x_2 - x_1)^2 + (y_2 - y_1)^2 + (z_2 - z_1)^2}$.
Theorem 15.1	A reflection is an isometry.
Theorem 15.2	A translation is an isometry.
Theorem 15.3	The resultant image determined by two successive reflections about parallel lines is a translation.
Theorem 15.4	A rotation is an isometry.
Theorem 15.5	The measure of the angle of rotation formed by two successive reflections (or by the product of two reflections) is twice the measure of the non-obtuse angle between the two lines of symmetry.
Theorem 15.6	The transformation defined by adding a constant to the coordinates of each point is a translation.
Theorem 15.7	The transformation defined by multiplying each coordinate of each point by a constant is a dilation.

Postulates, Theorems and Corollaries

Table of Roots and Powers

No.	Sq.	Sq. Root	Cube	Cu. Root	No.	Sq.	Sq. Root	Cube	Cu. Root
1	1	1.000	1	1.000	51	2,601	7.141	132,651	3.708
2	4	1.414	8	1.260	52	2,704	7.211	140,608	3.733
3	9	1.732	27	1.442	53	2,809	7.280	148,877	3.756
4	16	2.000	64	1.587	54	2,916	7.348	157,564	3.780
5	25	2.236	125	1.710	55	3,025	7.416	166,375	3.803
6	36	2.449	216	1.817	56	3,136	7.483	175,616	3.826
7	49	2.646	343	1.913	57	3,249	7.550	185,193	3.849
8	64	2.828	512	2.000	58	3,364	7.616	195,112	3.871
9	81	3.000	729	2.080	59	3,481	7.681	205,379	3.893
10	100	3.162	1,000	2.154	60	3,600	7.746	216,000	3.915
11	121	3.317	1,331	2.224	61	3,721	7.810	226,981	3.936
12	144	3.464	1,728	2.289	62	3,844	7.874	238,328	3.958
13	169	3.606	2,197	2.351	63	3,969	7.937	250,047	3.979
14	196	3.742	2,744	2.410	64	4,096	8.000	262,144	4.000
15	225	3.875	3,375	2.466	65	4,225	8.062	274,625	4.021
16	256	4.000	4,096	2.520	66	4,356	8.124	287,496	4.041
17	289	4.123	4,913	2.571	67	4,489	8.185	300,763	4.062
18	324	4.243	5,832	2.621	68	4,624	8.246	314,432	4.082
19	361	4.359	6,859	2.668	69	4,761	8.307	328,509	4.102
20	400	4.472	8,000	2.714	70	4,900	8.367	343,000	4.121
21	441	4.583	9,261	2.759	71	5,041	8.426	357,911	4.141
22	484	4.690	10,648	2.802	72	5,184	8.485	373,248	4.160
23	529	4.796	12,167	2.844	73	5,329	8.544	389,017	4.179
24	576	4.899	13,824	2.884	74	5,476	8.602	405,224	4.198
25	625	5.000	15,625	2.924	75	5,625	8.660	421,875	4.217
26	676	5.099	17,576	2.962	76	5,776	8.718	438,976	4.236
27	729	5.196	19,683	3.000	77	5,929	8.775	456,533	4.254
28	784	5.292	21,952	3.037	78	6,084	8.832	474,552	4.273
29	841	5.385	24,389	3.072	79	6,241	8.888	493,039	4.291
30	900	5.477	27,000	3.107	80	6,400	8.944	512,000	4.309
31	961	5.568	29,791	3.141	81	6,561	9.000	531,441	4.327
32	1,024	5.657	32,768	3.175	82	6,724	9.055	551,368	4.344
33	1,089	5.745	35,937	3.208	83	6,889	9.110	571,787	4.362
34	1,156	5.831	39,304	3.240	84	7,056	9.165	592,704	4.380
35	1,225	5.916	42,875	3.271	85	7,225	9.220	614,125	4.397
36	1,296	6.000	46,656	3.302	86	7,396	9.274	636,056	4.414
37	1,369	6.083	50,653	3.332	87	7,569	9.327	658,503	4.431
38	1,444	6.164	54,872	3.362	88	7,744	9.381	681,472	4.448
39	1,521	6.245	59,319	3.391	89	7,921	9.434	704,969	4.465
40	1,600	6.325	64,000	3.420	90	8,100	9.487	729,000	4.481
41	1,681	6.403	68,921	3.448	91	8,281	9.539	753,571	4.498
42	1,764	6.481	74,088	3.476	92	8,464	9.592	778,688	4.514
43	1,849	6.557	79,507	3.503	93	8,649	9.644	804,357	4.531
44	1,936	6.633	85,184	3.530	94	8,836	9.695	830,584	4.547
45	2,025	6.708	91,125	3.557	95	9,025	9.747	857,375	4.563
46	2,116	6.782	97,336	3.583	96	9,216	9.798	884,736	4.579
47	2,209	6.856	103,823	3.609	97	9,409	9.849	912,673	4.595
48	2,304	6.928	110,592	3.634	98	9,604	9.899	941,192	4.610
49	2,401	7.000	117,649	3.659	99	9,801	9.950	970,299	4.626
50	2,500	7.071	125,000	3.684	100	10,000	10.000	1,000,000	4.642

Table

Trigonometric Ratios

Angle Measure	Sin	Cos	Tan	Angle Measure	Sin	Cos	Tan
0°	0.000	1.000	0.000	46°	.7193	.6947	1.036
1°	.0175	.9998	.0175	47°	.7314	.6820	1.072
2°	.0349	.9994	.0349	48°	.7431	.6691	1.111
3°	.0523	.9986	.0524	49°	.7547	.6561	1.150
4°	.0698	.9976	.0699	50°	.7660	.6428	1.192
5°	.0872	.9962	.0875	51°	.7771	.6293	1.235
6°	.1045	.9945	.1051	52°	.7880	.6157	1.280
7°	.1219	.9925	.1228	53°	.7986	.6018	1.327
8°	.1392	.9903	.1405	54°	.8090	.5878	1.376
9°	.1564	.9877	.1584	55°	.8192	.5736	1.428
10°	.1736	.9848	.1763	56°	.8290	.5592	1.483
11°	.1908	.9816	.1944	57°	.8387	.5446	1.540
12°	.2079	.9781	.2126	58°	.8480	.5299	1.600
13°	.2250	.9744	.2309	59°	.8572	.5150	1.664
14°	.2419	.9703	.2493	60°	.8660	.5000	1.732
15°	.2588	.9659	.2679	61°	.8746	.4848	1.804
16°	.2756	.9613	.2867	62°	.8829	.4695	1.881
17°	.2924	.9563	.3057	63°	.8910	.4540	1.963
18°	.3090	.9511	.3249	64°	.8988	.4384	2.050
19°	.3256	.9455	.3443	65°	.9063	.4226	2.145
20°	.3420	.9397	.3640	66°	.9135	.4067	2.246
21°	.3584	.9336	.3839	67°	.9205	.3907	2.356
22°	.3746	.9272	.4040	68°	.9272	.3746	2.475
23°	.3907	.9205	.4245	69°	.9336	.3584	2.605
24°	.4067	.9135	.4452	70°	.9397	.3420	2.747
25°	.4226	.9063	.4663	71°	.9455	.3256	2.904
26°	.4384	.8988	.4877	72°	.9511	.3090	3.077
27°	.4540	.8910	.5095	73°	.9563	.2924	3.270
28°	.4695	.8829	.5317	74°	.9613	.2756	3.487
29°	.4848	.8746	.5543	75°	.9659	.2588	3.732
30°	.5000	.8660	.5774	76°	.9703	.2419	4.010
31°	.5150	.8572	.6009	77°	.9744	.2250	4.331
32°	.5299	.8480	.6249	78°	.9781	.2079	4.704
33°	.5446	.8387	.6494	79°	.9816	.1908	5.145
34°	.5592	.8290	.6745	80°	.9848	.1736	5.671
35°	.5736	.8192	.7002	81°	.9877	.1564	6.314
36°	.5878	.8090	.7265	82°	.9903	.1392	7.115
37°	.6018	.7986	.7536	83°	.9925	.1219	8.144
38°	.6157	.7880	.7813	84°	.9945	.1045	9.514
39°	.6293	.7771	.8098	85°	.9962	.0872	11.43
40°	.6428	.7660	.8391	86°	.9976	.0698	14.30
41°	.6561	.7547	.8693	87°	.9986	.0523	19.08
42°	.6691	.7431	.9004	88°	.9994	.0349	28.64
43°	.6820	.7314	.9325	89°	.9998	.0175	57.29
44°	.6947	.7193	.9657	90°	1.000	0.000	
45°	.7071	.7071	1.000				

Glossary

acute angle: An angle with a measure less than 90. (p. 16)

acute triangle: A triangle with three acute angles. (p. 139)

adjacent angles: Two coplanar angles with one common side and a common vertex, but no common interior points. (p. 21)

adjacent sides of a polygon: Sides which intersect at a vertex. (p. 225)

alternate interior/exterior angles: Interior/Exterior, nonadjacent angles which lie on opposite sides of a transversal. (pp. 97, 98)

altitude of a prism: Any segment perpendicular to the planes containing the two bases with endpoints in these planes. (p. 564)

altitude of a triangle: A segment from a vertex of the triangle perpendicular to the opposite side. (p. 194)

altitude of a trapezoid: A perpendicular segment from any point on one base to the other base. (p. 286)

angle: Two rays with a common endpoint. The rays are the *sides*, the endpoint is the *vertex*. (p. 15)

angle bisector: A segment, line, ray, or plane which divides an angle into two congruent angles. (p. 23)

apothem of a regular polygon: The length of a perpendicular segment from the center of the polygon to a side. (p. 499)

arc: An unbroken part of a circle measured by referring to an angle whose vertex is the center of the circle. (p. 432)

area of a circle: The limit of the areas of inscribed regular polygons as the number of sides increases indefinitely. (p. 512)

area of a polygon: The number of square units in the region bounded. (p. 478)

base angles of an isosceles triangle: The angles of an isosceles triangle opposite the legs or congruent sides. (p. 177)

base of an isosceles triangle: The side opposite the vertex angle in an isosceles triangle. (p. 177)

bases of a prism: Two congruent polygonal faces that lie in parallel planes. (p. 560)

bisector of a segment: A line, ray, segment, or plane that intersects a segment at its midpoint. (p. 10)

center of a regular polygon: The common center of its inscribed and circumscribed circles. (p. 498)

central angle: An angle whose vertex is the center of the circle. (p. 432)

chord: A segment within a circle with endpoints on the circle. (p. 416)

circle: The set of all points in a plane that are a given distance from a fixed point in that plane. (p. 415)

circumcenter: The intersection of perpendicular bisectors of sides of a triangle. (p. 542)

circumscribed circle: A circle in which a polygon is inscribed. (p. 423)

circumscribed polygon: A polygon whose sides are tangent to a circle. (p. 427)

collinear: Points contained within the same line. (p. 2)

complementary angles: Two angles with measures whose sum is 90. (p. 28)

concentric circles: Coplanar circles of different radii with a common center. (p. 438)

concurrent lines: Lines that converge, or intersect, in one common point. (p. 541)

conditional: A statement that can be written in the form "If p, then q." (p. 63)

cone: A pyramid-like object whose base is a circle and whose lateral surface is made up of an infinite number of segments between the circle and a vertex. (p. 576)

congruent angles: Angles that have the same measure. (p. 17)

congruent arcs: Arcs that have the same degree measure. (p. 438)

congruent circles: Circles that have congruent radii. (p. 415)

congruent quadrilaterals: Quadrilaterals with congruent corresponding sides. (p. 269)

congruent segments: Segments that have the same length. (p. 10)

congruent triangles: Triangle with congruent, corresponding sides and angles. (p. 143)

conjunction: If p and q are both statements, the statement "p and q" is their conjunction. (p. 32)

converse: The statement formed by interchanging the hypothesis and conclusion of a conditional. (p. 109)

convex: A polygon is convex if a segment joining any two interior points of the polygon is in the interior of the polygon. (p. 227)

coordinates: A corresponding ordered pair (x,y) for every point in a plane. (p. 379)

coplanar: Points within the same plane. (p. 2)

corollary: A theorem whose proof follows from another theorem in a few steps. (p. 71)

corresponding angles: Angles which lie on the same side of a transversal. (p. 98)

cosine: In right $\triangle ABC$ with acute $\angle A$, cosine $\angle A = \dfrac{\text{length of adjacent leg}}{\text{length of hypotenuse}}$ (p. 366)

cylinder: A cylinder consists of two congruent and parallel circular bases whose lateral surface is the infinite number of segments connecting the circles. (p. 561)

deductive reasoning: Inferring from general principles to prove a statement. (p. 67)

diagonal: A segment which joins two non-consecutive vertices. (p. 226)

diameter: A chord that contains the center of a circle. (p. 416)

dilation: A transformation which produces an image similar to the original. (p. 622)

disjunction: The logical union "p or q," where p and q are statements. (p. 33)

equiangular: Has congruent angles. (p. 139)

equidistant: Equally distant from. (p. 11)

equilateral: With congruent sides. (p. 139)

exterior angle: An angle that is adjacent and supplementary to one of the angles of a triangle. (p. 121)

geometric mean: The geometric mean of two positive numbers, a and b, is the positive number m, such that $\dfrac{a}{m} = \dfrac{m}{b}$. (p. 305)

geometric transformation: Matching the points of a figure with the points of a second figure called the *image*. (p. 599)

hypotenuse: The side opposite the right angle in a right triangle. (p. 188)

hypothesis: The hypothesis is the "if" part of a conditional statement. (p. 63)

indirect proof: A proof which assumes that the desired conclusion is not true and shows that this assumption leads to a contradiction. (p. 106)

inductive reasoning: Generalizations based on repeated observations. (p. 68)

inscribed angle: An angle whose vertex lies on a circle and whose sides contain chords of the circle. (p. 442)

inscribed circle: A circle about which a polygon is circumscribed. (p. 427)

inscribed polygon: A polygon whose sides are chords of a circle. (p. 423)

intersection: A set of points contained in two or more intersecting figures. (p. 3)

intercepted arc: The arc $\overset{\frown}{AB}$ is called the intercepted arc of inscribed angle ACB. (p. 442)

inverse: The statement formed by negating the hypothesis and the conclusion. (p. 128)

isometry: A transformation that is distance-preserving. The distance between two points is the same as the distance between the corresponding images of these points. (p. 600)

isosceles trapezoid: A trapezoid with congruent legs. (p. 290)

isosceles triangle: A triangle with at least two congruent sides. (p. 139)

kite: A quadrilateral in which one diagonal is the perpendicular bisector of the other. (p. 489)

lateral faces of a prism: The faces of a prism that are formed by parallelograms. (p. 560)

lateral surface: The curved surface between the bases of a geometric figure. (p. 561)

legs of an isosceles triangle: The two congruent sides of an isosceles triangle. (p. 177)

legs of a right triangle: The sides forming the right angle of a right triangle. (p. 188)

linear equation: An equation whose graph is a line in which the set of points has the coordinates that satisfy the given equation. (p. 393)

linear pair: Two adjacent angles whose outer rays are opposite rays. (p. 26)

line of symmetry: The reflection or mirror line of a reflection transformation. (p. 600)

locus: The set of all points that satisfy a given set of conditions. (p. 529)

major arc: The major arc $\overset{\frown}{ACB}$ consists of points A and B and all points of the circle in the *exterior* of central angle AOB. (p. 433)

means: b and c are the means of the proportion $\frac{a}{b} = \frac{c}{d}$ ($b \neq 0$, $d \neq 0$). (p. 303)

median of a trapezoid: A segment that joins the midpoints of the legs. (p. 286)

midpoint of an arc: A point M is the midpoint of $\overset{\frown}{AMB}$ if $\overset{\frown}{AM} \cong \overset{\frown}{MB}$. (p. 439)

midpoint of a segment: The point at which a segment is divided into two congruent segments. (p. 10)

minor arc: The minor arc, $\overset{\frown}{AB}$, of central angle AOB consists of points A, B, and all points on the circle that lie in the interior of the central angle. (p. 432)

oblique cylinder: A cylinder that is not a right cylinder. (p. 561)

oblique prism: A prism that is not a right prism. (p. 560)

obtuse angle: An angle with a measure greater than 90. (p. 16)

obtuse triangle: A triangle with one obtuse angle. (p. 139)

opposite rays: If point O is between points A and B on $\overleftrightarrow{AB}$, then $\overrightarrow{OA}$ and $\overrightarrow{OB}$ are called opposite rays. (p. 26)

opposite sides of a quadrilateral: The nonadjacent sides of a quadrilateral. (p. 241)

parallel lines: Coplanar lines that do not intersect. (p. 93)

parallelogram: A quadrilateral with both pairs of opposite sides parallel. (p. 241)

parallel planes: Planes that do not intersect. (p. 130)

perimeter of a triangle: The sum of the lengths of its three sides. (p. 178)

perpendicular: Two lines which intersect to form a right angle. (p. 27)

perpendicular bisector: A perpendicular bisector of a side is perpendicular to that side at its midpoint. (p. 199)

polygon: The union of three or more coplanar segments such that each endpoint is shared by exactly two segments; segments intersect only at their endpoints; and intersecting segments are noncollinear. (p. 225)

postulate: A statement that is accepted without proof. (p. 5)

prism: A polyhedron with two congruent polygonal faces, called the bases, in parallel planes. (p. 560)

proportion: An equation of the form $\frac{a}{b} = \frac{c}{d}$ ($b \neq 0$, $d \neq 0$). (p. 303)

pyramid: A polyhedron composed of a polygonal region, called the base, and triangular regions, called the lateral faces which intersect at a vertex. (p. 574)

quadrilateral: A quadrilateral is a polygon with four sides. (p. 241)

radius: In a circle, a radius is a segment from the center to any point on the circle. (plural: radii) (p. 415)

radius of a regular polygon: A radius of the regular polygon is a segment from the center to a vertex of the polygon. (p. 499)

ratio: A ratio is a comparison of two numbers by division. The ratio of a to b ($b \neq 0$) may be written a to b, $\frac{a}{b}$, or $a{:}b$. (p. 303)

ray: $\overrightarrow{XY}$ consists of $\overline{XY}$ and all points P such that Y is between X and P. (p. 7)

rectangle: A parallelogram with four right angles. (p. 273)

reflection: In a plane, a reflection about line l is a transformation which maps each point P into a point P' such that (1) if P is on l, then $P' = P$, and (2) if P is not on l, then l is the perpendicular bisector of $\overline{PP'}$. (p. 599)

regular polygon: A convex polygon that is both equilateral and equiangular. (p. 227)

regular polyhedron: A polyhedron in which all the faces are congruent regular polygons. (p. 557)

remote interior angle: An interior angle that is not adjacent to the given exterior angle. (p. 121)

rhombus: A parallelogram with four congruent sides. (p. 273)

right angle: An angle with a measure of 90. (p. 16)

right cylinder: A cylinder in which a segment connecting a pair of corresponding points of the bases is perpendicular to the bases. (p. 561)

right prism: A prism in which the lateral edges are perpendicular to the bases. (p. 560)

right triangle: A triangle with one right angle. (p. 139)

rotation: A rotation $R_{P,m}$ of point A about point P through an angle of measure m is a transformation which *maps* A into its *rotation image* A' such that $PA = PA'$ and m $\angle APA' = m$. (p. 613)

rotational symmetry: A figure has rotational symmetry if there is a rotation in which the figure and its image coincide under the rotation. (p. 615)

scalene triangle: A triangle with no congruent sides. (p. 139)

secant: A line that intersects the circle in two points. (p. 417)

sector of a circle: The region bounded by an arc of a circle and two radii whose endpoints are endpoints of that arc. (p. 516)

segment: Segment $\overline{AB}$ is the set of points consisting of points A, B, and all points between A and B. (p. 6)

similar polygons: Polygons in which corresponding angles are congruent and the ratios of the lengths of corresponding sides are equal. (p. 308)

similar triangles: Triangles in which corresponding angles are congruent and corresponding sides are proportional. (p. 313)

sine: The sine of an acute angle of a right triangle is the ratio of the leg opposite the angle to the length of the hypotenuse. (p. 360)

skew lines: Noncoplanar lines. (p. 93)

slope: The slope of a segment $\overline{P_1P_2}$, with endpoints $P_1(x_1,y_1)$ and $P_2(x_2,y_2)$, is the ratio $\dfrac{y_2 - y_1}{x_2 - x_1}$ $(x_2 \neq x_1)$. (p. 388)

space: The set of all points. (p. 2)

sphere: The set of all points in space at a given distance from a given point called the center. (p. 584)

square: A rectangle with four congruent sides. (p. 273)

standard form: The standard form of an equation of a line is an equation of the form $ax + by = c$, where a and b are not both zero, and where $a > 0$. (p. 393)

straight angle: An angle with a measure of 180. (p. 16)

supplementary angles: Two angles with measures whose sum is 180. Each angle is called a *supplement* of the other. (p. 27)

symmetric figures: Figures that are their own mirror images about a line or about a plane. (p. 601)

tangent: The tangent of an angle is the ratio of the length of the opposite leg to the length of the adjacent leg. (p. 366)

tangent circles: Two coplanar circles which are tangent to the same line at the same point. (p. 428)

tangent segment: A segment that contains a point of tangency and another point of a tangent line to a circle. (p. 425)

tangent to a circle: A line that is coplanar with the circle and intersects the circle in exactly one point. (p. 417)

theorem: A statement that has been proved true. (p. 27)

translation: A translation in a plane from A to A' is a transformation which maps any point P into a point P' such that $\overline{PP'} \cong \overline{AA'}$ and $\overline{PP'} \parallel \overline{AA'}$. (p. 607)

transversal: A line, ray, or segment that intersects two or more coplanar lines, rays, or segments, each at a different point. (p. 97)

trapezoid: A quadrilateral with exactly one pair of parallel sides. (p. 285)

triangle: A figure formed by three segments joining three noncollinear points. (p. 115)

vertex: The common endpoint of the sides of an angle. (p. 15)

vertex angle of an isosceles triangle: The angle formed by the legs of an isosceles triangle. (p. 177)

vertical angles: Two nonadjacent angles formed by intersecting lines. (p. 77)

volume of a prism: The product of the area of a base and the length of an altitude. ($V = Bh$, where B = area of base and h = altitude.) (p. 569)

x-axis: The horizontal line in a plane. (p. 379)

y-axis: The vertical line in a plane. (p. 379)

y-intercept: The point at which the line intersects the y-axis. (p. 394)

Additional Answers

1.

Statement	Reason
1. $\overline{AC} \cong \overline{CE}$, $\overline{AB} \cong \overline{DE}$, $\angle 1 \cong \angle 2$, $\overline{BG} \cong \overline{DF}$	1. Given
2. $AC - AB = CE - DE$	2. Equations may be subtracted.
3. $AC - AB = BC$, $CE - DE = CD$	3. Seg Add Post
4. $BC = CD$	4. Sub
5. $\triangle BGC \cong \triangle DFC$	5. SAS

2.

Statement	Reason
1. $\angle 3 \cong \angle 4$, $\angle 5 \cong \angle 6$, C is the midpt of $\overline{BD}$	1. Given
2. $\angle 1$ and $\angle 3$ are supp, $\angle 2$ and $\angle 4$ are supp	2. If the outer rays of 2 adj $\angle$s form a st $\angle$, then the $\angle$s are supp.
3. $\angle 1 \cong \angle 2$	3. Supp $\angle$s of $\cong$ $\angle$s are $\cong$.
4. $\overline{BC} \cong \overline{DC}$	4. Def of midpt
5. $\triangle BGC \cong \triangle DFC$	5. ASA

3.

Statement	Reason
1. $\overline{AB} \cong \overline{CD}$, $\angle 3 \cong \angle 2$, $\overline{EC} \cong \overline{FB}$	1. Given
2. $AB + BC = CD + BC$	2. Add Prop of Eq
3. $AB + BC = AC$, $CD + BC = BD$	3. Seg Add Post
4. $AC = BD$	4. Sub
5. $\triangle AEC \cong \triangle DFB$	5. SAS

4.

Statement	Reason
1. $\angle 1 \cong \angle 4$, $\overline{EC} \cong \overline{FB}$, $\angle E \cong \angle F$	1. Given
2. $\angle 2$ and $\angle 1$ are supp, $\angle 3$ and $\angle 4$ are supp	2. If the outer rays of 2 adj $\angle$s form a st $\angle$, then $\angle$s are supp.
3. $\angle 2 \cong \angle 3$	3. Supp $\angle$s of $\cong$ $\angle$s are $\cong$.
4. $\triangle AEC \cong \triangle DFB$	4. ASA

5.

Statement	Reason
1. $\overline{AG} \cong \overline{DG}$, $\overline{FG} \cong \overline{EG}$, $\overline{AB} \cong \overline{DC}$, $\angle A \cong \angle D$	1. Given
2. $AG - FG = DG - EG$	2. Equations may be subt.
3. $AG - FG = AF$, $DG - EG = DE$	3. Seg Add Post
4. $AF = DE$	4. Sub
5. $\triangle FAB \cong \triangle EDC$	5. SAS

6.

Statement	Reason
1. $\angle 1 \cong \angle 2$, $\overline{FB} \perp \overline{AD}$, $\overline{EC} \perp \overline{AD}$, $\overline{FB} \cong \overline{EC}$	1. Given
2. $\angle 1$ and $\angle 3$ are supp, $\angle 2$ and $\angle 4$ are supp	2. If the outer rays of 2 adj $\angle$s form a st $\angle$, then the $\angle$s are supp.
3. $\angle 3 \cong \angle 4$	3. Supp $\angle$s of $\cong$ $\angle$s are $\cong$.
4. m $\angle FBA = 90$, m $\angle ECD = 90$	4. Def of $\perp$
5. m $\angle FBA =$ m $\angle ECD$	5. Sub
6. $\triangle FAB \cong \triangle EDC$	6. ASA

7.

Statement	Reason
1. $\overline{RP} \cong \overline{US}$, $\overline{RQ} \cong \overline{UT}$, $\overline{QW} \cong \overline{TV}$, $\overline{RP} \parallel \overline{US}$	1. Given
2. $RP - RQ = US - UT$	2. Equations may be subtracted.
3. $RP - RQ = PQ$, $US - UT = ST$	3. Seg Add Post
4. $PQ = ST$	4. Sub
5. $\angle Q \cong \angle T$	5. If lines are $\parallel$, then alt int $\angle$s are $\cong$.
6. $\triangle PQW \cong \triangle STV$	6. SAS

8.

Statement	Reason
1. $\angle 1 \cong \angle 2$, $\overline{GB} \cong \overline{FC}$, $\overline{AC} \cong \overline{DB}$	1. Given
2. $\angle 1$ and $\angle ABG$ are supp, $\angle 2$ and $\angle DCF$ are supp	2. If the outer rays of 2 adj $\angle$s form a st $\angle$, then the $\angle$s are supp.
3. $\angle ABG \cong \angle DCF$	3. Supp $\angle$s of $\cong$ $\angle$s are $\cong$.
4. $AC - CB = DB - CB$	4. Subt Prop of Eq
5. $AC - CB = AB$, $DB - CB = DC$	5. Seg Add Post
6. $AB = DC$	6. Sub
7. $\triangle BAG \cong \triangle CDF$	7. SAS

9.

Statement	Reason
1. $\overline{AD} \cong \overline{CB}$, $\overline{CE} \parallel \overline{BF}$, $\angle 1$ and $\angle 2$ are supp	1. Given
2. $\angle 1$ and $\angle 3$ are supp	2. If the outer rays of 2 adj $\angle$s form a st $\angle$, then the $\angle$s are supp.
3. $\angle 2 \cong \angle 3$	3. Supp $\angle$s of the same $\angle$ are $\cong$.
4. $AD - BD = CB - BD$	4. Subt Prop of Eq
5. $AD - BD = AB$, $CB - BD = CD$	5. Seg Add Post
6. $AB = CD$	6. Sub
7. $\angle DCE \cong \angle ABF$	7. If lines are $\parallel$, then corr $\angle$s are $\cong$.
8. $\triangle CED \cong \triangle BFA$	8. ASA

10.

Statement	Reason
1. $\overline{RV} \perp \overline{PT}$, $\angle 1 \cong \angle 2$, $\angle 3 \cong \angle 4$, R is the midpt of $\overline{QS}$	1. Given
2. $\angle WRQ$ and $\angle 1$ are comp, $\angle URS$ and $\angle 2$ are comp	2. If the outer rays of 2 adj acute $\angle$s are $\perp$, then the $\angle$s are comp.
3. $\angle WRQ \cong \angle URS$	3. Comp $\angle$s of $\cong$ $\angle$s are $\cong$.
4. $\angle WQR$ and $\angle 3$ are supp, $\angle USR$ and $\angle 4$ are supp	4. If the outer rays of 2 adj $\angle$s form a st $\angle$, then the $\angle$s are supp.
5. $\angle WQR \cong \angle USR$	5. Supp $\angle$s of $\cong$ $\angle$s are $\cong$.
6. $\overline{QR} \cong \overline{SR}$	6. Def of midpt
7. $\triangle QRW \cong \triangle SRU$	7. ASA

649

11.

Statement	Reason
1. $\overline{PQ} \cong \overline{RS}$, $\overline{WP} \perp \overline{PS}$, $\overline{SX} \perp \overline{PS}$, $\angle 3 \cong \angle 4$, $\angle 1 \cong \angle 2$	1. Given
2. $PQ + QR = RS + QR$	2. Add Prop of Eq
3. $PQ + QR = PR$, $RS + QR = QS$	3. Seg Add Post
4. $PR = QS$	4. Sub
5. $\angle 3$ and $\angle UPR$ are comp, $\angle 4$ and $\angle TSQ$ are comp	5. If the outer rays of 2 adj acute $\angle$s are $\perp$, then the $\angle$s are comp.
6. $\angle UPR \cong \angle TSQ$	6. Comp $\angle$s of $\cong \angle$s are $\cong$.
7. $\angle 1$ and $\angle TQR$ are supp, $\angle 2$ and $\angle URQ$ are supp	7. If the outer rays of 2 adj $\angle$s form a st $\angle$, then the $\angle$s are supp.
8. $\angle TQR \cong \angle URQ$	8. Supp $\angle$s of $\cong \angle$s are $\cong$.
9. $\triangle PUR \cong \triangle STQ$	9. ASA

12.

Statement	Reason
1. $\overline{EB}$ bis $\angle FBD$, $\overline{EB} \perp \overline{AC}$, $\angle 3 \cong \angle 4$, $\overline{FB} \cong \overline{DB}$	1. Given
2. $\angle FBE \cong \angle DBE$	2. Def of $\angle$ bis
3. $\angle FBA$ and $\angle FBE$ are comp, $\angle DBE$ and $\angle DBC$ are comp	3. If the outer rays of 2 adj acute $\angle$s are $\perp$, then the $\angle$s are comp.
4. $\angle FBA \cong \angle DBC$	4. Comp $\angle$s of $\cong \angle$s are $\cong$.
5. $\angle AFB$ and $\angle 3$ are supp, $\angle CDB$ and $\angle 4$ are supp	5. If the outer rays of 2 adj $\angle$s form a st $\angle$, then the $\angle$s are supp.
6. $\angle AFB \cong \angle CDB$	6. Supp $\angle$s of $\cong \angle$s are $\cong$.
7. $\triangle FAB \cong \triangle DCB$	7. ASA

13.

Statement	Reason
1. $\overline{AB} \cong \overline{DC}$, $\overline{AE} \cong \overline{DF}$, $\overline{CE} \cong \overline{BF}$	1. Given
2. $AB + BC = DC + BC$	2. Add Prop of Eq
3. $AB + BC = AC$, $DC + BC = BD$	3. Seg Add Post
4. $AC = BD$	4. Sub
5. $\triangle ACE \cong \triangle DBF$	5. SSS

3.

Statement	Reason
1. $\overline{AC}$ and $\overline{DE}$ bis each other	1. Given
2. $\overline{AB} \cong \overline{CB}$, $\overline{DB} \cong \overline{EB}$	2. Def of seg bis
3. $\angle DBA \cong \angle EBC$	3. Vert $\angle$s are $\cong$.
4. $\triangle BAD \cong \triangle BCE$	4. SAS
5. $\angle A \cong \angle C$	5. CPCTC
6. $\overline{AD} \parallel \overline{CE}$	6. If alt int $\angle$s are $\cong$, then lines are $\parallel$.

4.

Statement	Reason
1. $\overline{AD} \parallel \overline{CE}$, B is the midpt of $\overline{AC}$	1. Given
2. $\angle A \cong \angle C$	2. If lines are $\parallel$, then alt int $\angle$s are $\cong$.
3. $\overline{BA} \cong \overline{BC}$	3. Def of midpt
4. $\angle DBA \cong \angle EBC$	4. Vert $\angle$s are $\cong$.
5. $\triangle BAD \cong \triangle BCE$	5. ASA
6. $\overline{DB} \cong \overline{EB}$	6. CPCTC
7. B is the midpt of $\overline{DE}$	7. Def of midpt

5.

Statement	Reason
1. $\overline{FS} \cong \overline{UQ}$, $\overline{FP} \parallel \overline{UR}$, $\overline{FP} \cong \overline{UR}$	1. Given
2. $FS + SQ = UQ + SQ$	2. Add Prop of Eq
3. $FS + SQ = FQ$, $UQ + SQ = US$	3. Seg Add Post
4. $FQ = US$	4. Sub
5. $\angle F \cong \angle U$	5. If lines are $\parallel$, then alt int $\angle$s are $\cong$.
6. $\triangle PFQ \cong \triangle RUS$	6. SAS
7. $\angle PQF \cong \angle RSU$	7. CPCTC
8. $\overline{PQ} \parallel \overline{RS}$	8. If alt int $\angle$s are $\cong$, then lines are $\parallel$.

6.

Statement	Reason
1. $\overline{FP} \parallel \overline{UR}$, $\overline{PQ} \parallel \overline{RS}$, $\overline{PQ} \cong \overline{RS}$	1. Given
2. $\angle RUS \cong \angle PFQ$, $\angle RSU \cong \angle PQF$	2. If lines are $\parallel$, then alt int $\angle$s are $\cong$.
3. $\triangle PQF \cong \triangle RSU$	3. AAS
4. $\overline{FQ} \cong \overline{US}$	4. CPCTC

7.

Statement	Reason
1. $\overline{TQ} \perp \overline{PR}$, $\angle 1 \cong \angle 2$, $\overline{TQ} \perp \overline{US}$	1. Given
2. $\angle TQU$ and $\angle 1$ are comp, $\angle TQS$ and $\angle 2$ are comp	2. If the outer rays of 2 adj acute $\angle$s form a st $\angle$, then the $\angle$s are comp.
3. $\angle TQU \cong \angle TQS$	3. Comp $\angle$s of $\cong \angle$s are $\cong$.
4. $\overline{TQ} \cong \overline{TQ}$	4. Reflex Prop of Congr
5. $m\angle UTQ = 90$, $m\angle STQ = 90$	5. Def of $\perp$
6. $m\angle UTQ = m\angle STQ$	6. Sub
7. $\triangle UTQ \cong \triangle STQ$	7. ASA
8. $\overline{UQ} \cong \overline{SQ}$	8. CPCTC

8.

Statement	Reason
1. $\overline{TQ}$ bis $\angle UQS$, $\overline{UQ} \cong \overline{SQ}$	1. Given
2. $\angle UQT \cong \angle SQT$	2. Def of $\angle$ bis
3. $\overline{TQ} \cong \overline{TQ}$	3. Reflex Prop of Congr
4. $\triangle TUQ \cong \triangle TSQ$	4. SAS
5. $\angle STQ \cong \angle UTQ$	5. CPCTC
6. $m\angle STQ + m\angle UTQ = 180$	6. If the outer rays of 2 adj $\angle$s form a st $\angle$, then the sum of their meas is 180.
7. $m\angle STQ + m\angle STQ = 180$	7. Sub
8. $m\angle STQ = 90$	8. Div Prop of Eq
9. $\overline{TQ} \perp \overline{US}$	9. Def of $\perp$

9.

Statement	Reason
1. $\overline{CB} \perp \overline{AB}$, $\overline{CB} \perp \overline{BH}$, $\overline{DB} \cong \overline{IB}$, $\angle 1 \cong \angle 2$	1. Given
2. $\angle CBD$ and $\angle 1$ are comp, $\angle CBI$ and $\angle 2$ are comp	2. If the outer rays of 2 adj acute $\angle$s are $\perp$, then the $\angle$s are comp.
3. $\angle CBD \cong \angle CBI$	3. Comp $\angle$s of $\cong \angle$s are $\cong$.
4. $\overline{CB} \cong \overline{CB}$	4. Reflex Prop of Congr
5. $\triangle DBC \cong \triangle IBC$	5. SAS
6. $\angle 5 \cong \angle 6$	6. CPCTC

650

10.

Statement	Reason
1. $\overline{BC} \perp \overline{EC}$, $\overline{BC} \perp \overline{GC}$, $\overline{EC} \cong \overline{GC}$, $\overline{ED} \cong \overline{GI}$, $\angle 3 \cong \angle 4$	1. Given
2. m$\angle ECB$ = 90, m$\angle GCB$ = 90	2. Def of $\perp$
3. m$\angle ECB$ = m$\angle GCB$	3. Sub
4. $\angle 5$ and $\angle 3$ are supp, $\angle 6$ and $\angle 4$ are supp	4. If the outer rays of 2 adj $\angle$s form a st $\angle$, then the $\angle$s are supp.
5. $\angle 5 \cong \angle 6$	5. Supp $\angle$s of $\cong \angle$s are $\cong$.
6. $EC - ED = GC - GI$	6. Equations may be subtracted.
7. $EC - ED = DC$, $GC - GI = IC$	7. Seg Add Post
8. $DC = IC$	8. Sub
9. $\triangle DBC \cong \triangle IBC$	9. ASA
10. $\overline{BD} \cong \overline{BI}$	10. CPCTC

PAGE 181

25.

Statement	Reason
1. $\overline{BC} \cong \overline{AC}$, $\overrightarrow{CE}$ bis $\angle DCA$	1. Given
2. m$\angle A$ = m$\angle B$	2. $\angle$s opp $\cong$ sides are $\cong$.
3. m$\angle A$ + m$\angle B$ = m$\angle DCA$	3. Meas of ext $\angle$ = sum of meas of 2 remote int $\angle$s.
4. m$\angle B$ + m$\angle B$ = m$\angle DCA$, or 2 $\times$ m$\angle B$ = m$\angle DCA$	4. Sub
5. m$\angle B = \frac{1}{2}$(m$\angle DCA$)	5. Div Prop Eq
6. m$\angle DCE = \frac{1}{2}$(m$\angle DCA$)	6. Def of $\angle$ bis
7. m$\angle B$ = m$\angle DCE$	7. Sub
8. $\overline{CE} \parallel \overline{AB}$	8. If corr $\angle$s are $\cong$, then lines are $\parallel$.

26.

Statement	Reason
1. $\overline{BA} \perp$ every line in $\mathcal{M}$ through A, $\angle ACB \cong \angle ADB$	1. Given
2. m$\angle BAC$ = 90, m$\angle BAD$ = 90	2. Def of $\perp$
3. m$\angle BAC$ = m$\angle BAD$	3. Sub
4. $\overline{AB} \cong \overline{AB}$	4. Reflex Prop of Congr
5. $\triangle ABC \cong \triangle ABD$	5. AAS
6. $\overline{CB} \cong \overline{DB}$	6. CPCTC
7. $\triangle BCD$ is isos	7. Def of isos $\triangle$

27.

Statement	Reason
1. $\overline{AE} \perp \overline{CD}$, E is midpt of $\overline{CD}$, $\overline{BA} \perp$ all lines in $\mathcal{M}$ through A	1. Given
2. m$\angle AED$ = 90, m$\angle AEC$ = 90	2. Def of $\perp$
3. m$\angle AED$ = m$\angle AEC$	3. Sub
4. $\overline{CE} \cong \overline{ED}$	4. Def of midpt
5. $\overline{AE} \cong \overline{AE}$	5. Reflex Prop of Congr
6. $\triangle CEA \cong \triangle DEA$	6. SAS
7. $\overline{AC} \cong \overline{AD}$	7. CPCTC
8. m$\angle BAD$ = 90, m$\angle BAC$ = 90	8. Def of $\perp$
9. m$\angle BAD$ = m$\angle BAC$	9. Sub
10. $\overline{BA} \cong \overline{BA}$	10. Reflex Prop of Congr
11. $\triangle CAB \cong \triangle DAB$	11. SAS
12. $\overline{BC} \cong \overline{BD}$	12. CPCTC

PAGE 187

6.

Statement	Reason
1. $\overline{HG}$ and $\overline{DE}$ bis each other	1. Given
2. $\overline{HB} \cong \overline{GB}$, $\overline{DB} \cong \overline{EB}$	2. Def of seg bis
3. $\angle 1 \cong \angle 2$	3. Vert $\angle$s are $\cong$.
4. $\triangle DBH \cong \triangle EBG$	4. SAS
5. $\overline{HD} \cong \overline{GE}$	5. CPCTC

7.

Statement	Reason
1. B is the midpt of $\overline{AC}$, $\overline{AD} \perp \overline{AC}$, $\overline{EC} \perp \overline{AC}$	1. Given
2. $\overline{AB} \cong \overline{CB}$	2. Def midpt
3. $\angle DBA \cong \angle EBC$	3. Vert $\angle$s are $\cong$.
4. m$\angle DAB$ = 90, m$\angle ECB$ = 90	4. Def of $\perp$
5. m$\angle DAB$ = m$\angle ECB$	5. Sub
6. $\triangle DAB \cong \triangle ECB$	6. ASA
7. $\overline{DB} \cong \overline{EB}$	7. CPCTC
8. B is midpt of $\overline{DE}$	8. Def of midpt

8.

Statement	Reason
1. $\angle 1 \cong \angle 3$, $\angle ABC \cong \angle DCB$	1. Given
2. $\overline{BC} \cong \overline{BC}$	2. Reflex Prop of Congr
3. $\triangle ABC \cong \triangle DCB$	3. ASA
4. $\overline{AB} \cong \overline{DC}$	4. CPCTC

9.

Statement	Reason
1. $\angle A \cong \angle D$, $\overline{AE} \cong \overline{DE}$	1. Given
2. $\angle 5 \cong \angle 6$	2. Vert $\angle$s are $\cong$.
3. $\triangle ABE \cong \triangle DCE$	3. ASA
4. $\overline{AB} \cong \overline{DC}$	4. CPCTC

10.

Statement	Reason
1. $\overline{AB} \cong \overline{DC}$, $\angle 1 \cong \angle 3$, $\angle 2 \cong \angle 4$	1. Given
2. m$\angle 1$ + m$\angle 2$ = m$\angle 3$ + m$\angle 4$	2. Equations may be added.
3. m$\angle 1$ + m$\angle 2$ = m$\angle ABC$, m$\angle 3$ + m$\angle 4$ = m$\angle DCB$	3. Angle Add Post
4. m$\angle ABC$ = m$\angle DCB$	4. Sub
5. $\triangle ABC \cong \triangle DCB$	5. AAS
6. $\overline{AC} \cong \overline{DB}$	6. CPCTC

11.

Statement	Reason
1. $\overrightarrow{PR}$ bis $\angle SPT$, $\angle 3 \cong \angle 4$	1. Given
2. $\angle 9 \cong \angle 10$	2. Def of $\angle$ bis
3. $\overline{PR} \cong \overline{PR}$	3. Reflex Prop of Congr
4. $\triangle PSR \cong \triangle PTR$	4. ASA
5. $\overline{PS} \cong \overline{PT}$	5. CPCTC
6. $\overline{PQ} \cong \overline{PQ}$	6. Reflex Prop of Congr
7. $\triangle PSQ \cong \triangle PTQ$	7. SAS
8. $\overline{QS} \cong \overline{QT}$	8. CPCTC

12.

Statement	Reason
1. $\overline{PR}$ bis $\angle SRT$, $\overline{SR} \cong \overline{TR}$	1. Given
2. $\angle 3 \cong \angle 4$	2. Def of $\angle$ bis
3. $\overline{PR} \cong \overline{PR}$	3. Reflex Prop of Congr
4. $\triangle PSR \cong \triangle PTR$	4. SAS
5. $\overline{PS} \cong \overline{PT}$, $\angle 9 \cong \angle 10$	5. CPCTC
6. $\overline{PQ} \cong \overline{PQ}$	6. Reflex Prop of Congr
7. $\triangle PSQ \cong \triangle PTQ$	7. SAS
8. $\angle 1 \cong \angle 6$	8. CPCTC

13.

Statement	Reason
1. $\overline{PS} \cong \overline{PT}$, $\overline{SR} \cong \overline{TR}$	1. Given
2. $\overline{PR} \cong \overline{PR}$	2. Reflex Prop of Congr
3. $\triangle PSR \cong \triangle PTR$	3. SSS
4. $\angle 3 \cong \angle 4$	4. CPCTC
5. $\overline{RQ} \cong \overline{RQ}$	5. Reflex Prop of Congr
6. $\triangle RSQ \cong \triangle RTQ$	6. SAS
7. $\angle 2 \cong \angle 5$	7. CPCTC

14.

Statement	Reason
1. $\overline{AD}$ and $\overline{CF}$ bis each other	1. Given
2. $\overline{AG} \cong \overline{DG}$, $\overline{CG} \cong \overline{FG}$	2. Def of seg bis
3. $\angle AGC \cong \angle DGF$	3. Vert $\angle$s are $\cong$.
4. $\triangle AGC \cong \triangle DGF$	4. SAS
5. $\angle 1 \cong \angle 3$	5. Vert $\angle$s are $\cong$.
6. $\angle BAG \cong \angle EDG$	6. CPCTC
7. $\triangle BAG \cong \triangle EDG$	7. ASA
8. $\angle 5 \cong \angle 6$	8. CPCTC

15.

Statement	Reason
1. $\overline{AD}$ bis $\overline{BE}$, $\angle 5 \cong \angle 6$	1. Given
2. $\overline{BG} \cong \overline{EG}$	2. Def of seg bis
3. $\angle 1 \cong \angle 3$	3. Vert $\angle$s are $\cong$.
4. $\triangle ABG \cong \triangle DEG$	4. ASA
5. $\overline{AG} \cong \overline{DG}$, $\angle CAG \cong \angle FDG$	5. CPCTC
6. $\angle AGC \cong \angle DGF$	6. Vert $\angle$s are $\cong$.
7. $\triangle CAG \cong \triangle FDG$	7. ASA
8. $\overline{AC} \cong \overline{DF}$	8. CPCTC

16. Student flowcharts may vary. See page 667 for flowchart.

17.

Statement	Reason
1. $\overline{AC} \parallel \overline{FD}$, $\overline{BC} \cong \overline{EF}$	1. Given
2. $\angle F \cong \angle C$	2. If lines are $\parallel$, then alt int $\angle$s are $\cong$.
3. $\angle 2 \cong \angle 4$	3. Vert $\angle$s are $\cong$.
4. $\triangle BCG \cong \triangle EFG$	4. AAS
5. $\overline{CG} \cong \overline{FG}$	5. CPCTC
6. $\angle AGC \cong \angle DGF$	6. Vert $\angle$s are $\cong$.
7. $\triangle ACG \cong \triangle DGF$	7. ASA
8. $\overline{AC} \cong \overline{DF}$	8. CPCTC

18.

Statement	Reason
1. $\triangle AED$ and $\triangle BCD$ are equil, D is midpt of $\overline{EC}$	1. Given
2. $AE = ED$, $BC = CD$	2. Def of equil $\triangle$
3. $ED = CD$	3. Def midpt
4. $AE = BC$	4. Sub
5. $EC = CE$	5. Reflex Prop of Eq
6. m $\angle AED = 60$, m $\angle BCD = 60$	6. In an equil $\triangle$, each $\angle$ has meas of 60.
7. m $\angle AED = $ m $\angle BCD$	7. Sub
8. $\triangle AEC \cong \triangle BCE$	8. SAS
9. $\angle ACE \cong \angle BEC$	9. CPCTC
10. $\overline{XE} \cong \overline{XC}$ ($XE = XC$)	10. Sides opp $\cong$ $\angle$s are $\cong$

19.

Statement	Reason
1. $AX = BX$, $CX = EX$	1. Given
2. m $\angle BAX = $ m $\angle ABX$, m $\angle CEX = $ m $\angle ECX$	2. $\angle$s opp $\cong$ sides are $\cong$.
3. m $\angle AXB = $ m $\angle CXE$	3. Vert $\angle$s are $\cong$.
4. m $\angle BAX + $ m $\angle ABX + $ m $\angle AXB = 180$, m $\angle CEX + $ m $\angle ECX + $ m $\angle CXE = 180$	4. Sum of meas of all $\angle$s of $\triangle = 180$.
5. m $\angle BAX + $ m $\angle ABX + $ m $\angle AXB = $ m $\angle CEX + $ m $\angle ECX + $ m $\angle CXE$	5. Sub
6. $2 \times $ m $\angle BAX + $ m $\angle AXB = 2 \times $ m $\angle ECX + $ m $\angle CXE$	6. Sub
7. $2 \times $ m $\angle BAX + $ m $\angle AXB = 2 \times $ m $\angle ECX + $ m $\angle AXB$	7. Sub
8. $2 \times $ m $\angle BAX = 2 \times $ m $\angle CEX$	8. Subt Prop of Eq
9. m $\angle BAX = $ m $\angle ECX$	9. Div Prop of Eq
10. $\overline{AB} \parallel \overline{EC}$	10. If alt int $\angle$s are $\cong$, then lines are $\parallel$.

Brainteaser

$\triangle DEF$ is equil. $\overline{AN} \cong \overline{NC} \cong \overline{CM} \cong \overline{MB} \cong \overline{BP} \cong \overline{PA}$ since each are $\frac{1}{2} \times$ length of a side of $\triangle ABC$. $\angle A \cong \angle B \cong \angle C$, so by SAS, $\triangle PAN \cong \triangle NCM \cong \triangle MBP$. By CPCTC, $\overline{PN} \cong \overline{NM} \cong \overline{MP}$; therefore, $\triangle PMN$ is equil. Same reasoning is used to prove each successive int $\triangle$ equil.

9.

Statement	Reason
1. $\overline{PQ} \perp \overline{PS}$, $\overline{SR} \perp \overline{SP}$, $\overline{PT} \cong \overline{ST}$	1. Given
2. m∠QPS = 90, m∠RSP = 90	2. Def of ⊥
3. m∠QPS = m∠RSP	3. Sub
4. ∠1 ≅ ∠2	4. ∠s opp ≅ sides are ≅.
5. $\overline{PS} \cong \overline{PS}$	5. Reflex Prop of Congr
6. △PSR ≅ △SPQ	6. ASA
7. ∠Q ≅ ∠R	7. CPCTC

10.

Statement	Reason
1. $\overline{PQ} \perp \overline{PS}$, $\overline{SR} \perp \overline{SP}$, $\overline{PR} \cong \overline{SQ}$	1. Given
2. △PSR and △SPQ are rt △s.	2. Def of rt △
3. $\overline{PS} \cong \overline{PS}$	3. Reflex Prop of Congr
4. △PSR ≅ △SPQ	4. HL
5. ∠1 ≅ ∠2	5. CPCTC
6. $\overline{PT} \cong \overline{ST}$	6. Sides opp ≅ ∠s are ≅.

11.

Statement	Reason
1. $\overline{PQ} \perp \overline{PS}$, $\overline{SR} \perp \overline{SP}$, ∠3 ≅ ∠4	1. Given
2. m∠QPS = 90, m∠RSP = 90	2. Def of ⊥
3. m∠QPS = m∠RSP	3. Sub
4. ∠1 and ∠3 are comp, ∠2 and ∠4 are comp.	4. If the outer rays of 2 adj acute ∠s are ⊥, then ∠s are comp.
5. ∠1 ≅ ∠2	5. comp ∠s of ≅ ∠s are ≅.
6. $\overline{PS} \cong \overline{PS}$	6. Reflex Prop of Congr
7. △QPS ≅ △RSP	7. ASA
8. ∠Q ≅ ∠R	8. CPCTC

12.

Statement	Reason
1. $\overline{AB} \perp \overline{AE}$, $\overline{CB} \perp \overline{CD}$, $\overline{AB} \cong \overline{CB}$, ∠1 ≅ ∠2	1. Given
2. △AEB and △CDB are rt △s.	2. Def of rt △
3. $\overline{EB} \cong \overline{DB}$	3. Sides opp ≅ ∠s are ≅.
4. △AEB ≅ △CDB	4. HL
5. $\overline{AE} \cong \overline{CD}$	5. CPCTC

13.

Statement	Reason
1. $\overline{AB} \perp \overline{AE}$, $\overline{CB} \perp \overline{CD}$, $\overline{AB} \cong \overline{CB}$, ∠ABD ≅ ∠EBC	1. Given
2. m∠BAE = 90, m∠BCD = 90	2. Def of ⊥
3. m∠BAE = m∠BCD	3. Sub
4. m∠ABD − m∠4 = m∠EBC − m∠4	4. Subt Prop of Eq
5. m∠ABD − m∠4 = m∠3, m∠EBC − m∠4 = m∠5	5. Angle Add Post
6. m∠3 = m∠5	6. Sub
7. △ABE ≅ △CBD	7. ASA
8. $\overline{EB} \cong \overline{BD}$	8. CPCTC
9. △EBD is isos.	9. Def of isos △

Using rt △s ABC and DEF for Exercises 14–16:

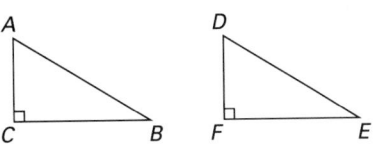

14.

Statement	Reason
1. $\overline{AB} \cong \overline{DE}$, ∠A ≅ ∠D, ∠C is a rt ∠, ∠F is a rt ∠.	1. Given
2. ∠C ≅ ∠F	2. All right ∠s are ≅.
3. △ABC ≅ △DEF	3. AAS

15.

Statement	Reason
1. $\overline{AC} \cong \overline{DF}$, $\overline{BC} \cong \overline{EF}$, ∠C is a rt ∠, ∠F is a rt ∠.	1. Given
2. ∠C ≅ ∠F	2. All right ∠s are ≅.
3. △ABC ≅ △DEF	3. SAS

16. Leg-Acute Angle Method. Two rt △s are ≅ if a leg and an acute ∠ of one are ≅ to a leg and an acute ∠ of the other.

(1) The leg is incl between the acute ∠ and the rt ∠. ∠A ≅ ∠D, $\overline{AC} \cong \overline{DF}$, ∠C ≅ ∠F (Given); △ABC ≅ △DEF (ASA)

(2) The leg is not incl between acute ∠ and rt ∠. ∠A ≅ ∠D, ∠C ≅ ∠F, $\overline{BC} \cong \overline{EF}$ (Given); △ABC ≅ △DEF (AAS)

17. Student flowcharts may vary. See page 667 for flowchart.

18.

Statement	Reason
1. $\overline{PR} \perp \overline{SQ}$, $\overline{QT} \perp \overline{SP}$, $\overline{SP} \cong \overline{SQ}$	1. Given
2. m∠QRP = 90, m∠PTQ = 90	2. Def of ⊥
3. m∠PTQ = m∠QRP	3. Sub
4. ∠SPQ ≅ ∠SQP	4. ∠s opp ≅ sides are ≅.
5. $\overline{PQ} \cong \overline{QP}$	5. Reflex Prop of Congr
6. △TPQ ≅ △RQP	6. AAS
7. $\overline{TP} \cong \overline{RQ}$	7. CPCTC
8. ∠TUP ≅ ∠RUQ	8. Vert ∠s are ≅.
9. △TUP ≅ △RUQ	9. AAS
10. $\overline{UP} \cong \overline{UQ}$	10. CPCTC
11. △UPQ is isos.	11. Def of isos △

19. Student flowcharts may vary. See page 667 for flowchart.

20.

Statement	Reason
1. $\overline{PR} \perp \overline{SQ}$, $\overline{QT} \perp \overline{SP}$, ∠3 ≅ ∠4	1. Given
2. m∠PTQ = 90, m∠QRP = 90	2. Def of ⊥
3. ∠PTQ ≅ ∠QRP	3. Sub
4. $\overline{PQ} \cong \overline{QP}$	4. Reflex Prop of Congr
5. △PTQ ≅ △QRP	5. AAS
6. ∠TPQ ≅ ∠RQP	6. CPCTC
7. △SPQ is isos.	7. Def isos △

21.

Statement	Reason
1. $\overline{PR} \perp \overline{SQ}$, $\overline{QT} \perp \overline{SP}$, $\overline{UT} \cong \overline{UR}$	1. Given
2. m∠UTP = 90, m∠URQ = 90	2. Def of ⊥
3. m∠UTP = m∠URQ	3. Sub
4. ∠TUP ≅ ∠RUQ	4. Vert ∠s are ≅.
5. △TUP ≅ △RUQ	5. ASA
6. $\overline{UP} \cong \overline{UQ}$	6. CPCTC
7. UP + UR = UQ + UT	7. Add Prop of Eq
8. UP + UR = PR, UQ + UT = QT	8. Seg Add Post
9. PR = QT	9. Sub
10. ∠S ≅ ∠S	10. Reflex Prop of Congr
11. m∠SRP = 90, m∠STQ = 90	11. Def of ⊥
12. m∠SRP = m∠STQ	12. Sub
13. △SRP ≅ △STQ	13. AAS
14. $\overline{SR} \cong \overline{ST}$	14. CPCTC

Midchapter Review

5.

Statement	Reason
1. $\overline{QS} \perp \overline{PS}$, $\overline{QR} \perp \overline{PR}$, ∠3 ≅ ∠4	1. Given
2. m∠QSP = 90, m∠QRP = 90	2. Def of ⊥
3. m∠QSP = m∠QRP	3. Sub
4. $\overline{QP} \cong \overline{QP}$	4. Reflex Prop of Cong
5. △QSP ≅ △QRP	5. AAS

6.

Statement	Reason
1. $\overline{SP} \cong \overline{RP}$ m∠QSP = 90, m∠QRP = 90	1. Given
2. $\overline{QS} \perp \overline{SP}$, $\overline{QR} \perp \overline{RP}$	2. Def of ⊥
3. $\overline{QP} \cong \overline{QP}$	3. Reflex Prop of Cong
4. △QSP ≅ △QRP	4. HL
5. ∠3 ≅ ∠4	5. CPCTC

7.

Statement	Reason
1. $\overline{QP}$ bis ∠SQR, ∠3 ≅ ∠4, $\overline{ST} \cong \overline{RT}$	1. Given
2. ∠SQP ≅ ∠RQP	2. Def ∠ bis
3. $\overline{QP} \cong \overline{QP}$	3. Reflex Prop of Congr
4. △QSP ≅ △QRP	4. ASA
5. $\overline{QS} \cong \overline{QR}$	5. CPCTC
6. $\overline{QT} \cong \overline{QT}$	6. Reflex Prop of Congr
7. △QST ≅ △QRT	7. SAS
8. ∠1 ≅ ∠2	8. CPCTC

PAGE 198

16.

Statement	Reason
1. $\overline{DB}$ is an alt from the vertex ∠ of isos △ADC.	1. Given
2. ∠DBA and ∠DBC are rt ∠s.	2. Def of alt
3. ∠DBA ≅ ∠DBC	3. All rt ∠s are ≅.
4. $\overline{AD} \cong \overline{CD}$	4. Def of isos △
5. ∠A ≅ ∠C	5. ∠s opp ≅ sides are ≅.
6. △ABD ≅ △CBD	6. AAS
7. $\overline{BE} \cong \overline{BF}$	7. Corr medians of ≅ △s are ≅.

17. Student flowcharts may vary. See page 667 for flowchart.

18.

Statement	Reason
1. $\overline{DB}$ is a median from vertex ∠ of isos △ADC, ∠1 ≅ ∠2	1. Given
2. $\overline{AB} \cong \overline{CB}$	2. Def of median
3. ∠A ≅ ∠C	3. Base ∠s of isos △ are ≅.
4. △ABE ≅ △CBF	4. ASA
5. $\overline{BE} \cong \overline{BF}$	5. CPCTC

19. True.

Statement	Reason
1. △ABC with pt D on $\overline{BC}$, $\overline{AD}$ is alt, $\overline{AD}$ bis ∠BAC	1. Given
2. ∠BAD ≅ ∠CAD	2. Def ∠ bis
3. ∠ADB and ∠ADC are rt ∠s.	3. Def of alt
4. ∠ADB ≅ ∠ADC	4. All rt ∠s are ≅.
5. $\overline{AD} \cong \overline{AD}$	5. Reflex Prop of Congr
6. △ADB ≅ △ADC	6. ASA
7. $\overline{AB} \cong \overline{AC}$	8. CPCTC
9. △ABC is isos.	9. Def of isos △

20. True.

Statement	Reason
1. △ABC with pts M and N on $\overline{AC}$ and $\overline{AB}$, respectively, △ABC is isos with AB = AC, $\overline{BM}$ and $\overline{CN}$ are medians.	1. Given
2. $\frac{1}{2}AB = \frac{1}{2}AC$	2. Mult Prop of Eq
3. $BN = \frac{1}{2}AB$, $CM = \frac{1}{2}AC$	3. Def of median
4. BM = CN	4. Sub
5. ∠ABC ≅ ∠ACB	5. ∠opp ≅ sides are ≅.
6. $\overline{BC} \cong \overline{CB}$	6. Reflex Prop of Congr
7. △NBC ≅ △MCB	7. SAS
8. $\overline{BM} \cong \overline{CN}$	8. CPCTC

21. False, except in the special case of an equil △. Isos △ABC with AB = AC, AB = BC, and ∠ bis $\overline{BM}$ (Given). Assume that $\overline{BM}$ is also an alt. Then △BMA ≅ △BMC by ASA. So ∠A ≅ ∠C by ASA; hence △ABC is equil with AB = BC. But this contradicts the given information, so the assumption that $\overline{BM}$ is an alt is false.

PAGE 251

18.

Statement	Reason
1. ▱ABDE, $\overline{FE}$ ≅ $\overline{CB}$	1. Given
2. ∠DEA ≅ ∠DBA	2. Opp ∠s of ▱ are ≅.
3. ∠FEA and ∠DEA are supp, ∠CBD and ∠DBA are supp.	3. If outer rays of 2 adj ∠s form a st ∠, then ∠s are supp.
4. ∠CBD ≅ ∠FEA	4. Supp ∠s of ≅ ∠s are ≅.
5. $\overline{AE}$ ≅ $\overline{DB}$	5. Opp sides of ▱ are ≅.
6. △AEF ≅ △DBC	6. SAS

19.

Statement	Reason
1. $\overline{AF}$ ∥ $\overline{CD}$, △AFE ≅ △DCB	1. Given
2. $\overline{AF}$ ≅ $\overline{CD}$, $\overline{AE}$ ≅ $\overline{DB}$, $\overline{FE}$ ≅ $\overline{CB}$	2. CPCTC
3. ACDF is ▱.	3. If 1 pair opp sides are both ∥ and ≅, then quad is ▱.
4. FD = AC	4. Def of ▱
5. FD − FE = AC − CB	5. Equations may be subtracted.
6. FD − FE = ED, AC − CB = AB	6. Seg Add Post
7. ED = AB ($\overline{ED}$ ≅ $\overline{AB}$)	7. Sub
8. ABDE is ▱.	8. If both pairs of opp sides are ≅, then quad is ▱.

20.

Statement	Reason
1. △GHJ and △IHJ are equil.	1. Given
2. $\overline{GH}$ ≅ $\overline{GJ}$ ≅ $\overline{HJ}$, $\overline{IJ}$ ≅ $\overline{IH}$ ≅ $\overline{HJ}$	2. Def of equil △.
2. $\overline{GH}$ ≅ $\overline{GJ}$ $\overline{IJ}$ ≅ $\overline{IH}$	3. Trans Prop of Congr
4. GHIJ is ▱.	4. If 2 pair opp sides are ≅, then quad is ▱.

21. True

Statement	Reason
1. Quad ABCD, ∠A and ∠D are supp, ∠D and ∠C are supp	1. Given
2. $\overline{AB}$ ∥ $\overline{CD}$, $\overline{BC}$ ∥ $\overline{AD}$	2. If int ∠s on same side of transv are supp, then lines are ∥.
3. ABCD is a ▱.	3. Def of ▱

22. False

23. False

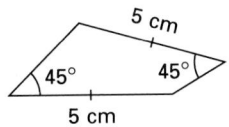

24. True

Statement	Reason
1. Quad ABCD, AB + BC = K, BC + CD = K, CD + DA = K	1. Given
2. AB + BC = BC + CD, BC + CD = CD + DA	2. Sub
3. AB = CD, BC = DA	3. Subt Prop of Eq
4. ABCD is ▱.	4. If 2 pair opp sides are ≅, then quad is ▱.

25. True

Statement	Reason
1. Quad ABCD, m ∠A + m ∠B = K, m ∠B + m ∠C = K, m ∠C + m ∠D = K	1. Given
2. m ∠A + m ∠B = m ∠B + m ∠C, m ∠B + m ∠C = m ∠C + m ∠D	2. Sub
3. m ∠A = m ∠C, m ∠B = m ∠D	3. Subt Prop of Eq
4. ABCD is ▱.	4. If both pair opp ∠s are ≅, then quad is ▱.

PAGE 255

20.

Statement	Reason
1. U is midpt of $\overline{RS}$, V is midpt of $\overline{ST}$, $\overline{UV}$ ≅ $\overline{SV}$	1. Given
2. $VU = \frac{1}{2}RT$	2. Midseg Thm
3. $VS = \frac{1}{2}TS$	3. def of midpt
4. $\frac{1}{2}RT = \frac{1}{2}TS$	4. Sub
5. RT = TS	5. Mult Prop of Eq
6. △RST is isos.	6. Def of isos △

21.

Statement	Reason
1. Quad ABCD, E, F, G, H are midpts of $\overline{AB}$, $\overline{BC}$, $\overline{DC}$, $\overline{DA}$, respectively.	1. Given
2. $\overline{GF}$ ∥ $\overline{DB}$, $GF = \frac{1}{2}DB$, $\overline{HE}$ ∥ $\overline{DB}$, $HE = \frac{1}{2}DB$	2. Midseg Thm
3. $\overline{HE}$ ∥ $\overline{GF}$	3. 2 lines ∥ to the same line are ∥.
4. HE = GF	4. Sub
5. EFGH is ▱.	5. If same pair opp sides are ∥ and ≅, then quad is ▱.
6. $\overline{HF}$ and $\overline{GE}$ bis each other	6. Diags of ▱ bis each other.

22.

Statement	Reason
1. H, I, and J are the midpts of $\overline{DF}$, $\overline{EF}$, and $\overline{GF}$, respectively.	1. Given
2. $HI = \frac{1}{2}DE$, $JI = \frac{1}{2}GE$, $HJ = \frac{1}{2}DG$	2. Def of Midpt
3. Construct $\overline{KL}$ such that K is midpt of $\overline{DG}$ and L is midpt of $\overline{DE}$.	3. Between any two pts exactly one line can be drawn.
4. $DL = \frac{1}{2}DE$, $KL = \frac{1}{2}GE$, $DK = \frac{1}{2}DG$	4. Midseg Thm
5. $HI = DL$, $JI = KL$, $HJ = DK$	5. Sub
6. $\triangle JHI \cong \triangle KDL$	6. SSS
7. $\angle JHI \cong \angle KDL$	7. CPCTC

23.

Statement	Reason
1. H, I, J, K, and L are midpts of $\overline{DF}$, $\overline{EF}$, $\overline{GF}$, $\overline{DG}$, and $\overline{DE}$, respectively.	1. Given
2. $JH = \frac{1}{2}DG$, $HI = \frac{1}{2}DE$, $JI = \frac{1}{2}GE$, $KL = \frac{1}{2}GE$	2. Midseg Thm
3. $DK = \frac{1}{2}DG$, $DL = \frac{1}{2}DE$	3. Def of midpt
4. $JH = DK$, $HI = DL$, $JI = KL$	4. Sub
5. $\triangle JHI \cong \triangle KDL$	5. SSS

6.

Statement	Reason
1. $\triangle GEF \cong \triangle CED$, $AF \cong BD$	1. Given
2. $\overline{FG} \cong \overline{DC}$	2. $FG = DC$
3. $\overline{GE} \cong \overline{CE}$	3. $GE = CE$
4. $\overline{FE} \cong \overline{DE}$	4. $FE = DE$
5. $\angle F \cong \angle D$, $\angle FGE \cong \angle DCE$	5. CPCTC
6. $FE + CE = DE + GE$	6. Equations may be added.
7. $FE + CE = FC$, $DE + GE = GD$	7. Seg Add Post
8. $FC = GD$	8. Sub
9. $AF - FG = BD - DC$	9. Equations may be subtracted.
10. $AF - FG = AG$, $BD - DC = BC$	10. Seg Add Post
11. $AG = BC$	11. Sub
12. $\angle FGE$ and $\angle AGE$ are supp. $\angle DCE$ and $\angle BCE$ are supp	12. If outer rays of 2 adj $\angle$s form st $\angle$, then $\angle$s are supp.
13. $\angle AGE \cong \angle BCE$	13. Supp $\angle$s of $\cong \angle$s are $\cong$.
14. Quad $ABDG \cong$ quad $BAFC$	14. SASAS

7.

Statement	Reason
1. $\overline{TX} \cong \overline{WU}$, $\triangle TZY \cong \triangle WZV$	1. Given
2. $\overline{YZ} \cong \overline{VZ}$, $\overline{TZ} \cong \overline{WZ}$, $\overline{TY} \cong \overline{WV}$, $\angle TYZ \cong \angle WVZ$	2. CPCTC
3. $TX - TY = WU - WV$	3. Equations may be subtracted.
4. $TX - TY = YX$, $WU - WV = VU$	4. Seg Add Post
5. $YX = VU$	5. Sub
6. $\angle XYZ$ and $\angle TYZ$ are supp, $\angle UVZ$ and $\angle WVZ$ are supp.	6. If outer rays of 2 adj $\angle$s form st $\angle$, then $\angle$s are supp.
7. $\angle XYZ \cong \angle UVZ$	7. Supp of $\cong \angle$s are $\cong$.
8. $\angle YZW \cong \angle VZT$	8. Vert $\angle$s are $\cong$.
9. Quad $TUVZ \cong$ quad $WXYZ$	9. SASAS

8.

Statement	Reason
1. $\triangle TUW \cong \triangle WXT$, $\overline{TY} \cong \overline{XY}$, $\overline{UV} \cong \overline{WV}$, $\overline{TZ} \cong \overline{WZ}$	1. Given
2. $\angle ZWX \cong \angle ZTU$, $\overline{WX} \cong \overline{TU}$, $\angle WXY \cong \angle TUV$, $\overline{XT} \cong \overline{UW}$	2. CPCTC
3. $XY = \frac{1}{2}XT$, $UV = \frac{1}{2}UW$	3. Def of midpt
4. $XY = \frac{1}{2}XT$, $UV = \frac{1}{2}XT$	4. Sub
5. $XY = UV$	5. Mult Prop of Eq

9.

Statement	Reason
1. $\square TUWX$, $\overline{YV}$ and $\overline{TW}$ bis each other at Z.	1. Given
2. $\overline{TU} \cong \overline{XW}$	2. Opp sides of $\square$ are $\cong$.
3. $\overline{TU} \parallel \overline{XW}$	3. Def of $\square$
4. $\angle XWZ \cong \angle UTZ$	4. If lines are $\parallel$, then alt int $\angle$s are $\cong$.
5. $\overline{WZ} \cong \overline{TZ}$, $\overline{YZ} \cong \overline{VZ}$	5. Def of seg bis
6. $\angle WZY \cong \angle TZV$	6. Vert $\angle$s are $\cong$.
7. Quad $TUVZ \cong$ quad $WXYZ$	7. SASAS

10.

Statement	Reason
1. $\overline{AB} \cong \overline{EF}$, $\overline{BC} \cong \overline{FG}$, $\overline{CD} \cong \overline{GH}$, $\angle B \cong \angle F$, $\angle C \cong \angle G$	1. Given
2. Draw $\overline{AC}$, $\overline{EG}$	2. 2 pts determine a line.
3. $\triangle ABC \cong \triangle EFG$	3. SAS
4. $\overline{AC} \cong \overline{EG}$, $\angle 1 \cong \angle 5$, $\angle 2 \cong \angle 6$	4. CPCTC
5. $m\angle 2 + m\angle 3 = m\angle BCD$, $m\angle 6 + m\angle 7 = m\angle FGH$	5. $\angle$ Add Post
6. $m\angle 2 + m\angle 3 = m\angle 6 + m\angle 7$	6. Sub
7. $m\angle 3 = m\angle 7$	7. Equations may be subtracted.
8. $\triangle ACD \cong \triangle EGH$	8. SAS
9. $\overline{AD} \cong \overline{EH}$, $\angle D \cong \angle H$, $\angle 4 \cong \angle 8$	9. CPCTC
10. $m\angle 4 + m\angle 1 = m\angle 8 + m\angle 5$	10. Equations may be added.
11. $m\angle 4 + m\angle 1 = m\angle BAD$, $m\angle 8 + m\angle 5 = m\angle FEH$	11. Angle Add Post
12. $m\angle BAD = m\angle FEH$	12. Sub
13. Quad $ABCD \cong$ quad $EFGH$	13. Def of $\cong$ quads

11.

Statement	Reason
1. $\angle F \cong \angle J$, $\angle G \cong$ $\angle K$, $\angle H \cong \angle L$, $\overline{FG}$ $\cong \overline{JK}$, $\overline{GH} \cong \overline{KL}$	1. Given
2. Draw $\overline{HF}$, $\overline{LJ}$.	2. 2 pts determine a line.
3. $\triangle HFG \cong \triangle LJK$	3. SAS
4. m $\angle GHF =$ m $\angle KLJ$, m $\angle GFH =$ m $\angle KJL$, $\overline{HF} \cong \overline{LJ}$	4. CPCTC
5. m $\angle GHF +$ m $\angle EHF =$ m $\angle EHG$, m $\angle KLJ +$ m $\angle ILJ$ $=$ m $\angle ILK$, m $\angle GFH +$ m $\angle EFH$ $=$ m $\angle EFG$, m $\angle KJL +$ m $\angle IJL =$ m $\angle IJK$	5. Angle Add Post
6. m $\angle GHF +$ m $\angle EHF =$ m $\angle KLJ$ $+$ m $\angle ILJ$, m $\angle GFH +$ m $\angle EFH$ $=$ m $\angle KJL +$ m $\angle IJL$	6. Sub
7. m $\angle EHF =$ m $\angle ILJ$, m $\angle EFH =$ m $\angle IJL$	7. Equations may be subtracted.
8. $\triangle HEF \cong \triangle LIJ$	8. ASA
9. $\overline{HE} \cong \overline{LI}$	9. CPCTC
10. Quad $EFGH \cong$ quad $IJKL$	10. SASAS

12.

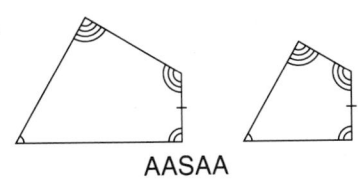

AASAA

AASAA and ASSAS are *not* congruence patterns. Counterexamples:

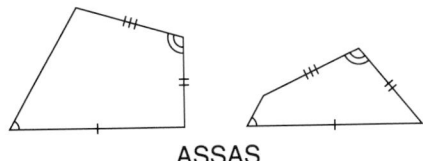

ASSAS

SSAAS, AASSA and SSASS *are* congruence patterns.

Sample proof (AASSA):

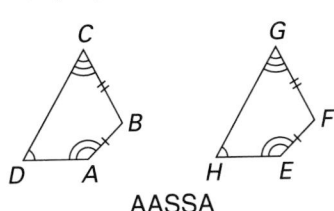

AASSA

12.

Statement	Reason
1. $\overline{AB} \cong \overline{EF}$, $\overline{BC} \cong$ $\overline{FG}$, $\angle D \cong \angle H$, $\angle A$ $\cong \angle E$, $\angle C \cong \angle G$	1. Given
2. Draw $\overline{AC}$ and $\overline{EG}$	2. 2 pts determine a line.
3. m $\angle DAC +$ m $\angle DCA +$ m $\angle D =$ 180, m $\angle HEG +$ m $\angle HGE +$ m $\angle H =$ 180	3. Sum of $\angle$ meas in $\triangle$ $= 180$.
4. m $\angle DAC +$ m $\angle DCA +$ m $\angle D =$ m $\angle HEG +$ m $\angle HGE +$ m $\angle H$	4. Sub
5. m $\angle DAC +$ m $\angle DCA =$ m $\angle HEG$ $+$ m $\angle HGE$	5. Equations may be subtracted.
6. m $\angle DAC +$ m $\angle BAC =$ m $\angle DAB$, m $\angle HEG +$ m $\angle DCA =$ m $\angle HEF$, m $\angle DCA +$ m $\angle ACB =$ m $\angle C$, m $\angle HGE +$ m $\angle EGF$ $=$ m $\angle FGA$	6. Angle Add Post
7. m $\angle DAC +$ m $\angle BAC =$ m $\angle HEG$ $+$ m $\angle DCA$, m $\angle DCA +$ m $\angle ACB$ $=$ m $\angle HGE +$ m $\angle EGF$	7. Sub
8. m $\angle DAC +$ m $\angle BAC +$ m $\angle DCA$ $+$ m $\angle ACB =$ m $\angle HEG +$ m $\angle DCA$ $+$ m $\angle HGE +$ m $\angle EGF$	8. Equations may be added.
9. m $\angle BAC +$ m $\angle ACB =$ m $\angle DCA$ $+$ m $\angle EGF$	9. Equations may be subtracted.
10. m $\angle B +$ m $\angle BAC$ $+$ m $\angle ACB =$ 180, m $\angle F +$ m $\angle DCA$ $+$ m $\angle EGF =$ 180	10. Sum of $\angle$ meas in $\triangle$ $= 180$.
11. m $\angle B =$ m $\angle F$	11. Equations may be subtracted.
12. Quad $ABCD \cong$ quad $EFGH$	12. ASASA

19. Sufficient

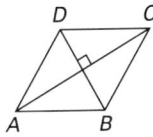

Statement	Reason
1. Quad $ABCD$, $\overline{BD}$ and $\overline{AC}$ bis each other, $\overline{BD} \perp \overline{AC}$	1. Given
2. $ABCD$ is $\square$.	2. If diags bis each other, then quad is $\square$.
3. $ABCD$ is rhom.	3. $\square$ with $\perp$ diags is rhom.

20. Not sufficient

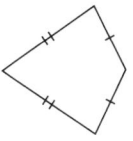

21. Not sufficient

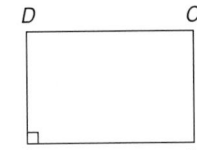

22. Not sufficient

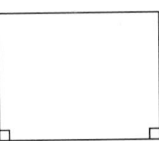

23. Sufficient

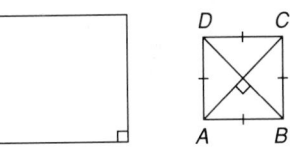

Statement	Reason
1. Rect $ABCD$ with $\overline{AC} \perp$ $\overline{DB}$	1. Given
2. $ABCD$ is $\square$.	2. Def rect
3. $ABCD$ is rhom.	3. $\square$ with $\perp$ diags is rhom.
4. $\overline{AB} \cong \overline{BC} \cong$ $\overline{CD} \cong \overline{DA}$	4. Def rhom
5. $ABCD$ is sq.	5. Def sq

24. Sufficient

Statement	Reason
1. Rhom $ABCD$ $\overline{AC}$ $\cong \overline{BD}$	1. Given
2. $ABCD$ is $\square$.	2. Def of rhom
3. $ABCD$ is rect.	3. $\square$ with $\cong$ diags is rect.
4. $\overline{AB} \cong \overline{BC} \cong$ $\overline{CD} \cong \overline{DA}$	4. Def rhom
5. $ABCD$ is sq.	5. Def sq

657

25. A *necessary* condition is true of all quads that are rects, but not enough to prove quad is a rectangle. A *sufficient* condition is enough for quad to be rectangle.

26.

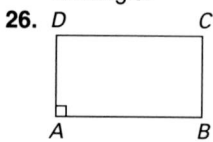

Statement	Reason
1. ▱ABCD, m ∠A = 90	1. Given
2. m ∠B = 90, m ∠D = 90	2. Consec ∠s of ▱ are supp.
3. m ∠C = 90	3. Opp ∠s of ▱ are ≅.
4. ▱ABCD is rect.	4. Def rect

27.

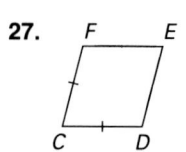

Statement	Reason
1. ▱ CDEF, $\overline{CD} \cong \overline{CF}$	1. Given
2. $\overline{CD} \cong \overline{FE}$, $\overline{CF} \cong \overline{DE}$	2. Opp sides of ▱ are ≅.
3. $\overline{CD} \cong \overline{DE} \cong \overline{FE} \cong \overline{CF}$	3. Trans Prop of Congr
4. ▱CDEF is rhom.	4. Def of rhom

28.

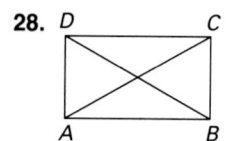

Statement	Reason
1. ▱ ABCD, $\overline{AC} \cong \overline{DB}$	1. Given
2. $\overline{DA} \cong \overline{CB}$	2. Opp sides of ▱ are ≅.
3. $\overline{AB} \cong \overline{AB}$.	3. Reflex Prop of Congr
4. △DAB ≅ △CBA	4. SSS
5. m ∠DAB = m ∠CBA	5. CPCTC
6. m ∠DAB + m ∠CBA = 180	6. Consec ∠s of ▱ are supp.
7. 2 × m ∠CBA = 180	7. Sub
8. m ∠CBA = 90	8. Div Prop of Eq
9. ▱ ABCD is rect.	9. ▱ with 1 rt ∠ is rect.

29.

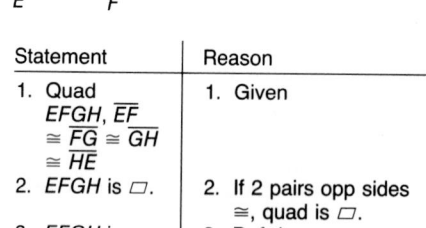

Statement	Reason
1. ▱ ABCD, $\overline{DB}$ bis opp ∠s	1. Given
2. ∠1 ≅ ∠2, ∠3 ≅ ∠4	2. Def ∠ bis
3. $\overline{BD} \cong \overline{BD}$	3. Reflex Prop of Congr
4. △ABD ≅ △CBD	4. ASA
5. $\overline{AB} \cong \overline{CB}$	5. CPCTC
6. ▱ABCD is rhom.	6. ▱ with adj ≅ sides is rhom.

30.

Statement	Reason
1. Quad EFGH, $\overline{EF} \cong \overline{FG} \cong \overline{GH} \cong \overline{HE}$	1. Given
2. EFGH is ▱.	2. If 2 pairs opp sides ≅, quad is ▱.
3. EFGH is rhom.	3. Def rhom

31. Sufficient

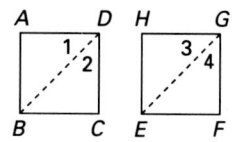

Statement	Reason
1. Sq ABCD and sq EFGH, $\overline{BD} \cong \overline{EG}$	1. Given
2. m ∠A = 90, m ∠ADC = 90, m ∠H = 90, m ∠HGF = 90, m ∠ABC = 90, m ∠HEF = 90	2. Def sq
3. m ∠1 = m ∠2, m ∠3 = m ∠4	3. Diags of rhom bis opp ∠s.
4. m ∠1 + m ∠2 = m ∠ADC, m ∠3 + m ∠4 = m ∠HGF	4. Angle Add Post
5. 2 (m ∠1) = 90, 2 (m ∠3) = 90	5. Sub
6. m ∠1 = 45, m ∠3 = 45	6. Div Prop of Eq
7. △ABD ≅ △HEG	7. AAS
8. $\overline{AD} \cong \overline{HG}$, $\overline{AB} \cong \overline{HE}$	8. CPCTC
9. ABCD ≅ EFGH	9. ASASA

32. Not sufficient

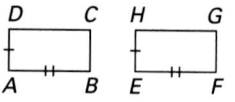

33. Sufficient

Statement	Reason
1. Rects ABCD and EFGH, $\overline{DA} \cong \overline{HE}$, $\overline{AB} \cong \overline{EF}$	1. Given
2. m ∠D = 90, m ∠A = 90, m ∠B = 90, m ∠H = 90, m ∠E = 90, m ∠F = 90	2. Def rect
3. m ∠D = m ∠H, m ∠A = m ∠E, m ∠B = m ∠F	3. Sub
4. ABCD ≅ EFGH	4. ASASA

34. Sufficient

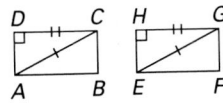

Statement	Reason
1. Rect *ABCD* and *EFGH*, $\overline{AC} \cong \overline{EG}$, $\overline{DC} \cong \overline{HG}$	1. Given
2. m ∠*HGF* = 90, m ∠*H* = 90, m ∠*HEF* = 90, m ∠*DCB* = 90, m ∠*D* = 90, m ∠*DAB* = 90	2. Def rect
3. m ∠*HGF* = m ∠*DCB*, m ∠*H* = m ∠*D*, m ∠*HEF* = m ∠*DAB*	3. Sub
4. △*DAC* ≅ △*HEG*	4. HL
5. $\overline{DA} \cong \overline{HE}$	5. CPCTC
6. *ABCD* ≅ *EFGH*	6. ASASA

Midchapter Review

1.

Statement	Reason
1. Rects *ABCD* and *WXYZ*, $\overline{AB} \cong \overline{WX}$	1. Given
2. $\overline{DC} \cong \overline{AB}$, $\overline{ZY} \cong \overline{WX}$	2. Opp sides of rect are ≅.
3. $\overline{AB} \cong \overline{CD} \cong \overline{WX} \cong \overline{ZY}$	3. Trans Prop of Congr
4. All ∠s are rt ∠s.	4. All ∠s of rect are rt ∠s.
5. $\overline{AD} \parallel \overline{WZ} \parallel \overline{BC} \parallel \overline{XY}$	5. 2 lines ⊥ same line are ∥.
6. $\overline{AX} \parallel \overline{DY}$	6. Opp sides of rect are ∥.
7. $\overline{AD} \cong \overline{WZ} \cong \overline{BC} \cong \overline{XY}$	7. ∥ lines are equidist at all pts.
8. Quad *ABCD* ≅ quad *WXYZ*	8. ASASA

PAGE 289

24.

Statement	Reason
1. Trap *ABCD* with median $\overline{GF}$, $\overline{AC}$ intersects $\overline{GF}$ at *H*, $\overline{DE} \perp \overline{AC}$ at *H*	1. Given
2. *G* is midpt. $\overline{AD}$	2. Def of median
3. $\overline{DC} \parallel \overline{GF} \parallel \overline{AB}$	3. Median of trap is ∥ to bases.
4. $\overline{HD} \cong \overline{HE}$	4. If 3 ∥ lines cut ≅ segs on 1 transv, they cut ≅ segs on every transv.
5. ∠*DHA* ≅ ∠*EHA*	5. Rt ∠s are ≅.
6. $\overline{AH} \cong \overline{AH}$	6. Reflex Prop of Congr
7. △*AHD* ≅ △*AHE*	7. SAS
8. ∠*DAH* ≅ ∠*EAH*	8. CPCTC
9. ∠*EAH* ≅ ∠*GHA*	9. Alt int ∠s of ∥ lines are ≅.
10. ∠*DAH* ≅ ∠*GHA*	10. Trans Prop of Congr
11. $\overline{GA} \cong \overline{GH}$	11. Sides opp ≅ ∠s of △ are ≅.
12. △*AHG* is isos.	12. Def isos △

25. Counterexample

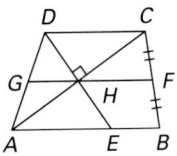

Mixed Review

1.

Statement	Reason
1. Isos △*ABC* with $\overline{AC} \cong \overline{BC}$	1. Given
2. Draw seg from midpt *M* of $\overline{AB}$ to *C*.	2. 2 pts determine line.
3. $\overline{AM} \cong \overline{MB}$	3. Def midpt
4. $\overline{CM} \cong \overline{CM}$	4. Reflex Prop of Congr
5. △*AMC* ≅ △*BMC*	5. SSS
6. ∠*A* ≅ ∠*B*	6. CPCTC

2.

Statement	Reason
1. ▱ *ABCD*	1. Given
2. Draw $\overline{DB}$.	2. 2 pts determine line.
3. $\overline{DB} \cong \overline{DB}$	3. Reflex Prop of Congr
4. $\overline{AB} \cong \overline{CD}$, $\overline{DA} \cong \overline{BC}$	4. Opp sides of ▱ are ≅.
5. △*ABD* ≅ △*CDB*.	5. SSS
6. ∠*A* ≅ ∠*C*	6. CPCTC

PAGE 317

21.

Statement	Reason
1. Rt △*ABC*, $\overline{ED} \perp \overline{AB}$, $\overline{BE}$ bis ∠*ABC*	1. Given
2. m ∠*EDB* = 90	2. Def of ⊥
3. m ∠*ABC* = 90	3. Def rt △
4. m ∠*CBE* + m ∠*DBE* = m ∠*ABC*	4. Angle Add Post
5. m ∠*CBE* + m ∠*DBE* = 90	5. Sub
6. m ∠*CBE* = m ∠*DBE*	6. Def of ∠ bis
7. 2 × m ∠*DBE* = 90	7. Sub
8. m ∠*DBE* = 45	8. Div Prop of Eq
9. m ∠*DBE* + m ∠*EDB* + m ∠*DEB* = 180	9. Sum of meas of ∠s of △ = 180.
10. 45 + 90 + m ∠*DEB* = 180	10. Sub
11. m ∠*DEB* = 45	11. Subt Prop of Eq
12. m ∠*DBE* = m ∠*DEB*	12. Sub
13. *ED* = *BD* ($\overline{ED} \cong \overline{BD}$)	13. Sides opp ≅ ∠s of △ are ≅.
14. ∠*A* ≅ ∠*A*	14. Reflex Prop of Congr
15. m ∠*EDA* = 90	15. Def of ⊥
16. m ∠*EDA* = m ∠*ABC*	16. Sub
17. △*ADE* ~ △*ABC*	17. AA~
18. $\frac{AE}{AC} = \frac{DE}{BC}$	18. Def of ~ △s
19. $\frac{AE}{AC} = \frac{BD}{BC}$	19. Sub

22.

Statement	Reason
1. Rt $\triangle ABC$, $\overline{ED} \perp \overline{AB}$, $\angle DEB \cong \angle EBA$	1. Given
2. m $\angle ABC$ = 90	2. Def rt $\triangle$
3. m $\angle EDA$ = 90	3. Def of $\perp$
4. m $\angle ABC$ = m $\angle EDA$	4. Sub
5. $\angle A \cong \angle A$	5. Reflex Prop of Congr
6. $\triangle ADE \sim \triangle ABC$	6. AA$\sim$
7. $ED = BD$ ($\overline{ED} \cong \overline{BD}$)	7. Sides opp $\cong \angle$s of $\triangle$ are $\cong$.
8. $\frac{ED}{BC} = \frac{AD}{AB}$	8. Def of $\sim \triangle$s
9. $AD = AB - BD$	9. Seg Add Post
10. $AD = AB - ED$	10. Sub
11. $\frac{ED}{BC} = \frac{AB - ED}{AB}$	11. Sub
12. $\frac{ED}{BC} = \frac{AB}{AB} - \frac{ED}{AB}$	12. Distr Prop
13. $\frac{ED}{BC} = 1 - \frac{ED}{AB}$	13. A nonzero number div by itself = 1.
14. $\frac{ED}{BC} + \frac{ED}{AB} = 1$	14. Add Prop of Eq
15. $\frac{1}{ED}(\frac{ED}{BC} + \frac{ED}{AB}) = \frac{1}{ED} \cdot 1$	15. Mult Prop of Eq
16. $\frac{1}{BD} + \frac{1}{AB} = \frac{1}{ED}$	16. Distr Prop

23.

Statement	Reason
1. $\triangle ABC \cong \triangle DEF$	1. Given
2. $\angle A \cong \angle D$, $\angle B \cong \angle E$, $\angle C \cong \angle F$, $AB = DE$, $BC = EF$, $CA = FD$	2. Def of $\cong \triangle$s
3. $\frac{AB}{AB} = \frac{BC}{BC} = \frac{CA}{CA}$	3. A nonzero number div by itself = 1.
4. $\frac{AB}{DE} = \frac{BC}{EF} = \frac{CA}{FD}$	4. Sub
5. $\triangle ABC \sim \triangle DEF$	5. Def of $\sim \triangle$s

24. There are 2 possibilities.

First:

Statement	Reason
1. $\triangle ABC \sim \triangle DEF$, $\triangle DEF \sim \triangle GHI$	1. Given
2. m $\angle A =$ m $\angle D$, m $\angle D =$ m $\angle G$, m $\angle B =$ m $\angle E$, m $\angle E =$ m $\angle H$, m $\angle C =$ m $\angle F$, m $\angle F =$ m $\angle I$	2. Def of $\sim \triangle$s
3. m $\angle A =$ m $\angle G$, m $\angle B =$ m $\angle H$, m $\angle C =$ m $\angle I$	3. Sub
4. $\triangle ABC \sim \triangle GHI$	4. AA$\sim$

Second:

Statement	Reason
1. $\triangle ABC \cong \triangle DEF$	1. Given
2. $\frac{AB}{DE} = \frac{BC}{EF} = \frac{CA}{FD}$, $\frac{DE}{GH} = \frac{EF}{HI} = \frac{FD}{IG}$	2. Def of $\sim \triangle$s
3. $\frac{AB}{BC} = \frac{DE}{EF}$, $\frac{BC}{CA} = \frac{EF}{FD}$, $\frac{CA}{AB} = \frac{FD}{DE}$, $\frac{DE}{EF} = \frac{GH}{HI}$, $\frac{EF}{FD} = \frac{HI}{IG}$, $\frac{FD}{DE} = \frac{GH}{IG}$	3. If $\frac{a}{b} = \frac{c}{d}$, then $\frac{a}{c} = \frac{b}{d}$
4. $\frac{AB}{BC} = \frac{GH}{HI}$, $\frac{BC}{CA} = \frac{HI}{IG}$, $\frac{AB}{CA} = \frac{GH}{IG}$	4. Sub
5. $\frac{AB}{GH} = \frac{BC}{HI}$, $\frac{BC}{GH} = \frac{CA}{IG}$, $\frac{AB}{GH} = \frac{CA}{IG}$	5. If $\frac{a}{b} = \frac{c}{d}$, then $\frac{a}{c} = \frac{b}{d}$
6. $\frac{AB}{GH} = \frac{BC}{HI} = \frac{CA}{IG}$	6. Sub
7. $\triangle ABC \sim \triangle GHI$	7. Def of $\sim \triangle$s

25.

Statement	Reason
1. $\square XYZW$	1. Given
2. $\overline{WZ} \parallel \overline{XY}$	2. Opp sides of $\square$ are $\parallel$.
3. $\angle 1 \cong \angle 3$, $\angle 2 \cong \angle 4$,	3. If lines are $\parallel$, then alt int $\angle$s are $\cong$.
4. $\triangle XRQ \sim \triangle ZPQ$	4. AA$\sim$
5. $\frac{XQ}{ZQ} = \frac{RQ}{PQ}$	5. Def of $\sim \triangle$s
6. $XQ \cdot PQ = ZQ \cdot RQ$	6. If proport, then prod of means = prod of extremes.

21.

Statement	Reason
1. $\triangle ABC$ with D between A and C, and E between B and C, $\frac{AD}{DC} = \frac{BE}{EC}$	1. Given
2. Draw a line through D $\parallel$ to $\overline{AB}$.	2. Parallel Post
3. Label the inters of this line and $\overline{CB}$ F so that $\overline{DF} \parallel \overline{AB}$.	3. 2 lines inters in 1 pt.
4. $\frac{DC}{AC} = \frac{FC}{BC}$	4. Triangle Proport Thm
5. $\frac{AD + DC}{DC} = \frac{BE + EC}{EC}$	5. If $\frac{a}{b} = \frac{c}{d}$, then $\frac{a + b}{b} = \frac{c + d}{d}$
6. $AD + DC = AC$, $BE + EC = BC$	6. Seg Add Post
7. $\frac{AC}{DC} = \frac{BC}{EC}$	7. Sub
8. $\frac{DC}{AC} = \frac{EC}{BC}$	8. If $\frac{a}{b} = \frac{c}{d}$, then $\frac{b}{a} = \frac{d}{c}$
9. $\frac{FC}{BC} = \frac{EC}{BC}$	9. Sub
10. $FC = EC$	10. Mult Prop of Eq
11. Pt F is pt E.	11. Ruler Post
12. $\overleftrightarrow{DE}$ and $\overleftrightarrow{DF}$ are same line.	12. 2 pts determine a line.
13. $\overline{DE} \parallel \overline{AB}$	13. Sub

25. $\overleftrightarrow{AB} \parallel \overleftrightarrow{CD} \parallel \overleftrightarrow{EF}$, with transv l and m (Given); There are 2 possibilities. See page 661 for proofs.

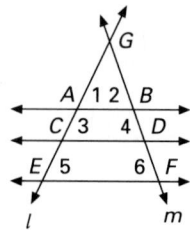

First:

Statement	Reason
1. l and m intersect at G.	1. 2 lines inters in at most 1 pt.
2. $m\angle 1 = m\angle 3$, $m\angle 2 = m\angle 4$, $m\angle 3 = m\angle 5$, $m\angle 4 = m\angle 6$	2. If 2 ∥ lines are inters by a trans, then corr ∠s are ≅.
3. $\triangle GAB \sim \triangle GCD$, $\triangle GCD \sim \triangle GEF$	3. AA~
4. $\triangle GAB \sim \triangle GCD \sim \triangle GEF$	4. Trans Prop of ~
5. $\frac{EG}{CG} = \frac{FG}{DG}$	5. Def of ~ △s
6. $\frac{EG - CG}{CG} = \frac{FG - DG}{DG}$	6. If $\frac{a}{b} = \frac{c}{d}$, then $\frac{a-b}{b} = \frac{c-d}{d}$
7. $EG - CG = EC$, $FG - DG = FD$	7. Seg Add Post
8. $\frac{EC}{CG} = \frac{FD}{DG}$	8. Sub
9. $\frac{EC}{FD} = \frac{CG}{DG}$	9. If $\frac{a}{b} = \frac{c}{d}$, then $\frac{a}{c} = \frac{b}{d}$
10. $\frac{CG}{AG} = \frac{DG}{BG}$	10. Def of ~ △s
11. $\frac{CG - AG}{AG} = \frac{DG - BG}{BG}$	11. If $\frac{a}{b} = \frac{c}{d}$, then $\frac{a-b}{b} = \frac{c-d}{d}$
12. $CG - AG = CA$, $DG - BG = BD$	12. Seg Add Post
13. $\frac{CA}{AG} = \frac{BD}{BG}$	13. Sub
14. $\frac{AG}{CA} = \frac{BG}{BD}$	14. If $\frac{a}{b} = \frac{c}{d}$, then $\frac{b}{a} = \frac{d}{c}$
15. $\frac{AG + CA}{CA} = \frac{BG + BD}{BD}$	15. If $\frac{a}{b} = \frac{c}{d}$, then $\frac{a+b}{b} = \frac{c+d}{d}$
16. $AG + CA = CG$, $BG + BD = GD$	16. Angle Add Post
17. $\frac{CG}{CA} = \frac{GD}{DB}$	17. Sub
18. $\frac{CG}{DG} = \frac{CA}{DB}$	18. If $\frac{a}{b} = \frac{c}{d}$, then $\frac{a}{c} = \frac{b}{d}$
19. $\frac{EC}{FD} = \frac{CA}{DB}$	19. Sub
20. $\frac{EC}{CA} = \frac{FD}{DB}$	20. If $\frac{a}{b} = \frac{c}{d}$, then $\frac{a}{c} = \frac{b}{d}$

```
   A  1  2  B
 ◄──┼──┼──►
  C  3  4  D
 ◄──┼──┼──►
  E  5  6  F
 ◄──┼──┼──►
    l    m
```

Second:

Statement	Reason
1. $l \parallel m$	1. Parallel Post
2. $ABDC$, $CDFE$ are ▱s.	2. Def of ▱
3. $AC = BD$, $CE = DF$ ($\overline{AC} \cong \overline{BD}$, $\overline{CE} \cong \overline{DF}$)	3. Opp sides of ▱ are ≅.
4. $\frac{AC}{AC} = \frac{CE}{CE} = 1$	4. Any nonzero number div by itself = 1.
5. $\frac{AC}{BD} = \frac{CE}{DF}$	5. Sub
6. $\frac{AC}{CE} = \frac{BD}{DF}$	6. If $\frac{a}{b} = \frac{c}{d}$, then $\frac{a}{c} = \frac{b}{d}$

26.

Statement	Reason
1. $\overline{DE} \parallel \overline{AB}$, $\overline{FE} \parallel \overline{CB}$	1. Given
2. $\frac{AD}{DG} = \frac{BE}{EG}$, $\frac{BE}{EG} = \frac{CF}{FG}$	2. Triangle Proport Thm
3. $\frac{AD}{DG} = \frac{CF}{FG}$	3. Sub
4. $\overline{FD} \parallel \overline{CA}$	4. Parallel Post

27.

Statement	Reason
1. $\overline{FD} \parallel \overline{CA}$, $\overline{DE} \parallel \overline{AB}$	1. Given
2. $\frac{GF}{FC} = \frac{GD}{DA}$, $\frac{GD}{DA} = \frac{GE}{EB}$	2. Triangle Proport Thm
3. $\frac{GF}{FC} = \frac{GE}{EB}$	3. Sub

28.

Statement	Reason
1. $\triangle EFG$ with midseg $\overline{HI}$ containing midpts H and I of $\overline{EG}$ and $\overline{FG}$, respectively	1. Given
2. $GH = EH$, $GI = FI$	2. Def of midpt
3. $\frac{GH}{GH} = \frac{GI}{GI} = 1$	3. Any nonzero number div by itself = 1.
4. $\frac{GH}{EH} = \frac{GI}{FI}$	4. Sub
5. $\overline{HI} \parallel \overline{EF}$	5. Thm 8.6
6. $\angle GHI \cong \angle GEF$, $\angle GIH \cong \angle GFE$	6. If 2 ∥ lines are inters by transv, then corr ∠s are ≅.
7. $\triangle GHI \sim \triangle GEF$	7. AA~
8. $GH = \frac{1}{2}GE$	8. Def of midpt
9. $\frac{GH}{GE} = \frac{1}{2}$	9. Mult Prop of Eq
10. $\frac{HI}{EF} = \frac{1}{2}$	10. Corr sides of ~ △s are proport.
11. $HI = \frac{1}{2}EF$	11. Mult Prop of Eq

Mixed Review

1. For all real numbers a and b, if $a = b$, then $b = a$.
2. For any segments $\overline{AB}$, $\overline{CD}$ and $\overline{PQ}$, if $\overline{AB} \cong \overline{CD}$ and $\overline{CD} \cong \overline{PQ}$, then $\overline{AB} \cong \overline{PQ}$. For any angles A, B and C, if $\angle A \cong \angle B$ and $\angle B \cong \angle C$, then $\angle A \cong \angle C$.
3. Concave pentagon

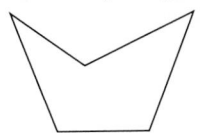

4. Convex pentagon

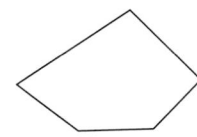

PAGE 327

19.

Statement	Reason
1. $\triangle ABC$, $\triangle EDF$ such that $\overleftrightarrow{ED} \perp \overline{AB}$, $\overline{DF} \perp \overline{BC}$, $\overline{FE} \perp \overline{CA}$	1. Given
2. $\angle DJH \cong \angle BJK$	2. Vert ∠s are ≅.
3. $\angle DHJ \cong \angle BKJ$	3. Rt ∠s are ≅.
4. $\triangle DHJ \sim \triangle BKJ$	4. AA~
5. $\angle JDH \cong \angle B$	5. Def ~ △s
6. $\angle EKL \cong \angle AGL$	6. Rt ∠s are ≅.
7. $\angle KLE \cong \angle GLA$	7. Vert ∠s are ≅.
8. $\triangle EKL \sim \triangle AGL$	8. AA~
9. $\angle KEL \cong \angle A$	9. Def of ~ △s
10. $\triangle ABC \sim \triangle EDF$	10. AA~

661

20.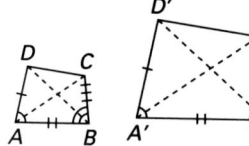

Statement	Reason
1. Quads $ABCD$ and $A'B'C'D'$ with $\frac{AD}{A'D'}$ $= \frac{AB}{A'B'} = \frac{BC}{B'C'}$ and $\angle A \cong \angle A'$, $\angle B \cong \angle B$	1. Given
2. $\triangle ABC \sim \triangle A'B'C'$	2. SAS
3. m $\angle CAB =$ m $\angle C'A'B'$, m $\angle ACB$ $=$ m $\angle A'C'B'$, $\frac{AC}{A'C'}$ $= \frac{AD}{A'D'}$	3. Def $\sim \triangle$s
4. m $\angle A -$ m $\angle CAB$ $=$ m $\angle A' -$ m $\angle C'A'B'$	4. Equations may be subtracted.
5. m $\angle A -$ m $\angle CAB$ $=$ m $\angle DAC$, m $\angle A' -$ m $\angle C'A'B'$ $=$ m $\angle D'A'C'$	5. Angle Add Post
6. m $\angle DAC =$ m $\angle D'A'C'$	6. Sub
7. $\triangle ADC \sim \triangle A'D'C'$	7. SAS$\sim$
8. m $\angle ACD =$ m $\angle A'C'D'$, m $\angle D =$ m $\angle D'$, $\frac{CD}{C'D'} = \frac{AD}{A'D'}$	8. Def $\sim \triangle$s
9. m $\angle ACD +$ m $\angle ACB =$ m $\angle A'C'D' +$ m $\angle A'C'B'$	9. Equations may be added.
10. m $\angle ACD +$ m $\angle ACB =$ m $\angle C$, m $\angle A'C'D' +$ m $\angle A'C'B' =$ m $\angle C'$	10. Angle Add Post
11. m $\angle C =$ m $\angle C'$	11. Sub
12. $\frac{AD}{A'D'} = \frac{AB}{A'B'} = \frac{BC}{B'C'}$ $= \frac{CD}{C'D'}$	12. Sub
13. Quad $ABCD \sim$ quad $A'B'C'D'$	13. All corr $\angle$s are $\cong$ and all corr sides are proport.

21.

Statement	Reason
1. Quads $ABCD$ and $A'B'C'D'$ with m $\angle A =$ m $\angle A'$, m $\angle D =$ m $\angle D'$, m $\angle B =$ m $\angle B'$, $\frac{AD}{A'D'}$ $= \frac{AB}{A'B'}$	1. Given
2. $\triangle ADB \sim \triangle A'D'B'$	2. SAS$\sim$
3. m $\angle ADB =$ m $\angle A'D'B'$, m $\angle ABD$ $=$ m $\angle A'B'D'$, $\frac{DB}{D'B'}$ $= \frac{AD}{A'D'}$	3. Def $\sim \triangle$s
4. m $\angle D -$ m $\angle ADB$ $=$ m $\angle D' -$ m $\angle A'D'B'$, m $\angle B -$ m $\angle ABD =$ m $\angle B'$ $-$ m $\angle A'B'D'$	4. Equations may be subtracted.
5. m $\angle D -$ m $\angle ADB$ $=$ m $\angle D' -$ m $\angle ADB =$ m $\angle CDB$, m $\angle D' -$ m $\angle A'D'B' =$ m $\angle C'D'B'$	5. Angle Add Post
6. m $\angle CDB =$ m $\angle C'D'B'$	6. Sub
7. $\triangle CDB \sim$ $\triangle C'D'B'$, $\triangle CBD$ $\sim \triangle C'B'D'$	7. AA$\sim$
8. $\frac{DC}{D'C'} = \frac{CB}{C'B'} = \frac{DB}{D'B'}$, m $\angle C =$ m $\angle C'$	8. Def $\sim \triangle$s
9. $\frac{AD}{A'D'} = \frac{AB}{A'B'} = \frac{BC}{B'C'}$ $= \frac{DC}{D'C'}$	9. Sub
10. Quad $ABCD \sim$ quad $A'B'C'D'$	10. All corr $\angle$s are $\cong$ and all corr sides are proport.

PAGE 446

32.

Statement	Reason
1. Inscr trap $ABCD$ with bases $\overline{AB}$ and $\overline{DC}$	1. Given
2. $\overline{AB} \parallel \overline{DC}$	2. Bases of trap are $\parallel$.
3. m$\overset{\frown}{AD} =$ m$\overset{\frown}{BC}$	3. 2 arcs of $\odot$ included between $\parallel$ chords are $\cong$.
4. m$\overset{\frown}{AB} =$ m$\overset{\frown}{AB}$	4. Reflex Prop of Eq
5. m$\overset{\frown}{AD} +$ m$\overset{\frown}{AB} =$ m$\overset{\frown}{BC} +$ m$\overset{\frown}{AB}$	5. Equations may be added.
6. m$\overset{\frown}{DAB} =$ m$\overset{\frown}{CBA}$	6. Arc Add Post
7. $\overline{DB} \cong \overline{AC}$	7. If arcs are $\cong$, corr chords are $\cong$.

Algebra Review

1. ◄─────⊕─── -5 0 5   8. ◄──⊕─────── -5 0 5

2. ◄━━━━━━⊕─── -5 0 5   9. ◄───⊕─⊕──── -5 0 5

3. ◄──⊕━━━━━━ -5 0 5   10. ◄───⊕─⊕──── -5 0 5

4. ◄──⊕─────── -5 0 5   11. ◄─────⊕━━━━ -5 0 5

5. ◄━━━━━━━━━⊕ -5 0 5   12. ◄────────⊕── -5 0 5

6. ◄─────⊕──── -5 0 5   13. ◄━━━━━━━━⊕── -5 0 5

7. ◄──⊕━━━━━━ -5 0 5   14. ◄─────⊕──── -5 0 5

33.

Statement	Reason
1. m $\angle 1 = \frac{1}{2}$ m$\overarc{CD}$, diam $\overline{BD}$	1. Given
2. $\frac{1}{2}$ m$\overarc{AD}$ = m $\angle 1$	2. Meas inscr $\angle = \frac{1}{2}$ meas intercepted arc.
3. $\frac{1}{2}$ m$\overarc{CD}$ = $\frac{1}{2}$ m$\overarc{AD}$	3. Sub
4. m$\overarc{CD}$ = m$\overarc{AD}$	4. Mult Prop of Eq
5. m$\overarc{BCD}$ = m$\overarc{BAD}$	5. Semicircles have meas of 180.
6. m$\overarc{BCD}$ − m$\overarc{CD}$ = m$\overarc{BAD}$ − m$\overarc{AD}$	6. Equations may be subtr.
7. m$\overarc{BC}$ = m$\overarc{BA}$	7. Arc Add Post

34.

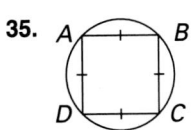

Statement	Reason
1. Inscr $\square$ ABCD	1. Given
2. m $\angle A$ + m $\angle C$ = 180	2. Opp $\angle$s of inscr quad are supp.
3. m $\angle A$ = m $\angle C$	3. Opp $\angle$s of $\square$ are $\cong$.
4. 2 (m $\angle A$) = 180	4. Sub
5. m $\angle A$ = 90	5. Div Prop of Eq
6. ABCD is a rect.	6. $\square$ with 1 rt $\angle$ is rect.

35.

Statement	Reason
1. Inscr rhombus ABCD	1. Given
2. Prove rhombus ABCD is rect.	2. See answer to Ex 34.
3. $\overline{AB} \cong \overline{BC} \cong \overline{CD} \cong \overline{DA}$	3. Def of rhombus
4. ABCD is square.	4. Def square

36.

Statement	Reason
1. Inscr quad ABCD, with $\overline{AB} \cong \overline{CD}$	1. Given
2. m$\overarc{AB}$ = m$\overarc{CD}$.	2. If chords are $\cong$, corr arcs are $\cong$.
3. m$\overarc{BC}$ = m$\overarc{BC}$	3. Reflex Prop of Eq
4. m$\overarc{AB}$ + m$\overarc{BC}$ = m$\overarc{CD}$ + m$\overarc{BC}$	4. Equations may be added.
5. m$\overarc{AB}$ + m$\overarc{BC}$ = m$\overarc{ABC}$, m$\overarc{BC}$ + m$\overarc{CD}$ = m$\overarc{BCD}$	5. Arc Add Post
6. m$\overarc{ABC}$ = m$\overarc{BCD}$	6. Sub
7. m $\angle D$ = m $\angle A$	7. If inscr $\angle$s intercept $\cong$ arcs, they are $\cong$.
8. m $\angle B$ + m $\angle D$ = 180	8. Opp $\angle$s of inscr quad are supp.
9. m $\angle B$ + m $\angle A$ = 180	9. Sub
10. $\overline{BC} \parallel \overline{AD}$	10. Thm 3.2

37.

Statement	Reason
1. Inscr equil $\triangle ABC$, R and S are midpts of $\overarc{AB}$ and $\overarc{BC}$ resp	1. Given
2. m $\angle A$ = m $\angle B$ = m $\angle C$ = 60	2. Def equilat $\triangle$
3. m$\overarc{AB}$ = m$\overarc{BC}$ = m$\overarc{CA}$ = 120	3. Meas inscr $\angle = \frac{1}{2}$ meas intercepted arc.
4. m$\overarc{CAR}$ = m$\overarc{ACS}$ = 180	4. Arc Add Post
5. m $\angle ARS$ = m $\angle RSC$ = 90	5. Meas inscr $\angle = \frac{1}{2}$ meas intercepted arc.
6. m $\angle RAC$ = m $\angle SCA$ = 90	6. In inscr quad, opp $\angle$s are supp.
7. ARSC is rect.	7. Def of rect

38.

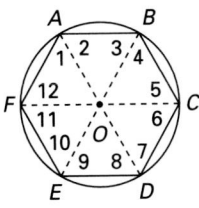

Statement	Reason
1. Equilateral hexagon ABCDEF inscr in $\odot O$	1. Given
2. AB = BC = CD = DE = EF = FA	2. Def equil
3. OA = OB = OC = OD = OE = OF	3. Radii of $\odot$ are $\cong$.
4. $\triangle AOB \cong \triangle BOC \cong \triangle COD \cong \triangle DOE \cong \triangle EOF \cong \triangle FOA$	4. SSS
5. $\angle 2 \cong \angle 3$, $\angle 4 \cong \angle 5$, $\angle 6 \cong \angle 7$, $\angle 8 \cong \angle 9$, $\angle 10 \cong \angle 11$, $\angle 12 \cong \angle 1$	5. If 2 sides of $\triangle \cong$, $\angle$s opp $\cong$.
6. $\angle 1 \cong \angle 2$, $\angle 3 \cong \angle 4$, $\angle 5 \cong \angle 6$, $\angle 7 \cong \angle 8$, $\angle 9 \cong \angle 10$, $\angle 11 \cong \angle 12$	6. CPCTC
7. $\angle 1 \cong \angle 2 \cong \angle 3 \cong \angle 4 \cong \angle 5 \cong \angle 6 \cong \angle 7 \cong \angle 8 \cong \angle 9 \cong \angle 10 \cong \angle 11 \cong \angle 12$	7. Trans Prop of Congr
8. m $\angle 1$ + m $\angle 2$ = m $\angle 3$ + m $\angle 4$ = m $\angle 5$ + m $\angle 6$ = m $\angle 7$ + m $\angle 8$ = m $\angle 9$ + m $\angle 10$ = m $\angle 11$ + m $\angle 12$	8. Add Prop of Eq
9. m $\angle A$ = m $\angle B$ = m $\angle C$ = m $\angle D$ = m $\angle E$ = m $\angle F$	9. Angle Add Post
10. Hexagon ABCDEF is regular.	10. Equiangular and equilat polygons are regular.

This result extends to all equilat n-gons, since all consist of $\cong$ isos triangles that form $\cong$ int angles of the polygon.

663

14.

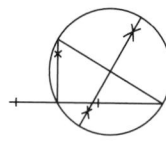

Construct the midpt of the hyp of the triangle. Using this pt as ctr, and radius equal to $\frac{1}{2}$ length of hyp, construct a ⊙. The meas of the arc formed by the hyp is 180, since the inscribed angle meas 90. Thus, the hyp forms 2 semicircles and is a diam.

15. (a) $\overline{OT}$ is a diam of ⊙M because it has endpts on the ⊙ and contains ctr M.
(b) ∠OAT is rt because an angle inscribed in a semicircle is a rt ∠.
(c) $\overline{TA}$ is ⊥ to ⊙O at pt A because ∠OAT is rt.

16. (a) The hexagon is equil because arcs of equal measure are marked off on the ⊙. If arcs are ≅, then corr chords are ≅.
(b) The hexagon is equiangular because it consists of 6 ≅, isos △s, the angle meas of which are all ≅. Thus, the int angles of the hexagon, each consisting of two such 60 deg ∠s, are ≅.

17. Construct a seg $\overline{OM}$ 13 units in length. Construct C, the midpt of $\overline{OM}$. From O draw a ⊙ of radius 3 and from M draw a ⊙ of radius 8. From M draw an auxiliary circle with a radius of 8 − 3, or 5. Use CO as the length to swing 2 arcs that intersect the auxiliary circle at pts P and Q. Draw $\overrightarrow{MP}$ and $\overrightarrow{MQ}$ intersecting ⊙M at A and B. From pts A and B, swing arcs of length PO intersecting ⊙O at pts D and E, respectively. Draw tans $\overline{AD}$ and $\overline{BE}$, which should measure 12 units in length by Pythagorean Thm.

18.

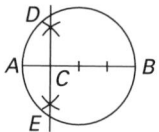

Check students constructions.

Statement	Reason
1. $CD = \sqrt{3}$: △ADB is rt	1. An ∠ inscribed in a semicircle is a rt ∠.
2. $AC = 1$, $CB = 3$, $\overline{DC} \perp \overline{AB}$	2. Given
3. $(DC)^2 = AC \cdot CB$	3. Square of altitude to hyp is product of lengths of seg.
4. $DC^2 = 1 \cdot 3 = 3$	4. Sub
5. $DC = \sqrt{3}$	5. Def of square

Application

1. Left figure: Mark 6 arcs of equal measure on a ⊙. Open compass to distance between two such pts. From any pt, swing arc intersecting ctr of ⊙ and ⊙ at two pts. From pt opposite the first pt, swing another arc. Shade area between arcs.

Right figure: Mark 6 arcs of equal measure. Swing arcs as in procedure above from each of 6 pts on ⊙.

19. Part I.

Statement	Reason
1. P lies on the ⊥ bis of $\overline{AB}$.	1. Given
2. $\overline{AD} \cong \overline{BD}$	2. Def of bis
3. ∠PDA and ∠PDE are rt ∠s.	3. Def of ⊥
4. ∠PDA ≅ ∠PDE	4. All rt ∠s are ≅.
5. $\overline{PD} \cong \overline{PD}$	5. Reflex
6. △PDA ≅ △PDB	6. SAS
7. $PA = PB$	7. CPCTC

Part II.

Statement	Reason
1. $PA = PB$	1. Given
2. Draw $\overrightarrow{PD}$ ⊥ $\overline{PB}$.	2. There is exactly one ⊥ from a pt to a line.
3. $\overline{PD} \cong \overline{PD}$	3. Reflex
4. △PDA ≅ △PDB	4. HL
5. $AD = BD$	5. CPCTC
6. $\overline{PD}$ is ⊥ bis of AB.	6. Def of ⊥ bis

20. (Ref to illus in text)

Statement	Reason
1. $\overrightarrow{OC}$ bis ∠AOB, P is on $\overrightarrow{OC}$, $\overline{PD} \perp \overline{OA}$, $\overline{PE} \perp \overline{OB}$	1. Given
2. ∠PDO and ∠PEO are rt ∠s.	2. Def of ⊥
3. ∠PDO ≅ ∠PEO	3. ⊥s form ≅ rt ∠s.
4. ∠DOP ≅ ∠EOP	4. Def of ∠ bis
5. $\overline{PO} \cong \overline{PO}$	5. Reflex
6. △PDO ≅ △PEO	6. AAS
7. $PD \cong PE$	7. CPCTC

21. (Ref to illus in text)

Statement	Reason
1. P is in the int of ∠AOB, $PD = PE$, $\overline{PD} \perp \overline{OA}$, $\overline{PE} \perp \overline{OB}$	1. Def of distance
2. ∠ODP and ∠OEP are rt ∠s	2. Def of ⊥
3. $\overline{OP} \cong \overline{OP}$	3. Reflex
4. △DOP and △EOP are rt △s.	4. Def of rt △.
5. △POD ≅ △POE	5. HL
6. ∠POD ≅ ∠POE	6. CPCTC
7. $\overline{OP}$ is the bis of ∠AOB.	7. Def of ∠ bis

22. Part I.

Statement	Reason
1. $\odot P$, $\overline{AB}$ and $\overline{CD}$ are $\cong$ chords of $\odot P$	1. Given
2. Draw $\overline{M_1P} \perp \overline{AB}$, $\overline{M_2P} \perp \overline{CD}$.	2. From pt to line, just 1 $\perp$ can be drawn.
3. $\overline{M_1P}$ bis $\overline{AB}$, $\overline{M_2P}$ bis $\overline{CD}$	3. If a line contains the center of a $\odot$ and is $\perp$ a chord, then it bis chord.
4. M_1 is midpt of $\overline{AB}$, M_2 is midpt of $\overline{CD}$	4. Def of bis
5. $\overline{M_1P} \cong \overline{M_2P}$	5. $\cong$ chords are equidistant from center of $\odot$.
6. M_1, M_2 lie on $\odot P_m$	6. Def of $\odot$
7. $\odot P_m$, $\odot P$ are concentric.	7. Def of concentric

Part II.

Statement	Reason
1. $\odot P$, $\odot P_c$, M_1, M_2 lie on P_c	1. Given
2. Draw $\overline{AB}$, $\overline{CD}$ tan to $\odot P_c$ at M_1, M_2, respectively.	2. Unique line tan at a given pt on a $\odot$
3. Draw $\overline{M_1P} \perp \overline{AB}$, $M_2P \perp \overline{CD}$	3. Radius is $\perp$ to tan at pt of tangency.
4. $\overline{AM_1} \cong \overline{BM_1}$, $\overline{CM_2} \cong \overline{DM_2}$	4. If a line contains the ctr of a $\odot$ and is $\perp$ to a chord, then it bis the chord.
5. M_1, M_2 midpts of $\overline{AB}$, $\overline{CD}$, respectively.	5. Def of bis
6. $PM_1 = PM_2$	6. Def of $\odot$
7. $\overline{AB} \cong \overline{CD}$	7. Chords that are equidistant from the ctr of a $\odot$ are $\cong$.

23. Part I.

Statement	Reason
1. $\odot P$, V lies on $\odot P$, $\overline{AB}$ diam of $\odot P$	1. Given
2. Draw $\overline{VA}$, $\overline{VB}$	2. 2 pts determine a line.
3. $\triangle AVB$ is a rt $\triangle$.	3. An $\angle$ inscr in a semicircle is a rt $\triangle$.

Part II.

Statement	Reason
1. $\odot P$ $\overline{AB}$ a diam of $\odot P$, $\overline{AB}$ a hyp of $\triangle AVB$	1. Given
2. Draw $\overline{CV}$, $\overline{AV}$ so that X lies on $\overline{BV}$.	2. $\parallel$ Post
3. $\angle PXB$ is a rt $\angle$.	3. Corr formed by $\parallel$ lines are $\cong$.
4. $BP = AP$	4. Radii are $\cong$.
5. $BX = VX$	5. Line $\parallel$ base of a $\triangle$ cuts off seg proportional to the sides.
6. $PX = PX$	6. Reflex
7. $\triangle BXP \cong \triangle VXP$	7. Rt $\angle$s are $\cong$.
8. $\triangle BXP \cong \triangle VXP$	8. SAS
9. $PV = PB$	9. CPCTC
10. PV is a radius of $\odot P$	10. Radii are $\cong$.
11. V is on $\odot P$.	11. Def of $\odot$

Mixed Review

1. Given $\overline{AB}$, swing equal arcs from A and B and find the intersection, P and Q. Draw $\overleftrightarrow{PQ}$, the $\perp$ bis of $\overline{AB}$.

2. Given obtuse $\angle XOY$, swing equal arcs from O to $\overrightarrow{OX}$ and $\overrightarrow{OY}$. From the pts of intersection C and D, swing equal arcs. Label the pt of intersection E. Draw $\overrightarrow{OE}$, the $\angle$ bis.

3. From pt A, not on line l, swing an arc. Label the pts of intersection with l, as B and C. Swing equal arcs from B and C, intersecting at D. Draw $\overline{AD}$, the $\perp$ to l.

20.

Statement	Reason
1. $\triangle ABC$ with alt $\overline{AD}$, $\overline{BE}$, and $\overline{CF}$	1. Given
2. Through C draw $\overleftrightarrow{GH} \parallel \overline{AB}$; through B draw $\overleftrightarrow{HI} \parallel \overline{BC}$, through A draw $\overleftrightarrow{GI} \parallel \overline{BC}$	2. $\parallel$ Post
3. $ABCG$ and $ABHC$ are $\square$s.	3. Def of $\square$
4. $\overline{GC} \cong \overline{AB}$ and $\overline{AB} \cong \overline{CH}$	4. Opp sides of a $\square$ are $\cong$.
5. $\overline{GC} \cong \overline{CH}$	5. Sub
6. C is midpt of $\overline{GH}$.	6. Def of midpt
7. $\overline{CF} \perp \overline{AB}$	7. Def of alt
8. $\overline{CF} \perp \overline{GH}$	8. If a line is $\perp$ to one of two $\parallel$ lines, it is $\perp$ to the other.
9. $\overleftrightarrow{CF}$ is the $\perp$ bis of $\overline{GH}$.	9. Def of $\perp$ bis
10. Similarly $\overleftrightarrow{AD}$ is the bis of $\overline{GI}$; $\overline{BE}$ is the $\perp$ bis of $\overline{HI}$	10. Repeat the steps above.
11. $\overline{AD}$, $\overline{BE}$, and $\overline{CF}$ are cncrnt.	11. The $\perp$ bis of the sides of a $\triangle$ are cncrnt.

21. Given: $\triangle ABC$ with meds $\overline{AD}$, $\overline{BE}$, $\overline{CF}$. Prove: $\overline{AD}$, $\overline{BE}$, $\overline{CF}$ are concrnt at P, $AP = \frac{2}{3}AD$, $BP = \frac{2}{3}BE$, $CP = \frac{2}{3}CF$.

Statement	Reason
1. $\triangle ABC$ with meds $\overline{AD}$, $\overline{BE}$, and $\overline{CF}$	1. Given
2. $\overline{AD}$, $\overline{BE}$ inters at P, $AP = \frac{2}{3}AD$, $BP = \frac{2}{3}BE$	2. Thm 13.6
3. $\overline{AD}$, $\overline{CF}$ inters at Q, $AQ = \frac{2}{3}AD$, $CQ = \frac{2}{3}CF$	3. Thm 13.6
4. $AP = AQ$	4. Sub
5. $P = Q$	5. Ruler Post
6. $\overline{AD}$, $\overline{BE}$, $\overline{CF}$ are cncrnt.	6. Def of cncrnt
7. $CP = \frac{2}{3}CF$	7. Sub

22. Given: Equil $\triangle ABC$, with meds $\overline{AE}$, $\overline{BF}$, $\overline{CD}$ Prove: $\triangle$s AEB, AEC, BFA, BFC, CDA, CDB $\cong$.

Statement	Reason
1. $AB = BC = CA$	1. Def of equil $\triangle$
2. E, F, D are midpts.	2. Def med
3. $AD = DB$, $BE = EC$, $CF = FA$	3. Def midpt
4. $AD = \frac{1}{2} AB$, $BE = \frac{1}{2} BC$, $CF = \frac{1}{2} CA$	4. Def midpt
5. $\frac{1}{2} AB = \frac{1}{2} BC = \frac{1}{2} AC$	5. Div Prop Eq
6. $AD = BE = CF$	6. Sub
7. $AD = DB$, $BE = EC$, $CF = FA$	7. Subst
8. $AE = BF = CD$	8. Med of equil $\triangle \cong$.
9. $\triangle$s AEB, AEC, CDA, CDB, BFC, BFA $\cong$	9. SSS, Trans $\cong$

23. Given: Equil $\triangle ABC$, $\overline{AE}$, $\overline{BF}$. $\overline{CD}$ are alt inters at P. Prove: P is incenter, circumctr, orthoctr, centroid.

Statement	Reason
1. P is orthoctr.	1. Def orthoctr
2. $\overline{AE}$, $\overline{BF}$, $\overline{CD}$ are med.	2. Alt of equil $\triangle$ are med.
3. P is centroid.	3. Def of centroid
4. $\overline{AE}$, $\overline{BF}$, $\overline{CD}$ are $\angle$ bis.	4. Med of equil $\triangle$ is bis vert $\angle$.
5. P is inctr.	5. Def of inctr
6. $\overline{AE}$, $\overline{BF}$, $\overline{CD}$ are $\perp$ bis of sides.	6. Alt of equil $\triangle$ are $\perp$ bis of sides.
7. P is circumctr.	7. Def of circumctr

Mixed Review

1. Line $\parallel$ to the two given lines and halfway between them.

2. Line that is $\perp$ bis of seg having 2 pts as endpts.

Brainteaser

(a) The "cone" from the Sun would have ht = dist from Earth to Sun. Part of cone from disk to pavement is extreme bottom of cone and would be, for all practical purposes, a cylinder.

(b) Circle or ellipse

(c) Parallelograms of different shapes

(d) Triangles

11. $d = \sqrt{(x_1 - x_2)^2 + (y_1 - y_2)^2}$; $d' = \sqrt{[(x_1 + 1) - (x_2 + 1)]^2 + [(y_1 + 2) - (y_2 + 2)]^2} = \sqrt{(x_1 - x_2)^2 + (y_1 - y_2)^2}$

12. $d = \sqrt{(x_1 - x_2)^2 + (y_1 - y_2)^2}$; $d' = \sqrt{[(x_1 - 4) - (x_2 - 4)]^2 + [(y_1 + 5) - (y_2 + 5)]^2} = \sqrt{(x_1 - x_2)^2 + (y_1 - y_2)^2}$

13. $d = \sqrt{(x_1 - x_2)^2 + (y_1 - y_2)^2}$; $d' = \sqrt{[(x_1 + 6) - (x_2 + 6)]^2 + [(y_1 - 7) - (y_2 - 7)]^2} = \sqrt{(x_1 - x_2)^2 + (y_1 - y_2)^2}$

14. $d = \sqrt{(x_1 - x_2)^2 + (y_1 - y_2)^2}$; $d' = \sqrt{[(x_1 - 9) - (x_2 - 9)]^2 + [(y_1 - 7) - (y_2 - 7)]^2} = \sqrt{(x_1 - x_2)^2 + (y_1 - y_2)^2}$

15. $d = \sqrt{(x_1 - x_2)^2 + (y_1 - y_2)^2}$; $d' = \sqrt{(y_1 - y_2)^2 + (x_1 - x_2)^2} = \sqrt{(x_1 - x_2)^2 + (y_1 - y_2)^2}$

16. $d = \sqrt{(x_1 - x_2)^2 + (y_1 - y_2)^2}$; $d' = \sqrt{[-y_1 - (-y_2)]^2 + (x_1 - x_2)^2} = \sqrt{(y_2 - y_1)^2 + (x_1 - x_2)^2} = \sqrt{(y_1 - y_2)^2 + (x_1 - x_2)^2} = \sqrt{(x_1 - x_2) + (y_1 - y_2)}$

Mixed Review

1.

p	q	$p \lor q$	$(p \lor q) \to p$
T	T	T	T
T	F	T	T
F	T	T	F
F	F	F	T

2.

p	q	$p \land q$	$(p \land q) \to p$
T	T	T	T
T	F	F	T
F	T	F	T
F	F	F	T

3.

p	q	$p \to q$	$(p \to q) \land p$	$[(p \to q) \land p] \to q$
T	T	T	T	T
T	F	F	F	T
F	T	T	F	T
F	F	T	F	T

Application

Slide poles (trans) until the bottom of the pole is at the hole. Rotate poles counterclockwise 90 around the base of the pole to stand them up. Rotate the lights clockwise 90 for those on near side; counterclockwise for those on far side.

16.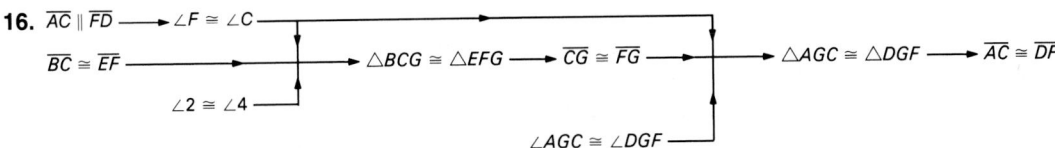

17.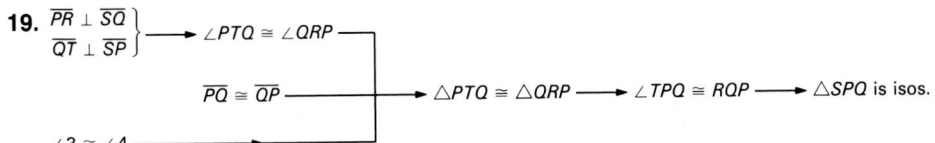

19.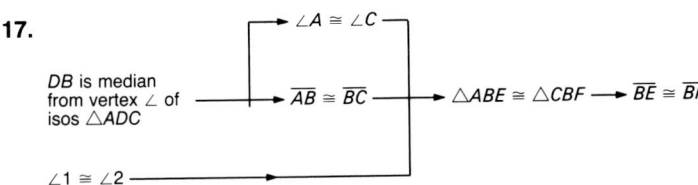

17.

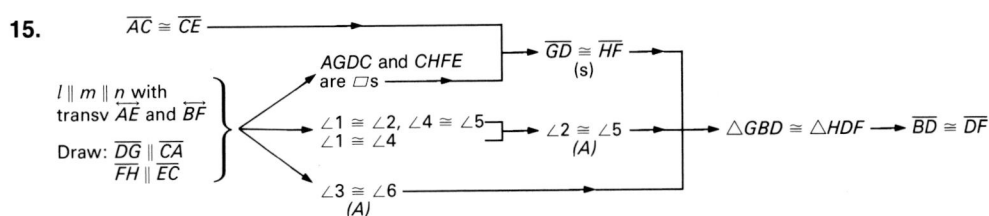

15.

$\overline{AC} \cong \overline{CE}$

$l \parallel m \parallel n$ with
transv $\overrightarrow{AE}$ and $\overrightarrow{BF}$

Draw: $\overline{DG} \parallel \overline{CA}$
$\overline{FH} \parallel \overline{EC}$

AGDC and CHFE
are ▱s

$\overline{GD} \cong \overline{HF}$
(s)

$\angle 1 \cong \angle 2, \angle 4 \cong \angle 5$
$\angle 1 \cong \angle 4$

$\angle 2 \cong \angle 5$
(A)

$\angle 3 \cong \angle 6$
(A)

$\triangle GBD \cong \triangle HDF \longrightarrow \overline{BD} \cong \overline{DF}$

Index

Boldfaced numerals indicate the pages that contain definitions.

AA Similarity Postulate,
314
AAS Theorem, 162–163, 171
Absolute value, 4, 38
Acute angle(s), 16, 39
Acute triangle(s), **139,** 170, 351, 374
Addition Property of Equality, 48
Adjacent angle(s), 20, **21**
Adjacent side(s) of polygons, 225
Agriculture, ancient, 68
Algebra Review(s)
  absolute value, 4
  division of fractions, 219
  division of polynomial by monomial,
    272
  fractional equations, 114, 235, 387
  graphing the solution set of
    inequality, 327
  linear equations, 47
  multiplication of fractions, 219
  rationalization of denominator, 353
  solution of inequality, 209, 588
  two equations with two variables, 166
Alternate exterior angle(s),
  98–99, 134
Alternate interior angle(s),
  97–101, 110–112, 134
Alternate Interior Angles Postulate,
  100–101
Altitude(s), **194**
  of isosceles triangle, 195, 200
  of right triangles, 339–342, 374
  of similar triangles, 328, 333
  of trapezoids, 286
  of triangles, 193–198
Angle Addition Postulate,
  21–22, 27, 57, 86
Angle bisector(s), 22–23
Angle of depression, 370, 375
Angle of repose, 152
Angle(s), **15**–19
  acute, 16, 39
  adjacent, 20, 21–25
  alternate exterior, 98–99
  alternate interior, 97–101,
    110–112, 134
Angle Addition Postulate,
  21–22, 27, 57–58, 86
  base, 177

bisectors of, 22–23
classification of, 16, 39
complementary, 20, **28**–30, 39, 72,
  87
congruence of, 17
construction of congruent angles,
  17
corresponding, 98–99, 110, 134
exterior of, 21
exterior of triangle, 121–124, 135
interior of, 21
of isosceles trapezoids,
  290–291, 299
linear pair of, 26
measure of, 28–29
measurement of, 16–17
nonadjacent, 77
obtuse, 16, 39
of polygons, 225, 230–239
proofs of theorems about,
  71–75
remote interior of triangle, 121–125,
  134–135
right, 16, 39, 73
sides of, 15
straight, 16, 26, 27, 39
supplementary, 20, 26–31, 39,
  71–72, 87, 110, 134
symbol of, 15
transversals and, 97–99
of triangle, 115–120, 135
vertex of, 15
vertical, 76–81, 87
Application(s)
  angles in sand piles, 152
  astronomical units, 31
  baseball diamond, 349
  conditions for quadrilaterals, 283
  congruent relationships, 204
  electric circuits, 35
  height of tree, 317
  light rays, 125
  mirrors, 239
  pitch of roof, 307
  ratio, 307
  soccer ball, 235
  surveying, 25
  tiling, 235
  tilt of the earth, 56

vector methods, 402
Arc(s)
  compass used in drawing of, 11
Area, 474–525
ASA Postulate, 154, 163, 171
Assumption(s)
  information from diagrams that can
    be assumed, 45
Astronomical units, 31
Auxiliary line(s), 109

Base angle(s)
  of isosceles trapezoids,
    290–291, 299
  of isosceles triangle, 177
Base(s)
  of isosceles triangle, 177
Bearing, 25
Biconditional, **127,** 135
Bisector(s), **199**
  of angles, **22**–23
  of angles of similar triangles, 329,
    333
  perpendicular, 199–203
  of segments, 10, 38
Box, drawing of, 94–95
Brainteaser(s)
  area of rectangle, 369, 482
  congruent triangles, 169
  construction of parallelogram, 14
  indirect proof, 108
  overlapping triangles, 187
  parallelograms, 246
  proof, 70
  squares, 297
  triangle inequalities, 214

Calculator(s)
  Exercises, 214, 233, 312, 321, 347,
    364–365, 387, 485–486, 567–568,
    581–582
  trigonometric ratios using, 365
Circles, 414–473
Closure, (See side column for each
  lesson.)
College Prep Test(s), 41, 89, 137,
  173, 223, 265, 301, 335, 377

Collinear points, 2
Common Error Analysis, 2, 6, 11, 22, 27, 45, 65, 78, 94, 102, 123, 128, 131, 140, 148, 154, 159, 162, 178, 185, 195, 201, 206, 212, 232, 238, 243, 259, 270, 275, 287, 305, 310, 320, 340, 345, 351, 367, 381, 385, 390, 405, 416, 433, 443, 448, 467, 485, 495, 506, 512, 531, 543, 548, 561, 565, 570, 575, 600, 614, 619
Compass use in drawing arcs, 11
Complementary angle(s)
  computer activity for, 20
  congruence of, 72, 87
  definition of, 28, 39
  measure of, 28–29
  proofs of theorems about, 72
Computer Investigation(s)
  altitudes and medians of triangle, 193
  angles of triangle, 120
  congruence of triangles, 146
  parallelograms, 240
  right triangles, 359
  similar triangles, 322
  supplementary and complementary angles, 20
  necessary conditions, 284
  vertical angles, 76
Concave polygon(s), **227,** 262
Conclusion(s), **63**
  conditional statements and, 63, 86
  drawing conclusions, 43–47, 86
Conditional statement(s), 63–66, 86, 126–29
Conditional statement(s), **63**
  biconditional of, 127
  contrapositive of, 128
  converse of, 109, 126–27
  inverse of, 128
  truth table for, 64
Congruence
  AAS Theorem, 162–163, 171
  of alternate interior angle(s), 110–113
  of angles, 17
  ASA Postulate, 154, 163, 171
  complementary angles and, 72, 87
  in complex figures, 158–160
  construction of congruent segments, 11
  corresponding altitudes of congruent triangles, 196
  corresponding medians of congruent triangles, 195–196

corresponding parts of congruent triangles, 167–169
  division of segment into three congruent segments, 260
  of isosceles triangles, 177–181, 204
  overlapping triangles, 184–187
  of parallelograms, 241–244, 263
  proof of parallel lines and, 167–168
  of quadrilaterals, 269–272, 298
  Reflexive Property of Congruence, 50, 184–185
  right angles and, 73
  of right triangles, 188–192
  SAS Postulate, 147–152, 163, 171
  of segments, 10–14
  SSS Postulate, 153, 163, 171
  supplementary angles and, 71–72, 87
  symbol for, 10
  Symmetric Property of Congruence, 50
  Third Angle Theorem, 161–162
  Transitive Property of Congruence, 50
  transversals of three parallel lines, 258–259, 263
  of triangles, 143–172
  vertical angles, 87
Conjunction(s), **32**–34, **39**
  graphing of, 36–37, 39
  truth table for, 32
Connections, Math, (See side column for each lesson.)
Consecutive angle(s)
  of parallelograms, 242–243, 244, 263
  of polygons, 225
Construction industry, 204
Construction(s)
  angle bisectors, 23
  congruent angles, 17
  division of segment into three congruent segments, 260
  of midpoint of segment, 12
  parallel lines, 100
  of parallelogram, 14
  perpendicular lines, 200–201
  of point equidistant from two points, 11
  proportional segments, 329–330
  quadrilaterals, 295–297
  rectangles, 295
  rhombus, 295–296

Contrapositive, **128,** 135
Converse(s), **109,** 126–27, 135
Convex polygon(s), **227,** 230–231, 236–237, 262
Cooperative Learning, (See *Group Projects.*)
Coordinate(s)
  of midpoint, 12, 38
  of point on number line, 5
Coplanar line(s), 93
Coplanar point(s), 2
Corresponding angle(s), 98–99, 110, 134
Corresponding median(s), 195–196
Corresponding Parts of Congruent Triangles are Congruent (CPCTC), 167
Cosine, 366–369, 375
Counterexample, 64
CPCTC, 167
Critical Thinking Questions, (See side column for each lesson.)

Decagon(s), 226
Deductive reasoning, 67, 87
Diagonal(s)
  of isosceles trapezoids, 291–292, 299
  of parallelograms, 241–244, 263
  of polygons, 226
  of quadrilaterals, 248, 263
  of rectangle, 274
  of rhombus, 274–275
Diagram(s)
  drawing of box, 94–95
  information that can be assumed, 45
  problem solving and, 62
Disjunction(s), **33**–34, **39**
  graphing of, 36–37, 39
  truth table for, 33
Distance, **5**–9
  between parallel lines, 258
  between two points, 5–9, 38
  Ruler Postulate, 5–6
Division
  of fractions, 219
  of polynomial by monomial, 272
Division Property of Equality, 48
Dodecagon(s), 226

Earth, tilt of, 56
Egyptians, ancient, 68
Electric circuit(s), 35
Endpoint of rays, **7**

Enrichment, (See side column for each lesson.)
Equality
  properties of, 48–52
  Subtraction Property of Equality, 48, 57–58
Equiangular polygon(s), 227, 262
Equiangular triangle(s), 139, 170, 177–178
Equidistant, 11
Equilateral polygon(s), 227, 262
Equilateral triangle(s), 139, 170, 177–178
Error Analysis, (See *Common Error Analysis*.)
Exterior Angle Inequality Theorem, 205
Exterior Angle Theorem, 121–124
Exterior angle(s)
  of polygons, 236–239, 262
  of triangles, 121–124, 135
Extreme(s), 303

Flow diagram(s), 185
Focus on Reading, 3, 18, 65, 73, 123, 141, 164, 202, 217, 275, 325, 357, 367, 381, 417, 435, 463, 476, 544, 558, 623
Fractional equation(s), 114, 235
Fraction(s)
  division of, 219
  multiplication of, 219

Generalization(s), 68
Geometric mean, 305, 332
Golden Ratio, 312
Graphing
  conjunctions, 36–37
  disjunctions, 36–37
    solution set of inequality, 327
Group Projects, xiv, 42, 138, 224, 268, 302, 414

Heptagon(s), 226
Hexagon(s), 226
HL Theorem, (See *Hypotenuse-Leg Theorem*)
Hypotenuse, **188**
  altitude to, 339–343
  of 45–45–90 triangle, 354
  length of, 210–211

of 30–60–90 triangle, 355
Hypotenuse-Leg Theorem, 188–190
Hypothesis, **63**
  conditional statements and, 63

Indirect proof, 82, 106–108, 135
Inductive reasoning, 67–68, 87
Inequalities, **205**
  Exterior Angle Inequality Theorem, 205
  graphing solution set of, 327
  SAS Inequality Theorem, 215–219, 221
  SSS Inequality Theorem, 216–219, 221
  Triangle Inequality Theorem, 210–214, 221
  in triangles, 205–219, 221
  for two triangles, 215–219
Interior angle(s) of polygons, 230–234
Intersection
  of line and plane, 3, 83, 87
  of lines, 3
  nonadjacent angles and, 77
  of two lines, 83, 84, 87
  of two parallel planes and a third plane, 131
  of two planes, 83, 87
Inverse, **128**, 135
Investigations, xiv, 42, 92, 138, 176, 224, 268, 302, 338, 378, 414, 474, 528, 554, 598
Isosceles trapezoid(s), **290**
  base angles of, 290–291, 299
  definition of, 290
  diagonals of, 291–292, 299
Isosceles triangle(s), **139**, 170, 177–181, 220
  altitude of, 195, 200
  base angles of, 177
  base of, 177
  congruent relationships in, 204
  leg of, 177
  median of, 200
  perpendicular bisector of, 199
  vertex angle of, 177

Law of Reflection of Light, 125
Light, reflection of, 125
Linear equation(s), 47
Linear pair of angles, 26
Line(s), **1**–4

auxiliary, 109
coplanar, 93
intersection of, 3, 83, 84, 87
naming of, 38
parallel, 93–96, 100–105, 134, 167–168
parallel to many lines, 257–261, 263
parallel to planes, 130
in planes, 83, 84, 87
points contained in, 82, 87
representation of, 1
skew, 93–96, 134
Logically equivalent statement(s), 128

Manipulatives, 92, 176, 378, 474, 528, 554, 598
Math Connections, (See side column for each lesson.)
Means, 303
Median(s), 194, 199
  corresponding, 195–196
  of isosceles triangle, 200
  of similar triangles, 328, 333
  of trapezoids, 286–289, 299
  of triangles, 193–198, 199
Midpoint Formula, 12–13
Midpoint(s), **10**
  construction of midpoint of segment, 11
  coordinates of, 12, 38
  formula for, 12–13
  of segments, 10–14, 38
Midsegment Theorem, 252–255, 263
Motivator, (See side column for each lesson.)
Multiplication of fractions, 219
Multiplication Property of Equality, 48

Necessary condition(s), 273–277
  for rectangles, 274, 275, 298
  for rhombus, 274, 275, 298
  for squares, 274, 275, 298
Negation, **126**, 135
Nonadjacent angle(s), 77
Nonadjacent side(s)
  of polygons, 225
Nonagon(s), 226
Noncollinear point(s), 2, 82, 84, 87

Nonconsecutive angle(s)
  of polygons, 225
Noncoplanar point(s), 2, 82

Obtuse angle(s), 16, 39
Obtuse triangle(s), 139, 170, 351, 374
Octagon(s), 226
Opposite ray(s), 26
Overlapping triangles, 182–187

Parallel line(s), 93–96
Parallel line(s)
  congruent triangles and,
    167–168
  construction of, 100
  distance between two parallel lines,
    258
  lines parallel to many lines, 257–261,
    263
  perpendicular lines and,
    102–103, 112, 134
  proofs of, 167–168
  in same plane, 112
  theorems pertaining to,
    100–105, 134
  transversals and, 100–102, 110–
    112, 134
Parallel plane(s), **130**
  intersection by a third plane, 131
Parallel Postulate, 109–14
Parallel ray(s), 94
Parallel segment(s), 94
Parallelogram(s), **241, 273**
  computer investigations, 240
  consecutive angles of,
    242–243, 244, 263
  construction of, 14
  diagonals of, 241–244, 263, 296–
    297
  proofs for, 247–251, 263
  properties of, 241–246, 263
  sufficient conditions for proving, 247–
    251, 263
Pentagon(s), 226
Perimeter of triangles, 178
Perpendicular symbol, 27
Perpendicular bisector(s),
  199–203
Perpendicular line(s), **27**
  Perpendicular line(s)
  construction of, 200–201
  parallel lines and, 102–103, 112, 134
  right angles and, 79

Perpendicular ray(s), 28
Physics, 125
Pi, **511**
  estimating, 509
Plane(s), **1**–4
  intersections, 3, 83, 87
  line and point contained in, 82–84,
    87
  parallel, 130–32
  parallel to lines, 130
  points contained in, 82–84, 87
  representation of, 1
  two intersecting lines in, 84, 87
Point(s), **1**–4
  collinear, 2
  construction equidistant from two
    points, 11
  coordinate of on number line, 5
  coplanar, 2
  distance between, 5–6, 38
  interior point of angle, 21
  in line, 82, 87
  noncollinear, 2
  noncoplanar, 2
  in plane, 82, 83, 84, 87
  representation of, 1
  in space, 82
Polygon(s), **225**
  adjacent sides of, 225
  classification of, 226
  concave, 227, 262
  consecutive angles of, 225
  convex, 227, 230–231, 236–237,
    262
  diagonal of, 226
  equiangular, 227, 262
  equilateral, 227, 262
  exterior angles of, 236–239, 262
  exterior of, 226
  interior angles of, 230–234
  interior of, 226
  nonadjacent sides of, 225
  nonconsecutive angles of, 225
  regular, 227, 232, 237, 262
  side of, 225
  similar, 308–311, 332
  sum of measures of exterior angles
    of, 236–239, 262
  sum of measures of interior angles
    of, 230–234, 262
  vertex of, 225
Polynomial(s)
  division of, 272
Postulate(s), **5.** See also (names).

Premises, 67
Problem solving
  drawing a diagram, 62
  formula for finding golden ratio,
    312
  making a table, 256
  more than one way, 487
  using an alternate approach, 133
Projects, Group, xiv, 42, 138, 224,
  268, 302, 414
Proof(s), **27**
  of complex figures, 57–61
  flow diagram of, 185
  indirect proof, 82, 106–108, 135
  introduction to, 48–52
  paragraph form of, 195
  for parallel lines, 100–105, 167–168
  Segment Addition Postulate and,
    57–59, 86
  Subtraction Property of Equality and,
    57–58
  of theorems about angles,
    71–75
  two-column form for, 49
  working backwards in developing, 58
  writing of, 53–56, 86
Proportion, 303–307, 314, 332
  extremes of, 303
  means of, 303
Protractor
  measurement of angles, 16
Protractor Postulate, 16–17
Pyramid, **574**
Pythagorean Theorem,
  344–348
  converse of, 350–353

Quadrilateral(s), 241
  conditions for, 283
  congruence of, 269–272, 298
  construction of, 295–297
  diagonals of, 248, 263
  number of sides, 226
  proofs for parallelograms, 247–251,
    263
  sum of measures of interior angles
    of, 231

Ratio, 303–307
Ray(s), **7**–9
  opposite rays, 26
  parallel, 94
  skew, 94

Reading, Focus on, 3, 18, 65, 73, 123, 141, 164, 202, 217, 275, 325, 357, 367, 381, 417, 435, 463, 476, 544, 558, 623

Real numbers
  properties of, 48–51

Reasoning. See *Deductive; Inductive*

Rectangle(s), **273, 298**
  construction of, 295
  diagonals of, 274
  necessary conditions for, 274, 275, 278, 298
  sufficient conditions for, 278–280, 299

Reflections, **599**

Reflexive Property, 49

Reflexive Property of Congruence, 50, 184–185

Regular polygon(s), **227,** 232, 237, 262

Remote interior angle(s) of triangles, 121–124, 135

Resources, Teaching, (See side column for each lesson.)

Rhombus, **273, 298**
  construction of, 295–296
  diagonals of, 274–275
  necessary conditions for, 274, 275, 298
  sufficient conditions for, 278–280, 299

Right angle(s), **16,** 39
  congruence of, 73
  perpendicular lines and, 79
  symbol of, 16

Right triangle(s), **139, 170**
  altitudes of, 339–342, 374
  angles of, 141
  computer investigations, 359
  congruence of, 188–192
  converse of Pythagorean Theorem, 350–353
  cosine of, 366–367, 375
  45–45–90 triangle, 354, 355, 375
  hypotenuse of, 188–190, 210–211, 339–342, 354–355
  legs of, 188
  Pythagorean Theorem, 344–348
  similarity properties of, 339–343, 374
  sine of, 360–367, 375
  tangent of, 366–367, 375
  30–60–90 triangle, 355–357

trigonometric ratios for, 360–373, 375

Ruler Postulate, 5–6

SAS Inequality Theorem, 215–219, 221

SAS Postulate, 148–149, 163, 171

SAS Similarity Theorem, 323–327, 333

Scalene triangle(s), 139, 170, 207

Segment Addition Postulate, 6–7, 39, 57–58, 86

Segment(s), **6–9**
  bisector of, **10,** 38
  congruence of, 10–14
  construction of congruent segments, 11
  construction of midpoint of, 11
  construction of proportional segments, 329–330, 333
  midpoint of, 10, 11, 38
  parallel, 94
  Segment Addition Postulate, 6–7, 39, 57–58, 86
  in similar triangles, 328–331
  skew, 94
  straightedge used in drawing of, 11

Similar polygon(s), 308–311, 332

Similar triangle(s), **313, 332**
  AA Similarity Postulate, 314–315
  altitudes of, 328, 333
  angle bisectors of, 329, 333
  computer investigations, 322
  and congruent triangles, 313–315
  medians of, 328, 333
  SAS Similarity Theorem, 323, 333
  segments in, 328–331, 333
  SSS Similarity Theorem, 324–327
  Transitive Property of Triangle Similarity, 313

Similarity
  AA Similarity Postulate, 314–315
  computer investigations, 322
  of polygons, 308–311, 332
  right triangles, 339–343, 374
  SAS Similarity Theorem, 323–327, 333
  SSS Similarity Theorem, 324–327, 333
  Transitive Property of Triangle Similarity, 313

of triangles, 313–317, 322–331

Sine, 360–365, 370–373

Skew line(s), 93–96, 134

Skew ray(s), 94

Skew segment(s), 94

Solving Problems, 62, 133, 256, 312, 487

Space, **2,** 82

Sphere, **417**

Square(s), **273, 298**
  necessary conditions for, 274, 275, 298
  sufficient conditions for, 280, 299

SSS Inequality Theorem, 216–219, 221

SSS Postulate, 153, 163, 171

SSS Similarity Theorem, 324–327, 333

Straight angle(s)
  measure of, 26, 39
  supplementary angles and, 27
  symbol of, 16

Straightedge
  use in drawing segments, 11

Strategies, Problem Solving, 62, 133, 256, 312, 487

Substitution Property, 49

Subtraction Property of Equality, 48, 57–58

Sufficient condition(s)
  for rectangles, 278–280, 299
  for rhombus, 278–280, 299
  for squares, 278–280, 299

Supplementary angle(s), **27, 39**
  computer activity for, 20
  congruence of, 71–72, 87
  measure of, 28–29
  parallel lines intersected by transversal, 110–112, 134
  proofs of theorems about, 71–75

Surveying, 25

Symmetric Property, 49

Symmetric Property of Congruence, 50

Tangent, 366–367, 375

Teaching Resources, (See side column for each lesson.)

Technology, xiiiA, 41A, 91A, 137A, 175A, 223A, 267A, 301A, 337A, 338, 377A, 413A, 473A, 527A, 553A

Theorem(s), 27. See also *(specific names)*

Third Angle Theorem, 161–162

Transformations, 598–629

Transitive Property, 49
Transitive Property of Congruence, 50
Transitive Property of Triangle
  Similarity, 313
Transversal(s), **97**
  parallel lines and, 100–102, 110–
    112, 134
  and special angle relationships, 97–
    99
  of three parallel lines,
    258–259, 263
Trapezoid(s), **285, 299**
  altitudes of, 286
  computer investigations, 284
  isosceles, 290–294, 299
  median of, 286–289, 299
Triangle Inequality Theorem, 210–214,
  221
Triangle Proportionality Theorem,
  318–321, 333
Triangle Proportionality Theorem
  converse of, 319, 333
Triangle(s), **115**
  AAS Theorem, 162–163, 171
  acute, 139, 170, 351, 374
  altitudes of, 193–198
  angle bisectors of, 199
  angles of, 115–120, 135
  ASA Postulate, 154, 163, 171
  classifications of, 139–142, 170
  computer investigations, 120
  congruence of, 143–169
  corresponding altitudes of congruent
    triangles, 196

corresponding medians of congruent
  triangles, 195–196
corresponding parts of congruent,
  167–169
equiangular, 139, 170, 177–178
equilateral, 139, 170, 177–178
exterior angle of, 121–125, 135
45–45–90 triangle, 354, 355, 375
inequalities in, 205–219, 221
isosceles, 139, 170, 177–181, 199–
  200, 204, 220
medians of, 193–198, 199
midsegment of, 252–255, 263
obtuse, 139, 170, 351, 374
overlapping, 182–187
perimeter of, 178
remote interior angles of,
  121–125, 135
right, 139, 141, 170, 339–375
SAS Postulate, 148–149, 163, 171
scalene, 139, 170, 207
sides of, 115
similar, 313–317, 322–331, 332
SSS Postulate, 153, 163, 171
sum of measures of angles of, 116–
  117, 135
Third Angle Theorem, 161–162
30–60–90 triangle, 355–356, 375
vertex of, 115
Trichotomy Property, 206
Trigonometric ratio(s)
  for right triangles, 360–375
  and word problems, 370–373
Trigonometry, 360

Truth table(s)
  conditional statement and its
    converse, 127
  for conditional statements, 64
  for conjunctions, 32
  for disjunctions, 33

Vertex
  of angle, 15
  of polygons, 225
  of triangle, 115
Vertex angle(s)
  of isosceles triangle, 177
Vertical angle(s), **77**
  computer investigation, 76
  congruence of, 78
Vertical Angles Theorem,
  78–79

Writing math (partial listing), 30(WE
  17, 30), 81(WE 19), 85(WE 6),
  104(WE 11), 108(WE 10), 197(CE
  7–8), 218(WE 11), 229(WE 26),
  239(WE 15), 282(WE 25), 311(WE
  18), 353(WE 23), 364(WE 32),
  406(WE 8–9), 419(WE 28),
  450(WE 32), 496(WE 24), 537(WE
  30), 550(WE 19), 559(WE 17),
  604(WE 16, 18–19), 610(WE 9, 12)